ENGINEERING MATHEMATICS WITH MATLAB®

ENGINEERING MATHEMATICS WITH MATLAB®

Won Y. Yang
Young K. Choi
Jaekwon Kim
Man Cheol Kim
H. Jin Kim
Taeho Im

CRC Press
Taylor & Francis Group
Boca Raton London New York

CRC P i i i f h

Limits of Liability and Disclaimer of Warranty of Software

The reader is expressly warned to consider and adopt all safety precautions that might be indicated by the activities herein and to avoid all potential hazards. By following the instructions contained herein, the reader willingly assumes all risks in connection with such instructions.

The authors and publisher of this book have used their best efforts and knowledge in preparing this book as well as developing the computer programs in it. However, they make no warranty of any kind, expressed or implied, with regard to the programs or the documentation contained in this book. Accordingly, they shall not be liable for any incidental or consequential damages in connection with, or arising out of, the readers' use of, or reliance upon, the material in this book.

Questions about the contents of this book can be mailed to wyyang53@hanmail.net.
Program files in this book can be down-loaded from the following website:
https://wyyang53.wixsite.com/mysite/publications

MATLAB® and Simulink® are registered trademarks of The MathWorks, Inc. and are used with permission. The MathWorks does not warrant the accuracy of the text or exercises in this book. This book's use or discussion of MATLAB® and Simulink® does not constitute endorsement or sponsorship by The MathWorks of a particular pedagogical approach or particular use of the MATLAB® and Simulink®. For MATLAB® and Simulink® product information, please contact:

The MathWorks, Inc.
3 Apple Hill Drive
Natick, MA 01760-2098, USA
☎: 508-647-7000, Fax: 508-647-7001
E-mail: info@mathworks.com
Web: www.mathworks.com

CRC Press
Taylor & Francis Group
6000 Broken Sound Parkway NW, Suite 300
Boca Raton, FL 33487-2742

Printed on acid-free paper
Version Date: 20171028

International Standard Book Number-13: 978-1-138-05933-7 (Hardback)

Visit the Taylor & Francis Web site at
http://www.taylorandfrancis.com

and the CRC Press Web site at
http://www.crcpress.com

To our parents and families
who love and support us
and
to our teachers and students
who enriched our knowledge

Contents

Preface

This book has been authored to help the students studying basic mathematical stuff that can be met up with in the course of engineering under the name of 'engineering mathematics'. They are supposed to have taken at least two freshman-level courses, one on differentiation/integration and one on linear algebra (vector and matrix). One more course on the basic level of MATLAB is recommended.

It is not the aim of this book to provide any foundation in the mathematical theory since the authors are not so knowledgeable as to do it. The first aim of this book is to help the readers understand the concepts, techniques, terminologies, and equations appearing in the existing books on engineering mathematics while using MATLAB software for computation that would be time-consuming, tedious, and error-prone with no computing device. Needless to say, the readers are recommended to learn some basic usage of MATLAB software that is available from the MATLAB help manual or the on-line documents at the website <http://www.mathworks.com/>. However, they are not required to be so good at MATLAB since most programs in this book have been composed carefully and completely so that they can be understood in connection with related/referred equations and algorithms. The readers are expected to get used to the MATLAB software while trying to modify/use the MATLAB codes in this book for solving the end-of-chapter problems or their own problems. The second and main aim of this book is to make even a novice at both MATLAB and mathematics become acquainted, at least comfortable, with MATLAB as well as mathematics while running the MATLAB codes on his/her computer, trying to understand what is going on among the equations, and making an interpretation of derived results or formulas. Is it too much to expect that a novice will become interested in engineering mathematics and simultaneously, fall in love with MATLAB, which is a universal language for engineers and scientists, after having read this book through? Is it just the authors' imagination that the readers would think of this book describing and explaining many concepts in MATLAB rather than in English? In any case, the authors have no intention to hide their hope that this book will be one of the all-the-time-reserved books in most libraries and can be found always on the desks of most engineers who are interested in developing their insight and getting an overview rather than repeating impractical computations when studying engineering mathematics. The features of this book can be summarized as follows:

1. This book presents more MATLAB programs for studying engineering mathematics than any existent books with the same or similar titles as an approach to explain most things using MATLAB and figures rather than English and equations.
2. Most MATLAB programs are presented in a complete form so that the readers can run them instantly with no programming skill and focus on understanding the mathematical manipulation process and making interpretations of the results obtained by running the programs.
3. The solutions for most examples are plotted for visualization using the related MATLAB programs because graphs are not only easier to understand than numeric values or literal expressions but also helpful in grasping the concept.

4. Authors never think that this book can replace the existent books made by the great authors. They neither expect that this book can take the place of the MATLAB manual. Instead, this book aims to bridge the gap between MATLAB software and theory, equations appearing in the field of engineering mathematics so that the readers can feel free to utilize the MATLAB software for studying mathematical stuff and become much more interested in engineering mathematics than before reading this book.

The contents of this book are derived from the works of many (known or unknown) great scientists, scholars, and researchers, all of whom are deeply appreciated. We would like to thank the reviewers for their valuable comments and suggestions, which contribute to enriching this book. We also thank the people of the School of Electronic & Electrical Engineering, Chung-Ang University for giving us an academic environment. Without affections and supports of our families and friends, this book could not be written. Special thanks should be given to the graduate students, Seungjoon Lee, Changhyeon Kim, and Wonchul Kim at the department of Aerospace Engineering in Seoul National University for their invaluable help in correction. We gratefully acknowledge the editorial and production staff (including Ms. Aastha Sharma, Ms. Michele Dimont, and Shikha Garg) of CRC Press (Taylor & Francis Group) for their kind, efficient, and encouraging guidance.

Program files can be downloaded from <https://wyyang53.wixsite.com/mysite/publications>. Any questions, comments, and suggestions regarding this book are welcome and they should be mailed to wyyang53@hanmail.net.

Won Y. Yang et al.

CHAPTER 1

Vectors and Matrices

CHAPTER OUTLINE

1.1 VECTORS

1.1.1 Geometry with Vector

Above of all, we need to distinguish between a scalar and a vector. A *scalar* is a real number with only magnitude. Length, mass, and density are examples of scalar quantities. A *vector* is a quantity with magnitude and direction that can be represented graphically as an arrow, i.e., a directed line segment (Fig. 1.1(a)) and algebraically as an ordered pair/triple $\mathbf{a}=[a_x\ a_y]^T/[a_x\ a_y\ a_z]^T$ (^{T}: transpose) of two/three real numbers in a 2D/3D Cartesian coordinate system. A 3D vector $[a_x\ a_y\ a_z]^T$ has a direction from the origin [0, 0, 0] to the point $[a_x, a_y, a_z]$ and a magnitude $\|a\|=\sqrt{a_x^2+a_y^2+a_z^2}$ equal to the length of the arrow or the distance from the origin to the point $[a_x, a_y, a_z]$. To make sure that $\mathbf{a}=[a_x\ a_y\ a_z]^T$ is a 3D vector rather than a set of coordinates of a geometric point in a 3D rectangular coordinate system, we sometimes write it as

$$\mathbf{a}=[a_x\quad a_y\quad a_z]^T=\begin{bmatrix}a_x\\ a_y\\ a_z\end{bmatrix}=a_x\mathbf{u}_x+a_y\mathbf{u}_y+a_z\mathbf{u}_z \tag{1.1.1}$$

where $\mathbf{u}_x$, $\mathbf{u}_y$, and $\mathbf{u}_z$ are the unit vectors along $+x$-, $+y$-, and $+z$-axes, respectively:

$$\mathbf{u}_x=\begin{bmatrix}1\\0\\0\end{bmatrix},\quad \mathbf{u}_y=\begin{bmatrix}0\\1\\0\end{bmatrix},\quad \mathbf{u}_z=\begin{bmatrix}0\\0\\1\end{bmatrix} \tag{1.1.2}$$

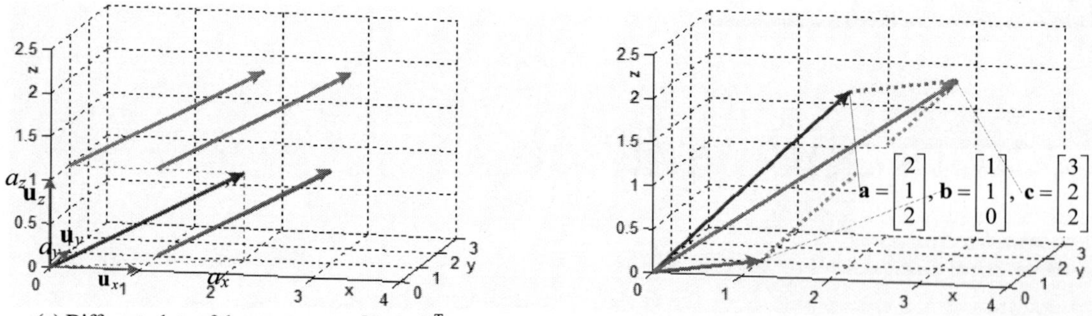

(a) Different plots of the same vector $[2 \ \ 1 \ \ 1]^T$

(b) The sum of two vectors by the parallelogram law

Figure 1.1 Plots of a vector and the sum of two vectors

Force, velocity, and acceleration are examples of vector quantities. Note that the vectors are most often written in a column vector rather than in a row vector.

Fig. 1.1(a) shows the three unit vectors $\mathbf{u}_x$, $\mathbf{u}_y$, $\mathbf{u}_z$ (each along the $+x$-, $+y$-, and $+z$-axes that are orthogonal) and different parallel arrows (with the same length and direction), which denote an identical vector $[2 \ \ 1 \ \ 1]^T$ wherever they may start and end. Fig. 1.1(b) shows the *parallelogram law* for vector addition, that the sum of two vectors is represented by the diagonal of the parallelogram having the two arrows representing the two vectors (with an identical starting point) as sides.

1.1.2 Dot Product

The *dot* (*scalar*, or *inner*) *product* of two vectors $\mathbf{a}=a_x\mathbf{u}_x+a_y\mathbf{u}_y+a_z\mathbf{u}_z$ and $\mathbf{b}=b_x\mathbf{u}_x+b_y\mathbf{u}_y+b_z\mathbf{u}_z$ is defined as

$$\mathbf{a}\bullet\mathbf{b}=\|\mathbf{a}\|\,\|\mathbf{b}\|\cos\theta_{\mathbf{ab}}=a_xb_x+a_yb_y+a_zb_z=\mathbf{a}^{*T}\mathbf{b} \ \ (^*\text{: conjugate, }^T\text{: transpose, } \|\cdot\|\text{ : norm}) \tag{1.1.3}$$

where $\theta_{\mathbf{ab}}$ is the angle between the two vectors. The second equality can be proved by using the distributive law for the dot product and the orthnormality among the unit vectors $\mathbf{u}_x$, $\mathbf{u}_y$, and $\mathbf{u}_z$:

$$\begin{bmatrix}\mathbf{u}_x & \mathbf{u}_y & \mathbf{u}_z\end{bmatrix}^T\begin{bmatrix}\mathbf{u}_x & \mathbf{u}_y & \mathbf{u}_z\end{bmatrix}=\begin{bmatrix}\mathbf{u}_x^T\\ \mathbf{u}_y^T\\ \mathbf{u}_z^T\end{bmatrix}\begin{bmatrix}\mathbf{u}_x & \mathbf{u}_y & \mathbf{u}_z\end{bmatrix}=\begin{bmatrix}\mathbf{u}_x^T\mathbf{u}_x & \mathbf{u}_x^T\mathbf{u}_y & \mathbf{u}_x^T\mathbf{u}_z\\ \mathbf{u}_y^T\mathbf{u}_x & \mathbf{u}_y^T\mathbf{u}_y & \mathbf{u}_y^T\mathbf{u}_z\\ \mathbf{u}_z^T\mathbf{u}_x & \mathbf{u}_z^T\mathbf{u}_y & \mathbf{u}_z^T\mathbf{u}_z\end{bmatrix}=\begin{bmatrix}1&0&0\\0&1&0\\0&0&1\end{bmatrix} \tag{1.1.4a}$$

$$\rightarrow\begin{cases}\mathbf{u}_x\bullet\mathbf{u}_x\overset{(1.1.3)}{=}\|\mathbf{u}_x\|\|\mathbf{u}_x\|\cos 0^\circ=1,\ \ \mathbf{u}_y\bullet\mathbf{u}_y=1,\ \ \mathbf{u}_z\bullet\mathbf{u}_z=1\\ \mathbf{u}_x\bullet\mathbf{u}_y=\mathbf{u}_y\bullet\mathbf{u}_x=\|\mathbf{u}_x\|\,\|\mathbf{u}_y\|\cos 90^\circ=0,\ \ \mathbf{u}_y\bullet\mathbf{u}_z=\mathbf{u}_z\bullet\mathbf{u}_y=0,\ \ \mathbf{u}_z\bullet\mathbf{u}_x=\mathbf{u}_x\bullet\mathbf{u}_z=0\end{cases} \tag{1.1.4b}$$

(Proof)

$$\mathbf{a}\bullet\mathbf{b}=(a_x\mathbf{u}_x+a_y\mathbf{u}_y+a_z\mathbf{u}_z)\bullet(b_x\mathbf{u}_x+b_y\mathbf{u}_y+b_z\mathbf{u}_z)\overset{(1.1.4)}{=}a_xb_x+a_yb_y+a_zb_z$$

Note that the magnitude, called the (Euclidean) *norm* or l^2-norm, of a vector $\mathbf{a}$ can be found from the dot product of $\mathbf{a}$ and itself as

$$\|\mathbf{a}\|=\sqrt{\|\mathbf{a}\|^2}=\sqrt{\mathbf{a}\bullet\mathbf{a}}=\sqrt{a_x^2+a_y^2+a_z^2} \tag{1.1.5}$$

Note also that two vectors are *orthogonal* or *perpendicular* if and only if their dot product is zero:

$$\mathbf{a}\bullet\mathbf{b}=\|\mathbf{a}\|\,\|\mathbf{b}\|\cos\theta_{\mathbf{ab}}=0 \ \Leftrightarrow\ \theta_{\mathbf{ab}}=\frac{\pi}{2} \ \text{ denoted by } \mathbf{a}\perp\mathbf{b} \tag{1.1.6}$$

1.1.2.1 The Angle between Two Nonzero Vectors

The angle between two nonzero vectors **a** and **b** can be found from their dot product as

$$\cos\theta_{\mathbf{ab}} \overset{(1.1.3)}{=} \frac{\mathbf{a}\cdot\mathbf{b}}{\|\mathbf{a}\|\|\mathbf{b}\|} \overset{(1.1.3)}{=} \frac{a_x b_x + a_y b_y + a_z b_z}{\|\mathbf{a}\|\|\mathbf{b}\|} \rightarrow \theta_{\mathbf{ab}} = \cos^{-1}\left(\frac{a_x b_x + a_y b_y + a_z b_z}{\|\mathbf{a}\|\|\mathbf{b}\|}\right) \tag{1.1.7}$$

1.1.2.2 The Law of Cosines (Cosine Formula or Cosine Rule)

To use Eq. (1.1.3) for proving the law of cosines, consider a triangle with three sides represented by the vectors $\mathbf{a}=[a_x\ a_y\ a_z]^T$, $\mathbf{b}=[b_x\ b_y\ b_z]^T$, and $\mathbf{c}=[c_x\ c_y\ c_z]^T=[a_x-b_x\ a_y-b_y\ a_z-b_z]^T$, respectively, as depicted in Fig. 1.2. Then we can derive the law of cosines as

$$\begin{aligned}\|\mathbf{c}\|^2 &= c_x^2 + c_y^2 + c_z^2 = (a_x-b_x)^2 + (a_y-b_y)^2 + (a_z-b_z)^2 \\ &= a_x^2 + a_y^2 + a_z^2 + b_x^2 + b_y^2 + b_z^2 - 2(a_x b_x + a_y b_y + a_z b_z) \underset{(1.1.3)}{\overset{(1.1.5)}{=}} \|\mathbf{a}\|^2 + \|\mathbf{b}\|^2 - 2\|\mathbf{a}\|\|\mathbf{b}\|\cos\theta_{\mathbf{ab}}\end{aligned} \tag{1.1.8}$$

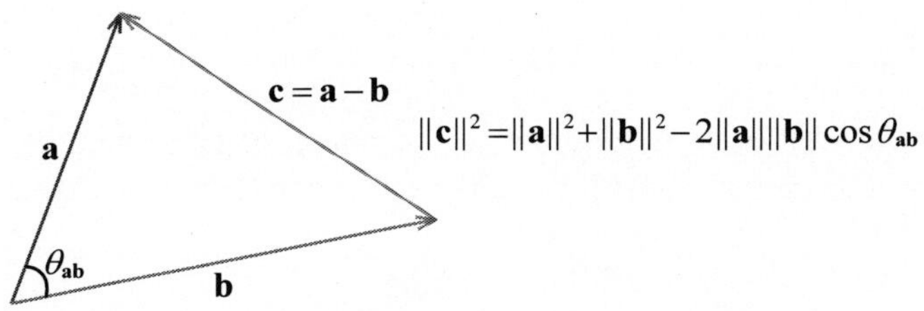

Figure 1.2 Dot product and the law of cosines

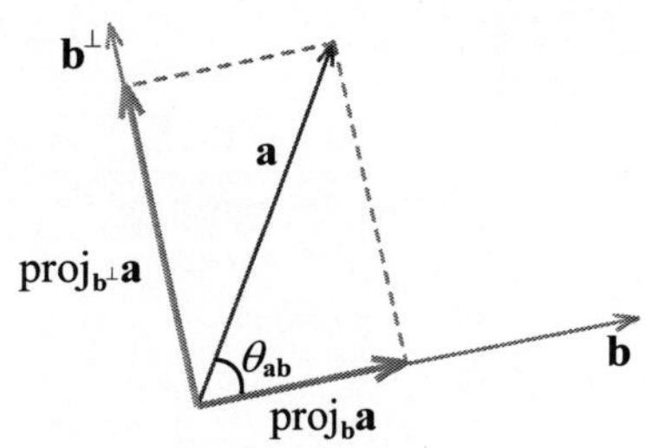

Figure 1.3 Projections of **a** on $\mathbf{b}^\perp$

1.1.2.3 Projections of Vector a onto Vectors b and $\mathbf{b}^\perp$

As depicted in Fig. 1.3, the magnitude of the component of a vector **a** along another vector **b** is $\|\mathbf{a}\|\cos\theta_{\mathbf{ab}}$ and can be found from the dot product of the two vectors as

$$\text{comp}_{\mathbf{b}}\mathbf{a} = \|\mathbf{a}\|\cos\theta_{\mathbf{ab}} \overset{(1.1.3)}{=} \frac{\mathbf{a}\cdot\mathbf{b}}{\|\mathbf{b}\|} = \mathbf{a}\cdot\frac{\mathbf{b}}{\|\mathbf{b}\|} \tag{1.1.9}$$

We can multiply this with the unit vector in the direction of **b**, i.e., **b**/||**b**|| to find the *projection* of **a** on **b** as

$$\text{proj}_{\mathbf{b}}\mathbf{a} = (\text{comp}_{\mathbf{b}}\mathbf{a})\frac{\mathbf{b}}{\|\mathbf{b}\|} \overset{(1.1.9)}{=} \left(\mathbf{a}\cdot\frac{\mathbf{b}}{\|\mathbf{b}\|}\right)\frac{\mathbf{b}}{\|\mathbf{b}\|} = (\mathbf{a}\cdot\mathbf{b})\frac{\mathbf{b}}{\|\mathbf{b}\|^2} \tag{1.1.10}$$

We can subtract this projection $\text{proj}_{\mathbf{b}}\mathbf{a}$ from **a** to find the projection of **a** on $\mathbf{b}^\perp$ as

$$\text{proj}_{\mathbf{b}^\perp}\mathbf{a} = \mathbf{a} - \text{proj}_{\mathbf{b}}\mathbf{a} \overset{(1.1.10)}{=} \mathbf{a} - (\mathbf{a}\cdot\mathbf{b})\frac{\mathbf{b}}{\|\mathbf{b}\|^2} \tag{1.1.11}$$

where $\mathbf{b}^\perp$ is the vector of magnitude ||**b**|| and direction perpendicular to **b** (see Fig. 1.3).

1.1.2.4 Point-Plane Distance

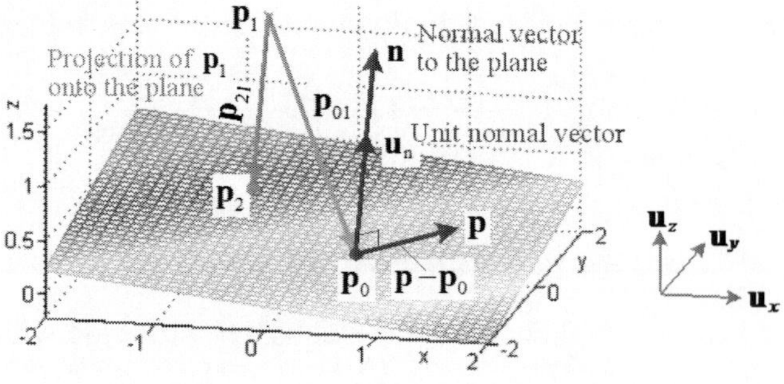

Figure 1.4 Distance from a point to a plane

Consider a point $\mathbf{p}_1=[x_1\ \ y_1\ \ z_1]^T$ and a plane described by

$$ax+by+cz+d=0 \tag{1.1.12}$$

as depicted in Fig. 1.4. To find the formula for the distance from the point $\mathbf{p}_1$ to the plane, take an arbitrary point $\mathbf{p}_0=[x_0\ \ y_0\ \ z_0]^T$ on the plane and a *normal vector*

$$\mathbf{n}=[a\ \ b\ \ c]^T \tag{1.1.13}$$

Why is $\mathbf{n}$ a normal vector to the plane? Because it is perpendicular to the plane, or equivalently, to any vector $\mathbf{p}_0-\mathbf{p}=[x_0-x\ \ y_0-y\ \ z_0-z]^T$ on the plane where the x-, y-, and z-coordinates of $\mathbf{p}_0$ and $\mathbf{p}=[x\ \ y\ \ z]^T$ satisfy the equation for the plane:

$$\mathbf{n}\cdot(\mathbf{p}-\mathbf{p}_0)=\begin{bmatrix}a\\b\\c\end{bmatrix}\cdot\begin{bmatrix}x-x_0\\y-y_0\\z-z_0\end{bmatrix}\overset{(1.1.3)}{=}a(x-x_0)+b(y-y_0)+c(z-z_0)\overset{(1.1.12)}{=}-d-(-d)=0 \tag{1.1.14}$$

Note from Eq. (1.1.6) that $\mathbf{n}$ and $(\mathbf{p}_0-\mathbf{p})$ are perpendicular (or orthogonal) to each other. Fig. 1.4 suggests that the distance from $\mathbf{p}_1$ to the plane is just the magnitude of the component of the vector $\mathbf{p}_{01}$ (from $\mathbf{p}_1$ to $\mathbf{p}_0$) along the normal vector $\mathbf{n}$, which can be found from Eq. (1.1.9) as

$$\|\mathbf{p}_{21}\|=|\mathrm{comp}_{\mathbf{n}}\mathbf{p}_{01}|\overset{(1.1.9)}{=}\left|\mathbf{p}_{01}\cdot\frac{\mathbf{n}}{\|\mathbf{n}\|}\right|\overset{(1.1.13)}{=}\frac{1}{\sqrt{a^2+b^2+c^2}}\left|\begin{bmatrix}x_0-x_1\\y_0-y_1\\z_0-z_1\end{bmatrix}\cdot\begin{bmatrix}a\\b\\c\end{bmatrix}\right|$$

$$=\left|\frac{a(x_0-x_1)+b(y_0-y_1)+c(z_0-z_1)}{\sqrt{a^2+b^2+c^2}}\right|\overset{(1.1.12)}{=}\left|\frac{-d-ax_1-by_1-cz_1}{\sqrt{a^2+b^2+c^2}}\right|=\left|\frac{ax_1+by_1+cz_1+d}{\sqrt{a^2+b^2+c^2}}\right| \tag{1.1.15}$$

Furthermore, the shortest vector $\mathbf{p}_{21}$ from $\mathbf{p}_1$ to the plane can be found as the projection of the vector $\mathbf{p}_{01}$ from $\mathbf{p}_1$ to $\mathbf{p}_0$ (the arbitrary point on the plane) onto the normal vector $\mathbf{n}$:

$$\mathbf{p}_{21}=\mathrm{proj}_{\mathbf{n}}\mathbf{p}_{01}\overset{(1.1.10)}{=}(\mathrm{comp}_{\mathbf{n}}\mathbf{p}_{01})\frac{\mathbf{n}}{\|\mathbf{n}\|}\overset{(1.1.15)}{=}-\frac{ax_1+by_1+cz_1+d}{\sqrt{a^2+b^2+c^2}}\frac{1}{\sqrt{a^2+b^2+c^2}}\begin{bmatrix}a\\b\\c\end{bmatrix}=-\frac{ax_1+by_1+cz_1+d}{a^2+b^2+c^2}\begin{bmatrix}a\\b\\c\end{bmatrix} \tag{1.1.16}$$

The projection $\mathbf{p}_2$ of the point $\mathbf{p}_1$ onto the plane can be found by adding $\mathbf{p}_{21}$ to $\mathbf{p}_1$ as

$$\mathbf{p}_2=\mathbf{p}_1+\mathbf{p}_{21}\overset{(1.1.16)}{=}\begin{bmatrix}x_1\\y_1\\z_1\end{bmatrix}-\frac{ax_1+by_1+cz_1+d}{a^2+b^2+c^2}\begin{bmatrix}a\\b\\c\end{bmatrix} \tag{1.1.17}$$

```
%em01f04.m
clear, clf % clear the workspace memory, clear the current figure window
x=[-2:0.1:2]; y=[-2:0.1:2]; [X,Y]=meshgrid(x,y); % x- and y-grids
a=1; b=-1; c=9; d=2;  Z=(-d-a*X-b*Y)/c;  % Plane equation (1.1.12)
n=[a b c].';  % Normal vector to the plane
mesh(X,Y,Z), hold on
p1=[-0.5 0.5 2].'; % A point as a column vector
% An arbitrary point on the plane
p0=zeros(3,1); p0(1)=p1(1)+1; p0(2)=p1(2)-1;
p0(3)=(-d-a*p0(1)-b*p0(2))/c; % fix p0 on the plane satisfying Eq.(1.1.12)
p01=p0-p1; % The vector from p1 to p0
un=n/norm(n); % The unit normal vector
p21=dot(p01,un)*un % Eq.(1.1.15): Projection of p01 on n
plot3(p0(1),p0(2),p0(3),'o'), plot3(p1(1),p1(2),p1(3),'rx') %Points p0, p1
quiver3(p0(1),p0(2),p0(3),un(1),un(2),un(3),0) % un from p0
quiver3(p1(1),p1(2),p1(3),p01(1),p01(2),p01(3),0) % p01 from p0
quiver3(p1(1),p1(2),p1(3),p21(1),p21(2),p21(3),0) % p21 from p1
p2=p1+p21; % Projection point p2 of p1 onto the plane: Eq.(1.1.17)
plot3(p2(1),p2(2),p2(3),'bv'), axis('equal'), grid on
xlabel('x'), ylabel('y'), zlabel('z'), view(12,15), shg
```

Fig. 1.4 has been plotted by running the above MATLAB script "em01f04.m" with a point $\mathbf{p}_1=[-0.5\ \ 0.5\ \ 2]^T$ and a plane equation $x-y+9z+2=0$, which uses the MATLAB function '`dot()`' to compute the dot product. Note that the MATLAB function '`quiver3()`' (with scale factor set to 1 for no scaling) has been used to draw 3D vectors. See Appendix E.10.

1.1.2.5 Cauchy–Schwarz Inequality

We can derive the Cauchy–Schwarz inequality by taking the absolute values of both sides of Eq. (1.1.3) as

$$|\mathbf{a}\bullet\mathbf{b}| \overset{(1.1.3)}{=} \|\mathbf{a}\|\|\mathbf{b}\||\cos\theta_{\mathbf{ab}}| \le \|\mathbf{a}\|\|\mathbf{b}\| \rightarrow |\mathbf{a}\bullet\mathbf{b}| \le \|\mathbf{a}\|\|\mathbf{b}\| \tag{1.1.18}$$

where the equality holds if and only if the two vectors **a** and **b** are parallel, i.e., **a**=k**b** (k: a scalar).

1.1.2.6 Triangle Inequality

Referring to Fig. 1.5, we can use the Cauchy–Schwartz inequality to verify the triangle inequality saying that the length of a side of a triangle is not larger than the sum of the lengths of the other two sides:

$$\begin{aligned}\|\mathbf{c}\|^2 &= \|\mathbf{a}-\mathbf{b}\|^2 \overset{(1.1.5)}{=} (\mathbf{a}-\mathbf{b})\bullet(\mathbf{a}-\mathbf{b}) = \mathbf{a}\bullet\mathbf{a}-2\mathbf{a}\bullet\mathbf{b}+\mathbf{b}\bullet\mathbf{b}\\ &\le \|\mathbf{a}\|^2+2|\mathbf{a}\bullet\mathbf{b}|+\|\mathbf{b}\|^2 \overset{(1.1.18)}{\le} \|\mathbf{a}\|^2+2\|\mathbf{a}\|\|\mathbf{b}\|+\|\mathbf{b}\|^2=(\|\mathbf{a}\|+\|\mathbf{b}\|)^2\\ ; \quad &\|\mathbf{c}\| = \|\mathbf{a}-\mathbf{b}\| \le \|\mathbf{a}\|+\|\mathbf{b}\|\end{aligned} \tag{1.1.19}$$

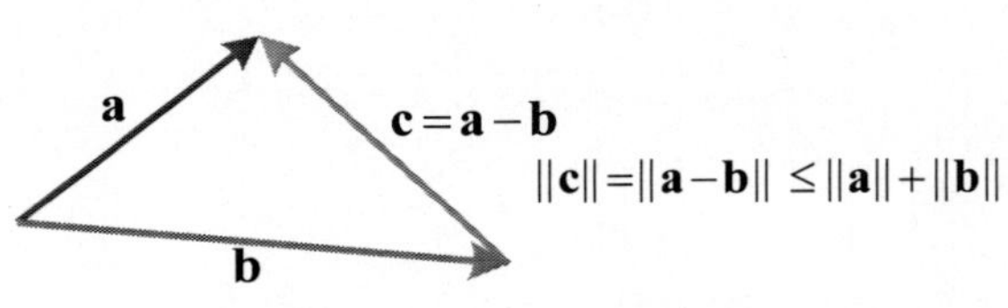

Figure 1.5 Triangle inequality

1.1.3 Cross Product

(a) (b)

Figure 1.6 Direction of cross product

The *cross* (*vector*, or *outer*) *product* of two vectors $\mathbf{a}=a_x\mathbf{u}_x+a_y\mathbf{u}_y+a_z\mathbf{u}_z$ and $\mathbf{b}=b_x\mathbf{u}_x+b_y\mathbf{u}_y+b_z\mathbf{u}_z$ is defined as

$$\mathbf{a}\times\mathbf{b} = \|\mathbf{a}\|\|\mathbf{b}\|\sin\theta_{\mathbf{ab}}\,\mathbf{u}_n = (a_y b_z - a_z b_y)\,\mathbf{u}_x + (a_z b_x - a_x b_z)\mathbf{u}_y + (a_x b_y - a_y b_x)\mathbf{u}_z$$

$$= \begin{vmatrix} \mathbf{u}_x & \mathbf{u}_y & \mathbf{u}_z \\ a_x & a_y & a_z \\ b_x & b_y & b_z \end{vmatrix} = \begin{bmatrix} a_y b_z - a_z b_y \\ a_z b_x - a_x b_z \\ a_x b_y - a_y b_x \end{bmatrix} \tag{1.1.20}$$

where $|C|$ is the determinant (Section 1.2.3) of matrix C and $\mathbf{u}_n$ is the unit normal vector to the plane spanned by **a** and **b**. (A space is said to be *spanned* by a set of vectors if any vector in the space can be expressed by a linear combination of the vectors belonging to the set.) Note that the direction of **a×b** is given by the right-hand rule in the sense that the thumb points in the direction of **a×b** if the forefinger (index finger) and middle finger point in the direction of **a** and **b**, respectively, as shown in Fig. 1.6.

The second equality of Eq. (1.1.20) can be proved by verifying that both sides are equal in terms of the magnitude and direction (perpendicular to both **a** and **b**):

$$\|\mathbf{a}\|^2\|\mathbf{b}\|^2\sin^2\theta_{\mathbf{ab}}\|\mathbf{u}_n\|^2 = \|\mathbf{a}\|^2\|\mathbf{b}\|^2(1-\cos^2\theta_{\mathbf{ab}}) = \|\mathbf{a}\|^2\|\mathbf{b}\|^2 - (\|\mathbf{a}\|\|\mathbf{b}\|\cos\theta_{\mathbf{ab}})^2$$

$$\underset{(1.1.5)}{\overset{(1.1.3)}{=}} (a_x^2+a_y^2+a_z^2)(b_x^2+b_y^2+b_z^2) - (a_x b_x + a_y b_y + a_z b_z)^2$$

$$= (a_y b_z - a_z b_y)^2 + (a_z b_x - a_x b_z)^2 + (a_x b_y - a_y b_x)^2 \tag{1.1.21}$$

$$\mathbf{a}\bullet(\mathbf{a}\times\mathbf{b}) = \begin{bmatrix} a_x \\ a_y \\ a_z \end{bmatrix} \bullet \left(\begin{bmatrix} a_x \\ a_y \\ a_z \end{bmatrix} \times \begin{bmatrix} b_x \\ b_y \\ b_z \end{bmatrix} \right) = \begin{bmatrix} a_x \\ a_y \\ a_z \end{bmatrix} \bullet \begin{bmatrix} a_y b_z - a_z b_y \\ a_z b_x - a_x b_z \\ a_x b_y - a_y b_x \end{bmatrix}$$

$$= a_x(a_y b_z - a_z b_y) + a_y(a_z b_x - a_x b_z) + a_z(a_x b_y - a_y b_x) = 0 \rightarrow \mathbf{a} \perp (\mathbf{a}\times\mathbf{b}) \tag{1.1.22a}$$

$$\mathbf{b}\bullet(\mathbf{a}\times\mathbf{b}) = \begin{bmatrix} b_x \\ b_y \\ b_z \end{bmatrix} \bullet \left(\begin{bmatrix} a_x \\ a_y \\ a_z \end{bmatrix} \times \begin{bmatrix} b_x \\ b_y \\ b_z \end{bmatrix} \right) = \begin{bmatrix} b_x \\ b_y \\ b_z \end{bmatrix} \bullet \begin{bmatrix} a_y b_z - a_z b_y \\ a_z b_x - a_x b_z \\ a_x b_y - a_y b_x \end{bmatrix}$$

$$= b_x(a_y b_z - a_z b_y) + b_y(a_z b_x - a_x b_z) + b_z(a_x b_y - a_y b_x) = 0 \rightarrow \mathbf{b} \perp (\mathbf{a}\times\mathbf{b}) \tag{1.1.22b}$$

Alternatively, we can use the relations among the unit vectors $\mathbf{u}_x$, $\mathbf{u}_y$, and $\mathbf{u}_z$:

$$\mathbf{u}_x \times \mathbf{u}_y \overset{(1.1.20),\,\theta=\pi/2}{=} \mathbf{u}_z \quad (1.1.23\text{a}), \qquad \mathbf{u}_y \times \mathbf{u}_x = -\mathbf{u}_z \quad (1.1.24\text{a})$$

$$\mathbf{u}_y \times \mathbf{u}_z = \mathbf{u}_x \quad (1.1.23\text{b}), \qquad \mathbf{u}_z \times \mathbf{u}_y = -\mathbf{u}_x \quad (1.1.24\text{b})$$

$$\mathbf{u}_z \times \mathbf{u}_x = \mathbf{u}_y \quad (1.1.23\text{c}), \qquad \mathbf{u}_x \times \mathbf{u}_z = -\mathbf{u}_y \quad (1.1.24\text{c})$$

$$\mathbf{u}_x \times \mathbf{u}_x = \mathbf{u}_y \times \mathbf{u}_y = \mathbf{u}_z \times \mathbf{u}_z \overset{(1.1.20),\,\theta=0}{=} 0 \tag{1.1.25}$$

to prove the second equality of Eq. (1.1.20) as

$$\mathbf{a} \times \mathbf{b} = (a_x\mathbf{u}_x + a_y\mathbf{u}_y + a_z\mathbf{u}_z) \times (b_x\mathbf{u}_x + b_y\mathbf{u}_y + b_z\mathbf{u}_z)$$
$$\overset{(1.1.23,24,25)}{=} (a_y b_z - a_z b_y)\,\mathbf{u}_x + (a_z b_x - a_x b_z)\mathbf{u}_y + (a_x b_y - a_y b_x)\mathbf{u}_z$$

1.1.3.1 Triple Scalar Product and Triple Vector Product

We can prove the equality among three *triple scalar products*

$$(\mathbf{a} \times \mathbf{b}) \bullet \mathbf{c} = (\mathbf{b} \times \mathbf{c}) \bullet \mathbf{a} = (\mathbf{c} \times \mathbf{a}) \bullet \mathbf{b} \tag{1.1.26}$$

as follows:

$$(\mathbf{a} \times \mathbf{b}) \bullet \mathbf{c} \overset{(1.1.20)}{=} (c_x\mathbf{u}_x + c_y\mathbf{u}_y + c_z\mathbf{u}_z) \bullet \left\{(a_y b_z - a_z b_y)\mathbf{u}_x + (a_z b_x - a_x b_z)\mathbf{u}_y + (a_x b_y - a_y b_x)\mathbf{u}_z\right\}$$
$$\overset{(1.1.3)}{=} c_x(a_y b_z - a_z b_y) + c_y(a_z b_x - a_x b_z) + c_z(a_x b_y - a_y b_x) \tag{1.1.27b}$$

$$(\mathbf{b} \times \mathbf{c}) \bullet \mathbf{a} \overset{(1.1.20)}{=} (a_x\mathbf{u}_x + a_y\mathbf{u}_y + a_z\mathbf{u}_z) \bullet \left\{(b_y c_z - b_z c_y)\mathbf{u}_x + (b_z c_x - b_x c_z)\mathbf{u}_y + (b_x c_y - b_y c_x)\mathbf{u}_z\right\}$$
$$\overset{(1.1.3)}{=} a_x(b_y c_z - b_z c_y) + a_y(b_z c_x - b_x c_z) + a_z(b_x c_y - b_y c_x) \tag{1.1.27a}$$

$$(\mathbf{b} \times \mathbf{c}) \bullet \mathbf{a} \overset{(1.1.20)}{=} (a_x\mathbf{u}_x + a_y\mathbf{u}_y + a_z\mathbf{u}_z) \bullet \left\{(b_y c_z - b_z c_y)\mathbf{u}_x + (b_z c_x - b_x c_z)\mathbf{u}_y + (b_x c_y - b_y c_x)\mathbf{u}_z\right\}$$
$$\overset{(1.1.3)}{=} b_x(c_y a_z - c_z a_y) + b_y(c_z a_x - c_x a_z) + b_z(c_x a_y - c_y a_x) \tag{1.1.27c}$$

Also, we can prove the equality for a *triple vector product*

$$\mathbf{a} \times (\mathbf{b} \times \mathbf{c}) = (\mathbf{c} \bullet \mathbf{a})\,\mathbf{b} - (\mathbf{b} \bullet \mathbf{a})\mathbf{c} \tag{1.1.28}$$

as follows:

$$\mathbf{a} \times (\mathbf{b} \times \mathbf{c}) \overset{(1.1.20)}{=} \begin{bmatrix} a_x \\ a_y \\ a_z \end{bmatrix} \times \begin{bmatrix} b_y c_z - b_z c_y \\ b_z c_x - b_x c_z \\ b_x c_y - b_y c_x \end{bmatrix} = \begin{bmatrix} a_y(b_x c_y - b_y c_x) - a_z(b_z c_x - b_x c_z) \\ a_z(b_y c_z - b_z c_y) - a_x(b_x c_y - b_y c_x) \\ a_x(b_z c_x - b_x c_z) - a_y(b_y c_z - b_z c_y) \end{bmatrix} \tag{1.1.29a}$$

$$\mathbf{b}(\mathbf{a} \bullet \mathbf{c}) - \mathbf{c}(\mathbf{a} \bullet \mathbf{b}) \overset{(1.1.3)}{=} \begin{bmatrix} b_x \\ b_y \\ b_z \end{bmatrix} (a_x c_x + a_y c_y + a_z c_z) - \begin{bmatrix} c_x \\ c_y \\ c_z \end{bmatrix} (a_x b_x + a_y b_y + a_z b_z) \tag{1.1.29b}$$

1.1.3.2 Jacobi Identity

We can use Eq. (1.1.28) for the triple vector product to prove the *Jacobi identity*:

$$\mathbf{a}\times(\mathbf{b}\times\mathbf{c})+\mathbf{b}\times(\mathbf{c}\times\mathbf{a})+\mathbf{c}\times(\mathbf{a}\times\mathbf{b})=\mathbf{0} \tag{1.1.30a}$$

$$(\mathbf{a}\times\mathbf{b})\times\mathbf{c}+(\mathbf{b}\times\mathbf{c})\times\mathbf{a}+(\mathbf{c}\times\mathbf{a})\times\mathbf{b}=\mathbf{0} \tag{1.1.30b}$$

1.1.3.3 Areas and Volume with Cross Product

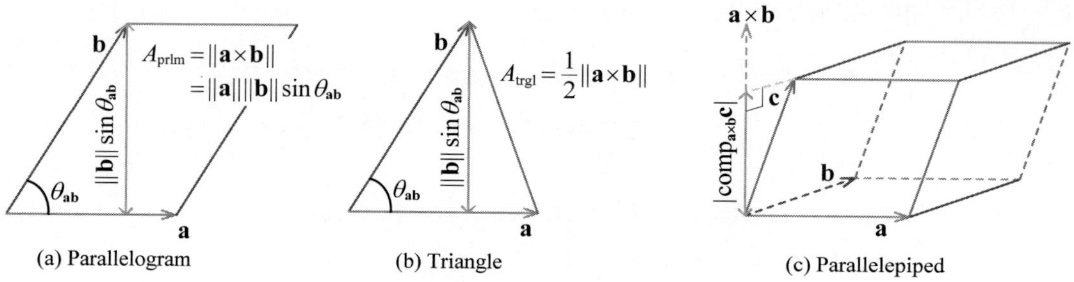

Figure 1.7 Areas and volume with cross product

The area of a parallelogram with its sides denoted by two nonparallel vectors **a** and **b** can be computed from the magnitude of the cross product of **a** and **b** (see Fig. 1.7(a)):

$$A_{\text{prlm}} = \|\mathbf{a}\times\mathbf{b}\| = \|\mathbf{a}\|\|\mathbf{b}\|\sin\theta_{\mathbf{ab}} \tag{1.1.31}$$

The area of a triangle with its sides denoted by two nonparallel vectors **a** and **b** can also be computed from the magnitude of the cross product of **a** and **b** (see Fig. 1.7(b)):

$$A_{\text{trgl}} = \frac{1}{2}\|\mathbf{a}\times\mathbf{b}\| = \frac{1}{2}\|\mathbf{a}\|\|\mathbf{b}\|\sin\theta_{\mathbf{ab}} \tag{1.1.32}$$

The volume of a parallelepiped with its sides denoted by three nonparallel vectors **a**, **b**, and **c** can be computed from the magnitude of the *triple scalar product* or *box product* of **a**, **b**, and **c**:

$$|(\mathbf{a}\times\mathbf{b})\cdot\mathbf{c}| = |\mathbf{c}\cdot(\mathbf{a}\times\mathbf{b})| = \|\mathbf{a}\times\mathbf{b}\|\left|\mathbf{c}\cdot\frac{\mathbf{a}\times\mathbf{b}}{\|\mathbf{a}\times\mathbf{b}\|}\right| \overset{(1.1.9)}{=} \|\mathbf{a}\times\mathbf{b}\|\,|\text{comp}_{\mathbf{a}\times\mathbf{b}}\mathbf{c}|$$
$$= (\text{base parallelogram area})(\text{height}) \tag{1.1.33}$$

Why is $\text{comp}_{\mathbf{a}\times\mathbf{b}}\mathbf{c}$ the height of the parallelepiped? Because it is the magnitude of the component of **c** along the normal vector **a**×**b** to the base plane spanned by **a** and **b** (see Fig. 1.7(c)).

1.1.3.4 Coplanar Vectors

The necessary and sufficient condition for three vectors **a**, **b**, and **c** to be *coplanar*, i.e., in the same plane, is

$$(\mathbf{a}\times\mathbf{b})\cdot\mathbf{c} = 0 \tag{1.1.34}$$

This forces the third vector to be perpendicular to the normal vector to the plane spanned by the other two vectors.

1.1.4 Lines and Planes

1.1.4.1 Vector Equation of a Line

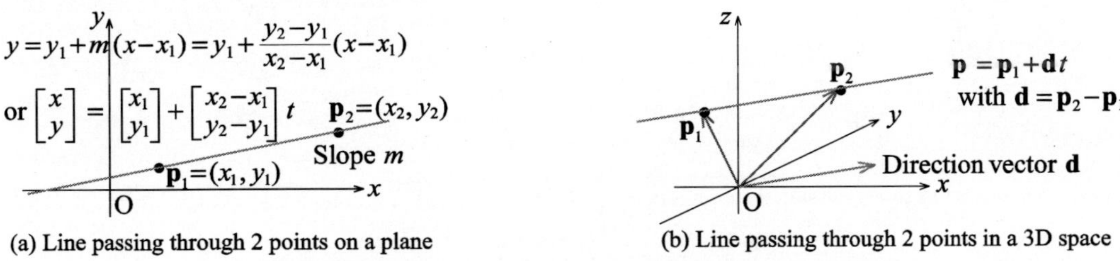

Figure 1.8 Vector equation of a line with direction vector **d**

The *vector equation* for the line with *direction vector* **d** passing through a point $\mathbf{p}_1$ is

$$\mathbf{p} = \mathbf{p}_1 + \mathbf{d}t \quad \text{with direction vector } \mathbf{d} \tag{1.1.35}$$

Accordingly, the vector equation for the line through any two distinct points $\mathbf{p}_1$ and $\mathbf{p}_2$ is

$$\mathbf{p} = \mathbf{p}_1 + \mathbf{d}t = \mathbf{p}_1 + (\mathbf{p}_2 - \mathbf{p}_1)t \quad \text{with direction vector } \mathbf{d} = \mathbf{p}_2 - \mathbf{p}_1 \tag{1.1.36}$$

This describes any point **p** on the line with an appropriate real value of parameter t (see Fig. 1.8). For example, $\mathbf{p}_1$ and $\mathbf{p}_2$ are described by Eq. (1.1.36) with t=0 and 1, respectively. Eq. (1.1.36) can also be used to describe a line segment from $\mathbf{p}_0 = \mathbf{p}_1 + \mathbf{d}t_0$ to $\mathbf{p}_f = \mathbf{p}_1 + \mathbf{d}t_f$ with an appropriate closed interval $[t_0, t_f]$.

Eq. (1.1.35) can be written in the form of *parametric equations* as

$$x = x_1 + d_x t, \; y = y_1 + d_y t, \; z = z_1 + d_z t \tag{1.1.37a}$$

or in the form of *symmetric equations* (with the parameter t eliminated) as

$$\frac{x - x_1}{d_x} = \frac{y - y_1}{d_y} = \frac{z - z_1}{d_z} \tag{1.1.37b}$$

if the three components of the direction vector $\mathbf{d} = [d_x \; d_y \; d_z]^T$ are all nonzero. If one of the three components of **d** is zero, say $d_x = 0$, it should be written as

$$x = x_1, \; \frac{y - y_1}{d_y} = \frac{z - z_1}{d_z} \tag{1.1.38}$$

The interested readers who want to feel the vector equation of a line through a soft experiment are advised to run the following MATLAB script "em01f08.m":

```
%em01f08.m
p1=[-1 -1 1.5].'; p2=[1 0 2].'; % Two points
plot3(p1(1),p1(2),p1(3),'o', p2(1),p2(2),p2(3),'o'), hold on
d=p2-p1; % Direction vector
t=[-1 2]; % Parametric vector
p=repmat(p1,1,length(t))+(p2-p1)*t; % Vector equation Eq.(1.1.36)
plot3(p(1,:),p(2,:),p(3,:),'r') % Plot the line in a 3D space
O=zeros(3,1); arrow3(p1,O), arrow3(p2,O) % Arrows of p1 and p2 from Origin
view(12,15), grid on
```

1.1.4.2 Vector Equation of a Plane

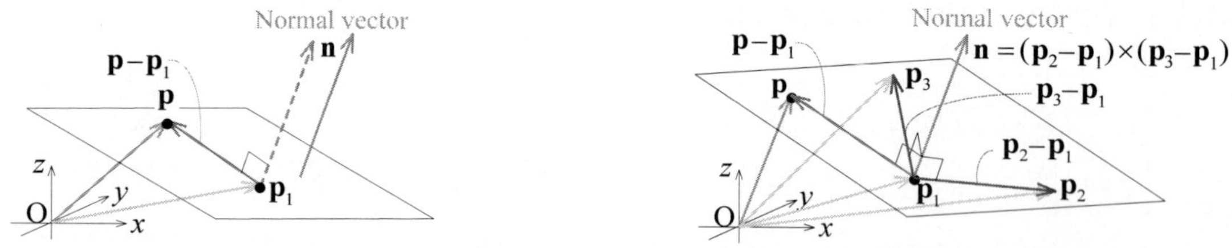

(a) Plane with normal vector **n** passing through a point $\mathbf{p}_1$ (b) Plane containing three points $\mathbf{p}_1$, $\mathbf{p}_2$, and $\mathbf{p}_3$

Figure 1.9 Vector equation of a plane

The *vector* and *Cartesian equations* of a plane with normal vector $\mathbf{n}=[a\ \ b\ \ c]^T$ passing through a point $\mathbf{p}_1=[x_1\ \ y_1\ \ z_1]^T$ (Fig. 1.9(a)) are

$$\mathbf{n}\bullet(\mathbf{p}-\mathbf{p}_1)=0 \tag{1.1.39a}$$

and

$$a(x-x_1)+b(y-y_1)+c(z-z_1)=0 \tag{1.1.39b}$$

This implies that a line from $\mathbf{p}_1$ to any point $\mathbf{p}=[x\ \ y\ \ z]^T$ on the plane is perpendicular to the normal vector **n**.

How do we write the equation for a plane containing some three points $\mathbf{p}_1$, $\mathbf{p}_2$, and $\mathbf{p}_3$ (Fig. 1.9(b))? It can be obtained from Eq. (1.1.39a) with the normal vector **n** constructed by the cross product of the two vectors $(\mathbf{p}_2-\mathbf{p}_1)$ and $(\mathbf{p}_3-\mathbf{p}_1)$ on the plane:

$$\{(\mathbf{p}_2-\mathbf{p}_1)\times(\mathbf{p}_3-\mathbf{p}_1)\}\bullet(\mathbf{p}-\mathbf{p}_1)=0 \tag{1.1.40}$$

(Example 1.1) Finding a Normal Vector to a Plane

We can use the cross product to find a normal vector to the plane containing three points $\mathbf{p}_1=[0\ \ 0\ \ -2/9]^T$, $\mathbf{p}_2=[1\ \ 0\ \ -1/3]^T$, and $\mathbf{p}_3=[0\ \ 1\ \ -1/9]^T$ as

$$\mathbf{n}=(\mathbf{p}_2-\mathbf{p}_1)\times(\mathbf{p}_3-\mathbf{p}_1)=\begin{bmatrix}1\\0\\-1/9\end{bmatrix}\times\begin{bmatrix}0\\1\\1/9\end{bmatrix}=\begin{bmatrix}0\cdot(1/9)-(-1/9)\cdot 1\\(-1/9)\cdot 1-0\cdot(1/9)\\1\cdot 1-0\cdot 0\end{bmatrix}=\begin{bmatrix}1/9\\-1/9\\1\end{bmatrix} \tag{E1.1.1}$$

Alternatively, we first can find the equation $ax+by+cz+d=0$ for the plane containing the three points $\mathbf{p}_1$, $\mathbf{p}_2$, and $\mathbf{p}_3$ by substituting the coordinates of the three points into the plane equation:

$$\begin{bmatrix}0&0&-2/9\\1&0&-1/3\\0&1&-1/9\end{bmatrix}\begin{bmatrix}a\\b\\c\end{bmatrix}=-\begin{bmatrix}d\\d\\d\end{bmatrix};\ \begin{bmatrix}a\\b\\c\end{bmatrix}=-\begin{bmatrix}0&0&2/9\\1&0&-1/3\\1&-1&0\end{bmatrix}^{-1}\begin{bmatrix}d\\d\\d\end{bmatrix}=-\begin{bmatrix}-3/2&1&0\\-1/2&0&1\\-9/2&0&0\end{bmatrix}\begin{bmatrix}d\\d\\d\end{bmatrix}=\begin{bmatrix}0.5d\\-0.5d\\4.5d\end{bmatrix} \tag{E1.1.2}$$

and then write a normal vector from the coefficients as

$$\mathbf{n}=\begin{bmatrix}0.5d\\-0.5d\\4.5d\end{bmatrix} \tag{E1.1.3}$$

Note that the two vectors obtained as Eqs. (E1.1.1) and (E1.1.3) are virtually equal (with possibly different lengths and opposite directions) as normal vectors to the plane since they are parallel.

1.1.4.3 Intersection of Two Planes

To find the vector equation for the intersection line of two planes P_a and P_b, we determine an intersection point $\mathbf{p}_1$ of the two planes and find the direction vector $\mathbf{d}$ of the line of intersection from the cross product of the normal vectors $\mathbf{n}_a$ and $\mathbf{n}_b$ to the two planes. Then we can use Eq. (1.1.35) to write the vector equation of the line of intersection as

$$\mathbf{p} \overset{(1.1.35)}{=} \mathbf{p}_1 + \mathbf{d}\,t = \mathbf{p}_1 + (\mathbf{n}_a \times \mathbf{n}_b)\,t \tag{1.1.41}$$

(Example 1.2) Finding the Line of Intersection of Two Planes

Find the line of intersection of the two planes described by

$$2x - 3y + 4z = 1 \tag{E1.2.1a}$$

$$x - y - z = 5 \tag{E1.2.1b}$$

We first find an intersection point $\mathbf{p}_1$ of the two planes by setting one of the x-, y-, and z-coordinates to zero, say $z=0$, and solving the two equations as

$$\left.\begin{matrix} 2x_1 - 3y_1 = 1 \\ x_1 - y_1 = 5 \end{matrix}\right\} \rightarrow \begin{bmatrix} 2 & -3 \\ 1 & -1 \end{bmatrix}\begin{bmatrix} x_1 \\ y_1 \end{bmatrix} = \begin{bmatrix} 1 \\ 5 \end{bmatrix};\ \begin{bmatrix} x_1 \\ y_1 \end{bmatrix} = \begin{bmatrix} 2 & -3 \\ 1 & -1 \end{bmatrix}^{-1}\begin{bmatrix} 1 \\ 5 \end{bmatrix} = \begin{bmatrix} -1 & 3 \\ -1 & 2 \end{bmatrix}\begin{bmatrix} 1 \\ 5 \end{bmatrix} = \begin{bmatrix} 14 \\ 9 \end{bmatrix} \tag{E1.2.2}$$

which yields an intersection point $\mathbf{p}_1 = [14 \;\; 9 \;\; 0]^T$. Then we find the direction vector $\mathbf{d}$ of the line of intersection from the cross product of the normal vectors $\mathbf{n}_a = [2 \; -3 \;\; 4]^T$ and $\mathbf{n}_b = [1 \; -1 \; -1]^T$ to the two planes as

$$\mathbf{d} = \mathbf{n}_a \times \mathbf{n}_b = \begin{bmatrix} 2 \\ -3 \\ 4 \end{bmatrix} \times \begin{bmatrix} 1 \\ -1 \\ -1 \end{bmatrix} = \begin{bmatrix} 7 \\ 6 \\ 1 \end{bmatrix} \tag{E1.2.3}$$

Thus we can write the vector equation of the line of intersection as

$$\mathbf{p} \overset{(1.1.41)}{=} \mathbf{p}_1 + \mathbf{d}\,t \overset{(E1.2.3)}{=} \begin{bmatrix} 14 \\ 9 \\ 0 \end{bmatrix} + \begin{bmatrix} 7 \\ 6 \\ 1 \end{bmatrix} t \tag{E1.2.4}$$

Alternatively, we can solve the two equations for x and y with $z = t$ (like a given constant) to get

$$\left.\begin{matrix} 2x - 3y = 1 - 4t \\ x - y = 5 + t \end{matrix}\right\} \rightarrow \begin{bmatrix} 2 & -3 \\ 1 & -1 \end{bmatrix}\begin{bmatrix} x \\ y \end{bmatrix} = \begin{bmatrix} 1-4t \\ 5+t \end{bmatrix};\ \begin{bmatrix} x \\ y \end{bmatrix} = \begin{bmatrix} -1 & 3 \\ -1 & 2 \end{bmatrix}\begin{bmatrix} 1-4t \\ 5+t \end{bmatrix} = \begin{bmatrix} 14+7t \\ 9+6t \end{bmatrix} \tag{E1.2.5}$$

which is the same result $\{x=14+7t,\ y=9+6t,\ z=t\}$ as Eq. (E1.2.4). For the same purpose, we can also run the following statements to use the MATLAB function '`intersect_planes()`' listed below:

```
>>planes=[2 -3 4 -1; 1 -1 -1 -5]; p1=[14; 9; 0]; Kg=1;
>>[d,p]=intersect_planes(planes,Kg), [d,p]=intersect_planes(planes,p1,Kg)
    d = 7              p =  0
        6                  -3
        1                  -2
```

This means that an equation for the line of intersection between the two planes is $\{x=7t,\ y=-3+6t,\ z=-2+t\}$. Does it seem to differ from Eq. (E1.2.4)? Then replace t by $t+2$ to see their equivalence. The MATLAB function '`intersect_planes(planes,Kg)`' with Kg=1 gives the plot of planes/lines as shown in Fig. 1.10.

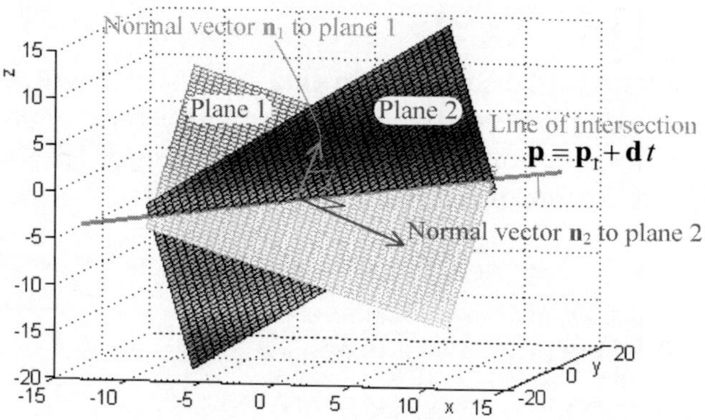

Figure 1.10 Line of intersection between two planes

```
function [d,p]=intersect_planes(planes,Kg)
% This computes the line of intersection of 2 planes.
% Inputs : planes=[a1 b1 c1 d1; a2 b2 c2 d2]
%          Kg=1/0  to or not to plot the graph
% Outputs: d = Direction vector of the line of intersection
%          p = An intersection point of the 2 planes
%Copyleft: Won Y. Yang, wyyang53@hanmail.net, CAU for academic use only
A=planes(:,1:3);  b=-planes(:,4);
RankA=rank(A); RankP=rank(planes);
if RankA==1
  if RankP==1, error('The 2 planes coincide (redundancy)!');
   else  error('The 2 planes are parallel! (inconsistency)');
  end
end
d=cross(planes(1,1:3),planes(2,1:3)).'; % Direction vector Eq.(1.1.41)
p=A\b; % An intersection point
if nargin<2, Kg=0; end
if (Kg>0|nargout==0)&abs(A(1,3)*A(2,3))>1e-8 % To plot the planes/lines
  x=[p(1)-10:0.5:p(1)+10]; % x-grid
  y=[p(2)-10:0.5:p(2)+10]; % y-grid
  [X,Y]=meshgrid(x,y); % x/y-grid matrix
  Z1=(b(1)*ones(size(X))-(A(1,1)*X+A(1,2)*Y))/A(1,3); % Plane 1
  Z2=(b(2)*ones(size(X))-(A(2,1)*X+A(2,2)*Y))/A(2,3); % Plane 2
  clf, mesh(X,Y,Z1), hold on, surf(X,Y,Z2)
  t0=-2; tf=2; t=[t0 tf];  % t-parameter vector
  % Only two end points are enough for plot3() to draw a straight line.
  iline=repmat(p,1,2)+d*t; % Vector equation of the line of intersection
  plot3(iline(1,:),iline(2,:),iline(3,:),'r'), grid on
end
% Examples of usage:
% planes=[2 -3  4 -1; 1 -1 -1 -5]; [d,p]=intersect_planes(planes,1)
```

If you don't have time or interest to learn about the graphic stuff, i.e., how to graph planes using '`mesh()`' or '`surf()`', never mind about what is going on in the latter part of the above MATLAB function '`intersect_planes()`' starting from the 18th line, which is executed only when the second input argument Kg is given as a positive number. Also, if you don't know how to tell whether a singular set of linear equations corresponds to the underdetermined case (redundancy: infinitely many solutions) or the overdetermined case (inconsistency: no exact solution) based on the rank of the coefficient matrix, put off trying to understand the part from the 10th line to the 15th line till Section 1.3.

1.1.4.4 Intersection of a Plane and a Line

Referring to Fig. 1.11, the intersection point **p** of a plane with normal vector $\mathbf{n}=[a\ b\ c]^T$ containing a point $\mathbf{p}_1=[x_1\ y_1\ z_1]^T$ and a line with direction vector $\mathbf{d}=[d_x\ d_y\ d_z]^T$ passing through a point $\mathbf{p}_2=[x_2\ y_2\ z_2]^T$ off the plane can be found by solving the set of two vector equations of the plane and line:

$$\mathbf{n}\bullet(\mathbf{p}-\mathbf{p}_1)\overset{(1.1.39a)}{=}0 \tag{1.1.42}$$

$$\mathbf{p}\overset{(1.1.35)}{=}\mathbf{p}_2+\mathbf{d}\,t \tag{1.1.43}$$

Substituting Eq. (1.1.43) into Eq. (1.1.42) yields

$$\mathbf{n}\bullet(\mathbf{p}_2+\mathbf{d}\,t-\mathbf{p}_1)\underset{(1.1.43)}{\overset{(1.1.42)}{=}}0;\quad t=\frac{\mathbf{n}\bullet(\mathbf{p}_1-\mathbf{p}_2)}{\mathbf{n}\bullet\mathbf{d}}=\frac{a(x_1-x_2)+b(y_1-y_2)+c(z_1-z_2)}{ad_x+bd_y+cd_z} \tag{1.1.44}$$

where it is assumed that $\mathbf{n}\bullet\mathbf{d}\neq 0$, which implies that the direction vector of the line is not perpendicular to the normal vector of the plane so that the line does not run parallel to the plane. If $\mathbf{n}\bullet\mathbf{d}=0$, the line is parallel to the plane so that there are infinitely many intersections (on the entire line) or there is no intersection depending on whether the numerator of Eq. (1.1.44) is also zero (when $\mathbf{p}_2$ is on the plane) or not. We can find the intersection point by substituting Eq. (1.1.44) into Eq. (1.1.43).

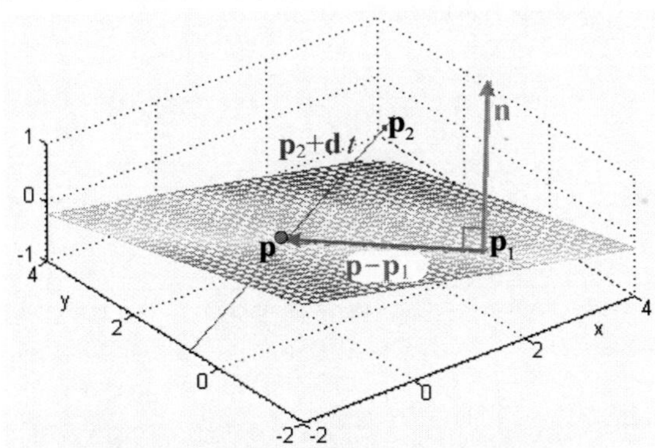

Figure 1.11 Intersection of a plane and a line

(Example 1.3) Finding the Intersection of a Plane and a Line

Find the intersection point of a plane described by

$$x+y+5z=1 \tag{E1.3.1}$$

and a line described by

$$\mathbf{p}\overset{(1.1.35)}{=}\mathbf{p}_2+\mathbf{d}\,t;\quad \begin{bmatrix}x\\y\\z\end{bmatrix}=\begin{bmatrix}1\\0\\2\end{bmatrix}+\begin{bmatrix}-1\\1\\-2\end{bmatrix}t \tag{E1.3.2}$$

With a point $\mathbf{p}_1=[2\ \ -1\ \ 0]^T$ on the plane (as a trivial solution of Eq. (E1.3.1)), let us use Eq. (1.1.44) to get

$$t\overset{(1.1.44)}{=}\frac{\mathbf{n}\bullet(\mathbf{p}_1-\mathbf{p}_2)}{\mathbf{n}\bullet\mathbf{d}}=\frac{1(x_1-x_2)+1(y_1-y_2)+5(z_1-z_2)}{1\cdot(-1)+1\cdot 1+5\cdot(-2)}=\frac{1(2-1)+1(-1-0)+5(0-2)}{-10}=1 \tag{E1.3.3}$$

Substituting this value of t into Eq. (E1.3.2) yields the intersection point as

$$\mathbf{p} \underset{(E1.3.3)}{\overset{(E1.3.2)}{=}} \begin{bmatrix} 1 \\ 0 \\ 2 \end{bmatrix} + \begin{bmatrix} -1 \\ 1 \\ -2 \end{bmatrix} \cdot 1 = \begin{bmatrix} 0 \\ 1 \\ 0 \end{bmatrix} \tag{E1.3.4}$$

We can type the following statements in the MATLAB Command Window to get the same result:

```
>>n=[1 1 5].'; p1=[2 -1 0].'; % Normal vector and a point of the plane
>>d=[-1 1 -2].'; p2=[1 0 2].'; % Direction vector and a point of the line
>>t=dot(n,p1-p2)/dot(n,d); % Eq.(E1.3.3)
>>p=p2+d*t  % Intersection point by Eq.(E1.3.2)
```

1.1.4.5 Point-Line Distance

Referring to Fig. 1.12, the (shortest) distance between a point $\mathbf{p}_1=[x_1\ y_1\ z_1]^T$ and a line $\mathbf{p}_2+\mathbf{d}t$ (with $\mathbf{p}_2=[x_2\ y_2\ z_2]^T$) can be found as the length (norm) of the cross product of the difference vector $\mathbf{p}_1-\mathbf{p}_2$ and the (unit) direction vector $\mathbf{d}/||\mathbf{d}||$:

$$\left\| \frac{(\mathbf{p}_1-\mathbf{p}_2)\times\mathbf{d}}{||\mathbf{d}||} \right\| = ||\mathbf{p}_1-\mathbf{p}_2||\,|\sin\theta| \tag{1.1.45}$$

Also, the point $\mathbf{p}$ (on the line) closest to the point $\mathbf{p}_1$ can be found by taking the projection of the difference vector $\mathbf{p}_1-\mathbf{p}_2$ onto the direction vector $\mathbf{d}$:

$$\text{proj}_{\mathbf{d}}(\mathbf{p}_1-\mathbf{p}_2) \overset{(1.1.10)}{=} ((\mathbf{p}_1-\mathbf{p}_2)\bullet\mathbf{d})\frac{\mathbf{d}}{||\mathbf{d}||^2} \tag{1.1.46}$$

This process to find the distance between a point and a line has been cast into the following MATLAB function '`distance_point_line()`'.

```
function [distance,p_nearest]=distance_point_line(p1,p2,d)
% To find the shortest distance between a point p1 and a line p2+d*t
%  and the point (on the line) closest to p1
distance = norm(cross(p1-p2,d))/norm(d); % Eq.(1.1.45)
p_nearest = p2 + proj_a_onto_b(p1-p2,d); % Eq.(1.1.46)
```

```
function pab=proj_a_onto_b(a,b)
% Projection of vector a onto vector b
pab = dot(a,b)*b/norm(b)^2;  % Eq.(1.1.10)
```

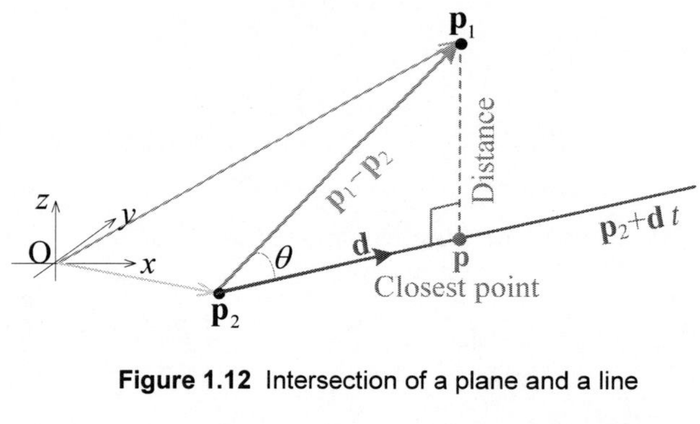

Figure 1.12 Intersection of a plane and a line

1.1.4.6 Line-Line Distance

How can we find the (shortest) distance between two lines, each described by the following vector equations?

$$\begin{aligned}&\text{Line 1: } \mathbf{p}=\mathbf{p}_1+\mathbf{d}_1 t_1\\&\text{Line 2: } \mathbf{p}=\mathbf{p}_2+\mathbf{d}_2 t_2\end{aligned} \tag{1.1.47}$$

The vector for measuring the distance between two lines is perpendicular to both lines (each with direction vector $\mathbf{d}_1$ and $\mathbf{d}_2$) and accordingly, it can be obtained from the projection of the difference vector $\mathbf{p}_1-\mathbf{p}_2$ onto the cross vector $(\mathbf{d}_1\times\mathbf{d}_2)$ of $\mathbf{d}_1$ and $\mathbf{d}_2$:

$$\text{Distance vector}: \mathbf{d}=\text{proj}_{\mathbf{d}_1\times\mathbf{d}_2}(\mathbf{p}_1-\mathbf{p}_2)\overset{(1.1.10)}{=}(\mathbf{p}_1-\mathbf{p}_2)\bullet(\mathbf{d}_1\times\mathbf{d}_2)\frac{\mathbf{d}_1\times\mathbf{d}_2}{\|\mathbf{d}_1\times\mathbf{d}_2\|^2} \tag{1.1.48}$$

and its length (norm)

$$\text{Distance: } d=\|\mathbf{d}\|=\left\|\text{proj}_{\mathbf{d}_1\times\mathbf{d}_2}(\mathbf{p}_1-\mathbf{p}_2)\right\| \tag{1.1.49}$$

is the distance between the two lines. What if $\mathbf{d}_1\times\mathbf{d}_2=\mathbf{0}$? It implies that the two lines are parallel and possibly, they coincide. In this case, the problem should be converted into a point-to-line distance problem, say, to find the distance between point $\mathbf{p}_2$ and line 1.

If the two lines are not parallel, the closest points $\mathbf{p}_a$ and $\mathbf{p}_b$ (each on line 1 and line 2, respectively) can be found in the following two steps:

(Step 1) Move line 2 by the distance vector $\mathbf{d}$ and find the intersection of the moved line 2 with line 1 by solving

$$\mathbf{p}_1+\mathbf{d}_1 t_1=(\mathbf{p}_2+\mathbf{d})+\mathbf{d}_2 t_2\ ;\ \begin{bmatrix}\mathbf{d}_1 & -\mathbf{d}_2\end{bmatrix}\begin{bmatrix}t_1\\t_2\end{bmatrix}=\mathbf{p}_2+\mathbf{d}-\mathbf{p}_1 \tag{1.1.50}$$

$$;\qquad \begin{bmatrix}t_1\\t_2\end{bmatrix}=\begin{bmatrix}\mathbf{d}_1 & -\mathbf{d}_2\end{bmatrix}\backslash(\mathbf{p}_2+\mathbf{d}-\mathbf{p}_1) \tag{1.1.51}$$

for t_1 and t_2, and substitute t_1 into Eq. (1.1.47a) to get the intersection of the moved line 2 with line 1, which must be $\mathbf{p}_a$, i.e., one of the closest points on line 1. Note that Eq. (1.1.50) with a 3×2 (rectangular) coefficient matrix $[\mathbf{d}_1\ -\mathbf{d}_2]$ is a set of three linear simultaneous equations in two unknowns t_1 and t_2, and it can be solved by using the backslash(\) operator (see Section 1.3) like Eq. (1.1.51). If the rank of the coefficient matrix is 1, Eq. (1.1.50) has infinitely many solutions. But such an undetermined case never happens because it corresponds to the case of "parallel lines" that has been dealt with separately as a point-to-line distance problem above.

(Step 2) Subtract $\mathbf{d}$ from $\mathbf{p}_a$ to get $\mathbf{p}_b=\mathbf{p}_a-\mathbf{d}$, which is another one of the closest points (on line 2).

$$\mathbf{p}_a=\mathbf{p}_1+\mathbf{d}_1 t_1,\qquad \mathbf{p}_b=\mathbf{p}_a-\mathbf{d} \tag{1.1.52}$$

This process to find the distance between two lines and their closest points has been cast into the following MATLAB function: '`intersect_lines()`'. To try it, type the following statements at the MATLAB prompt, which will yield Fig. 1.13.

```
>>p1=[-1;-1;-1]; d1=[1;0;1]; p2=[1;1;-1]; d2=[-1;0;1]; fig=1;
  clf, [p_int,dist,dist_vector]=intersect_lines(p1,d1,p2,d2,fig)
```

```
function [p_int,distance,dist_vector]=intersect_lines(p1,d1,p2,d2,fig)
% finds the shortest distance between 2 lines p1 + d1*t1 and p2 + d2*t2
%  and the closest points on each line (possibly intersection point)
% Output:
%   p_int = 3x2 matrix containing 2 point vectors closest to each other
%   distance = Distance between the 2 lines (the two closest points)
%   dist_vector = Distance vector connecting the two closest points
%Copyleft: Won Y. Yang, wyyang53@hanmail.net, CAU for academic use only
if nargin<5, fig = 0; end  % To get the picture or not
tt = [-10 10]; EPS = 1e-6;
% To find the shortest distance between the two lines
d1_cross_d2 = cross(d1,d2);
norm_d1_cross_d2 = norm(d1_cross_d2); % Denominator of Eq.(1.1.48)
if norm_d1_cross_d2>EPS % If the two lines are not parallel,
  % find the distance vector and the distance between the two lines
  dist_vector = proj_a_onto_b(p1-p2,d1_cross_d2); % Eq.(1.1.48)
  distance = norm(dist_vector); % Distance % Eq.(1.1.49)
  if distance>EPS % If the two lines do not cross each other,
    % then find the 2 points (each on the 2 lines) closest to each other
    [p_int(:,1),distance0] = intersect_lines(p1,d1,p2+dist_vector,d2);
    p_int(:,2) = p_int(:,1) - dist_vector;
    if fig>0
      plot3(p1(1)+d1(1)*tt,p1(2)+d1(2)*tt,p1(3)+d1(3)*tt,'b'), hold on
      plot3(p2(1)+d2(1)*tt,p2(2)+d2(2)*tt,p2(3)+d2(3)*tt,'r')
      plot3(p_int(1,:),p_int(2,:),p_int(3,:),'r-o'), grid on
    end
   else % If the two lines cross each other at some point
    distance = 0;  dist_vector = zeros(3,1);
    % Find the intersection point on line 1
    A = [d1 -d2]; % 3x2 Coefficient matrix
    t = A\[p2-p1]; % Eq.(1.1.51)
    p_int = p1 + d1*t(1); % Intersection point Eq.(1.1.52)
  end
else % If the two lines are parallel
  [distance,p_int,dist_vector] = distance_point_line(p2,p1,d1);
  if distance<eps, p_int = 'The 2 lines coincide.';
   else  p_int = 'The 2 lines are parallel.';
  end
end
```

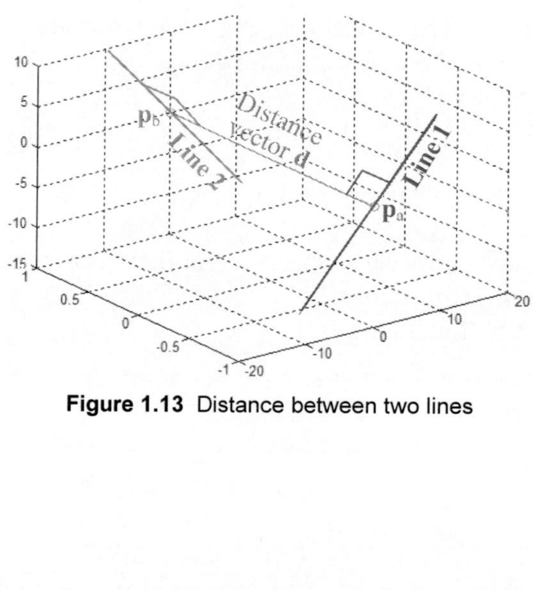

Figure 1.13 Distance between two lines

1.1.5 Vector Space

In this section, we generalize the space of vector from 3-D space R^3 (with *x*-/*y*-/*z*-coordinates) into *N*-D space R^N (that can no longer be visualized for $N>3$) where an *N*-D vector (or more simply an *N*-vector) is expressed as an ordered *N*-tuple $\mathbf{x}=[x_1, x_2, \ldots, x_N]^T$ where x_n is called the *n*th component of the vector **x**. Why do we use the numbers {1, 2, 3, ...} instead of $\{x, y, z, \ldots\}$ as a subscript for denoting each component of a vector? Because we may run out of letters for high-dimensional vectors. The concepts of vector addition, scalar addition, dot product, and so on carry over to R^N:

$$\text{Vector addition: } \mathbf{x}+\mathbf{y}=[x_1+y_1 \;\; x_2+y_2 \;\; \cdots \;\; x_N+y_N]^T \tag{1.1.53}$$

$$\text{Multiplication by scalar: } a\mathbf{x}=[ax_1 \;\; ax_2 \;\; \cdots \;\; ax_N]^T \tag{1.1.54}$$

$$\text{Dot product: } \mathbf{x}\bullet\mathbf{y}=x_1y_1+x_2y_2+\cdots+x_Ny_N \tag{1.1.55}$$

A set *V* of vectors is said to be a *vector space* or *linear space* if it satisfies the following properties:

Vector addition	Closure under addition: $\mathbf{x}+\mathbf{y}\in V$ for any $\mathbf{x},\mathbf{y}\in V$	(1.1.56a)
	Commutative law: $\mathbf{x}+\mathbf{y}=\mathbf{y}+\mathbf{x}$ for any $\mathbf{x},\mathbf{y}\in V$	(1.1.56b)
	Associative law: $\mathbf{x}+(\mathbf{y}+\mathbf{z})=(\mathbf{x}+\mathbf{y})+\mathbf{z}$ for any $\mathbf{x},\mathbf{y},\mathbf{z}\in V$	(1.1.56c)
	Zero vector: There exists a unique vector $\mathbf{0}$ such that $\mathbf{x}+\mathbf{0}=\mathbf{x}\;\forall\;\mathbf{x}\in V$	(1.1.56d)
	Additive inverse: $\exists$ a negative vector $-\mathbf{x}$ such that $\mathbf{x}+(-\mathbf{x})=\mathbf{0}\;\forall\;\mathbf{x}\in V$	(1.1.56e)
Scalar multiplication	Closure under scalar multiplication: $a\mathbf{x}\in V$ for all $\mathbf{x}\in V$	(1.1.56f)
	Distributive law: $a(\mathbf{x}+\mathbf{y})=a\mathbf{x}+a\mathbf{y}$	(1.1.56g)
	Distributive law: $(a_1+a_2)\mathbf{x}=a_1\mathbf{x}+a_2\mathbf{x}$	(1.1.56h)
	Associative law: $a_1(a_2\mathbf{x})=(a_1a_2)\mathbf{x}$	(1.1.56i)
	Multiplicative identity: $1\mathbf{x}=\mathbf{x}$	(1.1.56j)

A subset *W* of a vector space *V* is called a *subspace* of *V* if it is also a vector space with vector addition and scalar multiplication. A set of vectors $\{\mathbf{x}_1, \mathbf{x}_2, \ldots, \mathbf{x}_N\}$ is said to be *linearly independent* if their linear combination becomes zero only for all zero coefficients, i.e.,

$$a_1\mathbf{x}_1+a_2\mathbf{x}_2+\cdots+a_N\mathbf{x}_N=\mathbf{0} \text{ only for } a_1=a_2=\cdots=a_N=0 \tag{1.1.57}$$

Otherwise, it is said to be *linearly dependent*. The set of all the linear combinations of vectors $\mathbf{x}_1$, $\mathbf{x}_2$, ..., and $\mathbf{x}_N$ is called the *span* of the vectors and denoted by Span($\mathbf{x}_1, \mathbf{x}_2, \ldots, \mathbf{x}_N$). If every vector in a vector space *V* can be expressed as a linear combination of vectors $\mathbf{x}_1$, $\mathbf{x}_2$, ..., and $\mathbf{x}_N$, the vectors are said to *span V*, or equivalently, *V* is said to be *spanned* by the vectors and the set $\{\mathbf{x}_1, \mathbf{x}_2, \ldots, \mathbf{x}_N\}$ is called a *spanning set* for *V*. If every vector in a vector space *V* can be (uniquely) expressed as a linear combination of linearly independent vectors $\mathbf{x}_1$, $\mathbf{x}_2$, ..., and $\mathbf{x}_N$, the set $\{\mathbf{x}_1, \mathbf{x}_2, \ldots, \mathbf{x}_N\}$ is referred to as a *basis* for the vector space *V* and the number of vectors constituting the basis is called the *dimension* of *V*.

For example, let us consider the following three sets of vectors in the 3-D vector space R^3:

$$\{\mathbf{u}_x,\mathbf{u}_y,\mathbf{u}_z\},\ \{\mathbf{u}_x,\ \mathbf{u}_x+\mathbf{u}_y,\ \mathbf{u}_x+\mathbf{u}_y+\mathbf{u}_z\},\text{ and }\{\mathbf{u}_x,\mathbf{u}_y,\mathbf{u}_z,\ \mathbf{u}_x+\mathbf{u}_y,\ \mathbf{u}_x+\mathbf{u}_y+\mathbf{u}_z\}$$

While each of these three sets may well be called a spanning set for R^3, only the first two sets are linearly independent, so that each of them can be called a basis for R^3.

1.1.6 Coordinate Systems

A coordinate system is a system that uses an ordered 2/3-tuple of real numbers to uniquely specify the position of a point in 2/3-D space. Fig. 1.14.1 shows the (orthogonal) unit vectors of a Cartesian (rectangular) coordinate system. Figs. 1.14.2/1.14.3 show the unit vectors of the cylindrical/spherical coordinate systems and their relationships with Cartesian unit vectors, respectively.

Referring to Figs. 1.14.2(a)–(b) or Fig. 1.15.1(a), the relationship between the cylindrical unit vectors and the Cartesian unit vectors can be written as

$$\begin{bmatrix}\mathbf{u}_\rho^T\\ \mathbf{u}_\phi^T\\ \mathbf{u}_z^T\end{bmatrix}=\overbrace{\begin{bmatrix}\cos\phi & \sin\phi & 0\\ -\sin\phi & \cos\phi & 0\\ 0 & 0 & 1\end{bmatrix}}^{T_{\rm CR}}\begin{bmatrix}\mathbf{u}_x^T\\ \mathbf{u}_y^T\\ \mathbf{u}_z^T\end{bmatrix} \text{ or equivalently, } \begin{bmatrix}\mathbf{u}_x^T\\ \mathbf{u}_y^T\\ \mathbf{u}_z^T\end{bmatrix}=\overbrace{\begin{bmatrix}\cos\phi & -\sin\phi & 0\\ \sin\phi & \cos\phi & 0\\ 0 & 0 & 1\end{bmatrix}}^{T_{\rm RC}=T_{\rm CR}^T=T_{\rm CR}^{-1}}\begin{bmatrix}\mathbf{u}_\rho^T\\ \mathbf{u}_\phi^T\\ \mathbf{u}_z^T\end{bmatrix} \quad (1.1.58a,b)$$

where

$$\begin{cases}x=\rho\cos\phi\\ y=\rho\sin\phi\\ z\end{cases} \leftrightarrow \begin{cases}\rho=\sqrt{x^2+y^2}\\ \phi=\tan^{-1}(y/x)\\ z\end{cases} \quad (1.1.58c,d)$$

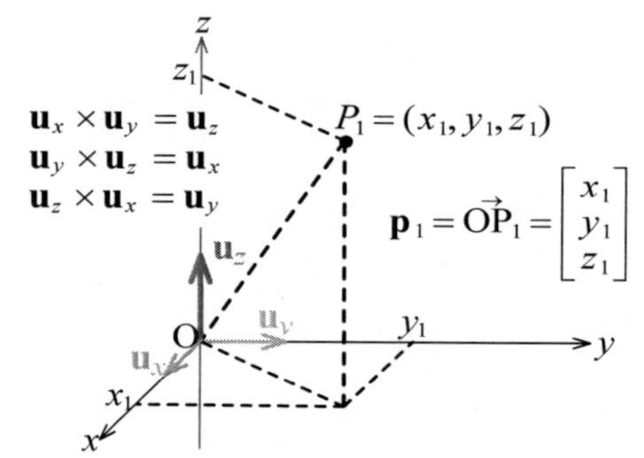

Figure 1.14.1 Cartesian coordinate system

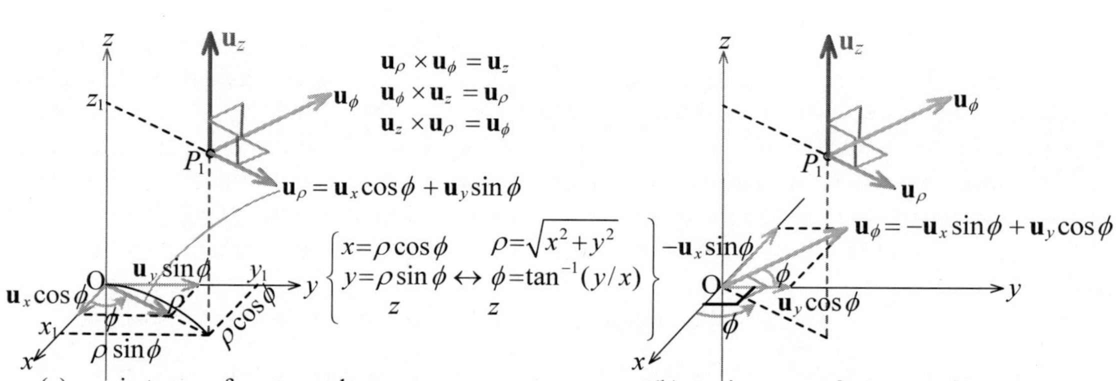

(a) $\mathbf{u}_\rho$ in terms of $\mathbf{u}_x$, $\mathbf{u}_y$, and $\mathbf{u}_z$ (b) $\mathbf{u}_\phi$ in terms of $\mathbf{u}_x$, $\mathbf{u}_y$, and $\mathbf{u}_z$

Figure 1.14.2 Cylindrical coordinate system

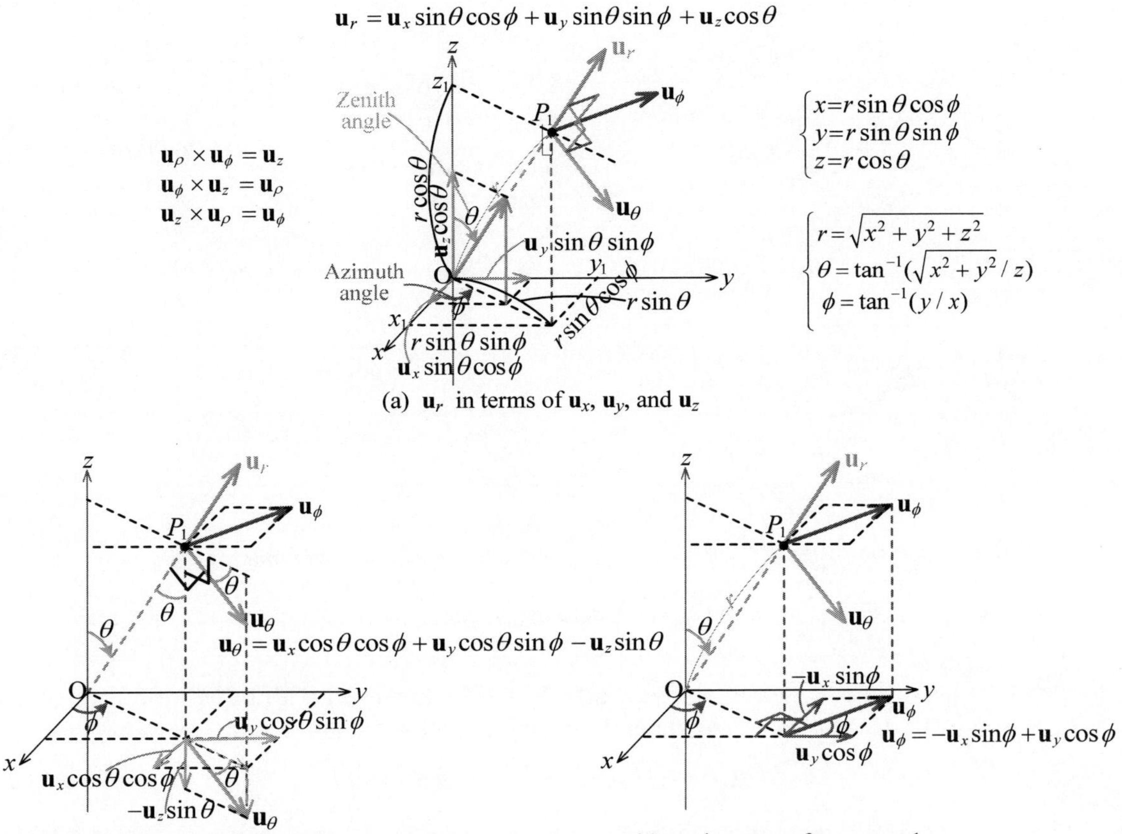

(a) $\mathbf{u}_r$ in terms of $\mathbf{u}_x$, $\mathbf{u}_y$, and $\mathbf{u}_z$

(b) $\mathbf{u}_\theta$ in terms of $\mathbf{u}_x$, $\mathbf{u}_y$, and $\mathbf{u}_z$

(c) $\mathbf{u}_\phi$ in terms of $\mathbf{u}_x$, $\mathbf{u}_y$, and $\mathbf{u}_z$

Figure 1.14.3 Spherical coordinate system

Also, referring to Figs. 1.14.3(a)–(c) or Fig. 1.15.2(a), the relationship between the spherical unit vectors and the Cartesian unit vectors can be written as

$$\begin{bmatrix}\mathbf{u}_r^T\\ \mathbf{u}_\theta^T\\ \mathbf{u}_\phi^T\end{bmatrix} = \overbrace{\begin{bmatrix}\sin\theta\cos\phi & \sin\theta\sin\phi & \cos\theta\\ \cos\theta\cos\phi & \cos\theta\sin\phi & -\sin\theta\\ -\sin\phi & \cos\phi & 0\end{bmatrix}}^{T_{SR}}\begin{bmatrix}\mathbf{u}_x^T\\ \mathbf{u}_y^T\\ \mathbf{u}_z^T\end{bmatrix} \tag{1.1.59a}$$

$$\text{or equivalently,}\quad \begin{bmatrix}\mathbf{u}_x^T\\ \mathbf{u}_y^T\\ \mathbf{u}_z^T\end{bmatrix} = \overbrace{\begin{bmatrix}\sin\theta\cos\phi & \cos\theta\cos\phi & -\sin\phi\\ \sin\theta\sin\phi & \cos\theta\sin\phi & \cos\phi\\ \cos\theta & -\sin\theta & 0\end{bmatrix}}^{T_{RS}=T_{SR}^T=T_{SR}^{-1}}\begin{bmatrix}\mathbf{u}_r^T\\ \mathbf{u}_\theta^T\\ \mathbf{u}_\phi^T\end{bmatrix} \tag{1.1.59b}$$

where

$$\begin{cases}x=r\sin\theta\cos\phi\\ y=r\sin\theta\sin\phi\\ z=r\cos\theta\end{cases} \leftrightarrow \begin{cases}r=\sqrt{x^2+y^2+z^2}\\ \theta=\tan^{-1}(\sqrt{x^2+y^2}/z)\\ \phi=\tan^{-1}(y/x)\end{cases} \tag{1.1.59c,d}$$

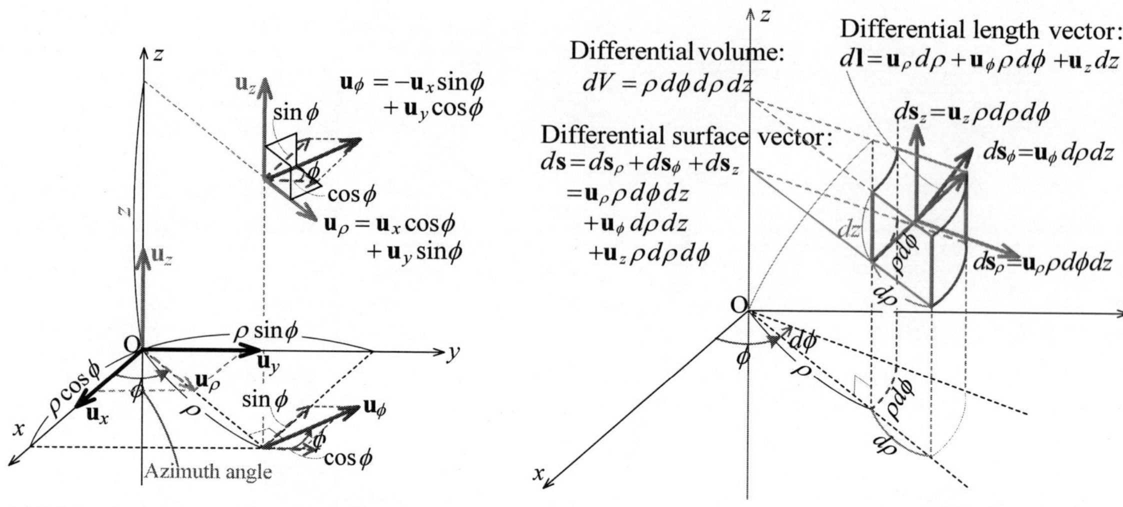

(a) Relationship between cylindrical/Cartesian coordinate systems (b) Differential surface/length vectors and differential volume

Figure 1.15.1 Differential volume and surface/length vectors in the cylindrical coordinate system

On the other hand, Fig. 1.15.1(b) shows a differential hexahedron with three infinitesimal sides, each expressed by $\mathbf{u}_\rho d\rho$, $\mathbf{u}_\phi \rho d\phi$, and $\mathbf{u}_z dz$ in the cylindrical coordinate system, whose volume is

$$\textit{Differential volume:}\quad dV = (\rho d\phi) d\rho dz \tag{1.1.60a}$$

The *differential surface vector* for the differential hexahedron in the cylindrical coordinate system is defined as the sum of three surface vectors $d\mathbf{s}_\rho$, $d\mathbf{s}_\phi$, and $d\mathbf{s}_z$, each representing the face of the differential hexahedron perpendicular to the ρ-, ϕ-, and z-axis, respectively:

$$\textit{Differential surface vector:}\, d\mathbf{s} = d\mathbf{s}_\rho + d\mathbf{s}_\phi + d\mathbf{s}_z = \mathbf{u}_\rho(\rho d\phi)dz + \mathbf{u}_\phi d\rho dz + \mathbf{u}_z(\rho d\phi)d\rho \tag{1.1.60b}$$

Fig. 1.15.1(b) also shows that a differential displacement resulting from infinitesimal movements along the three axes in the cylindrical coordinate system can be expressed by

$$\textit{Differential length vector:}\quad d\mathbf{l} = \mathbf{u}_\rho d\rho + \mathbf{u}_\phi \rho d\phi + \mathbf{u}_z dz \tag{1.1.60c}$$

Fig. 1.15.2(b) shows a differential hexahedron with three infinitesimal sides, each expressed by $\mathbf{u}_r dr$, $\mathbf{u}_\theta r d\theta$, and $\mathbf{u}_\phi r \sin\theta d\phi$ in the spherical coordinate system, whose volume is

$$\textit{Differential volume:}\quad dV = dr(rd\theta)(r\sin\theta d\phi) = r^2 \sin\theta\, dr d\theta d\phi \tag{1.1.61a}$$

The *differential surface vector* for the differential hexahedron in the spherical coordinate system is defined as the sum of three surface vectors $d\mathbf{s}_r$, $d\mathbf{s}_\theta$, and $d\mathbf{s}_\phi$, each representing the face of the differential hexahedron perpendicular to the r-, θ-, and ϕ-axis, respectively:

$$d\mathbf{s} = d\mathbf{s}_r + d\mathbf{s}_\theta + d\mathbf{s}_\phi = (rd\theta)(r\sin\theta d\phi)\mathbf{u}_r + (r\sin\theta d\phi)dr\,\mathbf{u}_\theta + (rd\theta)dr\,\mathbf{u}_\phi \tag{1.1.61b}$$

Fig. 1.15.2(b) also shows that a differential displacement resulting from infinitesimal movements along the three axes in the spherical coordinate system can be expressed by

$$\textit{Differential length vector:}\quad d\mathbf{l} = \mathbf{u}_r dr + \mathbf{u}_\theta r d\theta + \mathbf{u}_\phi r\sin\theta d\phi \tag{1.1.61c}$$

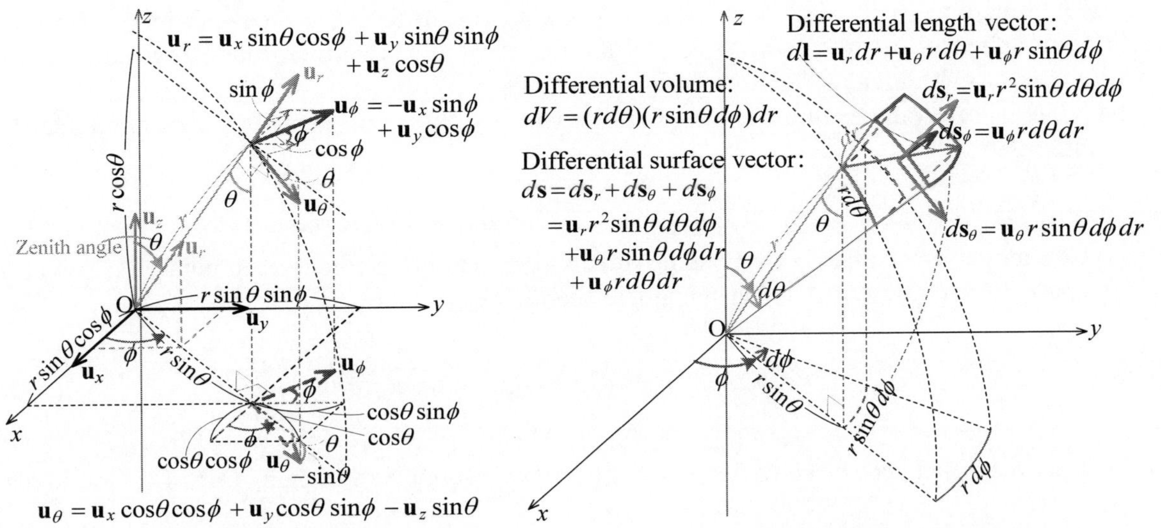

(a) Relationship with spherical/Cartesian coordinate systems (b) Differential surface/length vectors and differential volume

Figure 1.15.2 Differential volume and surface/length vectors in the spherical coordinate system

Table 1.1 Properties of and transformations among the three coordinate systems

	Cartesian $\{x, y, z\}$	Cylindrical $\{\rho,\phi,z\}$	Spherical $\{r,\theta,\phi\}$
Cartesian $\{x, y, z\}$	$\mathbf{u}_x\times\mathbf{u}_y=\mathbf{u}_z$ $\mathbf{u}_y\times\mathbf{u}_z=\mathbf{u}_x$ $\mathbf{u}_z\times\mathbf{u}_x=\mathbf{u}_y$	$\begin{cases}x=\rho\cos\phi\\ y=\rho\sin\phi\\ z\end{cases}$ [x,y,z]=pol2cart(ϕ,ρ,z)	$\begin{cases}x=r\sin\theta\cos\phi\\ y=r\sin\theta\sin\phi\\ z=r\cos\theta\end{cases}$ [x,y,z]=sph2cart($\phi,\frac{\pi}{2}-\theta,r$)
Cylindrical $\{\rho, \phi, z\}$	$\begin{cases}\rho=\sqrt{x^2+y^2}\\ \phi=\tan^{-1}(y/x)\\ z\end{cases}$ [ϕ,ρ,z]=cart2pol(x,y,z)	$\mathbf{u}_\rho\times\mathbf{u}_\phi=\mathbf{u}_z$ $\mathbf{u}_\phi\times\mathbf{u}_z=\mathbf{u}_\rho$ $\mathbf{u}_z\times\mathbf{u}_\rho=\mathbf{u}_\phi$	$\begin{bmatrix}\mathbf{u}_x^T\\ \mathbf{u}_y^T\\ \mathbf{u}_z^T\end{bmatrix}=\begin{bmatrix}\cos\phi & -\sin\phi & 0\\ \sin\phi & \cos\phi & 0\\ 0 & 0 & 1\end{bmatrix}\begin{bmatrix}\mathbf{u}_\rho^T\\ \mathbf{u}_\phi^T\\ \mathbf{u}_z^T\end{bmatrix}$
Spherical $\{r, \theta, \phi\}$	$\begin{cases}r=\sqrt{x^2+y^2+z^2}\\ \theta=\tan^{-1}(\sqrt{x^2+y^2}/z)\\ \phi=\tan^{-1}(y/x)\end{cases}$ [$\phi,\frac{\pi}{2}-\theta,r$]=cart2sph(x,y,z)	$\begin{bmatrix}\mathbf{u}_x^T\\ \mathbf{u}_y^T\\ \mathbf{u}_z^T\end{bmatrix}=\begin{bmatrix}\sin\theta\cos\phi\\ \sin\theta\sin\phi\\ \cos\theta\end{bmatrix}$	$\mathbf{u}_r\times\mathbf{u}_\theta=\mathbf{u}_\phi$ $\mathbf{u}_\theta\times\mathbf{u}_\phi=\mathbf{u}_r$ $\mathbf{u}_\phi\times\mathbf{u}_r=\mathbf{u}_\theta$ $\begin{matrix}\cos\theta\cos\phi & -\sin\phi\\ \cos\theta\sin\phi & \cos\phi\\ -\sin\theta & 0\end{matrix}\Bigg]\begin{bmatrix}\mathbf{u}_r^T\\ \mathbf{u}_\theta^T\\ \mathbf{u}_\phi^T\end{bmatrix}$
Differential length vector	$d\mathbf{l}=dx\,\mathbf{u}_x+dy\,\mathbf{u}_y+dz\,\mathbf{u}_z$	$d\mathbf{l}=d\rho\,\mathbf{u}_\rho+\rho d\phi\,\mathbf{u}_\phi+dz\,\mathbf{u}_z$	$d\mathbf{l}=dr\,\mathbf{u}_r+rd\theta\,\mathbf{u}_\theta+r\sin\theta d\phi\,\mathbf{u}_\phi$
Differential surface vector	$d\mathbf{s}=(dydz)\mathbf{u}_x+(dzdx)\mathbf{u}_y+(dxdy)\mathbf{u}_z$	$d\mathbf{s}=(\rho d\phi)dz\,\mathbf{u}_\rho+d\rho dz\,\mathbf{u}_\phi+(\rho d\phi)d\rho\,\mathbf{u}_z$	$d\mathbf{s}=(rd\theta)(r\sin\theta d\phi)\mathbf{u}_r+(r\sin\theta d\phi)dr\,\mathbf{u}_\theta+(rd\theta)dr\,\mathbf{u}_\phi$
Differential volume	$dV=dx\,dy\,dz$	$dV=(\rho d\phi)d\rho dz$	$dV=(rd\theta)(r\sin\theta d\phi)dr$

Table 1.1, together with Figs. 1.14 and 1.15, shows the differential elements in the Cartesian/cylindrical/spherical coordinate systems and the transformations between the Cartesian coordinate system and the cylindrical/spherical coordinate systems.

Now, let us see some examples to get familiar with the three coordinate systems and understand their interrelation.

(Example 1.4) Cylindrical/Spherical-to-Cartesian Coordinate Conversions

(a) Consider a vector field, corresponding to an assignment of vectors to each point in a 2-D or 3-D space, which is described in the cylindrical coordinate system as

$$\mathbf{H}_\mathrm{a}(\rho,\phi,z)= 0\mathbf{u}_\rho+\frac{1}{\rho}\mathbf{u}_\phi+0\mathbf{u}_z=[\mathbf{u}_\rho\ \mathbf{u}_\phi\ \mathbf{u}_z]\begin{bmatrix}0\\1/\rho\\0\end{bmatrix} \tag{E1.4.1}$$

This vector field can be converted into the Cartesian coordinate system by using Eq. (1.1.58) as

$$\mathbf{H}_\mathrm{a}(x,y,z)=[\mathbf{u}_\rho\ \mathbf{u}_\phi\ \mathbf{u}_z]\begin{bmatrix}0\\1/\rho\\0\end{bmatrix}\overset{(1.1.58\mathrm{a})}{=}\left(\overbrace{\begin{bmatrix}\cos\phi & \sin\phi & 0\\-\sin\phi & \cos\phi & 0\\0 & 0 & 1\end{bmatrix}}^{T_\mathrm{CR}}\begin{bmatrix}\mathbf{u}_x^\mathrm{T}\\\mathbf{u}_y^\mathrm{T}\\\mathbf{u}_z^\mathrm{T}\end{bmatrix}\right)^\mathrm{T}\begin{bmatrix}0\\1/\rho\\0\end{bmatrix}$$

$$=[\mathbf{u}_x\ \mathbf{u}_y\ \mathbf{u}_z]\begin{bmatrix}\cos\phi & -\sin\phi & 0\\\sin\phi & \cos\phi & 0\\0 & 0 & 1\end{bmatrix}\begin{bmatrix}0\\1/\rho\\0\end{bmatrix}=[\mathbf{u}_x\ \mathbf{u}_y\ \mathbf{u}_z]\begin{bmatrix}(-\sin\phi)/\rho\\(\cos\phi)/\rho\\0\end{bmatrix}$$

$$\overset{(1.1.58\mathrm{d})}{=}\frac{-y}{x^2+y^2}\mathbf{u}_x+\frac{x}{x^2+y^2}\mathbf{u}_y \tag{E1.4.2}$$

This cylindrical-to-Cartesian coordinate conversion can be performed by using the MATLAB function '`cyl2cart_v()`' (see Section E.7 in Appendix E) as

```
>>syms x y z r th ph;
  Ha_cyl = [0; 1/r; 0]; % Eq.(E1.4.1)
  Ha_cart = cyl2cart_v(Ha_cyl); % Eq.(E1.4.2)
  pretty(simplify(Ha_cart.')) % transpose, make simple and pretty
```

Running these MATLAB statements yields

```
+-                        -+
|       y         x         |
|  - -------, -------, 0    |
|     2    2   2    2       |
|    x  + y   x  + y        |
+-                        -+
```

The following MATLAB script, "em01e04a.m," can be run to plot the vector field $\mathbf{H}_\mathrm{a}$ at several points as shown in Fig. 1.16(a). Note that the direction of $\mathbf{u}_\phi$ at a point $\mathbf{p}$ is the direction along which the (counterclockwise) azimuth angle ϕ of the *position* (*location* or *radius*) *vector* $\mathbf{p}$ (from the origin to the point $\mathbf{p}$) w.r.t. the $+x$-axis on the xy-plane increases, as illustrated in Fig. 1.16(a). Note also that the direction of $\mathbf{u}_\rho$ at a point $\mathbf{p}$ is the direction along which the radian distance ρ of the point $\mathbf{p}$ from the origin on the xy-plane increases. It is interesting that $\mathbf{u}_\rho$ and $\mathbf{u}_\phi$ at a point $\mathbf{p}=[x\ y\ z]^\mathrm{T}$ vary with $\phi=\tan^{-1}(y/x)$ while $\mathbf{u}_z$ is identical at any point.

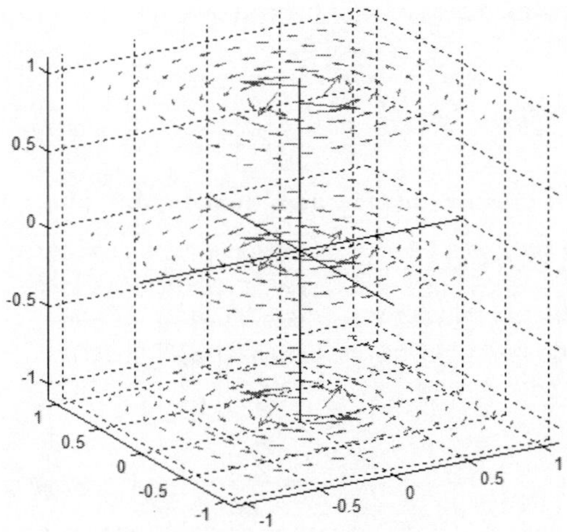

(a) $\mathbf{H}_a = (1/\rho)\,\mathbf{u}_\phi$

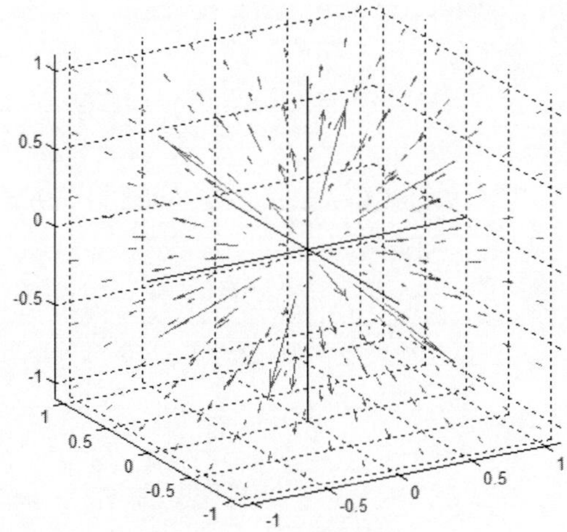

(b) $\mathbf{H}_b = (1/r^2)\,\mathbf{u}_r$

Figure 1.16 Plots of the vector functions given in Example 1.4(a) and (b)

```
%em01e04a.m
clear, clf
syms x y z r th ph;
Ha_cyl=[0; 1/r; 0]; % Eq.(E1.4.1)
Ha_cart = cyl2cart_v(Ha_cyl); pretty(simplify(Ha_cart)) % Eq.(E1.4.2)
Ha_cyl1 = cart2cyl_v(Ha_cart); pretty(simplify(Ha_cyl1)) % Eq.(E1.4.1)
% Grid points along x-/y-/z-axes
xx=[-0.9:0.2:0.9]; yy=[-0.9:0.2:0.9]; zz=[-1:1:1];
[x,y,z] = meshgrid(xx,yy,zz);
Hax = eval(Ha_cart(1)); % x-component of vector field Ha
Hay = eval(Ha_cart(2)); % y-component of vector field Ha
Haz = zeros(size(x));   % z-component of vector field Ha
% Here, zero component has been treated specially.
scale=2; quiver3(x,y,z, Hax,Hay,Haz, scale);
axis([-1 1 -1 1 -1 1]); axis('equal')
view(-30,20)
```

```
%em01e04b.m
clear
syms x y z r th ph;
Hb_sph=[1/r^2; 0; 0]; % Eq.(E1.4.3)
Hb_cart = sph2cart_v(Hb_sph); pretty(simplify(Hb_cart)) % Eq.(E1.4.4)
Hb_sph1 = cart2sph_v(Hb_cart); pretty(simplify(Hb_sph1)) % Eq.(E1.4.3)
% Grid points along x-/y-/z-axes
xx=[-1:0.4:1]; yy=[-1:0.4:1]; zz=[-1:0.4:1];
[x,y,z] = meshgrid(xx,yy,zz);
Hbx = eval(Hb_cart(1)); % x-component of vector field F
Hby = eval(Hb_cart(2)); % y-component of vector field F
Hbz = eval(Hb_cart(3));  % z-component of vector field F
scale=2; quiver3(x,y,z, Hbx,Hby,Hbz, scale);
axis([-1 1 -1 1 -1 1]); axis('equal')
view(-30,20)
```

(b) Consider a vector field, which is described in the spherical coordinate system as

$$\mathbf{H}_b(r,\theta,\phi)=\frac{1}{r^2}\mathbf{u}_r+0\mathbf{u}_\theta+0\mathbf{u}_\phi=[\mathbf{u}_r\ \mathbf{u}_\theta\ \mathbf{u}_\phi]\begin{bmatrix}1/r^2\\0\\0\end{bmatrix} \tag{E1.4.3}$$

This vector field can be converted into the Cartesian coordinate system by using Eq. (1.1.59) as

$$\mathbf{H}_b(x,y,z)=[\mathbf{u}_r\ \mathbf{u}_\theta\ \mathbf{u}_\phi]\begin{bmatrix}1/r^2\\0\\0\end{bmatrix}\overset{(1.1.59a)}{=}\left(\overset{T_{SR}}{\begin{bmatrix}\sin\theta\cos\phi & \sin\theta\sin\phi & \cos\theta\\ \cos\theta\cos\phi & \cos\theta\sin\phi & -\sin\theta\\ -\sin\phi & \cos\phi & 0\end{bmatrix}}\begin{bmatrix}\mathbf{u}_x^T\\ \mathbf{u}_y^T\\ \mathbf{u}_z^T\end{bmatrix}\right)^T\begin{bmatrix}1/r^2\\0\\0\end{bmatrix}$$

$$=[\mathbf{u}_x\ \mathbf{u}_y\ \mathbf{u}_z]\begin{bmatrix}\sin\theta\cos\phi & \cos\theta\cos\phi & -\sin\phi\\ \sin\theta\sin\phi & \cos\theta\sin\phi & \cos\phi\\ \cos\theta & -\sin\theta & 0\end{bmatrix}\begin{bmatrix}1/r^2\\0\\0\end{bmatrix}=[\mathbf{u}_x\ \mathbf{u}_y\ \mathbf{u}_z]\begin{bmatrix}\sin\theta\cos\phi/r^2\\ \sin\theta\sin\phi/r^2\\ \cos\theta/r^2\end{bmatrix}$$

$$\overset{(1.1.59d)}{=}\frac{x}{(x^2+y^2+z^2)^{3/2}}\mathbf{u}_x+\frac{y}{(x^2+y^2+z^2)^{3/2}}\mathbf{u}_y+\frac{z}{(x^2+y^2+z^2)^{3/2}}\mathbf{u}_z \tag{E1.4.4}$$

This spherical-to-Cartesian coordinate conversion can be performed by using the MATLAB function '`sph2cart_v()`' (see Section E.7 in Appendix E) as

```
>>syms x y z r th ph;
  Hb_sph = [1/r^2; 0; 0]; % Eq.(E1.4.3)
  Hb_cart = sph2cart_v(Hb_sph); % Eq.(E1.4.4)
  pretty(simplify(Hb_cart.')) % transpose, make simple and pretty
```

Running these MATLAB statements yields

```
+-                                                                    -+
|           x                         y                        z        |
|  -------------------,  -------------------,  -------------------   |
|    2      2      2 3/2    2      2      2 3/2    2      2      2 3/2 |
|  (x  + y  + z )          (x  + y  + z )          (x  + y  + z )      |
+-                                                                    -+
```

The above MATLAB script "em01e04b.m" can be run to plot the vector field $\mathbf{H}_b$ at several points as shown in Fig. 1.16(b). Note that the direction of $\mathbf{u}_r$ at a point **p** is the direction along which the radial distance r of the point **p** from the origin increases, as illustrated in Fig. 1.16(b). Note also that the direction of $\mathbf{u}_\theta$ at a point **p** is the direction along which the angle of the position vector **p** from the $+z$-axis increases and that the direction of $\mathbf{u}_\phi$ at a point **p** is the direction along which the (counterclockwise) azimuth angle of the projection of the position vector **p** onto the xy-plane w.r.t. the $+x$-axis increases. It is interesting that $\mathbf{u}_r$, $\mathbf{u}_\phi$, and $\mathbf{u}_\theta$ at a point $\mathbf{p}=[x\ y\ z]^T$ vary with $\phi=\tan^{-1}(y/x)$ and $\theta=\tan^{-1}(\sqrt{x^2+y^2}/z)$.

If the value of the vector field $\mathbf{H}_b(r,\theta,\phi)$ at a point represented by its spherical coordinate (r,θ,ϕ), say, $(1,\pi/2,\pi/2)$, is wanted in the Cartesian coordinate system, you can use '`sph2cart_v()`' as

```
>>r=1; th=pi/2; ph=pi/2; eval(sph2cart_v(Hb_sph,[r th ph]))
  [x,y,z]=sph2cart(ph,pi/2-th,r); eval(Hb_cart) % Alternatively,
```

Here, it is not the zenith angle θ but the elevation angle $\pi/2-\theta$ that is put as the 2nd input argument of the MATLAB function '`sph2cart()`' for the spherical-to-Cartesian conversion.

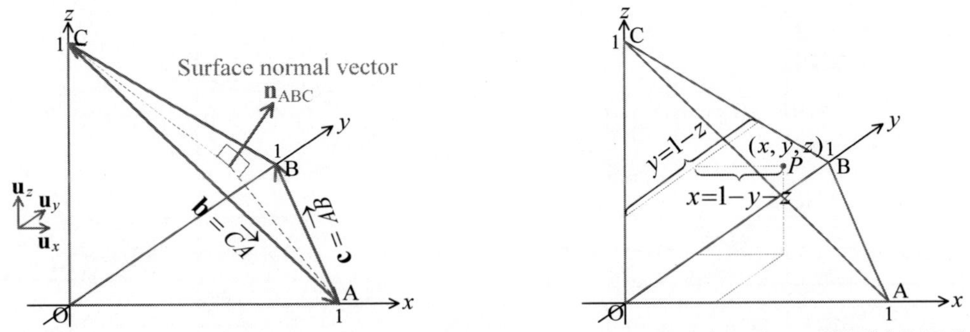

(a) To find the length of **b** and the area of ΔABC (b) To find the volume of Solid OABC

Figure 1.17 Single/double/triple integrals in the Cartesian coordinate system

(Example 1.5) Single/Double/Triple Integrals in the Cartesian Cooordinate System

Consider the solid figure in Fig. 1.17.

(a) Referring to Fig. 1.17(a), let us find the length of the vector **b** by integrating the dot product of the unit vector $\mathbf{u}_b$ along **b** and the Cartesian differential length vector $d\mathbf{l}$ as

$$l_{\mathbf{b}} = \int_{(x,z)=(0,1)}^{(x,z)=(1,0)} \mathbf{u}_{\mathrm{b}} \bullet d\mathbf{l} \overset{\text{Table1.1}}{\underset{\text{Cartesian}}{=}} \int_{(x,z)=(0,1)}^{(x,z)=(1,0)} \frac{1}{\sqrt{2}}(\mathbf{u}_x - \mathbf{u}_z)\bullet(\mathbf{u}_x\,dx + \mathbf{u}_y\,dy + \mathbf{u}_z\,dz)$$

$$= \frac{1}{\sqrt{2}}\left(\int_0^1 1dx - \int_1^0 1dz\right) = \sqrt{2} \tag{E1.5.1}$$

(b) Referring to Fig. 1.17(a), let us find the area of the triangle ΔABC by double-integrating the dot product of the unit normal vector $\mathbf{u}_{n,\Delta ABC}$ to ΔABC and the Cartesian differential surface vector $d\mathbf{s}$ as

$$A = \iint_S \mathbf{u}_{n,\Delta ABC} \bullet d\mathbf{s} \overset{\text{Table1.1}}{\underset{\text{Cartesian}}{=}} \iint_A \frac{1}{\sqrt{3}}(\mathbf{u}_x + \mathbf{u}_y + \mathbf{u}_z)\bullet(dy\,dz\,\mathbf{u}_x + dx\,dz\,\mathbf{u}_y + dx\,dy\,\mathbf{u}_z)$$

$$= \frac{1}{\sqrt{3}}\left(\int_0^1\int_0^{1-z} 1dy\,dz + \int_0^1\int_0^{1-x} 1\,dz\,dx + \int_0^1\int_0^{1-y} 1\,dx\,dy\right) = \frac{1}{\sqrt{3}}\left(\int_0^1 (1-z)\,dz + \int_0^1 (1-x)\,dx + \int_0^1 (1-y)\,dy\right)$$

$$= \frac{1}{\sqrt{3}}\left\{\left(z - \frac{z^2}{2}\right)\Big|_0^1 + \left(x - \frac{x^2}{2}\right)\Big|_0^1 + \left(y - \frac{y^2}{2}\right)\Big|_0^1\right\} = \frac{\sqrt{3}}{2} \tag{E1.5.2}$$

where the normal vector $\mathbf{n}_{\Delta ABC}$ of ΔABC can be obtained from the plane equation

$$x + y + z - 1 = 0 \tag{E1.5.3}$$

or the cross product of $\mathbf{b} = \overrightarrow{CA} = [1\ \ 0\ \ -1]^T$ and $\mathbf{c} = \overrightarrow{AB} = [-1\ \ 1\ \ 0]^T$.

(c) Referring to Fig. 1.17(b), let us find the volume of the tetrahedron (or triangular pyramid) OABC by using the triple integral (based on the Cartesian differential volume dV) as

$$V_{\mathrm{OABC}} = \iiint_V 1dV \overset{\text{Table1.1}}{\underset{\text{Cartesian}}{=}} \int_0^1\int_0^{1-z}\int_0^{1-y-z} 1dx\,dy\,dz = \int_0^1\int_0^{1-z}(1-y-z)\,dy\,dz = \int_0^1\left((1-z)y - \frac{y^2}{2}\right)\Big|_0^{1-z} dz$$

$$= \int_0^1\left((1-z)^2 - \frac{(1-z)^2}{2}\right)dz = \int_0^1 \frac{(1-z)^2}{2}\,dz = -\frac{(1-z)^3}{6}\Big|_0^1 = \frac{1}{6} \tag{E1.5.3}$$

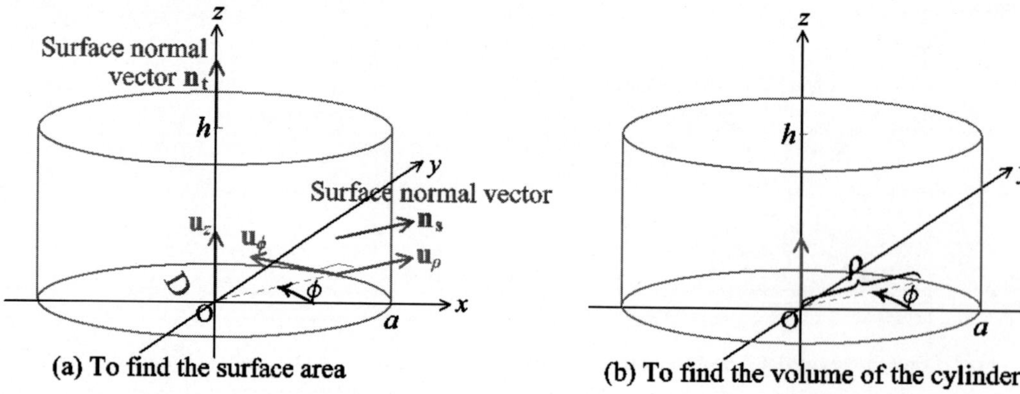

Figure 1.18 Single/double/triple integrals in the cylindrical coordinate system

(Example 1.6) Single/Double/Triple Integrals in the Cylindrical Cooordinate System

Consider the cylindrical object in Fig. 1.18.

(a) Referring to Fig. 1.18(a), let us find the circumference of the circle of radius a (centered at the origin O) by integrating the dot product of the unit vector $\mathbf{u}_\phi$ along ϕ and the cylindrical differential length vector $d\mathbf{l}$ as

$$l=\int_{(\rho,\phi)=(a,0)}^{(\rho,\phi)=(a,2\pi)}\mathbf{u}_\phi\bullet d\mathbf{l}=\int_{(\rho,\phi)=(a,0)}^{(\rho,\phi)=(a,2\pi)}\mathbf{u}_\phi\bullet(d\rho\mathbf{u}_\rho+\rho d\phi\mathbf{u}_\phi+dz\mathbf{u}_z)=\int_0^{2\pi}a\,d\phi=2\pi a \quad \text{(E1.6.1)}$$

(b) Referring to Fig. 1.18(a), let us find the area of the disk D of radius a by double-integrating the dot product of the unit normal vector $\mathbf{u}_z$ to the disk and the cylindrical differential surface vector $d\mathbf{s}$ as

$$A=\iint_S\mathbf{u}_z\bullet d\mathbf{s}\overset{\text{Table1.1}}{\underset{\text{cylindrical}}{=}}\iint_A\mathbf{u}_z\bullet\{(\rho d\phi)dz\,\mathbf{u}_\rho+d\rho dz\,\mathbf{u}_\phi+(\rho d\phi)d\rho\,\mathbf{u}_z\}$$

$$=\int_0^a\int_0^{2\pi}\rho\,d\phi d\rho=2\pi\int_0^a\rho\,d\rho=2\pi\frac{\rho^2}{2}\bigg|_0^a=\pi a^2 \quad \text{(E1.6.2)}$$

(c) Referring to Fig. 1.18(b), let us find the volume of the cylinder by using the triple integral (based on the cylindrical differential volume dV) as

$$V=\iiint_V 1dV\overset{\text{Table1.1}}{\underset{\text{cylindrical}}{=}}\int_0^h\int_0^a\int_0^{2\pi}1(\rho d\phi)d\rho dz=\int_0^h\int_0^a 2\pi\,\rho d\rho dz=\pi a^2 h \quad \text{(E1.6.3)}$$

(Example 1.7) Double/Triple Integrals in the Spherical Cooordinate System

Consider the spherical object in Fig. 1.19.

(a) Referring to Fig. 1.19, let us find the surface area of the sphere of radius a by double-integrating the dot product of the unit normal vector $\mathbf{u}_r$ to the spherical surface and the spherical differential surface vector $d\mathbf{s}$ as

$$A=\iint_S\mathbf{u}_r\bullet d\mathbf{s}\overset{\text{Table1.1}}{\underset{\text{spherical}}{=}}\iint_A\mathbf{u}_r\bullet\{(rd\theta)(r\sin\theta d\phi)\mathbf{u}_r+(r\sin\theta d\phi)dr\,\mathbf{u}_\theta+(rd\theta)dr\,\mathbf{u}_\phi\}$$

$$\overset{r=a}{=}\int_0^{2\pi}\int_0^\pi a^2\sin\theta d\theta d\phi=a^2\int_0^{2\pi}(-\cos\theta)\big|_0^\pi d\phi=a^2\int_0^{2\pi}2d\phi=4\pi a^2 \quad \text{(E1.7.1)}$$

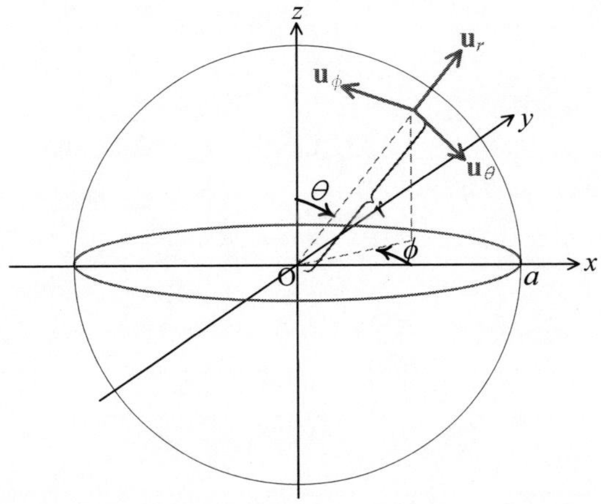

Figure 1.19 Double/triple integrals in the spherical coordinate system

(b) Referring to Fig. 1.19, let us find the volume of the sphere by using the triple integral (based on the cylindrical differential volume dV) as

$$V=\iiint_V 1dV \overset{\text{Table1.1}}{\underset{\text{spherical}}{=}} \int_0^a\int_0^{2\pi}\int_0^{\pi} 1\,(rd\theta)(r\sin\theta d\phi)dr=\int_0^a r^2\int_0^{2\pi}(-\cos\theta)\Big|_0^{\pi}d\phi dr$$

$$=\int_0^a r^2\int_0^{2\pi}2d\phi dr=4\pi\frac{r^3}{3}\Bigg|_0^a=\frac{4}{3}\pi a^3 \qquad \text{(E1.7.2)}$$

(Example 1.8) Magnetic Field by a Wire of Length $2L$ and with Current I

Consider a wire of length $2L$ (ranging from $z=-L$ to L) and with current density I in the $+z$-direction as described in Fig. 1.20.1. Find the magnetic field (produced by the current segment) at an observation point **p**, which is ρ[m] away from the wire at $z=0$ on the x-y plane.

Referring to Fig. 1.20.1 where $d\mathbf{l}$ is a differential conductor element of length dl directed along $+z$-direction and $\mathbf{r}$ is the vector from the differential conductor element $d\mathbf{l}=dl\,\mathbf{u}_z$ to the observation point **p**, we apply the Biot-Savart law to compute the magnetic field intensity as

$$\mathbf{H(r)} \overset{\text{Biot-Savart law}}{=} \frac{1}{4\pi}\int\frac{Id\mathbf{l}\times\mathbf{r}}{r^3} \text{ with } r=\|\mathbf{r}\|,\ d\mathbf{l}=dl\,\mathbf{u}_z,\ \mathbf{r}=r\cos\theta\,\mathbf{u}_\rho-r\sin\theta\,\mathbf{u}_z \qquad \text{(E1.8.1)}$$

$$;\ \mathbf{H}=\frac{1}{4\pi}\int_{-L}^{L}\frac{I\,dl\,\mathbf{u}_z\times(r\cos\theta\,\mathbf{u}_\rho-r\sin\theta\,\mathbf{u}_z)}{r^3} \overset{\text{Table1.1}}{=} \frac{1}{4\pi}\int_{-L}^{L}\frac{I\,dl\cos\theta}{r^2}\mathbf{u}_\phi=\frac{1}{4\pi}\int_{-\tan^{-1}(L/\rho)}^{\tan^{-1}(L/\rho)}\frac{I\,r\,d\theta}{r^2}\mathbf{u}_\phi$$

$$\overset{r=\rho\sec\theta}{=}\frac{1}{4\pi}\int_{-\tan^{-1}(L/\rho)}^{\tan^{-1}(L/\rho)}\frac{I\,d\theta}{\rho\sec\theta}\mathbf{u}_\phi=\frac{I}{4\pi\rho}\int_{-\tan^{-1}(L/\rho)}^{\tan^{-1}(L/\rho)}\cos\theta d\theta\,\mathbf{u}_\phi=\frac{I}{4\pi\rho}\sin\theta\Big|_{-\tan^{-1}(L/\rho)}^{\tan^{-1}(L/\rho)}\mathbf{u}_\phi$$

$$=\frac{LI}{2\pi\rho\sqrt{\rho^2+L^2}}\mathbf{u}_\phi \overset{(1.1.58a)}{=} \frac{LI}{2\pi\rho\sqrt{\rho^2+L^2}}(-\sin\phi\,\mathbf{u}_x+\cos\phi\,\mathbf{u}_y) \qquad \text{(E1.8.2)}$$

Note that with the assumption that $L>>\rho$, this becomes

$$\mathbf{H}=\frac{I}{2\pi\rho}\mathbf{u}_\phi \qquad \text{(E1.8.3)}$$

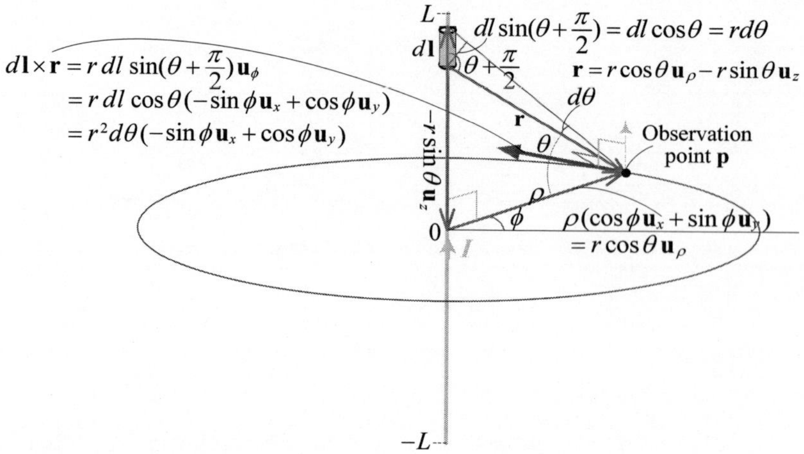

Figure 1.20.1 Magnetic field intensity produced by a 2*L*[m]-length line conducting current *I*

```
function Hs=H_by_line_current(ps,I,L)
% Magnetic field intensity at ps=[p1x p1y p1z; p2x p2y p2z; ... ]
% produced by a 2L[m]-length wire along the z-axis carrying current I
dth=pi/360;
for np=1:size(ps,1)
   p=ps(np,:); % For each point p
   rho=norm(p(1:2));
   ph=atan2(p(2),p(1));
   if nargin<3
     ths=[-179:179]*dth; L=1e10;
    else
     ths1=[dth:dth:atan2(L,rho)];
     ths=[-fliplr(ths1) 0 ths1];
   end
   for nth=1:length(ths)
      th=ths(nth); % Differential elevation angle
      pth=[0 0 rho*tan(th)]; % Position of differential wire
      r=p-pth; % Vector from differential wire to observation point p
      rm=norm(r); % Distance between differential wire and p
      dl=[0 0 rm*dth/cos(th)]; % Direction of differential wire
      dH(nth,:)=(I*cross(dl,r)/(4*pi*rm^3))*(rm>=1e-6); % Eq.(E1.8.1)
   end
   H=mu*sum(dH); Hs(np,:)=H;
   H1=L*I/2/pi/rho/sqrt(rho^2+L^2)*[-sin(ph) cos(ph) 0]; % Eq.(E1.8.2)
   discrepancy=norm(H-H1)/norm(H1);
   if discrepancy>1e-2, fprintf('%12.4e ',H,H1); fprintf('\n');  end
end
```

```
%show_H_by_line_current.m
% plots the magnetic flux densities at points from a circular current
clear, clf
plot3([0 0],[0 0],[-10 10],'Linewidth',2)
grid on, hold on
ps=[3 3 0; 3 -3 0;-3 3 0;-3 -3 0; 2 2 0; 2 -2 0;-2 2 0;-2 -2 0];
I=10; L=1; Hs=H_by_line_current(ps,I,L);
for np=1:size(ps,1)
   arrow3(Hs(np,:)*50,ps(np,:)) % Plot the magnetic flux at each point p
end
```

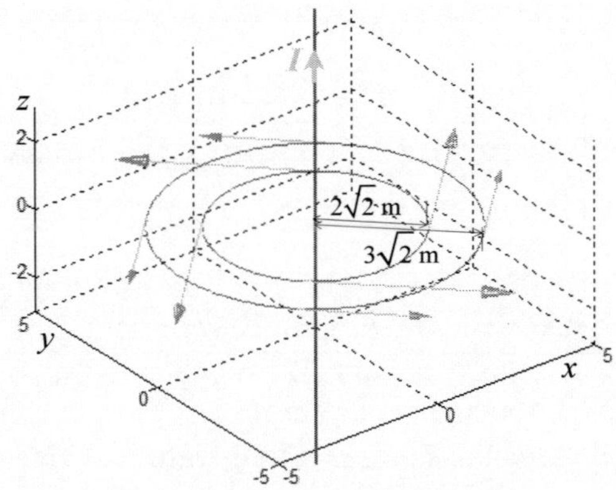

Figure 1.20.2 Magnetic field intensity produced by a 2[m]-length line conducting current $I = 10$A

With the above MATLAB function 'H_by_line_current()' and script "show_H_by_line_current.m" saved in your computer, you can type the following statement at the MATLAB prompt

```
>>show_H_by_line_current
```

to plot the magnetic field vectors at the eight points:

{[3 3 0], [3 -3 0], [-3 3 0], [-3 -3 0], [2 2 0], [2 -2 0], [-2 2 0], [-2 -2 0]}

as depicted in Fig. 1.20.2.

1.1.7 Gram–Schmidt Orthogonalization

Given a set of N linearly independent $M(\geq N)$-dimensional vectors $\{\mathbf{v}_n,\ n=1:N\}$, the Gram–Schmidt orthogonalization process generates a set of orthonormal vectors $\{\mathbf{q}_n,\ n=1:N\}$ that spans the vector space formed by $\mathbf{v}_n$'s, called the orthonormal basis of the vector space, as follows:

$$\begin{aligned}
&\mathbf{q}_1 = \frac{\mathbf{v}_1}{\|\mathbf{v}_1\|}\\
&\mathbf{q}'_2 = \mathbf{v}_2 - \mathrm{proj}_{\mathbf{q}_1}\mathbf{v}_2 \overset{(1.1.10)}{=} \mathbf{v}_2 - (\mathbf{v}_2\bullet\mathbf{q}_1)\frac{\mathbf{q}_1}{\|\mathbf{q}_1\|^2} \overset{\text{normalize}}{\rightarrow} \mathbf{q}_2 = \frac{\mathbf{q}'_2}{\|\mathbf{q}'_2\|} \qquad (1.1.62)\\
&\mathbf{q}'_3 = \mathbf{v}_3 - \mathrm{proj}_{\mathbf{q}_1}\mathbf{v}_3 - \mathrm{proj}_{\mathbf{q}_2}\mathbf{v}_3 \overset{(1.1.10)}{=} \mathbf{v}_3 - (\mathbf{v}_3\bullet\mathbf{q}_1)\frac{\mathbf{q}_1}{\|\mathbf{q}_1\|^2} - (\mathbf{v}_3\bullet\mathbf{q}_2)\frac{\mathbf{q}_2}{\|\mathbf{q}_2\|^2} \overset{\text{normalize}}{\rightarrow} \mathbf{q}_3 = \frac{\mathbf{q}'_3}{\|\mathbf{q}'_3\|}\\
&\qquad\cdots\cdots
\end{aligned}$$

where the projection was defined in Section 1.1.2.3 and the orthonormality is defined by Eq. (1.2.30):

$$Q^{\mathrm{T}}Q = I \text{ (Identity matrix)} \qquad (1.1.63)$$

This algorithm is listed in the following box:

for n = 1 : N

$$\mathbf{q}_n = \mathbf{v}_n - \sum_{k=1}^{n-1}\mathrm{proj}_{\mathbf{q}_k}\mathbf{v}_n \rightarrow \mathbf{q}_n = \mathbf{q}_n/\|\mathbf{q}_n\| \qquad (1.1.64)$$

end

$Q = [\mathbf{q}_1\ \mathbf{q}_2 \cdots \mathbf{q}_N]$: a set of orthonormal vectors

In connection with the QR decomposition (Section 1.2.10.3), note that Eq. (1.1.62) can be written as

$$\begin{bmatrix} \mathbf{q}_1 & \mathbf{q}_2 & \mathbf{q}_3 \end{bmatrix} \begin{bmatrix} \|\mathbf{v}_1\| & (\mathbf{v}_2\bullet\mathbf{q}_1)/\|\mathbf{q}_1\|^2 & (\mathbf{v}_3\bullet\mathbf{q}_1)/\|\mathbf{q}_1\|^2 \\ 0 & \|\mathbf{q}_2'\| & (\mathbf{v}_3\bullet\mathbf{q}_2)/\|\mathbf{q}_2\|^2 \\ 0 & 0 & \|\mathbf{q}_3'\| \end{bmatrix} = \begin{bmatrix} \mathbf{v}_1 & \mathbf{v}_2 & \mathbf{v}_3 \end{bmatrix};\ QR = A \quad (1.1.65)$$

This Gram-Schmidt process has been cast into the following MATLAB function, '[Q,R]= Gram_Schmidt(A)', which takes an $M\times N$ matrix A consisting of N linearly independent $M(\geq N)$-dimensional vectors and generates a unitary $M\times N$ matrix Q, consisting of N orthonormal column vectors (such that $Q^{*T}Q = I$), together with an upper triangular matrix R such that $QR = A$, where the superscripts * and ^{T} denote the complex conjugate and transpose, respectively. (See Section 1.2.7.2 for the definition of *unitary* matrix.)

```
function [Q,R]=Gram_Schmidt(A)
% Gram-Schmidt orthonormalization of the columns of MxN matrix A
%  on the assumption that the columns of A are linearly independent.
% Q = MxN matrix whose columns form an orthonormal basis
%     for the column space of A.
% R = An invertible upper triangular matrix R so that A = Q*R.
% Warning: For a more stable algorithm, use [Q,R] = qr(A,0).
[M,N] = size(A);
v1 = A(:,1);
R(1,1) = norm(v1);
Q(:,1) = v1/R(1,1);
for n=2:min(M,N)
  vn = A(:,n); % Take each column as a vector
   R(1:n-1,n) = Q(:,1:n-1)'*vn; % components of vn onto qk's Eq.(1.1.65)
   qn = vn - Q(:,1:n-1)*R(:,n); % Eq.(1.1.64) with Projection Q*(Q'*vn)
   R(n,n) = norm(qn); Q(:,n) = qn/R(n,n); % Normalize
end
```

(Example 1.9) Gram–Schmidt Orthogonalization

Use the Gram-Schmidt process to find an orthonormal basis for a vector space R^3 spanned by

$$\mathbf{v}_1 = \begin{bmatrix} 1 \\ 1 \\ 1 \end{bmatrix},\ \mathbf{v}_2 = \begin{bmatrix} 1 \\ 2 \\ 2 \end{bmatrix},\ \text{and}\ \mathbf{v}_3 = \begin{bmatrix} 1 \\ 1 \\ 0 \end{bmatrix} \quad (\text{E1.9.1})$$

Following the steps of Eq. (1.1.62), we get

$$\mathbf{q}_1 = \frac{1}{\|\mathbf{v}_1\|}\mathbf{v}_1 = \frac{1}{\sqrt{3}}\begin{bmatrix} 1 \\ 1 \\ 1 \end{bmatrix} \quad (\text{E1.9.2})$$

$$\mathbf{q}_2 = \mathbf{v}_2 - \mathrm{proj}_{\mathbf{q}_1}\mathbf{v}_2 \overset{(1.1.10)}{=} \mathbf{v}_2 - (\mathbf{v}_2\bullet\mathbf{q}_1)\frac{\mathbf{q}_1}{\|\mathbf{q}_1\|^2} = \begin{bmatrix} 1 \\ 2 \\ 2 \end{bmatrix} - \left(\begin{bmatrix} 1 \\ 2 \\ 2 \end{bmatrix}\bullet\begin{bmatrix} 1 \\ 1 \\ 1 \end{bmatrix}\right)\frac{1}{3}\begin{bmatrix} 1 \\ 1 \\ 1 \end{bmatrix} = \begin{bmatrix} -2/3 \\ 1/3 \\ 1/3 \end{bmatrix}$$

$$\overset{\text{normalize}}{\rightarrow}\ \mathbf{q}_2 = \frac{\mathbf{q}_2}{\|\mathbf{q}_2\|} = \frac{1}{\sqrt{6}/3}\begin{bmatrix} -2/3 \\ 1/3 \\ 1/3 \end{bmatrix} = \frac{1}{\sqrt{6}}\begin{bmatrix} -2 \\ 1 \\ 1 \end{bmatrix} \quad (\text{E1.9.3})$$

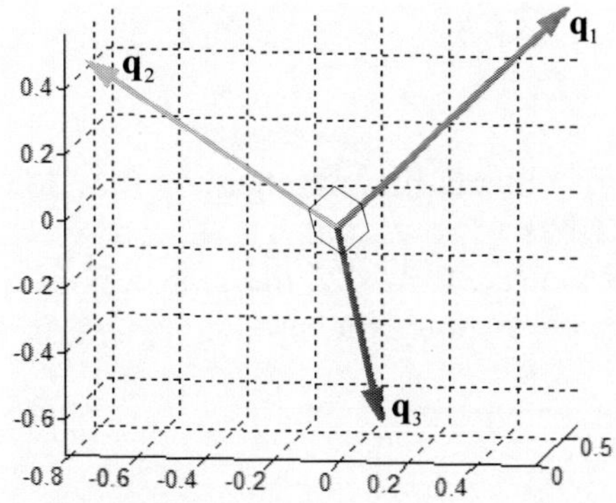

Figure 1.21 Three basis vectors obtained using the Gram-Schmidt orthogonalization in Example 1.9

$$\mathbf{q}_3=\mathbf{v}_3-\mathrm{proj}_{\mathbf{q}_1}\mathbf{v}_3-\mathrm{proj}_{\mathbf{q}_2}\mathbf{v}_3 \overset{(1.1.10)}{=} \mathbf{v}_3-(\mathbf{v}_3\bullet\mathbf{q}_1)\frac{\mathbf{q}_1}{||\mathbf{q}_1||^2}-(\mathbf{v}_3\bullet\mathbf{q}_2)\frac{\mathbf{q}_2}{||\mathbf{q}_2||^2}$$

$$=\begin{bmatrix}1\\1\\0\end{bmatrix}-\left(\begin{bmatrix}1\\1\\0\end{bmatrix}\bullet\begin{bmatrix}1\\1\\1\end{bmatrix}\right)\frac{1}{3}\begin{bmatrix}1\\1\\1\end{bmatrix}-\left(\begin{bmatrix}1\\1\\0\end{bmatrix}\bullet\begin{bmatrix}-2\\1\\1\end{bmatrix}\right)\frac{1}{6}\begin{bmatrix}-2\\1\\1\end{bmatrix}=\begin{bmatrix}1\\1\\0\end{bmatrix}-\begin{bmatrix}2/3\\2/3\\2/3\end{bmatrix}+\begin{bmatrix}-2/6\\1/6\\1/6\end{bmatrix}=\begin{bmatrix}0\\1/2\\-1/2\end{bmatrix}$$

$$\overset{\text{normalize}}{\rightarrow}\quad \mathbf{q}_3=\frac{\mathbf{q}_3}{||\mathbf{q}_3||}=\frac{1}{\sqrt{2}/2}\begin{bmatrix}0\\1/2\\-1/2\end{bmatrix}=\frac{1}{\sqrt{2}}\begin{bmatrix}0\\1\\-1\end{bmatrix} \tag{E1.9.4}$$

These computations can be done by typing the following statements at the MATLAB prompt:

```
>>A=[1  1  1; 1  2  1; 1  2  0];
  [Q,R]=Gram_Schmidt(A);  Q
```

This will yield

```
 Q =     0.5774    -0.8165    -0.0000
         0.5774     0.4082     0.7071
         0.5774     0.4082    -0.7071    % Eqs.(E1.9.2),(E1.9.3),(E1.9.4)
 R =     1.7321     2.8868     1.1547
              0     0.8165    -0.4082
              0          0     0.7071
```

The three column vectors constituting Q are a set of orthonormal basis vectors for the 3D space R^3 spanned by the three vectors $\mathbf{v}_1$, $\mathbf{v}_2$, and $\mathbf{v}_3$, as depicted in Fig. 1.21. If Fig. 1.21 does not make you sure that the column vector $\mathbf{q}_n$'s of $Q=[\mathbf{q}_1\ \ \mathbf{q}_2\ \ \mathbf{q}_3]$ are mutually orthogonal, how about checking if the orthonormality condition (1.1.63) holds, or aligning the viewing angle of the graph with $\mathbf{q}_3$ by running the following MATLAB statements?

```
>>Q'*Q % Is Q orthonormal by satisfying Eq.(1.1.63)?
>>hold on, grid on;
  S=0; % No scaling
  quiver3(0,0,0,Q(1,1),Q(2,1),Q(3,1),S,'Color','r','Linewidth',2);
  quiver3(0,0,0,Q(1,2),Q(2,2),Q(3,2),S,'Color','g','Linewidth',2);
  quiver3(0,0,0,Q(1,3),Q(2,3),Q(3,3),S,'Color','b','Linewidth',2);
  axis('equal'), xlabel('x'), ylabel('y'), zlabel('z')
  shg; pause
  % To align the viewing angle with q3
  view(Q(:,3)) % Now, what do you see?
```

1.2 MATRICES

1.2.1 Matrix Algebra

A matrix is a rectangular array of objects like numbers or functions, called *entries* or *elements*, in rows and columns. If a matrix A has M rows and N columns, we say that the size of A is M by N (written $M\times N$) or A is an $M\times N$ matrix. If $M=N$, the matrix is said to be *square*. An $M\times N$ matrix may be written as $[a_{mn}]_{M\times N}$ where a_{mn} denotes the m-n entry, i.e., the entry in row m and column n.

1.2.1.1 Addition and Subtraction

$$A\pm B=\begin{bmatrix} a_{11} & a_{12} & \bullet & a_{1N} \\ a_{21} & a_{22} & \bullet & a_{2N} \\ \bullet & \bullet & \bullet & \bullet \\ a_{M1} & a_{M2} & \bullet & a_{MN} \end{bmatrix} \pm \begin{bmatrix} b_{11} & b_{12} & \bullet & b_{1N} \\ b_{21} & b_{22} & \bullet & b_{2N} \\ \bullet & \bullet & \bullet & \bullet \\ b_{M1} & b_{M2} & \bullet & b_{MN} \end{bmatrix} = \begin{bmatrix} c_{11} & c_{12} & \bullet & c_{1N} \\ c_{21} & c_{22} & \bullet & c_{2N} \\ \bullet & \bullet & \bullet & \bullet \\ c_{M1} & c_{M2} & \bullet & c_{MN} \end{bmatrix} = C \tag{1.2.1}$$

$$\text{with } a_{mn}\pm b_{mn}=c_{mn}$$

1.2.1.2 Multiplication

$$AB=\begin{bmatrix} a_{11} & a_{12} & \bullet & a_{1K} \\ a_{21} & a_{22} & \bullet & a_{2K} \\ \bullet & \bullet & \bullet & \bullet \\ a_{M1} & a_{M2} & \bullet & a_{MK} \end{bmatrix} \begin{bmatrix} b_{11} & b_{12} & \bullet & b_{1N} \\ b_{21} & b_{22} & \bullet & b_{2N} \\ \bullet & \bullet & \bullet & \bullet \\ b_{K1} & b_{K2} & \bullet & b_{KN} \end{bmatrix} = \begin{bmatrix} c_{11} & c_{12} & \bullet & c_{1N} \\ c_{21} & c_{22} & \bullet & c_{2N} \\ \bullet & \bullet & \bullet & \bullet \\ c_{M1} & c_{M2} & \bullet & c_{MN} \end{bmatrix} = C \tag{1.2.2}$$

$$\text{with } c_{mn}=\sum\nolimits_{k=1}^{K} a_{mk}b_{kn}$$

(Note) For this multiplication to be done, the number of columns of A must equal the number of rows of B.
(Note) Note that the commutative law does not hold for the matrix multiplication, i.e., $AB\neq BA$.

1.2.2 Rank and Row/Column Spaces

The *rank* of an $M\times N$ matrix is the number of linearly independent rows/columns, and it can be determined using the echelon form of the matrix (Sec. 1.2.7.5). If it equals $\min(M,N)$, then the matrix is said to be of *maximal* or *full rank*; otherwise, the matrix is said to be *rank-deficient* or to have *rank-deficiency*. The MATLAB statement to obtain the rank of a matrix A is '`rank(A)`'. For example, consider the following three matrices:

$$A_1=\begin{bmatrix} 1 & -2 & 3 \\ 2 & -4 & 5 \\ 1 & 2 & 1 \end{bmatrix},\quad A_2=\begin{bmatrix} 1 & 1 & 0 \\ 0 & 1 & 1 \\ -2 & -5 & -3 \end{bmatrix},\quad A_3=\begin{bmatrix} 1 & -2 & 3 \\ 2 & -4 & 6 \\ -3 & 6 & -9 \end{bmatrix}$$

The matrix A_1 is of full rank (3) because all the rows/columns are independent in the sense that no row/column can be expressed as a linear combination of the other rows/columns. However, A_2 is of rank 2 because the first two rows are independent and the third row is the sum of −2 times the first row and −3 times the second row. In fact, the rank-deficient matrix A_2 has a dependency not only among the rows but also among the columns ($A_2(:,2)-A_2(:,3)=A_2(:,1)$), although it may not easily be seen. What about A_3? It is of rank 1 because the second and third rows are a scalar (2 and −3) times the first row, respectively. What would the MATLAB function '`rank()`' say about the ranks of A_1, A_2, and A_3? It is recommended that interested readers to run the following statements:

```
>>A1=[1 -2 3;2 -4 5;1 2 1]; A2=[1 -2 3;2 -4 6;1 2 -1];
>>A3=[1 -2 3;2 -4 6;-3 6 -9];
>>[rank(A1) rank(A2) rank(A3)]
```

The *row space* of an $M\times N$ matrix, denoted by $\mathcal{R}(A)$, is the space spanned by the row vectors, i.e., the set of all possible linear combinations of row vectors of A that can be expressed by $A^{\mathrm{T}}\boldsymbol{\alpha}$, with an M-dimensional column vector $\boldsymbol{\alpha}$, where

$$A^{\mathrm{T}}\boldsymbol{\alpha}=\begin{bmatrix} a_{11} & \cdots & a_{M1} \\ \vdots & \ddots & \vdots \\ a_{1N} & \cdots & a_{MN}\end{bmatrix}\begin{bmatrix}\alpha_1 \\ \vdots \\ \alpha_M\end{bmatrix}=\alpha_1\begin{bmatrix}a_{11} \\ \vdots \\ a_{1N}\end{bmatrix}+\cdots+\alpha_M\begin{bmatrix}a_{M1} \\ \vdots \\ a_{MN}\end{bmatrix} \tag{1.2.3}$$

On the other hand, the *null space* of the matrix A, denoted by $\mathcal{N}(A)$, is the space orthogonal to the row space, i.e., the set of all possible linear combinations of the N-dimensional vector $\mathbf{n}$'s satisfying

$$A\mathbf{n}=\mathbf{0} \tag{1.2.4}$$

(Example 1.10) Row Space and Null Space

Consider a matrix

$$A=\begin{bmatrix}1 & 3 & 2 \\ 1 & 2 & 1\end{bmatrix} \tag{E1.10.1}$$

Any vector in the row space of A can be expressed by Eq. (1.2.3) as

$$\mathbf{r}=A^{\mathrm{T}}\boldsymbol{\alpha}\,;\quad \begin{bmatrix}r_x \\ r_y \\ r_z\end{bmatrix}=\begin{bmatrix}1 & 1 \\ 3 & 2 \\ 2 & 1\end{bmatrix}\begin{bmatrix}\alpha_1 \\ \alpha_2\end{bmatrix} \tag{E1.10.2}$$

To plot the row space for visualization, we rewrite Eq. (E1.10.2) as an expression of r_z in terms of r_x and r_y:

$$r_z=[2\ \ 1]\begin{bmatrix}\alpha_1 \\ \alpha_2\end{bmatrix}=[2\ \ 1]\begin{bmatrix}1 & 1 \\ 3 & 2\end{bmatrix}^{-1}\begin{bmatrix}r_x \\ r_y\end{bmatrix}=[2\ \ 1]\begin{bmatrix}-2 & 1 \\ 3 & -1\end{bmatrix}\begin{bmatrix}r_x \\ r_y\end{bmatrix}=[-1\ \ 1]\begin{bmatrix}r_x \\ r_y\end{bmatrix}=-r_x+r_y \tag{E1.10.3}$$

Also, we express the null space, i.e., the set of all possible linear combinations of the 3-D vector $\mathbf{n}$'s satisfying Eq. (1.2.4), as

$$A\mathbf{n}=\begin{bmatrix}1 & 3 & 2 \\ 1 & 2 & 1\end{bmatrix}\begin{bmatrix}n_x \\ n_y \\ n_z\end{bmatrix}\overset{(1.2.4)}{=}\mathbf{0}\ \overset{n_x=t}{\rightarrow}\ \begin{bmatrix}3 & 2 \\ 2 & 1\end{bmatrix}\begin{bmatrix}n_y \\ n_z\end{bmatrix}=\begin{bmatrix}-t \\ -t\end{bmatrix};\ \begin{bmatrix}n_y \\ n_z\end{bmatrix}=\begin{bmatrix}3 & 2 \\ 2 & 1\end{bmatrix}^{-1}\begin{bmatrix}-t \\ -t\end{bmatrix}=\begin{bmatrix}1 & -2 \\ -2 & 3\end{bmatrix}\begin{bmatrix}t \\ t\end{bmatrix}=\begin{bmatrix}-t \\ t\end{bmatrix}$$

$$;\ n_x=t,\ n_y=-t,\ n_z=t\ \text{(parametric equation) or }\ \mathbf{n}=\mathbf{0}+[1\ \ -1\ \ 1]^{\mathrm{T}}t\ \text{(vector equation)} \tag{E1.10.4}$$

In fact, this is the intersection line of two planes, each described by $n_x+3n_y+2n_z=0$ and $n_x+2n_y+n_z=0$, respectively, which can also be obtained by using the MATLAB function '`intersect_planes()`' (Section 1.1.4.3) as

```
>>[d,p]=intersect_planes([1 3 2 0; 1 2 1 0])
  d =   -1          p =    0
         1                 0
        -1                 0
```

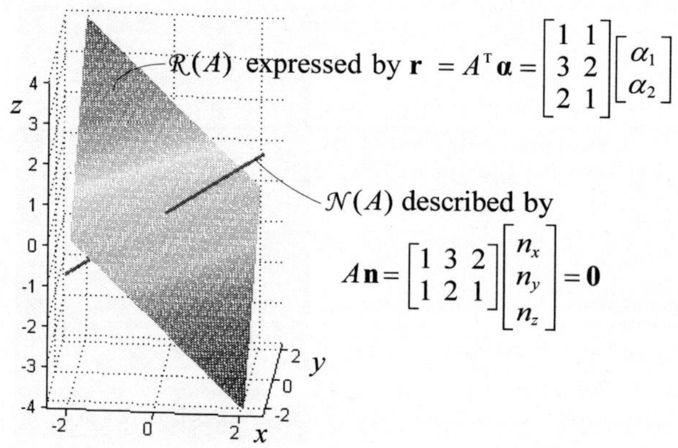

Figure 1.22 Row space and null space of a matrix A given in Example 1.10

Based on these two equations (E1.10.3) and (E1.10.4), we compose the following MATLAB script "em01f22.m" and run it to get Fig. 1.22, which shows the row space and null space of A orthogonal (perpendicular) to each other.

```
%em01f22.m
% shows the row and null spaces of a matrix given in Example 1.10
x=[-2:0.05:2]; y=[-2:0.05:2]; [X,Y]=meshgrid(x,y); % X-Y grids for graph
rZ=-X+Y; % Eq.(E1.10.3) for Row space
mesh(X,Y,rZ), hold on
t=[-2.5 2.5];
nx=t; ny=-t; nz=t; % Eq.(E1.10.4) for Null space
plot3(nx,ny,nz, 'Linewidth',2)
axis('equal'), view(5,20)
```

1.2.3 Determinant and Trace

The *determinant* of a $K\times K$ (square) matrix $A=[a_{mn}]$ is defined by

$$\det(A)=|A|=\sum_{k=0}^{K} a_{kn}(-1)^{k+n}M_{kn} \text{ or } \sum_{k=0}^{K} a_{mk}(-1)^{m+k}M_{mk} \tag{1.2.5}$$

for any fixed $1\le n\le K$ or $1\le m\le K$

where the *minor* M_{kn} is the determinant of the $(K-1)\times(K-1)$ (minor) matrix formed by removing the kth row and the nth column from A, and $A_{kn}=(-1)^{k+n}M_{kn}$ is called the *cofactor* of a_{kn}.

Especially, the determinants of a 2×2 matrix $A_{2\times 2}$ and a 3×3 matrix $A_{3\times 3}$ are

$$\det(A_{2\times 2})=\begin{vmatrix} a_{11} & a_{12} \\ a_{21} & a_{22}\end{vmatrix}=\sum_{k=0}^{2} a_{kn}(-1)^{k+n}M_{kn}=a_{11}a_{22}-a_{12}a_{21} \tag{1.2.6}$$

$$\det(A_{3\times 3})=\begin{vmatrix} a_{11} & a_{12} & a_{13} \\ a_{21} & a_{22} & a_{23} \\ a_{31} & a_{32} & a_{33}\end{vmatrix}=a_{11}\begin{vmatrix} a_{22} & a_{23} \\ a_{32} & a_{33}\end{vmatrix}-a_{12}\begin{vmatrix} a_{21} & a_{23} \\ a_{31} & a_{33}\end{vmatrix}+a_{13}\begin{vmatrix} a_{21} & a_{22} \\ a_{31} & a_{32}\end{vmatrix}$$

$$=a_{11}(a_{22}a_{33}-a_{23}a_{32})-a_{12}(a_{21}a_{33}-a_{23}a_{31})+a_{13}(a_{21}a_{32}-a_{22}a_{31}) \tag{1.2.7}$$

Note the following about the determinant of a $K\times K$ (square) matrix $A=[a_{mn}]$:

(1) $\det(A^{\mathrm{T}}) = \det(A)$ (1.2.8a)

(2) $\det(AB) = \det(A)\det(B)$ (1.2.8b)

(3) $\det(A^{-1}) = 1/\det(A)$ (1.2.8c)

(4) $\det(cA) = c^K \det(A)$ (1.2.8d)

(5) If any row/column of a square matrix A has all zero entries or is the same as another row/column multiplied by a real number, then $\det(A) = 0$.

(6) If any row/column of a square matrix A is multiplied by a real number c, it increases the determinant by c times. However, even if any row/column of a square matrix A is multiplied by a real number and added to another row/column, it makes no difference to the determinant.

(7) If any two rows/columns of a square matrix A are switched (interchanged), it reverses the sign of the determinant.

(8) If the determinant of a matrix is zero, the matrix is said to be *singular*.

(9) The determinant of a matrix equals the product of all the eigenvalues (see Section 1.2.4) of the matrix.

(10) If A is upper/lower triangular having only zeros below/above the diagonal in each column, its determinant is the product of the diagonal elements:

$$\det(A) = \prod_{k=1}^{K} a_{kk} \tag{1.2.9}$$

(11) The MATLAB statement to get the determinant of a given matrix A is '`det(A)`'.

The *trace* of a $K\times K$ (square) matrix $A=[a_{mn}]$ is the sum of all the entries on the main diagonal:

$$\mathrm{tr}(A) = \sum_{k=1}^{K} a_{kk} \tag{1.2.10}$$

Note the following about the trace:

(1) $\mathrm{tr}(A^{\mathrm{T}}) = \mathrm{tr}(A)$ (1.2.11a)

(2) $\mathrm{tr}(A+B) = \mathrm{tr}(A) + \mathrm{tr}(B)$ (1.2.11b)

(3) $\mathrm{tr}(AB) = \sum_{k=1}^{K}\sum_{m=1}^{K} a_{km} b_{mk} = \sum_{k=1}^{K}\sum_{m=1}^{K} b_{km} a_{mk} = \mathrm{tr}(BA)$: *Cyclic property* (1.2.11c)

(4) $\mathrm{tr}(cA) = c\,\mathrm{tr}(A)$ (1.2.11d)

(5) $\mathrm{tr}(A) = \mathrm{tr}(P^{-1}AP)$: The trace of a matrix is invariant under similarity transformation. (1.2.11e)

(6) $\mathrm{tr}(A^{*\mathrm{T}}A) = \mathrm{tr}(AA^{*\mathrm{T}}) = \sum_{k=1}^{K}\sum_{m=1}^{K} |a_{mk}|^2$ (*: conjugate, T: transpose) (1.2.11f)

(7) The MATLAB statement to get the trace of a given matrix A is '`trace(A)`'.

1.2.4 Eigenvalues and Eigenvectors

The *eigenvalue* or *characteristic value* and its corresponding *eigenvector* or *characteristic vector* of an $N\times N$ matrix A are defined to be a scalar λ and a nonzero vector $\mathbf{v}$ satisfying the eigenpair equation

$$A\mathbf{v} = \lambda\mathbf{v} \Leftrightarrow (A-\lambda I)\mathbf{v} = \mathbf{0} \;\; (\text{with } \mathbf{v} \neq \mathbf{0}) \tag{1.2.12}$$

respectively, where I denotes the $N \times N$ *identity matrix* having ones on its diagonal and zeros elsewhere. Note that $(\lambda, \mathbf{v})$ is called an *eigenpair* and there are N eigenpairs for an $N \times N$ matrix A.

The eigenvalues of a matrix A can be obtained as the roots of the *characteristic equation* in λ:

$$|A - \lambda I| = 0 \tag{1.2.13}$$

and the eigenvector corresponding to an eigenvalue λ_i can be obtained by substituting λ_i into Eq. (1.2.12) and solving it for $\mathbf{v}$.

Note the following about the eigenvalue/eigenvector:

(1) If a square matrix A consisting of only real entries has a complex eignevalue $\lambda = \alpha + j\beta$, its conjugate $\lambda^* = \alpha - j\beta$ is also an eigenvalue of A. If $\mathbf{v}$ is an eigenvector corresponding to λ, its conjugate $\mathbf{v}^*$ is an eigenvector corresponding to λ^*.

(2) Upper/lower triangular or diagonal matrices have all their eigenvalues on their diagonal.

(3) If A is real and symmetric, all the eigenvalues are real-valued and the eigenvectors corresponding to distinct eigenvalues of A are orthogonal in the sense that their dot product is zero.

(4) If A is symmetric and positive definite, all the eigenvalues are real and positive.

(5) If $\mathbf{v}$ is an eigenvector of A, so is $c\,\mathbf{v}$ for any nonzero scalar c.

(6) The MATLAB statement to get the eigenpairs of a given matrix A is `[V,LAMBDA]=eig(A)`, where the output arguments of '`eig()`', V and LAMDA, are a *modal matrix* with eigenvectors in each of its columns and another diagonal matrix with eigenvalues on its diagonal, respectively.

(Example 1.11) Eigenvalues and Eigenvectors of a Matrix

Consider a matrix

$$A = \begin{bmatrix} 0 & 2 \\ 1 & 1 \end{bmatrix} \tag{E1.11.1}$$

Let us find the eigenvalues and eigenvectors of the matrix. First, we use Eq. (1.2.13) to find the eigenvalues as

$$|A - \lambda I| = \left| \begin{bmatrix} -\lambda & 2 \\ 1 & 1-\lambda \end{bmatrix} \right| = \lambda^2 - \lambda - 2 = (\lambda + 1)(\lambda - 2) = 0 \tag{E1.11.2}$$

$$; \qquad \lambda_1 = -1, \quad \lambda_2 = 2 \tag{E1.11.3}$$

and then, we use Eq. (1.2.12) to get the corresponding eigenvectors as

$$[A - \lambda_1 I]\mathbf{v}_1 = \begin{bmatrix} 1 & 2 \\ 1 & 2 \end{bmatrix} \begin{bmatrix} v_{11} \\ v_{21} \end{bmatrix} = \begin{bmatrix} v_{11} + 2v_{21} \\ v_{11} + 2v_{21} \end{bmatrix} = \begin{bmatrix} 0 \\ 0 \end{bmatrix}; \; v_{11} = -2v_{21}; \; \mathbf{v}_1 = \begin{bmatrix} v_{11} \\ v_{21} \end{bmatrix} = \frac{1}{\sqrt{5}} \begin{bmatrix} -2 \\ 1 \end{bmatrix} \tag{E1.11.4a}$$

$$[A - \lambda_2 I]\mathbf{v}_2 = \begin{bmatrix} -2 & 2 \\ 1 & -1 \end{bmatrix} \begin{bmatrix} v_{12} \\ v_{22} \end{bmatrix} = \begin{bmatrix} -2v_{12} + 2v_{22} \\ v_{12} - v_{22} \end{bmatrix} = \begin{bmatrix} 0 \\ 0 \end{bmatrix}; \; v_{12} = v_{22}; \; \mathbf{v}_2 = \begin{bmatrix} v_{12} \\ v_{22} \end{bmatrix} = \begin{bmatrix} 1/\sqrt{2} \\ 1/\sqrt{2} \end{bmatrix} \tag{E1.11.4b}$$

where we have chosen the values of v_{11}, v_{12}, v_{21}, and v_{22} so that the eigenvectors can be normalized with unit magnitude (norm) as '`eig()`' returns.

You can save time and effort by using the MATLAB software. To this aim, compose the following MATLAB script, save it as an M-file named "em01e11.m,"

```
%em01e11.m
% to get the eigenvalues an eigenvectors of a matrix A.
A=[0 2; 1 1];
[V,L]=eig(A) % V= modal matrix composed of eigenvectors
% L= diagonal matrix with eigenvalues on its diagonal
e=eig(A) % An alternative to get just eigenvalues
ch_eq=poly(A) % Characteristic eq. (E1.11.2) in unknown eigenvalues
roots(ch_eq) % Roots of (E1.11.2)
L1=V^-1*A*V % diagonalize A through similarity transformation
     % into a diagonal matrix having the eigenvalues on diagonal.
```

and enter the name of the M-file at the MATLAB prompt:

```
>>em01e11
```

Then you can see the outputs as

```
V  = -0.8944   -0.7071       L = -1     0       e = -1       ans =  2
      0.4472   -0.7071            0     2            2             -1
L1 = -1.0000    0.0000
           0    2.0000
```

Note that the differences of the sign and magnitude of the eigenvectors do not matter. Here, the same eigenvalues have been computed three times to illustrate different usages of '`eig()`' and make the readers become familiar with using MATLAB. The last one shows that a modal matrix (consisting of eigenvectors) of a matrix can be used to diagonalize the matrix (see Section 1.2.6).

(Example 1.12) Repeated Eigenvalues, Generalized Eigenvectors, and Jordan Canonical Form

(a) Find the eigenvalues and eigenvectors of a matrix

$$A_1 = \begin{bmatrix} 1 & 0 & 1 \\ 0 & 1 & 0 \\ 0 & 0 & 0 \end{bmatrix} \tag{E1.12.1}$$

The eigenvalues are found by using Eq. (1.2.13) as

$$|A_1 - \lambda I| = \begin{vmatrix} 1-\lambda & 0 & 1 \\ 0 & 1-\lambda & 0 \\ 0 & 0 & -\lambda \end{vmatrix} = -\lambda(1-\lambda)^2 = 0\,;\ \lambda = 1,\,1,\text{ and }0 \tag{E1.12.2}$$

The eigenvectors corresponding to the repeated eigenvalue λ=1 of (*algebraic*) *multiplicity* 2 are found as follows:

$$[A_1 - \lambda I]_{\lambda=1}\mathbf{v} = \begin{bmatrix} 0 & 0 & 1 \\ 0 & 0 & 0 \\ 0 & 0 & -1 \end{bmatrix}\begin{bmatrix} v_1 \\ v_2 \\ v_3 \end{bmatrix} = \begin{bmatrix} 0 \\ 0 \\ 0 \end{bmatrix} \overset{v_3=0}{\rightarrow} \mathbf{v}_1 = \begin{bmatrix} 1 \\ 0 \\ 0 \end{bmatrix},\ \mathbf{v}_2 = \begin{bmatrix} 0 \\ 1 \\ 0 \end{bmatrix} \tag{E1.12.3}$$

This equation holds for any values of v_1 and v_2 as long as $v_3=0$, and therefore, we could determine the two orthogonal eigenvectors $\mathbf{v}_1$ and $\mathbf{v}_2$ corresponding to the eigenvalue $\lambda=1$, just like that. The (Gram-Schmidt) orthogonal process would have been needed to find orthogonal eigenvectors if they could not easily be determined. Here, the eigenvalue $\lambda=1$ is said to be of *geometric multiplicity* m_g=2 (that is defined as *the number of linearly independent eigenvectors corres-*

ponding to the eigenvalue), which can be determined as m_g=dim(A_1) − rank(A_1−λI)=3−1. On the other hand, the eigenvector corresponding to the eigenvalue λ_3=1 is found as follows:

$$[A_1 - \lambda I]_{\lambda=0}\mathbf{v}_3 = \begin{bmatrix} 1 & 0 & 1 \\ 0 & 1 & 0 \\ 0 & 0 & 0 \end{bmatrix}\begin{bmatrix} v_{13} \\ v_{23} \\ v_{33} \end{bmatrix} = \begin{bmatrix} 0 \\ 0 \\ 0 \end{bmatrix} \underset{v_2=0}{\overset{v_1+v_3=0}{\longrightarrow}} \mathbf{v}_3 = \begin{bmatrix} 1/\sqrt{2} \\ 0 \\ -1/\sqrt{2} \end{bmatrix} \tag{E1.12.4}$$

This is supported by the results of using the MATLAB function '`eig()`':

```
>>A1=[1 0 1;0 1 0;0 0 0];   [V,L]=eig(A1)
  V =   1.0000         0    -0.7071       L =       1      0      0
             0    1.0000          0                 0      1      0
             0         0     0.7071                 0      0      0
```

(b) Find the eigenvalues and eigenvectors of a matrix

$$A_2 = \begin{bmatrix} 0 & 1 & 0 \\ 0 & 0 & 1 \\ -2 & -5 & -4 \end{bmatrix} \tag{E1.12.5}$$

The eigenvalues are found by using Eq. (1.2.13) as

$$|A_2 - \lambda I| = \begin{vmatrix} -\lambda & 1 & 0 \\ 0 & -\lambda & 1 \\ -2 & -5 & -4-\lambda \end{vmatrix} = -(\lambda+1)^2(\lambda+2) = 0\ ;\ \lambda = -1,\ -1,\ \text{and}\ -2 \tag{E1.12.6}$$

The eigenvectors corresponding to the eigenvalue λ=−1 of (algebraic) multiplicity m_a=2 are found as follows:

$$[A_2 - \lambda I]_{\lambda=-1}\mathbf{v}_1 = \begin{bmatrix} 1 & 1 & 0 \\ 0 & 1 & 1 \\ -2 & -5 & -3 \end{bmatrix}\begin{bmatrix} v_{11} \\ v_{21} \\ v_{31} \end{bmatrix} = \begin{bmatrix} 0 \\ 0 \\ 0 \end{bmatrix} \rightarrow \mathbf{v}_1 = \begin{bmatrix} 1 \\ -1 \\ 1 \end{bmatrix},\quad \mathbf{v}_2 = ? \tag{E1.12.7}$$

where the second eigenvector $\mathbf{v}_2$ independent of $\mathbf{v}_1$ could not be determined (even by using '`eig()`'). Here, the eigenvalue λ=−1 is of geometric multiplicity m_g=dim(A_2) − rank(A_2 − λI) =3−2=1. The matrices having an eigenvalue of geometric multiplicity m_g less than the algebraic multiplicity m_a are said to be *defective* where r=m_a−m_g is called the *degree of deficiency* of the eigenvalue. For an eigenvalue with degree of deficiency r, r *generalized eigenvectors* can be determined starting from $\mathbf{v}_k$, where the (k+1)th generalized eigenvector $\mathbf{v}_{k+1}$ is generated from Eq. (1.2.12) with the zero vector replaced by the kth eigenvector $\mathbf{v}_k$ on its RHS:

$$[A_2 - \lambda I]_{\lambda=-1}\mathbf{v}_{k+1} = \mathbf{v}_k \tag{E1.12.8}$$

Accordingly, the second generalized eigenvector independent of $\mathbf{v}_2$ can be determined as

$$(A_2 - \lambda_1 I)\mathbf{v}_2 = \mathbf{v}_1;\ \begin{bmatrix} 1 & 1 & 0 \\ 0 & 1 & 1 \\ -2 & -5 & -3 \end{bmatrix}\begin{bmatrix} v_{12} \\ v_{22} \\ v_{32} \end{bmatrix} = \begin{bmatrix} 1 \\ -1 \\ 1 \end{bmatrix};\ \mathbf{v}_2 = \begin{bmatrix} v_{12} \\ v_{22} \\ v_{32} \end{bmatrix} = \begin{bmatrix} 0 \\ 1 \\ -2 \end{bmatrix} \tag{E1.12.9}$$

On the other hand, the eigenvector corresponding to the eigenvalue λ_3=−2 is found as follows:

$$[A_2 - \lambda I]_{\lambda=-2}\mathbf{v}_3 = \begin{bmatrix} 2 & 1 & 0 \\ 0 & 2 & 1 \\ -2 & -5 & -2 \end{bmatrix}\begin{bmatrix} v_{13} \\ v_{23} \\ v_{33} \end{bmatrix} = \begin{bmatrix} 0 \\ 0 \\ 0 \end{bmatrix} \rightarrow \mathbf{v}_3 = \begin{bmatrix} 1 \\ -2 \\ 4 \end{bmatrix} \tag{E1.12.10}$$

Combining these results yields the modal matrix and *Jordan canonical* (or *normal*) *form* of A as

$$V = \begin{bmatrix} \mathbf{v}_1 & \mathbf{v}_2 & \mathbf{v}_3 \end{bmatrix} = \begin{bmatrix} 1 & 0 & 1 \\ -1 & 1 & -2 \\ 1 & -2 & 4 \end{bmatrix}, \quad V^{-1}AV \overset{(1.2.25)}{=} \begin{bmatrix} \lambda_1=-1 & 1 & 0 \\ 0 & \lambda_1=-1 & 0 \\ 0 & 0 & \lambda_3=-2 \end{bmatrix} \tag{E1.12.11}$$

It is good to use '`jordan()`' rather than '`eig()`' to find the generalized eigenvectors:

```
>>A2=[0 1 0;0 0 1;-2 -5 -4]; [V,J]=jordan(A2)
    V =    1     2     0        J =  -2     0     0
          -2    -2     2              0    -1     1
           4     2    -4              0     0    -1
```

(Example 1.13) Complex Eigenvalues/Eigenvectors

Find the eigenvalues and eigenvectors of the matrix

$$A = \begin{bmatrix} 2 & 1 \\ -2 & 0 \end{bmatrix} \tag{E1.13.1}$$

The eigenvalues are found by using Eq. (1.2.13) as

$$|A-\lambda I| = \begin{vmatrix} 2-\lambda & 1 \\ -2 & -\lambda \end{vmatrix} = \lambda^2 - 2\lambda + 2 = (\lambda-1)^2 + 1 = 0\,;\ \lambda = 1 \pm j \tag{E1.13.2}$$

The eigenvectors corresponding to the eigenvalues $\lambda = 1 \pm j$ are found as follows:

$$[A-\lambda I]_{\lambda=1\pm j}\mathbf{v} \overset{(1.2.12)}{=} \begin{bmatrix} 0 \\ 0 \end{bmatrix} \overset{(E1.13.2)}{\rightarrow} \begin{bmatrix} 1\mp j & 1 \\ -2 & -1\mp j \end{bmatrix} \begin{bmatrix} v_1 \\ v_2 \end{bmatrix} = \begin{bmatrix} 0 \\ 0 \end{bmatrix} \rightarrow \mathbf{v}_1,\ \mathbf{v}_2 = \begin{bmatrix} 1 \\ -1\pm j \end{bmatrix} \tag{E1.13.3}$$

```
>>A=[2 1;-2 0]; [V,L]=eig(A)
   V = -0.4082 - 0.4082i    -0.4082 + 0.4082i        L = 1 + 1 i    0
        0.8165               0.8165                       0         1 - 1 i
```

1.2.5 Inverse of a Matrix

The *inverse* matrix of an $N \times N$ (square) matrix $A=[a_{mn}]$ is denoted by A^{-1} and defined to be a matrix that is pre-multiplied/post-multiplied by A to form an identity matrix, i.e., satisfies

$$AA^{-1} = A^{-1}A = I \text{ (Identity matrix)} \tag{1.2.14}$$

The inverse matrix A^{-1} can be computed as

$$A^{-1} = \frac{1}{\det(A)}\text{adj}(A) = \frac{1}{\det(A)} \begin{bmatrix} \alpha_{11} & \cdots & \alpha_{1N} \\ \vdots & \ddots & \vdots \\ \alpha_{N1} & \cdots & \alpha_{NN} \end{bmatrix} \quad \text{with } \alpha_{mn} = (-1)^{n+m} M_{nm} \tag{1.2.15}$$

where adj(A) is called the *adjoint* or *adjugate matrix* of A, and M_{nm} and $(-1)^{n+m}M_{nm}$ are the minor and cofactor of a_{mn}, respectively.

Note that a square matrix A is *invertible/nonsingular* if and only if

- the determinant of A is nonzero, or equivalently,
- no eigenvalue of A is zero, or equivalently,
- the rows (and the columns) of A are linearly independent.

Note also the following about the inverse of a matrix:

(1) $(A^{-1})^{-1} = A$ if A is nonsingular. (1.2.16a)

(2) $(AB)^{-1} = B^{-1}A^{-1}$ if A and B are nonsingular. (1.2.16b)

(3) $(A^{T})^{-1} = (A^{-1})^{T}$ if A is nonsingular where the superscript ^{T} denotes the transpose. (1.2.16c)

(4) The MATLAB statement to get the inverse of a given matrix A is '`inv(A)`'.

(**Example 1.14**) Finding the Inverse of a Matrix

(a) Find the inverse of

$$A = \begin{bmatrix} a_{11} & a_{12} & a_{13} \\ a_{21} & a_{22} & a_{23} \\ a_{31} & a_{32} & a_{33} \end{bmatrix} \tag{E1.14.1}$$

We use Eq. (1.2.15) together with Eq. (1.2.7) to get the inverse of A as

$$A^{-1} \overset{(1.2.15)}{=} \frac{1}{\det(A)} \begin{bmatrix} (-1)^{1+1}M_{11} & (-1)^{2+1}M_{21} & (-1)^{3+1}M_{31} \\ (-1)^{1+2}M_{12} & (-1)^{2+2}M_{22} & (-1)^{3+2}M_{32} \\ (-1)^{1+3}M_{13} & (-1)^{2+3}M_{23} & (-1)^{3+3}M_{33} \end{bmatrix}$$

$$\overset{(1.2.7)}{=} \frac{1}{\Delta} \begin{bmatrix} \begin{vmatrix} a_{22} & a_{23} \\ a_{32} & a_{33} \end{vmatrix} & -\begin{vmatrix} a_{12} & a_{13} \\ a_{32} & a_{33} \end{vmatrix} & \begin{vmatrix} a_{12} & a_{13} \\ a_{22} & a_{23} \end{vmatrix} \\ -\begin{vmatrix} a_{21} & a_{23} \\ a_{31} & a_{33} \end{vmatrix} & \begin{vmatrix} a_{11} & a_{13} \\ a_{31} & a_{33} \end{vmatrix} & -\begin{vmatrix} a_{11} & a_{13} \\ a_{21} & a_{23} \end{vmatrix} \\ \begin{vmatrix} a_{21} & a_{22} \\ a_{31} & a_{32} \end{vmatrix} & -\begin{vmatrix} a_{11} & a_{12} \\ a_{31} & a_{32} \end{vmatrix} & \begin{vmatrix} a_{11} & a_{12} \\ a_{21} & a_{22} \end{vmatrix} \end{bmatrix} = \frac{1}{\Delta} \begin{bmatrix} \begin{pmatrix} a_{22}a_{33} \\ -a_{23}a_{32} \end{pmatrix} & \begin{pmatrix} -a_{12}a_{33} \\ +a_{13}a_{32} \end{pmatrix} & \begin{pmatrix} a_{12}a_{23} \\ -a_{13}a_{22} \end{pmatrix} \\ \begin{pmatrix} -a_{21}a_{33} \\ +a_{23}a_{31} \end{pmatrix} & \begin{pmatrix} a_{11}a_{33} \\ -a_{13}a_{31} \end{pmatrix} & \begin{pmatrix} -a_{11}a_{23} \\ +a_{13}a_{21} \end{pmatrix} \\ \begin{pmatrix} a_{21}a_{32} \\ -a_{22}a_{31} \end{pmatrix} & \begin{pmatrix} -a_{11}a_{32} \\ +a_{12}a_{31} \end{pmatrix} & \begin{pmatrix} a_{11}a_{22} \\ -a_{12}a_{21} \end{pmatrix} \end{bmatrix} \tag{E1.14.2}$$

with $\Delta = a_{11}(a_{22}a_{33} - a_{23}a_{32}) - a_{12}(a_{21}a_{33} - a_{23}a_{31}) + a_{13}(a_{21}a_{32} - a_{22}a_{31})$

Check this result by entering the following statements at the MATLAB prompt:

```
>>syms a11 a12 a13 a21 a22 a23 a31 a32 a33
  A=[a11 a12 a13; a21 a22 a23; a31 a32 a33];
  inv(A)
```

(b) Find the inverse of

$$A = \begin{bmatrix} a_{11} & a_{12} \\ a_{21} & a_{22} \end{bmatrix} \tag{E1.14.3}$$

We use Eq. (1.2.15) together with Eq. (1.2.6) to get the inverse of A as

$$A^{-1} \overset{(1.2.15)}{=} \frac{1}{\det(A)} \begin{bmatrix} (-1)^{1+1}M_{11} & (-1)^{2+1}M_{21} \\ (-1)^{1+2}M_{12} & (-1)^{2+2}M_{22} \end{bmatrix} \overset{(1.2.6)}{=} \frac{1}{a_{11}a_{22} - a_{12}a_{21}} \begin{bmatrix} a_{22} & -a_{12} \\ -a_{21} & a_{11} \end{bmatrix} \tag{E1.14.4}$$

(c) Find the inverse of

$$A = \begin{bmatrix} 1 & -2 \\ -2 & 4 \end{bmatrix} \tag{E1.14.5}$$

We use Eq. (E1.14.4) to get the inverse of A as

$$A^{-1} \overset{\text{(E1.14.4)}}{=} \frac{1}{a_{11}a_{22}-a_{12}a_{21}}\begin{bmatrix} a_{22} & -a_{12} \\ -a_{21} & a_{11} \end{bmatrix} = \frac{1}{1\cdot 4-(-2)(-2)}\begin{bmatrix} 4 & 2 \\ 2 & 1 \end{bmatrix} = \frac{1}{0}\begin{bmatrix} 4 & 2 \\ 2 & 1 \end{bmatrix} \quad \text{(E1.14.6)}$$

What is happening here? We cannot find A^{-1}. Why? Because it does not exist. Why? Because its determinant in the denominator is zero, or equivalently, one of the row/columns is equal to the other row/column multiplied by some constant ($c=-2$). How will the MATLAB command '`inv()`' deal with this singular case? Let us try:

```
>>A=[1 -2; -2 4]; inv(A)
```

Then you can see the following output, which is quite reasonable:

```
Warning: Matrix is singular to working precision.
ans =  Inf   Inf
       Inf   Inf
```

Now, to raise the issue of ill-conditioned matrix, let us consider the inverses of

$$A_1 = \begin{bmatrix} 1 & -2 \\ -2 & 4+10^{-10} \end{bmatrix},\ A_2 = \begin{bmatrix} 1 & -2 \\ -2 & 4+10^{-15} \end{bmatrix},\ \text{and } A_3 = \begin{bmatrix} 1 & -2 \\ -2 & 4+10^{-16} \end{bmatrix} \quad (1.2.17)$$

We use Eq. (E1.14.4) to get the inverse of A as

$$A_1^{-1} \overset{\text{(E1.14.4)}}{\approx} \frac{1}{10^{-10}}\begin{bmatrix} 4 & 2 \\ 2 & 1 \end{bmatrix} = 10^{10}\begin{bmatrix} 4 & 2 \\ 2 & 1 \end{bmatrix},\ A_2^{-1} \approx 10^{15}\begin{bmatrix} 4 & 2 \\ 2 & 1 \end{bmatrix},\ \text{and } A_3^{-1} \approx 10^{16}\begin{bmatrix} 4 & 2 \\ 2 & 1 \end{bmatrix} \quad (1.2.18)$$

How will the MATLAB function '`inv()`' deal with these near-singular cases? Let us run the following block of statements:

```
A1=[1 -2; -2 4+1e-10];   A2=[1 -2; -2 4+1e-15];   A3=[1 -2; -2 4+1e-16];
A1_inv=inv(A1), A2_inv=inv(A2), A3_inv=inv(A3)
```

Then you can see the outputs as

```
A1_inv =
   1.0e+010 *
      4.0000      2.0000
      2.0000      1.0000

Warning: Matrix is close to singular or badly scaled.
         Results may be inaccurate. RCOND = 2.467162e-017.

A2_inv =
   1.0e+015 *
      4.5036      2.2518
      2.2518      1.1259

Warning: Matrix is singular to working precision.
A3_inv =
      Inf    Inf
      Inf    Inf
```

You might have been surprised to see that the numerical results for A_2^{-1} and A_3^{-1} are not the same as the corresponding analytical ones given in Eq. (1.2.17). However, it is quite natural from the point of view of digital computers with finite wordlength that have difficulty in telling a very small number from zero. To computers, whether a number is zero or not is a matter of degree, i.e., how close the number is to zero. They will decide a number is zero unless its magnitude is less than the smallest positive number that they can represent in the finite-wordlength number representation system adopted by the software. In this context, the singularity of a matrix is also a matter of degree to computers. Here, a question may arise in your mind:

"Isn't there any quantitative criterion to measure the near-singularity or ill-conditionedness of a matrix?"

Yes, there is a criterion to measure how ill-conditioned a matrix is. It is the *condition number* defined as

$$\text{cond}(A_{M\times N}) = \|A\|\,\|A^{-1}\| \quad \text{with } \|A\| = \|A\|_2 = \underset{1\le n\le N}{\text{Max}}\,\sigma_n(A) = \underset{1\le n\le N}{\text{Max}}\left\{\sqrt{\lambda_n(A^{*\text{T}}A)}\right\} \tag{1.2.19}$$

that can be computed using the MATLAB command '`cond(A)`' where $\sigma_n(A)=\sqrt{\lambda_n(A^{*\text{T}}A)}$ is the square root of the nth eigenvalue of $A^{*\text{T}}A$ and is called a *singular value* of matrix A. The singular values $\{\sigma_n, 1\le n\le N\}$ in the descending order of magnitude can be computed using '`svd(A)`' (see Section 1.2.10.4). cond(A) is equal to the ratio between the largest and smallest singular values of A. Matrix A is said to be *ill-conditioned/well-conditioned* if cond(A) is large/small. It may be more convenient to use the *reciprocal condition number* defined as

$$\text{rcond}(A_{M\times N}) = \frac{1}{\|A\|_1\|A^{-1}\|_1} \quad \text{with } \|A\|_1 = \underset{1\le n\le N}{\text{Max}}\left\{\textstyle\sum_{m=1}^{M}|a_{mn}|\right\} \tag{1.2.20}$$

that can be computed using 'rcond(A)', which is large (close to one) and small (close to zero) for *well-conditioned* and *ill-conditioned* matrices, respectively. Let us try using '`cond()`' and '`rcond()`' for a matrix:

```
>>A4=[1 2 -1;-1 2 1;3 4 5]; S=svd(A4)
>>cond(A4), norm(A4)*norm(inv(A4)), S(1)/S(3)
>>rcond(A4), 1/(norm(A4,1)*norm(inv(A4),1))
```

We can also use '`rcond()`' to find the reciprocal condition number of the above three near-singular matrices:

```
>>rcond(A1), rcond(A2), rcond(A3)
   ans = 2.7778e-012,  ans = 2.4672e-017,  ans = 0
```

1.2.6 Similarity Transformation and Diagonalization

Pre-multiplying a matrix A by P^{-1} and post-multiplying it by P makes a *similarity transformation*:

$$A \xrightarrow{\text{Similarity transformation}} P^{-1}AP \tag{1.2.21}$$

Note the following about the effect of similarity transformation on eigenvalues/eigenvectors:

(1) The characteristic equation (1.2.13) producing eigenvalues is invariant under a similarity transformation:

$$|P^{-1}AP-\lambda I| = |P^{-1}AP-P^{-1}\lambda IP| = |P^{-1}||A-\lambda I||P| = |A-\lambda I| \tag{1.2.22}$$

and consequently, eigenvalues are not changed by a similarity transformation.

(2) Substituting $\mathbf{v} = P\mathbf{w}$ into Eq. (1.2.12) yields

$$A\mathbf{v} \overset{(1.2.12)}{=} \lambda\mathbf{v} \overset{\mathbf{v}=P\mathbf{w}}{\rightarrow} AP\mathbf{w} = \lambda P\mathbf{w} = P\lambda\mathbf{w}; \quad [P^{-1}AP]\,\mathbf{w} = \lambda\mathbf{w} \tag{1.2.23}$$

This implies that the matrix $P^{-1}AP$ obtained by a similarity transformation has $\mathbf{w} = P^{-1}\mathbf{v}$ as its eigenvector if $\mathbf{v}$ is an eigenvector of the matrix A.

In order to understand the diagonalization of a matrix into a diagonal matrix (having its eigenvalues on the main diagonal) through a similarity transformation, we have to know the following theorem:

> **[Theorem 1.1]** Distinct Eigenvalues and Independent Eigenvectors
>
> If the eigenvalues of a matrix A are all distinct, i.e., different from each other, then the corresponding eigenvectors are independent of each other and, consequently, the *modal matrix* composed of the eigenvectors as columns is nonsingular.

Now, for an $N \times N$ matrix A whose eigenvalues are all distinct, let us put all of the eigenpair equations (1.2.12) for each eigenvalue–eigenvector pair together to write

$$A\,[\mathbf{v}_1\ \mathbf{v}_2 \cdots \mathbf{v}_N] = [\mathbf{v}_1\ \mathbf{v}_2 \cdots \mathbf{v}_N] \begin{bmatrix} \lambda_1 & 0 & \cdot & 0 \\ 0 & \lambda_2 & \cdot & 0 \\ \cdot & \cdot & \cdot & \cdot \\ 0 & 0 & \cdot & \lambda_N \end{bmatrix}; \quad AV = V\Lambda \tag{1.2.24}$$

Then, noting that the modal matrix V is nonsingular and invertible by Theorem 1.1, we can pre-multiply the above equation by the inverse modal matrix V^{-1} to get

$$V^{-1}AV = V^{-1}V\Lambda \equiv \Lambda \tag{1.2.25}$$

This implies that the modal matrix V composed of the eigenvectors of a matrix A is a similarity transformation matrix that can be used for converting the matrix A into a diagonal matrix having its eigenvalues on the main diagonal. Here is an example to illustrate how a matrix can be diagonalized using a similarity transformation with the model matrix.

(Example 1.15) Diagonalization Using Modal Matrix

Consider the following matrix, which was dealt with in Example 1.11:

$$A = \begin{bmatrix} 0 & 2 \\ 1 & 1 \end{bmatrix} \tag{E1.15.1}$$

We can use the eigenvectors (E1.11.4) (obtained in Example 1.11) to construct the modal matrix V as

$$V = [\mathbf{v}_1\ \ \mathbf{v}_2] = \begin{bmatrix} -2/\sqrt{5} & 1/\sqrt{2} \\ 1/\sqrt{5} & 1/\sqrt{2} \end{bmatrix} \tag{E1.15.2}$$

and use this matrix V to make a similarity transformation of the matrix A as

$$V^{-1}AV=\begin{bmatrix}-2/\sqrt{5} & 1/\sqrt{2}\\ 1/\sqrt{5} & 1/\sqrt{2}\end{bmatrix}^{-1}\begin{bmatrix}0 & 2\\ 1 & 1\end{bmatrix}\begin{bmatrix}-2/\sqrt{5} & 1/\sqrt{2}\\ 1/\sqrt{5} & 1/\sqrt{2}\end{bmatrix}=\begin{bmatrix}-\sqrt{5}/3 & \sqrt{5}/3\\ \sqrt{2}/3 & 2\sqrt{2}/3\end{bmatrix}\begin{bmatrix}2/\sqrt{5} & \sqrt{2}\\ -1/\sqrt{5} & \sqrt{2}\end{bmatrix}=\begin{bmatrix}-1 & 0\\ 0 & 2\end{bmatrix}. \tag{E1.15.3}$$

This is a diagonal matrix having the eigenvalues on its main diagonal. This diagonalization process can be done by typing the following block of statements at the MATLAB prompt:

```
A=[0 2; 1 1]; % Eq.(E1.15.1)
[V,L]=eig(A); % Modal matrix V (Eq.(E1.15.2)) and Eigenvalue matrix L
L1=V^-1*A*V   % Eq.(E1.15.3)
```

1.2.7 Special Matrices

1.2.7.1 Transpose and Symmetric/Hermitian Matrix

The *transpose* of an $M\times N$ matrix A, denoted by A^{T}, is an $N\times M$ matrix whose rows/columns are the columns/rows of A, respectively. Therefore, taking the transpose of a matrix A means the following:

$$A=\begin{bmatrix}a_{11} & a_{12} & \bullet & a_{1N}\\ a_{21} & a_{22} & \bullet & a_{2N}\\ \bullet & \bullet & \bullet & \bullet\\ a_{M1} & a_{M2} & \bullet & a_{MN}\end{bmatrix}\overset{\text{Transpose}}{\rightarrow}\begin{bmatrix}a_{11} & a_{21} & \bullet & a_{M1}\\ a_{12} & a_{22} & \bullet & a_{M2}\\ \bullet & \bullet & \bullet & \bullet\\ a_{1N} & a_{2N} & \bullet & a_{MN}\end{bmatrix}=A^{\mathrm{T}} \tag{1.2.26}$$

A square matrix A is said to be *symmetric*, if it is equal to its transpose, i.e.,

$$A^{\mathrm{T}}=A \tag{1.2.27}$$

A complex-valued matrix is said to be *Hermitian* or *self-adjoint* if it is equal to its *tranjugate* (transposed conjugate or conjugate transpose), i.e.,

$A^{*\mathrm{T}}$ (denoted by A^{H} or A^{*} in many references) $=A$ where the superscript * means the conjugate. (1.2.28)

Note the following properties of transpose:

(1) $(A^{\mathrm{T}})^{\mathrm{T}}=A$ (1.2.29a)
(2) $(A+B)^{\mathrm{T}}=A^{\mathrm{T}}+B^{\mathrm{T}}$ (1.2.29b)
(3) $(AB)^{\mathrm{T}}=B^{\mathrm{T}}A^{\mathrm{T}}$ (1.2.29c)
(4) If A is real-symmetric/Hermitian, all the eigenvalues are real-valued and the eigenvectors corresponding to distinct eigenvalues of A are orthogonal/unitary in the sense that their dot product is zero.
(5) The MATLAB statements to get the transpose and conjugate transpose of a given matrix A are `A.'` (dot-apostrophe operator) and `A'` (apostrophe operator), respectively.

1.2.7.2 Orthogonal/Unitary Matrix

A nonsingular (square) matrix A is said to be *orthogonal* if its row/column vectors form an orthonormal set so that its transpose is equal to its inverse, i.e.,

$$A^{\mathrm{T}}A\equiv I\ ;\quad A^{\mathrm{T}}\equiv A^{-1} \tag{1.2.30}$$

A complex-valued (square) matrix is said to be *unitary* if its conjugate transpose is equal to its inverse, i.e.,

$$A^{*\mathrm{T}}A \equiv I\ ;\quad A^{*\mathrm{T}} \equiv A^{-1} \tag{1.2.31}$$

Note the following properties of an orthogonal/unitary matrix:

(1) The magnitude (absolute value) of every eigenvalue is one. (1.2.32a)

(2) The product of two orthogonal/unitary matrices is also orthogonal/unitary:

$$(AB)^{*\mathrm{T}}(AB) = B^{*\mathrm{T}}(A^{*\mathrm{T}}A)B \equiv I \tag{1.2.32b}$$

(3) $|A| = 1$ or -1. (1.2.32c)

1.2.7.3 Permutation Matrix

A matrix P having only one nonzero element of value 1 in each row and column is called a *permutation matrix* and has the following properties:

(1) Pre-multiplication/post-multiplication of a matrix A by a permutation matrix P, i.e., PA or AP yields a row/column permutation of the matrix A, respectively. For example, we have

$$\text{Row permutation with } P: PA = \begin{bmatrix} 0 & 0 & 1 \\ 1 & 0 & 0 \\ 0 & 1 & 0 \end{bmatrix} \begin{bmatrix} a_{11} & a_{12} & a_{13} \\ a_{21} & a_{22} & a_{23} \\ a_{31} & a_{32} & a_{33} \end{bmatrix} = \begin{bmatrix} a_{31} & a_{32} & a_{33} \\ a_{11} & a_{12} & a_{13} \\ a_{21} & a_{22} & a_{23} \end{bmatrix} \tag{1.2.33a}$$

$$\text{Column permutation with } P: AP = \begin{bmatrix} a_{11} & a_{12} & a_{13} \\ a_{21} & a_{22} & a_{23} \\ a_{31} & a_{32} & a_{33} \end{bmatrix} \begin{bmatrix} 0 & 0 & 1 \\ 1 & 0 & 0 \\ 0 & 1 & 0 \end{bmatrix} = \begin{bmatrix} a_{12} & a_{13} & a_{11} \\ a_{22} & a_{23} & a_{21} \\ a_{32} & a_{33} & a_{31} \end{bmatrix} \tag{1.2.33b}$$

(2) A permutation matrix P is orthogonal, i.e., $P^{\mathrm{T}}P=I$, or equivalently, $P^{\mathrm{T}}=P^{-1}$. For example, we have

$$P^{\mathrm{T}}P = \begin{bmatrix} 0 & 1 & 0 \\ 0 & 0 & 1 \\ 1 & 0 & 0 \end{bmatrix} \begin{bmatrix} 0 & 0 & 1 \\ 1 & 0 & 0 \\ 0 & 1 & 0 \end{bmatrix} = \begin{bmatrix} 1 & 0 & 0 \\ 0 & 1 & 0 \\ 0 & 0 & 1 \end{bmatrix}. \tag{1.2.34}$$

1.2.7.4 Rotation Matrix

There are three basic 3-D rotation matrices that can be used to rotate a vector by θ about the x-, y-, and z-axes:

$$R_x(\theta) = \begin{bmatrix} 1 & 0 & 0 \\ 0 & \cos\theta & -\sin\theta \\ 0 & \sin\theta & \cos\theta \end{bmatrix},\ R_y(\theta) = \begin{bmatrix} \cos\theta & 0 & \sin\theta \\ 0 & 1 & 0 \\ -\sin\theta & 0 & \cos\theta \end{bmatrix},\ R_z(\theta) = \begin{bmatrix} \cos\theta & -\sin\theta & 0 \\ \sin\theta & \cos\theta & 0 \\ 0 & 0 & 1 \end{bmatrix} \tag{1.2.35a,b,c}$$

To see how these matrices rotate a vector, let us multiply $R_z(\pi/4)$ with a vector $\mathbf{a}=[2\ \ 1\ \ 0]^{\mathrm{T}}$ on the xy-plane:

$$\mathbf{b} = R_z(\tfrac{\pi}{4})\mathbf{a} = \begin{bmatrix} \cos(\pi/4) & -\sin(\pi/4) & 0 \\ \sin(\pi/4) & \cos(\pi/4) & 0 \\ 0 & 0 & 1 \end{bmatrix} \begin{bmatrix} 2 \\ 1 \\ 0 \end{bmatrix} = \begin{bmatrix} 2\cos(\pi/4)-\sin(\pi/4) \\ 2\sin(\pi/4)+\cos(\pi/4) \\ 0 \end{bmatrix} = \begin{bmatrix} 1/\sqrt{2} \\ 3/\sqrt{2} \\ 0 \end{bmatrix} \tag{1.2.36}$$

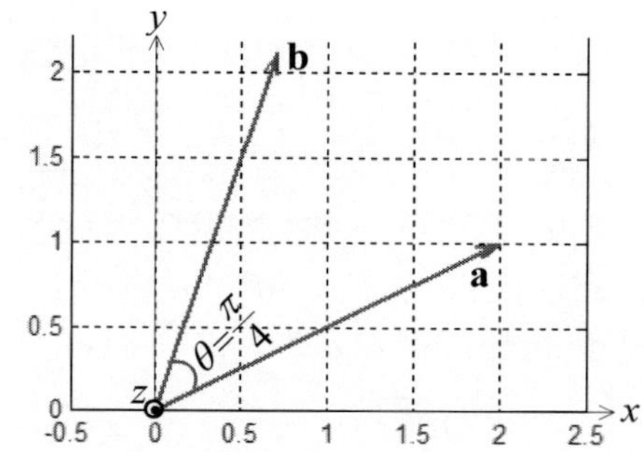

Figure 1.23 Effect of multiplying a vector **a** by a rotation matrix $R_z(\theta)$ with $\theta=\pi/4$

Fig. 1.23 shows the vectors **a** and **b**, each before and after being multiplied by $R_z(\pi/4)$, respectively, where **b** turns out to have been rotated from **a** by $\pi/4$ in the right-hand direction around the $+z$-axis.

Rotation matrices have the following properties:

(1) $R_a^{\mathrm{T}}(\theta)=R_a^{-1}(\theta)$ (orthogonality) (1.2.37a)
(2) $\det(R_a(\theta))=1$ (1.2.37b)
(3) $R_a(\theta_1)\,R_a(\theta_2)=R_a(\theta_1+\theta_2)$ (1.2.37c)
(4) The eigenvalues of $R_a(\theta)$ are $\{1,\cos\theta\pm j\sin\theta\}$ with $j=\sqrt{-1}$. (1.2.37d)
(5) The trace, i.e., the sum of the diagonal elements of $R_z(\theta)$ is $1+2\cos\theta$. (1.2.37e)

Now we consider the rotation matrix R_u for rotating a vector by θ about an arbitrary axis $\mathbf{u}=[u_x\ u_y\ u_z]^{\mathrm{T}}$ (passing through the origin):

$$R_u(\theta)=\begin{bmatrix}\cos\theta+u_x^2(1-\cos\theta) & u_xu_y(1-\cos\theta)-u_z\sin\theta & u_xu_z(1-\cos\theta)+u_y\sin\theta\\ u_yu_x(1-\cos\theta)+u_z\sin\theta & \cos\theta+u_y^2(1-\cos\theta) & u_yu_z(1-\cos\theta)-u_x\sin\theta\\ u_zu_x(1-\cos\theta)-u_y\sin\theta & u_zu_y(1-\cos\theta)+u_x\sin\theta & \cos\theta+u_z^2(1-\cos\theta)\end{bmatrix}\quad(1.2.38)$$

Since any vector along the axis of rotation is rotation-invariant, we can write

$$R_u(\theta)\mathbf{u}_a=\mathbf{u}_a;\quad (R_u(\theta)-1\cdot I)\mathbf{u}_a=\mathbf{0}\quad(1.2.39)$$

Compared to the eigenpair equation (1.2.12), this equation implies that the unit vector $\mathbf{u}_a$ along the axis of rotation is nothing but the eigenvector of the rotation matrix $R_u(\theta)$ corresponding to the eigenvalue $\lambda_1=1$ (Eq. (1.2.37d)). Even if it is complicated and time-consuming to solve Eq. (1.2.39) for $\mathbf{u}_a=[u_{a,x}\ u_{a,y}\ u_{a,z}]^{\mathrm{T}}$ as a symbolic expression, it is simple to insist that $\mathbf{u}=[u_x\ u_y\ u_z]^{\mathrm{T}}$ is the solution to the equation by substituting **u** for $\mathbf{u}_a$ into Eq. (1.2.39) and verifying the equality. This verification can be done by running the following MATLAB script:

```
%show_rotation_matrix.m
% To show that [ux;uy;uz] is the rotation axis satisfying Eq.(1.2.39)
syms ux uy th
uz=sqrt(1-ux^2-uy^2); u=[ux; uy; uz]; c=cos(th); s=sin(th);
R=[c+ux^2*(1-c) ux*uy*(1-c)-uz*s  ux*uz*(1-c)+uy*s;
   uy*ux*(1-c)+uz*s  c+uy^2*(1-c) uy*uz*(1-c)-ux*s;
   uz*ux*(1-c)-uy*s  uz*uy*(1-c)+ux*s c+uz^2*(1-c)]; % Eq.(1.2.38)
simplify([R*u-u R.'*R]), simplify(trace(R)) % Eqs.(1.2.39,37a,37e)
```

which yields

```
ans =  0        1, 0, 0              ans = 2*cos(th) + 1
       0        0, 1, 0
       0        0, 0, 1
```

Thus the vector $\mathbf{u}=[u_x\ u_y\ u_z]^T$ has turned out to be the unit vector along the axis of rotation. Now, how do we determine the angle of rotation for a given rotation matrix like Eq. (1.2.38)? In light of the fact that the trace of a rotation matrix is $1+2\cos\theta$ (Eq. (1.2.37e)), the rotation angle θ can be found as

$$\theta = \cos^{-1}\left\{\frac{1}{2}(\text{trace}(R_u(\theta))-1)\right\} \tag{1.2.40}$$

1.2.7.5 Row Echelon Form

A matrix is said to be of *row echelon form* if

- each nonzero row having at least one nonzero element has a 1 as its first nonzero element, and
- the leading 1 in a row is in a column to the right of the leading 1 in the upper row.
- All-zero rows are below the rows that have at least one nonzero element.

A matrix is said to be of *reduced row echelon form* if it satisfies the above conditions and additionally, each column containing a leading 1 has no other nonzero elements.

Any matrix, singular or rectangular, can be transformed into this form through the Gaussian elimination procedure, i.e., a series of elementary row operations (like row switching, row multiplication, and row addition/subtraction), or equivalently, by using the MATLAB function '`rref()`'. For example, we have

$$A = \begin{bmatrix} \boxed{0\ 0\ 1\ \ 3} \\ \boxed{2\ 4\ 0\ -8} \\ \boxed{1\ 2\ 1\ -1} \end{bmatrix} \underset{\text{switching}}{\overset{\text{row}}{\rightarrow}} \begin{bmatrix} \boxed{2\ 4\ 0\ -8} \\ \boxed{0\ 0\ 1\ \ 3} \\ \boxed{1\ 2\ 1\ -1} \end{bmatrix} \underset{\substack{\text{row} \\ \text{subtraction}}}{\overset{\substack{\text{row} \\ \text{division}}}{\rightarrow}} \begin{bmatrix} 1 & 2 & 0 & -4 \\ 0 & 0 & 1 & 3 \\ 0 & 0 & 1 & 3 \end{bmatrix} \underset{\text{subtraction}}{\overset{\text{row}}{\rightarrow}} \begin{bmatrix} \textcircled{1} & 2 & 0 & -4 \\ 0 & 0 & \textcircled{1} & 3 \\ 0 & 0 & 0 & 0 \end{bmatrix} = \text{rref}(A) \tag{1.2.41}$$

This can be obtained by running the following MATLAB statements:

```
>>A=[0 0 1 3; 2 4 0 -8; 1 2 1 -1];   rref(A)
   ans =      1        2        0       -4
              0        0        1        3
              0        0        0        0
```

Once the row echelon form is obtained, it is easy to compute the rank, the determinant, and even the inverse of the matrix as long as the matrix is invertible.

1.2.8 Positive Definiteness

A square matrix A is said to be *positive definite* if

$$\mathbf{x}^{*T} A\, \mathbf{x} > 0 \quad \text{for any nonzero vector } \mathbf{x} \tag{1.2.42}$$

A square matrix A is said to be *positive semi-definite* if

$$\mathbf{x}^{*T} A\, \mathbf{x} \geq 0 \quad \text{for any nonzero vector } \mathbf{x} \tag{1.2.43}$$

Note the following properties of a positive definite matrix A:

- A is nonsingular and all of its eigenvalues are positive, and
- the inverse of A is also positive definite.

There are similar definitions for *negative definiteness* and *negative semi-definiteness*.

Note the following properties, which can be used to determine if a matrix is positive (semi-) definite or not. A square matrix is *positive definite* if and only if

i) every diagonal element is positive and
ii) every leading principal minor matrix has positive determinant.

On the other hand, a square matrix is *positive semi-definite* if and only if

i) every diagonal element is nonnegative and
ii) every principal minor matrix has nonnegative determinant.

Note also that the *principal minor*s are the sub-matrices taking the diagonal elements from the diagonal of the matrix A. For example, the principal minors of a 3×3 matrix $A=[a_{mn}]$ are

$$a_{11},\ a_{22},\ a_{33},\quad \begin{bmatrix} a_{11} & a_{12} \\ a_{21} & a_{22} \end{bmatrix}, \begin{bmatrix} a_{22} & a_{23} \\ a_{32} & a_{33} \end{bmatrix}, \begin{bmatrix} a_{11} & a_{13} \\ a_{31} & a_{33} \end{bmatrix}, \begin{bmatrix} a_{11} & a_{12} & a_{13} \\ a_{21} & a_{22} & a_{23} \\ a_{31} & a_{32} & a_{33} \end{bmatrix}$$

among which the leading ones are

$$a_{11},\quad \begin{bmatrix} a_{11} & a_{12} \\ a_{21} & a_{22} \end{bmatrix}, \begin{bmatrix} a_{11} & a_{12} & a_{13} \\ a_{21} & a_{22} & a_{23} \\ a_{31} & a_{32} & a_{33} \end{bmatrix}$$

1.2.9 Matrix Inversion Lemma

[*Matrix Inversion Lemma*]

Let A, C, and $[C^{-1}+DA^{-1}B]$ be well-defined with non-singularity as well as compatible dimensions. Then we have

$$[A+BCD]^{-1} = A^{-1} - A^{-1}B[C^{-1}+DA^{-1}B]^{-1}DA^{-1} \tag{1.2.44}$$

(Proof)

We will show that post-multiplying Eq. (1.2.44) by $[A+BCD]$ yields an identity matrix.

$$\begin{aligned}
&[A^{-1} - A^{-1}B[C^{-1}+DA^{-1}B]^{-1}DA^{-1}][A+BCD] \\
&= I + A^{-1}BCD - A^{-1}B[C^{-1}+DA^{-1}B]^{-1}D - A^{-1}B[C^{-1}+DA^{-1}B]^{-1}DA^{-1}BCD \\
&= I + A^{-1}BCD - A^{-1}B[C^{-1}+DA^{-1}B]^{-1}C^{-1}CD - A^{-1}B[C^{-1}+DA^{-1}B]^{-1}DA^{-1}BCD \\
&= I + A^{-1}BCD - A^{-1}B[C^{-1}+DA^{-1}B]^{-1}[C^{-1}+DA^{-1}B]CD \\
&= I + A^{-1}BCD - A^{-1}BCD \equiv I
\end{aligned}$$

1.2.10 LU, Cholesky, QR, and Singular Value Decompositions

1.2.10.1 LU Decomposition (Factorization)

LU decomposition (*factorization*) of a nonsingular (square) matrix A means expressing the matrix as the multiplication of a lower triangular matrix L and an upper triangular matrix U, where a lower/upper triangular matrix is a matrix having no non-zero elements above/below the diagonal. For the case where some row switching operations are needed like in the Gauss elimination (Section 1.3.4), we include a permutation matrix P representing the necessary row switching operation(s) (Section 1.2.7.3) to write the LU decomposition as

$$PA = LU; \quad A = P^{-1}LU \overset{(1.2.34)}{=} P^{T}LU \tag{1.2.45}$$

LU decomposition procedure has been cast into the MATLAB function '`lu()`'.

Here is an example of LU decomposition:

$$\begin{bmatrix} 1 & 2 & 5 \\ 0.2 & 1.6 & 7.4 \\ 0.5 & 4 & 8.5 \end{bmatrix} = \begin{bmatrix} 1 & 0 & 0 \\ 0 & 0 & 1 \\ 0 & 1 & 0 \end{bmatrix}^{T} \begin{bmatrix} 1 & 0 & 0 \\ 0.5 & 1 & 0 \\ 0.2 & 0.4 & 1 \end{bmatrix} \begin{bmatrix} 1 & 2 & 5 \\ 0 & 3 & 6 \\ 0 & 0 & 4 \end{bmatrix} \tag{1.2.46}$$

This can be obtained by using '`lu()`' as follows:

```
>>A=[1 2 5;0.2 1.6 7.4; 0.5 4 8.5];
>>[L,U,P]=lu(A) % LU decomposition
  L = 1.0000        0         0      U =  1    2    5     P = 1    0    0
      0.5000   1.0000         0           0    3    6         0    0    1
      0.2000   0.4000    1.0000           0    0    4         0    1    0
>>P'*L*U-A % A zero or nearly-zero matrix as an evidence of A=P'*L*U
  ans =  0     0     0
         0     0     0
         0     0     0
```

1.2.10.2 Cholesky Decomposition (Factorization)

Cholesky decomposition factors a positive definite symmetric/Hermitian matrix A into an upper triangular matrix premultiplied by its conjugate transpose as

$$A = U^{*T}U \quad (U: \text{an upper triangular matrix}) \tag{1.2.47}$$

Cholesky decomposition can be performed using the MATLAB function '`chol()`'.

Here is an illustration of its usage:

```
>>A=[2 3 4; 3 5 6; 4 6 9]; % A positive definite symmetric matrix
>>U=chol(A) % Cholesky decomposition
  U = 1.4142      2.1213      2.8284
      0           0.7071      0.0000
      0           0           1.0000
>>U'*U-A % A (virtual) zero matrix as an evidence of A=U'*U
  ans =  0     0     0
         0     0     0
         0     0     0
```

1.2.10.3 QR Decomposition (Factorization)

QR (orthogonal-right-triangular) decomposition is used to express a square or rectangular matrix A as the product of an orthogonal (unitary) matrix Q and a right (or upper) triangular matrix R as

$$A = QR \tag{1.2.48}$$

where $Q^{*T}Q = I$. QR decomposition can be performed by using the MATLAB function '`qr()`' or '`Gram_Schmidt()`' (Section 1.1.7).

Here is an illustration of the usage of '`qr()`' for QR decomposition:

```
>>A=[2 3 4; 3 5 6; 4 6 9]; % A positive definite symmetric matrix
>>[Q,R]=qr(A) % QR decomposition
     Q =  -0.3714    0.2491   -0.8944    R =  -5.3852   -8.3563  -11.5131
          -0.5571   -0.8305   -0.0000               0   -0.4152    0.4983
          -0.7428    0.4983    0.4472               0         0    0.4472
```

1.2.10.4 SVD (Singular Value Decomposition)

SVD (singular value decomposition) is used to express an $M\times N$ matrix A in the following form:

$$A = USW^{*T} \tag{1.2.49}$$

where U (called the *left-singular matrix*) is an orthogonal (unitary) $M\times M$ matrix, W (called the *right-singular matrix*) an orthogonal (unitary) $N\times N$ matrix, and S a real diagonal $M\times N$ matrix having the *singular values* (s_{mn}'s) of A (the square roots of the eigenvalues of $A^{*T}A$) in decreasing order on its diagonal. Note that we have

$$A^{*T}A\,W = (USW^{*T})^{*T}USW^{*T}W \overset{(1.2.29c)}{=} WS^{*T}U^{*T}US = W(S^{*T}S) = W[s_{mm}^2] \tag{1.2.50a}$$

$$AA^{*T}U = USW^{*T}(USW^{*T})^{*T}U = USW^{*T}WS^{*T}U^{*T}U \overset{(1.2.31)}{=} U[s_{mm}^2] \tag{1.2.50b}$$

This implies that s_{mm}^2's are the common eigenvalues of $A^{*T}A$ and AA^{*T} and that the columns of W/U, each called the *right/left singular vectors* of A, are the same as the eigenvectors of $A^{*T}A$ / AA^{*T}, respectively. Note that the number of nonzero singular values equals the rank of A (see Corollary 6.2 of [K-1]). SVD can be performed by using the MATLAB function '`svd()`'.

Here is an illustration of the usage of '`svd()`' for SVD:

```
>>A=[1 2;2 3;3 5]; % A rectangular matrix
>>[U,S,W]=svd(A) % Singular Value Decomposition
    U= -0.3092  0.7557 -0.5774   S= 7.2071       0   W= 0.5184  -0.8552
       -0.4998 -0.6456 -0.5774        0     0.2403      0.8552   0.5184
       -0.8090  0.1100  0.5774        0          0
>>smm=sqrt(eig(A'*A)), err=U*S*W'-A  % to check if the result is right
    smm=  0.2403         err= 1.0e-015*  -0.2220  -0.2220
          7.2071                               0        0
                                          0.4441        0
```

1.2.11 Geometrical Meaning of Eigenvalues/Eigenvectors

To realize the geometrical meaning of an eigenvalue/eigenvector, consider a premultiplication of a vector $\mathbf{x}$ with a matrix A (dealt with in Example 1.15) to make another vector $\mathbf{y}$:

$$\mathbf{y} = A\mathbf{x} = \begin{bmatrix} 0 & 2 \\ 1 & 1 \end{bmatrix}\begin{bmatrix} 1 \\ 0 \end{bmatrix} = \begin{bmatrix} 0 \\ 1 \end{bmatrix} \tag{1.2.51}$$

Here, we can run the following MATLAB statements:

```
>>A=[0  2; 1  1]; [V,L]=eig(A),  [U,S,W]=svd(A)
```

to get

```
  V =  -0.8944  -0.7071   L =     -1        0
        0.4472  -0.7071            0        2
  U =  -0.8507  -0.5257   S =  2.2882       0   W =  -0.2298   0.9732
       -0.5257   0.8507            0   0.8740       -0.9732  -0.2298
```

This means that the *eigendecomposition* (ED) and SVD of the matrix A are

$$A \overset{\text{ED}}{=} V\Lambda V^{-1} \text{ with } V = [\mathbf{v}_1 \;\; \mathbf{v}_2] \overset{\text{(E1.15.2')}}{=} \begin{bmatrix} -2/\sqrt{5} & -1/\sqrt{2} \\ 1/\sqrt{5} & -1/\sqrt{2} \end{bmatrix},\; \Lambda = \begin{bmatrix} \lambda_1{=}-1 & 0 \\ 0 & \lambda_2{=}2 \end{bmatrix} \tag{1.2.52}$$

$$A \overset{\text{SVD}}{=} USW^{-1} = \begin{bmatrix} -0.8507 & -0.5257 \\ -0.5257 & 0.8507 \end{bmatrix}\begin{bmatrix} s_1{=}2.2882 & 0 \\ 0 & s_2{=}0.8740 \end{bmatrix}\begin{bmatrix} -0.2298 & 0.9732 \\ -0.9732 & -0.2298 \end{bmatrix}^{-1} \tag{1.2.53}$$

where the matrix V of the SVD has been replaced by W to keep it from being mixed up with the modal matrix V.

With the component-wise expression of the vector $\mathbf{x}=[1 \;\; 0]^T$ in terms of the eigenvector $\mathbf{v}_i$'s:

$$\mathbf{x} = VV^{-1}\mathbf{x} = V\boldsymbol{\alpha}_V \text{ with } \boldsymbol{\alpha}_V = V^{-1}\mathbf{x} = \begin{bmatrix} \alpha_{V,1} = -0.7454 \\ \alpha_{V,2} = -0.4714 \end{bmatrix} \tag{1.2.54}$$

```
>>x=[1; 0]; alpha_V=V\x
     alpha_V =    -0.7454
                  -0.4714
```

we can rewrite Eq. (1.2.51) as

$$\mathbf{y} = A\mathbf{x} \underset{(1.2.54)}{\overset{(1.2.52)}{=}} V\Lambda V^{-1}V\boldsymbol{\alpha}_V = V\Lambda\boldsymbol{\alpha}_V = [\mathbf{v}_1 \;\; \mathbf{v}_2]\begin{bmatrix} \lambda_1{=}-1 & 0 \\ 0 & \lambda_2{=}2 \end{bmatrix}\begin{bmatrix} \alpha_{V,1} \\ \alpha_{V,2} \end{bmatrix} = [\mathbf{v}_1 \;\; \mathbf{v}_2]\begin{bmatrix} \lambda_1\alpha_{V,1} \\ \lambda_2\alpha_{V,2} \end{bmatrix} \tag{1.2.55}$$

This implies that premultiplying a vector $\mathbf{x}$ with a matrix A amounts to multiplying each component $\alpha_{V,i}$ of $\mathbf{x}$ (along each eigenvector $\mathbf{v}_i$) with λ_i.

Now, with the component-wise expression of the vector $\mathbf{x}=[1 \;\; 0]^T$ in terms of the right singular vectors, i.e., the column vector $\mathbf{w}_i$'s of W:

$$\mathbf{x} = WW^{-1}\mathbf{x} = W\boldsymbol{\alpha}_W \text{ with } \boldsymbol{\alpha}_W = W^{-1}\mathbf{x} = \begin{bmatrix} \alpha_{W,1} = -0.2298 \\ \alpha_{W,2} = 0.9732 \end{bmatrix} \tag{1.2.56}$$

we can rewrite Eq. (1.2.51) as

$$\mathbf{y} = A\mathbf{x} \underset{(1.2.56)}{\overset{(1.2.53)}{=}} USW^{-1}W\boldsymbol{\alpha}_W = US\boldsymbol{\alpha}_W = [\mathbf{u}_1 \;\; \mathbf{u}_2]\begin{bmatrix} s_1 & 0 \\ 0 & s_2 \end{bmatrix}\begin{bmatrix} \alpha_{W,1} \\ \alpha_{W,2} \end{bmatrix} = [\mathbf{u}_1 \;\; \mathbf{u}_2]\begin{bmatrix} s_1\alpha_{W,1} \\ s_2\alpha_{W,2} \end{bmatrix} \tag{1.2.57}$$

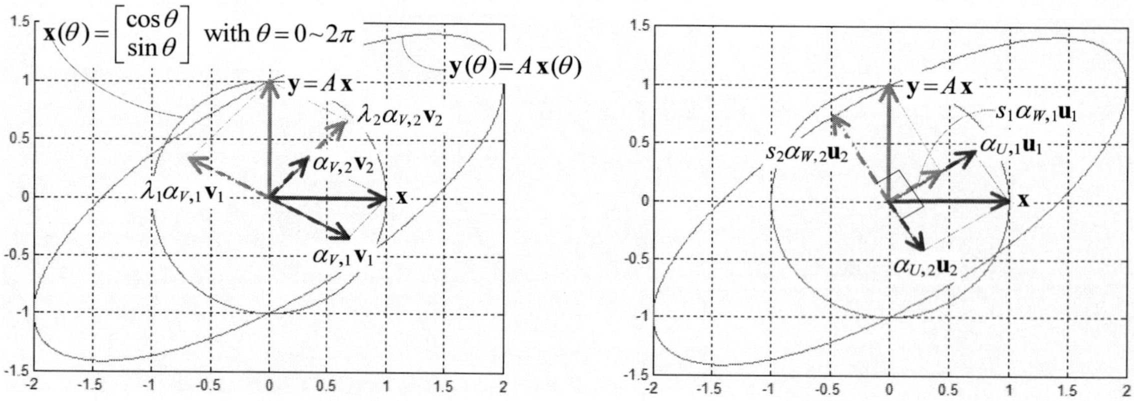

(a) Physical meaning of eigenvalues/eigenvectors (b) Physical meaning of singular values/vectors

Figure 1.24 Geometrical meanings of eigenvalues/eigenvectors and singular values/vectors

This implies that the length of $A\mathbf{x}$ along each left singular vector $\mathbf{u}_i$ is equal to the length $\alpha_{W,i}$ of $\mathbf{x}$ along the corresponding right singular vector $\mathbf{w}_i$ times the corresponding singular value s_i.

Note that if the matrix is Hermitian, the ED (1.2.52) and the SVD (1.2.53) are the same although the orders of the eigenvalues (singular values) and corresponding eigenvectors (singular vectors) may differ. See Fig. 1.24 (made availabe by typing 'eigshow(A)' at the MATLAB prompt) for a visual understanding of the above explanation about the geometrical meaning of eigenvalues/eigenvectors. It is interesting that the left singular vectors $\mathbf{u}_1/\mathbf{u}_2$ corresponding to the larger/smaller singular values s_1/s_2 are along the major/minor axes of the ellipse of $\mathbf{y}=A\mathbf{x}$ with $||\mathbf{x}||=1$, respectively, where the lengths of the semi-major/minor axes are s_1/s_2, respectively.

To see an application of ED, let us consider the computation of the Kth power of an $N\times N$ matrix A, which is diagonalizable in the sense that it can be written in *eigendecomposition* (or *spectral decomposition* or *Jordan normal/canonical*) form as

$$A=V\Lambda W^{*\mathrm{T}}=\begin{bmatrix}\mathbf{v}_1 & \bullet & \mathbf{v}_N\end{bmatrix}\begin{bmatrix}\lambda_1 & 0 & 0\\ 0 & \bullet & 0\\ 0 & 0 & \lambda_N\end{bmatrix}\begin{bmatrix}\mathbf{w}_1^{*\mathrm{T}}\\ \bullet\\ \mathbf{w}_N^{*\mathrm{T}}\end{bmatrix}=\sum_{i=1}^{N}\lambda_i\,\mathbf{v}_i\mathbf{w}_i^{*\mathrm{T}} \tag{1.2.58}$$

Here, $\mathbf{v}_i$'s/$\mathbf{w}_i$'s, each called the *right/left eigenvector* corresponding to the eigenvalue λ_i of A, are the column vectors of the *right/left modal matrices* V/W such that

$$AV=V\Lambda;\ \begin{bmatrix}\mathbf{a}_1 & \bullet & \mathbf{a}_N\end{bmatrix}\begin{bmatrix}\mathbf{v}_1 & \bullet & \mathbf{v}_N\end{bmatrix}=\begin{bmatrix}\mathbf{v}_1 & \bullet & \mathbf{v}_N\end{bmatrix}\begin{bmatrix}\lambda_1 & 0 & 0\\ 0 & \bullet & 0\\ 0 & 0 & \lambda_N\end{bmatrix} \tag{1.2.59a}$$

$$W^{*\mathrm{T}}A=\Lambda W^{*\mathrm{T}};\ \begin{bmatrix}\mathbf{w}_1^{*\mathrm{T}}\\ \bullet\\ \mathbf{w}_N^{*\mathrm{T}}\end{bmatrix}\begin{bmatrix}\mathbf{a}_1 & \bullet & \mathbf{a}_N\end{bmatrix}=\begin{bmatrix}\lambda_1 & 0 & 0\\ 0 & \bullet & 0\\ 0 & 0 & \lambda_N\end{bmatrix}\begin{bmatrix}\mathbf{w}_1^{*\mathrm{T}}\\ \bullet\\ \mathbf{w}_N^{*\mathrm{T}}\end{bmatrix} \tag{1.2.59b}$$

where such matrices V and W can be obtained by using the MATLAB function '`eig()`' as

```
>>[V,L,W]=eig(A)
```

The right eigenvectors are usually referred to as the eigenvectors with no modifier, and if every eigenvalue is real-valued, then the left modal matrix of A is equal to the right modal matrix of $A^{*\mathrm{T}}$. If the matrix A is Hermitian (symmetric) (Section 1.2.7.1), the right modal matrix V and left modal matrix W are equal to each other and unitary (see Theorem 7.14.4 of [K-2], called the spectral theorem) so that the ED of A can be written as

$$A = V\Lambda W^{*T} = V\Lambda V^{*T} = V\Lambda V^{-1} \tag{1.2.60}$$

$$\because V\Lambda V^{-1} \overset{(1.2.25)}{=} A \overset{(1.2.58)}{=} V \overset{\text{real}}{\Lambda} W^{*T} \underset{(1.2.28)}{\overset{\text{Hermitian}}{=}} A^{*T} \overset{(1.2.29c)}{=} W \overset{\text{real}}{\Lambda} V^{*T};\quad W = V \text{ and } V^{*T} \overset{\text{unitary}}{=} V^{-1}$$

In this case the inverse and the Kth power of the Hermitian matrix A become easy to compute as

$$A^{-1} \overset{(1.2.60)}{=} (V\Lambda V^{-1})^{-1} \overset{(1.2.16b)}{=} V\Lambda^{-1}V^{-1} = V\begin{bmatrix} \lambda_1^{-1} & 0 & 0 \\ 0 & \bullet & 0 \\ 0 & 0 & \lambda_N^{-1} \end{bmatrix} V^{*T} \tag{1.2.61}$$

$$A^K \overset{(1.2.60)}{=} \overbrace{(V\Lambda V^{-1})(V\Lambda V^{-1})\cdots(V\Lambda V^{-1})}^{K\text{ times}} = V\Lambda^K V^{-1} = V\begin{bmatrix} \lambda_1^K & 0 & 0 \\ 0 & \bullet & 0 \\ 0 & 0 & \lambda_N^K \end{bmatrix} V^{*T} \tag{1.2.62}$$

The ED of A can also be used to transform any quadratic form in N variables $\{x_1,\ldots, x_N\}$ into an equivalent quadratic form with no cross terms. For example,

$$\begin{aligned} g(\mathbf{x}) &= a_{11}x_1^2 + a_{22}x_2^2 + a_{33}x_3^2 + 2a_{12}x_1x_2 + 2a_{13}x_1x_3 + 2a_{23}x_2x_3 \\ &= \mathbf{x}^T A\,\mathbf{x} = \begin{bmatrix} x_1 & x_2 & x_3 \end{bmatrix}\begin{bmatrix} a_{11} & a_{12} & a_{13} \\ a_{12} & a_{22} & a_{23} \\ a_{13} & a_{23} & a_{33} \end{bmatrix}\begin{bmatrix} x_1 \\ x_2 \\ x_3 \end{bmatrix} \end{aligned} \tag{1.2.63a}$$

$$\rightarrow h(\mathbf{y}) = \lambda_1 y_1^2 + \lambda_2 y_2^2 + \lambda_3 y_3^2 = \mathbf{y}^T\Lambda\mathbf{y} = \begin{bmatrix} y_1 & y_2 & y_3 \end{bmatrix}\begin{bmatrix} \lambda_1 & 0 & 0 \\ 0 & \lambda_2 & 0 \\ 0 & 0 & \lambda_3 \end{bmatrix}\begin{bmatrix} y_1 \\ y_2 \\ y_3 \end{bmatrix} \text{ with } \mathbf{y} = V^T\mathbf{x} \tag{1.2.63b}$$

(Example 1.16) Geometrical Meaning of Eigenvalues/Eigenvectors

(a) Consider a quadratic (objective) function

$$f(\mathbf{x}) = \mathbf{x}^T A\mathbf{x} = \begin{bmatrix} x_1 & x_2 \end{bmatrix}\begin{bmatrix} 5 & 3 \\ 3 & 5 \end{bmatrix}\begin{bmatrix} x_1 \\ x_2 \end{bmatrix} \tag{E1.16.1}$$

The eigenvalues and eigenvectors of the coefficient matrix A can be found by using the MATLAB function '`eig()`' as

```
>>A=[5 3; 3 5]; [V,L]=eig(A)
  V =  -0.7071     0.7071        L =   2           0
        0.7071     0.7071              0           8
```

$$\mathbf{v}_a = \begin{bmatrix} -1/\sqrt{2} \\ 1/\sqrt{2} \end{bmatrix},\ \lambda_a = 2 \text{ and } \mathbf{v}_b = \begin{bmatrix} 1/\sqrt{2} \\ 1/\sqrt{2} \end{bmatrix},\ \lambda_b = 8 \tag{E1.16.2}$$

The following MATLAB script, "em01e16a.m," can be run to plot the contours of the above function as Fig. 1.25(a), which shows the eigenvectors $\mathbf{v}_a$ and $\mathbf{v}_b$ where the values of the objective function (E1.16.1) at $\mathbf{x}=\mathbf{v}_a$ and $\mathbf{v}_b$ are

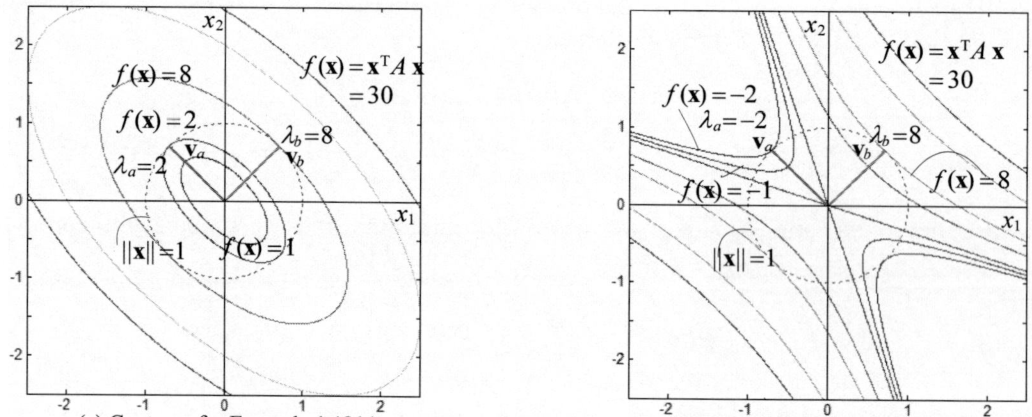

(a) Contours for Example 1.12(a) (b) Contours for Example 1.12(b)

Figure 1.25 Contours of $f(\mathbf{x})=\mathbf{x}^{\mathrm{T}}A\mathbf{x}$ and eigenvectors of A

```
%em01e16a.m
A=[5 3; 3 5]; % Coefficient matrix
fx=@(x)x'*A*x; % Objective function (E1.16.1)
x1=[-2.5:0.01:2.5]; x2=[-2.5:0.01:2.5];  [X1,X2]=meshgrid(x1,x2);
for m=1:length(x1)
   for n=1:length(x2), FX(n,m)=fx([X1(n,m) X2(n,m)].');  end
end
contour(X1,X2,FX,[1 2 8 20 30]), hold on
plot(exp(j*[0:0.01:2*pi]),':') % Plot the unit circle for x'*x=1
axis('equal'), axis([-2.5 2.5 -2.5 2.5])
[V,L]=eig(A), va=V(:,1), vb=V(:,2)
quiver(0,0,va(1),va(2)), quiver(0,0,vb(1),vb(2))
fa=fx(va), fb=fx(vb)
```

$$f(\mathbf{v}_a)=\mathbf{v}_a^{\mathrm{T}}A\mathbf{v}_a=\mathbf{v}_a^{\mathrm{T}}\lambda_a\mathbf{v}_a=\lambda_a\|\mathbf{v}_a\|^2=\lambda_a=2 \tag{E1.16.3a}$$

$$f(\mathbf{v}_b)=\mathbf{v}_b^{\mathrm{T}}A\mathbf{v}_b=\mathbf{v}_b^{\mathrm{T}}\lambda_b\mathbf{v}_b=\lambda_b\|\mathbf{v}_b\|^2=\lambda_b=8 \tag{E1.16.3b}$$

This implies that for a quadratic (objective) function $f(\mathbf{x})=\mathbf{x}^{\mathrm{T}}A\,\mathbf{x}$ with a real symmetric coefficient matrix A, the eigenvalues of A are the values of $f(\mathbf{x})$ increasing from zero when $\mathbf{x}$ moves from the origin to the unit circle along the direction of the corresponding eigenvectors. Note that the values of $f(\mathbf{x})$ on the unit circle (for unit-magnitude vectors) are between the maximum and minimum eigenvalues.

(b) Consider a quadratic (objective) function

$$f(\mathbf{x})=\mathbf{x}^{\mathrm{T}}A\,\mathbf{x}=\begin{bmatrix}x_1 & x_2\end{bmatrix}\begin{bmatrix}3 & 5\\ 5 & 3\end{bmatrix}\begin{bmatrix}x_1\\ x_2\end{bmatrix} \tag{E1.16.4}$$

The eigenvalues and eigenvectors of the coefficient matrix A can be found as

$$\mathbf{v}_a=\begin{bmatrix}-1/\sqrt{2}\\ 1/\sqrt{2}\end{bmatrix},\ \lambda_a=-2 \ \text{ and }\ \mathbf{v}_b=\begin{bmatrix}1/\sqrt{2}\\ 1/\sqrt{2}\end{bmatrix},\ \lambda_b=8 \tag{E1.16.5}$$

where the contours of the above function together with the eigenvectors of A are shown in Fig. 1.25(b). It can be seen that as $\mathbf{x}$ moves from the origin to the unit circle along the direction of the eigenvectors $\mathbf{v}_a$ and $\mathbf{v}_b$, the objective function $f(\mathbf{x})$ increases by $\lambda_a=-2$ and $\lambda_b=8$, respectively. You may be interested in modifying the above MATLAB script "em01e16a.m" so that it can be run to yield Fig. 1.25(b).

1.3 SYSTEMS OF LINEAR EQUATIONS

In this section, we discuss the solution of a system of equations:

$$\begin{aligned} a_{11}x_1 + a_{12}x_2 + \cdots + a_{1N}x_N &= b_1 \\ a_{21}x_1 + a_{22}x_2 + \cdots + a_{2N}x_N &= b_2 \\ \cdots\cdots &= \,. \\ a_{M1}x_1 + a_{M2}x_2 + \cdots + a_{MN}x_N &= b_M \end{aligned} \tag{1.3.1a}$$

which can be written in a compact form by using matrix-vector notations as

$$A_{M\times N}\,\mathbf{x} = \mathbf{b} \tag{1.3.1b}$$

where

$$A_{M\times N} = \begin{bmatrix} a_{11} & a_{12} & \cdot\ \cdot & a_{1N} \\ a_{21} & a_{22} & \cdot\ \cdot & a_{2N} \\ \cdot & \cdot & \cdot\ \cdot & \cdot \\ a_{M1} & a_{M1} & \cdot\ \cdot & a_{MN} \end{bmatrix}, \quad \mathbf{x} = \begin{bmatrix} x_1 \\ x_2 \\ \cdot \\ x_N \end{bmatrix}, \quad \text{and } \mathbf{b} = \begin{bmatrix} b_1 \\ b_2 \\ \cdot \\ b_M \end{bmatrix}$$

We will deal with the three cases:

(i) The case where the number (M) of equations and the number (N) of unknowns are equal ($M=N$) so that the coefficient matrix $A_{M\times N}$ is square.

(ii) The case where the number (M) of equations is smaller than the number (N) of unknowns ($M<N$) so that we might have to find the minimum-norm solution among infinitely many solutions.

(iii) The case where the number of equations is greater than the number of unknowns ($M>N$) so that there might exist no exact solution, and we must find a solution based on a global error minimization, like the LSE (least squares error) solution.

1.3.1 Nonsingular Case ($M = N$)

If the number (M) of equations and the number (N) of unknowns are equal ($M=N$), then the coefficient matrix A is square so that the solution can be written as

$$\mathbf{x} = A^{-1}\,\mathbf{b} \tag{1.3.2}$$

so long as the matrix A is not singular. Here are the MATLAB statements for this job:

```
>>A=[1 2;3 4]; b=[-1;-1];
>>x=A^-1*b % or, x=inv(A)*b
   x =   1.0000
        -1.0000
```

What if A is square, but singular?

```
>>A=[1 2;2 4]; b=[-1;-1];
>>x=A^-1*b
   Warning: Matrix is singular to working precision.
    x = -Inf
        -Inf
```

This is the case where some or all of the rows of the coefficient matrix A are dependent on other rows and so the rank of A is deficient, which implies that there are some equations equivalent (*redundant*) to or *inconsistent* with other equations. If we remove the dependent rows until all the (remaining) rows are independent of each other so that A is of full rank (equal to M), it leads to the case of $M<N$, which will be dealt with in the next section.

1.3.2 Underdetermined Case ($M < N$) — Minimum-Norm Solution

If the number (M) of equations is less than the number (N) of unknowns, the solution is not unique, but numerous. Suppose the M rows of the coefficient matrix A are independent. Then, any N-dimensional vector can be decomposed into two components

$$\mathbf{x} = \mathbf{x}^+ + \mathbf{x}^- \tag{1.3.3}$$

where the one is in the row space $\mathcal{R}(A)$ of A that can be expressed as a linear combination of the M row vectors

$$\mathbf{x}^+ \overset{(1.2.3)}{=} A^{\mathrm{T}}\boldsymbol{\alpha} \tag{1.3.4}$$

and the other is in the null space $\mathcal{N}(A)$ orthogonal (perpendicular) to the row space so that

$$A\,\mathbf{x}^- \overset{(1.2.4)}{=} \mathbf{0} \tag{1.3.5}$$

Substituting the arbitrary N-dimensional vector representation (1.3.3) into Eq. (1.3.1b) yields

$$A_{M\times N}\mathbf{x} \overset{(1.3.1b)}{=} \mathbf{b} \overset{(1.3.3)}{\rightarrow} A(\mathbf{x}^+ + \mathbf{x}^-) \overset{(1.3.4)}{=} AA^{\mathrm{T}}\boldsymbol{\alpha} + A\mathbf{x}^- \overset{(1.3.5)}{=} AA^{\mathrm{T}}\boldsymbol{\alpha} = \mathbf{b} \tag{1.3.6}$$

Since AA^{T} is supposedly a nonsingular $M\times M$ matrix resulting from multiplying an $M\times N$ matrix by an $N\times M$ matrix, we can solve this equation for $\boldsymbol{\alpha}$ to get

$$\boldsymbol{\alpha}^{\mathrm{o}} \overset{(1.3.6)}{=} [AA^{\mathrm{T}}]^{-1}\mathbf{b} \tag{1.3.7}$$

Substituting this equation (1.3.7) into Eq. (1.3.4) yields

$$\mathbf{x}^{\mathrm{o}+} \overset{(1.3.4)}{=} A^{\mathrm{T}}\boldsymbol{\alpha}^{\mathrm{o}} \overset{(1.3.7)}{=} A^{\mathrm{T}}[AA^{\mathrm{T}}]^{-1}\mathbf{b} \tag{1.3.8}$$

This satisfies Eq. (1.3.1) and thus qualifies as its solution. However, it is far from being unique because the sum of $\mathbf{x}^{\mathrm{o}+}$ and any vector $\mathbf{x}^-$ belonging to the null space satisfying Eq. (1.3.5) still satisfies Eq. (1.3.1) (as seen from Eq. (1.3.6)), yielding infinitely many solutions.

Based on the principle that any one of the two perpendicular legs is shorter than the hypotenuse in a right-angled triangle, Eq. (1.3.8) is believed to represent the *minimum-norm solution*. Note that the projections of a solution $\mathbf{x}^{\mathrm{o}}$ of Eq. (1.3.1) (satisfying $A\mathbf{x}^{\mathrm{o}} = \mathbf{b}$) onto the row and null spaces of A are

$$\mathbf{x}^{\mathrm{o}+} \overset{(1.3.8)}{=} A^{\mathrm{T}}[AA^{\mathrm{T}}]^{-1}\mathbf{b} = A^{\mathrm{T}}[AA^{\mathrm{T}}]^{-1}A\mathbf{x}^{\mathrm{o}} \tag{1.3.9}$$

and

$$\mathbf{x}^{\mathrm{o}-} \overset{(1.3.3)}{=} \mathbf{x}^{\mathrm{o}} - \mathbf{x}^{\mathrm{o}+} \overset{(1.3.9)}{=} \mathbf{x}^{\mathrm{o}} - A^{\mathrm{T}}[AA^{\mathrm{T}}]^{-1}A\mathbf{x}^{\mathrm{o}} = \left[I - A^{\mathrm{T}}[AA^{\mathrm{T}}]^{-1}A\right]\mathbf{x}^{\mathrm{o}}, \tag{1.3.10}$$

respectively. Note also that the matrix $A^T[AA^T]^{-1}$ is called the *right pseudo* (*generalized*) *inverse* of A in the sense that multiplying A on the right by it yields the identity matrix.

The MATLAB function 'pinv()' or slash(/) operator can be used to find the minimum-norm solution (1.3.8) to the system of linear equations (1.3.1):

```
>>A=[1  2]; b=3;
>>x=pinv(A)*b % x=A'*(A*A')^-1*b or eye(size(A,2))/A*b by Eq.(1.3.9)
   x = 0.6000
       1.2000
```

(**Example 1.17**) Minimum-Norm Solution

Consider the problem of solving the equation

$$[1\ \ 2]\begin{bmatrix} x_1 \\ x_2 \end{bmatrix} = 3;\ A\mathbf{x} = \mathbf{b} \text{ with } A = [1\ \ 2] \text{ and } \mathbf{b} = 3 \tag{E1.17.1}$$

This has infinitely many solutions and any $\mathbf{x} = [x_1\ \ x_2]^T$ satisfying this equation, or equivalently,

$$x_1 + 2x_2 = 3;\ \ x_2 = -\frac{1}{2}x_1 + \frac{3}{2} \tag{E1.17.2}$$

is a qualified solution. This equation (E1.17.2) describes the solution space as depicted in Fig. 1.26.

On the other hand, any vector in the row space of the coefficient matrix A can be expressed by Eq. (1.3.4) as

$$\mathbf{x}^+ = A^T\boldsymbol{\alpha} = \begin{bmatrix} 1 \\ 2 \end{bmatrix}\alpha\ \ (\alpha \text{ is a scalar, since } M = 1) \tag{E1.17.3}$$

and any vector in the null space of A can be expressed by Eq. (1.3.5) as

$$A\mathbf{x}^- = [1\ \ 2]\begin{bmatrix} x_1^- \\ x_2^- \end{bmatrix} = 0;\ \ x_2^- = -\frac{1}{2}x_1^- \tag{E1.17.4}$$

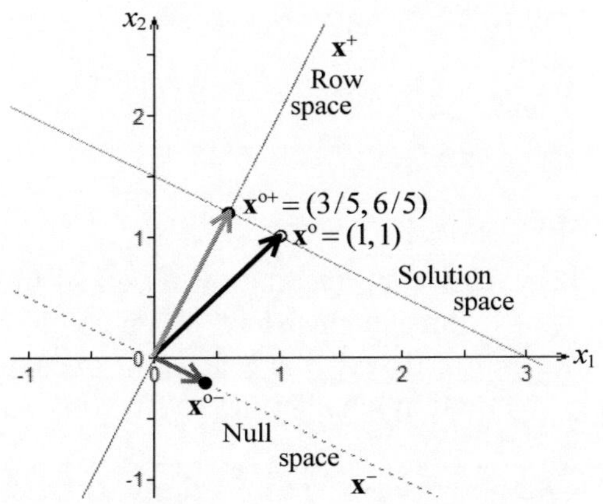

Figure 1.26 A minimum-norm solution

We use Eq. (1.3.8) to obtain the minimum-norm solution

$$\mathbf{x}^{o+} = A^T[AA^T]^{-1}\mathbf{b} = \begin{bmatrix}1\\2\end{bmatrix}\left(\begin{bmatrix}1 & 2\end{bmatrix}\begin{bmatrix}1\\2\end{bmatrix}\right)^{-1} 3 = \frac{3}{5}\begin{bmatrix}1\\2\end{bmatrix} = \begin{bmatrix}0.6\\1.2\end{bmatrix} \tag{E1.17.5}$$

Note from Fig. 1.26 that the minimum-norm solution $\mathbf{x}^{o+}$ is the intersection of the solution space and the row space and is the closest to the origin among the vectors in the solution space, being worthy of the modifier "minimum-norm."

1.3.3 Overdetermined Case ($M > N$) — Least Squares Error Solution

If the number (M) of (independent) equations is greater than the number (N) of unknowns, there exists no solution strictly satisfying all the equations. Thus we try to find the *LS* (*least squares*) or *LSE* (*least squares error*) solution minimizing the norm of the (inevitable) error vector

$$\mathbf{e} = A\mathbf{x} - \mathbf{b} \tag{1.3.11}$$

Then, our problem is to minimize the objective function

$$J = \frac{1}{2}\|\mathbf{e}\|^2 = \frac{1}{2}\|A\mathbf{x}-\mathbf{b}\|^2 = \frac{1}{2}[A\mathbf{x}-\mathbf{b}]^T[A\mathbf{x}-\mathbf{b}] \tag{1.3.12}$$

whose solution can be obtained by setting the derivative of this function (1.3.12) w.r.t. $\mathbf{x}$ to zero:

$$\frac{\partial}{\partial \mathbf{x}}J = A^T[A\,\mathbf{x}-\mathbf{b}] = A^TA\,\mathbf{x} - A^T\mathbf{b} = \mathbf{0};\quad A^TA\,\mathbf{x} = A^T\mathbf{b}$$

$$;\qquad \mathbf{x}^o = [A^TA]^{-1}A^T\mathbf{b} \tag{1.3.13}$$

This is called the LS or LSE solution. Note that the matrix A having more rows than columns ($M>N$) does not have its inverse, but has its *left pseudo* (*generalized*) *inverse* $[A^TA]^{-1}A^T$ as long as A is not rank-deficient, i.e., all of its columns are independent of each other.

The MATLAB function '`pinv()`' or backslash(\) operator can be used to find the LSE solution (1.3.13) to the system of linear equations (1.3.1):

```
>>A=[1; 2]; b=[2.1; 3.9];
>>x=pinv(A)*b % A\b or x=(A'*A)^-1*A'*b by Eq.(1.3.13)
   x = 1.9800
```

[Remark 1.1] MATLAB Function '`pinv()`'

(1) The power of '`pinv()`' is beyond our imagination. Even in the case of $M \le N$, 'pinv(A)*b' returns the LSE solution if Ax=b is inconsistent. Even in the case of $M \ge N$, 'pinv(A)*b' returns the minimum-norm solution if Ax=b is redundant. Actually, a set of equations can be solved by using the '`pinv()`' function whether the coefficient matrix A is square or not.

```
>>A1=[1 2 3;2 4 6]; b1=[1.9;4.1]; x1=pinv(A1)*b1, norm(A1*x1-b1) %incons
>>A2=[1 2;2 4;3 6]; b2=[3; 6; 9]; x2=pinv(A2)*b2, norm(x2) % redundant
     x1 =  0.1443      ans = 0.1342       x2 =   0.6000      ans = 1.3416
           0.2886                                1.2000
           0.4329
```

(2) The MATLAB statement '`pinv(A)`' returns a matrix X of the same dimension as A^T such that $AXA=A$ and $XAX=X$. It can be used to get the right/left pseudo (generalized) inverse $A^T[AA^T]^{-1}$ $/[A^TA]^{-1}A^T$ for a matrix A given as its input argument, depending on whether the number (M) of rows is smaller or greater than the number (N) of columns, so long as the matrix is of full rank, i.e., rank(A)=min(M, N) (see Section 1.2.2). Note that $A^T[AA^T]^{-1}/\,[A^TA]^{-1}A^T$ is called the right/left inverse since it is multiplied onto the right/left side of A to yield the identity matrix. However, you should be careful when using '`pinv(A)`' to find the right/left inverse of a rank-deficient matrix A because its output is no longer the right/left inverse, which does not even exist for rank-deficient matrices.

(**Example 1.18**) LSE and Minimum-Norm Solutions

(a) Consider the problem of solving the equation

$$\begin{bmatrix}1&2&3\\2&4&6\end{bmatrix}\begin{bmatrix}x_1\\x_2\\x_3\end{bmatrix}=\begin{bmatrix}1.9\\4.1\end{bmatrix};\ A\mathbf{x}=\mathbf{b}\text{ with }A=\begin{bmatrix}1&2&3\\2&4&6\end{bmatrix}\text{ and }\mathbf{b}=\begin{bmatrix}1.9\\4.1\end{bmatrix} \tag{E1.18.1}$$

Unlike the usual case of $M=2<3=N$, having fewer equations than unknowns, the two equations are inconsistent, so that we cannot directly apply the minimum-norm solution (1.3.8). First, let us formulate this problem as a kind of LSE problem to minimize the sum of the squared residual errors:

$$J(x_1,x_2,x_3)=(x_1+2x_2+3x_3-1.9)^2+(2x_1+4x_2+6x_3-4.1)^2 \tag{E1.18.2}$$

This objective function can be minimized by setting its partial derivatives w.r.t. x_1, x_2, and x_3 to zero:

$$\begin{cases}\dfrac{\partial J}{\partial x_1}=2(x_1+2x_2+3x_3-1.9)+4(2x_1+4x_2+6x_3-4.1)=0\\ \dfrac{\partial J}{\partial x_2}=4(x_1+2x_2+3x_3-1.9)+8(2x_1+4x_2+6x_3-4.1)=0\\ \dfrac{\partial J}{\partial x_3}=6(x_1+2x_2+3x_3-1.9)+12(2x_1+4x_2+6x_3-4.1)=0\end{cases} \tag{E1.18.3}$$

However, these three equations happen to be identical and accordingly, we find the minimum-norm solution to one of them:

$$\begin{bmatrix}10&20&30\end{bmatrix}\begin{bmatrix}x_1\\x_2\\x_3\end{bmatrix}=20.2;\quad A_1\mathbf{x}=\mathbf{b}_1$$

$$;\ \mathbf{x}\overset{(1.3.8)}{=}A_1^T[A_1A_1^T]^{-1}\mathbf{b}_1=\begin{bmatrix}0.1443\\0.2886\\0.4329\end{bmatrix} \tag{E1.18.4}$$

This can be obtained by using '`pinv()`' as

```
>>A1=[1 2 3;2 4 6]; b1=[1.9;4.1]; x1=pinv(A1)*b1
   x1 =  0.1443
         0.2886
         0.4329
```

(b) Consider the problem of solving the equation

$$\begin{bmatrix} 1 & 2 \\ 2 & 4 \\ 3 & 6 \end{bmatrix} \begin{bmatrix} x_1 \\ x_2 \end{bmatrix} = \begin{bmatrix} 3 \\ 6 \\ 9 \end{bmatrix} \tag{E1.18.5}$$

Unlike the usual case of $M=3>2=N$, having more equations than unknowns, the three equations are identical. That is why we remove two (redundant) equations and solve only one equation with two variables by using the minimum-norm solution (1.3.8) as

$$[1 \;\; 2]\begin{bmatrix} x_1 \\ x_2 \end{bmatrix} = 3; \; \begin{bmatrix} x_1 \\ x_2 \end{bmatrix} \underset{\text{Minimum-norm}}{\overset{(1.3.8)}{=}} A^T[AA^T]^{-1}\mathbf{b} = \begin{bmatrix} 1 \\ 2 \end{bmatrix}\left[[1 \;\; 2]\begin{bmatrix} 1 \\ 2 \end{bmatrix}\right]^{-1} 3 = \begin{bmatrix} 0.6 \\ 1.2 \end{bmatrix} \tag{E1.18.6}$$

This can be obtained by using '`pinv()`' as

```
>>A2=[1 2;2 4;3 6]; b2=[3; 6; 9]; x2=pinv(A2)*b2
  x2 =      0.6000
            1.2000
```

[Remark 1.2] Backslash(\) and Slash(/) Operator

Overall, the MATLAB statement '`A\B`' (with matrices A and B of compatible dimensions) returns a solution X to the equation A*X = B. Let us denote the row and column dimensions of the coefficient matrix A by M and N, respectively.

(1) If the matrix A is square and upper/lower-triangular in the sense that all of its elements below/above the diagonal are zero, then '`A\B`' finds the solution by applying the backward/forward substitution method:

$$\begin{bmatrix} a_{11} & a_{12} & a_{13} \\ 0 & a_{22} & a_{23} \\ 0 & 0 & a_{33} \end{bmatrix} \begin{bmatrix} x_1 \\ x_2 \\ x_3 \end{bmatrix} = \begin{bmatrix} b_1 \\ b_2 \\ b_3 \end{bmatrix}; \; \begin{bmatrix} x_1 \\ x_2 \\ x_3 \end{bmatrix} \underset{\text{substitution}}{\overset{\text{Backward}}{=}} \uparrow \begin{bmatrix} (b_1 - a_{12}x_2 - a_{13}x_3)/a_{11} \\ (b_2 - a_{23}x_3)/a_{22} \\ b_3/a_{33} \end{bmatrix} \tag{1.3.14a}$$

$$\begin{bmatrix} a_{11} & 0 & 0 \\ a_{21} & a_{22} & 0 \\ a_{31} & a_{32} & a_{33} \end{bmatrix} \begin{bmatrix} x_1 \\ x_2 \\ x_3 \end{bmatrix} = \begin{bmatrix} b_1 \\ b_2 \\ b_3 \end{bmatrix}; \; \begin{bmatrix} x_1 \\ x_2 \\ x_3 \end{bmatrix} \underset{\text{substitution}}{\overset{\text{Forward}}{=}} \downarrow \begin{bmatrix} b_1/a_{11} \\ (b_2 - a_{21}x_1)/a_{22} \\ (b_3 - a_{32}x_2 - a_{31}x_1)/a_{33} \end{bmatrix} \tag{1.3.14b}$$

(2) If matrix A is square, symmetric (Hermitian), and positive definite, then '`A\B`' finds the solution by using Cholesky factorization (Section 1.2.10.2).

(3) If matrix A is square and has no special feature, then '`A\B`' finds the solution by using LU decomposition (Section 1.2.10.1).

$$A\mathbf{x} = P^T LU\mathbf{x} = \mathbf{b} \;\text{ with }\; A \overset{(1.2.45)}{=} P^T LU = P^T \begin{bmatrix} 1 & 0 & 0 \\ l_{21} & 1 & 0 \\ l_{31} & l_{32} & 1 \end{bmatrix} \begin{bmatrix} u_{11} & u_{12} & u_{13} \\ 0 & u_{22} & u_{23} \\ 0 & 0 & u_{33} \end{bmatrix}$$

$$; \;\; \mathbf{x} = U^{-1}L^{-1}P\mathbf{b} = \underset{\text{substitution}}{\text{Backward}}\left(U, \underset{\text{substitution}}{\text{Forward}}(L, \text{Permutation}(\mathbf{b}))\right) \tag{1.3.15}$$

(4) If matrix A is rectangular, then '`A\B`' uses QR decomposition (Section 1.2.10.3) to find a solution:

$$A\mathbf{x} = \mathbf{b} \text{ with } A \overset{(1.2.48)}{=} QR = Q\begin{bmatrix} r_{11} & r_{12} & r_{13} \\ 0 & r_{22} & r_{23} \\ 0 & 0 & r_{33} \end{bmatrix}$$

$$; \quad \mathbf{x} = R^{-1}Q^{-1}\mathbf{b} = \underset{\text{substitution}}{\text{Backward}}\left(R, Q^{\mathrm{T}}\mathbf{b}\right) \tag{1.3.16}$$

If A is rectangular and of full rank with rank(A)=min(M,N), it will be the LS (least-squares) solution (Eq. (1.3.13)) for M>N (overdetermined case) and one of the many solutions that is not always the same as the minimum-norm solution (Eq. (1.3.8)) for M<N (underdetermined case). But for the case where A is rectangular and has rank deficiency, what 'A\B' gives us may be useless. Therefore, you must pay attention to the warning message about rank deficiency, which might tell you not to count on the dead-end solution made by the backslash(\) operator. To find an alternative for the case of rank deficiency, you can resort to the SVD.

(5) The MATLAB statement '`B/A`' using the slash operator is equivalent to MATLAB statement '`(A'\B')'`' using the backslash operator.

1.3.4 Gauss(ian) Elimination

For simplicity, we assume that the coefficient matrix A in Eq. (1.3.1) is a nonsingular 3×3 matrix with $M=N=3$. Then we can rewrite the equation as

$$a_{11}x_1 + a_{12}x_2 + a_{13}x_3 = b_1 \tag{1.3.17-1}$$

$$a_{21}x_1 + a_{22}x_2 + a_{23}x_3 = b_2 \tag{1.3.17-2}$$

$$a_{31}x_1 + a_{32}x_2 + a_{33}x_3 = b_3 \tag{1.3.17-3}$$

First, to remove the x_1 terms from Eqs. (1.3.17-m) (with $m\neq1$), we subtract (1.3.17-1) $\times a_{m1}/a_{11}$ from each of them to get

$$a_{11}^{(0)}x_1 + a_{12}^{(0)}x_2 + a_{13}^{(0)}x_3 = b_1^{(0)} \tag{1.3.18-1}$$

$$a_{22}^{(1)}x_2 + a_{23}^{(1)}x_3 = b_2^{(1)} \tag{1.3.18-2}$$

$$a_{32}^{(1)}x_2 + a_{33}^{(1)}x_3 = b_3^{(1)} \tag{1.3.18-3}$$

with

$$a_{mn}^{(0)} = a_{mn},\; b_m^{(0)} = b_m \quad \text{for } m,n = 1,2,3 \tag{1.3.19a}$$

$$a_{mn}^{(1)} = a_{mn}^{(0)} - \frac{a_{m1}^{(0)}}{a_{11}^{(0)}}a_{1n}^{(0)},\; b_m^{(1)} = b_m^{(0)} - \frac{a_{m1}^{(0)}}{a_{11}^{(0)}}b_1^{(0)} = b_m \quad \text{for } m,n = 2,3 \tag{1.3.19b}$$

We call this work "doing *ERO*s (*elementary row operations*) at a_{11}" and the center element a_{11} a "pivot (element)".

Next, to remove the x_2 term from Eq. (1.3.18-m) (with $m\neq1,2$), we subtract (1.3.18-2) $\times a_{m2}^{(1)}/a_{22}^{(1)}$ (m=3) from it to get

$$a_{11}^{(0)}x_1 + a_{12}^{(0)}x_2 + a_{13}^{(0)}x_3 = b_1^{(0)} \tag{1.3.20-1}$$

$$a_{22}^{(1)}x_2 + a_{23}^{(1)}x_3 = b_2^{(1)} \tag{1.3.20-2}$$

$$a_{33}^{(2)}x_3 = b_3^{(2)} \tag{1.3.20-3}$$

with

$$a_{mn}^{(2)} = a_{mn}^{(1)} - \frac{a_{m2}^{(1)}}{a_{22}^{(1)}}a_{2n}^{(1)},\ \ b_m^{(2)} = b_m^{(1)} - \frac{a_{m2}^{(1)}}{a_{22}^{(1)}}b_2^{(1)} \ \text{ for } m,n=3 \tag{1.3.21a,b}$$

We call this procedure "*Gauss(ian) (forward) elimination*" and can generalize the updating formula (1.3.19)/(1.3.21) as

$$a_{mn}^{(k)} = a_{mn}^{(k-1)} - \frac{a_{mk}^{(k-1)}}{a_{kk}^{(k-1)}}a_{kn}^{(k-1)} \ \text{ for } m,n=k+1,\,k+2,\cdots,M \tag{1.3.22a}$$

$$b_m^{(k)} = b_m^{(k-1)} - \frac{a_{mk}^{(k-1)}}{a_{kk}^{(k-1)}}b_k^{(k-1)} \ \text{ for } m=k+1,\,k+2,\cdots,M \tag{1.3.22b}$$

where $a_{mk}^{(k-1)}/\,a_{kk}^{(k-1)}$ is called the "*Gauss multiplier*".

After having the triangular matrix-vector equation like Eq. (1.3.20), we can solve the last equation (1.3.20-3) first to get

$$x_3 = \frac{b_3^{(2)}}{a_{33}^{(2)}} \tag{1.3.23a}$$

and then substitute this result into Eq. (1.3.20-2) to get

$$x_2 = \frac{b_2^{(1)} - a_{23}^{(1)}x_3}{a_{22}^{(1)}} \tag{1.3.23b}$$

Successively, we substitute Eqs. (1.3.23a,b) into Eq. (1.3.20-1) to get

$$x_1 = \frac{b_1^{(0)} - \sum_{n=2}^{3} a_{1n}^{(0)}x_n}{a_{11}^{(0)}} \tag{1.3.23c}$$

We call this procedure "*backward substitution*" and can generalize the solution formula (1.3.23) as

$$x_m = \frac{b_m^{(m-1)} - \sum_{n=m+1}^{M} a_{mn}^{(m-1)}x_n}{a_{mm}^{(m-1)}} \ \text{ for } m=M,M-1,\cdots,1 \tag{1.3.24}$$

Overall, the Gauss elimination procedure consists of two steps, i.e., forward elimination and backward substitution. Noting that

- this procedure has nothing to do with the specific values of the unknown variable x_m's and involves only the coefficients, and
- the formulas (1.3.22a) on the coefficient matrix A and (1.3.22b) on the RHS (right-hand side) vector $\mathbf{b}$ conform with each other,

we will augment A with $\mathbf{b}$ (in the (N+1)th column) and put the formulas (1.3.22a,b) together into one framework when programming the Gauss forward elimination procedure.

Now, to realize the necessity of row switching for forward elimination, consider the following example:

$$\begin{bmatrix} 0 & 1 & 1 \\ 2 & -1 & -1 \\ 1 & 1 & -1 \end{bmatrix}\begin{bmatrix} x_1 \\ x_2 \\ x_3 \end{bmatrix} = \begin{bmatrix} b_1=2 \\ b_2=0 \\ b_3=1 \end{bmatrix} \tag{1.3.25}$$

We construct an augmented matrix by combining the coefficient matrix and the RHS vector as

$$\begin{bmatrix} a_{11} & a_{12} & a_{13} & b_1 \\ a_{21} & a_{22} & a_{23} & b_2 \\ a_{31} & a_{32} & a_{33} & b_3 \end{bmatrix} = \begin{bmatrix} \boxed{0} & 1 & 1 & 2 \\ \textcircled{2} & -1 & -1 & 0 \\ 1 & 1 & -1 & 1 \end{bmatrix} \begin{matrix} : r_1 \\ : r_2 \\ : r_3 \end{matrix} \tag{1.3.26}$$

and apply the Gauss elimination procedure.

In the stage of forward elimination, we want to do EROs at a_{11}, but $a_{11}= 0$ cannot be used as the pivot because it is zero. Thus we switch the 1st row and the 2nd row having the element of the largest absolute value in the 1st column:

$$\begin{bmatrix} a_{11}^{(1)} & a_{12}^{(1)} & a_{13}^{(1)} & b_1^{(1)} \\ a_{21}^{(1)} & a_{22}^{(1)} & a_{23}^{(1)} & b_2^{(1)} \\ a_{31}^{(1)} & a_{32}^{(1)} & a_{33}^{(1)} & b_3^{(1)} \end{bmatrix} = \begin{bmatrix} \textcircled{2} & -1 & -1 & 0 \\ 0 & 1 & 1 & 2 \\ 1 & 1 & -1 & 1 \end{bmatrix} \begin{matrix} : r_1^{(1)} \\ : r_2^{(1)} \\ : r_3^{(1)} \end{matrix} \tag{1.3.27a}$$

Then we do EROs at $a_{11}^{(1)} = 2$ by applying Eq. (1.3.19) to write

$$\begin{matrix} r_1^{(1)} \\ r_2^{(1)} - a_{21}^{(1)}/a_{11}^{(1)} \times r_1^{(1)} \\ r_3^{(1)} - a_{31}^{(1)}/a_{11}^{(1)} \times r_1^{(1)} \end{matrix} \begin{bmatrix} a_{11}^{(2)} & a_{12}^{(2)} & a_{13}^{(2)} & b_1^{(2)} \\ a_{21}^{(2)} & a_{22}^{(2)} & a_{23}^{(2)} & b_2^{(2)} \\ a_{31}^{(2)} & a_{32}^{(2)} & a_{33}^{(2)} & b_3^{(2)} \end{bmatrix} = \begin{bmatrix} 2 & -1 & -1 & 0 \\ 0 & \underline{1} & 1 & 2 \\ 0 & \textcircled{3/2} & -1/2 & 1 \end{bmatrix} \begin{matrix} : r_1^{(2)} \\ : r_2^{(2)} \\ : r_3^{(2)} \end{matrix} \tag{1.3.27b}$$

Here, instead of doing EROs at $a_{22}^{(2)}$, we switch the 2nd row and the 3rd row having the element of the largest absolute value among the elements not above $a_{22}^{(2)}$ in the 2nd column to write

$$\begin{bmatrix} a_{11}^{(3)} & a_{12}^{(3)} & a_{13}^{(3)} & b_1^{(3)} \\ a_{21}^{(3)} & a_{22}^{(3)} & a_{23}^{(3)} & b_2^{(3)} \\ a_{31}^{(3)} & a_{32}^{(3)} & a_{33}^{(3)} & b_3^{(3)} \end{bmatrix} = \begin{bmatrix} 2 & \underline{-1} & -1 & 0 \\ 0 & \textcircled{3/2} & -1/2 & 1 \\ 0 & 1 & 1 & 2 \end{bmatrix} \begin{matrix} : r_1^{(3)} \\ : r_2^{(3)} \\ : r_3^{(3)} \end{matrix} \tag{1.3.27c}$$

Then we do EROs at $a_{22}^{(3)}$ by applying Eq. (1.3.21), more generally, Eq. (1.3.22), to get

$$\begin{matrix} r_1^{(3)} \\ r_2^{(3)} \\ r_3^{(3)} - a_{32}^{(3)}/a_{22}^{(3)} \times r_2^{(3)} \end{matrix} \begin{bmatrix} a_{11}^{(4)} & a_{12}^{(4)} & a_{13}^{(4)} & b_1^{(4)} \\ a_{21}^{(4)} & a_{22}^{(4)} & a_{23}^{(4)} & b_2^{(4)} \\ a_{31}^{(4)} & a_{32}^{(4)} & a_{33}^{(4)} & b_3^{(4)} \end{bmatrix} = \begin{bmatrix} 2 & -1 & -1 & 0 \\ 0 & 3/2 & -1/2 & 1 \\ 0 & 0 & 4/3 & 4/3 \end{bmatrix} \begin{matrix} : r_1^{(4)} \\ : r_2^{(4)} \\ : r_3^{(4)} \end{matrix} \tag{1.3.27d}$$

Since this is an upper-triangularized form, we perform the backward substitution (Eq. (1.3.23), or more generally, Eq. (1.3.24)) to get the final solution as

$$\left.\begin{aligned} x_3 &= \frac{b_3^{(4)}}{a_{33}^{(4)}} = \frac{4/3}{4/3} = 1 \\ x_2 &= \frac{b_2^{(4)} - a_{23}^{(4)} x_3}{a_{22}^{(4)}} = \frac{1-(-1/2)\times 1}{3/2} = 1 \\ x_1 &= \frac{b_1^{(4)} - \sum_{n=2}^{3} a_{1n}^{(4)} x_n}{a_{11}^{(4)}} = \frac{0-(-1)\times 1-(-1)\times 1}{2} = 1 \end{aligned}\right\} \rightarrow \begin{bmatrix} x_3 \\ x_2 \\ x_1 \end{bmatrix} = \begin{bmatrix} 1 \\ 1 \\ 1 \end{bmatrix} \tag{1.3.28}$$

Note that the strategy to select the entry with the largest absolute value among $\{a_{kk}, a_{k(k+1)}, \dots\}$ in the kth column as the pivot even through row switching is called *partial pivoting.*

Let us consider another system of equations:

$$\begin{bmatrix} 1 & 0 & 1 \\ 1 & 1 & 1 \\ 1 & -1 & 1 \end{bmatrix} \begin{bmatrix} x_1 \\ x_2 \\ x_3 \end{bmatrix} = \begin{bmatrix} b_1=2 \\ b_2=3 \\ b_3=1 \end{bmatrix} \tag{1.3.29}$$

We construct an augmented matrix by combining the coefficient matrix and the RHS vector as

$$\begin{bmatrix} a_{11} & a_{12} & a_{13} & b_1 \\ a_{21} & a_{22} & a_{23} & b_2 \\ a_{31} & a_{32} & a_{33} & b_3 \end{bmatrix} = \begin{bmatrix} 1 & 0 & 1 & 2 \\ 1 & 1 & 1 & 3 \\ 1 & -1 & 1 & 1 \end{bmatrix} \begin{matrix} :r_1 \\ :r_2 \\ :r_3 \end{matrix} \tag{1.3.30}$$

and apply the Gauss elimination procedure as follows:

$$\begin{matrix} r_1^{(0)} \\ r_2^{(0)} - a_{21}^{(0)}/a_{11}^{(0)} \times r_1^{(0)} \\ r_3^{(0)} - a_{31}^{(0)}/a_{11}^{(0)} \times r_1^{(0)} \end{matrix} \begin{bmatrix} a_{11}^{(1)} & a_{12}^{(1)} & a_{13}^{(1)} & b_1^{(1)} \\ a_{21}^{(1)} & a_{22}^{(1)} & a_{23}^{(1)} & b_2^{(1)} \\ a_{31}^{(1)} & a_{32}^{(1)} & a_{33}^{(1)} & b_3^{(1)} \end{bmatrix} = \begin{bmatrix} 1 & 0 & 1 & 2 \\ 0 & 1 & 0 & 1 \\ 0 & -1 & 0 & -1 \end{bmatrix} \begin{matrix} :r_1^{(1)} \\ :r_2^{(1)} \\ :r_3^{(1)} \end{matrix} \tag{1.3.31a}$$

$$\begin{matrix} r_1^{(1)} \\ r_2^{(1)} \\ r_3^{(1)} - a_{32}^{(1)}/a_{22}^{(1)} \times r_2^{(1)} \end{matrix} \begin{bmatrix} a_{11}^{(2)} & a_{12}^{(2)} & a_{13}^{(2)} & b_1^{(2)} \\ a_{21}^{(2)} & a_{22}^{(2)} & a_{23}^{(2)} & b_2^{(2)} \\ a_{31}^{(2)} & a_{32}^{(2)} & a_{33}^{(2)} & b_3^{(2)} \end{bmatrix} = \begin{bmatrix} 1 & 0 & 1 & 2 \\ 0 & 1 & 0 & 1 \\ 0 & 0 & \boxed{0} & 0 \end{bmatrix} \begin{matrix} :r_1^{(2)} \\ :r_2^{(2)} \\ :r_3^{(2)} \end{matrix} \tag{1.3.31b}$$

Since this is an upper-triangularized form, we are at the stage of backward substitution. However, $a_{33}^{(2)}$, which is supposed to be the denominator in Eq. (1.3.24), is zero. We may face such a weird situation of zero-division even during the forward elimination process where the pivot is zero, and besides, we may not find any (nonzero) element below it in the same column and on its right in the same row except for the RHS element. In this case, we cannot go further. This implies that some or all rows of the coefficient matrix A are dependent on others, corresponding to the case of redundancy (infinitely many solutions) or inconsistency (no exact solution). Noting that the RHS element of the zero row in Eq. (1.3.31b) is also zero, we should declare a case of redundancy and may have to be satisfied with one of the infinitely many solutions being the RHS vector as

$$[x_1 \;\; x_2 \;\; x_3] = [b_1^{(2)} \;\; b_2^{(2)} \;\; b_3^{(2)}] = [2 \;\; 1 \;\; 0] \tag{1.3.32}$$

Furthermore, if we remove the all-zero row(s), the problem can be treated as an underdetermined case handled in Section 1.3.2. Note that, if the RHS element were not zero, we would have to declare a case of inconsistency, as will be illustrated below.

Suppose that $b_1=1$ in Eq. (1.3.29). Then, the Gauss elimination would have proceeded as follows:

$$\begin{bmatrix} 1 & 0 & 1 & 1 \\ 1 & 1 & 1 & 3 \\ 1 & -1 & 1 & 1 \end{bmatrix} \rightarrow \begin{matrix} r_1 \\ r_2-r_1 \\ r_3-r_1 \end{matrix} \begin{bmatrix} 1 & 0 & 1 & 1 \\ 0 & 1 & 0 & 2 \\ 0 & -1 & 0 & 0 \end{bmatrix} \rightarrow \begin{matrix} r_1 \\ r_2 \\ r_3+r_2 \end{matrix} \begin{bmatrix} 1 & 0 & 1 & 1 \\ 0 & 1 & 0 & 2 \\ 0 & 0 & 0 & 2 \end{bmatrix} \tag{1.3.33}$$

This ended up with an all-zero row except the non-zero RHS element, corresponding to the case of inconsistency. Thus we must declare a case of "no exact solution" and try with the LSE solution.

The Gauss elimination algorithm with partial pivoting has been cast into the following MATLAB function '`Gauss_elimination()`', which can be used to solve a system of linear equations with a nonsingular coefficient matrix. For example, to solve Eq. (1.3.25), type the following statements at the MATLAB prompt:

```
>>A=[0 1 1; 2 -1 -1; 1 1 -1]; b=[2 0 1]'; % Eq.(1.3.25)
>>x=Gauss_elimination(A,b)
>>x1=A\b % for comparison with the result of backslash operation
```

```
function x=Gauss_elimination(A,B)
% This function solves Ax=B by Gauss elimination algorithm.
% The sizes of matrices A,B are supposed to be NAxNA and NAxNB.
NA=size(A,2);   [NB1,NB]=size(B);
if NB1~=NA, error('A and B must have compatible dimension'); end
N=NA+NB; AB=[A(1:NA,1:NA) B(1:NA,1:NB)]; % Augmented matrix
for k=1:NA
   % Partial Pivoting at AB(k,k)
   [akx,kx]=max(abs(AB(k:NA,k)));
   if akx<eps, error('Singular matrix'); end
   if kx>1 % Row change if necessary
     mx=k+kx-1; % Index of row to be exchanged with row k
     tmp_row=AB(k,k:N); AB(k,k:N)=AB(mx,k:N); AB(mx,k:N)=tmp_row;
   end
   % Gauss forward elimination
   for m=k+1:NA
      AB(m,k+1:N)=AB(m,k+1:N)-AB(m,k)/AB(k,k)*AB(k,k+1:N); % Eq.(1.3.22)
      AB(m,k) =0;
   end
end
% Backward substitution for a upper-triangular matrix eqation
% having all the diagonal elements equal to one
x(NA,:)= AB(NA,NA+1:N);
for m=NA-1:-1:1
    x(m,:)= AB(m,NA+1:N)-AB(m,m+1:NA)*x(m+1:NA,:); % Eq.(1.3.24)
end
```

1.3.5 RLS (Recursive Least Squares) Algorithm

In this section, we will see the so-called *RLSE* (*recursive least squares estimation*) algorithm, which is a recursive method to compute the LSE solution. Suppose we know that the theoretical relationship between the temperature t[°] and the resistance $r\,[\Omega]$ of a resistor is

$$c_1 t + c_2 = r \tag{1.3.34}$$

where we have many pairs of experimental data $\{(t_1,r_1), (t_2,r_2), \ldots, (t_k,r_k)\}$ collected up to time k. Since the above equation cannot be satisfied for all the data with any value of the parameters c_1 and c_2, we should try to get the parameter estimates that are optimal in some sense. This corresponds to the overdetermined case dealt with in Section 1.3.3, and can be formulated as an LSE problem where we must solve a set of linear equations:

$$A_k \mathbf{x}_k \approx \mathbf{b}_k, \text{ where } A_k = \begin{bmatrix} t_1 & 1 \\ t_2 & 1 \\ \cdot & \cdot \\ t_k & 1 \end{bmatrix}, \ \mathbf{x}_k = \begin{bmatrix} c_{1,k} \\ c_{2,k} \end{bmatrix}, \text{ and } \mathbf{b}_k = \begin{bmatrix} r_1 \\ r_2 \\ \cdot \\ r_k \end{bmatrix} \tag{1.3.35}$$

We can use Eq. (1.3.13) to get the LSE solution as

$$\mathbf{x}_k \overset{(1.3.13)}{=} [A_k^{\mathrm{T}} A_k]^{-1} A_k^{\mathrm{T}} \mathbf{b}_k \tag{1.3.36}$$

Now, we are given an additional pair of data (t_{k+1}, r_{k+1}) and must find the new parameter estimate

$$\mathbf{x}_{k+1} \overset{(1.3.13)}{=} [A_{k+1}^{\mathrm{T}} A_{k+1}]^{-1} A_{k+1}^{\mathrm{T}} \mathbf{b}_{k+1} \tag{1.3.37}$$

$$\text{with } A_{k+1} = \begin{bmatrix} t_1 & 1 \\ \cdot & \cdot \\ t_k & 1 \\ t_{k+1} & 1 \end{bmatrix}, \ \mathbf{x}_{k+1} = \begin{bmatrix} c_{1,k+1} \\ c_{2,k+1} \end{bmatrix}, \text{ and } \mathbf{b}_{k+1} = \begin{bmatrix} r_1 \\ \cdot \\ r_k \\ r_{k+1} \end{bmatrix}$$

How do we compute this? If we discard the previous estimate $\mathbf{x}_k$ and make direct use of Eq. (1.3.37) to compute a new estimate $\mathbf{x}_{k+1}$ every time a new data pair is available, the size of matrix A will get bigger and bigger as the data pile up, eventually defying any powerful computer in this world.

How about updating the previous estimate by just adding a correction term based on the new data to get a new estimate? This is the basic idea of the RLSE algorithm, which we are going to study. To get the idea, let us define the following notations:

$$A_{k+1} = \begin{bmatrix} A_k \\ \mathbf{a}_{k+1}^{\mathrm{T}} \end{bmatrix}, \ \mathbf{a}_{k+1} = \begin{bmatrix} t_{k+1} \\ 1 \end{bmatrix}, \ \mathbf{b}_{k+1} = \begin{bmatrix} \mathbf{b}_k \\ r_{k+1} \end{bmatrix}, \text{ and } P_k = [A_k^{\mathrm{T}} A_k]^{-1} \tag{1.3.38}$$

and see how the inverse matrix P_{k+1} is to be updated on the arrival of new data (t_{k+1}, r_{k+1}):

$$P_{k+1} = [A_{k+1}^{\mathrm{T}} A_{k+1}]^{-1} = \left[[A_k^{\mathrm{T}} \ \ \mathbf{a}_{k+1}] \begin{bmatrix} A_k \\ \mathbf{a}_{k+1}^{\mathrm{T}} \end{bmatrix} \right]^{-1} = \left[A_k^{\mathrm{T}} A_k + \mathbf{a}_{k+1} \mathbf{a}_{k+1}^{\mathrm{T}} \right]^{-1} = \left[P_k^{-1} + \mathbf{a}_{k+1} \mathbf{a}_{k+1}^{\mathrm{T}} \right]^{-1} \tag{1.3.39}$$

(Matrix Inversion Lemma (1.2.44) in Section 1.2.9)

$$; \qquad P_{k+1} = P_k - P_k \mathbf{a}_{k+1} [\mathbf{a}_{k+1}^{\mathrm{T}} P_k \mathbf{a}_{k+1} + 1]^{-1} \mathbf{a}_{k+1}^{\mathrm{T}} P_k \tag{1.3.40}$$

It is interesting that $[\mathbf{a}_{k+1}^{\mathrm{T}} P_k \mathbf{a}_{k+1} + 1]$ is nothing but a scalar and thus we do not need to compute the matrix inverse thanks to the matrix inversion lemma (Eq. (1.2.44)). It is much better in the computational aspect to use this recursive formula (1.3.40) than to compute $P_{k+1} = [A_{k+1}^{\mathrm{T}} A_{k+1}]^{-1}$ directly. Eq. (1.3.37) can also be written in recursive form as

$$\mathbf{x}_{k+1} \overset{(1.3.37,39)}{=} P_{k+1} A_{k+1}^{\mathrm{T}} \mathbf{b}_{k+1} \overset{(1.3.38)}{=} P_{k+1} [A_k^{\mathrm{T}} \ \ \mathbf{a}_{k+1}] \begin{bmatrix} \mathbf{b}_k \\ r_{k+1} \end{bmatrix} = P_{k+1} [A_k^{\mathrm{T}} \mathbf{b}_k + \mathbf{a}_{k+1} r_{k+1}]$$

$$\overset{(1.3.35)}{=} P_{k+1} [A_k^{\mathrm{T}} A_k \mathbf{x}_k + \mathbf{a}_{k+1} r_{k+1}] \overset{(1.3.38)}{=} P_{k+1} [(A_{k+1}^{\mathrm{T}} A_{k+1} - \mathbf{a}_{k+1} \mathbf{a}_{k+1}^{\mathrm{T}}) \mathbf{x}_k + \mathbf{a}_{k+1} r_{k+1}]$$

$$\overset{(1.3.39)}{=} P_{k+1}[P_{k+1}^{-1}\mathbf{x}_k - \mathbf{a}_{k+1}\mathbf{a}_{k+1}^{\mathrm{T}}\mathbf{x}_k + \mathbf{a}_{k+1}r_{k+1}]$$

$$; \qquad \mathbf{x}_{k+1} = \mathbf{x}_k + P_{k+1}\mathbf{a}_{k+1}(r_{k+1} - \mathbf{a}_{k+1}^{\mathrm{T}}\mathbf{x}_k) \qquad (1.3.41)$$

Eq. (1.3.40) can be used to rewrite the gain matrix $P_{k+1}\mathbf{a}_{k+1}$ multiplied by the "error" $(r_{k+1}-\mathbf{a}_{k+1}^{\mathrm{T}}\mathbf{x}_k)$ to make the correction term on the RHS of Eq. (1.3.41) as

$$K_{k+1} = P_{k+1}\mathbf{a}_{k+1} \overset{(1.3.40)}{=} [P_k - P_k\mathbf{a}_{k+1}[\mathbf{a}_{k+1}^{\mathrm{T}}P_k\mathbf{a}_{k+1}+1]^{-1}\mathbf{a}_{k+1}^{\mathrm{T}}P_k]\mathbf{a}_{k+1}$$

$$= P_k\mathbf{a}_{k+1}[I - [\mathbf{a}_{k+1}^{\mathrm{T}}P_k\mathbf{a}_{k+1}+1]^{-1}\mathbf{a}_{k+1}^{\mathrm{T}}P_k\mathbf{a}_{k+1}]$$

$$= P_k\mathbf{a}_{k+1}[\mathbf{a}_{k+1}^{\mathrm{T}}P_k\mathbf{a}_{k+1}+1]^{-1}\{[\mathbf{a}_{k+1}^{\mathrm{T}}P_k\mathbf{a}_{k+1}+1] - \mathbf{a}_{k+1}^{\mathrm{T}}P_k\mathbf{a}_{k+1}\}$$

$$; \qquad K_{k+1} = P_k\mathbf{a}_{k+1}[\mathbf{a}_{k+1}^{\mathrm{T}}P_k\mathbf{a}_{k+1}+1]^{-1} \qquad (1.3.42)$$

We can substitute this back into Eq. (1.3.40) to write it as

$$P_{k+1} = P_k - K_{k+1}\mathbf{a}_{k+1}^{\mathrm{T}}P_k \qquad (1.3.43)$$

The following MATLAB function '`rlse_online()`' implements this RLSE (recursive least squares estimation) algorithm that updates the parameter estimates by using Eqs. (1.3.42), (1.3.41), and (1.3.43).

```
function [x,K,P]=RLSE_online(a_k,b_k,x,P)
% One step of RLSE (Recursive Least Squares Estimation) algorithm
a_k=a_k(:); b_k=b_k(:); % To make sure them column vectors
K = P*a_k/(a_k'*P*a_k+1); % Eq.(1.3.42)
x = x +K*(b_k-a_k'*x); % Eq.(1.3.41)
P = P -K*a_k'*P;   % Eq.(1.3.43)
```

Noting that

- the matrix $[A_k^{\mathrm{T}}A_k]$ as well as $\mathbf{b}_k$ consists of information and is a nonnegative matrix, and
- as valuable information data accumulate, $[A_k^{\mathrm{T}}A_k]$ will get larger, or equivalently, $P_k=[A_k^{\mathrm{T}}A_k]^{-1}$ will get smaller, and consequently, the gain matrix K_k will get smaller,

one could understand that P_k is initialized to a very large identity matrix, since no information is available in the beginning. Since a large/small P_k makes the correction term on the RHS of Eq. (1.3.41) large/small, the RLSE algorithm becomes more conservative and reluctant to learn from new data as the data pile up, while it is willing to make use of new data for updating the estimates when it is hungry for information in the beginning.

Let us apply the RLSE algorithm to estimate the parameter $[c_1^o \;\; c_2^o]=[-0.5 \;\; 1]$ of the following relation between temperature t and resistance r where $n(t)$ denotes a measurement error or noise:

$$r(t) = -0.5t + 1 + n(t) \;\text{ with } n(t)\text{: measurement error or noise} \qquad (1.3.44)$$

For this purpose, we run the following MATLAB script "do_RLSE.m," which updates the estimate $[c_1^e \;\; c_2^e]$ of the parameter $[c_1 \;\; c_2]$ in the relation

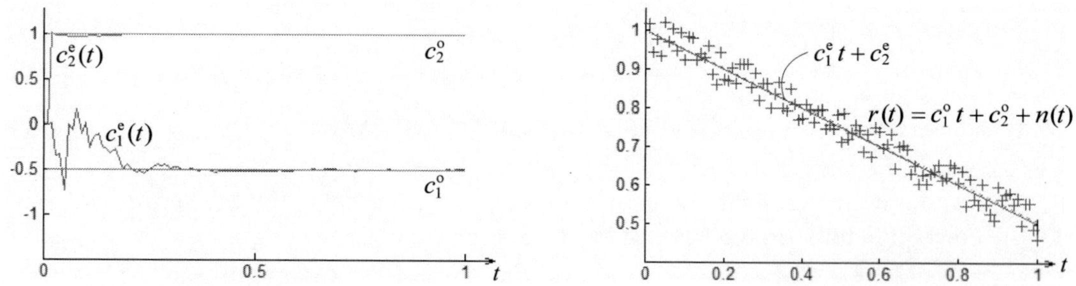

(a) Parameter estimates updated on-line (b) Data and curve fitting

Figure 1.27 Example of applying the RLSE algorithm for parameter estimation

```
%do_RLSE.m
clear, clf
xo=[-0.5  1]'; % True value of the parameters
NA=length(xo); % Dimension of the parameter vector
kmax=100; % Maximum iteration number
t=0.01*[1:kmax]; % Temperature vector
xk_on =zeros(NA,kmax+1); % Initial estimate of the parameters
P=1000*eye(NA,NA); % Initial error covariance matrix, which is large.
for k=1:kmax
   A(k,:)=[t(k) 1];
   r(k)=A(k,:)*xo + 0.1*(rand-0.5); % Eq.(1.3.44)
   [xk_on(:,k+1),K,P] = RLSE_online(A(k,:),r(k),xk_on(:,k),P);
end
b=r(:); x_off=A\b % Off-line estimate
x1 =(A'*A)^-1*A'*b % To compare with Eq.(1.3.13)
x_on=xk_on(:,end) % Last on-line estimate
norm(x_on-x_off)/norm(x_off) % Discrepancy between parameter estimates
subplot(221) % in the 1st one of 2x2 panes (numbered row-wise)
plot(t,xk_on(:,1:end-1).', t,xo(1)*ones(size(t)), t,xo(2)*ones(size(t)))
title('Parameter estimates updated on-line and True parameter values')
subplot(222) % in the 2nd one of 2x2 panes on the screen
plot(t,b,'b+', t,A*xo,'b', t,A*x_on,'r:')
title('r(t) and Curvefitting results with the parameter estimates')
```

$$c_1 t + c_2 = [t \quad 1]\begin{bmatrix} c_1 \\ c_2 \end{bmatrix} = r(t) \tag{1.3.45}$$

every time new data $[t \;\; r(t)]$ arrives, as depicted in Fig. 1.27(a). Fig. 1.27(b) shows the result of curvefitting the data with the last parameter estimate, say $[c_1^e(\text{kmax}) \;\; c_2^e(\text{kmax})]=[-0.4863 \;\; 0.9952]$ (kmax: the number for the last data or equivalently, the total number of data), which is close to the true value $[c_1^o \;\; c_2^o]=[-0.5 \;\; 1]$, together with the true relation $r(t)=-0.5t+1$ with no measurement error or noise.

Problems

1.1 Area of Parallelogram

Consider the parallelogram with four vertices where the coordinates of three vertices are given as (1, -1, 1), (2, 0, 1), and (3, -1, 1).

(a) Find one of the three possible points as the fourth vertex of the parallelogram.
(b) Find the area of the parallelogram.
(c) Find the volume of the parallelepiped with eight vertices including the 3 points and (2, -1, 2).

1.2 Intersection of Two Planes

Consider the two planes described by

$$2x - y + 3z = 4 \quad \text{(P1.2.1a)}$$

$$3x - y - z = 1 \quad \text{(P1.2.1b)}$$

Referring to Example 1.2, find the line of intersection betwen the planes and plot the line together with the two planes as in Fig. 1.10.

1.3 Intersection of Two Planes

Consider Example 1.2 where the vector equation of the line of intersection of the two planes

$$2x - 3y + 4z = 1 \quad \text{(E1.2.1a)}$$

$$x - y - z = 5 \quad \text{(E1.2.1b)}$$

was found as

$$\mathbf{p} \overset{(1.1.41)}{=} \mathbf{p}_1 + \mathbf{d}\,t = \begin{bmatrix} 14 \\ 9 \\ 0 \end{bmatrix} + \begin{bmatrix} 7 \\ 6 \\ 1 \end{bmatrix} t \quad \text{(E1.2.4)}$$

(a) Noting that the two planes have a line of intersection so that there are infinitely many points shared by them, we may want to see that the position of point $\mathbf{p}_1$ does not matter so long as it is shared by the two planes. To this aim, solve the set of plane equations with y=0 instead of z=0 to find a point of intersection and go on with that point and the same direction vector to find the vector equation of the line of intersection.
(b) Find the angle of intersection of the two planes from the angle between their normal vectors. Also, use the MATLAB function '`intersect_planes()`' (Example 1.2) with Kg=1 to get Fig. 1.10 and type

```
>>axis('equal')
  view(atan2(d(2),d(1))*180/pi-90,-5)
```

at the MATLAB prompt, which will produce Fig. P1.3. Can you see the angle of intersection of the two planes from the figure? Note that the two input arguments of '`view()`' are the azimuth/elevation angles (in degrees) of view w.r.t. the $-y$-/ρ-axes, respectively. How about typing the following statement?

```
>>view(d)
```

Can you see the angle more clearly?

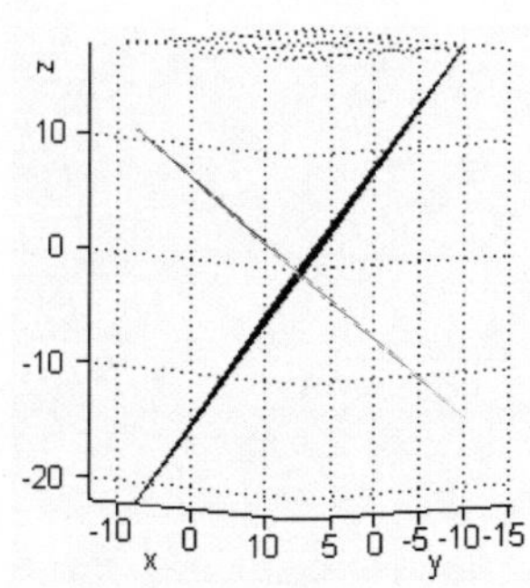

Figure P1.3

1.4 Intersection of a Plane and a Line

Consider three points $\mathbf{p}_1=(1, 1, 1)$, $\mathbf{p}_2=(2, 2, 1)$, and $\mathbf{p}_3=(3, 1, 2)$.

(a) Find the equation for the plane containing the three points.

(b) Find the equation of the line that passes through the point $\mathbf{p}_4=(-1,-5, 9)$ and is normal to the plane.

(c) Find the intersection point $\mathbf{p}_i$ of the line (obtained in (b)) and the plane (obtained in (a)).

(d) Find the distance from the point $\mathbf{p}_4=(-1,-5, 9)$ to the intersection point $\mathbf{p}_i$ obtained in (c). Check if it conforms with what is obtained from Eq. (1.1.15).

(e) Referring to the MATLAB script "em01f04.m" (Section 1.1.2.4), complete the following script "em01p04.m" (including the equation numbers in the comments) and run it to plot the plane, the line, the point $\mathbf{p}_4$, and the intersection point as depicted in Fig. P1.4.

```
%em01p04.m
% Intersection of a plane and a line normal to the plane
syms x y z
p1=[1;1;1]; p2=[2;2;1]; p3=[3;1;2]; p=[x;y;z];
dot(cross((p2-p1),(p3-p1)),p-p1) % Eq.(1.1.??)
% From the result of running the above statements
a=1; b=-1; c=??; d1=2; % Coefficients of plane equation
n=[? b c].'; % Normal vector to the plane
d=n;  % Direction vector of the line
p4=[?? -5  ?].'; % A point off the plane
t_i=dot(n,p1-p?)/???(?,d) % Eq.(1.1.??)
p_i = p4 + n*t_i  % Intersection point (1.1.??)
norm(p_i-p4) % Distance between p4 and the intersection point p_i
distance=abs([a ? c]*p4+d1)/sqrt(a^2+?^2+c^2) % Eq.(1.1.??)
% To plot the plane
x=p_i(1)+[-5:5]; y=p_i(2)+[-5:5]; % Grid vectors along x,y-axes
[X,Y]=meshgrid(x,y); % X,Y-grid matrices
Z=(-d1-?*X-b*?)/?; % Plane equation
mesh(X,Y,Z), hold on % Plot the plane
t=[0 4];  p = p4*ones(size(t)) + n*t; % Line equation (1.1.??)
plot3(p(1,:),p(2,:),p(3,:)), hold on % Plot the line
plot3(p4(1),p4(2),p4(3),'o')
plot3(p_i(1),p_i(2),p_i(3),'rx'), view(40,25), grid on
xlabel('x'), ylabel('y'), zlabel('z'), axis('equal'), shg
```

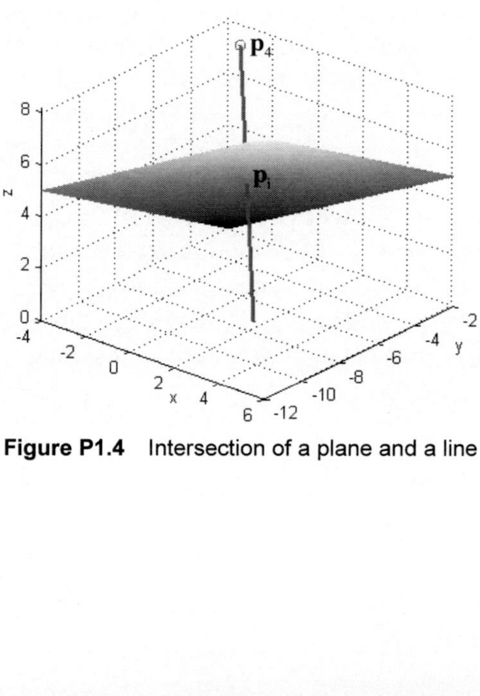

Figure P1.4 Intersection of a plane and a line

1.5 Magnetic Field by a Ring of Radius a Carrying Current I

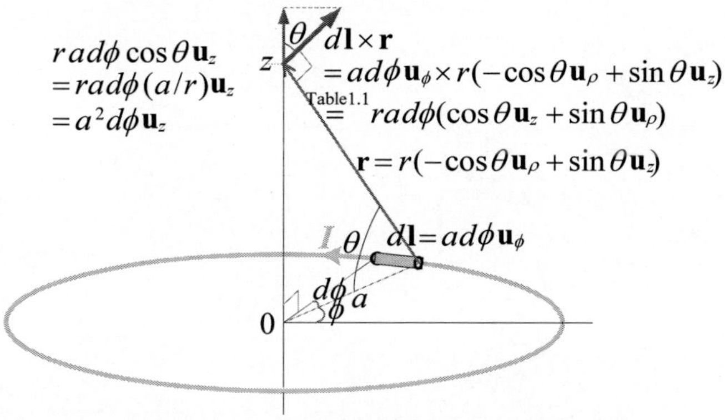

Figure P1.5.1 Magnetic field intensity produced by a ring with radius a and current I

Consider a ring with radius a and current I in the $+\mathbf{u}_\phi$ (counter-clockwise) direction on the x-y plane as described in Fig. P1.5.1. Let us find the magnetic field (produced by the current ring) at some point [0, 0, z] on the z-axis.

Referring to Fig. P1.5.1, we apply the Biot-Savart law to compute the magnetic field intensity as

$$\mathbf{H} \overset{\text{Biot-Savart law}}{=} \frac{1}{4\pi}\int \frac{I d\mathbf{l}\times\mathbf{r}}{\|\mathbf{r}\|^3} = \frac{1}{4\pi}\int_0^{2\pi} \frac{I a^2}{r^3} d\phi \mathbf{u}_z = \frac{I a^2}{4\pi r^3} 2\pi \mathbf{u}_z = \frac{I a^2}{2(a^2+z^2)^{3/2}} \mathbf{u}_z \quad \text{(P1.5.1)}$$

With the following MATLAB function "H_by_circular_current()" and script "show_H_by_circular _current.m" (listed below) saved in your computer, you can type the following statements at the MATLAB prompt to plot the magnetic field intensities at the nineteen points: {[0 0 0], [0 0 0.5], [0 0 -0.5], [-1 0 0.1], [0 -1 0.1], [1 0 0.1], [0 1 0.1], [-1 0 -0.1], [0 -1 -0.1], [1 0 -0.1], [0 1 -0.1], [-0.9 0 0], [0 -0.9 0], [0.9 0 0], [0 0.9 0], [-1.1 0 0], [0 -1.1 0], [1.1 0 0], [0 1.1 0]}, as depicted in Fig. P1.5.2.

```
>>show_H_by_circular_current
>>[ps Bs]
  %px       py       pz          Hx        Hy        Hz
    0        0        0           0         0    0.5000
    0        0   0.5000     -0.0000    0.0000    0.3578
  .. .. .. .. .. .. .. ..     .. .. .. .. .. .. .. .. ..
```

```
function Hs=H_by_circular_current(ps,a,I)
% Magnetic field intensity at ps by a ring with radius a and current I
dph=pi/180; phs=[0:359]*dph; adph=a*dph; adph1=adph/10;
for np=1:size(ps,1)
  p=ps(np,:); % For each point p
  for nph=1:length(phs)
    ph=phs(nph); pph=a*[cos(ph) sin(ph) 0];
    r=p-pph; rm=norm(r);  dl=adph*[-sin(ph) cos(ph) 0];
    dH(nph,:)=(I*cross(dl,r)/(4*pi*rm^3))*(rm>=adph1); %Biot-Savart law
  end
  H=sum(dH); Hs(np,:)=H;
  if norm(p(1:2))<eps % if the point is virtually on the z-axis
    H1=I*a^2/2/(a^2+p(3)^2)^(3/2)*[0 0 1]; e=norm(H-H1); % Eq.(P1.5.1)
    if e/norm(H1)>1e-2, fprintf('%13.4e',H,H1); fprintf('\n');  end
  end
end
```

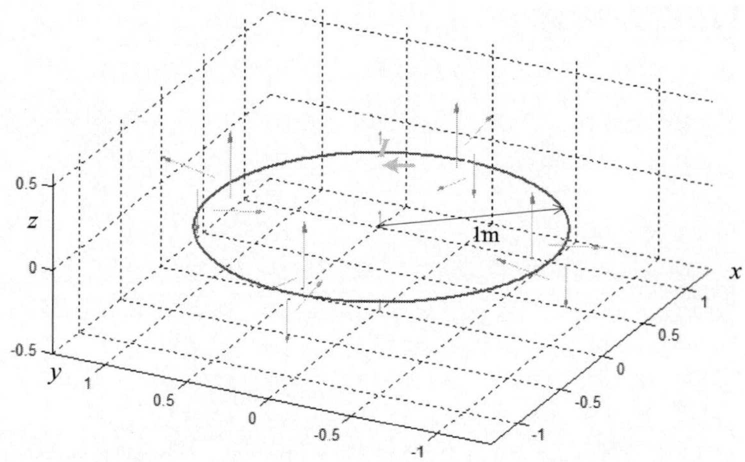

Figure P1.5.2 Magnetic field intensity produced by a 1[m]-radius ring conducting current I=1A

```
%show_H_by_circular_current.m
% plots the magnetic field at points from a circular current
dth=pi/180; ths=[0:359]*dth; N=length(ths); a=1; I=1;
plot3(a*cos(ths),a*sin(ths),zeros(1,N),'Linewidth',2), grid on, hold on
z1=0.1*a; z2=-z1; a1=a-0.1; a2=a+0.1;
ps=[0 0 0; 0 0 0.5*a; 0 0 -0.5*a;
     -a 0 z1; 0 -a z1; a 0 z1; 0 a z1;
    -a 0 z2; 0 -a z2; a 0 z2; 0 a z2;
    -a1 0 0; 0 -a1 0; a1 0 0; 0 a1 0;
     -a2 0 0; 0 -a2 0; a2 0 0; 0 a2 0];
Hs=H_by_circular_current(ps,a,I);
% Magnetic field at each point p
for np=1:size(ps,1), arrow3(Hs(np,:)/5,ps(np,:)); end
```

Note that the closed-loop formula (P1.5.1) is not applicable for the points off the z-axis while the above function 'H_by_circular_current()' can be used for any point in the 3-dimensional space.

1.6 Coordinate Conversion

(a) Use Eqs. (1.1.59b,c) to make the rectangular-to-spherical coordinate conversion of

$$x\mathbf{u}_x + y\mathbf{u}_y + z\mathbf{u}_z \tag{P1.6.1}$$

as

$$\begin{aligned}[x\ y\ z]\begin{bmatrix}\mathbf{u}_x^T\\ \mathbf{u}_y^T\\ \mathbf{u}_z^T\end{bmatrix} &\overset{(1.1.59b)}{\underset{(1.1.59c)}{=}} [r\sin\theta\cos\phi\ \ r\sin\theta\sin\phi\ \ r\cos\theta]\begin{bmatrix}\sin\theta\cos\phi & \cos\theta\cos\phi & -\sin\phi\\ \sin\theta\sin\phi & \cos\theta\sin\phi & \cos\phi\\ \cos\theta & -\sin\theta & 0\end{bmatrix}\begin{bmatrix}\mathbf{u}_r^T\\ \mathbf{u}_\theta^T\\ \mathbf{u}_\phi^T\end{bmatrix}\\ &= (r\sin\theta\cos\phi\sin\theta\cos\phi + r\sin\theta\sin\phi\sin\theta\sin\phi + r\cos\theta\cos\theta)\mathbf{u}_r^T\\ &\quad + (r\sin\theta\cos\phi\cos\theta\cos\phi + r\sin\theta\sin\phi\cos\theta\sin\phi - r\cos\theta\sin\theta)\mathbf{u}_\theta^T\\ &\quad + (-r\sin\theta\cos\phi\sin\phi + r\sin\theta\sin\phi\cos\phi)\mathbf{u}_\phi^T\\ &= \boxed{\qquad\qquad\qquad\qquad} \end{aligned} \tag{P1.6.2}$$

(b) Use Eqs. (1.1.59a,c) to make the spherical-to-rectangular coordinate conversion of

$$r\mathbf{u}_r \text{ with } r=\sqrt{x^2+y^2+z^2} \tag{P1.6.3}$$

as

$$r\mathbf{u}_r \rightarrow r\mathbf{u}_r^{\mathrm{T}} \overset{(1.1.59a.1)}{=} r\sin\theta\cos\phi\,\mathbf{u}_x^{\mathrm{T}} + r\sin\theta\sin\phi\,\mathbf{u}_y^{\mathrm{T}} + r\cos\theta\,\mathbf{u}_z^{\mathrm{T}}$$

$$\overset{(1.1.59c)}{\rightarrow} \boxed{} \tag{P1.6.4}$$

1.7 Coordinate Conversion for Easier Computation of Triple Integral

Consider the following triple integral of a three-variable function

$$f(x,y,z)=x^2+y^2+z^2 \tag{P1.7.1}$$

over the upper unit hemisphere centered at the origin. It can be evaluated in the rectangular coordinate system as

$$\begin{aligned}
&\int_{-1}^{1}\int_{-\sqrt{1-x^2}}^{\sqrt{1-x^2}}\int_0^{\sqrt{1-x^2-y^2}}(x^2+y^2+z^2)\,dz\,dy\,dx \underset{dxdy=\rho d\rho d\phi}{\overset{x^2+y^2=\rho^2}{=}} \int_0^{2\pi}\int_0^1\int_0^{\sqrt{1-\rho^2}}(\rho^2+z^2)\,dz\,\rho d\rho\,d\phi \\
&=\int_0^{2\pi}\int_0^1 \rho(\boxed{}3)\Big|_0^{\sqrt{1-\rho^2}} d\rho d\phi = \int_0^{2\pi}\int_0^1 \rho\{\rho^2\sqrt{1-\rho^2}+(1-\rho^2)^{3/2}/3\}\,d\rho d\phi \\
&=2\pi\left\{\int_0^1(-2\rho)(1-\rho^2)^{1/2}(-\rho^2/2)\,d\rho+\frac{1}{3}\int_0^1\rho(1-\rho^2)^{3/2}d\rho\right\} \\
&\overset{(C.38)}{=} 2\pi\left\{\frac{2}{3}(1-\rho^2)^{3/2}(-\rho^2/2)\Big|_0^1+\frac{2}{3}\int_0^1(1-\rho^2)^{3/2}\boxed{}\,d\rho+\frac{1}{3}\int_0^1\rho(1-\rho^2)^{3/2}d\rho\right\} \\
&=2\pi\left\{-\frac{1}{5}(1-\rho^2)^{5/2}\Big|_0^1\right\}=\frac{2\pi}{5}
\end{aligned} \tag{P1.7.2}$$

```
>>syms x y z r th ph
>>int(int(int(x^2+y^2+z^2,z,0,sqrt(1-x^2-?^2)), ...
            y,-sqrt(1-?^2),sqrt(1-y^2)),x,-?,?)
```

(a) Evaluate the integral in the cylindrical coordinate system:

$$\iiint_V(\rho^2+z^2)\,dV \overset{(1.1.60a)}{=} \int_0^{2\pi}\int_0^1\int_0^{\sqrt{1-\rho^2}}(\rho^2+z^2)\boxed{}\,dz\,d\rho\,d\phi=\cdots=\frac{2\pi}{5} \tag{P1.7.3}$$

```
>>int(int(int(r*(r^2+z^2),z,0,sqrt(1-?^2)),?,0,?),ph,0,2*pi)
```

(b) Evaluate the integral in the spherical coordinate system:

$$\begin{aligned}
&\iiint_V r^2\,dV \overset{(1.1.61a)}{=} \int_0^{2\pi}\int_0^{\pi/2}\int_0^1 r^2\boxed{}\,dr\,d\theta\,d\phi=\frac{1}{5}\int_0^{2\pi}\int_0^{\pi/2}\sin\theta\,d\theta\,d\phi \\
&=\frac{1}{5}\int_0^{2\pi}(\boxed{})\Big|_0^{\pi/2}=\frac{2\pi}{5}
\end{aligned} \tag{P1.7.4}$$

```
>>int(int(int(r^4*sin(th),r,0,1),??,0,pi/2),ph,0,2*pi)
```

1.8 Cauchy-Schwartz Inequality for Definite Integrals

Let $a(t)$ and $b(t)$ be real functions continuous on a closed interval $[t_1, t_2]$. Then we have the so-called *Cauchy-Schwarz inequality for definite integrals*:

$$\left|\int_{t_1}^{t_2} a(\tau)b(\tau)\,d\tau\right| \le \left(\int_{t_1}^{t_2} a^2(\tau)\,d\tau\right)^{1/2}\left(\int_{t_1}^{t_2} b^2(\tau)\,d\tau\right)^{1/2} \tag{P1.8.1}$$

where the equality holds if and only if $a(t)=x\,b(t)$ for a real number x.

(a) Prove Inequality (P1.8.1) by writing a quadratic polynomial in real variable x as

$$\begin{aligned} f(x) &= \int_{t_1}^{t_2} \{a(\tau)x+b(\tau)\}^2 d\tau = \int_{t_1}^{t_2} \left\{a^2(\tau)x^2+2a(\tau)b(\tau)x+b^2(\tau)\right\} d\tau \\ &= \left(\int_{t_1}^{t_2} a^2(\tau)\,d\tau\right)x^2 + 2\left(\int_{t_1}^{t_2} a(\tau)b(\tau)\,d\tau\right)x + \left(\int_{t_1}^{t_2} b^2(\tau)\,d\tau\right) = Ax^2+Bx+C \ge 0 \;\forall x \end{aligned} \tag{P1.8.2}$$

and claiming that its discriminant $D=B^2-4AC$ is less than or equal to zero because the value of the quadratic polynomial $f(x)$ is nonnegative for any x so that $f(x)=0$ cannot have two distinct real roots.

(b) Prove Inequality (P1.8.1) by defining the inner product for any pair of real functions $a(t)$ and $b(t)$ continuous on a closed interval $[t_1, t_2]$ as

$$<a(t),b(t)> = \int_{t_1}^{t_2} a(\tau)b(\tau)\,d\tau \tag{P1.8.3}$$

and invoking the Cauchy-Schwarz inequality (1.1.18) where the norm of a real function is computed as

$$||a(t)|| = \int_{t_1}^{t_2} a^2(\tau)\,d\tau \tag{P1.8.4}$$

1.9 Gram-Schmidt Process to Find a Set of Orthogonal Continuous Functions

With the inner product defined as Eq. (P1.8.3), two real functions $a(t)$ and $b(t)$ continuous on a closed interval $[t_1, t_2]$ are orthogonal to each other if

$$<a(t),b(t)> = \int_{t_1}^{t_2} a(\tau)b(\tau)\,d\tau = 0 \tag{P1.9.1}$$

(a) Referring to Eq. (1.1.62), use the Gram-Schmidt process to find a set of orthonormal functions spanning the vector space that is formed by the three polynomial functions $p_1=1$, $p_2=t$, and $p_3=t^2$ defined on $t=[-1,1]$ as follows:

$$q_1 = \frac{p_1}{||p_1||} = \frac{1}{\left(\int_{-1}^{1} 1^2\,dt\right)^{1/2}} = \boxed{} \tag{P1.9.2a}$$

$$q_2' = p_2 - \mathrm{proj}_{q_1} p_2 \overset{(1.1.10)}{=} p_2 - <p_2,q_1>\frac{q_1}{||q_1||^2} = t - \int_{-1}^{1} t\frac{1}{\sqrt{2}}\,dt\,\frac{1/\sqrt{2}}{1^2} = t$$

$$\overset{\text{normalize}}{\rightarrow}\; q_2 = \frac{q_2'}{||q_2'||} = \frac{t}{||t||} = \frac{t}{\left(\int_{-1}^{1} t^2\,dt\right)^{1/2}} = \boxed{} \tag{P1.9.2b}$$

$$q_3' = p_3 - \mathrm{proj}_{q_1} p_3 - \mathrm{proj}_{q_2} p_3 \overset{(1.1.10)}{=} p_3 - <p_3,q_1>\frac{q_1}{||q_1||^2} - <p_3,q_2>\frac{q_2}{||q_2||^2}$$

```
function [Q,L]=Gram_Schmidt_t(A,interval)
% Gram-Schmidt orthonormalization of the rows of MxN matrix A
%  assuming that the rows of A are linearly independent polynomials
% Q: MxN matrix whose rows form an orthonormal basis for row space of A.
% L: An invertible lower triangular matrix such that A = L*Q.
if nargin<2, interval=[-1 1]; end
[M,N] = size(A);
v1 = A(1,:);
L(1,1) = norm_t(v1,interval); Q(1,:) = v1/L(1,1);
for n=2:min(M,N)
   vn = A(n,:); % Take each row of A as a polynomial
   for m=1:n-1, L(n,m) = dot_t(vn,Q(m,:),interval); end
   qn = vn - L(n,:)*Q(1:n-1,:); % Eq.(1.1.64) with Projection Q*(Q'*vn)
   L(n,n) = norm_t(qn,interval); Q(n,:) = qn/L(n,n); % Normalize
end

function norm_p = norm_t(p,interval)
% Find the norm of a polynomial.
if nargin<2, interval=[-1 1]; end
norm_p = sqrt(dot_t(p,p,interval));

function p3 = dot_t(p1,p2,interval)
% Find the inner product of two polynomials
%  described by their coefficient vectors in descending order.
if nargin<3, interval=[-1 1]; end
p1 = p1(:).'; p2 = p2(:).';  p3 = polyint(conv(p1,p2));
p3 = polyval(p3,interval(2))-polyval(p3,interval(1));
```

$$=t^2-\int_{-1}^{1}t^2\frac{1}{\sqrt{2}}dt\frac{1/\sqrt{2}}{1^2}-\int_{-1}^{1}t^2\frac{t}{\sqrt{2/3}}dt\frac{t/\sqrt{2/3}}{1^2}=t^2-\frac{1}{3}$$

$$\xrightarrow{\text{normalize}} q_3=\frac{q_3'}{\|q_3'\|}=\frac{t^2-1/3}{\left(\int_{-1}^{1}(t^2-1/3)^2\,dt\right)^{1/2}}=\frac{t^2-1/3}{(2/5-4/9+2/9)^{1/2}}=\boxed{} \tag{P1.9.2c}$$

(b) The Gram-Schmidt process to find a set of orthonormal functions spanning the vector space that is formed by a given set of polynomial functions defined on $[t_1, t_2]$ has been cast into the above MATLAB function '`Gram_Schmidt_t()`', like '`Gram_Schmidt()`' (Section 1.7). Complete and run the following script "em01p09.m" to see if it works well for part (a). Also see if the Legendre polynomials (Eq. (3.3.37)) are obtained by dividing the orthonormal polynomials (obtained as each row of Q) by their values at t=1 so that their values at t=1 can be one. Note that in MATLAB, a polynomial can be represented by a numeric vector p containing its coefficients in descending order and its value at t0 can be found by running the MATLAB function '`polyval(p,t0)`'. Note also that the derivative/integral of polynomial p can be found using '`polyder(p)`'/ '`polyint(p)`' and the product of two polynomials p1 and p2 can be obtained using '`conv(p1,p2)`' (see Section E.1).

```
%em01p09.m
N=3; A=fliplr(eye(N)); % Three polynomial functions {1,t,t^2}
Q=Gram_Schmidt_t(?,[-1 1])
Q1=[0 0 1/sqrt(2); 0 1/sqrt(2/3) 0; sqrt(45/8)*[1 0 -1/3]] % (P1.9.2)
for n=1:N, Q2(n,:)=Q(n,:)/polyval(Q(n,:),1); end
Q2 % Does this conform with Eq.(3.3.37)? Also, for N=4,5,...?
```

1.10 Eigenvalues and *Gershgorin's Theorem*

Let $A = (a_{mn})$ be a complex $N \times N$ (square) matrix. Then every eigenvalue of the matrix lies within at least one of the *Gershgorin discs* that are defined as the discs each centered at a_{mm} and with radius R_{mr} where R_{mr} is defined as

$$R_{mr} = \sum_{n \neq m}^{N} |a_{mn}| \qquad \text{(P1.10.1)}$$

This can be proved by manipulating the component of Eq. (1.2.12) corresponding to $\mathrm{Max}_{1 \le n \le N} |v_{ni}|$ of the maximum absolute element of the eigenvector $\mathbf{v}_i$:

$$A\,\mathbf{v}_i \overset{(1.2.12)}{=} \lambda_i \mathbf{v}_i \overset{\text{the } m\text{th one}}{\rightarrow} \begin{bmatrix} a_{m1} & a_{m2} & \cdots & a_{mN} \end{bmatrix} \begin{bmatrix} v_{1i} \\ v_{2i} \\ \vdots \\ v_{Ni} \end{bmatrix} = \lambda_i v_{mi} \quad \text{with } m = \underset{1 \le n \le N}{\mathrm{ArgMax}} |v_{ni}| \qquad \text{(P1.10.2)}$$

$$; \quad |(\lambda_i - a_{mm}) v_{mi}| = \left| \sum_{n \neq m}^{N} a_{mn} v_{ni} \right|$$

$$; \quad |\lambda_i - a_{mm}| \le \left| \sum_{n \neq m}^{N} a_{mn} \right| \frac{|v_{ni}|}{|v_{mi}|} \le \sum_{n \neq m}^{N} |a_{mn}| \overset{(\text{P1.10.1})}{=} R_{mr} \qquad \text{(P1.10.3)}$$

Note that the radius, which is the absolute row sum, can be replaced by the absolute column sum:

$$R_{mc} = \sum_{n \neq m}^{N} |a_{nm}| \qquad \text{(P1.10.4)}$$

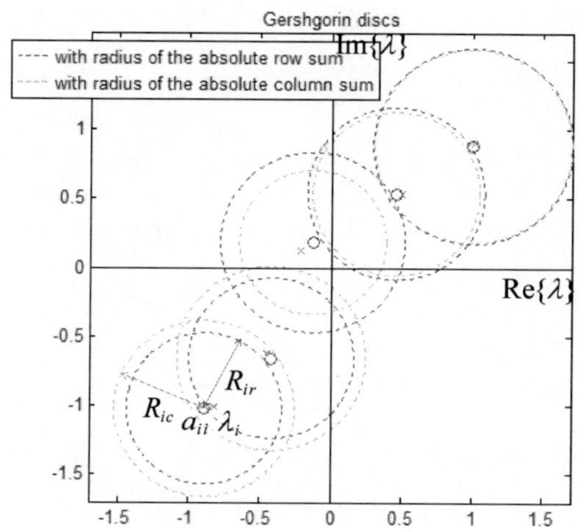

Figure P1.10 Gershgorin discs

Complete the following MATLAB function '`Gershdisc()`' and run the MATLAB script "em01p10.m" to use '`Gershdisc()`' for plotting the Gershgorin discs together with the eigenvalues of a partially random matrix A as depicted in Fig. P1.10.

```
%em01p10.m
A=diag([-1:0.5:1]+j*[-1:0.5:1])+0.4*(rand(5,5)-0.4+j*(rand(5,5)-0.5))
Gershdisc(A)
```

```
function [c,r]=Gershdisc(A)
% This plots the Gershgorin discs for the matrix A
[Nr,Nc]=size(A); N=360; ths=[0:N]*(2*pi/N); expjths=exp(j*ths);
Lams = eig(A); % Eigenvalues
for m=1:Nr
  c(m) = A(m,m); % Centers of each Gershgorin discs
  r1(m) = sum(abs(A(m,:))) - abs(A(m,?)); % Absolute row sum
  r2(m) = sum(abs(A(?,m))) - abs(A(m,m)); % Absolute column sum
  plot(c(m)+r1(m)*expjths,'b:'), hold on % Plot Gershgorin discs
  plot(c(m)+r2(m)*expjths,'g:') % Plot Gershgorin discs
  plot(real(c(m)),imag(c(m)),'o', real(Lams(m)),imag(Lams(m)),'rx')
end
axis equal;
```

1.11 *Companion Matrix* and Its Characteristic Equation

Consider the following forms of companion matrix:

$$A_{c1} = \begin{bmatrix} -a_1 & -a_2 & -a_3 & -a_4 & -a_5 \\ 1 & 0 & 0 & 0 & 0 \\ 0 & 1 & 0 & 0 & 0 \\ 0 & 0 & 1 & 0 & 0 \\ 0 & 0 & 0 & 1 & 0 \end{bmatrix} \quad \text{or} \quad A_{c2} = \begin{bmatrix} 0 & 0 & 0 & 0 & -a_5 \\ 1 & 0 & 0 & 0 & -a_4 \\ 0 & 1 & 0 & 0 & -a_3 \\ 0 & 0 & 1 & 0 & -a_2 \\ 0 & 0 & 0 & 1 & -a_1 \end{bmatrix} \tag{P1.11.1}$$

The characteristic equations of these matrices (that are used to find the eigenvalues) can be obtained by using Eq. (1.2.13) as

$$|xI - A_{c1}| = 0 \quad \text{or} \quad |xI - A_{c2}| = 0 \;:\; A_5(x) = x^5 + a_1 x^4 + \cdots + a_4 x + a_5 \tag{P1.11.2}$$

Try running the following MATLAB statements and see the results to get familiar with the characteristic equation of the companion matrices:

```
>>a=[1  -3  -23  51  94 -120]; % A 5th-degree polynomial
>>Ac1=[-a(2:6); eye(4) zeros(4,1)]; % The companion matrix
>>Ac1=compan(a) % Alternatively
>>syms x; det(x*eye(size(Ac1))-Ac1) % Eq.(P1.11.2)
   ans = x^5 - 3*x^4 - 23*x^3 + 51*x^2 + 94*x - 120
>>char_eq=poly(Ac1) % Characteristic equation of matrix Ac1
   char_eq =  1.0000   -3.0000  -23.0000   51.0000   94.0000 -120.0000
>>Lam=roots(char_eq).' % Eigenvalues of matrix A
   x =  5.0000   -4.0000   -2.0000    3.0000    1.0000
>>Lam=eig(Ac1).' % Eigenvalues of matrix A
   x =  5.0000   -4.0000   -2.0000    3.0000    1.0000
```

Noting that the matrices given in Eq. (P1.11.1) are called the (*Frobenius*) *companion matrices* of the monic polynomial (with unit leading coefficient), show that the characteristic equation of another companion matrix A_{c2} is also the same as Eq. (P1.11.2).

1.12 Power Iterations for Finding Eigenvalues/Eigenvectors

Suppose all of the eigenvalues of an $M \times N$ matrix A are distinct with magnitudes

$$|\lambda_1| > |\lambda_2| \ge |\lambda_3| \ge \cdots \ge |\lambda_N| \tag{P1.12.1}$$

(a) Scaled Power Method

The dominant eigenvalue λ_1 with the largest magnitude and its corresponding eigenvector $\mathbf{v}_1$ can be found by the following algorithm called the *scaled power method*:

(Step 1) Determine an initial vector $\mathbf{x}_0$ that has some nonzero component in the direction of $\mathbf{v}_1$.

(Step 2) With the iteration number $k=0$, divide $\mathbf{x}_k$ by its largest component (in absolute value) for normalization (scaling) and premultiply the normalized vector by the matrix A:

$$\mathbf{x}_{k+1} = A\frac{\mathbf{x}_k}{\|\mathbf{x}_k\|_\infty} \quad \text{with } \|\mathbf{x}\|_\infty = \underset{1\le n\le N}{\text{Max}}|x_n| \tag{P1.12.2}$$

(Step 3) Repeat Step 2 with $k \to k+1$, until $\|\mathbf{x}_{k+1} - \mathbf{x}_k\| < \varepsilon$ (some threshold value).

The convergence of this algorithm can be shown by expressing the initial vector in terms of the eigenvectors of A:

$$\mathbf{x}_0 = \alpha_1\mathbf{v}_1 + \alpha_2\mathbf{v}_2 + \cdots + \alpha_N\mathbf{v}_N, \tag{P1.12.3}$$

premultiplying Eq. (P1.12.3) by A, substituting $A\mathbf{v}_n \overset{(1.2.12)}{=} \lambda_n\mathbf{v}_n$ to write

$$\begin{aligned}\mathbf{x}_1 &= A\mathbf{x}_0 = \alpha_1\lambda_1\mathbf{v}_1 + \alpha_2\lambda_2\mathbf{v}_2 + \cdots + \alpha_N\lambda_N\mathbf{v}_N \\ &= \lambda_1\left(\alpha_1\mathbf{v}_1 + \alpha_2\frac{\lambda_2}{\lambda_1}\mathbf{v}_2 + \cdots + \alpha_N\frac{\lambda_N}{\lambda_1}\mathbf{v}_N\right),\end{aligned} \tag{P1.12.4}$$

and repeating this multiplication to obtain

$$\mathbf{x}_k = A^k\mathbf{x}_0 = \lambda_1^k\left\{\alpha_1\mathbf{v}_1 + \alpha_2\left(\frac{\lambda_2}{\lambda_1}\right)^k\mathbf{v}_2 + \cdots + \alpha_N\left(\frac{\lambda_N}{\lambda_1}\right)^k\mathbf{v}_N\right\} \rightarrow \lambda_1^k\alpha_1\mathbf{v}_1, \tag{P1.12.5}$$

This will converge to the eigenvector corresponding to the eigenvalue λ_1 with the largest magnitude as long as $\alpha_1 \neq 0$. Besides, since we keep scaling before premultiplication at every iteration, the component of the limit vector must be $|\lambda_1|$:

$$\mathbf{x}_{k+1} = A\frac{\mathbf{x}_k}{\|\mathbf{x}_k\|_\infty} \rightarrow A\frac{\mathbf{v}_1}{\|\mathbf{v}_1\|_\infty} \overset{(1.2.12)}{=} \lambda_1\frac{\mathbf{v}_1}{\|\mathbf{v}_1\|_\infty} \tag{P1.12.6}$$

Complete the following MATLAB function '`eig_power()`' to implement this algorithm and use it to find the largest eigenvalue and its corresponding eigenvector of

$$A = \begin{bmatrix} 2 & 0 & 1 \\ 0 & -2 & 0 \\ 1 & 0 & 2 \end{bmatrix} \tag{P1.12.7}$$

where the eigenvalue can be found by using the Rayleigh quotient (P1.25.2). Does the result conform with that obtained using the MATLAB built-in function '`eig()`'?

```
function [lambda,v,k]=eig_power(A,x,EPS,MaxIter)
% The power method to find the largest eigenvalue (lambda) and
%  the corresponding eigenvector (v) of a matrix A.
if nargin<4, MaxIter=100; end % Maximum number of iterations
if nargin<3, EPS=1e-8; end % Threshold on the difference
N=size(A,2);
if nargin<2, x=[1:N]; end % Initial vector
x=x(:);  xm=max(abs(x));
for k=1:MaxIter
   x0=x;  x0m=xm;
   x = ?*?/xm; % Eq.(P1.12.2)
   [xm,m]=max(abs(x)); % norm(x,inf);
   if norm(abs(x)-abs(x0))<EPS&abs(xm-x0m)<EPS, break; end
end
v= x/norm(x); % Eigenvector
lambda= v'*A*v/(v'*v); % Rayleigh quotient - largest eigenvalue
if k==MaxIter, disp('Warning: you may have to increase MaxIter'); end
```

(b) Inverse power method

We can multiply both sides of Eq. (1.2.12) by $A^{-1}\lambda^{-1}$ to write

$$A\mathbf{v} \overset{(1.2.12)}{=} \lambda\mathbf{v} \overset{\times A^{-1}\lambda^{-1}}{\rightarrow} A^{-1}\mathbf{v} = \lambda^{-1}\mathbf{v}, \tag{P1.12.8}$$

This implies that A^{-1} has the eigenvalues that are the reciprocals of the eigenvalues of A, but the same eigenvectors. Thus we can apply the scaled power method to A^{-1} and take the inverse of the component of the limit as the smallest (magnitude) eigenvalue λ_N. This algorithm is called the *inverse power method* and it works only in cases where the matrix A is nonsingular and so has no zero eigenvalue. Based on this idea, use the MATLAB function '`eig_power()`' to find the smallest eigenvalue and its corresponding eigenvector of the matrix A given as Eq. (P1.12.7).

(c) Shifted inverse power method

We can subtract $s\mathbf{v}$ (s: a number that does not happen to coincide with any eigenvalue) from both sides of Eq. (1.2.12) to write

$$A\mathbf{v} \overset{(1.2.12)}{=} \lambda\mathbf{v} \overset{-s\mathbf{v}}{\rightarrow} [A-sI]\mathbf{v} = (\lambda-s)\mathbf{v} \tag{P1.12.9}$$

This implies that $(\lambda-s)$ is the eigenvalue of $[A-sI]$. Thus, we can apply the inverse power method for $[A-sI]$ to get its smallest magnitude eigenvalue (λ_k-s) with $\text{Min}_{1\le i\le N}|\lambda_i-s|$ and add the shift s to it to obtain the eigenvalue of A that is closest to the number s. This algorithm is called the *shifted inverse power method*. Based on this idea, use the MATLAB function '`eig_power()`' to find the eigenvalue closest to -3 and its corresponding eigenvector of the matrix A given as Eq. (P1.12.7).

(d) Using power method to find the M largest eigenvalues of a Hermitian (symmetric) matrix

Noting that the modal matrix V of a Hermitian matrix A, which consists of the eigenvectors, is unitary (see property (4) in Section 1.2.7.1) so that

$$A_{N\times N}V \overset{(1.2.24)}{=} V\Lambda \rightarrow AVV^{*\text{T}} = V\Lambda V^{*\text{T}}$$

$$\underset{\text{unitary}}{\overset{VV^{*\text{T}}=I}{\rightarrow}} A = V\Lambda V^{*\text{T}} = \begin{bmatrix} \mathbf{v}_1 & \cdot & \mathbf{v}_N \end{bmatrix} \begin{bmatrix} \lambda_1 & 0 & 0 \\ 0 & \cdot & 0 \\ 0 & 0 & \lambda_N \end{bmatrix} \begin{bmatrix} \mathbf{v}_1^{*\text{T}} \\ \cdot \\ \mathbf{v}_N^{*\text{T}} \end{bmatrix} = \sum_{i=1}^{N} \lambda_i \mathbf{v}_i \mathbf{v}_i^{*\text{T}}, \tag{P1.12.10}$$

we can use the power method to find not only the largest eigenvalue (and its corresponding eigenvector) but also as many largest eigenvalues and their corresponding eigenvectors of a Hermitian/symmetric matrix A as we want as follows:

1. Use the power method to find the largest eigenvalue λ_1 and its corresponding eigenvector $\mathbf{v}_1$ of the matrix A.
2. Use the power method to find the largest eigenvalue λ_2 and its corresponding eigenvector $\mathbf{v}_2$ of the matrix

$$A_1 = A(I-\mathbf{v}_1\mathbf{v}_1^{*\text{T}}) = A - A\mathbf{v}_1\mathbf{v}_1^{*\text{T}} \overset{(1.2.12)}{=} A - \lambda_1\mathbf{v}_1\mathbf{v}_1^{*\text{T}} \tag{P1.12.11}$$

which is the next largest eigenvalue and its corresponding eigenvector of the matrix A.

Complete the following MATLAB function '`eigs_power()`' (implementing this idea) and run the script file "test_eigs_power.m" to find the M=3 largest eigenvalues and their corresponding eigenvectors of a random 5×5 Hermitian matrix. Do the results agree with those obtained by running the following MATLAB statement?

```
function [lambdas,vs,ks]=eigs_power(A,M,x,EPS,MaxIter)
% Uses the power method to find the M largest eigenvalue lambdas
%   and their corresponding eigenvectors vs of a Hermitian matrix A
% Input:
%   A       = A Hermitian matrix
%   M       = Number of largest eigenvalues/eigenvectors to be found
%   x       = Initial guess of the eigenvector
%   EPS     = Threshold for judging the convergence
%   MaxIter = Maximum iteration number
% Output:
%   lambdas = M Largest eigenvalues
%   vs      = Eigenvectors corresponding to the eigenvalues lambdas
%   ks      = Numbers of iterations performed
if nargin<5, MaxIter=100; end % Maximum number of iterations
if nargin<4, EPS=1e-6; end % Threshold on the difference
N=size(A,2);
if nargin<3, x=[1:N]; end % Initial vector
lambdas=zeros(1,M); vs = zeros(N,M); ks=zeros(1,M);
[lambdas(1),vs(:,1),ks(1)] = eig_power(A);
I_vv = eye(N,N); % NxN Identity matrix
for n=2:M
   I_vv = I_vv - vs(:,n-1)*??(:,n-1)'; % Eq.(P1.12.11)
   [lambdas(n),vs(:,n),ks(n)] = eig_power(?*I_vv,x,EPS,MaxIter);
end
```

```
%test_eigs_power.m
N=5; A0=rand(N,N)-0.5 + (rand(N,N)-0.5)*i;
A=A0*A0'; % An arbitrary 5x5 Hermitian matrix
M=3; [lambdas,vs,ks]=eigs_power(A,M);
[lambdas,idx]=sort(lambdas); vs=vs(:,idx);
[V0,L0]=eigs(A,M); lambdas0=diag(L0).';
[lambdas0,idx0]=sort(lambdas0); vs0=V0(:,idx0);
Discrepancy_eigenvalues = norm(lambdas-lambdas0)/norm(lambdas0)
Discrepancy_eigenvectors = norm(abs(vs)-abs(vs0))/norm(vs0)
```

```
>>[V,L]=eigs(A,M)
```

1.13 Eigenvalues/Eigenvectors

Find the eigenvalues and eigenvectors of a matrix

$$A=\begin{bmatrix}3 & 0 & 0\\ 2 & 1 & 0\\ 2 & -4 & 2\end{bmatrix} \tag{P1.13.1}$$

1.14 Repeated Eigenvalues, Generalized Eigenvectors, and Jordan Canonical (Normal) Form

Find the eigenvalues and eigenvectors of the following matrices and check the validity through their eigendecompositions.

(a)
$$A_1=\begin{bmatrix}0 & 1 & 1\\ 1 & 0 & 1\\ 1 & 1 & 0\end{bmatrix} \tag{P1.14.1}$$

(b)
$$A_2=\begin{bmatrix}1 & -4 & 0\\ 1 & 0 & 2\\ 0 & 2 & 1\end{bmatrix} \tag{P1.14.2}$$

(c)
$$A_3 = \begin{bmatrix} 1 & 0 & 0 \\ 2 & 1 & 2 \\ 1 & 0 & 0 \end{bmatrix} \tag{P1.14.3}$$

(d)
$$A_4 = \begin{bmatrix} 0 & 1 & 0 \\ 0 & 0 & 1 \\ 0 & -1 & -2 \end{bmatrix} \tag{P1.14.4}$$

1.15 Eigenvalues/Eigenvectors of a Symmetric Tri-diagonal Toeplitz Matrix

Consider the following $N \times N$ symmetric tri-diagonal *Toeplitz* or *diagonal-constant matrix* (having constants along negative-sloping diagonals) as

$$\begin{bmatrix} a & b & 0 & \cdot & 0 & 0 \\ b & a & b & \cdot & 0 & 0 \\ 0 & b & a & \cdot & 0 & 0 \\ \cdot & \cdot & \cdot & \cdot & \cdot & \cdot \\ 0 & 0 & 0 & \cdot & a & b \\ 0 & 0 & 0 & \cdot & b & a \end{bmatrix} \tag{P1.15.1}$$

(a) Verify that the eigenvalues and eigenvectors of this matrix are as follows, with N=3 for convenience.

$$\lambda_n = a + 2b\cos\left(\frac{n\pi}{N+1}\right) \quad \text{for } n = 1 \text{ to } N \tag{P1.15.2}$$

$$\mathbf{v}_n = \sqrt{\frac{2}{N+1}}\left[\sin\left(\frac{n\pi}{N+1}\right) \;\; \sin\left(\frac{2n\pi}{N+1}\right) \;\; \cdots \;\; \sin\left(\frac{N\,n\pi}{N+1}\right)\right]^T \tag{P1.15.3}$$

(b) Letting N=3, a=2, and b=1, find the eigenvalues/eigenvectors of the above matrix by using (P1.15.2,3) and by using the MATLAB function '`eig()`' for crosscheck.

1.16 Eigenvalues/Eigenvectors of a Circulant Matrix

Consider the following $N \times N$ *circulant matrix* (in which every row is a right cyclic shift of the row above it) as

$$\begin{bmatrix} h(0) & h(N-1) & h(N-2) & \cdot & h(1) \\ h(1) & h(0) & h(N-1) & \cdot & h(2) \\ h(2) & h(1) & h(0) & \cdot & h(3) \\ \cdot & \cdot & \cdot & \cdot & \cdot \\ h(N-1) & h(N-2) & h(N-3) & \cdot & h(0) \end{bmatrix} \tag{P1.16.1}$$

(a) Verify that the eigenvalues and eigenvectors of this matrix are as follows, with N=4 for convenience.

$$\lambda_n = h(0) + h(N-1)e^{j2\pi n/N} + h(N-2)e^{j2\pi 2n/N} + \cdots + h(1)e^{j2\pi(N-1)n/N} \tag{P1.16.2}$$

$$\mathbf{v}_n = \left[1 \;\; e^{j2\pi n/N} \;\; e^{j2\pi 2n/N} \;\; \cdots \;\; e^{j2\pi(N-1)n/N}\right]^T, \quad \text{for } n = 0 \text{ to } N-1 \tag{P1.16.3}$$

(b) Letting N=4, $h(0)$=2, $h(3)$=$h(1)$=1, and $h(2)$=0, find the eigenvalues/eigenvectors of the above matrix by using (P1.16.2,3) and by using the MATLAB function '`eig()`'. Do they agree? Do they satisfy Eq. (1.2.12)?

1.17 Repetitive Matrix-Vector Multiplications Using Eigendecomposition

Consider a vector $\mathbf{x} = [1 \quad -1]^T$.

(a) Find $\mathbf{y} = A^{20}\mathbf{x}$ with

$$A = \begin{bmatrix} 0 & 2 \\ 1 & 1 \end{bmatrix} \tag{P1.17.1}$$

If you need the eigenvalues/eigenvectors of A, make use of the results obtained in Example 1.11.

(b) Find $\mathbf{y} = A^{20}\mathbf{x}$ with

$$A = \begin{bmatrix} 0.68 & -0.24 \\ -0.24 & 0.82 \end{bmatrix} \tag{P1.17.2}$$

(c) Based on the results of (a) and (b), discuss the effect of the magnitudes of the eigenvalues on the convergence or divergence.

1.18 Computation of Matrix Exponential Using Eigendecomposition

Let the *matrix exponential* of a matrix A be defined as

$$\phi = e^A = I + A + \frac{1}{2!}A^2 + \frac{1}{3!}A^3 + \cdots \tag{P1.18.1}$$

Note that this is implemented by the MATLAB command '`expm()`'.

(a) Show that the matrix exponentials of

$$\Lambda_a = \begin{bmatrix} \lambda_1 & 1 & 0 \\ 0 & \lambda_1 & 1 \\ 0 & 0 & \lambda_1 \end{bmatrix} \quad \text{and} \quad \Lambda_b = \begin{bmatrix} \lambda_1 & 0 & 0 \\ 0 & \lambda_1 & 0 \\ 0 & 0 & \lambda_1 \end{bmatrix} \tag{P1.18.2}$$

are

$$e^{\Lambda_a} = \begin{bmatrix} e^{\lambda_1} & e^{\lambda_1} & e^{\lambda_1}/2 \\ 0 & e^{\lambda_1} & e^{\lambda_1} \\ 0 & 0 & e^{\lambda_1} \end{bmatrix} \quad \text{and} \quad e^{\Lambda_b} = \begin{bmatrix} e^{\lambda_1} & 0 & 0 \\ 0 & e^{\lambda_1} & 0 \\ 0 & 0 & e^{\lambda_1} \end{bmatrix} \tag{P1.18.3}$$

```
syms L
A=[L 1 0;0 L 1;0 0 L];
eye(size(A))+A+A^2/2+A^3/6
expm(A)
A=[L 0 0;0 L 0;0 0 L];
eye(size(A))+A+A^2/2+A^3/6
expm(A)
```

(b) Show that the *matrix exponential* can be computed by using the eigendecomposition $A = V\Lambda V^{-1}$ (Λ: Jordan form) as

$$e^{V\Lambda V^{-1}} = I + V\Lambda V^{-1} + \frac{1}{2!}(V\Lambda V^{-1})^2 + \frac{1}{3!}(V\Lambda V^{-1})^3 + \cdots = Ve^{\Lambda}V^{-1} \tag{P1.18.4}$$

where

$$e^{\Lambda_c} = \begin{bmatrix} e^{\lambda_1} & e^{\lambda_1} & e^{\lambda_1}/2 & 0 \\ 0 & e^{\lambda_1} & e^{\lambda_1} & 0 \\ 0 & 0 & e^{\lambda_1} & 0 \\ 0 & 0 & 0 & e^{\lambda_4} \end{bmatrix} \quad \text{or} \quad e^{\Lambda_d} = \begin{bmatrix} e^{\lambda_1} & 0 & 0 & 0 \\ 0 & e^{\lambda_2} & 0 & 0 \\ 0 & 0 & e^{\lambda_3} & 0 \\ 0 & 0 & 0 & e^{\lambda_4} \end{bmatrix} \tag{P1.18.5}$$

depending on whether the Jordan form is of the form

$$\Lambda_c = \begin{bmatrix} \lambda_1 & 1 & 0 & 0 \\ 0 & \lambda_1 & 1 & 0 \\ 0 & 0 & \lambda_1 & 0 \\ 0 & 0 & 0 & \lambda_4 \end{bmatrix} \quad \text{or} \quad \Lambda_d = \begin{bmatrix} \lambda_1 & 0 & 0 & 0 \\ 0 & \lambda_2 & 0 & 0 \\ 0 & 0 & \lambda_3 & 0 \\ 0 & 0 & 0 & \lambda_4 \end{bmatrix} \tag{P1.18.6}$$

1.19 Diagonalization of Covariance Matrix for Aligning the Data Orientation

Consider the four points $\mathbf{x}^{(1)}=[0 \quad -1]^T$, $\mathbf{x}^{(2)}=[-1 \quad 0]^T$, $\mathbf{x}^{(3)}=[2 \quad 3]^T$, and $\mathbf{x}^{(4)}=[3 \quad 2]^T$ plotted as square boxes on the x_1x_2-plane in Fig. P1.19(a).

(a) Find the covariance matrix $C_\mathbf{x}$ of the vectors representing the K=4 points defined as

$$C_\mathbf{x} = \frac{1}{K}\sum_{k=1}^{K}[\mathbf{x}^{(k)}-\mathbf{m}_\mathbf{x}][\mathbf{x}^{(k)}-\mathbf{m}_\mathbf{x}]^T = \begin{bmatrix} 2.5 & 2 \\ 2 & 2.5 \end{bmatrix} \text{ with } \mathbf{m}_\mathbf{x}=\frac{1}{K}\sum_{k=1}^{K}\mathbf{x}^{(k)} \tag{P1.19.1}$$

where the mean vector represents the center of the points.

(b) Check that the eigendecomposition of $C_\mathbf{x}$ is

$$C_\mathbf{x} = V\Lambda V^{-1} = \begin{bmatrix} \mathbf{v}_1 & \mathbf{v}_2 \end{bmatrix}\begin{bmatrix} \lambda_1 & 0 \\ 0 & \lambda_2 \end{bmatrix}\begin{bmatrix} \mathbf{v}_1 & \mathbf{v}_2 \end{bmatrix}^{-1} = \begin{bmatrix} -1/\sqrt{2} & 1/\sqrt{2} \\ 1/\sqrt{2} & 1/\sqrt{2} \end{bmatrix}\begin{bmatrix} 0.5 & 0 \\ 0 & 4.5 \end{bmatrix}\begin{bmatrix} -1/\sqrt{2} & 1/\sqrt{2} \\ 1/\sqrt{2} & 1/\sqrt{2} \end{bmatrix} \tag{P1.19.2}$$

where the modal matrix V has the eigenvectors in its columns and the Jordan matrix Λ has the eigenvalues on its diagonal.

(c) Transform $\{\mathbf{x}^{(1)}, \mathbf{x}^{(2)}, \mathbf{x}^{(3)}, \mathbf{x}^{(4)}\}$ into $\mathbf{y}=V^{-1}(\mathbf{x}-m_\mathbf{x})$ and check that the covariance matrix of $\mathbf{y}$'s is

$$C_\mathbf{y} = \frac{1}{K}\sum_{k=1}^{K}[\mathbf{y}^{(k)}-\mathbf{m}_\mathbf{y}][\mathbf{y}^{(k)}-\mathbf{m}_\mathbf{y}]^T = \Lambda = \begin{bmatrix} 0.5 & 0 \\ 0 & 4.5 \end{bmatrix} \text{ with } \mathbf{m}_\mathbf{y}=\frac{1}{K}\sum_{k=1}^{K}\mathbf{y}^{(k)} \tag{P1.19.3}$$

(d) Noting from Fig. P1.19(a) that the transformation of vector $\mathbf{x}$ into $\mathbf{y}=V^{-1}(\mathbf{x}-m_\mathbf{x})$ (through the inverse modal matrix of the covariance matrix $C_\mathbf{x}$) amounts to the alignment of $\mathbf{x}$ with the principal axes, tell which one of the smaller and larger axes of distribution (variances) of $\mathbf{x}/\mathbf{y}$ corresponds to the smaller or larger eigenvalues of $C_\mathbf{x}/C_\mathbf{y}$, where the difference among the eigenvalues determines how oblong the overall shape of the distribution is and the samples have a larger variance along the larger axis.

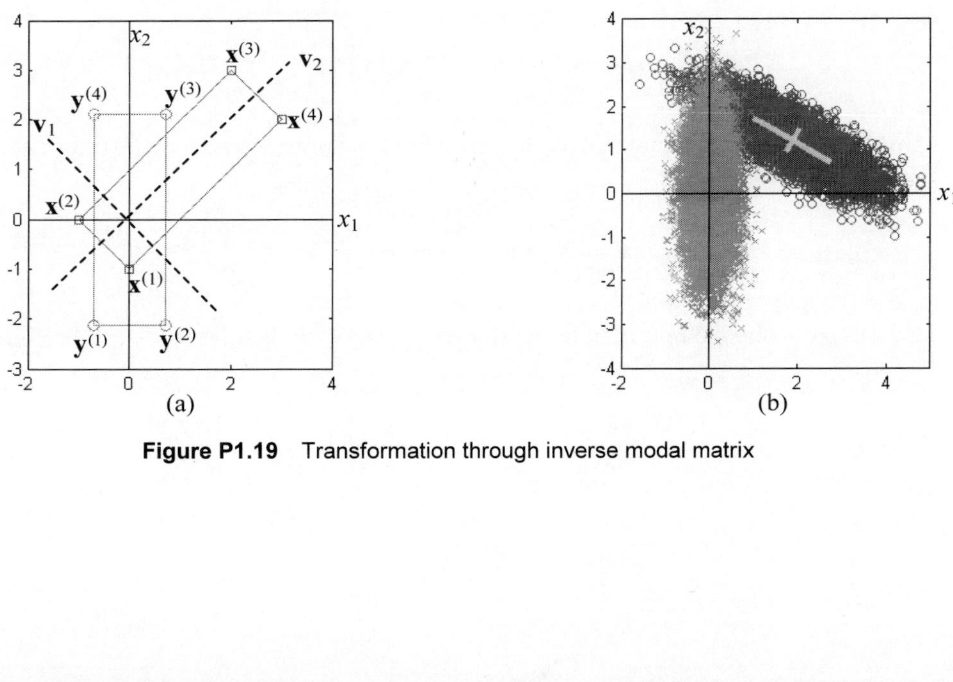

Figure P1.19 Transformation through inverse modal matrix

```
%em01p19e.m
% Diagonalize of covariance matrix for aligning data orientation
K = 10000; % Number of random data points
mx0 = [2 -1]; sgmx0 = [0.3 1]; % Means and Square root variances
x0(1,:)=mx0(1)+sgmx0(1)*randn(1,K); %Gaussian data of mean 1, var 1
x0(2,:)=mx0(2)+sgmx0(2)*randn(1,K); %Gaussian data of mean 1, var 0.3^2
% Rotation matrix Eq.(1.2.35)
R=@(th)[cos(th) -sin(th); sin(th) cos(th)];
th = 60*pi/180;
x = R(th)*x0; % Rotate the data points by 60[degree]=pi/3[rad]
subplot(222), plot(x(1,:),x(2,:),'o'), hold on
mx=mean(x,2); % Row-wise mean
x_m = x-repmat(mx,1,K); % Repetition of mx for dimension compatibility
Cx = x_m*x_?'/K % Covariance matrix Eq.(P1.19.1)
[V,L] = ei?(Cx); % Eigendecomposition
[R(th) V] % Rotation matrix and Modal matrix
sgmx = sqrt([L(1,1) L(2,2)]); % Square root eigenvalues
[sgmx0; sgmx] % Square root variance and Square root eigenvalues
for n=1:2 % To plot the axes of x with eigenvector directions
  plot(mx(1)+[-V(1,n) V(1,n)]*sgmx(n),mx(2)+[-V(2,n) V(2,n)]*sgmx(n),'g')
end
y=V^-?*x_m; atan(V(2,1)/V(1,1))*180/pi
plot(y(1,:),y(2,:),'rx') % Plot the transformed data y
```

(e) Now, do the same for many more data points as was done for the 4 data points in (a)–(c) by completing and running the above script "em01p19e.m", which is supposed to generate $K=10^4$ data points in the $2\times K$ matrix $\mathbf{x}$ containing x_1/x_2-values in its first/second rows, and align them with the principal axes as depicted in Fig. P1.19(b). Find the modal matrix V consisting of the eigenvectors of the covariance matrix $C_{\mathbf{x}}=(\mathbf{x}-m_{\mathbf{x}})(\mathbf{x}-m_{\mathbf{x}})^T/K$. Regarding V as a 2-D rotation matrix

$$V=\begin{bmatrix} v_{11} & v_{12} \\ v_{21} & v_{22} \end{bmatrix} \sim \begin{bmatrix} \cos\theta & -\sin\theta \\ \sin\theta & \cos\theta \end{bmatrix} = R(\theta) \text{ with } \theta=\tan^{-1}\frac{v_{21}}{v_{11}}, \quad \text{(P1.19.4)}$$

find the rotation angle θ.

1.20 Diagonalization of Coefficient Matrix for Rotating the Graph Plane

(a) Consider the following equation of an ellipse:

$$5x_1^2+6x_1x_2+5x_2^2=32\,;\ \begin{bmatrix} x_1 & x_2 \end{bmatrix}\begin{bmatrix} 5 & 3 \\ 3 & 5 \end{bmatrix}\begin{bmatrix} x_1 \\ x_2 \end{bmatrix}=\mathbf{x}^T A\,\mathbf{x}=32 \quad \text{(P1.20.1)}$$

Check that by the change of variables (based on the eigendecomposition of the coefficient matrix $A=V\Lambda V^{-1}$),

$$\mathbf{x}=V\mathbf{y},\text{ i.e., } \begin{bmatrix} x_1 \\ x_2 \end{bmatrix}=\begin{bmatrix} 1/\sqrt{2} & 1/\sqrt{2} \\ -1/\sqrt{2} & 1/\sqrt{2} \end{bmatrix}\begin{bmatrix} y_1 \\ y_2 \end{bmatrix} \quad \text{(P1.20.2)}$$

Eq. (P1.20.1) can be rewritten in the standard form with the diagonalized coefficient matrix:

$$\mathbf{y}^T\Lambda\mathbf{y}=\begin{bmatrix} y_1 & y_2 \end{bmatrix}\begin{bmatrix} \lambda_1=2 & 0 \\ 0 & \lambda_2=8 \end{bmatrix}\begin{bmatrix} y_1 \\ y_2 \end{bmatrix}=2y_1^2+8y_2^2=32 \quad \text{(P1.20.3)}$$

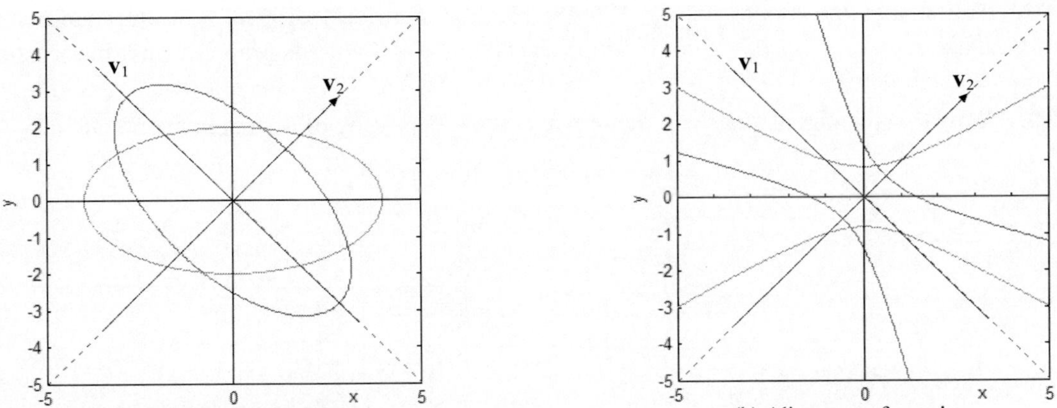

(a) Alignment of an ellipse (b) Alignment of a conic

Figure P1.20 Alignments of ellipse and conic through diagonalization of the coefficient matrices

```
%em01p20.m
% Diagonalization of coefficient matrix for rotating the graph plane
disp('(a)')
A=[5 3; 3 5];
subplot(121), ezplot('5*x^2+6*x*y+5*y^2-??'), hold on
[V,L]=eig(A); [V L]  % V=[-0.7071 0.7071; 0.7071 0.7071], L=[2 0;0 8];
[A V*L*V']
t=[-10 10];
plot(V(1,1)*t,V(2,1)*t,'k:', V(1,2)*t,V(2,2)*t,'k:')
ezplot('?*x^2+8*y^2-32')
axis('equal'), axis([-5 5 -5 5])
disp('(b)')
A=[1 2; 2 1];
subplot(122), ezplot('x^2+??x*y+y^2-2'), hold on
[V,L]=eig(A); [V L] % V=[-0.7071 0.7071; 0.7071 0.7071], L=[-1 0;0 3];
[A V*L*V']
plot(V(1,1)*t,V(?,1)*t,'k:', V(1,?)*t,V(2,2)*t,'k:')
ezplot('-x^2+??y^2-2'), axis('equal'), axis([-5 5 -5 5]), shg
```

Determine from Fig. P1.20(a) how the major/minor (principal) axes of the ellipse are related to the eigenvectors corresponding with the smaller/larger eigenvalues.

(b) Consider the following equation of a conic:

$$x_1^2+4x_1x_2+x_2^2=2\,;\ \begin{bmatrix}x_1 & x_2\end{bmatrix}\begin{bmatrix}1 & 2\\ 2 & 1\end{bmatrix}\begin{bmatrix}x_1\\ x_2\end{bmatrix}=\mathbf{x}^T A\mathbf{x}=2 \tag{P1.20.4}$$

Check that by the change of variables (based on the eigendecomposition of the coefficient matrix $A=V\Lambda V^{-1}$)

$$\mathbf{x}=V\mathbf{y},\ \text{i.e.,}\ \begin{bmatrix}x_1\\ x_2\end{bmatrix}=\begin{bmatrix}1/\sqrt{2} & 1/\sqrt{2}\\ -1/\sqrt{2} & 1/\sqrt{2}\end{bmatrix}\begin{bmatrix}y_1\\ y_2\end{bmatrix} \tag{P1.20.5}$$

Eq. (P1.20.4) can be rewritten in the standard form with the diagonalized coefficient matrix:

$$\mathbf{y}^T\Lambda\mathbf{y}=\begin{bmatrix}y_1 & y_2\end{bmatrix}\begin{bmatrix}\lambda_1{=}-1 & 0\\ 0 & \lambda_2{=}3\end{bmatrix}\begin{bmatrix}y_1\\ y_2\end{bmatrix}=-y_1^2+3y_2^2=2 \tag{P1.20.6}$$

(c) Noting that the modal matrix V's are rotation matrices, discuss how the standard form of equation of an ellipse or a conic should be changed to rotate its graph by an arbitrary angle. You can complete the above script "em01p20.m" to plot Fig. P1.20 for reference.

(d) Show that the eigenvalues and the modal matrix consisting of the eigenvectors of

$$K=\begin{bmatrix}\sigma_{\mathbf{x}}^2 & \sigma_{\mathbf{xy}}\\ \sigma_{\mathbf{xy}} & \sigma_{\mathbf{y}}^2\end{bmatrix}=\begin{bmatrix}1 & 0.5\\ 0.5 & 1\end{bmatrix} \quad \text{(P1.20.7)}$$

are

$$\begin{bmatrix}\lambda_1\\ \lambda_2\end{bmatrix}=\frac{1}{2}\begin{bmatrix}\sigma_{\mathbf{x}}^2+\sigma_{\mathbf{y}}^2+\sqrt{(\sigma_{\mathbf{x}}^2-\sigma_{\mathbf{y}}^2)^2+4\sigma_{\mathbf{xy}}^4}\\ \sigma_{\mathbf{x}}^2+\sigma_{\mathbf{y}}^2-\sqrt{(\sigma_{\mathbf{x}}^2-\sigma_{\mathbf{y}}^2)^2+4\sigma_{\mathbf{xy}}^4}\end{bmatrix} \quad \text{(P1.20.8)}$$

$$V=\begin{bmatrix}\cos\theta & -\sin\theta\\ \sin\theta & \cos\theta\end{bmatrix} \text{ with } \theta=\frac{1}{2}\tan^{-1}\frac{2\sigma_{\mathbf{xy}}}{\sigma_{\mathbf{x}}^2-\sigma_{\mathbf{y}}^2} \quad \text{(P1.20.9)}$$

1.21 Successive 3-D Rotations Described by a Concatenation of Rotation Matrices

A series of 3-D rotations can be described by concatenating the basic rotation matrices. Sketch how a vector $\mathbf{p}=[0\ \ 0\ \ 1]^T$ will be changed as a result of the three successive rotations described by

$$R_y(\frac{\pi}{4})R_z(\frac{\pi}{2})R_x(\frac{\pi}{4})=\begin{bmatrix}\cos\frac{\pi}{4} & 0 & \sin\frac{\pi}{4}\\ 0 & 1 & 0\\ -\sin\frac{\pi}{4} & 0 & \cos\frac{\pi}{4}\end{bmatrix}\begin{bmatrix}\cos\frac{\pi}{2} & -\sin\frac{\pi}{2} & 0\\ \sin\frac{\pi}{2} & \cos\frac{\pi}{2} & 0\\ 0 & 0 & 1\end{bmatrix}\begin{bmatrix}1 & 0 & 0\\ 0 & \cos\frac{\pi}{4} & -\sin\frac{\pi}{4}\\ 0 & \sin\frac{\pi}{4} & \cos\frac{\pi}{4}\end{bmatrix} \quad \text{(P1.21.1)}$$

Complete and run the following script, "em01p21.m," to plot the rotated vectors for reference.

```
%em01p21.m
Rx=@(th)[1 0 0; 0 ???(th) -sin(th); 0 sin(th) cos(th)]; %Eq.(1.2.35a)
Ry=@(th)[cos(th) 0 sin(th); 0 1 0; -???(th) 0 cos(th)]; %Eq.(1.2.35b)
Rz=@(th)[cos(??) -sin(th) 0; sin(th) cos(th) 0; 0 0 ?]; %Eq.(1.2.35c)
p=[0;0;1]; quiver3(0,0,0,p(1),p(2),p(3)), hold on
axis('equal'), axis([-1 1 -1 1 -0.1 1]), view(30,20), shg, pause
p1=Rx(pi/4)*p, quiver3(0,0,0,p1(1),p1(2),p1(3)), pause
p2=Rz(pi/2)*p1, quiver3(0,0,0,p2(1),p2(2),p2(3)), pause
p3=Ry(pi/4)*p2, quiver3(0,0,0,p3(1),p3(2),p3(3))
Ry(????)*Rz(pi/2)*Rx(pi/4)
xlabel('x'), ylabel('y'), zlabel('z')
```

1.22 Rotation w.r.t an Arbitrary Axis

(a) The rotation by θ with respect to an arbitrary axis $\mathbf{u}=[u_x\ \ u_y\ \ u_z]^T$ passing through the origin can be done systematically by taking the following steps, where the two angles ϕ_1 and ϕ_2 are defined in Fig. P1.22(a) as

$$\phi_1=\tan^{-1}\frac{u_y}{u_x} \quad\text{and}\quad \phi_2=\tan^{-1}\frac{\sqrt{u_x^2+u_y^2}}{u_z} \quad \text{(P1.22.1)}$$

(Step 1) Rotate the object (together with $\mathbf{u}$) by $-\phi_1$ about the z-axis and then by $-\phi_2$ about the y-axis so that the rotated $\mathbf{u}$ can be aligned with the z-axis.

(Step 2) Rotate the object by θ about the z-axis.

(Step 3) Rotate the object (together with $\mathbf{u}$) by ϕ_2 about the y-axis and then by ϕ_1 about the z-axis so that $\mathbf{u}$ can come back to its original posture.

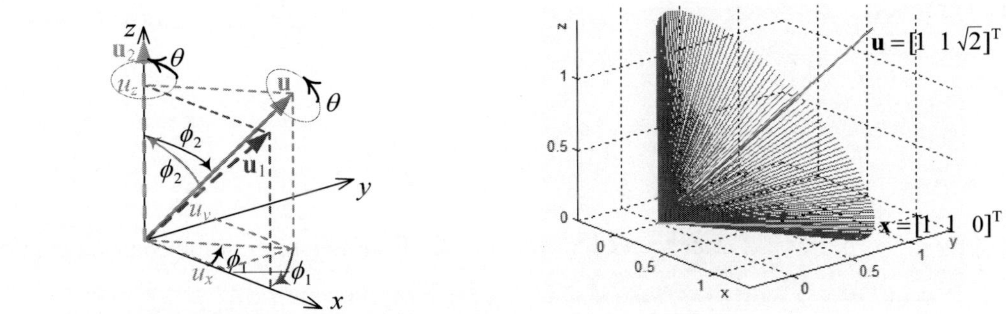

(a) Alignment of a rotation axis with the z-axis (b) Locus of a rotated vector **x** w.r.t **u**

Figure P1.22 The locus of the vectors produced by rotating **x** w.r.t. **u**

These steps can be described as a concatenation of rotations about the principal axes (Eqs. (1.2.35a,b,c)):

$$R_u(\theta) = R_z(\phi_1)R_y(\phi_2)R_z(\theta)R_y(-\phi_2)R_z(-\phi_1) \tag{P1.22.2}$$

Complete and run the following MATLAB script, "test_rotation_matrix1.m," to check that this yields the rotation matrix R_u defined by Eq. (1.2.38).

```
%test_rotation_matrix1.m
syms ph1 ph2 th
ux=sin(ph2)*cos(ph1); uy=sin(ph2)*sin(ph1); uz=cos(ph2);
Rx=@(th)[1 0 0;0 cos(th) -sin(th);0 sin(th) cos(th)]; %Eq.(1.2.35a)
Ry=@(th)[cos(th) 0 sin(th);0 1 0;-sin(th) 0 cos(th)]; %Eq.(1.2.35b)
Rz=@(th)[cos(th) -sin(th) 0;sin(th) cos(th) 0;0 0 1]; %Eq.(1.2.35c)
R1=Rz(ph1)*Ry(ph?)*Rz(th)*Ry(?ph2)*Rz(-ph?);
c=cos(th); s=sin(th); % Cosine/Sine of rotation angle
Ra=@(ux,uy,uz,th)[c+ux^2*(1-c) ux*uy*(1-c)-uz*s ux*uz*(1-c)+uy*s;
   uy*ux*(1-c)+uz*s c+uy^2*(1-c) uy*uz*(1-c)-ux*s;
   uz*ux*(1-c)-uy*s uz*uy*(1-c)+ux*s c+uz^2*(1-c)]; % Eq.(1.2.38)
simplify(R1-Ra(ux,uy,uz,th))
```

(b) Complete the following MATLAB function, '`rotate_by_th_wrt_u(x,th,u)`', so that it can be used to rotate a vector x by an angle th w.r.t. an axis u. Then save the function as an M-file named "rotate_by_th_wrt_u.m" and run the following script, "test_rotation_matrix2.m," to plot the locus of the vector **y**'s that are produced by rotating $\mathbf{x}=[1\ \ 1\ \ 0]^T$ by (0:0.04:2π)[rad] w.r.t. $\mathbf{u}=[1\ \ 1\ \ \sqrt{2}]^T$, as depicted in Fig. P1.22(b).

```
%test_rotation_matrix2.m
x=[1;1;0]; u=[1;1;sqrt(2)];
plot3([0 u(1)],[0 u(2)],[0 u(3)],'LineWidth',2,'Color','r'), hold on
for th=[0:0.04:2*pi]
   y=rotate_by_th_wrt_u(x,th,u); plot3([0 y(1)],[0 y(2)],[0 y(3)])
end
axis([0 1.5 0 1.5 0 1.5]), axis('equal')
```

```
function y=rotate_by_th_wrt_u(x,th,u)
u=u/norm(u); ux=u(1); uy=u(2); uz=u(3); % Rotation axis
c=cos(th); s=sin(th); % Rotation angle
R=[c+ux^2*(1-c) u?*uy*(1-c)-uz*s ux*uz*(1-?)+uy*s;
   uy*ux*(1-c)+uz*s c+u?^2*(1-c) uy*uz*(1-c)-ux*?;
   uz*ux*(1-c)-uy*s u?*uy*(1-c)+ux*s c+uz^2*(1-c)]; % Eq.(1.2.38)
y=R*x;
```

1.23 LU Decomposition, QR Decomposition, and SVD for Solving a Set of Linear Equations

Consider the following system of linear equations:

$$A\mathbf{x}=\mathbf{b};\quad \begin{bmatrix}1 & -2 & 3\\ 1 & 0 & 4\\ 0 & 2 & 5\end{bmatrix}\begin{bmatrix}x_1\\ x_2\\ x_3\end{bmatrix}=\begin{bmatrix}2\\ 7\\ 9\end{bmatrix} \tag{P1.23.1}$$

(a) '`[L,U,P]=lu(A)`' can be used to find the LU decomposition of the coefficient matrix A as

$$A=P^{\mathrm{T}}LU;\quad \begin{bmatrix}1 & -2 & 3\\ 1 & 0 & 4\\ 0 & 2 & 5\end{bmatrix}=\begin{bmatrix}1 & 0 & 0\\ 0 & 1 & 0\\ 0 & 0 & 1\end{bmatrix}\begin{bmatrix}1 & 0 & 0\\ 1 & 1 & 0\\ 0 & 1 & 1\end{bmatrix}\begin{bmatrix}1 & -2 & 3\\ 0 & 2 & 1\\ 0 & 0 & 4\end{bmatrix} \tag{P1.23.2}$$

Solve Eq. (P1.23.1) by exploiting the LU structure as

$$P^{\mathrm{T}}LU\ \mathbf{x}=\mathbf{b};\ LU\mathbf{x}=P\mathbf{b};\ U\mathbf{x}=L^{-1}P\mathbf{b};\ \mathbf{x}=U^{-1}L^{-1}P\mathbf{b} \tag{P1.23.3}$$

so that no matrix inversion is needed, since the premultiplication of a vector by L^{-1} and U^{-1} can be performed by the forward and backward substitutions, respectively. It can be done by running the following MATLAB statements if you saved the M-files '`forsubst()`' and '`backsubst()`' (listed below) on your PC:

```
>>A=[1 -2  3; 1  0  4; 0  2  5];  b=[2; 7; 9];
  [L,U,P]=lu(A);
  x=backsubst(U,forsubst(L,P*b))
```

```
function x=forsubst(L,B)
% Forward substitution for a lower-triangular matrix equation Lx=B
N=min(size(L)); x(1,:)=B(1,:)/L(1,1);
for m=2:N
   x(m,:)=(B(m,:)-L(m,1:m-1)*x(1:m-1,:))/L(m,m);
end
```

```
function x=backsubst(U,B)
% Backward substitution for a upper-triangular matrix equation Ux=B
N=min(size(U)); x(N,:)=B(N,:)/U(N,N);
for m=N-1:-1:1
   x(m,:)=(B(m,:)-U(m,m+1:N)*x(m+1:N,:))/U(m,m);
end
```

(b) '`[Q,R,P]=qr(A)`' can be used to find the QR decomposition of the coefficient matrix A as

$$A=QRP^{\mathrm{T}}$$

$$;\begin{bmatrix}1 & -2 & 3\\ 1 & 0 & 4\\ 0 & 2 & 5\end{bmatrix}=\begin{bmatrix}-0.4243 & 0.8083 & 0.4082\\ -0.5657 & 0.1155 & -0.8165\\ -0.7071 & -0.5774 & 0.4082\end{bmatrix}\begin{bmatrix}-7.0711 & -0.5657 & -0.9899\\ 0 & -2.7713 & 0.9238\\ 0 & 0 & -0.4082\end{bmatrix}\begin{bmatrix}0 & 0 & 1\\ 0 & 1 & 0\\ 1 & 0 & 0\end{bmatrix} \tag{P1.23.4}$$

Solve Eq. (P1.23.1) by exploiting the QR structure as

$$QRP^{\mathrm{T}}\mathbf{x}=\mathbf{b};\ RP^{\mathrm{T}}\mathbf{x}=Q^{*\mathrm{T}}\mathbf{b};\ P^{\mathrm{T}}\mathbf{x}=R^{-1}Q^{*\mathrm{T}}\mathbf{b};\ \mathbf{x}=PR^{-1}Q^{*\mathrm{T}}\mathbf{b} \tag{P1.23.5}$$

so that no matrix inversion is needed. It can be done by running the following MATLAB statements:

```
>>[Q,R,P]=qr(A); x=P*backsubst(R,Q'*b)
```

Note that the LSE (least-squares error) solution for an overdetermined set of linear equations is easy to get with the QR decomposition since

$$\mathbf{x} \overset{(1.3.13)}{=} [A^{\mathrm{H}}A]^{-1}A^{\mathrm{H}}\mathbf{b} = \left[(QRP^{\mathrm{T}})^{\mathrm{H}}QRP^{\mathrm{T}}\right]^{-1}(QRP^{\mathrm{T}})^{\mathrm{H}}\mathbf{b}$$

$$= [PR^{\mathrm{H}}Q^{\mathrm{H}}QRP^{\mathrm{T}}]^{-1}PR^{\mathrm{H}}Q^{\mathrm{H}}\mathbf{b} = P(R^{\mathrm{H}}R)^{-1}R^{\mathrm{H}}Q^{\mathrm{H}}\mathbf{b} \overset{R^{\mathrm{H}}R \text{ is of full rank}}{=} PR^{-1}Q^{\mathrm{H}}\mathbf{b} \quad \text{(P1.23.6)}$$

where the superscript ^{H} denotes the Hermitian (conjugate) transpose operation like $^{*\mathrm{T}}$.

(c) '`[U,S,V]=svd(A)`' can be used to find the SVD decomposition of the coefficient matrix A as

$$A = USV^{\mathrm{H}}; \quad \begin{bmatrix} 1 & -2 & 3 \\ 1 & 0 & 4 \\ 0 & 2 & 5 \end{bmatrix} = \quad \text{(P1.23.7)}$$

$$\begin{bmatrix} -0.4087 & 0.7950 & 0.4482 \\ -0.5702 & 0.1610 & -0.8056 \\ -0.7126 & -0.5849 & 0.3875 \end{bmatrix} \begin{bmatrix} 7.1633 & 0 & 0 \\ 0 & 2.9225 & 0 \\ 0 & 0 & 0.3821 \end{bmatrix} \begin{bmatrix} -0.1367 & 0.3271 & -0.9350 \\ -0.0848 & -0.9443 & -0.3180 \\ -0.9870 & 0.0359 & 0.1568 \end{bmatrix}$$

Solve Eq. (P1.23.1) by exploiting the USV structure (with unitary/diagonal/unitary matrices $U/S/V$) as

$$USV^{\mathrm{H}}\mathbf{x} = \mathbf{b}; \; SV^{\mathrm{H}}\mathbf{x} = U^{\mathrm{H}}\mathbf{b}; \; V^{\mathrm{H}}\mathbf{x} = S^{-1}U^{\mathrm{H}}\mathbf{b}; \; \mathbf{x} = VS^{-1}U^{\mathrm{H}}\mathbf{b} \quad \text{(P1.23.8)}$$

This needs no matrix inversion because the inverse matrix S^{-1} of the diagonal matrix $S = [s_{mm}]$ is $S^{-1} = [s_{mm}^{-1}]$, which can be obtained by just taking the reciprocal of each diagonal element if every s_{mm} is nonzero. It can be done by running the following MATLAB statements:

```
>>[U,S,V]=svd(A); x=V*diag(1./diag(S))*U'*b
```

Note that the SVD can be used to find a minimum-norm solution to a (possibly) rank-deficient least squares problem with rank$(A) < \min(M,N)$ (M,N: the numbers of rows/columns of A), which minimizes both $\|\mathbf{x}\|$ and $\|A\mathbf{x} - \mathbf{b}\|$. The minimum-norm least squares solution can be found as

$$\mathbf{x} = V(:,1:K)S(1:K,1:K)^{-1}U(:,1:K)^{\mathrm{H}}\mathbf{b} \text{ with } K = \text{rank}(A) \quad \text{(P1.23.9)}$$

```
>>sing=diag(S); sing=sing(find(sing>=1e-9)); %Nonzero singular values
  K=length(sing); % Rank of A possibly rank-deficient matrix
  x_svd = V(:,1:K)*diag(1./sing)*U(:,1:K)'*b; % SVD solution (P1.23.9)
```

1.24 Systems of Linear Equations with Rectangular or Rank-Deficient Coefficient Matrices

Consider the following sets of linear equations:

$$\text{(a)} \begin{bmatrix} 2 & -1 \\ 1 & -5 \\ -3 & 6 \end{bmatrix} \begin{bmatrix} x_1 \\ x_2 \end{bmatrix} = \begin{bmatrix} 5 \\ 10 \\ -14 \end{bmatrix} \quad \text{(b)} \begin{bmatrix} 2 & -1 & 1 \\ 1 & -5 & 2 \\ -3 & 6 & -3 \end{bmatrix} \begin{bmatrix} x_1 \\ x_2 \\ x_3 \end{bmatrix} = \begin{bmatrix} 7 \\ 17 \\ -24 \end{bmatrix} \quad \text{(c)} \begin{bmatrix} 2 & 1 & -3 \\ -1 & -5 & 1 \end{bmatrix} \begin{bmatrix} x_1 \\ x_2 \\ x_3 \end{bmatrix} = \begin{bmatrix} -9 \\ 12 \end{bmatrix} \quad \text{(P1.24.1)}$$

$$\text{rank}(A) = 2 = \min(M,N) \qquad \text{rank}(A) = 2 < 3 = \min(M,N) \qquad \text{rank}(A) = 2 = \min(M,N)$$

Among the above three sets of equations, the coefficient matrix of Eq. (P1.24.1b) is of deficient rank, while those of Eqs. (P1.24.1a,c) are of full rank. Since Eq. (P1.24.1a) with rank([A **b**]) =rank(A)+1 > min(M,N) is inconsistent, we should find the LSE (least-squares error) solution to it. Since Eq. (P1.24.1b) with rank([A **b**])=rank(A)=2 < min(M,N) is redundant, it has infinitely many solutions. Eq. (P1.24.1c) has also infinitely many solutions since it has more unknowns than equations. Referring to the idea presented in Problem 1.23, complete and run the following MATLAB script, "em01p24.m," to solve each of the three sets of equations by using the QR method, the SVD method, the backslash operator, and '`pinv()`'. Fill Table P1.24 with the results of running the script. Which ones among the four methods give the minimum-norm/LSE solution for the redundant/undetermined case, seemingly unaffected by the rank deficiency of A?

Table P1.24 Solutions of (P1.24.1a,b,c) obtained using QR decomposition, SVD, \, and `pinv()`

Equation		(a) Inconsistent (No exact solutions)	(b) Redundant (Many solutions)	(c) Underdetermined (Many solutions)
QR solution Eq. (P1.23.6)	**x**	[ ]	[0.2821 -1.2821 5.1536]	[0 -1.929 2.357]
	$\|\|\mathbf{x}\|\|$		5.3182	3.0456
	$\|\|A\mathbf{x}-\mathbf{b}\|\|$		1.7764×10^{-15}	8.8818×10^{-15}
SVD solution Eq. (P1.23.9)	**x**	[1.5185 -1.6296]	[ ]	[-1.166 -1.845 1.608]
	$\|\|\mathbf{x}\|\|$	2.2275		2.7109
	$\|\|A\mathbf{x}-\mathbf{b}\|\|$	0.5774	$\times 10^{-15}$	3.5527×10^{-15}
Backslash solution A\b	**x**	[1.5185 -1.6296]	[NaN NaN NaN]	[ ]
	$\|\|\mathbf{x}\|\|$	2.2275	NaN	
	$\|\|A\mathbf{x}-\mathbf{b}\|\|$	0.5774	NaN	$\times 10^{-15}$
pinv(A)*b	**x**	[1.5185 -1.6296]	[1.546 -2.546 1.364]	[-1.166 -1.845 1.608]
	$\|\|\mathbf{x}\|\|$	2.2275	3.2753	2.7109
	$\|\|A\mathbf{x}-\mathbf{b}\|\|$	0.5774	6.4047×10^{-15}	3.5527×10^{-15}

```
%em01p24.m
EPS=1e-12;
format short e
for k=1:3
  if k==1, A=[2 -1; 1 -?; -3 6]; b =[5; 10; -14]; %Eq.(P1.24.1a)
   elseif k==2, A=[2 -1 1;1 -5 2; -3 6 -3]; b=[7; 17;-??]; %Eq.(P1.24.1b)
   else A= [2 1 -3; -1 -5 ?]; b= [-9; 12]; %Eq.(P1.24.1c)
  end
  [Q,R,P] = qr(A); % Q*R = A*P
  x_qr = backsubst(R,?'*b); % permutated QR solution
  [M,N] = size(A);
  if N>M, x_qr = [x_qr; zeros(N-M,1)]; end % possibly augmented
  x_qr = P*x_qr; % depermutated QR solution (P1.23.?)
  [U,S,V] = svd(A); % SVD decomposition
  sing = diag(S); % Singular values
  sing = sing(find(sing>=EPS)); % Choose only nonzero eigenvalues
  K = length(sing); % Rank of A possibly rank-deficient
  x_svd = V(:,1:K)*diag(1./sing)*?(:,1:K)'*b; % SVD solution (P1.23.?)
  x_bs = A\b; % Solution using the backslash operator
  x_pinv = pinv(A)*b; % Solution using pinv()
  disp('    x_qr        x_svd        x_bs        x_pinv')
  solutions=[x_qr x_svd x_bs x_pinv]
  norms= [norm(x_qr) norm(x_svd) norm(x_bs) norm(x_pinv)]
  errs= [norm(A*x_qr-b) norm(A*x_svd-b) norm(A*x_bs-b) norm(A*x_pinv-b)]
end
format short
```

1.25 Rayleigh Quotient as an LS (Least Squares) Solution

Suppose we have a guess of an eigenvector $\mathbf{v}_1$ of a matrix A. How can we find the corresponding eigenvalue λ_1 such that the following approximate equality holds?

$$A\mathbf{v}_1 \approx \lambda_1 \mathbf{v}_1 \;\rightarrow\; \mathbf{v}_1\lambda_1 - A\mathbf{v}_1 \approx \mathbf{0} \tag{P1.25.1}$$

The LS solution (1.3.13) with $A=\mathbf{v}_1$ and $\mathbf{b}=A\mathbf{v}_1$ can be applied to get an optimal estimate for the eigenvalue

$$\lambda_1=[A^{\mathrm{T}}A]^{-1}A^{\mathrm{T}}\mathbf{b} \underset{\mathbf{b}=A\mathbf{v}_1}{\overset{A=\mathbf{v}_1}{=}} (\mathbf{v}_1^{\mathrm{T}}\mathbf{v}_1)^{-1}\mathbf{v}_1^{\mathrm{T}}A\mathbf{v}_1=\frac{\mathbf{v}_1^{\mathrm{T}}A\mathbf{v}_1}{\mathbf{v}_1^{\mathrm{T}}\mathbf{v}_1}: \text{Rayleigh quotient} \tag{P1.25.2}$$

which is referred to as the *Rayleigh quotient*.

The *rayleigh quotient iteration* (*RQI*) is an algorithm using the Rayleigh quotient to find the eigenvector close to a given approximate eigenvector, which has been cast into the following MATLAB function: '`Rayleigh_quotient_iteration()`'. Complete the MATLAB function and use it to find the eigenvector of

$$A=\begin{bmatrix} 2 & 0 & 1 \\ 0 & -2 & 0 \\ 1 & 0 & 2 \end{bmatrix} \tag{P1.25.3}$$

closest to $[1.6 \quad -0.2 \quad 1.1]^{\mathrm{T}}$.

```
function
   [v,lambda,k]=Rayleigh_quotient_iteration(A,v0,lambda0,EPS,MaxIter)
% Rayleigh quotient iteration (RQI) to find the eigenvector v
%   closest to a given vector v0
% Input:
%   A       = A square matrix
%   v0      = Initial guess of eigenvector
%   lambda0 = Initial guess of the corresponding eigenvalue
%   EPS     = Threshold for judging the convergence
%   MaxIter = Maximum iteration number
% Output:
%   lambda  = Eigenvalue lambda largest/smallest/closest to x
%   v       = Eigenvector corresponding to the eigenvalue lambda
%   k       = Number of iteration performed
if nargin<5, MaxIter=100; end
if nargin<4, EPS=1e-8; end
v0=v0(:);
if nargin<3
  lambda0=v0'*A*v0;
end
I=eye(size(A)); v=v0; lambda=lambda0;
for k=1:MaxIter
   w = (A-lambda*I)\v; %w = AI*v; Shifted inverse power iteration
   v = w/norm(w);
   lambda = v'*?*?; % Eq.(P1.25.2) with v'*v=1
   if abs(lambda-lambda0)/abs(lambda)<EPS, break; end
   lambda0 = lambda;
end
```

1.26 Gauss-Jordan Elimination with Partial Pivoting

To solve a system of linear equations $A\mathbf{x} = \mathbf{b}$, Gauss elimination performs forward elimination to make the coefficient matrix A an upper-triangular matrix U and then performs backward substitution to get the solution $\mathbf{x}$, as explained in Sec. 1.3.4. Gauss-Jordan elimination performs forward/backward elimination to make the coefficient matrix A an identity matrix I and then takes the RHS as the solution. The general formula applicable for Gauss-Jordan elimination is the same as Eq. (1.3.22), except that the index set is $m \neq k$, i.e., all the numbers from $m=1$ to $m=M$ except $m=k$.

(a) Complete and use the MATLAB function '`Gauss_Jordan()`' to solve Eq. (1.3.25). Explain what operations have been done in the intermediate steps (1)–(5) shown below:

$$\begin{bmatrix} 0 & 1 & 1 & 2 \\ 2 & -1 & -1 & 0 \\ 1 & 1 & -1 & 1 \end{bmatrix} \overset{(1)}{\rightarrow} \begin{bmatrix} 2 & -1 & -1 & 0 \\ 0 & 1 & 1 & 2 \\ 1 & 1 & -1 & 1 \end{bmatrix} \overset{(2)}{\rightarrow} \begin{bmatrix} 1 & -1/2 & -1/2 & 0 \\ 0 & 1 & 1 & 2 \\ 0 & 3/2 & -1/2 & 1 \end{bmatrix}$$
$$\overset{(3)}{\rightarrow} \begin{bmatrix} 1 & -1/2 & -1/2 & 0 \\ 0 & 3/2 & -1/2 & 1 \\ 0 & 1 & 1 & 2 \end{bmatrix} \overset{(4)}{\rightarrow} \begin{bmatrix} 1 & 0 & -2/3 & 1/3 \\ 0 & 1 & -1/3 & 2/3 \\ 0 & 0 & 4/3 & 4/3 \end{bmatrix} \overset{(5)}{\rightarrow} \begin{bmatrix} 1 & 0 & 0 & 1 \\ 0 & 1 & 0 & 1 \\ 0 & 0 & 1 & 1 \end{bmatrix} \tag{P1.26.2}$$

```
function  x=Gauss_Jordan(A,B)
%This function solves A*x=B by Gauss-Jordan elimination algorithm.
%The sizes of given matrices A,B are supposed to be NAxNA and NAxNB.
[NA1,NA]=size(A);  [NB1,NB]=size(B);
if NA1~=NA|NB1~=NA
  error('A and B must have compatible dimension');
end
N=NA+NB;
AB=[A(1:NA,1:NA) B(1:NA,1:NB)]; % Augmented matrix
for k=1:NA
   %Partial Pivoting at AB(k,k)
   [akx,kx]= max(abs(AB(k:NA,k)));
   if akx<eps,  error('Singular coefficient matrix');  end
   if kx>1 % Row change if necessary
     mx=k+kx-1;
     tmp_row =AB(k,k:N);
     AB(k,k:N)=AB(mx,k:N);
     AB(mx,k:N)=tmp_row;
   end
   AB % To check the intermediate result stepwise
   % Gauss-Jordan elimination
   AB(k,k+1:N)=AB(k,k+1:N)/AB(k,k);
   AB(k,k)=1;
   for m=1:NA
     if m??? % if m is not equal to k
       AB(m,k+1:N)=AB(m,k+1:N)-AB(m,k)*AB(k,k+1:N);
       AB(m,k)=0;
     end
   end
   AB % To check the intermediate result stepwise
end
x=AB(1:NA,NA+1:N);
```

(b) How is the solution X of the following equation related to the coefficient matrix A?

$$AX = B; \quad \begin{bmatrix} 0 & 1 & 1 \\ 2 & -1 & -1 \\ 1 & 1 & -1 \end{bmatrix} X = \begin{bmatrix} 1 & 0 & 0 \\ 0 & 1 & 0 \\ 0 & 0 & 1 \end{bmatrix} \tag{P1.26.3}$$

Use the above MATLAB function '`Gauss_Jordan()`' to find the solution:

```
>>A=[0 1 1; 2 -1 -1; 1 1 -1]; B=eye(3); X=Gauss_Jordan(A,B)
```

1.27 Application of Eigendecomposition for DoA (Degree of Arrival) Estimation

Fig. P1.27.1 shows a UCA (uniform circular array) having N_e=4 antenna elements equally spaced on a circle of radius R, which is impinged on by N_s=2 signals $s_1(n)$ and $s_2(n)$ with DoAs (ϕ_1=60°, θ_1) and (ϕ_2=210°, θ_2), respectively, where the DoA (ϕ, θ) of an incident signal is a pair of azimuth and zenith angles. The output signals measured by each element of the UCA at time n can be written as

$$\underset{\mathbf{y}(n)}{\begin{bmatrix} y_1(n) \\ y_2(n) \\ \cdot \\ y_{N_e}(n) \end{bmatrix}} = \underset{A:\, N_e \times N_s \text{ matrix}}{\begin{bmatrix} e^{j\beta R \sin\theta_1 \cos(\phi_1 - 2\pi 0/N_e)} & e^{j\beta R \sin\theta_{Ns} \cos(\phi_{Ns} - 2\pi 0/N_e)} \\ e^{j\beta R \sin\theta_1 \cos(\phi_1 - 2\pi 1/N_e)} & e^{j\beta R \sin\theta_{Ns} \cos(\phi_{Ns} - 2\pi 1/N_e)} \\ \cdot & \cdot \\ e^{j\beta R \sin\theta_1 \cos(\phi_1 - 2\pi (N_e-1)N_e)} & e^{j\beta R \sin\theta_{Ns} \cos(\phi_{Ns} - 2\pi (N_e-1)N_e)} \end{bmatrix}} \underset{\mathbf{s}(n)}{\begin{bmatrix} s_1(n) \\ s_{N_s}(n) \end{bmatrix}} + \underset{\mathbf{w}(n)}{\begin{bmatrix} w_1(n) \\ w_2(n) \\ \cdot \\ w_{N_e}(n) \end{bmatrix}} \tag{P1.27.1}$$

$$; \quad \mathbf{y}(n) = \underset{A:\, N_e \times N_s \text{ steering matrix}}{\left[\mathbf{a}(\phi_1,\theta_1)\ \mathbf{a}(\phi_{Ns},\theta_{Ns})\right]} \mathbf{s}(n) + \mathbf{w}(n) \text{ with } \underset{\text{Steering vector}}{\mathbf{a}(\phi,\theta)} = \begin{bmatrix} e^{j\beta R \sin\theta \cos(\phi - 2\pi 0/N_e)} \\ e^{j\beta R \sin\theta \cos(\phi - 2\pi 1/N_e)} \\ \cdot \\ e^{j\beta R \sin\theta \cos(\phi - 2\pi (N_e-1)N_e)} \end{bmatrix} \tag{P1.27.2}$$

where $w_m(n)$ denotes a complex noise with normal distribution $N(0, 0.5)$ of mean 0 and variance 0.5 (see Chapter 10 for "normal distriution"). For simplicity, we suppose that the two signals are

$$s_1(n) = \sin(\frac{\pi}{8}n) \quad \text{and} \quad s_2(n) = \cos(\frac{\pi}{12}n), \tag{P1.27.3a,b}$$

while the zenith angles of all the received signals are θ=90°, the wave propagation factor is β=2π/λ=1 (λ: wavelength), and the radius of the UCA is R=1. To estimate the DoAs, we construct the signal correlation matrix R_y from {$\mathbf{y}(n)$, n=1:N_d=200} and use the MATLAB command '`[V,Λ]=eig(`R_y`)`' to find its eigenvectors and eigenvalues as

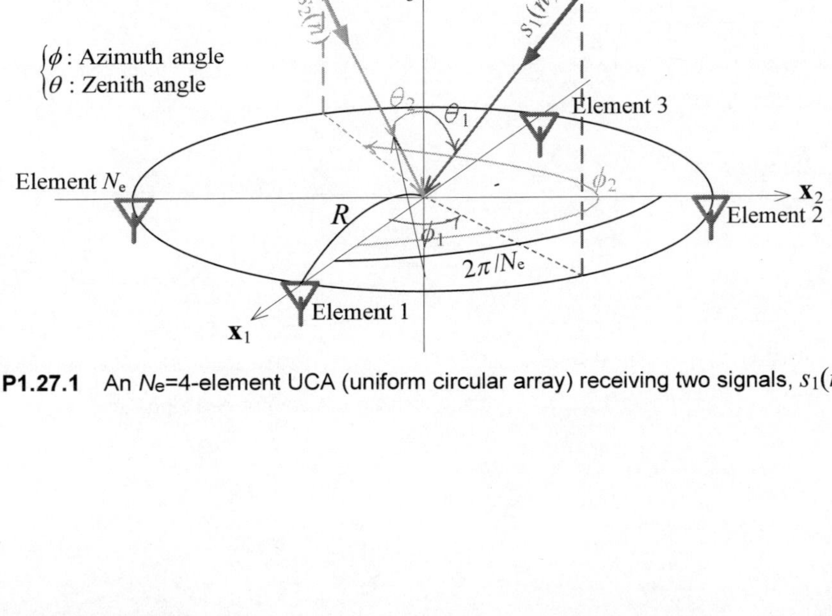

Figure P1.27.1 An N_e=4-element UCA (uniform circular array) receiving two signals, $s_1(n)$ and $s_2(n)$

$$R_{\mathbf{y}} = \frac{YY^{*\mathrm{T}}}{200} = \begin{bmatrix} 1.3728 & 0.87-j0.3237 & 0.1237-j0.0103 & 0.213-j0.0463 \\ 0.87+j0.3237 & 1.4144 & 0.3024+j0.0456 & 0.2842+j0.0811 \\ 0.1237+j0.0103 & 0.3024-j0.0456 & 1.6276 & 1.0259+j0.3658 \\ 0.213+j0.0463 & 0.2842-j0.0811 & 1.0259-j0.3658 & 1.6152 \end{bmatrix} \text{ with } Y = \begin{bmatrix} y_1(1) & \cdots & y_1(200) \\ y_2(1) & \cdots & y_2(200) \\ y_3(1) & \cdots & y_3(200) \\ y_4(1) & \cdots & y_4(200) \end{bmatrix} \tag{P1.27.4}$$

$$\rightarrow V = \begin{bmatrix} -0.1044-j0.7008 & -0.0251-j0.0441 & -0.6035-j0.058 & 0.3575-j0.008 \\ -0.1354+j0.6837 & -0.0834+j0.0774 & -0.5268-j0.2475 & 0.378+j0.1399 \\ 0.0723-j0.0241 & -0.6465-j0.2482 & 0.3686+j0.1976 & 0.556+j0.1749 \\ -0.0802 & 0.7106 & 0.3446 & 0.6082 \end{bmatrix}, \Lambda = \begin{bmatrix} 0.4614 & 0 & 0 & 0 \\ 0 & 0.5249 & 0 & 0 \\ 0 & 0 & 2.0644 & 0 \\ 0 & 0 & 0 & 2.9793 \end{bmatrix} \tag{P1.27.5}$$

Even if the specific values of $\mathbf{y}(n)$, $R_{\mathbf{y}}$, and its eigenvalues/eigenvectors vary with the noise $\mathbf{w}(n)$, (the last) two eigenvalues λ_3=2.0644 and λ_4=2.9793 are conspicuously larger than the other ones, λ_1=0.4614 and λ_2=0.5249, so that their corresponding eigenvectors $\mathbf{v}_3$ and $\mathbf{v}_4$ can be regarded as spanning the signal subspace, while the eigenvectors $\mathbf{v}_1$ and $\mathbf{v}_2$ corresponding to the two smaller eigenvalues λ_1 and λ_2 can be regarded as spanning the noise subspace.

(a) One way to estimate the azimuth angles of the two signals is to find the local maxima of the spatial spectrum based on the signal subspace

$$P_{\mathrm{s}}(\phi) = \mathbf{a}^{*\mathrm{T}}(\phi) S S^{*\mathrm{T}} \mathbf{a}(\phi) \text{ with } S = [V(:,3)\ V(:,4)] \tag{P1.27.6}$$

which yields the azimuth angle estimates around the true values 60° and 210° (Fig. P1.27.2(a1)). An alternative is to find the local maxima of the spatial spectrum based on the noise subspace

$$P_{\mathrm{n}}(\phi) = \frac{1}{\mathbf{a}^{*\mathrm{T}}(\phi) N N^{*\mathrm{T}} \mathbf{a}(\phi)} \text{ with } N = [V(:,1)\ V(:,2)] \tag{P1.27.7}$$

which also yields the azimuth angle estimate $\hat{\phi}$'s around the true values 60° and 210° (Fig. P1.27.2(a2)). This estimation method is referred to as the *MUSIC* (*multiple sIgnal classification*) algorithm. Which one of these two spatial spectra, P_{s} or P_{n}, looks better in terms of selectivity? To carry out this process, complete and run the MATLAB script "DoA_estimation_using_ eigendecomposition.m" to get the plots like Figs. P1.27.2(a1)&(a2).

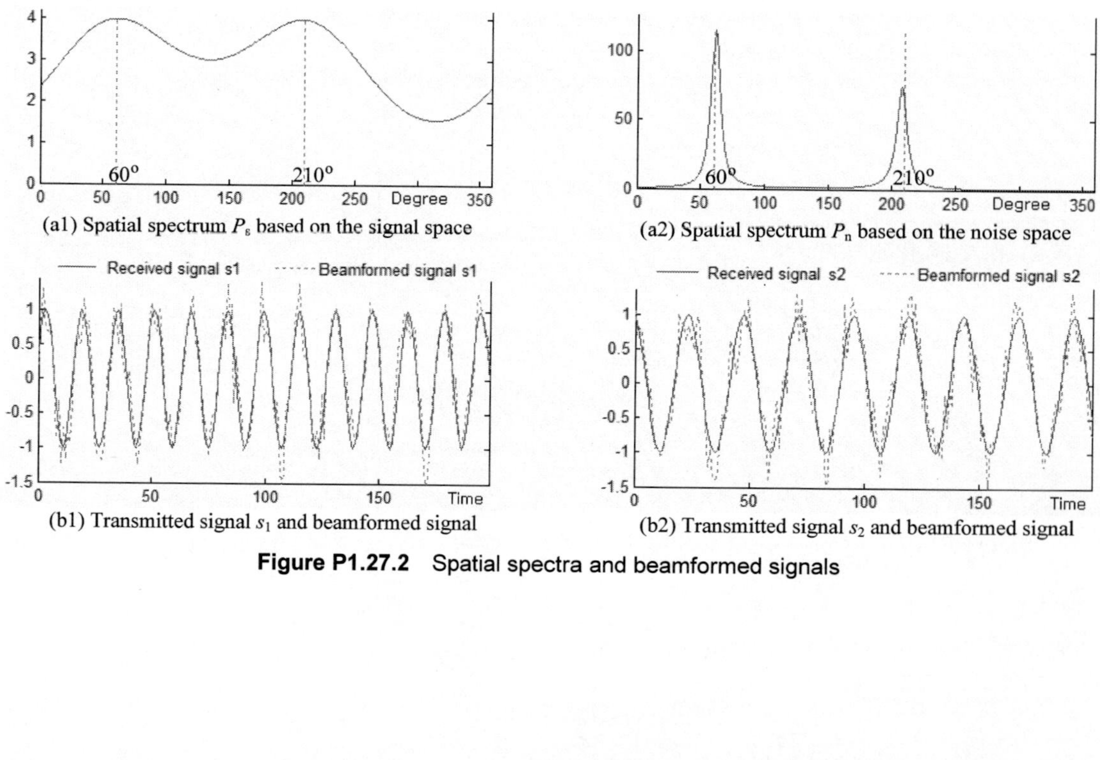

(a1) Spatial spectrum P_{s} based on the signal space

(a2) Spatial spectrum P_{n} based on the noise space

(b1) Transmitted signal s_1 and beamformed signal

(b2) Transmitted signal s_2 and beamformed signal

Figure P1.27.2 Spatial spectra and beamformed signals

```
%DoA_estimation_using_eigendecomposition.m
% Use the eigendecomposition to estimate the DoA (Degree of Arrival)
% with UCA (uniform circular array)
clear, clf
b=1; R=1;  % Wave propagation factor 2*pi/wavelength
Ne=4; Ns=2;  % Number of antenna elements and Number of signals
% True DoAs (azimuth angles) in degree/radian
phs_deg = [60 210]; phs = phs_deg*pi/180;
Nd=200; % Number of data used for DoA estimation
% Steering function for azimuth angle ph
a = @(ph,th)exp(j*b*R*sin(th)*cos(ph-[0:Ne-1]'*2*pi/Ne));
for m=1:Ns
   A(:,m) = a(phs(m),pi/2); % LxM steering matrix
end
for n=1:Nd  % Make a noisy received signal with some directivity
   s(:,n) = [sin(pi/8*n); cos(pi/12*n)]; % Received signal
   y(:,n) = A*s(:,n)+0.5*(randn(Ne,1)+j*randn(Ne,1)); % Eq.(P1.27.1)
end
Ry = y*y'/Nd; % UCA output signal correlation matrix  Eq.(P1.27.4)
[V,Lam]=eig(Ry) % Eigendecomposition of Ry: Eq.(P1.27.5)
% partition modal matrix V into Noise/Signal subspace matrices N,S
N = V(:,1:Ne-Ns);      % Noise sub-eigenspace matrix
S = V(:,Ne-Ns+1:Ne); % Signal sub-eigenspace matrix
phph = 0:359; % Whole coverage of azimuth angle
for i=1:length(phph) % To construct the spatial spectra
   phi_rad = phph(i)*pi/180; a_phi = a(phi_rad,pi/2);
   Ps(i)=real(a_phi'*S*?'*a_phi);    % Eq.(P1.27.6)
   Pn(i)=1/real(a_phi'*?*N'*a_phi); % Eq.(P1.27.7)
end
spsm = max(Ps);  spnm = max(Pn);
subplot(221), plot(phph,Ps) % plot the spatial spectrum based on S
subplot(222), plot(phph,Pn) % plot the spatial spectrum based on N
[lmaxima,indices] = localmax(Pn,[],false); % Find the local maxima
ph_est = phph(indices) % Estimates of azimuth angles
```

```
%try_beamforming.m
for m=1:Ns % NexNs steering matrix with DoA estimates
   Ae(:,m) = a(ph_est(m)*pi/180,pi/2); % Eq.(P1.27.8)
end
W = pinv(Ae); % (Ae'*Ae)^-1*Ae' : NsxNe beamformer gain matrix
sh = real(?*y); % Eq.(P1.27.8): Beamformed signal
nn=[0:Nd-1]; % Time vector
subplot(223), plot(nn,s(1,:), nn,sh(1,:),'r:')
legend('Received signal s1','Beamformed signal s1')
subplot(224), plot(nn,s(2,:), nn,sh(2,:),'r:')
legend('Received signal s2','Beamformed signal s2')
```

(b) To reconstruct the received signals s_1 and s_2 from the UCA output signal vector **y**, beamforming with the estimated DoA is needed, and it can be done by premultiplying the UCA output signal vector **y** by the left pseudo inverse of the (estimated) steering matrix $A_e(\hat{\boldsymbol{\phi}})$ as

$$\hat{\mathbf{s}}(n) = \overset{N_s\times N_e \text{ matrix}}{W(\hat{\boldsymbol{\phi}})}\mathbf{y}(n) = \left[A_e(\hat{\boldsymbol{\phi}})^{*T} A_e(\hat{\boldsymbol{\phi}})\right]^{-1} A_e(\hat{\boldsymbol{\phi}})^{*T}\mathbf{y}(n) \text{ with } A_e(\hat{\boldsymbol{\phi}}) = \overset{N_e\times N_s \text{ matrix}}{\left[\mathbf{a}(\hat{\phi}_1)\ \mathbf{a}(\hat{\phi}_2)\right]} \quad \text{(P1.27.8)}$$

Together with the received signals $s_1(n)$ and $s_2(n)$, their reconstructions $\hat{s}_1(n)$ and $\hat{s}_2(n)$ obtained through beamforming like this are shown in Figs. P1.27.2(b1) and (b2), respectively.

To simulate this receive-beamforming process, complete and run the above MATLAB script "try_beamforming.m" to get the plots like Figs. P1.27.2(b1) and (b2).

1.28 SVD and *Principal Component Analysis* (*PCA*)

PCA is a technique that can be used to simplify/compress data consisting of many values of several variables by reducing the dimensionality of the data, i.e., by finding a smaller set of (new) variables (called principal components) that retain most of the information about the data.

Let a set of N (real-valued) data be observed for each of M ($<N$) variables (properties) $\{x_1,\ldots,x_M\}$ so that an $N\times M$ data matrix X can be formed. PCA on the data matrix X is performed step by step as follows:

(Step 1) Detrend the data matrix by subtracting the mean of each column $\mathbf{x}_m$ from every piece of data in the column to get a detrended data matrix $\underline{X}$, each column of which has zero mean.

(Step 2) Find the SVD of $\underline{X}$:

$$\underline{X} = USW^{\mathrm{T}} = \overset{N\times N}{\begin{bmatrix}\mathbf{u}_1 & \mathbf{u}_2 & \cdots & \mathbf{u}_N\end{bmatrix}} \overset{N\times M}{\begin{bmatrix} s_1 & 0 & 0 & 0 \\ 0 & s_2 & 0 & 0 \\ 0 & 0 & \ddots & 0 \\ 0 & 0 & 0 & s_M \\ 0 & 0 & \cdots & 0 \\ 0 & 0 & \cdots & 0 \end{bmatrix}} \overset{M\times M}{\begin{bmatrix}\mathbf{w}_1^{\mathrm{T}} \\ \mathbf{w}_2^{\mathrm{T}} \\ \vdots \\ \mathbf{w}_M^{\mathrm{T}}\end{bmatrix}} = \sum_{m=1}^{M} s_m \mathbf{u}_m \mathbf{w}_m^{\mathrm{T}} \tag{P1.28.1}$$

with the singular values sorted in descending order:

$$s_1 \ge s_2 \ge \cdots \ge s_M \ge 0 \tag{P1.28.2}$$

(Step 3) Take the partial sum up to the pth principal components of (P1.28.1) (corresponding to the largest $p(<M)$ s_m's) to get the compressed data Y_{1p} (called the score matrix of size $N\times p$):

$$Y_{1p} = \sum_{m=1}^{p} s_m \mathbf{u}_m = \begin{bmatrix}\mathbf{u}_1 & \mathbf{u}_2 & \cdots & \mathbf{u}_p\end{bmatrix} \begin{bmatrix} s_1 & 0 & 0 & 0 \\ 0 & s_2 & 0 & 0 \\ 0 & 0 & \ddots & 0 \\ 0 & 0 & 0 & s_p \end{bmatrix} = U(:,1:p)S(1:p,1:p) \tag{P1.28.3}$$

This can also be obtained by post-multiplying $\underline{X}$ by the coefficients $W(:,1:p)$

$$Y_{1p} = \underline{X}\begin{bmatrix}\mathbf{w}_1 & \mathbf{w}_2 & \cdots & \mathbf{w}_p\end{bmatrix} = \underline{X}W(:,1:p) \tag{P1.28.4}$$

since

$$\underline{X}W = USW^{\mathrm{T}}W \underset{W^{\mathrm{T}}W=I}{\overset{W:\text{ orthonormal}}{=}} US \tag{P1.28.5}$$

(Step 4) If necessary, the original data can be reconstructed by postmultiplying this (compressed) data matrix Y_{1p} by W_{1p}^{T}:

$$\hat{\underline{X}}_p = Y_{1p}W_{1p}^{\mathrm{T}} \text{ with } W_{1p}^{\mathrm{T}} = \overset{p\times M}{\begin{bmatrix}\mathbf{w}_1^{\mathrm{T}} \\ \mathbf{w}_2^{\mathrm{T}} \\ \vdots \\ \mathbf{w}_p^{\mathrm{T}}\end{bmatrix}} = W(:,1:p)^{\mathrm{T}} \tag{P1.28.6}$$

Note that since

$$\underline{X}^{\mathrm{T}}\underline{X} = (USW^{\mathrm{T}})^{\mathrm{T}}USW^{\mathrm{T}} = WS^{\mathrm{T}}U^{\mathrm{T}}USW^{\mathrm{T}} = WS^2W^{\mathrm{T}} \tag{P1.28.7}$$

the "right-singular" matrix W of $\underline{X}$ and S^2 equals the modal matrix (consisting of the eigenvectors) and the "eigenvalue-diagonal" matrix (having the eigenvalues on its diagonal) of $\underline{X}^T\underline{X}$, respectively. That is why the PCA problem can be solved by using the eigendecomposition (Section 1.2.11) of $\underline{X}^T\underline{X}$ as well as by using the SVD (Section 1.2.10.4) of the data matrix $\underline{X}$.

(a) Complete and run the MATLAB script "do_PCA.m" to try the above PCA for a 10×3 data matrix and see that the *Frobenius norm* of the reconstruction error matrix

$$||E_p||_F^2 = \sum_{i=1}^{N}\sum_{j=1}^{N} E_p^2(i,j) \overset{(1.2.11f)}{=} \text{trace}\{E_p E_p^T\} = \sum_{m=p+1}^{M} s_m^2\, \mathbf{u}_m \mathbf{w}_m^T \mathbf{w}_m \mathbf{u}_m^T = \sum_{m=p+1}^{M} s_m^2 \quad \text{(P1.28.8)}$$

$$\left(\because\ E_p = \hat{\underline{X}}_p - \underline{X} \overset{(P1.28.6)}{=} Y_{1p} W_{1p}^T - \underline{X} \underset{(P1.28.1)}{\overset{(P1.28.4)}{=}} \sum_{m=1}^{p} s_m \mathbf{u}_m \mathbf{w}_m^T - \sum_{m=1}^{M} s_m \mathbf{u}_m \mathbf{w}_m^T = -\sum_{m=p+1}^{M} s_m \mathbf{u}_m \mathbf{w}_m^T \right)$$

becomes smaller as the number p of principal components increases the more noticeably, the larger s_{p+1} is. From the above equations, find the notations for Xbh, Xb, rec_err, and the three output arguments {coefs, scores, variances} of the MATLAB built-in function '`princomp()`' or '`pca()`' (available in MATLAB 2015a).

```
%do_PCA.m
N=10; % Data size
X(:,[1 3])=randn(N,2); X(:,2)=X(:,1)/2+0.1*randn(N,1);
[N,M] = size(X); Xm=repmat(mean(X),N,1);
Xb = X-Xm; % Demean the data
[U,S,W] = svd(??); % SVD
rec_err(1)=trace(Xb'*Xb); % Actually, no reconstruction
discrepancy0=rec_err(1)-norm(Xb,'fro')^2 % Square of Frobenius norm?
for p=1:min(M,14)
   W1p=W(:,1:p); % Transformation matrix
   Y1p = U(:,1:p)*?(1:p,1:p); % Data compression (P1.28.3)
   Y1p1 = Xb*???; % Data compression (P1.28.4)
   Discrepancy_between_Y1p_and_Y1p1 = norm(Y1p-Y1p1)
   Xbh = ???*W1p'; % Data reconstruction (P1.28.6)
   er = trace((Xbh-Xb)*(Xbh-Xb)'); % Reconstruction error (P1.28.?)
   rec_err(p+1,:) = er;
end
% PCA analysis using MATLAB function princomp()/pca() if available
[coefs,scores,variances] = princomp(Xb);
% Full reconstruction error
full_rec_err_pca = trace((scores*coefs'-Xb)*(scores*coefs'-Xb)')
% To see the relationship among scores, Y1p, coefs, and W
[scores Y1p], [coefs W(:,1:p)]
ss=diag(S).^2; % Squares of singular values
% Relationship among s^2, variance, and reconstruction error
[ss(1:p) variances(1:p)*(N-1) -diff(rec_err(1:p+1)) rec_err(1:p)]
```

(b) To get a real feeling about using the PCA for data compression, make a data matrix X from a black and white image like '1245.bmp' and try the PCA for it by inserting the following statements in appropriate places and running the modified script:

```
X=imread('1245','bmp');
subplot(351), imshow(X)
subplot(3,5,p+1), imshow(Xbh+Xm)
```

Then, in addition to the original image, you will see images $\{\hat{\underline{X}}_p,\ p=1\sim14\}$ of quality improving as p increases from 1 to 14. Which one does the reconstruction error decrease by less than 0.1×10^3 from? Does it conform with your inspection of the images?

CHAPTER 2

Vector Calculus

CHAPTER OUTLINE

2.1 DERIVATIVES

The *derivative* of a function $y=f(x)$ of one variable x is defined as

$$\frac{d}{dx}f(x) = \lim_{\Delta x\to 0}\frac{f(x+\Delta x)-f(x)}{\Delta x} \tag{2.1.1}$$

This is often replaced by one of the following approximations:

$$\text{Forward difference:} \quad \frac{f(x+h)-f(x)}{h} \tag{2.1.2a}$$

$$\text{Backward difference:} \quad \frac{f(x)-f(x-h)}{h} \tag{2.1.2b}$$

$$\text{Central difference:} \quad \frac{f(x+h)-f(x-h)}{2h} \tag{2.1.2c}$$

The *partial derivatives* of a multi-variable function, say, $f(x, y)$ of two variables x and y, with respect to x and y, are defined as

$$\frac{\partial}{\partial x}f(x,y) = \lim_{\Delta x\to 0}\frac{f(x+\Delta x,y)-f(x,y)}{\Delta x} \tag{2.1.3a}$$

$$\frac{\partial}{\partial y}f(x,y) = \lim_{\Delta y\to 0}\frac{f(x,y+\Delta y)-f(x,y)}{\Delta y} \tag{2.1.3b}$$

<Chain Rule>

Let $y=f(u)$ be a differentiable function of a variable u where $u=g(x)$ is a function of another variable x. Then the *chain rule* states that the derivative of y w.r.t. x is

$$\frac{dy}{dx}=\frac{dy}{du}\frac{du}{dx} \tag{2.1.4}$$

If $z=f(u,v)$ is a differentiable function of two variables u and v where $u=g(x,y)$ and $v=h(x,y)$ are functions of two other variables x and y and they have continuous partial derivatives, then the *chain rule* states that the partial derivatives of z w.r.t. x and y are

$$\frac{\partial z}{\partial x}=\frac{\partial z}{\partial u}\frac{\partial u}{\partial x}+\frac{\partial z}{\partial v}\frac{\partial v}{\partial x} \text{ and } \frac{\partial z}{\partial y}=\frac{\partial z}{\partial u}\frac{\partial u}{\partial y}+\frac{\partial z}{\partial v}\frac{\partial v}{\partial y}, \tag{2.1.5}$$

<Gradient>

The (1D) *gradient* of a one-variable function $f(x)$ can be defined as

$$\nabla f(x)=\mathbf{u}_x\frac{\partial f}{\partial x} \tag{2.1.6}$$

The (2D) *gradient* of a two-variable function $f(x,y)$ is defined as

$$\nabla f(x,y)=\mathbf{u}_x\frac{\partial f}{\partial x}+\mathbf{u}_y\frac{\partial f}{\partial y} \tag{2.1.7}$$

Likewise, the (3D) *gradient* of a three-variable function $f(x,y,z)$ is defined as

$$\nabla f(x,y,z)=\mathbf{u}_x\frac{\partial f}{\partial x}+\mathbf{u}_y\frac{\partial f}{\partial y}+\mathbf{u}_z\frac{\partial f}{\partial z} \tag{2.1.8}$$

where ∇ (called "del" or "nabla") denotes the *vector differential operator* defined (in the 3D Cartesian coordinate system) as

$$\nabla=\mathbf{u}_x\frac{\partial}{\partial x}+\mathbf{u}_y\frac{\partial}{\partial y}+\mathbf{u}_z\frac{\partial}{\partial z} \quad (\mathbf{u}_x, \mathbf{u}_y, \mathbf{u}_z\text{: unit vectors along the } x\text{-, } y\text{-, } z\text{-axes, respectively}) \tag{2.1.9}$$

Note that the gradient points to the direction in which a point should move to increase the value of the function with the maximum rate.

<Directional Derivative>

The *directional derivative* of $f(x,y,z)$ in the direction of a unit vector $\mathbf{u}=\alpha_x\mathbf{u}_x+\alpha_y\mathbf{u}_y+\alpha_z\mathbf{u}_z$ with $\|\mathbf{u}\|=1$ is the dot product of the gradient and the unit vector:

$$\nabla_{\mathbf{u}}f(x,y,z)=\lim_{t\to 0}\frac{f(x+t\alpha_x,y+t\alpha_y,z+t\alpha_z)-f(x,y,z)}{t}$$

$$\overset{\text{Chain rule}}{=}\frac{\partial f}{\partial x}\frac{\partial(x+t\alpha_x)}{\partial t}+\frac{\partial f}{\partial y}\frac{\partial(y+t\alpha_y)}{\partial t}+\frac{\partial f}{\partial z}\frac{\partial(z+t\alpha_z)}{\partial t}=\nabla f(x,y,z)\bullet\mathbf{u} \tag{2.1.10}$$

Note that the magnitude of the directional derivative is maximized when $\mathbf{u}$ is aligned with ∇f.

<Tangent Line/Plane>

The equation of the *tangent line* of a curve described by $y=f(x)$ at a point (x_1, y_1) can be written as

$$y = y_1 + \left.\frac{\partial f}{\partial x}\right|_{x=x_1}(x-x_1) \tag{2.1.11}$$

The equation of the *tangent line* of a (level) curve described by $f(x, y)=c$ at a point (x_1, y_1) can be written as

$$\left.\frac{\partial f}{\partial x}\right|_{(x,y)=(x_1,y_1)}(x-x_1) + \left.\frac{\partial f}{\partial y}\right|_{(x,y)=(x_1,y_1)}(y-y_1) = 0 \tag{2.1.12}$$

Likewise, the equation of the *tangent plane* of a (level) surface described by $f(x,y,z)=c$ at a point (x_1, y_1, z_1) can be written as

$$\left.\frac{\partial f}{\partial x}\right|_{(x,y,z)=(x_1,y_1,z_1)}(x-x_1) + \left.\frac{\partial f}{\partial y}\right|_{(x,y,z)=(x_1,y_1,z_1)}(y-y_1) + \left.\frac{\partial f}{\partial z}\right|_{(x,y,z)=(x_1,y_1,z_1)}(z-z_1) = 0 \tag{2.1.13}$$

Eqs. (2.1.12) and (2.1.13) can be derived from Eq. (1.1.39) with the gradient as a normal vector to the tangent plane. Eq. (2.1.11) is identical to Eq. (2.1.12) with $f(x)-y=0$ in place of $f(x,y)=c$.

To realize the geometric interpretations of the gradient and tangent plane, let us see the following example:

(Example 2.1) Gradient and Tangent Line/Plane

(a) Consider a level curve described by

$$f(x,y) = x^2 - y = 4 \tag{E2.1a.1}$$

We can apply Eq. (2.1.7) to get the *gradient* of $f(x,y)=x^2-y$ at $\mathbf{p}_1=[2\ \ 0]^T$:

$$\nabla f(x,y) \overset{(2.1.7)}{=} \mathbf{u}_x\frac{\partial f}{\partial x} + \mathbf{u}_y\frac{\partial f}{\partial y} = 2x\,\mathbf{u}_x - \mathbf{u}_y \overset{(x,y)=(2,0)}{=} 4\mathbf{u}_x - \mathbf{u}_y \tag{E2.1a.2}$$

We can use Eq. (2.1.12) to get the equation of the *tangent line* to the level curve at the point $p_1=(2,0)$ (belonging to the level curve), which must be perpendicular to the gradient, as

$$\left.\frac{\partial f}{\partial x}\right|_{(x,y)=(x_1,y_1)}(x-x_1) + \left.\frac{\partial f}{\partial y}\right|_{(x,y)=(x_1,y_1)}(y-y_1) \overset{(2.1.12)}{=} 0;\ 4(x-2)-(y-0)=0;\ y=4x-8 \tag{E2.1a.3}$$

Fig. 2.1(a) shows the gradient vector and tangent line at $p_1=(2,0)$. Fig. 2.1(b) shows the contours for $f(x,y)=x^2-y=c$ (with $c=-1,0,\cdots,5$) and the gradient vectors (perpendicular to the contours) at many grid points as obtained using the MATLAB function '`gradient()`'.

(b) Consider a surface described by

$$f(x,y,z) = x^2 - y - 4z = 0 \tag{E2.1b.1}$$

We can apply Eq. (2.1.8) to get the *gradient* of $f(x,y,z)=x^2-y-4z$ at $\mathbf{p}_1=[2\ \ 0\ \ 1]^T$:

$$\nabla f(x,y,z) = \mathbf{u}_x\frac{\partial f}{\partial x} + \mathbf{u}_y\frac{\partial f}{\partial y} + \mathbf{u}_z\frac{\partial f}{\partial z} \overset{(E2.1b.1)}{=} 2x\,\mathbf{u}_x - \mathbf{u}_y - 4\mathbf{u}_z \underset{=(2,0,1)}{\overset{(x,y,z)}{=}} 4\mathbf{u}_x - \mathbf{u}_y - 4\mathbf{u}_z \tag{E2.1b.2}$$

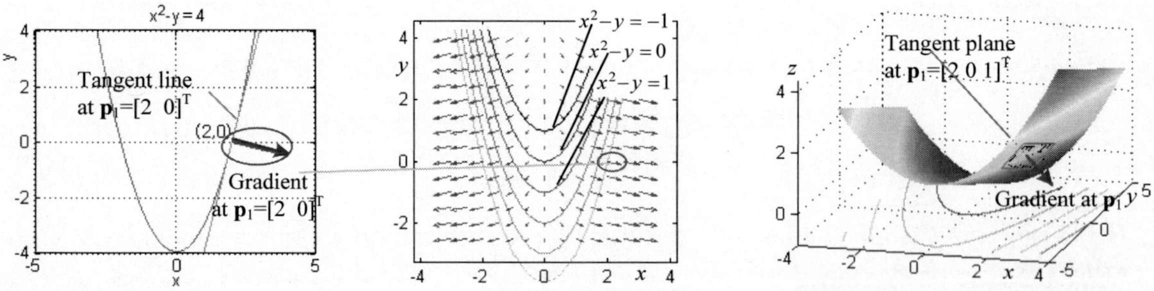

(a) Gradient and tangent plane to $x^2-y=4$ (b) Gradients on the contour (c) Gradient and tangent plane to $x^2-y-4z=0$

Figure 2.1 Example of gradient and tangent line/plane

```
%em02e01.m
% Gradients and Tangent Planes for Example 2.1 (Fig.2.1)
clear, clf
% (a) Tangent line to a level curve x^2-y=4
p1=[2; 0]; % An arbitrary point on the curve
grad_f=[2*p1(1); -1];  % Gradient at p1 by Eq.(E2.1a.2)
subplot(131)
x=p1(1)+[-1 1]; % Domain on which to plot the tangent line
y=-grad_f(1)/grad_f(2)*(x-p1(1))+p1(2); % Tangent line Eq.(E2.1a.3)
ezplot('x^2-y=4'), hold on   % Level curve
plot(x,y,'r') % Tangent line
quiver(p1(1),p1(2),grad_f(1),grad_f(2),0.5,'linewidth',2)
axis('equal'), grid on
% (b) Using gradient()
xs=[-4:0.05:4]; ys=[-4:0.05:6];
[Xs,Ys]=meshgrid(xs,ys); % Grids for surface
Fs = Xs.^2-Ys; % Level surface
subplot(132)
contour(xs,ys,Fs,[-1 0 1 2 3 4 5]), hold on
dx=0.5; dy=0.5;
x=[-3:dx:3]; y=[-2.5:dy:4]; [X,Y]=meshgrid(x,y);
F = X.^2-Y; % Level surface where (0-(-2.5))/0.5+1=6,((2-(-3))/0.5+1=11
[dFx,dFy] = gradient(F,dx,dy);
fprintf(' Gradient at (%3.1f,%3.1f) is [%3.1f %3.1f]\n',...
    X(6,11),Y(6,11),dFx(6,11),dFy(6,11))
quiver(x,y,dFx,dFy), axis('equal')
% (c) Tangent plane to a level surface x^2-y-4*z=0
p1=[2; 0; 1]; % An arbitrary point on the surface
grad_f=[2*p1(1);-1;-4];  % Gradient at p1 by Eq.(E2.1b.2)
xs=[-3:0.02:4]; ys=[-4:0.02:4];
[Xs,Ys]=meshgrid(xs,ys); % Grids for surface
Zs = (Xs.^2-Ys)/4; % Level surface
subplot(133)
x=[1.6:0.2:2.4]; y=[-2:0.5:0.2];
[Xp,Yp]=meshgrid(x,y); % Grids for tangent plane
Zp=(grad_f'*p1-grad_f(1)*Xp-grad_f(2)*Yp)/grad_f(3); % Eq.(E2.1b.3)
meshc(Xs,Ys,Zs), hold on   % Level surface
surf(Xp,Yp,Zp) % Tangent plane
quiver3(p1(1),p1(2),p1(3),grad_f(1),grad_f(2),grad_f(3),0.2)
axis('equal'), view(15,15)
```

We can use Eq. (2.1.13) to get the equation of the *tangent plane* to the level surface at the point $p_1=(2,0,1)$ (belonging to the surface), which must be perpendicular to the gradient, as

$$\left.\frac{\partial f}{\partial x}\right|_{(x,y,z)=(x_1,y_1,z_1)}(x-x_1)+\left.\frac{\partial f}{\partial y}\right|_{(x,y,z)=(x_1,y_1,z_1)}(y-y_1)+\left.\frac{\partial f}{\partial z}\right|_{(x,y,z)=(x_1,y_1,z_1)}(z-z_1)=0$$

$$;\,2x_1(x-x_1)-(y-y_1)-4(z-z_1)=4(x-2)-(y-0)-4(z-1)=0;\ 4x-y-4z-4=0 \quad \text{(E2.1b.3)}$$

Fig. 2.1(c) shows the gradient vector (Eq. (E2.1b.2)) and tangent plane (Eq. (E2.1b.3)) where the gradient points to the direction in which a point should move to increase the value of the function. It was obtained by running the above MATLAB script "em02e01.m" where '[dFx,dFy]= gradient(F,dx,dy)' is used to compute $\partial f/\partial x$ and $\partial f/\partial y$ at each point on the grid specified by (X,Y) as follows:

$$\mathrm{X}=\begin{bmatrix}\ddots & \vdots & \vdots & \vdots & \iddots\\ \cdots & 1.5 & 2.0 & 2.5 & \cdots\\ \cdots & 1.5 & \boxed{2.0} & 2.5 & \cdots\\ \cdots & 1.5 & 2.0 & 2.5 & \cdots\\ \iddots & \vdots & \vdots & \vdots & \ddots\end{bmatrix},\quad \mathrm{Y}=\begin{bmatrix}\ddots & \vdots & \vdots & \vdots & \iddots\\ \cdots & -0.5 & -0.5 & -0.5 & \cdots\\ \cdots & 0.0 & \boxed{0.0} & 0.0 & \cdots\\ \cdots & 0.5 & 0.5 & 0.5 & \cdots\\ \iddots & \vdots & \vdots & \vdots & \ddots\end{bmatrix},\quad \mathrm{Z}=\begin{bmatrix}\ddots & \vdots & \vdots & \vdots & \vdots & \vdots & \iddots\\ \cdots & 2.0 & 3.25 & 5.0 & 7.25 & 10.0 & \cdots\\ \cdots & 1.5 & 2.75 & \boxed{4.5} & \boxed{6.75} & 9.5 & \cdots\\ \cdots & 1.0 & \boxed{2.25} & 4.0 & \boxed{6.25} & 9.0 & \cdots\\ \cdots & 0.5 & 1.75 & \boxed{3.5} & 5.75 & 8.5 & \cdots\\ \cdots & 0.0 & 1.25 & 3.0 & 5.25 & 8.0 & \cdots\\ \iddots & \vdots & \vdots & \vdots & \vdots & \vdots & \ddots\end{bmatrix}$$

$$\mathrm{dFx}=\begin{bmatrix}\ddots & \vdots & \vdots & \vdots & \iddots\\ \cdots & \frac{4.5-1.5}{2\times0.5}=3 & \frac{6.75-2.75}{2\times0.5}=4 & \frac{9.5-4.5}{2\times0.5}=5 & \cdots\\ \cdots & \frac{4.0-1.0}{2\times0.5}=3 & \frac{\boxed{6.25}-\boxed{2.25}}{2\times0.5}=\boxed{4} & \frac{9.0-4.0}{2\times0.5}=5 & \cdots\\ \cdots & \frac{3.5-0.5}{2\times0.5}=3 & \frac{5.75-1.75}{2\times0.5}=4 & \frac{8.5-3.5}{2\times0.5}=5 & \cdots\\ \iddots & \vdots & \vdots & \vdots & \ddots\end{bmatrix},\quad \mathrm{dFy}=\begin{bmatrix}\ddots & \vdots & \vdots & \vdots & \iddots\\ \cdots & \frac{2.25-3.25}{2\times0.5}=-1 & \frac{4.0-5.0}{2\times0.5}=-1 & \frac{6.25-7.25}{2\times0.5}=-1 & \cdots\\ \cdots & \frac{1.74-2.75}{2\times0.5}=-1 & \frac{\boxed{3.5}-\boxed{4.5}}{2\times0.5}=\boxed{-1} & \frac{5.75-6.75}{2\times0.5}=-1 & \cdots\\ \cdots & \frac{1.25-2.25}{2\times0.5}=-1 & \frac{3.0-4.0}{2\times0.5}=-1 & \frac{5.25-6.25}{2\times0.5}=-1 & \cdots\\ \iddots & \vdots & \vdots & \vdots & \ddots\end{bmatrix}$$

(Example 2.2) (Negative) Gradient of an Electric Potential to Get the Electric Field

The electric potential V formed by two positive/negative point charges $+Q/-Q$ (of equal magnitude and opposite polarity), each located at $(0, +d/2, 0)$ and $(0, -d/2, 0)$, respectively, is

$$V=\frac{Q}{4\pi\varepsilon_0}\left(\frac{1}{R_1}-\frac{1}{R_2}\right)=\frac{Q}{4\pi\varepsilon_0}\left(\frac{1}{\sqrt{x^2+(y-d/2)^2}}-\frac{1}{\sqrt{x^2+(y+d/2)^2}}\right) \quad \text{(E2.2.1)}$$

The electric field $\mathbf{E}$ can be obtained as the negative gradient of V:

$$\mathbf{E}=-\nabla V(x,y)\overset{(2.1.7)}{=}-\mathbf{u}_x\frac{\partial V}{\partial x}-\mathbf{u}_y\frac{\partial V}{\partial y}$$

$$\overset{(E2.2.1)}{=}\frac{Q}{4\pi\varepsilon_0}\left\{\left(\frac{x}{(x^2+(y-d/2)^2)^{3/2}}-\frac{x}{(x^2+(y+d/2)^2)^{3/2}}\right)\mathbf{u}_x+\left(\frac{y-d/2}{(x^2+(y-d/2)^2)^{3/2}}-\frac{y+d/2}{(x^2+(y+d/2)^2)^{3/2}}\right)\mathbf{u}_y\right\} \quad \text{(E2.2.2)}$$

The MATLAB program "em02e02.m" has been run to plot $V(x, y)$ and $\mathbf{E}$ as depicted in Figs. 2.2(a) and (b), respectively.

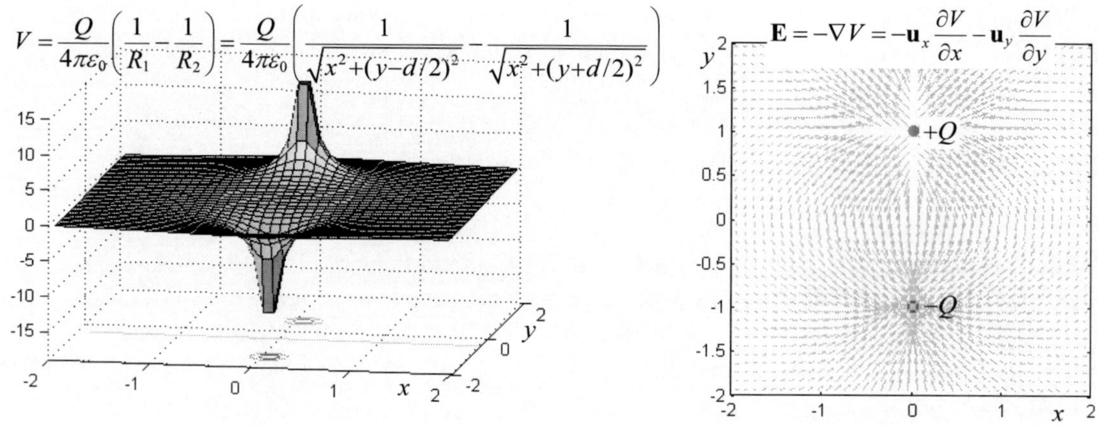

(a) Electric potential V (b) Electric field **E**= -grad V

Figure 2.2 Electric potential and field produced by an electric dipole

```
%em02e02.m: Example 2.2 (Fig. 2.2)
clear, clf
Vxy=@(x,y,Q,d)Q./sqrt(x.^2+(y-d/2).^2)-Q./sqrt(x.^2+(y+d/2).^2);%Eq.(E2.2.1)
dx=0.1; dy=0.1; x=[-2+dx/2:dx:2]; y=[-2+dx/2:dy:2];
[X,Y]=meshgrid(x,y); % Grid matrices
Q=1; d=2; V=Vxy(X,Y,Q,d); % Electric potential Eq.(E2.2.1)
R1 = sqrt(X.^2+(Y-d/2).^2)+1e-8; R2 = sqrt(X.^2+(Y+d/2).^2)+1e-8;
Ex = X./R1.^3-X./R2.^3; Ey = (Y-d/2)./R1.^3-(Y+d/2)./R2.^3; % Eq.(E2.2.2)
a=1; % To saturate the length of arrow describing the electric field
Ex = sign(Ex).*(1-exp(-abs(Ex)/a)); Ey = sign(Ey).*(1-exp(-abs(Ey)/a));
subplot(121), surfc(X,Y,V), view(10,15)
subplot(122), quiver(X,Y,Ex,Ey,2,'g')
```

2.2 VECTOR FUNCTIONS

2D and 3D *vector(-valued) functions* are written as

$$\mathbf{p}(t) = x(t)\mathbf{u}_x + y(t)\mathbf{u}_y \tag{2.2.1a}$$

and

$$\mathbf{p}(t) = x(t)\mathbf{u}_x + y(t)\mathbf{u}_y + z(t)\mathbf{u}_z, \tag{2.2.1b}$$

respectively. They can be drawn as 2D and 3D curves, possibly with an arrowhead indicating the moving direction of the point $\mathbf{p}(t)$ as the parameter t increases.

The *tangent line* to the curve (representing $\mathbf{p}(t)$) at a point $\mathbf{p}(t_1)$ is the line through the point $\mathbf{p}(t_1)$ parallel to the derivative, called a *tangent vector*,

$$\mathbf{p}'(t) = \frac{d\mathbf{p}(t)}{dt} \overset{(2.2.1a)}{=} \frac{dx(t)}{dt}\mathbf{u}_x + \frac{dy(t)}{dt}\mathbf{u}_y = x'(t)\mathbf{u}_x + y'(t)\mathbf{u}_y \tag{2.2.2a}$$

or

$$\mathbf{p}'(t) = \frac{d\mathbf{p}(t)}{dt} \overset{(2.2.1b)}{=} \frac{dx(t)}{dt}\mathbf{u}_x + \frac{dy(t)}{dt}\mathbf{u}_y + \frac{dz(t)}{dt}\mathbf{u}_z = x'(t)\mathbf{u}_x + y'(t)\mathbf{u}_y + z'(t)\mathbf{u}_z \quad (2.2.2b)$$

The *arc length* of the curve (representing $\mathbf{p}(t)$) from $\mathbf{p}(t_0)$ and $\mathbf{p}(t_1)$ can be obtained by integrating the magnitude of the derivative (2.2.2) over $[t_0, t_1]$ as

$$l = \int_{t_0}^{t_1} \|\mathbf{p}'(t)\|\,dt = \int_{t_0}^{t_1} \sqrt{(x'(t))^2 + (y'(t))^2 + (z'(t))^2}\,dt \quad (2.2.3)$$

only when $\mathbf{p}(t)$ and the corresponding curve are *smooth* in the sense that the component functions $x(t)$, $y(t)$, and $z(t)$ have continuous first derivatives that are not all zero for the open interval (t_0, t_1). Note that the vector function $\mathbf{p}(t)$ can be parameterized as $\mathbf{p}(l)$ in terms of the arc length

$$l = \int_0^t \|\mathbf{p}'(\tau)\|\,d\tau = \int_0^t \sqrt{(x'(\tau))^2 + (y'(\tau))^2 + (z'(\tau))^2}\,d\tau = h(t) \quad (2.2.4)$$

by substituting the inverse function $t=h^{-1}(l)$ of $l=h(t)$ for t into Eq. (2.2.1), as illustrated in Example 2.3(c).

Before ending this section, let us note the following rules about the differentiation of differentiable vector functions $\mathbf{p}(t)$, $\mathbf{p}_1(t)$, and $\mathbf{p}_2(t)$ where $q(t)$ is a differentiable scalar function:

$$\frac{d}{dt}\big(q(t)\mathbf{p}(t)\big) = \left(\frac{d}{dt}q(t)\right)\mathbf{p}(t) + q(t)\left(\frac{d}{dt}\mathbf{p}(t)\right) = q'(t)\mathbf{p}(t) + q(t)\mathbf{p}'(t) \quad (2.2.5a)$$

$$\frac{d}{dt}\big(\mathbf{p}_1(t)+\mathbf{p}_2(t)\big) = \frac{d}{dt}\mathbf{p}_1(t) + \frac{d}{dt}\mathbf{p}_2(t) = \mathbf{p}_1'(t) + \mathbf{p}_2'(t) \quad (2.2.5b)$$

$$\frac{d}{dt}\big(\mathbf{p}_1(t)\cdot\mathbf{p}_2(t)\big) = \left(\frac{d}{dt}\mathbf{p}_1(t)\right)\cdot\mathbf{p}_2(t) + \mathbf{p}_1(t)\cdot\left(\frac{d}{dt}\mathbf{p}_2(t)\right) = \mathbf{p}_1'(t)\cdot\mathbf{p}_2(t) + \mathbf{p}_1(t)\cdot\mathbf{p}_2'(t) \quad (2.2.5c)$$

$$\frac{d}{dt}\big(\mathbf{p}_1(t)\times\mathbf{p}_2(t)\big) = \left(\frac{d}{dt}\mathbf{p}_1(t)\right)\times\mathbf{p}_2(t) + \mathbf{p}_1(t)\times\left(\frac{d}{dt}\mathbf{p}_2(t)\right) = \mathbf{p}_1'(t)\times\mathbf{p}_2(t) + \mathbf{p}_1(t)\times\mathbf{p}_2'(t) \quad (2.2.5d)$$

(Example 2.3) Tangent Line and Arc Length of a Circular Helix

Consider a 3D vector function

$$\mathbf{p}(t) = t\,\mathbf{u}_x + \cos(2\pi t)\,\mathbf{u}_y + \sin(2\pi t)\,\mathbf{u}_z \quad (E2.3.1)$$

Its graph is the circular helix in Fig. 2.3 that is plotted by running the following MATLAB statements:

```
>>t=[0:0.01:2]; % Time vector
>>plot3(t,cos(2*pi*t),sin(2*pi*t)), view(20,10), grid on
```

(a) We use Eq. (1.1.35) to find the *tangent line* of the helix at $\mathbf{p}(1)=[1\ \ 1\ \ 0]^T$ (corresponding to $t=1$) as

$$\mathbf{p} \overset{(1.1.35)}{=} \mathbf{p}(1)+\mathbf{d}t = \mathbf{p}(1) + \mathbf{p}'(t)\big|_{t=1} t$$

$$\overset{(E2.3.1)}{=} \begin{bmatrix} t \\ \cos(2\pi t) \\ \sin(2\pi t) \end{bmatrix}\Bigg|_{t=1} + \begin{bmatrix} 1 \\ -2\pi\sin(2\pi t) \\ 2\pi\cos(2\pi t) \end{bmatrix}\Bigg|_{t=1} t = \begin{bmatrix} 1 \\ 1 \\ 0 \end{bmatrix} + \begin{bmatrix} 1 \\ 0 \\ 2\pi \end{bmatrix} t = \begin{bmatrix} 1+t \\ 1 \\ 2\pi t \end{bmatrix} \quad (E2.3.2)$$

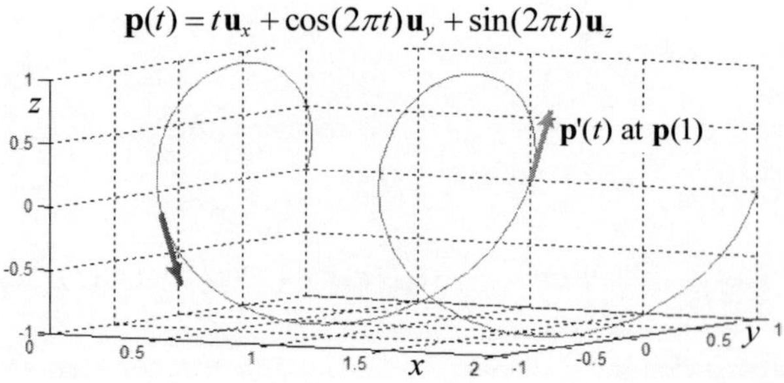

Figure 2.3 Circular helix for Example 2.3

This tangent line for a time segment of length 0.2 can be plotted as a red arrow by running the following MATLAB statements:

```
>>hold on, ts=0.2;
>>quiver3(1,1,0,1*ts,0*ts,2*pi*ts,'r') % Eq.(E2.3.2)
```

(b) Let us find the *arc length* of the helix for time segment [0, 1] by using Eq. (2.2.3) as

$$l \overset{(2.2.3)}{=} \int_0^1 \|\mathbf{p}'(t)\|\,dt = \int_0^1 \sqrt{(x'(t))^2+(y'(t))^2+(z'(t))^2}\,dt$$
$$= \int_0^1 \sqrt{1+(-2\pi\sin 2\pi t)^2+(2\pi\cos 2\pi t)^2}\,dt = \int_0^1 \sqrt{1+4\pi^2}\,dt = \sqrt{1+4\pi^2} = 6.3623 \qquad \text{(E2.3.3)}$$

```
>>dt=0.0001; t=[0:dt:1].';
>>dp=diff([t cos(2*pi*t) sin(2*pi*t)]); % Derivative of vector (E2.3.1)
>>length_arc=sum(sqrt(sum(dp.^2,2))) % 6.3623
>>f=@(t)sqrt(1+(-2*pi*sin(2*pi*t)).^2+(2*pi*cos(2*pi*t)).^2); % Or
>>l=quadl(f,0,1) % (E2.3.3) -> 6.3623
```

(c) Let us parameterize $\mathbf{p}(t)$ in terms of the arc length l defined by Eq. (2.2.4):

$$l = \int_0^t \|\mathbf{p}'(t)\|\,dt = \int_0^t \sqrt{1+(-2\pi\sin 2\pi t)^2+(2\pi\cos 2\pi t)^2}\,d\tau = \sqrt{1+4\pi^2}\,t \qquad \text{(E2.3.4)}$$

We can substitute the inverse function

$$t = l/\sqrt{1+4\pi^2} \qquad \text{(E2.3.5)}$$

into Eq. (E2.3.1):

$$\mathbf{p}(l) = \frac{l}{\sqrt{1+4\pi^2}}\mathbf{u}_x + \cos\left(\frac{2\pi l}{\sqrt{1+4\pi^2}}\right)\mathbf{u}_y + \sin\left(\frac{2\pi l}{\sqrt{1+4\pi^2}}\right)\mathbf{u}_z \qquad \text{(E2.3.6)}$$

Differentiating this w.r.t. l yields the *unit tangent vector* at a point that is l[m] distant from $\mathbf{p}(0)$ along the curve:

$$\mathbf{p}'(l) = \frac{d}{dl}\mathbf{p}(l) = \frac{1}{\sqrt{1+4\pi^2}}\left(\mathbf{u}_x - 2\pi\sin\left(\frac{2\pi l}{\sqrt{1+4\pi^2}}\right)\mathbf{u}_y + 2\pi\cos\left(\frac{2\pi l}{\sqrt{1+4\pi^2}}\right)\mathbf{u}_z\right) \qquad \text{(E2.3.7)}$$

2.3 VELOCITY AND ACCELERATION

If the position of a particle or a body at time t is described by

$$\mathbf{p}(t) = x(t)\mathbf{u}_x + y(t)\mathbf{u}_y + z(t)\mathbf{u}_z \tag{2.3.1}$$

its first and second derivatives

$$\mathbf{v}(t) = \frac{d}{dt}\mathbf{p}(t) \overset{(2.3.1)}{=} \frac{dx(t)}{dt}\mathbf{u}_x + \frac{dy(t)}{dt}\mathbf{u}_y + \frac{dz(t)}{dt}\mathbf{u}_z = x'(t)\mathbf{u}_x + y'(t)\mathbf{u}_y + z'(t)\mathbf{u}_z \tag{2.3.2}$$

$$\mathbf{a}(t) = \frac{d^2\mathbf{p}(t)}{dt^2} \overset{(2.3.1)}{=} \frac{d^2x(t)}{dt^2}\mathbf{u}_x + \frac{d^2y(t)}{dt^2}\mathbf{u}_y + \frac{d^2z(t)}{dt^2}\mathbf{u}_z = x''(t)\mathbf{u}_x + y''(t)\mathbf{u}_y + z''(t)\mathbf{u}_z \tag{2.3.3}$$

are called the *velocity* and *acceleration* of the particle, respectively. The magnitude of the velocity

$$\|\mathbf{v}(t)\| \overset{(2.3.2)}{=} \|\mathbf{p}'(t)\| = \sqrt{\left(\frac{dx(t)}{dt}\right)^2 + \left(\frac{dy(t)}{dt}\right)^2 + \left(\frac{dz(t)}{dt}\right)^2} = \sqrt{(x'(t))^2 + (y'(t))^2 + (z'(t))^2} \tag{2.3.4}$$

is called the *speed* and is equal to the derivative of the arc length (2.2.4):

$$v(t) = \frac{dl}{dt} \overset{(2.2.4)}{=} \frac{d}{dt}\int_0^t \|\mathbf{p}'(t)\|\,dt = \|\mathbf{p}'(t)\| \tag{2.3.5}$$

Note that if the speed is constant so that

$$\|\mathbf{v}\|^2 = \mathbf{v} \bullet \mathbf{v} = c^2 \tag{2.3.6}$$

then we can differentiate this w.r.t. t to write

$$\frac{d}{dt}(\mathbf{v}(t) \bullet \mathbf{v}(t)) \overset{(2.2.5c)}{=} \left(\frac{d}{dt}\mathbf{v}(t)\right) \bullet \mathbf{v}(t) + \mathbf{v}(t) \bullet \left(\frac{d}{dt}\mathbf{v}(t)\right) = 2\mathbf{v}(t) \bullet \left(\frac{d}{dt}\mathbf{v}(t)\right) = 0$$

$$; \quad \mathbf{v}(t) \cdot \mathbf{a}(t) = 0 \tag{2.3.7}$$

This implies that the velocity vector and acceleration vector of a particle moving with a constant speed are perpendicular to each other. For a particle moving with a constant speed $R\omega$ along a circle (on the xy-plane) with radius R, its position vector is described by

$$\mathbf{p}(t) = R\cos(\omega t)\mathbf{u}_x + R\sin(\omega t)\mathbf{u}_y \tag{2.3.8}$$

Thus we can use Eqs. (2.3.2) and (2.3.3) to get the velocity and (centripetal) acceleration vectors as

$$\mathbf{v}(t) \overset{(2.3.2)}{=} \frac{d}{dt}\mathbf{p}(t) \overset{(2.3.8)}{=} -R\omega\sin(\omega t)\mathbf{u}_x + R\omega\cos(\omega t)\mathbf{u}_y \tag{2.3.9}$$

$$\mathbf{a}(t) \overset{(2.3.3)}{=} \frac{d\mathbf{v}(t)}{dt} \overset{(2.3.9)}{=} -R\omega^2\cos(\omega t)\mathbf{u}_x - R\omega^2\sin(\omega t)\mathbf{u}_y \overset{(2.3.8)}{=} -\omega^2\mathbf{p}(t) \tag{2.3.10}$$

where $\mathbf{v}(t)$ and $\mathbf{a}(t)$ turn out to be perpendicular to each other:

$$\mathbf{v}(t)\cdot\mathbf{a}(t)=\left(-R\omega\sin(\omega t)\mathbf{u}_x+R\omega\cos(\omega t)\mathbf{u}_y\right)\cdot\left(-R\omega^2\cos(\omega t)\mathbf{u}_x-R\omega^2\sin(\omega t)\mathbf{u}_y\right)=0 \quad (2.3.11)$$

On the other hand, we can integrate the acceleration $\mathbf{a}(t)$ successively to get the velocity $\mathbf{v}(t)$ and position vector $\mathbf{p}(t)$:

$$\mathbf{v}(t)=\int_{-\infty}^{t}\mathbf{a}(t)\,dt=\int_{-\infty}^{t_0}\mathbf{a}(t)\,dt+\int_{t_0}^{t}\mathbf{a}(t)\,dt=\mathbf{v}(t_0)+\int_{t_0}^{t}\mathbf{a}(t)\,dt \quad (2.3.12)$$

$$\mathbf{p}(t)=\int_{-\infty}^{t}\mathbf{v}(t)\,dt=\int_{-\infty}^{t_0}\mathbf{v}(t)\,dt+\int_{t_0}^{t}\mathbf{v}(t)\,dt=\mathbf{p}(t_0)+\int_{t_0}^{t}\mathbf{v}(t)\,dt \quad (2.3.13)$$

(Example 2.4) Trajectory of a Shell

Find the trajectory of a shell fired with initial speed v_0=250[m/s] and elevation angle θ_0=30° at the initial position z_0=2[m] high from ground level till it reaches the ground level where the gravitational acceleration is g=9.8[m/s^2]. Also find the range and impact speed of the shell crashing into the ground.

First, with

$$\mathbf{p}(0)=z_0\mathbf{u}_z,\ \mathbf{v}(0)=v_0\cos\theta_0\,\mathbf{u}_x+v_0\sin\theta_0\,\mathbf{u}_z,\ \text{and}\ \mathbf{a}(t)=-g\mathbf{u}_z=-9.8\mathbf{u}_z \quad (E2.4.1)$$

we can use Eq. (2.3.12) to find thc *velocity* as

$$\mathbf{v}(t)\overset{(2.3.12)}{=}\mathbf{v}(0)+\int_0^t\mathbf{a}(t)\,dt\overset{(E2.4.1)}{=}v_0\cos\theta_0\,\mathbf{u}_x+(v_0\sin\theta_0-gt)\mathbf{u}_z \quad (E2.4.2)$$

Subsequently, we use Eq. (2.3.13) to find the *position (trajectory)* as

$$\mathbf{p}(t)\overset{(2.3.13)}{=}\mathbf{p}(0)+\int_0^t\mathbf{v}(t)\,dt\overset{(E2.4.2)}{=}z_0\mathbf{u}_z+v_0t\cos\theta_0\,\mathbf{u}_x+(v_0t\sin\theta_0-\frac{1}{2}gt^2)\mathbf{u}_z \quad (E2.4.3)$$

The landing time t_f of the shell can be found by setting the z-component of $\mathbf{p}(t_f)$ to zero as

$$z_f=-\frac{1}{2}gt_f^2+(v_0\sin\theta_0)t_f+z_0=0$$

$$;\ t_f=\frac{v_0\sin\theta_0\pm\sqrt{v_0^2\sin^2\theta_0+4z_0g/2}}{g}=\frac{125\pm\sqrt{125^2+4\times9.8}}{9.8}=25.5[\text{s}] \quad (E2.4.4)$$

```
>>z0=2; v0=250; th=30*pi/180; g=9.8; tf=roots([-g/2  v0*sin(th)  z0])
   tf =    25.5262
           -0.0160 % irrelevant root
```

This landing time can be substituted into Eqs. (E2.4.3) and (E2.4.2) to yield the *range* and *impact speed* as

$$\mathbf{p}(t_f)\overset{(E2.4.3)}{=}v_0t_f\cos\theta_0\mathbf{u}_x+(z_0+v_0t_f\sin\theta_0-\frac{1}{2}gt_f^2)\mathbf{u}_z\overset{(E2.4.4)}{=}5527\mathbf{u}_x[\text{m}] \quad (E2.4.5)$$

$$\mathbf{v}(t_f)\overset{(E2.4.2)}{=}v_0\cos\theta_0\mathbf{u}_x+(v_0\sin\theta_0-gt_f)\mathbf{u}_z;\ \ v(t_f)=\|\mathbf{v}(t_f)\|=250[\text{m/s}] \quad (E2.4.6)$$

```
>>tf=tf(1); xf=v0*cos(th)*tf, vf=norm([v0*cos(th) v0*sin(th)-g*tf])
   xf =    5.5266e+003   % Range
   vf = 250.0784         % Impact speed
```

Also, we can run the following block of MATLAB statements to plot the (vertical) z-component and speed versus the (horizontal) x-component of the trajectory for t=[0, t_f], as depicted in Fig. 2.4:

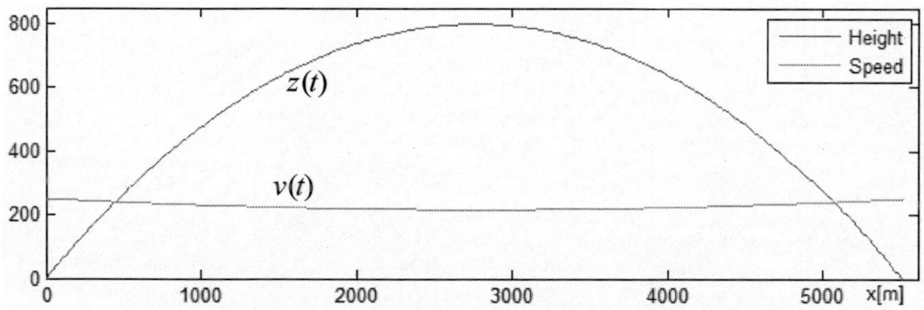

Figure 2.4 Trajectory and speed of the shell for Example 2.4

```
t=[0:0.1:tf+0.1]; % Time vector
x=v0*cos(th)*t;
z=z0+(v0*sin(th)*t-g/2*t.^2); % x/z-components
v=sqrt((v0*cos(th))^2+(v0*sin(th)-g*t).^2); % Speed
plot(x,z, x,v,'r'), xlabel('x[m]'), legend('Height','Speed')
```

Now, as the amount by which a smooth curve representing $\mathbf{p}(t)$ deviates from being flat or straight, we define the *curvature* of the curve at a point as the derivative of the unit tangent vector $\mathbf{u}_\mathrm{T}$ w.r.t. the arc length parameter l:

$$\kappa = \left\| \frac{d\mathbf{u}_\mathrm{T}}{dl} \right\| = \left\| \frac{d\mathbf{u}_\mathrm{T}/dt}{dl/dt} \right\| = \left\| \frac{\mathbf{u}_\mathrm{T}'(t)}{l'(t)} \right\| = \frac{\|\mathbf{u}_\mathrm{T}'(t)\|}{\|\mathbf{p}'(t)\|=v(t)} \tag{2.3.14}$$

where the *unit tangent vector* $\mathbf{u}_\mathrm{T}$ is defined as

$$\mathbf{u}_\mathrm{T} = \frac{\mathbf{v}(t)}{\|\mathbf{v}(t)\|=v(t)} = \frac{\mathbf{p}'(t)}{\|\mathbf{p}'(t)\|} \underset{(2.2.4)}{\overset{(2.2.2)}{\equiv}} \frac{d\mathbf{p}(t)/dt}{dl(t)/dt} = \frac{d\mathbf{p}(l)}{dl} \tag{2.3.15}$$

The curvature defined by Eq. (2.3.14) describes how abruptly/smoothly a curve changes its direction w.r.t. arc length at a point $\mathbf{p}(t)$ on the curve. Also, the *radius of curvature* is defined as the reciprocal of the curvature:

$$r_\kappa = \frac{1}{\kappa} \tag{2.3.16}$$

The circle through a point $\mathbf{p}(t)$ with radius r_κ that best fits the curve is called the *circle of curvature* or *osculating circle* and its center is the *center of curvature*.

(Example 2.5) Curvature of an Ellipse

Find the values of curvature and the circles of curvature of the ellipse represented by

$$\mathbf{p}(t) = 2\cos(\pi t)\mathbf{u}_x + \sin(\pi t)\mathbf{u}_y \tag{E2.5.1}$$

at two points (2, 0) and (0, 1), each corresponding to $t=0$ and 1/2, respectively.

We apply Eq. (2.3.15) for Eq. (E2.5.1) to find the *unit tangent vector*

$$\mathbf{u}_\mathrm{T} \overset{(2.3.15)}{=} \frac{\mathbf{p}'(t)}{\|\mathbf{p}'(t)\|} \overset{(E2.5.1)}{=} \frac{-2\pi\sin(\pi t)\mathbf{u}_x + \pi\cos(\pi t)\mathbf{u}_y}{\sqrt{4\pi^2\sin^2(\pi t) + \pi^2\cos^2(\pi t)}} = \begin{cases} \mathbf{u}_y & \text{at } t=0 \\ -\mathbf{u}_x & \text{at } t=1/2 \end{cases} \tag{E2.5.2}$$

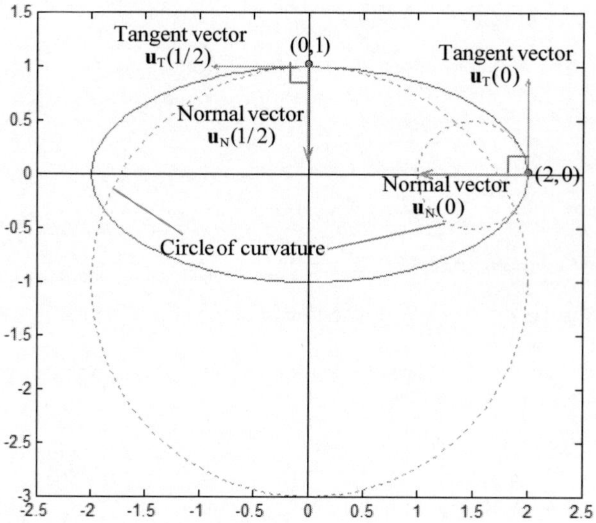

Figure 2.5 Unit tangent vectors and circles of curvature at two points (2, 0) and (0, 1) for Example 2.5

```
%em02e05.m
% Curvature
clear, clf
t=0:0.01:2; pit=pi*t; cpt=cos(pit); spt=sin(pit); % Time vector, ...
p=@(t)[2*cos(pi*t); sin(pi*t)]; pt=p(t); % Eq.(E2.5.1)
plot(pt(1,:),pt(2,:), 1.5+cpt/2,spt/2,'r:', 2*cpt,2*spt-1,'r:'), hold on
p1=p(0); p2=p(1/2); % Two observation points
uT1=[0; 1]; uT2=[-1; 0]; % Unit tangent vectors (E2.5.2)
quiver(p1(1),p1(2),uT1(1),uT1(2),'r')
quiver(p2(1),p2(2),uT2(1),uT2(2),'r')
axis('equal'), axis([-2.5 2.5 -3 1.5])
```

Then we use Eq. (2.3.14) to find the *curvature*

$$\kappa(t) \overset{(2.3.14)}{=} \frac{\|\mathbf{u}_T'(t)\|}{\|\mathbf{p}'(t)\|} \overset{(E2.5.2)}{=} \frac{\|-2\pi^2\cos(\pi t)\mathbf{u}_x - \pi^2\sin(\pi t)\mathbf{u}_y\|}{4\pi^2\sin^2(\pi t)+\pi^2\cos^2(\pi t)} = \begin{cases} 2 & \text{at } t=0 \\ 1/2 & \text{at } t=1/2 \end{cases} \tag{E2.5.3}$$

We can run the above MATLAB script "em02e05.m" to plot the two circles of curvature together with the ellipse as depicted in Fig. 2.5, where the two circles are represented by

$$(x-1.5)^2 + y^2 = (1/2)^2 \text{ with radius } r_{\kappa 1} = 1/\kappa(0) = 1/2 \text{ passing through } (2,0) \tag{E2.5.4a}$$

$$x^2 + (y+1)^2 = 2^2 \text{ with radius } r_{\kappa 2} = 1/\kappa(1/2) = 2 \text{ passing through } (0,1) \tag{E2.5.4b}$$

Here, just like Eq. (2.3.7), noting that the magnitude of the unit tangent vector $\mathbf{u}_T(t)$ is constant so that

$$\|\mathbf{u}_T\|^2 = \mathbf{u}_T \bullet \mathbf{u}_T = 1$$

we can differentiate this w.r.t. t to get

$$\mathbf{u}_T(t) \bullet \mathbf{u}_T'(t) = 0 \tag{2.3.17}$$

This implies that $\mathbf{u}_T(t)$ and $\mathbf{u}_T'(t)$ are perpendicular to each other. Thus we can define the *unit normal vector* to the curve, which is perpendicular to the tangent vector $\mathbf{u}_T(t)$, as

$$\mathbf{u}_N(t) = \frac{\mathbf{u}_T'(t)}{||\mathbf{u}_T'(t)||} \tag{2.3.18}$$

Also, from Eq. (2.3.15), we can write the velocity as

$$\mathbf{v}(t) = v(t)\mathbf{u}_T(t) \tag{2.3.19}$$

Differentiating this w.r.t. t via Eq. (2.2.5a) yields an expression of the acceleration as

$$\mathbf{a}(t) = \frac{d}{dt}\mathbf{v}(t) \overset{(2.3.19)}{=} \frac{d}{dt}(v(t)\mathbf{u}_T(t)) \overset{(2.2.5a)}{=} v'(t)\mathbf{u}_T(t) + v(t)\mathbf{u}_T'(t) \overset{(2.3.18)}{=} v'(t)\mathbf{u}_T(t) + v(t)||\mathbf{u}_T'(t)||\mathbf{u}_N(t)$$

$$\overset{(2.3.14)}{=} v'(t)\mathbf{u}_T(t) + \kappa(t)v^2(t)\mathbf{u}_N(t) = a_T\mathbf{u}_T(t) + a_N\mathbf{u}_N(t) \text{ with } a_T = v'(t) \text{ and } a_N = \kappa(t)v^2(t) \tag{2.3.20}$$

Here, a_T and a_N, called the *tangential* and *normal* (or *centripetal*) *components* of the acceleration, reflect the changes of velocity vector $\mathbf{v}(t)$ in magnitude and direction, respectively.

In conjunction with the unit tangent vector $\mathbf{u}_T(t)$ and the unit normal vector $\mathbf{u}_N(t)$, we define a third unit vector, called the *binormal vector*, as

$$\mathbf{u}_B = \mathbf{u}_T \times \mathbf{u}_N \tag{2.3.21}$$

The three vectors $\mathbf{u}_T$, $\mathbf{u}_N$, and $\mathbf{u}_B$ form a right-handed coordinate system of mutually orthogonal vectors, called the *Frenet-Serret frame* or *moving trihedral*. The plane spanned by $\mathbf{u}_T$ and $\mathbf{u}_N$, the plane spanned by $\mathbf{u}_N$ and $\mathbf{u}_B$, and the plane spanned by $\mathbf{u}_B$ and $\mathbf{u}_T$ are called the *osculating plane*, the *normal plane*, and the *rectifying plane*, respectively. They are illustrated in Fig. 2.6.

<Frenet–Serret formulas>

Before ending this section, it is good to see the *Frenet–Serret formulas*, which can be written in matrix form as

$$\left.\begin{matrix} d\mathbf{u}_T/dl = \kappa\mathbf{u}_N & \text{(a)} \\ d\mathbf{u}_N/dl = -\kappa\mathbf{u}_T + \tau\mathbf{u}_B & \text{(b)} \\ d\mathbf{u}_B/dl = -\tau\mathbf{u}_N & \text{(c)} \end{matrix}\right\} \rightarrow \begin{bmatrix} d\mathbf{u}_T/dl \\ d\mathbf{u}_N/dl \\ d\mathbf{u}_B/dl \end{bmatrix} = \begin{bmatrix} 0 & \kappa & 0 \\ -\kappa & 0 & \tau \\ 0 & -\tau & 0 \end{bmatrix} \begin{bmatrix} \mathbf{u}_T \\ \mathbf{u}_N \\ \mathbf{u}_B \end{bmatrix} \tag{2.3.22}$$

where $\tau = -(d\mathbf{u}_B/dl)\bullet\mathbf{u}_N$ is the negative proportionality constant between $d\mathbf{u}_B/dl$ and $\mathbf{u}_N$, called the *torsion* of the curve, which measures how much a curve with the moving trihedral twists (from being planar) as the parameter t increases.

(Proof)

First, we multiply Eqs. (2.3.14) and (2.3.18) to get Eq. (2.3.22a):

$$\kappa\mathbf{u}_N \underset{(2.3.18)}{\overset{(2.3.14)}{=}} \frac{||\mathbf{u}_T'||}{||\mathbf{p}'|| = v} \frac{\mathbf{u}_T'}{||\mathbf{u}_T'||} = \frac{\mathbf{u}_T' = d\mathbf{u}_T/dt}{v = dl/dt} = \frac{d\mathbf{u}_T}{dl} \tag{2.3.23}$$

Then, as in Eq. (2.3.17), note that

$$\mathbf{u}_B\bullet\frac{d\mathbf{u}_B}{dl} = 0 \tag{2.3.24}$$

just because $\mathbf{u}_B$ is also a unit vector of constant magnitude. Now, differentiating $\mathbf{u}_B=\mathbf{u}_T\times\mathbf{u}_N$ (Eq. (2.3.21)) w.r.t. the arc length parameter l and using Eq. (2.2.5d) yields

$$\frac{d\mathbf{u}_B}{dl} \underset{(2.2.5d)}{\overset{(2.3.21)}{=}} \frac{d\mathbf{u}_T}{dl}\times\mathbf{u}_N + \mathbf{u}_T\times\frac{d\mathbf{u}_N}{dl} \overset{(2.3.23)}{=} \kappa(t)\mathbf{u}_N(t)\times\mathbf{u}_N + \mathbf{u}_T\times\frac{d\mathbf{u}_N}{dl} = \mathbf{u}_T\times\frac{d\mathbf{u}_N}{dl} \tag{2.3.25}$$

Eqs. (2.3.24) and (2.3.25) imply that $d\mathbf{u}_B/dl$ is perpendicular to both $\mathbf{u}_B$ and $\mathbf{u}_T$ so that it must be in the direction of $\mathbf{u}_N$. That is why we can write Eq. (2.3.22c):

$$\frac{d\mathbf{u}_B}{dl} = -\tau\,\mathbf{u}_N \text{ with torsion } \tau = -\frac{d\mathbf{u}_B}{dl}\bullet\mathbf{u}_N \tag{2.3.26}$$

Now, differentiating $\mathbf{u}_N=\mathbf{u}_B\times\mathbf{u}_T$ w.r.t. l and using Eq. (2.2.5d) yields

$$\frac{d\mathbf{u}_N}{dl} \underset{(2.2.5d)}{=} \frac{d\mathbf{u}_B}{dl}\times\mathbf{u}_T + \mathbf{u}_B\times\frac{d\mathbf{u}_T}{dl} \underset{(2.3.23)}{\overset{(2.3.26)}{=}} -\tau\mathbf{u}_N\times\mathbf{u}_T + \mathbf{u}_B\times\kappa\mathbf{u}_N = \tau\mathbf{u}_B - \kappa\mathbf{u}_T \tag{2.3.27}$$

(Example 2.6) Tangent, Normal, Binormal Vectors, and Torsion of a Circular Helix

Find the unit tangent, normal, binormal vectors, and torsion of the curve represented by

$$\mathbf{p}(t) = t\,\mathbf{u}_x + \cos(\pi t)\,\mathbf{u}_y + \sin(\pi t)\,\mathbf{u}_z \tag{E2.6.1}$$

at two points (0, 1, 0) and (1/2, 0, 1), each corresponding to $t=0$ and 1/2, respectively (see Fig. 2.6).

We apply Eq. (2.3.15) for Eq. (E2.6.1) to find the *unit tangent vector:*

$$\mathbf{u}_T(t) \overset{(2.3.15)}{=} \frac{\mathbf{p}'(t)}{\|\mathbf{p}'(t)\|} \overset{(E2.6.1)}{=} \frac{\mathbf{u}_x - \pi\sin(\pi t)\mathbf{u}_y + \pi\cos(\pi t)\mathbf{u}_z}{\sqrt{1+\pi^2}} = \begin{cases} (\mathbf{u}_x+\pi\mathbf{u}_z)/\sqrt{1+\pi^2} & \text{at } t=0 \\ (\mathbf{u}_x-\pi\mathbf{u}_y)/\sqrt{1+\pi^2} & \text{at } t=1/2 \end{cases} \tag{E2.6.2}$$

Then we use Eq. (2.3.18) to find *the unit normal vector*:

$$\mathbf{u}_N(t) \overset{(2.3.18)}{=} \frac{\mathbf{u}_T'(t)}{\|\mathbf{u}_T'(t)\|} \overset{(E2.6.2)}{=} -\cos(\pi t)\mathbf{u}_y - \sin(\pi t)\mathbf{u}_z = \begin{cases} -\mathbf{u}_y & \text{at } t=0 \\ -\mathbf{u}_z & \text{at } t=1/2 \end{cases} \tag{E2.6.3}$$

We also use Eq. (2.3.21) to find the *binormal vector*:

$$\mathbf{u}_B \overset{(2.3.21)}{=} \mathbf{u}_T\times\mathbf{u}_N \underset{(E2.6.3)}{\overset{(E2.6.2)}{=}} \frac{1}{\sqrt{1+\pi^2}}\begin{bmatrix} 1 \\ -\pi\sin(\pi t) \\ \pi\cos(\pi t)\end{bmatrix}\times\begin{bmatrix} 0 \\ -\cos(\pi t) \\ -\sin(\pi t)\end{bmatrix} \overset{(1.1.20)}{=} \frac{1}{\sqrt{1+\pi^2}}\begin{bmatrix} \pi \\ \sin(\pi t) \\ -\cos(\pi t)\end{bmatrix}$$

$$= \begin{cases} (\pi\mathbf{u}_x-\mathbf{u}_z)/\sqrt{1+\pi^2} & \text{at } t=0 \\ (\pi\mathbf{u}_x+\mathbf{u}_y)/\sqrt{1+\pi^2} & \text{at } t=1/2 \end{cases} \tag{E2.6.4}$$

Now, to find the torsion of the curve using Eq. (2.3.26), we compute

$$\frac{d\mathbf{u}_B}{dl} = \frac{d\mathbf{u}_B/dt}{dl/dt=v(t)=\|\mathbf{p}'(t)\|} \underset{(E2.6.1)}{\overset{(E2.6.4)}{=}} \frac{1}{1+\pi^2}\begin{bmatrix} 0 \\ \pi\cos(\pi t) \\ \pi\sin(\pi t)\end{bmatrix} \overset{(E2.6.3)}{=} -\frac{\pi}{1+\pi^2}\mathbf{u}_N \tag{E2.6.5}$$

Matching this with Eq. (2.3.26) yields the *torsion* as

$$\tau(t) = -\frac{d\mathbf{u}_B}{dl}\bullet\mathbf{u}_N = \frac{\pi}{1+\pi^2} \text{ for all } t \tag{E2.6.6}$$

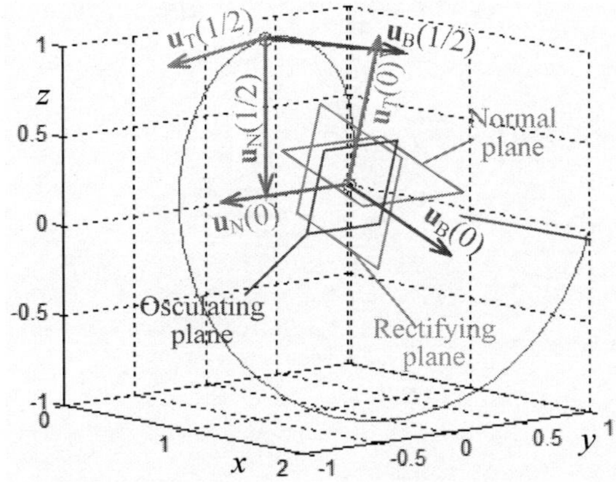

Figure 2.6 Illustration of tangent, normal, binormal vectors, and moving trihedral for Example 2.6

2.4 DIVERGENCE AND CURL

The *divergence* of a vector field **F** (described by a vector function of two or three variables) at an observation point is defined as

$$\text{div}\,\mathbf{F} = \lim_{\Delta V \to 0} \frac{1}{\Delta V} \oiint_{\Delta S} \mathbf{F} \bullet d\mathbf{s} \tag{2.4.1}$$

where $d\mathbf{s}=\mathbf{u}_n ds$ ($\mathbf{u}_n$: the outward unit normal vector to a differential surface of area ds) is a differential surface vector outward from a differential volume ΔV (containing the observation point) and $\oiint_{\Delta S}\mathbf{F}\bullet d\mathbf{s}$ is the surface integral of **F** over the closed boundary surface ΔS surrounding the volume ΔV. This is the limit of the net outward flux of **F** per unit volume as the volume goes to zero.

The *curl* of a vector field **F** at an observation point is defined as

$$\text{curl}\,\mathbf{F} = \mathbf{u}_n \lim_{\Delta S \to 0} \frac{1}{\Delta S} \oint_{\Delta C} \mathbf{F} \bullet d\mathbf{l} \tag{2.4.2}$$

where $\oint_{\Delta C}\mathbf{F}\bullet d\mathbf{l}$ is the line integral of **F** along the closed boundary line ΔC of a differential planar surface (containing the observation point) having area ΔS and unit normal vector $\mathbf{u}_n$, which is the axis of rotation determined by the right-hand rule. This is the limit of the net circulation (rotation) of **F** per unit area as the area goes to zero.

From the definitions of divergence and curl, do they look difficult to deal with? Don't be afraid. There are useful formulas for computing the gradient, divergence, curl, and Laplacian (the divergence of the gradient) of a vector field, depending on which one of the three coordinate systems the vector field is represented in:

(1) Cartesian coordinate system with $f(x,y,z)$ and $\mathbf{F}=F_x\mathbf{u}_x+F_y\mathbf{u}_y+F_z\mathbf{u}_z$

$$\text{Gradient: } \nabla f = \frac{\partial f}{\partial x}\mathbf{u}_x + \frac{\partial f}{\partial y}\mathbf{u}_y + \frac{\partial f}{\partial z}\mathbf{u}_z \tag{2.4.3a}$$

$$\text{Divergence: } \nabla \bullet \mathbf{F} = \frac{\partial F_x}{\partial x} + \frac{\partial F_y}{\partial y} + \frac{\partial F_z}{\partial z} \tag{2.4.4a}$$

$$\text{Curl: } \nabla\times\mathbf{F} = \left(\frac{\partial F_z}{\partial y}-\frac{\partial F_y}{\partial z}\right)\mathbf{u}_x + \left(\frac{\partial F_x}{\partial z}-\frac{\partial F_z}{\partial x}\right)\mathbf{u}_y + \left(\frac{\partial F_y}{\partial x}-\frac{\partial F_x}{\partial y}\right)\mathbf{u}_z \tag{2.4.5a}$$

$$\text{Laplacian: } \nabla\bullet\nabla f = \nabla^2 f = \frac{\partial^2 f}{\partial x^2}+\frac{\partial^2 f}{\partial y^2}+\frac{\partial^2 f}{\partial z^2} \tag{2.4.6a}$$

(2) Cylindrical coordinate system with $f(\rho,\phi,z)$ and $\mathbf{F}=F_\rho\mathbf{u}_x+F_\phi\mathbf{u}_\phi+F_z\mathbf{u}_z$

$$\text{Gradient: } \nabla f = \frac{\partial f}{\partial\rho}\mathbf{u}_\rho+\frac{1}{\rho}\frac{\partial f}{\partial\phi}\mathbf{u}_\phi+\frac{\partial f}{\partial z}\mathbf{u}_z \tag{2.4.3b}$$

$$\text{Divergence: } \nabla\bullet\mathbf{F} = \frac{1}{\rho}\frac{\partial(\rho F_\rho)}{\partial\rho}+\frac{1}{\rho}\frac{\partial F_\phi}{\partial\phi}+\frac{\partial F_z}{\partial z} \tag{2.4.4b}$$

$$\text{Curl: } \nabla\times\mathbf{F} = \left(\frac{1}{\rho}\frac{\partial F_z}{\partial\phi}-\frac{\partial F_\phi}{\partial z}\right)\mathbf{u}_\rho + \left(\frac{\partial F_\rho}{\partial z}-\frac{\partial F_z}{\partial\rho}\right)\mathbf{u}_\phi + \frac{1}{\rho}\left(\frac{\partial(\rho F_\phi)}{\partial\rho}-\frac{\partial F_\rho}{\partial\phi}\right)\mathbf{u}_z \tag{2.4.5b}$$

$$\text{Laplacian: } \nabla\bullet\nabla f = \nabla^2 f = \frac{1}{\rho}\frac{\partial(\rho\,\partial f/\partial\rho)}{\partial\rho}+\frac{1}{\rho^2}\frac{\partial^2 f}{\partial\phi^2}+\frac{\partial^2 f}{\partial z^2} \tag{2.4.6b}$$

(3) Spherical coordinate system with $f(r,\theta,\ \phi)$ and $\mathbf{F}=F_r\mathbf{u}_r+F_\theta\mathbf{u}_\theta+F_\phi\mathbf{u}_\phi$

$$\text{Gradient: } \nabla f = \frac{\partial f}{\partial r}\mathbf{u}_r+\frac{1}{r}\frac{\partial f}{\partial\theta}\mathbf{u}_\theta+\frac{1}{r\sin\theta}\frac{\partial f}{\partial\phi}\mathbf{u}_\phi \tag{2.4.3c}$$

$$\text{Divergence: } \nabla\bullet\mathbf{F} = \frac{1}{r^2}\frac{\partial(r^2F_r)}{\partial r}+\frac{1}{r\sin\theta}\frac{\partial(F_\theta\sin\theta)}{\partial\theta}+\frac{1}{r\sin\theta}\frac{\partial F_\phi}{\partial\phi} \tag{2.4.4c}$$

$$\text{Curl: } \nabla\times\mathbf{F} = \frac{1}{r\sin\theta}\left(\frac{\partial(F_\phi\sin\theta)}{\partial\theta}-\frac{\partial F_\theta}{\partial\phi}\right)\mathbf{u}_r + \frac{1}{r}\left(\frac{1}{\sin\theta}\frac{\partial F_r}{\partial\phi}-\frac{\partial(rF_\phi)}{\partial r}\right)\mathbf{u}_\theta + \frac{1}{r}\left(\frac{\partial(rF_\theta)}{\partial r}-\frac{\partial F_r}{\partial\theta}\right)\mathbf{u}_\phi \tag{2.4.5c}$$

$$\text{Laplacian: } \nabla\bullet\nabla f = \nabla^2 f = \frac{1}{r^2}\frac{\partial(r^2\,\partial f/\partial r)}{\partial r}+\frac{1}{r^2\sin\theta}\frac{\partial(\sin\theta\,\partial f/\partial\theta)}{\partial\theta}+\frac{1}{r^2\sin^2\theta}\frac{\partial^2 f}{\partial\phi^2} \tag{2.4.6c}$$

These formulas have been cast into the following MATLAB functions:

```
function divF=diva(F,coords)
%find the analytical divergence of a given symbolic 3D vector function F
% coords = 'car'/'cyl'/'sph'
% Copyleft: Won Y. Yang, wyyang53@hanmail.net, CAU for academic use only
if nargin<2, coords='ca'; end  % Cartesian coordinate system by default
switch lower(coords(1:2)) % Depending on which coordinate system to use
  case 'cy', syms r ph z; coord='cylindrical';
             divF=diff(r*F(1),r)/r+diff(F(2),ph)/r+diff(F(3),z); %(2.4.4b)
  case 'sp', syms r th ph; coord='spherical';
             divF = diff(r^2*F(1),r)/r^2 +diff(F(2)*sin(th),th)/r/sin(th)
                    +diff(F(3),ph)/r/sin(th); %(2.4.4c)
  otherwise, syms x y z; coord='Cartesian';
             divF = diff(F(1),x)+diff(F(2),y)+diff(F(3),z); %(2.4.4a)
end
if nargout<1
  fprintf(['Div(F) in ' coord ' coordinate is']); pretty(simplify(DivF))
end
```

```
function curlF=curla(F,coords)
% To find the analytical curl of a given symbolic 3D vector function F
% F=[Fx Fy Fz]|[Frho Fph Fz]|[Fr Fth Fph]/coords= 'car'|'cyl'|'sph'
% Copyleft: Won Y. Yang, wyyang53@hanmail.net, CAU for academic use only
if nargin<2, coords='ca'; end % Cartesian coordinate system by default
switch lower(coords(1:2)) % Depending on which coordinate system to use
  case 'cy', syms r ph z; coord='cylindrical'; %(2.4.5b)
             curlF = [diff(F(3),ph)/r-diff(F(2),z);
                      diff(F(1),z)-diff(F(3),r);
                      (diff(r*F(2),r)-diff(F(1),ph))/r];
  case 'sp', syms r th ph; coord='spherical'; %(2.4.5c)
             curlF = [(diff(F(3)*sin(th),th)-diff(F(2),ph))/r/sin(th);
                       (diff(F(1),th)/sin(ph)-diff(r*F(3),r))/r;
                       (diff(r*F(2),r)-diff(F(1),th))/r];
  otherwise, syms x y z; coord='Cartesian'; %(2.4.5a)
             curlF=[diff(F(3),y)-diff(F(2),z);
                 diff(F(1),z)-diff(F(3),x); diff(F(2),x)-diff(F(1),y)];
end
if nargout<1
  fprintf(['Curl(F) in ' coord ' coordinate is']); pretty(simplify(CurlF))
end
```

```
function Df=gradienta(f,coords)
% find the analytical gradient of a given symbolic function f
% coords = 'car'/'cyl'/'sph'
if nargin<2, coords='ca'; end % Cartesian coordinate system by default
switch lower(coords(1:2)) % Depending on which coordinate system to use
  case 'cy', syms r ph z;  coord='cylindrical';
             Df=[diff(f,r); diff(f,ph)/r; diff(f,z)]; %(2.4.3b)
  case 'sp', syms r th ph;  coord='spherical';
             Df=[diff(f,r); diff(f,th)/r; diff(f,ph)/r/sin(th)]; %(2.4.3c)
  otherwise, syms x y z;  coord='Cartesian';
             Df=[diff(f,x); diff(f,y); diff(f,z)]; %(2.4.3a)
end
if nargout<1
 fprintf(['Gradient of f in ' coord ' coordinate is']); pretty(simplify(Df))
end
```

```
function Lapf=Laplaciana(f,coords)
% find the analytical Laplacian of a given symbolic function f
% coords = 'car'/'cyl'/'sph'
if nargin<2, coords='ca'; end % Cartesian coordinate system by default
switch lower(coords(1:2)) % Depending on which coordinate system to use
  case 'cy', syms r ph z;  coord='cylindrical';
             lapf = (diff(r*diff(f,r),r)/r +diff(diff(f,ph),ph))/r^2 ...
                       +diff(diff(f,z),z); %(2.4.6b)
  case 'sp', syms r th ph; coord='spherical';
             lapf = (diff(r^2*diff(f,r),r)/r^2
                    +diff(sin(th)*diff(f,th),th))/r^2/sin(th) ...
                     +diff(diff(f,ph),ph)/r^2/sin(th)^2; %(2.4.6c)
  otherwise, syms x y z; coord='Cartesian';
             lapf = diff(diff(f,x),x) +diff(diff(f,y),y) ...
                    +diff(diff(f,z),z); %(2.4.6a)
end
if nargout<1
  fprintf(['Laplacian(f) in ' coord ' coordinate is']);
  pretty(simplify(lapf))
end
```

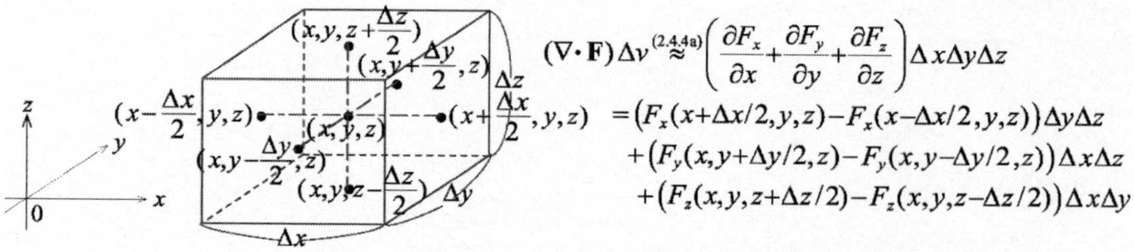

Figure 2.7 Divergence of **F** and flow of flux in a differential volume

Here, to grasp the physical meaning of the divergence, let us multiply Eq. (2.4.4a) with a differential volume $\Delta v = \Delta x \Delta y \Delta z$:

$$(\nabla \bullet \mathbf{F})\Delta v \overset{(2.4.4a)}{\approx} \left(\frac{\partial F_x}{\partial x}+\frac{\partial F_y}{\partial y}+\frac{\partial F_z}{\partial z}\right)\Delta x \Delta y \Delta z = \big(F_x(x+\Delta x/2,y,z)-F_x(x-\Delta x/2,y,z)\big)\Delta y \Delta z$$
$$+\big(F_y(x,y+\Delta y/2,z)-F_y(x,y-\Delta y/2,z)\big)\Delta x \Delta z+\big(F_z(x,y,z+\Delta z/2)-F_z(x,y,z-\Delta z/2)\big)\Delta x \Delta y \quad (2.4.7)$$

This is the sum of the flows of the flux described by the 3D vector $\mathbf{F}=F_x\,\mathbf{u}_x+F_y\,\mathbf{u}_y+F_z\,\mathbf{u}_z$ along the x, y, and z-axes in a differential rectangular box of dimensions Δx, Δy, and Δz as described in Fig. 2.7. This implies that the divergence of **F**, denoted by $\nabla \bullet \mathbf{F}$, is a measure of the divergence of the vector field **F** away from each point.

Also, to grasp the physical meaning of the curl computed by Eq. (2.4.5a), let us consider an object rotating with constant angular speed ω[rad/s] around a line as depicted in Fig. 2.8. Note that the direction of the angular velocity $\boldsymbol{\omega}=\omega_x\mathbf{u}_x+\omega_y\mathbf{u}_y+\omega_z\mathbf{u}_z$, as specified by the right-hand rule, is the direction in which a right-handed screw rotated in the direction of rotation of the object will move. That is why we can write the (tangential) linear velocity **v** as the cross product of the angular velocity $\boldsymbol{\omega}$ and the position vector **r**:

$$\mathbf{v}=\boldsymbol{\omega}\times\mathbf{r}=\begin{bmatrix}\omega_x\\ \omega_y\\ \omega_z\end{bmatrix}\times\begin{bmatrix}x\\ y\\ z\end{bmatrix}=\begin{bmatrix}\omega_y z-\omega_z y\\ \omega_z x-\omega_x z\\ \omega_x y-\omega_y x\end{bmatrix} \quad (2.4.8)$$

On the other hand, noting that ω_x, ω_y, and ω_z are constant, we can take the curl of this linear velocity **v** to write

$$\nabla\times\mathbf{v}=\begin{bmatrix}\partial/\partial x\\ \partial/\partial y\\ \partial/\partial z\end{bmatrix}\times\begin{bmatrix}\omega_y z-\omega_z y\\ \omega_z x-\omega_x z\\ \omega_x y-\omega_y x\end{bmatrix}=\begin{bmatrix}2\omega_x\\ 2\omega_y\\ 2\omega_z\end{bmatrix}=2\boldsymbol{\omega} \quad (2.4.9)$$

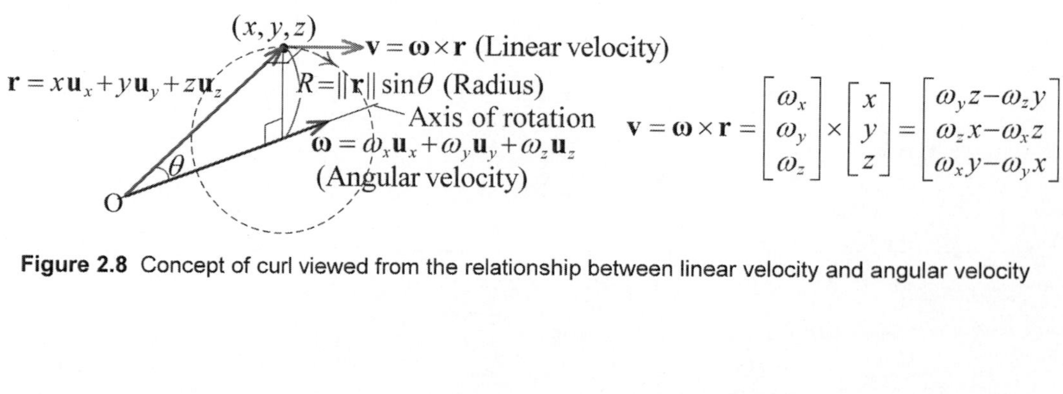

Figure 2.8 Concept of curl viewed from the relationship between linear velocity and angular velocity

(Example 2.7) Divergence and Curl

Find the divergence and curl of the vector fields

$$\mathbf{F}_1(x,y,z) = x^3\mathbf{u}_x + y^3\mathbf{u}_y + 0\mathbf{u}_z \quad \text{(E2.7.1a)} \qquad \mathbf{F}_2(x,y,z) = -y^3\mathbf{u}_x + x^3\mathbf{u}_y + 0\mathbf{u}_z \quad \text{(E2.7.2a)}$$

that are depicted in Figs. 2.9(a1) and (a2), respectively.

We can use Eq. (2.4.4a) to get the divergences of $\mathbf{F}_1$ and $\mathbf{F}_2$:

$$\nabla\bullet\mathbf{F}_1 \overset{(2.4.4a)}{=} \frac{\partial F_x}{\partial x} + \frac{\partial F_y}{\partial y} + \frac{\partial F_z}{\partial z} \overset{(E2.7.1a)}{=} 3(x^2+y^2) \,\text{(E2.7.1b)} \quad \nabla\bullet\mathbf{F}_2 \overset{(2.4.4a)}{=} \frac{\partial F_x}{\partial x} + \frac{\partial F_y}{\partial y} + \frac{\partial F_z}{\partial z} \overset{(E2.7.2a)}{=} 0 \,\text{(E2.7.2b)}$$

```
%em02e07.m
% to plot the vector fields F = x^3 ux + y^3 uy and F = -y^3 ux + x^3 uy.
clear, figure(1), clf
dx=0.5; dy=0.5; xx=-2:dx:2; yy=-2:dy:2;   [X,Y]=meshgrid(xx,yy);
syms x y z ux uy uz;   sxy=sqrt(x^2+y^2);
for iter=0:1
   if iter==0, FX=X.^3; FY=Y.^3;   Fx=x^3; Fy=y^3; Fz=0; % (E2.7.1a)
               div_F_analytical = 3*(X.^2+Y.^2); % (E2.7.1b)
               curl_F_analytical = zeros(size(X)); % (E2.7.1c)
    else FX=-Y.^3; FY=X.^3;   Fx=-y^3; Fy=x^3; Fz=0; % (E2.7.2a)
         div_F_analytical = zeros(size(X)); % (E2.7.2b)
         curl_F_analytical = 3*(X.^2+Y.^2); % (E2.7.2c)
   end
   div_F = divergence(X,Y,FX,FY);   curl_F = curl(X,Y,FX,FY);
   figure(1)
   subplot(2,3,1+iter*3), quiver(X,Y,FX,FY), axis image
   subplot(2,3,2+iter*3), imagesc(xx,yy,div_F), axis image, colorbar
   subplot(2,3,3+iter*3), imagesc(xx,yy,curl_F), axis image, colorbar
   DivF=diva([Fx Fy Fz]); fprintf('Div_F'); pretty(simplify(DivF))
   CurlF=curla([Fx Fy Fz]); fprintf('Curl_F'); pretty(simplify(CurlF))
end
```

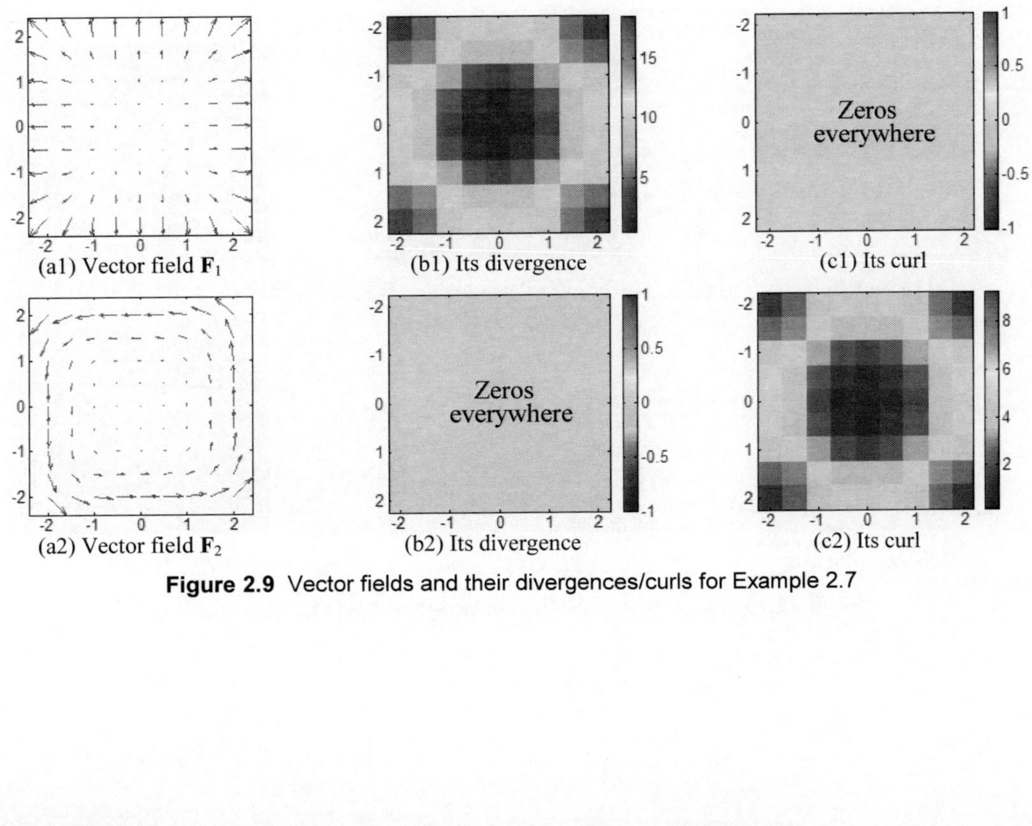

Figure 2.9 Vector fields and their divergences/curls for Example 2.7

We can also use Eq. (2.4.5a) to get the curls of $\mathbf{F}_1$ and $\mathbf{F}_2$:

$$\nabla\times\mathbf{F}_1 \overset{(2.4.5a)}{=} \left(\frac{\partial F_z}{\partial y}-\frac{\partial F_y}{\partial z}\right)\mathbf{u}_x+\left(\frac{\partial F_x}{\partial z}-\frac{\partial F_z}{\partial x}\right)\mathbf{u}_y+\left(\frac{\partial F_y}{\partial x}-\frac{\partial F_x}{\partial y}\right)\mathbf{u}_z=\mathbf{0} \quad \text{(E2.7.1c)}$$

$$\nabla\times\mathbf{F}_2 \overset{(2.4.5a)}{=} \left(\frac{\partial F_z}{\partial y}-\frac{\partial F_y}{\partial z}\right)\mathbf{u}_x+\left(\frac{\partial F_x}{\partial z}-\frac{\partial F_z}{\partial x}\right)\mathbf{u}_y+\left(\frac{\partial F_y}{\partial x}-\frac{\partial F_x}{\partial y}\right)\mathbf{u}_z=\left(3x^2+3y^2\right)\mathbf{u}_z \quad \text{(E2.7.2c)}$$

The above MATLAB script "em02e07.m" can be run to yield Fig. 2.9, which shows the two vector fields $\mathbf{F}_1$ and $\mathbf{F}_2$ together with their divergences and curls.

(Example 2.8) Divergence and Curl

Find the divergence and curl of the vector fields

$$\mathbf{F}_1(x,y,z)=\frac{x}{\sqrt{x^2+y^2}}\mathbf{u}_x+\frac{y}{\sqrt{x^2+y^2}}\mathbf{u}_y \quad \text{(E2.8.1a)} \quad \mathbf{F}_2(x,y,z)=-\frac{y}{\sqrt{x^2+y^2}}\mathbf{u}_x+\frac{x}{\sqrt{x^2+y^2}}\mathbf{u}_y \quad \text{(E2.8.2a)}$$

that are depicted in Fig. 2.10(a1) and (a2), respectively.

We can use Eq. (2.4.4a) to get the divergences of $\mathbf{F}_1$ and $\mathbf{F}_2$:

$$\nabla\cdot\mathbf{F}_1 \overset{(2.4.4a)}{=} \frac{\partial F_x}{\partial x}+\frac{\partial F_y}{\partial y}+\frac{\partial F_z}{\partial z}=\frac{2(x^2+y^2)^{1/2}-x^2(x^2+y^2)^{-1/2}-y^2(x^2+y^2)^{-1/2}}{x^2+y^2}=\frac{1}{(x^2+y^2)^{1/2}} \quad \text{(E2.8.1b)}$$

$$\nabla\cdot\mathbf{F}_2 \overset{(2.4.4a)}{=} \frac{\partial F_x}{\partial x}+\frac{\partial F_y}{\partial y}+\frac{\partial F_z}{\partial z}=\frac{xy(x^2+y^2)^{-1/2}-xy(x^2+y^2)^{-1/2}}{x^2+y^2}=0 \quad \text{(E2.8.2b)}$$

We can also use Eq. (2.4.5a) to get the curls of $\mathbf{F}_1$ and $\mathbf{F}_2$:

$$\nabla\times\mathbf{F}_1 \overset{(2.4.5a)}{=} \left(\frac{\partial F_z}{\partial y}-\frac{\partial F_y}{\partial z}\right)\mathbf{u}_x+\cdots+\left(\frac{\partial F_y}{\partial x}-\frac{\partial F_x}{\partial y}\right)\mathbf{u}_z=\frac{-xy(x^2+y^2)^{-1/2}+xy(x^2+y^2)^{-1/2}}{x^2+y^2}=\mathbf{0} \quad \text{(E2.8.1c)}$$

$$\nabla\times\mathbf{F}_2 \overset{(2.4.5a)}{=} \frac{(x^2+y^2)^{1/2}-x^2(x^2+y^2)^{-1/2}+(x^2+y^2)^{1/2}-y^2(x^2+y^2)^{-1/2}}{x^2+y^2}\mathbf{u}_z=\frac{1}{(x^2+y^2)^{1/2}}\mathbf{u}_z \quad \text{(E2.8.2c)}$$

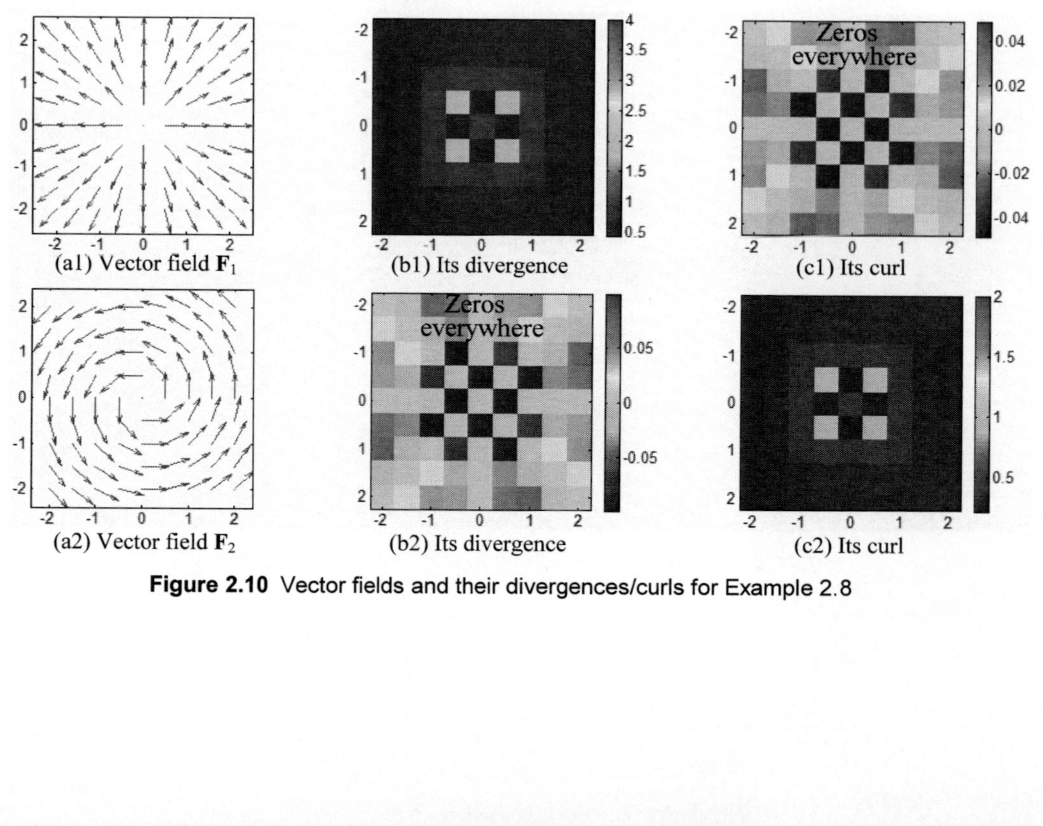

Figure 2.10 Vector fields and their divergences/curls for Example 2.8

The MATLAB script "em02e08.m," similar to "em02e07.m," can be run to yield Fig. 2.10, which shows the two vector fields $\mathbf{F}_1$ and $\mathbf{F}_2$ together with their divergences and curls. Note the following:

(1) Why does the divergence in Fig. 2.9(b1) increase, in contrast with that in Fig. 2.10(b1), as the observation point moves away from the origin? Because the outward vectors in Fig. 2.9(a1) increase in magnitude as the observation point moves away from the origin, while those in Fig. 2.10(a1) do not.

(2) Why does the curl in Fig. 2.9(c2) increase, in contrast with that in Fig. 2.10(c2), as the observation point moves away from the origin? Because the circulating vectors in Fig. 2.9(a2) increase in magnitude as the observation point moves away from the origin, while those in Fig. 2.10(a2) do not.

(Example 2.9) Divergence and Curl of an Electric Field

A sphere (of radius a) carrying a total charge Q[Coulomb] that is uniformly distributed over the volume of the sphere produces an electric field at a point r[m] away from the center of the sphere as

$$\mathbf{E}=\begin{cases} \overset{\text{spherical}}{\dfrac{rQ}{4\pi\varepsilon_0 a^3}\mathbf{u}_r} = \overset{\text{Cartesian (rectangular)}}{\dfrac{Q}{4\pi\varepsilon_0 a^3}(x\mathbf{u}_x+y\mathbf{u}_y+z\mathbf{u}_z)} & \text{for } r\le a \text{ (inside the sphere)} \\ \dfrac{Q}{4\pi\varepsilon_0 r^2}\mathbf{u}_r = \dfrac{Q}{4\pi\varepsilon_0 r^3}(x\mathbf{u}_x+y\mathbf{u}_y+z\mathbf{u}_z) & \text{for } r>a \text{ (outside the sphere)} \end{cases} \tag{E2.9.1}$$

where $\mathbf{u}_r$ is the unit vector in the direction of increasing the radial distance r from the sphere's center. Noting that $E_\theta=0$ and $E_\phi=0$, we can use Eq. (2.4.4c) to find the divergence of $\mathbf{E}$ as

$$\nabla\bullet\mathbf{E} \overset{(2.4.4c)}{=} \frac{1}{r^2}\frac{\partial(r^2E_r)}{\partial r}+\frac{1}{r\sin\theta}\frac{\partial(\not{E}_\theta\sin\theta)}{\partial\theta}+\frac{1}{r\sin\theta}\frac{\partial \not{E}_\phi}{\partial\phi}$$

$$=\begin{cases} \dfrac{1}{r^2}\dfrac{\partial}{\partial r}\left(\dfrac{r^3Q}{4\pi\varepsilon_0 a^3}\right)=\dfrac{3}{r^2}\dfrac{r^2Q}{4\pi\varepsilon_0 a^3}=\dfrac{1}{\varepsilon_0}\dfrac{Q}{4\pi a^3/3}=\dfrac{\rho_v}{\varepsilon_0} & \text{for } r\le a \\ \dfrac{1}{r^2}\dfrac{\partial}{\partial r}\left(\dfrac{r^2Q}{4\pi\varepsilon_0 r^2}\right)=\dfrac{1}{r^2}\dfrac{\partial}{\partial r}\left(\dfrac{Q}{4\pi\varepsilon_0}\right)=0 & \text{for } r>a \end{cases} \tag{E2.9.2}$$

Also, noting that E_r is not related to ϕ or θ, we can use Eq. (2.4.5c) to find the curl of $\mathbf{E}$ as

$$\nabla\times\mathbf{E} \overset{(2.4.5c)}{=} \frac{1}{r\sin\theta}\left(\frac{\partial(\not{E}_\phi\sin\theta)}{\partial\theta}-\frac{\partial\not{E}_\theta}{\partial\phi}\right)\mathbf{u}_r+\frac{1}{r}\left(\frac{1}{\sin\theta}\frac{\partial \not{E}_r}{\not{\partial}\phi}-\frac{\partial(r\not{E}_\phi)}{\partial r}\right)\mathbf{u}_\theta+\frac{1}{r}\left(\frac{\partial(r\not{E}_\theta)}{\partial r}-\frac{\partial\not{E}_r}{\not{\partial}\theta}\right)\mathbf{u}_\phi=\mathbf{0} \tag{E2.9.3}$$

From these results on the divergence and curl, we may say that the electric field is divergent, but irrotational (not circulant) (see Fig. 2.11).

We can use the Cartesian coordinate system to get the same result even if it is not recommended for this problem due to its complexity. To do this, we use Eq. (2.4.4a) to find the divergence of $\mathbf{E}$ as

$$\nabla\bullet\mathbf{E}=\frac{\partial E_x}{\partial x}+\frac{\partial E_y}{\partial y}+\frac{\partial E_z}{\partial z}$$

$$=\begin{cases} \dfrac{Q}{4\pi\varepsilon_0 a^3}(1+1+1)=\dfrac{1}{\varepsilon_0}\dfrac{Q}{4\pi a^3/3}=\dfrac{\rho_v}{\varepsilon_0} & \text{for } r\le a \\ \dfrac{(x^2+y^2+z^2)^{1/2}}{4\pi\varepsilon_0(x^2+y^2+z^2)^3}\left(x^2+y^2+z^2-3x^2+x^2+y^2+z^2-3y^2+x^2+y^2+z^2-3z^2\right)=0 & \text{for } r>a \end{cases} \tag{E2.9.4}$$

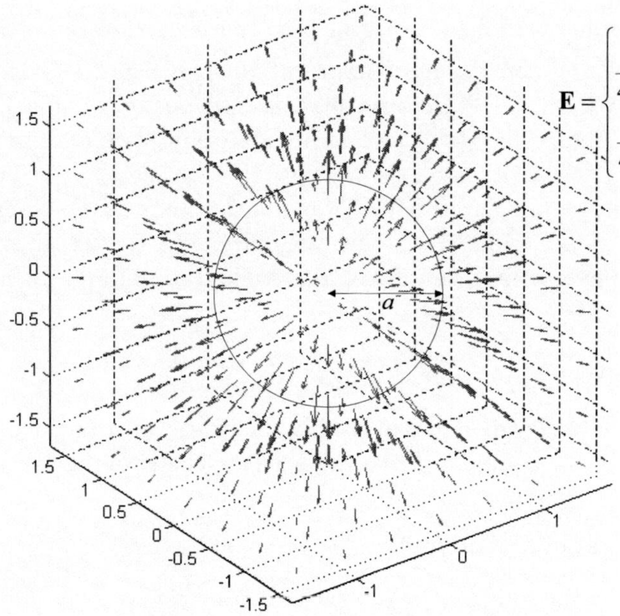

$$\mathbf{E}=\begin{cases}\overset{\text{spherical}}{\dfrac{rQ}{4\pi\varepsilon_0 a^3}\mathbf{u}_r}=\overset{\text{Cartesian (rectangular)}}{\dfrac{Q}{4\pi\varepsilon_0 a^3}(x\mathbf{u}_x+y\mathbf{u}_y+z\mathbf{u}_z)} & \text{for } r\le a\\ \dfrac{Q}{4\pi\varepsilon_0 r^2}\mathbf{u}_r=\dfrac{Q}{4\pi\varepsilon_0 r^3}(x\mathbf{u}_x+y\mathbf{u}_y+z\mathbf{u}_z) & \text{for } r>a\end{cases}$$

Figure 2.11 Electric field produced by a sphere of radius *a* carrying a uniformly distributed charge (Example 2.9)

Also, we can use Eq. (2.4.5a) to find the curl of **E** (in the Cartesian coordinate system) as

$$\nabla\times\mathbf{E}=\left(\frac{\partial E_z}{\partial y}-\frac{\partial E_y}{\partial z}\right)\mathbf{u}_x+\left(\frac{\partial E_x}{\partial z}-\frac{\partial E_z}{\partial x}\right)\mathbf{u}_y+\left(\frac{\partial E_y}{\partial x}-\frac{\partial E_x}{\partial y}\right)\mathbf{u}_z$$

$$=\left(\frac{-2y(3/2)(x^2+y^2+z^2)^{1/2}z}{4\pi\varepsilon_0(x^2+y^2+z^2)^3}+\frac{2z(3/2)(x^2+y^2+z^2)^{1/2}y}{4\pi\varepsilon_0(x^2+y^2+z^2)^3}\right)\mathbf{u}_x+0\mathbf{u}_y+0\mathbf{u}_z=\mathbf{0} \qquad \text{(E2.9.5)}$$

Note that Eqs. (E2.9.4) and (E2.9.5) obtained in the Cartesian coordinate system agree with Eqs. (E2.9.2) and (E2.9.3) obtained in the spherical coordinate system, respectively.

These computations can be done by using '`diva()`' and '`curla()`' as follows:

```
>>syms r Q e0 a;
>>Ea=[r*Q/4/pi/e0/a^3 0 0]; % Eq.(2.9.1a)
>>Eb=[Q/4/pi/e0/r^2 0 0]; % Eq.(2.9.1b)
>>diva(Ea,'sp'); diva(Eb,'sp'); % Eq.(2.9.2)
    Divergence(F)     Divergence(F)
        3 Q
     ----------            0
            3
     4 pi a  e0

>>curla(Ea,'sp'); curla(Eb,'sp'); % Eq.(2.9.3)
    Curl(F)           Curl(F)
    +-    -+          +-    -+
    |  0   |          |  0   |
    |      |          |      |
    |  0   |          |  0   |
    |      |          |      |
    |  0   |          |  0   |
    +-    -+          +-    -+
```

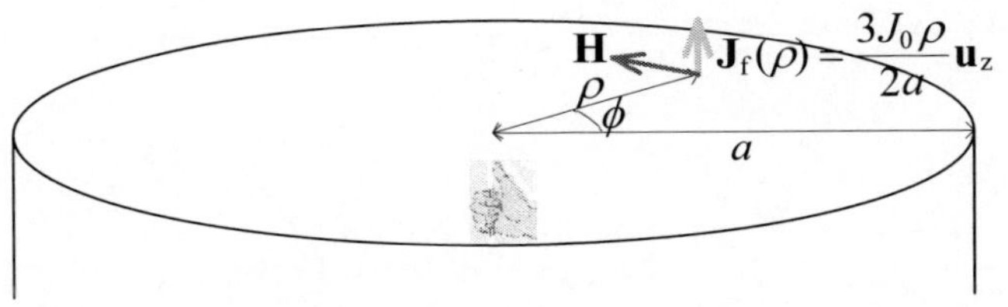

Figure 2.12 Magnetic field produced by a long wire of radius a having current density $J_f(\rho)=3J_0\rho/2a$

(Example 2.10) Curl and Divergence of a Magnetic Field

A long straight wire (of radius a) having current density $J_f(\rho)=3J_0\rho/2a$[C/m²] increasing linearly with radial distance ρ from the center axis of the wire produces a magnetic field at a point ρ[m] away from the axis center as

$$\mathbf{H}=\begin{cases}\dfrac{J_0\,\rho^2}{2a}\mathbf{u}_\phi=\dfrac{J_0\,\rho^2}{2a}(-\sin\phi\mathbf{u}_x+\cos\phi\mathbf{u}_y) & \text{for } \rho\le a \quad \text{(E2.10.1a)}\\ \dfrac{J_0\,a^2}{2\rho}\mathbf{u}_\phi=\dfrac{J_0\,a^2}{2\rho}(-\sin\phi\mathbf{u}_x+\cos\phi\mathbf{u}_y) & \text{for } \rho>a \quad \text{(E2.10.1b)}\end{cases}$$

where $\mathbf{u}_\phi$ is the unit vector in the direction of increasing the azimuth angle ϕ from the $+x$-axis (see Fig. 2.12).

Noting that $H_\rho=0$, $H_\phi=J_0\rho^2/2a$, or $J_0a^2/2\rho$, and $H_z=0$, we can use Eq. (2.4.4b) to find the divergence of **H** as

$$\nabla\cdot\mathbf{H}\overset{(2.4.4b)}{=}\frac{1}{\rho}\frac{\partial(\rho H_\rho)}{\partial\rho}+\frac{1}{\rho}\frac{\partial H_\phi}{\partial\phi}+\frac{\partial H_z}{\partial z}\begin{cases}\overset{(E2.10.1a)}{=}\dfrac{1}{\rho}\dfrac{\partial H_\phi}{\partial\phi}=\dfrac{J_0}{\rho}\dfrac{\partial}{\partial\phi}\left(\dfrac{\rho^2}{2a}\right)=0 & \text{for } \rho\le a\\ \overset{(E2.10.1b)}{=}\dfrac{1}{\rho}\dfrac{\partial H_\phi}{\partial\phi}=\dfrac{J_0}{\rho}\dfrac{\partial}{\partial\phi}\left(\dfrac{a^2}{2\rho}\right)=0 & \text{for } \rho>a\end{cases} \quad \text{(E2.10.2)}$$

Also, we can use Eq. (2.4.5b) to find the curl of **H** as

$$\nabla\times\mathbf{H}\overset{(2.4.5b)}{=}\left(\frac{1}{\rho}\frac{\partial H_z}{\partial\phi}-\frac{\partial H_\phi}{\partial z}\right)\mathbf{u}_\rho+\left(\frac{\partial H_\rho}{\partial z}-\frac{\partial H_z}{\partial\rho}\right)\mathbf{u}_\phi+\frac{1}{\rho}\left(\frac{\partial(\rho H_\phi)}{\partial\rho}-\frac{\partial H_\rho}{\partial\phi}\right)\mathbf{u}_z$$

$$\begin{cases}\overset{(E2.10.1a)}{=}\dfrac{1}{\rho}\dfrac{\partial(\rho H_\phi)}{\partial\rho}\mathbf{u}_z=\dfrac{J_0}{\rho}\dfrac{\partial}{\partial\rho}\left(\dfrac{\rho^3}{2a}\right)\mathbf{u}_z=\dfrac{J_0}{\rho}\dfrac{3\rho^2}{2a}\mathbf{u}_z=\dfrac{3J_0\rho}{2a}\mathbf{u}_z & \text{for } \rho\le a\\ \overset{(E2.10.1b)}{=}\dfrac{1}{\rho}\dfrac{\partial(\rho H_\phi)}{\partial\rho}\mathbf{u}_z=\dfrac{J_0}{\rho}\dfrac{\partial}{\partial\rho}\left(\dfrac{a^2}{2}\right)\mathbf{u}_z=\mathbf{0} & \text{for } \rho>a\end{cases} \quad \text{(E2.10.3)}$$

From these results on the divergence and curl, we may say that the magnetic field is circulant (only inside the wire), but not divergent anywhere.

These computations can be done by using '`diva()`' and '`curla()`' as follows:

```
>>syms r J0 a;
>>Ha=[0 J0*r^2/2/a 0]; % Eq.(2.10.1a)
>>Hb=[0 J0*a^2/r 0]; % Eq.(2.10.1b)
>>diva(Ha,'cy'); diva(Hb,'cy'); % Eq.(2.10.2)
>>curla(Ha,'cy'); curla(Hb,'cy'); % Eq.(2.10.3)
```

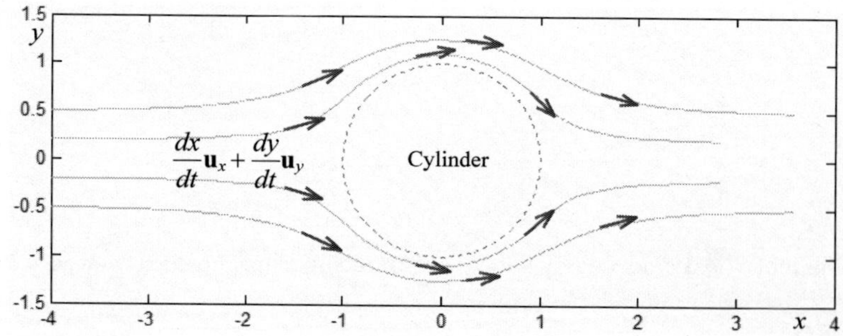

Figure 2.13 Velocity vector field of an ideal fluid around a cylinder for Example 2.11

(Example 2.11) Curl and Divergence of a Velocity Vector Field

Find the divergence and curl of a velocity vector field

$$\mathbf{v}(t) = v_x \mathbf{u}_x + v_y \mathbf{u}_y = \frac{dx}{dt}\mathbf{u}_x + \frac{dy}{dt}\mathbf{u}_y = V_0\left(1 - \frac{x^2-y^2}{(x^2+y^2)^2}\right)\mathbf{u}_x + V_0\frac{-2xy}{(x^2+y^2)^2}\mathbf{u}_y \quad \text{(E2.11.1)}$$

which is depicted in Fig. 2.13. Both of them turn out to be zero as follows:

$$\nabla \bullet \mathbf{v} \overset{(2.4.4a)}{=} \frac{\partial v_x}{\partial x} + \frac{\partial v_y}{\partial y} = V_0 \frac{\partial}{\partial x}\left(\frac{x^4+2x^2y^2+y^4-x^2+y^2}{(x^2+y^2)^2}\right) + V_0 \frac{\partial}{\partial y}\frac{-2xy}{(x^2+y^2)^2}$$

$$= V_0\left(\frac{2x(2x^2+2y^2-1)(x^2+y^2)-4x(x^4+2x^2y^2+y^4-x^2+y^2)}{(x^2+y^2)^3} + \frac{-2x(x^2+y^2)+8xy^2}{(x^2+y^2)^3}\right)$$

$$= V_0\frac{2x(2x^2+2y^2-2)(x^2+y^2)-4x(x^4+2x^2y^2+y^4-x^2-y^2)}{(x^2+y^2)^3} = 0 \quad \text{(E2.11.2)}$$

$$\nabla \times \mathbf{v} \overset{(2.4.5a)}{=} \left(\frac{\partial v_z}{\partial y} - \frac{\partial v_y}{\partial z}\right)\mathbf{u}_x + \left(\frac{\partial v_x}{\partial z} - \frac{\partial v_z}{\partial x}\right)\mathbf{u}_y + \left(\frac{\partial v_y}{\partial x} - \frac{\partial v_x}{\partial y}\right)\mathbf{u}_z$$

$$= \left(\frac{-2y(x^2+y^2)+8x^2y}{(x^2+y^2)^3} - \frac{2y(2x^2+2y^2+1)(x^2+y^2)-4y(x^4+2x^2y^2+y^4-x^2+y^2)}{(x^2+y^2)^3}\right)\mathbf{u}_z = \mathbf{0} \quad \text{(E2.11.3)}$$

Note that with

$$x^2+y^2=\rho^2, \quad \frac{x}{(x^2+y^2)^{1/2}}=\cos\phi, \quad \frac{y}{(x^2+y^2)^{1/2}}=\sin\phi \quad \text{(E2.11.4a)}$$

and

$$\begin{bmatrix}\mathbf{u}_x \\ \mathbf{u}_y\end{bmatrix} \overset{\text{Table 1.1}}{=} \begin{bmatrix}\cos\phi & -\sin\phi \\ \sin\phi & \cos\phi\end{bmatrix}\begin{bmatrix}\mathbf{u}_\rho \\ \mathbf{u}_\phi\end{bmatrix}, \quad \text{(E2.11.4b)}$$

we can rewrite Eq. (E2.11.1) in the cylindrical (polar) coordinate system as

$$\mathbf{v}(t) = V_0\left(1 - \frac{\cos^2\phi - \sin^2\phi}{\rho^2}\right)(\cos\phi\,\mathbf{u}_\rho - \sin\phi\,\mathbf{u}_\phi) - V_0\frac{2\cos\phi\sin\phi}{\rho^2}(\sin\phi\,\mathbf{u}_\rho + \cos\phi\,\mathbf{u}_\phi)$$

$$= V_0\left(1 - \frac{1}{\rho^2}\right)\cos\phi\,\mathbf{u}_\rho - V_0\left(1 + \frac{1}{\rho^2}\right)\sin\phi\,\mathbf{u}_\phi \quad \text{(E2.11.5)}$$

Once the vector field equation is written in the cylindrical (polar) coordinate system, it is easy to use Eq. (2.4.4b) to find the divergence of $\mathbf{v}$ as

$$\nabla\bullet\mathbf{v}\overset{(2.4.4b)}{=}\frac{1}{\rho}\frac{\partial(\rho H_\rho)}{\partial\rho}+\frac{1}{\rho}\frac{\partial H_\phi}{\partial\phi}+\frac{\partial H_z}{\partial z}=\frac{V_0}{\rho}\left\{\frac{\partial}{\partial\rho}\left(\rho-\frac{1}{\rho}\right)\cos\phi-\frac{\partial}{\partial\phi}\left(1+\frac{1}{\rho^2}\right)\sin\phi\right\}$$

$$=\frac{V_0}{\rho}\left\{\left(1+\frac{1}{\rho^2}\right)\cos\phi-\left(1+\frac{1}{\rho^2}\right)\cos\phi\right\}=0 \tag{E2.11.6}$$

Also, we can use Eq. (2.4.5b) to find the curl of $\mathbf{v}$ as

$$\nabla\times\mathbf{v}\overset{(2.4.5b)}{=}\left(\frac{1}{\rho}\frac{\partial H_z}{\partial\phi}-\frac{\partial H_\phi}{\partial z}\right)\mathbf{u}_\rho+\left(\frac{\partial H_\rho}{\partial z}-\frac{\partial H_z}{\partial\rho}\right)\mathbf{u}_\phi+\frac{1}{\rho}\left(\frac{\partial(\rho H_\phi)}{\partial\rho}-\frac{\partial H_\rho}{\partial\phi}\right)\mathbf{u}_z$$

$$=\frac{V_0}{\rho}\left\{-\frac{\partial}{\partial\rho}\left(\rho+\frac{1}{\rho}\right)\sin\phi-\frac{\partial}{\partial\phi}\left(1-\frac{1}{\rho^2}\right)\cos\phi\right\}\mathbf{u}_z$$

$$=\frac{V_0}{\rho}\left\{-\left(1-\frac{1}{\rho^2}\right)\sin\phi+\left(1-\frac{1}{\rho^2}\right)\sin\phi\right\}\mathbf{u}_z=\mathbf{0} \tag{E2.11.7}$$

These computations can be done by using '`diva()`' and '`curla()`' as follows:

```
>>syms r ph V0;
>>v=V0*[(1-1/r^2)*cos(ph); -(1+1/r^2)*sin(ph); 0]; % Eq.(2.11.5)
>>diva(v,'cy'); curla(v,'cy');
```

Many useful theorems and identities on divergence and curl are listed in Table 2.1. Note that Eq. (2.4.13) can be verified as

$$\nabla\bullet(a\mathbf{A})=\frac{\partial(aA_x)}{\partial x}+\frac{\partial(aA_y)}{\partial y}+\frac{\partial(aA_z)}{\partial z}=\frac{\partial a}{\partial x}A_x+a\frac{\partial A_x}{\partial x}+\frac{\partial a}{\partial y}A_y+a\frac{\partial A_y}{\partial y}+\frac{\partial a}{\partial z}A_z+a\frac{\partial A_z}{\partial z}$$

$$=\begin{bmatrix}\partial a/\partial x\\ \partial a/\partial y\\ \partial a/\partial z\end{bmatrix}\bullet\begin{bmatrix}A_x\\ A_y\\ A_z\end{bmatrix}+a\begin{bmatrix}\partial/\partial x\\ \partial/\partial y\\ \partial/\partial z\end{bmatrix}\bullet\begin{bmatrix}A_x\\ A_y\\ A_z\end{bmatrix}=\mathbf{A}\bullet\nabla a+a\nabla\bullet\mathbf{A}$$

Eq. (2.4.15) can be verified as

$$\nabla\times(a\mathbf{A})=\begin{bmatrix}\partial/\partial x\\ \partial/\partial y\\ \partial/\partial z\end{bmatrix}\times\begin{bmatrix}aA_x\\ aA_y\\ aA_z\end{bmatrix}=\begin{bmatrix}\partial(aA_z)/\partial y-\partial(aA_y)/\partial z\\ \partial(aA_x)/\partial z-\partial(aA_z)/\partial x\\ \partial(aA_y)/\partial x-\partial(aA_x)/\partial y\end{bmatrix}$$

$$=\begin{bmatrix}(\partial a/\partial y)A_z-(\partial a/\partial z)A_y\\ (\partial a/\partial y)A_x-(\partial a/\partial z)A_z\\ (\partial a/\partial y)A_y-(\partial a/\partial z)A_x\end{bmatrix}+\begin{bmatrix}a\partial A_z/\partial y-a\partial A_y/\partial z\\ a\partial A_x/\partial z-a\partial A_z/\partial x\\ a\partial A_y/\partial x-a\partial A_x/\partial y\end{bmatrix}$$

$$=\begin{bmatrix}\partial a/\partial x\\ \partial a/\partial y\\ \partial a/\partial z\end{bmatrix}\times\begin{bmatrix}A_x\\ A_y\\ A_z\end{bmatrix}+a\begin{bmatrix}\partial/\partial x\\ \partial/\partial y\\ \partial/\partial z\end{bmatrix}\times\begin{bmatrix}A_x\\ A_y\\ A_z\end{bmatrix}=\nabla a\times\mathbf{A}+a\nabla\times\mathbf{A}$$

Eq. (2.4.16) can be verified as

$$\text{LHS} = \nabla \bullet (\mathbf{A} \times \mathbf{B})$$

$$= \begin{bmatrix} \partial/\partial x \\ \partial/\partial y \\ \partial/\partial z \end{bmatrix} \bullet \left(\begin{bmatrix} A_x \\ A_y \\ A_z \end{bmatrix} \times \begin{bmatrix} B_x \\ B_y \\ B_z \end{bmatrix} \right) \overset{(1.1.20)}{=} \begin{bmatrix} \partial/\partial x \\ \partial/\partial y \\ \partial/\partial z \end{bmatrix} \bullet \begin{bmatrix} A_y B_z - A_z B_y \\ A_z B_x - A_x B_z \\ A_x B_y - A_y B_x \end{bmatrix}$$

$$= \frac{\partial (A_y B_z - A_z B_y)}{\partial x} + \frac{\partial (A_z B_x - A_x B_z)}{\partial y} + \frac{\partial (A_x B_y - A_y B_x)}{\partial z}$$

$$\text{RHS} = \mathbf{B} \bullet (\nabla \times \mathbf{A}) - \mathbf{A} \bullet (\nabla \times \mathbf{B})$$

$$\overset{(1.1.20)}{=} \begin{bmatrix} B_x \\ B_y \\ B_z \end{bmatrix} \bullet \left(\begin{bmatrix} \partial/\partial x \\ \partial/\partial y \\ \partial/\partial z \end{bmatrix} \times \begin{bmatrix} A_x \\ A_y \\ A_z \end{bmatrix} \right) - \begin{bmatrix} A_x \\ A_y \\ A_z \end{bmatrix} \bullet \left(\begin{bmatrix} \partial/\partial x \\ \partial/\partial y \\ \partial/\partial z \end{bmatrix} \times \begin{bmatrix} B_x \\ B_y \\ B_z \end{bmatrix} \right)$$

$$= B_x \left(\frac{\partial A_z}{\partial y} - \frac{\partial A_y}{\partial z} \right) + B_y \left(\frac{\partial A_x}{\partial z} - \frac{\partial A_z}{\partial x} \right) + B_z \left(\frac{\partial A_y}{\partial x} - \frac{\partial A_x}{\partial y} \right)$$

$$- A_x \left(\frac{\partial B_z}{\partial y} - \frac{\partial B_y}{\partial z} \right) - A_y \left(\frac{\partial B_x}{\partial z} - \frac{\partial B_z}{\partial x} \right) - A_z \left(\frac{\partial B_y}{\partial x} - \frac{\partial B_x}{\partial y} \right)$$

Eq. (2.4.17) can be verified as

$$\text{LHS} = \nabla (\mathbf{A} \bullet \mathbf{B}) = \begin{bmatrix} \partial (A_x B_x + A_y B_y + A_z B_z)/\partial x \\ \partial (A_x B_x + A_y B_y + A_z B_z)/\partial y \\ \partial (A_x B_x + A_y B_y + A_z B_z)/\partial z \end{bmatrix}$$

$$= \begin{bmatrix} A_x \partial B_x/\partial x + B_x \partial A_x/\partial x + A_y \partial B_y/\partial x + B_y \partial A_y/\partial x + A_z \partial B_z/\partial x + B_z \partial A_z/\partial x \\ A_x \partial B_x/\partial y + B_x \partial A_x/\partial y + A_y \partial B_y/\partial y + B_y \partial A_y/\partial y + A_z \partial B_z/\partial y + B_z \partial A_z/\partial y \\ A_x \partial B_x/\partial z + B_x \partial A_x/\partial z + A_y \partial B_y/\partial z + B_y \partial A_y/\partial z + A_z \partial B_z/\partial z + B_z \partial A_z/\partial z \end{bmatrix}$$

$$\text{RHS} = (\mathbf{A} \bullet \nabla)\mathbf{B} + (\mathbf{B} \bullet \nabla)\mathbf{A} + \mathbf{A} \times (\nabla \times \mathbf{B}) + \mathbf{B} \times (\nabla \times \mathbf{A})$$

$$\overset{(1.1.20)}{=} A_x \frac{\partial}{\partial x} \begin{bmatrix} B_x \\ B_y \\ B_z \end{bmatrix} + A_y \frac{\partial}{\partial y} \begin{bmatrix} B_x \\ B_y \\ B_z \end{bmatrix} + A_z \frac{\partial}{\partial z} \begin{bmatrix} B_x \\ B_y \\ B_z \end{bmatrix} + B_x \frac{\partial}{\partial x} \begin{bmatrix} A_x \\ A_y \\ A_z \end{bmatrix} + B_y \frac{\partial}{\partial y} \begin{bmatrix} A_x \\ A_y \\ A_z \end{bmatrix} + B_z \frac{\partial}{\partial z} \begin{bmatrix} A_x \\ A_y \\ A_z \end{bmatrix}$$

$$+ \begin{bmatrix} A_x \\ A_y \\ A_z \end{bmatrix} \times \begin{bmatrix} \partial B_z/\partial y - \partial B_y/\partial z \\ \partial B_x/\partial z - \partial B_z/\partial x \\ \partial B_y/\partial x - \partial B_x/\partial y \end{bmatrix} + \begin{bmatrix} B_x \\ B_y \\ B_z \end{bmatrix} \times \begin{bmatrix} \partial A_z/\partial y - \partial A_y/\partial z \\ \partial A_x/\partial z - \partial A_z/\partial x \\ \partial A_y/\partial x - \partial A_x/\partial y \end{bmatrix}$$

$$\overset{(1.1.20)}{=} A_x \frac{\partial}{\partial x} \begin{bmatrix} B_x \\ B_y \\ B_z \end{bmatrix} + A_y \frac{\partial}{\partial y} \begin{bmatrix} B_x \\ B_y \\ B_z \end{bmatrix} + A_z \frac{\partial}{\partial z} \begin{bmatrix} B_x \\ B_y \\ B_z \end{bmatrix} + B_x \frac{\partial}{\partial x} \begin{bmatrix} A_x \\ A_y \\ A_z \end{bmatrix} + B_y \frac{\partial}{\partial y} \begin{bmatrix} A_x \\ A_y \\ A_z \end{bmatrix} + B_z \frac{\partial}{\partial z} \begin{bmatrix} A_x \\ A_y \\ A_z \end{bmatrix}$$

$$+ \begin{bmatrix} A_y(\partial B_y/\partial x - \partial B_x/\partial y) - A_z(\partial B_x/\partial z - \partial B_z/\partial x) + B_y(\partial A_y/\partial x - \partial A_x/\partial y) - B_z(\partial A_x/\partial z - \partial A_z/\partial x) \\ A_z(\partial B_z/\partial y - \partial B_y/\partial z) - A_x(\partial B_y/\partial x - \partial B_x/\partial y) + B_z(\partial A_z/\partial y - \partial A_y/\partial z) - B_x(\partial A_y/\partial x - \partial A_x/\partial y) \\ A_x(\partial B_x/\partial z - \partial B_z/\partial x) - A_y(\partial B_z/\partial y - \partial B_y/\partial z) + B_x(\partial A_x/\partial z - \partial A_z/\partial x) - B_y(\partial A_z/\partial y - \partial A_y/\partial z) \end{bmatrix}$$

Eq. (2.4.18) can be verified as

$$\nabla\times(\mathbf{A}\times\mathbf{B}) \overset{(2.1.9)}{=} \begin{bmatrix}\partial/\partial x\\ \partial/\partial y\\ \partial/\partial z\end{bmatrix}\times\begin{bmatrix}A_x\\ A_y\\ A_z\end{bmatrix}\times\begin{bmatrix}B_x\\ B_y\\ B_z\end{bmatrix} \overset{(1.1.20)}{=} \begin{bmatrix}\partial/\partial x\\ \partial/\partial y\\ \partial/\partial z\end{bmatrix}\times\begin{bmatrix}A_yB_z-A_zB_y\\ A_zB_x-A_xB_z\\ A_xB_y-A_yB_x\end{bmatrix}$$

$$=\left(\frac{\partial(A_xB_y-A_yB_x)}{\partial y}-\frac{\partial(A_zB_x-A_xB_z)}{\partial z}\right)\mathbf{u}_x+\left(\frac{\partial(A_yB_z-A_zB_y)}{\partial z}-\frac{\partial(A_xB_y-A_yB_x)}{\partial x}\right)\mathbf{u}_y$$

$$+\left(\frac{\partial(A_zB_x-A_xB_z)}{\partial x}-\frac{\partial(A_yB_z-A_zB_y)}{\partial y}\right)\mathbf{u}_z$$

$$\mathbf{A}(\nabla\cdot\mathbf{B})-\mathbf{B}(\nabla\cdot\mathbf{A})+(\mathbf{B}\cdot\nabla)\mathbf{A}-(\mathbf{A}\cdot\nabla)\mathbf{B}$$

$$\underset{(1.1.3)}{\overset{(2.1.9)}{=}}(A_x\mathbf{u}_x+A_y\mathbf{u}_y+A_z\mathbf{u}_z)\left(\frac{\partial B_x}{\partial x}+\frac{\partial B_y}{\partial y}+\frac{\partial B_z}{\partial z}\right)-(B_x\mathbf{u}_x+B_y\mathbf{u}_y+B_z\mathbf{u}_z)\left(\frac{\partial A_x}{\partial x}+\frac{\partial A_y}{\partial y}+\frac{\partial A_z}{\partial z}\right)$$

$$+\left(B_x\frac{\partial}{\partial x}+B_y\frac{\partial}{\partial y}+B_z\frac{\partial}{\partial z}\right)(A_x\mathbf{u}_x+A_y\mathbf{u}_y+A_z\mathbf{u}_z)-\left(A_x\frac{\partial}{\partial x}+A_y\frac{\partial}{\partial y}+A_z\frac{\partial}{\partial z}\right)(B_x\mathbf{u}_x+B_y\mathbf{u}_y+B_z\mathbf{u}_z)$$

$$=\left(A_x\frac{\partial B_y}{\partial y}+\frac{\partial A_x}{\partial y}B_y-A_y\frac{\partial B_x}{\partial y}-\frac{\partial A_y}{\partial y}B_x-A_z\frac{\partial B_x}{\partial z}-\frac{\partial A_z}{\partial z}B_x+A_x\frac{\partial B_z}{\partial z}-\frac{\partial A_x}{\partial z}B_z\right)\mathbf{u}_x$$

$$+\left(A_y\frac{\partial B_z}{\partial z}+\frac{\partial A_y}{\partial z}B_z-A_z\frac{\partial B_y}{\partial z}-\frac{\partial A_z}{\partial z}B_y-A_x\frac{\partial B_y}{\partial x}-\frac{\partial A_x}{\partial x}B_y+A_y\frac{\partial B_x}{\partial x}-\frac{\partial A_y}{\partial x}B_x\right)\mathbf{u}_y$$

$$+\left(A_z\frac{\partial B_x}{\partial x}+\frac{\partial A_z}{\partial x}B_x-A_x\frac{\partial B_z}{\partial x}-\frac{\partial A_x}{\partial x}B_z-A_y\frac{\partial B_z}{\partial y}-\frac{\partial A_y}{\partial y}B_z+A_z\frac{\partial B_y}{\partial y}-\frac{\partial A_z}{\partial y}B_y\right)\mathbf{u}_y$$

Eq. (2.4.20) can be verified as

$$\nabla(\nabla\cdot\mathbf{A})\underset{(2.4.4a)}{\overset{(2.1.9)}{=}}\frac{\partial}{\partial x}\left(\frac{\partial A_x}{\partial x}+\frac{\partial A_y}{\partial y}+\frac{\partial A_z}{\partial z}\right)\mathbf{u}_x+\frac{\partial}{\partial y}\left(\frac{\partial A_x}{\partial x}+\frac{\partial A_y}{\partial y}+\frac{\partial A_z}{\partial z}\right)\mathbf{u}_y+\frac{\partial}{\partial z}\left(\frac{\partial A_x}{\partial x}+\frac{\partial A_y}{\partial y}+\frac{\partial A_z}{\partial z}\right)\mathbf{u}_z$$

$$\nabla\times(\nabla\times\mathbf{A})\overset{(2.4.5a)}{=}\left(\left(\frac{\partial^2 A_y}{\partial y\partial x}-\frac{\partial^2 A_x}{\partial y^2}\right)-\left(\frac{\partial^2 A_x}{\partial z^2}-\frac{\partial^2 A_z}{\partial z\partial x}\right)\right)\mathbf{u}_x+\left(\left(\frac{\partial^2 A_z}{\partial z\partial y}-\frac{\partial^2 A_y}{\partial z^2}\right)-\left(\frac{\partial^2 A_y}{\partial x^2}-\frac{\partial^2 A_x}{\partial x\partial y}\right)\right)\mathbf{u}_y$$

$$+\left(\left(\frac{\partial^2 A_x}{\partial x\partial z}-\frac{\partial^2 A_z}{\partial x^2}\right)-\left(\frac{\partial^2 A_z}{\partial y^2}-\frac{\partial^2 A_y}{\partial y\partial z}\right)\right)\mathbf{u}_z$$

$$;\nabla(\nabla\cdot\mathbf{A})-\nabla\times(\nabla\times\mathbf{A})=\left(\frac{\partial^2 A_x}{\partial x^2}+\frac{\partial^2 A_y}{\partial x\partial y}+\frac{\partial^2 A_z}{\partial x\partial z}-\frac{\partial^2 A_y}{\partial y\partial x}+\frac{\partial^2 A_x}{\partial y^2}+\frac{\partial^2 A_x}{\partial z^2}-\frac{\partial^2 A_z}{\partial z\partial x}\right)\mathbf{u}_x$$

$$+\left(\frac{\partial^2 A_x}{\partial x\partial y}+\frac{\partial^2 A_y}{\partial y^2}+\frac{\partial^2 A_z}{\partial y\partial z}-\frac{\partial^2 A_z}{\partial z\partial y}+\frac{\partial^2 A_y}{\partial z^2}+\frac{\partial^2 A_y}{\partial x^2}-\frac{\partial^2 A_x}{\partial x\partial y}\right)\mathbf{u}_y$$

$$+\left(\frac{\partial^2 A_x}{\partial x\partial z}+\frac{\partial^2 A_y}{\partial y\partial z}+\frac{\partial^2 A_z}{\partial z^2}-\frac{\partial^2 A_x}{\partial x\partial z}+\frac{\partial^2 A_z}{\partial x^2}+\frac{\partial^2 A_z}{\partial y^2}-\frac{\partial^2 A_y}{\partial y\partial z}\right)\mathbf{u}_z$$

$$=\left(\frac{\partial^2}{\partial x^2}+\frac{\partial^2}{\partial y^2}+\frac{\partial^2}{\partial z^2}\right)(A_x\mathbf{u}_x+A_y\mathbf{u}_y+A_z\mathbf{u}_z)=\nabla^2\mathbf{A}$$

Also, Eqs. (2.4.21) an (2.4.22) can be verified as

$$\nabla\cdot(\nabla\times\mathbf{A}) \overset{(2.4.4a)}{\underset{(2.4.5a)}{=}} \left(\mathbf{u}_x\frac{\partial}{\partial x}+\mathbf{u}_y\frac{\partial}{\partial y}+\mathbf{u}_z\frac{\partial}{\partial z}\right)\cdot\left\{\left(\frac{\partial A_z}{\partial y}-\frac{\partial A_y}{\partial z}\right)\mathbf{u}_x+\left(\frac{\partial A_x}{\partial z}-\frac{\partial A_z}{\partial x}\right)\mathbf{u}_y+\left(\frac{\partial A_y}{\partial x}-\frac{\partial A_x}{\partial y}\right)\mathbf{u}_z\right\}$$

$$\overset{(1.1.3)}{=} \frac{\partial}{\partial x}\left(\frac{\partial A_z}{\partial y}-\frac{\partial A_y}{\partial z}\right)+\frac{\partial}{\partial y}\left(\frac{\partial A_x}{\partial z}-\frac{\partial A_z}{\partial x}\right)+\frac{\partial}{\partial z}\left(\frac{\partial A_y}{\partial x}-\frac{\partial A_x}{\partial y}\right)=0$$

$$\nabla\times\nabla a \overset{(2.1.9)}{=} \nabla\times\left(\mathbf{u}_x\frac{\partial a}{\partial x}+\mathbf{u}_y\frac{\partial a}{\partial y}+\mathbf{u}_z\frac{\partial a}{\partial z}\right)$$

$$\overset{(2.1.9)}{=} \begin{bmatrix}\partial/\partial x\\ \partial/\partial y\\ \partial/\partial z\end{bmatrix}\times\begin{bmatrix}\partial a/\partial x\\ \partial a/\partial y\\ \partial a/\partial z\end{bmatrix} \overset{(1.1.20)}{=} \left(\frac{\partial^2 a}{\partial y\partial z}-\frac{\partial^2 a}{\partial z\partial y}\right)\mathbf{u}_x+\left(\frac{\partial^2 a}{\partial z\partial x}-\frac{\partial^2 a}{\partial x\partial z}\right)\mathbf{u}_y+\left(\frac{\partial^2 a}{\partial x\partial y}-\frac{\partial^2 a}{\partial y\partial x}\right)\mathbf{u}_z=0$$

Table 2.1 Identities and Theorems on Divergence and Curl

Vector Identities:

$$\mathbf{A}\times\mathbf{B}=-\mathbf{B}\times\mathbf{A} \tag{2.4.10}$$

$$(\mathbf{A}\times\mathbf{B})\cdot\mathbf{C}=(\mathbf{B}\times\mathbf{C})\cdot\mathbf{A}=(\mathbf{C}\times\mathbf{A})\cdot\mathbf{B} \tag{2.4.11}$$

$$\mathbf{A}\times(\mathbf{B}\times\mathbf{C})=(\mathbf{C}\cdot\mathbf{A})\mathbf{B}-(\mathbf{B}\cdot\mathbf{A})\mathbf{C} \tag{2.4.12}$$

$$\nabla\cdot(a\mathbf{A})=\mathbf{A}\cdot\nabla a+a\nabla\cdot\mathbf{A} \tag{2.4.13}$$

$$\nabla(ab)=a\nabla b+b\nabla a \tag{2.4.14}$$

$$\nabla\times(a\mathbf{A})=\nabla a\times\mathbf{A}+a\nabla\times\mathbf{A} \tag{2.4.15}$$

$$\nabla\cdot(\mathbf{A}\times\mathbf{B})=\mathbf{B}\cdot(\nabla\times\mathbf{A})-\mathbf{A}\cdot(\nabla\times\mathbf{B}) \tag{2.4.16}$$

$$\nabla(\mathbf{A}\cdot\mathbf{B})=(\mathbf{A}\cdot\nabla)\mathbf{B}+(\mathbf{B}\cdot\nabla)\mathbf{A}+\mathbf{A}\times(\nabla\times\mathbf{B})+\mathbf{B}\times(\nabla\times\mathbf{A}) \tag{2.4.17}$$

$$\nabla\times(\mathbf{A}\times\mathbf{B})=\mathbf{A}(\nabla\cdot\mathbf{B})-\mathbf{B}(\nabla\cdot\mathbf{A})+(\mathbf{B}\cdot\nabla)\mathbf{A}-(\mathbf{A}\cdot\nabla)\mathbf{B} \tag{2.4.18}$$

$$\nabla\cdot\nabla a=\nabla^2 a=\frac{\partial^2 a}{\partial x^2}+\frac{\partial^2 a}{\partial y^2}+\frac{\partial^2 a}{\partial z^2}:\ \text{Laplacian of a scalar function} \tag{2.4.19}$$

$$\nabla^2\mathbf{A}=\left(\frac{\partial^2}{\partial x^2}+\frac{\partial^2}{\partial y^2}+\frac{\partial^2}{\partial z^2}\right)(A_x\mathbf{u}_x+A_y\mathbf{u}_y+A_z\mathbf{u}_z):\ \text{Laplacian of a vector function} \tag{2.4.20}$$

$$=\nabla(\nabla\cdot\mathbf{A})-\nabla\times(\nabla\times\mathbf{A})$$

$$\nabla\cdot(\nabla\times\mathbf{A})=0:\ \text{Null identity I} \tag{2.4.21}$$

$$\nabla\times\nabla a=\mathbf{0}\quad:\ \text{Null identity II} \tag{2.4.22}$$

Stokes' Theorem: $\oint\!\!\int_S \nabla\times\mathbf{A}\cdot d\mathbf{s}=\oint_C \mathbf{A}\cdot d\mathbf{l}$ (2.4.23)

$\iiint_V \nabla\times\mathbf{A}\,dv=-\oint\!\!\int_S \mathbf{A}\times d\mathbf{s}$ (To be discussed in Sec. 2.9) (2.4.24)

$\iiint_V \nabla a\,dv=\oint\!\!\int_S a\,d\mathbf{s}$ (2.4.25)

$\oint\!\!\int_S \nabla a\times d\mathbf{s}=-\oint_C a\,d\mathbf{l}$ (2.4.26)

Divergence Theorem: $\iiint_V \nabla\cdot\mathbf{A}\,dv=\oint\!\!\int_S \mathbf{A}\cdot d\mathbf{s}$ (see Sec. 2.11) (2.4.27)

2.5 LINE INTEGRALS AND PATH INDEPENDENCE

2.5.1 Line Integrals

2.5.1.1 Line (Path or Curve) Integral of Scalar Function on Plane

<Line Integral of a Scalar Function *f*(*x*,*y*) along a Curve *C* Defined Parametrically>

The integrations of a scalar function $f(x,y)$ along a smooth curve C parameterized by $x=x(t)$ and $y=y(t)$ for $a\le t\le b$ w.r.t. x, y, and arc length l are carried out w.r.t. t as follows:

$$\int_C f(x,y)\,dx=\int_a^b f(x(t),y(t))\,x'(t)\,dt \tag{2.5.1a}$$

$$\int_C f(x,y)\,dy=\int_a^b f(x(t),y(t))\,y'(t)\,dt \tag{2.5.1b}$$

$$\int_C f(x,y)\,dl=\int_a^b f(x(t),y(t))\,\sqrt{(x'(t))^2+(y'(t))^2}\,dt \tag{2.5.1c}$$

(**Example 2.12**) Line Integrals on a Plane

Evaluate the following three integrals

$$\text{(a) }\int_C x^2y\,dx \qquad \text{(b) }\int_C x^2y\,dy \qquad \text{(c) }\int_C x^2y\,dl \tag{E2.12.1}$$

on the quarter-circle C parameterized by $2\cos t\,\mathbf{u}_x+2\sin t\,\mathbf{u}_y$ for $0\le t\le \pi/2$.

(a) $\int_C x^2y\,dx \overset{(2.5.1a)}{=} \int_a^b f(x(t),y(t))\,x'(t)\,dt=\int_0^{\pi/2}\{(2\cos t)^2 2\sin t\}(-2\sin t)\,dt$

$$\overset{(C.16)}{=}-4\int_0^{\pi/2}\sin^2 2t\,dt \overset{(C.14)}{=}-2\int_0^{\pi/2}(1-\cos 4t)\,dt \overset{(C.36)}{=}-2\left(\frac{\pi}{2}-\frac{1}{4}\sin 4t\Big|_0^{\pi/2}\right)=-\pi \tag{E2.12.2}$$

(b) $\int_C x^2y\,dy \overset{(2.5.1b)}{=} \int_a^b f(x(t),y(t))\,y'(t)\,dt=\int_0^{\pi/2}(2\cos t)^2(2\sin t)\,(2\cos t)\,dt$

$$=16\int_0^{\pi/2}\cos^3 t\,(\sin t\,dt) \overset{(C.34)}{=}-4\cos^4 t\Big|_0^{\pi/2}=4 \tag{E2.12.3}$$

(c) $\int_C x^2y\,dl \overset{(2.5.1c)}{=} \int_a^b f(x(t),y(t))\,\sqrt{(x'(t))^2+(y'(t))^2}\,dt=\int_0^{\pi/2}(2\cos t)^2(2\sin t)\sqrt{(-2\sin t)^2+(2\cos t)^2}\,dt$

$$=16\int_0^{\pi/2}\cos^2 t\,\sin t\,dt \overset{(C.34)}{=}-\frac{16}{3}\cos^3 t\Big|_0^{\pi/2}=\frac{16}{3} \tag{E2.12.4}$$

<Line Integral of a Scalar Function *f*(*x*,*y*) along a Curve *C* Defined by an Explicit Function>

The integrations of a scalar function $f(x,y)$ along a smooth curve C defined by an explicit function $y=g(x)$ for $a\le t\le b$ w.r.t. x, y, and arc length l are carried out w.r.t. x as follows:

$$\int_C f(x,y)\,dx=\int_a^b f(x,g(x))\,dx \tag{2.5.2a}$$

$$\int_C f(x,y)\,dy=\int_a^b f(x,g(x))\,g'(x)\,dx \tag{2.5.2b}$$

$$\int_C f(x,y)\,dl=\int_a^b f(x,g(x))\,\sqrt{1+(g'(x))^2}\,dx \tag{2.5.2c}$$

(Example 2.13) Line Integrals on a Plane

Evaluate the integrals

$$\oint_{C_1\text{-}C_2\text{-}C_3} -y\,dx + x\,dy \qquad \text{(E2.13.1a)}$$

$$\int_{C_2} 1\,dl \qquad \text{(E2.13.1b)}$$

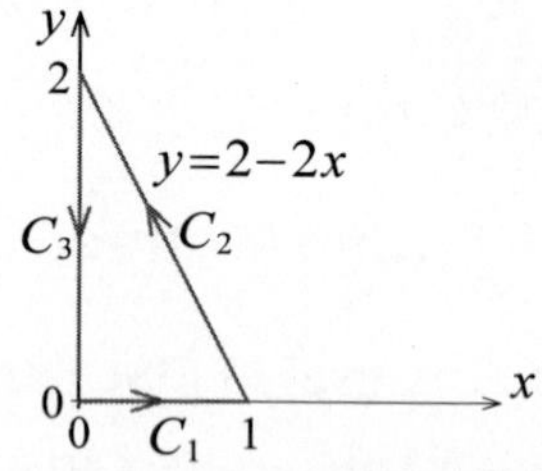

Figure 2.14 Closed curves as the paths of integration for Example 2.13

where the curves C_1, C_2, and C_3 are shown in Fig. 2.14.

(a) The integral on the piecewise-smooth closed curve C_1-C_2-C_3 can be obtained as the sum of three integrals each on C_1, C_2, and C_3:

$$\int_{C_1} -y\,dx + x\,dy \overset{y=0,\,dy=0}{=} \int_0^1 0\,dx + \int_0^0 x\,dy = 0 \qquad \text{(E2.13.2a)}$$

$$\int_{C_2} -y\,dx + x\,dy \overset{y=2-2x,\,dy=-2dx}{=} \int_1^0 (-2+2x)\,dx + \int_1^0 x(-2dx) = -2x + x^2 - x^2\Big|_1^0 = 2 \qquad \text{(E2.13.2b)}$$

$$\int_{C_3} -y\,dx + x\,dy \overset{x=0,\,dx=0}{=} \int_0^0 -y\,dx + \int_2^0 0\,dy = 0 \qquad \text{(E2.13.2c)}$$

$$;\quad \oint_{C_1\text{-}C_2\text{-}C_3} -y\,dx + x\,dy = \text{Sum of Eqs. (E2.13.2a,b,c)} = 0 + 2 + 0 = 2 \qquad \text{(E2.13.3)}$$

This illustrates that the line integral of $(-y\,dx + x\,dy)$ around a closed curve equals twice the area of the region enclosed by the curve given as the path of integration.

(b) We can find the line integral (E2.13.1b) w.r.t. arc length by using Eq. (2.5.1c) with $f(x, y) = 1$, $x=1-t$, and $y=2t$ for $0 \le t \le 1$:

$$\int_{C_2} 1\,dl \underset{x=1-t,\,y=2t \text{ for } 0\le t\le 1}{\overset{(2.5.1c)\text{ with } f(x,y)=1}{=}} \int_0^1 \sqrt{(x'(t))^2 + (y'(t))^2}\,dt = \int_0^1 \sqrt{(-1)^2 + 2^2}\,dt = \sqrt{5} \qquad \text{(E2.13.4)}$$

This illustrates that the line integral of $f(x,y) = 1$ w.r.t. arc length around a curve equals the length of the curve given as the path of integration.

2.5.1.2 Line (Path or Curve) Integral of Scalar Function in Space

The integrations of a scalar function $f(x, y, z)$ along a smooth curve C parameterized by $x=x(t)$, $y=y(t)$, and $z=z(t)$ for $a \le t \le b$ w.r.t. x, y, z, and arc length l are carried out w.r.t. t as follows:

$$\int_C f(x,y,z)\,dx = \int_a^b f(x(t), y(t), z(t))\, x'(t)\,dt \qquad \text{(2.5.3a)}$$

$$\int_C f(x,y,z)\,dy = \int_a^b f(x(t), y(t), z(t))\, y'(t)\,dt \qquad \text{(2.5.3b)}$$

$$\int_C f(x,y,z)\,dz = \int_a^b f(x(t), y(t), z(t))\, z'(t)\,dt \qquad \text{(2.5.3c)}$$

$$\int_C f(x,y,z)\,dl = \int_a^b f(x(t), y(t), z(t)) \sqrt{(x'(t))^2 + (y'(t))^2 + (z'(t))^2}\,dt \qquad \text{(2.5.3d)}$$

(Example 2.14) Line Integral of a Scalar Function in 3D Space

Evaluate the following integrals w.r.t. arc length l :

$$\int_{C_1} \frac{2}{x^2+y^2+z^2} dl, \ \int_{C_2} \frac{2z}{(x^2+y^2+z^2)^{3/2}} dl, \text{ and } \int_{C_3} \frac{2x}{(x^2+y^2+z^2)^{3/2}} dl \tag{E2.14.1}$$

where the three curves C_1, C_2, and C_3, depicted in Fig. 2.15, are parameterized by

$$(1-t)\mathbf{u}_x+(1-t)\mathbf{u}_z \text{ for } 0\le t\le \frac{1}{2},\ \mathbf{u}_x+(1-t)\mathbf{u}_z \text{ for } 0\le t\le \frac{1}{2}, \text{ and } (1-t)\mathbf{u}_x+\frac{1}{2}\mathbf{u}_z \text{ for } 0\le t\le \frac{1}{2}, \tag{E2.14.2}$$

respectively.

The integrals are computed as follows:

$$\int_{C_1} \frac{2}{x^2+y^2+z^2} dl \overset{(2.5.3d)}{=} \int_0^{0.5} \frac{2\sqrt{(x'(t))^2+(y'(t))^2+(z'(t))^2}}{x^2+y^2+z^2} dt = \int_0^{0.5} \frac{2\sqrt{(-1)^2+0^2+(-1)^2}}{(1-t)^2+0^2+(1-t)^2} dt$$
$$= \int_0^{0.5} \frac{2\sqrt{2}}{2(1-t)^2} dt = \left.\frac{\sqrt{2}}{1-t}\right|_0^{0.5} = \sqrt{2} \tag{E2.14.3}$$

$$\int_{C_2} \frac{2z}{(x^2+y^2+z^2)^{3/2}} dl \overset{(2.5.3d)}{=} \int_0^{0.5} \frac{2(1-t)\sqrt{0^2+0^2+(-1)^2}}{(1^2+0^2+(1-t)^2)^{3/2}} dt \overset{1-t=\tan\theta}{=} \int_{\pi/4}^{\tan^{-1}0.5} \frac{2\tan\theta}{(1+\tan^2\theta)^{3/2}}(-\sec^2\theta)\, d\theta$$
$$= -\int_{\pi/4}^{\tan^{-1}0.5} 2\sin\theta\, d\theta = 2\cos\theta\Big|_{\pi/4}^{\tan^{-1}0.5} = 2\left(\frac{2}{\sqrt{5}} - \frac{1}{\sqrt{2}}\right) \tag{E2.14.4}$$

$$\int_{C_3} \frac{2x}{(x^2+y^2+z^2)^{3/2}} dl \overset{(2.5.3d)}{=} \int_0^{0.5} \frac{2(1-t)\sqrt{(-1)^2+0^2+0^2}}{((1-t)^2+0^2+0.5^2)^{3/2}} dt = \int_0^{0.5} \frac{2(1-t)}{((1-t)^2+0.5^2)^{3/2}} dt$$
$$\overset{1-t=0.5\tan\theta}{=} \int_{\tan^{-1}2}^{\pi/4} \frac{2\times 0.5\tan\theta}{0.5^3(1+\tan^2\theta)^{3/2}}(-0.5\sec^2\theta)\, d\theta \underset{\text{Eq.(E2.14.4)}}{\overset{\text{like}}{=}} 4\cos\theta\Big|_{\tan^{-1}2}^{\pi/4} = 4\left(\frac{1}{\sqrt{2}} - \frac{1}{\sqrt{5}}\right) \tag{E2.14.5}$$

```
>>syms x y z t; I=int_line(2*x/(x^2+y^2+z^2)^(3/2),[1-t 0 1/2],0,1/2)
```

Note that the value of Eq. (E2.14.3) is easily obtained in the spherical coordinate system:

$$\int_{C_1} \frac{2}{x^2+y^2+z^2} dl \underset{dl=dr}{\overset{(x^2+y^2+z^2)^{1/2}=r}{=}} \int_{C_1} \frac{2}{r^2} dr = -\left.\frac{2}{r}\right|_{0.5\sqrt{2}}^{\sqrt{2}} = -\left(\frac{2}{\sqrt{2}} - \frac{4}{\sqrt{2}}\right) = \sqrt{2} \tag{E2.14.6}$$

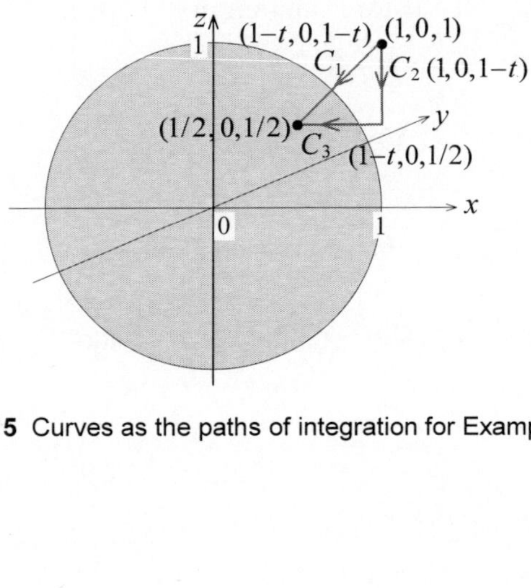

Figure 2.15 Curves as the paths of integration for Example 2.14

(Example 2.15) Line Integral of a Scalar Function in 3D Space

Evaluate the following integral w.r.t. arc length l:

$$\int_C 8xyz\,dl \tag{E2.15.1}$$

where the curve C is parameterized by

$$t\,\mathbf{u}_x + \cos t\,\mathbf{u}_y + \sin t\,\mathbf{u}_z \quad \text{for} \quad 0 \le t \le \pi \tag{E2.15.2}$$

The integral is computed as follows:

$$\begin{aligned}
\int_C 8xyz\,dl &\overset{(2.5.3d)}{=} \int_0^\pi 8xyz\sqrt{(x'(t))^2+(y'(t))^2+(z'(t))^2}\,dt \\
&\overset{(E2.15.2)}{=} \int_0^\pi 8t\cos t\sin t\sqrt{1^2+(-\sin t)^2+(\cos t)^2}\,dt \\
&\overset{(C.16)}{=} 4\sqrt{2}\int_0^\pi t\sin 2t\,dt \underset{\text{by parts (C.38)}}{\overset{\text{Integration}}{=}} 2\sqrt{2}\,(-t\cos 2t)\Big|_0^\pi + 2\sqrt{2}\int_0^\pi \cos 2t\,dt \\
&= -2\sqrt{2}\pi + \sqrt{2}(\sin 2t)\Big|_0^\pi = -2\sqrt{2}\pi
\end{aligned} \tag{E2.15.3}$$

2.5.1.3 Line Integral of Vector Function in Space

The integration of a vector function

$$\mathbf{F}(x,y,z) = F_x(x,y,z)\mathbf{u}_x + F_y(x,y,z)\mathbf{u}_y + F_z(x,y,z)\mathbf{u}_z \tag{2.5.4}$$

along a smooth curve C parameterized by $x=x(t)$, $y=y(t)$, and $z=z(t)$ for $a \le t \le b$ can be computed as

$$\begin{aligned}
\int_C \mathbf{F}(x,y,z)\bullet d\mathbf{l} &= \int_C (F_x(x,y,z)\mathbf{u}_x + F_y(x,y,z)\mathbf{u}_y + F_z(x,y,z)\mathbf{u}_z)\bullet(dx(t)\mathbf{u}_x + dy(t)\mathbf{u}_y + dz(t)\mathbf{u}_z) \\
&= \int_a^b \{F_x(x(t),y(t),z(t))\frac{dx}{dt} + F_y(x(t),y(t),z(t))\frac{dy}{dt} + F_z(x(t),y(t),z(t))\frac{dz}{dt}\}dt
\end{aligned} \tag{2.5.5}$$

where $d\mathbf{l}$ is the differential tangent vector of the curve.

(Example 2.16) Line Integral of a Vector Function in 3D Space

Evaluate the line integrals of a vector function:

$$\int_{C_i} \frac{2(x\mathbf{u}_x + y\mathbf{u}_y + z\mathbf{u}_z)}{(x^2+y^2+z^2)^{3/2}}\bullet d\mathbf{l} \tag{E2.16.1}$$

along the three curves C_1, C_2, and C_3 that are depicted in Fig. 2.15 and parameterized by

$$(1-t)\mathbf{u}_x + (1-t)\mathbf{u}_z \text{ for } 0 \le t \le \frac{1}{2},\ \mathbf{u}_x + (1-t)\mathbf{u}_z \text{ for } 0 \le t \le \frac{1}{2},\ \text{and } (1-t)\mathbf{u}_x + \frac{1}{2}\mathbf{u}_z \text{ for } 0 \le t \le \frac{1}{2}, \tag{E2.16.2}$$

respectively.

The integrals are computed as follows:

$$\int_{C_1}\frac{2(x\mathbf{u}_x+y\mathbf{u}_y+z\mathbf{u}_z)}{(x^2+y^2+z^2)^{3/2}}\bullet d\mathbf{l}\overset{(2.5.5)}{=}\int_0^{1/2}\frac{2((1-t)\mathbf{u}_x+0\mathbf{u}_y+(1-t)\mathbf{u}_z)}{((1-t)^2+0^2+(1-t)^2)^{3/2}}\bullet((1-t)'\mathbf{u}_x+(1-t)'\mathbf{u}_z)\,dt$$

$$=\int_0^{1/2}\frac{2((1-t)\mathbf{u}_x+0\mathbf{u}_y+(1-t)\mathbf{u}_z)}{((1-t)^2+0^2+(1-t)^2)^{3/2}}\bullet(-\mathbf{u}_x-\mathbf{u}_z)\,dt=\int_0^{1/2}\frac{2\cdot 2(t-1)}{2^{3/2}(1-t)^3}dt=\left.\frac{\sqrt{2}}{t-1}\right|_0^{1/2}=-\sqrt{2}\quad\text{(E2.16.3)}$$

$$\int_{C_2}\frac{2(x\mathbf{u}_x+y\mathbf{u}_y+z\mathbf{u}_z)}{(x^2+y^2+z^2)^{3/2}}\bullet d\mathbf{l}\overset{(2.5.5)}{=}\int_0^{1/2}\frac{2(\mathbf{u}_x+0\mathbf{u}_y+(1-t)\mathbf{u}_z)}{(1^2+0^2+(1-t)^2)^{3/2}}\bullet((1-t)'\mathbf{u}_z)\,dt=\int_0^{1/2}\frac{2(t-1)}{(1^2+(1-t)^2)^{3/2}}dt$$

$$\overset{1-t=\tan\theta}{=}\int_{\pi/4}^{\tan^{-1}0.5}\frac{-2\tan\theta}{(1+\tan^2\theta)^{3/2}}(-\sec^2\theta)\,d\theta\overset{\text{(E2.14.4)}}{=}-2\left(\frac{2}{\sqrt{5}}-\frac{1}{\sqrt{2}}\right)\quad\text{(E2.16.4)}$$

$$\int_{C_3}\frac{2(x\mathbf{u}_x+y\mathbf{u}_y+z\mathbf{u}_z)}{(x^2+y^2+z^2)^{3/2}}\bullet d\mathbf{l}\overset{(2.5.5)}{=}\int_0^{1/2}\frac{2((1-t)\mathbf{u}_x+0\mathbf{u}_y+0.5\mathbf{u}_z)}{((1-t)^2+0^2+0.5^2)^{3/2}}\bullet((1-t)'\mathbf{u}_x)dt=\int_0^{1/2}\frac{2(t-1)}{((1-t)^2+0.5^2)^{3/2}}dt$$

$$\overset{1-t=0.5\tan\theta}{=}\int_{\tan^{-1}2}^{\pi/4}\frac{-2\times 0.5\tan\theta}{0.5^3(1+\tan^2\theta)^{3/2}}(-0.5\sec^2\theta)\,d\theta\overset{\text{(E2.14.5)}}{=}-4\left(\frac{1}{\sqrt{2}}-\frac{1}{\sqrt{5}}\right)\quad\text{(E2.16.5)}$$

```
>> syms x y z t; int_line(2*[x y z]/(x^2+y^2+z^2)^(3/2),[1-t 0 1/2],0,1/2)
```

Note that the value of Eq. (E2.16.3) is easily obtained in the spherical coordinate system:

$$\int_{C_1}\frac{2(x\mathbf{u}_x+y\mathbf{u}_y+z\mathbf{u}_z)}{(x^2+y^2+z^2)^{3/2}}\bullet d\mathbf{l}\underset{d\mathbf{l}=-d\mathbf{r}}{\overset{(x^2+y^2+z^2)^{1/2}=r}{=}}\int_{C_1}\frac{2\mathbf{r}}{r^3}\bullet(-d\mathbf{r})=-\int_{0.5\sqrt{2}}^{\sqrt{2}}\frac{2}{r^2}dr=\left.\frac{2}{r}\right|_{0.5\sqrt{2}}^{\sqrt{2}}=-\frac{2}{\sqrt{2}}=-\sqrt{2}\quad\text{(E2.16.6)}$$

(Example 2.17) Line Integral of a Vector Function

Evaluate the line integrals of a vector function:

$$\int_C\frac{-y\mathbf{u}_x+x\mathbf{u}_y}{x^2+y^2}\bullet d\mathbf{l}\quad\text{(E2.17.1)}$$

along the unit circle C_1 and C_2 (Fig. 2.16(a)), each parameterized by

$$\begin{cases}\cos t\,\mathbf{u}_x+\sin t\,\mathbf{u}_y\ \text{(countercloclwise)} & \text{for } 0\le t\le 2\pi\\ \cos t\,\mathbf{u}_x-\sin t\,\mathbf{u}_y\ \text{(clockwise)} & \text{for } 0\le t\le 2\pi\end{cases}\quad\text{(E2.17.2)}$$

The integrals are computed as follows:

$$\int_{C_1}\frac{-y\mathbf{u}_x+x\mathbf{u}_y}{x^2+y^2}\bullet d\mathbf{l}\underset{\text{(E2.17.2a)}}{\overset{(2.5.5)}{=}}\int_0^{2\pi}\frac{-\sin t\,\mathbf{u}_x+\cos t\,\mathbf{u}_y}{\cos^2 t+\sin^2 t}\bullet(-\sin t\,\mathbf{u}_x+\cos t\,\mathbf{u}_y)\,dt=\int_0^{2\pi}1\,dt=2\pi\quad\text{(E2.17.3)}$$

$$\int_{C_2}\frac{-y\mathbf{u}_x+x\mathbf{u}_y}{x^2+y^2}\bullet d\mathbf{l}\underset{\text{(E2.17.2b)}}{\overset{(2.5.5)}{=}}\int_0^{2\pi}\frac{\sin t\,\mathbf{u}_x+\cos t\,\mathbf{u}_y}{\cos^2 t+\sin^2 t}\bullet(-\sin t\,\mathbf{u}_x-\cos t\,\mathbf{u}_y)\,dt=\int_0^{2\pi}(-1)\,dt=-2\pi\quad\text{(E2.17.4)}$$

This illustrates that if the direction of the integration path is reversed, the sign of a line integral along a curve is reversed:

$$\int_{-C}\mathbf{F}(x,y,z)\bullet d\mathbf{l}=-\int_C\mathbf{F}(x,y,z)\bullet d\mathbf{l}\quad\text{(E2.17.5)}$$

2.5.2 Path Independence

A vector function **G** is said to be *conservative* if it can be written as the gradient of a scalar function, say, $f(x,y,z)$, called a *potential function* for **G**:

$$\mathbf{G}(x,y,z)=G_x(x,y,z)\mathbf{u}_x+G_y(x,y,z)\mathbf{u}_y+G_z(x,y,z)\mathbf{u}_z=\frac{\partial f}{\partial x}\mathbf{u}_x+\frac{\partial f}{\partial y}\mathbf{u}_y+\frac{\partial f}{\partial z}\mathbf{u}_z=\nabla f(x,y,z) \quad (2.5.6)$$

A conservative vector field is called a gradient vector field. How can we tell whether a vector field is conservative or not? The criterion for conservative vector fields is that G_x, G_y, G_z, and their first partial derivatives are continuous, and besides, the curl of the vector function **G** is zero in an open *simply-connected* region R (that is connected and does not contain any hole at which **G** is singular):

$$\nabla\times\mathbf{G}=\left(\frac{\partial G_z}{\partial y}-\frac{\partial G_y}{\partial z}\right)\mathbf{u}_x+\left(\frac{\partial G_x}{\partial z}-\frac{\partial G_z}{\partial x}\right)\mathbf{u}_y+\left(\frac{\partial G_y}{\partial x}-\frac{\partial G_x}{\partial y}\right)\mathbf{u}_z=\mathbf{0}:\text{irrotational} \quad (2.5.7)$$

Note that $\nabla\times\nabla f \overset{(2.4.22)}{=} 0$ and that in the case of 2D vector function, the criterion (2.5.7) becomes

$$\frac{\partial G_y}{\partial x}=\frac{\partial G_x}{\partial y} \quad (2.5.8)$$

On the other hand, a line integral from one point to another point is said to be *path independent* or *independent of path* if its value depends only on the two end points, but not on the path of integration.

According to the *fundamental theorem for line integrals*, if and only if **G** is a conservative vector field $\mathbf{G}=\nabla f$ with a scalar potential function f, then its line integral is path independent, i.e.,

$$\int_C \mathbf{G}(x,y,z)\bullet d\mathbf{l}=\int_A^B \nabla f(x,y,z)\bullet d\mathbf{l}=f(B)-f(A) \text{ with } A=\mathbf{l}(a) \text{ and } B=\mathbf{l}(b) \quad (2.5.9)$$

where C is an integration path parameterized by $\mathbf{l}(t)=x(t)\mathbf{u}_x+y(t)\mathbf{u}_y+z(t)\mathbf{u}_z$ for $a\le t\le b$ in an open simply connected region R. In other words, for a conservative vector field **G** over an open simply connected region R, its line integral around any simple closed curve C in R, called the *circulation* of **G** around C, is zero:

$$\int_C \mathbf{G}(x,y,z)\bullet d\mathbf{l}=0 \text{ for any simple closed curve } C \quad (2.5.10)$$

Note that a region R is said to be *open* if it has no boundary, and is said to be *closed* if the region it includes all of its boundary, if any. A region R is said to be *connected* if any two points in R can be connected with a path lying completely in R. A curve C is said to be *simple* if it does not cross itself.

(**Example 2.18**) Path Independence of Line Integrals

Determine the path independence of the following line integrals in connection with whether or not the integrand vector functions are conservative:

$$\text{(a)} \int_C \mathbf{G}(x,y)\bullet d\mathbf{l}=\int_C \{G_x(x,y)\mathbf{u}_x+G_y(x,y)\mathbf{u}_y\}\bullet d\mathbf{l}=\int_C \frac{-y\mathbf{u}_x+x\mathbf{u}_y}{x^2+y^2}\bullet d\mathbf{l} \quad \text{(E2.18a.1)}$$

The vector function $\mathbf{G}(x, y)$ satisfies the criterion (2.5.8) for 2D conservative vector fields:

$$\frac{\partial G_y}{\partial x}=\frac{(x^2+y^2)-x\cdot 2x}{(x^2+y^2)^2}=\frac{y^2-x^2}{(x^2+y^2)^2}=\frac{-(x^2+y^2)+y\cdot 2y}{(x^2+y^2)^2}=\frac{\partial G_x}{\partial y} \quad \text{(E2.18a.2)}$$

However, it is a fact that the line integral of $\mathbf{G}(x, y)$ is not path-independent, since the line integral along the unit circle (which is a closed curve) is not zero, as found in Example 2.17. How can this seeming contradiction be explained? The problem is the hole of singularity of $\mathbf{G}$ at (0, 0), which keeps the open domain D (containing the path of integration) from being simply connected (see Fig. 2.16(a)).

(b) $$\int_C \mathbf{G}(x,y)\bullet d\mathbf{l} = \int_C \{G_x(x,y)\mathbf{u}_x + G_y(x,y)\mathbf{u}_y\}\bullet d\mathbf{l} = \int_C \frac{-y\mathbf{u}_x + x\mathbf{u}_y}{x^2}\bullet d\mathbf{l} \quad \text{(E2.18b.1)}$$

The vector function $\mathbf{G}(x, y)$ satisfies the criterion (2.5.8) for 2D conservative vector fields:

$$\frac{\partial G_y}{\partial x} = \frac{-1}{x^2} = \frac{\partial G_x}{\partial y} \quad \text{(E2.18b.2)}$$

over the two simply connected regions $\{x>0\}$and $\{x<0\}$ separated by $x=0$ (see Fig. 2.16(b)). The scalar potential function f such that $\nabla f = \mathbf{G}$ can be found as

$$f(x,y) = \frac{y}{x} + C \text{ with } \frac{\partial f}{\partial x} = -\frac{y}{x^2} \text{ and } \frac{\partial f}{\partial y} = \frac{1}{x} \quad \text{(E2.18b.3)}$$

Let us evaluate the integral over two paths:

$$\{C_1: \mathbf{u}_x + (1-t)\mathbf{u}_y \text{ for } 0 \le t \le 2\} \text{ and } \begin{Bmatrix} C_2: (1+t)\mathbf{u}_x + \mathbf{u}_y \text{ for } 0 \le t \le 1 \\ C_3: 2\mathbf{u}_x + (1-t)\mathbf{u}_y \text{ for } 0 \le t \le 2 \\ C_4: (2-t)\mathbf{u}_x - \mathbf{u}_y \text{ for } 0 \le t \le 1 \end{Bmatrix} \quad \text{(E2.18b.4)}$$

$$\int_{C_1} \frac{-y\mathbf{u}_x + x\mathbf{u}_y}{x^2}\bullet d\mathbf{l} = \int_0^2 \frac{-(1-t)\mathbf{u}_x + \mathbf{u}_y}{1^2}\bullet\{(1-t)'\mathbf{u}_y\}dt = \int_0^2 (-1)\,dt = -2 \quad \text{(E2.18b.5)}$$

$$\int_{C_2} \frac{-y\mathbf{u}_x + x\mathbf{u}_y}{x^2}\bullet d\mathbf{l} = \int_0^1 \frac{-\mathbf{u}_x + (1+t)\mathbf{u}_y}{(1+t)^2}\bullet\{(1+t)'\mathbf{u}_x\}dt = \int_0^1 \frac{-1}{(1+t)^2}dt = \frac{1}{1+t}\Big|_0^1 = -\frac{1}{2} \quad \text{(E2.18b.6)}$$

$$\int_{C_3} \frac{-y\mathbf{u}_x + x\mathbf{u}_y}{x^2}\bullet d\mathbf{l} = \int_0^2 \frac{-(1-t)\mathbf{u}_x + 2\mathbf{u}_y}{2^2}\bullet\{(1-t)'\mathbf{u}_y\}dt = \int_0^2 \frac{-1}{2}dt = -1 \quad \text{(E2.18b.7)}$$

$$\int_{C_4} \frac{-y\mathbf{u}_x + x\mathbf{u}_y}{x^2}\bullet d\mathbf{l} = \int_0^1 \frac{\mathbf{u}_x + (2-t)\mathbf{u}_y}{(2-t)^2}\bullet\{(2-t)'\mathbf{u}_x\}dt = \int_0^1 \frac{-1}{(2-t)^2}dt = \frac{1}{t-2}\Big|_0^1 = -\frac{1}{2} \quad \text{(E2.18b.8)}$$

$$\int_{C_2\text{-}C_3\text{-}C_4} \frac{-y\mathbf{u}_x + x\mathbf{u}_y}{x^2}\bullet d\mathbf{l} = \text{Sum of Eqs. (E2.18.8b6,7,8)} = -\frac{1}{2} - 1 - \frac{1}{2} = -2 \quad \text{(E2.18b.9)}$$

The equality of the values of the line integral over the two paths supports that the line integral of $\mathbf{G}(x, y)$ is path-independent.

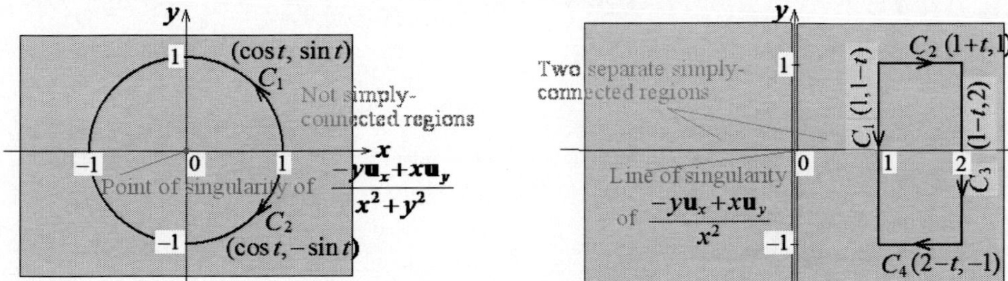

(a) Paths of integration for Example 2.17(a)

(b) Paths of integration for Example 2.17(b)

Figure 2.16 Paths of integration for Example 2.18(a) and (b)

2.6 DOUBLE INTEGRALS

(a) Type I region of integration (b) Type II region of integration

Figure 2.17 Regions of integration for double (2D) integral in Cartesian coordinates

The double or 2D integral of a function $f(x,y)$ over a region R of integration on the xy-plane can be computed as one of the following two iterated integrals depending on the shape of R (see Fig. 2.17):

$$\iint_R f(x,y)dA = \int_a^b \int_{g_1(x)}^{g_2(x)} f(x,y)dydx \text{ for } R=\{a\le x\le b,\ g_1(x)\le y\le g_2(x)\}: \text{Type I } (y\text{-simple}) \quad (2.6.1a)$$

$$\iint_R f(x,y)dA = \int_c^d \int_{h_1(y)}^{h_2(y)} f(x,y)dxdy \text{ for } R=\{h_1(y)\le x\le h_2(x),\ c\le y\le d\}: \text{Type II } (x\text{-simple}) \quad (2.6.1b)$$

In general, the value of the double integral is the volume of a 3D object with R and $f(x,y)$ as the base plane and height, respectively (Fig. 2.18(a)). However, if we let $f(x,y)=1\ \forall(x,y)\in R$, it is equal to the area of the integration region R (Fig. 2.18(b)).

(a) Double integral to get the volume of a 3D object (b) Double integral to get the area of a 2D region

Figure 2.18 Geometric interpretation of double (2D) integral

The double or 2D integral of a function $f(\rho,\phi)$ over a region R of integration on the $\rho\phi$-plane can be computed as one of the following two iterated integrals, depending on the shape of R: (Fig. 2.19):

$$\iint_R f(\rho,\phi)dA = \int_\alpha^\beta \int_{g_1(\phi)}^{g_2(\phi)} f(\rho,\phi)\,\rho d\rho d\phi \text{ for } R=\{\alpha\le\phi\le\beta,\ g_1(\phi)\le\rho\le g_2(\phi)\}: \text{Type I} \quad (2.6.2a)$$

$$\iint_R f(\rho,\phi)dA = \int_a^b \int_{h_1(\rho)}^{h_2(\rho)} f(\rho,\phi)\,\rho d\phi d\rho \text{ for } R=\{h_1(\rho)\le\phi\le h_2(\rho),\ a\le\rho\le b\}: \text{Type II} \quad (2.6.2b)$$

(a) Type I region of integration (b) Type II region of integration

Figure 2.19 Regions of integration for double (2D) integral in polar coordinates

<Center of Mass and Moments of Laminas with Variable Density>

For a lamina with a variable density $\rho(x,y)$ that is continuous over a region R, its mass m and the x/y-coordinates $(\bar{x}, \bar{y})$ of its center of mass are computed as follows:

$$\text{Mass: } m = \iint_R \rho(x,y)\,dA \tag{2.6.3}$$

$$\text{Center of mass: } (\bar{x}, \bar{y}) = \left(\frac{M_y}{m}, \frac{M_x}{m}\right) \tag{2.6.4}$$

with

$$M_x = \iint_R y\rho(x,y)\,dA \quad \text{and} \quad M_y = \iint_R x\rho(x,y)\,dA \tag{2.6.5}$$

where M_x and M_y are called the (first) moments of the lamina about the x- and y-axes, respectively.
Also, the moments of inertia about the x-axis, y-axis, and origin are defined as follows:

$$I_x = \iint_R y^2\rho(x,y)\,dA,\; I_y = \iint_R x^2\rho(x,y)\,dA, \text{ and } I_o = I_x + I_y = \iint_R (x^2+y^2)\rho(x,y)\,dA \tag{2.6.6}$$

<Cartesian-to-Polar Coordinate Conversion>

Many double integrals of integrand functions depending on x^2+y^2 can be easier to evaluate in the polar coordinate system than in the Cartesian coordinate system. Those double integrals in the Cartesian coordinate system can be converted into versions within polar coordinates as follows:

$$\iint_R f(x,y)\,dA \overset{x=\rho\cos\phi,\ y=\rho\sin\phi}{\longrightarrow} \int_{\phi_1}^{\phi_2}\int_{\rho_1(\phi)}^{\rho_2(\phi)} f(\rho\cos\phi, \rho\sin\phi)\,\rho\,d\rho\,d\phi \tag{2.6.7}$$

(Example 2.19) Double Integral

(a) Find the area of the figure consisting of a square and two half circles depicted in Fig. 2.20(a).
To this end, we set $f(x,y)=1$ and use Eq. (2.6.1a) to find the area as

$$\iint_R f(x,y)\,dA \overset{(2.6.1a)}{=} \int_{-1}^{1}\int_{-1-\sqrt{1-x^2}}^{1+\sqrt{1-x^2}} 1\,dy\,dx = \int_{-1}^{1} 2(1+\sqrt{1-x^2})\,dx \underset{dx=-\sin\phi d\phi}{\overset{x=\cos\phi}{=}} 4 + 2\int_{\pi}^{0} \sin\phi(-\sin\phi)\,d\phi$$

$$\overset{(C.14)}{=} 4 - \int_{\pi}^{0} (1-\cos 2\phi)\,d\phi = 4 + \pi = 7.1416 \tag{E2.19.1}$$

```
>>fxya=@(x,y)ones(size(x)); % f(x,y)=1
>>c=@(x)-1-sqrt(1-x.^2);  d=@(x)1+sqrt(1-x.^2);
>>Area=quad2d(fxya,-1,1,c,d)
   Area =     7.1416
```

(b) Find the volume of the 3D object consisting of a half cylinder and two quarter spheres depicted in Fig. 2.20(b).
To this end, we set $f(x,y)$ as

$$f(x,y) = \begin{cases} \sqrt{1-x^2} & \text{for } -1 \le x \le 1,\ |y| \le 1 \\ \sqrt{1-x^2-(|y|-1)^2} & \text{for } -1 \le x \le 1,\ |y| > 1 \\ 0 & \text{elsewhere} \end{cases} \tag{E2.19.2}$$

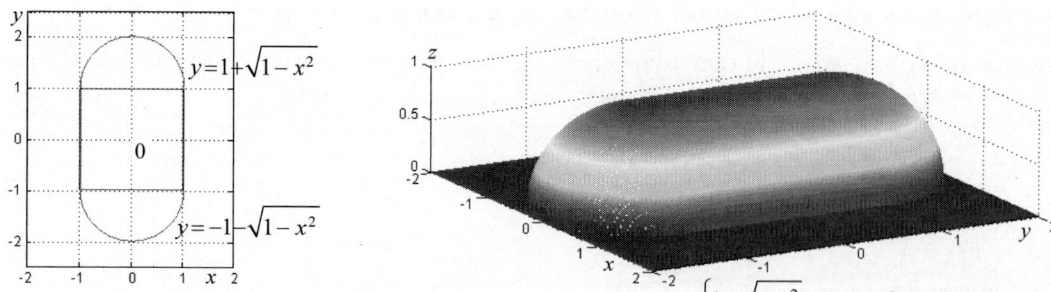

(a) Area with $f(x,y)=1$ (b) Volume with $f(x,y)=\begin{cases}1-\sqrt{1-x^2} & \text{for } |y|\le 1\\ 1-\sqrt{1-x^2-(|y|-1)^2} & \text{for } |y|>1\end{cases}$

Figure 2.20 Double integral to get an area or a volume

Noting that the upper and lower quarter spheres are symmetric, we use Eq. (2.6.1a) and Cartesian-to-polar coordinate conversion to find the volume as

$$\iint_R f(x,y)\,dA \overset{(2.6.1a)}{=} 2\int_{-1}^{1}\int_{-1-\sqrt{1-x^2}}^{-1}\sqrt{1-x^2-(|y|-1)^2}\,dy\,dx+\int_{-1}^{1}\int_{-1}^{1}\sqrt{1-x^2}\,dy\,dx$$

$$\underset{x=\cos\theta}{\overset{x^2+(|y|-1)^2\to r^2,\ dxdy\to rdrd\theta}{=}} 2\int_0^{\pi}\int_0^1\sqrt{1-r^2}\,r\,dr\,d\theta+2\int_0^1\int_{\pi}^{0}\sin\theta(-\sin\theta)\,d\theta\,dy$$

$$\underset{(C.14)}{\overset{(C.34)}{=}} -\frac{2}{3}\pi(1-r^2)^{3/2}\Big|_0^1-\int_{\pi}^{0}(1-\cos 2\theta)\,d\theta=\frac{2}{3}\pi+\pi=5.2360 \tag{E2.19.3}$$

```
>>fxyb=@(x,y)(-1<=x&x<1).*((abs(y)<=1).*sqrt(max(1-x.^2,0))...
    +(abs(y)>1).*sqrt(max(1-x.^2-(abs(y)-1).^2,0)));
>>c=@(x)-1-sqrt(1-x.^2); d=@(x)1+sqrt(1-x.^2);
>>Volume=quad2d(fxyb,-1,1,c,d)
   Volume =     5.2360
```

<Change of Variables in a Double Integral>

We can change the variable of integration in a definite integral as

$$\int_a^b f(x)\,dx \overset{x=x(u)}{\rightarrow} \int_{\alpha}^{\beta} f(x(u))\frac{dx}{du}du \tag{2.6.8}$$

by substituting $x(u)$ for x and replacing the lower/upper limits a/b of the interval of integration with α/β such that $x(\alpha)=a$ and $x(\beta)=b$, where $x(u)$ is supposed to be continuous and have a continuous derivative for $\alpha\le u\le\beta$. Likewise, we can change the variables of integration in a double integral as

$$\iint_{R_{xy}} f(x,y)\,dx\,dy \overset{x=x(u,v),\ y=y(u,v)}{\rightarrow} \iint_{R_{uv}} f(x(u,v),y(u,v))\,|J|\,du\,dv \tag{2.6.9}$$

by substituting $x(u,v)$ and $y(u,v)$ for x and y, respectively, where J is the *Jacobian* matrix defined as

$$J=\frac{\partial(x,y)}{\partial(u,v)}=\begin{bmatrix}\partial x/\partial u & \partial x/\partial v\\ \partial y/\partial u & \partial y/\partial v\end{bmatrix} \tag{2.6.10}$$

Note that the *Jacobian* (*determinant*), i.e., the determinant of the Jacobian matrix for Cartesian-to-polar coordinate conversion, is

$$|J| = \left|\frac{\partial(x,y)}{\partial(\rho,\phi)}\right| = \left\|\begin{bmatrix} \frac{\partial}{\partial r}(\rho\cos\phi) & \frac{\partial}{\partial\theta}(\rho\cos\phi) \\ \frac{\partial}{\partial r}(\rho\sin\phi) & \frac{\partial}{\partial\theta}(\rho\sin\phi) \end{bmatrix}\right\| = \left\|\begin{bmatrix} \cos\phi & -\rho\sin\phi \\ \sin\phi & \rho\cos\phi \end{bmatrix}\right\| = \rho \tag{2.6.11}$$

Thus we can substitute $x=x(\rho,\phi)=\rho\cos\phi$ and $y=y(\rho,\phi)=\rho\sin\phi$ together with $|J|=\rho$ into Eq. (2.6.9) to obtain the double integration formula (2.6.7) in polar coordinates.

(**Example 2.20**) Double Integral

Let us evaluate the following double integral

$$\iint_R (x+y)\, dx\,dy \tag{E2.20.1}$$

over the regions R_1 and R_2 depicted in Fig. 2.21.

(a) $\iint_{R_1}(x+y)\,dx\,dy \overset{(2.6.1b)}{=} \int_{-1}^{1}\int_{-(1-|y|)}^{1-|y|}(x+y)\,dx\,dy = \int_{-1}^{1}\left(\frac{1}{2}x^2+xy\right)\Big|_{-(1-|y|)}^{1-|y|}dy = 2\int_{-1}^{1}(1-|y|)y\,dy$

$$= 2\int_0^1(1-y)y\,dy + 2\int_{-1}^{0}(1+y)y\,dy = \left(y^2-\frac{2}{3}y^3\right)\Big|_0^1 + \left(y^2+\frac{2}{3}y^3\right)\Big|_{-1}^{0} = 0 \tag{E2.20.2}$$

If we change the variables of integration by $x=(u+v)/2$ and $y=(u-v)/2$ with Jacobian determinant $|J| = (\partial x/\partial u)(\partial y/\partial v) - (\partial x/\partial v)(\partial y/\partial u) = (1/2)(-1/2) - (1/2)(1/2) = -1/2$, it may be easier to compute:

$$\iint_{R_1}(x+y)\,dx\,dy \underset{(2.6.9),\,|J|=-1/2}{\overset{x=(u+v)/2,\,y=(u-v)/2}{\longrightarrow}} \int_{-1}^{1}\int_{-1}^{1} u\left(-\frac{1}{2}\right)du\,dv = -\frac{1}{4}\int_{-1}^{1}u^2\Big|_{-1}^{1}dv = 0 \tag{E2.20.3}$$

(b) $\iint_{R_2}(x+y)\,dx\,dy \overset{(2.6.1b)}{=} \int_{-1}^{1}\int_{-\sqrt{1-y^2}}^{\sqrt{1-y^2}}(x+y)\,dx\,dy = \int_{-1}^{1}\left(\frac{1}{2}x^2+xy\right)\Big|_{-\sqrt{1-y^2}}^{\sqrt{1-y^2}}dy$

$$= \int_{-1}^{1}\sqrt{1-y^2}\,2y\,dy = -\frac{2}{3}(1-y^2)^{3/2}\Big|_{-1}^{1} = 0 \tag{E2.20.4}$$

If we change the variables of integration by $x=r\cos\theta$ and $y=r\sin\theta$ with Jacobian determinant $|J|=r$ (Eq. (2.6.11)) corresponding to the Cartesian-to-polar coordinate conversion, it may be easier to compute:

$$\iint_R(x+y)\,dx\,dy \underset{(2.6.9),\,|J|=\rho}{\overset{x=\rho\cos\phi,\,y=\rho\sin\phi}{\longrightarrow}} \int_0^{2\pi}\int_0^1 \rho(\cos\phi+\sin\phi)\,\rho\,d\rho\,d\phi$$

$$= \int_0^1\rho^2 d\rho\int_0^{2\pi}(\cos\phi+\sin\phi)\,d\phi = 0 \tag{E2.20.5}$$

Figure 2.21 Regions of integration for the double integral of Examples 2.20 and 2.21

(Example 2.21) Double Integral

Let us evaluate the following double integral

$$\iint_R e^{x^2+y^2} dA \tag{E2.21.1}$$

over the region R of the unit disk depicted in Fig. 2.21(b).

(a) This integral is difficult to deal with analytically in the Cartesian coordinate system:

$$\iint_R e^{x^2+y^2} dA \overset{(2.6.1a)}{=} \int_{-1}^{1} \int_{-\sqrt{1-x^2}}^{\sqrt{1-x^2}} e^{x^2+y^2} dy\, dx \tag{E2.21.2}$$

That is why we try with the numerical double integrator '`dblquad()`':

```
>>fxy=@(x,y)(x.^2+y.^2<=1).*exp(x.^2+y.^2);
>>dblquad(fxy,-1,1,-1,1)
   ans   = 5.3981
```

(b) This integral is easy to deal with in the polar coordinate system:

$$\iint_R e^{x^2+y^2} dA \overset{(2.6.7)}{=} \int_0^{2\pi} \int_0^1 e^{\rho^2} \rho\, d\rho\, d\phi = \frac{1}{2}\int_0^{2\pi} e^{\rho^2}\Big|_0^1 d\phi = \pi(e-1) \tag{E2.21.3}$$

2.7 GREEN'S THEOREM

Green's theorem gives the relationship between a line integral around a simple closed 2D curve C and a double integral over the planar region R bounded by C.

[Green's Theorem]

Let C be a piecewise-smooth, simple closed curve enclosing a 2D region R (on the xy-plane) in the counter-clockwise (positive) direction, i.e., to the left. Also, let P and Q be functions of (x, y) *defined on* R and have continuous partial derivatives. Then

$$\oint_C P dx + Q dy = \iint_R \left(\frac{\partial Q}{\partial x} - \frac{\partial P}{\partial y} \right) dA \tag{2.7.1}$$

(Proof)

Here, assuming that the integration region R is of type I, i.e., $R=\{a \le x \le b,\ g_1(x) \le y \le g_2(x)\}$ and also is of type II, i.e., $R=\{h_1(y) \le x \le h_2(y),\ c \le y \le d\}$ as depicted in Fig. 2.22, we are going to verify the following two equalities:

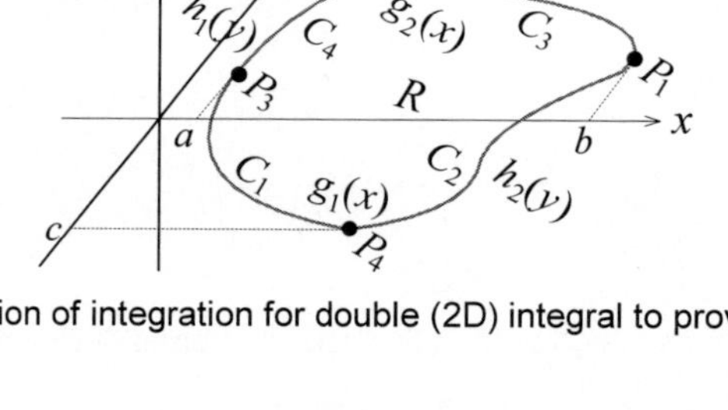

Figure 2.22 Region of integration for double (2D) integral to prove Green's theorem

$$-\iint_R \frac{\partial P}{\partial y} dA = \oint_C P dx \qquad (2.7.2a) \qquad\qquad \iint_R \frac{\partial Q}{\partial x} dA = \oint_C Q\,dy \qquad (2.7.2b)$$

$$-\iint_R \frac{\partial P}{\partial y} dA = -\int_a^b \int_{g_1(x)}^{g_2(x)} \frac{\partial P}{\partial y} dy dx = -\int_a^b \left(P(x,g_2(x)) - P(x,g_1(x))\right) dx$$

$$= \int_a^b P(x,g_1(x))\,dx + \int_b^a P(x,g_2(x))\,dx = \oint_{C_1\text{-}C_2} Pdx + \oint_{C_3\text{-}C_4} Pdx \overset{\text{OK}}{\underset{(2.7.2a)}{=}} \oint_C P\,dx$$

$$\iint_R \frac{\partial Q}{\partial x} dA = \int_c^d \int_{h_1(y)}^{h_2(y)} \frac{\partial Q}{\partial x} dx dy = \int_c^d \left(Q(h_2(y),y) - Q(h_1(x),y)\right) dy$$

$$= \int_c^d Q(h_2(y),y)\,dy + \int_d^c Q(h_1(x),y)\,dy = \oint_{C_2\text{-}C_3} Qdy + \oint_{C_4\text{-}C_1} Qdy \overset{\text{OK}}{\underset{(2.7.2b)}{=}} \oint_C Q\,dy$$

This proof can be extended to the case with a more general shape of region by decomposing the region into a number of subregions and adding the results for each subregion.(Q.E.D)

(**Example 2.22**) Line Integral and Double Integral — Green's Theorem

Let us evaluate the following line integral

$$\oint_C -y\,dx + x\,dy \qquad \text{(E2.22.1)}$$

around the piecewise-smooth, simple closed curve C_a consisting of the four straight line segments C_1, C_2, C_3, and C_4 depicted in Fig. 2.23(a).

$$\int_{C_1} -y\,dx + x dy = \int_{-2}^{2} -(-2)\,dx + \int_{-2}^{-2} x\,dy = 8 \qquad \text{(E2.22.2a)}$$

$$\int_{C_2} -y\,dx + x dy = \int_{2}^{2} (-y)\,dx + \int_{-2}^{2} 2\,dy = 8 \qquad \text{(E2.22.2b)}$$

$$\int_{C_3} -y\,dx + x\,dy = \int_{2}^{-2} (-2)\,dx + \int_{2}^{2} x\,dy = 8 \qquad \text{(E2.22.2c)}$$

$$\int_{C_4} -y\,dx + x\,dy = \int_{-2}^{-2} (-y) dx + \int_{2}^{-2} (-2) dy = 8 \qquad \text{(E2.22.2d)}$$

Summing up the values of these four integrals yields the line integral around C_a=C_1-C_2-C_3-C_4 as

$$\oint_{C_a = \cup_{i=1}^4 C_i} (-y\,dx + x\,dy) = \sum_{i=1}^{4} \oint_{C_i} (\text{the same}) \overset{\text{(E2.22.2a,b,c,d)}}{=} 8+8+8+8 = 32 \qquad \text{(E2.22.3)}$$

(a) A closed curve C1-C2-C3-C- and another one C_b

(b) A multiply connected region R (with a hole viewed as the union of two simply connected regions R_1 and R_1

Figure 2.23 Regions of integration and closed curves enclosing them for the double integral of Example 2.22

Now, let us use Green's theorem to recompute the integral:

$$\oint_C -y\,dx + x\,dy = \oint_C P dx + Q dy \overset{(2.7.1)}{=} \iint_R \left(\frac{\partial Q}{\partial x} - \frac{\partial P}{\partial y}\right) dA$$
$$= \iint_R (1-(-1))\,dA = \int_{-2}^{2}\int_{-2}^{2} 2\,dx\,dy = 2\cdot 4\cdot 4 = 32 \tag{E2.22.4}$$

This is identical to Eq. (E2.22.3), which implies that Green's theorem works. Note that each term of the line integral (E2.22.1) yields the area of region R enclosed by the path C of integration.

(Example 2.23) Seeming Failure of Green's Theorem

Consider the following line integral

$$\oint_C \frac{-y}{x^2+y^2}dx + \frac{x}{x^2+y^2}dy \tag{E2.23.1}$$

(a) Evaluate the line integral around the piecewise-smooth, simple closed curve C_a consisting of the four straight line segments C_1, C_2, C_3, and C_4 depicted in Fig. 2.23(a).

$$\int_{C_1} \frac{-y}{x^2+y^2}dx + \frac{x}{x^2+y^2}dy = \int_{-2}^{2} \frac{-(-2)}{x^2+(-2)^2}dx + \int_{-2}^{-2} \frac{x}{x^2+(-2)^2}dy \overset{x=2\tan\theta}{=} \int_{-\pi/4}^{\pi/4} 1\,d\theta = \frac{\pi}{2} \tag{E2.23.2a}$$

$$\int_{C_2} \frac{-y}{x^2+y^2}dx + \frac{x}{x^2+y^2}dy = \int_{2}^{2} \frac{-y}{2^2+y^2}dx + \int_{-2}^{2} \frac{2}{2^2+y^2}dy \overset{y=2\tan\theta}{=} \int_{-\pi/4}^{\pi/4} 1\,d\theta = \frac{\pi}{2} \tag{E2.23.2b}$$

$$\int_{C_3} \frac{-y}{x^2+y^2}dx + \frac{x}{x^2+y^2}dy = \int_{2}^{-2} \frac{-2}{x^2+2^2}dx + \int_{2}^{2} \frac{x}{x^2+2^2}dy \overset{x=2\tan\theta}{=} \int_{\pi/4}^{-\pi/4} -1\,d\theta = \frac{\pi}{2} \tag{E2.23.2c}$$

$$\int_{C_4} \frac{-y}{x^2+y^2}dx + \frac{x}{x^2+y^2}dy = \int_{-2}^{-2} \frac{-y}{(-2)^2+y^2}dx + \int_{2}^{-2} \frac{-2}{(-2)^2+y^2}dy \overset{y=2\tan\theta}{=} \int_{\pi/4}^{-\pi/4} -1\,d\theta = \frac{\pi}{2} \tag{E2.23.2d}$$

Summing up the values of these four integrals yields the line integral around C_a=C_1-C_2-C_3-C_4 as

$$\oint_{C_a=\cup_{i=1}^4 C_i} \left(\frac{-y}{x^2+y^2}dx + \frac{x}{x^2+y^2}dy\right) = \sum_{i=1}^{4} \oint_{C_i}(\text{the same}) \overset{(\text{E2.23.2a,b,c,d})}{=} \frac{\pi}{2}+\frac{\pi}{2}+\frac{\pi}{2}+\frac{\pi}{2} = 2\pi \tag{E2.23.3}$$

(b) Evaluate the line integral around the smooth, simple closed curve C_b, which is enclosing the hole of the unit circle (described by $x^2+y^2=1$) and the (shaded) region R each in the clockwise/counter-clockwise direction (to the right/left) as depicted in Fig. 2.23(a).

$$\oint_{C_b} \frac{-y}{x^2+y^2}dx + \frac{x}{x^2+y^2}dy \underset{x=\cos t,\, y=\sin t}{\overset{x^2+y^2=1}{=}} \int_{2\pi}^{0} -\sin t(-\sin t)\,dt + \int_{2\pi}^{0} \cos t \cos t\,dt = -2\pi \tag{E2.23.4}$$

(c) To check whether Green's theorem works for the line integral (E2.23.1), let us compute the double integral on the RHS of Eq. (2.7.1):

$$\oint_C \frac{-y}{x^2+y^2}dx + \frac{x}{x^2+y^2}dy = \oint_C P dx + Q dy \overset{(2.7.1)}{=} \iint_R \left(\frac{\partial Q}{\partial x} - \frac{\partial P}{\partial y}\right) dA$$
$$= \iint_R \left(\frac{(x^2+y^2)-x\cdot 2x}{(x^2+y^2)^2} + \frac{(x^2+y^2)-y\cdot 2y}{(x^2+y^2)^2}\right) dA = \iint_R \frac{0}{(x^2+y^2)^2} dA = 0 \tag{E2.23.5}$$

This is not equal to Eq. (E2.23.3), which is the value of the line integral around C_a. Green's theorem seems to have failed for this line integral, making us doubtful of the theorem. Why has such a weird thing happened? It is due to the existence of the hole $(x, y) = (0, 0)$ in the region R of integration at which P, Q, $\partial P/\partial y$, and $\partial Q/\partial x$ are not continuous, which makes R multiply connected, not simply connected. However, if we regard the region R as the union of two simply connected subregions R_1 and R_2 as depicted in Fig. 2.23(b), then Green's theorem is applicable for the line integral around the simple closed path consisting of the line segments C_1-C_{22}-C_{62}-C_{b2}-C_{52}-C_{42} and C_3-C_{41}-C_{51}-C_{b1}-C_{61}-C_{21}, or equivalently, around the two simple closed curves C_1-C_2-C_3-C_4 and C_b enclosing the (shaded) region R in the counter-clockwise direction (Fig. 2.23(a)) where the two pairs of line segments C_{51}-C_{52} and C_{61}-C_{62} are cancelled because they are pairwisely reversed. Note that the sum of Eqs. (E2.23.3) and (E2.23.4) equals Eq. (E2.23.5), which implies that the Green's theorem works fine for the line integral (E2.23.1).

(Example 2.24) Line Integral and Double Integral — Green's Theorem

Let us evaluate the following line integral:

$$\oint_C -y\,dx + x\,dy \tag{E2.24.1}$$

around the simple closed curve C, called a cardioid (from the Greek καρδία, "heart"), which is depicted in Fig. 2.24 and described by

$$r = 2(1-\cos\theta) \ \text{ for } 0 \le \theta \le 2\pi \tag{E2.24.2}$$

We apply the Cartesian-to-polar coordinate conversion with $x = r\cos\theta$, $y = r\sin\theta$, $dx = dr\cos\theta + r(-\sin\theta)d\theta$, and $dy = dr\sin\theta + r\cos\theta d\theta$ for Eq. (E2.24.1) to rewrite and compute it as

$$\int_0^{2\pi} r^2(\sin^2\theta+\cos^2\theta)\,d\theta + \int_0^4(-r\sin\theta\cos\theta + r\cos\theta\sin\theta)\,dr = \int_0^{2\pi} r^2 d\theta = \int_0^{2\pi} 2^2(1-\cos\theta)^2\,d\theta$$

$$= \int_0^{2\pi} 4(1-2\cos\theta+\cos^2\theta)d\theta \overset{(C.15)}{=} \int_0^{2\pi}(4-8\cos\theta+2+2\cos 2\theta)d\theta = 12\pi \tag{E2.24.3}$$

Now, let us use Green's theorem to recompute the integral:

$$\oint_C -y\,dx + x\,dy = \oint_C P dx + Q dy \overset{(2.7.1)}{=} \iint_R \left(\frac{\partial Q}{\partial x} - \frac{\partial P}{\partial y}\right) dA = \iint_R (1-(-1))\,dA$$

$$= \int_0^{2\pi}\int_0^{2(1-\cos\theta)} 2r\,dr\,d\theta = \int_0^{2\pi} r^2\Big|_0^{2(1-\cos\theta)} d\theta = \int_0^{2\pi} 2^2(1-\cos\theta)^2 d\theta \overset{\text{like (E2.24.3)}}{=} 12\pi \tag{E2.24.4}$$

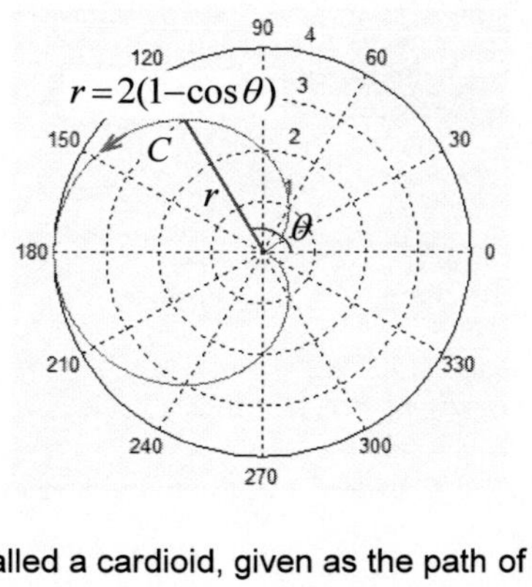

Figure 2.24 Closed curve, called a cardioid, given as the path of integration for Example 2.24

This can be obtained by running the following MATLAB statements:

```
>>syms r th
>>int(int(2*r,r,0,2*(1-cos(th))),th,0,2*pi)
   ans   = 12*pi
```

This is identical to Eq. (E2.24.3), which implies that Green's theorem works.

2.8 SURFACE INTEGRALS

<Double Integral to Get a Surface Area>

Fig. 2.25(a) shows that the two sides of the differential tangent plane ΔT_n of a surface (described by a scalar function $f(x,y)$) at (x_n, y_n) are given by

$$\mathbf{v}_n = \Delta x\,\mathbf{u}_x + \Delta x \left.\frac{\partial f(x,y)}{\partial x}\right|_{x=x_n,y=y_n} \mathbf{u}_z \quad (2.8.1a) \qquad \mathbf{w}_n = \Delta y\,\mathbf{u}_x + \Delta y \left.\frac{\partial f(x,y)}{\partial y}\right|_{x=x_n,y=y_n} \mathbf{u}_z \quad (2.8.1b)$$

Noting from Eq. (1.1.31) that the (rectangular) area of the differential tangent plane ΔT_n is the magnitude of the cross product of the two side vectors $\mathbf{u}_n$ and $\mathbf{w}_n$:

$$A(\Delta T_n) = \|\mathbf{v}_n \times \mathbf{w}_n\| = \left\| \begin{bmatrix} \Delta x \\ 0 \\ \Delta x(\partial f(x,y)/\partial x) \end{bmatrix} \times \begin{bmatrix} 0 \\ \Delta y \\ \Delta y(\partial f(x,y)/\partial y) \end{bmatrix} \right\| = \Delta x \Delta y \left\| \begin{bmatrix} -\partial f(x,y)/\partial x \\ -\partial f(x,y)/\partial y \\ 1 \end{bmatrix} \right\|_{x=x_n,y=y_n}$$

$$= \sqrt{1+(\partial f/\partial x|_{x=x_n,y=y_n})^2+(\partial f/\partial y|_{x=x_n,y=y_n})^2}\,\Delta x \Delta y \qquad (2.8.2)$$

we can take the double integral of this to find the area of the surface described by $f(x,y)$ over a closed region R (on the xy-plane) as

$$\text{Surface area: } A(S) = \iint_S 1\,dS = \iint_R \sqrt{1+(\partial f/\partial x)^2+(\partial f/\partial y)^2}\,dA \qquad (2.8.3)$$

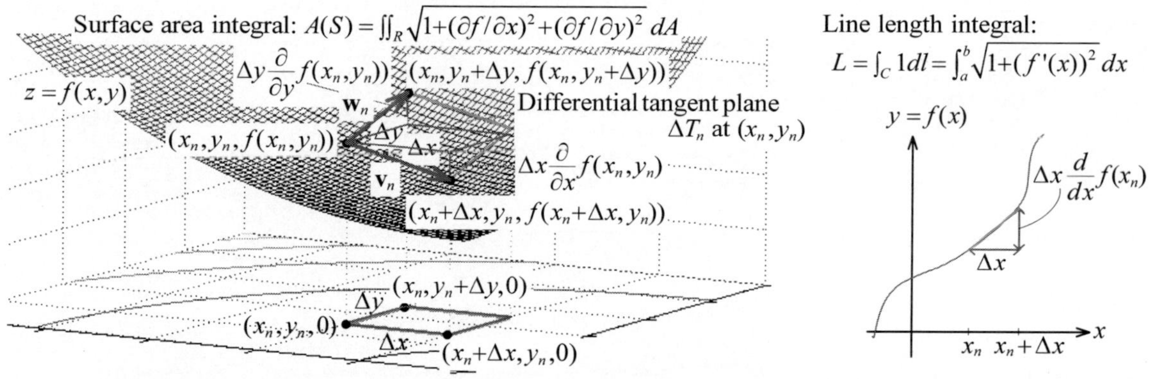

(a) Double integral to get a surface area - differential tangent plane (b) Line integral to get an arc length

Figure 2.25 Double integral to get the area of a surface and line integral to get the length of a curve where $f(x,y)$ is assumed to have continuous first partial derivatives $\partial f/\partial x$ and $\partial f/\partial y$. This is similar to the fact that the length of a smooth curve C defined by an explicit function $y=f(x)$ for $a\le x\le b$ can be obtained as the line integral

$$\text{Curve length: } L=\int_C 1\,dl=\int_a^b\sqrt{1+(f'(x))^2}\,dx \tag{2.8.4}$$

w.r.t. arc length l, as depicted in Fig. 2.25(b).

<*Surface Integral* **of a Scalar Function**>

Let $G(x,y,z)$ be a scalar function defined over a spatial region R containing a surface S described by $z=f(x,y)$, where $G, f, \partial f/\partial x$, and $\partial f/\partial y$ are assumed to be continuous through the region R (on the xy-plane). Then the surface integral of G over S can be computed as

$$\iint_S G(x,y,z)\,dS=\iint_{R(x,y)}G(x,y,f(x,y))\sqrt{1+(\partial f/\partial x)^2+(\partial f/\partial y)^2}\,dA \tag{2.8.5a}$$

If the surface S is described by $y=g(x,z)$ where G, g, $\partial g/\partial x$, and $\partial g/\partial z$ are assumed to be continuous through the region R (on the xz-plane), the surface integral of G over S can be computed as

$$\iint_S G(x,y,z)\,dS=\iint_{R(x,z)}G(x,g(x,z),z)\sqrt{1+(\partial g/\partial x)^2+(\partial g/\partial z)^2}\,dA \tag{2.8.5b}$$

If the surface S is described by $x=h(y,z)$ where G, h, $\partial h/\partial y$, and $\partial h/\partial z$ are assumed to be continuous through the region R (on the yz-plane), the surface integral of G over S can be computed as

$$\iint_S G(x,y,z)\,dS=\iint_{R(y,z)}G(h(y,z),y,z)\sqrt{1+(\partial h/\partial y)^2+(\partial h/\partial z)^2}\,dA \tag{2.8.5c}$$

If the surface S is parameterized by a position vector $\mathbf{s}(v,w)=x(v,w)\mathbf{u}_x+y(v,w)\mathbf{u}_y+z(v,w)\mathbf{u}_z$, the surface integral of G over S can be computed as

$$\iint_S G(x,y,z)\,dS=\iint_{R(v,w)}G(x(v,w),y(v,w),z(v,w))\left\|\frac{\partial\mathbf{s}}{\partial v}\times\frac{\partial\mathbf{s}}{\partial w}\right\|dv\,dw \tag{2.8.6}$$

where

$$\frac{\partial\mathbf{s}}{\partial v}=\frac{\partial}{\partial v}x(v,w)\mathbf{u}_x+\frac{\partial}{\partial v}y(v,w)\mathbf{u}_y+\frac{\partial}{\partial v}z(v,w)\mathbf{u}_z \tag{2.8.7a}$$

$$\frac{\partial\mathbf{s}}{\partial w}=\frac{\partial}{\partial w}x(v,w)\mathbf{u}_x+\frac{\partial}{\partial w}y(v,w)\mathbf{u}_y+\frac{\partial}{\partial w}z(v,w)\mathbf{u}_z \tag{2.8.7b}$$

<*Surface Integral* **of a Vector Function**>

Let $\mathbf{G}(x,y,z)$ be a vector (field) function defined over a spatial region R containing a smooth surface S described by $f(x,y,z)=0$. The flux or flow of G passing through S can be expressed as the surface integral:

$$\text{Flux}=\iint_S(\mathbf{G}\cdot\mathbf{u}_n)\,dS=\iint_S\mathbf{G}\cdot(\mathbf{u}_n dS)=\iint_S\mathbf{G}\cdot d\mathbf{s}\ \text{ with } d\mathbf{s}=\mathbf{u}_n dS \tag{2.8.8}$$

where $\mathbf{u}_n$ is a unit normal vector to dS given by

$$\mathbf{u}_n=\pm\frac{\nabla f}{\|\nabla f\|}\ \text{(with sign +/– for outward/inward direction)} \tag{2.8.9}$$

(a) Torus - orientable (b) Möbius strip - non-orientable

Figure 2.26 Examples of orientable (two-sided) surface and non-orientable (one-sided) surface

and the surface S is assumed to be not only smooth but also *orientable* or *two-sided* in the sense that no 2D figure can be moved (continuously) on the surface and back to where it started so that it looks like its own mirror image reflected w.r.t. S. Figs. 2.26(a)/(b) show examples of orientable/non-orientable surfaces, respectively, where (b) having only one side is called the *Möbius strip* or *twisted cylinder*. Note that if the surface S is piecewise-smooth, the flux can be expressed as the sum of the surface integrals over the several smooth surfaces.

If the surface S is parameterized by a position vector $\mathbf{s}(v,w)=x(v,w)\mathbf{u}_x+y(v,w)\mathbf{u}_y+z(v,w)\mathbf{u}_z$, the surface integral of $\mathbf{G}$ over S can be computed as

$$\text{Flux}=\iint_S \mathbf{G}(x,y,z)\bullet d\mathbf{s}=\pm\iint_{R(v,w)}\mathbf{G}(x(v,w),y(v,w),z(v,w))\bullet\left(\frac{\partial\mathbf{s}}{\partial v}\times\frac{\partial\mathbf{s}}{\partial w}\right)dv\,dw \tag{2.8.10}$$

To realize how this formula can be used, let us compute the surface integral of a vector function

$$G(x,y,z)=x\,\mathbf{u}_x \tag{2.8.11}$$

over the three surfaces S_1, S_2, and S_3, each of which is depicted in Fig. 2.27(a), (b), and (c), respectively.

(1) Surface integral of $\mathbf{G}$ over S_1

Noting that S_1 is parameterized by $\mathbf{s}(v,w)=1\,\mathbf{u}_x+v\,\mathbf{u}_y+w\,\mathbf{u}_z$ with $0\le v\le 1$ and $0\le w\le 1$, we can compute

$$\iint_{S_1}\mathbf{G}\bullet d\mathbf{s}\overset{(2.8.10)}{=}\pm\iint_{R(v,w)}(x\mathbf{u}_x)\bullet\left(\frac{\partial\mathbf{s}}{\partial v}\times\frac{\partial\mathbf{s}}{\partial w}\right)dv\,dw\underset{\substack{y=v\\z=w}}{\overset{x=1}{=}}\int_0^1\!\!\int_0^1\begin{bmatrix}1\\0\\0\end{bmatrix}\bullet\left(\begin{bmatrix}0\\1\\0\end{bmatrix}\times\begin{bmatrix}0\\0\\1\end{bmatrix}\right)dv\,dw=\int_0^1\!\!\int_0^1 1\,dv\,dw=1$$

Alternatively, noting that a unit normal vector to the surface S_1 described by $x=1$ or $f(x,y,z)=x-1=0$ is $\mathbf{n}=\nabla f/||\nabla f||\overset{(2.1.8)}{=}\mathbf{u}_x$, we can directly use Eq. (2.8.8) to compute

$$\iint_{S_1}\mathbf{G}\bullet d\mathbf{s}\overset{(2.8.8)}{=}\iint_{S_1}(\mathbf{G}\bullet\mathbf{u}_n)dS=\int_0^1\!\!\int_0^1(x\mathbf{u}_x)\bullet\mathbf{u}_x\,dy\,dz\overset{x=1}{=}\int_0^1\!\!\int_0^1 1\,dy\,dz=1$$

(2) Surface integral of $\mathbf{G}$ over S_2

Noting that S_2 is parameterized by $\mathbf{s}(v,w)=v\,\mathbf{u}_x+1\,\mathbf{u}_y+w\,\mathbf{u}_z$ with $0\le v\le 1$ and $0\le w\le 1$, we can compute

$$\iint_{S_2}\mathbf{G}\bullet d\mathbf{s}\overset{(2.8.10)}{=}\pm\iint_{R(v,w)}(x\mathbf{u}_x)\bullet\left(\frac{\partial\mathbf{s}}{\partial v}\times\frac{\partial\mathbf{s}}{\partial w}\right)dv\,dw\underset{\substack{y=1\\z=w}}{\overset{x=v}{=}}\int_0^1\!\!\int_0^1\begin{bmatrix}v\\0\\0\end{bmatrix}\bullet\left(\begin{bmatrix}1\\0\\0\end{bmatrix}\times\begin{bmatrix}0\\0\\1\end{bmatrix}\right)dv\,dw=\int_0^1\!\!\int_0^1 0\,dv\,dw=0$$

(3) Surface integral of $\mathbf{G}$ over S_3

This can be obtained just like the surface integral of $\mathbf{G}$ over S_2.

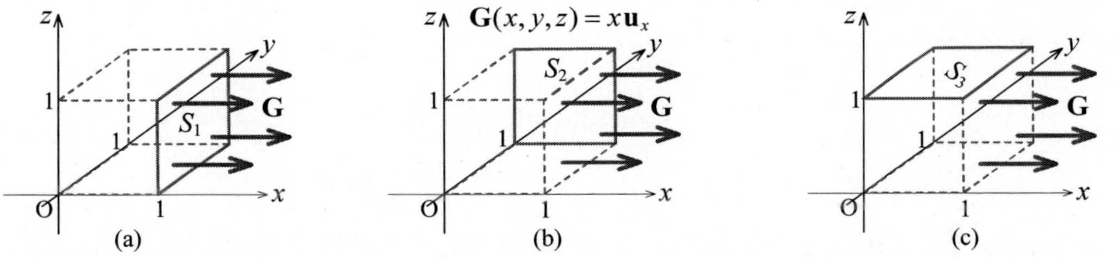

Figure 2.27 Surfaces for the surface integral of a vector function $G(x,y,z)=x\mathbf{u}_x$

(Example 2.25) Surface Area Integral

Use the surface area integral formula (2.8.3) to find the surface area of the unit hemisphere (with radius 1) (see Fig. 2.28(a)).

Noting that the surface S of the upper unit hemisphere is described by

$$z=f(x,y)=\sqrt{1-x^2-y^2} \tag{E2.25.1}$$

we use Eq. (2.8.3) or (2.8.5a) with $G(x,y,z)=1$ to find the area of the surface as

$$A(S)\overset{(2.8.3)}{=}\iint_R\sqrt{1+(\frac{\partial f}{\partial x})^2+(\frac{\partial f}{\partial y})^2}\,dA=\iint_R\sqrt{1+\frac{x^2}{1-x^2-y^2}+\frac{y^2}{1-x^2-y^2}}\,dA=\int_{-1}^{1}\int_{-\sqrt{1-y^2}}^{\sqrt{1-y^2}}\frac{1}{\sqrt{1-x^2-y^2}}\,dx\,dy$$

$$\overset{\text{Cartesian-to-polar}}{=}\int_0^{2\pi}\int_0^1\frac{1}{\sqrt{1-\rho^2}}\rho\,d\rho\,d\phi=-\int_0^{2\pi}\sqrt{1-\rho^2}\Big|_0^1 d\phi=2\pi \tag{E2.25.2}$$

Alternatively, we can parameterize the surface S of the upper unit hemisphere in terms of θ and ϕ in the spherical coordinate system as

$$\mathbf{s}(\theta,\phi)=r\mathbf{u}_r\underset{r=1}{\overset{(1.1.59a)}{=}}\sin\theta\cos\phi\,\mathbf{u}_x+\sin\theta\sin\phi\,\mathbf{u}_y+\cos\theta\,\mathbf{u}_z \tag{E2.25.3}$$

and then, use Eq. (2.8.6) with $G(x,y,z)=1$ to find the area of S as

$$A(S)\underset{G(x,y,z)=1}{\overset{(2.8.6)}{=}}\iint_R\left\|\frac{\partial\mathbf{s}}{\partial\theta}\times\frac{\partial\mathbf{s}}{\partial\phi}\right\|dA=\int_0^{2\pi}\int_0^{\pi/2}\left\|\begin{bmatrix}\cos\theta\cos\phi\\ \cos\theta\sin\phi\\ -\sin\theta\end{bmatrix}\times\begin{bmatrix}-\sin\theta\sin\phi\\ \sin\theta\cos\phi\\ 0\end{bmatrix}\right\|d\theta\,d\phi=\int_0^{2\pi}\int_0^{\pi/2}\left\|\begin{bmatrix}\sin^2\theta\cos\phi\\ \sin^2\theta\sin\phi\\ \cos\theta\sin\theta\end{bmatrix}\right\|d\theta\,d\phi$$

$$=\int_0^{2\pi}\int_0^{\pi/2}\sqrt{\sin^4\theta+\cos^2\theta\sin^2\theta}\,d\theta d\phi=\int_0^{2\pi}\int_0^{\pi/2}\sin\theta\,d\theta d\phi=\int_0^{2\pi}(-\cos\theta)\Big|_0^{\pi/2}d\phi=2\pi \tag{E2.25.4}$$

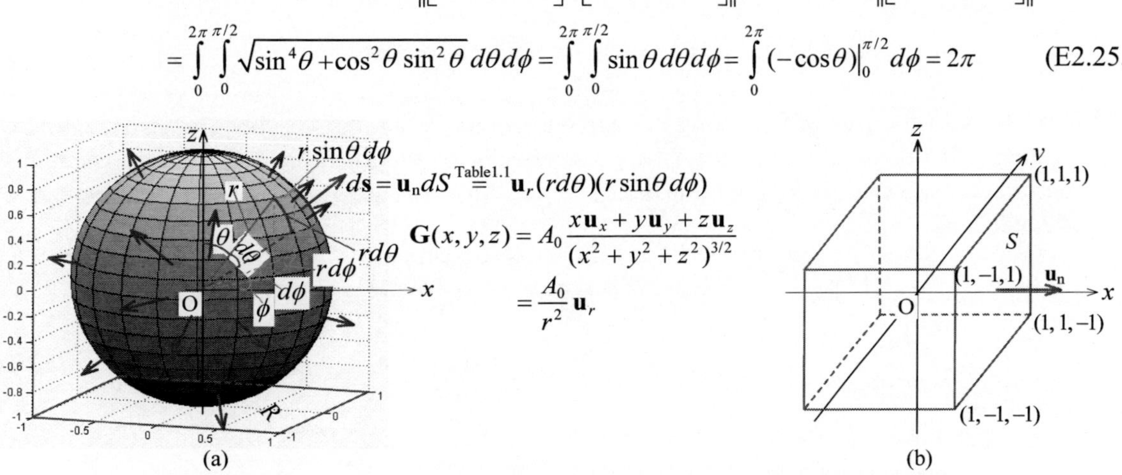

Figure 2.28 Surfaces for the surface integrals in Examples 2.25 and 2.26

(Example 2.26) Surface Integral of a Vector Function

Consider a vector field given by

$$\mathbf{G}(x,y,z)=\frac{x\mathbf{u}_x+y\mathbf{u}_y+z\mathbf{u}_z}{(x^2+y^2+z^2)^{3/2}} \text{ or equivalently, } \mathbf{G}(r,\theta,\phi)=\frac{1}{r^3}\mathbf{r}=\frac{1}{r^2}\mathbf{u}_r \tag{E2.26.1}$$

where $\mathbf{r}=x\mathbf{u}_x+y\mathbf{u}_y+z\mathbf{u}_z$ and $r=\|\mathbf{r}\|=\sqrt{x^2+y^2+z^2}$

(a) Find the outward flux of **G** passing through the surface of the unit sphere depicted in Fig. 2.28(a). Noting that the surface *S* of the unit sphere and its outward normal vector $\mathbf{u_n}$ are described by

$$f(x,y,z)=x^2+y^2+z^2-1=0 \rightarrow \mathbf{u}_\mathrm{n} \overset{(2.8.9)}{=} \frac{\nabla f}{\|\nabla f\|}=\frac{x\mathbf{u}_x+y\mathbf{u}_y+z\mathbf{u}_z}{\sqrt{x^2+y^2+z^2}}=\mathbf{u}_r \tag{E2.26.2}$$

we use Eq. (2.8.8) to find the flux of **G** through *S* as

$$\iint_S \mathbf{G}\bullet d\mathbf{s} \underset{(1.1.61b)}{\overset{(E2.26.1),(E2.26.2)}{=}} \int_0^{2\pi}\int_0^{\pi}\frac{\mathbf{u}_r}{\cancel{r^2}}\bullet\mathbf{u}_r\cancel{r^2}\sin\theta\, d\theta\, d\phi$$

$$=\int_0^{2\pi}\int_0^{\pi}\sin\theta\, d\theta d\phi=\int_0^{2\pi}(-\cos\theta)\Big|_0^{\pi}\, d\phi=4\pi \tag{E2.26.3}$$

Alternatively, with the surface *S* of the unit hemisphere parameterized in terms of θ and ϕ in the spherical coordinate system as

$$\mathbf{s}(\theta,\phi)=r\mathbf{u}_r \underset{r=1}{\overset{(1.1.59a)}{=}} \sin\theta\cos\phi\,\mathbf{u}_x+\sin\theta\sin\phi\,\mathbf{u}_y+\cos\theta\,\mathbf{u}_z \tag{E2.26.4}$$

we may use Eq. (2.8.10) to find the surface integral of **G** as

$$\text{Flux} \overset{(2.8.10)}{=} \iint_R \mathbf{G}\bullet\left(\frac{\partial\mathbf{s}}{\partial\theta}\times\frac{\partial\mathbf{s}}{\partial\phi}\right)d\theta\, d\phi \text{ (with the direction of } \left(\frac{\partial\mathbf{s}}{\partial\theta}\times\frac{\partial\mathbf{s}}{\partial\phi}\right)=\mathbf{u}_\theta\times\mathbf{u}_\phi \overset{\text{Table1.1}}{=} \mathbf{u}_r\text{: outward)}$$

$$\underset{(E2.26.4)}{\overset{(E2.26.1)}{=}} \int_0^{2\pi}\int_0^{\pi}\frac{1}{1^3}\begin{bmatrix}x{=}\sin\theta\cos\phi\\ y{=}\sin\theta\sin\phi\\ z{=}\cos\theta\end{bmatrix}\bullet\left(\begin{bmatrix}\cos\theta\cos\phi\\ \cos\theta\sin\phi\\ -\sin\theta\end{bmatrix}\times\begin{bmatrix}-\sin\theta\sin\phi\\ \sin\theta\cos\phi\\ 0\end{bmatrix}\right)d\theta\, d\phi$$

$$=\int_0^{2\pi}\int_0^{\pi}1\begin{bmatrix}\sin\theta\cos\phi\\ \sin\theta\sin\phi\\ \cos\theta\end{bmatrix}\bullet\begin{bmatrix}\sin^2\theta\cos\phi\\ \sin^2\theta\sin\phi\\ \cos\theta\sin\theta\end{bmatrix}d\theta\, d\phi=\int_0^{2\pi}\int_0^{\pi}(\sin^3\theta+\cos^2\theta\sin\theta)d\theta d\phi$$

$$=\int_0^{2\pi}\int_0^{\pi}\sin\theta\, d\theta d\phi=\int_0^{2\pi}(-\cos\theta)\Big|_0^{\pi}\, d\phi=\int_0^{2\pi}2\,d\phi=4\pi \tag{E2.26.5}$$

(b) Find the outward flux of **G** passing through one of the six faces of the cube shown in Fig. 2.28(b). Noting that the surface *S* of the face perpendicular to the *x*-axis and its normal vector $\mathbf{u}_\mathrm{n}$ are described by $f(x,y,z)=x-1=0$ and $\mathbf{u}_\mathrm{n}=\nabla f/\|\nabla f\|=\mathbf{u}_x$, we use Eq. (2.8.8) to find the flux of **G** through *S* as

$$\text{Flux} \overset{(2.8.8)}{=} \iint_S(\mathbf{G}\bullet\mathbf{u}_\mathrm{n})dS=\int_{-1}^{1}\int_{-1}^{1}\frac{x\mathbf{u}_x+y\mathbf{u}_y+z\mathbf{u}_z}{(x^2+y^2+z^2)^{3/2}}\Bigg|_{x=1}\bullet\mathbf{u}_x\,dydz$$

$$=\int_{-1}^{1}\int_{-1}^{1}\frac{1}{(1+y^2+z^2)^{3/2}}dy\,dz \tag{E2.26.6}$$

```
>>dblquad(@(x,y)(1+x.^2+y.^2).^(-3/2),-1,1,-1,1)
   ans = 2.0944   %=4*pi/6: How is this related with Eq.(E2.26.3)?
```

(Example 2.27) Surface Integral of Scalar Functions

(a) $\iint_S 8xy\,dS$ where S is the portion of the plane $z=2(1-x-y)$ in the first octant (Fig. 2.29(a)).
First, noting that the surface S is described by $z=f(x,y)=2(1-x-y)$, we use Eq. (2.8.5a):

$$\iint_S G(x,y,z)\,dS \overset{(2.8.5a)}{=} \iint_{R(x,y)} 8xy\sqrt{1+(\partial f/\partial x)^2+(\partial f/\partial y)^2}\,dA = \int_0^1\int_0^{1-x} 8xy\sqrt{1+(-2)^2+(-2)^2}\,dy\,dx$$

$$=12\int_0^1 xy^2\Big|_0^{1-x}dx = 12\int_0^1 x(1-x)^2dx = (6x^2-8x^3+3x^4)\Big|_0^1 = 1 \quad \text{(E2.27a.1)}$$

Also, with the surface S described by $x=h(y,z)=1-y-z/2$, we can use Eq. (2.8.5c)

$$\iint_S G(x,y,z)\,dS \overset{(2.8.5c)}{=} \iint_{R(y,z)} 8xy\sqrt{1+(\frac{\partial h}{\partial y})^2+(\frac{\partial h}{\partial z})^2}\,dA = \int_0^1\int_0^{2(1-y)} 8(1-y-\frac{z}{2})y\sqrt{1+(-1)^2+(-\frac{1}{2})^2}\,dz\,dy$$

$$=12\int_0^1 \{(1-y)yz-\frac{1}{4}yz^2\}\Big|_0^{2(1-y)}dy = 12\int_0^1\{2(1-y)^2y-y(1-y)^2\}dy=1 \quad \text{(E2.27a.2)}$$

Alternatively, we can parameterize the surface S by

$$\mathbf{s}(x,y)=x\,\mathbf{u}_x+y\,\mathbf{u}_y+2(1-x-y)\mathbf{u}_z \text{ with } x\ge 0,\ y\ge 0, \text{ and } x+y\le 1 \quad \text{(E2.27a.3)}$$

and use Eq. (2.8.6):

$$\iint_S G(x,y,z)\,dS \overset{(2.8.6)}{=} \iint_{R(x,y)} 8xy\left\|\frac{\partial \mathbf{s}}{\partial x}\times\frac{\partial \mathbf{s}}{\partial y}\right\|dx\,dy \overset{(E2.27a.3)}{=} 8\int_0^1\int_0^{1-y} xy\left\|\begin{bmatrix}1\\0\\-2\end{bmatrix}\times\begin{bmatrix}0\\1\\-2\end{bmatrix}\right\|dx\,dy$$

$$=8\int_0^1\int_0^{1-y} xy\sqrt{2^2+2^2+1^2}\,dx\,dy = 24\int_0^1\int_0^{1-y} xy\,dx\,dy = 12\int_0^1 x^2y\Big|_0^{1-y}dy = 1 \quad \text{(E2.27a.4)}$$

(b) $\iint_S (x+y+z)\,dS$ where S is the part of the cylinder $x^2+y^2=1$ between $z=0$ and $z=1$ (Fig. 2.29(b)).
We can parameterize the surface S by

$$\mathbf{s}(\phi,z) \underset{\rho=1}{\overset{(1.1.58c)}{=}} \cos\phi\,\mathbf{u}_x+\sin\phi\,\mathbf{u}_y+z\,\mathbf{u}_z \text{ with } 0\le\phi\le 2\pi \text{ and } 0\le z\le 1 \quad \text{(E2.27b.1)}$$

and use Eq. (2.8.6):

$$\iint_S G(x,y,z)\,dS \underset{(E2.27b.1)}{\overset{(2.8.6)}{=}} \iint_{R(\phi,z)} \begin{pmatrix}x+\\y+z\end{pmatrix}\Bigg|_{\substack{x=\cos\phi\\y=\sin\phi}} \left\|\frac{\partial \mathbf{s}}{\partial \phi}\times\frac{\partial \mathbf{s}}{\partial z}\right\|d\phi\,dz = \int_0^1\int_0^{2\pi}\begin{pmatrix}\cos\phi+\\\sin\phi+z\end{pmatrix}\left\|\begin{bmatrix}-\sin\phi\\\cos\phi\\0\end{bmatrix}\times\begin{bmatrix}0\\0\\1\end{bmatrix}\right\|d\phi\,dz$$

$$=\int_0^1\int_0^{2\pi}(\cos\phi+\sin\phi+z)\sqrt{\cos^2\phi+\sin^2\phi+0^2}\,d\phi\,dz = \int_0^1 2\pi z\,dz = \pi z^2\Big|_0^1 = \pi \quad \text{(E2.27b.2)}$$

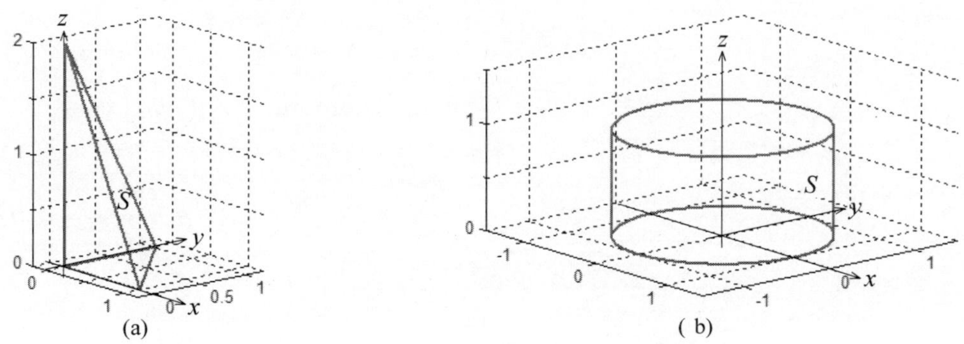

Figure 2.29 Surfaces for the surface integrals in Example 2.27

(Example 2.28) Surface Integral of a Vector Function

Consider a vector field given by

$$\mathbf{G}(x,y,z)=A_0\frac{x\mathbf{u}_x+y\mathbf{u}_y}{(x^2+y^2)^{3/2}}=A_0\frac{\mathbf{u}_\rho}{\rho^2}\ \text{with}\ \mathbf{u}_\rho=\frac{x\mathbf{u}_x+y\mathbf{u}_y}{\sqrt{x^2+y^2}} \tag{E2.28.1}$$

Find the flux of **G** passing through the cylindrical surface that is depicted in Fig. 2.30.

Noting that the cylindrical surface S and its (outward) unit normal vector $\mathbf{u}_n$ are described by

$$f(x,y,z)=x^2+y^2-1=0 \tag{E2.28.2a}$$

$$\mathbf{u}_n=\frac{\nabla f}{||\nabla f||}=\frac{x\mathbf{u}_x+y\mathbf{u}_y}{\sqrt{x^2+y^2}}=\mathbf{u}_\rho \tag{E2.28.2b}$$

we use Eq. (2.8.8) to find the flux of **G** through S as

$$\text{Flux}\overset{(2.8.8)}{=}\iint_S(\mathbf{G}\bullet\mathbf{u}_n)\,dS\underset{(E2.28.2b)}{\overset{(E2.28.1)}{=}}\int_0^1\int_0^{2\pi}A_0\frac{x\mathbf{u}_x+y\mathbf{u}_y}{(x^2+y^2)^{3/2}}\bullet\frac{x\mathbf{u}_x+y\mathbf{u}_y}{\sqrt{x^2+y^2}}\rho\,d\phi\,dz$$

$$=\int_0^1\int_0^{2\pi}\frac{A_0}{\rho}\,d\phi\,dz=2\frac{A_0}{\rho}\pi\overset{\rho=1}{=}2A_0\pi \tag{E2.28.3}$$

Alternatively, with the cylindrical surface S parameterized in terms of ϕ and z in the cylindrical coordinate system as

$$\mathbf{s}(\phi,z)\overset{(1.1.58c)}{=}\rho\cos\phi\,\mathbf{u}_x+\rho\sin\phi\,\mathbf{u}_y+z\mathbf{u}_z\ \text{with}\ \rho=1 \tag{E2.28.4}$$

we can use Eq. (2.8.10) to find the surface integral of **G** as

$$\text{Flux}\overset{(2.8.10)}{=}\iint_R\mathbf{G}\bullet\left(\frac{\partial\mathbf{s}}{\partial\phi}\times\frac{\partial\mathbf{s}}{\partial z}\right)d\phi\,dz\ (\text{with the direction of}\left(\frac{\partial\mathbf{s}}{\partial\phi}\times\frac{\partial\mathbf{s}}{\partial z}\right)=\mathbf{u}_\phi\times\mathbf{u}_z\overset{\text{Table1.1}}{=}\mathbf{u}_\rho:\text{outward})$$

$$\underset{(E2.28.4)}{\overset{(E2.28.1)}{=}}\int_0^1\int_0^{2\pi}\frac{A_0}{\rho^3}\begin{bmatrix}x=\rho\cos\phi\\ y=\rho\sin\phi\\ 0\end{bmatrix}\bullet\left(\begin{bmatrix}-\rho\sin\phi\\ \rho\cos\phi\\ 0\end{bmatrix}\times\begin{bmatrix}0\\0\\1\end{bmatrix}\right)d\phi\,dz$$

$$=\int_0^1\int_0^{2\pi}\frac{A_0}{\rho}\begin{bmatrix}\cos\phi\\ \sin\phi\\ 0\end{bmatrix}\bullet\begin{bmatrix}\cos\phi\\ \sin\phi\\ 0\end{bmatrix}d\phi\,dz=\int_0^1\int_0^{2\pi}\frac{A_0}{\rho}\,d\phi\,dz=2\frac{A_0}{\rho}\pi\overset{\rho=1}{=}2A_0\pi \tag{E2.28.5}$$

```
>>syms x y z v w;
>>I=int_surf([x y 0]/(x^2+y^2)^(3/2),[cos(v) sin(v) w],[0 2*pi],[0 1],0)
```

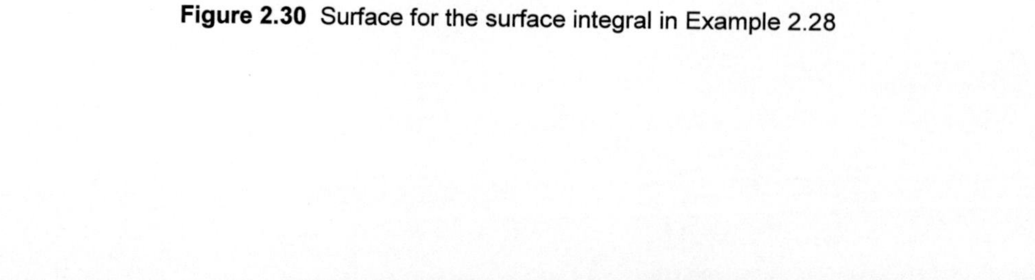

Figure 2.30 Surface for the surface integral in Example 2.28

2.9 STOKES' THEOREM

Just like Green's theorem, Stokes' theorem also gives the relationship between the line integral of a vector function around a simple closed curve C and the surface integral over the region R bounded by C, but in 3D space.

[Stokes' Theorem]

Let C be a piecewise-smooth, simple closed curve enclosing an orientable piecewise-smooth surface S (in 3D space) in the counter-clockwise (positive) direction, i.e., to the left. Also, let $\mathbf{F}(x, y, z)=F_x(x, y, z)\mathbf{u}_x+F_y(x, y, z)\mathbf{u}_y+F_z(x, y, z)\mathbf{u}_z$ be a continuous vector function which has continuous first partial derivatives in a 3D region containing S. Then

$$\oint_C \mathbf{F}\cdot d\mathbf{l}=\oint_C \mathbf{F}\cdot\mathbf{u}_T\,dl=\iint_S(\nabla\times\mathbf{F})\cdot d\mathbf{s} \text{ with } d\mathbf{s}=\mathbf{u}_n\,dS=(n_x\mathbf{u}_x+n_y\mathbf{u}_y+n_z\mathbf{u}_z)dS \quad (2.9.1a)$$

$$;\oint_C\{F_x\,dx+F_y\,dy+F_z\,dz\}=\iint_R\left\{\left(\frac{\partial F_z}{\partial y}-\frac{\partial F_y}{\partial z}\right)dy\,dz+\left(\frac{\partial F_x}{\partial z}-\frac{\partial F_z}{\partial x}\right)dz\,dx+\left(\frac{\partial F_y}{\partial x}-\frac{\partial F_x}{\partial y}\right)dx\,dy\right\} \quad (2.9.1b)$$

where $\mathbf{u}_T$ and $\mathbf{u}_n$ denote the unit tangent vector along C and the unit normal vector (determined by the right-hand rule from the direction of C) to a differential surface dS, respectively.

Stokes' theorem states that the total amount of circulation along a surface is equal to the amount of circulation on its boundary. Since it is nothing but a generalization of Green's theorem into a 3D vector form, let us leave the proof to other references like [K-2], [Z-1], ... and instead, approximate each side of Eq. (2.9.1a) by

$$\sum_i \oint_{\Delta C_i}\mathbf{F}\cdot d\mathbf{l}=\sum_i \nabla\times\mathbf{F}\cdot\Delta\mathbf{s}_i \text{ with } \Delta\mathbf{s}=\mathbf{u}_n\Delta S, \quad (2.9.2)$$

and visualize it as depicted in Fig. 2.31. Fig. 2.31 states that the sum of $\nabla\times\mathbf{F}$ on the RHS of Eq. (2.9.2) for all subsurfaces $\{\Delta S_i\}$ becomes the flux of $\mathbf{F}$ through the whole surface S made of ΔS_i's where the line integrals along a line segment shared by two adjacent small closed curve ΔC_i's are of opposite signs, so that the sum of line integrals on the LHS of Eq. (2.9.2) for all small curve ΔC_i's becomes the line integral of $\mathbf{F}$ along the large closed curve C, enclosing the whole surface S. In addition, recall the definition (2.4.2) of curl:

$$\text{curl }\mathbf{F}=\nabla\times\mathbf{F}\overset{(2.4.2)}{=}\mathbf{u}_n\lim_{\Delta S\to 0}\frac{1}{\Delta S}\oint_{\Delta C}\mathbf{F}\cdot d\mathbf{l}\;\rightarrow(\nabla\times\mathbf{F})\cdot\mathbf{u}_n\Delta S\approx\oint_{\Delta C}\mathbf{F}\cdot d\mathbf{l} \quad (2.9.3)$$

Don't these statements help you to believe the validity and realize the meaning of Stokes' theorem?

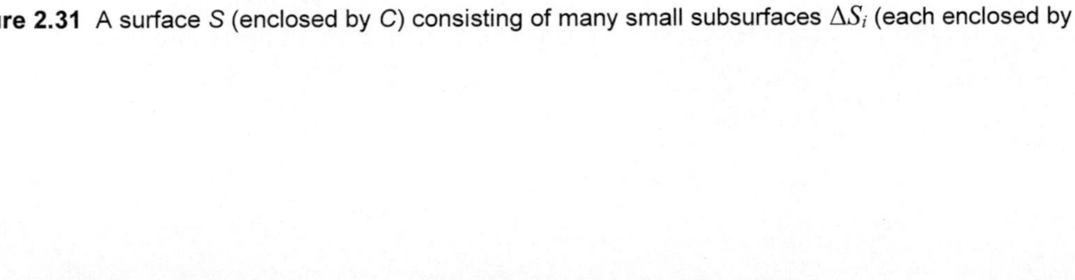

Figure 2.31 A surface S (enclosed by C) consisting of many small subsurfaces ΔS_i (each enclosed by ΔC_i)

<Curl of $\mathbf{F}=\mathbf{u}_\phi/\rho$ and 2D Unit Impulse (Dirac Delta) Function>

Consider the vector field that was dealt with in Example 2.17:

$$\mathbf{F}(x,y,z)=\frac{-y}{x^2+y^2}\mathbf{u}_x+\frac{x}{x^2+y^2}\mathbf{u}_y,\ \text{or equivalently,}\ \mathbf{F}(\rho,\phi,z)=\frac{1}{\rho}\mathbf{u}_\phi \tag{2.9.4}$$

where

$$\rho=\sqrt{x^2+y^2}$$

For this vector field, let us see if Stokes' theorem (2.9.1) holds with the surface S of the unit disk enclosed by the (counterclockwise) unit circle C on the xy-plane. Since the LHS (line integral) of Eq. (2.9.1a) with Eq. (2.9.4) was obtained as Eq. (E2.17.3) in Example 2.17:

$$\oint_C \mathbf{F}\bullet d\mathbf{l}\overset{(2.9.4)}{=}\oint_C\left(\frac{-y}{x^2+y^2}\mathbf{u}_x+\frac{x}{x^2+y^2}\mathbf{u}_y\right)\bullet d\mathbf{l}\overset{(\text{E2.17.3})}{=}2\pi \tag{2.9.5}$$

we have only to find the RHS (surface integral) of Eq. (2.9.1a) where the unit normal vector of the surface S is $\mathbf{u}_\text{n}=\mathbf{u}_z$:

$$\oiint_S(\nabla\times\mathbf{F})\bullet d\mathbf{s}\overset{(2.4.5\text{a})}{=}\iint_S\left\{\left(\frac{\partial F_z}{\partial y}-\frac{\partial F_y}{\partial z}\right)\mathbf{u}_x+\left(\frac{\partial F_x}{\partial z}-\frac{\partial F_z}{\partial x}\right)\mathbf{u}_y+\left(\frac{\partial F_y}{\partial x}-\frac{\partial F_x}{\partial y}\right)\mathbf{u}_z\right\}\bullet\mathbf{u}_\text{n}dS$$

$$\underset{\mathbf{u}_\text{n}=\mathbf{u}_z}{\overset{(2.9.4)}{=}}\iint_S\left(\frac{\partial}{\partial x}\frac{x}{x^2+y^2}-\frac{\partial}{\partial y}\frac{-y}{x^2+y^2}\right)\mathbf{u}_z\bullet\mathbf{u}_z dS=\iint_S\left(\frac{(x^2+y^2)-2x^2}{(x^2+y^2)^2}+\frac{(x^2+y^2)-2y^2}{(x^2+y^2)^2}\right)dS=0 \tag{2.9.6}$$

Actually, the line/surface integrals in the LHS/RHS of Eq. (2.9.1a) with (2.9.4) are easier to compute in the cylindrical coordinate system as

$$\oint_C\mathbf{F}\bullet d\mathbf{l}\underset{(1.1.60\text{c})}{\overset{(2.9.4)}{=}}\int_0^{2\pi}\frac{1}{\rho}\mathbf{u}_\phi\bullet\mathbf{u}_\phi\rho\,d\phi=2\pi \tag{2.9.7}$$

$$\oiint_S(\nabla\times\mathbf{F})\bullet d\mathbf{s}\overset{(2.4.5\text{b})}{=}\iint_S\left\{\left(\cancel{\frac{1}{\rho}\frac{\partial F_z}{\partial\phi}}-\cancel{\frac{\partial F_\phi}{\partial z}}\right)\mathbf{u}_\rho+\left(\cancel{\frac{\partial F_\rho}{\partial z}}-\cancel{\frac{\partial F_z}{\partial\rho}}\right)\mathbf{u}_\phi+\frac{1}{\rho}\left(\frac{\partial(\rho F_\phi)}{\partial\rho}-\cancel{\frac{\partial F_\rho}{\partial\phi}}\right)\mathbf{u}_z\right\}\bullet\mathbf{u}_\text{n}dS$$

$$\underset{\mathbf{u}_\text{n}=\mathbf{u}_z}{\overset{(2.9.4)}{=}}\iint_S\frac{1}{\rho}\frac{\partial(\rho/\rho)}{\partial\rho}\mathbf{u}_z\bullet\mathbf{u}_z dS=\iint_S 0\,dS=0 \tag{2.9.8}$$

Contrary to our expectations, Eqs. (2.9.5)/(2.9.7) and (2.9.6)/(2.9.8) are not equal. Is this a counterexample to disprove Stokes' theorem? No way. The truth is that we have overlooked the singularity of the vector field $\mathbf{F}=\mathbf{u}_\phi/\rho$ at $\rho=0$ where $\mathbf{F}$ blows up so that it cannot satisfy the premise of Stokes' theorem that $\mathbf{F}$ must be continuous in a 3D region containing the surface S. This seeming contradiction can be resolved by letting

$$\text{curl}\left(\frac{-y}{x^2+y^2}\mathbf{u}_x+\frac{x}{x^2+y^2}\mathbf{u}_y\right)=\text{curl}\left(\frac{\mathbf{u}_\phi}{\rho}\right)=\nabla\times\frac{\mathbf{u}_\phi}{\rho}=2\pi\delta^2(x,y)\,\mathbf{u}_z=2\pi\delta(x)\,\delta(y)\,\mathbf{u}_z \tag{2.9.9}$$

where $\delta^2(x,y)=\delta(x)\delta(y)$ is called the *2D Dirac delta function*, which is a generalization of the 1D Dirac delta or unit impulse function $\delta(t)$, to be introduced in Section 4.1.2, and satisfies

$$\iint_S\delta^2(x,y)dS=\iint_S\delta(x)\,\delta(y)\,dS=1\ \text{for any surface}\ S\ \text{containing a point}\ (0,0,z) \tag{2.9.10}$$

To get familiar with Stokes' theorem, see the following examples:

(Example 2.29) Stokes' Theorem

Show the validity of Stokes' theorem (2.9.1) for a vector field

$$\mathbf{F}(x,y,z)=z\,\mathbf{u}_x+x\,\mathbf{u}_y+y\,\mathbf{u}_z \tag{E2.29.1}$$

with the surface S (Fig. 2.32) of the partial plane $z=g(x,y)=2-x-y$ in the first octant described by

$$f(x,y,z)=x+y+z-2=0 \ \text{ for } 0\le z\le 2 \tag{E2.29.2}$$

Noting that the curve C enclosing the triangular surface consists of the three line segments C_1, C_2, and C_3, each of which is parameterized by

$$\mathbf{p}_1(t)=(2-t)\,\mathbf{u}_x+t\,\mathbf{u}_y+0\,\mathbf{u}_z \ \text{ for } t=0\sim 2 \tag{E2.29.3a}$$

$$\mathbf{p}_2(t)=0\,\mathbf{u}_x+(2-t)\,\mathbf{u}_y+t\,\mathbf{u}_z \ \text{ for } t=0\sim 2 \tag{E2.29.3b}$$

$$\mathbf{p}_3(t)=t\,\mathbf{u}_x+0\,\mathbf{u}_y+(2-t)\,\mathbf{u}_z \ \text{ for } t=0\sim 2 \tag{E2.29.3c}$$

the *LHS* of Eq. (2.9.1a) can be computed as

$$\oint_C \mathbf{F}\cdot d\mathbf{l}=\oint_C \mathbf{F}\cdot\mathbf{u}_T\,dl \overset{\text{(E2.29.1)}}{\underset{\text{(2.3.15)}}{=}} \oint_C (z\,\mathbf{u}_x+x\mathbf{u}_y+y\mathbf{u}_z)\cdot d\mathbf{p} \overset{\text{(E2.29.3a,b,c)}}{=} \int_0^2((2-t)\mathbf{u}_y+t\mathbf{u}_z)\cdot(-\mathbf{u}_x+\mathbf{u}_y)\,dt$$
$$+\int_0^2(t\mathbf{u}_x+(2-t)\mathbf{u}_z)\cdot(-\mathbf{u}_y+\mathbf{u}_z)\,dt+\int_0^2((2-t)\mathbf{u}_x+t\mathbf{u}_y)\cdot(\mathbf{u}_x-\mathbf{u}_z)\,dt$$
$$=\int_0^2(2-t)dt+\int_0^2(2-t)dt+\int_0^2(2-t)\,dt=3(2t-\frac{1}{2}t^2)\Big|_0^2=6 \tag{E2.29.4}$$

Noting that the surface S is described by $f(x,y,z)=x+y+z-2=0$, we can evaluate the *RHS* of Eq. (2.9.1a) with (E2.29.1) as

$$\iint_S(\nabla\times\mathbf{F})\cdot d\mathbf{s} \overset{\text{(2.4.5a)}}{=} \iint_S\left\{\left(\frac{\partial F_z}{\partial y}-\frac{\partial F_y}{\partial z}\right)\mathbf{u}_x+\left(\frac{\partial F_x}{\partial z}-\frac{\partial F_z}{\partial x}\right)\mathbf{u}_y+\left(\frac{\partial F_y}{\partial x}-\frac{\partial F_x}{\partial y}\right)\mathbf{u}_z\right\}\cdot\mathbf{u}_n dS$$
$$\overset{\text{(E2.29.1)}}{\underset{\text{(2.8.9)}}{=}} \iint_S\{(1-0)\mathbf{u}_x+(1-0)\mathbf{u}_y+(1-0)\mathbf{u}_z\}\cdot\frac{\nabla f=1\mathbf{u}_x+1\mathbf{u}_y+1\mathbf{u}_z}{\|\nabla f\|=\sqrt{3}}\,dS=\iint_S\sqrt{3}\,dS$$
$$\overset{\text{(2.8.5a)}}{\underset{z=f(x,y)=2-x-y}{=}} \int_0^2\int_0^{2-x}\sqrt{3}\sqrt{1+(-1)^2+(-1)^2}\,dydx=\int_0^2 3(2-x)\,dx=(6x-\frac{3}{2}x^2)\Big|_0^2=6 \tag{E2.29.5}$$

Also, the *RHS* of Eq. (2.9.1b) can be used directly:

$$\iint_R\left\{\left(\frac{\partial F_z}{\partial y}-\frac{\partial F_y}{\partial z}\right)dydz+\left(\frac{\partial F_x}{\partial z}-\frac{\partial F_z}{\partial x}\right)dzdx+\left(\frac{\partial F_y}{\partial x}-\frac{\partial F_x}{\partial y}\right)dxdy\right\}=\iint_R 1dydz+\iint_R 1dzdx+\iint_R 1dxdy$$
$$=\text{The sum of the areas of the triangular regions } R(y,z), R(z,x), \text{ and } R(x,y)=3\cdot\frac{1}{2}2^2=6 \tag{E2.29.6}$$

Alternatively, with the triangular surface S parameterized by

$$\mathbf{s}(x,y)=x\,\mathbf{u}_x+y\,\mathbf{u}_y+(2-x-y)\,\mathbf{u}_z \tag{E2.29.7}$$

we may use Eq. (2.8.6) to compute the above surface integral as

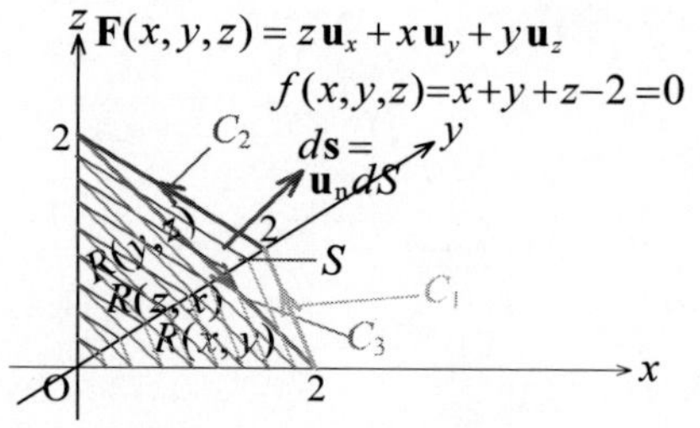

Figure 2.32 A curve and a surface for Example 2.29

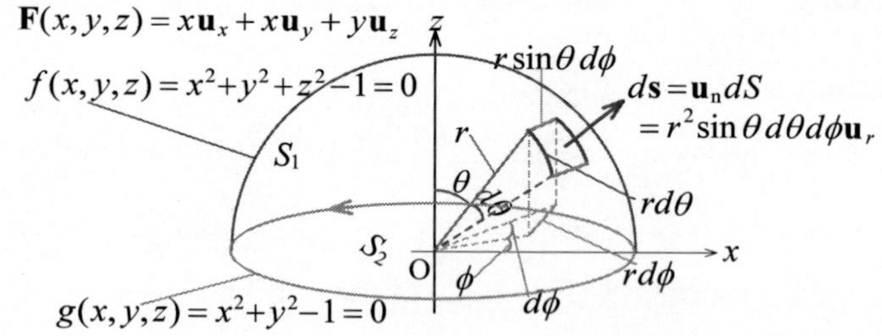

Figure 2.33 A curve and a surface for Example 2.30

$$\iint_S \sqrt{3}\,dS \underset{\mathbf{s}=x\mathbf{u}_x+y\mathbf{u}_y+(2-x-y)\mathbf{u}_z}{\overset{(2.8.6)}{=}} \iint_R \sqrt{3}\left\|\frac{\partial \mathbf{s}}{\partial x}\times\frac{\partial \mathbf{s}}{\partial y}\right\| dA = \iint_R \sqrt{3}\left\|\begin{bmatrix}1\\0\\-1\end{bmatrix}\times\begin{bmatrix}0\\1\\-1\end{bmatrix}\right\| dA = \iint_R \sqrt{3}\left\|\begin{bmatrix}1\\1\\1\end{bmatrix}\right\| dA = \iint_R 3\,dA$$

$$= 3\cdot(\text{Area of the triangular region } R(x,y)) = 3\cdot\frac{1}{2}2^2 = 6 \tag{E2.29.8}$$

(Example 2.30) Stokes' Theorem

Show the validity of Stokes' theorem (2.9.1) for a vector field

$$\mathbf{F}(x,y,z)=x\,\mathbf{u}_x+x\,\mathbf{u}_y+y\,\mathbf{u}_z \tag{E2.30.1}$$

with the hemispherical surface S_1 (Fig. 2.33) described by

$$f(x,y,z)=x^2+y^2+z^2-1=0 \quad \text{for } z\ge 0 \tag{E2.30.2}$$

that is enclosed by the unit circle C parameterized as

$$\mathbf{p}(t)=\cos t\,\mathbf{u}_x+\sin t\,\mathbf{u}_y \quad \text{for } t=0\sim 2\pi \tag{E2.30.3}$$

The *LHS* of Eq. (2.9.1a) can be computed as

$$\oint_C \mathbf{F}\bullet d\mathbf{l}=\oint_C \mathbf{F}\bullet\mathbf{u}_T\,dl \overset{(2.3.15)}{=} \int_0^{2\pi}\mathbf{F}\bullet d\mathbf{p} \underset{(E2.30.3)}{\overset{(E2.30.1)}{=}} \int_0^{2\pi}(\cos t\,\mathbf{u}_x+\cos t\,\mathbf{u}_y+\sin t\,\mathbf{u}_z)\bullet(-\sin t\,\mathbf{u}_x+\cos t\,\mathbf{u}_y)\,dt$$

$$=\int_0^{2\pi}\left(-\frac{1}{2}\sin 2t+\cos^2 t\right)dt=\int_0^{2\pi}\left(-\frac{1}{2}\sin 2t+\frac{1}{2}(1+\cos 2t)\right)dt=\pi \tag{E2.30.4}$$

The *RHS* of Eq. (2.9.1a) can be computed as

$$\iint_{S_1}(\nabla\times\mathbf{F})\bullet d\mathbf{s} \overset{(2.4.5a)}{=} \iint_{S_1}\left\{\left(\frac{\partial F_z}{\partial y}-\frac{\partial F_y}{\partial z}\right)\mathbf{u}_x+\left(\frac{\partial F_x}{\partial z}-\frac{\partial F_z}{\partial x}\right)\mathbf{u}_y+\left(\frac{\partial F_y}{\partial x}-\frac{\partial F_x}{\partial y}\right)\mathbf{u}_z\right\}\bullet d\mathbf{s}$$

$$\underset{(1.1.61b)}{\overset{(E2.30.1)}{=}} \int_0^{2\pi}\int_0^{\pi/2}\{(1-0)\mathbf{u}_x+(0-0)\mathbf{u}_y+(1-0)\mathbf{u}_z\}\bullet\mathbf{u}_r\,r^2\sin\theta\,d\theta\,d\phi$$

$$\underset{(1.1.59a)}{\overset{r=1}{=}} \int_0^{2\pi}\int_0^{\pi/2}(\sin\theta\cos\phi+\cos\theta)\sin\theta\,d\theta\,d\phi=\int_0^{2\pi}\int_0^{\pi/2}\left(\sin^2\theta\cos\phi+\frac{1}{2}\sin 2\theta\right)d\theta\,d\phi$$

$$=\int_0^{\pi/2}\sin^2\theta\,d\theta\int_0^{2\pi}\cos\phi\,d\phi-\frac{1}{4}\cos 2\theta\Big|_0^{\pi/2}\int_0^{2\pi}d\phi=\pi \tag{E2.30.5}$$

The validity of Stokes' theorem (2.9.1) is assured since Eqs. (E2.30.4) and (E2.30.5) are equal. In fact, the RHS of Eq. (2.9.1) is the same for different surfaces as long as they are enclosed by the same closed curve. Also, it becomes zero for any closed surface with no boundary, like a sphere or a torus.

(Example 2.30) Stokes' Theorem (*Continued*)

Now, considering the circular disk S_2 (which is the base of the hemisphere) in Fig. 2.33 as a surface (with the direction of $-\mathbf{u}_z$) enclosed by the unit circle C, let us compute the *RHS* of Eq. (2.9.1b):

$$\iint_{S_2}(\nabla\times\mathbf{F})\bullet d\mathbf{s} \overset{(2.4.5a)}{=} \iint_{S_2}\left\{\left(\frac{\partial F_z}{\partial y}-\frac{\partial F_y}{\partial z}\right)\mathbf{u}_x+\left(\frac{\partial F_x}{\partial z}-\frac{\partial F_z}{\partial x}\right)\mathbf{u}_y+\left(\frac{\partial F_y}{\partial x}-\frac{\partial F_x}{\partial y}\right)\mathbf{u}_z\right\}\bullet d\mathbf{s}$$

$$\underset{\text{Table1.1}}{\overset{(E2.30.1)}{=}} \int_{-1}^{1}\int_{-\sqrt{1-x^2}}^{\sqrt{1-x^2}}\{(1-0)\mathbf{u}_x+(0-0)\mathbf{u}_y+(1-0)\mathbf{u}_z\}\bullet(-\mathbf{u}_z dydx)=-\pi \qquad \text{(E2.30.6)}$$

This differs from Eq. (2.30.4) (in the sign). Why is that? Because the direction of normal vector to the surface S_2 was set opposite to that determined by the right-hand rule from the direction of the closed curve C enclosing S_2, which is the integration path for the line integral (E2.30.4).

It may be interesting to use the MATLAB function 'curla()' for computing the curl:

```
>>syms x y z; F_cart=[x; x; y];   % Eq.(E2.30.1)
>>curlF_cart=curla(F_cart,'ca');  pretty(simplify(curlF_cart))
   +-   -+
   |  1  |
   |     |
   |  0  |
   |     |
   |  1  |
   +-   -+
```

(Example 2.31) Stokes' Theorem

Show the validity of Stokes' theorem (2.9.1) for a vector field

$$\mathbf{F}(x,y,z)=\frac{x}{x^2+y^2}\mathbf{u}_z=\frac{\cos\phi}{\rho}\mathbf{u}_z \qquad \text{(E2.31.1)}$$

with the hemicylindrical surface S (Fig. 2.34(a)) described by

$$f(x,y,z)=x^2+y^2-1=\rho^2-1=0 \;\text{ for } 0\le z\le 1 \qquad \text{(E2.31.2)}$$

that is enclosed by a curve C=C_1-C_2-C_3-C_4, each piece of which is parameterized by

$$\mathbf{p}_1(t)=-1\,\mathbf{u}_x+0\,\mathbf{u}_y+t\,\mathbf{u}_z \;\text{ or }\; 1\,\mathbf{u}_\rho+0\,\mathbf{u}_\phi+t\,\mathbf{u}_z \;\text{ for } t=0\sim 1 \qquad \text{(E2.31.3a)}$$

$$\mathbf{p}_2(t)=\cos(\pi-t)\,\mathbf{u}_x+\sin(\pi-t)\,\mathbf{u}_y+1\,\mathbf{u}_z \;\text{ or }\; 1\,\mathbf{u}_\rho+0\,\mathbf{u}_\phi+1\,\mathbf{u}_z \;\text{ for } t=0\sim\pi \qquad \text{(E2.31.3b)}$$

$$\mathbf{p}_3(t)=1\,\mathbf{u}_x+0\,\mathbf{u}_y+(1-t)\,\mathbf{u}_z \;\text{ or }\; 1\,\mathbf{u}_\rho+0\,\mathbf{u}_\phi+(1-t)\,\mathbf{u}_z \;\text{ for } t=0\sim 1 \qquad \text{(E2.31.3c)}$$

$$\mathbf{p}_4(t)=\cos t\,\mathbf{u}_x+\sin t\,\mathbf{u}_y+0\,\mathbf{u}_z \;\text{ or }\; 1\,\mathbf{u}_\rho+0\,\mathbf{u}_\phi+0\,\mathbf{u}_z \;\text{ for } t=0\sim\pi \qquad \text{(E2.31.3d)}$$

The LHS of Eq. (2.9.1a) is computed as

$$\oint_C\mathbf{F}\bullet d\mathbf{l}=\oint_C\mathbf{F}\bullet\mathbf{u}_T\,dl \overset{(2.3.15)}{=} \int_0^{2\pi}\mathbf{F}\bullet d\mathbf{p}$$

$$\underset{(E2.31.3)}{\overset{(E2.31.1)}{=}} \int_0^1\frac{\cos\pi}{1}\mathbf{u}_z\bullet\mathbf{u}_z\,dt+\int_0^\pi\frac{\cos(\pi-t)}{1}\mathbf{u}_z\bullet\mathbf{0}\,dt+\int_0^t\frac{\cos 0}{1}\mathbf{u}_z\bullet(-\mathbf{u}_z)\,dt+\int_0^\pi\frac{\cos t}{1}\mathbf{u}_z\bullet\mathbf{0}\,dt$$

$$=\int_0^1(-1)dt+\int_0^1(-1)\,dt=-2 \qquad \text{(E2.31.4)}$$

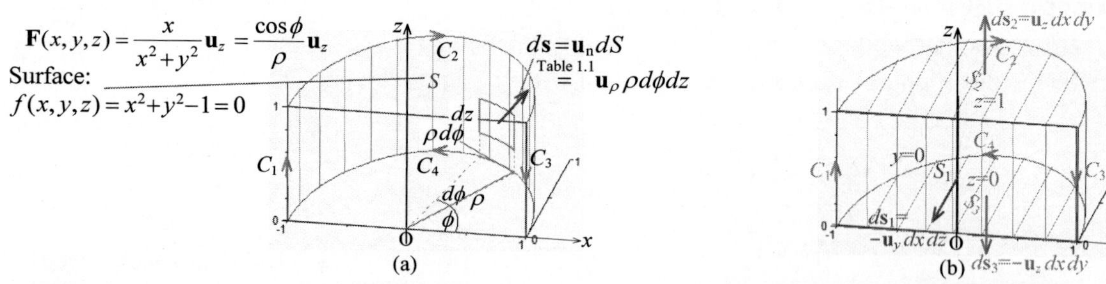

Figure 2.34 A hemicylindrical surface S and the closed curve C enclosing the surface for Example 2.31

The RHS of Eq. (2.9.1a) for the hemicylindrical surface S enclosed by $C=C_1\text{-}C_2\text{-}C_3\text{-}C_4$ is computed as

$$\iint_S (\nabla\times\mathbf{F})\bullet d\mathbf{s} \overset{(2.4.5b)}{=} \iint_S \left\{\left(\frac{1}{\rho}\frac{\partial F_z}{\partial\phi}-\frac{\partial F_\phi}{\partial z}\right)\mathbf{u}_\rho+\left(\frac{\partial F_\rho}{\partial z}-\frac{\partial F_z}{\partial\rho}\right)\mathbf{u}_\phi+\frac{1}{\rho}\left(\frac{\partial(\rho F_\phi)}{\partial\rho}-\frac{\partial F_\rho}{\partial\phi}\right)\mathbf{u}_z\right\}\bullet d\mathbf{s}$$

$$\underset{(1.1.60b)}{\overset{(E2.31.1)}{=}} \int_0^1\int_0^\pi\left(-\frac{1}{\rho^2}\sin\phi\,\mathbf{u}_\rho+\frac{\cos\phi}{\rho^2}\mathbf{u}_\phi\right)\bullet\mathbf{u}_\rho\,\rho\,d\phi\,dz \overset{\rho=1}{=} \int_0^1 \cos\phi\Big|_0^\pi dz=-2 \quad \text{(E2.31.5)}$$

The validity of Stokes' theorem (2.9.1) is assured since Eqs. (E2.31.4) and (E2.31.5) are equal.

It may be interesting to use the MATLAB function '`curla()`' for computing the curl:

```
>>syms r ph z; F_cyl=[0; 0; cos(ph)/r]; % Eq.(E2.31.1)
>>curlF_cyl=curla(F_cyl,'cy');  pretty(simplify(curlF_cyl))
  +-              -+
  |    sin(ph)     |
  |  - -------     |
  |        2       |
  |       r        |
  |                |
  |    cos(ph)     |
  |    -------     |
  |        2       |
  |       r        |
  |                |
  |       0        |
  +-              -+
```

Now, for another surface $S_{123}=\cup_{i=1}^{3}S_i$ consisting of the rectangular surface S_1 (described by $y=0$ with the direction of $-\mathbf{u}_y$) and two semi-circular disks S_2 and S_3 (each with the direction of $\pm\mathbf{u}_z$), let us compute the *RHS* of Eq. (2.9.1b) as

$$\iint_{S_{123}} (\nabla\times\mathbf{F})\bullet d\mathbf{s} \overset{(2.4.5a)}{=} \iint_{S_{123}} \left\{\left(\frac{\partial F_z}{\partial y}-\frac{\partial F_y}{\partial z}\right)\mathbf{u}_x+\left(\frac{\partial F_x}{\partial z}-\frac{\partial F_z}{\partial x}\right)\mathbf{u}_y+\left(\frac{\partial F_y}{\partial x}-\frac{\partial F_x}{\partial y}\right)\mathbf{u}_z\right\}\bullet d\mathbf{s}$$

$$\underset{\text{Table1.1}}{\overset{(E2.31.1)}{=}} \int_0^1\int_{-1}^1\left(\frac{-2xy}{(x^2+y^2)^2}\mathbf{u}_x-\frac{(x^2+y^2)-2x^2}{(x^2+y^2)^2}\mathbf{u}_y\right)\bullet(-\mathbf{u}_y)dx\,dz$$

$$+\int_{-1}^1\int_0^{\sqrt{1-x^2}}\left(\frac{-2xy}{(x^2+y^2)^2}\mathbf{u}_x-\frac{(x^2+y^2)-2x^2}{(x^2+y^2)^2}\mathbf{u}_y\right)\bullet\mathbf{u}_z\,dy\,dx$$

$$+\int_{-1}^1\int_0^{\sqrt{1-x^2}}\left(\frac{-2xy}{(x^2+y^2)^2}\mathbf{u}_x-\frac{(x^2+y^2)-2x^2}{(x^2+y^2)^2}\mathbf{u}_y\right)\bullet(-\mathbf{u}_z)dy\,dx$$

$$=\int_0^1 \frac{x}{x^2+y^2}\bigg|_{-1}^{1}\bigg|_{y=0} dz=2 \quad \text{(E2.31.6)}$$

This differs from Eq. (2.31.4) (in the sign). Why is that?

2.10 TRIPLE INTEGRALS

<Triple Integrals in Cartesian Coordinates>

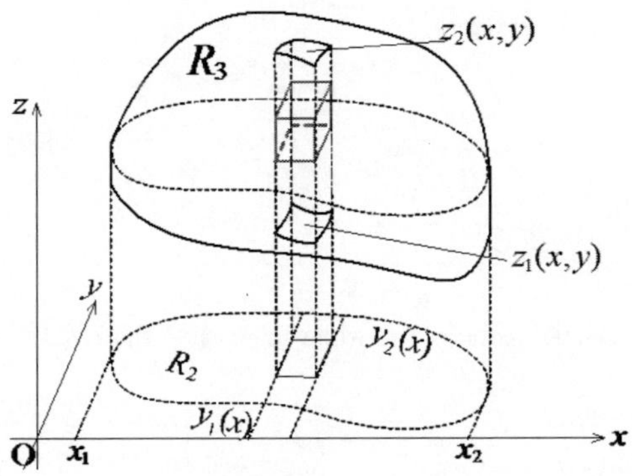

Figure 2.35 Geometric illustration of the triple integral in the Cartesian coordinate system

The triple integral of a three-variable scalar function $G(x,y,z)$ over a closed 3D region R_3 in space can be computed as the following iterated integral in the Cartesian coordinate system:

$$\iiint_{R_3} G(x,y,z)dV = \iint_{R_2} \int_{z_1(x,y)}^{z_2(x,y)} G(x,y,z)dz\,dA = \int_{x_1}^{x_2} \int_{y_1(x)}^{y_2(x)} \int_{z_1(x,y)}^{z_2(x,y)} G(x,y,z)\,dz\,dy\,dx \quad (2.10.1)$$

This can be done if the 3D region R_3 is z-simple in the sense that it intersects any line parallel to the z-axis in a single line element, and the projection R_2 of R_3 onto the xy-plane is a Type-I region (like Fig. 2.17(a)) so that it is y-simple, as illustrated in Fig. 2.35. Depending on the geometric shape of the region R_3, the order of integration can be selected among the following:

$$\begin{array}{ll} \text{(a)} \int_{x_1}^{x_2} \int_{y_1(x)}^{y_2(x)} \int_{z_1(x,y)}^{z_2(x,y)} G(x,y,z)\,dz\,dy\,dx & \text{(b)} \int_{y_1}^{y_2} \int_{x_1(y)}^{x_2(y)} \int_{z_1(x,y)}^{z_2(x,y)} G(x,y,z)\,dz\,dx\,dy \\ \text{(c)} \int_{x_1}^{x_2} \int_{z_1(x)}^{z_2(x)} \int_{y_1(x,z)}^{y_2(x,z)} G(x,y,z)\,dy\,dz\,dx & \text{(d)} \int_{z_1}^{z_2} \int_{x_1(z)}^{x_2(z)} \int_{y_1(x,z)}^{y_2(x,z)} G(x,y,z)\,dy\,dx\,dz \\ \text{(e)} \int_{y_1}^{y_2} \int_{z_1(y)}^{z_2(y)} \int_{x_1(y,z)}^{x_2(y,z)} G(x,y,z)\,dx\,dz\,dy & \text{(f)} \int_{z_2}^{z_2} \int_{y_1(z)}^{y_2(z)} \int_{x_1(y,z)}^{x_2(y,z)} G(x,y,z)\,dx\,dy\,dz \end{array} \quad (2.10.2)$$

Note that the triple integrals with $G(x,y,z)=1$ and $G(x,y,z)=\rho(x,y,z)$ (3D density) yield the volume and mass of an object described by the 3D region R_3, respectively:

$$\text{Volume: } \iiint_{R_3} dV \quad (2.10.3)$$

$$\text{Mass: } \iiint_{R_3} \rho(x,y,z)\,dV \quad (2.10.4)$$

<Triple Integrals in Cylindrical Coordinates>

The triple integral of a three-variable scalar function $G(\rho,\phi,z)$ over a closed 3D region R_3 in space can be computed as the following iterated integral in the cylindrical coordinate system:

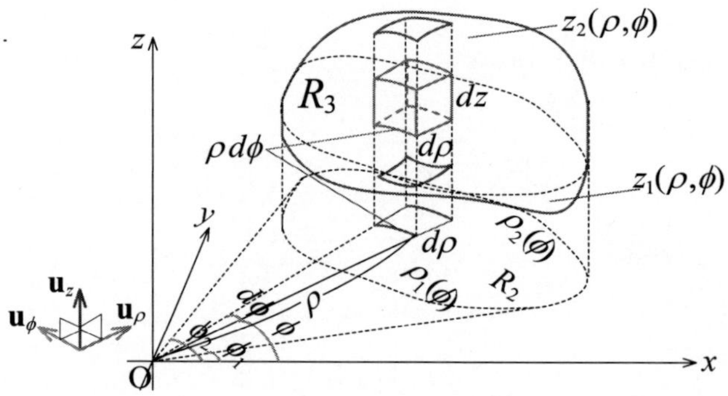

Figure 2.36 Geometric illustration of the triple integral in the cylindrical coordinate system

$$\iiint_{R_3} G(\rho,\phi,z)\,dV = \iint_{R_2}\int_{z_1(\rho,\phi)}^{z_2(\rho,\phi)} G(\rho,\phi,z)\,dz\,dA = \int_{\phi_1}^{\phi_2}\int_{\rho_1(x)}^{\rho_2(x)}\int_{z_1(\rho,\phi)}^{z_2(\rho,\phi)} G(\rho,\phi,z)\rho\,dz\,d\rho\,d\phi \quad (2.10.5)$$

if the 3D region R_3 is z-simple and the projection R_2 of R_3 onto the xy-plane is a Type-I region (like Fig. 2.19(a)) so that it is ρ-simple, as illustrated in Fig. 2.36. Depending on the geometric shape of the region R_3, the order of integration can be changed.

<Triple Integrals in Spherical Coordinates>

The triple integral of a three-variable scalar function $G(r,\theta,\phi)$ over a closed 3D region R_3 in space can be computed as the following iterated integral in the spherical coordinate system:

$$\iiint_{R_3} G(r,\theta,\phi)\,dV = \iint_{R_2}\int_{r_1(\theta,\phi)}^{r_2(\theta,\phi)} G(r,\theta,\phi)\,dr\,dA \overset{(1.1.61a)}{=} \int_{\phi_1}^{\phi_2}\int_{\theta_1(\phi)}^{\theta_2(\phi)}\int_{r_1(\theta,\phi)}^{r_2(\theta,\phi)} G(r,\theta,\phi)\,r^2\sin\theta\,dr\,d\theta\,d\phi \quad (2.10.6)$$

if the 3D region R_3 is *r-simple* in the sense that it intersects any radial line in a single line element, and the projection R_2 of R_3 onto the $\theta\phi$-plane is θ-simple, as illustrated in Fig. 2.37. Depending on the geometric shape of the region R_3, the order of integration can be changed.

Note that the formulas for the conversions among rectangular (Cartesian), cylindrical, and spherical coordinate systems are listed in Table 1.1 (Section 1.1.6).

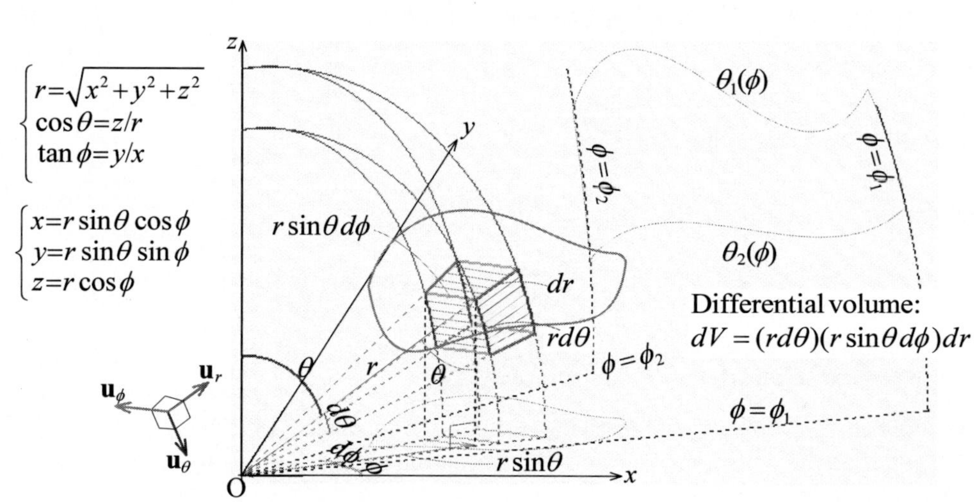

Figure 2.37 Geometric illustration of the triple integral in the spherical coordinate system

<Change of Variables in a Triple Integral>

As in Eq. (2.6.9) for changing the variables of integration in a double integral, we can change the variables of integration in a triple integral as

$$\iiint_{R_{xyz}} G(x,y,z)\,dx\,dy\,dz \underset{z=z(u,v,w)}{\overset{\substack{x=x(u,v,w)\\ y=y(u,v,w)}}{\longrightarrow}} \iiint_{R_{uvw}} f(x(u,v,w),y(u,v,w),z(u,v,w))\,|J|\,du\,dv\,dw \tag{2.10.7}$$

by substituting $x\,(u,v,w)$, $y\,(u,v,w)$, and $z\,(u,v,w)$ for x, y, and z, respectively, where J is the *Jacobian* matrix defined as

$$J = \frac{\partial(x,y,z)}{\partial(u,v,w)} = \begin{bmatrix} \partial x/\partial u & \partial x/\partial v & \partial x/\partial w \\ \partial y/\partial u & \partial y/\partial v & \partial y/\partial w \\ \partial z/\partial u & \partial z/\partial v & \partial z/\partial w \end{bmatrix} \tag{2.10.8}$$

Note that the *Jacobian* (*determinant*), i.e., the determinant of the Jacobian matrix for Cartesian-to-spherical coordinate conversion, is

$$|J| = \left|\frac{\partial(x,y,z)}{\partial(r,\theta,\phi)}\right| \overset{\text{Table 1.1}}{=} \left\|\begin{matrix} \frac{\partial}{\partial r}(r\sin\theta\cos\phi) & \frac{\partial}{\partial\theta}(r\sin\theta\cos\phi) & \frac{\partial}{\partial\phi}(r\sin\theta\cos\phi) \\ \frac{\partial}{\partial r}(r\sin\theta\sin\phi) & \frac{\partial}{\partial\theta}(r\sin\theta\sin\phi) & \frac{\partial}{\partial\phi}(r\sin\theta\sin\phi) \\ \frac{\partial}{\partial r}(r\cos\theta) & \frac{\partial}{\partial\theta}(r\cos\theta) & \frac{\partial}{\partial\phi}(r\cos\theta) \end{matrix}\right\|$$

$$= \left\|\begin{matrix} \sin\theta\cos\phi & r\cos\theta\cos\phi & -r\sin\theta\sin\phi \\ \sin\theta\sin\phi & r\cos\theta\sin\phi & r\sin\theta\cos\phi \\ \cos\theta & -r\sin\theta & 0 \end{matrix}\right\| = \begin{matrix} r^2\sin^2\theta\,\cos\phi\sin\theta\cos\phi \\ +r^2\cos\theta\sin\theta\cos\phi\cos\theta\cos\phi \\ +r^2\sin\theta\,\sin\phi(\sin^2\theta\,\sin\phi+\cos^2\theta\,\sin\phi) \end{matrix}$$

$$= r^2\cos^2\phi\,\sin\theta + r^2\sin\theta\sin^2\phi = r^2\sin\theta \tag{2.10.9}$$

Thus, we can apply the substitutions $x=x(r,\theta,\phi)=r\sin\theta\cos\phi$, $y=y(r,\theta,\phi)=r\sin\theta\sin\phi$, and $z=z(r,\theta,\phi)=r\cos\theta$ together with $|J|=r^2\sin\theta$ to obtain the triple integration formula (2.10.6) in spherical coordinates.

(**Example 2.32**) Triple Integral

Find the volume of the triangular pyramid (tetrahedron) depicted in Fig. 2.38.

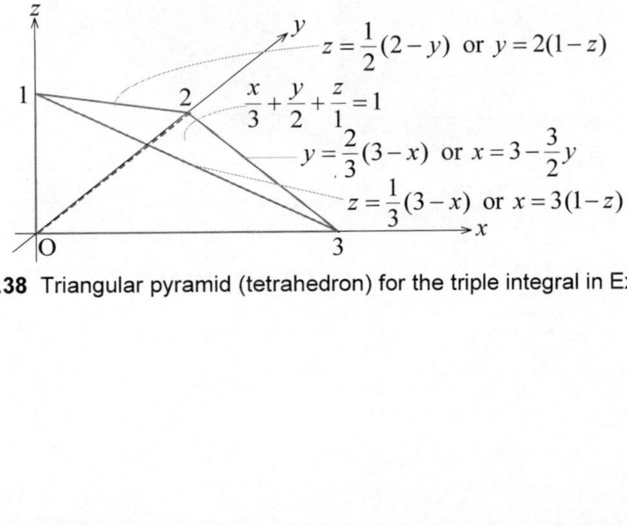

Figure 2.38 Triangular pyramid (tetrahedron) for the triple integral in Example 2.32

Noting that the equation of the oblique plane is

$$\frac{x}{3}+\frac{y}{2}+\frac{z}{1}=1 \tag{E2.32.1}$$

we can use Eq. 2.10.2(a) with $G(x,y,z)=1$ to get the volume V:

$$V=\iiint_{R_3} G(x,y,z)\,dV \overset{(2.10.2a)}{\underset{G=1}{=}} \int_0^3\int_0^{2-2x/3}\int_0^{1-x/3-y/2} 1\,dz\,dy\,dx=\int_0^3\int_0^{2-2x/3}(1-\frac{x}{3}-\frac{y}{2})\,dy\,dx$$
$$=\int_0^3 ((1-\frac{1}{3}x)y-\frac{1}{4}y^2)\Big|_0^{2-2x/3}dx=\int_0^3((1-\frac{1}{3}x)(2-\frac{2}{3}x)-\frac{1}{4}(2-\frac{2}{3}x)^2)\,dx$$
$$=\int_0^3(1-\frac{2}{3}x+\frac{1}{9}x^2)\,dx=(x-\frac{1}{3}x^2+\frac{1}{27}x^3)\Big|_0^3=1 \tag{E2.32.2}$$

```
>>syms x y z
>>V=int(int(int(1,z,0,1-x/3-y/2),y,0,2-2*x/3),x,0,3)   % Eq.(E2.32.2)
   V = 1  % Symbolic computation
>>F=@(x,y,z)(z<=1-x/3-y/2).*(x>=0).*(y>=0); % Alternatively, triplequad()
>>V=triplequad(F,0,3,0,2,0,1)
   V = 1.0000  % Numerical computation
```

(Example 2.33) Triple Integral

Find the volume of the object depicted in Fig. 2.39.
Considering the shape of the object, we use Eq. 2.10.2(d) with $G(x,y,z)=1$ to get the volume V:

$$V=\iiint_{R_3} G(x,y,z)\,dV \overset{(2.10.2d)}{\underset{G=1}{=}} \int_{-1}^1\int_{-\sqrt{1-z^2}}^{\sqrt{1-z^2}}\int_{x^2+z^2}^{4-2(x^2+z^2)} 1\,dy\,dx\,dz=\int_{-1}^1\int_{-\sqrt{1-z^2}}^{\sqrt{1-z^2}}(4-3x^2-3z^2)\,dx\,dz$$
$$=\int_{-1}^1 x(4-x^2-3z^2)\Big|_{-\sqrt{1-z^2}}^{\sqrt{1-z^2}}dz=2\int_{-1}^1\sqrt{1-z^2}\{4-(1-z^2)-3z^2\}\,dz$$
$$=2\int_{-1}^1\{(1-z^2)^{1/2}+2(1-z^2)^{3/2}\}\,dz \overset{z=\sin\theta}{\underset{dz=\cos\theta d\theta}{=}} 2\int_{-\pi/2}^{\pi/2}(\cos^2\theta+2\cos^4\theta)\,d\theta$$
$$\overset{(C.15)}{=}\int_{-\pi/2}^{\pi/2}((1+\cos 2\theta)+(1+\cos 2\theta)^2)\,d\theta=\int_{-\pi/2}^{\pi/2}(2+3\cos 2\theta+\cos^2 2\theta)\,d\theta$$
$$\overset{(C.15)}{=}\int_{-\pi/2}^{\pi/2}\Big(2+3\cos 2\theta+\frac{1}{2}(1+\cos 4\theta)\Big)\,d\theta=\frac{5}{2}\pi \tag{E2.33.1}$$

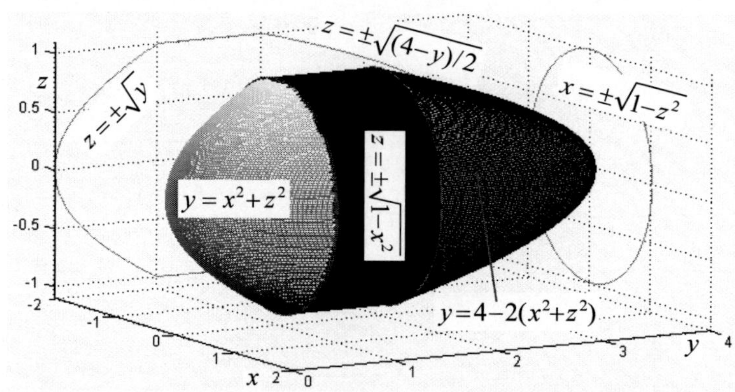

Figure 2.39 Object for the triple integral in Example 2.33

With the substitution of variables $x=\rho\cos\phi$ and $z=\rho\sin\phi$ (Eq. (1.1.58c)), we can use the triple integral formula (2.10.5) in cylindrical coordinates to get the volume V more easily:

$$V=\iiint_{R_3} G(x,y,z)dV \overset{(2.10.2d)}{\underset{G=1}{=}} \int_{-1}^{1}\int_{-\sqrt{1-z^2}}^{\sqrt{1-z^2}}\int_{x^2+z^2}^{4-2(x^2+z^2)} 1\,dy\,dx\,dz \underset{x=\rho\cos\phi,z=\rho\sin\phi}{\overset{(2.10.5)}{=}} \int_0^{2\pi}\int_0^1\int_{\rho^2}^{4-2\rho^2} 1\,\rho\,dy\,d\rho\,d\phi$$

$$=\int_0^{2\pi}\int_0^1 (4-3\rho^2)\,\rho\,d\rho\,d\phi=\int_0^{2\pi}\left(2\rho^2-\frac{3}{4}\rho^4\right)\Big|_0^1 d\phi=\int_0^{2\pi}\frac{5}{4}d\phi=\frac{5}{2}\pi \quad \text{(E2.33.2)}$$

```
>>syms x y z r
>>V=int(int(int(1,y,x^2+z^2,4-2*(x^2+z^2)),x,-sqrt(1-z^2),sqrt(1-z^2)),...
      z,-1,1)   % Eq.(E2.33.1)
>>V=int(int(int(r,y,r^2,4-2*r^2),r,0,1),z,0,2*pi) % Eq.(E2.33.2)
   V = (5*pi)/2
```

(Example 2.34) Triple Integral

Find the triple integral of

$$G(x,y,z)=x^2+y^2+z^2 \quad \text{(E2.34.1)}$$

over the region R specified as a sphere of radius 1 centered at $(0,0,1)$, i.e.,

$$R=\{(x,y,z)\,|\,x^2+y^2+(z-1)^2\le 1\} \quad \text{(E2.34.2)}$$

Noting that this spherical region is x-, y-, and z-simple, let us use Eq. (2.10.2d):

$$\iiint_R G(x,y,z)dV \overset{(2.10.2d)}{\underset{G=1}{=}} \int_0^2\int_{-\sqrt{1-(z-1)^2}}^{\sqrt{1-(z-1)^2}}\int_{-\sqrt{1-(x^2+(z-1)^2)}}^{\sqrt{1-(x^2+(z-1)^2)}}(x^2+y^2+z^2)\,dy\,dx\,dz$$

$$\overset{z-1=z'}{=}\int_{-1}^{1}\int_{-\sqrt{1-z^2}}^{\sqrt{1-z^2}}\int_{-\sqrt{1-(x^2+z^2)}}^{\sqrt{1-(x^2+z^2)}}\{x^2+y^2+(z+1)^2\}\,dy\,dx\,dz$$

$$=\int_{-1}^{1}\int_{-\sqrt{1-z^2}}^{\sqrt{1-z^2}}\left(x^2y+\frac{1}{3}y^3+(z+1)^2y\right)\Big|_{-\sqrt{1-(x^2+z^2)}}^{\sqrt{1-(x^2+z^2)}}dx\,dz=\cdots \quad \text{(E2.34.3)}$$

This triple integral is formidable to compute in Cartesian coordinates, while it can be computed numerically as

```
>>F=@(x,y,z)(x.^2+y.^2+z.^2).*(x.^2+y.^2+(z-1).^2<=1); % Eq.(E2.34.1,2)
>>V=triplequad(F,-1,1,-1,1,-1,1) % Eq.(E2.34.3)
   V = 6.7021 % 32/15*pi
```

Alternatively, with the substitution of variables $x=r\sin\theta\cos\phi$, $y=r\sin\theta\sin\phi$, and $z=1+r\cos\theta$ (Eq. (1.1.59c)) so that the new region of integration can be centered at the origin $(0,0,0)$, we can use the triple integral formula (2.10.6) in spherical coordinates:

$$\iiint_R G(x,y,z)dV \overset{(2.10.6)}{=} \int_0^{2\pi}\int_0^1\int_0^{\pi}(x^2+y^2+z^2)\Big|_{\substack{x=r\sin\theta\cos\phi\\ y=r\sin\theta\sin\phi\\ z=1+r\cos\theta}} r^2\sin\theta\,d\theta\,dr\,d\phi$$

$$=\int_0^{2\pi}\int_0^1\int_0^{\pi} r^2(r^2+2r\cos\theta+1)\sin\theta\,d\theta\,dr\,d\phi=\int_0^{2\pi}\int_0^1-\left((r^4+r^2)\cos\theta+\frac{1}{2}r^3\cos 2\theta\right)\Big|_0^{\pi}dr\,d\phi$$

$$=2\int_0^{2\pi}\int_0^1(r^4+r^2)\,dr\,d\phi=2\int_0^{2\pi}\left(\frac{1}{5}r^5+\frac{1}{3}r^3\right)\Big|_0^1 d\phi=2\int_0^{2\pi}\frac{8}{15}d\phi=\frac{32}{15}\pi \quad \text{(E2.34.4)}$$

This is much easier to compute than (E2.34.3), yielding the same result with it.

2.11 DIVERGENCE THEOREM

Just like Stokes' theorem, the divergence or Gauss' theorem can also be viewed as a generalization of Green's theorem, presenting the relationship between the surface integral of a vector function around a closed surface S and the volume integral over the region R bounded by S.

[Divergence or **Gauss' Theorem]**

Let R be a closed 3D region bounded by a piecewise-smooth surface S that is oriented outward from R. Also, let $\mathbf{G}(x,y,z) = G_x(x,y,z)\,\mathbf{u}_x + G_y(x,y,z)\,\mathbf{u}_y + G_z(x,y,z)\,\mathbf{u}_z$ be a continuous vector function that has continuous first partial derivatives in a 3D region containing R. Then

$$\oiint_S \mathbf{G}\cdot d\mathbf{s} = \oiint_S \mathbf{G}\cdot\mathbf{u}_n\,dS = \iiint_R \operatorname{div}\mathbf{G}\,dV \tag{2.11.1a}$$

$$\iint_S G_x\,dydz + G_y\,dzdx + G_z\,dxdy = \iiint_R \left(\frac{\partial G_x}{\partial x} + \frac{\partial G_y}{\partial y} + \frac{\partial G_z}{\partial z}\right) dV \tag{2.11.1b}$$

where $\mathbf{u}_n$ denotes the unit outward normal vector to a differential surface dS.

Divergence theorem states that the total amount of divergence over a volume is equal to the amount of flux flowing out through its boundary surface. Since it is nothing but another generalization of Green's theorem into a 3D vector form, let us leave the proof to other references like [K-2], [Z-1], ... and instead, approximate each side of Eq. (2.11.1a) as

$$\sum_i \oiint_{\Delta S_i} \mathbf{G}\cdot\mathbf{u}_n\,\Delta S_i = \sum_i \nabla\cdot\mathbf{G}\,\Delta V_i \tag{2.11.2}$$

and visualize it as depicted in Fig. 2.40. Fig. 2.40 states that the sum of $\nabla\cdot\mathbf{G}$ on the RHS of Eq. (2.11.2) for all subvolumes $\{\Delta V_i\}$ becomes the flux of $\mathbf{G}$ through the whole surface S made of ΔS_i's, where the (outward) normal vector $\mathbf{u}_n$'s to a surface segment shared by two adjacent small volume ΔV_i's are of opposite directions so, that the sum of volume integrals of $\nabla\cdot\mathbf{G}$ on the RHS of Eq. (2.11.2) for all small volume ΔV_i's becomes the surface integral of $\mathbf{G}$ over the large closed surface S, enclosing the whole volume V. In addition, recall the definition (2.4.1) of divergence:

$$\operatorname{div}\mathbf{G} = \nabla\cdot\mathbf{G} \overset{(2.4.1)}{=} \lim_{\Delta V\to 0}\frac{1}{\Delta V}\oiint_{\Delta S}\mathbf{G}\cdot d\mathbf{s} \rightarrow (\nabla\cdot\mathbf{G})\Delta V \approx \oiint_{\Delta S}\mathbf{G}\cdot\Delta\mathbf{s} = \oiint_{\Delta S}\mathbf{G}\cdot\mathbf{u}_n\,\Delta S \tag{2.11.3}$$

Don't these statements help you to believe the validity and realize the meaning of the divergence theorem? To help consolidate this intuition about divergence theorem, see the following examples:

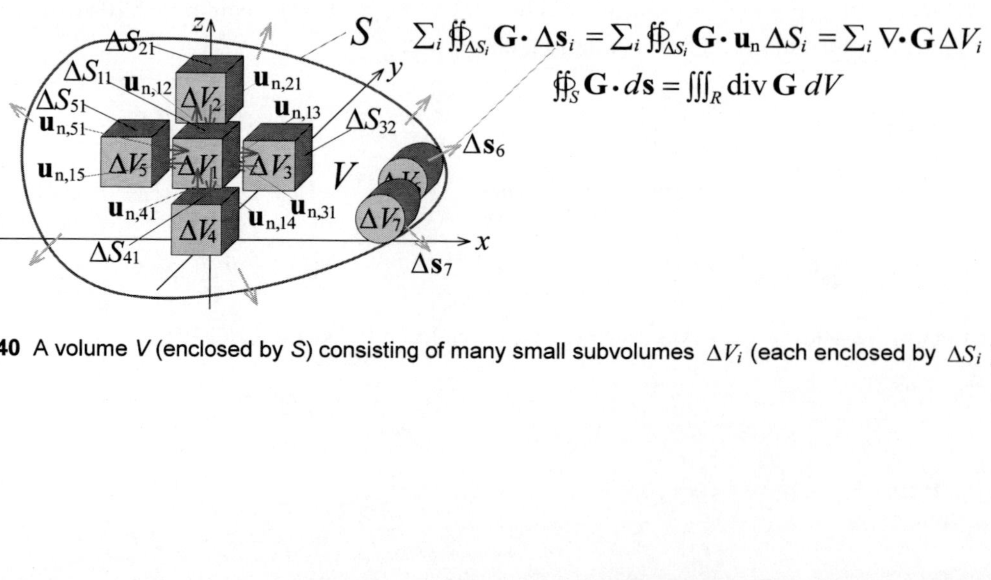

Figure 2.40 A volume V (enclosed by S) consisting of many small subvolumes ΔV_i (each enclosed by ΔS_i)

<Divergence of $\mathbf{G}=\mathbf{u}_r/r^2$ and 3D Unit Impulse (Dirac Delta) Function>

Consider the vector field that was dealt with in Example 2.26:

$$\mathbf{G}(x,y,z)=\frac{x\mathbf{u}_x+y\mathbf{u}_y+z\mathbf{u}_z}{(x^2+y^2+z^2)^{3/2}} \text{ or equivalently, } \mathbf{G}(r,\theta,\phi)=\frac{1}{r^3}\mathbf{r}=\frac{1}{r^2}\mathbf{u}_r \tag{2.11.4}$$

where

$$\mathbf{r}=x\mathbf{u}_x+y\mathbf{u}_y+z\mathbf{u}_z \text{ and } r=\|\mathbf{r}\|=\sqrt{x^2+y^2+z^2}$$

For this vector field, let us see if the divergence theorem (2.11.1) holds with the unit spherical region that is depicted in Fig. 2.28(a). Since the LHS (surface integral) of the divergence theorem (2.11.1) was obtained as Eq. (E2.26.3) or (E2.26.5):

$$\iint_S \mathbf{G}\bullet d\mathbf{s} \overset{\text{(E2.26.3)}}{=} 4\pi, \tag{2.11.5}$$

we have only to find the RHS (volume integral) of Eq. (2.11.1):

$$\begin{aligned}\iiint_R \operatorname{div}\mathbf{G}\,dV &\overset{\text{(2.4.4c)}}{=} \iiint_R \left(\frac{1}{r^2}\frac{\partial(r^2G_r)}{\partial r}+\frac{1}{r\sin\theta}\frac{\partial(G_\theta\sin\theta)}{\partial\phi}+\frac{1}{r\sin\theta}\frac{\partial G_\phi}{\partial\phi}\right)dV\\ &\overset{\text{(2.11.4)}}{=} \iiint_R \left(\frac{1}{r^2}\frac{\partial(r^2/r^2)}{\partial r}\right)dV=0\end{aligned} \tag{2.11.6}$$

Contrary to our expectations, the LHS and RHS of Eq. (2.11.1) for Eq. (2.11.4) do not conform with each other. Is this a counterexample to disprove the divergence theorem? No way. The truth is that we have overlooked the singularity of the vector field $\mathbf{G}=\mathbf{u}_r/r^2$ at $r=0$ where $\mathbf{G}$ blows up, and for the divergence theorem to hold, we must have

$$\iiint_R \operatorname{div}\mathbf{G}\,dV \overset{\text{(2.11.4)}}{=} \iiint_R \nabla\bullet\frac{\mathbf{u}_r}{r^2}\,dV=4\pi \tag{2.11.7}$$

or equivalently,

$$\operatorname{div}\left(\frac{\mathbf{u}_r}{r^2}\right)=\nabla\bullet\frac{\mathbf{u}_r}{r^2} \underset{\text{gradient}}{\overset{\text{(2.4.3c)}}{=}} \nabla\bullet\nabla\left(-\frac{1}{r}\right)=-\nabla^2\left(\frac{1}{r}\right)=4\pi\delta^3(\mathbf{r}) \tag{2.11.8}$$

where $\delta^3(\mathbf{r})$ is the *3D Dirac delta function* (Section 1.5.3 of [G-1]), which is a generalization of the 1D Dirac delta or unit impulse function $\delta(t)$, to be introduced in Section 4.1.2:

$$\delta^3(\mathbf{r})=\delta(x)\delta(y)\delta(z) \text{ with } \mathbf{r}=x\mathbf{u}_x+y\mathbf{u}_y+z\mathbf{u}_z \tag{2.11.9}$$

Like the properties (Eqs. (4.1.7) and (4.1.13)) of $\delta(t)$, the 3D Dirac delta function $\delta^3(\mathbf{r})$ has the following properties:

$$\iiint_{\text{all space}} \delta^3(\mathbf{r})dV=\iiint_{R_{xyz}} \delta(x)\delta(y)\delta(z)\,dx\,dy\,dz=1 \tag{2.11.10}$$

$$\text{Sampling property: } \iiint_{\text{all space}} f(\mathbf{r})\delta^3(\mathbf{r}-\mathbf{r}_1)dV=f(\mathbf{r}_1) \tag{2.11.11}$$

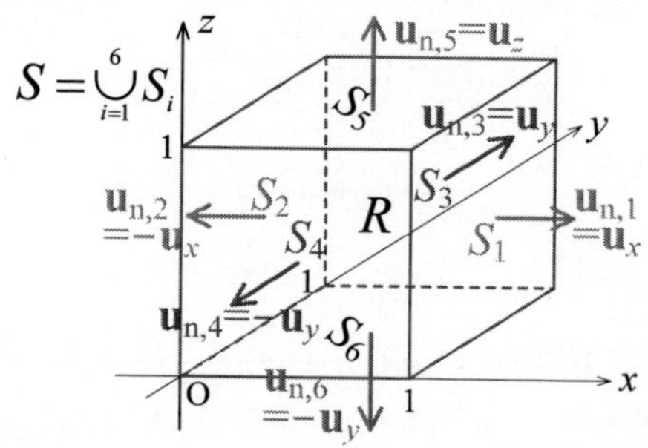

Figure 2.41 Rectangular region for Example 2.35

(Example 2.35) Divergence Theorem

Show the validity of the divergence theorem (2.11.1) for a vector field

$$\mathbf{G}(x,y,z) = x^2\mathbf{u}_x + 2xy\,\mathbf{u}_y + 2xz\,\mathbf{u}_z \tag{E2.35.1}$$

with the rectangular region that is depicted in Fig. 2.41:

$$R = \{0 \le x \le 1,\ 0 \le y \le 1,\ 0 \le z \le 1\} \tag{E2.35.2}$$

The LHS of divergence theorem (2.11.1) for Eq. (E2.35.1) with the closed surface S enclosing the rectangular region R is

$$\begin{aligned}
&\iint_{S_1} G_x\mathbf{u}_x\cdot\mathbf{u}_x\,dydz + \iint_{S_2} G_x\mathbf{u}_x\cdot(-\mathbf{u}_x)dydz + \iint_{S_3} G_y\mathbf{u}_y\cdot\mathbf{u}_y\,dzdx + \iint_{S_4} G_y\mathbf{u}_y\cdot(-\mathbf{u}_y)dzdx\\
&+\iint_{S_5} G_z\mathbf{u}_z\cdot\mathbf{u}_z\,dxdy + \iint_{S_6} G_z\mathbf{u}_z\cdot(-\mathbf{u}_z)dxdy\\
&\overset{\text{(E2.35.1)}}{=} \int_0^1\int_0^1 (x^2|_{x=1} - x^2|_{x=0})\,dydz + \int_0^1\int_0^1 (2xy|_{y=1} - 2xy|_{y=0})\,dzdx\\
&\qquad + \int_0^1\int_0^1 (2xz|_{z=1} - 2xz|_{z=0})\,dxdy\\
&= \int_0^1 y|_0^1\,dz + \int_0^1 2x\,z|_0^1\,dx + \int_0^1 x^2|_0^1\,dy\\
&= \int_0^1 1dz + \int_0^1 2xdx + \int_0^1 1dy = z|_0^1 + x^2|_0^1 + y|_0^1 = 3
\end{aligned} \tag{E2.35.3}$$

On the other hand, the RHS of divergence theorem (2.11.1) for Eq. (E2.35.1) with the rectangular region R enclosed by the closed surface S is

$$\begin{aligned}
\iiint_R\left(\frac{\partial G_x}{\partial x}+\frac{\partial G_y}{\partial y}+\frac{\partial G_z}{\partial z}\right)dV &\overset{\text{(E2.35.1)}}{=} \int_0^1\int_0^1\int_0^1 (2x+2x+2x)\,dx\,dydz\\
&= \int_0^1\int_0^1 3x^2|_0^1\,dydz = \int_0^1\int_0^1 3\,dydz = \int_0^1 3y|_0^1\,dz = \int_0^1 3dz = 3z|_0^1 = 3
\end{aligned} \tag{E2.35.4}$$

```
>>syms x y z; G=[x^2; 2*x*y; 2*x*z]; % Eq.(E2.35.1)
>>RHS=int(int(int(diva(G),x,0,1),y,0,1),z,0,1)
```

Since Eqs. (E2.35.3) and (E2.35.4), which are the LHS and RHS of divergence theorem (2.11.1), respectively, are equal, the validity of the divergence theorem has been checked.

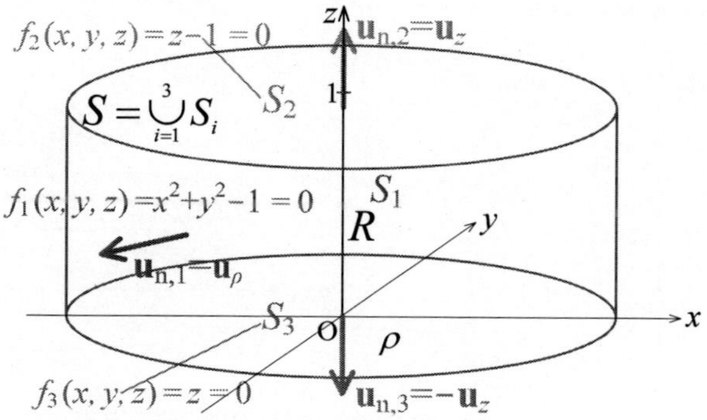

Figure 2.42 Cylindrical region for Example 2.36

(Example 2.36) Divergence Theorem

Show the validity of the divergence theorem (2.11.1) for a vector field

$$\mathbf{G}(x, y, z) = x^2\mathbf{u}_x + xy\,\mathbf{u}_y + 2z\,\mathbf{u}_z \tag{E2.36.1}$$

with the cylindrical region that is depicted in Fig. 2.42:

$$R = \{x^2+y^2 \le 1,\ 0 \le z \le 1\} \tag{E2.36.2}$$

First, note that the cylindrical surface S_1 is described by

$$f(x, y, z) = x^2 + y^2 - 1 = 0 \tag{E2.36.3}$$

and accordingly, its normal vector is

$$\mathbf{u}_{n,1} \overset{(2.8.9)}{=} \frac{\nabla f}{||\nabla f||} \overset{(2.4.3a)}{=} \frac{\frac{\partial f}{\partial x}\mathbf{u}_x + \frac{\partial f}{\partial y}\mathbf{u}_y + \frac{\partial f}{\partial z}\mathbf{u}_z}{\sqrt{(\frac{\partial f}{\partial x})^2 + (\frac{\partial f}{\partial y})^2 + (\frac{\partial f}{\partial z})^2}} \overset{(E2.36.3)}{=} \frac{2x\,\mathbf{u}_x + 2y\,\mathbf{u}_y}{\sqrt{(2x)^2+(2y)^2}} \overset{x^2+y^2=1}{=} x\,\mathbf{u}_x + y\,\mathbf{u}_y \tag{E2.36.4}$$

Note also from Eq. (2.8.5b) that the differential surface area dS of the cylindrical surface S_1 is related to the corresponding regional area dA on the xz-plane as

$$y = g(x, z) = \pm\sqrt{1-x^2} \overset{(2.8.5b)}{\rightarrow} dS = \sqrt{1+(\partial g/\partial x)^2+(\partial g/\partial z)^2}\,dA = \sqrt{1+\frac{x^2}{1-x^2}}\,dA = \frac{1}{\sqrt{1-x^2}}\,dA \tag{E2.36.5}$$

Thus, the LHS of divergence theorem (2.11.1) for Eq. (E2.36.1) with the closed surface S enclosing the cylindrical region R can be computed as

$$\begin{aligned}
\oiint_S \mathbf{G}\bullet d\mathbf{s} &= \oiint_S \mathbf{G}\bullet\mathbf{u}_n\,dS \overset{(E2.36.1)}{=} \iint_{S_1} (x^2\mathbf{u}_x+xy\mathbf{u}_y+2z\mathbf{u}_z)\bullet(x\mathbf{u}_x+y\mathbf{u}_y)\,dS \\
&+ \iint_{S_2} (x^2\mathbf{u}_x+xy\mathbf{u}_y+2z\mathbf{u}_z)|_{z=1}\bullet\mathbf{u}_z\,dS + \iint_{S_3} \cancel{(x^2\mathbf{u}_x+xy\mathbf{u}_y+2z\mathbf{u}_z)|_{z=0}\bullet(-\mathbf{u}_z)}\,dS \\
&\overset{(E2.36.5)}{=} \int_0^1\int_{-1}^1 (x^3+xy^2)|_{x^2+y^2=1}\frac{1}{\sqrt{1-x^2}}\,dx\,dz + \iint_{S_2} 2\,dS = \int_0^1\int_{-1}^1 \frac{2x}{\sqrt{1-x^2}}\,dx\,dz + 2\pi \\
&= -2\int_0^1 \sqrt{1-x^2}\Big|_{-1}^{1}\,dz + 2\pi = 2\pi
\end{aligned} \tag{E2.36.6}$$

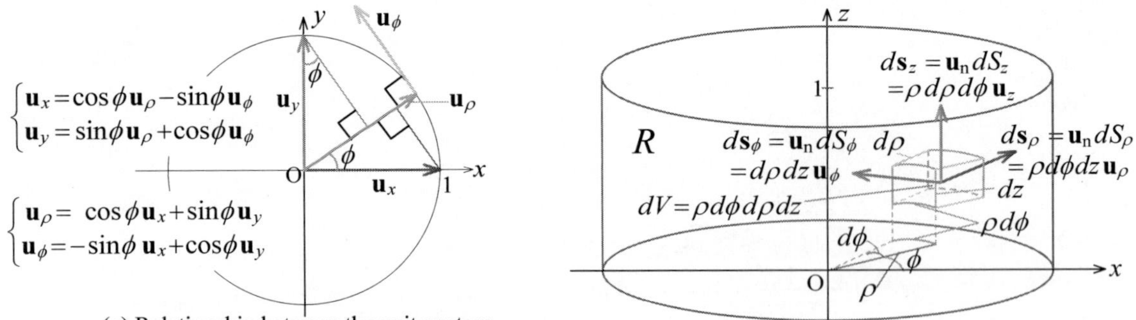

(a) Relationship between the unit vectors in Cartesian/cylindrical coordinates

(b) Cylindrical region R enclosed by a surface S

Figure 2.43 Cylindrical region for Example 2.35

On the other hand, the RHS of divergence theorem (2.11.1) for Eq. (E2.36.1) with the cylindrical region R is

$$\iiint_R\left(\frac{\partial G_x}{\partial x}+\frac{\partial G_y}{\partial y}+\frac{\partial G_z}{\partial z}\right)dV \overset{\text{(E2.36.1)}}{=} \int_0^1\int_{-1}^1\int_{-\sqrt{1-y^2}}^{\sqrt{1-y^2}}(2x+x+2)\,dx\,dy\,dz=\int_0^1\int_{-1}^1\left(\frac{3}{2}x^2+2x\right)\Big|_{-\sqrt{1-y^2}}^{\sqrt{1-y^2}}dy\,dz$$

$$=\int_0^1\int_{-1}^1 4\sqrt{1-y^2}\,dy\,dz \underset{dy=\cos\theta}{\overset{y=\sin\theta}{=}} 4\int_0^1\int_{-\pi/2}^{\pi/2}\cos^2\theta\,d\theta\,dz=2\int_0^1\int_{-\pi/2}^{\pi/2}(1+2\cos 2\theta)d\theta\,dz=2\pi \quad \text{(E2.36.7)}$$

```
>>syms x y z; G=[x^2; x*y; 2*z];
>>RHS=int(int(int(diva(G),x,-sqrt(1-y^2),sqrt(1-y^2)),y,-1,1),z,0,1)
```

Since Eqs. (E2.36.6) and (E2.36.7), which are the LHS and RHS of divergence theorem (2.11.1), respectively, are equal, the validity of the divergence theorem has been checked.

Alternatively, since the cylindrical region R is easier to handle in cylindrical coordinates, let us make the Cartesian-to-cylindrical coordinate conversion of the vector function $\mathbf{G}(x,y,z)$ via the change of variables $x=\rho\cos\phi$, $y=\rho\sin\phi$, $\mathbf{u}_x=\cos\phi\,\mathbf{u}_\rho-\sin\phi\,\mathbf{u}_\phi$, and $\mathbf{u}_y=\sin\phi\,\mathbf{u}_\rho+\cos\phi\,\mathbf{u}_\phi$ (see Fig. 2.43(a) or Table 1.1):

$$\mathbf{G}(x,y,z)=x^2\mathbf{u}_x+xy\,\mathbf{u}_y+2z\mathbf{u}_z$$

$$\underset{\text{(1.1.58b)}}{\overset{\text{(1.1.58c)}}{\longrightarrow}}\mathbf{G}(\rho,\phi,z)=\rho^2\cos^2\phi(\cos\phi\,\mathbf{u}_\rho-\sin\phi\,\mathbf{u}_\phi)+\rho^2\cos\phi\sin\phi(\sin\phi\,\mathbf{u}_\rho+\cos\phi\,\mathbf{u}_\phi)+2z\mathbf{u}_z$$

$$=\rho^2\cos\phi\,\mathbf{u}_\rho+2z\mathbf{u}_z \quad \text{(E2.36.8)}$$

Then, referring to Fig. 2.43(b), the LHS of divergence theorem (2.11.1) for Eq. (E2.36.8) with the cylindrical region R enclosed by the closed surface S can be computed in cylindrical coordinates as

$$\iint_S G_\rho\mathbf{u}_\rho\cdot d\mathbf{s}_\rho+G_\phi\mathbf{u}_\phi\cdot d\mathbf{s}_\phi+G_z\mathbf{u}_z\cdot d\mathbf{s}_z \underset{\text{(1.1.60b)}}{\overset{\text{Fig.2.43(b)}}{=}} \iint_S G_\rho\,\rho\,d\phi\,dz+G_\phi\,dz\,d\rho+G_z\,\rho\,d\rho\,d\phi$$

$$\overset{\text{(E2.36.8)}}{=}\int_0^1\int_0^{2\pi}\rho^2\cos\phi\big|_{\rho=1}\rho\,d\phi\,dz+\int_0^1\int_0^1 0\,dz\,d\rho+\int_0^{2\pi}\int_0^1 2\rho\left(z\big|_{z=1}-z\big|_{z=0}\right)d\rho\,d\phi$$

$$=\int_0^1\sin\phi\big|_0^{2\pi}dz+\int_0^{2\pi}\rho^2\big|_0^1 d\phi=2\pi \quad \text{(E2.36.9)}$$

The RHS of divergence theorem (2.11.1) for Eq. (E2.36.8) with the cylindrical region R enclosed by the closed surface S is

$$\iiint_R\overset{\text{Eq. (2.4.4b)}}{\left(\frac{1}{\rho}\frac{\partial(\rho G_\rho)}{\partial\rho}+\frac{1}{\rho}\frac{\partial G_\phi}{\partial\phi}+\frac{\partial G_z}{\partial z}\right)}dV \underset{\text{Fig.2.43(b),(1.1.60a)}}{\overset{\text{(E2.36.8)}}{=}} \int_0^1\int_0^{2\pi}\int_0^1(3\rho\cos\phi+2)\rho\,d\rho\,d\phi\,dz=2\pi \quad \text{(E2.36.10)}$$

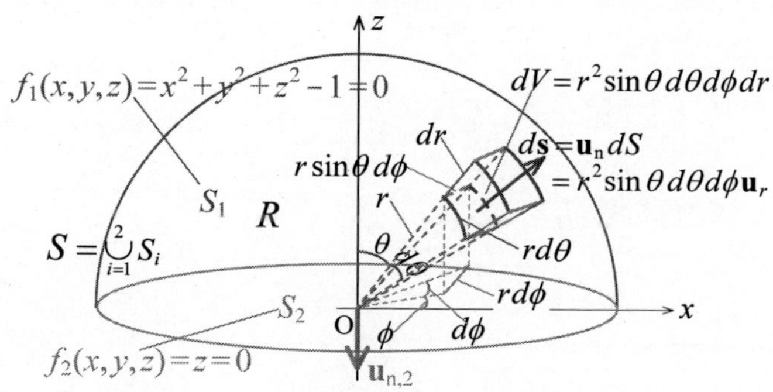

Figure 2.44 A hemispherical surface S and the unit circle C enclosing the surface for Example 2.37

(Example 2.37) Divergence Theorem

Show the validity of divergence theorem (2.11.1) for a vector field

$$\mathbf{G}(x,y,z) = x\mathbf{u}_x + y\mathbf{u}_y + z\mathbf{u}_z \tag{E2.37.1}$$

with the hemispherical region that is depicted in Fig. 2.44:

$$R=\{x^2+y^2+z^2\le 1,\ 0\le z\le 1\} \tag{E2.37.2}$$

First, note that the hemispherical surface S_1 is described by

$$f_1(x,y,z) = x^2 + y^2 + z^2 - 1 = 0 \tag{E2.37.3}$$

and accordingly, its normal vector is

$$\mathbf{u}_n \overset{(2.8.9)}{=} \frac{\nabla f}{||\nabla f||} \overset{(E2.37.3)}{=} \frac{2x\mathbf{u}_x+2y\mathbf{u}_y+2x\mathbf{u}_z}{\sqrt{(2x)^2+(2y)^2+(2z)^2}} \overset{x^2+y^2+z^2=1}{=} x\mathbf{u}_x+y\mathbf{u}_y+z\mathbf{u}_z \tag{E2.37.4}$$

Note also from Eq. (2.8.5a) that the differential surface area dS of the hemispherical surface S_1 is related to the corresponding regional area dA on the xy-plane as

$$z=g(x,y)=\pm\sqrt{1-x^2-y^2} \overset{(2.8.5a)}{\rightarrow} dS=\sqrt{1+(\frac{\partial g}{\partial x})^2+(\frac{\partial g}{\partial y})^2}\, dA=\sqrt{1+\frac{x^2+y^2}{1-x^2-y^2}}\, dA=\frac{1}{\sqrt{1-x^2-y^2}}\, dA \tag{E2.37.5}$$

Thus, the LHS of divergence theorem (2.11.1) for Eq. (E2.37.1) with the closed surface S enclosing the hemispherical region R can be computed as

$$\begin{aligned}\oiint_S \mathbf{G}\cdot d\mathbf{s} &= \oiint_S \mathbf{G}\cdot\mathbf{u}_n\, dS \overset{(E2.37.1)}{\underset{(E2.37.4)}{=}} \iint_{S_1}(x\mathbf{u}_x+y\mathbf{u}_y+z\mathbf{u}_z)\cdot(x\mathbf{u}_x+y\mathbf{u}_y+z\mathbf{u}_z)\, dS \\ &+ \iint_{S_2}(x\mathbf{u}_x+y\mathbf{u}_y+z\mathbf{u}_z)|_{z=0}\cdot(-\mathbf{u}_z)\, dS \\ &\overset{(E2.37.5)}{=} \int_0^1\int_{-\sqrt{1-y^2}}^{\sqrt{1-y^2}}(x^2+y^2+z^2)|_{x^2+y^2+z^2=1}\frac{1}{\sqrt{1-x^2-y^2}}\, dxdy \\ &\overset{\text{Rectangular-to-polar}}{=} \int_0^{2\pi}\int_0^1\frac{1}{\sqrt{1-r^2}}\, r\, drd\theta=\int_0^{2\pi}(-\sqrt{1-r^2})|_0^1\, d\theta=\int_0^{2\pi}1\, d\theta=2\pi\end{aligned} \tag{E2.37.6}$$

```
>>syms x y z; Gxy=1/sqrt(1-x^2-y^2);
>>LHS=int(int(Gxy,x,-sqrt(1-y^2),sqrt(1-y^2)),y,-1,1) % Eq.(E2.37.6)
   2*pi
```

On the other hand, the RHS of divergence theorem (2.11.1) for Eq. (E2.37.1) with the hemispherical region R is

$$\iiint_R \left(\frac{\partial G_x}{\partial x} + \frac{\partial G_y}{\partial y} + \frac{\partial G_z}{\partial z} \right) dV \overset{\text{(E2.37.1)}}{=} \int_0^1 \int_{-\sqrt{1-z^2}}^{\sqrt{1-z^2}} \int_{-\sqrt{1-y^2-z^2}}^{\sqrt{1-y^2-z^2}} (1+1+1)\,dx\,dy\,dz = \int_0^1 \int_{-\sqrt{1-z^2}}^{\sqrt{1-z^2}} 6\sqrt{1-y^2-z^2}\,dy\,dz$$

$$\overset{\text{Rectangular-to-polar}}{=} \int_0^{\pi} \int_0^1 6\sqrt{1-r^2}\,r\,dr\,d\theta = -\int_0^{\pi} 2(1-r^2)^{3/2}\Big|_0^1 d\theta = \int_0^{\pi} 2\,d\theta = 2\pi \qquad \text{(E2.37.7)}$$

```
>>syms x y z; Gca=[x; y; z]; % Eq.(R2.37.1)
>>RHS=int(int(int(diva(Gca),x,-sqrt(1-y^2-z^2),sqrt(1-y^2-z^2)), ...
      y,-sqrt(1-z^2),sqrt(1-z^2)),z,0,1)  % Eq.(E2.37.7)
   2*pi
```

Since Eqs. (E2.37.6) and (E2.37.7), which are the LHS and RHS of divergence theorem (2.11.1), respectively, are equal, the validity of the divergence theorem has been checked.

Now, since the spherical region R is easier to deal with in spherical coordinates, let us make the Cartesian-to-spherical coordinate conversion of the vector function $G(x, y, z)$ via the change of variables $x = r\sin\theta\cos\phi$, $y = r\sin\theta\sin\phi$, and $z = r\cos\theta$, $\mathbf{u}_x = \sin\theta\cos\phi\,\mathbf{u}_r + \cos\theta\cos\phi\,\mathbf{u}_\theta - \sin\phi\,\mathbf{u}_\phi$, $\mathbf{u}_y = \sin\theta\sin\phi\,\mathbf{u}_r + \cos\theta\sin\phi\,\mathbf{u}_\theta + \cos\phi\,\mathbf{u}_\phi$, and $\mathbf{u}_z = \cos\theta\,\mathbf{u}_r - \sin\theta\,\mathbf{u}_\theta$ (Eqs. (1.1.59b,c)),

$$\mathbf{G}(x, y, z) = x\mathbf{u}_x + y\mathbf{u}_y + z\mathbf{u}_z$$

$$\underset{\text{(1.1.59b)}}{\overset{\text{(1.1.59c)}}{\longrightarrow}} \mathbf{G}(r, \theta, \phi) = r\sin\theta\cos\phi(\sin\theta\cos\phi\,\mathbf{u}_r + \cos\theta\cos\phi\,\mathbf{u}_\theta - \sin\phi\,\mathbf{u}_\phi)$$

$$+ r\sin\theta\sin\phi(\sin\phi\sin\theta\,\mathbf{u}_r + \sin\phi\cos\theta\,\mathbf{u}_\theta + \cos\phi\,\mathbf{u}_\phi) + r\cos\theta(\cos\theta\,\mathbf{u}_r - \sin\theta\,\mathbf{u}_\theta) = r\,\mathbf{u}_r \qquad \text{(E2.37.8)}$$

```
>>Gsp=simplify(cart2sph_v(Gca)) % Cartesian-to-Spherical conversion
  Gsp = r
        0
        0
```

Then, with the normal vector $\mathbf{u}_r$ to the hemispherical surface S_1 ($r=1$) and the normal vector $-\mathbf{u}_z$ to the unit disk surface S_2, the LHS of divergence theorem (2.11.1) for Eq. (E2.37.8) with the hemispherical region R can be computed in spherical coordinates as

$$\oiint_S \mathbf{G} \bullet d\mathbf{s} = \oiint_S \mathbf{G} \bullet \mathbf{u}_n\,dS \underset{r=1}{\overset{\text{(E2.37.8)}}{=}} \overbrace{\iint_{S_1} \mathbf{u}_r \bullet \mathbf{u}_r\,dS}^{\text{hemispherical surface}} + \overbrace{\iint_{S_2} \mathbf{u}_r \bullet (-\mathbf{u}_z)\,dS}^{\text{base disk}}$$

$$\underset{\text{(1.1.59b)}}{\overset{\text{(1.1.61b)}}{=}} \int_0^{2\pi} \int_0^{\pi/2} r^2\Big|_{r=1} \sin\theta\,d\theta\,d\phi + \iint_{S_2} \mathbf{u}_r \bullet (-\cos\theta\,\mathbf{u}_r + \sin\theta\,\mathbf{u}_\theta)\Big|_{\theta=\pi/2} dS$$

$$= \int_0^{2\pi} (-\cos\theta)\Big|_0^{\pi/2} d\phi + 0 = 2\pi \qquad \text{(E2.37.9)}$$

```
>>syms r th ph; RHS=int(int(sin(th),th,0,pi/2),ph,0,2*pi)
  2*pi
```

The RHS of divergence theorem (2.11.1) for Eq. (E2.37.8) with the hemispherical region R enclosed by the closed surface S is

$$\iiint_R \left(\frac{1}{r^2}\frac{\partial(r^2 G_r)}{\partial r} + \frac{1}{r\sin\theta}\frac{\partial(G_\theta \sin\theta)}{\partial\phi} + \frac{1}{r\sin\theta}\frac{\partial G_\phi}{\partial\phi} \right)_{\text{(2.4.4c)}} dV \underset{\text{(2.10.6)}}{\overset{\text{(E2.37.8)}}{=}} \int_0^{2\pi} \int_0^{\pi/2} \int_0^1 (3) r^2 \sin\theta\,dr\,d\theta\,d\phi = 2\pi$$

$$\text{(E2.37.10)}$$

```
>>RHS=int(int(int(diva(Gsp,'sp')*r^2*sin(th),r,0,1),th,0,pi/2),ph,0,2*pi)
  2*pi
```

(Example 2.38) Volume via Surface Integral Based on the Divergence Theorem

Based on the divergence theorem (2.11.1), use the surface integral to find the volume of the object depicted in Fig. 2.39 or 2.45, which was dealt with in Example 2.33.

To use the surface integral (LHS of Eq. (2.11.1)) for finding a volume, we need to find a vector function **G** such that div(**G**)=1. Even if there are many choices such as $\mathbf{G}=x\,\mathbf{u}_x,\ y\,\mathbf{u}_y,\ z\,\mathbf{u}_z,\ \ldots$, we select

$$\mathbf{G} = y\mathbf{u}_y \tag{E2.38.1}$$

to exploit the cylindrical symmetry of the surface/volume. Then we substitute this into the LHS of Eq. (2.11.1) to write the surface integral and compute it as

$$\begin{aligned} V &\overset{(2.11.1)}{=} \oint\!\!\!\oint_S \mathbf{G}\cdot d\mathbf{s} = \iint_{S_1} y\mathbf{u}_y\cdot d\mathbf{s}_1 + \iint_{S_2} y\mathbf{u}_y\cdot\mathbf{u}_\rho\, dS_2 + \iint_{S_3} y\mathbf{u}_y\cdot d\mathbf{s}_3 \\ &= -\int_{-1}^{1}\int_{-\sqrt{1-z^2}}^{\sqrt{1-z^2}} (x^2+z^2)\,dx\,dz + \int_{-1}^{1}\int_{-\sqrt{1-z^2}}^{\sqrt{1-z^2}} \{4-2(x^2+z^2)\}\,dx\,dz \\ &\overset{(2.6.7)}{=} -\int_0^{2\pi}\int_0^1 \rho^2\rho\,d\rho\,d\phi + \int_0^{2\pi}\int_0^1 (4-2\rho^2)\rho\,d\rho\,d\phi = 4\frac{2\pi}{2} - 3\frac{2\pi}{4} = \frac{5}{2}\pi \end{aligned} \tag{E2.38.2}$$

This gives us the same result as obtained using the volume integral in Example 2.33, but with less complex computation.

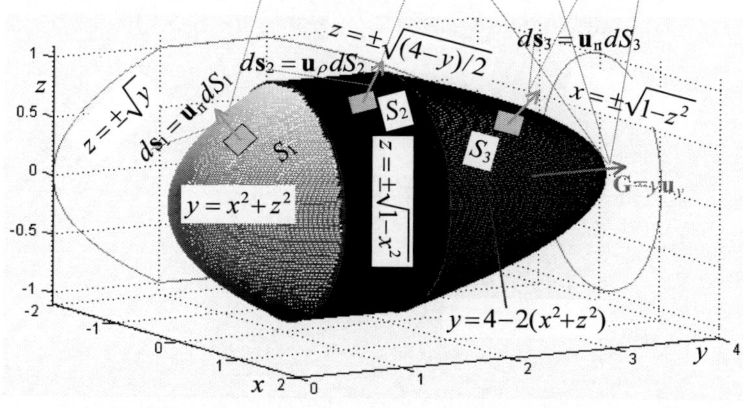

Figure 2.45 Object for the surface integral in Example 2.38

(Example 2.39) Volume via Surface Integral Based on the Divergence Theorem

Based on the divergence theorem (2.11.1), use the surface integral to find the volume of a sphere with radius r_1.

As in Example 2.38, to use the surface integral (LHS of Eq. (2.11.1)) for finding a volume, we need to find a vector function **G** such that div(**G**)=1. Even if there are many choices such as $\mathbf{G}=x\,\mathbf{u}_x,\ y\,\mathbf{u}_y,\ z\,\mathbf{u}_z,\ \ldots$, we select

$$G = \frac{1}{3}\left(x\mathbf{u}_x + y\mathbf{u}_y + z\mathbf{u}_z\right) = \frac{1}{3}r\,\mathbf{u}_r \text{ with } r = \sqrt{x^2+y^2+z^2} \tag{E2.39.1}$$

to exploit the spherical symmetry of the surface/volume. Then we substitute this into the LHS of Eq. (2.11.1) to write the surface integral (with the spherical surface S_1 of radius r_1) and compute it as

$$V \overset{(2.11.1)}{=} \oiint_S \mathbf{G} \cdot d\mathbf{s} = \iint_{S_1} \frac{1}{3} r\mathbf{u}_r \cdot \mathbf{u}_r \, dS$$

$$\overset{(1.1.61b)}{=} \frac{1}{3}\int_0^{2\pi}\int_0^{\pi} r^3\Big|_{r=r_1} \sin\theta \, d\theta \, d\phi = \frac{4}{3}\pi r_1^3 \quad \text{(E2.39.2)}$$

(Example 2.40) Heat Equation with the Divergence Theorem

Let the temperature distribution function in a region R be denoted by $u(x,y,z,t)$ at point (x,y,z) and time t. Then the total heat (energy) in R is

$$H = \iiint_R c d \, u \, dV \, [\mathrm{J}] \quad \text{(E2.40.1)}$$

where c[J/kg/°K] (J: Joule, °K(elvin): the unit of absolute thermodynamic temperature) and d[kg/m^3] are the specific heat and density (mass per unit volume) of the material, respectively. On the other hand, Fourier's law says that the heat flux **J**, i.e., the rate of heat transfer through unit area in a homogeneous material, is proportional to the negative gradient of the temperature distribution function $u(x,y,z,t)$ at point (x,y,z) and time t, as described by

$$\mathbf{J} = -k\nabla u \, [\mathrm{W/m^2}] \quad \text{(E2.40.2)}$$

where the minus sign stems from the fact that the heat flows in the direction of decreasing temperature (from warm to cold). Also, the heat energy conservation law says that the time rate of decrease of heat H in the region R is equal to the outward surface integral of the heat flux over the boundary surface S of R:

$$-\frac{\partial H}{\partial t} = \oiint_S \mathbf{J} \cdot d\mathbf{s} \, [\mathrm{W}] \quad \text{(E2.40.3)}$$

Substituting Eqs. (E2.40.1) and (E2.40.2) into both sides yields

$$\text{LHS: } -\frac{\partial H}{\partial t} \overset{(E2.40.1)}{=} -\frac{\partial}{\partial t}\iiint_R c d \, u \, dV = -cd\iiint_R \frac{\partial u}{\partial t} dV \quad \text{(E2.40.4)}$$

$$\text{RHS: } \oiint_S \mathbf{J} \cdot d\mathbf{s} \overset{(E2.40.2)}{=} \oiint_S (-k\nabla u) \cdot d\mathbf{s} \underset{\text{divergence thm}}{\overset{(2.11.1a)}{=}} -k\iiint_R \mathrm{div}\,(\nabla u) dV = -k\iiint_R \nabla \cdot \nabla u \, dV$$

$$= -k\iiint_R \nabla^2 u \, dV \quad \text{(E2.40.5)}$$

Substituting these equations into both sides of Eq. (E2.40.3) and taking the triple integral off the equation yields

$$a^2\nabla^2 u = \frac{\partial u}{\partial t} \text{ with } a^2 = \frac{k}{cd} \text{: thermal diffusivity} \quad \text{(E2.40.6)}$$

This is called the *heat* (*diffusion*) *equation*, which will be dealt with extensively in Chapter 7.

Problems

2.1 Gradient and Tangent Plane

Consider a function

$$f(x,y,z) = x^2+y^2+2z^2-4 \tag{P2.1.1}$$

(a) Find the gradient vector of the function at $\mathbf{p}_1=(-1,-1,1)$.

(b) Find the tangent plane of the function at $\mathbf{p}_1=(-1,-1,1)$.

(c) Complete and run the MATLAB script "em02p01.m" to plot the gradient vector and tangent plane found in (a) and (b) together with the graph of $f(x,y,z)=0$ as shown in Fig. P2.1.

```
%em02p01.m
% Gradient and Tangent Plane for Problem 2.1 (Fig.P2.1)
%                f(x,y,z)=x^2+y^2+2*z^2-4       (P2.1.1)
clear, clf
syms x y z;
f=@(x,y,z)x^2+?^2+?*z^2-4; % Eq.(P2.1.1)
dx=0.005; dy=0.005; xs=[-2:dx:2]; ys=[-2:dy:2];
[X,Y]=meshgrid(xs,ys); % Grids for the graph
Z2=(4-X.^2-Y.^2)/2; [ix,iy]=find(Z2<0); % Z^2 such that f(X,Y,Z)=0
Z = sqrt(max(Z2,0)); % To take only real numbers
for m=1:length(ix), Z(ix(m),iy(m))=nan; end %imaginary number ->NaN
p=[-1; -1; ?]; % A given point on the graph
gradf=gradienta(f) %Gradient of (P2.1.1)
gradfp=subs(gradf,{x,y,z},{p(1),p(2),p(3)}); gradfp=double(gradfp);
fprintf('Gradient vector at %6.2f,%6.2f,%6.2f): %6.2fux+%6.2fuy
     +%6.2fuz\n', p,gradfp);
x=p(1)+[-0.2 0.2]; y=p(2)+[-0.2 0.2];
[Xp,Yp]=meshgrid(x,y); % Grids for tangent plane
fprintf('Tangent plane equation: %6.2fx+%6.2fy+%6.2fz=%6.2f\n',
         gradfp,gradfp'*p);
Zp=(gradfp'*p-gradfp(1)*Xp-gradfp(2)*Yp)/gradfp(?); % Eq.(2.1.13)
subplot(221), hold on
meshc(X,Y,Z), hold on, mesh(X,Y,-Z) % Plot the graph
surf(Xp,Yp,Zp,'Linewidth',2) % Plot the tangent plane
quiver3(p(1),p(2),p(3),gradfp(1),gradfp(2),gradfp(3),0.25) %Gradient
axis('equal'), axis([-2 2 -2 2 -1 1.5]), grid on, view(68,20)
```

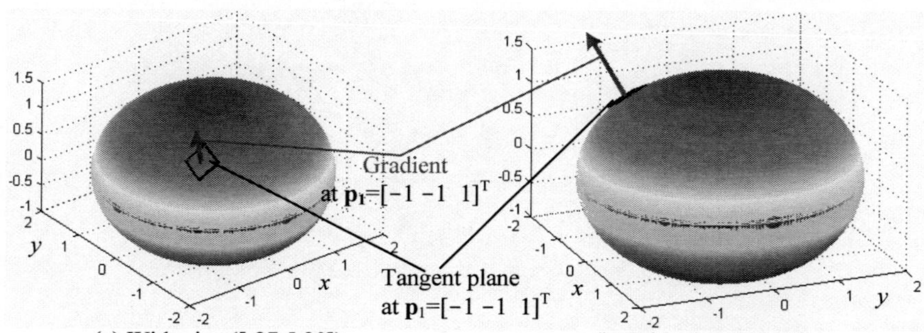

(a) With view([-37.5 30]) (b) With view([68 20])

Figure P2.1 Graph for Problem 2.1

2.2 Gradient and Tangent Plane

Consider a function

$$f(x,y,z)=x^2+2y^2-z \tag{P2.2.1}$$

(a) Find the gradient vector of the function at $\mathbf{p}_1=(1, 1, 3)$.

(b) Find the tangent plane of the function at $\mathbf{p}_1=(1, 1, 3)$.

(c) Complete and run the MATLAB script "em02p02.m" to plot the gradient vector and tangent plane found in (a) and (b) together with the graph of $f(x,y,z)=0$ as shown in Fig. P2.2.

```
%em02p02.m
% Gradient and Tangent Plane for Problem 2.2 (Fig.P2.2)
clear, clf
syms x y z;
f=@(x,y,z)x^2+?^2-z; % Eq.(P2.2.1)
dx=0.05; dy=0.05; xs=[-3:dx:3]; ys=[-2:dy:2];
[X,Y]=meshgrid(xs,ys); % Grids for the graph
Z = X.^2 + ?*Y.^2; % Eq.(P2.2.1)
meshc(X,Y,Z), hold on % Plot the graph
p1=[1; 1; 3]; % A given point on the graph
gradf=gradienta(f) %Gradient of (P2.2.1)
gradfp=subs(gradf,{x,y,z},{p(1),p(2),p(3)}); gradfp=double(gradfp);
% Plot the gradient vector at p1 in the 3D space
quiver3(p1(1),p1(2),p1(3),gradfp(1),gradfp(2),gradfp(3),0.2)
% Plot the gradient vector at p1 on the xy-plane
quiver3(p1(1),p1(2),0,gradfp(1),gradfp(2),0,0.2,'linewidth',2)
fprintf(' Gradient at (%3.1f,%3.1f) is [%3.1f %3.1f]\n',
             p1(1:2),gradfp(1:2))
% To plot the tangent plane
x=p1(1)+[-0.2 0.2]; y=p1(2)+[-0.1 0.1];
[Xp,Yp]=meshgrid(x,y); % Grids for the tangent plane
Zp=(gradfp'*p1-gradfp(1)*Xp-gradfp(2)*Yp)/gradfp(3); % Eq.(2.1.13)
surf(Xp,Yp,Zp,'Linewidth',2) % Plot the tangent plane
view(100,10), axis('equal'), axis([-4 4 -2 2 0 5]), grid on
set(gcf,'color','white'), shg
```

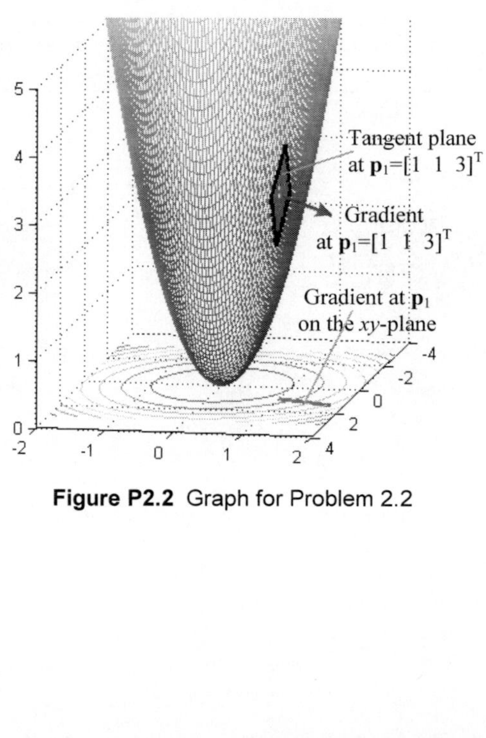

Figure P2.2 Graph for Problem 2.2

2.3 Tangent, Normal, and Binormal Vectors

Consider a 3D vector function

$$\mathbf{p}(t) = 3\cos t\,\mathbf{u}_x + 3\sin t\,\mathbf{u}_y + 4t\,\mathbf{u}_z \qquad \text{(P2.3.1)}$$

(a) Find the arc length of the helix for time segment $[0, \pi]$.
(b) Find the tangent line of the helix at $t=\pi$.
(c) Parameterize $\mathbf{p}(t)$ in terms of the arc length l defined by Eq. (2.2.4).
(d) Find the unit tangent vector $\mathbf{u}_T$ at a point that is l[m] distant from $\mathbf{p}(0)$ along the curve.
(e) Find the curvature κ of the helix.
(f) Find the unit normal vector $\mathbf{u}_N$ of the helix.
(g) Find the unit binormal vector $\mathbf{u}_B$ of the helix.
(h) Find the torsion τ of the helix.
(i) Complete the MATLAB script "em02p03.m" together with the function '`uTNBKTor()`' and run it to graph the helix (for $t=[0, 2\pi]$) with the moving trihedral (Frenet–Serret frame) consisting of $\mathbf{u}_T$, $\mathbf{u}_N$, and $\mathbf{u}_B$ at $t=\pi$ as in Fig. P2.3.

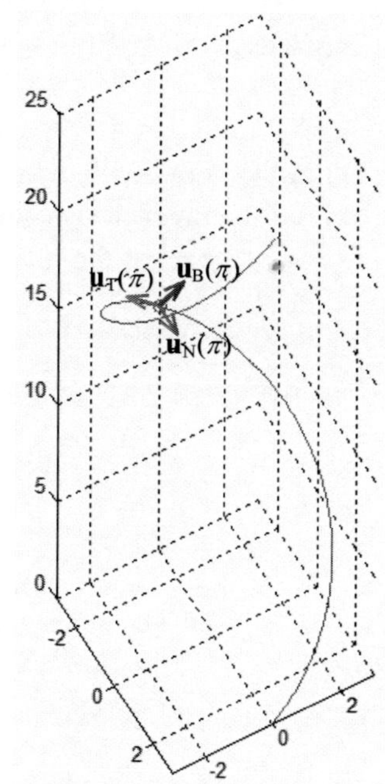

Figure P2.3 Moving trihedral for Problem 2.3

```
%em02p03.m
% Tangent, Normal, Binormal vectors
clear, clf
syms t; p=[3*cos(t); 3*sin(t); 4*t];
[v,a,uT,uN,uB,kappa,tau]=uTNBKTor(p)
t1=pi; uT1=subs(uT,t,t1), uN1=subs(uN,t,t1), uB1=subs(uB,t,t1)
t=0:0.01:2*pi; % Time vector and Differential time
p=@(t)[3*cos(t); 3*sin(t); 4*t]; pt=p(t); % Eq.(P2.3.1)
plot3(pt(1,:),pt(2,:),pt(3,:)), grid on, hold on
p1=p(t1); plot3(p1(1),p1(2),p1(3),'o') % An observation point
quiver3(p1(1),p1(2),p1(3),uT1(1),uT1(2),uT1(3),'r','Linewidth',2)
quiver3(p1(1),p1(2),p1(3),uN1(1),uN1(2),uN1(3),'g','Linewidth',2)
quiver3(p1(1),p1(2),p1(3),uB1(1),uB1(2),uB1(3),'b','Linewidth',2)
view(60,60), axis([-2.5 2.5 -2.5 2.5 0 20])
axis('equal'), set(gcf,'color','white'), shg
```

```
function [v,a,uT,uN,uB,kappa,tau]=uTNBKTor(p)
% p=[px(t); py(t); pz(t)] : A 3-D position vector function in t
% uT|uN|uB : Unit tangent|normal|binormal vectors
% kappa: Curvature, tau: Torsion
v=diff(p); a=diff(v); % Eqs.(2.3.9),(2.3.10)
dp=v; norm_dp=simplify(sqrt(dp(1)^2+dp(2)^2+dp(3)^2));
uT=simplify(dp/norm_dp); % Eq.(2.3.15)
duT=diff(uT); norm_duT=simplify(sqrt(duT(1)^2+duT(2)^2+duT(3)^2));
uN=simplify(duT/norm_duT); % Eq.(2.3.18)
uB=simplify(cross(uT,uN)); % Eq.(2.3.21)
duB=simplify(diff(uB));
tau=simplify(-duB'*uN*uN)/norm_dp); % Eq.(2.3.26)
kappa=simplify(norm_duT/norm_dp); % Eq.(2.3.14)
```

2.4 Gradient, Divergence, Laplacian, and Curl

Consider a two-variable scalar function

$$f(x,y) = \ln(x^2+y^2) = \ln\rho^2 \text{ with } \rho=\sqrt{x^2+y^2} \tag{P2.4.1}$$

(a) Find the gradient vector **G** of $f(x,y)$.
(b) Find the divergence of the gradient, i.e., the Laplacian of $f(x, y)$.
(c) Find the curl of the gradient of $f(x, y)$.

2.5 Gradient, Divergence, Laplacian, and Curl

Consider a two-variable scalar function

$$f(x,y) = \tan^{-1}\frac{y}{x} = \phi \tag{P2.5.1}$$

where

$$v=\tan u \rightarrow \frac{dv}{du} = \sec^2 u = 1+\tan^2 u = 1+v^2;\ u=\tan^{-1} v \rightarrow \frac{du}{dv} = \frac{1}{dv/du} = \frac{1}{1+v^2} \tag{P2.5.2}$$

$$\frac{\partial}{\partial x}(\tan^{-1}\frac{y}{x}) \overset{y/x=v}{=} \frac{\partial}{\partial x}(\tan^{-1} v) = \frac{\partial}{\partial v}(\tan^{-1} v)\frac{\partial v}{\partial x} \overset{\text{(P2.5.2)}}{=} \frac{1}{1+v^2}\frac{\partial v}{\partial x} \overset{v=y/x}{=} \frac{-y/x^2}{1+(y/x)^2} = \frac{-y}{x^2+y^2} \tag{P2.5.3a}$$

$$\frac{\partial}{\partial y}(\tan^{-1}\frac{y}{x}) \overset{y/x=v}{=} \frac{\partial}{\partial y}(\tan^{-1} v) = \frac{\partial}{\partial v}(\tan^{-1} v)\frac{\partial v}{\partial y} \overset{\text{(P2.5.2)}}{=} \frac{1}{1+v^2}\frac{\partial v}{\partial y} \overset{v=y/x}{=} \frac{1/x}{1+(y/x)^2} = \frac{x}{x^2+y^2} \tag{P2.5.3b}$$

```
>>syms x y; diff(atan(y/x)), diff(atan(y/x),y)
   ans = -y/(x^2*(y^2/x^2 + 1)),   ans = 1/(x*(y^2/x^2 + 1))
```

(a) Find the gradient vector **G** of $f(x,y)$.
(b) Find the divergence of the gradient, i.e., the Laplacian of $f(x,y)$.
(c) Find the curl of the gradient of $f(x,y)$.

```
>>f=atan(y/x); G=gradienta(f); diva(G), curla(G)
```

2.6 Gradient, Divergence, Laplacian, and Curl

Consider a three-variable scalar function

$$f(x,y,z) = \frac{c}{r} = \frac{c}{\sqrt{x^2+y^2+z^2}} \tag{P2.6.1}$$

(a) Find the gradient vector **G** of $f(x,y,z)$.
(b) Find the divergence of the gradient, i.e., the Laplacian of $f(x,y,z)$.
(c) Find the curl of the gradient of $f(x,y,z)$.

2.7 Magnetic Field by a Coaxial Cable Carrying Current I

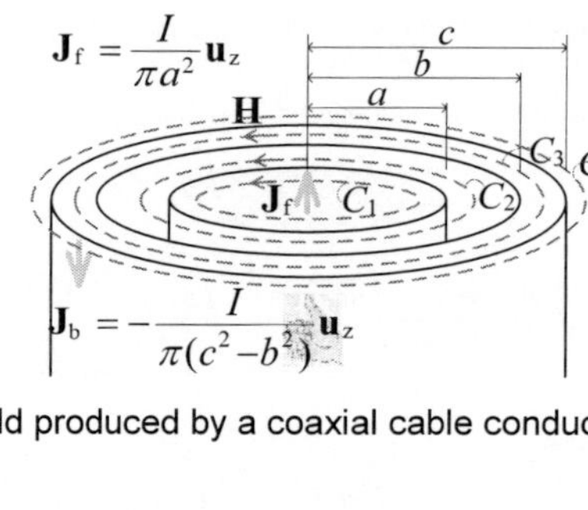

Figure P2.7 Magnetic field produced by a coaxial cable conducting current density $J_f = I/\pi a^2$

Consider a coaxial cable where the inner cable of radius a carries current density $J_f=I/\pi a^2$ in the $+z$ direction and the outer cable of inner/outer radii b/c carries current density $J_b=I/\pi(c^2-b^2)$ in the $+z$ direction, respectively, as shown in Fig. P2.7. It produces the magnetic field at a point ρ[m] away from the center of the cable axis as

$$\mathbf{H}=\begin{cases}\dfrac{I\rho}{2\pi a^2}\mathbf{u}_\phi & \text{for } \rho\le a\\ \dfrac{I}{2\pi\rho}\mathbf{u}_\phi & \text{for } a<\rho\le b\\ \dfrac{(c^2-\rho^2)I}{2\pi\rho(c^2-b^2)}\mathbf{u}_\phi & \text{for } b<\rho\le c\\ \mathbf{0} & \text{for } \rho>c\end{cases} \tag{P2.7.1}$$

(a) Use Eq. (2.4.5b) to find the curl of the magnetic field as follows:

(i) For $\rho\le a$,

$$\nabla\times\mathbf{H}\overset{\text{(P2.7.1a)}}{=}\nabla\times\frac{I\rho}{2\pi a^2}\mathbf{u}_\phi=\left(-\frac{\partial H_\phi}{\partial z}\right)\mathbf{u}_\rho+\frac{1}{\rho}\left(\frac{\partial}{\partial\rho}(\rho H_\phi)\right)\mathbf{u}_z=\frac{1}{\rho}\frac{\partial}{\partial\rho}\left(\rho\frac{I\rho}{2\pi a^2}\right)\mathbf{u}_z$$
$$=\boxed{} \tag{P2.7.2}$$

(ii) For $a<\rho\le b$,

$$\nabla\times\mathbf{H}\overset{\text{(P2.7.1b)}}{=}\nabla\times\frac{I}{2\pi\rho}\mathbf{u}_\phi=\left(-\frac{\partial H_\phi}{\partial z}\right)\mathbf{u}_\rho+\frac{1}{\rho}\frac{\partial}{\partial\rho}(\rho H_\phi)\mathbf{u}_z=\frac{1}{\rho}\frac{\partial}{\partial\rho}\left(\rho\frac{I}{2\pi\rho}\right)\mathbf{u}_z=\boxed{} \tag{P2.7.3}$$

(iii) For $b<\rho\le c$,

$$\nabla\times\mathbf{H}\overset{\text{(P2.7.1c)}}{=}\nabla\times\frac{(c^2-\rho^2)I}{2\pi\rho(c^2-b^2)}\mathbf{u}_\phi=\left(-\frac{\partial H_\phi}{\partial z}\right)\mathbf{u}_\rho+\frac{1}{\rho}\left(\frac{\partial}{\partial\rho}(\rho H_\phi)\right)\mathbf{u}_z$$
$$=\frac{1}{\rho}\frac{\partial}{\partial\rho}\left(\rho\frac{(c^2-\rho^2)I}{2\pi\rho(c^2-b^2)}\right)\mathbf{u}_z=\frac{I}{2\pi\rho}\frac{\partial}{\partial\rho}\left(\frac{c^2-\rho^2}{c^2-b^2}\right)\mathbf{u}_z=\boxed{} \tag{P2.7.4}$$

(iv) For $\rho>c$ where the current density is zero,

$$\nabla\times\mathbf{H}\overset{\text{(P2.7.1d)}}{=}\nabla\times\mathbf{0}=\mathbf{0} \tag{P2.7.5}$$

(b) The same result can be obtained by using the definition (2.4.5a) of the cross product w.r.t. the Cartesian coordinate system where

$$\mathbf{u}_\phi\overset{\text{(1.1.58a)}}{=}-\sin\phi\,\mathbf{u}_x+\cos\phi\,\mathbf{u}_y=\frac{-y}{\sqrt{x^2+y^2}}\mathbf{u}_x+\frac{x}{\sqrt{x^2+y^2}}\mathbf{u}_y \tag{P2.7.6}$$

(i) For $\rho\le a$,

$$\nabla\times\mathbf{H}\overset{\text{(P2.7.1a)}}{=}\nabla\times\frac{I\rho}{2\pi a^2}\mathbf{u}_\phi\overset{\text{(P2.7.6)}}{=}\nabla\times\frac{I\sqrt{x^2+y^2}}{2\pi a^2}\left(\frac{-y}{\sqrt{x^2+y^2}}\mathbf{u}_x+\frac{x}{\sqrt{x^2+y^2}}\mathbf{u}_y\right)$$
$$=\nabla\times\frac{I}{2\pi a^2}\left(-y\,\mathbf{u}_x+x\,\mathbf{u}_y\right)=\left(-\frac{\partial H_y}{\partial z}\right)\mathbf{u}_x+\left(\frac{\partial H_x}{\partial z}\right)\mathbf{u}_y+\left(\frac{\partial H_y}{\partial x}-\frac{\partial H_x}{\partial y}\right)\mathbf{u}_z$$
$$=\boxed{} \tag{P2.7.7}$$

(ii) For $a < \rho \le b$,

$$\nabla\times\mathbf{H} \overset{\text{(P2.7.1b)}}{=} \nabla\times\frac{I}{2\pi\rho}\mathbf{u}_\phi \overset{\text{(P2.7.6)}}{=} \nabla\times\frac{I}{2\pi\sqrt{x^2+y^2}}\left(\frac{-y}{\sqrt{x^2+y^2}}\mathbf{u}_x+\frac{x}{\sqrt{x^2+y^2}}\mathbf{u}_y\right)$$

$$= \nabla\times\frac{I}{2\pi}\left(\frac{-y}{x^2+y^2}\mathbf{u}_x+\frac{x}{x^2+y^2}\mathbf{u}_y\right)=\left(\frac{\partial H_y}{\partial x}-\frac{\partial H_x}{\partial y}\right)\mathbf{u}_z$$

$$= \boxed{} \tag{P2.7.8}$$

(iii) For $b < \rho \le c$,

$$\nabla\times\mathbf{H} \overset{\text{(P2.7.1c)}}{=} \nabla\times\frac{(c^2-\rho^2)I}{2\pi\rho(c^2-b^2)}\mathbf{u}_\phi \overset{\text{(P2.7.6)}}{=} \nabla\times\frac{(c^2-x^2-y^2)I}{2\pi(c^2-b^2)}\left(\frac{-y}{x^2+y^2}\mathbf{u}_x+\frac{x}{x^2+y^2}\mathbf{u}_y\right)$$

$$=\left(\frac{\partial H_y}{\partial x}-\frac{\partial H_x}{\partial y}\right)\mathbf{u}_z=\frac{I}{2\pi(c^2-b^2)}\left(\frac{\begin{array}{c}(x^2+y^2)(\cancel{c^2}-\cancel{3}x^2-y^2)\\ -\cancel{2x^2}(\cancel{c^2}-x^2-y^2)\end{array}}{(x^2+y^2)^2}+\frac{\begin{array}{c}(x^2+y^2)(\cancel{c^2}-\cancel{3}y^2-x^2)\\ -\cancel{2y^2}(\cancel{c^2}-x^2-y^2)\end{array}}{(x^2+y^2)^2}\right)\mathbf{u}_z$$

$$= \boxed{} \tag{P2.7.9}$$

(c) Complete the MATLAB program "show_H_by_coaxial_cable.m" for any test point and check if the discrepancy between the two values of $\nabla\times\mathbf{H}$, one computed numerically using Eq. (2.4.5a) and the other (Eq. (P2.7.2~5)) computed analytically, is close to zero.

```
%show_H_by_coaxial_cable.m
I=1; a=1; b=2; c=3; h=1e-6; h2=2*h; uz=[0 0 1];
J=@(x,a,b,c,I)(x<=a)*I/pi/a^2-(b<x&x<=c)*I/pi/(c^2-b^2); % Current density
p=[-0.6 -0.6 0]; rho=norm(p(1:2)); % A test point and its radial distance
H=H_cable(p,a,b,c,I);
dHdx=H_cable(p+[h 0 0],a,b,c,I)-H_cable(p+[-h 0 0],a,b,c,I);
dHdy=H_cable(p+[0 h 0],a,b,c,I)-H_cable(p+[0 -h 0],a,b,c,I);
dHdz=H_cable(p+[0 0 h],a,b,c,I)-H_cable(p+[0 0 -h],a,b,c,I);
curl_H=[dHdy(3)-dHdz(2) dHdz(1)-dHdx(?) dHdx(2)-dHdy(?)]/h2; % Eq.(2.4.5a)
discrepancy=norm(curl_H-J(rho,a,b,c,I)*uz) % Error from Eq.(P2.7.2~5)
```

```
function H=H_cable(p,a,b,c,I)
% Magnetic Field at p by a coaxial cable with radii a/b/c and current I
rho=norm(p(1:2)); phi=atan2(p(2),p(1));
uphi=[-sin(phi) cos(phi) 0]; % Eq.(P2.7.6)
H=(rho<=a)*rho/a^2+(a<rho&rho<=b)/rho...
      +(c^2-???^2)/(?^2-b^2)/rho*(b<rho&rho<=c);
H=H*I/2/pi*uphi; % Eq.(P2.7.1)
```

2.8 Curl

Find the divergences and curls of the following vector fields:

$$\mathbf{F}_{\text{A}}(x,y)=x\mathbf{u}_x+y\mathbf{u}_y=\rho(\cos\phi\,\mathbf{u}_x+\sin\phi\,\mathbf{u}_y)=\rho\mathbf{u}_\rho \tag{P2.8A}$$

$$\mathbf{F}_{\text{B}}(x,y)=\frac{x}{\sqrt{x^2+y^2}}\mathbf{u}_x+\frac{y}{\sqrt{x^2+y^2}}\mathbf{u}_y=\mathbf{u}_\rho,\ \mathbf{F}_{\text{C}}(x,y)=\frac{x}{x^2+y^2}\mathbf{u}_x+\frac{y}{x^2+y^2}\mathbf{u}_y=\frac{1}{\rho}\mathbf{u}_\rho \tag{P2.8B,C}$$

2.9 Divergence and Curl

Find the divergences and curls of the following vector fields:

$$\mathbf{F}_A(x,y) = -y\mathbf{u}_x + x\mathbf{u}_y = \rho(-\sin\phi\,\mathbf{u}_x + \cos\phi\,\mathbf{u}_y) = \rho\mathbf{u}_\phi \quad \text{(P2.9A)}$$

$$\mathbf{F}_B(x,y) = \frac{-y}{\sqrt{x^2+y^2}}\mathbf{u}_x + \frac{x}{\sqrt{x^2+y^2}}\mathbf{u}_y = \mathbf{u}_\phi \quad \text{(P2.9B)}$$

$$\mathbf{F}_C(x,y) = \frac{-y}{x^2+y^2}\mathbf{u}_x + \frac{x}{x^2+y^2}\mathbf{u}_y = \frac{1}{\rho}\mathbf{u}_\phi \quad \text{(P2.9C)}$$

2.10 Line Integrals

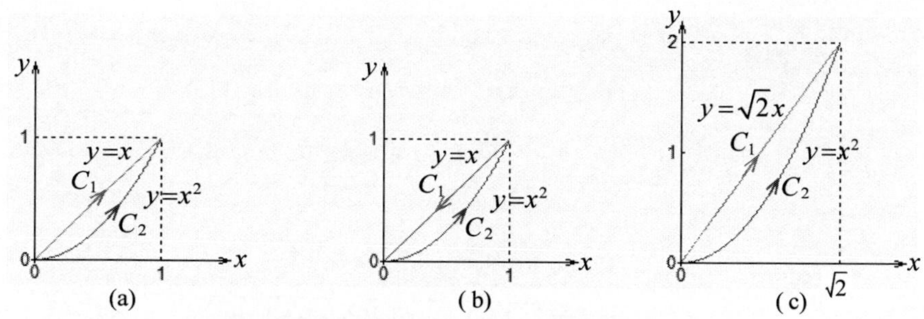

Figure P2.10 Paths of integration for Problem 2.10

(a) Evaluate the following integral twice

$$\int_{C_i} (12xy\,dx + 2x\,dy), \quad \text{(P2.10a)}$$

once on the straight line C_1 from (0, 0) to (1, 1) and once on the parabola C_2 that is given as an explicit function $y=x^2$ for $0 \le x \le 1$ (Fig. P2.10(a)).

(b) Evaluate the integral

$$\int_C (-y\,dx + x\,dy), \quad \text{(P2.10b)}$$

on the closed path C consisting of a line C_1 and a parabola C_2 depicted in Fig. P2.10(b).

(c) Evaluate the following integral twice

$$\int_{C_i} 6x\,dl, \quad \text{(P2.10c)}$$

once on the straight line C_1 from (0, 0) to $(\sqrt{2}, 2)$ and once on the parabola C_2 that is given as an explicit function $y = x^2$ for $0 \le x \le \sqrt{2}$ or parameterized by $t\mathbf{u}_x + t^2\mathbf{u}_y$ for $0 \le t \le \sqrt{2}$ (Fig. P2.10(c)).

2.11 Line Integrals

Evaluate the following integrals:

$$\int_C (1+x+y)dx, \quad \int_C (1+x+y)dy, \quad \int_C (1+x+y)dl \quad \text{(P2.11a,b,c)}$$

on a semicircular path C parameterized by

$$\mathbf{p}(t) = \cos t\,\mathbf{u}_x + \sin t\,\mathbf{u}_z \text{ for } 0 \le t \le \pi \quad \text{(P2.11.1)}$$

2.12 Line Integral w.r.t. Arc Length

Evaluate the integral

$$\int_C 4xyz\,dl \tag{P2.12.1}$$

on the helix C that is parameterized by

$$\mathbf{p}(t) = \cos t\,\mathbf{u}_x + \sin t\,\mathbf{u}_y + t\,\mathbf{u}_z \text{ for } 0 \le t \le \pi \tag{P2.12.2}$$

2.13 Line Integrals and Path Independence

Evaluate the following integrals twice, once on the straight line C_1 from (1, 0, 0) to (−1, 0, 1) parameterized by

$$\mathbf{p}(t) = (1-2t)\mathbf{u}_x + t\,\mathbf{u}_z \text{ for } 0 \le t \le 1 \tag{P2.13.1}$$

and once on the helix C_2 that is parameterized by

$$\mathbf{p}(t) = \cos \pi t\,\mathbf{u}_x + \sin \pi t\,\mathbf{u}_y + t\,\mathbf{u}_z \text{ for } 0 \le t \le 1 \tag{P2.13.2}$$

(a)
$$\int_{C_i} \frac{2(x\mathbf{u}_x + y\mathbf{u}_y + z\mathbf{u}_z)}{(x^2+y^2+z^2)^{3/2}} \bullet d\mathbf{l} \tag{P2.13a}$$

(b)
$$\int_{C_i} (y\mathbf{u}_x + x\mathbf{u}_y + z\mathbf{u}_z) \bullet d\mathbf{l} \tag{P2.13b}$$

(c)
$$\int_{C_i} (y\mathbf{u}_x - x\mathbf{u}_y + z\mathbf{u}_z) \bullet d\mathbf{l} \tag{P2.13c}$$

(d) Determine whether or not each of the above three integrals (a), (b), and (c) is path-independent and tell why it is so.

2.14 Double Integrals

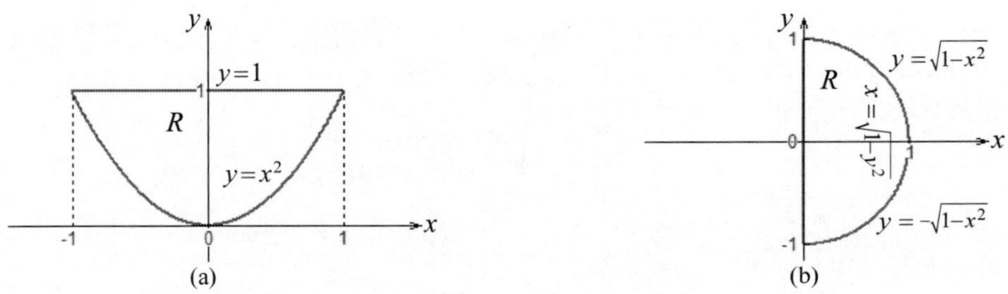

Figure P2.14 Regions for the double integrals in Problem 2.14

Evaluate the following integrals:

(a) $\iint_R 1\,dA$ over the region depicted in Fig. P2.14(a).

```
>>fxya=@(x,y)ones(size(x));   c=@(x)(x.^2);   d=@(x)ones(size(x));
  Area=quad2d(fxya,-1,1,c,d)
```

(b) $\iint_R x\,dA$ over the region depicted in Fig. P2.14(a).

(c) $\iint_R y\,dA$ over the region depicted in Fig. P2.14(a).

(d) $\iint_R x^2 dA$ over the region depicted in Fig. P2.14(a).

(e) $\iint_R y^2 dA$ over the region depicted in Fig. P2.14(a).

(f) $\iint_R 1dA$ over the region depicted in Fig. P2.14(b).

```
>>fxyf=@(x,y)ones(size(x));   c=@(x)0*x; d=@(x)sqrt(1-x.^2);
  Area=quad2d(fxyf,-1,1,c,d) % Area = 355/226 ~ pi/2
```

(g) $\iint_R x\,dA$ over the region depicted in Fig. P2.14(b).

```
>>fxyg=@(x,y)y;   c=@(x)0*x; d=@(x)sqrt(1-x.^2);
  mx=quad2d(fxyg,-1,1,c,d) % x-coordinate of centroid (center of mass)
```

(h) $\iint_R y\,dA$ over the region depicted in Fig. P2.14(b).

(i) $\iint_R x^2dA$ over the region depicted in Fig. P2.14(b).

(j) $\iint_R y^2dA$ over the region depicted in Fig. P2.14(b).

2.15 Double Integral to Get the Volume of a Solid

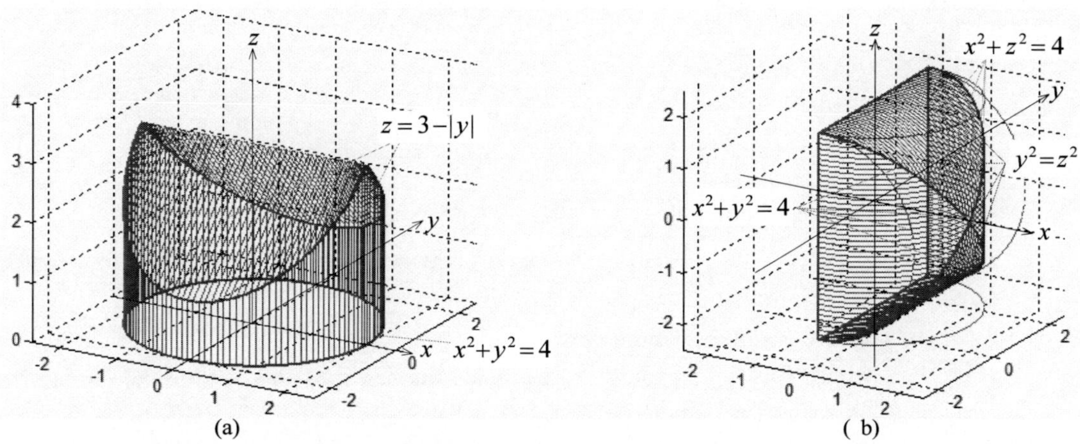

Figure P2.15 3D solids for Problem 2.15

(a) Use the double integral to find the volume of the solid depicted in Fig. P2.15(a).
(b) Use the double integral to find the volume of the solid depicted in Fig. P2.15(b).
(c) Among the following MATLAB scripts, choose the one(s) that can be used to solve each of the above problems and run them to check if they work well.

```
%em02p15_1.m
fxy=@(x,y)(3-y); c=@(x)0*x; d=@(x)sqrt(4-x.^2);
Integral=2*quad2d(fxy,-2,2,c,d)
```

```
%em02p15_2.m
syms x y; V=2*int(int(3-y,y,0,sqrt(4-x^2)),x,-2,2)
```

```
%em02p15_3.m
fxy=@(x,y)(3-x); c=@(x)0*x; d=@(x)sqrt(4-x.^2);
Integral=4*quad2d(fxy,0,2,c,d)
```

```
%em02p15_4.m
syms x y; V=4*int(int(3-y,x,0,sqrt(4-y^2)),y,0,2)
```

```
%em02p15_5.m
fxy=@(x,y)sqrt(4-x.^2);   c=@(x)0*x; d=@(x)sqrt(4-x.^2);
V=4*quad2d(fxy,0,2,c,d)
```

```
%em02p15_6.m
syms x y; V=4*int(int(sqrt(4-x^2),y,0,sqrt(4-x^2)),x,0,2)
```

```
%em02p15_7.m
fyz=@(z,y)sqrt(4-z.^2);   c=@(z)0*z; d=@(z)z;
V=8*quad2d(fyz,0,2,c,d)
```

```
%em02p15_8.m
syms y z; V=8*int(int(sqrt(4-z^2),y,0,z),z,0,2)
```

2.16 Double Integrals in Polar Coordinates

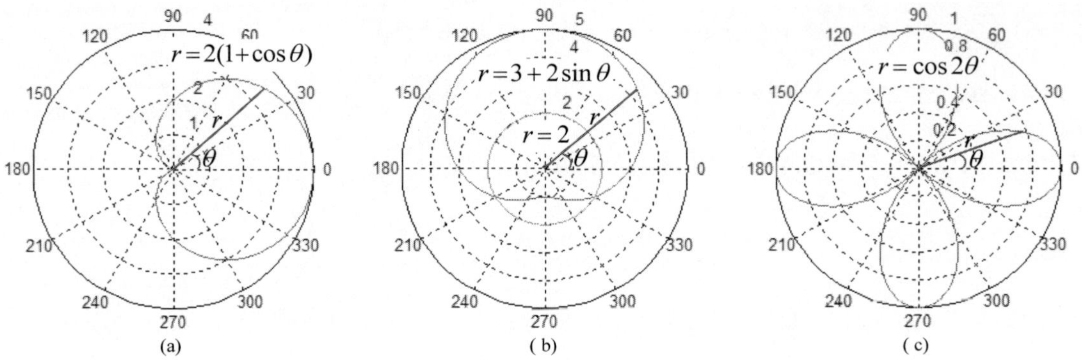

Figure P2.16 Regions for the double integrals in Problem 2.16

Evaluate the following double integrals:

(a) $I=\iint_R 8xy\,dA$ where R is the part of the region between the two circles of radius 1 and radius 2 centered at the origin that lies in the first quadrant.

(b) $I=\iint_R 1dA$ where R is the cardioid depicted in Fig. P2.16(a).

(c) $I=\iint_R 1dA$ where R is the region that lies inside the cardioid and outside the circle of radius 2 centered at the origin, as depicted in Fig. P2.16(b).

(d) $I=\iint_R 1dA$ where R is the region that lies inside the four leaves depicted in Fig. P2.16(c).

2.17 Double Integrals in Polar Coordinates

Evaluate the following double integrals:

(a) $I=\iint_R 3\sqrt{1-x^2-y^2}\,dA$ where R is the disk of radius 1/2 centered at (0, 1/2) as depicted in Fig. P2.17(a). It seems that, noting that the circular boundary of R can be represented by

$$(x-\frac{1}{2})^2+y^2=\frac{1}{4};\ x^2+y^2=x;\ r^2=r\cos\theta;\ r=\cos\theta\,, \tag{P2.17a.1}$$

we could compute the integral as

$$I=\iint_R 3\sqrt{1-x^2-y^2}\,dA \overset{(2.6.7)}{=} 3\int_{-\pi/2}^{\pi/2}\int_0^{\cos\theta}\sqrt{1-r^2}\,r\,dr\,d\theta=-\int_{-\pi/2}^{\pi/2}(1-r^2)^{3/2}\Big|_0^{\cos\theta}d\theta$$
$$=-\int_{-\pi/2}^{\pi/2}(\sin^3\theta-1)\,d\theta=\int_{-\pi/2}^{\pi/2}(1-(1-\cos^2\theta)\sin\theta)\,d\theta=\cdots \tag{P2.17a.2}$$

or

$$I=\iint_R 3\sqrt{1-x^2-y^2}\,dA \overset{(2.6.7)}{=} 6\int_0^{\pi/2}\int_0^{\cos\theta}\sqrt{1-r^2}\,r\,dr\,d\theta=\cdots \tag{P2.17a.3}$$

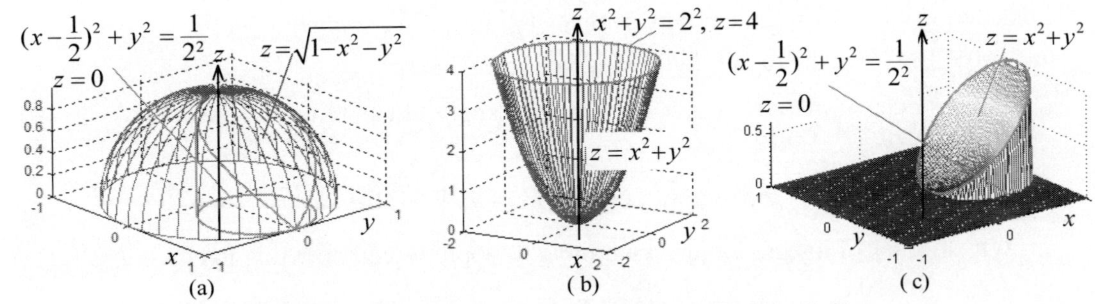

Figure P2.17 Double integrals to get volumes in Problem 2.17

Do both ways yield the same results? If not, choose the right one and tell what is wrong with the wrong one. Refer to the following MATLAB statements that can be used to do the computations:

```
>>fxy=@(x,y)sqrt(1-x.^2-y.^2);
>>c=@(x)-sqrt(x-x.^2); d=@(x)sqrt(x-x.^2);
>>Ixy=3*quad2d(fxy,0,1,c,d) % Numerical computation in Cartesian
>>ftr=@(t,r)r.*sqrt(1-r.^2);  c=@(t)0*t; d=@(t)cos(t);
>>Itr=6*quad2d(ftr,0,pi/2,c,d) % Numerical computation in polar
>>syms r t;
>>I=6*int(int(r*sqrt(1-r^2),r,0,cos(t)),t,0,pi/2)%Symbolic computation
```

(b) $V=\iint_R(4-x^2-y^2)\,dA$ where R is the cardioid depicted in Fig. P2.17(b).

(c) $V=\iint_R(x^2+y^2)\,dA$ where R is the disk of radius 1/2 centered at (0, 1/2) as depicted in Fig. P2.17(c).

(d) $V=\iint_R(\sqrt{2-x^2-y^2}-(x^2+y^2))\,dA$ where R is the disk bounded by the intersection of $z=\sqrt{2-x^2-y^2}$ and $z=x^2+y^2$.

```
>>ftr=@(t,r)r.*(sqrt(2-r.^2)-r.^2);
>>V=quad2d(ftr,0,2*pi,0,1) % Numerical computation
>>syms r t;
>>V=int(int(r*(sqrt(2-r^2)-r^2),r,0,1),t,0,2*pi) %Symbolic computation
```

(e) $V=\iint_R(1-\frac{1}{a}\sqrt{x^2+y^2})\,dA$ where R is the disk of radius a centered at the origin (0, 0).

(f) $V=\iint_R\sqrt{a^2-x^2-y^2}\,dA$ where R is the disk of radius a centered at the origin (0, 0).

2.18 Double Integral and the Order of Integration

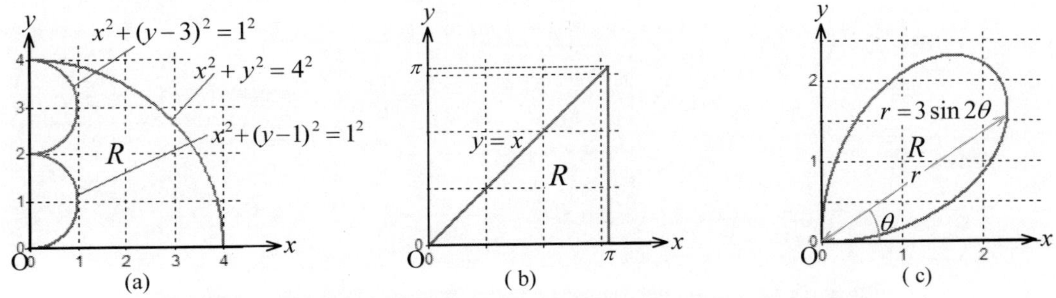

Figure P2.18 Regions for the double integrals in Problem 2.18

Considering the order of integration, evaluate the following double integrals.

(a) $I = \iint_R 1dA$ over the region R depicted in Fig. P2.18(a).

Which order of integration are you going to apply to compute this integral, $I = \iint_R () dy\, dx$, or $I = \iint_R () dx\, dy$? Why?

(b) $I = \iint_R \frac{\sin x}{x} dA$ over the region R depicted in Fig. P2.18(b).

Which order of integration are you going to apply to compute this integral, $I = \iint_R () dy\, dx$, or $I = \iint_R () dx\, dy$?

(c) $I = \iint_R r dA$ over the region R depicted in Fig. P2.18(c).

2.19 Green's Theorem — Line Integral and Double Integral

Evaluate the following line integrals twice, once without using Green's theorem and once using Green's theorem.

(a) $I = \oint_C -y\,dx + x\,dy$ where C is the boundary of the ellipse (Fig. P2.19(a)), oriented in the counterclockwise direction, that can be parameterized by

$$\mathbf{p}(t) = a\cos t\,\mathbf{u}_x + b\sin t\,\mathbf{u}_y,\ 0 \le t \le 2\pi \qquad \text{(P2.19a.1)}$$

(b) $I = \oint_C xy\,dx + x\,dy$ where C is the boundary of the rectangle with vertices at (0,0), (3,0), (3,2), and (0,2), oriented in the counterclockwise direction, that consists of four curves $C_1 = \{0 \le x \le 3, y = 0\}$, $C_2 = \{x = 3, 0 \le y \le 2\}$, $C_3 = \{3 \ge x \ge 0, y = 2\}$, and $C_4 = \{x = 0, 2 \ge y \ge 0\}$ (Fig. P2.19(b)).

(c) $I = \oint_C (\cos y)\,dx + (\sin x)\,dy$ where C is the boundary of the triangle with vertices at (0,0), $(0, \pi/2)$, and $(\pi/2, \pi/2)$, oriented in the counterclockwise direction, that consists of three curves $C_1 = \{0 \le x \le \pi/2, y = 0\}$, $C_2 = \{x = \pi/2, 0 \le y \le \pi/2\}$, and $C_3 = \{\pi/2 \ge x = y \ge 0\}$ (Fig. P2.19(c)).

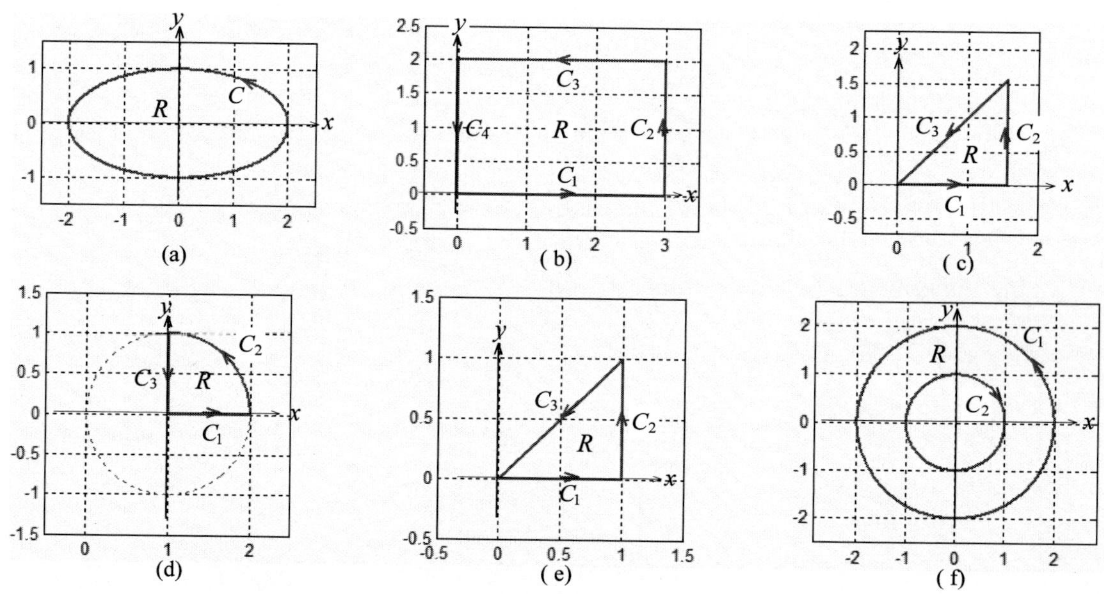

Figure P2.19 Closed curves as the paths of integration for Problem 2.19

(d) $I = \oint_C \frac{-y}{x^2+y^2}dx + \frac{x}{x^2+y^2}dy$ where C is a quarter of the unit circle that is centered at (1, 0), left-bounded by x=1, and lower-bounded by y=0 (Fig. P2.19(d)). Does Green's theorem hold? If so, what is the difference between this case and the case of Example 2.23 where Green's theorem does not hold for the same integral?

(e) $I = \oint_C \frac{-y}{\sqrt{x^2+y^2}}dx + \frac{x}{\sqrt{x^2+y^2}}dy$ where C is a triangle with vertices at (0, 0), (0, 1), and (1, 1) (Fig. P2.19(e)).

(f) $I = \oint_C \frac{-y}{\sqrt{x^2+y^2}}dx + \frac{x}{\sqrt{x^2+y^2}}dy$ where C consists of a circle C_1 of radius 2 oriented in the counterclockwise direction, and a circle C_2 of radius 1 oriented in the clockwise direction (Fig. P2.19(f)).

(g) $I = \oint_C -(y^3/3)\,dx + y^2x\,dy$ where C is the ellipse oriented in the counterclockwise direction (Fig. P2.19(a)).

2.20 Surface Integral

Use the surface area integral formula (2.8.3) to find the areas of the surfaces shown in Fig. P2.20.

(a) The surface area of the part of the plane z=2(1−x−y) in the first octant shown in Fig. P2.20(a).

(b) The upper surface area of the circular cone shown in Fig. P2.20(b).

(c) The upper surface area of the cylindrical segment, which is a part of a cylinder cut by the plane $z = \sqrt{3/2}(\sqrt{2} - x - y)$ as shown in Fig. P2.20(c).

(d) The lower surface area of the cone $z = \sqrt{x^2+y^2}$ cut by the plane z=1.

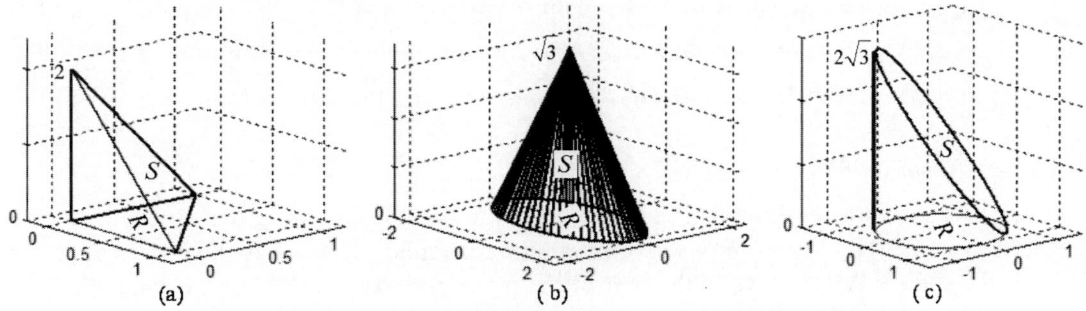

Figure P2.20 Surfaces for the surface integrals in Problem 2.20

2.21 Surface Integral Formulas

To show that each of the surface integral formulas (2.8.5a,b,c) is nothing but a special case of the more general surface integral formula (2.8.6), we can derive (2.8.5a) from (2.8.6). To this end, let the surface be parameterized by

$$\mathbf{s}(x, y) = x\,\mathbf{u}_x + y\,\mathbf{u}_y + f(x, y)\,\mathbf{u}_z \qquad \text{(P2.21.1)}$$

and start from Eqs. (2.8.6) with (x, y) in place of (v, w).

2.22 Surface Integral of Scalar Functions

Use the surface integral formula (2.8.5) or (2.8.6) to find the following scalar functions.

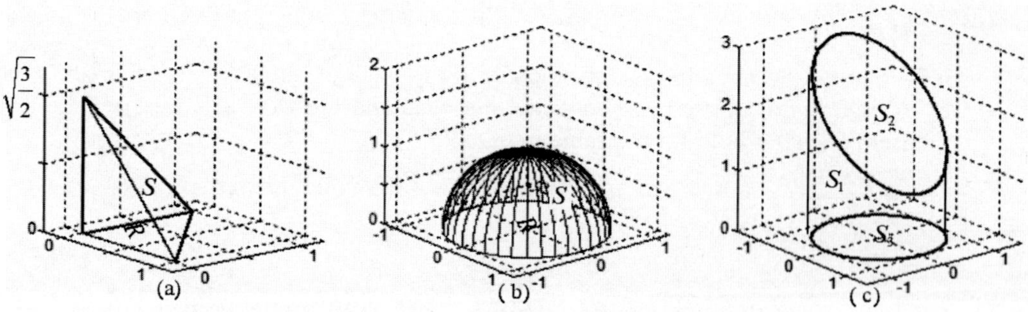

Figure P2.22 Surfaces for the surface integrals in Problem 2.22

(a) $G(x,y,z)=2x+y$ over the triangular surface S with vertices at (1,0,0), (0,1,0), and $(\sqrt{3/2},0,0)$, which is the part of the plane $z=\sqrt{3/2}(1-x-y)$ in the first octant (Fig. P2.21(a)), which can be parameterized by

$$\mathbf{s}(x,y)=x\,\mathbf{u}_x+y\,\mathbf{u}_y+\sqrt{3/2}\,(1-x-y)\mathbf{u}_z \text{ with } x,y\ge 0 \text{ and } 0\le x+y\le 1 \quad \text{(P2.22a.1)}$$

(b) $G(x,y,z)=z$ over the hemispherical surface S, which is the upper half of a unit sphere (Fig. P2.22(b)) and can be parameterized by

$$\mathbf{s}(\phi,\theta)=\sin\theta\cos\phi\,\mathbf{u}_x+\sin\theta\sin\phi\,\mathbf{u}_y+\cos\theta\,\mathbf{u}_z \quad \text{(P2.22b.1)}$$

(c) $G(x,y,z)=3x+6z$ over the surface S, which consists of the side S_1, the top S_2, and the bottom S_3 of the cylindrical segment where S_2 belongs to the plane $z=2-x$ as depicted in Fig. P2.22(c). Note that each of the three partial surfaces S_1, S_2, and S_3 can be parameterized as follows:

$$\mathbf{s}(\phi,z)=\cos\phi\,\mathbf{u}_x+\sin\phi\,\mathbf{u}_y+z\,\mathbf{u}_z \text{ with } 0\le\phi\le 2\pi \text{ and } 0\le z\le 2-\cos\phi \quad \text{(P2.22c.1)}$$

$$\mathbf{s}(\rho,\phi)=\rho\cos\phi\,\mathbf{u}_x+\rho\sin\phi\,\mathbf{u}_y+(2-\rho\cos\phi)\mathbf{u}_z \text{ with } 0\le\rho\le 1 \text{ and } 0\le\phi\le 2\pi \quad \text{(P2.22c.2)}$$

$$\mathbf{s}(\rho,\phi)=\rho\cos\phi\,\mathbf{u}_x+\rho\sin\phi\,\mathbf{u}_y+0\,\mathbf{u}_z \text{ with } 0\le\rho\le 1 \text{ and } 0\le\phi\le 2\pi \quad \text{(P2.22c.3)}$$

2.23 Surface Integral of Vector Functions

Use the surface integral formula (2.8.8) or (2.8.10) to find the following flux.

(a) $\mathbf{G}(x,y,z)=y\,\mathbf{u}_x+(z-y)\mathbf{u}_y+x\,\mathbf{u}_z$ through the triangular surface S with vertices at (1, 0, 0), (0, 1, 0), and (0, 0, 1), which is the part of the plane $z=1-x-y$ in the first octant (Fig. P2.23(a)). Note that the surface S is parameterized by

$$\mathbf{s}(x,y)=x\,\mathbf{u}_x+y\,\mathbf{u}_y+(1-x-y)\mathbf{u}_z \text{ with } 0\le x+y\le 1 \quad \text{(P2.23a.1)}$$

(b) $\mathbf{G}(x,y,z)=x\,\mathbf{u}_x+y\,\mathbf{u}_y+z\,\mathbf{u}_z$ through the paraboloidal surface S, which is described by $S=$ $S=\{(x,y,z)\,|\,0\le z=1-x^2-y^2\le 1\}$ (Fig. P2.23(b)). Note that the surface S is parameterized by

$$\mathbf{s}(x,y)=x\,\mathbf{u}_x+y\,\mathbf{u}_y+(1-x^2-y^2)\mathbf{u}_z \text{ with } 0\le x^2+y^2\le 1 \quad \text{(P2.23b.1)}$$

(c) $\mathbf{G}(x,y,z)=0\,\mathbf{u}_x+y\,\mathbf{u}_y-z\,\mathbf{u}_z$ through the paraboloidal surface $S_1=\{(x,y,z)\,|\,0\le y=x^2+z^2\le 1\}$ and the disk $S_2=\{(x,y,z)\,|\,y=1,\ x^2+z^2\le 1\}$ corresponding to the cap of the paraboloid (Fig. P2.23(c)). Note that the surface S_1 is parameterized by

$$\mathbf{s}(x,z)=x\,\mathbf{u}_x+(x^2+z^2)\mathbf{u}_y+z\,\mathbf{u}_z \quad \text{(P2.23c.1)}$$

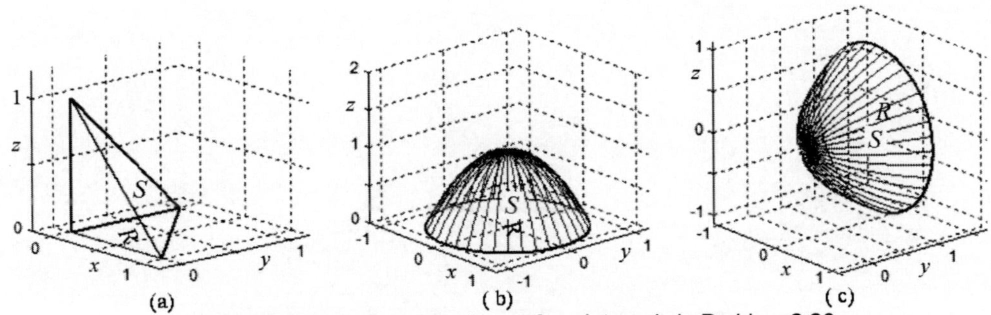

Figure P2.23 Surfaces for the surface integrals in Problem 2.23

(d) $\mathbf{G}(x,y,z)=y\,\mathbf{u}_x+x\,\mathbf{u}_y+z\,\mathbf{u}_z$ through the partial cylindrical surface S, which is parameterized by

$$\mathbf{s}(\phi,z)=\cos\phi\,\mathbf{u}_x+\sin\phi\,\mathbf{u}_y+z\mathbf{u}_z \text{ with } 0\le\phi\le\pi/2 \text{ and } 0\le z\le 2 \tag{P2.23d.1}$$

(e) $\mathbf{G}(x,y,z)=y\,\mathbf{u}_x-x\,\mathbf{u}_y+z\,\mathbf{u}_z$ through the conic surface $S=\{(x,y,z)\,|\,0\le z=\sqrt{x^2+y^2}\le 3\}$, which can be parameterized by

$$\mathbf{s}(\rho,\phi)=\rho\cos\phi\,\mathbf{u}_x+\rho\sin\phi\,\mathbf{u}_y+\rho\mathbf{u}_z \text{ with } 0\le\phi\le 2\pi \text{ and } 0\le\rho\le 3 \tag{P2.23e.1}$$

(f) $\mathbf{G}(x,y,z)=y\,\mathbf{u}_x-x\,\mathbf{u}_y+z\,\mathbf{u}_z$ through the spherical surface $S=\{(x,y,z)\,|\,x^2+y^2+z^2=1\}$, which can be parameterized by

$$\mathbf{s}(\theta,\phi)=\sin\theta\cos\phi\,\mathbf{u}_x+\sin\theta\sin\phi\,\mathbf{u}_y+\cos\theta\,\mathbf{u}_z \text{ with } 0\le\theta\le\pi \text{ and } 0\le\phi\le 2\pi \tag{P2.23f.1}$$

2.24 Green's Theorem and Stokes' Theorem

To show that Green's theorem is just a special case of Stokes' theorem, plug $F_x=P$, $F_y=Q$, and $F_z=0$ into Eq. (2.9.1b) (Stokes' theorem) to get Eq. (2.7.1) (Green's theorem).

$$\oint_C\{F_x\,dx+F_y\,dy+F_z\,dz\}\overset{(2.9.1b)}{=}\iint_R\left\{\left(\frac{\partial F_z}{\partial y}-\frac{\partial F_y}{\partial z}\right)dy\,dz+\left(\frac{\partial F_x}{\partial z}-\frac{\partial F_z}{\partial x}\right)dz\,dx+\left(\frac{\partial F_y}{\partial x}-\frac{\partial F_x}{\partial y}\right)dx\,dy\right\} \tag{P2.24.1}$$

2.25 Stokes' Theorem

Check the validity of Stokes' theorem (2.9.1) for the following vector fields with given closed curves enclosing some surfaces. If it does not seem to hold, explain why.

$$\oint_C\mathbf{F}\cdot d\mathbf{l}=\oint_C\mathbf{F}\cdot\mathbf{u}_T\,dl\overset{(2.9.1a)}{=}\iint_S(\nabla\times\mathbf{F})\cdot d\mathbf{s} \text{ with } d\mathbf{s}=\mathbf{u}_n\,dS \tag{P2.25.1}$$

(a) $\mathbf{F}(x,y,z)=z^2\,\mathbf{u}_x+y^2\,\mathbf{u}_y+x\,\mathbf{u}_z$ (P2.25a.1)
where $C=C_1$-C_2-C_3 is the piecewise linear path enclosing the triangle S with vertices at (1, 0, 0), (0, 1, 0), and (0, 0, 1) (Fig. P2.25(a)), which is the part of the plane $z=1-x-y$ in the first octant. Note that the surface S is described by $f(x,y,z)=x+y+z-1=0$.

(b) $\mathbf{F}(x,y,z)=-y^2\mathbf{u}_x+x\,\mathbf{u}_y+z^2\mathbf{u}_z$ (P2.25b.1)
where C is the circle of radius 2 in the xy-plane, which encloses the surface S_1 of the cone (Fig. P2.25(b)) described by

$$f(x,y,z)=\sqrt{x^2+y^2}+z-2=\rho+z-2=0 \text{ for } 0\le z\le 2 \tag{P2.25b.2}$$

Note that the conical surface S_1 can be represented by

$$\mathbf{s}(\rho,z)=\rho\cos\phi\,\mathbf{u}_x+\rho\sin\phi\,\mathbf{u}_y+(2-\rho)\mathbf{u}_z \tag{P2.25b.3}$$

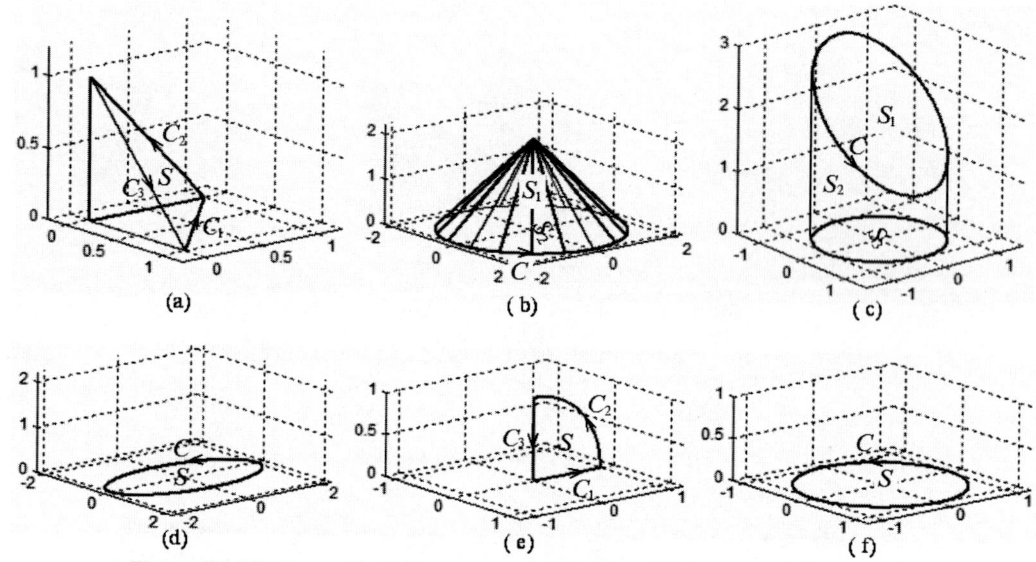

Figure P2.25 Curves/Surfaces for the line/surface integrals in Problem 2.25

How about the surface integral over the circular surface S_2 on the RHS of Eq. (2.9.1a)?

(c) $\mathbf{F}(x,y,z)=z\,\mathbf{u}_x+x\,\mathbf{u}_y+y\,\mathbf{u}_z$ (P2.25c.1)

where C is the curve enclosing the (elliptical) surface S_1, which is the intersection of the plane $f(x)=x+z-2=0$ and the cylinder $x^2+y^2=1$ (Fig. P2.25(c)).

(d) $\mathbf{F}(x,y,z)=x^2\mathbf{u}_x+2x\,\mathbf{u}_y+z^2\mathbf{u}_z$ (P2.25d.1)

where C is the ellipse in the xy-plane, which is described by $x^2+(y/2)^2=1$ (Fig. P2.25(d)).

(e) $\mathbf{F}(x,y,z)=4(x^2+z^2)\,\mathbf{u}_x-4y\,\mathbf{u}_y+4(x-y)\mathbf{u}_z$ (P2.25e.1)

where $C=C_1$-C_2-C_3 is the curve enclosing the quarter of the unit disk S centered at the origin in the first quadrant of the yz-plane (Fig. P2.25(e)).

(f) $\mathbf{F}(x,y,z)=\dfrac{-y}{x^2+y^2}\mathbf{u}_x+\dfrac{x}{x^2+y^2}\mathbf{u}_y$ (P2.25f.1)

where C is the unit circle enclosing the unit disk S centered at the origin in the xy-plane (Fig. P2.25(f)).

2.26 Triple Integral and Order of Integration

With the order of integration as in Eq. (2.10.2c) or (2.10.2f), evaluate the triple integral to find the volume of the triangular pyramid in Fig. 2.38.

2.27 Triple Integrals

Evaluate the following triple integrals.

(a1) $V=\iiint_{R_3} 1\,dV$ over the solid shown in Fig. P2.27(a) to find its volume.

<Hint>

```
>>syms x y z
>>V=int(int(int(1,z,0,sqrt(1-x^2)),y,x,3-2*x),x,0,1)
```

(a2) $m=\iiint_{R_3}\delta(x,y,z)dV$ with density function $\delta(x,y,z)=80z$ over the solid shown in Fig. P2.27(a) to find its mass.

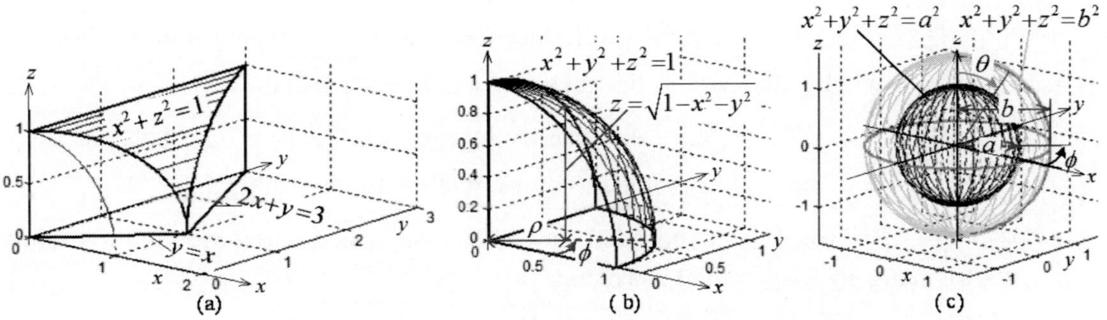

Figure P2.27 Solids for the triple integrals in Problem 2.27

(a3) $M_{xy}=\iiint_{R_3} z\delta(x,y,z)dV$ with density function $\delta(x,y,z)=80z$ over the solid shown in Fig. P2.27(a) to find its moment about the xy-plane.

(a4) $M_{xz}=\iiint_{R_3} y\,\delta(x,y,z)dV$ with density function $\delta(x,y,z)=80z$ over the solid shown in Fig. P2.27(a) to find its moment about the xz-plane.

(a5) $M_{yz}=\iiint_{R_3} x\delta(x,y,z)dV$ with density function $\delta(x,y,z)=80z$ over the solid shown in Fig. P2.27(a) to find its moment about the yz-plane.

(a6) $I_x=\iiint_{R_3} (y^2+z^2)\delta(x,y,z)\,dV$ with density function $\delta(x,y,z)=80z$ over the solid shown in Fig. P2.27(a) to find its moment of inertia about the x-axis.

(a7) $I_y=\iiint_{R_3} (x^2+z^2)\delta(x,y,z)\,dV$ with density function $\delta(x,y,z)=80z$ over the solid shown in Fig. P2.27(a) to find its moment of inertia about the y-axis.

(a8) $I_z=\iiint_{R_3} (x^2+y^2)\delta(x,y,z)\,dV$ with density function $\delta(x,y,z)=80z$ over the solid shown in Fig. P2.27(a) to find its moment of inertia about the z-axis.

(b1) $V=\iiint_{R_3} 1\,dV$ over the solid shown in Fig. P2.27(b) to find its volume.

<Hint>

```
>>syms x y z r ph
>>V=int(int(int(1,z,0,sqrt(1-x^2-y^2)),y,0,sqrt(1-x^2)),x,0,1)
>>V=int(int(int(r,z,0,sqrt(1-r^2)),r,0,1),ph,0,pi/2)
```

(b2) $m=\iiint_{R_3}\delta(x,y,z)dV$ with density function $\delta(x,y,z)=480\rho=480\sqrt{x^2+y^2}$ over the solid shown in Fig. P2.27(b) to find its mass.

(b3) $M_{xy}=\iiint_{R_3} z\delta(x,y,z)dV$ with density function $\delta(x,y,z)=480\rho=480\sqrt{x^2+y^2}$ over the solid shown in Fig. P2.27(b) to find its moment about the xy-plane.

(b4) $M_{xz}=\iiint_{R_3} y\,\delta(x,y,z)dV$ with density function $\delta(x,y,z)=480\rho=480\sqrt{x^2+y^2}$ over the solid shown in Fig. P2.27(b) to find its moment about the xz-plane.

(b5) $M_{yz}=\iiint_{R_3} x\delta(x,y,z)dV$ with density function $\delta(x,y,z)=480\rho=480\sqrt{x^2+y^2}$ over the solid shown in Fig. P2.27(b) to find its moment about the yz-plane.

(b6) $I_x=\iiint_{R_3} (y^2+z^2)\delta(x,y,z)\,dV$ with density function $\delta(x,y,z)=480\rho=480\sqrt{x^2+y^2}$ over the solid shown in Fig. P2.27(b) to find its moment of inertia about the x-axis.

(b7) $I_y = \iiint_{R_3}(x^2+z^2)\delta(x,y,z)\,dV$ with density function $\delta(x,y,z)=480\rho=480\sqrt{x^2+y^2}$ over the solid shown in Fig. P2.27(b) to find its moment of inertia about the y-axis.

(b8) $I_z = \iiint_{R_3}(x^2+y^2)\delta(x,y,z)\,dV$ with density function $\delta(x,y,z)=480\rho=480\sqrt{x^2+y^2}$ over the solid shown in Fig. P2.27(b) to find its moment of inertia about the z-axis.

(c1) $V=\iiint_{R_3} 1\,dV$ over the solid bounded between the two spheres each of radius a=1 and b=4 as shown in Fig. P2.27(c) to find its volume.

<Hint>

```
>>syms x y z r ph th
>>V=int(int(int(r^2*sin(th),r,1,4),th,0,pi),ph,0,2*pi)
```

(c2) $M_{xy}=\iiint_{R_3} z\delta(x,y,z)dV$ with density function $\delta(x,y,z)=2$ over the solid shown in Fig. P2.27(c) to find its moment about the xy-plane.

(c3) $M_{xz}=\iiint_{R_3} y\delta(x,y,z)dV$ with density function $\delta(x,y,z)=2$ over the solid shown in Fig. P2.27(c) to find its moment about the xz-plane.

(c4) $M_{yz}=\iiint_{R_3} x\delta(x,y,z)dV$ with density function $\delta(x,y,z)=2$ over the solid shown in Fig. P2.27(c) to find its moment about the yz-plane.

(c5) $I_x = \iiint_{R_3}(y^2+z^2)\delta(x,y,z)\,dV$ with density function $\delta(x,y,z)=2$ over the solid shown in Fig. P2.27(c) to find its moment of inertia about the x-axis.

(c6) $I_y = \iiint_{R_3}(x^2+z^2)\delta(x,y,z)\,dV$ with density function $\delta(x,y,z)=2$ over the solid shown in Fig. P2.27(c) to find its moment of inertia about the y-axis.

(c7) $I_z = \iiint_{R_3}(x^2+y^2)\delta(x,y,z)\,dV$ with density function $\delta(x,y,z)=2$ over the solid shown in Fig. P2.27(c) to find its moment of inertia about the z-axis.

2.28 Divergence Theorem

Check the validity of the divergence theorem (2.11.1) for the following vector fields where

$$\oiint_S \mathbf{G}\cdot d\mathbf{s} = \oiint_S \mathbf{G}\cdot\mathbf{u}_n\,dS = \iiint_R \mathrm{div}\,\mathbf{G}\,dV \qquad (2.11.1a)$$

$$\iint_S G_x\,dydz+G_y\,dzdx+G_z\,dxdy = \iiint_R\left(\frac{\partial G_x}{\partial x}+\frac{\partial G_y}{\partial y}+\frac{\partial G_z}{\partial z}\right)dV \qquad (2.11.1b)$$

(a) $\mathbf{G}(x,y,z)=x\,\mathbf{u}_x+y\,\mathbf{u}_y+z\,\mathbf{u}_z$ (P2.28a.1)

where R is the region bounded by the parabolic surface $S_1=\{0\le z=x^2+y^2\le 1\}$ and the partial plane $S_2=\{x^2+y^2\le 1, z=1\}$.

(b) $\mathbf{G}(x,y,z)=(x-y)\mathbf{u}_x+(x+y)\mathbf{u}_y+z\,\mathbf{u}_z$ P2.28b.1)

where R is the region bounded by the cylindrical surface $S_1=\{x^2+y^2=1\}$, the top lateral surface $S_2=\{z=1\}$, and the bottom surface $S_3=\{z=-1\}$.

(c) $\mathbf{G}(x,y,z)=2yz\,\mathbf{u}_y$ (P2.28c.1)

where R is the region bounded by the hemispherical surface $S_1=\{x^2+y^2+z^2=1\}$ and the bottom surface $S_2=\{z=0\}$.

2.29 Divergence Theorem

(a) Use the divergence theorem (2.11.1) with $\mathbf{G}(x,y,z)=x\,\mathbf{u}_x+y\,\mathbf{u}_y+z\,\mathbf{u}_z$ to show that the volume of a body with boundary surface S can be found using the surface integral as

$$V=\frac{1}{3}\oiint_S(x\mathbf{u}_x+y\mathbf{u}_y+z\mathbf{u}_z)\bullet d\mathbf{s}=\frac{1}{3}\oiint_S r\mathbf{u}_r\bullet\mathbf{u}_n dS \text{ with } r=\sqrt{x^2+y^2+z^2} \quad \text{(P2.29a.1)}$$

(b) Use Eq. (P2.29a.1) to find the volume of a sphere of radius a from its surface area $A(S)=4\pi a^2$.

(c) Use Eq. (P2.29a.1) to find the volume of a cylinder with radius a and height h where its lateral surface area (of the side) is $A(S_1)=2\pi ha$ and its base surface area (of the top or bottom) is $A(S_2)=\pi a^2$.

2.30 Divergence Theorem

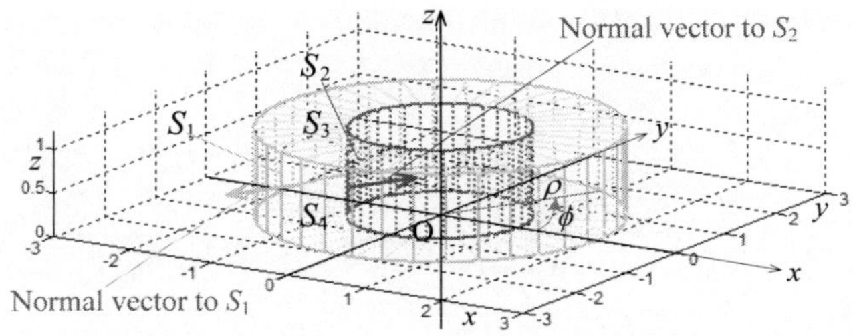

Figure P2.30 Surface S enclosing a wheel-shaped region for Problem 2.30

Consider a vector field given by

$$\mathbf{G}(x,y,z)=\frac{x\mathbf{u}_x+y\mathbf{u}_y}{(x^2+y^2)^{3/2}}=\frac{1}{\rho^2}\mathbf{u}_\rho \text{ with } \mathbf{u}_\rho=\frac{x\mathbf{u}_x+y\mathbf{u}_y}{\sqrt{x^2+y^2}} \quad \text{(P2.30.1)}$$

and the closed surface S that consists of four subspaces S_1, S_2, S_3, and S_4 as depicted in Fig. P2.30 where S_1/S_2 denote the cylindrical surfaces each of radius 2/1, respectively, and S_3/S_4 denote the top/bottom annular rings of the same size, respectively. Find the flux of **G** passing through the closed surface S in the following two ways.

(a) Noting that the outward unit normal vectors to the cylindrical surfaces S_1 and S_2 are $\mathbf{u}_\rho$ and $-\mathbf{u}_\rho$, respectively, use Eq. (E2.28.3) to find the flux of **G** passing through S.

(b) Use the divergence theorem (2.11.1) to find the flux of **G** passing through S enclosing the wheel-shaped region R.

2.31 Stokes' Theorem, Divergence Theorem, and Vector Identities for Maxwell's Equations

Coulomb's law states that the electric field **E** at a target point $\mathbf{r}=(x,y,z)$ induced by a volume of stationary charge density q[Coulomb/m^3] at $\mathbf{r}'=(x',y',z')$ is

$$\mathbf{E}(\mathbf{r})=\frac{1}{4\pi\varepsilon}\iiint_R\frac{q(\mathbf{r}')(\mathbf{r}-\mathbf{r}')}{r^3}dV=\frac{1}{4\pi\varepsilon}\iiint_R\frac{q(\mathbf{r}')\mathbf{u}_{r-r'}}{r^2}dV \text{ with } r=\|\mathbf{r}-\mathbf{r}'\|,\ \varepsilon\text{: permittivity} \quad \text{(P2.31.1)}$$

where $\mathbf{u}_{r\text{-}r'}$ is the unit vector having the same direction with $(\mathbf{r}-\mathbf{r}')$, and $q(\mathbf{r}')$ is the charge density at $\mathbf{r}'$. On the other hand, the *Biot–Savart law* mentioned in Example 1.8 gives the magnetic field **B** at a target point $\mathbf{r}=(x,y,z)$ induced by a volume of many small current elements $\{Id\mathbf{l}$ at $\mathbf{r}'=(x',y',z')\}$ as

$$\mathbf{B}(\mathbf{r})=\mu\mathbf{H}(\mathbf{r})=\frac{\mu}{4\pi}\int\frac{\mathbf{J}(\mathbf{r}')\times(\mathbf{r}-\mathbf{r}')}{r^3}Adl=\frac{\mu}{4\pi}\iiint_R\frac{\mathbf{J}(\mathbf{r}')\times\mathbf{u}_{r-r'}}{r^2}dV \text{ with } r=\|\mathbf{r}-\mathbf{r}'\|,\ \mu\text{: permeability} \quad \text{(P2.31.2)}$$

where $\mathbf{J}(\mathbf{r}')$ is the current density vector (at $\mathbf{r}'$) with the magnitude of I/A[Ampere/m^2] (A: the cross-sectional area of conductor carrying current I) and the same direction with $d\mathbf{l}$. Fill in the following parentheses or boxes with appropriate words/equations, or related equation numbers.

(a) Gauss' law, saying that the electric field through any closed surface S is proportional to the charge Q[Coulomb] in the volume enclosed by S, can be written as

$$\varepsilon \oiint_S \mathbf{E}\bullet d\mathbf{s} = \varepsilon \oiint_S \mathbf{E}\bullet \mathbf{u}_n dS = Q = \iiint_R q\, dV \;(\mathbf{E}\text{: electric field, } q\text{:charge density}) \quad \text{(P2.31.3)}$$

Applying the (   ) theorem on the LHS, we can rewrite this equation as

$$\varepsilon \iiint_R \operatorname{div}\mathbf{E}\, dV = \iiint_R q\, dV \;\rightarrow\; \operatorname{div}\mathbf{E} = \nabla\bullet\mathbf{E} = \boxed{} \quad \text{(P2.31.4)}$$

(b) Gauss' law (in differential form) can also be derived by taking the divergence of Coulomb's law (P2.31.1):

$$\nabla\bullet\mathbf{E}(\mathbf{r}) \overset{\text{(P2.31.1)}}{=} \frac{1}{4\pi\varepsilon}\iiint_R \left(\nabla\bullet\frac{\mathbf{u}_{r-r'}}{r^2}\right) q(\mathbf{r}')\, dV$$
$$\overset{\boxed{}}{=} \frac{1}{4\pi\varepsilon}\iiint_R 4\pi\delta^3(\mathbf{r}-\mathbf{r}')q(\mathbf{r}')\,dV \overset{\text{(2.11.11)}}{=} \frac{q(\mathbf{r})}{\varepsilon} \quad \text{(P2.31.5)}$$

(c) Taking the curl of the electric field $\mathbf{E}$ (P2.31.1) yields

$$\nabla\times\mathbf{E}(\mathbf{r}) \overset{\text{(P2.31.1)}}{=} \frac{1}{4\pi\varepsilon}\iiint_R \left(\nabla\times\frac{\mathbf{u}_{r-r'}}{r^2}\right) q(\mathbf{r}')\,dV \underset{\text{(E2.9.3)}}{\overset{\text{(2.4.5c)}}{=}} \boxed{} \;\forall\, \mathbf{r} \quad \text{(P2.31.6)}$$

where we have applied $\nabla\times(\mathbf{u}_r/r^2)=\mathbf{0}$ (Eq. (E2.9.3)). This implies that the curl (infinitesimal rotation) of an electrostatic field (associated with static electric charges) is zero.

If we take the surface integral of Eq. (P2.31.6) over the surface S enclosed by a closed curve C and apply the (   ) theorem, we can obtain

$$\oint_C \mathbf{E}\bullet d\mathbf{l} \overset{\boxed{}}{=} \iint_S (\nabla\times\mathbf{E})\bullet d\mathbf{s} \overset{\text{(P2.31.6)}}{=} 0 \text{ for any closed curve (loop) } C \quad \text{(P2.31.7)}$$

Comparing Eq. (P2.31.6) to $\nabla\times\nabla a=0$ (Eq. (2.4.22) in Table 2.1 in Section 2.4) gives us a suggestion that the electric field $\mathbf{E}$ can be the gradient of a scalar function, say V:

$$\mathbf{E} = -\nabla V \quad \text{(P2.31.8)}$$

where the scalar function V is called the *electric potential.*

(d) Taking the divergence of the magnetic field $\mathbf{B}$ (Eq. (P2.31.2)) yields (see Table 2.1)

$$\nabla\bullet\mathbf{B} \overset{\text{(P2.31.2)}}{=} \frac{\mu}{4\pi}\iiint_R \nabla\bullet\frac{\mathbf{J}(\mathbf{r}')\times\mathbf{u}_{r-r'}}{r^2}\,dV$$
$$\overset{\boxed{}}{=} \frac{\mu}{4\pi}\iiint_R \left\{\frac{\mathbf{u}_{r-r'}}{r^2}\bullet(\nabla\times\mathbf{J}) - \mathbf{J}\bullet\left(\nabla\times\frac{\mathbf{u}_{r-r'}}{r^2}\right)\right\} dV \underset{\text{(E2.9.3)}}{\overset{\nabla\times\mathbf{J}=0}{=}} \boxed{} \quad \text{(P2.31.9)}$$

where we have applied $\nabla\times\mathbf{J}=\mathbf{0}$ (because the current density vector $\mathbf{J}$ does not depend on the position $\mathbf{r}$ of the target point) and $\nabla\times(\mathbf{u}_r/r^2)=\mathbf{0}$ (Eq. (E2.9.3)). In contrast with Eq. (P2.31.5), this implies that there are no magnetic sources (like monopoles producing magnetic flux) that can be compared to electric charges (producing electric flux).

Comparing Eq. (P2.31.9) to $\nabla\bullet(\nabla\times\mathbf{A})=0$ (Eq. (2.4.21) in Table 2.1) gives us a suggestion that the magnetic field $\mathbf{B}$ can be the curl of a vector function, say, $\mathbf{A}$:

$$\mathbf{B} = \nabla \times \mathbf{A} \tag{P2.31.10}$$

where the vector function **A** is called the *magnetic vector potential*.

(e) Taking the curl of the magnetostatic field **B** (Eq. (P2.31.2)) yields (see Table 2.1 in Section 2.4)

$$\nabla \times \mathbf{B}(\mathbf{r}) \overset{\text{(P2.31.2)}}{=} \frac{\mu}{4\pi}\iiint_R \nabla \times \frac{\mathbf{J}(\mathbf{r}')\times \mathbf{u}_{r-r'}}{r^2}\,dV$$

$$\overset{\boxed{\quad\quad}}{=} \frac{\mu}{4\pi}\iiint_R \left\{ \mathbf{J}(\mathbf{r}')\left(\nabla\bullet\frac{\mathbf{u}_{r-r'}}{r^2}\right) - \frac{\mathbf{u}_{r-r'}}{r^2}\nabla\bullet\mathbf{J}(\mathbf{r}') + \left(\frac{\mathbf{u}_{r-r'}}{r^2}\bullet\nabla\right)\mathbf{J}(\mathbf{r}') - (\mathbf{J}(\mathbf{r}')\bullet\nabla)\frac{\mathbf{u}_{r-r'}}{r^2} \right\} dV$$

$$; \qquad \nabla \times \mathbf{B}(\mathbf{r}) \overset{(2.11.8)}{=} \frac{\mu}{4\pi}\iiint_R \{\mathbf{J}(\mathbf{r}')4\pi\delta^3(\mathbf{r}-\mathbf{r}')\}\,dV \overset{\boxed{\quad\quad}}{=} \mu\mathbf{J}(\mathbf{r}) \tag{P2.31.11}$$

where two terms involving derivatives of $\mathbf{J}(\mathbf{r}')$ (not dependent on **r**) are zero and another term is also zero (for some reason that cannot be explained in this book) so that all the three terms have been dropped out as shown by the three arrows. This equation is called Ampere's law in differential form. Taking the surface integral of both sides of Eq. (P2.31.11) and using Stokes' theorem yields Ampere's law in integral form:

$$\oint_C \mathbf{B}\bullet d\mathbf{l} \overset{\boxed{\quad\quad}}{=} \iint_S (\nabla\times\mathbf{B})\bullet d\mathbf{s} \overset{\text{(P2.31.11)}}{=} \iint_S \mu\mathbf{J}(\mathbf{r})\bullet d\mathbf{s} = \mu I_{\text{enc}} \tag{P2.31.12}$$

where I_{enc} is the total current passing through the surface enclosed by a closed curve C.

(f) Faraday's law, saying that the electromotive force equals the rate of change of the magnetic flux Φ linking a circuit C, can be written in integral form as

$$\oint_C \mathbf{E}\bullet d\mathbf{l} = -\frac{\partial\Phi}{\partial t} = -\frac{\partial}{\partial t}\iint_S \mathbf{B}\bullet d\mathbf{s} = \iint_S\left(-\frac{\partial}{\partial t}\mathbf{B}\right)\bullet d\mathbf{s} \quad (S\text{: surface enclosed by } C) \tag{P2.31.13}$$

Applying the (   ) theorem on the LHS, we can rewrite this equation as

$$\oint_C \mathbf{E}\bullet d\mathbf{l} = \iint_S (\nabla\times\mathbf{E})\bullet d\mathbf{s} \tag{P2.31.14}$$

Matching the RHSs of these two equations yields *Faraday's law* in differential form:

$$\text{curl}(\mathbf{E}) = \nabla\times\mathbf{E} = -\frac{\partial}{\partial t}\mathbf{B} \tag{P2.31.15}$$

(g) Noting that the current density **J** is the sum of free (conduction) current density $\mathbf{J}_\text{f}=\sigma\mathbf{E}$ (σ: conductivity) (due to the flow of electrons or positive holes) and displacement current density $\mathbf{J}_\text{d}=\varepsilon\partial\mathbf{E}/\partial t$ (ε: permittivity) (due to the time variation of electric field **E**), we can rewrite Eq. (P2.31.11) as

$$\nabla\times\mathbf{B} \overset{\text{(P2.31.11)}}{=} \mu\mathbf{J} = \mu(\mathbf{J}_\text{f}+\mathbf{J}_\text{d}) = \mu\left(\sigma\mathbf{E}+\varepsilon\frac{\partial\mathbf{E}}{\partial t}\right) \tag{P2.31.16}$$

Taking the curl of both sides of this equation and using Eqs. (2.4.20), (P2.31.9), and (P2.31.15) yields

$$\nabla\times(\nabla\times\mathbf{B}) \overset{\boxed{\quad\quad}}{=} \nabla(\nabla\bullet\mathbf{B}) - \nabla^2\mathbf{B} \overset{\text{(P2.31.9)}}{=} -\nabla^2\mathbf{B} \tag{P2.31.17a}$$

$$\mu\nabla\times\left(\sigma\mathbf{E}+\varepsilon\frac{d\mathbf{E}}{dt}\right) = \mu\sigma\nabla\times\mathbf{E} + \mu\varepsilon\frac{\partial}{\partial t}(\nabla\times\mathbf{E}) \overset{\text{(P2.31.15)}}{=} -\mu\sigma\frac{\partial\mathbf{B}}{\partial t} - \mu\varepsilon\frac{\partial^2\mathbf{B}}{\partial t^2} \tag{P2.31.17b}$$

Equating the RHSs of these two equations yields the Maxwell's equation for the magnetic field **B**:

$$\nabla^2\mathbf{B} = \mu\sigma\frac{\partial \mathbf{B}}{\partial t} + \mu\varepsilon\frac{\partial^2 \mathbf{B}}{\partial t^2} \tag{P2.31.18}$$

(h) Similarly, taking the curl of both sides of Eq. (P2.31.15) and using Eqs. (2.4.20) and (P2.31.16) yields

$$\nabla\times(\nabla\times\mathbf{E}) \overset{\boxed{}}{=} \nabla(\nabla\cdot\mathbf{E}) - \nabla^2\mathbf{E} \overset{\nabla\cdot\mathbf{E}=q/\varepsilon=0}{=} -\nabla^2\mathbf{E} \tag{P2.31.19a}$$

$$\nabla\times\left(-\frac{\partial \mathbf{B}}{\partial t}\right) = -\frac{\partial}{\partial t}(\nabla\times\mathbf{B}) \overset{(P2.31.16)}{=} -\mu\sigma\frac{\partial \mathbf{E}}{\partial t} - \mu\varepsilon\frac{\partial^2 \mathbf{E}}{\partial t^2} \tag{P2.31.19b}$$

where the space charge density q has been assumed to be zero. Equating the RHSs of these two equations yields the Maxwell's equation for the electric field $\mathbf{E}$:

$$\nabla^2\mathbf{E} = \mu\sigma\frac{\partial \mathbf{E}}{\partial t} + \mu\varepsilon\frac{\partial^2 \mathbf{E}}{\partial t^2} \tag{P2.31.20}$$

(i) We can also use the (   ) theorem to write the current I leaving a volume V as the surface integral of the current density J over the surface S enclosing V:

$$I = \iiint_R \nabla\cdot\mathbf{J}\,dV \overset{\boxed{}}{=} \oiint_S \mathbf{J}\cdot d\mathbf{s} \;\; (\mathbf{J}\text{: current density vector}[\text{A/m}^2]) \tag{P2.31.21}$$

According to the charge conservation law, this must be equal to the reduction rate of the charge inside V:

$$\iiint_R \nabla\cdot\mathbf{J}\,dV \overset{\text{Charge conservation}}{=} -\frac{d}{dt}\iiint_R q\;dV \;\; (q\text{: charge density}[\text{C/m}^3]) \tag{P2.31.22}$$

Equating the RHSs of these two equations yields

$$\nabla\cdot\mathbf{J} = -\frac{dq}{dt} \tag{P2.31.23}$$

This is called the *continuity equation*.

CHAPTER 3

Ordinary Differential Equations

CHAPTER OUTLINE

An *ordinary differential equation* (*ODE*) is an equation consisting of one or more derivatives of a dependent variable with respect to only one independent variable. In contrast, a *partial differential equation* (*PDE*) is an equation consisting of partial derivatives of an unknown function of more than one variable. The *order* of a differential equation (DE) is the order of the highest derivative. A DE is said to be *linear* or *nonlinear* depending on whether it is linear in the unknown variables/functions or not. A linear DE is said to be *homogeneous* or *nonhomogenous* depending on whether it consists of only the sum of terms involving the unknown variables/functions or not. For example,

$$y' + 2y = 0 \quad : \text{homogeneous, first-order linear ODE}$$

$$xy'' + 2y' + y^2 = e^{-x} : \text{nonhomogeneous, second-order nonlinear ODE}$$

$$\frac{\partial^2 u}{\partial x^2} + \frac{\partial^2 u}{\partial y^2} = f(x, y) : \text{nonhomogeneous, second-order linear PDE}$$

where

$$y' = \frac{dy}{dx},\ y'' = \frac{d^2 y}{dx^2},\ y^{(n)} = \frac{d^n y}{dx^n}$$

Also, a DE is said to be in *implicit* or *explicit* form if it is written as

$$F(x, y', y'', \cdots, y^{(n)}) = 0 \quad \text{or} \quad y^{(n)} = F(x, y', y'', \cdots, y^{(n-1)}),$$

respectively.

3.1 FIRST-ORDER DIFFERENTIAL EQUATIONS

3.1.1 Separable Equations

A first-order DE is said to be *separable* if can be written as

$$g(y)\,dy = f(x)\,dx \tag{3.1.1}$$

where the variables x and y are separated so that each of them appears on each side of the DE. The solution to such a DE can be found by integrating both of its sides as

$$\int g(y)\,dy = \int f(x)\,dx + c \tag{3.1.2}$$

where c is the constant of integration.

Some first-order DEs can be made separable after a change of variables, although they are not separable as they are. One such class of DE is of the form

$$y' = g\left(\frac{y}{x}\right) \tag{3.1.3}$$

By substitution of variables

$$y{=}ux \ \text{ with } \ y' = u + u'x \ \left(\text{i.e., } \frac{dy}{dx} \overset{\text{Product rule}}{=} u\frac{dx}{dx} + \frac{du}{dx}x\right), \tag{3.1.4}$$

this DE can be rewritten as

$$u + \frac{du}{dx}x = g(u); \ \ \frac{du}{dx}x = g(u) - u; \ \ \frac{dx}{x} = \frac{du}{g(u)-u} \tag{3.1.5}$$

where the variables x and u are separated so that each of them appears on each side of the DE. The solution to such a DE can be found by integrating both of its sides and substituting $u{=}y/x$.

(**Example 3.1**) Separable DE

Solve

$$y' + 2xy = 0 \ \text{ with } \ y(0) = 2 \tag{E3.1.1}$$

By separating variables, we rewrite this DE as

$$\frac{dy}{dx} = -2xy; \ \ \frac{dy}{y} = -2x\,dx \tag{E3.1.2}$$

Integrating both sides and using the initial condition $y(0)$=2 yields

$$\ln|y| = -x^2 + c \ ; \ \ |y| = e^{-x^2+c} = c_1 e^{-x^2} \overset{y(0)=2}{\rightarrow} |y| = 2e^{-x^2} \tag{E3.1.3}$$

We can use the MATLAB built-in function '`dsolve()`' with t in place of the independent variable x or with x (other than t by default) specified as an independent variable:

```
>>yt=dsolve('Dy+2*t*y=0','y(0)=2'), yx=dsolve('Dy+2*x*y=0','y(0)=2','x')
```

```
%em03e01.m
% To solve a DE (Eq.(E3.1.1)) y'+2xy=0 with y(0)=2
clear, clf
x0=0; xf=2; xspan=[x0 xf]; y0=2; % Solution interval and Initial value
% Plot of direction field for dydx=4-2y
xgrid=x0:(xf-x0)/20:xf;  ygrid=0:0.1:2; % Grids of x and y
[X,Y]=meshgrid(xgrid,ygrid); % Meshgrids of (x,y)
pX=ones(size(X)); pY=-2*X.*Y.*pX; % dy=(-2xy)dx
quiver(X,Y,pX,pY,3), hold on % plot dy vs. dx (directional/slope field)
% Numerical solution to y'=-2xy with y(0)=y0
dy=@(x,y)-2*x*y; % DE to solve
[xx,yy]=ode45(dy,xspan,y0); % Numerical solution
% Animated show of the numerical solution
for n=1:length(xx),  plot(xx(n),yy(n),'rx'), pause(0.1);  end
% Symbolic solution just like an analytic solution
yt=dsolve('Dy+2*t*y=0','y(0)=2'), yx=dsolve('Dy+2*x*y=0','y(0)=2','x')
% Conversion of the symbolic solution to numeric data
t=xx; yt_=eval(yt);
% To compute the discrepancy between the numerical/analytical solutions
discrepancy = norm(yy-yt_)/norm(yt_)
plot(t,yt_,'g') % To plot the analytical solution
```

to get the equivalent result:

```
yt = 2/exp(t^2),     yx = 2/exp(x^2)
```

On the other hand, we can use the function handle operator @ or the '`inline()`' function to define the DE (E3.1.1) (having the independent/dependent variable as its first/second input argument)

```
>>dy=@(x,y)-2*x*y; % Function handle for the DE to solve
>>dy=inline('-2*x*y','x','y'); % Inline function for the DE to solve
```

and then use '`ode45()`', one of the several MATLAB built-in numerical DE solvers, with the interval xspan=(0, 2) and the initial value y(0)=2 to get the numerical solution:

```
>>xspan=[0 2]; y0=2; [xx,yy]=ode45(dy,xspan,y0);
```

To compare the analytical solution with the numerical solution and plot it, we use the '`eval()`' command to evaluate the symbolic solution yst with t=xx or ysx with x=xx where xx is the first output argument of '`ode45()`':

```
>>t=xx; yt_=eval(yt); % x=xx; yx_=eval(yx); Alternatovely
>>plot(xx,yy, t,yt_,'g:')
```

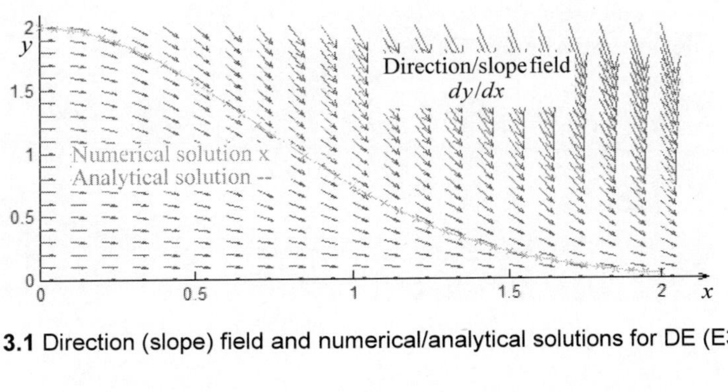

Figure 3.1 Direction (slope) field and numerical/analytical solutions for DE (E3.1.1)

All these MATLAB statements together with the plotting commands are listed in the above script file "em03e01.m," which has been run to yield Fig. 3.1, showing the directional field as well as the analytical/numerical solutions.

(Example 3.2) Separable DE

Solve

$$yy'+4x=0,\ \text{i.e.,}\ y\frac{dy}{dx}+4x=0\ \text{with}\ y(0)=2\sqrt{2} \tag{E3.2.1}$$

By separating variables, we rewrite this DE as

$$y\,dy=-4x\,dx \tag{E3.2.2}$$

Integrating both sides yields

$$\frac{1}{2}y^2=-2x^2+c;\ x^2+\frac{1}{4}y^2=c_1 \overset{y(0)=2\sqrt{2}}{\rightarrow} x^2+\frac{1}{4}y^2=2 \tag{E3.2.3}$$

We can use the MATLAB function '`dsolve()`' (with t in place of the independent variable x):

```
>>y=dsolve('y*Dy+4*t=0','y(0)=2*sqrt(2)')
```

to get

```
    y =    2^(1/2)*(4 - 2*t^2)^(1/2)
```

which means

$$y=\sqrt{2}\sqrt{4-2x^2} \tag{E3.2.4}$$

This does not differ from Eq. (E3.2.3) since squaring both sides of Eq. (E3.2.4) results in Eq. (E3.2.3).

(Example 3.3) DE Reducible to Separable Form

Solve

$$2xyy'-y^2+x^2=0\quad\text{with}\ y(1)=1 \tag{E3.3.1}$$

We divide both sides of this equation by x^2 and make a substitution of variables (Eq. (3.1.4)) to write it as

$$2\frac{y}{x}y'-\left(\frac{y}{x}\right)^2+1=0 \overset{(3.1.4)}{\rightarrow} 2u(u+u'x)-u^2+1=2uu'x+u^2+1=0 \tag{E3.3.2}$$

By separating variables, we rewrite this DE as

$$2u\frac{du}{dx}x+u^2+1=0;\ \ 2u\,x\,du+(u^2+1)dx=0;\ \ \frac{2udu}{u^2+1}=-\frac{dx}{x} \tag{E3.3.3}$$

Integrating both sides yields

$$\ln(u^2+1)=-\ln|x|+\ln c;\ u^2+1=\frac{c}{x} \overset{u=y/x}{\rightarrow} \left(\frac{y}{x}\right)^2+1=\frac{c}{x} \overset{y(1)=1}{\rightarrow} (x-1)^2+y^2=1^2 \tag{E3.3.4}$$

3.1.2 Exact Differential Equations and Integrating Factors

A first-order differential expression $M(x,y)\,dx+N(x,y)\,dy$ is called an *exact differential form* in a domain D of the xy-plane if it corresponds to the (total or exact) differential of a function $F(x,y)$, called the *potential function*, i.e.,

$$M(x,y)dx+N(x,y)dy=\frac{\partial F}{\partial x}dx+\frac{\partial F}{\partial y}dy=dF(x,y),\text{ i.e., }M(x,y)=\frac{\partial F}{\partial x}\text{ and }N(x,y)=\frac{\partial F}{\partial y}\quad(3.1.6)$$

Setting an exact differential form to zero yields an exact DE:

$$M(x,y)dx+N(x,y)dy=dF(x,y)=0\ \text{ with }M(x,y)=\frac{\partial F}{\partial x}\text{ and }N(x,y)=\frac{\partial F}{\partial y}\qquad(3.1.7)$$

Here, we have the necessary and sufficient condition for an exact DE:

$$M(x,y)=\frac{\partial F}{\partial x}\quad\text{and}\quad N(x,y)=\frac{\partial F}{\partial y}\qquad(3.1.8\text{a,b})$$

$$\Leftrightarrow\quad\frac{\partial M}{\partial y}=\frac{\partial^2 F}{\partial y\partial x}\equiv\frac{\partial^2 F}{\partial x\partial y}=\frac{\partial N}{\partial x};\quad\frac{\partial M}{\partial y}=\frac{\partial N}{\partial x}\qquad(3.1.9)$$

We can integrate both sides of this equation to obtain its solution in implicit form as

$$F(x,y)=c\qquad(3.1.10)$$

For some simple exact DEs like

$$y\,dx+x\,dy=0\qquad(3.1.11)$$

it may be easy to find $F(x,y)$ (by inspection) so that we can write the solution as

$$x\,y=c\qquad(3.1.12)$$

However, we usually need to take the following procedure:

(Step 1) Regarding y as a constant, integrate $M(x,y)=\partial F/\partial x$ w.r.t. x:

$$F(x,y)=\int M(x,y)\,dx+G(y)\ \text{ with }G(y)\text{: the "constant" of integration}\qquad(3.1.13)$$

(Step 2) Substitute Eq. (3.1.13) into Eq. (3.1.8b) to write

$$\frac{\partial}{\partial y}F(x,y)=\frac{\partial}{\partial y}\int M(x,y)\,dx+\frac{\partial}{\partial y}G(y)=N(x,y)$$

$$;\quad\frac{\partial}{\partial y}G(y)=N(x,y)-\frac{\partial}{\partial y}\int M(x,y)\,dx\qquad(3.1.14)$$

(Step 3) Integrate Eq. (3.1.14) to find $G(y)$ and substitute it into Eq. (3.1.13).

An alternative procedure can be started from integrating $N(x,y)$ w.r.t. y with x regarded as a constant.

(Example 3.4) Exact DE

Solve

$$\frac{dy}{dx} + \frac{y^2 - 2x}{2xy + 1} = 0 \text{ with } y(0) = 1 \tag{E3.4.1}$$

We can multiply both sides with $(2xy+1)dx$ to rewrite this equation as

$$Mdx + Ndy = (y^2 - 2x)dx + (2xy + 1)dy = 0 \tag{E3.4.2}$$

Noting that this DE is exact since it satisfies Eq. (3.1.9):

$$\frac{\partial M}{\partial y} = 2y = \frac{\partial N}{\partial x} \tag{E3.4.3}$$

we use Eq. (3.1.13) to write

$$F(x, y) \overset{(3.1.13)}{=} \int M(x,y)dx + G(y) = \int (y^2 - 2x)dx + G(y) = xy^2 - x^2 + G(y) \tag{E3.4.4}$$

and substitute this into Eq. (3.1.8b) to write

$$\frac{\partial}{\partial y}F(x, y) \overset{(3.1.8b)}{=} N(x, y);\ 2xy + \frac{\partial}{\partial y}G(y) = 2xy + 1\ ;\ \frac{\partial}{\partial y}G(y) = 1$$

$$;\qquad G(y) = y \tag{E3.4.5}$$

We substitute this into Eq. (E3.4.4) and then plug the result into Eq. (3.1.10) to get the solution as

$$F(x, y) \overset{(3.1.10)}{=} c \underset{(E3.4.5)}{\overset{(E3.4.4)}{\longrightarrow}} xy^2 - x^2 + y = c = 1 \tag{E3.4.6}$$

where $c=1$ has been obtained by using the initial condition $y(0)=1$.

Interested readers should run the MATLAB script "em03e04.m" to graph the analytical solution (E3.4.6) together with the numerical solution obtained using '`ode45()`'.

```
%em03e04.m
ezplot('x*y^2-x^2+y-1=0'), % plot the analytical solution (E3.4.6)
x0=0; xf=5; xspan=[x0 xf]; y0=1; % Solution interval and Initial value
dy=@(x,y)(2*x-y^2)/(2*x*y+1); % DE (E3.4.1) to solve
[xx,yy]=ode45(dy,xspan,y0); % Numerical solution
hold on, plot(xx,yy,'r:') % plot the numerical solution
Numerical_error=error_DE_sol(dy,yy,xx) % To measure the error of the sol
```

```
function Numerical_error=error_DE_sol(dx,xn,tt)
% To measure the 'error' of a given numerical solution for a DE
%Input:   dx = Differential equation as an inline function or in an M-file
%         xn = Numerical solution,    tt = Time vector
%Output: Numerical_error = Error between dxdt and dxdt computed from xn
dxdt_approximate= ...
 (xn(3:end,:)-xn(1:end-2,:))./repmat(tt(3:end)-tt(1:end-2),1,size(xn,2));
for n=1:length(tt)-2
   tmp=feval(dx,tt(n+1),xn(n+1,:)); dxdt(n,:)=tmp(:).';
end
Numerical_error=norm(dxdt-dxdt_approximate)/norm(dxdt);
```

(Example 3.5) Exact DE

Solve

$$\frac{dy}{dx}=\frac{-2xy/(x^2+y^2)^2}{1+(y^2-x^2)/(x^2+y^2)^2} \text{ with } y(-2)=1 \tag{E3.5.1}$$

This can be written as

$$Mdx+Ndy=\frac{2xy}{(x^2+y^2)^2}dx+\left(1+\frac{y^2-x^2}{(x^2+y^2)^2}\right)dy=0 \tag{E3.5.2}$$

Noting that this DE is exact since it satisfies Eq. (3.1.9):

$$\begin{aligned}\frac{\partial M}{\partial y}&=\frac{2x(x^2+y^2)^2-8xy^2(x^2+y^2)}{(x^2+y^2)^4}=\frac{2x(x^2-3y^2)}{(x^2+y^2)^3}\\&=\frac{-2x(x^2+y^2)^2-(y^2-x^2)4x(x^2+y^2)}{(x^2+y^2)^4}=\frac{\partial N}{\partial x}\end{aligned} \tag{E3.5.3}$$

we use Eq. (3.1.13) to write

$$F(x,y)\overset{(3.1.13)}{=}\int M(x,y)dx+G(y)=\int\frac{2xy}{(x^2+y^2)^2}dx+G(y)=\frac{-y}{x^2+y^2}+G(y) \tag{E3.5.4}$$

and substitute this into Eq. (3.1.8b) to write

$$\frac{\partial}{\partial y}F(x,y)\overset{(3.1.8b)}{=}N(x,y);\quad \frac{-(x^2+y^2)+2y^2}{(x^2+y^2)^2}+\frac{\partial}{\partial y}G(y)=\frac{y^2-x^2}{(x^2+y^2)^2}+1;\quad \frac{\partial}{\partial y}G(y)=1$$

$$;\qquad G(y)=y \tag{E3.5.5}$$

We substitute this into Eq. (E3.5.4) and then the result into Eq. (3.1.10) to get the solution as

$$F(x,y)\overset{(3.1.10)}{=}c\ \underset{(E3.5.5)}{\overset{(E3.5.4)}{\longrightarrow}}\ \frac{-y}{x^2+y^2}+y=c\overset{y(-2)=1}{=}0.8 \tag{E3.5.6}$$

where $c=0.8$ has been obtained by using the initial condition $y(-2)=1$.

Interested readers should to run the MATLAB script "em03e05.m" to graph the analytical solution (E3.5.6) together with the numerical solution obtained using '`ode45()`' where x0 (the left bound of the solution interval) has been set to −2 to implement the initial condition $y(-2)=1$ as if it were $y(0)=1$.

```
%em03e05.m
clear, clf
ezplot('-y/(x^2+y^2)+y=0.8') % plot the analytical solution (E3.5.6)
x0=-2; xf=2; xspan=[x0 xf]; y0=1; % Solution interval and Initial value
dy=@(x,y)-2*x*y/((x^2+y^2)^2+y^2-x^2); % DE (E3.5.1)
[xx,yy]=ode45(dy,xspan,y0); % Numerical solution
hold on, plot(xx,yy,'r:') % plot the numerical solution
Numerical_error=error_DE_sol(dy,yy,xx) % To measure the error of the sol
```

Some nonexact DE can be converted into an exact one by multiplying it with an *integrating factor* $\mu(x,y)$ as

$$\mu(x,y)M(x,y)dx+\mu(x,y)N(x,y)dy=0 \tag{3.1.15}$$

The necessary and sufficient condition for this DE to be exact is

$$\frac{\partial}{\partial y}\left(\mu(x,y)M(x,y)\right)\overset{(3.1.9)}{=}\frac{\partial}{\partial x}\left(\mu(x,y)N(x,y)\right) \tag{3.1.16a}$$

$$;\quad \frac{\partial\mu(x,y)}{\partial y}M(x,y)+\mu(x,y)\frac{\partial M(x,y)}{\partial y}=\frac{\partial\mu(x,y)}{\partial x}N(x,y)+\mu(x,y)\frac{\partial N(x,y)}{\partial x} \tag{3.1.16b}$$

$$;\quad \frac{\partial\mu(x,y)}{\partial x}N(x,y)-\frac{\partial\mu(x,y)}{\partial y}M(x,y)=\left(\frac{\partial M(x,y)}{\partial y}-\frac{\partial N(x,y)}{\partial x}\right)\mu(x,y) \tag{3.1.16c}$$

This is a PDE, which is not easy to solve for the integrating factor $\mu(x,y)$. However, it can be simplified if $\mu(x,y)$ is a function of a single variable x or y. Let $\mu(x,y)$ be $\mu(x)$ depending only on x. Then the condition (3.1.16c) becomes

$$\frac{d\mu(x)}{dx}N(x,y)=\left(\frac{\partial M(x,y)}{\partial y}-\frac{\partial N(x,y)}{\partial x}\right)\mu(x)$$

$$;\quad \frac{1}{\mu(x)}d\mu(x)=\frac{M_y-N_x}{N}dx \text{ with } M_y=\frac{\partial M(x,y)}{\partial y},\ N_x=\frac{\partial N(x,y)}{\partial x},\text{ and } N=N(x,y) \tag{3.1.17}$$

If $(M_y-N_x)/N$ in the RHS of this equation is a function of x alone, then an integratng factor can be found as

$$\ln|\mu(x)|=\int\frac{M_y-N_x}{N}dx;\quad \mu(x)=e^{\int\frac{M_y-N_x}{N}dx} \tag{3.1.18}$$

Now, let $\mu(x,y)$ be $\mu(y)$ depending only on y. Then the condition (3.1.16c) becomes

$$-\frac{d\mu(y)}{dy}M(x,y)=\left(\frac{\partial M(x,y)}{\partial y}-\frac{\partial N(x,y)}{\partial x}\right)\mu(y)$$

$$;\quad \frac{1}{\mu(y)}d\mu(y)=\frac{N_x-M_y}{M}dy \tag{3.1.19}$$

If the RHS of this equation, i.e., $(N_x-M_y)/M$, is a function of y alone, then an integrating factor can be found as

$$\ln|\mu(y)|=\int\frac{N_x-M_y}{M}dy;\quad \mu(y)=e^{\int\frac{N_x-M_y}{M}dy} \tag{3.1.20}$$

Depending on the case, we multiply the given non-exact DE by the integrating factor (3.1.18) or (3.1.20) to make it exact and then apply the procedure to find $F(x,y)$ such that

$$\mu(x,y)M(x,y)=\frac{\partial F}{\partial x}\ \text{ and }\ \mu(x,y)N(x,y)=\frac{\partial F}{\partial y} \tag{3.1.21}$$

(Example 3.6) Conversion of Non-Exact DE to Exact DE Using Integrating Factor

Solve

$$\frac{dy}{dx} = -\frac{y^2 + 3xy}{x(x+y)} \text{ with } y(1) = 2 \tag{E3.6.1}$$

This can be written as

$$Mdx + Ndy = (y^2 + 3xy)dx + x(x+y)dy = 0 \tag{E3.6.2}$$

Note that this DE is not exact as it is since it does not satisfy Eq. (3.1.9). That is why we try to find an integrating factor by using (3.1.18) or (3.1.20):

$$\frac{M_y - N_x}{N} = \frac{(2y+3x)-(2x+y)}{x(x+y)} = \frac{1}{x} \tag{E3.6.3}$$

Since this is a function of x alone, we can use Eq. (3.1.18) to determine an integrating factor as

$$\mu(x) \overset{(3.1.18)}{=} e^{\int \frac{M_y - N_x}{N} dx} \overset{(E3.6.3)}{=} e^{\int \frac{1}{x} dx} = e^{\ln x} = x \tag{E3.6.4}$$

We multiply Eq. (E3.6.2) by this integrating factor (E3.6.4) to make it exact as

$$\mu Mdx + \mu Ndy = x(y^2 + 3xy)dx + x^2(x+y)dy = 0 \tag{E3.6.5}$$

Then we use Eq. (3.1.13) to write

$$F(x,y) \overset{(3.1.13)}{=} \int \mu M\, dx + G(y) = \int x(y^2 + 3xy)dx + G(y) = \frac{1}{2}x^2y^2 + x^3y + G(y) \tag{E3.6.6}$$

and substitute this into Eq. (3.1.21b) to write

$$\frac{\partial}{\partial y}F(x,y) \overset{(3.1.21b)}{=} \mu N; \;\; x^2y + x^3 + \frac{\partial}{\partial y}G(y) = x^2(x+y); \;\; \frac{\partial}{\partial y}G(y) = 0\; ; \;\; G(y) = c_1 \tag{E3.6.7}$$

We substitute this into Eq. (E3.6.6) and then plug the result into Eq. (3.1.10) to get the solution as

$$F(x,y) \overset{(3.1.10)}{=} c \underset{(E3.6.7)}{\overset{(E3.6.6)}{\rightarrow}} \frac{1}{2}x^2y^2 + x^3y = c \overset{y(1)=2}{=} 4 \tag{E3.6.8}$$

where c=4 has been obtained by using the initial condition $y(1)$=2.

Interested readers should to run the MATLAB script "em03e06.m" to graph the analytical solution (E3.6.8) together with the numerical solution obtained using '`ode45()`' where x0 (the left bound of the solution interval) has been set to 1 to implement the initial condition $y(1)=2$.

```
%em03e06.m
clear, clf
ezplot('x^2*y^2/2+x^3*y=4') % plot the analytical solution (E3.6.8)
x0=1; xf=5; xspan=[x0 xf]; % Solution interval
y0=2; % Initial value y(1)=2
dy=@(x,y)-(y^2+3*x*y)/x/(x+y); % DE (E3.6.1)
[xx,yy]=ode45(dy,xspan,y0);
hold on, plot(xx,yy,'r:'), shg % plot the numerical solution
```

3.1.3 Linear First-Order Differential Equations

Consider the following first-order DE:

$$y'+a(x)y=f(x);\quad \frac{dy}{dx}+a(x)y=f(x) \text{ with } y(0)=y_0 \tag{3.1.22}$$

Since the coefficient a is only a function of the independent variable x, but not of the dependent variable y, the DE is said to be *linear*; if the coefficient a were dependent on y (and possibly x), the DE would have to be called *nonlinear*. If the input or forcing function $f(x)$ on the RHS (right-hand side) is zero, the linear DE is said to be *homogeneous*; otherwise, it is called *nonhomogeneous*.

Eq. (3.1.22) is not an exact DE, as can be seen by rewriting it as

$$(a(x)y-f(x))dx+dy=0;\ Mdx+Ndy=0 \text{ with } M=a(x)y-f(x) \text{ and } N=1;\ \frac{\partial M}{\partial y}\neq\frac{\partial N}{\partial x}$$

However, fortunately, $(M_y-N_x)/N=a(x)$ is a function of x alone so that we can make it an exact DE by multiplying both its sides with *integrating factor* $\mu(x)\overset{(3.1.18)}{=}\exp\{\int((M_y-N_x)/N)dx\}=e^{\int a(x)dx}$ as

$$e^{\int a(x)dx}\frac{dy}{dx}+e^{\int a(x)dx}a(x)y=e^{\int a(x)dx}f(x);\ \frac{d}{dx}\left(e^{\int a(x)dx}y\right)=e^{\int a(x)dx}f(x) \tag{3.1.23}$$

integrate both sides, and then divide both sides by the integrating factor to get the solution:

$$e^{\int a(x)dx}y=c+\int e^{\int a(x)dx}f(x)dx;\ y(x)=ce^{-\int a(x)dx}+e^{-\int a(x)dx}\int e^{\int a(x)dx}f(x)dx \tag{3.1.24}$$

The constant c is to be found from the initial condition $y(0)=y_0$.

(Example 3.7) Linear First-Order DE with Constant Coefficient

Solve

$$\frac{dy}{dx}+2y=4 \text{ with } y(0)=1 \tag{E3.7.1}$$

We use Eq. (3.1.24) to get

$$y(x)\overset{(3.1.24)}{=}ce^{-\int a(x)dx}+e^{-\int a(x)dx}\int e^{\int a(x)dx}f(x)dx\underset{f(x)=4}{\overset{a(x)=2}{=}}ce^{-2x}+e^{-2x}\int e^{2x}4dx=ce^{-2x}+2 \tag{E3.7.2}$$

and find the constant c from the initial condition $y(0)=1$ as

$$y(0)\underset{x=0}{\overset{(E3.7.2)}{=}}c+2=1;\ c=-1 \tag{E3.7.3}$$

Thus we substitute this into Eq. (E3.7.2) to write the solution as

$$y(x)\underset{(E3.7.3)}{\overset{(E3.7.2)}{=}}-e^{-2x}+2 \tag{E3.7.4}$$

We can run the MATLAB script "em03e07.m" to get Fig. 3.2 (showing a numerical solution obtained using an ODE solver '`ode45()`') and the symbolic solution (obtained using '`dsolve()`') just like the analytical solution:

```
ys   = 2-exp(-2*t) % with the independent variable t by default
```

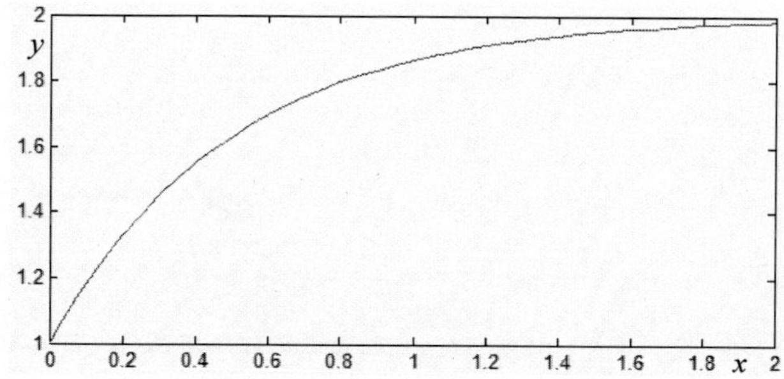

Figure 3.2 Numerical solution for DE (E3.7.1)

```
%em03e07.m
% To solve a DE (Eq.(E3.7.1)) y'+2y=4
clear, clf
x0=0; xf=2; xspan=[x0 xf]; y0=1; % Solution interval and Initial value
dy=@(x,y)4-2*y; % DE to solve
% Numerical solution to y'=-2y+4 with y(0)=y0
[xx,yy]=ode45(dy,xspan,y0);
plot(xx,yy) % plot the numerical solution
% Symbolic solution just like an analytic solution
ys=dsolve('Dy+2*y=4','y(0)=1','x')
x=xx; ys_=eval(ys);
discrepancy = norm(yy-ys_)/norm(ys_)
```

(Example 3.8) Linear First-Order DE with Variable Constant

Solve

$$x\frac{dy}{dx} - 2y = x^4 e^{-x} \text{ with } y(-1) = 1 \tag{E3.8.1}$$

First we divide this by x to write it in standard form as

$$\frac{dy}{dx} - \frac{2}{x}y = x^3 e^{-x} \tag{E3.8.2}$$

Now we use Eq. (3.1.24) to get

$$\begin{aligned} y(x) &\overset{(3.1.24)}{=} ce^{-\int a(x)dx} + e^{-\int a(x)dx}\int e^{\int a(x)dx} f(x)dx \underset{f(x)=x^3e^{-x}}{\overset{a(x)=-2/x}{=}} ce^{\int (2/x)dx} + e^{\int(2/x)dx}\int e^{-\int(2/x)dx} x^3 e^{-x}dx \\ &= ce^{2\ln x} + e^{2\ln x}\int e^{-2\ln x}x^3e^{-x}dx = cx^2 + x^2\int xe^{-x}dx \overset{\text{Integration by parts}}{=} cx^2 + x^2\left(-xe^{-x} + \int e^{-x}dx\right) \\ &= cx^2 + x^2(-xe^{-x} - e^{-x}) = cx^2 - x^2(x+1)e^{-x} \end{aligned} \tag{E3.8.3}$$

and find the constant c from the initial condition $y(-1)=1$ as

$$y(-1) \underset{x=-1}{\overset{(E3.8.3)}{=}} c(-1)^2 = c = 1; \quad c = 1 \tag{E3.8.4}$$

Then we substitute this into Eq. (E3.8.3) to write the solution as

$$y(x) \underset{(E3.8.4)}{\overset{(E3.8.3)}{=}} x^2 - x^2(x+1)e^{-x} \tag{E3.8.5}$$

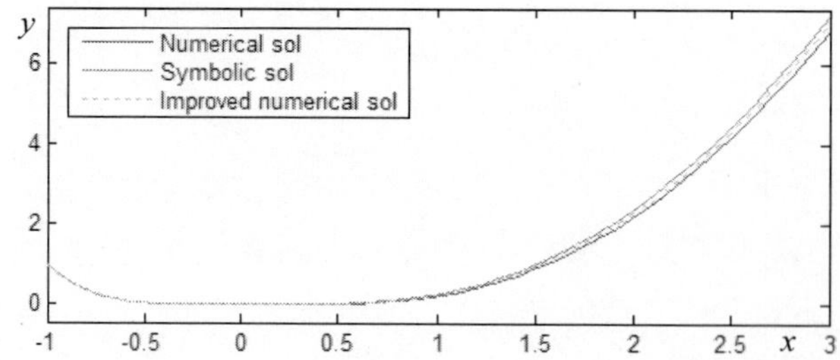

Figure 3.3 Plot of the numerical and symbolic solutions for DE (E3.8.1)

We can run the MATLAB script "em03e08.m" to get Fig. 3.3 (showing a numerical solution obtained using '`ode45()`') and the symbolic solution (obtained using '`dsolve()`') just like the analytical solution:

```
ys   = x^2 - (x^2*(x+1))/exp(x) % agrees with Eq.(E3.8.5)
```

Additionally, this symbolic solution (ys) has been plotted for the set (xx) of x-values (between x0=−1 and xf=3) that was obtained as the first output argument of the ODE solver '`ode45()`'. Further, to improve the accuracy of the numerical solution, '`ode45()`' has been used once again with the optional parameters, like the relative/absolute error tolerances that are set to $10^{-6}/10^{-8}$ by using '`odeset()`'.

```
%em03e08.m
% To solve a DE (Eq.(E3.8.1)) y'-2/x*y=x^3*exp(-x)
x0=-1; xf=3; xspan=[x0 xf]; y0=1; % Solution interval and Initial value
% Numerical solution with y(-1)=1
dy=@(x,y)2/x*y+x^3*exp(-x); % DE to solve
[xx,yy]=ode45(dy,xspan,y0);
plot(xx,yy);  hold on  % Plot the numerical solution
% Symbolic solution just like an analytic solution
ys=dsolve('Dy-2*y/x=x^3*exp(-x)','y(-1)=1','x')
x=xx; ys_=eval(ys);
plot(xx,ys_,'r') % Plot the symbolic solution
% Another numerical solution with reduced error
options=odeset('RelTol',1e-6,'AbsTol',1e-8); % Set relative/absolute error
[xx1,yy1]=ode45(dy,xspan,y0,options); % Expected to have smaller error
plot(xx1,yy1,'g:'), shg % Plot the improved (?) numerical solution
legend('Numerical sol','Symbolic sol','Improved numerical sol')
```

(Example 3.9) Linear First-Order DE with Discontinuous Forcing Function

Find the continuous-time solution $y(t)$, for $t \geq 0$, of

$$\frac{dy}{dt}+2y=f(t)=\begin{cases}2 & \text{for } 0\leq t<1\\ 0 & \text{elsewhere}\end{cases} \quad \text{with } y(0)=0 \tag{E3.9.1}$$

First, for $0 \leq t < 1$, we use Eq. (3.1.24) with $a(t)=2$ and $f(t)=2$ to get

$$y(t)\overset{(3.1.24)}{\underset{x=t}{=}}ce^{-\int a(t)dt}+e^{-\int a(t)dt}\int e^{\int a(t)dt}f(t)dt\overset{a(t)=2}{\underset{f(t)=2}{=}}ce^{-2t}+e^{-2t}\int e^{2t}2dt=ce^{-2t}+1 \tag{E3.9.2}$$

and find the constant c from the initial condition $y(0)=0$ as

$$y(0)\overset{(E3.9.2)}{\underset{t=0}{=}}c+1=0\;;\; c=-1 \tag{E3.9.3}$$

Then we substitute this into Eq. (E3.9.2) to write the solution for $0 \le t < 1$ as

$$y(t) \underset{(E3.9.3)}{\overset{(E3.9.2)}{=}} 1 - e^{-2t} \quad \text{for } 0 \le t < 1 \tag{E3.9.4}$$

Second, for $t \ge 1$, we use Eq. (3.1.3) with $f(t) = 0$ to get

$$y(t) \underset{x=t}{\overset{(3.1.3)}{=}} ce^{-\int a(t)\,dt} + e^{-\int a(t)\,dt} \int e^{\int a(t)\,dt} f(t)\,dt \underset{f(t)=0}{\overset{a(t)=2}{=}} ce^{-2t} \tag{E3.9.5}$$

and find the constant c from the initial condition $y(1) \underset{t=1}{\overset{(E3.9.4)}{=}} 1 - e^{-2}$ as

$$y(0) \underset{t=1}{\overset{(E3.9.5)}{=}} ce^{-2} = 1 - e^{-2}\,; \quad c = e^2 - 1 \tag{E3.9.6}$$

Then we substitute this into Eq. (E3.9.5) to write the solution for $t \ge 1$ as

$$y(t) \underset{(E3.9.6)}{\overset{(E3.9.5)}{=}} (e^2 - 1)e^{-2t} \quad \text{for } t \ge 1 \tag{E3.9.7}$$

Now we combine Eqs. (E3.9.4) and (E3.9.7) to write the total solution as

$$y(t) \underset{(E3.9.7)}{\overset{(E3.9.4)}{=}} \begin{cases} 1 - e^{-2t} & \text{for } 0 \le t < 1 \\ (e^2 - 1)e^{-2t} & \text{for } t \ge 1 \end{cases} \tag{E3.9.8}$$

We can run the following MATLAB script "em03e09.m" to get Fig. 3.4, which shows the plots of a numerical solution (obtained using '`ode45()`') and the analytical solution (Eq. (E3.9.8)) for the set (tt) of t-values (between t0=0 and tf=3) that was obtained as the first output argument of the ODE solver '`ode45()`'. Note that the differential equation 'dy' can be defined as a function handle (by using @) or an inline function (by using '`inline()`'). It can also be saved in a separate M-file named, say, "em03e09_dy.m."

Readers who want to make the numerical solution more accurate, i.e., closer to the analytical solution, are recommended to do as was done in the MATLAB program "em03e08.m" for Example 3.8 where the ODE solver '`ode45()`' was used with optional parameters set for smaller error tolerances by using '`odeset()`'.

```
%em03e09.m
% To solve a DE (Eq.(E3.9.1)) y'+2y=4 for 0<=t<1 and 0 for t>=1
clear, clf
t0=0; tf=3; tspan=[t0 tf]; y0=0; % Solution interval and Initial value
% Numerical solution to the DE with the initial condition y(0)=0
dy=@(t,y)-2*y+2*((0<=t)&(t<1)); % Eq.(E3.9.1))
[tt,yy]=ode45(dy,tspan,y0); % with DE described as a function handle
% Alternatively, you can define the DE in an M-file as below.
[tt,yy]=ode45(@em03e09_dy,tspan,y0); % with DE described in an M-file
plot(tt,yy), hold on  % Plot the numerical solution
% The analytical solution obtained as Eq.(E3.9.8)
ya=@(t)(1-exp(-2*t)).*(t<1)+(exp(2)-1)*exp(-2*t).*(t>=1);
plot(tt,ya(tt),'r:')  % Plot the analytical solution
legend('Numerical sol','Analytical sol')
```

```
function dy=em03e09_dy(t,y)
dy=-2*y+2*((0<=t)&(t<1)); % Eq.(E3.9.1)
```

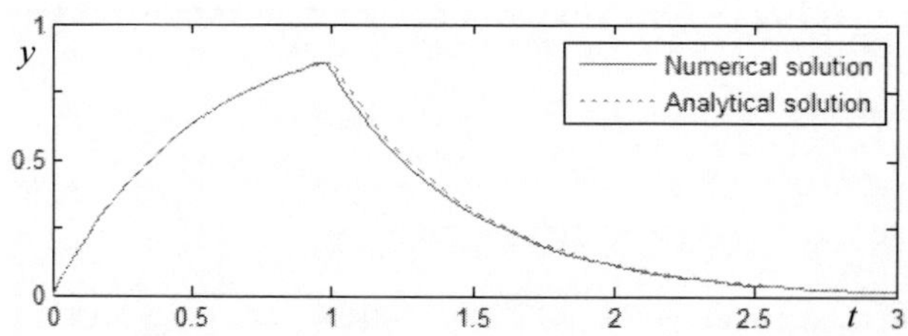

Figure 3.4 Plot of the numerical and analytical solutions for DE (E3.9.1)

3.1.4 Nonlinear First-Order Differential Equations

As an example of a nonlinear first-order DE, consider the logistic equation that can be used as a mathematical model for a population dynamics:

$$y' - ay + by^2 = 0\,;\quad \frac{dy}{dt} = -by\left(y - \frac{a}{b}\right) \text{ with } y(0) = y_0 \tag{3.1.25}$$

For separation of variables, we can multiply both sides with $-dt/by(y-a/b)$ to write

$$-\frac{(1/b)dy}{y(y-a/b)} = dt \overset{\text{Partial fractions}}{\rightarrow} \frac{1/a}{y}dy - \frac{1/a}{y-a/b}dy = dt \tag{3.1.26}$$

so that the two variables y and t have been separated. Integrating both sides gives

$$\ln|y| - \ln\left|y - \frac{a}{b}\right| = a(t+c) \rightarrow \ln\left|\frac{y}{y-a/b}\right| = a(t+c)$$

$$\overset{0<y<a/b}{\rightarrow} \frac{y}{y-a/b} = -e^{a(t+c)} = -c_1 e^{at}$$

$$;\ y = -c_1 e^{at}(y - a/b) \rightarrow y(t) = \frac{(a/b)c_1 e^{at}}{1+c_1 e^{at}} = \frac{a/b}{1+e^{-at}/c_1} \tag{3.1.27}$$

Now we find the constant c_1 of integration from the initial condition $y(0)=y_0$ as

$$y_0 \underset{t=0}{\overset{(3.1.27)}{=}} \frac{(a/b)c_1}{1+c_1};\quad c_1 = \frac{y_0}{a/b - y_0} = \frac{by_0}{a - by_0} \tag{3.1.28}$$

where we assumed $y_0 \neq a/b$. Then we plug this into Eq. (3.1.28) to write the solution as

$$y \overset{(3.1.27)}{=} \frac{a/b}{1+e^{-at}/c_1} \overset{(3.1.28)}{=} \frac{a/b}{1+\dfrac{a-by_0}{by_0}e^{-at}} = \frac{ay_0}{by_0+(a-by_0)e^{-at}} \tag{3.1.29}$$

This analytical procedure does not seem to be easy. That is why we try using MATLAB in the following example.

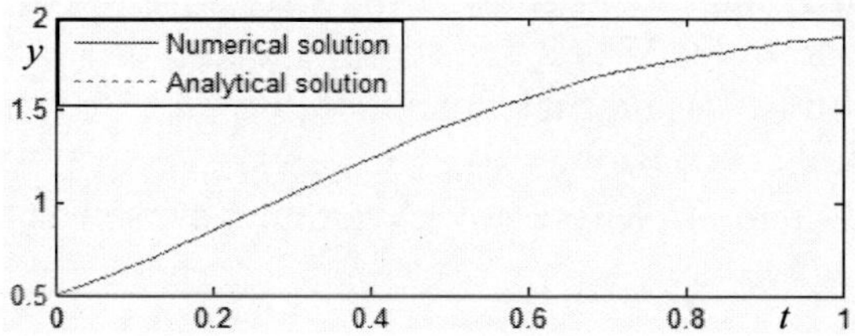

Figure 3.5 Plot of the numerical and analytical solutions for DE (E3.10.1)

(**Example 3.10**) Nonlinear First-Order DE

Find the solution $y(t)$ (for t=0~1[s]) of

$$y'-4y+2y^2=0;\ \frac{dy}{dt}=-2y(y-2) \text{ with } y(0)=0.5 \tag{E3.10.1}$$

Since this corresponds to the logistic equation (3.1.25) with a=4, b=2, and y_0=0.5, we can use Eq. (3.1.29) to write the analytical solution as

$$y \overset{(3.1.29)}{=} \frac{ay_0}{by_0+(a-by_0)e^{-at}} = \frac{2}{1+3e^{-4t}} \tag{E3.10.2}$$

We can run the MATLAB script "em03e10.m" to get the symbolic solution (obtained using '`dsolve()`') and Fig. 3.5, which shows the plots of $y(t)$ together with a numerical solution obtained using '`ode45()`' for the time interval [0,1].

```
%em03e10.m
% To solve a DE (Eq.(E3.10.1)) y'-4y+2y^2=0 with y(0)=0.5 for 0<=t<1
t0=0; tf=1; tspan=[t0 tf]; y0=0.5; % Solution interval and Initial value
dy=@(t,y)4*y-2*y.^2; % Eq.(E3.10.1)
[tt,yy]=ode45(dy,tspan,y0); % with DE defined as a function handle
ys=dsolve('Dy-4*y+2*y^2=0','y(0)=0.5') % Symbolic solution
ya=@(t)2./(1+3*exp(-4*t)); % Analytical solution Eq.(E3.10.2)
plot(tt,yy, tt,ya(tt),'r:') % Plot the numerical/analytical solutions
legend('Numerical solution','Analytical solution')
```

3.1.5 Systems of First-Order Differential Equations

Consider a system of linear first-order DE:

$$\begin{aligned}\frac{dx_1}{dt}&=a_{11}x_1+a_{12}x_2+f_1(t)\\ \frac{dx_2}{dt}&=a_{21}x_1+a_{22}x_2+f_2(t)\end{aligned} \quad \text{with } x_1(0)=x_{10} \text{ and } x_2(0)=x_{20} \tag{3.1.30}$$

We may apply the method of elimination, which eliminates other unknown functions than one to make a higher-order DE in a single unknown function as follows:

$$\begin{aligned} x_1'' &\overset{(3.1.30a)}{=} a_{11}x_1'+a_{12}x_2'+f_1'(t) \overset{(3.1.30b)}{=} a_{11}x_1'+a_{12}(a_{21}x_1+a_{22}x_2+f_2(t))+f_1'(t)\\ &= a_{11}x_1'+a_{12}a_{21}x_1+a_{22}a_{12}x_2+a_{12}f_2(t)+f_1'(t)\\ &\overset{(3.1.30a)}{=} a_{11}x_1'+a_{12}a_{21}x_1+a_{22}(x_1'-a_{11}x_1-f_1(t))+a_{12}f_2(t)+f_1'(t)\end{aligned}$$

$$;\quad x_1''-(a_{11}+a_{22})x_1'+(a_{11}a_{22}-a_{12}a_{21})x_1=a_{12}f_2(t)-a_{22}(t)f_1(t)+f_1'(t) \tag{3.1.31}$$

where $x_1'(0)$ and $x_2'(0)$ can be obtained by substituting $x_1(0)$, $x_2(0)$, $f_1(0)$, and $f_2(0)$ into Eq. (3.1.30). Then we can solve this second-order DE (by using any method of solving higher-order DEs to be discussed in Section 3.2) to find $x_1(t)$ and substitute it into Eq. (3.1.30a) to find $x_2(t)$.

On the other hand, the numerical DE solvers like '`ode45()`', '`ode23()`', '`ode23s()`', etc., work for systems of first-order DEs in multiple (unknown) variables as well as for DEs in a single variable. Let us see some examples.

(Example 3.11) System of Linear First-Order DEs

Find the solution $x(t)$ and $y(t)$ (for t=0~5[s]) of

$$\frac{dx}{dt} = -x + 2y - 1 \qquad \text{with } x(0) = 1 \text{ and } y(0) = 2 \qquad \text{(E3.11.1)}$$
$$\frac{dy}{dt} = -0.5x - y + 1.5$$

First, let us use the symbolic DE solver '`dsolve()`'

```
>>sol=dsolve('Dx+x-2*y+1=0','Dy+0.5*x+y-1.5=0','x(0)=1','y(0)=2');
>>[sol.x; sol.y]
```

to get

```
ans =        i/exp(t*(i + 1)) - i*exp(t*(i - 1)) + 1
       exp(t*(i - 1))/2 + 1/(2*exp(t*(i + 1))) + 1
```

where sol.x and sol.y, as members of the structure 'sol', are supposed to carry the solutions for $x(t)$ and $y(t)$, respectively. What is meant by the output given by MATLAB (with i standing for the imaginary unit)? It means the following analytical solution (with j standing for the imaginary unit):

$$x(t) = je^{-(j+1)t} - je^{(j-1)t} + 1 = e^{-t}\left(j(\cos t - j\sin t) - j(\cos t + j\sin t)\right) + 1 = 1 + 2e^{-t}\sin t$$
$$y(t) = \frac{1}{2}e^{(j-1)t} + \frac{1}{2}e^{-(j+1)t} + 1 = \frac{1}{2}e^{-t}\left((\cos t + j\sin t) + (\cos t - j\sin t)\right) + 1 = 1 + e^{-t}\cos t \qquad \text{(E3.11.2)}$$

Now, to use the numerical DE solver '`ode45()`', we write Eq. (E3.11.1) in state equation form as

$$\begin{bmatrix} x_1' \\ x_2' \end{bmatrix} = \begin{bmatrix} -x_1 + 2x_2 - 1 \\ -0.5x_1 - x_2 + 1.5 \end{bmatrix} \text{ with } \begin{bmatrix} x_1(0) \\ x_2(0) \end{bmatrix} = \begin{bmatrix} x(0) \\ y(0) \end{bmatrix} = \begin{bmatrix} 1 \\ 2 \end{bmatrix} \qquad \text{(E3.11.3)}$$

and build it as a function handle

```
>>dx=@(t,x) [-x(1)+2*x(2)-1; -0.5*x(1)-x(2)+1.5];
```

or in an M-file named, say 'em03e11_dx.m' as listed in the following box with a dotted borderline. Then, with the time span tspan=[0 5] and initial value x0=[1; 2], the numerical solution can be obtained by running

```
>>[tt,xn]=ode45(dx,tspan,x0); %with DE defined as a function handle 'dx'
```

or

```
>>[tt,xn1]=ode45('em03e11_dx',tspan,x0); % with DE defined in an M-file
```

These statements, together with those for plotting the solutions, are listed in a script file named "em03e11.m," as shown in the solid-line box below. The script file can be run to yield Fig. 3.6, where the numerical and analytical solutions are so close that they cannot be distinguished from each other.

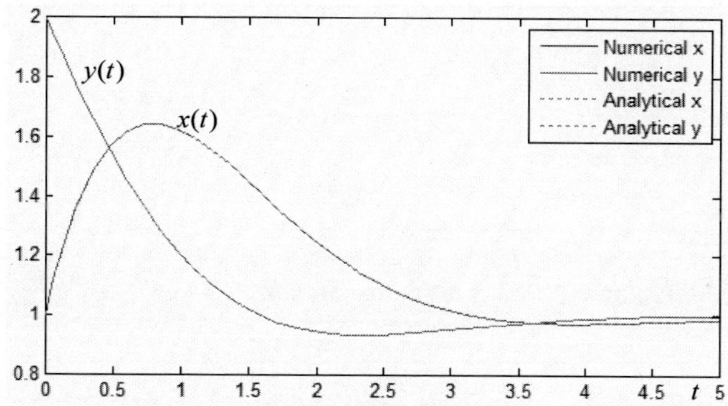

Figure 3.6 Plot of the numerical and analytical solutions for DE (E3.11.1)

```
%em03e11.m
% To solve the DE (E3.11.1) for t=[0,5]
t0=0; tf=5; tspan=[t0 tf]; % Solution interval (Time span)
x0=[1; 2]; % Initial values of x(t) and y(t)
dx=@(t,x)[-x(1)+2*x(2)-1; -0.5*x(1)-x(2)+1.5]; % Eq.(E3.11.3)
[tt,xn]=ode45(dx,tspan,x0); % with DE defined as a function handle
plot(tt,xn), hold on  % Plot the numerical solution
% To get the symbolic solution if possible
xs=dsolve('Dx+x-2*y+1=0','Dy+0.5*x+y-1.5=0','x(0)=1','y(0)=2');
t=tt; xs_=real(eval([xs.x xs.y]));
plot(tt,xs_,'r:')  % Plot the symbolic solution
legend('Numerical x','Numerical y','Analytical x','Analytical y')
% With the DE defined in an M-file named 'em03e11_dx.m'
[tt,xn1]=ode45('em03e11_dx',tspan,x0);
discrepancy=norm(xn-xn1)
```

```
function dx=em03e11_dx(t,x)
dx=[-x(1)+2*x(2)-1; -0.5*x(1)-x(2)+1.5]; %Eq.(E3.11.3)
```

(Example 3.12) System of Nonlinear First-Order DEs

Find the solution $x(t)$ and $y(t)$ (for t=0~20[s]) of

$$\frac{dx}{dt}=-0.2x+0.1xy, \quad \frac{dy}{dt}=4x-xy \quad \text{with } x(0)=3 \text{ and } y(0)=2 \tag{E3.12.1}$$

To use the numerical DE solver '`ode45()`', we write Eq. (E3.12.1) in state equation form as

$$\begin{bmatrix} x_1' \\ x_2' \end{bmatrix}=\begin{bmatrix} -0.2x_1+0.1x_1x_2 \\ 4x_1-x_1x_2 \end{bmatrix} \text{ with } \begin{bmatrix} x_1(0) \\ x_2(0) \end{bmatrix}=\begin{bmatrix} 3 \\ 2 \end{bmatrix} \tag{E3.12.2}$$

and build it as a function handle:

```
>>dx=@(t,x)[-0.2*x(1)+0.1*x(1)*x(2); 4*x(1)-x(1)*x(2)];
```

Then, with the time span tspan=[0 20] and initial value x0=[3; 2], the numerical solution can be obtained by running

```
>>tspan=[0 20]; x0=3; 2];
  [tt,xn]=ode45(dx,tspan,x0); %with DE defined as a function handle 'dx'
```

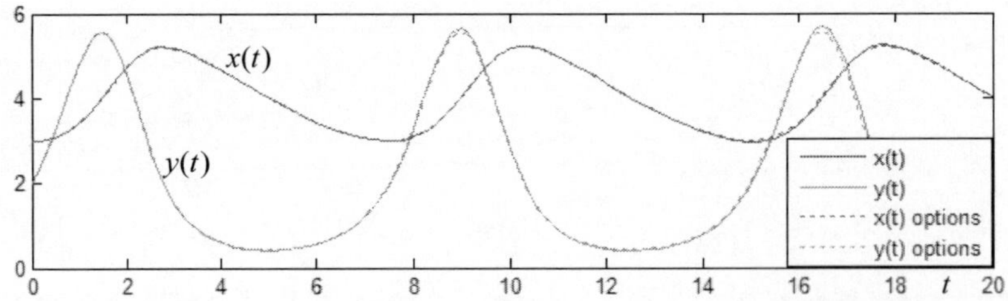

Figure 3.7 Plot of the numerical solutions for DE (E3.12.1)

```
%em03e12.m
% To solve the DE (E3.12.1) for t=[0,20]
tspan=[0 20]; x0=[3; 2]; %Solution interval and Initial value
dx=@(t,x)[-0.2*x(1)+0.1*x(1)*x(2);4*x(2)-x(1)*x(2)]; % Eq.(E3.12.2)
[tt,xn]=ode45(dx,tspan,x0); % Using ode45()
Numerical_error=error_DE_sol(dx,xn,tt)
options=odeset('RelTol',1e-6,'AbsTol',[1e-6 1e-6]); % For better accuracy
[tt2,xn2]=ode45(dx,tspan,x0,options);
Numerical_error2=error_DE_sol(dx,xn2,tt2)
plot(tt,xn(:,1), tt,xn(:,2),'r') % Plot the numerical solution
hold on, plot(tt2,xn2(:,1),':', tt2,xn2(:,2),'r:') % Plot the solution 2
legend('x(t)','y(t)')
```

Here, don't you wonder how closely the numerical solution satisfies the DE? It can be measured from the error between the derivatives $\mathbf{x}'(t)$ (computed from DE) and $(\mathbf{x}(t+\Delta t)-\mathbf{x}(t-\Delta t))/2/\Delta t$ (computed from the numerical solution), which can be found by using another MATLAB function '`error_DE_sol()`' listed in the box below Example 3.4.

```
>>Numerical_error=error_DE_sol(dx,xn,tt)
```

Now, don't you wonder if the error can be decreased to improve the numerical solution? It may be possible if we use the DE solver with an optional 4th input argument 'options' to specify smaller relative/absolute error tolerances that can be constructed by using the '`odeset()`' command, as illustrated below:

```
>>options=odeset('RelTol',1e-6,'AbsTol',[1e-6 1e-6]);
>>[tt2,xn2]=ode45(dx,tspan,x0,options);
```

These statements, together with those for plotting the solutions, are listed in a script file named "em03e12.m," as shown in the above solid-line box, which can be run to yield Fig. 3.7.

(Example 3.13) System of Nonlinear First-Order DEs in Cartesian/Polar Coordinate Systems

Find the four solutions $(x(t), y(t))$ (for t=0~8[s]) of

$$\frac{dx}{dt} = V_0\left(1 - \frac{x^2-y^2}{(x^2+y^2)^2}\right)$$
$$\frac{dy}{dt} = V_0\frac{-2xy}{(x^2+y^2)^2} \tag{E3.13.1}$$

with V_0=1 and four different initial conditions $(x(0), y(0))$=(−4, 0.5), (−4, 0.2), (−4, −0.2), and (−4, −0.5), and graph them together on the xy-plane.

To use the numerical DE solver '`ode45()`', we write Eq. (E3.13.1) in state equation form as

$$\begin{bmatrix} x_1' \\ x_2' \end{bmatrix} = \begin{bmatrix} V_0\{1-(x_1^2-x_2^2)/(x_1^2+x_2^2)^2\} \\ V_0\{-2x_1x_2/(x_1^2+x_2^2)^2\} \end{bmatrix} \text{ with } \begin{bmatrix} x_1(0) \\ x_2(0) \end{bmatrix} = \begin{bmatrix} -4 \\ 0.5 \end{bmatrix}, \begin{bmatrix} -4 \\ 0.2 \end{bmatrix}, \begin{bmatrix} -4 \\ -0.2 \end{bmatrix}, \text{ or } \begin{bmatrix} -4 \\ -0.5 \end{bmatrix} \quad \text{(E3.13.2)}$$

and build it as a function handle named 'dx' as done in the 7–8th lines of the following script, "em03e13.m." Then, with the time span tspan=[0 8] and initial value x0 chosen from the matrix x0s (having the four initial values as its column vectors in the 1415th lines) depending on the iteration number i of the 'for' loop, the corresponding numerical solutions are obtained by using '`ode45()`' in the 18th line. In addition to solving the DE, the script also has several statements for plotting the solution curves as in Fig. 3.8, which plots the trajectories of four corks moving from the four different initial positions in a fluid with the velocity vector field described by the DE (E3.13.1), where there is a cylinder of radius 1 centered at (0, 0). In addition, it shows the velocity vectors $(dx/dt, dy/dt)$ at the mid time t=4, implying that the flow of the fluid close to the cylinder is delayed due to the cylinder.

```
%em03e13.m
% To solve the DE (E3.13.2) for the position of a cork for t=[0,8]
%   x'= V0*(1-(x^2-y^2)/(x^2+y^2)^2);
%   y'= -2*V0*x*y/(x^2+y^2)^2;
clear, clf
V0=1; %+x-directional velocity of the fluid far from the cylinder
dx=@(t,x) [V0*(1-(x(1)^2-x(2)^2)/(x(1)^2+x(2)^2)^2);
             -2*V0*x(1)*x(2)/(x(1)^2+x(2)^2)^2]; % DE (E3.13.2)
% Plot the cylinder as a circle.
plot(exp(j*pi/180*[0:359]),'r:'), hold on
axis('equal'), axis([-4 4 -1.5 1.5])
title('Position (x(t),y(t)) of cork flowing in fluid around cylinder')
t0=0; tf=8; tspan=[t0 tf]; % Solution interval (Time span)
x0s=[-4 -4 -4 -4;
     0.5 0.2 -0.2 -0.5]; % Four initial positions of a small cork
for i=1:size(x0s,2)
   x0=x0s(:,i); % An initial position
   [tt,xn]=ode45(dx,tspan,x0); % Solve the DE dx
   plot(xn(:,1),xn(:,2),'g') % Plot the numerical solutions on xy-plane
   xm=xn(round(end/2),1); % x-position of the cork at the mid time
   ym=xn(round(end/2),2); % y-position of the cork at the mid time
   % To plot the velocity vector at the mid time
   quiver(xm,ym,...
          V0*(1-(xm^2-ym^2)/(xm^2+ym^2)^2),-V0*2*xm*ym/(xm^2+ym^2)^2,0.2)
end
```

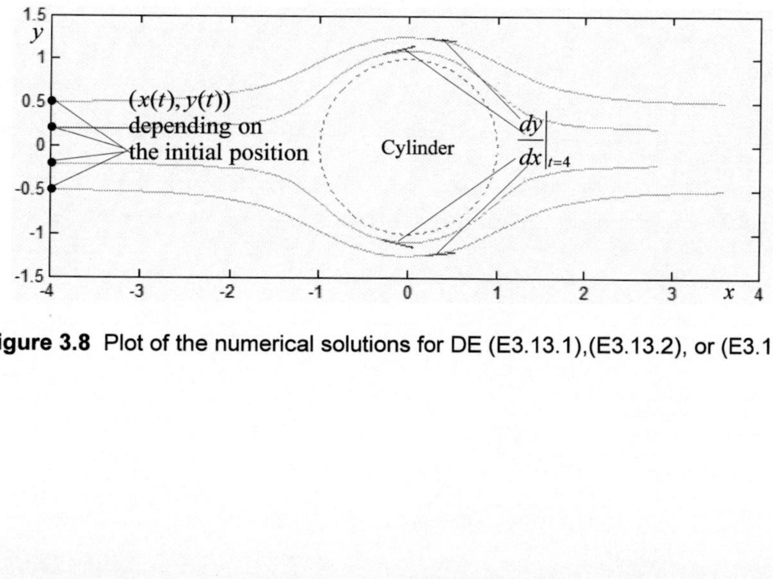

Figure 3.8 Plot of the numerical solutions for DE (E3.13.1),(E3.13.2), or (E3.13.4)

Note that dividing the second eq. (called Eq. (E3.13.1b)) by the first eq. (called Eq. (E3.13.1a)) of Eq. (E3.13.1) yields Eq. (E3.5.1), whose solution was found as Eq. (E3.5.6):

$$\frac{-y}{x^2+y^2}+y=c \text{ where } c \text{ is determined from a given point } (x_0,y_0) \tag{E3.5.6}$$

This problem can be made simpler if we use

$$\begin{cases} \dfrac{dr}{dt}=\dfrac{d}{dt}(x^2+y^2)^{1/2}=\dfrac{dx}{dt}\dfrac{d}{dx}(x^2+y^2)^{1/2}+\dfrac{dy}{dt}\dfrac{d}{dy}(x^2+y^2)^{1/2}=\dfrac{1}{r}\left(x\dfrac{dx}{dt}+y\dfrac{dy}{dt}\right) & \text{(E3.13.3a)} \\ \tan\theta=\dfrac{y}{x};\; d\theta\sec^2\theta=dx\dfrac{-y}{x^2}+dy\dfrac{1}{x};\; \dfrac{d\theta}{dt}=\dfrac{x^2}{r^2}\left(\dfrac{-y}{x^2}\dfrac{dx}{dt}+\dfrac{1}{x}\dfrac{dy}{dt}\right)=\dfrac{1}{r^2}\left(-y\dfrac{dx}{dt}+x\dfrac{dy}{dt}\right) & \text{(E3.13.3b)} \end{cases}$$

to reformulate the DE in the polar (2D cylindrical) coordinate system as

$$\frac{dr}{dt}\overset{\text{(E3.13.3a)}}{=}\frac{1}{r}\left(x\frac{dx}{dt}+y\frac{dy}{dt}\right)\overset{\text{(E3.13.1)}}{=}\frac{V_0}{r}\left\{x\left(1-\frac{x^2-y^2}{(x^2+y^2)^2}\right)+y\frac{-2xy}{(x^2+y^2)^2}\right\}$$

$$=\frac{V_0}{r^5}x\left\{(x^2+y^2)^2-(x^2+y^2)\right\}=V_0\cos\theta(1-r^{-2}) \tag{E3.13.4a}$$

$$\frac{d\theta}{dt}\overset{\text{(E3.13.3b)}}{=}\frac{1}{r^2}\left(-y\frac{dx}{dt}+x\frac{dy}{dt}\right)\overset{\text{(E3.13.1)}}{=}\frac{V_0}{r^2}\left\{-y\left(1-\frac{x^2-y^2}{(x^2+y^2)^2}\right)+x\frac{-2xy}{(x^2+y^2)^2}\right\}$$

$$=\frac{V_0}{r^6}\left\{-y(x^2+y^2)^2-y(x^2+y^2)\right\}=-\frac{V_0}{r}\sin\theta(1+r^{-2}) \tag{E3.13.4b}$$

This system of DEs (in the polar coordinate system) is defined as a function handle named 'drth' in the 6–7th lines of the script "em03e13b.m," which performs the same job as in the script "em03e13.m," but in polar coordinates. Note that the MATLAB function '`cart2pol()`' is used (in the 12th line) to convert the Cartesian coordinates of the four initial positions into the polar ones.

```
%em03e13b.m
% To solve the DE (E3.13.4) for the position of a cork for t=[0,8]
%   r'= V0*cos(th)*(1-r^-2);
%   th'=-V0/r*sin(th)*(1+r^-2);
V0=1; % +x-directional velocity of the fluid at x=-inf
drth=@(t,rth)[V0*cos(rth(2))*(1-rth(1)^-2);
                  -V0/rth(1)*sin(rth(2))*(1+rth(1)^-2)]; % DE (E3.13.3)
plot(exp(j*pi/180*[0:359]),'r:'), hold on
axis('equal'); axis([-4 4 -1.5 1.5]);
t0=0; tf=8; tspan=[t0 tf]; % Solution interval (Time span)
x0s=[-4 -4 -4 -4; 0.5 0.2 -0.2 -0.5]; % Initial positions of the cork
[ths,rs]=cart2pol(x0s(1,:),x0s(2,:)); % Cartesian-to-polar conversion
rth0s=[rs; ths]; % Initial positions of cork in polar coordinate system
for i=1:size(x0s,2)
   rth0=rth0s(:,i); % An initial position
   [tt,rthn]=ode45(drth,tspan,rth0); % with the DE drth
   plot(rthn(:,1).*exp(j*rthn(:,2)),'g') % Plot the numerical solutions
end
h=ezplot('-y/(x^2+y^2)+y=0.5*(1-1/((-4)^2+0.5^2))'); % Eq.(E3.5.6)
set(h,'Color','m','Linestyle',':') % To change the line properties
```

(Example 3.14) System of Linear First-Order DEs with Parameters

Find the solution $i(t)$ and $v(t)$ (for t=0~0.1[s]) of

$$L\frac{di(t)}{dt} + v(t) = V_{\rm m}\sin\omega t$$
$$C\frac{dv(t)}{dt} + \frac{v(t)}{R} = i(t) \tag{E3.14.1}$$

with R=1 [kΩ], L=1[mH], C=1[mF], $V_{\rm m}$=5[V], ω=120π[rad/s], and the initial conditions ($i(0)$, $v(0)$) =(1[A], 2[V]).

To use the numerical DE solver '`ode45()`', we write Eq. (E3.14.1) in state equation form as

$$\begin{bmatrix} x_1' \\ x_2' \end{bmatrix} = \begin{bmatrix} (V_{\rm m}\sin\omega t - x_2)/L \\ x_1/C - x_2/RC \end{bmatrix} \text{ with } \begin{bmatrix} x_1(0) \\ x_2(0) \end{bmatrix} = \begin{bmatrix} i(0) \\ v(0) \end{bmatrix} = \begin{bmatrix} 1 \\ 2 \end{bmatrix} \tag{E3.14.2}$$

and build it as a function handle 'dx' as listed in the 8–9th lines of the MATLAB script "em03e14.m," which has been run to yield Fig. 3.9. Note that the first and second columns of the numerical solution matrix 'xn' are interpreted as $i(t)$ and $v(t)$, respectively, in the 13th line.

```
%em03e14.m
% To solve the DE (E3.14.1) for i(t) and v(t) for t=[0,0.1s]
%  L di(t)/dt + v(t) = Vm sin(120pi t)
%  C dv(t)/dt + v(t)/R = i(t)
clear, clf
R=1000; L=0.001; C=0.001; % Parameter values
Vm=5; w=120*pi; % Amplitude and frequency of the input voltage vi(t)
dx=@(t,x) [(Vm*sin(w*t)-x(2))/L;
           x(1)/C-x(2)/R/C]; % DE (E3.14.2)
t0=0; tf=0.1; tspan=[t0 tf]; % Solution interval (Time span)
x0=[1; 2];   % Initial condition [i(0); v(0)]
[tt,xn]=ode45(dx,tspan,x0); % with the DE dx
it=xn(:,1); vt=xn(:,2); % Interprete the numerical solutions as i and v
plot(tt,it, tt,vt,'r') % Plot the current i(t) and voltage v(t)
hold on, plot(tt,Vm*sin(w*tt),'g') % Plot the input voltage
xlabel('t[s]')
legend('i(t)','v(t)','v_i (t)=Vm*sin(wt)')
title('v_i (t), i(t), and v(t) of an LRC circuit')
```

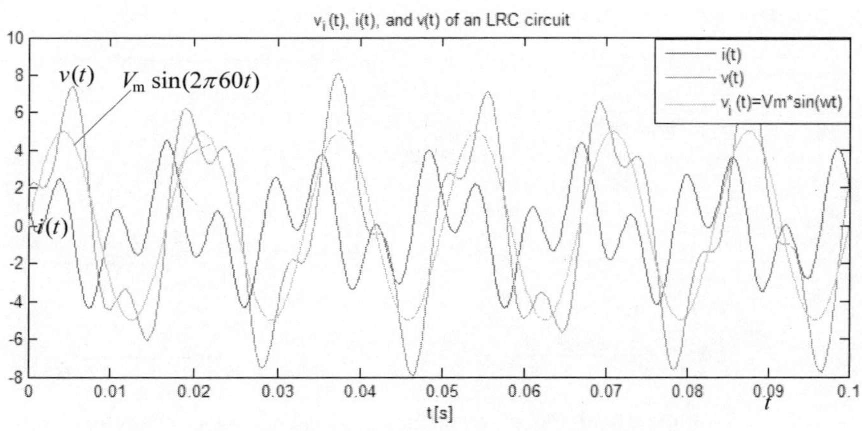

Figure 3.9 Plot of the numerical solutions for DE (E3.14.1)

(Example 3.15) System of Nonlinear First-Order DEs with Parameters

Find the solution ($\theta(t)$, $\omega(t)=\theta'(t)$) (for t = 0~20[s]) of

$$\begin{bmatrix}\theta' \\ \omega'\end{bmatrix}=\begin{bmatrix}\omega \\ -k\sin\theta\end{bmatrix} \text{ with } \begin{bmatrix}\theta(0) \\ \omega(0)\end{bmatrix}=\begin{bmatrix}0 \\ 1.9\sqrt{k}\end{bmatrix}, \begin{bmatrix}0 \\ 2\sqrt{k}\end{bmatrix}, \text{ or } \begin{bmatrix}0 \\ 2.1\sqrt{k}\end{bmatrix} \quad \text{(E3.15.1)}$$

and plot it in the $\theta\omega$-plane (called the phase plane) where k=4.

This DE has been defined as a function handle in the 3rd line of the MATLAB script "em03e15.m," which can be run to yield the numerical solutions on the phase plane as depicted in Fig. 3.10. Interested readers may set the value of 'which_to_animate' (in the 8th line) to 1, 2, or 3 and run the script to enjoy the animation for the selected case. Which value of $\omega(0)$ results in continuing rotation?

```
%em03e15.m
k=4; sqk=sqrt(k);
dx=@(t,x)[x(2); -k*sin(x(1))]; % Eq.(3.15.1)
t0=0; tf=20; tspan=[t0 tf]; % Solution interval (Time span)
x0s=[0 0 0; 1.9*sqk 2*sqk 2.1*sqk]; % A set of initial vectors
subplot(221)
gss='rgb'; % Specify the colors of graphs for the three cases
which_to_animate=1; % Select the case to be animated
for i=1:size(x0s,2)
   x0=x0s(:,i); % Initial condition
   [tt,xn]=ode45(dx,tspan,x0);
   plot(xn(:,1),xn(:,2),gss(i)), hold on
   % Save the solution for the selected case
   if i==which_to_animate,  tta=tt; xna=xn; ia=i;  end
end
plot([-10 10],[0 0],'k'), axis([-3.5 10 -5 5])
set(gca,'xtick',-pi:pi:3*pi)
set(gca,'xticklabel',{'-pi','0','pi','2*pi','3*pi'})
legend('w(0)=1.9*sqrt(k)','w(0)=2*sqrt(k)','w(0)=2.1*sqrt(k)')
% To animate the motion of the pendulum for the selected case
subplot(222)
for n=1:length(tta)-1
   xt=sin(xna(n,1)); yt=-cos(xna(n,1));
   plot([0 xt],[0 yt], xt,yt,'o', ...
           'Linewidth',2, 'Markersize',7, 'Color',gss(ia))
   axis('equal'), axis([-1.2 1.2 -1.2 1.2])
   dt=tta(n+1)-tta(n); % Interval between successive snapshot times
   pause(dt)
end
```

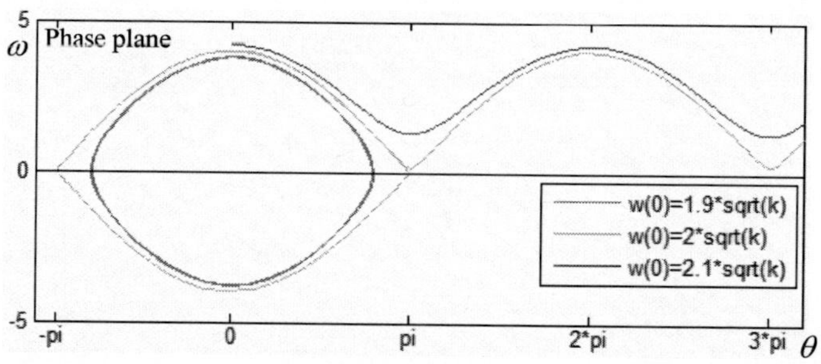

Figure 3.10 Plot of the numerical solutions for DE (E3.15.1)

3.2 HIGHER-ORDER DIFFERENTIAL EQUATIONS

3.2.1 Undetermined Coefficients

A set of functions $\{y_1(t), y_2(t),\ldots, y_N(t)\}$ is said to be *linearly independent* on an interval $I=(a, b)$ if

$$\sum_{i=1}^{N} c_i y_i(t) = 0 \ \forall t \in I \ \overset{\text{implies}}{\Rightarrow} \ c_i = 0 \ \forall i \tag{3.2.1}$$

Otherwise, the set of functions is said to be *linearly dependent*. In the case where each of the N functions has at least $(N-1)$ derivatives, the necessary and sufficient condition for their linear independence is

$$W(y_1, y_2, \cdots, y_N) = \begin{vmatrix} y_1 & y_2 & \cdots & y_N \\ y_1' & y_2' & \cdots & y_N' \\ \vdots & \vdots & \ddots & \vdots \\ y_1^{(N-1)} & y_2^{(N-1)} & \cdots & y_N^{(N-1)} \end{vmatrix} \neq 0 \ \forall x \in I \tag{3.2.2}$$

where $|\cdot|$ denotes the determinant, W is called the *Wronskian* of the functions, and $y^{(n)} = d^n y(t)/dt^n$.

Let $\{y_1(t), y_2(t), \ldots, y_N(t)\}$ be a (fundamental) set of N linearly independent solutions of a homogeneous linear Nth-order DE with constant coefficients:

$$y^{(N)}(t) + a_{N-1} y^{(N-1)}(t) + \cdots + a_1 y(t) = 0 \tag{3.2.3}$$

Then any linear combination of $y_1(t), y_2(t),\ldots,$ and $y_N(t)$, i.e.,

$$y_c(t) = \sum_{i=1}^{N} c_i y_i(t) \text{ with arbitrary constants } c_i,\ i=1:N \tag{3.2.4}$$

is called the *general solution* of the homogeneous DE. A fundamental set of solutions can be found as

$$\{e^{s_1 t}, te^{s_1 t}, \cdots, t^{M-1}e^{s_1 t}, e^{s_{M+1} t}, e^{s_{M+2} t}, \cdots, e^{s_N t}\} \tag{3.2.5}$$

from the roots $\{\overset{\text{a root of multiplicity } M}{s_1, s_1, \cdots, s_1}, \overset{(N-M)\text{ distinct roots}}{s_{M+1}, s_{M+2}, \cdots, s_N}\}$ of the corresponding *characteristic* or *auxiliary equation*:

$$s^N + a_{N-1} s^{N-1} + \cdots + a_1 = 0 \tag{3.2.6}$$

which has been assumed to have a root s_1 of multiplicity M and $(N-M)$ distinct roots. Note that if any root of the characteristic equation (3.2.6) with only real coefficients is a complex number $\sigma + j\omega$ $(j=\sqrt{-1})$, its conjugate $\sigma - j\omega$ is also a root and therefore, with the aid of Euler's formula (C.20), the solutions corresponding to the pair of complex roots can be written as

$$e^{\sigma t}\cos \omega t \text{ and } e^{\sigma t}\sin \omega t \tag{3.2.7}$$

instead of $e^{(\sigma+j\omega)t}$ and $e^{(\sigma-j\omega)t}$. Eq. (3.2.4) is called the *complementary solution* of the non-homogeneous DE

$$y^{(N)}(t) + a_{N-1} y^{(N-1)}(t) + \cdots + a_1 y(t) = f(t) \tag{3.2.8}$$

where $f(t)$ on the RHS is called the *input function*, *forcing function*, or excitation function, and $y(t)$ on the LHS is called the *output* or *response* of the linear system described by Eq. (3.2.8).

The *general solution* of this nonhomogeneous DE is the sum of the complementary solution $y_c(t)$ and a particular solution $y_p(t)$

$$y(t) = y_c(t) + y_p(t) \tag{3.2.9}$$

where the *particular solution* is a function satisfying the nonhomogeneous DE (3.2.8).

Here is the method of *undetermined coefficients*, also called "educated guessing," which can be used for solving only nonhomogeneous DEs with constant coefficients and certain types of forcing functions listed in the first columns of Table 3.1:

(Step 1) Write the *complementary solution* $y_c(t)$ as a linear combination of the functions belonging to the fundamental set of solutions (3.2.5) from the roots of the corresponding characteristic or auxiliary equation (3.2.6).

(Step 2) Write the form of particular solution $y_p(t)$ as a linear combination of certain types of trial functions (that are generated by repeated differentiations of the forcing function $f(t)$ as listed in Table 3.1), which depends mainly on the type of $f(t)$ on the RHS and partly on whether or not any term of the particular trial solution coincides with any term of $y_c(t)$ (see Remark 3.1).

(Step 3) Plugging $y_p(t)$ into the DE, find the values of its undetermined *coefficients*.

(Step 4) Determine the *coefficients* of $y_c(t)$ from the initial conditions and write the *complete solution* as Eq. (3.2.9).

[**Remark 3.1**] Proper Guess for the Form of a Particular Solution

There are two cases concerning the guess for a particular solution, to be dealt with in different ways:

(**Case I**) When no term in the trial function (listed in the second column of Table 3.1) corresponding to the forcing function $f(t)$ coincides with any term in the complementary solution $y_c(t)$, it is enough. If $f(t)$ is the sum of several terms in the first column, guess that the form of the particular solution is the sum of the corresponding terms in the second column with any duplicated ones removed.

(**Case II**) When one or more terms in the trial function corresponding to the forcing function $f(t)$ coincide with any term(s) in the complementary solution $y_c(t)$, multiply the trial function with t to make the guess of a particular solution (and possibly repetitively) until no duplicates appear, which is called the *modification rule*.

Table 3.1 Trial functions for a particular solution of nonhomogeneous DEs [W-2]

Type of the forcing function $f(t)$ on the RHS of nonhomogeneous DE	Trial functions for a particular solution
(1) Constant	k (a constant)
(2) t^M (Mth-degree polynomial)	$\sum_{i=0}^{M} k_i t^i$
(3) $e^{\sigma t}$	$k e^{\sigma t}$
(4) $\cos(\omega t+\theta)$ or $\sin(\omega t+\theta)$	$k_0\cos(\omega t+\theta) + l_0\sin(\omega t+\theta)$
(5) $t^M e^{\sigma t}$	$e^{\sigma t}\sum_{i=0}^{M} k_i t^i$
(6) $t^M\cos(\omega t+\theta)$ or $t^M\sin(\omega t+\theta)$	$\sum_{i=0}^{M} k_i t^i\cos(\omega t+\theta)+\sum_{i=0}^{M} l_i t^i\sin(\omega t+\theta)$
(7) $e^{\sigma t}\cos(\omega t+\theta)$ or $e^{\sigma t}\sin(\omega t+\theta)$	$e^{\sigma t}\{k_0\cos(\omega t+\theta) + l_0\sin(\omega t+\theta)\}$
(8) $t^M e^{\sigma t}\cos(\omega t+\theta)$ or $t^M e^{\sigma t}\sin(\omega t+\theta)$	$e^{\sigma t}\left\{\sum_{i=0}^{M} k_i t^i\cos(\omega t+\theta)+\sum_{i=0}^{M} l_i t^i\sin(\omega t+\theta)\right\}$

```
function ycis=complementary_sol(a)
% Find the terms constituting the complementary solution of a DE
%  with coefficient vector a
syms t
ycis = [];  EPS = 1e-4;
s = roots(a); N = length(s);
s = [sort(round(s/EPS)*EPS); NaN; NaN];
n = 1;
while n<=N
   mn = 1;
   sigma=quant(real(s(n)),EPS);   w=quant(imag(s(n)),EPS);
   if abs(w)<EPS  % Real roots
     % To find the multiplicity mn
     while abs(s(n+mn-1)-s(n+mn))<EPS, mn=mn+1; end
     tmp = exp(sigma*t);  ycis = [ycis tmp]; % Eq.(3.2.5)
     for m=2:mn, tmp = tmp*t;  ycis = [ycis tmp];  end
     n = n + mn;
    else   % Complex roots
     % To find the multiplicity mn
     mn1=mn; while abs(s(n+mn1-1)-s(n+mn1))<EPS, mn1=mn1+1; end
     mn2=mn+1; while abs(s(n+mn2-1)-s(n+mn2+1))<EPS, mn2=mn2+2; end
     mn = max(mn1,mn2/2);
     tmp = exp(sigma*t)*[cos(w*t) -sin(w*t)]; ycis = [ycis tmp];
     for m=2:mn, tmp = tmp*t;  ycis = [ycis tmp];  end  % Eq.(3.2.5)
     n = n + 2*mn;
   end
end
```

(**Example 3.16**) Nonhomogeneous Linear Second-Order DE with Constant Coefficients

Solve the following DE and plot its solution for t=0~10[sec]:

$$y''(t)+a_1y'(t)+a_0y(t) = y''(t)+3y'(t)+2y(t) = 10\sin t \text{ with } y(0)=0 \text{ and } y'(0)=-3 \quad \text{(E3.16.1)}$$

Based on the roots of the characteristic equation:

$$s^2+3s+2=0;\ (s+1)(s+2)=0;\ s=-1 \text{ and } -2 \quad \text{(E3.16.2)}$$

we write the *complementary solution* to Eq. (3.16.1) as

$$y_c(t) \underset{(3.2.5)\text{ with } s_1=-1 \text{ and } s_2=-2}{\overset{(3.2.4)}{=}} c_1e^{-t}+c_2e^{-2t} \quad \text{(E3.16.3)}$$

```
>>ycis=complementary_sol([1 3 2])
   ycis = [ exp(-2*t), exp(-t) ]
```

where it can easily be verified that each term e^{s_it} is the solution of the associated homogeneous DE, making the LHS of Eq. (E3.16.1) zero. On the other hand, from Table 3.1(4) (corresponding to the RHS of Eq. (E3.16.1)), we write the *particular solution* as

$$y_p(t) \underset{\omega=1,\,\theta=0}{\overset{\text{Table3.1(4)}}{=}} k_0\cos t + l_0\sin t \quad \text{(E3.16.4)}$$

Noting that no term of the trial $y_p(t)$ (Eq. (E3.16.4)) coincides with any term of $y_c(t)$ (Eq. (E3.16.3)), we plug this into Eq. (E3.16.1) for $y(t)$ to find its *coefficients* as

$$y_p''(t)+3y_p'(t)+2y_p(t)=(-k_0\cos t-l_0\sin t)+3(-k_0\sin t+l_0\cos t)+2(k_0\cos t+l_0\sin t)$$
$$=(k_0+3l_0)\cos t+(-3k_0+l_0)\sin t=10\sin t \quad \text{(E3.16.5a)}$$

$$;\quad \begin{bmatrix}1&3\\-3&1\end{bmatrix}\begin{bmatrix}k_0\\l_0\end{bmatrix}=\begin{bmatrix}0\\10\end{bmatrix};\ \begin{bmatrix}k_0\\l_0\end{bmatrix}=\begin{bmatrix}1&3\\-3&1\end{bmatrix}^{-1}\begin{bmatrix}0\\10\end{bmatrix}=\frac{1}{10}\begin{bmatrix}1&-3\\3&1\end{bmatrix}\begin{bmatrix}0\\10\end{bmatrix}=\begin{bmatrix}-3\\1\end{bmatrix} \quad \text{(E3.16.5b)}$$

Thus we use Eq. (3.2.9) to write the *general solution* of the nonhomogeneous DE (E3.16.1) as

$$y(t)\overset{(3.2.9)}{=}y_c(t)+y_p(t)\underset{\text{(E3.16.4) with (E3.16.5b)}}{\overset{\text{(E3.16.3)}}{=}}c_1e^{-t}+c_2e^{-2t}-3\cos t+\sin t \quad \text{(E3.16.6)}$$

Now we use the initial conditions to determine the *coefficients* c_1 and c_2 as

$$\left.\begin{aligned}y(0)&\overset{\text{(E3.16.6)}}{=}c_1e^{-t}+c_2e^{-2t}-3\cos t+\sin t\big|_{t=0}=c_1+c_2-3\overset{\text{(E3.16.1)}}{=}0\\ y'(0)&\overset{\text{(E3.16.6)}}{=}-c_1e^{-t}-2c_2e^{-2t}+3\sin t+\cos t\big|_{t=0}=-c_1-2c_2+1\overset{\text{(E3.16.1)}}{=}-3\end{aligned}\right\}$$

$$;\quad \begin{bmatrix}1&1\\1&2\end{bmatrix}\begin{bmatrix}c_1\\c_2\end{bmatrix}=\begin{bmatrix}3\\4\end{bmatrix};\ \begin{bmatrix}c_1\\c_2\end{bmatrix}=\begin{bmatrix}1&1\\1&2\end{bmatrix}^{-1}\begin{bmatrix}3\\4\end{bmatrix}=\begin{bmatrix}2&-1\\-1&1\end{bmatrix}\begin{bmatrix}3\\4\end{bmatrix}=\begin{bmatrix}2\\1\end{bmatrix} \quad \text{(E3.16.7)}$$

We substitute this into Eq. (E3.16.6) to write the *complete solution* as

$$y(t)\underset{\text{(E3.16.7)}}{\overset{\text{(E3.16.6)}}{=}}2e^{-t}+e^{-2t}-3\cos t+\sin t \quad \text{(E3.16.8)}$$

This procedure is called the method of undetermined coefficients. The same solution can also be obtained by using the symbolic DE solver '`dsolve()`' as

```
>>ys=dsolve('D2y+3*Dy+2*y=10*sin(t)','y(0)=0,Dy(0)=-3')
   ys = 2/exp(t) + 1/exp(2*t) - 3*cos(t) + sin(t) % agrees with (E3.16.8)
```

However, to use '`ode45()`' for finding a numerical solution, we should write the DE (E3.16.1) in state equation form as

$$\begin{bmatrix}y'\\y''\end{bmatrix}=\begin{bmatrix}0&1\\-a_0=-2&-a_1=-3\end{bmatrix}\begin{bmatrix}y\\y'\end{bmatrix}+\begin{bmatrix}0\\10\sin t\end{bmatrix}\ \text{with}\ \begin{bmatrix}y(0)\\y'(0)\end{bmatrix}=\begin{bmatrix}0\\-3\end{bmatrix}$$

$$;\quad \begin{bmatrix}y_1'\\y_2'\end{bmatrix}=\begin{bmatrix}0&1\\-2&-3\end{bmatrix}\begin{bmatrix}y_1\\y_2\end{bmatrix}+\begin{bmatrix}0\\10\sin t\end{bmatrix}\ \text{with}\ \begin{bmatrix}y_1(0)\\y_2(0)\end{bmatrix}=\begin{bmatrix}y(0)\\y'(0)\end{bmatrix}=\begin{bmatrix}0\\-3\end{bmatrix} \quad \text{(E3.16.9)}$$

where $y(t)$ and $y'(t)$ have been written as $y_1(t)$ and $y_2(t)$, respectively. This equation can be defined as a function handle by using the function handler @:

```
>>dy=@(t,y)[0 1; -2 -3]*y+[0; 10*sin(t)]; % DE to solve
```

To graph the numerical solution and analytical solution (Eq. (E3.16.8)) for comparison, we run the following script file "em03e16.m," which yields Fig. 3.11.

Figure 3.11 Numerical/analytical solutions for DE (E3.16.1)

```
%em03e16.m
% To solve a DE (Eq.(E3.16.1)) y"+3y'+2y=10*sin(t) with y(0)=0, y'(0)=-3
clear, clf
tspan=[0 10]; y0=[0;-3]; % Solution interval and Initial value
dy=@(t,y)[0 1; -2 -3]*y+[0; 10*sin(t)]; % DE to solve
[tt,yy]=ode45(dy,tspan,y0); % Numerical solution
% To get the analytical solution
ys=dsolve('D2y+3*Dy+2*y=10*sin(t)','y(0)=0,Dy(0)=-3') % Symbolic solution
t=tt; yst=eval(ys);
% To compute the discrepancy between the numerical/analytical solutions
discrepancy = norm(yy(:,1)-yst)/norm(yst)
plot(tt,yy, tt,yst,'g') % To plot the numerical/analytical solutions
legend('Numerical sol y(t)','Numerical sol y''(t)','Analytical sol y(t)')
```

(Example 3.17) Nonhomogeneous Linear Second-Order DE with Constant Coefficients

Solve the following DE and plot its solution for t=0~5[sec]:

$$y''(t)+3y'(t)+2y(t)=e^{-t} \text{ with } y(0)=3 \text{ and } y'(0)=-4 \tag{E3.17.1}$$

Since the LHS is the same as that of Eq. (E3.16.1), we can write the *complementary solution* as

$$y_c(t) \overset{\text{(E3.16.3)}}{=} c_1 e^{-t}+c_2 e^{-2t} \tag{E3.17.2}$$

From Table 3.1(3) (corresponding to the RHS of Eq. (E3.17.1)), we write the trial function for a *particular solution* as

$$y_p(t) \underset{\sigma=-1}{\overset{\text{Table3.1(3)}}{=}} k_0 e^{-t} \tag{E3.17.3}$$

If we plugged this into the DE (E3.17.1) for $y(t)$ to find the undetermined coefficient k_0:

$$y''(t)+3y'(t)+2y(t)=k_0e^{-t}-3k_0e^{-t}+2k_0e^{-t}=0 \overset{\text{impossible}}{=} e^{-t} \tag{E3.17.4}$$

we must have failed because this equation cannot be satisfied irrespective of the value of k_0. The underlying cause of this difficulty is the coincidence of the trial $y_p(t)$ (Eq. (E3.17.3)) with a term of $y_c(t)$ (Eq. (E3.17.2)). That is why we multiply the trial function (for the *particular solution*) by t, by the modification rule (Remark 3.1, Case II), to modify it as

$$y_p(t)=k_0te^{-t} \tag{E3.17.5}$$

and plug this into Eq. (E3.17.1) for $y(t)$ to find its *coefficient* k_0 as

$$y_p''(t)+3y_p'(t)+2y_p(t)=-2k_0e^{-t}+k_0te^{-t}+3(k_0e^{-t}-k_0te^{-t})+2k_0te^{-t}=k_0e^{-t}=e^{-t}$$

$$; \quad k_0=1 \tag{E3.17.6}$$

Thus we use Eq. (3.2.9) to write the *general solution* of the nonhomogeneous DE (E3.17.1) as

$$y(t) \overset{(3.2.9)}{=} y_c(t)+y_p(t) \underset{\text{(E3.17.5) with (E3.17.6)}}{\overset{\text{(E3.17.2)}}{=}} c_1e^{-t}+c_2e^{-2t}+te^{-t} \tag{E3.17.7}$$

Now we use the initial conditions with this general solution to determine the *coefficients* c_1 and c_2 as

$$\left.\begin{array}{l} y(0) \overset{(E3.17.7)}{=} c_1 e^{-t} + c_2 e^{-2t} + te^{-t}\Big|_{t=0} = c_1 + c_2 \overset{(E3.17.1)}{=} 3 \\ y'(0) \overset{(E3.17.7)}{=} -c_1 e^{-t} - 2c_2 e^{-2t} + e^{-t} - te^{-t}\Big|_{t=0} = -c_1 - 2c_2 + 1 \overset{(E3.17.1)}{=} -4 \end{array}\right\}$$

$$; \begin{bmatrix} 1 & 1 \\ 1 & 2 \end{bmatrix}\begin{bmatrix} c_1 \\ c_2 \end{bmatrix} = \begin{bmatrix} 3 \\ 5 \end{bmatrix}; \quad \begin{bmatrix} c_1 \\ c_2 \end{bmatrix} = \begin{bmatrix} 1 & 1 \\ 1 & 2 \end{bmatrix}^{-1}\begin{bmatrix} 3 \\ 5 \end{bmatrix} = \begin{bmatrix} 2 & -1 \\ -1 & 1 \end{bmatrix}\begin{bmatrix} 3 \\ 5 \end{bmatrix} = \begin{bmatrix} 1 \\ 2 \end{bmatrix} \tag{E3.17.8}$$

We plug this into Eq. (E3.17.7) to write the *complete solution* as

$$y(t) \underset{(E3.17.8)}{\overset{(E3.17.7)}{=}} e^{-t} + 2e^{-2t} + te^{-t} \tag{E3.17.9}$$

This solution can also be obtained by using the symbolic DE solver '`dsolve()`' as

```
>>ys=dsolve('D2y+3*Dy+2*y=exp(-t)','y(0)=3,Dy(0)=-4')
   ys = 1/exp(t) + 2/exp(2*t) + t/exp(t) % agrees with (E3.17.9)
```

However, to use '`ode45()`' for finding a numerical solution, we should write the DE (E3.17.1) in state equation form like Eq. (E3.16.9) as

$$\begin{bmatrix} y_1' \\ y_2' \end{bmatrix} = \begin{bmatrix} 0 & 1 \\ -2 & -3 \end{bmatrix}\begin{bmatrix} y_1 \\ y_2 \end{bmatrix} + \begin{bmatrix} 0 \\ e^{-t} \end{bmatrix} \text{ with } \begin{bmatrix} y_1(0) \\ y_2(0) \end{bmatrix} = \begin{bmatrix} y(0) \\ y'(0) \end{bmatrix} = \begin{bmatrix} 3 \\ -4 \end{bmatrix} \tag{E3.17.10}$$

where $y(t)$ and $y'(t)$ have been written as $y_1(t)$ and $y_2(t)$, respectively. This equation can be defined as a function handle or an inline function by using @ or the '`inline()`' function:

```
>>dy=@(t,y) [0 1; -2 -3]*y+[0;exp(-t)]; % DE to solve
```

To graph the numerical solution and analytical solution (Eq. (E3.17.9)) for comparison, we run the script file "em03e17.m," which yields Fig. 3.12.

```
%em03e17.m
% To solve a DE (Eq.(E3.17.1)) y"+3y'+2y=exp(-t) with y(0)=3, y'(0)=-4
tspan=[0 5]; y0=[3; -4]; % Solution interval and Initial value
dy=@(t,y) [0 1; -2 -3]*y+[0; exp(-t)]; % DE to solve
[tt,yy]=ode45(dy,tspan,y0); % Numerical solution
% To get the analytical solution
ys=dsolve('D2y+3*Dy+2*y=exp(-t)','y(0)=3,Dy(0)=-4') % Symbolic solution
t=tt; yst=eval(ys);
% To compute the discrepancy between the numerical/analytical solutions
discrepancy=norm(yy(:,1)-yst)/norm(yst)
plot(tt,yy, tt,yst,'g:') % To plot the numerical/analytical solutions
legend('Numerical sol y(t)','Numerical sol y''(t)','Analytical sol y(t)')
```

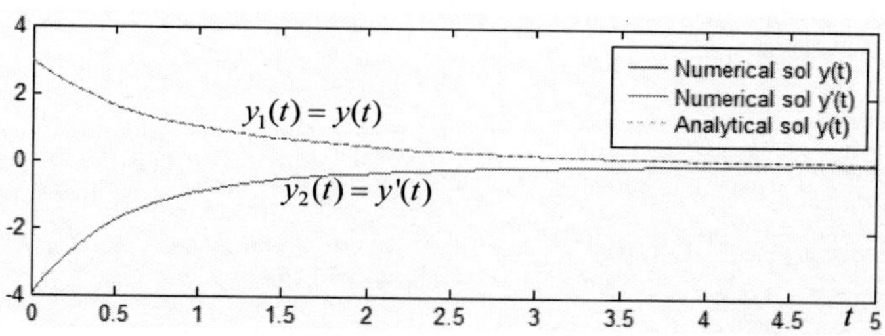

Figure 3.12 Numerical/analytical solutions for DE (E3.17.1)

(Example 3.18) Nonhomogeneous Linear Second-Order DE with Constant Coefficients

Solve the following DE and plot its solution for t=0~5[sec]:

$$y''(t)+2y'(t)+2y(t)=2 \text{ with } y(0)=2 \text{ and } y'(0)=1 \tag{E3.18.1}$$

Since the characteristic equation and its roots are

$$s^2+2s+2=(s+1)^2+1^2=0\,; \qquad s=-1\pm j \tag{E3.18.2}$$

we can write the *complementary solution* as

$$y_c(t) \underset{(3.2.5)}{\overset{(3.2.4)}{=}} c_1 e^{(-1+j)t}+c_1^* e^{(-1-j)t}=e^{-t}(d_1\cos t+d_2\sin t) \tag{E3.18.3}$$

```
>>ycis=complementary_sol([1 2 2])
  ycis = [ exp(-t)*cos(t), exp(-t)*sin(t) ]
```

From Table 3.1(1) (corresponding to the RHS of Eq. (E3.18.1)), we write the trial function for a *particular solution* as

$$y_p(t) \overset{\text{Table3.1(1)}}{=} k \tag{E3.18.4}$$

Noting that no term of Eq. (E3.18.4) coincides with any term of Eq. (E3.18.3), we plug this into Eq. (E3.18.1) for $y(t)$, to find the *coefficient* of the particular solution as

$$y_p''(t)+2y_p'(t)+2y_p(t)=2; \quad 2k=2; \quad k=1 \tag{E3.18.5}$$

Then we use Eq. (3.2.9) to write the *general solution* of the nonhomogeneous DE (E3.18.1) as

$$y(t) \overset{(3.2.9)}{=} y_c(t)+y_p(t) \underset{\text{(E3.18.4) with (E3.18.5)}}{\overset{\text{(E3.18.3)}}{=}} e^{-t}(d_1\cos t+d_2\sin t)+1 \tag{E3.18.6}$$

Now we use the initial conditions to determine the *coefficients* d_1 and d_2 as

$$\left.\begin{aligned} y(0) &\overset{\text{(E3.18.6)}}{=} e^{-t}(d_1\cos t+d_2\sin t)+1\big|_{t=0}=d_1+1 \overset{\text{(E3.18.1)}}{=} 2 \\ y'(0) &\overset{\text{(E3.18.6)}}{=} -e^{-t}(d_1\cos t+d_2\sin t)+e^{-t}(-d_1\sin t+d_2\cos t)\big|_{t=0}=-d_1+d_2 \overset{\text{(E3.18.1)}}{=} 1 \end{aligned}\right\}; \begin{aligned} d_1&=1 \\ d_2&=2 \end{aligned} \tag{E3.18.7}$$

We substitute this into Eq. (E3.18.6) to write the *complete solution* as

$$y(t) \underset{\text{(E3.18.7)}}{\overset{\text{(E3.18.6)}}{=}} e^{-t}(\cos t+2\sin t)+1 \tag{E3.18.8}$$

This solution can also be obtained by using the symbolic DE solver '`dsolve()`' as

```
>>ys=dsolve('D2y+2*Dy+2*y=2','y(0)=2,Dy(0)=1')
  ys = cos(t)/exp(t) + (2*sin(t))/exp(t) + 1 % agrees with (E3.18.8)
```

However, to use '`ode45()`' to find a numerical solution, we should write the DE (E3.18.1) in state equation form like Eq. (E3.16.9), as

$$\begin{bmatrix} y_1' \\ y_2' \end{bmatrix}=\begin{bmatrix} 0 & 1 \\ -2 & -2 \end{bmatrix}\begin{bmatrix} y_1 \\ y_2 \end{bmatrix}+\begin{bmatrix} 0 \\ 2 \end{bmatrix} \text{ with } \begin{bmatrix} y_1(0) \\ y_2(0) \end{bmatrix}=\begin{bmatrix} y(0) \\ y'(0) \end{bmatrix}=\begin{bmatrix} 2 \\ 1 \end{bmatrix} \tag{E3.18.9}$$

where $y(t)$ and $y'(t)$ have been written as $y_1(t)$ and $y_2(t)$, respectively. This equation can be defined as a function handle by using the function handle operator @:

```
>>dy=@(t,y)[0 1; -2 -2]*y+[0; 2]; % DE to solve
```

To graph the numerical solution and analytical solution (Eq. (E3.18.8)) for comparison, we run the script file "em03e18.m," which yields Fig. 3.13.

```
%em03e18.m
% To solve a DE (Eq.(E3.18.1)) y"+2y'+2y=2 with y(0)=2, y'(0)=1
clear, clf
t0=0; tf=5; tspan=[t0 tf]; y0=[2; 1];
dy=@(t,y)[0 1; -2 -2]*y+[0; 2]; % DE to solve
[tt,yy]=ode45(dy,tspan,y0); % Numerical solution
% To get the analytical solution
ys=dsolve('D2y+2*Dy+2*y=2','y(0)=2,Dy(0)=1') % Symbolic solution
t=tt; yst=eval(ys);
% To compute the discrepancy between the numerical/analytical solutions
discrepancy = norm(yy(:,1)-yst)/norm(yst)
plot(tt,yy, tt,yst,'g:') % To plot the numerical/analytical solutions
legend('Numerical sol y(t)','Numerical sol y''(t)','Analytical sol y(t)')
```

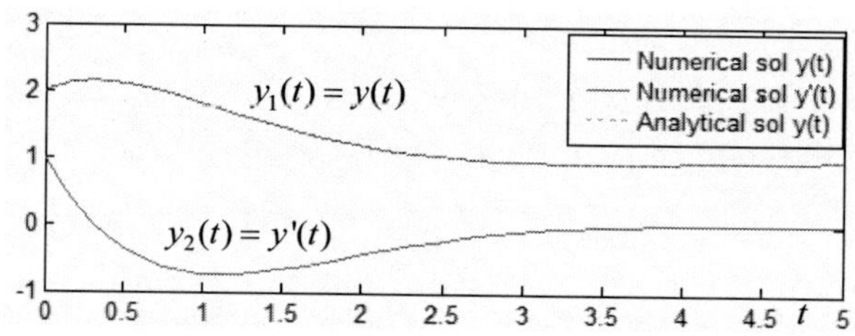

Figure 3.13 Numerical/analytical solutions for DE (E3.18.1)

(Example 3.19) Nonhomogeneous Linear Third-Order DE with Constant Coefficients

Solve the following DE and plot its solution for t=0~5[sec]:

$$y^{(3)}(t)+a_2y''(t)+a_1y'(t)+a_0y(t) = y^{(3)}(t)+4y''(t)+5y'(t)+2y(t)=2 \quad \text{(E3.19.1a)}$$

$$\text{with initial conditions } y(0)=5,\ y'(0)=-5, \text{ and } y''(0)=9 \quad \text{(E3.19.1b)}$$

Since the characteristic equation and its roots are

$$s^3+4s^2+5s+2=(s+1)^2(s+2)=0;\ s=-1, -1, \text{ and } -2 \quad \text{(E3.19.2)}$$

and $s=-1$ is of order 2, we can write the *complementary solution* as

$$y_c(t) \underset{(3.2.5) \text{ with } s_1=-1(M=2) \text{ and } s_2=-2}{\overset{(3.2.4)}{=}} (c_1+c_2t)e^{-t}+c_3e^{-2t} \quad \text{(E3.19.3)}$$

From Table 3.1(1) (corresponding to the RHS of Eq. (E3.19.1a)), we write the trial function for the *particular solution* as

$$y_p(t) \overset{\text{Table3.1(1)}}{=} k \quad \text{(E3.19.4)}$$

Noting that no term of Eq. (E3.19.4) coincides with any term of Eq. (E3.19.3), we plug this into Eq. (E3.19.1) for $y(t)$ to find the *coefficient* of the particular solution as

$$y_p^{(3)}(t)+4y_p''(t)+5y_p'(t)+2y_p(t)=2;\quad 2k=2;\quad k=1 \quad \text{(E3.19.5)}$$

Thus, we use Eq. (3.2.9) to write the *general solution* of the nonhomogeneous DE (E3.19.1a) as

$$y(t) \overset{(3.2.9)}{=} y_c(t) + y_p(t) \underset{(E3.19.4)\text{ with }(E3.19.5)}{\overset{(E3.19.3)}{=}} (c_1+c_2 t)e^{-t} + c_3 e^{-2t} + 1 \tag{E3.19.6}$$

Now we use the initial conditions to determine the *coefficients* c_1, c_2, and c_3, as

$$\left.\begin{aligned} y(0) &\overset{(E3.19.6)}{=} (c_1 + c_2 t)\,e^{-t} + c_3 e^{-2t} + 1\Big|_{t=0} = c_1 + c_3 + 1 \overset{(E3.19.1b)}{=} 5 \\ y'(0) &\overset{(E3.19.6)}{=} -c_1 e^{-t} + c_2 e^{-t} - c_2 t e^{-t} - 2c_3 e^{-2t}\Big|_{t=0} = -c_1 + c_2 - 2c_3 \overset{(E3.19.1b)}{=} -5 \\ y''(0) &\overset{(E3.19.6)}{=} c_1 e^{-t} - c_2 e^{-t} - c_2 e^{-t} + c_2 t e^{-t} + 4c_3 e^{-2t}\Big|_{t=0} = c_1 - 2c_2 + 4c_3 \overset{(E3.19.1b)}{=} 9 \end{aligned}\right\}$$

$$; \quad \begin{bmatrix} 1 & 0 & 1 \\ -1 & 1 & -2 \\ 1 & -2 & 4 \end{bmatrix}\begin{bmatrix} c_1 \\ c_2 \\ c_3 \end{bmatrix} = \begin{bmatrix} 4 \\ -5 \\ 9 \end{bmatrix}; \quad \begin{bmatrix} c_1 \\ c_2 \\ c_3 \end{bmatrix} = \begin{bmatrix} 1 & 0 & 1 \\ -1 & 1 & -2 \\ 1 & -2 & 4 \end{bmatrix}^{-1}\begin{bmatrix} 4 \\ -5 \\ 9 \end{bmatrix} = \begin{bmatrix} 1 \\ 2 \\ 3 \end{bmatrix} \tag{E3.19.7}$$

We substitute this into Eq. (E3.19.6) to write the *complete solution* as

$$y(t) \underset{(E3.19.7)}{\overset{(E3.19.6)}{=}} (1+2t)e^{-t} + 3e^{-2t} + 1 \tag{E3.19.8}$$

This solution can also be obtained by using the symbolic DE solver '`dsolve()`' as

```
>>ys=dsolve('D3y+4*D2y+5*Dy+2*y=2','y(0)=5,Dy(0)=-5,D2y(0)=9')
   ys = 1/exp(t) + 3/exp(2*t) + (2*t)/exp(t) + 1 % agrees with (E3.19.8)
```

However, to use '`ode45()`' to find a numerical solution, we should write the DE (E3.19.1) in state equation form like Eq. (E3.16.9) as

$$\begin{bmatrix} y_1' \\ y_2' \\ y_3' \end{bmatrix} = \begin{bmatrix} 0 & 1 & 0 \\ 0 & 0 & 1 \\ -2 & -5 & -4 \end{bmatrix}\begin{bmatrix} y_1 \\ y_2 \\ y_3 \end{bmatrix} + \begin{bmatrix} 0 \\ 0 \\ 2 \end{bmatrix} \text{ with } \begin{bmatrix} y_1(0) \\ y_2(0) \\ y_3(0) \end{bmatrix} = \begin{bmatrix} y(0) \\ y'(0) \\ y''(0) \end{bmatrix} = \begin{bmatrix} 5 \\ -5 \\ 9 \end{bmatrix} \tag{E3.19.9}$$

where $y(t)$, $y'(t)$, and $y''(t)$ have been written as $y_1(t)$, $y_2(t)$, and $y_3(t)$, respectively. This equation can be defined as a function handle by using the function handle operator @:

```
>>dy=@(t,y)[0 1 0; 0 0 1;-2 -5 -4]*y +[0;0;2]; % DE to solve
```

To graph the numerical solution and analytical solution (Eq. (E3.19.8)) for comparison, we run the script file "em03e19.m," which yields Fig. 3.14.

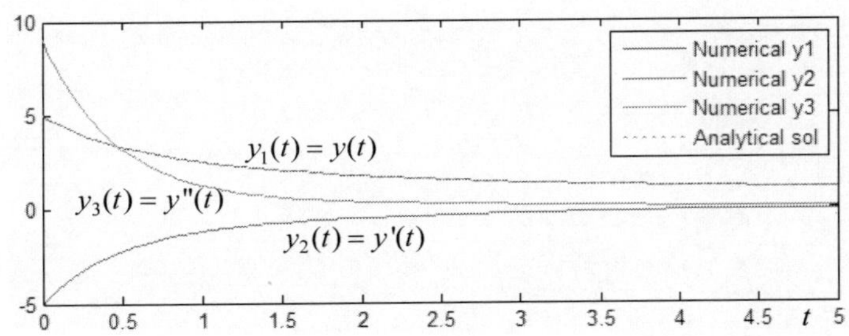

Figure 3.14 Numerical/analytical solutions for DE (E3.19.1)

```
%em03e19.m
% To solve a DE (Eq.(E3.19.1))
%    y"'+4y"+5y'+2y=2 with y(0)=5, y'(0)=-5, and y"(0)=9
t0=0; tf=5; tspan=[t0 tf]; y0=[5;-5;9];
dy=@(t,y)[0 1 0; 0 0 1;-2 -5 -4]*y+[0;0;2]; % DE to solve
[tt,yy]=ode45(dy,tspan,y0); % Numerical solution
% To get the analytical solution
ys=dsolve('D3y+4*D2y+5*Dy+2*y=2','y(0)=5,Dy(0)=-5,D2y(0)=9')%Symbolic sol
t=tt; yst=eval(ys);
% To plot the numerical/analytical solutions
plot(tt,yy, tt,yst,'r:')
legend('Numerical y1','Numerical y2','Numerical y3','Analytical sol')
```

3.2.2 Variation of Parameters

The method of variation of parameters differs from the method of undetermined coefficients in how to find the particular solution. Consider a linear Nth-order DE:

$$y^{(N)}(t) + a_{N-1}y^{(N-1)}(t) + \cdots + a_1 y(t) = f(t) \tag{3.2.10}$$

and let its *complementary solution* be

$$y_c(t) = \sum_{i=1}^{N} c_i y_i(t) \text{ with arbitrary constants } c_i,\ i=1:N \tag{3.2.11}$$

where $y_1(t)$, $y_2(t)$, ..., $y_N(t)$ are N linearly independent solutions (Eq. (3.2.5)) of the associated homogeneous DE. Then, instead of looking for trial functions, say, in Table 3.1, the method of variation of parameters uses the N linearly independent solutions (constituting the complementary solution $y_c(t)$) to write a *particular solution* as

$$y_p(t) = \sum_{i=1}^{N} k_i(t) y_i(t) \tag{3.2.12}$$

where the *coefficient function* $k_i(t)$'s can be found by integrating their first derivatives (with no constant of integration) that are the solutions to the following system of equations:

$$\begin{bmatrix} y_1(t) & y_2(t) & \cdots & y_N(t) \\ y_1'(t) & y_2'(t) & \cdots & y_N'(t) \\ \vdots & \vdots & \ddots & \vdots \\ y_1^{(N-1)}(t) & y_2^{(N-1)}(t) & \cdots & y_N^{(N-1)}(t) \end{bmatrix} \begin{bmatrix} k_1'(t) \\ k_2'(t) \\ \vdots \\ k_N'(t) \end{bmatrix} = \begin{bmatrix} 0 \\ 0 \\ \vdots \\ f(t) \end{bmatrix} \tag{3.2.13}$$

```
function yp=particular_sol(ycis,f)
% Use the method of variation of parameters to find a particular sol
%  to the linear Nth-order DE (driven by forcing function f)
%  given its fundamental set of complementary solutions ycis
% Input:  ycis = [yc1 yc2 ... ycN]: consisting of complementary sol
%         f    = Forcing function f(t)
% Output: yp = Particular solution
% where yp may contain terms of yc, which can be dropped by inspection.
if ischar(ycis),  ycis=sym(ycis);  end
N=length(ycis);  % Order of DE
zerof='[0;'; % Construct RHS vector of Eq.(3.2.13) starting from yi(t)'s
for n=2:N-1; zerof=[zerof '0;']; end
zerof=[zerof f ']']; % Complete RHS vector of Eq.(3.2.13) by 'f(t)]'
yp=ycis*int(Wronskian(ycis)\zerof); % solving Eq.(3.2.13)
```

```
function [Wr,W]=Wronskian(y)
% Input:  y  = String expression of a row vector consisting of yi(t)'s
%              belonging to the fundamental set of complementary sol.
% Output: Wr = Wronskian matrix
%         W  = Wronskian (the determinant of Wr)
if ischar(y), ycis=sym(y); end
N=length(y); % Dimension of y
Wr = y; for i=1:N-1, y=diff(y); Wr=[Wr; y]; end
W = det(Wr);
```

This method of variation of parameters for finding a particular solution to an Nth-order linear DE, given its fundamental set of complementary solutions, has been cast into the above MATLAB function '`particular_sol()`'. Note that the particular solution $y_p(t)$ returned by this function may contain some terms of the complementary solution $y_c(t)$, which can be dropped by inspection before adding $y_p(t)$ and $y_c(t)$ to write the general solution.

For example, consider the DE (E3.16.1) that was dealt with in Example 3.16:

$$y''(t)+3y'(t)+2y(t)=10\sin t$$

whose *complementary solution* is

$$y_c(t)=c_1y_1(t)+c_2y_2(t)\overset{(E3.16.3)}{=}c_1e^{-t}+c_2e^{-2t}$$

As stated above, the *coefficient functions* of a particular solution can be determined by solving Eq. (3.2.13):

$$\begin{bmatrix} y_1(t) & y_2(t) \\ y_1'(t) & y_2'(t) \end{bmatrix}\begin{bmatrix} k_1'(t) \\ k_2'(t) \end{bmatrix}\overset{(3.2.13)}{=}\begin{bmatrix} 0 \\ f(t) \end{bmatrix};\quad \begin{bmatrix} e^{-t} & e^{-2t} \\ -e^{-t} & -2e^{-2t} \end{bmatrix}\begin{bmatrix} k_1'(t) \\ k_2'(t) \end{bmatrix}=\begin{bmatrix} 0 \\ 10\sin t \end{bmatrix} \tag{3.2.14}$$

to get

$$\begin{bmatrix} k_1'(t) \\ k_2'(t) \end{bmatrix}=\begin{bmatrix} e^{-t} & e^{-2t} \\ -e^{-t} & -2e^{-2t} \end{bmatrix}^{-1}\begin{bmatrix} 0 \\ 10\sin t \end{bmatrix}=\frac{1}{-e^{-3t}}\begin{bmatrix} -2e^{-2t} & -e^{-2t} \\ e^{-t} & e^{-t} \end{bmatrix}\begin{bmatrix} 0 \\ 10\sin t \end{bmatrix}=\begin{bmatrix} 10e^{t}\sin t \\ -10e^{2t}\sin t \end{bmatrix} \tag{3.2.15}$$

and integrating this as

$$\begin{bmatrix} k_1(t) \\ k_2(t) \end{bmatrix}=\begin{bmatrix} 10\int e^{t}\sin t\,dt \\ -10\int e^{2t}\sin t\,dt \end{bmatrix}=\begin{bmatrix} 5e^{t}(\sin t-\cos t) \\ -2e^{2t}(2\sin t-\cos t) \end{bmatrix} \tag{3.2.16}$$

where

$$\int e^{t}\sin t\,dt\overset{(C.20)}{=}\mathrm{Im}\left\{\int e^{(1+j)t}dt\right\}=\mathrm{Im}\left\{\frac{1}{1+j}e^{(1+j)t}\right\}=\frac{1}{2}e^{t}\,\mathrm{Im}\{(1-j)(\cos t+j\sin t)\}$$

$$=\frac{1}{2}e^{t}(\sin t-\cos t)$$ `>>syms t; int(exp(t)*sin(t))`

$$\int e^{2t}\sin t\,dt\overset{(C.20)}{=}\mathrm{Im}\left\{\int e^{(2+j)t}dt\right\}=\mathrm{Im}\left\{\frac{1}{2+j}e^{(2+j)t}\right\}=\frac{1}{5}e^{2t}\,\mathrm{Im}\{(2-j)(\cos t+j\sin t)\}$$

$$=\frac{1}{5}e^{2t}(2\sin t-\cos t)$$ `>>syms t; int(exp(2*t)*sin(t))`

Now we substitute these coefficient functions into Eq. (3.2.12) to get the particular solution as

$$y_p(t) \overset{(3.2.12)}{=} \sum_{i=1}^{N=2} k_i(t)y_i(t) \overset{(3.2.16)}{=} 5e^t(\sin t - \cos t)e^{-t} - 2e^{2t}(2\sin t - \cos t)e^{-2t}$$
$$= \sin t - 3\cos t \tag{3.2.17}$$

It can also be obtained by using the MATLAB function '`particular_sol()`' as

```
>>ycis='[exp(-t) exp(-2*t)]'; f='10*sin(t)';
>>yp=particular_sol(ycis,f)
  yp = sin(t) -3*cos(t) % agrees with Eq.(3.2.17)
```

This agrees with Eq. (E3.16.4), with Eq. (3.16.5b) plugged in.

As another example, consider the DE (E3.17.1) that was dealt with in Example 3.17:

$$y''(t)+3y'(t)+2y(t) = e^{-t}$$

whose *complementary solution* is

$$y_c(t) = c_1y_1(t)+c_2y_2(t) \overset{(E3.16.3)}{=} c_1e^{-t}+c_2e^{-2t}$$

The *coefficient functions* of a particular solution can be determined by solving Eq. (3.2.13):

$$\begin{bmatrix} e^{-t} & e^{-2t} \\ -e^{-t} & -2e^{-2t} \end{bmatrix}\begin{bmatrix} k_1'(t) \\ k_2'(t) \end{bmatrix} \overset{(3.2.13)}{=} \begin{bmatrix} 0 \\ e^{-t} \end{bmatrix}; \quad \begin{bmatrix} k_1'(t) \\ k_2'(t) \end{bmatrix} = \begin{bmatrix} e^{-t} & e^{-2t} \\ -e^{-t} & -2e^{-2t} \end{bmatrix}^{-1}\begin{bmatrix} 0 \\ e^{-t} \end{bmatrix} = \begin{bmatrix} 1 \\ -e^{t} \end{bmatrix} \tag{3.2.18}$$

and integrating this as

$$\begin{bmatrix} k_1(t) \\ k_2(t) \end{bmatrix} = \begin{bmatrix} \int 1\,dt \\ -\int e^t dt \end{bmatrix} = \begin{bmatrix} t \\ -e^t \end{bmatrix} \tag{3.2.19}$$

Now we substitute these coefficient functions into Eq. (3.2.12) to get the *particular solution* as

$$y_p(t) \overset{(3.2.12)}{=} \sum_{i=1}^{N=2} k_i(t)y_i(t) \overset{(3.2.19)}{=} te^{-t} - e^te^{-2t} = te^{-t} - e^{-t} \tag{3.2.20}$$

It can also be obtained by using the MATLAB function '`particular_sol()`' as

```
>>ycis='[exp(-t) exp(-2*t)]'; f='exp(-t)';
>>yp=particular_sol(ycis,f)
  yp = t/exp(t) - 1/exp(t)   % agrees with Eq.(3.2.20)
```

Is this identical to Eq. (E3.17.5) with Eq. (E3.17.6), which is $y_p(t) = k_0te^{-t} \overset{k_0=1}{=} te^{-t}$? No, it contains an additional term $-e^{-t}$. However, there is no problem because the additional term is just one of the complementary solution terms, so that it can be absorbed into the general solution.

As another example, consider the DE (E3.18.1) that was dealt with in Example 3.18:

$$y''(t)+2y'(t)+2y(t) = 2$$

whose *complementary solution* is

$$y_c(t) \overset{(E3.18.3)}{=} d_1 e^{-t}\cos t + d_2 e^{-t}\sin t$$

The *coefficient functions* of a particular solution can be determined by solving Eq. (3.2.13) and integrating the solution as

$$\begin{bmatrix} e^{-t}\cos t & e^{-t}\sin t \\ -e^{-t}\begin{pmatrix}\cos t + \\ \sin t\end{pmatrix} & e^{-t}\begin{pmatrix}\cos t - \\ \sin t\end{pmatrix} \end{bmatrix} \begin{bmatrix} k_1'(t) \\ k_2'(t) \end{bmatrix} = \begin{bmatrix} 0 \\ 2 \end{bmatrix}; \begin{bmatrix} k_1'(t) \\ k_2'(t) \end{bmatrix} = \begin{bmatrix} -2e^t \sin t \\ 2e^t \cos t \end{bmatrix} \rightarrow \begin{bmatrix} k_1(t) \\ k_2(t) \end{bmatrix} = \begin{bmatrix} e^t(\cos t - \sin t) \\ e^t(\cos t + \sin t) \end{bmatrix} \tag{3.2.21}$$

Now we substitute these coefficient functions into Eq. (3.2.12) to get the *particular solution* as

$$y_p(t) \overset{(3.2.12)}{=} \sum_{i=1}^{N=2} k_i(t) y_i(t) \overset{(3.2.21)}{=} e^t(\cos t - \sin t) e^{-t}\cos t + e^t(\cos t + \sin t) e^{-t}\sin t = 1 \tag{3.2.22}$$

It can also be obtained by using the MATLAB function 'particular_sol()' as

```
>>ycis='[exp(-t)*cos(t) exp(-t)*sin(t)]'; f='2';
>>yp=particular_sol(ycis,f); simplify(yp)
  1  % identical to Eq.(3.2.22) and Eqs.(E3.18.4)&(E3.18.5)
```

3.2.3 Cauchy–Euler Equations

Any ODE with variable constants of the form

$$a_N t^N \frac{d^N y}{dt^N} + a_{N-1} t^{N-1} \frac{d^{N-1} y}{dt^{N-1}} + \cdots + a_1 t \frac{dy}{dt} + a_0 y = g(t) \tag{3.2.23}$$

where a_N, $a_{N-1}, \ldots$, a_0 are constants is called a Cauchy–Euler, an Euler–Cauchy equation, or an equidimensional equation. Note that the degree of t^n equals the order of derivative $d^n y/dt^n$ in each term on the LHS.

To solve this, we substitute $y = t^m$ as a trial solution

$$a_N m(m-1)(m-2)\cdots(m-N+1)t^m + \cdots + a_2 m(m-1)t^m + a_1 m t^m + a_0 t^m = 0 \tag{3.2.24}$$

to write the auxiliary equation (also called the indicial equation)

$$a_N m(m-1)(m-2)\cdots(m-N+1) + \cdots + a_2 m(m-1) + a_1 m + a_0 = 0 \tag{3.2.25}$$

The *complementary solution* of Eq. (3.2.23) depends on whether the roots of this equation are real and distinct, real and repeated, or complex (as a conjugate pair) as follows:

<**Case I**: Real and Distinct Roots $m_1, m_2, \ldots$>

$$y(t) = c_1 y_1(t) + c_2 y_2(t) + \cdots = c_1 t^{m_1} + c_2 t^{m_2} + \cdots \tag{3.2.26}$$

<**Case II**: Real and Repeated Roots $m_1, m_1, \ldots$ of multiplicity K>

$$y(t) = c_1 y_1(t) + c_2 y_2(t) + c_3 y_3(t) + \cdots + c_K y_K(t)$$

$$y(t) = c_1 y_1(t) + c_2 y_2(t) + \cdots = c_1 t^{m_1} + c_2 t^{m_2} + \cdots \quad (3.2.26) \tag{3.2.27}$$

<**Case III**: Conjugate Complex Roots $m_{1,2}=\sigma\pm j\omega$, possibly of multiplicity K>

$$y(t)=t^{\sigma}\left(c_{11}\cos(\omega\ln|t|)+c_{12}\sin(\omega\ln|t|)\right)+(\ln|t|)t^{\sigma}\left(c_{21}\cos(\omega\ln|t|)+c_{22}\sin(\omega\ln|t|)\right)+\cdots+(\ln|t|)^{K-1}t^{\sigma}\left(c_{K1}\cos(\omega\ln|t|)+c_{K2}\sin(\omega\ln|t|)\right) \tag{3.2.28}$$

The *particular solution* can be obtained by applying the method of variation of parameters (Section 3.2.2), i.e., Eqs. (3.2.12) and (3.2.13), with the set of terms $\{y_i(t)\}$ and $f(t)=g(t)/a_N t^N$, where $f(t)$ corresponds to the forcing function of the "normalized" DE that has been obtained by dividing Eq. (3.2.23) by its leading coefficient (of the highest-order term) so that the resulting leading coefficient becomes like that of Eq. (3.2.10). Note that the method of variation of parameters has been cast into the MATLAB function '`particular_sol()`' and also that the roots of the auxiliary equation (3.2.25) can be obtained by using the MATLAB function '`m_for_Cauchy_eq()`'. We can write the complete solution by adding the particular solution $y_p(t)$ to the complementary solution $y_c(t)$ (consisting of Eqs. (3.2.26)–(3.2.28)) and determining the coefficients of $y_c(t)$ from the initial conditions.

```
function [ms,aux_eq]=m_for_Cauchy_eq(a)
% Root m's of auxiliary eqn for Cauchy eqn with coefficient vector a
N=length(a);
aux_eq=a(1)*poly([0:N-2]); aux_eq(N)=aux_eq(N)+a(N);
for n=2:N-1, aux_eq(n:end)=aux_eq(n:end)+a(n)*poly([0:N-1-n]); end
ms=roots(aux_eq); % Roots of Eq.(3.2.25)
```

(Example 3.20) Homogeneous Third-Order Cauchy–Euler DE

Find the general solution of the following Cauchy–Euler DE:

$$t^3y'''(t)+8t^2y''(t)+18ty'(t)+8y(t)=0 \tag{E3.20.1}$$

Substituting $y=t^m$ into this equation yields the auxiliary equation

$$m(m-1)(m-2)+8m(m-1)+18m+8=m^3+5m^2+12m+8=(m+1)(m^2+4m+8)\overset{(3.2.25)}{=}0, \tag{E3.20.2}$$

whose solutions are

$$m=-1,\ -2\pm j2 \tag{E3.20.3}$$

These values of m can also be found by using the MATLAB function '`m_for_Cauchy_eq()`' as

```
>>m_for_Cauchy_eq([1 8 18 8])
   ans =   -2.0000 + 2.0000i
           -2.0000 - 2.0000i
           -1.0000
```

Thus the general solution of Eq. (E3.20.1) can be written as

$$y(t)\underset{(E3.20.3)}{\overset{(3.2.26,28)}{=}}c_1t^{-1}+t^{-2}\left(c_2\cos(2\ln|t|)+c_3\sin(2\ln|t|)\right) \tag{E3.20.4}$$

(Example 3.21) Nonhomogeneous Second-Order Cauchy–Euler DE (with Distinct Roots)

Find the general solution of the following Cauchy–Euler DE:

$$t^2y''(t)+4ty'(t)+2y(t)=g(t)=e^{-t}\ \text{with}\ y(1)=e^{-1}\ \text{and}\ y'(1)=1-3e^{-1} \tag{E3.21.1}$$

Substituting $y=t^m$ into this equation with $g(t)=0$ yields the auxiliary equation

$$m(m-1)+4m+2=m^2+3m+2=(m+1)(m+2)=0 \tag{E3.21.2}$$

whose solutions are

$$m=-1,\ -2 \tag{E3.21.3}$$

Thus the *complementary solution* of Eq. (E3.21.1) can be written as

$$y_c(t) \underset{(E3.21.3)}{\overset{(3.2.26)}{=}} c_1 t^{-1} + c_2 t^{-2} \tag{E3.21.4}$$

The *coefficient functions* of a particular solution can be determined by solving Eq. (3.2.13) (with $f(t)=g(t)/t^2=e^{-t}t^{-2}$) and integrating the solution as

$$\begin{bmatrix} t^{-1} & t^{-2} \\ -t^{-2} & -2t^{-3} \end{bmatrix}\begin{bmatrix} k_1'(t) \\ k_2'(t) \end{bmatrix}=\begin{bmatrix} 0 \\ e^{-t}t^{-2} \end{bmatrix};\ \begin{bmatrix} k_1'(t) \\ k_2'(t) \end{bmatrix}=\begin{bmatrix} 2t & t^2 \\ -t^2 & -t^3 \end{bmatrix}\begin{bmatrix} 0 \\ e^{-t}t^{-2} \end{bmatrix}=\begin{bmatrix} e^{-t} \\ -te^{-t} \end{bmatrix}$$

$$\rightarrow \begin{bmatrix} k_1(t) \\ k_2(t) \end{bmatrix}=\begin{bmatrix} \int e^{-t}dt \\ \int -te^{-t}\,dt \end{bmatrix}=\begin{bmatrix} -e^{-t} \\ (t+1)e^{-t} \end{bmatrix} \tag{E3.21.5}$$

We substitute these coefficient functions into Eq. (3.2.12) to get the *particular solution* as

$$y_p(t) \overset{(3.2.12)}{=} \sum_{i=1}^{N=2} k_i(t)y_i(t) \underset{(E3.21.5)}{\overset{(E3.21.4)}{=}} -e^{-t}t^{-1}+(t+1)e^{-t}t^{-2}=e^{-t}t^{-2} \tag{E3.21.6}$$

This particular solution can also be obtained by using the MATLAB function '`particular_sol()`' as

```
>>ycis='[t^-1 t^-2]'; f='t^-2*exp(-t)'; % Complementary solution terms, f(t)
>>yp=particular_sol(ycis,f); simplify(yp)
      ans = exp(-t)/t^2
```

Thus we add the complementary and particular solutions to write the *general solution* as

$$y(t)=y_c(t)+y_p(t) \overset{(E3.21.4,6)}{=} c_1t^{-1}+c_2t^{-2}+e^{-t}t^{-2} \tag{E3.21.7}$$

Now we use the initial conditions to determine the *coefficients* c_1 and c_2 as

$$\left.\begin{aligned} y(1) &\overset{(E3.21.7)}{=} c_1t^{-1}+c_2t^{-2}+e^{-t}t^{-2}\Big|_{t=1}=c_1+c_2+e^{-1} \overset{(E3.21.1)}{=} e^{-1} \\ y'(1) &\overset{(E3.21.7)}{=} -c_1t^{-2}-2c_2t^{-2}-e^{-t}t^{-2}-2e^{-t}t^{-3}\Big|_{t=1}=-c_1-2c_2-3e^{-1} \overset{(E3.21.1)}{=} 1-3e^{-1} \end{aligned}\right\};\ \begin{aligned} c_1&=1 \\ c_2&=-1 \end{aligned} \tag{E3.21.8}$$

We plug this into Eq. (E3.21.7) to write the *complete solution* as

$$y(t) \underset{(E3.21.8)}{\overset{(E3.21.7)}{=}} t^{-1}-t^{-2}+e^{-t}t^{-2} \tag{E3.17.9}$$

This solution can also be obtained by using the symbolic DE solver '`dsolve()`' as

```
>>ys=dsolve('t^2*D2y+4*t*Dy+2*y=exp(-t)','y(1)=exp(-1),Dy(1)=1-3*exp(-1)')
    ys = (t + exp(-t) - 1)/t^2
```

(Example 3.22) Nonhomogeneous Second-Order Cauchy–Euler DE (with Repeated Roots)

Find the general solution of the following Cauchy–Euler DE:

$$t^2y''(t)+3ty'(t)+y(t)=g(t)=4t \quad \text{with } y(1)=2 \text{ and } y'(1)=1 \tag{E3.22.1}$$

Substituting $y=t^m$ into this equation yields the auxiliary equation

$$m(m-1)+3m+1=m^2+2m+1=(m+1)^2=0 \tag{E3.22.2}$$

whose solutions are

$$m=-1,\,-1 \text{ (double roots)} \tag{E3.22.3}$$

Thus, the *complementary solution* of Eq. (E3.22.1) can be written as

$$y_c(t)\underset{\text{(E3.22.3)}}{\overset{\text{(3.2.27)}}{=}}c_1y_1(t)+c_2y_2(t)=c_1t^{-1}+c_2t^{-1}\ln|t| \tag{E3.22.4}$$

The *coefficient functions* of a particular solution can be determined by solving Eq. (3.2.13) (with $f(t)=g(t)/t^2=4t^{-1}$) and integrating the solution as

$$\begin{bmatrix} t^{-1} & t^{-1}\ln|t| \\ -t^{-2} & -t^{-2}\ln|t|+t^{-2}\end{bmatrix}\begin{bmatrix}k_1'(t)\\ k_2'(t)\end{bmatrix}=\begin{bmatrix}0\\ 4t^{-1}\end{bmatrix};\ \begin{bmatrix}k_1'(t)\\ k_2'(t)\end{bmatrix}=\begin{bmatrix}-t\ln|t|+t & -t^2\ln|t|\\ t & t^2\end{bmatrix}\begin{bmatrix}0\\ 4t^{-1}\end{bmatrix}=4\begin{bmatrix}-t\ln|t|\\ t\end{bmatrix}$$

$$\rightarrow \begin{bmatrix}k_1(t)\\ k_2(t)\end{bmatrix}=\begin{bmatrix}-\int 4t\ln|t|\,dt\\ \int 4t\,dt\end{bmatrix}\underset{\text{(C.34)}}{\overset{\text{(C.38)}}{=}}\begin{bmatrix}-2t^2\ln|t|+\int 2t\,dt\\ 2t^2\end{bmatrix}\overset{\text{(C.34)}}{=}\begin{bmatrix}t^2(1-2\ln|t|)\\ 2t^2\end{bmatrix} \tag{E3.22.5}$$

We substitute these coefficient functions into Eq. (3.2.12) to get the *particular solution* as

$$y_p(t)\overset{\text{(3.2.12)}}{=}\sum_{i=1}^{N=2}k_i(t)y_i(t)\underset{\text{(E3.22.5)}}{\overset{\text{(E3.22.4)}}{=}}t^2(1-2\ln|t|)t^{-1}+2t^2t^{-1}\ln|t|=t \tag{E3.22.6}$$

This particular solution can also be obtained by using the MATLAB function '`particular_sol()`' as

```
>>ycis='[t^-1 t^-1*log(t)]'; f='4*t^-1'; %Complementary solution terms, f(t)
>>yp=particular_sol(ycis,f); simplify(yp)
    ans = t
```

Thus we add the complementary and particular solutions to write the *general solution* as

$$y(t)\overset{\text{(3.2.9)}}{=}y_c(t)+y_p(t)\overset{\text{(E3.22.4,6)}}{=}c_1t^{-1}+c_2t^{-1}\ln|t|+t \tag{E3.22.7}$$

Now we use the initial conditions to determine the *coefficients* c_1 and c_2 as

$$\left.\begin{aligned} y(1)&\overset{\text{(E3.22.7)}}{=}c_1t^{-1}+c_2t^{-1}\ln|t|+t\Big|_{t=1}=c_1+1\overset{\text{(E3.22.1)}}{=}2\\ y'(1)&\overset{\text{(E3.22.7)}}{=}-c_1t^{-2}-c_2t^{-2}\ln|t|+c_2t^{-2}\Big|_{t=1}+1=-c_1+c_2+1\overset{\text{(E3.22.1)}}{=}1\end{aligned}\right\};\ \begin{aligned}c_1&=1\\ c_2&=1\end{aligned} \tag{E3.22.8}$$

We plug this into Eq. (E3.22.7) to write the *complete solution* as

$$y(t)\underset{\text{(E3.22.8)}}{\overset{\text{(E3.22.7)}}{=}}t^{-1}+t^{-1}\ln|t|+t \tag{E3.22.9}$$

This solution can also be obtained by using the symbolic DE solver '`dsolve()`' as

```
>>ys=dsolve('t^2*D2y+3*t*Dy+y=4*t','y(1)=2,Dy(1)=2')
   ys = (log(t) + t^2)/t + 1/t
```

3.2.4 Systems of Linear Differential Equations

Consider a linear 4th-order DE

$$y^{(4)}(t)+a_3y^{(3)}(t)+a_2y''(t)+a_1y'(t)+a_0y(t)=u(t) \tag{3.2.29}$$

with initial conditions $y(0)$, $y'(0)$, $y''(0)$, and $y^{(3)}(0)$

This can be written in state equation form as

$$\mathbf{x}'(t)=A\mathbf{x}(t)+\mathbf{b}u(t) \text{ with } \mathbf{x}(0)=\mathbf{x}_0=[y(0)\ y'(0)\ y''(0)\ y^{(3)}(0)]^T \tag{3.2.30a}$$

where

$$\mathbf{x}(t)=\begin{bmatrix}x_1(t)\\x_2(t)\\x_3(t)\\x_4(t)\end{bmatrix}=\begin{bmatrix}y(t)\\y'(t)\\y''(t)\\y^{(3)}(t)\end{bmatrix},\ \mathbf{x}'(t)=\begin{bmatrix}x_1'(t)\\x_2'(t)\\x_3'(t)\\x_4'(t)\end{bmatrix},\ A=\begin{bmatrix}0&1&0&0\\0&0&1&0\\0&0&0&1\\-a_0&-a_1&-a_2&-a_3\end{bmatrix},\text{ and }\mathbf{b}=\begin{bmatrix}0\\0\\0\\1\end{bmatrix} \tag{3.2.30b}$$

As will be discussed in more detail later, after the Laplace transform is introduced in Chapter 4, the solution of the state equation can be written as

$$\mathbf{x}(t)=\phi(t)\mathbf{x}(0)+\int_0^t\phi(t-\tau)\mathbf{b}u(\tau)d\tau, \tag{3.2.31}$$

Here, $\phi(t)$, called the *fundamental* or *state transition matrix* of the system described by the DE (3.2.30a), is defined in the form of a matrix exponential, i.e., the exponential of matrix At as

$$\phi(t)=e^{At}=I+At+\frac{1}{2!}(At)^2+\frac{1}{3!}(At)^3+\cdots=e^{V\Lambda V^{-1}t}=Ve^{\Lambda t}V^{-1}, \tag{3.2.32}$$

satisfying the associated homogeneous state equation

$$\phi'(t)=A\phi(t) \text{ with } \phi(0)=I \tag{3.2.33}$$

and $A=V\Lambda V^{-1}$ is the eigendecomposition of the coefficient matrix A. This is verified by substituting Eq. (3.2.32) into Eq. (3.2.33) as

$$\phi'(t)=A+A^2t+\frac{1}{2}A^3t^2+\cdots=A\left(I+At+\frac{1}{2}A^2t^2+\cdots\right)\overset{(3.2.32)}{=}A\phi(t) \tag{3.2.34}$$

(Q) Why is $\phi(t)=e^{At}$ called the state transition matrix?
(A) Because it is multiplied with $\mathbf{x}(0)$ to yield $\mathbf{x}(t)$ which is like transiting the state at a certain time to that at another time.

Depending on whether the coefficient matrix A is defective (see Example 1.12) or not, the matrix Λ having eigenvalues on its diagonal is of *Jordan canonical* (or *normal*) *form*

$$\Lambda_a = \begin{bmatrix} \lambda_1 & 1 & 0 & 0 \\ 0 & \lambda_1 & 1 & 0 \\ 0 & 0 & \lambda_1 & 0 \\ 0 & 0 & 0 & \lambda_4 \end{bmatrix} \quad \text{or} \quad \Lambda_b = \begin{bmatrix} \lambda_1 & 0 & 0 & 0 \\ 0 & \lambda_2 & 0 & 0 \\ 0 & 0 & \lambda_3 & 0 \\ 0 & 0 & 0 & \lambda_4 \end{bmatrix} \tag{3.2.35a,b}$$

and the corresponding fundamental matrix is

$$\phi(t) = Ve^{\Lambda_a t}V^{-1} = V\begin{bmatrix} e^{\lambda_1 t} & te^{\lambda_1 t} & t^2e^{\lambda_1 t}/2 & 0 \\ 0 & e^{\lambda_1 t} & e^{\lambda_1 t} & 0 \\ 0 & 0 & e^{\lambda_1 t} & 0 \\ 0 & 0 & 0 & e^{\lambda_4 t} \end{bmatrix}V^{-1} \quad \text{or} \quad Ve^{\Lambda_b t}V^{-1} = V\begin{bmatrix} e^{\lambda_1 t} & 0 & 0 & 0 \\ 0 & e^{\lambda_2 t} & 0 & 0 \\ 0 & 0 & e^{\lambda_3 t} & 0 \\ 0 & 0 & 0 & e^{\lambda_4 t} \end{bmatrix}V^{-1} \tag{3.2.36a,b}$$

Note that as the solution to the linear first-order DE

$$x'(t) = -a\,x(t) + f(t), \tag{3.2.37}$$

Eq. (3.1.24) will be

$$x(t) = e^{-at}x(0) + \int_0^t e^{-a(t-\tau)} f(\tau)\,d\tau \tag{3.2.38}$$

and it is nothing but Eq. (3.2.31) with $\phi(t)=e^{-at}$ and $\mathbf{b}u(t)=f(t)$.

If you know about the Laplace transform, which will be introduced in Chapter 4, you can get the fundamental matrix by taking the inverse Laplace transform of $[sI-A]^{-1}$ as

$$\phi(t) = \mathcal{L}^{-1}\left\{[sI-A]^{-1}\right\}, \tag{3.2.39}$$

because the Laplace transform of the fundamental matrix (3.2.32) is

$$\begin{aligned} \Phi(s) = \mathcal{L}\{\phi(t)\} &= s^{-1}I + As^{-2} + A^2s^{-3} + \cdots \\ &= s^{-1}(I + As^{-1} + A^2s^{-1} + \cdots) = s^{-1}[I - As^{-1}]^{-1} \end{aligned}$$

$$; \qquad \Phi(s) = [sI - A]^{-1} \tag{3.2.40}$$

Eq. (3.2.40) can also be obtained by taking the Laplace transform of Eq. (3.2.33) as

$$\mathcal{L}\{\phi'(t)\} \overset{\text{Table A.7(5)}}{=} s\Phi(s) - \phi(0) = A\Phi(s); \; [sI-A]\Phi(s) = \phi(0) = I$$

$$; \qquad \Phi(s) = [sI - A]^{-1} \tag{3.2.41}$$

Further, the solution (3.2.31) of the DE (3.2.30a) can be obtained by taking the inverse Laplace transform of Eq. (3.2.43) as

$$\mathbf{x}(t) \overset{(3.2.43)}{=} \mathcal{L}^{-1}\left\{[sI-A]^{-1}\mathbf{x}(0) + [sI-A]^{-1}\mathbf{b}U(s)\right\} \tag{3.2.42}$$

because the Laplace transform of the DE (3.2.30a) yields

$$\mathbf{x}'(t) \overset{(3.2.30a)}{=} A\,\mathbf{x}(t)+\mathbf{b}\,u(t) \underset{\text{Table A.7(5)}}{\overset{\mathcal{L}}{\rightarrow}} sX(s)-\mathbf{x}(0)=AX(s)+\mathbf{b}\,U(s)$$

$$;\qquad [sI-A]X(s)=\mathbf{x}(0)+\mathbf{b}\,U(s)$$

$$;\qquad X(s)=[sI-A]^{-1}\mathbf{x}(0)+[sI-A]^{-1}\mathbf{b}U(s) \tag{3.2.43}$$

(Example 3.23) Nonhomogeneous Linear Second-Order DE with Constant Coefficients

Solve the following DE that was dealt with in Example 3.18:

$$y''(t)+2y'(t)+2y(t)=2 \text{ with } y(0)=2 \text{ and } y'(0)=1 \tag{E3.23.1}$$

This can be written in state equation form as

$$\mathbf{x}'(t)\underset{(E3.23.1)}{\overset{(3.2.30a)}{=}}A\,\mathbf{x}(t)+\mathbf{b}\,u(t)\,;\ \begin{bmatrix}x_1'\\x_2'\end{bmatrix}=\begin{bmatrix}0&1\\-2&-2\end{bmatrix}\begin{bmatrix}x_1\\x_2\end{bmatrix}+\begin{bmatrix}0\\2\end{bmatrix} \text{ with } \begin{bmatrix}x_1(0)\\x_2(0)\end{bmatrix}=\begin{bmatrix}y(0)\\y'(0)\end{bmatrix}=\begin{bmatrix}2\\1\end{bmatrix} \tag{E3.23.2}$$

Note that the eigendecomposition of the coefficient matrix A is

$$A=\begin{bmatrix}0&1\\-2&-2\end{bmatrix}=\begin{bmatrix}-1-j&-1+j\\2&2\end{bmatrix}\begin{bmatrix}\lambda_1=-1+j&0\\0&\lambda_2=-1-j\end{bmatrix}\begin{bmatrix}j/2&(1+j)/4\\-j/2&(1-j)/4\end{bmatrix}=V\Lambda V^{-1} \tag{E3.23.3}$$

and accordingly, the fundamental matrix of the system described by Eq. (E3.23.2) is

$$\phi(t)\underset{(3.2.36b)}{\overset{(3.2.32)}{=}}Ve^{\Lambda t}V^{-1}=\begin{bmatrix}-1-j&-1+j\\2&2\end{bmatrix}\begin{bmatrix}e^{\lambda_1 t}=e^{(-1+j)t}&0\\0&e^{\lambda_2 t}=e^{(-1-j)t}\end{bmatrix}\begin{bmatrix}j/2&(1+j)/4\\-j/2&(1-j)/4\end{bmatrix}$$

$$=\begin{bmatrix}e^{-t}(\cos t+\sin t)&e^{-t}\sin t\\-2e^{-t}\sin t&e^{-t}(\cos t-\sin t)\end{bmatrix}. \tag{E3.23.4}$$

Now we use Eq. (3.2.31) together with Eq. (E3.23.4) to write the solution to Eq. (E3.23.2) as

$$\mathbf{x}(t)\underset{(E3.23.4)}{\overset{(3.2.31)}{=}}\begin{bmatrix}e^{-t}(\cos t+\sin t)&e^{-t}\sin t\\-2e^{-t}\sin t&e^{-t}(\cos t-\sin t)\end{bmatrix}\begin{bmatrix}2\\1\end{bmatrix}+\int_0^t\begin{bmatrix}2e^{-(t-\tau)}\sin(t-\tau)\\2e^{-(t-\tau)}(\cos(t-\tau)-\sin(t-\tau))\end{bmatrix}d\tau$$

$$=\begin{bmatrix}e^{-t}(2\cos t+3\sin t)\\e^{-t}(\cos t-5\sin t)\end{bmatrix}+\begin{bmatrix}e^{-(t-\tau)}(\cos(t-\tau)+\sin(t-\tau))\big|_0^t\\-2e^{-(t-\tau)}\sin(t-\tau)\big|_0^t\end{bmatrix}$$

$$=\begin{bmatrix}e^{-t}(2\cos t+3\sin t)\\e^{-t}(\cos t-5\sin t)\end{bmatrix}+\begin{bmatrix}1-e^{-t}(\cos t+\sin t)\\2e^{-t}\sin t\end{bmatrix}=\begin{bmatrix}1+e^{-t}(\cos t+2\sin t)\\e^{-t}(\cos t-3\sin t)\end{bmatrix} \tag{E3.23.5}$$

```
>>syms t tau; x0=[2; 1]; b=[0; 1];
>>phi=exp(-t)*[cos(t)+sin(t) sin(t); -2*sin(t) cos(t)-sin(t)];%Eq.(E3.23.4)
>>xt=phi*x0+int(subs(phi,t,t-tau)*b*2,tau,0,t) %Eq.(3.2.31)
   xt = exp(-t)*sin(t) + exp(-t)*(cos(t) + sin(t)) + 1
              exp(-t)*(cos(t) - sin(t)) - 2*exp(-t)*sin(t))
```

This gives the solution of the DE (E3.23.1), which agrees with Eq. (E3.18.8), as

$$y(t)=x_1(t)=1+e^{-t}(\cos t+2\sin t) \tag{E3.23.6}$$

Also, using the Laplace transform solution (3.2.43), we can get the same result:

$$X(s) \overset{(3.2.43)}{=} [sI-A]^{-1}(\mathbf{x}(0)+\mathbf{b}U(s)) = \begin{bmatrix} s & -1 \\ 2 & s+2 \end{bmatrix}^{-1}\left(\begin{bmatrix} 2 \\ 1 \end{bmatrix} + \begin{bmatrix} 0 \\ 1 \end{bmatrix}\frac{2}{s}\right) = \frac{1}{s(s^2+2s+2)}\begin{bmatrix} s+2 & 1 \\ -2 & s \end{bmatrix}\begin{bmatrix} 2s \\ s+2 \end{bmatrix}$$

$$= \frac{1}{s(s^2+2s+2)}\begin{bmatrix} 2s^2+5s+2 \\ s^2-2s \end{bmatrix} = \begin{bmatrix} 1/s+\{(s+1)+2\}/\{(s+1)^2+1^2\} \\ \{(s+1)-3\}/\{(s+1)^2+1^2\} \end{bmatrix} \quad \text{(E3.23.7)}$$

$$; \quad y(t) = x_1(t) = \mathcal{L}^{-1}\{X_1(s)\} = \mathcal{L}^{-1}\left\{\frac{1}{s} + \frac{(s+1)+2}{(s+1)^2+1^2}\right\} \overset{A.8(3),(12),(10)}{=} 1 + e^{-t}(\cos t + 2\sin t) \quad \text{(E3.23.8)}$$

We can ask MATLAB to help us with this computation:

```
>>syms s; A=[0 1; -2 -2]; b=[0; 1]; x0=[2; 1]; Us=2/s;
>>xt=ilaplace((s*eye(size(A))-A)\(x0+b*Us)) % Eq.(3.2.43)
        xt = (cos(t) + 2*sin(t))/exp(t) + 1   % Eq.(E3.23.8)
             (cos(t) - 3*sin(t))/exp(t)
```

(Example 3.24) Nonhomogeneous Linear Third-Order DE with Constant Coefficients

Solve the following DE that was dealt with in Example 3.19:

$$y^{(3)}(t)+a_2 y''(t)+a_1 y'(t)+a_0 y(t) = y^{(3)}(t)+4y''(t)+5y'(t)+2y(t) = 2 \quad \text{(E3.24.1a)}$$

$$\text{with initial conditions } y(0)=5,\ y'(0)=-5, \text{ and } y''(0)=9 \quad \text{(E3.24.1b)}$$

This can be written in state equation form as

$$\mathbf{x}'(t)=A\mathbf{x}(t)+\mathbf{b}u(t) \text{ with } A=\begin{bmatrix} 0 & 1 & 0 \\ 0 & 0 & 1 \\ -2 & -5 & -4 \end{bmatrix}, \mathbf{x}(t)=\begin{bmatrix} x_1(t) \\ x_2(t) \\ x_3(t) \end{bmatrix}=\begin{bmatrix} y(t) \\ y'(t) \\ y''(t) \end{bmatrix}, \mathbf{b}=\begin{bmatrix} 0 \\ 0 \\ 1 \end{bmatrix}, \text{ and } u(t)=2 \quad \text{(E3.24.2)}$$

The eigenvalues, modal matrix (consisting of eigenvectors and generalized one), and Jordan canonical form of the coefficient matrix A were obtained in Example 1.12(b) as

$$A = V\Lambda V^{-1} \overset{(E1.12.11)}{=} \overset{\mathbf{v}_1\ \ \mathbf{v}_2\ \ \mathbf{v}_3}{\begin{bmatrix} 1 & 0 & 1 \\ -1 & 1 & -2 \\ 1 & -2 & 4 \end{bmatrix}}\begin{bmatrix} \lambda_1=-1 & 1 & 0 \\ 0 & \lambda_1=-1 & 0 \\ 0 & 0 & \lambda_2=-2 \end{bmatrix}\begin{bmatrix} 0 & -2 & -1 \\ 2 & 3 & 1 \\ 1 & 2 & 1 \end{bmatrix} \quad \text{(E3.24.3)}$$

Accordingly, the fundamental matrix of the system described by Eq. (E3.24.2) is

$$\phi(t) \overset{(3.2.36a)}{=} Ve^{\Lambda t}V^{-1} \overset{(3.2.35a)}{=} \begin{bmatrix} 1 & 0 & 1 \\ -1 & 1 & -2 \\ 1 & -2 & 4 \end{bmatrix}\begin{bmatrix} e^{\lambda_1 t} & te^{\lambda_1 t} & 0 \\ 0 & e^{\lambda_1 t} & 0 \\ 0 & 0 & e^{\lambda_3 t} \end{bmatrix}\begin{bmatrix} 0 & -2 & -1 \\ 2 & 3 & 1 \\ 1 & 2 & 1 \end{bmatrix}$$

$$= \begin{bmatrix} e^{-2t}+2te^{-t} & 2e^{-2t}-2e^{-t}+3te^{-t} & e^{-2t}-e^{-t}+te^{-t} \\ 2e^{-t}-2e^{-2t}-2te^{-t} & 5e^{-t}-4e^{-2t}-3te^{-t} & 2e^{-t}-2e^{-2t}-te^{-t} \\ 4e^{-2t}-4e^{-t}+2te^{-t} & 8e^{-2t}-8e^{-t}+3te^{-t} & 4e^{-2t}-3e^{-t}+te^{-t} \end{bmatrix} \quad \text{(E3.24.4)}$$

```
>>A=[0 1 0; 0 0 1;-2 -5 4]; x0=[5;-5;9]; bu=[0;0;2]; syms t; % Eq.(E3.24.2)
>>V=[1 0 1;-1 1 -2;1 -2 4]; L=[-1 1 0;0 -1 0;0 0 -2]; % Eigendecomposition
>>phi=V*expm(L*t)*V^-1; phi_=expm(A*t) % Fundamental matrix (E3.24.4)
```

Thus, we use Eq. (3.2.31) to write the solution to Eq. (E3.24.2) as

$$\mathbf{x}(t) \overset{(3.2.31)}{=} \phi(t)\,\mathbf{x}(0) + \int_0^t \phi(t-\tau)\,\mathbf{b}\,u(\tau)\,d\tau = \begin{bmatrix} 1+e^{-t}+2te^{-t}+3e^{-2t} \\ e^{-t}-2te^{-t}-6e^{-2t} \\ -3e^{-t}+2te^{-t}+12e^{-2t} \end{bmatrix} \tag{E3.24.5}$$

which has been computed by running the following MATLAB statements:

```
>>syms t tau;   x0=[5;-5;9]; bu=[0;0;2];
>>xt=phi*x0+int(subs(phi,t,t-tau)*bu,tau,0,t) % Eq.(E3.24.5)
```

If we use the Laplace transform solution Eq. (3.2.43), we can get the above result directly in much less time:

```
>>syms s; A=[0 1 0;0 0 1;-2 -5 -4]; b=[0; 0; 1]; x0=[5;-5;9]; Us=2/s;
>>xt=ilaplace((s*eye(size(A))-A)\(x0+b*Us)); simplify(xt) % Eq.(3.2.43)
    ans =  exp(-t) + 3*exp(-2*t) + 2*t*exp(-t) + 1
                -exp(-2*t)*(2*t*exp(t) - exp(t) + 6)
               exp(-2*t)*(2*t*exp(t) - 3*exp(t) + 12)
```

This gives the solution of the DE (E3.24.1) as

$$y(t) = x_1(t) = 1 + e^{-t} + 2te^{-t} + 3e^{-2t}, \tag{E3.24.6}$$

which agrees with Eq. (E3.19.8).

3.3 SPECIAL SECOND-ORDER LINEAR ODES

3.3.1 Bessel's Equation [Chapter 7 of C-1][Chapter 12 of W-1][K-2]

The *Bessel's (differential) equation of order* $\nu \geq 0$ is

$$x^2 y'' + x y' + (x^2 - \nu^2) y = 0 \tag{3.3.1}$$

A solution of this equation can be found by substituting a power series for y:

$$y = \sum_{n=0}^{\infty} c_n x^{n+r} \text{ with } c_0 \neq 0 \tag{3.3.2}$$

as an assumed solution:

$$\sum_{n=0}^{\infty} c_n (n+r)(n+r-1) x^{n+r} + \sum_{n=0}^{\infty} c_n (n+r) x^{n+r} + \sum_{n=0}^{\infty} c_n x^{n+r+2} - \nu^2 \sum_{n=0}^{\infty} c_n x^{n+r}$$

$$= x^r \left\{ \sum_{n=0}^{\infty} c_n \left((n+r)^2 - \nu^2 \right) x^n + \sum_{n=0}^{\infty} c_n x^{n+2} \right\}$$

$$= x^r \left\{ c_0 \left(r^2 - \nu^2 \right) + c_1 \left((1+r)^2 - \nu^2 \right) x + \sum_{n=2}^{\infty} c_n \left((n+r)^2 - \nu^2 \right) x^n + \sum_{n=2}^{\infty} c_{n-2} x^n \right\} = 0 \tag{3.3.3}$$

This yields

$$c_0(r^2 - v^2) = 0: \text{ Indicial equation} \rightarrow \text{Indicial roots}: r = \pm v \tag{3.3.4a}$$

$$c_1\left\{(1+r)^2 - v^2\right\} = 0 \tag{3.3.4b}$$

$$c_n\left\{(n+r)^2 - v^2\right\} + c_{n-2} = 0 \quad \text{for } n \geq 2 \tag{3.3.4c}$$

Substituting one of the indicial roots of the indicial equation (3.3.4a), say $r=v$, into Eqs. (3.3.4b) and (3.3.4c) yields

$$(3.3.4b) \rightarrow c_1\left\{(1+r)^2 - v^2\right\} \overset{r=v}{=} c_1(1+2v) = 0 \overset{v \geq 0}{\rightarrow} c_1 = 0 \tag{3.3.6a}$$

$$(3.3.4c) \rightarrow c_n = \begin{cases} -\dfrac{c_{n-2}}{n(n+2v)} = -\dfrac{c_{n-2}}{2^2 m(m+v)} & \text{for } n=2m(\text{even}) \geq 2 \\ \dfrac{c_{n-2}}{n(n+2v)} = 0 \text{ since } c_1 = 0 & \text{for } n=2m+1(\text{odd}) \geq 3 \end{cases} \quad : \text{Recurrence relation} \tag{3.3.6b}$$

so that

$$c_{2m} = \frac{(-1)^m c_0}{2^{2m} m!\,(1+v)(2+v)\cdots(m+v)} \tag{3.3.7}$$

With a standard practical choice of c_0 as

$$c_0 = \frac{1}{2^v \Gamma(v+1)} \tag{3.3.8}$$

where we have used the *gamma function* $\Gamma(v) = \int_0^\infty t^{v-1} e^{-t} dt$ such that $\Gamma(v+1) = v\Gamma(v)$ and $\Gamma(n+1) = n!$ for integer n, Eq. (3.3.7) becomes

$$c_{2m} = \frac{(-1)^m}{2^{2m+v} m!\,\Gamma(m+v+1)} \tag{3.3.9}$$

We can substitute this into Eq. (3.3.2) to obtain a series solution of the Bessel's equation as

$$y = J_v(x) \overset{(3.3.2)}{=} \sum_{n=0}^{\infty} c_n x^{n+v} \overset{n=2m}{=} \sum_{m=0}^{\infty} c_{2m} x^{2m+v} \overset{(3.3.9)}{=} \sum_{m=0}^{\infty} \frac{(-1)^m}{m!\,\Gamma(m+v+1)} \left(\frac{x}{2}\right)^{2m+v} \tag{3.3.10a}$$

Substituting another of the indicial roots of the indicial equation (3.3.4a), $r=-v$, into Eqs. (3.3.4b) and (3.3.4c) would have yielded

$$y = J_{-v}(x) = \sum_{m=0}^{\infty} \frac{(-1)^m}{m!\,\Gamma(m-v+1)} \left(\frac{x}{2}\right)^{2m-v} \tag{3.3.10b}$$

Eqs. (3.3.10a) and (3.3.10b) are known as the *Bessel functions of the first kind*, *cylinder functions*, or *cylindrical harmonics of order v*.

The general solution of the Bessel's equation depends on whether its order v is an integer or not:

<**Case I**> If the order v is not an integer, the general solution of the Bessel's equation is

$$y = a_1 J_v(x) + a_2 J_{-v}(x) \quad \text{for all } x \neq 0 \tag{3.3.11}$$

<**Case II**> If the order v is an integer n, the two Bessel functions $J_n(x)$ and $J_{-n}(x)$ are linearly dependent:

$$J_{-n}(x) = (-1)^n J_n(x) \tag{3.3.12}$$

and the general solution of the Bessel's equation for any value of v is

$$y = a J_v(x) + b Y_v(x) \tag{3.3.13}$$

where $Y_v(x)$ is the *Bessel function of the second kind*, *Neumann function*, or *Weber function of order n*, which is defined as

$$Y_v(x) = \frac{1}{\sin v\pi}\left(J_v(x)\cos v\pi - J_{-v}(x)\right) \quad \forall\, v \tag{3.3.14a}$$

$$Y_n(x) = \lim_{v\to n} Y_v(x) \quad \text{for integer values of } n \tag{3.3.14b}$$

● Integral Representation of the Bessel Function of the First Kind

Another definition of the Bessel function is the *integral representation of the Bessel function of the first kind*:

$$J_n(x) = \frac{1}{2\pi}\int_{-\pi}^{\pi} e^{-j(n\tau - x\sin\tau)}\,d\tau = \frac{1}{\pi}\int_0^{\pi}\cos(n\tau - x\sin\tau)\,d\tau \;\; (j=\sqrt{-1}) \tag{3.3.15}$$

```
%plot_Bessel_ftn12.m
% Plot Bessel functions of the 1st/2nd kinds of order n=0:5
clear, clf
x=[1e-5:0.02:10]; % Domain
for n=1:6
   Jnx(n,:)=besselj(n-1,x); v=n-1+1e-6;  Ynx(n,:)=bessely(v,x);
   Ynx1(n,:)=(besselj(v,x)*cos(v*pi)-besselj(-v,x))/sin(v*pi); % Eq(3.3.14a)
end
[J,Jnx1]=besselj_my(5,x); % Bessel functions of upto order 5
discrepancy=norm(Jnx-Jnx1)/norm(Jnx) norm(Ynx-Ynx1)/norm(Ynx)]
subplot(211), plot(x,Jnx.'), axis([0 10 -0.5 1.1])
subplot(212), plot(x,Ynx1.'), axis([0 10 -1.5 0.6])
```

```
function [J,JJ]=besselj_my(N,x)
% Bessel function of the first kind of order n like besselj(n,x)
tmpn=ones(size(x));
for n=0:N
   tmp= tmpn;   JJ(n+1,:)= tmp;
   for m=1:100
      tmp=-tmp.*x.*x/4/m/(m+n); JJ(n+1,:)=JJ(n+1,:)+tmp; % Eq.(3.3.10a)
   end
   tmpn=tmpn.*x/2/(n+1);
end
J=JJ(N+1,:);
```

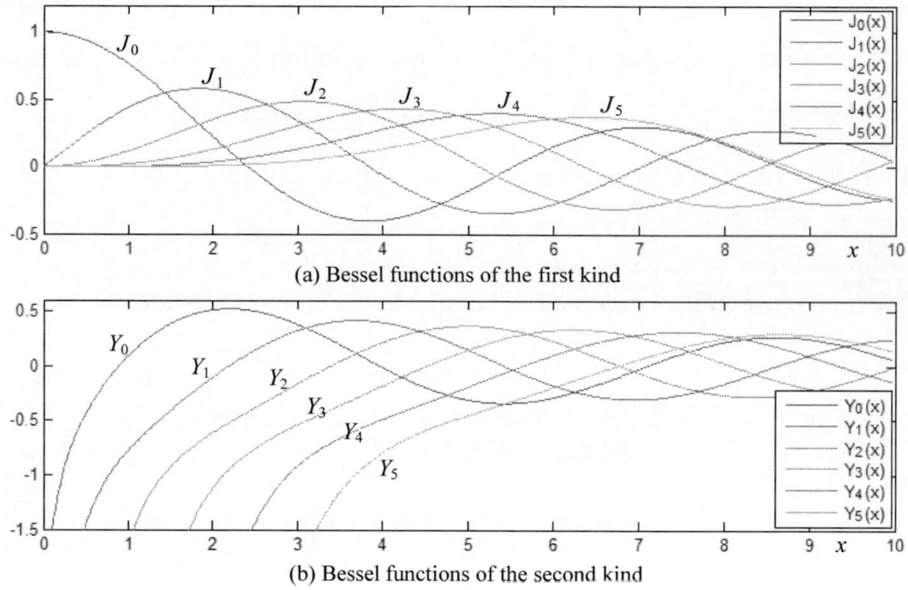

Figure 3.15 Bessel functions of the first and second kinds

Fig. 3.15 shows the graphs of Bessel functions of the first and second kinds of order n=0:5, which are obtained by running the above script "plot_Bessel_ftn12.m," where the Bessel functions of the first/second kinds, $J_v(x)/Y_v(x)$, are computed by using the MATLAB functions '`besselj()`'/ '`bessely()`', respectively.

Another set of two linearly independent solutions to Bessel's equation consists of the *Bessel functions of the third kind*, called the *Hankel functions*:

$$H_v^{(1)}(x) = J_v(x) + j\,Y_v(x) \quad (j = \sqrt{-1}) \tag{3.3.16a}$$

$$H_v^{(2)}(x) = J_v(x) - j\,Y_v(x) \tag{3.3.16b}$$

Running the script "get_Bessel_ftn3.m" yields

```
discrepancy =      1.0e-03 *
                     0.1497      0.1497
```

This implies that the values of the Hankel functions computed by using Eq. (3.3.16) and the MATLAB function '`besselh()`' agree with each other.

```
%get_Bessel_ftn3.m
% Bessel functions of the 3rd kind of order n=0:5
clear, clf
x=[1e-2:0.02:5]; % Domain
for n=1:6
   Hn1x(n,:)=besselh(n-1,1,x); Hn2x(n,:)=besselh(n-1,2,x);
   Jnx(n,:)=besselj(n-1,x);     Ynx(n,:)=bessely(n-1,x);
   Hn1x1(n,:)=Jnx(n,:)+i*Ynx(n,:); % Eq.(3.3.16a)
   Hn2x1(n,:)=Jnx(n,:)-i*Ynx(n,:); % Eq.(3.3.16b)
end
discrepancy=[norm(Hn1x-Hn1x1)/norm(Hn1x) norm(Hn2x-Hn2x1)/norm(Hn2x)]
```

● **Spherical Bessel Functions**

As linearly independent solutions to the following second-order ODE

$$x^2y'' + 2xy' + \left(x^2 - n(n+1)\right)y = 0 \quad (n\text{: integer}) \tag{3.3.17}$$

the *spherical Bessel functions* of the first/second/third kinds are defined by

$$j_n(x) = \sqrt{\frac{\pi}{2x}}J_{n+1/2}(x) = (-x)^n\left(\frac{1}{x}\frac{d}{dx}\right)^n \frac{\sin x}{x} \tag{3.3.18}$$

$$y_n(x) = \sqrt{\frac{\pi}{2x}}Y_{n+1/2}(x) = -(-x)^n\left(\frac{1}{x}\frac{d}{dx}\right)^n \frac{\cos x}{x} \tag{3.3.19}$$

$$h_v^{(1)}(x) = \sqrt{\frac{\pi}{2x}}H_{v+1/2}^{(1)}(x) = j_v(x) + j\,y_v(x) \quad (j = \sqrt{-1}) \tag{3.3.20a}$$

$$h_v^{(2)}(x) = \sqrt{\frac{\pi}{2x}}H_{v+1/2}^{(2)}(x) = j_v(x) - j\,y_v(x) \tag{3.3.20b}$$

● **Parametric Bessel's Equation**

Substituting λx and λdx for x and dx, respectively, into the Bessel's equation (3.3.1) yields the *parametric Bessel's equation* of order $v \ge 0$:

$$x^2y'' + xy' + (\lambda^2x^2 - v^2)y = 0 \tag{3.3.21}$$

A general solution of this equation can be obtained by substituting λr for x into Eq. (3.3.13) as

$$y = a(\lambda)J_v(\lambda x) + b(\lambda)Y_v(\lambda x) \tag{3.3.22}$$

● **Modified Bessel's Equation**

Substituting $j = \sqrt{-1}$ for λ into the parametric Bessel's equation (3.3.21) yields the *modified Bessel's equation of order* $v \ge 0$ as

$$x^2y'' + xy' - (x^2 + v^2)y = 0 \tag{3.3.23}$$

The general solution of the modified Bessel's equation depends on whether its order v is an integer or not:

<Case I> If the order v is not an integer, the general solution of the modified Bessel's equation is

$$y = a_1 I_v(x) + a_2 I_{-v}(x) \tag{3.3.24}$$

where

$$I_v(x) = j^{-v}J_v(jx) \overset{(3.3.10a)}{=} \sum_{m=0}^{\infty} \frac{1}{m!\,\Gamma(m+v+1)}\left(\frac{x}{2}\right)^{2m+v} \tag{3.3.25}$$

Eq. (3.3.25) is known as the *modified Bessel function of the first kind* of order v.

<Case II> If the order v is an integer n, the two modified Bessel functions $I_n(x)$ and $I_{-n}(x)$ are linearly dependent:

$$I_{-n}(x) = I_n(x) \tag{3.3.26}$$

and the general solution of the modified Bessel's equation for any value of v is

$$y = a_1 I_v(x) + a_2 K_v(x) \tag{3.3.27}$$

where $K_v(x)$ is the *modified Bessel function of the second kind* of order v, which is defined as

$$K_v(x) = \frac{\pi}{2\sin v\pi}\left(I_{-v}(x) - I_v(x)\right) \;\; \forall\, v \tag{3.3.28a}$$

$$K_n(x) = \lim_{v\to n} K_v(x) \;\; \text{for integer values of } n \tag{3.3.28b}$$

Fig. 3.16 shows the graphs of modified Bessel functions of the first and second kinds of order n=0:5, which are obtained by running the script "plot_modified_Bessel_ftn12.m" where the modified Bessel functions of the first/second kinds, $I_v(x)/K_v(x)$, are computed by using the MATLAB functions 'besseli()'/ 'besselk()', respectively.

```
%plot_modified_Bessel_ftn12.m
% Plot modified_Bessel functions of the 1st/2nd kinds of order n=0:5
clear, clf
x=[1e-5:0.02:6]; % Domain
for n=1:6
   v=n-1+1e-6; % To avoid division-by-zero
   Inx(n,:)=besseli(n-1,x);
   Knx(n,:)=besselk(v,x);
   Knx1(n,:)=pi/2/sin(v*pi)*(besseli(-v,x)-besseli(v,x)); % Eq.(3.3.28a)
end
discrepancy = norm(Knx-Knx1)/norm(Knx)
subplot(121), plot(x,Inx.'), axis([0 6 0 10])
legend('I_0 (x)','I_1 (x)','I_2 (x)','I_3 (x)','I_4 (x)','I_5 (x)')
subplot(122), plot(x,Knx.'), axis([0 6 0 10])
legend('K_0 (x)','K_1 (x)','K_2 (x)','K_3 (x)','K_4 (x)','K_5 (x)')
```

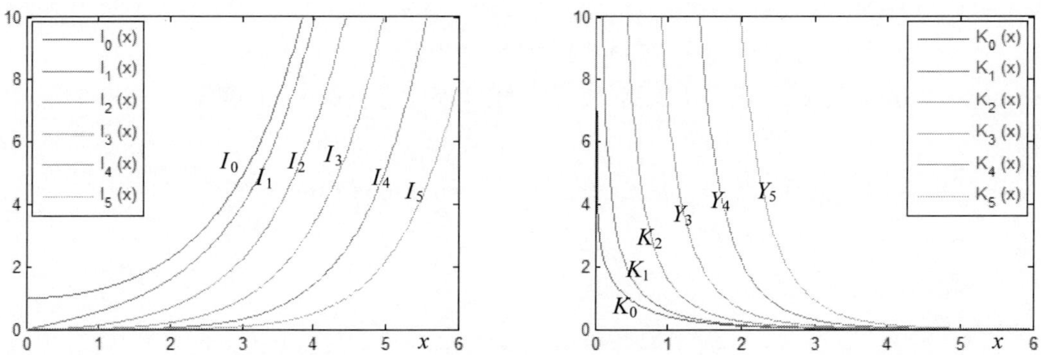

(a) Modified Bessel functions of the first kind (b) Modified Bessel functions of the second kind

Figure 3.16 Modified Bessel functions of the first and second kinds

3.3.2 Legendre's Equation [Chapter 5 of C-1]

The *Legendre's (differential) equation* is

$$(1-x^2)y''-2xy'+n(n+1)y=0 \text{ for } n>0 \text{ and } |x|<1 \tag{3.3.29}$$

A solution of this equation can be found by substituting a power series for y:

$$y=\sum_{m=0}^{\infty} c_m x^m \tag{3.3.30}$$

as an assumed solution:

$$(1-x^2)\sum_{m=2}^{\infty} c_m\, m(m-1)x^{m-2}-2x\sum_{m=1}^{\infty} c_m\, m x^{m-1}+n(n+1)\sum_{m=0}^{\infty} c_m x^m=0$$

$$;\sum_{m=0}^{\infty} c_{m+2}(m+2)(m+1)x^m-\sum_{m=2}^{\infty} c_m\, m(m-1)x^m-2\sum_{m=1}^{\infty} c_m m x^m+n(n+1)\sum_{m=0}^{\infty} c_m x^m=0 \tag{3.3.31}$$

This yields

$$2c_2+n(n+1)c_0=0: \quad \text{for } m=0 \text{(coefficients of } x^0) \tag{3.3.32a}$$

$$6c_3+\{n(n+1)-2\}c_1=0: \quad \text{for } m=1 \text{ (coefficients of } x^1) \tag{3.3.32b}$$

$$c_{m+2}(m+2)(m+1)-c_m\{m(m+1)-n(n+1)\}=0 \text{ for } m\geq 2$$

$$\rightarrow c_{m+2}=\frac{m(m+1)-n(n+1)}{(m+2)(m+1)}c_m=\frac{(m-n)(m+n+1)}{(m+2)(m+1)}c_m: \text{recurrence relation} \tag{3.3.32c}$$

Once c_0 and c_1 are somehow fixed, the remaining coefficients can be found successively by using the recurrence relation or recursion formula (3.3.32c):

$$c_2 \overset{(3.3.32a)}{=} -\frac{n(n+1)}{2!}c_0,\ c_4 \underset{m=2}{\overset{(3.3.32c)}{=}} -\frac{(n-2)(n+3)}{4\cdot 3}c_2=\frac{(n-2)n(n+1)(n+3)}{4!}c_0,\ \cdots \tag{3.3.33a}$$

$$c_3 \overset{(3.3.32b)}{=} -\frac{(n-1)(n+2)}{3!}c_1,\ c_5 \underset{m=3}{\overset{(3.3.32c)}{=}} -\frac{(n-3)(n+4)}{5\cdot 4}c_3=\frac{(n-3)(n-1)(n+2)(n+4)}{4!}c_1,\ \cdots \tag{3.3.33b}$$

Plugging these coefficient values into Eq. (3.3.30) yields a general solution for $|x|<1$:

$$y(x)=c_0 y_1(x)+c_1 y_2(x) \tag{3.3.34}$$

with

$$y_1(x)=1-\frac{n(n+1)}{2!}x^2+\frac{(n-2)n(n+1)(n+3)}{4!}x^4-\cdots \tag{3.3.35a}$$

$$y_2(x)=x-\frac{(n-1)(n+2)}{3!}x^3+\frac{(n-3)(n-1)(n+2)(n+4)}{5!}x^5-\cdots \tag{3.3.35b}$$

Since the coefficients obtained using Eq. (3.3.32c) are zero for even/odd $m>n$, either $y_1(x)$ or $y_2(x)$ becomes a *Legendre polynomial* of degree n depending on whether n is even or odd, where c_0 and c_1 are chosen differently depending on the order n:

$$c_0 = \begin{cases} 1 & \text{for } n=0 \\ (-1)^{n/2} \dfrac{1\cdot 3\cdot \cdots \cdot (n-1)}{2\cdot 4\cdot \cdots \cdot n} & \text{for } n=2,4,6,\cdots \end{cases} \tag{3.3.36a}$$

$$c_1 = \begin{cases} 1 & \text{for } n=1 \\ (-1)^{(n-1)/2} \dfrac{1\cdot 3\cdot \cdots \cdot n}{2\cdot 4\cdot \cdots \cdot (n-1)} & \text{for } n=3,5,7,\cdots \end{cases} \tag{3.3.36b}$$

The first several Legendre polynomial $P_n(x)$'s of degree n are

$$\begin{array}{l|l} P_0(x) = 1 & P_1(x) = x \\ P_2(x) = \dfrac{1}{2}(3x^2 - 1) & P_3(x) = \dfrac{1}{2}(5x^3 - 3x) \\ P_4(x) = \dfrac{1}{8}(35x^4 - 30x^2 + 3) & P_5(x) = \dfrac{1}{8}(63x^5 - 70x^3 + 15x) \end{array} \tag{3.3.37}$$

The Legendre polynomials can be generated by using the following three-term recurrence relation:

$$P_{n+1}(x) = \frac{(2n+1)x P_n(x) - n P_{n-1}(x)}{n+1} \quad \text{with } P_1(x) = x \text{ and } P_0(x) = 1 \tag{3.3.38}$$

or a *Rodrigues' formula*

$$P_n(x) = \frac{1}{2^n n!} \frac{d^n}{dx^n}(x^2 - 1)^n \quad \text{for } |x| < 1 \text{ and } n = 0, 1, 2, \cdots \tag{3.3.39}$$

or an explicit formula

$$P_n(x) = \sum_{m=0}^{\text{floor}(n/2)} (-1)^m \frac{(2n-2m)!}{2^n m!(n-m)!(n-2m)!} x^{n-2m} \quad \text{for } |x| < 1 \text{ and } n = 0, 1, 2, \cdots \tag{3.3.40}$$

```
%plot_Legendre_poly.m
% Plot Legendre polynomials of degree n=0:5
clear, clf
x=[-1:0.01:1]; % Domain
Ps=[zeros(1,5) 1; zeros(1,4) 1 0; zeros(4,6)]; % including P0 and P1
for n=1:6
   if n>2
     tmp = Ps(n-1,:); % P_(n-1): Legendre polynomial of degree n-1
     tmp = [tmp(2:end) 0]; % multiply P_(n-1) by x
     Ps(n,:)=((2*n-3)*tmp-(n-2)*Ps(n-2,:))/(n-1);  % Eq.(3.3.38)
   end
   P(n,:)=mfun('P',n-1,x);
   %tmp1=legendre(n-1,x); P(n,:)=tmp1(1,:); % This also works.
   P1(n,:)=polyval(Ps(n,:),x);
end
format rat, Ps, format short
plot(x,P.'), hold on, plot(x,P1.',':'), axis([-1 1 -1.2 1.2])
plot([-1 1],[0 0],'k', [0 0],[-1.2 1.2],'k')
```

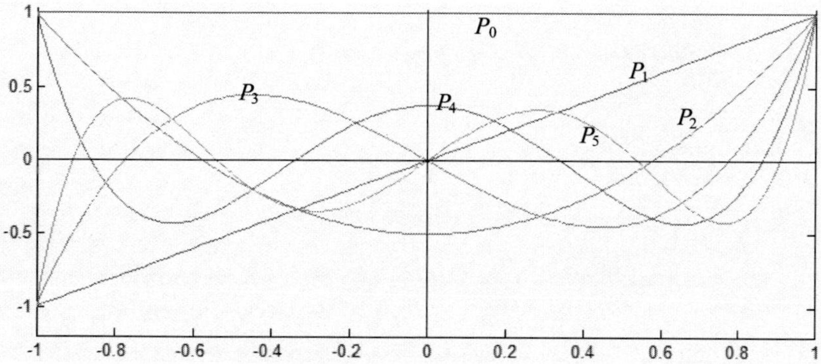

Figure 3.17 Legendre polynomials of degree n=0,1,2,3,4, and 5

Fig. 3.17 shows the graphs of Legendre polynomials for $n=0:5$ that are obtained by running the above script, "plot_Legendre_poly.m," where the Legendre polynomials are computed twice, once by using the MATLAB function '`legendre()`' and once by using the recurrence relation (3.3.38).

3.3.3 Chebyshev's Equation

The *Chebyshev's (differential) equation* of order n is

$$(1-x^2)y''-xy'+n^2y=0 \text{ for } n>0 \text{ and } |x|<1 \tag{3.3.41}$$

A solution of this equation can be found by substituting a power series for y:

$$y=\sum_{m=0}^{\infty} c_m x^m \tag{3.3.42}$$

as an assumed solution:

$$(1-x^2)\sum_{m=2}^{\infty} c_m\, m(m-1)x^{m-2} - x\sum_{m=1}^{\infty} c_m\, m x^{m-1} + n^2\sum_{m=0}^{\infty} c_m x^m = 0$$

$$;\ \sum_{m=0}^{\infty} c_{m+2}(m+2)(m+1)x^m - \sum_{m=2}^{\infty} c_m\, m(m-1)x^m - \sum_{m=1}^{\infty} c_m m x^m + n^2\sum_{m=0}^{\infty} c_m x^m = 0 \tag{3.3.43}$$

This yields

$$2c_2+n^2c_0=0: \quad \text{for } m=0 (\text{coefficients of } x^0) \tag{3.3.44a}$$

$$6c_3+(n^2-1)c_1=0: \text{ for } m=1 \text{ (coefficients of } x^1) \tag{3.3.44b}$$

$$c_{m+2}(m+2)(m+1)-c_m(m^2-n^2)=0 \text{ for } m\geq 2$$

$$\rightarrow c_{m+2}=\frac{m^2-n^2}{(m+2)(m+1)}c_m=\frac{(m-n)(m+n)}{(m+2)(m+1)}c_m : \text{recurrence relation} \tag{3.3.44c}$$

With the two series of coefficients, each generated by this recurrence equation, one starting $c_0=1$ & $c_1=0$ and one starting from $c_0=0$ & $c_1=1$, two independent solutions, called *Chebyshev polynomials*, can be written as

$$y_1(x) = 1 - \frac{n^2}{2!}x^2 + \frac{(n-2)n^2(n+2)}{4!}x^4 - \frac{(n-4)(n-2)n^2(n+2)(n+4)}{6!}x^6 + \cdots \quad (3.3.45a)$$

$$y_2(x) = x - \frac{(n-1)(n+1)}{3!}x^3 + \frac{(n-3)(n-1)(n+1)(n+3)}{5!}x^5 - \cdots \quad (3.3.45b)$$

constituting a general solution for $|x|<1$:

$$y(x) = c_0 y_1(x) + c_1 y_2(x) \quad (3.3.46)$$

Since the coefficients obtained using Eq. (3.3.44c) are zero for even/odd $m>n$, either $y_1(x)$ or $y_2(x)$ becomes a *Chebyshev polynomial* (*of the first kind*) of degree n depending on whether n is even or odd. The first several Chebyshev polynomial $T_n(x)$'s (of the first kind) of degree n are

$$T_0(x) = 1 \qquad T_1(x) = x$$
$$T_2(x) = 2x^2 - 1 \qquad T_3(x) = 4x^3 - 3x \quad (3.3.47)$$
$$T_4(x) = 8x^4 - 8x^2 + 1 \qquad T_5(x) = 16x^5 - 20x^3 + 5x$$

The Chebyshev polynomials of the first kind can be generated by using the following three-term recurrence relation:

$$T_{n+1}(x) = 2xT_n(x) - T_{n-1}(x) \text{ with } T_1(x) = x \text{ and } T_0(x) = 1 \quad (3.3.48)$$

or a Rodrigues-type formula

$$T_n(x) = (-1)^n \frac{n!}{(2n)!}\sqrt{1-x^2}\frac{d^n}{dx^n}(1-x^2)^{n-1/2} \text{ for } |x|<1 \text{ and } n = 0, 1, 2, \cdots \quad (3.3.49)$$

or an explicit formula

$$T_n(x) = \frac{n}{2}\sum_{m=0}^{\text{floor}(n/2)}(-1)^m \frac{(n-m-1)!}{m!(n-2m)!}(2x)^{n-2m} \text{ for } |x|<1 \text{ and } n = 0, 1, 2, \cdots \quad (3.3.50)$$

```
%plot_Chebyshev_poly.m
% Plot Chebyshev polynomials of degree n=0:5
clear, clf
x=[-1:0.01:1]; % Domain
Ts=[zeros(1,5) 1; zeros(1,4) 1 0; zeros(4,6)]; % including T0 and T1
for n=1:6
   if n>2
     tmp = Ts(n-1,:); % T_n-1: Chebyshev polynomial of degree n-1
     tmp = [tmp(2:end) 0]; % multiply T_(n-1) by x
     Ts(n,:)=2*tmp-Ts(n-2,:); % Eq.(3.3.48)
   end
   T(n,:)=mfun('T',n-1,x);
   T1(n,:)=polyval(Ts(n,:),x);
end
format rat, Ts, format short
plot(x,T.'), hold on, plot(x,T1.',':'), axis([-1 1 -1.2 1.2])
plot([-1 1],[0 0],'k', [0 0],[-1.2 1.2],'k')
```

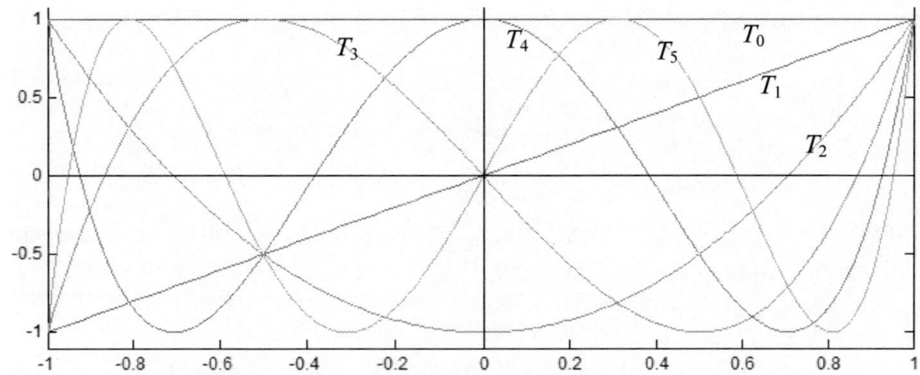

Figure 3.18 Chebyshev polynomials of degree n=0,1,2,3,4, and 5

Fig. 3.18 shows the graphs of (equiripple) Chebyshev polynomials for n=0:5, which are obtained by running the script "plot_Chebyshev_poly.m," where the Chebyshev polynomials are computed twice, once by using the MATLAB function '`mfun('T',n-1,x)`' and once by using the recurrence relation (3.3.48).

With $x=\cos\theta$, the *Chebyshev polynomials of the first kind* can be written as

$$T_n(x)=\cos(n\cos^{-1}x) \text{ for } |x|<1 \text{ and } n=0,1,2,\cdots \tag{3.3.51}$$

and their zeros, called the *Chebyshev nodes*, can be found as

$$T_n(x)=\cos(n\cos^{-1}x)\ ;\ n\cos^{-1}x=(2k+1)\frac{\pi}{2}\ ;\ x_k=\cos\frac{(2k+1)\pi}{2n} \tag{3.3.52}$$

How could we get Eq. (3.3.51) as an expression of Chebyshev polynomials? To find out how, we substistute $x=\cos\theta$ and $dx=-\sin\theta d\theta$ into the Chebyshev's equation (3.3.41) to rewrite it as

$$(1-x^2)y''-xy'+n^2y=\sin^2\theta\frac{d^2y}{dx^2}-\cos\theta\frac{dy}{dx}+n^2y=0\ \rightarrow\ \frac{d^2y}{d\theta^2}+n^2y=0 \tag{3.3.53}$$

since

$$\frac{dy}{d\theta}\underset{\text{Chain rule}}{\overset{(2.1.4)}{=}}\frac{dx}{d\theta}\frac{dy}{dx}=(-\sin\theta)\frac{dy}{dx} \tag{3.3.54a}$$

$$\frac{d^2y}{d\theta^2}=-\cos\theta\frac{dy}{dx}-\sin\theta\frac{d}{d\theta}\left(\frac{dy}{dx}\right)=-\cos\theta\frac{dy}{dx}-\sin\theta\frac{dx}{d\theta}\frac{d^2y}{dx^2}=-\cos\theta\frac{dy}{dx}+\sin^2\theta\frac{d^2y}{dx^2} \tag{3.3.54b}$$

Since the (homogeneous) second-order DE (3.3.53) has the characterstic roots $s=\pm jn$, we can use Eqs. (3.2.4) and (3.2.7) with σ=0, ω=n, and t=θ to write its general solution as

$$y(\theta)=A\cos n\theta+B\sin n\theta$$

$$;\ y(x)=A\cos(n\cos^{-1}x)+B\sin(n\cos^{-1}x)=AT_n(x)+BU_n(x) \tag{3.3.55}$$

where the first component function $T_n(x)$ is the Chebyshev polynomial of the first kind (see Eq. (3.3.52)) and the second component function

$$U_n(x)=\sin(n\cos^{-1}x) \tag{3.3.56}$$

is called the *Chebyshev polynomial of the second kind.*

Note that the Chebyshev polynomials are orthogonal w.r.t. the weighting function $1/\sqrt{1-x^2}$ in the sense that

$$\int_{-1}^{1}\frac{T_m(x)T_n(x)}{\sqrt{1-x^2}}dx=\begin{cases}0 & \text{for } m\neq n\\ \pi/2 & \text{for } m=n\neq 0\\ \pi & \text{for } m=n=0\end{cases} \tag{3.3.57}$$

This can be shown by substituting $x=\cos\theta$ and $dx=-\sin\theta d\theta=-\sqrt{1-x^2}d\theta$ and using the following equation of orthogonality among cosine functions:

$$\int_{-\pi}^{\pi}\cos m\theta\cos n\theta d\theta=\begin{cases}0 & \text{for } m\neq n\\ \pi/2 & \text{for } m=n\neq 0\\ \pi & \text{for } m=n=0\end{cases} \tag{3.3.58}$$

3.3.4 Hermite's Equation

The *Hermite's (differential) equation* of order n is

$$y''-2xy'+2ny=0 \text{ for } n>0 \tag{3.3.59}$$

A solution of this equation can be found by substituting a power series for y:

$$y=\sum_{m=0}^{\infty}c_m x^m \tag{3.3.60}$$

as an assumed solution:

$$\sum_{m=2}^{\infty}c_m\, m(m-1)x^{m-2}-2x\sum_{m=1}^{\infty}c_m\, m\, x^{m-1}+2n\sum_{m=0}^{\infty}c_m x^m=0$$

$$;\quad \sum_{m=0}^{\infty}c_{m+2}(m+2)(m+1)x^m-\sum_{m=0}^{\infty}c_m(2m-2n)x^m=0 \tag{3.3.61}$$

This yields

$$c_{m+2}=\frac{2(m-n)}{(m+2)(m+1)}c_m \text{ for } m=0,1,2,\cdots\text{: recurrence relation} \tag{3.3.62}$$

With two arbitrary coefficients c_0 and c_1, Eq. (3.3.60) yields a general solution as

$$y(x)=c_0y_1(x)+c_1y_2(x)=c_0\left\{1-\frac{2}{2!}nx^2+\frac{2^2}{4!}n(n-2)x^4-\frac{2^3}{6!}n(n-2)(n-4)x^6+\cdots\right\}$$

$$+c_1\left\{x-\frac{2}{3!}(n-1)x^3+\frac{2^2}{5!}(n-1)(n-3)x^5-\cdots\right\} \tag{3.3.63}$$

If the coefficients c_0 and c_1 are fixed as

$$\begin{cases}c_0=(-2)^{n/2}(n-1)(n-3)\cdots 1 \text{ and } c_1=0 & \text{for } n\text{: even}\\ c_0=0 \text{ and } c_1=-(-2)^{(n+1)/2}n(n-2)\cdots 1 & \text{for } n\text{:odd}\end{cases} \tag{3.3.64}$$

Eq. (3.3.63) becomes a *Hermite polynomial* $H_n(x)$ of degree n. The first several Hermite polynomials are

$$\begin{array}{l|l} H_0(x)=1 & H_1(x)=2x \\ H_2(x)=4x^2-2 & H_3(x)=8x^3-12x \\ H_4(x)=16x^4-48x^2+12 & H_5(x)=32x^5-160x^3+120x \end{array} \tag{3.3.65}$$

The Chebyshev polynomials of the first kind can be generated by using the following three-term recurrence relation:

$$H_{n+1}(x)=2xH_n(x)-2nH_{n-1}(x) \text{ with } H_1(x)=2x \text{ and } H_0(x)=1 \tag{3.3.66}$$

or a Rodrigues-type formula

$$H_n(x)=(-1)^n e^{x^2}\frac{d^n}{dx^n}e^{x^2} \text{ for } n=0,1,2,\cdots \tag{3.3.67}$$

or an explicit formula

$$H_n(x)=\sum_{m=0}^{\text{floor}(n/2)}(-1)^m\frac{n!}{m!(n-2m)!}(2x)^{n-2m} \text{ for } n=0,1,2,\cdots \tag{3.3.68}$$

```
%plot_Hermite_poly.m
% Plot Hermite polynomials of degree n=0:5
clear, clf
x=[-2:0.01:2]; % Domain
Hs=[zeros(1,5) 1; zeros(1,4) 2 0; zeros(4,6)]; % including H0 and H1
for n=1:5
   if n>2
     tmp = Hs(n-1,:); % H_n-1: Hermite polynomial of degree n-1
     tmp = [tmp(2:end) 0]; % multiply H_(n-1) by x
     Hs(n,:)=2*tmp-2*(n-2)*Hs(n-2,:); % Eq.(3.3.66)
   end
   H(n,:)=mfun('H',n-1,x);  H1(n,:)=polyval(Hs(n,:),x);
end
format rat, Hs, format short
plot(x,H.'), hold on, plot(x,H1.',':'), axis([-2 2 -40 80])
```

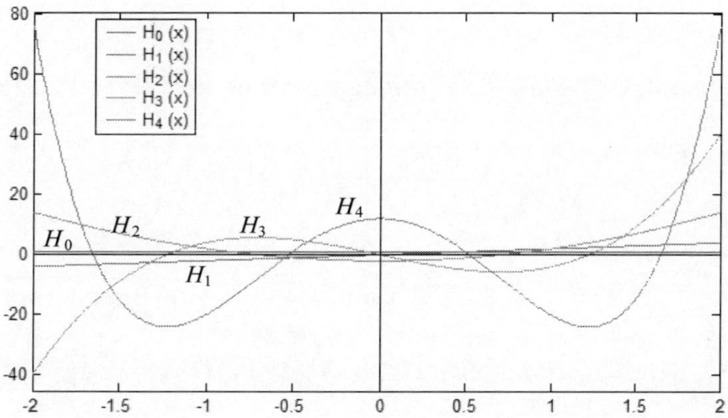

Figure 3.19 Hermite polynomials of degree n=0,1,2,3, and 4

Note that the Hermite polynomials are orthogonal w.r.t. the weighting function e^{-x^2} in the sense that

$$\int_{-\infty}^{\infty} H_m(x)H_n(x)e^{-x^2}dx = \begin{cases} 0 & \text{for } m\neq n \\ \sqrt{\pi}\,2^2 n! & \text{for } m=n \end{cases} \tag{3.3.69}$$

Fig. 3.19 shows the graphs of Hermite polynomials for $n=0{:}4$ that are obtained by running the script "plot_Hermite_poly.m," where the Hermite polynomials are computed twice, once by using the MATLAB function '`mfun('H',n-1,x)`' and once by using the recurrence relation (3.3.66).

3.3.5 Laguerre's Equation

The *Laguerre's (differential) equation* of order *n* is

$$xy'' + (1-x)y' + ny = 0 \ \text{ for } x \geq 0 \text{ and } n: \text{a nonnegative integer} \tag{3.3.70}$$

A solution of this equation can be found by substituting a power series for *y*:

$$y = \sum_{m=0}^{\infty} c_m x^m \tag{3.3.71}$$

as an assumed solution:

$$x\sum_{m=2}^{\infty} c_m\, m(m-1)x^{m-2} + (1-x)\sum_{m=1}^{\infty} c_m\, m x^{m-1} + n\sum_{m=0}^{\infty} c_m x^m = 0$$

$$;\ \sum_{m=1}^{\infty} c_{m+1}(m+1)m x^m + \sum_{m=0}^{\infty} c_{m+1}(m+1)x^m - \sum_{m=1}^{\infty} c_m m x^m + nc_0 + n\sum_{m=1}^{\infty} c_m x^m = 0 \tag{3.3.72}$$

$$;\ \sum_{m=1}^{\infty} \{c_{m+1}(m+1)^2 - c_m(m-n)\}x^m + c_1 + nc_0 = 0$$

This yields

$$c_{m+1} = \frac{m-n}{(m+1)^2}c_m \ \text{ for } m = 0,1,2,\cdots: \text{recurrence relation} \tag{3.3.73}$$

With an arbitrary coefficient c_0, Eq. (3.3.71) yields a general solution as

$$y(x) = c_0\left\{1 - nx + \frac{(1-n)n}{2^2}x^2 + \frac{(2-n)(1-n)n}{3^2 2^2}x^3 + \frac{(3-n)(2-n)(1-n)}{4^2 3^2 2^2}x^3 + \cdots\right\} \tag{3.3.74}$$

which becomes a *Laguerre polynomial* $L_n(x)$ of degree *n*. The first several Laguerre polynomials are

$$L_0(x) = 1 \qquad L_1(x) = -x+1$$

$$L_2(x) = \frac{1}{2!}(x^2 - 4x + 2) \qquad L_3(x) = \frac{1}{3!}(-x^3 + 9x^2 - 18x + 6) \tag{3.3.75}$$

The Laguerre polynomials can be generated by using the following three-term recurrence relation:

$$L_{n+1}(x) = \frac{1}{n+1}\{(1+2n-x)L_n(x) - nL_{n-1}(x)\} \ \text{ with } L_1(x) = 1-x \text{ and } L_0(x) = 1 \tag{3.3.76}$$

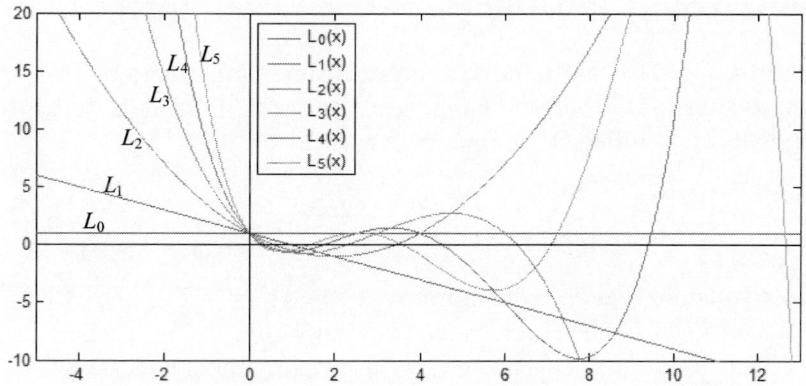

Figure 3.20 Laguerre polynomials of degree n=0,1,2,3, and 4

```
%plot_Laguerre_poly.m
% Plot Laguerre polynomials of degree n=0:5
clear, clf
x=[-2:0.01:2]; % Domain
Ls=[zeros(1,5) 1; zeros(1,4) -1 1; zeros(4,6)]; % including L0 and L1
for n=1:5
   if n>2
     tmp = Ls(n-1,:); % L_n-1: Laguerre polynomial of degree n-1
     tmp = [-tmp(2:end) 0] + (1+2*(n-2))*tmp;
     Ls(n,:)=(tmp-(n-2)*Ls(n-2,:))/(n-1); % Eq.(3.3.76)
   end
   L(n,:)=mfun('L',n-1,x);
   L1(n,:)=polyval(Ls(n,:),x);
end
format rat, Ls, format short
plot(x,L.')
```

or a Rodrigues-type formula

$$L_n(x) = e^x \frac{d^n}{dx^n}(x^n e^{-x}) \text{ for } n = 0, 1, 2, \cdots \tag{3.3.77}$$

or an explicit formula

$$L_n(x) = \sum_{m=0}^{n} (-1)^m \frac{(n!)^2}{(m!)^2 (n-m)!} x^m \text{ for } n = 0, 1, 2, \cdots \tag{3.3.78}$$

Note that the Laguerre polynomials are orthogonal w.r.t. the weighting function e^{-x} in the sense that

$$\int_0^\infty L_m(x) L_n(x) e^{-x} dx = \begin{cases} 0 & \text{for } m \neq n \\ 1 & \text{for } m = n \end{cases} \tag{3.3.79}$$

Fig. 3.20 shows the graphs of Laguerre polynomials for n=0:5 that are obtained by running the script "plot_Laguerre_poly.m," where the Laguerre polynomials are computed twice, once by using the MATLAB function '`mfun('L',n-1,x)`' and once by using the recurrence relation (3.3.76).

3.4 BOUNDARY VALUE PROBLEMS

A *boundary value problem* (*BVP*) is an *N*th-order differential equation with the BCs (boundary conditions) given as some of the values of dependent variable $y(t)$ and its derivative specified at the initial time t_0, and others specified at the final time t_f.

$$[\text{BVP}]_N:\ y^{(N)}(t) = f(t, y(t), y'(t), y^{(2)}(t), \cdots, y^{(N-1)}(t)) \quad \text{with the BCs } g(y(t_1),\ y'(t_2),\ \cdots,\ y^{(N-1)}(t_N)) = 0 \tag{3.4.1}$$

In some cases, some relations between the initial values and the final values may be given as a mixed-boundary condition instead of the initial/final values specified. This section illustrates how such second-order BVPs as Eq. (3.4.2) can be dealt with by using the symbolic DE solver '`dsolve()`' and the numerical BVP solver '`bvp4c()`':

$$[\text{BVP}]_2:\ y''(t)=f(t,y(t),y'(t)) \text{ with } y(t_0)=y_0 \text{ or } y'(t_0)=y_0 \text{ and } y(t_f)=y_f \text{ or } y'(t_f)=y_f \tag{3.4.2}$$

(Example 3.25) Linear Second-Order BVP

Solve the following homogeneous linear second-order BVP:

$$\frac{d^2y}{dt^2} + \omega^2 y = 0 \text{ with boundary conditions } y(0)=0 \text{ and } y(\frac{\pi}{2\omega})=1 \tag{E3.25.1}$$

Since the characteristic equation $s^2+\omega^2=0$ yields the complex eigenvalues $s=\pm j\omega$, we can use Eqs. (3.2.4) and (3.2.7) with $\sigma=0$ to write the general solution to this DE as

$$y(t) = c_1\cos(\omega t) + c_2\sin(\omega t) \tag{E3.25.2}$$

The coefficients can be determined from the BC (boundary condition):

$$y(0) = 0 \overset{\text{(E3.25.2)}}{\rightarrow} c_1\cos(0) + c_2\sin(0) = 0;\ c_1 = 0 \tag{E3.25.3a}$$

$$y(\frac{\pi}{2\omega}) = 1 \overset{\text{(E3.25.2)}}{\rightarrow} c_1\cos\left(\frac{\pi}{2}\right) + c_2\sin\left(\frac{\pi}{2}\right) \overset{\text{(E3.25.3a)}}{=} c_2\sin\left(\frac{\pi}{2}\right) = 1;\ c_2 = 1 \tag{E3.25.3b}$$

Thus, we can write the solution to Eq. (E3.25.1) as

$$y(t) \overset{\text{(E3.25.2)}}{=} c_1\cos(\omega t) + c_2\sin(\omega t) \underset{\text{(E3.25.3b)}}{\overset{\text{(E3.25.3a)}}{=}} \sin(\omega t) \tag{E3.25.4}$$

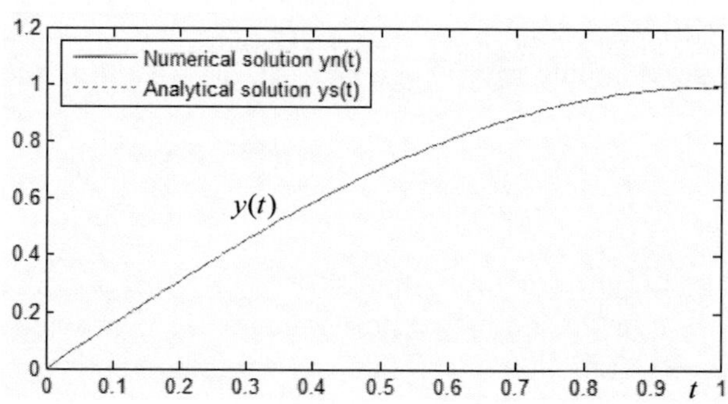

Figure 3.21 Plot of the numerical and analytical solutions for DE (E3.25.1)

```
%em03e25.m
clear, clf
dx=@(t,x)[0 1;-(pi/2)^2 0]*x; % ODE(E3.25.5) in state eq. form
fbc=@(x0,xf)[x0(1); xf(1)-1]; % Boundary condition
t0=0; tf=1; tt=[t0:0.01:tf]; % Solution time interval tt
w=pi/2;
solinit=bvpinit(linspace(t0,tf,5),[-1 1]); % Initial guess of solution
sol=bvp4c(dx,fbc,solinit); % Solution of the BVP
x=deval(sol,tt); ynt=x(1,:).'; % Evaluate the BVP solution at tt
% Using dsolve() to solve the DE (E3.25.1) given boundary values
ys=dsolve('D2y+(pi/2)^2*y=0','y(0)=0,y(1)=1') % Symbolic solution
t=tt; yst=eval(ys); % Numeric values of the symbolic solution
yat=sin(w*tt); % Analytical solution (E3.25.4)
discrepancy=norm(yst-yat)/norm(yat)
plot(tt,ynt, tt,yst,'r:'), axis([0 1 0 1.2])
legend('Numerical solution yn(t)','Analytical solution ys(t)'), shg
```

If the parameter ω is given as a numeric value like $\pi/2$, we can use '`dsolve()`' for the DE (E3.25.1) and '`bvp4c()`' for the DE written in state equation form like

$$\mathbf{x}'(t) = A\,\mathbf{x}(t);\quad \begin{bmatrix} x_1' \\ x_2' \end{bmatrix} = \begin{bmatrix} 0 & 1 \\ -(\pi/2)^2 & 0 \end{bmatrix} \begin{bmatrix} x_1 \\ x_2 \end{bmatrix} \text{ with } \begin{bmatrix} x_1(0) \\ x_1(1)-1 \end{bmatrix} = \begin{bmatrix} 0 \\ 0 \end{bmatrix} \tag{E3.25.5}$$

as illustrated in the above MATLAB script "em03e25.m." Running the script file yields Fig. 3.21, which shows that the numerical solution agrees with the analytical solution.

(Example 3.26) Linear Second-Order BVP

Solve the following homogeneous linear second-order BVP:

$$\frac{d^2y}{dt^2} - \gamma^2 y = 0 \text{ with boundary conditions } y(0)=0 \text{ and } y(1)=-1 \tag{E3.26.1}$$

Since the characteristic equation $s^2-\gamma^2=(s+\gamma)(s-\gamma)=0$ yields the eigenvalues $s=\pm\gamma$, we can use Eqs. (3.2.4) and (3.2.5) to write the general solution as

$$y(t) = c_1 e^{-\gamma t} + c_2 e^{\gamma t} \tag{E3.26.2}$$

The coefficients can be determined from the BC (boundary condition):

$$\left.\begin{aligned} y(0)=0 &\overset{\text{(E3.26.2)}}{\rightarrow} c_1 e^0 + c_2 e^0 = c_1 + c_2 = 0 \\ y(1)=-1 &\overset{\text{(E3.26.2)}}{\rightarrow} c_1 e^{-\gamma} + c_2 e^{\gamma} = -1 \end{aligned}\right\};\ \begin{bmatrix} 1 & 1 \\ e^{-\gamma} & e^{\gamma} \end{bmatrix} \begin{bmatrix} c_1 \\ c_2 \end{bmatrix} = \begin{bmatrix} 0 \\ -1 \end{bmatrix}$$

$$;\ \begin{bmatrix} c_1 \\ c_2 \end{bmatrix} = \begin{bmatrix} 1 & 1 \\ e^{-\gamma} & e^{\gamma} \end{bmatrix}^{-1} \begin{bmatrix} 0 \\ -1 \end{bmatrix} = \frac{1}{e^{\gamma}-e^{-\gamma}} \begin{bmatrix} e^{\gamma} & -1 \\ -e^{-\gamma} & 1 \end{bmatrix} \begin{bmatrix} 0 \\ -1 \end{bmatrix} = \frac{1}{e^{\gamma}-e^{-\gamma}} \begin{bmatrix} 1 \\ -1 \end{bmatrix} \tag{E3.26.3}$$

Thus, we can write the solution to Eq. (E3.26.1) as

$$y(t) \overset{\text{(E3.26.2)}}{=} c_1 e^{-\gamma t} + c_2 e^{\gamma t} \overset{\text{(E3.26.3)}}{=} \frac{1}{e^{\gamma}-e^{-\gamma}} \left(e^{-\gamma t} - e^{\gamma t}\right) \tag{E3.26.4}$$

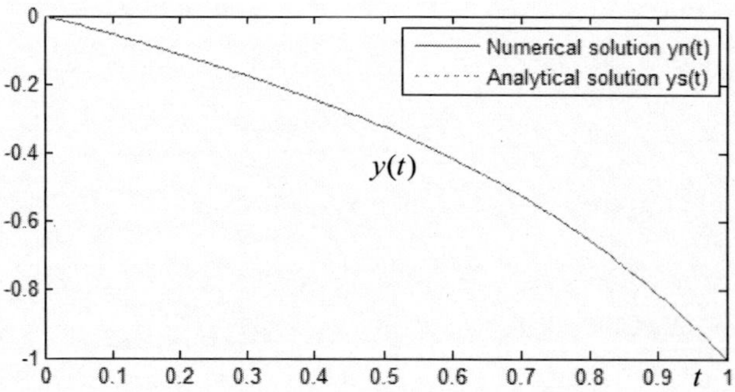

Figure 3.22 Plot of the numerical and analytical solutions for DE (E3.26.1)

If the parameter γ is given as a numeric value like 2, we can use '`dsolve()`' for the DE (E3.26.1) and '`bvp4c()`' for the DE written in state equation form like

$$\mathbf{x}'(t) = A\,\mathbf{x}(t); \quad \begin{bmatrix} x_1' \\ x_2' \end{bmatrix} = \begin{bmatrix} 0 & 1 \\ 2^2 & 0 \end{bmatrix} \begin{bmatrix} x_1 \\ x_2 \end{bmatrix} \text{ with } \begin{bmatrix} x_1(0) \\ x_1(1)+1 \end{bmatrix} = \begin{bmatrix} 0 \\ 0 \end{bmatrix} \tag{E3.26.5}$$

as illustrated in the MATLAB script "em03e26.m." Running the script file yields Fig. 3.22, which shows that the numerical solution agrees with the analytical solution.

```
%em03e26.m
clear, clf
dx=@(t,x)[0 1; 2^2 0]*x; % ODE (E3.26.5) in state eq. form
fbc=@(x0,xf)[x0(1); xf(1)+1]; % Boundary condition
t0=0; tf=1; tt=[t0:0.01:tf]; r=2; % Solution time interval tt
solinit=bvpinit(linspace(t0,tf,5),[0 2]); % Initial guess of solution
sol=bvp4c(dx,fbc,solinit); % Solution of the BVP
x=deval(sol,tt); ynt=x(1,:).'; % Evaluate the numerical solution at tt
% Using dsolve() to solve the DE (E3.26.1)
ys=dsolve('D2y-2^2*y=0','y(0)=0,y(1)=-1') % Symbolic solution
t=tt; yst=eval(ys); % Numeric values of the symbolic solution
yat=(exp(-tt)-exp(tt))/(exp(r)-exp(-r)); % Analytical solution (E3.26.4)
plot(tt,ynt, tt,yst,'r:')
legend('Numerical solution yn(t)','Analytical solution ys(t)'), shg
```

(Example 3.27) Linear Second-Order BVP with Variable Coefficients

Solve the following homogeneous linear second-order BVP:

$$y''(t) + \frac{2}{t}y'(t) - \frac{2}{t^2}y(t) = 0 \quad \text{with } y(1) = 5,\ y(2) = 3 \tag{E3.27.1}$$

We can use '`dsolve()`' for this DE as

```
>>ys=dsolve('D2y+2*(t*Dy-y)/t^2=0','y(1)=5, y(2)=3')
```

to get the analytical solution

```
ys =  t + 4/t^2
```

which implies

$$y(t) = t + \frac{4}{t^2} \tag{E3.27.2}$$

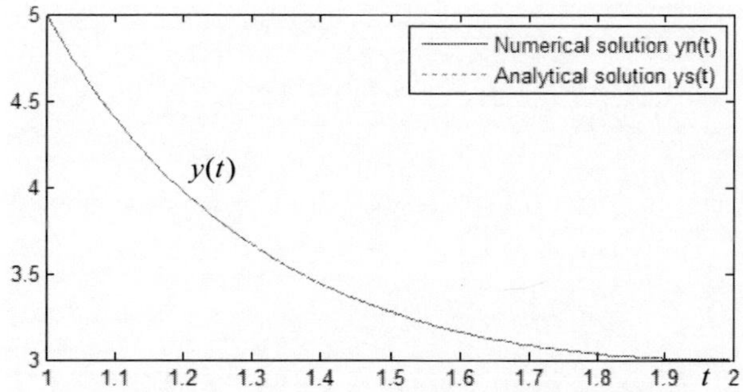

Figure 3.23 Plot of the numerical and analytical solutions for DE (E3.27.1)

Also, we can use '`bvp4c()`' for the DE written in state equation form like

$$\mathbf{x}'(t) = A\,\mathbf{x}(t); \quad \begin{bmatrix} x_1' \\ x_2' \end{bmatrix} = \begin{bmatrix} 0 & 1 \\ 2/t^2 & -2/t \end{bmatrix} \begin{bmatrix} x_1 \\ x_2 \end{bmatrix} \text{ with } \begin{bmatrix} x_1(1)-5 \\ x_1(2)-3 \end{bmatrix} = \begin{bmatrix} 0 \\ 0 \end{bmatrix} \tag{E3.27.3}$$

as illustrated in the MATLAB script "em03e27.m." Running the script file yields Fig. 3.23, which shows that the numerical solution agrees with the analytical solution.

```
%em03e27.m
clear, clf
t0=1; tf=2; tt=[t0:0.01:tf]; % Solution time interval tt
dx=@(t,x) [0 1; 2/t^2 -2/t]*x; % ODE(E3.27.3) in state eq. form
fbc=@(x0,xf) [x0(1)-5; xf(1)-3]; % Boundary condition
solinit=bvpinit(linspace(t0,tf,5),[1 10]); % Initial guess of solution
sol=bvp4c(dx,fbc,solinit,bvpset('RelTol',1e-4));
xnt=deval(sol,tt); % Evaluate the solution of DE.
ynt=xnt(1,:);
% Using 'dsolve()' to solve the DE (E3.27.1)
ys=dsolve('D2y+2/t*Dy-2/t^2*y=0','y(1)=5, y(2)=3') % Symbolic solution
t=tt; yst=eval(ys); % Numeric values of the symbolic solution
plot(tt,ynt, tt,yst,'r:') % To compare with analytical solution
legend('Numerical solution yn(t)','Analytical solution ys(t)'), shg
```

(Example 3.28) Nonlinear Second-Order BVP with Variable Coefficients

Solve the following homogeneous nonlinear second-order BVP:

$$y''(t) - 4t\,y(t)y'(t) - 2y^2(t) = 0 \text{ with } y(0) = \frac{1}{4} \text{ and } y'(1) = \frac{2}{9} \tag{E3.28.1}$$

We can substitute

$$y(t) = \frac{1}{4-t^2} \text{ with } y'(t) = \frac{2t}{(4-t^2)^2} = 2t\,y^2(t) \text{ and } y''(t) = 2y^2(t) + 4t\,y(t)y'(t) \tag{E3.28.2}$$

into the above DE to see that it is the solution satisfying the DE:

```
>> syms t; y=1/(4-t^2);
>>diff(diff(y,t),t)-4*t*y*diff(y)-2*y^2
    ans  =  0
```

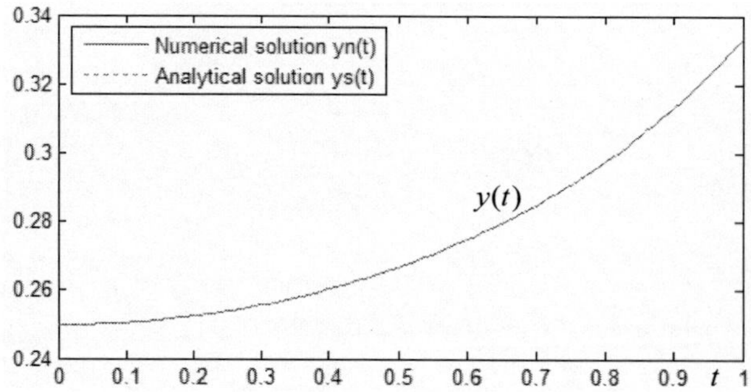

Figure 3.24 Plot of the numerical and analytical solutions for DE (E3.28.1)

We can use '`bvp4c()`' for the DE written in state equation form like

$$\mathbf{x}'(t) = A\,\mathbf{x}(t);\quad \begin{bmatrix} x_1' \\ x_2' \end{bmatrix} = \begin{bmatrix} x_2 \\ -4t\,x_1x_2 - 2x_1^2 \end{bmatrix} \text{ with } \begin{bmatrix} x_1(1) - 1/4 \\ x_2(2) - 2/9 \end{bmatrix} = \begin{bmatrix} 0 \\ 0 \end{bmatrix} \tag{E3.28.3}$$

as illustrated in the MATLAB script "em03e28.m." Running the script file yields Fig. 3.24, which shows that the numerical solution agrees with the analytical solution.

```
%em03e28.m
clear, clf
t0=0; tf=1; tt=[t0:0.01:tf]; % Solution time interval tt
dx=@(t,x)[x(2); 4*t*x(1)*x(2)+2*x(1)^2]; % Eq.(E3.28.3) in state eq. form
fbc=@(x0,xf)[x0(1)-1/4; xf(2)-2/9];
solinit=bvpinit(linspace(t0,tf,5),[0 1]); % Initial guess of solution
sol=bvp4c(dx,fbc,solinit); %,bvpset('RelTol',1e-4));
xnt=deval(sol,tt);
ynt=xnt(1,:);
syms t;
ys=1/(4-t^2); % Analytical solution (E3.28.2)
diff(diff(ys,t),t)-4*t*ys*diff(ys)-2*ys^2 % must be zero to satisfy the DE
t=tt; yst=eval(ys); % Numeric values of the symbolic solution
plot(tt,ynt, tt,yst,'r:') % To compare with analytical solution
legend('Numerical solution yn(t)','Analytical solution ys(t)'), shg
```

Now, we consider an important class of linear second-order DEs (differential equations), called the (regular) *Sturm–Liouville* (*SL*) *equation*, of the following form:

$$\frac{d}{dx}\left(p(x)\frac{dy}{dx}\right) + (q(x) - \lambda\, w(x))y = 0 \quad \text{for } x_0 \le x \le x_f \quad (\lambda : \text{not specified}) \tag{3.4.3}$$

subject to homogeneous mixed BCs (boundary conditions):

$$a_1 y(x_0) + a_2 y'(x_0) = 0 \quad (a_1 \text{ and } a_2 \text{ are not both zero}) \tag{3.4.4a}$$

$$b_1 y(x_f) + b_2 y'(x_f) = 0 \quad (b_1 \text{ and } b_2 \text{ are not both zero}) \tag{3.4.4b}$$

or periodic BCs:

$$y(x_0) = y(x_f) \text{ and } y'(x_0) = y'(x_f) \tag{3.4.5a,b}$$

where the functions $p(x)>0$, $p'(x)$, $q(x)$, and $w(x)>0$ are continuous on $[x_0, x_f]$. Any linear second-order DE including Bessel's equation (3.3.1) and Legendre's equation (3.3.29) dealt with in the previous section can be written in the form of Eq. (3.4.3) so that they belong to a class of SL equations. With the *SL* (*Sturm-Liouville*) *operator* defined as

$$L = \frac{1}{w(x)}\left\{\frac{d}{dx}\left(p(x)\frac{d}{dx}\right)+q(x)\right\} \tag{3.4.6}$$

the SL equation (with λ unspecified) can be written as

$$Ly = \lambda y \tag{3.4.7}$$

where λ and y satisfying this equation are called its eigenvalue and eigenfunction, respectively, compared to the definition of eigenvalue/eigenvector (Eq. (1.2.12): $A\mathbf{v}=\lambda\mathbf{v}$) of a matrix A. Note that the SL operator is *self-adjoint* in the sense that

$$<y_1, Ly_2>_w = <y_2, Ly_1>_w \tag{3.4.8}$$

where $<f, g>_w$ denotes the (w-)*weighted inner product* of two functions f and g defined as

$$<f, g>_w = \int_{x_0}^{x_f} f^*(x)g(x)w(x)dx \text{ (*: complex conjugate)} \tag{3.4.9}$$

and the *adjoint* $\boldsymbol{L}^*$ of operator $\boldsymbol{L}$ is an operator such that

$$<y_1, Ly_2>_w = <y_2, L^*y_1>_w \tag{3.4.10}$$

The eigenvalues/eigenfunctions, corresponding to the solution of the SL equation, have the following properties:

(SL-P1) The eigenvalues are real and bounded below (but not above) so that they can be ordered as

$$\lambda_0 < \lambda_1 < \lambda_2 < \cdots < \lambda_n < \cdots \tag{3.4.11}$$

This can be proved as follows:

$$<Ly, y>_w \overset{(3.4.8)}{=} <y, Ly>_w \underset{(3.4.7)}{\overset{(3.4.9)}{\rightarrow}} \lambda^*<y, y>_w = \lambda <y, y>_w \ ; \ \lambda^* = \lambda$$

(SL-P2) For each eigenvalue λ_n, there exists a unique corresponding eigenfunction y_n (except for nonzero multiples), which has $n-1$ zeros on the interval (x_0, x_f).

(SL-P3) Eigenfunctions y_m and y_n corresponding to different eigenvalues λ_m and λ_n are *orthogonal* (w.r.t. the weight function $w(x)$) in the sense that their weighted inner product is zero:

$$<y_m, y_n>_w = 0 \tag{3.4.12}$$

This can be proved as follows:

$$<Ly_m, y_n>_w \overset{(3.4.8)}{=} <y_m, Ly_n>_w \overset{(3.4.7)}{\rightarrow} \lambda_m <y_m, y_n>_w = \lambda_n <y_m, y_n>_w$$

$$\rightarrow (\lambda_m - \lambda_n)<y_m, y_n>_w = 0 \overset{\lambda_m-\lambda_n\neq 0}{\rightarrow} <y_m, y_n>_w = 0$$

(SL-P4) The set of eigenfunctions $\{y_n\}$ is *complete* in the sense that any piecewise smooth function $f(x)$ on $[x_0, x_f]$ can be represented by the weighted sum of the eigenfunctions that can be regarded as a generalized Fourier series representation (see Section 6.1.1):

$$f(x) = \sum_{n=0}^{\infty} c_n\, y_n \tag{3.4.13}$$

with coefficients

$$c_n = \frac{<f, y_n>_w}{<y_n, y_n>_w} \tag{3.4.14}$$

(SL-P5) Just like the Rayleigh quotient (P1.25.2) that is used to find the eigenvalue corresponding to an eigenvector, we can take the inner product of both sides of Eq. (3.4.7) for $y=y_n$

$$Ly_n \overset{(3.4.7)}{=} \lambda_n y_n \tag{3.4.15}$$

and y_n to get the *Rayleigh quotient* that is used to find the eigenvalue λ_n corresponding to an eigenfunction y_n:

$$\lambda_n = \frac{<Ly_n, y_n>_w}{<y_n, y_n>_w} = \frac{1}{<y_n, y_n>_w}\int_{x_0}^{x_f}\left\{\frac{d}{dx}\left(p(x)\frac{dy_n}{dx}\right)+q(x)y_n\right\}y_n\,dx$$

$$\overset{\text{Integration by parts}}{=} \frac{\left(py_n\dfrac{dy_n}{dx}\right)\Big|_{x_0}^{x_f} - \int_{x_0}^{x_f}\left\{p\left(\dfrac{dy_n}{dx}\right)^2 - q(x)y_n^2\right\}dx}{<y_n, y_n>_w} \tag{3.4.16}$$

This process has been cast into the following MATLAB function, '`eigenvalue_of_SLE()`'.

```
function lambda=eigenvalue_of_SLE(y,p,q,w,t0,tf)
% finds the eigenvalue of an SLE (Sturm-Liouville equation)
%   d{p(t)*dy/dt}/dt + {q(t)-lamda*w(t)}*y = 0   for t0<=t<=tf
%  corresponding to a given eigenfunction y(t).
dy=diff(y); pydy=p*y*dy;
num=subs(pydy,tf)-subs(pydy,t0)-int(p*dy^2-q*y^2,t0,tf);
yyw=dotf_w(y,y,w,t0,tf); % w-weighted dot product of y(t) and y(t)
lambda=eval(num/yyw); % Eq.(3.4.16)
```

(Example 3.29) SL (Sturm–Liouville) Problem

Solve the following linear second-order BVP:

$$(x\,y')' + \left(\frac{1}{x} - \lambda\frac{1}{x}\right)y = 0 \text{ with } y'(1) = 0 \text{ and } y'(2) = 0 \;\; (\lambda < 1\text{: not specified}) \tag{E3.29.1}$$

Since this equation can be rewritten as a homogeneous second-order Cauchy–Euler DE:

$$x^2 y''(x) + xy'(x) + (1-\lambda)\,y(x) = 0 \tag{E3.29.2}$$

we can make a substitution $y=x^m$ to write the auxiliary equation

$$m(m-1)+m+1-\lambda = m^2+1-\lambda = 0 \tag{E3.29.3}$$

whose solutions are

$$m=\pm j\sqrt{1-\lambda} \tag{E3.29.4}$$

Thus, the general solution of Eq. (E3.29.1) can be written as

$$y(x) \overset{(3.2.28),(E3.29.4)}{\underset{\sigma=0,\ \omega=\sqrt{1-\lambda},t=x}{=}} c_1\cos(\sqrt{1-\lambda}\ln|x|)+c_2\sin(\sqrt{1-\lambda}\ln|x|) \tag{E3.29.5}$$

Applying the boundary conditions to this yields

$$y'(1)\overset{(E3.29.5)}{=}-c_1\sqrt{1-\lambda}\frac{1}{x}\sin(\sqrt{1-\lambda}\ln|x|)+c_2\sqrt{1-\lambda}\frac{1}{x}\cos(\sqrt{1-\lambda}\ln|x|)\Big|_{x=1}=c_2\sqrt{1-\lambda}=0 \tag{E3.29.6a}$$
$$\rightarrow\ c_2=0$$

$$y'(2)\overset{(E3.29.5)}{=}-c_1\sqrt{1-\lambda}\frac{1}{x}\sin(\sqrt{1-\lambda}\ln|x|)+c_2\sqrt{1-\lambda}\frac{1}{x}\cos(\sqrt{1-\lambda}\ln|x|)\Big|_{x=2}$$
$$\overset{(P3.29.6a)}{=}-c_1\sqrt{1-\lambda}\frac{1}{2}\sin(\sqrt{1-\lambda}\ln 2)=0\ \rightarrow\ \sqrt{1-\lambda}\ln 2=n\pi;\ \sqrt{1-\lambda}=\frac{n\pi}{\ln 2} \tag{E3.29.6b}$$

Plugging Eqs. (E3.29.6a) and (E3.29.6b) into Eq. (E3.29.5) yields the eigenfunctions as

$$y_n(x)\overset{(E3.29.5)}{\underset{(E3.29.6)}{=}}c_1\cos(\frac{n\pi}{\ln 2}\ln|x|)\ \text{for}\ \lambda_n=1-\left(\frac{n\pi}{\ln 2}\right)^2 \tag{E3.29.7}$$

This solution y_1 with $\lambda_1=1-\{\pi/(\ln 2)\}^2$ can be obtained by using the symbolic DE solver '`dsolve()`' as follows:

```
>>y1=dsolve('t^2*D2y+t*Dy+(pi/log(2))^2*y=0','Dy(1)=0,Dy(2)=0')
   y1 = C16*cos((pi*log(t))/log(2))
```

Matching Eq. (E3.29.1) with Eq. (3.4.3), we can think of it as an SL equation with $p(x)=x$, $q(x)=1/x$, $x_0=1$, and $x_f=2$ so that Eq. (3.4.16) can be used to find the eigenvalue λ_1 if the eigenfunction y_1 has somehow been known:

```
>>yt=cos((pi*log(t))/log(2));
>>lambda1=1-(pi/log(2))^2 % from Eq.(E3.29.6b) with n=1
>>p=t; q=1/t; w=1/t; t0=1; tf=2;
>>eigenvalue_of_SLE(yt,p,q,w,t0,tf) % from Eq.(3.4.16) for y1
```

(Example 3.30) SL (Sturm–Liouville) Problem

Solve the following linear second-order BVP:

$$(x^2y')'-\lambda y/4=0\ \text{with}\ y(1)=0\ \text{and}\ y(2)=1\ (\lambda<-1:\text{not specified}) \tag{E3.30.1}$$

Since this equation can be rewritten as a homogeneous second-order Cauchy–Euler DE:

$$x^2y''(x)+2xy'(x)-\lambda y(x)/4=0 \tag{E3.30.2}$$

we can make a substitution $y=x^m$ to write the auxiliary equation

$$m(m-1)+2m-\lambda/4=m^2+m-\lambda/4=0 \tag{E3.30.3}$$

whose solutions are

$$m=\frac{-1\pm j\sqrt{-1-\lambda}}{2} \tag{E3.30.4}$$

Thus, the general solution of Eq. (E3.30.1) can be written as

$$y(x) \underset{\sigma=-1/2,\ \omega=\sqrt{-1-\lambda}/2,\ t=x}{\overset{(3.2.28),(E3.30.4)}{=}} c_1 x^{-1/2}\cos\left(\frac{\sqrt{-1-\lambda}}{2}\ln|x|\right) + c_2\, x^{-1/2}\sin\left(\frac{\sqrt{-1-\lambda}}{2}\ln|x|\right) \quad \text{(E3.30.5)}$$

Applying the boundary conditions to this yields

$$y(1) \overset{(E3.30.5)}{=} c_1 x^{-1/2}\cos(\frac{\sqrt{-1-\lambda}}{2}\ln|x|) + c_2\, x^{-1/2}\sin(\frac{\sqrt{-1-\lambda}}{2}\ln|x|)\Big|_{x=1} = c_1 = 0 \quad \text{(E3.30.6a)}$$

$$y(2) \overset{(E3.30.5)}{=} c_1 x^{-1/2}\cos(\frac{\sqrt{-1-\lambda}}{2}\ln|x|) + c_2\, x^{-1/2}\sin(\frac{\sqrt{-1-\lambda}}{2}\ln|x|)\Big|_{x=2}$$

$$\overset{(P3.30.6a)}{=} c_2\, 2^{-1/2}\sin(\frac{\sqrt{-1-\lambda}}{2}\ln 2) = 1 \;\rightarrow\; \frac{\sqrt{-1-\lambda}}{2}\ln 2 = \frac{n\pi}{2},\quad c_2 = 2^{1/2} \quad \text{(E3.30.6b)}$$

Plugging Eqs. (E3.30.6a) and (E3.30.6b) into Eq. (E3.30.5) yields the eigenfunctions as

$$y_n(x) \underset{(E3.30.6)}{\overset{(E3.30.5)}{=}} c_2\, x^{-1/2}\sin\left(\frac{\sqrt{-1-\lambda_n}}{2}\ln|x|\right) = c_2\, x^{-1/2}\sin(\frac{n\pi}{\ln 2}\ln|x|) \text{ for } \lambda_n = -1-\left(\frac{n\pi}{\ln 2}\right)^2 \quad \text{(E3.30.7)}$$

This solution y_1 with $\lambda_1 = -1 - \{\pi/(\ln 2)\}^2$ can be obtained by using the symbolic DE solver '`dsolve()`' as follows:

```
>>y1= dsolve('t^2*D2y+2*t*Dy+(1+(pi/log(2))^2)/4*y=0','y(1)=0,y(2)=1')
   y1 = (2^(1/2)*sin((pi*log(t))/(2*log(2))))/t^(1/2)
```

Interested readers should to run the MATLAB script "em03e30.m" to graph the analytical solution (E3.30.7) (n=1) together with the numerical solution obtained using '`bvp4c()`'.

```
%em03e30.m
clear, clf
lambda=-1-(pi/log(2))^2; c2=2^(1/2); % Eq.(E3.30.6b) with n=1
dy=@(x,y) [y(2); lambda/4*y(1)./x.^2-2*y(2)./x]; % Eq.(E3.30.2)
x0=1; xf=2; y0=0; yf=1; % Boundary condition
solinit=bvpinit(linspace(x0,xf,5),[0.5 1]); % Initial guess of solution
fbc=@(ya,yb) [ya(1)-y0; yb(1)-yf]; % BC function
sol=bvp4c(dy,fbc,solinit,bvpset('RelTol',1e-6));
N=1e3; xx=x0+[0:N]*(xf-x0)/N;
yyn=deval(sol,xx); % Evaluate the solution of DE.
residual_err=error_DE_sol(dy,yyn.',xx) % Error evaluation
yya=c2*xx.^(-1/2).*sin(sqrt(-1-lambda)/2*log(abs(xx))); % Eq.(E3.30.7)
plot(xx,yyn(1,:), xx,yya,'r:')
ys=dsolve('t^2*D2y+2*t*Dy+(1+(pi/log(2))^2)/4*y=0','y(1)=0,y(2)=1')
```

Problems

3.1 DEs That are Separable or Reducible to Separable Form

Find the solution of the following four DEs. Compose and run the MATLAB script (like "em03p01a.m" below) for each DE to plot $y(t)$ as in Fig. P3.1.

(a) $$\frac{dy}{dt} = y(1-y) \text{ with } y(0)=2 \quad \text{for } t=[0, 4] \tag{P3.1a}$$

(b) $$\frac{dy}{dt} = (2t+1)y \text{ with } y(0)=1 \quad \text{for } t=[0, 1] \tag{P3.1b}$$

(c) $$\frac{dy}{dt} = \frac{y^2-1}{t} \text{ with } y(1)=\frac{3}{2} \quad \text{for } t=[1, 2] \tag{P3.1c}$$

(d) $$\frac{dy}{dt} = \frac{ty^2}{\sqrt{1+t^2}} \text{ with } y(0)=-1 \quad \text{for } t=[0, 2] \tag{P3.1d}$$

```
%em03p01a.m
% To solve a DE (Eq.(P3.1a)) y'=y*(1-y) with y(0)=2
clear, clf
x0=0; xf=4; xspan=[x0 xf]; y0=2; % Solution interval and Initial value
dy=@(x,y)y*(1-y); % DE to solve
[xx,yy]=ode45(dy,xspan,y0); % Numerical solution
% Symbolic solution just like an analytic solution
ys=dsolve('Dy-y*(1-y)=0','y(0)=2'); pretty(simplify(ys))
t=xx; yst=eval(ys);
plot(xx,yy, xx,yst,'r:') % plot the numerical/analytical solutions
```

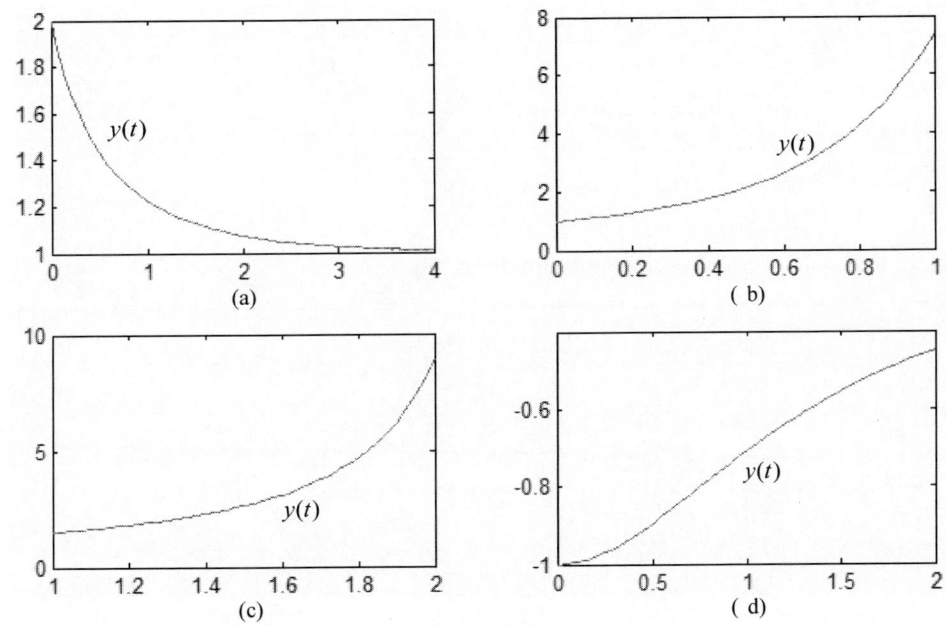

Figure P3.1

3.2 DE Reducible to Separable Form

Consider the following DE:

$$\frac{dy}{dx} = (x + y - 1)^2 \text{ with } y(0) = 1 \quad \text{for } x=[0, 1] \tag{P3.2.1}$$

With a substitution of variables: $u=x+y-1$ and $u'=1+y'$, this DE can be written in a separable form as

$$\frac{du}{dx} - 1 = u^2; \quad \frac{du}{u^2+1} = dx \tag{P3.2.2}$$

Starting from this form, find the solution to Eq. (P3.2.1) and plot it together with the numerical solution (obtained by using '`ode45()`') for $x=[0, 1]$. Check if it agrees with the solution obtained by completing and running the MATLAB script "em03p02.m."

```
%em03p02.m
% To solve DE (P3.2)
clear, clf
x0=0; xf=1; xspan=[x0 xf]; y0=1; % Solution interval and Initial value
dy=@(x,y)(x+y-1)^2; % DE to solve
[xx,yy]=?????(dy,xspan,y0); % Numerical solution
% To find the symbolic solution just like an analytic solution
ys=??????('Dy-(x+y-1)^2=0','y(0)=1','x')
% Conversion of symbolic solution to numeric data
x=xx; ysx=eval(ys);
plot(xx,yy, xx,ysx,'r:') % plot the numerical/analytical solutions
```

3.3 DEs That are Exact or Convertible to Exact Form

Solve the following DEs and graph their solutions for given intervals.

(a) $$\frac{dy}{dx} + \frac{6xy^2}{6x^2y+1} = 0 \text{ with } y(0) = 1 \quad \text{for } x=[0.01, 4] \tag{P3.3a}$$

(b) $$\frac{dy}{dx} = \frac{2x+y}{5-x} \text{ with } y(0) = 1 \quad \text{for } x=[0, 1] \tag{P3.3b}$$

(c) $$\frac{dy}{dx} = -\frac{xy}{x^2+y^2} \text{ with } y(0) = 2 \quad \text{for } x=[0, 4] \tag{P3.3c}$$

Refer to the MATLAB script "em03p03c.m" for (c).

```
%em03p03c.m
% To solve the DE (P3.3c)
clear, clf
ezplot('2*x^2*y^2+y^4-16'), hold on % True solution
x0=0; xf=4; xspan=[x0 xf]; y0=2; % Solution interval and Initial value
dy=@(x,y)-x*y/(x^2+y^2); % DE to solve
[xx,yy]=ode45(dy,xspan,y0); % Numerical solution
plot(xx,yy,'r:') % plot the numerical/analytical solutions
err1=error_DE_sol(dy,yy,xx) % How well Eq.(P3.3c) is satisfied
err2=norm(2*xx.^2.*yy.^2+yy.^4-16) % conforms with the true solution?
```

3.4 Linear First-Order DE

Consider

$$L\frac{di}{dt}+Ri=V_{\rm m}\cos\omega t \;\text{ with } i(0)=I_0 \tag{P3.4.1}$$

which can be written in standard form as

$$\frac{di}{dt}+\frac{R}{L}i=\frac{V_{\rm m}}{L}\cos\omega t \tag{P3.4.2}$$

Use Eq. (3.1.24) with the initial condition $i(0)=I_0$ to get

$$\begin{aligned} i(t) &= \left(I_0-\frac{RV_{\rm m}}{\square}\right)\square+\frac{V_{\rm m}}{\square}(R\cos\omega t+\omega L\sin\omega t) \\ &= \left(I_0-\frac{RV_{\rm m}}{\square}\right)\square+\frac{V_{\rm m}}{\square}\cos\left(\omega t-\tan^{-1}\frac{\omega L}{R}\right) \end{aligned} \tag{P3.4.3}$$

Note that the symbolic DE solver '`dsolve()`' can be used to get the same result:

```
>>i=dsolve('Dy+R/L*y=Vm/L*cos(w*t)','y(0)=I0','t'); % Eq.(P3.4.2)
>>pretty(i)

                                                         /                   R cos(t w) \
                                                      Vm | w sin(t w) + ---------- |
     /  R t \ /              R Vm              \           \                     L      /
  exp| - --- | | I0 - ---------------- | + --------------------------------
     \   L  / |           /  2        \ |                /  2              \
              |         2 | R       2 | |                | R            2 |
              |        L  | -- + w    | |              L | -- + w         |
              |           |  2        | |                |  2             |
              \           \ L         / /                \ L              /
```

3.5 Nonlinear First-Order DE

Complete and run the MATLAB script "em03p05.m" (below) to graph the numerical solution of the following DE for t=0~20[s] as in Fig. P3.5:

$$C\frac{dv(t)}{dt}+10^{-6}v^3(t)=\frac{1}{R}\left(V_{\rm s}+v_\delta\sin t-v(t)\right) \;\text{ with } v(0)=0 \tag{P3.5.1}$$

where $R=5\times10^5[\Omega]$, C=14[μF], V_s=6[V], and v_δ=1.4[V].

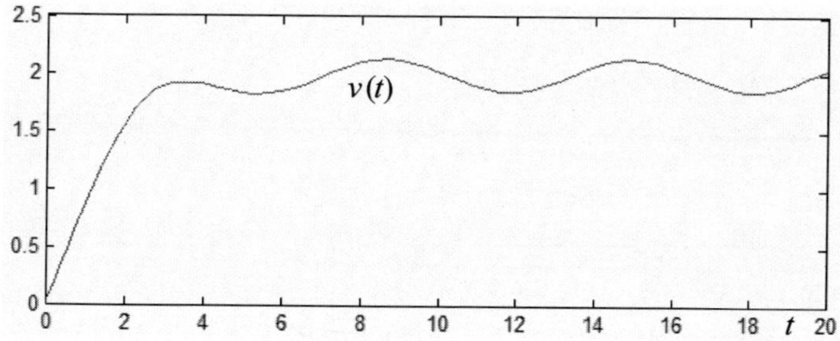

Figure P3.5

```
%em03p05.m
clear, clf
R=5e5; C=14e-6; Vs=6; vd=1.4;
dvdt=@(t,v)-1e-6/C*??? + (Vs+??*sin(t)-v)/?/C; % Eq.(P3.5.1)
t0=0; tf=20; tspan=[t0 tf]; v0=0; % Solution interval and Initial value
[tt,vv]=ode45(dvdt,tspan,v0); % Numerical solution
plot(tt,vv) % plot the numerical solutions
```

3.6 System of First-Order DEs

(a) Lorenz Equation — Turbulent Flow and Chaos

Consider the nonlinear state equation:

$$\begin{aligned} x_1'(t) &= \sigma(x_2(t)-x_1(t)) & & \sigma=10 \\ x_2'(t) &= (1+\lambda-x_3(t))x_1(t)-x_2(t) & \text{with } & \lambda=10\sim100 \\ x_3'(t) &= x_1(t)x_2(t)-\gamma\, x_3(t) & & \gamma=8/3 \end{aligned} \tag{P3.6a}$$

Complete the MATLAB script "em03p06a.m" to get the numerical solution of this DE, once with $\lambda=12$ and $[x_1(0)\ x_2(0)\ x_3(0)]=[-8\ -16\ 80]$, and three times with $\lambda=12, 27, 100$, and $[x_1(0)\ x_2(0)\ x_3(0)]=[-1\ 3\ 4]$. Run it to plot $x_3(t)$ vs. $x_1(t)$ for the time interval [0, 10] as in Fig. P3.6.1.

```
%em03p06a.m: Lorenz Attractor - (Caotic) Turbulent Flow & Caos
dxdt=@(t,x,sigma,lambda,r)[sigma*(x(2)-x(?));
                          (1+lambda-x(?)).*x(1)-x(2);
                          x(1).*x(?)-r*x(3)]; % Eq.(P3.6a)
sigma=10; lambdas=[12 12 27 ???]; r=8/3;
t0=0; tf=10; tspan=[t0 tf]; x0=[-8 ??? 80];
for i=1:length(lambdas)
  lambda=lambdas(i); if i>1, x0=[-1 3 4]; end
  [t,x]=ode23s(dxdt,tspan,x0,[],sigma,lambda,r); %Numerical solution
  subplot(220+i), plot(x(:,1),x(:,?)), xlabel('x_1'), ylabel('x_3')
end
```

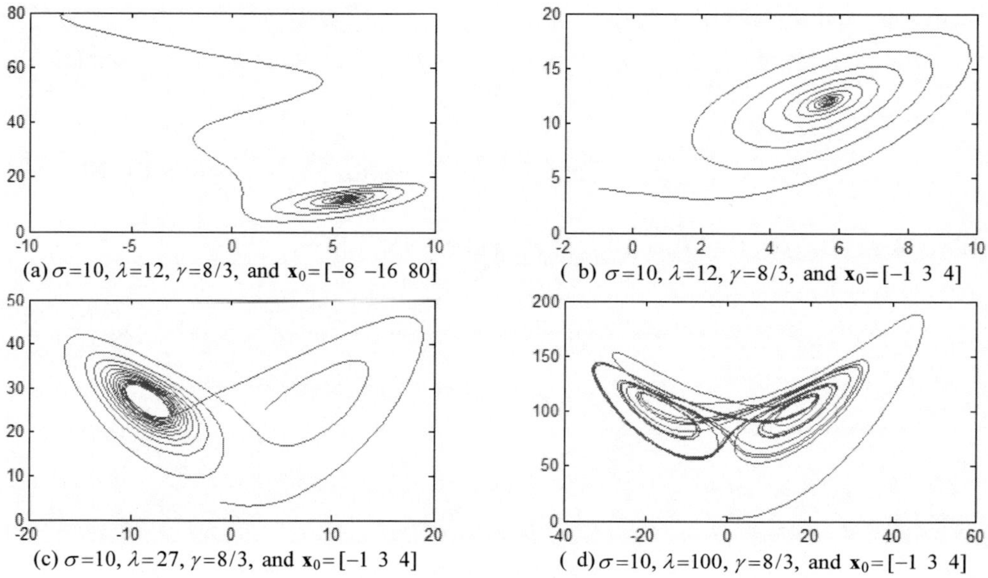

(a) $\sigma=10$, $\lambda=12$, $\gamma=8/3$, and $\mathbf{x}_0=[-8\ -16\ 80]$

(b) $\sigma=10$, $\lambda=12$, $\gamma=8/3$, and $\mathbf{x}_0=[-1\ 3\ 4]$

(c) $\sigma=10$, $\lambda=27$, $\gamma=8/3$, and $\mathbf{x}_0=[-1\ 3\ 4]$

(d) $\sigma=10$, $\lambda=100$, $\gamma=8/3$, and $\mathbf{x}_0=[-1\ 3\ 4]$

Figure P3.6.1

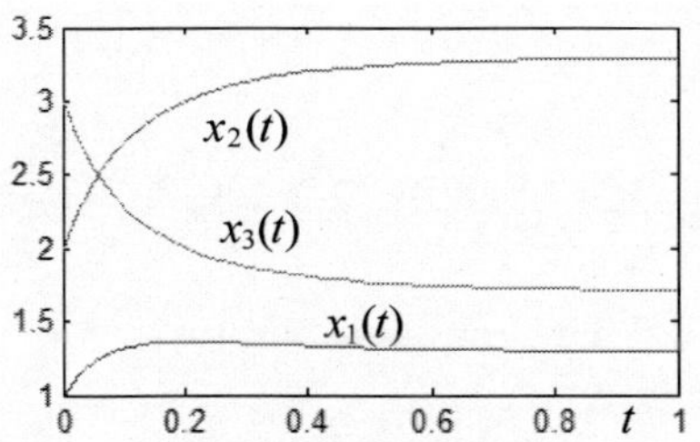

Figure P3.6.2

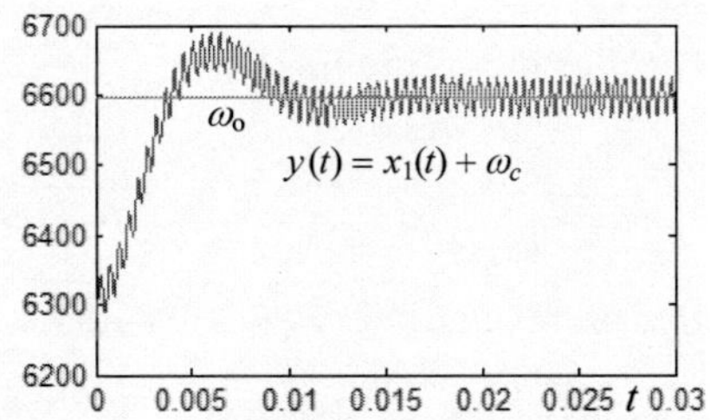

Figure P3.6.3

(b) Chemical Reactor

Consider the following nonlinear state equation:

$$\begin{aligned} x_1'(t) &= a(u_1 - x_1(t)) - bx_1(t)x_2(t) & & a = 5 \\ x_2'(t) &= a(u_2 - x_2(t)) - bx_1(t)x_2(t) & \text{with} \quad & b = 2 \\ x_3'(t) &= -ax_3(t) + bx_1(t)x_2(t) & & u_1 = 3, \; u_2 = 5 \end{aligned} \tag{P3.6b}$$

Complete the MATLAB script "em03p06b.m" to get the numerical solution of this DE with $[x_1(0) \; x_2(0) \; x_3(0)] = [1 \; 2 \; 3]$ and run it to plot $x_1(t)$, $x_2(t)$ and $x_3(t)$ for the time interval [0, 1] as in Fig. P3.6.2.

```
%em03p06b.m: Chemical Reactor
a=5; b=2; u1=3; u2=5;
dxdt=@(t,x)[a*(u1-x(?))-b*x(1).*x(2);
            a*(u2-x(?))-b*x(1).*x(2);
            -a*x(?)+b*x(1).*x(2)]; % Eq.(P3.6b)
t0=0; tf=1; x0=[1 2 3]; % Initial/final time and Initial values
[tt,xx]=ode45(dxdt,[t0 tf],x0); % Numerical solution
plot(tt,xx)
```

(c) Phase-Locked Loop (PLL)

Consider the following nonlinear state equation:

$$\begin{aligned} x_1'(t) &= \frac{au(t)\cos(x_2(t)) - x_1(t)}{\tau} & & a = 1500 \\ & & \text{with} \quad & \tau = 0.002 \\ x_2'(t) &= x_1(t) + \omega_c & & u(t) = \sin(\omega_o t) \end{aligned} \tag{P3.6c.1}$$

$$y(t) = x_1(t) + \omega_c \tag{P3.6c.2}$$

where ω_o=2100π[rad/s] and ω_c=2000π[rad/s]. Complete the above MATLAB script "em03p06c.m" to get the numerical solution of this DE with the initial condition $[x_1(0) \; x_2(0)] = [0 \; 0]$ and run it to plot $y(t)$ together with the constant ω_o for the time interval [0, 0.03] like Fig. P3.6.3.

```
%em03p06c.m: PLL(Phase-Locked Loop)
wo=2*pi*1050; wc=2*pi*1000; a=1500; tau=0.002; %4*pi/wc;
dxdt=@(t,x)[(a*sin(wo*t)*cos(x(?))-x(1))/???;
            x(?)+wc]; % Eq.(P3.6c.1)
t0=0; tf=0.03; x0=[0 0]; % Initial/final time and Initial values
[t,x]=ode45(dxdt,[t0 tf],x0); % Numerical solution
y = x(:,1)+wc; % Eq.(P3.6c.2)
plot(t,y, t,wo*ones(size(t)),'r')
```

3.7 Higher-Order Linear DE

Consider the following DE:

$$\frac{d^3}{dt^3}y(t)+2\frac{d^2}{dt^2}y(t)+\frac{d}{dt}y(t)=1 \quad \text{with } y(0)=3, y'(0)=2, \text{ and } y''(0)=-4 \tag{P3.7.1}$$

(a) Use the method of undetermined coefficients to find the complete solution.

```
>>ycis=complementary_sol([1 ? 1 ?])
```

(b) Use the method of variation of parameters to find the particular solution.

```
>>ycis='[1 exp(-t) t*exp(-t)]'; f='?';
>>yp=particular_sol(????,f)
```

(c) Use the symbolic DE solver '`dsolve()`' to find the complete solution.

(d) Noting that the DE can be written in state equation form as

$$\begin{aligned}\mathbf{x}'(t)&=A\mathbf{x}(t)+Bu(t)\\ y(t)&=C\mathbf{x}(t)\end{aligned} \quad \text{with } \mathbf{x}(t)=\begin{bmatrix} y(t)\\ y'(t)\\ y''(t)\end{bmatrix}, \ \mathbf{x}(0)=\begin{bmatrix} y(0)\\ y'(0)\\ y''(0)\end{bmatrix}=\begin{bmatrix} 3\\ 2\\ -4\end{bmatrix}, \text{ and } u(t)=1 \tag{P3.7.2a}$$

where

$$A=\begin{bmatrix} 0 & 1 & 0\\ 0 & 0 & 1\\ 0 & -1 & -2\end{bmatrix}, \ B=\begin{bmatrix} 0\\ 0\\ 1\end{bmatrix}, \text{ and } C=[1 \ \ 0 \ \ 0] \tag{P3.7.2b}$$

complete and run the MATLAB script "em03p07.m" to get the numerical and analytical solutions of the DE, and plot $y(t)$, $y'(t)$, and $y''(t)$ for the time interval [0, 5] as in Fig. P3.7.

```
%em03p07.m
% To solve the DE (P3.7.1).
A=[0 ? 0; 0 ? 1; 0 ?? -2]; B=[0;0;1]; C=[1 0 0]; u=1;
dx=@(t,x)A*x+B*u; % DE (P3.7.2)
t0=0; tf=5; x0=[3;2;??]; % Initial/final time and Initial state
[tt,xx]=ode45(??,[t0 tf],x0); % Numerical solution
% To get the analytical solution
ya=dsolve('???+2*D2y+Dy=1','y(0)=3,Dy(0)=2,D2y(0)=??')
t=tt; yat=eval(ya); % To evaluate the analytical solution for t=tt
% Compute the discrepancy between the numerical/analytical solutions
discrepancy = norm(xx(:,1)-yat)/norm(yat)
plot(??,xx, tt,???,'r:') % plot the numerical/analytical solutions
legend('Numerical y','Numerical y1','Numerical y2','Analytical y')
% Using the Laplace transform
syms s; y=ilaplace(C*(s*eye(size(A))-A)^-1*(x0+B/s)) % Eq.(4.4.5)
```

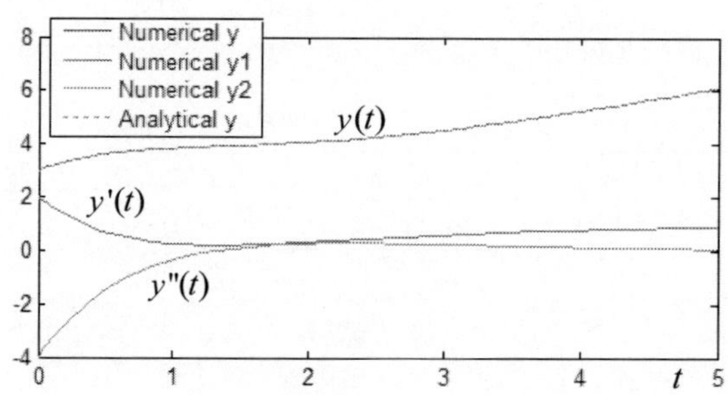

Figure P3.7 Numerical/analytical solutions for DE (P3.7.1)

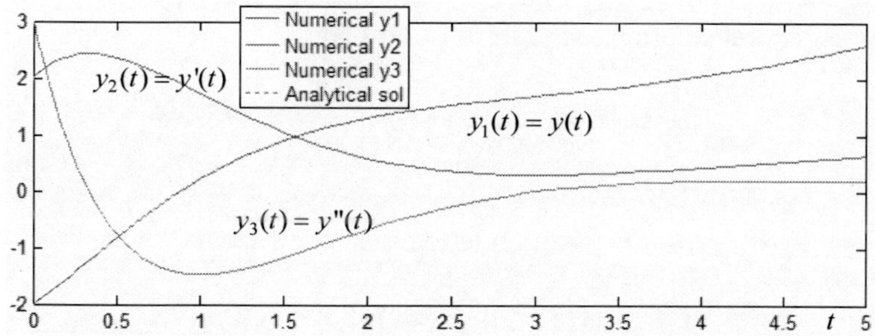

Figure P3.8 Numerical/analytical solutions for DE (P3.8.1)

3.8 Higher-Order Linear DE

Consider the following DE:

$$y^{(3)} + 3y'' + 3y' + y = t \text{ with } y(0)=-2, y'(0)=2, \text{ and } y''(0)=3 \tag{P3.8.1}$$

(a) Use the method of undetermined coefficients to find the complete solution.
(b) Use the method of variation of parameters to find the particular solution.
(c) Use the symbolic DE solver '`dsolve()`' to find the complete solution.
(d) Complete and run the MATLAB script "em03p08.m" to get the numerical and analytical solutions of the DE, and plot $y(t)$, $y'(t)$, and $y''(t)$ for the time interval [0, 5] as in Fig. P3.8.

```
%em03p08.m
A=[0 ? 0; 0 0 ?;-1 ?? -3]; B=[0;0;1]; C=[1 0 0];
dx=@(t,x)A*x+B*t; % DE to solve
t0=0; tf=5; x0=[-2; 2; 3]; % Initial/final time and Initial values
[tt,xx]=ode45(dy,[t0 tf],x0); % Numerical solution
ya=dsolve('D3y+3*D2y+3*Dy+y=t','y(0)=-2,Dy(0)=2,D2y(0)=3')
t=tt; yat=eval(ya); % evaluates the symbolic solution
plot(tt,xx, tt,yat,'r:')
legend('Numerical y1','Numerical y2','Numerical y3','Analytical sol')
sym s; y=ilaplace(C*(s*eye(size(A))-A)^-1*(x0+B/s^2)) % Eq.(4.4.5)
```

3.9 Cauchy–Euler DE

(a) Homogeneous Third-Order Cauchy–Euler DE
Find the solution to the third-order DE

$$t^2 \frac{d^2}{dt^2} y(t) + 2t \frac{d}{dt} y(t) - 2y(t) = 0 \text{ with } y(1) = 5 \text{ and } y(2) = 3, \tag{P3.9a}$$

(b) Homogeneous Third-Order Cauchy–Euler DE
Find the general solution to the third-order DE

$$t^3 \frac{d^3}{dt^3} y(t) + 6t^2 \frac{d^2}{dt^2} y(t) + 7t \frac{d}{dt} y(t) + y(t) = 0, \tag{P3.9b}$$

(c) Homogeneous Third-Order Cauchy–Euler DE
Find the general solution to the third-order DE

$$t^3 \frac{d^3}{dt^3} y(t) + 5t^2 \frac{d^2}{dt^2} y(t) + 7t \frac{d}{dt} y(t) + 8y(t) = 0, \tag{P3.9c}$$

(d) Homogeneous Third-Order Cauchy–Euler DE
Find the general solution to the third-order DE

$$t^4 \frac{d^4}{dt^4}y(t) + 10t^3 \frac{d^3}{dt^3}y(t) + 33t^2 \frac{d^2}{dt^2}y(t) + 39t\frac{d}{dt}y(t) + 25y(t) = 0\,, \qquad \text{(P3.9d)}$$

(e) Nonhomogeneous Third-Order Cauchy–Euler DE
Find the general solution (with undetermined coefficients) to the third-order DE

$$t^3 \frac{d^3}{dt^3}y(t) + t^2 \frac{d^2}{dt^2}y(t) - 2t\frac{d}{dt}y(t) + 2y(t) = g(t) = t^3e^t \qquad \text{(P3.9f)}$$

3.10 Nonlinear DEs

(a) Van der Pol Equation
Consider a second-order nonlinear DE

$$\frac{d^2}{dt^2}y(t) - \mu\,(1 - y^2(t))\frac{d}{dt}y(t) + y(t) = 0 \;\; \text{with } \mu = 2\,, \qquad \text{(P3.10a.1)}$$

Noting that the DE can be written in state equation form as

$$\begin{bmatrix} y'(t) \\ y''(t) \end{bmatrix} = \begin{bmatrix} y'(t) \\ -y(t) + \mu(1 - y^2(t))y'(t) \end{bmatrix} \qquad \text{(P3.10a.2)}$$

```
%em03p10a.m
u=2;
dx=@(t,x)[x(?); -x(?)+?*(1-x(?).^2).*x(2)];  % DE (P3.10a.2)
t0=0; tf=20; tspan=[t0 tf]; % Solution interval
x0s=[0.5 0; -1 ?]; % Two initial conditions
for i=1:size(x0s,2)
   x0=x0s(i,:); % Initial condition
   [tt,xx]=ode45(dx,tspan,x0); % Numerical solution
   subplot(219+2*i), plot(tt,xx)
   subplot(220+2*i), plot(xx(:,1),xx(:,2))
end
```

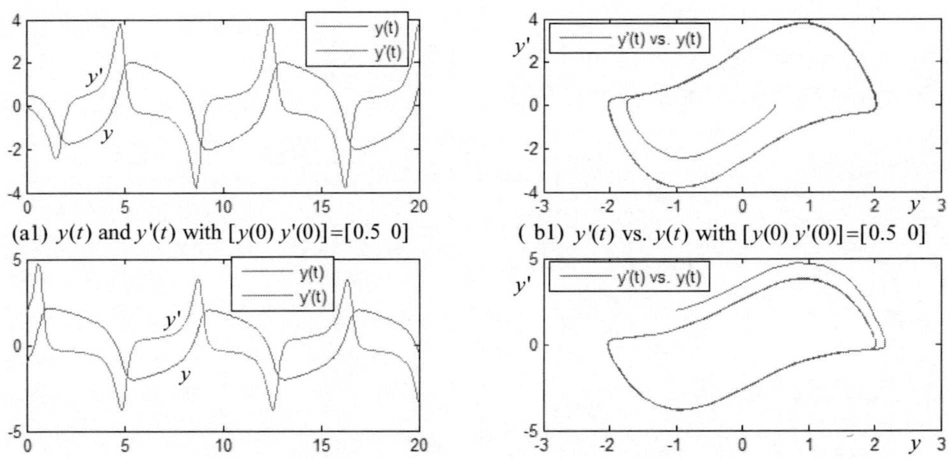

(a1) $y(t)$ and $y'(t)$ with $[y(0)\ y'(0)]=[0.5\ \ 0]$

(b1) $y'(t)$ vs. $y(t)$ with $[y(0)\ y'(0)]=[0.5\ \ 0]$

(a2) $y(t)$ and $y'(t)$ with $[y(0)\ y'(0)]=[-1\ \ 2]$

(b2) $y'(t)$ vs. $y(t)$ with $[y(0)\ y'(0)]=[-1\ \ 2]$

Figure P3.10.1

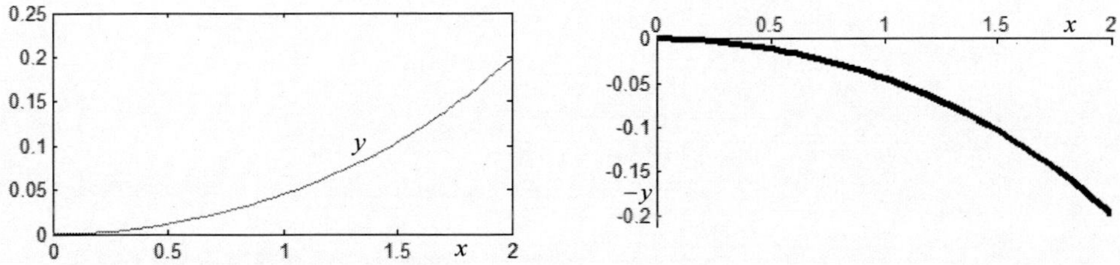

(a) The solution of DE (P3.10b) (b) Deflection of the cantilever beam

Figure P3.10.2

complete and run the above MATLAB script "em03p10a.m" to get the numerical solution of the DE twice, once with the initial condition $\mathbf{x}_0$=[0.5 0] and once with $\mathbf{x}_0$=[−1 2], and plot $y'(t)$ vs. $y(t)$ as well as $y(t)$ and $y'(t)$ vs. t for the time interval [0, 20] as in Fig. P3.10.1.

(b) Cantilever Beam – DE w.r.t. a Spatial Variable

Consider a nonlinear DE describing the vertical deflection y of a beam due to its own weight:

$$JE\frac{d^2y}{dx^2} = \rho g\left(1+\left(\frac{dy}{dx}\right)^2\right)^{3/2}\left\{x\left(x-\frac{L}{2}\right)+\frac{L^2}{2}\right\} \text{ with } [y(0)\ \ y'(0)] = [0\ \ 0] \qquad \text{(P3.10b)}$$

where g(gravitational acceleration)=9.8[m/s^2], JE(flexural rigidity)=2000[kg·m^3/s^2], ρ(linear density)=10[kg/m], and L(length)= 2[m]. Complete and run the MATLAB script "em03p10b.m" to get the numerical solution of this DE, and plot it for the interval [0, L] as depicted in Fig. P3.10.2(a) or (b).

```
%em03p10b.m: Cantilever
g=9.8; JE=2000; rho=10; L=2;
dy=@(x,y)[y(?); rho*g/JE*(1+y(?).^2).^1.5.*(x.*(x-L/2)+L^2/2)];
x0=0; xf=L; y0=[0 0]; % Left/Right boundaries and Boundary values
[xx,yy]=ode45(dy,[x0 xf],y0); % Numerical solution
plot(xx,yy(:,1))
```

(c) Motion of a Projectile

Consider the following DEs describing the motion of a projectile influenced by atmospheric drag proportional to the square of the velocity v:

$$\frac{dx}{dt} = v_x,\ \frac{dy}{dt} = v_y,\ \frac{dv_x}{dt} = -cvv_x \text{ and } \frac{dv_y}{dt} = -g - cvv_y \text{ with } v = \sqrt{v_x^2+v_y^2} \qquad \text{(P3.10c)}$$

where c(drag coefficient)=0.006[N·s^2/m^2] and g(gravitational acceleration)=9.8[m/s^2].

(c1) Complete and run the MATLAB script "em03p10c1.m" to get the numerical solution of this DE (with initial conditions $x(0)$=0, $y(0)$=0, $v(0)$=200m/s, and $\theta(0)$= $\tan^{-1}(v_y(0)/v_x(0))$=45°) and graph the solution $y(t)$ vs. $x(t)$ for t=0~12s as in Fig. P3.10.3.

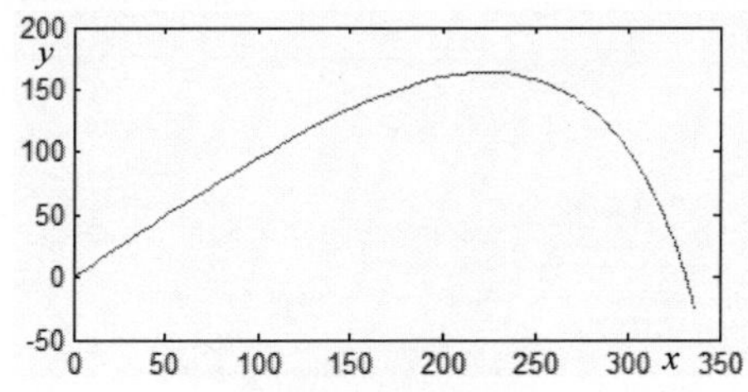

Figure P3.10.3

```
%em03p10c1.m: Motion of a Projectile
g=9.8; c=0.006; % Gravitational acceleration and Drag coefficient
dzdt=@(t,z)[z(3); z(4);
            -c*sqrt(z(3)^2+z(4)^2)*z(3);
            -g-c*sqrt(z(3)^2+z(4)^2)*z(4)]; % Eq.(P3.10c)
x0=0; y0=0; th0=deg2rad(45); v0=200;
z0 = [0; 0; v0*cos(th0); v0*sin(th0)]; % Initial state
t0=0; tf=12; tspan=[0 tf]; % Solution time interval
[tt,zz]=ode45(dzdt,tspan,z0); % Numerical solution
x=zz(:,1); y=zz(:,2); subplot(221), plot(x,y)
```

(c2) Run the MATLAB script "em03p10c2.m," which uses '`getframe()`' and '`movie()`' to make a series of movie frames for the projectile trajectory, and replay it.

```
%em03p10c2.m: Motion Picture (Animation) of a Projectile
for n=1:length(tt)
   plot(x(n),y(n),'ro'); xlabel('x[m]'); ylabel('y[m]');
   axis equal; axis([0,max(x),0,max(y)+10]);
   title('Projectile Trajectory'); M(n)=getframe;
end
movie(M)
```

3.11 Systems of Linear DEs

(a) DC Motor

Consider a linear DE describing the behavior of a DC motor system:

$$J\frac{d^2\theta(t)}{dt^2}+B\frac{d\theta(t)}{dt}=T(t)=K_T\, i(t)$$
$$L\frac{di(t)}{dt}+Ri(t)+K_b\frac{d\theta(t)}{dt}=v(t)=\begin{cases}10 & \text{for } 0\le t<D\\ 0 & \text{for } t\ge D\end{cases} \tag{P3.11a.1}$$

Convert this system of equations into a first-order vector DE, i.e., a state equation in the state vector $[\theta(t)\ \ \theta'(t)\ \ i(t)]$. Complete and run the MATLAB script "em03p11a.m," which uses '`ode45()`' to find the numerical solution of the DE with J=1[kg·m²], B=1[kg·m²/s], K_T=1[N·m/A], L=1[H], R=1[Ω], K_b=1[V/(rad/s)], and D=1[sec] for t=0~5[s] and plots the angular position $\theta(t)$, angular velocity $\theta'(t)$, and current $i(t)$ of the rotor as in Fig. P3.11(a). Identify the graphs of $\theta(t)$, $\theta'(t)$, and $i(t)$ among the three curves in that figure.

```
%em03p11a.m: to simulate a DC motor system
J=1; B=1; KT=1; Kb=1; R=1; L=1; Vm=10; D=1;
dx = @(t,x)[x(?); -B/?*x(2)+??/J*x(3);
            -Kb/?*x(2)-?/L*x(3)+??/L*(t<D)]; % Eq.(S3.11a.1)
t0=0; tf=5; x0-[0 0 0]; % Initial/Final time and initial value
% .........
```

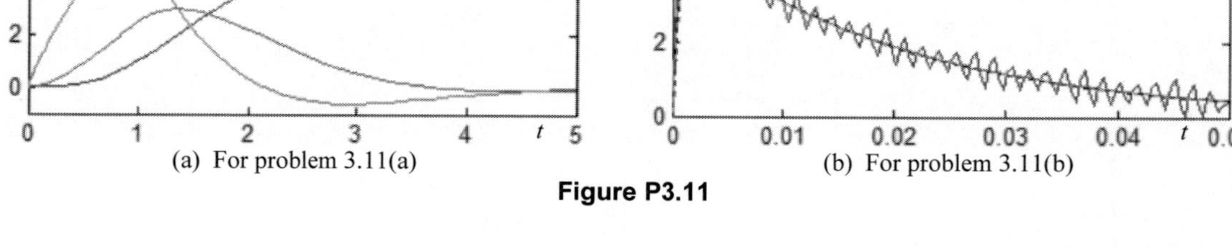

(a) For problem 3.11(a) (b) For problem 3.11(b)

Figure P3.11

(b) *RC* Circuit — A Stiff System

Consider the following DE in $i_1(t)$ and $i_2(t)$:

$$R_1 i_1(t) + \frac{1}{C_1}\int_{-\infty}^{t} i_1(\tau)d\tau + R_2(i_1(t) - i_2(t)) = v(t) = t$$
$$R_2(i_2(t) - i_1(t)) + \frac{1}{C_2}\int_{-\infty}^{t} i_2(\tau)\,d\tau = 0 \qquad \text{(P3.11b.1)}$$

To convert this system of DEs into a state equation, we differentiate both sides and rearrange them to get

$$(R_1 + R_2)\frac{di_1(t)}{dt} - R_2\frac{di_2(t)}{dt} + \frac{1}{C_1}i_1(t) = \frac{dv(t)}{dt} = 1$$
$$-R_2\frac{di_1(t)}{dt} + R_2\frac{di_2(t)}{dt} + \frac{1}{C_2}i_2(t) = 0 \qquad \text{(P3.11b.2)}$$

$$; \begin{bmatrix} R_1+R_2 & -R_2 \\ -R_2 & R_2 \end{bmatrix}\begin{bmatrix} i_1'(t) \\ i_2'(t) \end{bmatrix} = \begin{bmatrix} 1-i_1(t)/C_1 \\ -i_2(t)/C_2 \end{bmatrix} \qquad \text{(P3.11b.3)}$$

$$; \begin{bmatrix} i_1'(t) \\ i_2'(t) \end{bmatrix} = \begin{bmatrix} R_1+R_2 & -R_2 \\ -R_2 & R_2 \end{bmatrix}^{-1}\begin{bmatrix} 1-i_1(t)/C_1 \\ -i_2(t)/C_2 \end{bmatrix}$$
$$= \begin{bmatrix} -1/R_1C_1 & -1/R_1C_2 \\ -1/R_1C_1 & -(R_1+R_2)/R_1R_2C_2 \end{bmatrix}\begin{bmatrix} i_1(t) \\ i_2(t) \end{bmatrix} + \begin{bmatrix} 1/R_1 \\ 1/R_1 \end{bmatrix}u_s(t) \qquad \text{(P3.11b.4)}$$

where $u_s(t)$ denotes the unit step function whose value is 1 (one) $\forall\, t \geq 0$.

(i) Use the two DE solvers '`ode45()`' and '`ode23s()`' to get the numerical solution of the state equation with R_1=100[Ω], C_1=10[μF], R_2=1[kΩ], C_2=10[μF], and zero initial condition $i_1(0)=i_2(0)=0$, and plot $i_2(t)$ for t=0~0.05[s] as depicted in Fig. P3.11b.

```
%em03p11b1.m: RC Circuit - A Stiff System
R1=100; R2=1000; C1=1e-5; C2=1e-5;
dx=@(t,x)[(1-x(1)/C1-x(2)/C2)/R1;
          (1-x(1)/??)/R1-(R1+R2)/R1/??/C2*x(2)]; % Eq.(P3.11b.4)
t0=0; tf=0.05; x0=[0 0]; % Initial/Final time and Initial value
[tt,xx]=ode45(dx,[t0 tf],??); % Numerical solution
[tt1,xx1]=ode23s(??,[t0 tf],x0); % Hopefully with better accuracy
plot(tt,xx(:,2), tt1,xx1(:,2),'r')
legend('i2(t) from ode45()','i2(t) from ode23s()')
```

(ii) Using the MATLAB symbolic DE solver '`dsolve()`', find the analytical solution of the DE (P3.11b.2) and plot $i_2(t)$ for t=0~0.05[s]. Which of the two numerical solutions obtained in (i) is better in terms of closeness to the (true) analytical solution? You may refer to the following code.

```
syms R1 R2 C1 C2
i=dsolve('(R1+R2)*Di1-R2*D??+i1/C1=1',...
     '-R2*D??+R2*Di2+i2/C2','i1(0)=0','i2(0)=0'); %(P3.11b.2)
R1=100; R2=1000; C1=1e-5; C2=1e-5;
t0=0; tf=0.05; N=500; t=t0+(tf-t0)/N*[0:N]; % Time range
i2t=eval(i.i2);
hold on, plot(t,i2t,'m')
save t_i2t t i2t % save t and i2t in a data file named 't_i2t.mat'
```

(iii) If we apply the Laplace transform technique to solve this equation with zero initial condition $\mathbf{i}(0)=\mathbf{0}$, we can get

$$\begin{bmatrix} I_1(s) \\ I_2(s) \end{bmatrix} \overset{(3.2.43)}{=} [sI-A]^{-1}BU(s) = \begin{bmatrix} s+1/R_1C_1 & 1/R_1C_2 \\ 1/R_1C_2 & s+(R_1+R_2)/R_1R_2C_2 \end{bmatrix}^{-1} \begin{bmatrix} 1/R_1 \\ 1/R_1 \end{bmatrix} \frac{1}{s}$$

$$; \; I_2(s) = \frac{1/R_1}{s^2+(1/R_1C_1+(R_1+R_2)/R_1R_2C_2)s+1/R_1R_2C_1C_2} = \frac{1/100}{s^2+2100s+100000}$$

$$\cong \frac{1/100}{(s+2051.25)(s+48.75)} \cong \frac{1}{200250}\left(\frac{1}{s+48.75} - \frac{1}{s+2051.25}\right) \tag{P3.11b.5}$$

$$; \qquad i_2(t) \cong \frac{1}{200250}(e^{-48.75t} - e^{-2051.25t}) \tag{P3.11b.6}$$

where λ_1=−2051.25 and λ_2=−48.75 are the eigenvalues of the system matrix A in Eq. (P3.11b.4). Running a MATLAB script like "em03p110b3.m," plot this solution for t=0~ 0.05[s]. Also, find the measure of stiffness of the DE defined by

$$\eta(A) = \frac{\text{Max}\,\{|\text{Re}\,(\lambda_i)|\}}{\text{Min}\,\{|\text{Re}(\lambda_i)| \neq 0\}} \tag{P3.11b.7}$$

What exists between the failure/success of '`ode45()`'/'`ode23s()`' in (i)?

```
%em03p11b3.m
syms s; R1=100; R2=1e3; C1=1e-5; C2=1e-5;
A=[-1/R1/C1 -1/R1/C2;
   -1/R1/?? -(??+R2)/R1/R2/C2]; % Eq.(P3.11b.4)
BUs=[1/R1; 1/??]/s; Is=[s*eye(size(A))-A]^-1*Bus % Eq.(P3.11b.5)
i2=ilaplace(Is(2)) % Inverse Laplace transform Eq.(P3.11b.6)
t0=0; tf=0.05; N=500; t=t0+(tf-t0)/N*[0:N]; % Time range
i2t_i=eval(i2); % Eq.(P3.11b.6)
lambda=eig(A);
i2t=1/R1/(lambda(1)-lambda(?))*(exp(lambda(?)*t)-exp(lambda(2)*t));
hold on, plot(t,i2t_i, t,i2t,'r:')
lama=abs(real(lambda)); %Magnitudes of real parts of each eigenvalue
stiffness=max(lama)/min(lama(find(lama>1e-6))) % Eq.(P3.11b.7)
```

3.12 Bessel Functions

Verify the validity of the following identities for the Bessel function of the first kind:

(a) $e^{(z-z^{-1})x/2} = \sum_{n=-\infty}^{\infty} J_n(x)\,z^n$: *Generating function for the Bessel function* (P3.12.1)

where $e^{zx/2} \overset{(C.25)}{=} \sum_{l=0}^{\infty} \frac{(x/2)^l}{l!} z^l$ and $e^{-z^{-1}x/2} \overset{(C.25)}{=} \sum_{m=0}^{\infty} \frac{(-x/2)^m}{m!} z^{-m}$

(b) $v\,x^{v-1} J_v(x) + x^v J_v'(x) = \frac{d}{dx}\left(x^v J_v(x)\right) = x^v J_{v-1}(x)$: *Differential recurrence relation* (P3.12.2)

(c) $-v\,x^{-v-1} J_v(x) + x^{-v} J_v'(x) = \frac{d}{dx}\left(x^{-v} J_v(x)\right) = -x^{-v} J_{v+1}(x)$ (P3.12.3)

(d) $J_{v+1}(x) + J_{v-1}(x) = \frac{2v}{x} J_v(x)$ (P3.12.4)

(e) $J_{v-1}(x) - J_{v+1}(x) = 2\,J_v'(x)$ (P3.12.5)

3.13 Legendre Polynomials

Verify the validity of the following identities for the Legendre polynomial:

(a) $$\frac{1}{\sqrt{1-2xz+z^2}}=\sum_{n=0}^{\infty}P_n(x)\,z^n \quad : \textit{Generating function for the Legendre polynomial} \tag{P3.13.1}$$

(Hint) Note that the Taylor series expansion of $1/\sqrt{1-y}$ around $y=0$ is

$$f(y)=\frac{1}{\sqrt{1-y}}=(1-y)^{-1/2}\overset{(8.4.15)}{=}f(0)+f'(0)y+\frac{1}{2!}f''(0)y^2+\frac{1}{3!}f^{(3)}(0)y^3+\cdots$$
$$=1-\frac{1}{2}y+\frac{1}{2}\frac{3}{2}\frac{1}{2!}y^2-\frac{1}{2}\frac{3}{2}\frac{5}{2}\frac{1}{3!}y^3+\cdots$$

(b) $$\int_{-1}^{1}P_mP_n\,dx=\begin{cases}0 & \text{for } m\neq n\\ \dfrac{2}{2n+1} & \text{for } m=n\end{cases} \quad \text{(orthonality)} \tag{P3.13.2}$$

(b1) To show $\int_{-1}^{1}P_m\,P_n\,dx=0$ for $m\neq n$, rewrite the Legendre equation (3.3.21) as

$$(1-x^2)y''-2x\,y'+n(n+1)y=\left((1-x^2)y'\right)'+n(n+1)y=0 \tag{S3.13.2}$$

and plug $y=P_n(x)$ and $y=P_m(x)$ into this equation to write

$$\boxed{} \tag{S3.13.3a}$$
$$((1-x^2)P_m'(x))'+m(m+1)P_m(x)=0 \tag{S3.13.3b}$$

First, multiply Eq. (S3.13.3b) by P_n and integrate from $x=-1$ to 1 to get

$$\int_{-1}^{1}P_n((1-x^2)P_m'(x))'\,dx+m(m+1)\int_{-1}^{1}P_nP_m\,dx$$
$$=P_n(1-x^2)P_m'\Big|_{-1}^{1}-\int_{-1}^{1}P_n'(1-x^2)P_m'\,dx+m(m+1)\int_{-1}^{1}P_nP_m\,dx$$
$$=-\int_{-1}^{1}P_n'(1-x^2)P_m'\,dx+m(m+1)\int_{-1}^{1}P_nP_m\,dx=0 \tag{P3.13.4a}$$

Then, in the same way, multiply Eq. (S3.13.3a) by P_m and integrate from $x=-1$ to 1 to get

$$\boxed{} \tag{P3.13.4b}$$

Subtracting Eq. (P3.13.4a) from Eq. (P3.13.4b) yields

$$\left(m(m+1)-n(n+1)\right)\int_{-1}^{1}P_mP_n\,dx=0\overset{m\neq n}{\to}\int_{-1}^{1}P_mP_n\,dx=0 \tag{P3.13.5}$$

(b2) To show $\int_{-1}^{1}P_n^2\,dx=\dfrac{2}{2n+1}$, square the generating function (P3.13.1),

$$\frac{1}{1-2xz+z^2}=\sum_{m=0}^{\infty}P_m(x)\,z^m\sum_{n=0}^{\infty}P_n(x)\,z^n\underset{\text{(P3.13.5)}}{\overset{\text{Orthogonality}}{=}}\sum_{n=0}^{\infty}P_n^2(x)z^{2n} \tag{P3.13.6}$$

and integrate from $x=-1$ to 1 to write

$$\int_{-1}^{1}\frac{1}{1-2xz+z^2}dx=z^{2n}\int_{-1}^{1}\sum_{n=0}^{\infty}P_n^2(x)dx=z^{2n}\sum_{n=0}^{\infty}\int_{-1}^{1}P_n^2(x)dx \tag{P3.13.7}$$

where the LHS of this equation can be computed as

$$\begin{aligned}\int_{-1}^{1}\frac{1}{1-2xz+z^2}dx &= \frac{1}{-2z}\int_{-1}^{1}\frac{1}{x-(z+1/z)/2}dx = -\frac{1}{2z}\ln\left(x-(z+1/z)/2\right)\Big|_{-1}^{1}\\ &= -\frac{1}{2z}\ln\frac{1-(z+1/z)/2}{-1-(z+1/z)/2} = -\frac{1}{2z}\ln\frac{1-2z+z^2}{1+2z+z^2} = -\frac{1}{2z}\ln\left(\frac{1-z}{1+z}\right)^2\\ &= -\frac{1}{z}\ln\left(\frac{1-z}{1+z}\right) = \frac{1}{z}\ln\left(\frac{1+z}{1-z}\right) \overset{(8.4.24)}{=} \frac{2}{z}\left(z+\frac{z^3}{3}+\frac{z^5}{5}+\cdots\right) = \sum_{n=0}^{\infty}\frac{2}{2n+1}z^{2n}\end{aligned} \quad \text{(P3.13.8)}$$

Equate this with the RHS of Eq. (P3.13.7) to the desired equality:

$$\boxed{} \rightarrow \int_{-1}^{1} P_n^2\,dx = \frac{2}{2n+1} \quad \text{(P3.13.9)}$$

3.14 Bessel|Legendre|Chebyshev|Hermite|Laguerre's Equations

To get acquainted with Bessel|Legendre|Chebyshev|Hermite|Laguerre's equations or polynomials, check if the functions|polynomials numerically satisfy the corresponding DEs with small residual errors, respectively.

(a) To see if the Bessel function, $J_v(x)$ (Eq. (3.3.10a)), of order, say, $v=1/2$, numericallys atisfies the Bessel's equation (3.3.1):

$$x^2y''+xy'+(x^2-v^2)y=0 \text{ with } v=1/2 \text{ and } y(0)=0,\ y(1)=4 \quad \text{(P3.14.1)}$$

for $x=0\sim2$, run the following MATLAB statements:

```
>>xx=[0:1e-4:2]; % Solution interval
>>[y,yy,a,residual_error]=Bessel_eq(1/2,xx,1,4); residual_error
```

where the following two MATLAB functions are supposed to have been stored as m-files named "Bessel_eq.m" and "error_DE2_sol.m," respectively, on your PC. Is the relative residual error so small that $J_{1/2}(x)$ can be acknowledged as satisfying the Bessel's equation? The residual error may be reduced further by refining the grids in the solution interval. Note that the solution interval [0, 2] includes $x=0$ and $J_{-1/2}(x)$ has been excluded from the solution, since $J_{-v}(x)$ with a nonintger value of v is singular at $x=0$.

```
function [y,yy,a,residual_error]=Bessel_eq(v,xx,x1,y1)
% Numeric values yy of the sol. to Bessel DE of order v for xx:
%   x^2*y"(x) +x*y'(x) +(x^2-v^2)*y(x) = 0  such that y(x1)=y1
y=dsolve('t^2*D2y+t*Dy+(t^2-v^2)*y=0') % Analytical general sol
yy=besselj(v,xx); % Bessel function J_v of the first kind of order v
a=y1/besselj(v,x1); yy=a*yy; % so that y(x1)=y1
syms x; residual_error=error_DE2_sol([x.^2 x x.^2-v^2],yy,xx);
```

```
function residual_error=error_DE2_sol(a,yy,xx)
% Relative residual error of a numerical solution (yy,xx) of
%  a 2nd-order DE with its three coefficients in a.
dxx=diff(xx); syms x
dydx=diff(yy)./dxx; % 1st derivative
d2ydx2=diff(dydx)./dxx(1:end-1); % 2nd derivative
xx=xx(1:end-2); yy=yy(1:end-2); dydx=dydx(1:end-1); % adjust sizes
residual_errors=subs(a(1),x,xx).*d2ydx2 +subs(a(2),x,xx).*dydx ...
                  +subs(a(3),x,xx).*yy;
residual_error=eval(norm(residual_errors)/norm(yy));
```

(b) To see if the Bessel functions of the first/second kind, $J_v(x)$ (Eq. (3.3.10a)) and $Y_v(x)$ (Eq. (3.3.14a)), of order, say, v=2, numerically satisfy the Bessel's equation (3.3.1):

$$x^2y''+xy'+(x^2-v^2)y=0 \text{ with } v=2 \text{ and } y(1)=1,\ y(2)=3 \qquad \text{(P3.14.2)}$$

for x=1~2, run the following MATLAB statements:

```
>>xx=[1:1e-4:2]; % Solution interval
>>[y,yy,a,residual_error]=Bessel2_eq(2,xx,1,3); residual_error
>>yy(1), yy(end) % To see if the BC has been satisfied.
```

where the following MATLAB function is supposed to have been stored as an m-file named "Bessel2_eq.m" on your PC. Is the relative residual error so small that $J_2(x)$ and $Y_2(x)$ can be acknowledged as constituting the general solution of the Bessel equation? Have the BCs been satisfied? Note that the solution interval [1, 2] does not include x=0, at which $Y_v(x)$ is singular.

```
function [y,yy,a,residual_error]=Bessel2_eq(v,xx,y0,yf)
% Numeric values yy of the sol to Bessel's DE of order v (integer)
%  for xx=x0~xf.
%  x^2*y"(x) +x*y'(x) +(x^2-v^2)*y(x) = 0 s.t. y(x0)=y0 and y(xf)=yf
% where x0>0 so that singular point x=0 of Y_v(x) can be excluded.
y=dsolve('t^2*D2y+t*Dy+(t^2-v^2)*y=0') % Analytical general solution
yy(1,:)=besselj(v,xx); % Bessel function J_v of 1st kind of order v
yy(2,:)=bessely(v,xx); % Bessel function Y_v of 2nd kind of order v
a=yy(:,[1 end]).'\[y0; yf]; % so that y(x0)=y0 and y(xf)=yf.
yy=a.'*yy;
syms x; residual_error=error_DE2_sol([x.^2 x x.^2-v^2],yy,xx);
```

(c) To see if the modified Bessel functions of the first/second kind, $I_v(x)$ (Eq. (3.3.25a)) and $K_v(x)$ (Eq. (3.3.28a)), of order, say, v=3, numerically satisfy the modified Bessel's equation (3.3.23):

$$x^2y''+xy'-(x^2+v^2)y=0 \text{ with } v=3 \text{ and } y(1)=-1,\ y(2)=5 \qquad \text{(P3.14.3)}$$

for x=1~2, run the following MATLAB statements:

```
>>xx=[1:1e-4:2]; % Solution interval
>>[y,yy,a,residual_error]=Bessel_modified_eq(3,xx,-1,5);
>>residual_error
>>yy(1), yy(end) % To see if the BC has been satisfied.
```

where the following MATLAB function is supposed to have been stored as an m-file named "Bessel_modified_eq.m" on your PC. Is the relative residual error so small that $I_3(x)$ and $K_3(x)$ can be acknowledged as constituting the general solution of the modified Bessel's equation? Have the BCs been satisfied? Note that the solution interval [1, 2] does not include x=0, at which $K_3(x)$ is singular.

```
function [y,yy,a,residual_error]=Bessel_modified_eq(v,xx,y0,yf)
% Numeric values yy of the sol to the modified Bessel's DE
%           of order v (integer)   for xx=x0~xf.
%   x^2*y"(x) +x*y'(x) -(x^2+v^2)*y(x) = 0 s.t. y(x0)=y0 and y(xf)=yf
% where x0>0 so that singular point x=0 of K_v(x) can be excluded.
y=dsolve('t^2*D2y+t*Dy-(t^2+v^2)*y=0') % Analytical general solution
yy(1,:)=besseli(v,xx); % Bessel function I_v of 1st kind of order v
yy(2,:)=besselk(v,xx); % Bessel function K_v of 2nd kind of order v
a=yy(:,[1 end]).'\[y0; yf]; % so that y(x0)=y0 and y(xf)=yf.
yy=a.'*yy;
syms x; residual_error=error_DE2_sol([x.^2 x -x.^2-v^2],yy,xx);
```

(d) To see if the Legendre polynomial $P_n(x)$ (Eq. (3.3.37)) of degree, say, n=6, numerically satisfies the Legendre's equation (3.3.29):

$$(1-x^2)y''-2xy'+n(n+1)y=0 \text{ with } n=6 \text{ and } y(-1)=2,\ y(1)=7 \qquad \text{(P3.14.4)}$$

for $x=-1\sim1$, run the following MATLAB statements:

```
>>xx=[-1:1e-4:1]; % Solution interval
>>[y,yy,a,residual_error]=Legendre_eq(6,xx,2,7);
>>residual_error
>>yy(1), yy(end) % To see if the BC has been satisfied.
```

where the following MATLAB function is supposed to have been stored as an m-file named "Legendre_eq.m" on your PC. Is the relative error so small that $P_6(x)$ can be acknowledged as satisfying the Legendre's equation? Have the BCs been satisfied?

```
function [y,yy,a,residual_error]=Legendre_eq(n,xx,x1,y1)
% Numeric values yy of the sol to Legendre's DE of order n (integer)
%  for xx=-1~1.
%  (1-x^2)*y"(x) -2*x*y'(x) +n*(n+1)*y(x) = 0 s.t. y(x1)=y1
y=dsolve('(1-t^2)*D2y-2*t*Dy+n*(n+1)*y=0') % Analytical solution
yy=mfun('P',n,xx); % Legendre polynomial P_n(x) of degree n
a=y1/mfun('P',n,x1); yy=a*yy; % so that y(x1)=y1
syms x
residual_error=error_DE2_sol([1-x.^2 -2*x n*(n+1)],yy,xx);
```

(e) To see if the Chebyshev polynomial $T_n(x)$ (Eq. (3.3.47)) of degree, say, n=5, numerically satisfies the Chebyshev's equation (3.3.41):

$$(1-x^2)y''-xy'+n^2y=0 \text{ with } n=5 \text{ and } y(1)=4 \qquad \text{(P3.14.5)}$$

for $x=-1\sim1$, run the following MATLAB statements:

```
>>xx=[-1:1e-4:1]; % Solution interval
>>[y,yy,a,residual_error]=Chebyshev_eq(5,xx,1,4);
>>residual_error
>>yy(end) % To see if the BC has been satisfied.
```

where the following MATLAB function is supposed to have been stored as an m-file named "Legendre_eq.m" on your PC. Is the relative error so small that $P_6(x)$ can be acknowledged as satisfying the Legendre's equation? Has the BC been satisfied?

```
function [y,yy,a,residual_error]=Chebyshev_eq(n,xx,x1,y1)
% Numeric values yy of the sol to Chebyshev's DE of order n (integer)
%  for xx=-1~1.
%  (1-x^2)*y"(x) -x*y'(x) +n^2*y(x) = 0 s.t. y(x1)=y1
y=dsolve('(1-t^2)*D2y-t*Dy+n^2*y=0') % Analytical general solution
yy=mfun('T',n,xx); % Chebyshev polynomial T_n(x) of degree n
a=y1/mfun('T',n,x1); yy=a*yy; % so that y(x1)=y1
syms x
residual_error=error_DE2_sol([1-x.^2 -x n^2],yy,xx);
```

(f) To see if the Hermite polynomial $H_n(x)$ (Eq. (3.3.65)) of degree, say, n=4, numerically satisfies the Hermite's equation (3.3.59):

$$y''-2xy'+2ny=0 \text{ with } n=4 \text{ and } y(2)=3 \qquad \text{(P3.14.6)}$$

for $x=-2\sim2$, run the following MATLAB statements:

```
>>xx=[-2:1e-4:2]; % Solution interval
>>[y,yy,a,residual_error]=Hermite_eq(4,xx,2,3);
>>residual_error
>>yy(end) % To see if the BC has been satisfied.
```

where the following MATLAB function is supposed to have been stored as an m-file named "Hermite_eq.m" on your PC. Is the relative error so small that $H_4(x)$ can be acknowledged as satisfying the Hermite's equation? Has the BC been satisfied?

```
function [y,yy,a,residual_error]=Hermite_eq(n,xx,x1,y1)
% Numeric values yy of the sol to Hermite's DE of order n (integer)
%  for xx.
%  y"(x) -2*x*y'(x) +2*n*y(x) = 0 s.t. y(x1)=y1
y=dsolve('D2y-2*t*Dy+2*n*y=0') % Analytical general solution
yy=mfun('H',n,xx); % Hermite polynomial H_n(x) of degree n
a=y1/mfun('H',n,x1); yy=a*yy; % so that y(x1)=y1
syms x; residual_error=error_DE2_sol([1 -2*x 2*n],yy,xx);
```

(g) To see if the Laguerre polynomial $L_n(x)$ (Eq. (3.3.75)) of degree, say, $n=5$, numerically satisfies the Laguerre's equation (3.3.70):

$$x\,y''+(1-x)\,y'+n\,y=0 \quad \text{with } n=5 \text{ and } y(2)=1 \tag{P3.14.7}$$

for $x=-1\sim2$, run the following MATLAB statements:

```
>>xx=[-1:1e-4:2]; % Solution interval
>>[y,yy,a,residual_error]=Laguerre_eq(5,xx,2,1);
>>residual_error
>>yy(end) % To see if the BC has been satisfied.
```

where the following MATLAB function is supposed to have been stored as an m-file named "Laguerre_eq.m" on your PC. Is the relative error so small that $L_5(x)$ can be acknowledged as satisfying the Laguerre's equation? Has the BC been satisfied?

```
function [y,yy,a,residual_error]=Laguerre_eq(n,xx,x1,y1)
% Numeric values yy of the sol to Laguerre's DE of order n (integer)
%  for xx.
%         x*y"(x)+(1-x)*y'(x) +n*y(x) = 0 s.t. y(x1)=y1
y=dsolve('t*D2y+(1-t)*Dy+n*y=0') % Analytical general solution
yy=mfun('L',n,xx); % Laguerre polynomial L_n(x) of degree n
a=y1/mfun('L',n,x1); yy=a*yy; % so that y(x1)=y1
syms x; residual_error=error_DE2_sol([x 1-x n],yy,xx);
```

3.15 Power Series Solution of a Nonlinear Second-Order DE

Referring to Section 3.3, find a power series solution of the following 2nd-order DE:

$$(1+x^2)y''-6x\,y'+12\,y=0 \tag{P3.15.1}$$

by taking the following steps:

- As an assumed solution, subsubstitute a power series for y:

$$y=\sum_{m=0}^{\infty} c_m x^m \tag{P3.15.2}$$

to write

$$(1+x^2)\sum_{m=2}^{\infty} c_m\, m(m-1)x^{m-2} - 6x\sum_{m=1}^{\infty} c_m\, m\, x^{m-1} + 12\sum_{m=0}^{\infty} c_m x^m = 0$$

$$; \sum_{m=2}^{\infty}\{c_{m+2}(m+\square)(m+\square) + c_m\, m(m-\square)\}x^m - 6\sum_{m=2}^{\infty} c_m\, m\, x^m + 12\sum_{m=2}^{\infty} c_m\, x^m \qquad \text{(P3.15.3)}$$

$$+2c_2 + 3\cdot 2c_3\, x - 6c_1 x + 12(c_0 + c_1 x) = 0$$

This yields

$$c_2 = -6c_0,\ c_3 = -c_1,\ \ (m+\square)(m+\square)c_{m+2} + \{m(m-\square) - 6m + 12\}c_m = 0$$

$$; \quad c_{m+2} = -\boxed{\qquad\qquad\qquad}\, c_m \ \text{ for } m = 0, 1, 2, \cdots \qquad \text{(P3.15.4)}$$

Noting that $c_n = 0$ for all $n \geq 5$, write the general solution as

$$y = c_0\boxed{\qquad\quad} + c_1\boxed{\qquad\quad} \qquad \text{(P3.15.5)}$$

3.16 Boundary Value Problems

Referring to the MATLAB script "em03p16.m" below, use the BVP solver '`bvp4c()`' to get the numerical solutions of the following BVPs and graph them for $[t_0, t_f]/[x_0, x_f]$ as shown in Fig. P3.16.

(a) $y''(x) = y'(x) - y(x) + 3e^{2x} - 2\sin x$ with $y(0) = 5,\ y(2) = -10$ (P3.16a)

(b) $y''(t) = -2y(t) + \sin t$ with $y(0) = 0,\ y(1) = 0$ (P3.16b)

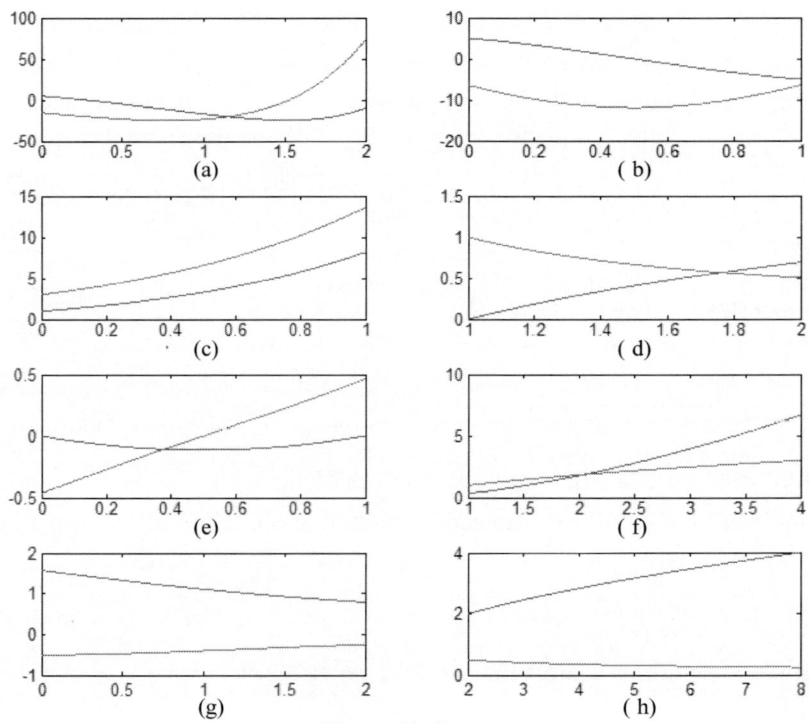

Figure P3.16

(c) $y''(t) = y'(t) + y(t) + e^x(1-2x)$ with $y(0)=0,\ y(1)=3e$ (P3.16c)

(d) $y''(t) + \frac{1}{t}\,y'(t) = 0$ with $y(1) = \ln 1,\ y(2) = \ln 2$ (P3.16d)

(e) $y'' - e^y = 0$ with the boundary cosndition $y(0) = 0,\ y(1) = 0$ (P3.16e)

(f) $y'' - \frac{2}{y'+1} = 0$ with the boundary condition $y(1) = \frac{1}{3},\ \ y(4) = \frac{20}{3}$ (P3.16f)

(g) $y'' = t(y')^2$ with the boundary condition $y(0) = \frac{\pi}{2},\ \ y(2) = \frac{\pi}{4}$ (P3.16g)

(h) $y'' + \frac{1}{y^2}\,y' = 0$ with the boundary condition $y(2) = 2,\ \ y'(8) = \frac{1}{4}$ (P3.16h)

```
%em03p16.m to solve the BVPs given in Problem 3.16
for itr=1:8
  switch itr
    case 1, %y"-y'+y=3*e^2t-2sin(t) with y(0)=5 & y(2)=-10
      df=@(t,y) [y(2); y(2)-y(1)+3*exp(2*t)-2*sin(t)];
      t0=0; tf=2; y0=5; yf=-10;
      solinit = bvpinit(linspace(t0,tf,5),[-10 5]); %[1 9]
      fbc=@(ya,yb)[ya(1)-5; ?????+10];
    case 2, %y"+4y=0 with y(0)=5 & y(1)=-5
      df=@(t,y)[y(2); -4*y(1)];
      %  . . . . . . . . . . . . . . . .
      % Fill this part with appropriate MATLAB statements
      %  . . . . . . . . . . . . . . . .
  end
  sol=bvp4c(df,fbc,solinit,bvpset('RelTol',1e-6)); %Numerical solution
  N=400; tt=t0+(tf-t0)/N*[0:N]; % Time range
  yy=deval(sol,tt); % evaluates the sol of a DE
  err=error_DE_sol(df,yy.',tt.') % Error evaluation
  subplot(420+max(min(itr,8),1)),
  plot(tt,yy) % Plot the solution graph one by one
end
```

CHAPTER 4

The Laplace Transform

CHAPTER OUTLINE

The Laplace transform was originally discovered by Leonhard Euler (1707–1783), the great 18th-century Swiss mathematician and physicist, but is named in honor of a French mathematician and astronomer, Pierre-Simon Laplace (1749–1827), who used the transform in his work on probability theory. He was such a genius, not only in mathematics, that the great mathematician Simeon Poisson (1781–1840) labeled him the Isaac Newton of France. As a genius in politics, he served three regimes in revolutionary France — the republic, the empire of Napoleon, and the Bourbon restoration, having been made a count by Napoleon and a marquis by Louis XVIII.

The Laplace transform is a very useful tool in solving differential equations and much further, it plays an important role in dealing with linear time-invariant systems.

4.1 DEFINITION OF THE LAPLACE TRANSFORM

The (unilateral or one-sided) *Laplace transform* is defined for a function $x(t)$ of a real variable t (often meaning the time) as

$$X(s) = \mathcal{L}\{x(t)\} = \int_{0-}^{\infty} x(t)e^{-st}\,dt \tag{4.1.1}$$

where s is a complex variable (parameter), the lower limit, 0_-(a negative infinitetsimal), of integration interval is the instant just before $t=0$ so that any point mass located at $t=0$ can be captured, and $x(t)$ is often assumed to be *causal* in the sense that it is zero for all $t<0$.

4.1.1 Laplace Transform of the Unit Step Function

The *unit* or *Heaviside step function* is defined as

$$u_s(t) = \begin{cases} 1 & \text{for } t \ge 0 \\ 0 & \text{for } t < 0 \end{cases} \tag{4.1.2}$$

which is depicted in Fig. 4.1(a). We can use Eq. (4.1.1) to find the Laplace transform of the unit step function as

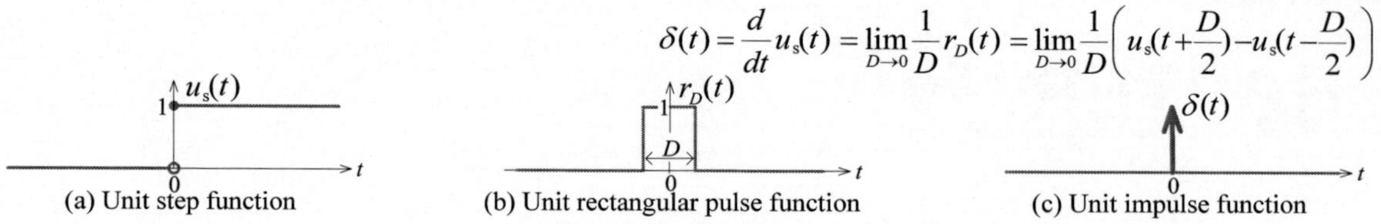

Figure 4.1 Basic functions

$$\mathcal{L}\{u_s(t)\} = \int_0^\infty u_s(t)e^{-st}dt = \int_0^\infty e^{-st}dt \overset{\text{(C.35)}}{=} \frac{1}{-s}e^{-st}\Big|_0^\infty = \frac{1}{-s}(0-1) = \frac{1}{s}$$

This Laplace transform pair is denoted by

$$u_s(t) \overset{\mathcal{L}}{\leftrightarrow} \frac{1}{s} \tag{4.1.3}$$

4.1.2 Laplace Transform of the Unit Impulse Function

The *unit impulse* or *Dirac delta function* can be defined as the limit of a unit-area rectangular pulse function $r_D(t)/D$ ($r_D(t)$: depicted in Fig. 4.1(b)) with pulsewidth $D\to 0$ (converging to zero) or simply, the time derivative of the unit step function:

$$\delta(t) = \lim_{D\to 0}\frac{1}{D}r_D(t) = \lim_{D\to 0}\frac{1}{D}(u_s(t+\frac{D}{2}) - u_s(t-\frac{D}{2})) = \frac{d}{dt}u_s(t) \tag{4.1.4}$$

is depicted in Fig. 4.1(c). We can use Eq. (4.1.1) to obtain the Laplace transform of the unit impulse function as

$$\mathcal{L}\{\delta(t)\} = \int_{0-}^\infty \delta(t)e^{-st}dt = e^{-st}\Big|_{t=0}\int_{0-}^\infty \delta(t)\,dt \overset{\text{unit area}}{=} 1$$

This Laplace transform pair is denoted by

$$\delta(t) \overset{\mathcal{L}}{\leftrightarrow} 1 \tag{4.1.5}$$

Eq. (4.1.4) implies that the integral of the unit impulse function is the unit step function:

$$\int_{-\infty}^t \delta(\tau)\,d\tau = u_s(t) \tag{4.1.6}$$

[Remark 4.1] Properties of the Unit Impulse Function $\delta(t)$

(1) The unit impulse function $\delta(t)$ has unity area around $t=0$, which means

$$\int_{-\infty}^{+\infty}\delta(\tau)\,d\tau = \int_{0^-}^{0^+}\delta(\tau)\,d\tau = 1. \tag{4.1.7}$$

$$\left(\because \int_{-\infty}^{0^+}\delta(\tau)\,d\tau - \int_{-\infty}^{0^-}\delta(\tau)\,d\tau \overset{(4.1.6)}{=} u_s(0^+) - u_s(0^-) \overset{(4.1.2)}{=} 1-0=1\right)$$

(2) The unit impulse function $\delta(t)$ is symmetric about $t=0$:

$$\delta(t)=\delta(-t) \tag{4.1.8}$$

(3) The convolution of a function $x(t)$ and the unit impulse function $\delta(t)$ makes the function itself:

$$x(t)*\delta(t) \overset{\text{definition of convolution integral}}{=} \int_{-\infty}^{+\infty} x(\tau)\delta(t-\tau)\,d\tau = x(t) \tag{4.1.9}$$

$$\left(\begin{array}{l} \because \int_{-\infty}^{+\infty} x(\tau)\delta(t-\tau)\,d\tau \overset{\delta(t-\tau)\neq 0 \text{ for only } \tau=t}{=} \int_{-\infty}^{+\infty} x(t)\delta(t-\tau)\,d\tau \overset{x(t) \text{ independent of } \tau}{=} x(t)\int_{-\infty}^{+\infty}\delta(t-\tau)\,d\tau \\ \overset{t-\tau\to t'}{=} x(t)\int_{t+\infty}^{t-\infty}\delta(t')(-dt') = x(t)\int_{t-\infty}^{t+\infty}\delta(t')\,dt' = x(t)\int_{-\infty}^{+\infty}\delta(\tau)\,d\tau \overset{(4.1.7)}{=} x(t) \end{array}\right)$$

What about the convolution of a time function $x(t)$ and a delayed unit impulse function $\delta(t-t_1)$? It becomes the delayed or shifted function $x(t-t_1)$:

$$x(t)*\delta(t-t_1)=\int_{-\infty}^{+\infty} x(\tau)\delta(t-\tau-t_1)\,d\tau = x(t-t_1) \tag{4.1.10}$$

What about the convolution of a delayed time function $x(t-t_2)$ and a delayed unit impulse function $\delta(t-t_1)$? It becomes another delayed time function $x(t-t_1-t_2)$:

$$x(t-t_2)*\delta(t-t_1)=\int_{-\infty}^{+\infty} x(\tau-t_2)\delta(t-\tau-t_1)\,d\tau = x(t-t_1-t_2) \tag{4.1.11}$$

If $x(t)*y(t)=z(t)$, we have

$$x(t-t_1)*y(t-t_2)=z(t-t_1-t_2) \tag{4.1.12}$$

However, with t replaced with $t-t_1$ on both sides, it does not hold, i.e.,

$$x(t-t_1)*y(t-t_1)\neq z(t-t_1)\text{, but } x(t-t_1)*y(t-t_1)=z(t-2t_1)$$

(4) The unit impulse function $\delta(t)$ has the *sampling* or *sifting property* that

$$\int_{-\infty}^{+\infty} x(t)\delta(t-t_1)\,dt \overset{\delta(t-t_1)\neq 0 \text{ for only } t=t_1}{=} x(t_1)\int_{-\infty}^{+\infty}\delta(t-t_1)\,dt \overset{(4.1.7)}{=} x(t_1) \tag{4.1.13}$$

This property enables us to sample or sift the sample value of a continuous-time signal $x(t)$ at a specific time $t=t_1$. It can also be used to model a discrete-time signal obtained from sampling a continuous-time signal.

[Remark 4.2] Several Models for the Unit Impulse Function $\delta(t)$

As depicted in Figs. 4.2(a)–(d), the unit impulse function can be modeled by the limit of various functions as follows:

$$\text{- } \delta(t)=\lim_{D\to 0^+}\frac{1}{D}\frac{\sin(\pi t/D)}{\pi t/D}=\lim_{D\to 0^+}\frac{1}{D}\operatorname{sinc}\left(\frac{t}{D}\right)\overset{\pi/D\to w}{=}\lim_{w\to\infty}\frac{w}{\pi}\frac{\sin(wt)}{wt} \tag{4.1.14a}$$

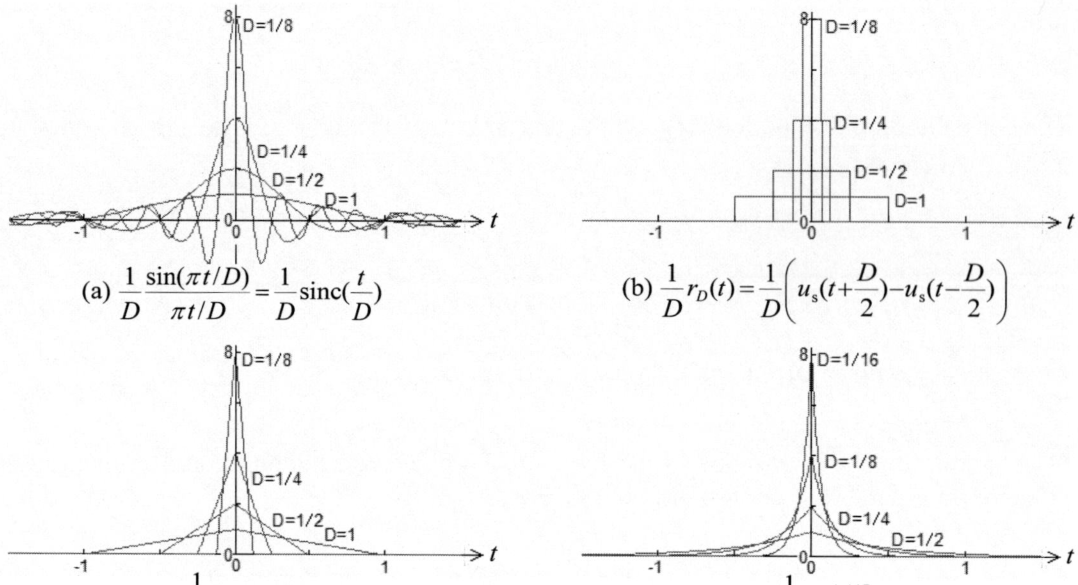

(a) $\frac{1}{D}\frac{\sin(\pi t/D)}{\pi t/D}=\frac{1}{D}\mathrm{sinc}(\frac{t}{D})$ (b) $\frac{1}{D}r_D(t)=\frac{1}{D}\left(u_s(t+\frac{D}{2})-u_s(t-\frac{D}{2})\right)$

(c) $\frac{1}{D}\lambda_D(t+D)$ (d) $\frac{1}{2D}e^{-|t|/D}$

Figure 4.2 Various models of the unit impulse function

$$- \ \delta(t)=\lim_{D\to 0^+}\frac{1}{D}r_D(t) \tag{4.1.14b}$$

$$- \ \delta(t)=\lim_{D\to 0^+}\frac{1}{D}\lambda_D(t+D) \tag{4.1.14c}$$

$$- \ \delta(t)=\lim_{D\to 0^+}\frac{1}{2D}e^{-|t|/D} \tag{4.1.14d}$$

where the *sinc function* sinc(t) and the *unit triangular function* $\lambda_D(t)$ are defined by

$$\mathrm{sinc}(t)=\frac{\sin(\pi t)}{\pi t} \quad \text{and} \quad \lambda_D(t)=\begin{cases}1-|t-D|/D & \text{for } |t-D|\le D\\ 0 & \text{elsewhere}\end{cases} \tag{4.1.15a,b}$$

4.1.3 Laplace Transform of the Ramp Function

The Laplace transform of the *unit ramp function* $tu_s(t)$ is obtained as

$$\mathcal{L}\{tu_s(t)\}\overset{(4.1.1)}{=}\int_0^\infty tu_s(t)e^{-st}dt=\int_0^\infty te^{-st}dt\overset{(C.38)}{=}\frac{t}{-s}e^{-st}\Big|_0^\infty+\frac{1}{s}\int_0^\infty e^{-st}dt\overset{(C.35)}{=}\frac{1}{s^2}$$

This Laplace transform pair is denoted by

$$tu_s(t)\ \overset{\mathcal{L}}{\leftrightarrow}\ \frac{1}{s^2} \tag{4.1.16}$$

```
>>syms t; laplace(dirac(t)), laplace(heaviside(t)), laplace(t)
   ans = 1,   ans = 1/s,   ans = 1/s^2
```

4.1.4 Laplace Transform of the Exponential Function

The Laplace transform of the one-sided (or right-sided) exponential function $e^{-at}u_s(t)$ is obtained as

$$\mathcal{L}\{e^{-at}u_s(t)\} \overset{(4.1.1)}{=} \int_0^\infty e^{-at}u_s(t)e^{-st}\,dt = \int_0^\infty e^{-at}e^{-st}\,dt = \int_0^\infty e^{-(s+a)t}\,dt \overset{(C.35)}{=} \frac{1}{s+a}$$

This Laplace transform pair is denoted by

$$e^{-at}u_s(t) \overset{\mathcal{L}}{\leftrightarrow} \frac{1}{s+a} \tag{4.1.17}$$

4.1.5 Laplace Transform of the Complex Exponential Function

Substituting $\sigma+j\omega$ for a into (4.1.17) yields the Laplace transform of the complex exponential function as

$$e^{-(\sigma+j\omega)t}u_s(t) \overset{(C.20)}{=} e^{-\sigma t}(\cos\omega t - j\sin\omega t)u_s(t) \overset{\mathcal{L}}{\leftrightarrow} \frac{1}{s+\sigma+j\omega} = \frac{s+\sigma}{(s+\sigma)^2+\omega^2} - j\frac{\omega}{(s+\sigma)^2+\omega^2}$$

$$e^{-\sigma t}\cos\omega t\,u_s(t) \overset{\mathcal{L}}{\leftrightarrow} \frac{s+\sigma}{(s+\sigma)^2+\omega^2} \tag{4.1.18}$$

$$e^{-\sigma t}\sin\omega t\,u_s(t) \overset{\mathcal{L}}{\leftrightarrow} \frac{\omega}{(s+\sigma)^2+\omega^2} \tag{4.1.19}$$

4.2 PROPERTIES OF THE LAPLACE TRANSFORM

Let the Laplace transforms of two functions $x(t)$ and $y(t)$ be $X(s)$ and $Y(s)$, respectively.

4.2.1 Linearity

The Laplace transform of a linear combination of $x(t)$ and $y(t)$ can be written as

$$\alpha x(t)+\beta y(t) \overset{\mathcal{L}}{\longleftrightarrow} \alpha X(s)+\beta Y(s) \tag{4.2.1}$$

4.2.2 Time Differentiation

The Laplace transform of the derivative of $x(t)$ w.r.t. t can be written as

$$x'(t) \overset{\mathcal{L}}{\longleftrightarrow} sX(s) - x(0) \tag{4.2.2}$$

(Proof)

$$\mathcal{L}\{x'(t)\} \overset{(4.1.1)}{=} \int_0^\infty \frac{dx}{dt}e^{-st}\,dt \overset{(C.38)}{=} x(t)e^{-st}\Big|_0^\infty - (-s)\int_0^\infty x(t)e^{-st}\,dt \overset{(4.1.1)}{=} sX(s)-x(0)$$

Repetitive application of the time differentiation property yields the Laplace transform of nth derivative of $x(t)$ w.r.t. t as

$$x^{(n)}(t) \xleftrightarrow{\mathcal{L}} s^n X(s) - s^{n-1}x(0) - s^{n-2}x'(0) - \cdots - x^{(n-1)}(0) \tag{4.2.3}$$

4.2.3 Time Integration

The Laplace transform of the integral of $x(t)$ w.r.t. t can be written as

$$\int_{-\infty}^{t} x(\tau)\,d\tau \xleftrightarrow{\mathcal{L}} \frac{1}{s}X(s) + \frac{1}{s}\int_{-\infty}^{0} x(\tau)\,d\tau \tag{4.2.4}$$

This can be derived by substituting $\int_{-\infty}^{t} x(\tau)d\tau$ and $x(t)$ for $x(t)$ and $x'(t)$, respectively, into Eq. (4.2.2) as

$$x(t) = \frac{d}{dt}\left\{\int_{-\infty}^{t} x(\tau)d\tau\right\} \xleftrightarrow{\mathcal{L}} X(s) = s\mathcal{L}\left\{\int_{-\infty}^{t} x(\tau)d\tau\right\} - \int_{-\infty}^{0} x(\tau)\,d\tau$$

Repetitive application of the time integration property yields the Laplace transform of nth integral of $x(t)$ w.r.t. t as

$$\int_{-\infty}^{t}\int_{-\infty}^{t}\cdots\int_{-\infty}^{t} x(\tau)\,d\tau^n \xleftrightarrow{\mathcal{L}} s^{-n}X(s) + s^{-n}\int_{-\infty}^{0} x(\tau)\,d\tau + \cdots + \int_{-\infty}^{0}\int_{-\infty}^{0}\cdots\int_{-\infty}^{0} x(\tau)\,d\tau^n \tag{4.2.5}$$

4.2.4 Time Shifting — Real Translation

The Laplace transform of a delayed function $x(t)$ can be obtained as follows:

$$\mathcal{L}\{x(t-t_1)\} \overset{(4.1.1)}{=} \int_0^{t_1} x(t-t_1)e^{-st}\,dt + \int_{t_1}^{\infty} x(t-t_1)e^{-st}\,dt$$

$$\overset{t-t_1=\tau}{=} e^{-st_1}\int_{-t_1}^{0} x(\tau)e^{-s\tau}\,d\tau + e^{-st_1}\int_0^{\infty} x(\tau)e^{-s\tau}\,d\tau$$

$$;x(t-t_1),\ t_1>0 \xleftrightarrow{\mathcal{L}} e^{-st_1}\left\{X(s) + \int_{-t_1}^{0} x(\tau)e^{-s\tau}\,d\tau\right\} \overset{\text{if } x(t)=0\ \forall\, t<0}{\rightarrow} e^{-st_1}X(s) \tag{4.2.6}$$

4.2.5 Frequency Shifting — Complex Translation

$$e^{s_1 t}x(t) \xleftrightarrow{\mathcal{L}} X(s-s_1) \tag{4.2.7}$$

4.2.6 Real Convolution

The (real) convolution of two (causal) functions $g(t)$ and $x(t)$ is defined as

$$g(t)*x(t) = \int_{-\infty}^{\infty} g(\tau)x(t-\tau)\,d\tau \underset{\text{causality}}{\overset{\text{if } g(t)=0 \text{ and } x(t)=0\ \forall\, t<0}{=}} \int_0^{t} g(\tau)\,x(t-\tau)\,d\tau \tag{4.2.8}$$

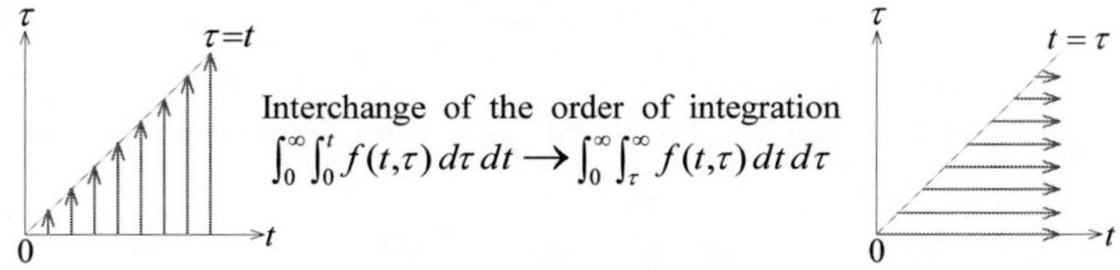

Figure 4.3 Interchange of the order of integration

The Laplace transform of the convolution $y(t) = g(t) * x(t)$ turns out to be the product of the Laplace transforms of the two functions as

$$y(t) = g(t) * x(t) \overset{\mathcal{L}}{\longleftrightarrow} Y(s) = G(s)X(s) \tag{4.2.9}$$

(Proof)

$$\mathcal{L}\{g(t)*x(t)\} \overset{(4.1.1)}{=} \int_0^\infty g(t) * x(t) e^{-st}\,dt \overset{(4.2.8)}{=} \int_0^\infty \left\{ \int_0^t g(\tau)x(t-\tau)\,d\tau \right\} e^{-st}\,dt$$

$$\overset{\text{Fig. 4.3}}{=} \int_0^\infty g(\tau)e^{-s\tau} \int_\tau^\infty x(t-\tau)e^{-s(t-\tau)}\,dt\,d\tau \overset{t-\tau=v}{=} \int_0^\infty g(\tau)e^{-s\tau} \left\{ \int_0^\infty x(v)e^{-sv}dv \right\} d\tau$$

$$\overset{(4.1.1)}{=} \int_0^\infty g(\tau)e^{-s\tau}d\tau\ X(s) \overset{(4.1.1)}{=} G(s)X(s)$$

This property is used to describe the input-output relationship of a continuous-time linear time-invariant (LTI) system having the input $x(t)$, the *impulse response* $g(t)$ (the output to an impulse input), and the output $y(t)$, where the Laplace transform of the impulse response, that is, $G(s) = \mathcal{L}\{g(t)\}$, is referred to as the *system function* or *transfer function* of the system.

4.2.7 Partial Differentiation

Given the Laplace transform of a function having a parameter a, that is, $\mathcal{L}\{x(t,a)\} = X(s,a)$, the Laplace transform of the derivative of the parameterized function $x(t, a)$ w.r.t. a equals the derivative of its Laplace transform $X(s, a)$ w.r.t. a:

$$\frac{\partial}{\partial a} x(t,a) \overset{\mathcal{L}}{\longleftrightarrow} \frac{\partial}{\partial a} X(s,a) \tag{4.2.10}$$

4.2.8 Complex Differentiation

The Laplace transform of $-tx(t)$ equals the derivative of $\mathcal{L}\{x(t)\} = X(s)$ w.r.t. s:

$$t\,x(t) \overset{\mathcal{L}}{\longleftrightarrow} -\frac{d}{ds} X(s) \tag{4.2.11}$$

This can be derived by differentiating Eq. (4.1.1) w.r.t. s.

(**Example 4.1**) Applying Partial/Complex Differentiation Property

To find the Laplace transform of $t\,e^{-at}u_s(t)$, we apply the partial differentiation property (4.2.10) to differentiate Eq. (4.1.17) w.r.t. a and multiply it by -1 to get

$$te^{-at}u_s(t) \overset{\mathcal{L}}{\longleftrightarrow} \frac{1}{(s+a)^2} \tag{E4.1.1}$$

Multiplying t by the left-hand side LHS and applying the complex differentiation property (4.2.11) repetitively, we get

$$t^2e^{-at}u_s(t) \overset{\mathcal{L}}{\longleftrightarrow} \frac{2!}{(s+a)^3} \tag{E4.1.2}$$

....

$$t^me^{-at}u_s(t) \overset{\mathcal{L}}{\longleftrightarrow} \frac{m!}{(s+a)^{m+1}} \tag{E4.1.3}$$

This Laplace transform pair is listed in Table A.10(8) (see Appendix A).

4.2.9 Initial Value Theorem (IVT)

We can get the initial value of $x(t)$ from its Laplace transform $\mathcal{L}\{x(t)\} = X(s)$ as

$$x(0) = \lim_{s\to\infty} sX(s) \tag{4.2.12}$$

This property, called the initial value theorem (IVT), can be derived by substituting $s=\infty$ into the time differentiation property (4.2.2) as

$$\lim_{s\to\infty} sX(s) - x(0) \overset{(4.2.2)}{=} \lim_{s\to\infty} \mathcal{L}\{x'(t)\} \overset{(4.1.1)}{=} \lim_{s\to\infty} \int_0^\infty x'(t)e^{-st}dt = 0$$

4.2.10 Final Value Theorem (FVT)

We can get the final value of $x(t)$ from its Laplace transform $\mathcal{L}\{x(t)\} = X(s)$ as

$$x(\infty) = \lim_{s\to 0} sX(s) \tag{4.2.13}$$

on the premise that $x(t)$ is convergent, or equivalently, all the poles of $X(s)$ are in the left-half plane (LHP) except for a simple pole at $s=0$. Note that $s=s_p$ is called a *pole* of $X(s)=Q(s)/P(s)$ if $P(s_p)=0$ and it is said to be a *multiple pole* of *order* (or *multiplicity*) N if the denominator $P(s)$ has $(s-s_p)^N$ $(N>1)$ as a factor where a pole of order $N=1$ is called a *simple pole*.

The property (4.2.13), called the final value theorem (FVT), can be derived by substituting $s=0$ into the time differentiation property (4.2.2) as

$$\lim_{s\to 0} sX(s) - x(0) \overset{(4.2.2)}{=} \lim_{s\to 0} \mathcal{L}\{x'(t)\} \overset{(4.1.1)}{=} \lim_{s\to 0} \int_0^\infty x'(t)e^{-st}dt = x(t)\big|_0^\infty = x(\infty) - x(0)$$

(Q1) Can we apply the FVT to find the final value of a sinusoidal signal $x(t)=\sin\omega t$, whose Laplace transform is

$$X(s) = \mathcal{L}\{x(t)\} = \frac{\omega}{s^2+\omega^2}\ ? \tag{4.2.14}$$

$$\sin\omega t\Big|_{t=\infty} = \lim_{s\to 0} sX(s) \overset{\text{Table 4.1(7)}}{=} \lim_{s\to 0} s\frac{\omega}{s^2+\omega^2} = 0 \tag{4.2.15}$$

(A1) No, because $X(s)$ has two simple poles at $s=\pm j\omega$ on the imaginary axis so that not all the poles of $X(s)$ are in the LHP. Note that $x(t)=\sin\omega t$ does not converge as $t\to\infty$, and therefore, its final value cannot be specified.

(Q2) Can we apply the FVT to find the final value of an exponentially decreasing sinusoidal signal $x(t)=e^{-t}\sin t$, whose Laplace transform is

$$X(s)=\mathcal{L}\{x(t)\}\overset{\text{Table 4.1(9)}}{=}\frac{1}{(s+1)^2+1^2}\,? \tag{4.2.16}$$

$$e^{-t}\sin\omega t\big|_{t=\infty}=\lim_{s\to 0} s\,X(s)=\lim_{s\to 0} s\frac{1}{(s+1)^2+1^2}=0 \tag{4.2.17}$$

(A2) Yes, because $X(s)$ has two simple poles at $s=-1\pm j$ so that all the poles of $X(s)$ are in the LHP.

(Q3) Can we apply the FVT to find the final value of an exponentially increasing sinusoidal signal $x(t)=e^{t}\sin t$, whose Laplace transform is

$$X(s)=\mathcal{L}\{x(t)\}\overset{\text{Table 4.1(9)}}{=}\frac{1}{(s-1)^2+1^2}\,? \tag{4.2.18}$$

$$e^{-t}\sin\omega t\big|_{t=\infty}=\lim_{s\to 0} sX(s)=\lim_{s\to 0} s\frac{1}{(s-1)^2+1^2}=0 \tag{4.2.19}$$

(A3) No, because $X(s)$ has two simple poles at $s=1\pm j$ so that not all the poles of $X(s)$ are in the LHP. Note that $x(t)=e^{t}\sin t$ diverges as $t\to\infty$ and therefore, its final value cannot be specified.

(Q4) Can we apply the FVT to find the final value of a unit step signal $x(t)=u_s(t)$, whose Laplace transform is

$$X(s)=\mathcal{L}\{x(t)\}\overset{\text{Table 4.1(3)}}{=}\frac{1}{s}\,? \tag{4.2.20}$$

$$u_s(t)\big|_{t=\infty}=\lim_{s\to 0} s\,X(s)=\lim_{s\to 0} s\frac{1}{s}=1 \tag{4.2.21}$$

(A4) Yes, because $X(s)$ has only a simple pole at $s=0$ even if not all poles of $X(s)$ are in the LHP.

(Q4) Can we apply the FVT to find the final value of a ramp signal $x(t)=tu_s(t)$, whose Laplace transform is

$$X(s)=\mathcal{L}\{x(t)\}\overset{\text{Table 4.1(4)}}{=}\frac{1}{s^2} \tag{4.2.22}$$

$$tu_s(t)\big|_{t=\infty}=\lim_{s\to 0} s\,X(s)\overset{\text{Table 4.1(9)}}{=}\lim_{s\to 0} s\frac{1}{s^2}=\infty \tag{4.2.23}$$

(A4) No, because $X(s)$ has a double pole at $s=0$. Note that $x(t)=tu_s(t)$ diverges as $t\to\infty$ and therefore, its final value cannot be specified.

Note that the imaginary axis $s=j\omega$ on the s-plane (the complex plane used for plotting the poles/zeros of a Laplace transform expression) can be considered as the boundary between the stable region (LHP) and the unstable region (RHP), since a time function $x(t)$ converges or diverges as $t\to\infty$ depending on whether or not all the poles of $X(s)$ are on the left side of the imaginary axis.

The Laplace transforms of basic functions and the properties of Laplace transform are listed in Tables 4.1/A.8 and 4.2/A.7, respectively.

Table 4.1 Laplace transforms of basic functions

$x(t)$	$X(s)$	$x(t)$	$X(s)$	$x(t)$	$X(s)$
(1) $\delta(t)$	1	(5) $e^{-at}u_s(t)$	$\frac{1}{s+a}$	(9) $e^{-at}\sin\omega t\, u_s(t)$	$\frac{\omega}{(s+a)^2+\omega^2}$
(2) $\delta(t-t_1)$	$e^{-t_1 s}$	(6) $t^m e^{-at}u_s(t)$	$\frac{m!}{(s+a)^{m+1}}$	(10) $e^{-at}\cos\omega t\, u_s(t)$	$\frac{s+a}{(s+a)^2+\omega^2}$
(3) $u_s(t)$	$\frac{1}{s}$	(7) $\sin\omega t\, u_s(t)$	$\frac{\omega}{s^2+\omega^2}$	(11) $e^{-at}\sinh bt\, u_s(t)$	$\frac{b}{(s-a)^2-b^2}$
(4) $t^m u_s(t)$	$\frac{m!}{s^{m+1}}$	(8) $\cos\omega t\, u_s(t)$	$\frac{s}{s^2+\omega^2}$	(12) $e^{-at}\cosh bt\, u_s(t)$	$\frac{s-a}{(s-a)^2-b^2}$

Table 4.2 Properties of Laplace transform

(0) Definition	$X(s)=\mathcal{L}\{x(t)\}=\int_0^\infty x(t)e^{-st}\,dt\,;\;\; x(t)\overset{\mathcal{L}}{\leftrightarrow}X(s)$
(1) Linearity	$\alpha\, x(t)+\beta\, y(t)\leftrightarrow \alpha\, X(s)+\beta\, Y(s)$
(2) Time shifting (Real translation)	$x(t-t_1)u_s(t-t_1),\; t_1>0 \leftrightarrow e^{-st_1}\left\{X(s)+\int_{-t_1}^{0}x(\tau)e^{-s\tau}d\tau\right\}$ $= e^{-st_1}X(s)$ for $x(t)=0\;\forall\, t<0$
(3) Frequency shifting (Complex translation)	$e^{s_1 t}x(t)\leftrightarrow X(s-s_1)$
(4) Real convolution	$g(t)*x(t)\leftrightarrow G(s)X(s)$
(5) Time derivative (Differentiation property)	$x'(t)=\frac{d}{dt}x(t)\leftrightarrow sX(s)-x(0)$
(6) Time integral (Integration property)	$\int_{-\infty}^{t}x(\tau)\,d\tau\leftrightarrow \frac{1}{s}X(s)+\frac{1}{s}\int_{-\infty}^{0}x(\tau)\,d\tau$
(7) Complex derivative	$t\,x(t)\leftrightarrow -\frac{d}{ds}X(s)$
(8) Complex convolution	$x(t)y(t)\leftrightarrow \frac{1}{2\pi j}\int_{\sigma_0-\infty}^{\sigma_0+\infty}X(v)Y(s-v)\,dv$
(9) Initial value theorem	$x(0)=\lim_{s\to\infty}sX(s)$
(10) Final value theorem	$x(\infty)=\lim_{s\to 0}sX(s)$

4.3 THE INVERSE LAPLACE TRANSFORM

Suppose an s-function $X(s)$ is given in the form of a rational function, i.e., the ratio of an Mth-degree polynomial $Q(s)$ to an Nth-degree polynomial $P(s)$ in s, and it is expanded into the form of partial fraction expansion (PFE) (with a multiple pole $s=p$ of order L and $N-L$ simple poles) as

$$X(s)=\frac{Q(s)}{P(s)}=\frac{b_M s^M+\dots+b_1 s+b_0}{a_N s^N+\dots+a_1 s+a_0}(M\le N)=\left(\sum_{n=1}^{N-L}\frac{r_n}{s-p_n}\right)+\frac{r_{N-L+1}}{s-p}+\cdots+\frac{r_N}{(s-p)^L}+K \quad (4.3.1)$$

where

$$r_n=(s-p_n)\left.\frac{Q(s)}{P(s)}\right|_{s=p_n},\quad n=1,2,\cdots,N-L \quad (4.3.2a)$$

$$r_{N-l}=\frac{1}{l!}\frac{d^l}{ds^l}\left\{(s-p)^L\frac{Q(s)}{P(s)}\right\}\Bigg|_{s=p},\quad l=0,1,\cdots,L-1 \quad (4.3.2b)$$

Then the inverse Laplace transform of $X(s)$ can be obtained as

$$x(t)=\mathcal{L}^{-1}\{X(s)\}=\left\{\sum_{n=1}^{N-L}r_n e^{p_n t}+r_{N-L+1}e^{pt}+r_{N-L+2}te^{pt}+\cdots+\frac{r_N}{(L-1)!}t^{L-1}e^{pt}\right\}u_s(t)+K\delta(t) \quad (4.3.3)$$

(Example 4.2) Inverse Laplace Transform

Let us find the inverse Laplace transform of the following s-functions.

(a) $$X(s)=\frac{3s^2+11s+11}{s^3+4s^2+5s+2}=\frac{3s^2+11s+11}{(s+1)^2(s+2)}=\frac{r_1}{s+2}+\frac{r_2}{s+1}+\frac{r_3}{(s+1)^2}\ \text{(with real poles)} \quad (E4.2.1)$$

We can use the formula (4.3.2a) to find the simple-pole coefficient r_1 as

$$r_1\overset{(4.3.2a)}{=}(s+2)X(s)\big|_{s=-2}\overset{(E4.2.1)}{=}(s+2)\left.\frac{3s^2+11s+11}{(s+1)^2(s+2)}\right|_{s=-2}=1 \quad (E4.2.2)$$

and can also use the formula (4.3.2b) to find the multiple-pole coefficients r_2 and r_3 as

$$r_3\overset{(4.3.2b)}{=}(s+1)^2X(s)\big|_{s=-1}\overset{(E4.2.1)}{=}\left.\frac{3s^2+11s+11}{s+2}\right|_{s=-1}=3 \quad (E4.2.3)$$

$$r_2=\frac{d}{ds}(s+1)^2X(s)\big|_{s=-1}\overset{(E4.2.1)}{=}\left.\frac{d}{ds}\frac{3s^2+11s+11}{s+2}\right|_{s=-1}\overset{(C.31)}{=}\left.\frac{(6s+11)(s+2)-(3s^2+11s+11)}{(s+2)^2}\right|_{s=-1}=2 \quad (E4.2.4)$$

Thus, the inverse Laplace transform can be written as

$$x(t)\underset{(4.3.3)}{\overset{(E4.2.1)}{=}}(r_1e^{-2t}+r_2e^{-t}+r_3te^{-t})u_s(t)\overset{(E4.2.2,3,4)}{=}(e^{-2t}+2e^{-t}+3te^{-t})u_s(t) \quad (E4.2.5)$$

```
>>Ns=[3 11 11]; Ds=[1 4 5 2];
>>[r,p,k]=residue(Ns,Ds), [A,B]=residue(r,p,k)
```

(b) $X(s)=\frac{s}{s^2+2s+2}=\frac{s}{(s+1)^2+1^2}$ (with complex conjugate poles at $s=-1\pm j$) (E4.2.6)

We may use the formula (4.3.2a) to find the coefficients of the PFE form:

$$X(s)=\frac{s}{s^2+2s+2}=\frac{r_1}{s+1-j}+\frac{r_2}{s+1+j} \tag{E4.2.7}$$

as

$$r_1\overset{(4.3.2a)}{=}(s+1-j)X(s)\Big|_{s=-1+j}\overset{(E4.2.6)}{=}\frac{s}{s+1+j}\Big|_{s=-1+j}=0.5(1+j)$$
$$r_2\overset{(4.3.2a)}{=}(s+1+j)X(s)\Big|_{s=-1-j}\overset{(E4.2.6)}{=}\frac{s}{s+1-j}\Big|_{s=-1-j}=0.5(1-j) \tag{E4.2.8}$$

Thus, we can write the inverse Laplace transform as

$$x(t)\underset{(4.3.3)}{\overset{(E4.2.7,8)}{=}}\{0.5(1+j)e^{(-1+j)t}+0.5(1-j)e^{(-1-j)t}\}u_s(t)\overset{(C.20)}{=}e^{-t}(\cos t-\sin t)u_s(t) \tag{E4.2.9}$$

In the case of complex conjugate poles, it may be simpler to equate $X(s)$ with the following form

$$X(s)=\frac{s}{s^2+2s+2}=\frac{A(s+1)}{(s+1)^2+1^2}+\frac{B\times 1}{(s+1)^2+1^2}\overset{\text{Common denominator}}{=}\frac{As+(A+B)}{(s+1)^2+1^2} \tag{E4.2.10}$$

and get the coefficients as $A=1$ and $B=-1$. Then we can directly find each term of the inverse Laplace transform from the Laplace transform (Table 4.1) and write the inverse Laplace transform as

$$X(s)\overset{(E4.2.10)}{=}\frac{s+1}{(s+1)^2+1^2}+\frac{(-1)\times 1}{(s+1)^2+1^2}\overset{\text{Table 4.1(9),(10)}}{\rightarrow}x(t)=(e^{-t}\cos t-e^{-t}\sin t)u_s(t) \tag{E4.2.11}$$

(c) Use of MATLAB for PFE (Partial Fraction Expansion) and Inverse Laplace Transform
We can use the MATLAB command '`residue()`' to get the PFE and '`ilaplace()`' to obtain the whole inverse Laplace transform. It should, however, be noted that '`ilaplace()`' might not work properly for high-degree rational functions.

```
>>Ns=[3 11 11]; Ds=[1 4 5 2]; [r,p,k]=residue(Ns,Ds); [r p],k % (E4.2.1)
   r = 1.0000      p = -2.0000     % (2) 1/(s-(-2))
       2.0000          -1.0000     % (4) 2/(s-(-1))
       3.0000          -1.0000     % (3) 3/(s-(-1))^2
   k = []

>>syms s, x=ilaplace((3*s^2+11*s+11)/(s^3+4*s^2+5*s+2))
   x = exp(-2*t)+3*t*exp(-t)+2*exp(-t)        % (E4.2.5)

>>Ns=[1 0]; Ds=[1 2 2]; [r,p,k]=residue(Ns,Ds); [r p],k   % (E4.2.6)
   r = 0.5000 + 0.5000i     p = -1.0000 + 1.0000i  % (8)  (0.5+0.5i)/(s+1-i)
       0.5000 - 0.5000i     p = -1.0000 - 1.0000i  % (8)  (0.5-0.5i)/(s+1+i)
   k = []

>>syms s, x= ilaplace(s/(s^2+2*s+2))
   x = exp(-t)*cos(t)-exp(-t)*sin(t)      % (E4.2.9) or (E4.2.11)
>>ilaplace(s/(s^4+10*s^3+9*s^2+6*s+1))  %?? Ns=[1 0]; Ds=[1 10 9 6 1];
```

4.4 USING THE LAPLACE TRANSFORM

We can use the Laplace transform to deal with a high-order linear ODE like

$$y^{(4)}(t)+a_3y^{(3)}(t)+a_2y''(t)+a_1y'(t)+a_0y(t)=u(t) \tag{4.4.1}$$

$$\text{with initial conditions } y(0),\ y'(0),\ y''(0), \text{ and } y^{(3)}(0)$$

in a very efficient way, by rewriting it in the form of state and output equations as

$$\mathbf{x}'(t)=A\,\mathbf{x}(t)+\mathbf{b}\,u(t) \text{ with } \mathbf{x}(0)=\mathbf{x}_0=[y(0)\ \ y'(0)\ \ y''(0)\ \ y^{(3)}(0)]^T \tag{4.4.2a}$$

$$y(t)=C\,\mathbf{x}(t) \tag{4.4.2b}$$

with

$$\mathbf{x}(t)=\begin{bmatrix}x_1(t)\\x_2(t)\\x_3(t)\\x_4(t)\end{bmatrix}=\begin{bmatrix}y(t)\\y'(t)\\y''(t)\\y^{(3)}(t)\end{bmatrix},\ A=\begin{bmatrix}0&1&0&0\\0&0&1&0\\0&0&0&1\\-a_0&-a_1&-a_2&-a_3\end{bmatrix},\ \mathbf{b}=\begin{bmatrix}0\\0\\0\\1\end{bmatrix},\text{ and } C=[1\ 0\ 0\ 0] \tag{4.4.3}$$

where $x(t)$ and A are referred to as the state vector and system matrix, respectively. Using the differentiation property (4.2.2), we take the Laplace transform of both sides of Eq. (4.4.2a):

$$sX(s)-\mathbf{x}(0)=A\,X(s)+\mathbf{b}U(s) \text{ with } X(s)=\mathcal{L}\{x(t)\},\ U(s)=\mathcal{L}\{u(t)\}$$

$$;\ \ [sI-A]X(s)=\mathbf{x}(0)+\mathbf{b}U(s);\quad X(s)=[sI-A]^{-1}[\mathbf{x}(0)+\mathbf{b}\,U(s)] \tag{4.4.4}$$

Then we take the Laplace transform of Eq. (4.4.2b), substitute Eq. (4.4.4) for $X(s)$ to get $Y(s)$, and take its inverse Laplace transform to get the t-domain solution $y(t)$ as done at the end of Example 3.23:

$$Y(s)\overset{(4.4.2b)}{=}CX(s)\overset{(4.4.4)}{=}C[sI-A]^{-1}[\mathbf{x}(0)+\mathbf{b}\,U(s)];\ \ y(t)=\mathcal{L}^{-1}\{Y(s)\} \tag{4.4.5}$$

(**Example 4.3**) Using the Laplace Transform to Solve a Linear DE

Let us solve the following DE:

$$\frac{d^2}{dt^2}y(t)+3\frac{d}{dt}y(t)+2y(t)=10u_s(t),\ \ t\ge 0 \tag{E4.3.1}$$

with the initial conditions $y(0)=1$ and $y'(0)=-2$.

We make use of the time differentiation properties (4.2.2) and (4.2.3) of Laplace transform to write the Laplace transform of both sides of Eq. (E4.3.1) as

$$s^2Y(s)-s\,y(0)-y'(0)+3(s\,Y(s)-y(0))+2Y(s)=\frac{10}{s}$$

$$;\ (s^2+3s+2)Y(s)=s\,y(0)+y'(0)+3y(0)+\frac{10}{s}=\frac{10}{s}+s+1 \tag{E4.3.2}$$

This algebraic equation can be solved for $Y(s)$ as

$$Y(s) = \frac{s(s+1)+10}{s(s^2+3s+2)} \tag{E4.3.3}$$

We expand this s-domain solution into partial fractions as

$$Y(s) = \frac{s(s+1)+10}{s(s^2+3s+2)} = \frac{r_1}{s} + \frac{r_2}{s+1} + \frac{r_3}{s+2} = \frac{5}{s} + \frac{-10}{s+1} + \frac{6}{s+2} \tag{E4.3.4}$$

where the coefficients are found by using the formula (4.3.2a) as

$$\begin{aligned} r_1 &\overset{(4.3.2a)}{=} sY(s)\big|_{s=0} \overset{(E4.3.3)}{=} \left.\frac{s^2+s+10}{s^2+3s+2}\right|_{s=0} = 5 \\ r_2 &\overset{(4.3.2a)}{=} (s+1)Y(s)\big|_{s=-1} \overset{(E4.3.3)}{=} \left.\frac{s^2+s+10}{s(s+2)}\right|_{s=-1} = -10 \\ r_3 &\overset{(4.3.2a)}{=} (s+2)Y(s)\big|_{s=-2} \overset{(E4.3.3)}{=} \left.\frac{s^2+s+10}{s(s+1)}\right|_{s=-2} = 6 \end{aligned} \tag{E4.3.5}$$

Thus, we can write the time-domain solution $y(t)$ as

$$y(t) = \mathcal{L}^{-1}\{Y(s)\} \underset{A.8(6)}{\overset{(E4.3.4)}{=}} (r_1 e^{-0t} + r_2 e^{-t} + r_3 e^{-2t})u_s(t) \overset{(E4.3.5)}{=} (5 - 10e^{-t} + 6e^{-2t})u_s(t) \tag{E4.3.6}$$

```
>>syms s, Ns=[1 1 10]; Ds=[1 3 2 0];
>>y=ilaplace(poly2sym(Ns,s)/poly2sym(Ds,s)) % Inverse Laplace transform
>>y1=ilaplace((s^2+s+10)/(s^3+3*s^2+2*s)) % Or, directly from Eq.(E4.3.3)
>>t=0:0.01:10; yt=eval(y); plot(t,yt), hold on % plot y(t) for t=[0,10]
>>[r,p,k]=residue(Ns,Ds) % Partial fraction expansion
>>yt1=real(r.'*exp(p*t)); plot(t,yt1,'r') % Another way to get y(t)
```

The symbolic DE solver '`dsolve()`' can also be used to get the same solution:

```
>>y=dsolve('D2y+3*Dy+2*y=10','y(0)=1,Dy(0)=-2') % DE (E4.3.1)
  y = 6/exp(2*t) - 10/exp(t) + 5    % Eq.(E4.3.6)
```

A much easier method for obtaining this solution is to write the linear DE in state equation form as

$$\mathbf{x}'(t) \underset{(E4.3.1)}{\overset{(4.4.2a)}{=}} A\mathbf{x}(t) + \mathbf{b}u(t);\ \begin{bmatrix} x_1' \\ x_2' \end{bmatrix} = \begin{bmatrix} 0 & 1 \\ -2 & -3 \end{bmatrix}\begin{bmatrix} x_1 \\ x_2 \end{bmatrix} + \begin{bmatrix} 0 \\ 10 \end{bmatrix} \text{ with } \begin{bmatrix} x_1(0) \\ x_2(0) \end{bmatrix} = \begin{bmatrix} y(0) \\ y'(0) \end{bmatrix} = \begin{bmatrix} 1 \\ -2 \end{bmatrix} \tag{E4.3.7}$$

and use the Laplace transform solution (4.4.5) to get the same result as Eq. (E4.3.3):

$$\begin{aligned} Y(s) &\overset{(4.4.5)}{=} CX(s) \overset{(4.4.4)}{=} C[sI-A]^{-1}(\mathbf{x}(0)+\mathbf{b}U(s)) = [1\ \ 0]\begin{bmatrix} s & -1 \\ 2 & s+3 \end{bmatrix}^{-1}\left(\begin{bmatrix} 1 \\ -2 \end{bmatrix} + \begin{bmatrix} 0 \\ 1 \end{bmatrix}\frac{10}{s}\right) \\ &= \frac{[1\ \ 0]}{s(s^2+3s+2)}\begin{bmatrix} s+3 & 1 \\ -2 & s \end{bmatrix}\begin{bmatrix} s \\ -2s+10 \end{bmatrix} = \frac{[1\ \ 0]}{s(s^2+3s+2)}\begin{bmatrix} s^2+s+10 \\ -2s^2+8s \end{bmatrix} = \frac{s^2+s+10}{s(s^2+3s+2)} \end{aligned} \tag{E4.3.8}$$

The inverse Laplace transform of this equation can be directly obtained by using 'ilaplace()' as done at the end of Examples 3.23 and 3.24:

```
>>syms s; A=[0 1; -2 -3]; b=[0; 1]; C=[1 0]; x0=[1; -2]; Us=10/s;
>>Ys=C*((s*eye(size(A))-A)\(x0+b*Us)) % Eq.(4.4.5) -> (E4.3.8)
>>y=ilaplace(Ys) % Inverse Laplace transform of Eq.(E4.3.8)
   y = 6/exp(2*t) - 10/exp(t) + 5    % Eq.(E4.3.6)
```

(Example 4.4) Using the Laplace Transform to Solve a Set of Linear DEs

Let us solve the following set of DEs:

$$\begin{aligned} x_1''+x_1'+2x_1+2x_2&=6 \\ 2x_1+x_2'+3x_2&=8 \end{aligned} \quad \text{with } x_1(0)=0,\ x_1'(0)=1, \text{ and } x_2(0)=0 \tag{E4.4.1}$$

We make use of the time differentiation properties (4.2.2) and (4.2.3) of Laplace transform to write the Laplace transforms of both sides of the two DEs as

$$s^2X_1(s)-sx_1(0)-x_1'(0)+sX_1(s)-x_1(0)+2X_1(s)+2X_2(s)=\frac{6}{s}$$
$$2X_1(s)+sX_2(s)-x_2(0)+3X_2(s)=\frac{8}{s}$$

$$;\ \begin{bmatrix} s^2+s+2 & 2 \\ 2 & s+3 \end{bmatrix}\begin{bmatrix} X_1(s) \\ X_2(s) \end{bmatrix}=\begin{bmatrix} sx_1(0)+x_1'(0)+x_1(0)+6/s \\ x_2(0)+8/s \end{bmatrix} \underset{x_2(0)=0}{\overset{x_1(0)=0,\,x_1'(0)=1}{=}} \begin{bmatrix} 1+6/s \\ 8/s \end{bmatrix} \tag{E4.4.2}$$

$$;\ \begin{bmatrix} X_1(s) \\ X_2(s) \end{bmatrix} \overset{(E4.4.2)}{=} \frac{1}{(s^2+s+2)(s+3)-4}\begin{bmatrix} s+3 & -2 \\ -2 & s^2+s+2 \end{bmatrix}\begin{bmatrix} 1+6/s \\ 8/s \end{bmatrix}=\frac{1}{s(s+1)^2(s+2)}\begin{bmatrix} s^2+9s+2 \\ 8s^2+6s+4 \end{bmatrix}$$

$$\overset{PFE}{=} \begin{bmatrix} 1/s-7/(s+1)+6/(s+1)^2+6/(s+2) \\ 2/s+10/(s+1)-6/(s+1)^2-12/(s+2) \end{bmatrix} \tag{E4.4.3}$$

MATLAB can be used to solve Eq. (E4.4.2) for $X_1(s)$ and $X_2(s)$ and find their inverse Laplace transforms as

```
>>syms s; x10=0; dx10=1; x20=0; % Initial values
>>Xs=[s^2+s+2 2; 2 s+3]\[s*x10+dx10+x10+6/s; x20+8/s] % solve Eq.(E4.4.2)
>>x=ilaplace(Xs) % Inverse Laplace transform
   x =  6/exp(2*t) - 7/exp(t) + (6*t)/exp(t) + 1
       10/exp(t) - 12/exp(2*t) - (6*t)/exp(t) + 2
```

Alternatively, we can rewrite the set of DEs (E4.4.1) in state equation form as

$$\begin{bmatrix} x_1' \\ x_2' \\ x_1'' \end{bmatrix}=\begin{bmatrix} 0 & 0 & 1 \\ -2 & -3 & 0 \\ -2 & -2 & -1 \end{bmatrix}\begin{bmatrix} x_1 \\ x_2 \\ x_1' \end{bmatrix}+\begin{bmatrix} 0 \\ 8 \\ 6 \end{bmatrix} \text{ with } \begin{bmatrix} x_1(0) \\ x_2(0) \\ x_1'(0) \end{bmatrix}=\begin{bmatrix} 0 \\ 0 \\ 1 \end{bmatrix} \tag{E4.4.4}$$

and use Eq. (4.4.4) to get the solution as

```
>>syms s; A=[0 0 1;-2 -3 0;-2 -2 -1]; b=[0;8;6]; x0=[0;0;1];
>>Xs=(s*eye(size(A))-A)\(x0+b/s); x=ilaplace(Xs) % Eq.(4.4.4)
```

If you want to make a two-column matrix with the values of $x_1(t)$ and $x_2(t)$ for a range of t in each column and plot it versus t, run the following MATLAB statements:

```
>>t=[0:0.01:10]; xt=eval(x(1:2)).'; plot(t,xt)
```

(Example 4.5) Linear DE with Discontinuous Forcing Function

Consider the following DE dealt with in Example 3.9:

$$\frac{dy}{dt}+2y=2(u_s(t)-u_s(t-1)) \text{ with } y(0)=0 \tag{E4.5.1}$$

Using the time differentiation property (Table A.7(5)), Table A.8(3), and the time shifting property (Table A.7(2)), we take the Laplace transform of this DE:

$$sY(s)-y(0)+2Y(s)=\frac{2}{s}(1-e^{-s}) \tag{E4.5.2}$$

$$;Y(s)=\frac{2}{s(s+2)}(1-e^{-s})=\left(\frac{1}{s}-\frac{1}{s+2}\right)-\left(\frac{1}{s}-\frac{1}{s+2}\right)e^{-s} \tag{E4.5.3}$$

Then, using Table A.8(3) and (6) together with the time differentiation property, we take the inverse Laplace transform of Eq. (E4.5.3) to write

$$y(t)=\mathcal{L}^{-1}\{Y(s)\}=(1-e^{-2t})u_s(t)-(1-e^{-2(t-1)})u_s(t-1) \tag{E4.5.4}$$

This is identical to Eq. (E3.9.8):

$$y(t)=\begin{cases}(1-e^{-2t}) & \text{for } 0\le t<1\\ (e^2-1)e^{-2t}=e^{-2(t-1)}-e^{-2t} & \text{for } t\ge 1\end{cases} \tag{E4.5.5}$$

We can run the following MATLAB statements to get the same result and plot it for t=0~3[s] as depicted in Fig. 3.4:

```
>>syms s; Ys=2/s/(s+2)*(1-exp(-s));
>>y=ilaplace(Ys) % Eq.(E4.5.4)
    y = 2*heaviside(t - 1)*(exp(2 - 2*t)/2 - 1/2) - 1/exp(2*t) + 1
>>t=[0:0.01:3]; yt=eval(y);  plot(t,yt)
```

(Example 4.6) Using the Laplace Transform to Compute a Convolution

Compute the convolution:

$$y(t)=g(t)*x(t)=e^{-2t}u_s(t)*2r_1(t)=\int_{-\infty}^{\infty}e^{-2(t-\tau)}u_s(t-\tau)\,2r_1(\tau)d\tau \tag{E4.6.1}$$

where $r_D(t)=u_s(t)-u_s(t-D)$ denotes the rectangular pulse function of duration [0, D].

Based on the real convolution property (Table 4.2(4)) of Laplace transform, we take the inverse transform of the product of $G(s)=\mathcal{L}\{y(t)\}$ and $X(s)=\mathcal{L}\{x(t)\}$ to get the convolution of $g(t)$ and $x(t)$:

$$\begin{aligned}y(t) &=\mathcal{L}^{-1}\{G(s)X(s)\}=\mathcal{L}^{-1}\left\{\mathcal{L}\{e^{-2t}u_s(t)\}\mathcal{L}\{2(u_s(t)-u_s(t-1))\}\right\}\\ &\underset{\text{Table 4.2(2)}}{\overset{\text{Table 4.1(3),(5)}}{=}}\mathcal{L}^{-1}\left\{\frac{1}{s+2}\left(\frac{2}{s}-\frac{2}{s}e^{-s}\right)\right\}\overset{\text{PFE}}{=}\mathcal{L}^{-1}\left\{\frac{1}{s}-\frac{1}{s+2}-\left(\frac{1}{s}-\frac{1}{s+2}\right)e^{-s}\right\}\\ &\underset{\text{Table 4.2(2)}}{\overset{\text{Table 4.1(3),(5)}}{=}}(1-e^{-2t})u_s(t)-(1-e^{-2(t-1)})u_s(t-1)\end{aligned} \tag{E4.6.2}$$

Without using the Laplace transform, we may compute the convolution in the time domain as

$$y(t)=g(t)*x(t)=e^{-2t}u_s(t)*2r_1(t)=\int_{-\infty}^{\infty}e^{-2(t-\tau)}u_s(t-\tau)\,2r_1(\tau)d\tau=2\int_0^1 e^{-2(t-\tau)}u_s(t-\tau)\,d\tau$$

$$=\begin{cases}2\int_0^t e^{-2(t-\tau)}u_s(t-\tau)\,d\tau = e^{-2(t-\tau)}\Big|_0^t=1-e^{-2t} & \text{for } t\le 1\\ 2\int_0^1 e^{-2(t-\tau)}u_s(t-\tau)\,d\tau = e^{-2(t-\tau)}\Big|_0^1=e^{-2(t-1)}-e^{-2t} & \text{for } t>1\end{cases} \tag{E4.6.3}$$

Which do you like, computing the convolution in the s-domain or in the t-domain? Are you still reluctant to use the Laplace transform for computing a convolution? Then, how about computing another convolution like this?

$$y(t)=g_1(t)*x(t)=e^{-t}\sin t\,u_s(t)*2r_1(t) \tag{E4.6.4}$$

Let the computer compute the convolution in both ways:

```
>>syms t tau s
>>y1=int(2*exp(-(t-tau))*sin(t-tau),tau,0,t) % for t<=1 in the t-domain
   y1 = 1 - sin(t)/exp(t) - cos(t)/exp(t)
>>y2=int(2*exp(-(t-tau))*sin(t-tau),tau,0,1) % for t>1 in the t-domain
   y2 = exp(1-t)*(cos(1-t)-sin(1-t)) - (cos(t)+sin(t))/exp(t)
>>y=ilaplace(2/((s+1)^2+1)/s*(1-exp(-s))) % for any t in the s-domain
   y = 2*heaviside(t-1)*((exp(1-t)*(cos(t-1)+sin(t-1)))/2-1/2)
            -(cos(t)+sin(t))/exp(t) + 1
>>pretty(y)
                        / exp(1-t)(cos(t-1)+sin(t-1))         \
   2 heaviside(t-1)  |  ---------------------------- - 1/2  |
                        \              2                      /
         - exp(-t) (cos(t) + sin(t)) + 1
```

Doesn't it seem that the convolution is much easier to deal with in the s-domain than in the t-domain, even if we do not perform hand calculations but use the MATLAB software for computations?

(Example 4.7) Using the Laplace Transform to Compute a Fundamental Matrix

Find the fundamental or state transition matrix $\phi(t)=e^{At}$ (defined by Eq. (3.2.32)) for the following DE that was dealt with in Example 3.23:

$$\mathbf{x}'(t)\overset{(E3.23.2)}{=}A\,\mathbf{x}(t)+\mathbf{b}\,u(t);\quad \begin{bmatrix}x_1'\\x_2'\end{bmatrix}=\begin{bmatrix}0&1\\-2&-2\end{bmatrix}\begin{bmatrix}x_1\\x_2\end{bmatrix}+\begin{bmatrix}0\\2\end{bmatrix} \tag{E4.7.1}$$

It was computed in the t-domain by using the eigendecomposition technique in Example 3.23 as

$$\phi(t)\overset{(E3.23.4)}{=}\begin{bmatrix}e^{-t}(\cos t+\sin t) & e^{-t}\sin t\\ -2e^{-t}\sin t & e^{-t}(\cos t-\sin t)\end{bmatrix} \tag{E4.7.2}$$

Here we use the Laplace transform technique based on Eq. (3.2.39):

$$\Phi(s)\overset{(3.2.40)}{=}[sI-A]^{-1}=\begin{bmatrix} s & -1 \\ 2 & s+2 \end{bmatrix}^{-1}=\frac{1}{s^2+2s+2}\begin{bmatrix} s+2 & 1 \\ -2 & s \end{bmatrix}=\frac{1}{(s+1)^2+1}\begin{bmatrix} s+1+1 & 1 \\ -2 & s+1-1 \end{bmatrix}$$

$$;\quad \phi(t)\overset{(3.2.39)}{=}\mathcal{L}^{-1}\{\Phi(s)\}=\begin{bmatrix} e^{-t}(\cos t+\sin t) & e^{-t}\sin t \\ -2e^{-t}\sin t & e^{-t}(\cos t-\sin t) \end{bmatrix}u_s(t) \tag{E4.7.3}$$

Also, we may let the computer compute $\phi(t)=e^{At}$ in the t-domain by using the exponential matrix function '`expm()`' and in the s-domain by using Eq. (3.2.39) with the Laplace transform technique:

```
>>syms t s;   A=[0 1;-2 -2];
>>phi=expm(A*t); pretty(phi);
      +-                                                    -+
      | exp(-t) (cos(t) + sin(t)),          exp(-t) sin(t)   |
      |      -2 exp(-t) sin(t),        exp(-t) (cos(t) - sin(t)) |
      +-                                                    -+
>>phi1=ilaplace((s*eye(size(A))-A)^-1) % Eq.(3.2.39)
    phi1 = [ (cos(t) + sin(t))/exp(t),                  sin(t)/exp(t)]
           [          -(2*sin(t))/exp(t),    (cos(t) - sin(t))/exp(t)]
```

Both of the results agree with Eq. (E3.23.4) or (E4.7.2).

(Example 4.8) Using the Laplace Transform for Differentiation/Integration

Use the Laplace transform to find the derivative and integral of

$$i(t) = 2e^{-t}\cos t + 4e^{-t}\sin t \tag{E4.8.1}$$

Noting that the Laplace transform of $i(t)$ is

$$I(s) = 2\frac{s+1}{(s+1)^2+1^2} + 4\frac{1}{(s+1)^2+1^2} = \frac{2s+6}{(s+1)^2+1^2} \tag{E4.8.2}$$

we can use the differentiation/integration properties (Table A.7(5)/(6)) of Laplace transform to find $i'(t)$ and $\int_{-\infty}^{\infty} i(\tau)d\tau$ as follows:

$$\mathcal{L}\{i'(t)\} \overset{\text{Table A.7(5)}}{=} sI(s)-i(0) \underset{\text{(E4.8.1)}}{\overset{\text{(E4.8.2)}}{=}} s\frac{2(s+3)}{(s+1)^2+1^2}-2=\frac{2s-4}{(s+1)^2+1^2} \tag{E4.8.3a}$$

$$;\quad i'(\tau) = \mathcal{L}^{-1}\left\{\frac{2(s+1)-6\times 1}{(s+1)^2+1^2}\right\} = (2e^{-t}\cos t - 6e^{-t}\sin t)u_s(t) \tag{E4.8.3b}$$

$$\mathcal{L}\left\{\int_0^t i(\tau)d\tau\right\} \overset{\text{Table A.7(6)}}{=} \frac{1}{s}I(s) \overset{\text{(E4.8.2)}}{=} \frac{2(s+3)}{s((s+1)^2+1^2)} = \frac{3}{s}-\frac{3s+4}{(s+1)^2+1^2} \tag{E4.8.4a}$$

$$;\quad \int_0^t i(\tau)d\tau = \mathcal{L}^{-1}\left\{\frac{3}{s}-\frac{3(s+1)+1}{(s+1)^2+1^2}\right\} = (3-3e^{-t}\cos t - e^{-t}\sin t)u_s(t) \tag{E4.8.4b}$$

How about computing $i'(t)$ and $\int_{-\infty}^{\infty} i(\tau)d\tau$ in the t-domain?

4.5 TRANSFER FUNCTION OF A CONTINUOUS-TIME SYSTEM

A single-input single-output (SISO) continuous-time (CT) system is a system that shows a CT signal as the output to a CT input. If a system has more than one input and more than one output, it is called a multiple-input multiple-output (MIMO) system. The input–output relationship of a SISO CT system with input $u(t)$ and output $y(t)$ is often described by a linear ODE like

$$y^{(3)}(t) + a_2 y''(t) + a_1 y'(t) + a_0 y(t) = u(t) \tag{4.5.1}$$
$$\text{with initial conditions } y(0),\ y'(0),\ \text{and } y''(0)$$

With zero initial conditions $y(0)=0$, $y'(0)=0$, and $y''(0)=0$, we can use the differentiation property of Laplace transform to write the Laplace transform of Eq. (4.5.1) as

$$s^3Y(s) + a_2s^2Y(s) + a_1sY(s) + a_0Y(s) = U(s) \tag{4.5.2}$$

which yields the transformed (s-domain) input-output relationship as

$$G(s) = \frac{Y(s)}{U(s)} = \frac{1}{s^3+a_2s^2+a_1s+a_0} \tag{4.5.3}$$

This is called the *transfer* or system function of the system, which is defined as the ratio of the transformed output $Y(s)=\mathcal{L}\{y(t)\}$ to the transformed input $U(s)=\mathcal{L}\{u(t)\}$.

The transfer function (4.5.3) can also be obtained from another Laplace transform solution (4.4.5) with zero initial condition $\mathbf{x}(0)=\mathbf{0}$:

$$Y(s) \overset{(4.4.5)}{=} C[sI-A]^{-1}\left[\mathbf{x}(0)+\mathbf{b}U(s)\right] \overset{\mathbf{x}(0)=0}{\rightarrow} Y(s) = C[sI-A]^{-1}\mathbf{b}U(s) \tag{4.5.4}$$

$$;\ G(s) = \frac{Y(s)}{U(s)} = C[sI-A]^{-1}\mathbf{b} \text{ with } C=[1\ 0\ 0],\ A=\begin{bmatrix} 0 & 1 & 0 \\ 0 & 0 & 1 \\ -a_0 & -a_1 & -a_2 \end{bmatrix},\ \text{and } \mathbf{b}=\begin{bmatrix} 0 \\ 0 \\ 1 \end{bmatrix}$$

$$= [1\ 0\ 0]\begin{bmatrix} s & -1 & 0 \\ 0 & s & -1 \\ a_0 & a_1 & s+a_2 \end{bmatrix}^{-1}\begin{bmatrix} 0 \\ 0 \\ 1 \end{bmatrix} = \frac{[1\ 0\ 0]}{s^3+a_2s^2+a_1s+a_0}\begin{bmatrix} s^2+a_2s+a_1s & s+a_2 & 1 \\ -a_0 & s^2+a_2s & s \\ -a_0s & -a_1s-a_0 & s^2 \end{bmatrix}\begin{bmatrix} 0 \\ 0 \\ 1 \end{bmatrix}$$

$$= \frac{1}{s^3+a_2s^2+a_1s+a_0} \tag{4.5.5}$$

```
>>syms s a2 a1 a0;
>>A=[0 1 0;0 0 1;-a0 -a1 -a2]; C=[1 0 0]; b=[0;0;1];
>>Gs=C*[s*eye(size(A))-A]^-1*b % Eq.(4.5.5)
   Gs = 1/(s^3 + a2*s^2 + a1*s + a0)
```

Here, if the transfer function is $G(s)=Q(s)/P(s)$, its denominator $P(s)$ is called the *characteristic polynomial*, and the equation $P(s)=0$, which is formed by equating the characteristic polynomial to zero, is called the *characteristic equation* of the system as well as the DE (4.5.1) describing the input–output relationship of the system (see Eq. (3.2.6)). This is because its solution, called the characteristic root(s), characterizes the essential property of the system. It is interesting to see that the characteristic equation of the system conforms with that of the system matrix A itself that has been defined by Eq. (1.2.13).

Once the transfer function $G(s)$ is found or given, the output $y(t)$ of the system to any Laplace transformable input $u(t)$ can be computed as

$$Y(s) = G(s)U(s); \quad y(t) = \mathcal{L}^{-1}\{Y(s)\} \tag{4.5.6}$$

How about the output of the system to an impulse input $u(t)=\delta(t)$ whose Laplace transform is $U(s)=\mathcal{L}\{\delta(t)\} \overset{\text{Table A.8(1)}}{=} 1$?

$$Y(s) \underset{U(s)=1}{\overset{(4.5.6)}{=}} G(s); \quad y(t) = g(t) = \mathcal{L}^{-1}\{G(s)\} \tag{4.5.7}$$

This implies that the transfer function of a CT system is the Laplace transform of the *impulse response*, which is defined as the output to an impulse input (with zero initial conditions).

(Example 4.9) Transfer Function, Impulse Response, and Step Response of a System

Find the transfer function and impulse response of a system whose input–output relationship is described by the following DE:

$$\frac{d^2}{dt^2}y(t) + 3\frac{d}{dt}y(t) + 2y(t) = u(t), \quad t \ge 0 \tag{E4.9.1}$$

where $u(t)$ and $y(t)$ denote the input and output of the system, respectively.

We can take the Laplace transform of Eq. (E4.9.1) with zero initial conditions $y(0)=0$ and $y'(0)=0$ to get the transfer function as

$$(s^2+3s+2)Y(s) = U(s) \rightarrow G(s) = \frac{Y(s)}{U(s)} = \frac{1}{s^2+3s+2} \tag{E4.9.2}$$

Then we can take the inverse Laplace transform of this transfer function to get the impulse response:

$$g(t) \overset{(4.5.7)}{=} \mathcal{L}^{-1}\{G(s)\} \overset{(E4.9.2)}{=} \mathcal{L}^{-1}\left\{\frac{1}{s^2+3s+2}\right\} \overset{\text{PFE}}{=} \mathcal{L}^{-1}\left\{\frac{1}{s+1} - \frac{1}{s+2}\right\} = (e^{-t} - e^{-2t})u_s(t) \tag{E4.9.3}$$

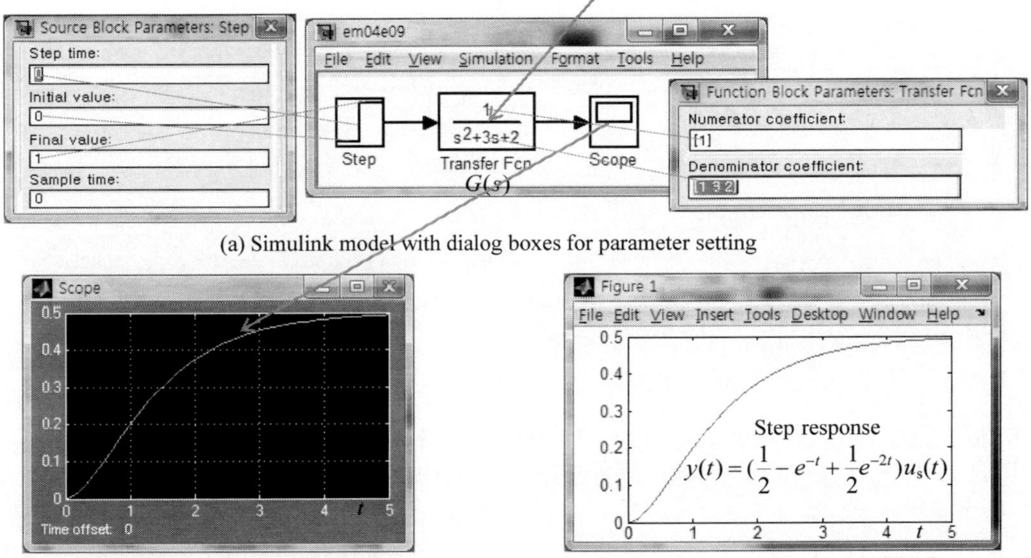

(a) Simulink model with dialog boxes for parameter setting

(b) Simulation result using Simulink

(c) Simulation result using MATLAB

Figure 4.4 Step response simulated using MATLAB® and Simulink® ("em04e09_sim.mdl")

Also, noting that the Laplace transform of the unit step function $u_s(t)$ is $U(s)=\mathcal{L}\{u_s(t)\}=1/s$, we can use Eq. (4.5.6) to find the *step response*, defined as the output to a unit step input $u(t)=u_s(t)$:

$$y_s(t)=\mathcal{L}^{-1}\{Y(s)\}\overset{(4.5.6)}{=}\mathcal{L}^{-1}\{G(s)U(s)\}=\mathcal{L}^{-1}\left\{G(s)\frac{1}{s}\right\}\overset{(E4.9.2)}{=}\mathcal{L}^{-1}\left\{\frac{1}{s(s^2+3s+2)}\right\}$$

$$\overset{\text{PFE}}{=}\mathcal{L}^{-1}\left\{\frac{1/2}{s}-\frac{1}{s+1}+\frac{1/2}{s+2}\right\}=\left(\frac{1}{2}-e^{-t}+\frac{1}{2}e^{-2t}\right)u_s(t) \tag{E4.9.4}$$

We can use MATLAB to compute the transfer function, impulse response, and step response:

```
>>a1=3; a0=2; A=[0 1;-a0 -a1]; C=[1 0]; b=[0;1];
>>syms s; Gs=C*[s*eye(size(A))-A]^-1*b % Eq.(4.5.5)
  Gs = 1/(s^2 + 3*s + 2) % Eq.(E4.9.2)
>>g=ilaplace(Gs) % Impulse response: Eq.(4.5.7)
  g = 1/exp(t) - 1/exp(2*t) % Eq.(E4.9.3)
>>Us=1/s; ys=ilaplace(Gs*Us) % Step response: Eq.(4.5.6)
  ys = 1/(2*exp(2*t)) - 1/exp(t) + 1/2 % Eq.(E4.9.4)
```

We can also compose a Simulink model "em04e09.mdl" as depicted in Fig. 4.4(a) and run it to get the step response of the system with the transfer function (E4.9.2), as depicted in Fig. 4.4(b). Fig. 4.3(c) showing the step response $y_s(t)$ has been obtained by running the following MATLAB statements:

```
>>t=[0:0.01:10]; yst=eval(ys); plot(t,yst) % Plot the step response
```

(Example 4.10) Sinusoidal Response and Resonance of a System

Find the responses of the systems whose input–output relationships are described by the following DEs, and graph them for comparison where their initial conditions are all zero, i.e., $y(0)=0$ and $y'(0)=0$, and their inputs are commonly described by a sinusoidal function $u(t)=4.01\sin t$:

System a: $$\frac{d^2}{dt^2}y(t)+2\frac{d}{dt}y(t)+2y(t)=u(t),\quad t\geq 0 \tag{E4.10.1a}$$

System b: $$\frac{d^2}{dt^2}y(t)+0.2\frac{d}{dt}y(t)+1.01y(t)=u(t),\quad t\geq 0 \tag{E4.10.1b}$$

System c: $$\frac{d^2}{dt^2}y(t)+y(t)=u(t),\quad t\geq 0 \tag{E4.10.1c}$$

Note that the Laplace transform of the common input is

$$U(s)=\mathcal{L}\{4.01\sin t\}=4.01/(s^2+1) \tag{E4.10.2}$$

(a) We can take the Laplace transform of Eq. (E4.10.1a) with zero initial conditions to get $Y_a(s)$:

$$(s^2+2s+2)Y_a(s)=\frac{4.01}{s^2+1}\rightarrow Y_a(s)=\frac{4.01}{(s^2+1)(s^2+2s+2)} \tag{E4.10.3a}$$

This can be expanded into partial fractions as

$$Y_a(s)=\frac{4.01}{(s^2+1)(s^2+2s+2)}=\frac{K_1(s+1)+K_2\times 1}{(s+1)^2+1^2}+\frac{K_3s+K_4\times 1}{s^2+1}$$
$$=\frac{1.604(s+1)+0.802\times 1}{(s+1)^2+1^2}+\frac{-1.604s+0.802\times 1}{s^2+1} \tag{E4.10.4a}$$

where the coefficients of each term of the pseudo PFE has been obtained as

$$\begin{cases}\text{Degree 3: } K_1+K_3=0\\ \text{Degree 2: } K_1+K_2+2K_3+K_4=0\\ \text{Degree 1: } K_1+2K_3+2K_4=0\\ \text{Degree 0: } K_1+K_2+2K_4=4.01\end{cases};\quad \begin{bmatrix}K_1\\K_2\\K_3\\K_4\end{bmatrix}=\begin{bmatrix}1&0&1&0\\1&1&2&1\\1&0&2&2\\1&1&0&2\end{bmatrix}^{-1}\begin{bmatrix}0\\0\\0\\4.01\end{bmatrix}=\begin{bmatrix}1.604\\0.802\\-1.604\\0.802\end{bmatrix}$$

Then the sinusoidal response $y_a(t)$ of System a is obtained as

$$y_a(t)=\mathcal{L}^{-1}\{Y_a(s)\}\overset{\text{(E4.10.4a)}}{=}0.802\{e^{-t}(2\cos t+\sin t)-2\cos t+\sin t\}u_s(t) \tag{E4.10.5a}$$

(b) We can take the Laplace transform of Eq. (E4.10.1b) with zero initial conditions to get $Y_b(s)$:

$$(s^2+0.2s+1.01)Y_b(s)=\frac{4.01}{s^2+1}\rightarrow Y_b(s)=\frac{4.01}{(s^2+1)(s^2+0.2s+1.01)} \tag{E4.10.3b}$$

and then get the sinusoidal response $y_b(t)$ of System b:

$$y_b(t)=\mathcal{L}^{-1}\{Y_b(s)\}\overset{\text{(E4.10.3b)}}{=}\mathcal{L}^{-1}\left\{\frac{4.01}{(s^2+1)(s^2+0.2s+1.01)}\right\}=\mathcal{L}^{-1}\left\{\frac{K_1(s+0.1)+K_2\times 1}{(s+0.1)^2+1^2}+\frac{K_3s+K_4\times 1}{s^2+1}\right\}$$
$$=\{e^{-0.1t}(20\cos t+\sin t)-20\cos t+\sin t\}u_s(t) \tag{E4.10.4b}$$

where the coefficients of each term of the pseudo PFE are obtained as

$$\frac{K_1(s+0.1)+K_2\times 1}{(s+0.1)^2+1^2}+\frac{K_3s+K_4\times 1}{s^2+1}=\frac{4.01}{(s^2+1)(s^2+0.2s+1.01)}$$

$$;\begin{cases}\text{Degree 3: } K_1+K_3=0\\ \text{Degree 2: } 0.1K_1+K_2+0.2K_3+K_4=0\\ \text{Degree 1: } K_1+1.01K_3+0.2K_4=0\\ \text{Degree 0: } 0.1K_1+K_2+1.01K_4=4.01\end{cases};\quad \begin{bmatrix}K_1\\K_2\\K_3\\K_4\end{bmatrix}=\begin{bmatrix}1&0&1&0\\0.1&1&0.2&1\\1&0&1.01&0.2\\0.1&1&0&1.01\end{bmatrix}^{-1}\begin{bmatrix}0\\0\\0\\4.01\end{bmatrix}=\begin{bmatrix}20\\1\\-20\\1\end{bmatrix} \tag{E4.10.5b}$$

(c) We can take the Laplace transform of Eq. (E4.10.1c) with zero initial conditions to get $Y_c(s)$:

$$(s^2+1)Y_c(s)=\frac{4.01}{s^2+1}\rightarrow Y_c(s)=\frac{4.01}{(s^2+1)^2} \tag{E4.10.3c}$$

and then get the sinusoidal response $y_c(t)$ of System c:

$$y_c(t)=\mathcal{L}^{-1}\{Y_c(s)\}\overset{\text{(E4.10.3c)}}{=}\mathcal{L}^{-1}\left\{\frac{4.01}{(s^2+1)^2}\right\}=\mathcal{L}^{-1}\left\{\frac{4.01/2}{s}\left(-\frac{d}{ds}\left(\frac{1}{s^2+1}\right)\right)\right\}\underset{(4.2.5)}{\overset{(4.2.11)}{=}}2.005\int_0^t \tau\sin\tau\, d\tau$$
$$=2.005\left(-t\cos t\Big|_0^t+2.005\int_0^t\cos\tau\, d\tau\right)=2.005(-t\cos t+\sin t)u_s(t) \tag{E4.10.4c}$$

(d) Fig. 4.5(a) shows the Simulink model "em04e10.mdl" for the simulation to get the sinusoidal responses of the above three systems. Figs. 4.5(b)/(c) show the sinusoidal responses of the three systems, each obtained using the Simulink model and running the MATLAB script "em04e10.m." Particular, $y_c(t)$ corresponds to the case of *resonance* where the frequency of the input signal conforms with the resonant or natural frequency of System c, $\omega_0=1$[rad/s], so that the output becomes infinitely large as $t\rightarrow\infty$, as could be anticipated from $t\cos t$ in Eq. (E4.10.4c).

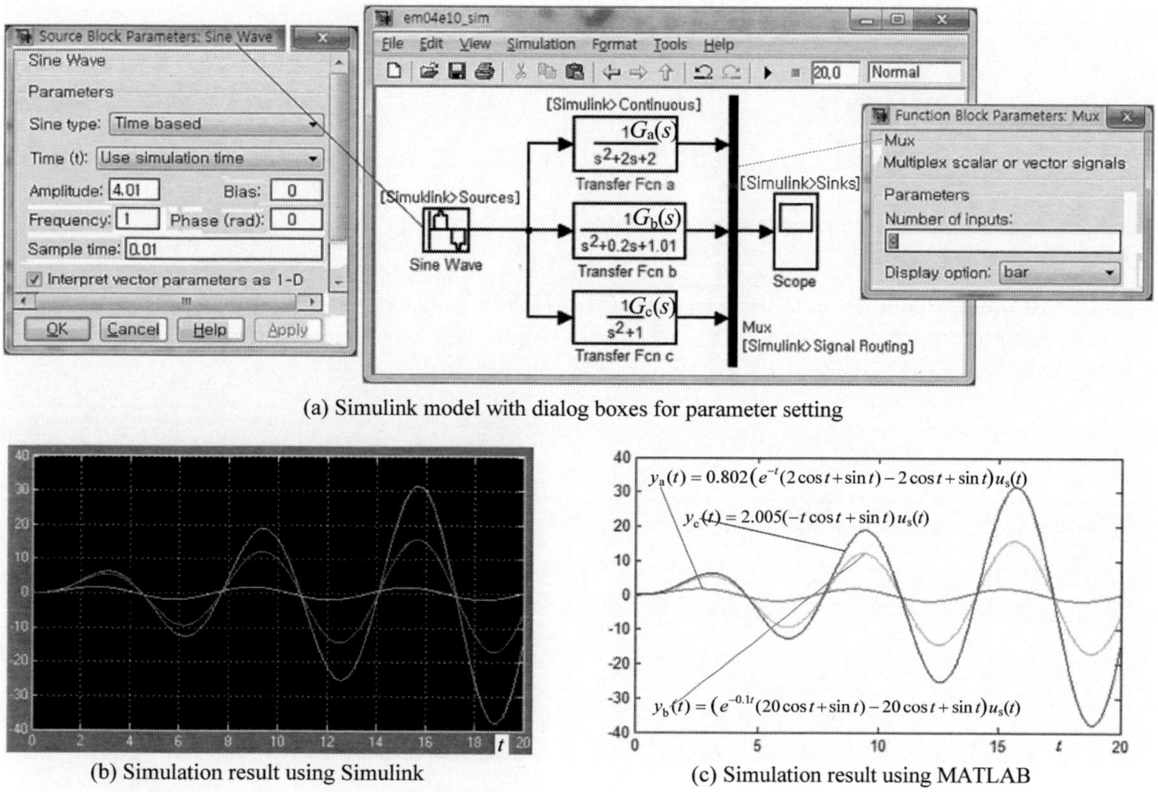

Figure 4.5 Sinusoidal responses simulated using MATLAB and Simulink ("em04e10.mdl")

```
%em04e10.m
% for Example 4.10
syms s
Us=4.01/(s^2+1); % The Laplace transform of the common sinusoidal input
t=[0:0.01:20]; % Time range to plot the outputs on
Gas=1/(s^2+2*s+2); Yas=Gas*Us; ya=ilaplace(Yas), yat=eval(ya);
Gbs=1/(s^2+0.2*s+1.01); Ybs=Gbs*Us; yb=ilaplace(Ybs), ybt=eval(yb);
Gcs=1/(s^2+1); Ycs=Gcs*Us;  yc=ilaplace(Ycs), yct=eval(yc);
plot(t,yat,'r', t,ybt,'g', t,yct)
```

Problems

4.1 Linear Second-Order DE

Consider the following DE for a series RLC circuit:

$$v_R(t)+v_L(t)+v_C(t)=R\,i(t)+L\frac{di(t)}{dt}+\frac{1}{C}\int_{-\infty}^{t}i(\tau)\,d\tau = v_{\rm i}(t) \tag{P4.1.1}$$

with the initial conditions $(i(0), v_C(0)=(1/C)\int_{-\infty}^{0}i(\tau)\,d\tau\,)=(I_0, V_0)=(1[\mathrm{A}], 0.5[\mathrm{V}])$. The differentiation/integration properties of Laplace transform (Table A.7(5)/(6)) can be used to take the Laplace transform of the DE as

$$R\,I(s)+L(s\,I(s)-i(0))+\frac{1}{C}\left(\frac{1}{s}I(s)+\frac{1}{s}\int_{-\infty}^{0}i(\tau)\,d\tau\right) = V_{\rm i}(s)$$

$$;\quad R\,I(s)+sL\,I(s)+\frac{1}{sC}I(s) = V_{\rm i}(s)+L\,i(0)-\frac{1}{s}v_C(0) \tag{P4.1.2}$$

Thus, the transformed solution of the DE is

$$I(s)=\frac{V_{\rm i}(s)+Li(0)-\frac{1}{s}v_C(0)}{R+sL+\frac{1}{sC}}=\frac{s(\frac{1}{L}V_{\rm i}(s)+i(0))-\frac{1}{L}v_C(0)}{s^2+\frac{R}{L}s+\frac{1}{LC}} \tag{P4.1.3}$$

Also, according to the voltage–current relationships of inductor L and capacitor C, the transformed voltages across L and C can be obtained from the current $I(s)$ as

$$v_L(t)=L\frac{di(t)}{dt}\ \underset{\text{Table A.7(5)}}{\overset{\mathcal{L}}{\rightarrow}}\ V_L(s)=L(sI(s)-i(0)) \tag{P4.1.4}$$

$$v_C(t)=\frac{1}{C}\int_{-\infty}^{0}i(\tau)\,d\tau\ \underset{\text{Table A.7(6)}}{\overset{\mathcal{L}}{\rightarrow}}\ V_C(s)=\frac{1}{sC}I(s)+\frac{1}{sC}\int_{-\infty}^{0}i(\tau)\,d\tau=\frac{1}{sC}I(s)+\frac{1}{s}v_C(0) \tag{P4.1.5}$$

Take the inverse Laplace transforms of $I(s)$ (Eq. (P4.1.3)), $V_L(s)$ (Eq. (P4.1.4)), and $V_C(s)$ (Eq. (P4.1.5)) with $v_{\rm i}(t)=2$[V] and the following values of parameters to get $i(t)$, $v_L(t)$, and $v_C(t)$. Plot $i(t)$, $v_L(t)$, and $v_C(t)$ for $t=0\sim10$[s] as depicted in Fig. P4.1.

(a) $R=3/2\,[\Omega]$, $L=1/2\,[\mathrm{H}]$, and $C=1\,[\mathrm{F}]$ $\rightarrow D=(R/L)^2-4/LC>0$ (overdamped case)

(b) $R=1\,[\Omega]$, $L=1/2\,[\mathrm{H}]$, and $C=2\,[\mathrm{F}]$ $\rightarrow D=(R/L)^2-4/LC=0$ (critically damped case)

(c) $R=1\,[\Omega]$, $L=1/2\,[\mathrm{H}]$, and $C=0.2[\mathrm{F}]$ $\rightarrow D=(R/L)^2-4/LC<0$ (underdamped case)

(d) $R=0\,[\Omega]$, $L=1/2\,[\mathrm{H}]$, and $C=1/2\,[\mathrm{F}]$ $\rightarrow D=(R/L)^2-4/LC<0$ (undamped case)

(e) Complete and run the MATLAB script "em04p01.m" to get the solutions, and graph them for all the cases given above as depicted in Fig. P4.1. Noting that $D=(R/L)^2-4/LC$ is the discriminant of the 2nd-degree polynomial equation formed by equating the denominator of $I(s)$ to zero, discuss how the sign of D determines whether the poles of $I(s)$ are located on the (negative) real axis on the s-plane or not and consequently, whether $i(t)$, $v_L(t)$, and $v_C(t)$ oscillate or not.

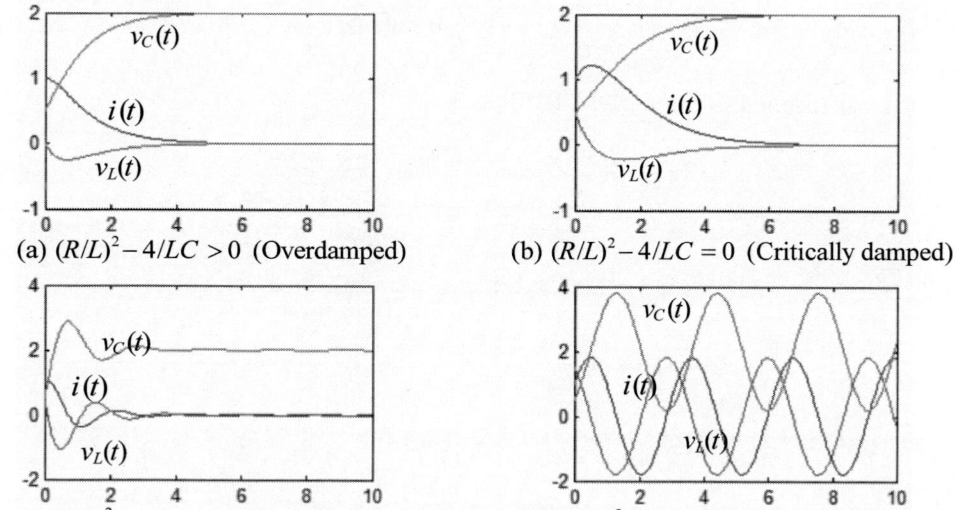

(a) $(R/L)^2-4/LC>0$ (Overdamped)

(b) $(R/L)^2-4/LC=0$ (Critically damped)

(c) $(R/L)^2-4/LC<0$ (Underdamped)

(d) $(R/L)^2-4/LC<0, R=0$ (Undamped)

Figure P4.1

```
%em04p01.m for Problem 4.1
syms s;
Vi=2; Vis=Vi/s; I0=1; V0=1/2; % Input voltage and Initial conditions
t0=0; tf=10; N=500; t=t0+(tf-t0)/N*[0:N]; % Time vector for t=[0,10]
for m=1:4
  if m==1, R=3/2; L=1/2; C=?; % (a)
    elseif m==2, R=?; L=1/2; C=2; % (b)
    elseif m==3, R=1; L=1/2; C=???; % (c)
    else R=?; L=1/2; C=1/2; % (d)
  end
  Is= ((s*???-V0)/L + ??*s)/(s^2+s*???+1/L/C); % Eq.(P4.1.3)
  i= ilaplace(Is) % the inverse Laplace transform i(t) of I(s)
  VLs= s*L*Is - L*I0; ??= ilaplace(VLs) % Inductor voltage Eq.(P4.1.4)
  VCs= Is/s/C + ??/s; vC= ilaplace(???) % Capacitor voltage Eq.(P4.1.5)
  it=eval(i); vLt=eval(vL); vCt=eval(vC);
  subplot(220+m), plot(t,it, t,vLt, t,vCt), hold on
end
```

(Note) The authors hope that this problem has made you happy with the Laplace transform and MATLAB or equivalent software. On the other hand, it may make you feel sorry for those who have not experienced the amazing usefulness and convenience of such tools.

4.2 Linear Second-Order DE

Consider the following DE for a parallel *RLC* circuit:

$$i_R(t)+i_L(t)+i_C(t)=\frac{v(t)}{R}+\frac{1}{L}\int_{-\infty}^{t}v(\tau)\,d\tau+C\frac{dv(t)}{dt}=i_i(t) \tag{P4.2.1}$$

with the initial conditions $(i_L(0)=(1/L)\int_{-\infty}^{0}v(\tau)\,d\tau,\ v(0))=(I_0,V_0)=(0[\mathrm{A}],0[\mathrm{V}])$. The differentiation/integration properties of the Laplace transform (Table 4.2(5)/(6)) can be used to transform the DE as

$$\frac{V(s)}{R}+\frac{1}{L}\left(\frac{1}{s}V(s)+\frac{1}{s}\int_{-\infty}^{0}v(\tau)\,d\tau\right)+C(sV(s)-V_0)=I_i(s)$$

$$; \left(\frac{1}{R}+\frac{1}{sL}+sC\right)V(s) = I_i(s)-\frac{1}{s}\left(\frac{1}{L}\int_{-\infty}^{0}v(\tau)d\tau\right)+CV_0 = I_i(s)-\frac{I_0}{s}+CV_0 \quad \text{(P4.2.2)}$$

Thus, the transformed solution of the DE is

$$V(s)=\frac{I_i(s)-I_0/s+CV_0}{1/R+sC+1/sL}=\frac{(sI_i(s)-I_0)/C+V_0 s}{s^2+s/RC+1/LC} \quad \text{(P4.2.3)}$$

Also, according to the voltage–current relationships of inductor L and capacitor C, the transformed currents through L and C can be obtained from the transformed voltage $V(s)$ as

$$i_C(t)=C\frac{dv(t)}{dt}\ \underset{\text{Table A.7(6)}}{\overset{\mathcal{L}}{\rightarrow}}\ I_C(s)=C(sV(s)-v(0)) \quad \text{(P4.2.4)}$$

$$i_L(t)=\frac{1}{L}\int_{-\infty}^{t}v(\tau)d\tau\ \underset{\text{Table A.7(6)}}{\overset{\mathcal{L}}{\rightarrow}}\ I_L(s)=\frac{1}{sL}V(s)+\frac{1}{sL}\int_{-\infty}^{0}v(\tau)d\tau=\frac{1}{sL}V(s)+\frac{1}{s}i_L(0) \quad \text{(P4.2.5)}$$

Take the inverse Laplace transforms of Eq. (P4.2.3) with $i_i(t)$=50sin2t[A] and the following values of parameters to get $v(t)$, where

$$i_i(t)=50\sin 2t\,u_s(t)\,[\text{A}]\ \underset{\text{Table A.7(7)}}{\overset{\mathcal{L}}{\rightarrow}}\ I_i(s)=\frac{50\times 2}{s^2+2^2},\ I_0=0\,[\text{A}],\ \text{and}\ V_0=0\,[\text{V}] \quad \text{(P4.2.6)}$$

Plot $v(t)$ for t = 0~20[s] as depicted in Fig. P4.2.

(a) R = 2/3[Ω], C = 1/2[F], L = 1[F] $\rightarrow D=(1/RC)^2-4/LC>0$ (overdamped)

(b) R = 1[Ω], C = 1/2[F], and L = 2[H] $\rightarrow D=(1/RC)^2-4/LC=0$ (critically damped)

(c) R = 1[Ω], C = 1/2[F], and L = 1[H] $\rightarrow D=(1/RC)^2-4/LC<0$ (underdamped)

(d) R = 1[Ω] (open), C = 1/2[F], and L = 2[H] $\rightarrow D=(1/RC)^2-4/LC<0$ (undamped)

(e) Complete and run the following MATLAB script "em04p02.m" get the solutions and graph them for all the cases given above as depicted in Fig. P4.2.

```
%em04p02.m
syms s
t0=0; tf=20; N=1000; t=t0+(tf-t0)/N*[0:N]; % Simulation interval
Iis= 50*2/(s^2+2^2); V0=0; I0=0; % Eq.(P4.2.6)
for m=1:4
   if m==1, R=2/3; C=1/2; L=1;
     elseif m==2, R=1; C=1/2; L=2;
     elseif m==3, R=1; C=1/2; L=1;
     else R=inf; C=1/2; L=2;
   end
   Vs=((s*???-I0)/C+??*s)/(s^2+s/R/C+1/L/C); % Eq.(P4.2.3)
   v=ilaplace(Vs) % Inverse Laplace transform
   vt=eval(v);
   subplot(220+m), plot(t,vt)
end
```

4.3 System of Linear First-Order DEs

Using the Laplace transform technique, solve the DE (E3.14.2):

$$\mathbf{x}'(t)=A\mathbf{x}(t)+\mathbf{b}u(t);\ \begin{bmatrix}x_1'\\x_2'\end{bmatrix}=\begin{bmatrix}0 & -1/L\\1/C & -1/RC\end{bmatrix}\begin{bmatrix}x_1\\x_2\end{bmatrix}+\begin{bmatrix}V_m/L\\0\end{bmatrix}\sin\omega t \text{ with } \begin{bmatrix}x_1(0)\\x_2(0)\end{bmatrix}=\begin{bmatrix}1\\2\end{bmatrix} \quad \text{(P4.3.1)}$$

where R=1[kΩ], L=1[mH], C=1[mF], V_m=5[V], and ω=120π[rad/s]. Graph the solution for t= 0~0.1[s].

You can use Eq. (4.4.4) to get the solution of Eq. (P4.3.1):

$$X(s) \overset{(4.4.4)}{=} [sI-A]^{-1}(\mathbf{x}(0)+\mathbf{b}U(s)) = \begin{bmatrix} s & 1/L \\ -1/C & \square \end{bmatrix}^{-1}\left(\begin{bmatrix}1\\2\end{bmatrix}+\begin{bmatrix}V_m/L\\0\end{bmatrix}\square\right)$$
$$= \frac{1}{s^2+s/RC+1/LC}\begin{bmatrix} \square & -1/L \\ 1/C & s \end{bmatrix}\left(\begin{bmatrix}1\\2\end{bmatrix}+\begin{bmatrix}V_m/L\\0\end{bmatrix}\square\right) \quad \text{(P4.3.2)}$$

Complete and run the MATLAB script "em04p03.m" to take the inverse Laplace transform of $X(s)$ to get $x_1(t)$ and $x_2(t)$ and graph them for t=0~0.1[s], as depicted in Fig. 3.9. If you meet up with any readability problem for $x_1(t)$ and $x_2(t)$, how about trying '`ilaplace_my()`' in place of '`ilaplace()`'?

```
%em04p03.m
syms s;
R=1000; L=0.001; C=?????; w=120*pi; Vm=5;
A=[0 -1/L; 1/C ??????]; b=[????; 0];
x0=[1; ?]; Us=?/(s^2+w^2);
Xs=(s*eye(size(A))-?)\(??+b*??) % Eq.(4.4.4)
x=ilaplace(Xs); % Inverse Laplace transform of Eq.(P4.3.2)
t0=0; tf=0.1; N=500; t=t0+(tf-t0)/N*[0:N].'; % Time vector in a column
xt=eval(x.');
plot(t,xt) % Plot x1(t) and x2(t)
x1=ilaplace_my(Xs(1)), x2=ilaplace_my(Xs(2))
discrepancy = norm([eval(x1) eval(x2)]-xt)
```

4.4 System of Linear First-Order DEs

Consider the following DE in $i_1(t)$ and $i_2(t)$, which was dealt with in Problem 3.11(b):

$$R_1 i_1(t)+\frac{1}{C_1}\int_{-\infty}^{t} i_1(\tau)d\tau + R_2(i_1(t)-i_2(t)) = v(t)=t$$
$$R_2(i_2(t)-i_1(t))+\frac{1}{C_2}\int_{-\infty}^{t} i_2(\tau)\,d\tau = 0 \quad \text{(P4.4.1)}$$

where $i_1(0)$=0 and $i_2(0)$=0. The differentiation/integration properties of Laplace transform (Table A.7(5)/(6)) can be used to transform this DE (with zero initial conditions) as

$$R_1 I_1(s)+\frac{1}{sC_1}I_1(s)+R_2(I_1(s)-I_2(s)) = V(s) = \square$$
$$R_2(I_2(s)-I_1(s))+\frac{1}{sC_2}I_2(s) = 0$$
$$;\ \begin{bmatrix} \square & -R_2 s \\ \square & R_2 s+1/C_2 \end{bmatrix}\begin{bmatrix}I_1(s)\\I_2(s)\end{bmatrix}=\begin{bmatrix}sV(s)\\0\end{bmatrix}=\begin{bmatrix}\square\\0\end{bmatrix} \quad \text{(P4.4.2)}$$

Complete and run the MATLAB script "em04p04.m" to solve this equation (R_1=100Ω, C_1=10μF, R_2=1kΩ, C_2=10μF) for $I_1(s)$ and $I_2(s)$, take the inverse Laplace transform of $I_2(s)$ to get $i_2(t)$, and plot it for t=0~0.05[s] as depicted in Fig. P3.11b.

```
%em04p04.m
syms s
R1=100; R2=1e3; C1=1e-5; C2=1e-5; Vs=1/s^2;
As=[(R1+R2)*s+1/C1  -????; -R2*s R2*s+????];
Is=As\[s*Vs; 0] % solve Eq.(P4.4.2)
i2=ilaplace(Is(2)); % Inverse Laplace transform of I2(s) to get i2(t)
t0=0; tf=0.05; N=500; t=t0+(tf-t0)/N*[0:N].';
i2t_1=eval(i2); t_1=t;
load t_i2t % Recall i2(t) obtained using 'dsolve()' in P3.11(b)
plot(t,i2t, t_1,i2t_1,'r:') % Plot the two i2(t)'s
```

4.5 Differential Equation, System Function, and Frequency Response

Consider a system whose input–output relationship is described by the following differential equation:

$$y''(t)+100y'(t)+100^2 y(t) = 100x(t) \text{ with } y(0)=y'(0)=0 \tag{P4.5.1}$$

where $x(t)$ and $y(t)$ denote the input and output of the system, respectively. Note that the *frequency response* $G(j\omega)$, obtained by substituting $s=j\omega$ into the transfer function $G(s)$, determines the steady-state response y_{ss} to a sinusoidal input $x(t)$ of radian frequency ω. The amplitude of y_{ss} is the product of $|G(j\omega)|$ and the amplitude of the input sinusoid where ω is the radian frequency of the input. The phase of y_{ss} is the sum of $\theta(\omega)=\angle G(j\omega)$ and the phase of the input sinusoid.

(a) Show that the transfer function of the system is

$$G(s)=\frac{Y(s)}{X(s)}=\frac{100}{s^2+100s+100^2} \tag{P4.5.2}$$

(b) Plot the magnitude of the *frequency response* $G(j\omega)=G(s)|_{s=j\omega}$ for ω=0:0.5:400[rad/s], which can be obtained by running the following MATLAB statements:

```
>>syms s; Gs=100/(s^2+100*s+100^2); % Transfer function (P4.5.2)
>>w=0:0.5:400; Gw=subs(Gs,s,j*w); % Frequency response for w=0~400
```

You can also use the MATLAB built-in function '`freqs()`' to find $G(j\omega)=G(s)|_{s=j\omega}$:

```
>>B=[100]; A=[1 100 100^2]; % Numerator/Denominator polynomials of Gs
>>GW=freqs(B,A,w); % An alternative to get the frequency response
```

Find the value of $G(\omega)$ at ω=0 and 100[rad/s] where $G(0)$ is referred to as the *DC* (*direct current*) *gain*.

(c) Find the step response, i.e., the output of the system to the unit step input $x(t)=u_s(t)$. Is the steady-state value of the step response equal to the DC gain obtained in (b) and the result of applying the FVT (final value theorem) (4.2.13)?

(d) Using the Laplace transform, show that the output to a sinusoidal input $x(t)=\sin(100t)u_s(t)$ can be found as

$$Y_{\sin}(s)=G(s)X(s)=G(s)\mathcal{L}\{\sin(100t)u_s(t)\}=\frac{100}{s^2+100s+100^2}\boxed{} \tag{P4.5.3}$$

$$;\ y_{\sin}(t)=\boxed{}\,u_s(t) \tag{P4.5.4}$$

$$\overset{t\to\infty}{\to} y_{\sin,ss}(t)=\boxed{}\sin(\boxed{}) \tag{P4.5.5}$$

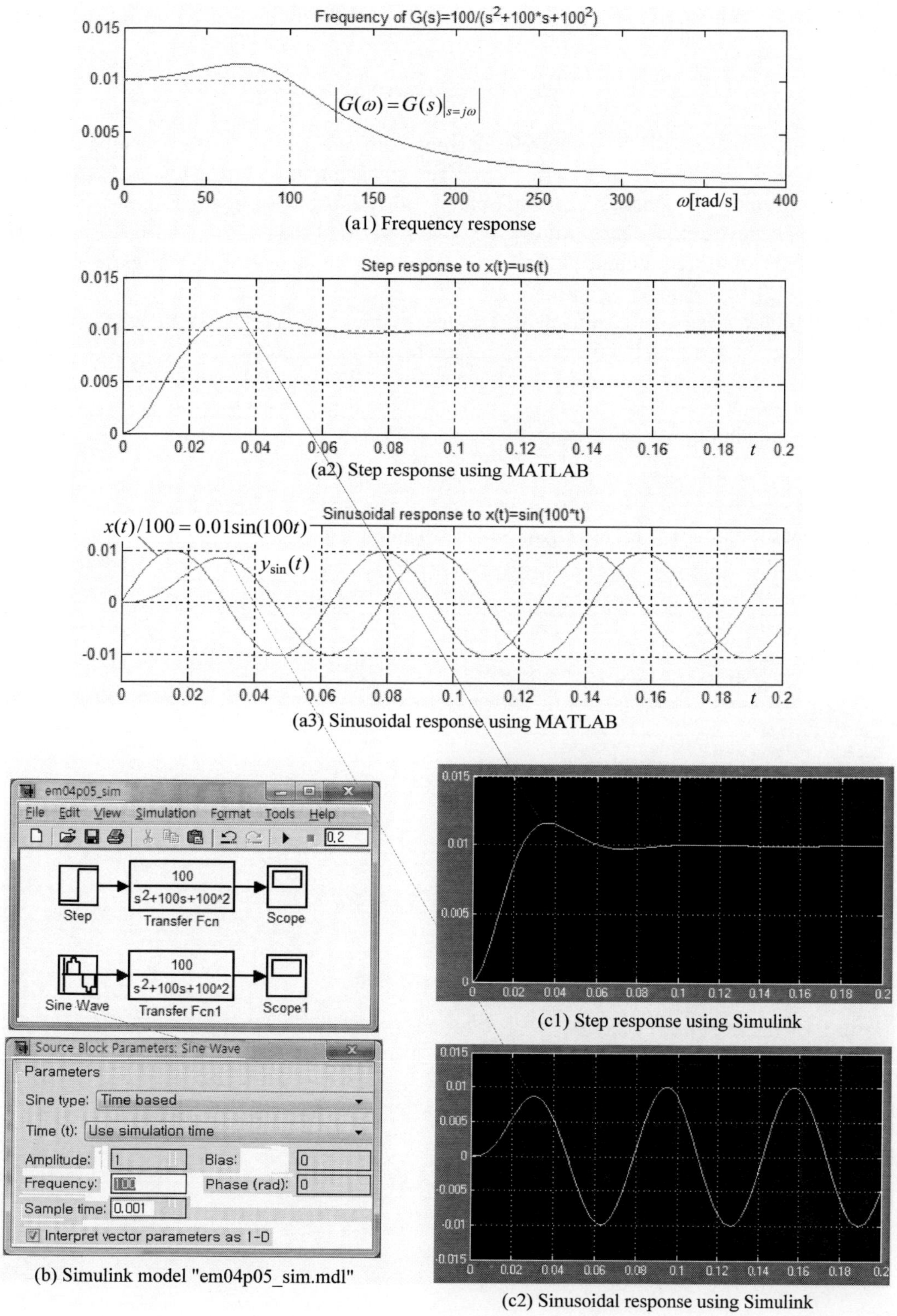

(a1) Frequency response

(a2) Step response using MATLAB

(a3) Sinusoidal response using MATLAB

(b) Simulink model "em04p05_sim.mdl"

(c1) Step response using Simulink

(c2) Sinusoidal response using Simulink

Figure P4.5 Frequency/step/sinusoidal responses simulated using MATLAB and Simulink ("em04p05_sim.mdl")

(i) Can you get the steady-state sinusoidal response by applying the FVT to Eq. (P4.5.3)?

(ii) Noting that the frequency response of the system at the frequency $\omega=100$[rad/s] of the input is

$$G(j\omega)\big|_{\omega=100} = -j0.01 = 0.01\angle -\frac{\pi}{2} \tag{P4.5.6}$$

what can you tell about the relationships between the amplitude/phase of the steady-state sinusoidal response and those of the sinusoidal input in terms of the frequency response magnitude and phase? The amplitude of the sinusoidal output is the (   ) of the frequency response magnitude $|G(j\omega)|$ and the amplitude of the sinusoidal input. The phase of the sinusoidal output is the (   ) of the frequency response phase $\angle G(j\omega)$ and the phase of the sinusoidal output.

(iii) Find the steady-state sinusoidal response to the input $x(t)=\cos(100t)u_s(t)$ by using not the Laplace transform but the concept of the frequency response.

(iv) Find the steady-state sinusoidal response to the input $x(t)=\sin(70.5t)u_s(t)$, where the frequency response $G(j\omega)\big|_{\omega=70.5} = 0.0067 - j0.0094 = 0.0115\angle -0.9511$ can be found by running the following statements:

```
>>syms s; Gs=100/(s^2+100*s+100^2);
>>w2=70.5; Gw2=subs(Gs,s,j*w2), abs(Gw2), angle(Gw2)
```

Does it conform with the result obtained using the Laplace transform as

```
>>y_sin2=ilaplace(Gs*w2/(s^2+w2^2)); pretty(y_sin2)
```

```
%em04p05.m
syms s t
Gs = 100/(s^2+100*s+100^2); % Transfer function
B=[100]; A=[1 100 100^2]; % Numerator/Denominator of Transfer function
disp('(b)')
w=[0:0.5:400]; % Frequency vector[rad/s]
Gw=freqs(B,A,w); Gw1=subs(Gs,s,j*w); % Frequency response
subplot(311), plot(w,abs(Gw))
ws=[0; 100]; % 2 frequency points
Freq_responses=freqs(B,A,ws)
DC_gain=Freq_responses(1)   % DC gain
disp('(c)') % Step response
y_step=ilaplace(Gs/s) % Step response using the Laplace transform
yss_step=double(subs(y_step,t,1e5)) % Steady-state value of y_step
tf=0.2; tt=tf/200*[0:200]; % Time vector
t=tt; yt_step=eval(y_step);
subplot(312), plot(tt,yt_step, tt,DC_gain*ones(size(tt)),':')
title('Step response to x(t)=us(t)')
disp('(d)') % Sinusoidal response
wi=100;  Xs=wi/(s^2+wi^2); % Input transform
y_sin=ilaplace(Gs*Xs) %Sinusoidal response using the Laplace transform
yt_sin=eval(y_sin);
% To get the numerical solution for the sinusoidal response
dx=@(t,x)[0 1; -100^2 -100]*x +[0;100]*sin(100*t);
x0=[0; 0]; % Initial state vector
[tt,xx]=ode45(dx,tt,x0); % Using the numerical DE solver
subplot(313), plot(tt,yt_sin, tt.',xx(:,1),'r:',
    tt,abs(Freq_responses(2))*ones(size(tt)),'k:')
title('Sinusoidal response to x(t)=sin(100*t)')
```

(e) Referring to the above MATLAB script "em04p05.m" and the Simulink model (block diagram) in Fig. P4.5(b), use MATLAB/Simulink to plot the frequency/step/sinusoidal responses of the system as depicted in Fig. P4.5(a1), (a2), (a3), (c1), and (c2).

CHAPTER 5

The Z-transform

CHAPTER OUTLINE

The z-transform is the discrete-time counterpart of the Laplace transform. It plays a vital role in the analysis and design of discrete-time linear time-invariant (LTI) systems that are described by difference equations. In this chapter we introduce the z-transform and discuss some of its properties.

5.1 DEFINITION OF THE Z-TRANSFORM[Y-2], Section 4.1

Consider a discrete-time (DT) sequence $x[n]$ that has possibly been obtained by sampling a continuous-time (CT) signal $x(t)$ every T seconds where T is called the *sampling period* or *interval*. The *bilateral* (or *two-sided*) and *unilateral* (or *one-sided*) *z-transforms* of $x[n]$ are defined as

$$\mathbb{X}[z] = Z\{x[n]\} = \sum_{n=-\infty}^{\infty} x[n]\, z^{-n} \quad (5.1.1a)$$

$$X[z] = \mathcal{Z}\{x[n]\} = \sum_{n=0}^{\infty} x[n]\, z^{-n} \quad (5.1.1b)$$

where z is a complex variable (parameter). For convenience, the z-transform relationship will sometimes be denoted as

$$x[n] \overset{Z}{\leftrightarrow} \mathbb{X}[z] \qquad\qquad x[n] \overset{\mathcal{Z}}{\leftrightarrow} X[z]$$

Note that the unilateral z-transform of $x[n]$ can be thought of as the bilateral z-transform of $x[n]u_s[n]$ ($u_s[n]=1/0$ for $n \geq 0 \,/\, n<0$: the *unit step sequence*) and therefore the two definitions will be identical for causal sequences that are zero for all $n<0$. The unilateral z-transform is particularly useful in analyzing causal systems described by linear constant-coefficient difference equations with initial conditions.

The z-transform does not necessarily exist for all values of z in the complex plane. For the z-transform to exist, the series in Eq. (5.1.1) must converge. For a given sequence $x[n]$, the domain of the z-plane within which the series converges is called the *region of convergence* (*ROC*) of the z-transform $\mathbb{X}[z]$ or $X[z]$. Thus, the specification of the z-transform requires both the algebraic expression in z and its ROC.

Note that a sequence $x[n]$ is called

- a *right-sided* sequence if $x[n]=0\ \forall\ n<n_0$ for some finite integer n_0,
- a *causal* sequence if $x[n]=0\ \forall\ n<n_0$ for some nonnegative integer $n_0\ge 0$,
- a *left-sided* sequence if $x[n]=0\ \forall\ n\ge n_0$ for some finite integer n_0, and
- an *anti-causal* sequence if $x[n]=0\ \forall\ n\ge n_0$ for some nonpositive integer $n_0\le 0$,

respectively.

(Example 5.1) The Z-transform of Exponential Sequences

(a) For a right-sided and causal sequence $x_1[n]=a^n u_s[n]$, we can use Eq. (5.1.1a) to get its bilateral z-transform as

$$\mathbb{X}_1[z]=Z\{x_1[n]\}=\sum_{n=-\infty}^{\infty}a^n u_s[n]z^{-n}=\sum_{n=0}^{\infty}a^n z^{-n}\overset{(C.23)}{=}\frac{1}{1-az^{-1}}=\frac{z}{z-a} \tag{E5.1.1}$$

This geometric sequence converges for $|az^{-1}|<1$, which implies that the ROC of $\mathbb{X}_1[z]$ is $\mathcal{R}_1=\{z:|z|>|a|\}$ (see Fig. 5.1(a)).

(b) For a left-sided and anti-causal sequence $x_2[n]=-b^n u_s[-n-1]$, we can use Eq. (5.1.1a) to get its bilateral z-transform as

$$\mathbb{X}_2[z]=Z\{x_2[n]\}=\sum_{n=-\infty}^{\infty}-b^n u_s[-n-1]z^{-n}=-\sum_{n=-\infty}^{-1}b^n z^{-n}\overset{n\to -n}{=}-\sum_{n=1}^{\infty}(b^{-1}z)^n\overset{(C.23)}{=}\frac{-b^{-1}z}{1-b^{-1}z}=\frac{z}{z-b} \tag{E5.1.2}$$

This geometric sequence converges for $|b^{-1}z|<1$, which implies that the ROC of $\mathbb{X}_2[z]$ is $\mathcal{R}_2=\{z:|z|<|b|\}$ (see Fig. 5.1(b)).

(c) For the two-sided sequence

$$y[n]=x_1[n]+x_2[n]=a^n u_s[n]-b^n u_s[-n-1] \tag{E5.1.3}$$

we can combine the above results of (a) and (b) to obtain its bilateral z-transform as

$$\mathbb{Y}[z]=\mathbb{X}_1[z]+\mathbb{X}_2[z]=\frac{z}{z-a}+\frac{z}{z-b} \tag{E5.1.4}$$

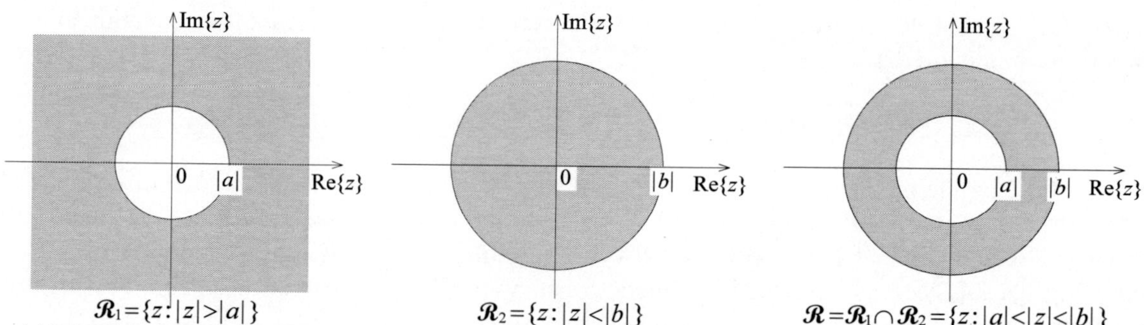

(a) ROC for a right-sided sequence (b) ROC for a left-sided sequence (c) ROC for a two-sided sequence

Figure 5.1 Three forms of ROC (region of convergence)

For this series to converge, both series must converge and therefore, its ROC is the intersection of the two ROCs $\mathcal{R}_1$ and $\mathcal{R}_2$: $\mathcal{R}=\mathcal{R}_1\cap\mathcal{R}_2=\{z:|a|<|z|<|b|\}$. This is an annular region (Fig. 5.1(c)) if $|b|>|a|$ and it will be an empty set if $|b|\le|a|$.

(Note) This example illustrates that different sequences may have the same *z*-transform, but with different ROCs. This implies that a *z*-transform expression may correspond to different sequences depending on its ROC.

[Remark 5.1] Region of Convergence (ROC)

(1) The ROC for $X[z]$ is an annular ring centered at the origin in the *z*-plane of the form

$$r^- < |z| < r^+ \tag{5.1.2}$$

where r^+ can be as large as ∞ (Fig. 5.1(a)) for causal sequences and r^- can be as small as zero (Fig. 5.1(b)) for anti-causal sequences.

(2) As illustrated in Example 5.1, the three different forms of ROC shown in Fig. 5.1 can be associated with the corresponding three different classes of DT sequences: $0<r^-<|z|$ for right-sided sequences, $|z|<r^+<\infty$ for left-sided sequences, and $r^-<|z|<r^+$ (annular ring) for two-sided sequences.

(Example 5.2) A Causal Sequence Having Multiple-Pole *Z*-transform

For a sum of two causal sequences

$$x[n] = x_1[n] + x_2[n] = 2u_s[n] - 2\left(\frac{1}{2}\right)^n u_s[n] \tag{E5.2.1}$$

we can use Eq. (E5.1.1) to get the *z*-transform for each sequence and combine the *z*-transforms as

$$\begin{aligned}\mathbb{X}[z] &= Z\{x[n]\} = \mathbb{X}_1[z] + \mathbb{X}_2[z] \\ &\overset{\text{(E5.1.1)}}{\underset{\text{with } a=1 \text{ and } a=1/2}{=}} 2\frac{z}{z-a}\bigg|_{a=1} - 2\frac{z}{z-a}\bigg|_{a=1/2} \\ &= \frac{2z}{z-1} - \frac{2z}{z-1/2} = \frac{z}{(z-1)(z-1/2)}\end{aligned} \tag{E5.2.2}$$

Since both $\mathbb{X}_1[z]=Z\{x_1[n]\}$ and $\mathbb{X}_2[z]=Z\{x_2[n]\}$ must exist for this *z*-transform to exist, the ROC is the intersection of the ROCs for the two *z*-transforms:

$$\mathcal{R}=\mathcal{R}_1\cap\mathcal{R}_2=\{z:|z|>1\}\cap\{z:|z|>1/2\}=\{z:|z|>1\} \tag{E5.2.3}$$

Fig. 5.2 shows the pole–zero pattern of the *z*-transform expression (E5.2.2) where pole and zero are marked with x and o, respectively. Note that the ROC of the *z*-transform $\mathbb{X}[z]$ of a right-sided sequence is the exterior of the circle that is centered at the origin and passes through the pole farthest from the origin in the *z*-plane.

Before working further with more examples, let us see the relationship between the *z*-transform and DTFT (discrete-time Fourier transform, see Chapter 6), which is summarized in the following remark:

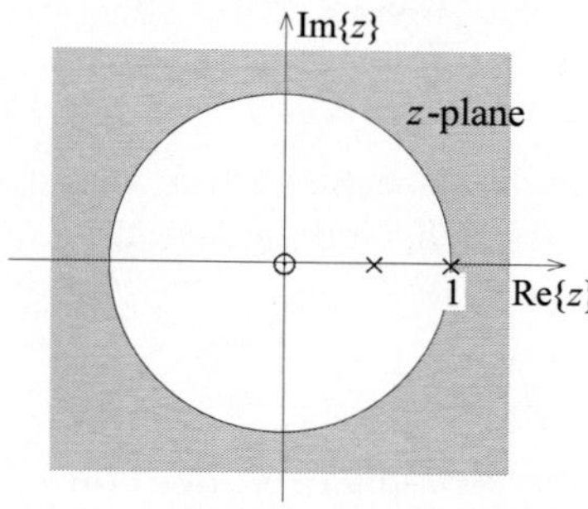

Figure 5.2 Pole–zero pattern of Eq. (E5.2.2)

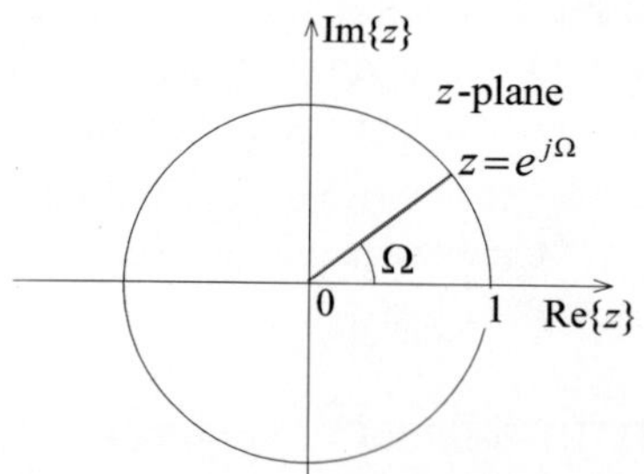

$$\mathcal{F}\{x[n]\} = X(j\Omega) = X[e^{j\Omega}] = (X[z] = Z\{x[n]\})\big|_{z=e^{j\Omega}}$$

Figure 5.3 The relationship between z-transform and DTFT

[Remark 5.2] The *Z*-transform and the DTFT (Discrete-Time Fourier Transform)

Substituting $z = e^{j\Omega}$ into the (bilateral) *z*-transform (5.1.1a) produces the DTFT defined by (6.3.1):

$$\mathcal{F}\{x[n]\} = X(j\Omega) \overset{(6.3.1)}{=} \sum_{n=-\infty}^{\infty} x[n]e^{-j\Omega n} = X[e^{j\Omega}] = X[z]\big|_{z=e^{j\Omega}} = Z\{x[n]\}\big|_{z=e^{j\Omega}} \quad (5.1.3)$$

This implies that the evaluation of the *z*-transform along the unit circle in the *z*-plane yields the DTFT. Therefore, in order for the DTFT to exist, the ROC of the *z*-transform must include the unit circle (see Fig. 5.3).

(Example 5.3) The *Z*-transform of a Complex Exponential Sequence

Let us consider a causal sequence

$$x[n] = e^{j\Omega_1 n}u_s[n] = \cos(\Omega_1 n)u_s[n] + j\sin(\Omega_1 n)u_s[n] \quad \text{(E5.3.1)}$$

We can find the *z*-transform of this exponential sequence by substituting $a = e^{j\Omega_1}$ into Eq. (E5.1.1) as

$$\mathbb{X}[z] = Z\{e^{j\Omega_1 n}u_s[n]\} \underset{a=e^{j\Omega_1}}{\overset{(E5.1.1)}{=}} \frac{z}{z - e^{j\Omega_1}} = \frac{z(z - e^{-j\Omega_1})}{(z - e^{j\Omega_1})(z - e^{-j\Omega_1})} \text{ with } |z| > |e^{\pm j\Omega_1}| = 1$$

$$;\ e^{j\Omega_1 n}u_s[n] \overset{(C.20)}{=} \cos(\Omega_1 n)u_s[n] + j\sin(\Omega_1 n)u_s[n] \overset{Z}{\leftrightarrow} \frac{z(z - \cos\Omega_1)}{z^2 - 2z\cos\Omega_1 + 1} + j\frac{z\sin\Omega_1}{z^2 - 2z\cos\Omega_1 + 1}$$

This implies the following two *z*-transform pairs:

$$\cos(\Omega_1 n)u_s[n] \overset{Z}{\leftrightarrow} \frac{z(z - \cos\Omega_1)}{z^2 - 2z\cos\Omega_1 + 1} = \frac{z(z - \cos\Omega_1)}{(z - \cos\Omega_1)^2 + \sin^2\Omega_1} \text{ with } \mathcal{R} = \{z : |z| > 1\} \quad \text{(E5.3.2)}$$

$$\sin(\Omega_1 n)u_s[n] \overset{Z}{\leftrightarrow} \frac{z\sin\Omega_1}{z^2 - 2z\cos\Omega_1 + 1} = \frac{z\sin\Omega_1}{(z - \cos\Omega_1)^2 + \sin^2\Omega_1} \text{ with } \mathcal{R} = \{z : |z| > 1\} \quad \text{(E5.3.3)}$$

(Example 5.4) The Z -transform of an Exponentially Decreasing Sinusoidal Sequence

For a causal sequence

$$x[n] = r^n \cos(\Omega_1 n) u_s[n] \overset{\text{(C.21)}}{=} \frac{1}{2} r^n (e^{j\Omega_1 n} + e^{-j\Omega_1 n}) u_s[n] \tag{E5.4.1}$$

we can use Eq. (E5.1.1) with $a = re^{\pm j\Omega_1}$ to find the z -transform of this sequence as

$$\mathbb{X}[z] \underset{a=re^{j\Omega_1} \text{ and } a=re^{-j\Omega_1}}{\overset{\text{(E5.1.1)}}{=}} \frac{1}{2}\left(\frac{z}{z-re^{j\Omega_1}} + \frac{z}{z-re^{-j\Omega_1}} \right) = \frac{z(z - r\cos\Omega_1)}{z^2 - 2zr\cos\Omega_1 + r^2} \tag{E5.4.2}$$

where the ROC is $\mathcal{R} = \{z : |z| > |re^{\pm j\Omega_1}| = |r|\}$.

In many cases, we deal with the z-transform that is expressed in the form of a rational function, i.e., a ratio of polynomials in z. For such a rational z-transform $\mathbb{X}[z] = \mathbb{Q}[z] / \mathbb{P}[z]$, we define its singularities, i.e., poles and zeros, as follows:

[Remark 5.3] Poles and Zeros of a Rational Z-transform Expression $\mathbb{X}[z] = \mathbb{Q}[z] / \mathbb{P}[z]$

(1) The roots of $\mathbb{Q}[z] = 0$, or the values of z at which $\mathbb{X}[z] = 0$ are called the *zeros* of $\mathbb{X}[z]$. The *poles* are the roots of $\mathbb{P}[z] = 0$, or the values of z for which $\mathbb{X}[z] = \infty$. In some cases, the number of poles is allegedly equal to that of zeros by drawing the singularities (possibly of multiple order) at $z=\infty$. For example, we may allege that the z-transform expression (E5.2.2)

$$\mathbb{X}[z] = \frac{z}{(z-1)(z-1/2)}$$

has two (finite) poles at $z=1$ and $1/2$, and also two zeros, one finite zero at $z=0$ and one zero at infinity.

(2) Notice that, for polynomials with real-valued coefficients, the complex-valued singularities must occur in complex-conjugate pairs.

The ROC of a z-transform $\mathbb{X}[z]$ can be described in terms of its poles as follows:

[Remark 5.4] Pole Locations and Region of Convergence (ROC)

(1) It is obvious that there can be no poles of a z-transform $\mathbb{X}[z]$ inside the ROC since the z-transform does not converge at its poles. In fact, the ROC is bounded by poles or infinity, as illustrated by Examples 5.1 and 5.2 (Fig. 5.2).

(2) For a right-sided sequence $x_a[n]$ and a left-sided sequence $x_b[n]$, the ROCs for their z-transforms $\mathbb{X}_a[z]$ and $\mathbb{X}_b[z]$ are outside/inside the outermost/innermost pole, respectively:

$$\mathcal{R}_1 = \{z : |z| > \text{the maximum of } |a_i|\text{'s}\} \quad (z = a_i\text{: the poles of } \mathbb{X}_a[z]) \tag{5.1.4a}$$

(as illustrated by Example 5.2), or

$$\mathcal{R}_2 = \{z : |z| < \text{the minimum of } |b_i|\text{'s}\} \quad (z = b_i\text{: the poles of } \mathbb{X}_b[z]) \tag{5.1.4b}$$

For a two-sided sequence $x[n] = x_a[n] + x_b[n]$, the ROC is an annular region given by

$$\mathcal{R} = \mathcal{R}_1 \cap \mathcal{R}_2 = \{z : \max_i |a_i| < |z| < \min_j |b_j|\} \tag{5.1.4c}$$

5.2 PROPERTIES OF THE Z-TRANSFORM[Y-2], Section 4.2

In this section we discuss the properties of the z-transform that are useful in obtaining z-transform pairs and in applying the transform for the analysis of discrete-time linear time-invariant systems. The properties are very similar to those of the Laplace transform (Chapter 4) and Fourier transform (Chapter 6).

Most of the properties are commonly applicable to the bilateral and unilateral z-transforms. There are, however, a few exceptions, such as the time shifting property, time reversal property, and initial/final value theorems. In particular, it is the time shifting property that makes the unilateral z-transform very useful for solving difference equations with initial conditions.

5.2.1 Linearity

Let the z-transforms of two time sequences $x[n]$ and $y[n]$ be $\mathcal{Z}\{x[n]\} = \mathbb{X}[z]$ with ROC $\mathcal{R}_x$ and $\mathcal{Z}\{y[n]\} = \mathbb{Y}[z]$ with ROC $\mathcal{R}_y$. Then we can express the z-transform of their linear combination in terms of $\mathbb{X}[z]$ and $\mathbb{Y}[z]$ as

$$\alpha x[n] + \beta y[n] \overset{Z}{\leftrightarrow} \alpha \mathbb{X}[z] + \beta \mathbb{Y}[z] \text{ with ROC } \mathcal{R} \supset (\mathcal{R}_1 \cap \mathcal{R}_2) \tag{5.2.1}$$

where the ROC is generally the intersection of the two ROCs $\mathcal{R}_x$ and $\mathcal{R}_y$, but it can be larger when any pole of $\mathbb{X}[z]$ and $\mathbb{Y}[z]$ is cancelled by the zeros resulting from the addition.

5.2.2 Time Shifting — Real Translation

Let the bilateral z-transform of $x[n]$ be $\mathcal{Z}\{x[n]\} = \mathbb{X}[z]$ with ROC $\mathcal{R}_x$. Then the bilateral z-transform of a time-shifted version $x[n-n_1]$ is

$$x[n-n_1] \overset{Z}{\leftrightarrow} z^{-n_1}\mathbb{X}[z], \text{ ROC } \mathcal{R}_x \text{ (possibly with } z=0 \text{ or } \infty \text{ added or removed)} \tag{5.2.2}$$

The addition/removal of $z=\infty$ to/from the ROC occurs when a causal/non-causal sequence is shifted to become a non-causal/causal sequence. Similarly, the addition/removal of the origin $z=0$ to/from the ROC is owing to the possible transition between anticausal and non-anticausal sequences resulting from the shift operation.

On the other hand, the time-shifting property of the unilateral z-transform is as follows:

<**Case 1**> $x[n-n_1]$ with $n_1 > 0$ (delayed sequence)

$$\mathcal{Z}\{x[n-n_1]\} = \sum_{n=0}^{\infty} x[n-n_1]\, z^{-n} = \sum_{n=0}^{n_1-1} x[n-n_1]\, z^{-n} + \sum_{n=n_1}^{\infty} x[n-n_1]\, z^{-n}$$

$$\overset{n-n_1=m,\; n=m+n_1}{=} \sum_{m=-n_1}^{-1} x[m]\, z^{-(m+n_1)} + \sum_{m=0}^{\infty} x[m]\, z^{-(m+n_1)} = z^{-n_1}\left(\sum_{m=0}^{\infty} x[m]\, z^{-m} + \sum_{m=-n_1}^{-1} x[m]\, z^{-m}\right)$$

$$; \qquad x[n-n_1] \overset{Z}{\leftrightarrow} z^{-n_1}\left(X[z] + \sum_{m=-n_1}^{-1} x[m]\, z^{-m}\right) \tag{5.2.3a}$$

where the second term in the RHS will disappear when $x[n]$ is causal.

<**Case 2**> $x[n+n_1]$ with $n_1>0$ (advanced sequence)

$$\mathcal{Z}\{x[n+n_1]\}=\sum_{n=0}^{\infty}x[n+n_1]z^{-n}\overset{n+n_1=m,\,n=m-n_1}{=}\sum_{m=n_1}^{\infty}x[m]z^{-(m-n_1)}$$

$$=z^{n_1}\left(\sum_{m=0}^{\infty}x[m]z^{-m}-\sum_{m=0}^{n_1-1}x[m]z^{-m}\right)$$

$$;\qquad x[n+n_1]\overset{Z}{\leftrightarrow}z^{n_1}\left(X[z]-\sum_{m=0}^{n_1-1}x[m]z^{-m}\right)\tag{5.2.3b}$$

(**Example 5.5**) Applying Linearity and Time Shifting Properties of the Z-transform

For a rectangular pulse sequence of duration N given by

$$x[n]=u_s[n]-u_s[n-N]\overset{n=\cdots-1\ \ 0\ \ 1\ \cdots N-1\ N\ \cdots}{=}\{\cdots 0\ \ 1\ \ 1\cdots\ 1\ \ 0\cdots\},\tag{E5.5.1}$$

we can use Eq. (E5.1.1) (with a=1) together with the linearity (5.2.1) and time-shifting property (5.2.2) to get its z-transform as

$$\mathbb{X}[z]\underset{(5.2.1)\,\&(5.2.2)}{\overset{(E5.1.1)\text{ with }a=1}{=}}\frac{z}{z-1}-z^{-N}\frac{z}{z-1}=\frac{z^N-1}{z^{N-1}(z-1)}\tag{E5.5.2}$$

Note that $\mathbb{X}[z]$ has a multiple pole of order $N-1$ at z=0 and $N-1$ zeros at $z=e^{j2m\pi/N}$ (for m=1, 2, ..., $N-1$) on the unit circle where the zero at z=1 (resulting from the addition) cancels the pole (of $\mathcal{Z}\{u_s[n]\}=z/(z-1)$) at that point. Due to this pole–zero cancellation at z=1, the ROC of $\mathbb{X}[z]$ becomes $|z|>0$ (the entire z-plane except the origin z=0), which is larger than that of $\mathcal{Z}\{u_s[n]\}=z/(z-1)$, i.e., $|z|>1$.

5.2.3 Frequency Shifting — Complex Translation

Let the bilateral z-transform of $x[n]$ be $\mathcal{Z}\{x[n]\}=\mathbb{X}[z]$ with ROC $\mathcal{R}_x=\{z:r_x^-<|z|<r_x^+\}$. Then we have the z-transform of a frequency-shifted or modulated version $z_1^n x[n]$ as

$$e^{j\Omega_1 n}x[n]\overset{Z}{\leftrightarrow}\mathbb{X}[e^{-j\Omega_1}z]\ \text{with ROC}\ \mathcal{R}_x=\{z:r_x^-<|z|<r_x^+\}\tag{5.2.4a}$$

$$z_1^n x[n]\overset{Z}{\leftrightarrow}\mathbb{X}[z/z_1]\ \text{with ROC}\ |z_1|\,\mathcal{R}_x=\{z:|z_1|\,r_x^-<|z|<|z_1|\,r_x^+\}\tag{5.2.4b}$$

5.2.4 Time Reversal

Let the bilateral z-transform of $x[n]$ be $\mathcal{Z}\{x[n]\}=\mathbb{X}[z]$ with ROC $\mathcal{R}_x=\{z:r_x^-<|z|<r_x^+\}$. Then we have the bilateral z-transform of a time-reversed version $x[-n]$ as

$$x[-n]\overset{Z}{\leftrightarrow}\mathbb{X}[z^{-1}]\ \text{with ROC}\ 1/\mathcal{R}_x=\{z:1/r_x^+<|z|<1/r_x^-\}\tag{5.2.5}$$

where if $z\in\mathcal{R}_x$, then $1/z\in 1/\mathcal{R}_x$.

5.2.5 Real Convolution

Let the z-transforms of two time sequences $g[n]$ and $x[n]$ be $Z\{g[n]\}=\mathbb{G}[z]$ with ROC $\mathcal{R}_g$, and $Z\{x[n]\}=\mathbb{X}[z]$ with ROC $\mathcal{R}_x$, respectively. Then we can express the z-transform of their convolution in terms of $\mathbb{G}[z]$ and $\mathbb{X}[z]$ as

$$y[n]=g[n]*x[n]=\sum_{m=-\infty}^{\infty} g[m]x[n-m] \overset{Z}{\leftrightarrow} \mathbb{Y}[z]=\mathbb{G}[z]\,\mathbb{X}[z] \text{ with ROC } \mathcal{R}\supset(\mathcal{R}_g\cap\mathcal{R}_x). \quad (5.2.6)$$

(Proof)

$$Z\{g[n]*x[n]\} \overset{(5.1.1a)}{=} \sum_{n=-\infty}^{\infty}(g[n]*x[n])\,z^{-n} = \sum_{n=-\infty}^{\infty}\left(\sum_{m=-\infty}^{\infty} g[m]x[n-m]\right)z^{-n}$$

$$=\sum_{m=-\infty}^{\infty}\left(g[m]\sum_{n=-\infty}^{\infty}x[n-m]\,z^{-(n-m)}z^{-m}\right)$$

$$\overset{n-m\to n,\ n\to n+m}{=} \sum_{m=-\infty}^{\infty}\left(g[m]z^{-m}\sum_{n=-\infty}^{\infty}x[n]\,z^{-n}\right) \overset{(5.1.1a)}{=} \mathbb{G}[z]\,\mathbb{X}[z]$$

This convolution property holds for causal sequences as well as non-causal sequences, which implies that the unilateral z-transform also has the same property. It is very useful for describing the input—output relationship of a discrete-time LTI (linear time-invariant) system with the input $x[n]$, output $y[n]$, and *impulse response* $g[n]$ (the output to a unit impulse input $\delta[n]$), where the z-transform of the impulse response, $Z\{g[n]\}=\mathbb{G}[z]$, is called the *system* or *transfer function* and the *unit impulse sequence* is defined by

$$\delta[n]=\begin{cases}1 & \text{for } n=0\\ 0 & \text{for } n\neq 0\end{cases} \quad (5.2.7)$$

Note that the unit impulse sequence is related to the unit step sequence as

$$\delta[n]=u_s[n]-u_s[n-1] \text{ and } u_s[n]=\sum_{m=-\infty}^{n}\delta[m]=\begin{cases}1 & \text{for } n\geq 0\\ 0 & \text{for } n<0\end{cases} \quad (5.2.8a,b)$$

5.2.6 Complex Convolution

Let the z-transforms of $x[n]$ and $y[n]$ be $Z\{x[n]\}=\mathbb{X}[z]$ with ROC $\mathcal{R}_x=\{z: r_x^- <|z|< r_x^+\}$, and $Z\{y[n]\}=\mathbb{Y}[z]$ with ROC $\mathcal{R}_y=\{z: r_y^- <|z|< r_y^+\}$, respectively. Then the z-transform of the product of the two sequences can be expressed in terms of $\mathbb{X}[z]$ and $\mathbb{Y}[z]$ as

$$x[n]y[n] \overset{Z}{\leftrightarrow} \frac{1}{2\pi j}\oint_{C_1}\mathbb{X}[z/v]\mathbb{Y}[v]v^{-1}dv \text{ with ROC } \mathcal{R}=\{z: r_x^- r_y^- <|z|< r_x^+ r_y^+\} \quad (5.2.9a)$$

$$x[n]y[n] \overset{Z}{\leftrightarrow} \frac{1}{2\pi j}\oint_{C_2}\mathbb{X}[v]\mathbb{Y}[z/v]v^{-1}dv \text{ with ROC } \mathcal{R}=\{z: r_x^- r_y^- <|z|< r_x^+ r_y^+\} \quad (5.2.9b)$$

where $\oint_{C_i}$ means the complex integral along a closed contour C_i within the intersection of the ROCs of $\mathbb{X}[z/v]$ and $\mathbb{Y}[v]$, or $\mathbb{X}[v]$, and $\mathbb{Y}[z/v]$ (see [O-2], Sec. 2.3.9 for its proof).

5.2.7 Complex Differentiation

Let the z-transform of $x[n]$ be $Z\{x[n]\} = \mathbb{X}[z]$ with ROC $\mathcal{R}_x$. Then the z-transform of $nx[n]$ can be expressed in terms of $\mathbb{X}[z]$ as

$$nx[n] \overset{Z}{\leftrightarrow} -z\frac{d}{dz}\mathbb{X}[z] \text{ with ROC } \mathcal{R}_x \tag{5.2.10}$$

This can be proved by differentiating the definition of the z-transform (5.1.1) w.r.t. z.

5.2.8 Partial Differentiation

If a z-transform pair is given in the form of $Z\{x(nT,a)\} = \mathbb{X}(z,a)$ with a parameter a, we can differentiate it w.r.t. a to obtain another z-transform pair as

$$\frac{\partial}{\partial a}x(nT,a) \overset{Z}{\leftrightarrow} \frac{\partial}{\partial a}X(z,a) \tag{5.2.11}$$

(**Example 5.6**) Complex Differentiation and Partial Differentiation

To find the z-transform of $y[n] = n^2a^n u_s[n]$, we can apply the complex differentiation property (5.2.10) twice for Eq. (E5.1.1) to write

$$na^n u_s[n] \overset{Z}{\leftrightarrow} -z\frac{d}{dz}\left(\frac{z}{z-a}\right) \overset{(C.31)}{=} -z\frac{(z-a)-z}{(z-a)^2} = \frac{az}{(z-a)^2} \tag{E5.6.1}$$

$$n^2a^n u_s[n] \overset{Z}{\leftrightarrow} -z\frac{d}{dz}\left(\frac{az}{(z-a)^2}\right) \overset{(C.31)}{=} -z\frac{a(z-a)^2-2az(z-a)}{(z-a)^4} = \frac{az(z+a)}{(z-a)^3} \tag{E5.6.2}$$

Alternatively, we can apply the partial differentiation property for Eq. (E5.1.1) to get the same results:

$$na^{n-1} u_s[n] \overset{Z}{\leftrightarrow} \frac{\partial}{\partial a}\left(\frac{z}{z-a}\right) \overset{(C.31)}{=} \frac{z}{(z-a)^2} \tag{E5.6.3}$$

$$n^2a^{n-1} u_s[n] \overset{Z}{\leftrightarrow} \frac{\partial}{\partial a}\left(\frac{az}{(z-a)^2}\right) \overset{(C.31)}{=} \frac{z(z-a)^2+2az(z-a)}{(z-a)^4} = \frac{z(z+a)}{(z-a)^3} \tag{E5.6.4}$$

Multiplying both sides of these equations by a yields Eqs. (E5.6.1) and (E5.6.2), which are listed in Table A.9(10) and (12) (see Appendix A).

5.2.9 Initial Value Theorem

For a causal sequence $x[n]$ such that $x[n] = 0 \ \forall n < 0$, we can get its initial value $x[0]$ from its unilateral z-transform as

$$x[0] = \lim_{z\to\infty} X[z] \tag{5.2.12}$$

This can easily be shown by substituting $z=\infty$ into the definition (5.1.1b) of the unilateral z-transform.

5.2.10 Final Value Theorem

For a causal sequence $x[n]$ such that $x[n]=0\ \forall\, n<0$, we can get its final value $x[\infty]$ from its unilateral z-transform as

$$x[\infty] = \lim_{z\to 1}(z-1)X[z] = \lim_{z\to 1}(1-z^{-1})X[z] \tag{5.2.13}$$

This requires that $x[n]$ should converge, or equivalently, all the poles of its z-transform $X[z]$ should lie inside the unit circle, possibly with the exception of a simple pole at $z=1$.

(Proof)

We can use the definition (5.1.1b) of the unilateral z-transform to write

$$\mathcal{Z}\{x[n+1]-x[n]\} = \lim_{k\to\infty}\left(\sum_{n=0}^{k} x[n+1]z^{-n} - \sum_{n=0}^{k} x[n]z^{-n}\right)$$

$$= \lim_{k\to\infty}\left(\sum_{n=1}^{k} x[n]\,z^{-(n-1)} + x[k+1]z^{-k} - x[0] - \sum_{n=1}^{k} x[n]\,z^{-n}\right)$$

On the other hand, from the time shifting property (5.2.3b) of the unilateral z-transform, we can write

$$\mathcal{Z}\{x[n+1]-x[n]\} \overset{(5.2.1)}{=} \mathcal{Z}\{x[n+1]\} - \mathcal{Z}\{x[n]\} \overset{(5.2.3b)}{=} z(X[z]-x[0]) - X[z]$$

$$= (z-1)X[z] - z\,x[0]$$

Noting that these two equations are commonly the z-transform of $(x[n+1]-x[n])$, we can equate their RHSs and substitute $z=1$ to get the desired result (5.2.13).

5.3 THE INVERSE Z-TRANSFORM[Y-2], Section 4.3

In this section we consider how to find $x[n]$ whose z-transform is given as $X[z]$ with its ROC. From the complex variable theory, the *inverse z-transform* (*IZT*) formula can be derived (see [O-2], Sec. 2.2) as

$$x[n] = \frac{1}{2\pi j}\oint_C \mathbb{X}[z]\,z^{-n}dz \tag{5.3.1}$$

where $\oint_C$ means the complex integral along a closed contour C (within the ROC of $\mathbb{X}[z]$) encircling the origin of the z-plane in the counterclockwise direction. It is, however, difficult to directly evaluate this integral, and therefore, we resort to alternative procedures for finding the inverse z-transform, which are similar to those for finding the inverse Laplace transform.

Noting that almost every basic z-transform listed in Table A.9 has the complex variable z in the numerator, we apply the same procedure as for the inverse Laplace transform (Sec. 4.3) to get the partial fraction expansion (PFE) on $X[z]/z$, and then multiply both sides by z so that we can directly use the z-transform table to get the inverse z-transform.

More specifically, let $X[z]/z$ be a rational function as

$$\frac{X[z]}{z}=\frac{Q_1[z]}{P[z]}=\frac{b_M z^M+\ldots+b_1 z+b_0}{a_N z^N+\ldots+a_1 z+a_0} \tag{5.3.2}$$

where M and N are the degrees of the numerator and denominator polynomials, respectively. If $M \geq N$, we divide $Q_1[z]$ by $P[z]$, starting with the highest power of z, to produce the remainder polynomial $Q[z]$ of degree less than N, so that we can rewrite Eq. (5.3.2) as

$$\frac{X[z]}{z}=\frac{Q[z]}{P[z]}+c_{M-N}z^{M-N}+\ldots+c_1 z+c_0 \tag{5.3.3}$$

If $M<N$, we have only the first term on the RHS where $c_i=0$ for all i. Now, for the purpose of illustration, we assume that all the poles of $Q[z]/P[z]$ are simple except for one multiple pole of order L at $z=p$, so that we can write $Q[z]/P[z]$ in the following form:

$$\frac{Q[z]}{P[z]}=\left(\sum_{i=1}^{N-L}\frac{r_i}{z-p_i}\right)+\frac{r_{N-L+1}}{z-p}+\ldots+\frac{r_N}{(z-p)^L}+K \tag{5.3.4}$$

where

$$r_i=(z-p_i)\left.\frac{Q[z]}{P[z]}\right|_{z=p_i},\quad i=1,2,\ldots,N-L \tag{5.3.5a}$$

$$r_{N-l}=\frac{1}{l!}\frac{d^l}{dz^l}\left.\left\{(z-p)^L\frac{Q[z]}{P[z]}\right\}\right|_{z=p},\quad l=0,1,\ldots,L-1 \tag{5.3.5b}$$

Now, substituting Eq. (5.3.4) into Eq. (5.3.3), multiplying both sides by z, and using the z-transform Table A.9, we can obtain the inverse z-transform of $X[z]$ as

$$\begin{aligned}x[n]=&\left\{\sum_{i=1}^{N-L}r_i\,p_i^n+r_{N-L+1}\,p^n+r_{N-L+2}\,n\,p^{n-1}+\ldots+r_N\frac{n!}{(L-1)!(n-L+1)!}p^{n-L+1}\right\}u_s[n]\\&+\sum_{i=0}^{M-N}c_i\,\delta[n+i+1]\end{aligned} \tag{5.3.6}$$

(Example 5.7) IZT (Inverse Z-transform) by PFE (Partial Fraction Expansion)

(a) Let us find the inverse z-transform of

$$X[z]=\frac{z}{z^3-2z^2+(5/4)z-1/4}=\frac{z}{(z-1)(z-1/2)^2}\quad\text{with ROC } \mathcal{R}=\{z:|z|>1\} \tag{E5.7.1}$$

We first divide this by z and then expand it into partial fractions:

$$\frac{X[z]}{z}=\frac{r_1}{z-1}+\frac{r_2}{z-1/2}+\frac{r_3}{(z-1/2)^2}\overset{\text{(E5.7.3)}}{=}\frac{4}{z-1}-\frac{4}{z-1/2}-\frac{2}{(z-1/2)^2} \tag{E5.7.2}$$

where the coefficient of each term can be found by using Eq. (5.3.5) as

$$r_1 \overset{(5.3.5a)}{=} (z-1)\frac{X[z]}{z}\bigg|_{z=1} = 4 \tag{E5.7.3a}$$

$$r_2 \underset{l=1}{\overset{(5.3.5b)}{=}} \frac{d}{dz}\left((z-1/2)^2\frac{X[z]}{z}\right)\bigg|_{z=1/2} = \frac{d}{dz}\left(\frac{1}{z-1}\right)\bigg|_{z=1/2} = -\frac{1}{(z-1)^2}\bigg|_{z=1/2} = -4 \tag{E5.7.3b}$$

$$r_3 \underset{l=0}{\overset{(5.3.5b)}{=}} (z-1/2)^2\frac{X[z]}{z}\bigg|_{z=1/2} = \frac{1}{z-1}\bigg|_{z=1/2} = -2 \tag{E5.7.3c}$$

Now, moving the z (which we have saved for now) from the LHS back into the RHS yields

$$X[z] = 4\frac{z}{z-1} - 4\frac{z}{z-1/2} - 4\frac{(1/2)z}{(z-1/2)^2} \tag{E5.7.4}$$

Since every term of this expression has a z in its numerator, we can easily use Table A.9(3), (5), and (10) to write the inverse z-transform as

$$x[n] = \left(4 - 4(\frac{1}{2})^n - 4n(\frac{1}{2})^n\right)u_s[n] \tag{E5.7.5}$$

where the right-sided sequences are chosen over the left-sided ones, since the given ROC is not the inside but the outside of a circle.

We can use the MATLAB command '`residue()`' or '`residuez()`' to get the PFE, and '`iztrans()`' to obtain the whole inverse z-transform. It should, however, be noted that '`iztrans()`' might not work properly for high-degree rational functions.

```
>>Nz=[0 1]; Dz=poly([1 1/2 1/2]), [r,p,k]=residue(Nz,Dz);
>>[r p],k % Eq.(E5.7.2)
   r =   4.0000     p =   1.0000     % Eq.(E5.7.2a)   4/(z-1)
        -4.0000           0.5000     % Eq.(E5.7.2b) -4/(z-0.5)
        -2.0000           0.5000     % Eq.(E5.7.2c) -2/(z-0.5)^2
   k = []

>>syms z, x=iztrans(z/(z^3-2*z^2+1.25*z-0.25)) % Eq.(E5.7.1)
   x = 4-4*(1/2)^n-4*(1/2)^n*n    % Eq.(E5.7.5)
```

(b) Let us find the inverse z-transform of

$$X[z] = \frac{3z}{z^2-(1/4)z-1/8} = \frac{3z}{(z-1/2)(z+1/4)} \tag{E5.7.6}$$

with one of the following three ROCs:

$$\mathcal{R}_1 = \{z : |z| > \frac{1}{2}\},\ \mathcal{R}_2 = \{z : |z| < \frac{1}{4}\},\ \text{and}\ \mathcal{R}_3 = \{z : \frac{1}{4} < |z| < \frac{1}{2}\} \tag{E5.7.7}$$

We first divide this by z, expand it into partial fractions, and then move z back to the RHS:

$$\frac{X[z]}{z} = \frac{4}{z-1/2} - \frac{4}{z+1/4}\ ; \quad X[z] = 4\frac{z}{z-1/2} - 4\frac{z}{z+1/4} \tag{E5.7.8}$$

Now, depending on the ROC, we use Table A.9(5) or (6) to write the inverse z-transform as follows:

$$\text{(i)}\quad \mathcal{R}_1=\{z:|z|>\frac{1}{2}\}:\quad x[n]=4(\frac{1}{2})^n u_s[n]-4(-\frac{1}{4})^n u_s[n] \tag{E5.7.9a}$$

$$\text{(ii)}\quad \mathcal{R}_2=\{z:|z|<\frac{1}{4}\}:\quad x[n]=-4(\frac{1}{2})^n u_s[-n-1]+4(-\frac{1}{4})^n u_s[-n-1] \tag{E5.7.9b}$$

$$\text{(iii)}\quad \mathcal{R}_3=\{z:\frac{1}{4}<|z|<\frac{1}{2}\}:\quad x[n]=-4(\frac{1}{2})^n u_s[-n-1]-4(-\frac{1}{4})^n u_s[n] \tag{E5.7.9c}$$

```
>>syms z, x=iztrans(3*z/(z^2-0.25*z-1/8)) % Eq.(E5.7.6)
  x = 4*(1/2)^n-4*(-1/4)^n % Eq.(E5.7.9a) just a right-sided sequence
```

(Example 5.8) IZT by PFE

Let us find the inverse z-transform of

$$X[z]=\frac{2z^2}{z^2-z+1/2}=\frac{2z^2}{(z-0.5-j0.5)(z-0.5+j0.5)}\quad \text{with ROC } \mathcal{R}=\{z:|z|>1\} \tag{E5.8.1}$$

We first divide this by z and then expand it into partial fractions:

$$\frac{X[z]}{z}=\frac{r_R+jr_I}{z-0.5-j0.5}+\frac{r_R-jr_I}{z-0.5+j0.5}=\frac{1-j}{z-0.5-j0.5}+\frac{1+j}{z-0.5+j0.5} \tag{E5.8.2}$$

Now, moving the z from the LHS back into the RHS yields

$$X[z]=\frac{(1-j)z}{z-0.5-j0.5}+\frac{(1+j)z}{z-0.5+j0.5} \tag{E5.8.3}$$

Then we can use Table A.9(5) to write the inverse z-transform as

$$\begin{aligned} x[n] &= (1-j)(0.5+j0.5)^n u_s[n] + (1+j)(0.5-j0.5)^n u_s[n] \\ &= \sqrt{2}\left(e^{-j\pi/4}\sqrt{2}^{-n}e^{jn\pi/4}+e^{j\pi/4}\sqrt{2}^{-n}e^{-jn\pi/4}\right)u_s[n] \\ &= \sqrt{2}^{-n+3}\cos((n-1)\frac{\pi}{4})u_s[n] \end{aligned} \tag{E5.8.4}$$

As a nice alternative for $X[z]$ having complex conjugate poles like (E5.8.1), we can decompose it into the following form, which can be matched exactly with some element of the z-transform table:

$$\begin{aligned} X[z] &= \frac{2z^2}{z^2-z+1/2}=\frac{2z(z-1/2)}{(z-1/2)^2+(1/2)^2}+\frac{2(1/2)z}{(z-1/2)^2+(1/2)^2} \\ &= 2\frac{z(z-r\cos\Omega_1)}{(z-r\cos\Omega_1)^2+r^2\sin^2\Omega_1}+2\frac{zr\sin\Omega_1}{(z-r\cos\Omega_1)^2+r^2\sin^2\Omega_1} \end{aligned} \tag{E5.8.5}$$

where $r=1/\sqrt{2}$, $\cos\Omega_1=1/\sqrt{2}$, $\sin\Omega_1=1/\sqrt{2}$, and $\Omega_1=\pi/4$. Then we can use Table A.9(18) and (17) to obtain the same result as (E5.8.4):

$$x[n]=2\{r^n\cos(\Omega_1 n)+r^n\sin(\Omega_1 n)\}u_s[n]=2\left\{\sqrt{2}^{-n}\cos(\frac{\pi}{4}n)+\sqrt{2}^{-n}\sin(\frac{\pi}{4}n)\right\}u_s[n]$$

$$\overset{(C.18)}{=}\sqrt{2}^{-n+3}\cos\left((n-1)\frac{\pi}{4}\right)u_s[n]\overset{(C.2)}{=}\sqrt{2}^{-n+3}\sin\left((n+1)\frac{\pi}{4}\right)u_s[n] \tag{E5.8.6}$$

(Note) The IZT solution presented by '`iztrans()`' may not be elegant like this.

```
syms z; x=iztrans(2*z^2/(z^2-z+1/2)) % IZT of Eq.(E5.8.1)
n=1:10; xn=2.^(-(n-3)/2).*cos((n-1)*pi/4); % (E5.8.6)
stem(n,xn) % Plot the discrete-time sequence xn.
[r,p,k]=residue([2 0],[1 -1 1/2]) % PFE
xn1=real(r.'*[p(1).^n; p(2).^n]); % Eq.(E5.8.4) as an alternative
hold on, stem(n,xn1,'r') % Plot xn1 with the existing graph of xn kept.
```

5.4 USING THE Z-TRANSFORM

We can use the z-transform to deal with a high-order linear difference equation like

$$y[n+4]+a_3y[n+3]+a_2y[n+2]+a_1y[n+1]+a_0y[n]=u[n] \quad (5.4.1)$$
$$\text{with initial conditions } y[0], y[1], y[2], \text{ and } y[3]$$

in a very efficient way, by rewriting it in the form of state and output equations as

$$\mathbf{x}[n+1]=A\mathbf{x}[n]+\mathbf{b}u[n] \text{ with } \mathbf{x}[0]=\mathbf{x}_0=[y[0]\;\; y[1]\;\; y[2]\;\; y[3]]^T \quad (5.4.2a)$$

$$y[n]=C\,\mathbf{x}[n] \quad (5.4.2b)$$

with

$$\mathbf{x}[n]=\begin{bmatrix} x_1[n] \\ x_2[n] \\ x_3[n] \\ x_4[n] \end{bmatrix}=\begin{bmatrix} y[n] \\ y[n+1] \\ y[n+2] \\ y[n+3] \end{bmatrix},\; A=\begin{bmatrix} 0 & 1 & 0 & 0 \\ 0 & 0 & 1 & 0 \\ 0 & 0 & 0 & 1 \\ -a_0 & -a_1 & -a_2 & -a_3 \end{bmatrix},\; \mathbf{b}=\begin{bmatrix} 0 \\ 0 \\ 0 \\ 1 \end{bmatrix},\; \text{and } C=[1\;0\;0\;0] \quad (5.4.3)$$

where $x[n]$ and A are referred to as the state vector and system matrix, respectively. Using the time shifting property (5.2.3b), we take the z-transform of both sides of Eq. (5.4.2a):

$$z\,(X[z]-\mathbf{x}(0))=AX[z]+\mathbf{b}U[z] \text{ with } X[z]=\mathcal{Z}\{x[n]\},\;\; U[z]=\mathcal{Z}\{u[n]\}$$
$$;\; [zI-A]X[z]=z\mathbf{x}[0]+\mathbf{b}U[z];\quad X[z]=[zI-A]^{-1}[z\,\mathbf{x}[0]+\mathbf{b}U[z]] \quad (5.4.4)$$

Then we can take the z-transform of Eq. (5.4.2b), substitute Eq. (5.4.4) for $X[z]$ to get $Y[z]$, and take its inverse z-transform to get the t-domain solution $y[n]$:

$$Y[z]\overset{(5.4.2b)}{=}CX[z]\overset{(5.4.4)}{=}C[zI-A]^{-1}[z\mathbf{x}[0]+\mathbf{b}U[z]];\;\; y[n]=\mathcal{Z}^{-1}\{Y[z]\} \quad (5.4.5)$$

(Example 5.9) Using the Z-transform to Solve a Linear Difference Equation

Let us solve the following difference equation:

$$y[n+2]-\frac{1}{4}y[n+1]-\frac{1}{8}y[n]=5u_s[n],\;\; n\ge 0 \quad (E5.9.1)$$

with the initial conditions $y[0]=20$ and $y[1]=5$.

We make use of the time shifting property (5.2.3b) of the z-transform and Eq. (E5.1.1) to write the z-transform of both sides of Eq. (E5.9.1) as

$$z^2(Y[z]-y[0]-z^{-1}y[1])-\frac{1}{4}z(Y[z]-y[0])-\frac{1}{8}Y[z]=5\frac{z}{z-1}$$

$$;(z^2-\frac{1}{4}z-\frac{1}{8})Y[z]=\frac{5z}{z-1}+z^2y[0]+zy[1]-\frac{1}{4}zy[0]=\frac{5z}{z-1}+20z^2=5z\frac{4z^2-4z+1}{z-1} \quad \text{(E5.9.2)}$$

This algebraic equation can be solved for $Y[z]$ as

$$Y[z]=\frac{5z(4z^2-4z+1)}{(z-1)(z^2-z/4-1/8)}=\frac{5z(4z^2-4z+1)}{(z-1)(z-1/2)(z+1/4)} \quad \text{(E5.9.3)}$$

An alternative for obtaining this solution is to write the linear difference equation in state equation form as

$$\mathbf{x}[n+1]\overset{(5.4.2a)}{\underset{(E5.9.1)}{=}}A\,\mathbf{x}[n]+\mathbf{b}\,u[n]\;;\begin{bmatrix}x_1[n+1]\\x_2[n+1]\end{bmatrix}=\begin{bmatrix}0&1\\1/8&1/4\end{bmatrix}\begin{bmatrix}x_1[n]\\x_2[n]\end{bmatrix}+\begin{bmatrix}0\\5\end{bmatrix}\text{ with }\begin{bmatrix}x_1[0]\\x_2[0]\end{bmatrix}=\begin{bmatrix}y[0]\\y[1]\end{bmatrix}=\begin{bmatrix}20\\5\end{bmatrix} \quad \text{(E5.9.4)}$$

and use the z-transform solution (5.4.5) to get the same result as Eq. (E5.9.3):

$$Y[z]\overset{(5.4.5)}{=}CX[z]\overset{(5.4.4)}{=}C[zI-A]^{-1}(z\mathbf{x}[0]+\mathbf{b}U[z])=[1\;\;0]\begin{bmatrix}z&-1\\-1/8&z-1/4\end{bmatrix}^{-1}\left(\begin{bmatrix}20z\\5z\end{bmatrix}+\begin{bmatrix}0\\1\end{bmatrix}\frac{5z}{z-1}\right)$$

$$=\frac{[1\;\;0]}{(z-1)(z^2-z/4-1/8)}\begin{bmatrix}z-1/4&1\\1/8&z\end{bmatrix}\begin{bmatrix}20z(z-1)\\5z^2\end{bmatrix}=\frac{[1\;\;0]}{(z-1)(z^2-z/4-1/8)}\begin{bmatrix}5z(4z^2-4+1)\\z(5z^2+5z/2-5/2)\end{bmatrix}$$

$$=\frac{5z(4z^2-4z+1)}{(z-1)(z^2-z/4-1/8)} \quad \text{(E5.9.5)}$$

We expand this z-domain solution into partial fractions as

$$\frac{Y[z]}{z}\overset{(E5.9.3)}{\underset{\text{or}(E5.9.5)}{=}}\frac{5(4z^2-4z+1)}{(z-1)(z-1/2)(z+1/4)}=\frac{r_1}{z-1}+\frac{r_2}{z-1/2}+\frac{r_3}{z+1/4}=\frac{8}{z-1}+\frac{0}{z-1/2}+\frac{12}{z+1/4} \quad \text{(E5.9.6)}$$

where the coefficients are found by using the formula (5.3.5a) as

$$r_1\overset{(5.3.5a)}{=}(z-1)\frac{Y[z]}{z}\bigg|_{z=1}\overset{(E5.9.3)}{=}\frac{4z^2-4z+1}{(z-1)(z-1/2)(z+1/4)}\bigg|_{z=1}=8$$

$$r_2\overset{(5.3.5a)}{=}(z-\frac{1}{2})\frac{Y[z]}{z}\bigg|_{z=1/2}\overset{(E5.9.3)}{=}\frac{4z^2-4z+1}{(z-1)(z+1/4)}\bigg|_{z=1/2}=0 \quad \text{(E5.9.7)}$$

$$r_3\overset{(5.3.5a)}{=}(z+\frac{1}{4})\frac{Y[z]}{z}\bigg|_{z=-1/4}\overset{(E5.9.3)}{=}\frac{4z^2-4z+1}{(z-1)(z-1/2)}\bigg|_{z=-1/4}=12$$

This PFE (partial fraction expansion) can be obtained by using the MATLAB command 'residuez()':

```
>>syms z; Nz=5*[4 -4 1]; Dz=poly([1 1/2 -1/4]);
>>[r,p,k]=residue(Nz,Dz) % Partial fraction expansion of Eq.(E5.9.6)
   r =  8.0000          p =   1.0000          k =  0
             0                 0.5000
       12.0000                -0.2500
```

We multiply the PFE (E5.9.6) by z to write

$$Y[z] \overset{(E5.9.4)}{\underset{(E5.9.5)}{=}} \frac{8z}{z-1} + \frac{12z}{z+1/4} \tag{E5.9.8}$$

and use Eq. (E5.1.1) or Table A.8(6) to take the inverse z-transform of Eq. (E5.9.3) to write the t-domain solution $y[n]$ as

$$y[n] = \{8 + 12(-1/4)^n\}u_s[n] \tag{E5.9.9}$$

This inverse z-transform can be directly obtained by using the MATLAB command '`iztrans()`' as

```
>>y1=iztrans(5*z*(4*z^2-4*z+1)/(z-1)/(z^2-z/4-1/8)) % IZT of Eq.(E5.9.3)
    y = 12*(-1/4)^n + 8    % like Eq.(E5.9.9)
```

or alternatively,

```
>>syms z; Num=5*[4 -4 1]; Den=poly([1 1/2 -1/4]);
>>Nz=poly2sym(Num,z); Dz=poly2sym(Den,z) %Numerator/Denominator polynomial
>>y=iztrans(Nz/Dz) % IZT
```

This IZT can also be obtained by using Eq. (5.4.5) with Eq. (5.4.4) for the state equation (E5.9.4):

```
>>syms z;
>>A=[0 1; 1/8 1/4]; b=[0;1]; C=[1 0]; x0=[20;5]; Uz=5*z/(z-1); Eq.(E5.9.4)
>>Yz=C*((z*eye(size(A))-A)\(z*x0+b*Uz)) % Eq.(5.4.4) with Eq.(5.4.5)
>>y=iztrans(Yz) % IZT of Eq.(E5.9.5)
    y = 12*(-1/4)^n + 8    % like Eq.(E5.9.9)
```

If you want to plot $y[n]$, run the following MATLAB statements:

```
>>n=0:10; yn=eval(y); stem(n,yn) % plot y for n=0:10
```

5.5 TRANSFER FUNCTION OF A DISCRETE-TIME SYSTEM

The input–output relationship of a single-input single-output (SISO) discrete-time (DT) system with input $u[n]$ and output $y[n]$ is often described by a linear difference equation like

$$y[n+3] + a_2 y[n+2] + a_1 y[n+1] + a_0 y[n] = u[n] \tag{5.5.1}$$

with initial conditions $y[0]$, $y[1]$, and $y[2]$

With zero initial conditions $y[0]=0$, $y[1]=0$, and $y[2]=0$, we can use the time shifting property of the z-transform to write the z-transform of Eq. (5.5.1) as

$$z^3 Y[z] + a_2 z^2 Y[z] + a_1 z Y[z] + a_0 Y[z] = U[z] \tag{5.5.2}$$

which yields the transformed (z-domain) input-output relationship as

$$G[z] = \frac{Y[z]}{U[z]} = \frac{1}{z^3 + a_2 z^2 + a_1 z + a_0} \tag{5.5.3}$$

This is called the *transfer* or *system function* of the system, which is defined as the ratio of the transformed output $Y[z]=\mathcal{Z}\{y[n]\}$ to the transformed input $U[z]=\mathcal{Z}\{u[n]\}$.

Just like Eq. (4.5.4) for a CT system, the transfer function of the DT system can also be obtained from another z-transform solution (5.4.5) with zero initial condition $\mathbf{x}[0]=\mathbf{0}$:

$$Y[z]\overset{(5.4.5)}{=}C[zI-A]^{-1}\left[z\mathbf{x}[0]+\mathbf{b}U[z]\right]\overset{\mathbf{x}(0)=0}{\rightarrow}Y[z]=C[zI-A]^{-1}\mathbf{b}U[z] \tag{5.5.4}$$

$$;\ G[z]=\frac{Y[z]}{U[z]}=C[zI-A]^{-1}\mathbf{b}=[1\ \ 0\ \ 0]\begin{bmatrix} z & -1 & 0 \\ 0 & z & -1 \\ a_0 & a_1 & z+a_2 \end{bmatrix}^{-1}\begin{bmatrix} 0 \\ 0 \\ 1 \end{bmatrix}=\frac{1}{z^3+a_2z^2+a_1z+a_0} \tag{5.5.5}$$

Once the transfer function $G[z]$ is found or given, the output $y[n]$ of the system to any z-transformable input $u[n]$ can be computed as

$$Y[z]=G[z]U[z];\ \ y[n]=\mathcal{Z}^{-1}\{Y[z]\} \tag{5.5.6}$$

How about the output of the system to an impulse input $u[n]=\delta[n]$, whose z-transform is $U[z]=\mathcal{Z}\{\delta[n]\}\overset{\text{Table A.8(1)}}{=}1$?

$$Y[z]\underset{U[z]=1}{\overset{(5.5.6)}{=}}G[z];\ \ y[n]=g[n]=\mathcal{Z}^{-1}\{G[z]\} \tag{5.5.7}$$

This implies that the transfer function of a DT system is the z-transform of the *impulse response*, which is defined as the output to an impulse input (with zero initial conditions).

(Example 5.10) Transfer Function, Impulse Response, and Step Response of a DT System

Find the transfer function and impulse response of a DT system with input $u[n]$ and output $y[n]$, whose input–output relationship is described by the following difference equation:

$$y[n+2]-\frac{1}{4}y[n+1]-\frac{1}{8}y[n]=u[n],\ \ n\ge 0 \tag{E5.10.1}$$

We can take the z-transform of Eq. (E5.10.1) with zero initial conditions $y[0]=0$ and $y[1]=0$ to get the transfer function $G[z]$:

$$(z^2-\frac{1}{4}z-\frac{1}{8})Y[z]=U[z]\rightarrow G[z]=\frac{Y[z]}{U[z]}=\frac{1}{z^2-z/4-1/8} \tag{E5.10.2}$$

We can take the IZT of this transfer function to get the impulse response $g[n]$:

$$g[n]=\mathcal{Z}^{-1}\{G[z]\}=\mathcal{Z}^{-1}\left\{\frac{1}{z^2-z/4-1/8}\right\}\overset{\text{PFE}}{=}\mathcal{Z}^{-1}\left\{\frac{4/3}{z-1/2}-\frac{4/3}{z+1/4}\right\}$$
$$\underset{\text{Time shifting or A.7(2)}}{\overset{\text{(E5.1.1) or A.9(5)}}{=}}\frac{4}{3}\left((\frac{1}{2})^{n-1}-(-\frac{1}{4})^{n-1}\right)u_s[n-1] \tag{E5.10.3}$$

We can use the MATLAB command '`iztrans()`' to get the IZT of $G[z]$:

```
>>syms z; g1 = iztrans(1/(z^2-z/4-1/8)) % IZT of Eq.(E5.9.2)
  g1  =   (8*(1/2)^n)/3 + (16*(-1/4)^n)/3 - 8*kroneckerDelta(n,0)
```

which implies the sequence

$$g_1[n]\ =\ \frac{8}{3}(\frac{1}{2})^n+\frac{16}{3}(-\frac{1}{4})^n-8\delta[n] \tag{E5.10.4}$$

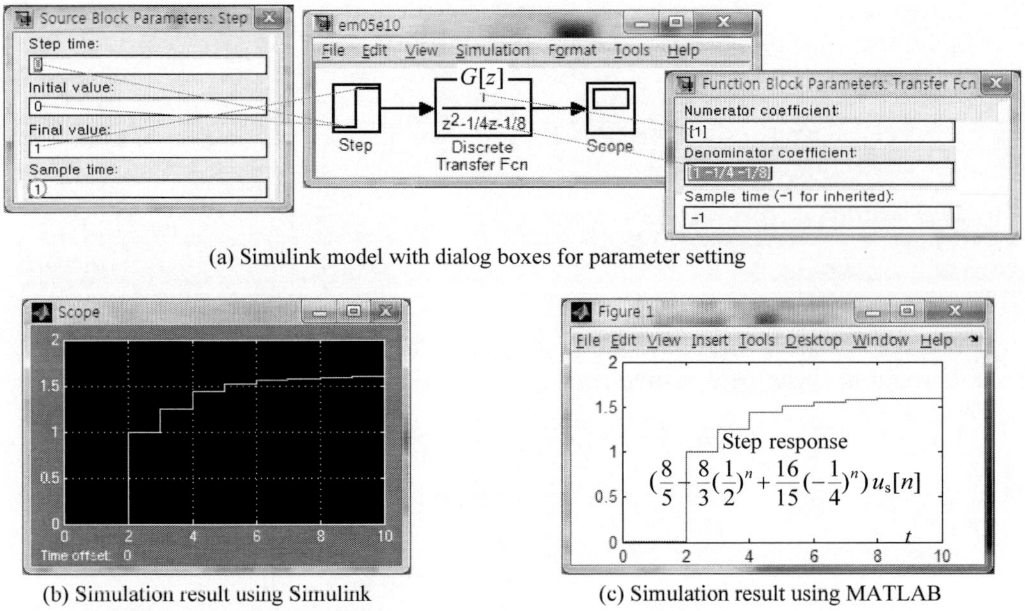

Figure 5.4 Step response simulated using MATLAB and Simulink ("em05e10_sim.mdl")

Although these two expressions (E5.10.3) and (E5.10.4) look different, they conform with each other, which can be checked by running the following MATLAB statements:

```
>>n=0:10; g1n=eval(g1) % Numeric value of Eq.(E5.10.4) for n=0:10
>>gn=4/3*((1/2).^(n-1)-(-1/4).^(n-1)).*(n>=1) % Eq.(E5.10.3) for n=0:10
>>discrepancy=norm(gn-g1n) % Are gn and g1n equal?
   discrepancy =   0
>>stem(n,gn), hold on, stem(n,g1n,'r.:') % plot Eqs.(E5.10.3),(E5.10.4)
```

Also, noting that the z-transform of the unit step sequence $u_s[n]$ is $U[z]=\mathcal{Z}\{u_s[z]\}=z/(z-1)$, we can use Eq. (5.5.6) to find the *step response*, defined as the output to a unit step input $u[n]=u_s[n]$:

$$y_s[n]=\mathcal{Z}^{-1}\{Y[z]\}\overset{(5.5.6)}{=}\mathcal{Z}^{-1}\{G[z]U[z]\}=\mathcal{Z}^{-1}\left\{G[z]\frac{z}{z-1}\right\}=\mathcal{Z}^{-1}\left\{\frac{z}{(z-1)(z^2-z/4-1/8)}\right\}$$

$$=\mathcal{Z}^{-1}\left\{\frac{(8/5)z}{z-1}-\frac{(8/3)z}{z-1/2}+\frac{(16/15)z}{z+1/4}\right\}\overset{(E5.1.1)\text{ or A.9(5)}}{=}\left(\frac{8}{5}-\frac{8}{3}\left(\frac{1}{2}\right)^n+\frac{16}{15}\left(-\frac{1}{4}\right)^n\right)u_s[n] \quad \text{(E5.10.5)}$$

We can use MATLAB to compute the transfer function and step response:

```
>>a1=-1/4; a0=-1/8; A=[0 1;-a0 -a1]; C=[1 0]; b=[0;1];
>>syms z; Gz=C*[z*eye(size(A))-A]^-1*b % Eq.(5.5.4)
   Gz = -8/(2*z - 8*z^2 + 1) % Eq.(E5.10.2): 1/(z^2-z/4-1/8)
>>Uz=z/(z-1); ys=iztrans(Gz*Uz) % Step response: Eq.(5.5.6)
   ys = (16*(-1/4)^n)/15 - (8*(1/2)^n)/3 + 8/5 % Eq.(E5.10.5)
```

We can also compose a Simulink model "em05e10.mdl," as depicted in Fig. 5.4(a), and run it to get the step response of the system with the transfer function (E5.10.2), as depicted in Fig. 5.4(b). Fig. 5.4(c) showing the step response $y_s[n]$ has been obtained by running the following MATLAB statements:

```
>>n=[0:10]; ysn=eval(ys); stairs(n,ysn) % Plot the step response
```

5.6 DIFFERENTIAL EQUATION AND DIFFERENCE EQUATION

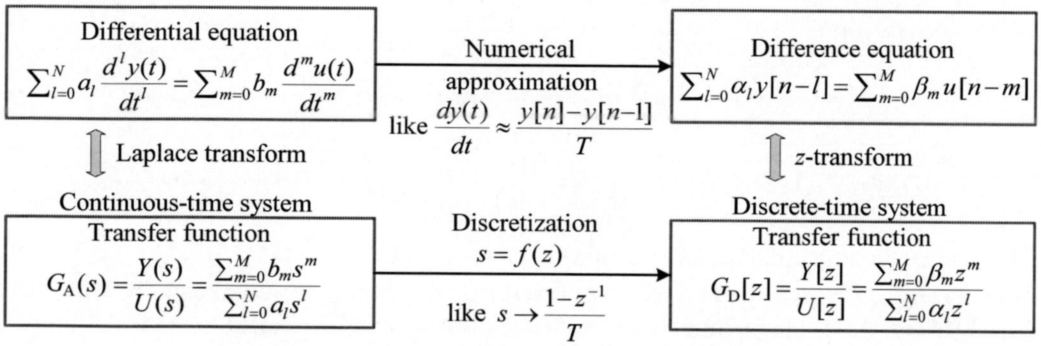

Figure 5.5 Numerical approximation and discretization

Most physical systems are CT (continuous-time) systems, whose input–output relationship and internal behavior are described by differential equations, while digital computers are DT (discrete-time) systems, whose input–output relationship and internal behavior are described by difference equations. Therefore, to use digital computers for the simulation or emulation of physical systems, we need to convert differential equations or CT systems into the "equivalent" difference equation or DT systems (see Fig. 5.5), although there can be no 'perfectly equivalent' difference equation that has the same output (solution) to any identical input as a differential equation. How can the continuous-to-discrete conversion, called *discretization*, be done? There are many numerical approximation methods that are useful for discretization, and one of them is the backward difference rule, which replaces the first derivative

$$y(t) = \frac{d}{dt} x(t) \quad \underset{\text{A.7(5)}}{\overset{\text{Laplace transform}}{\rightarrow}} \quad Y(s) = s\,X(s) \tag{5.6.1a}$$

by its backward difference approximation

$$y(nT) = \frac{x(nT)-x(nT-T)}{T} : y[n] = \frac{x[n]-x[n-1]}{T} \quad \underset{\text{A.7(5)}}{\overset{z\text{-transform}}{\rightarrow}} \quad Y[z] = \frac{1-z^{-1}}{T} X[z] \tag{5.6.1b}$$

where the initial values have been neglected since the input–output relationship is the focus. This numerical differentiation method leads to the *backward difference mapping* rule:

$$s \rightarrow \frac{1-z^{-1}}{T}, \text{ or equivalently, } \frac{1}{1-sT} \rightarrow z \tag{5.6.2}$$

(**Example 5.11**) Discretization of a CT System Using the Backward Difference Rule

Consider a CT system whose input–output relationship is described by the first-order DE

$$y'(t) + y(t) = u(t) \tag{E5.11.1}$$

We can take the Laplace transform of both sides (with zero initial conditions) to get the transfer function $G_A(s)$ as

$$sY(s) + Y(s) = U(s); \ (s+1)Y(s) = U(s); \ G_A(s) = \frac{Y(s)}{U(s)} = \frac{1}{s+1} \tag{E5.11.2}$$

(a) Convert the DE (E5.11.1) into the "equivalent" difference equation by using the backward difference rule (5.6.2). Then take the z-transform of the difference equation to find the transfer function $G_D[z]$ of the "equivalent" DT system whose input–output relationship is described by the difference equation.

$$\frac{y[n]-y[n-1]}{T}+y[n]=u[n]\,;\;\;(1+T)y[n]-y[n-1]=Tu[n] \tag{E5.11.3}$$

$$;(1+T)Y[z]-z^{-1}Y[z]=TU[z]\,;\;G_D[z]=\frac{Y[z]}{U[z]}=\frac{T}{1-z^{-1}+T}=\frac{1}{(1-z^{-1})/T+1}=\frac{Tz/(1+T)}{z-1/(1+T)} \tag{E5.11.4}$$

This can be directly obtained by substituting $s=(1-z^{-1})/T$ (Eq. (5.6.2)) into $G_A(s)$ (Eq. (E5.11.2)).

(b) Find the step responses of the CT system $G_A(s)$ (to a CT step input $u_s(t)$) and the DT system $G_D[z]$ (to a DT step input $u_s[n]$). Graph them with sampling interval $T=0.1$[s] for $t=0:T:5$[s].

Noting that $U(s)=\mathcal{L}\{u_s(t)\}=1/s$, we can get the step response of $G_A(s)$:

$$y_A(t)=\mathcal{L}^{-1}\left\{Y_A(s)\overset{(4.5.6)}{=}G_A(s)U(s)\overset{(E5.11.2)}{=}\frac{1}{(s+1)s}=\frac{1}{s}-\frac{1}{s+1}\right\}\overset{A.8(3),(6)}{=}(1-e^{-t})u_s(t) \tag{E5.11.5}$$

On the other hand, noting that $U[z]=\mathcal{Z}\{u_s[z]\}=z/(z-1)$, we can get the step response of $G_D[z]$:

$$y_D[n]=\mathcal{Z}^{-1}\left\{Y_D[z]\overset{(5.5.6)}{=}G_D[z]U[z]\overset{(E5.11.4)}{=}\frac{Tz/(1+T)}{z-1/(1+T)}\frac{z}{z-1}=\frac{z}{z-1}-\frac{z/(1+T)}{z-1/(1+T)}\right\}$$

$$\overset{A.9(3),(5)}{=}\left(1-(\frac{1}{1+T})^{n+1}\right)u_s[n] \tag{E5.11.6}$$

We can run the following MATLAB statements to compute the step responses of the CT/DT systems, and plot them as depicted in Fig. 5.6:

```
%em05e11.m
syms s z;
T=0.1; n=0:5/T; t=n*T; % Time range
GAs=1/(s+1); % Eq.(E5.11.2)
yA=ilaplace(GAs*1/s), yAt=eval(yA); % % Eq.(4.5.6) -> Eq.(E5.11.5)
GDz=T*z/(1+T)/(z-1/(1+T)); % Eq.(E5.11.4)
yD=iztrans(GDz*z/(z-1)), yDn=eval(yD); % Eq.(5.5.6) -> Eq.(E5.11.6)
plot(t,yAt); hold on; stem(t,yDn,'g'); % Plot the two step responses.
stem(t,1-1./(1+T).^(n+1),'r:x') % Plot Eq.(E5.11.6) as it is (for fun)
```

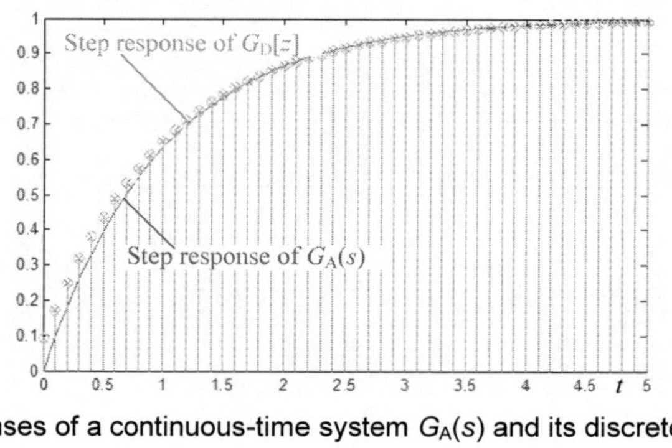

Figure 5.6 Step responses of a continuous-time system $G_A(s)$ and its discrete "equivalent" $G_D[z]$

Problems

5.1 *Z*-transform

(a) Find the bilateral z-transform $\mathbb{X}[z]$ (with its ROC) of

$$x[n] = a^{|n|} = a^n u_s[n] + a^{-n} u_s[-n-1] \quad \text{with } |a|<1 \tag{P5.1.1}$$

(b) Could you get the bilateral z-transform of

$$x[n] = 1 = u_s[n] + u_s[-n-1] \tag{P5.1.2}$$

If your answer is yes, find it. Otherwise, state why it is not possible.

(c) Find the bilateral z-transform $\mathbb{X}[z]$ (with its ROC) of

$$x[n] = u_s[-n] \tag{P5.1.3}$$

in the following three ways:

(i) Noting that $u_s[-n]$ is the one-sample delayed version of $u_s[-n-1]$, use the time shifting property (5.2.2).

(ii) Noting that $u_s[-n]$ is the time-reversal version of $u_s[n]$, use the time reversal property (5.2.5).

(iii) Noting that $u_s[-n]$ can be expressed as $u_s[-n-1]+\delta[n]$, use Table A.9(6) and (1).

5.2 Inverse *Z*-transform (IZT)

(a) Find the IZT $x[n]$ of

$$X[z] = \frac{z^2}{(z+1)^2} = z \times \frac{z}{(z+1)^2} = \frac{z}{z+1} \times \frac{z}{z+1} \quad \text{with ROC } \mathcal{R} = \{z : |z|>1\} \tag{P5.2.1}$$

in the following two different ways:

(i) Use the z-transform table A.9(10) and the time shifting property 5.2.3(b) or Table A.7(2).

(ii) Use the z-transform table A.9(5) and the convolution property 5.2.6 or Table A.7(4).

(b) Use the z-transform table A.9(17) and the time shifting property 5.2.3(a) or Table A.7(2) to find the IZT $x[n]$ of

$$X[z] = \frac{a}{z^2+a^2} = z^{-1} \times \frac{az}{z^2+a^2} \quad \text{with ROC } \mathcal{R} = \{z : |z| > a\} \tag{P5.2.2}$$

(c) Find the IZT $x[n]$ of

$$X[z] = \frac{2z-1}{(z-1)(z^2-z+1)} \quad \text{with ROC } \mathcal{R} = \{z : |z|>1\} \tag{P5.2.3}$$

(d) Find the IZT $x[n]$ of

$$X[z] = \frac{1}{(z-1)^2 z^2} = \frac{1}{(z-1)^2} - \frac{2}{z-1} + \frac{1}{z^2} + \frac{2}{z} \quad \text{with ROC } \mathcal{R} = \{z : |z|>1\} \tag{P5.2.4}$$

5.3 Inverse *Z*-transform, Causality, Convergence, and Stability

(a) Find the IZTs $g_{a1}[n]$, $g_{a2}[n]$, and $g_{a3}[n]$ of

$$G_a[z]=\frac{3(1-z^{-1})}{1-(5/2)z^{-1}+z^{-2}}=\frac{3z(z-1)}{(z-2)(z-1/2)} \tag{P5.3.1}$$

with one of the following ROCs:

(i) $\mathcal{R}_1=\{z:|z|>2\}$, (ii) $\mathcal{R}_2=\{z:\ |z|<1/2\}$, and (iii) $\mathcal{R}_3=\{z:1/2<|z|<2\}$

Then determine the causality and convergence (to zero) for each of $g_i[n]$'s, and fill in the related blanks in Table P5.3.

(b) Find the IZTs $g_{b1}[n]$, $g_{b2}[n]$, and $g_{b3}[n]$ of

$$G_b[z]=\frac{2z(z+1)}{z^2+(3/4)z+1/8}=\frac{2z(z+1)}{(z+1/4)(z+1/2)} \tag{P5.3.2}$$

with one of the following ROCs:

(i) $\mathcal{R}_1=\{z:|z|>1/2\}$, (ii) $\mathcal{R}_2=\{z:\ |z|<1/4\}$, and (iii) $\mathcal{R}_3=\{z:1/4<|z|<1/2\}$

Then determine the causality and convergence (to zero) for each of $g_i[n]$'s, and fill in the related blanks in Table P5.3.

Table P5.3 Pole locations and ROCs of *z*-transform expressions

	All poles inside the unit circle	The unit circle inside ROC	Causality	Convergence
(a)-(i)	No			No
(a)-(ii)			No	
(a)-(iii)		Yes		
(b)-(i)				Yes
(b)-(ii)			No	
(b)-(iii)	Yes	No		

5.4 To Solve a Difference Equation Using the *Z*-transform

Consider the following difference equation:

$$y[n]-\frac{1}{3}y[n-1]=x[n] \quad \text{with } x[n]=2^{-n}u_s[n] \tag{P5.4.1}$$

(a) Solve it for $y[n]$ twice, once with the initial condition $y[-1]=-3$ and once with $y[-1]=6$.

(b) Complete and run the MATLAB script “em05p04.m” to plot $y[n]$ (with $y[-1]=-3$) for $n=-1:8$.

```
%em05p04.m
syms z;
nn=[0:8]; x=2.^-nn; Xz=z/(z-???); % Input
y0=-3; y=iztrans(z/(z-???)*(Xz+??/3)) % Output
n=nn; yn=[y0 eval(y)];
yn2=[y0 subs(y,'n',nn)]; % Alternative to get y[n] for n=nn
yn1(1)=y0; % Initial value
for n=2:10 % To solve the difference equation (P5.4.1) in the time domain
   yn1(n) = yn1(n-1)/? +x(n-1); % Eq.(P5.4.1)
end
stem([-1 nn],yn), hold on, stem([-1 nn],yn1,'r:x')
```

5.5 Forward/Backward Solution of a Difference Equation by Iteration

Consider the following difference quation:

$$y[n]-\frac{1}{4}y[n-1]-\frac{1}{8}y[n-2]=3x[n-1] \quad \text{with } x[n]=\delta[n] \tag{P5.5.1}$$

```
%em05p05.m
% To solve a difference equation forward by iteration
ya(1)=0; ya(2)=0; % zero initial conditions
xa=zeros(1,10); xa(3)=1; % unit impulse input
for n=3:11, ya(n)= ya(n-1)/4 + ya(n-2)/? +?*xa(n-1);  end
nn=0:8; ya1= 4*(1/2).^nn - 4*(-1/4).^nn;
[ya(3:end); ya1]
% To solve a difference equation backward by iteration
yb(10)=0; yb(11)=0; % Zero initial conditions
xb= zeros(1,10); xb(9)=1; % Unit impulse input
for n=11:-1:3, yb(n-2)= -?*yb(n-1) + ?*yb(n) -24*xb(n-1); end
yb
nn=-8:0; yb1= -4*(1/2).^nn + 4*(-1/4).^nn;
[yb(1:end-2); yb1]
```

(a) Solve it forward with zero initial condition ($y[-2]=y[-1]=0$) to find $y_a[n]$ for $n=0:3$. To do this, you should rewrite the difference equation as

$$y_a[n]=\frac{1}{4}y_a[n-1]+\frac{1}{8}y_a[n-2]+3x[n-1] \quad \text{with } x[n]=\delta[n] \tag{P5.5.2}$$

(b) Solve it backward with zero initial condition ($y[1]=y[2]=0$) to find $y_b[n]$ for $n=-3:0$. To do this, you should rewrite the difference equation as

$$y_b[n-2]=-2y_b[n-1]+8y_b[n]-24x[n-1] \quad \text{with } x[n]=\delta[n] \tag{P5.5.3}$$

(c) Complete and run the above MATLAB script "em05p05.m" to get $y_a[n]$ for $n=0:3$ and $y_b[n]$ for $n=-3:0$.

5.6 Difference Equation, System Function, Impulse Response, and Frequency Response

Consider a system whose input–output relationship is described by the following difference quation:

$$y[n]-y[n-1]+0.5y[n-2]=x[n]+x[n-1] \quad \text{with } y[-2]=y[-1]=0 \tag{P5.6.1}$$

(a) Find the transfer or system function $G[z]=Y[z]/X[z]$.

(b) Plot the *frequency response* $G(\Omega)=G[e^{j\Omega}]=G[z]|_{z=e^{j\Omega}}$ for Ω=0:0.01:2π, which can be obtained by running the following MATLAB statements:

```
>>B=[1 1 0]; A=[1 -1 0.5]; W=0:0.01:2*pi; GW=freqz(B,A,W);
```

Find the value of $G[e^{j\Omega}]$ at $\Omega=0$, $\pi/2$, and π, or $e^{j\Omega}=1, j$, and -1 where $G[e^{j0}=1]$ is referred to as the *DC (direct current) gain*.

(c) Find the impulse response $g[n]$, i.e., the output of the system to the unit impulse input $u[n]=\delta[n]$ in two ways, one using the z-transform and the other using the iterative method (that is implemented by the MATLAB function '`filter()`').

(d) Find the step response, i.e., the output of the system to a unit step input $x[n]=u_s[n]$ by using the z-transform. Is the steady-state value of the step response equal to the DC gain obtained in (b) and the result of applying the final value theorem?

```
%em05p06.m
clear, clf
syms z;
Gz=z*(z+1)/(z^2-z+???);  % Transfer function
B=[1 1 0]; A=[1 ?? 0.5]; % Numerator/Denominator of Transfer function
disp('(b)')
Ws=[0; pi/2; pi]; % 3 frequency points
Freq_responses=freqz(B,A,Ws)
DC_gain= Freq_responses(1) % DC gain
disp('(c)') % Impulse response and frequency response
g= iztrans(??) % Impulse response as the inverse z-transform of G[z]
% To solve the difference equation by iteration
gn(1)=0; gn(2)=0; % Zero initial conditions
r=1/sqrt(2); W1=pi/4; nf=30;
xc=zeros(1,nf+3); xc(3)=1; % Unit impulse input
for n=0:nf  % Impulse response
   gn(n+3)= gn(n+2) - ???*gn(n+1) + xc(n+3) + xc(n+2); % Eq.(P5.6.1)
   gn_iztrans(n+1)= eval(g);
   gn_iztrans_hand(n+1)= r^n*(cos(W1*n)+3*sin(W1*n)); % Eq.(S5.6.4)
end
gn_filter=filter(B,A,xc(3:end));
W=0:0.01:2*pi; % Frequency vector
GW_DTFT= DTFT(gn,W,0); % Frequency response as the DTFT of g[n]
GW= freqz(B,A,W); % Frequency response by substituting z=e^jw into G[z]
discrepancy_GW= norm(GW-GW_DTFT)
subplot(311)
plot(W,abs(GW)), grid on
set(gca,'xtick',0:pi/2:2*pi,'xtickLabel',{'0','pi/2','pi','3pi/2','2pi'})
title('Frequency response of G[z]=z(z+1)/(z^2-z+0.5)')
disp('(d)') % Step response
Xz=z/(z-1); % The z-transform of the unit step input
y_step=iztrans(??*Gz) % Step response using the inverse z-transform
n=0:nf; % Time vector
yn_step_iztran=eval(y_step);
xd= ones(size(xd)); % Unit step input
yn_step = filter(B,A,xd);
yn_step_hand = 4-r.^n.*(3*cos(W1*n)-sin(W1*n));
discrepancy_yn1= norm(yn_step-yn_step_hand)
discrepancy_yn2= norm(yn_step-yn_step_iztran)
subplot(312)
stairs(n,yn_step), title('Step response')
yn_step(end)-DC_gain % Steady-state step response = DC gain?
disp('(e)') % Sinusoidal response
syms z n;
Xz=ztrans(sin(pi/2*n)); % Sinusoidal input
y_sin=iztrans(Xz*??) % Sinusoidal response
n=0:nf; % Time vector
yn_sin_iztran = eval(y_sin);
W2=Ws(2); xe= sin(W2*n); % Unit sinusoidal input
yn_sin = filter(B,A,xe);
yn_sin_hand = -1.2*cos(W2*n)-0.4*sin(??*n) ...
                  +r.^n.*(1.2*cos(??*n)+1.6* sin(W1*n));
discrepancy_yn1= norm(yn_sin-yn_sin_hand)
discrepancy_yn2= norm(yn_sin-yn_sin_iztran)
subplot(313)
stairs(n,yn_sin)
title('Sinusoidal response to x[n]=sin(pi/2*n)')
```

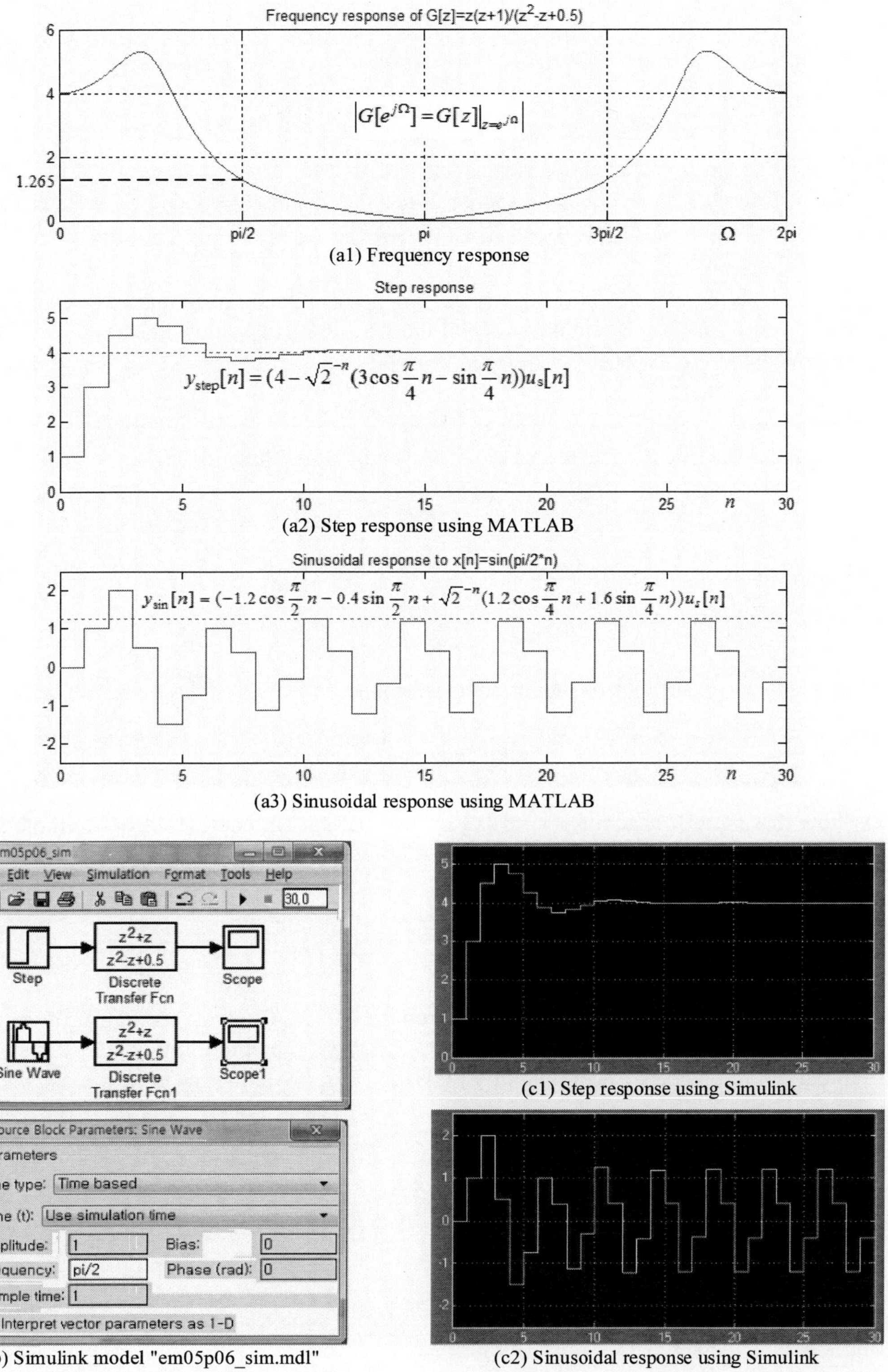

(a1) Frequency response

(a2) Step response using MATLAB

(a3) Sinusoidal response using MATLAB

(b) Simulink model "em05p06_sim.mdl"

(c1) Step response using Simulink

(c2) Sinusoidal response using Simulink

Figure P5.6 Frequency/step/sinusoidal responses simulated using MATLAB and Simulink ("em05p06_sim.mdl")

(e) Using the z-transform, show that the output to a sinusoidal input $x[n]=\sin(n\pi/2)u_s[n]$ is

$$y_{\sin}[n]=\left\{\sqrt{0.4^2+1.2^2}\,\sin\left(\frac{\pi}{2}n+\tan^{-1}\left(\frac{-1.2}{-0.4}\right)\right)+\sqrt{2}^{-n}(1.2\cos\frac{\pi}{4}n+1.6\sin\frac{\pi}{4}n)\right\}u_s[n] \quad \text{(P5.6.2)}$$

$$\overset{t\to\infty}{\to}\ y_{ss,\sin}(t)=0.01\sin(100t-\frac{\pi}{2}) \quad \text{(P4.5.2)}$$

How is the steady-state sinusoidal response related to the input in terms of the magnitude and phase? Note that the frequency response of the system at the frequency $\Omega=\pi/2$ of the input is

$$G[e^{j\Omega}]\big|_{\Omega=\pi/2}=-0.4-j1.2=1.265\angle-1.8925 \quad \text{(P5.6.3)}$$

(f) Referring to the above MATLAB script "em05p06.m" and the Simulink block diagram in Fig. P5.6(b), use MATLAB/Simulink to plot the frequency/step/sinusoidal responses of the system as depicted in Fig. P5.6(a1), (a2), (a3), (c1), and (c2).

5.7 Discretization of a Continuous-Time (CT) System — Bilinear Transformation (BLT)

By the trapezoidal rule (or Tustin's method), the integral of a function $x(t)$

$$y(t)=\int_0^t x(\tau)d\tau:\ Y(s)=\frac{1}{s}X(s) \quad \text{(P5.7.1a)}$$

can be approximated by

$$y(nT)=y(nT-T)+\frac{T}{2}(x(nT)+x(nT-T)):\ Y[z]=\frac{T}{2}\frac{1+z^{-1}}{1-z^{-1}}X[z] \quad \text{(P5.7.1b)}$$

This leads to one of the discretization methods, called the *bilinear transformation* rule:

$$s\to\frac{2}{T}\frac{1-z^{-1}}{1+z^{-1}},\ \text{or equivalently,}\ \frac{1+sT/2}{1-sT/2}\to z \quad \text{(P5.7.3)}$$

(a) Show that the CT system $G_A(s)=1/(s+1)$ (Eq. (E5.11.2)), considered in Example 5.11, can be discretized by using the bilinear transformation rule as

$$G_{Db}[z]=\frac{1}{s+1}\bigg|_{s=\frac{2(1-z^{-1})}{T(1+z^{-1})}}=\frac{1}{\frac{2(1-z^{-1})}{T(1+z^{-1})}+1}=\frac{T(z+1)/(2+T)}{z-\frac{1-T/2}{1+T/2}} \quad \text{(P5.7.4)}$$

(b) Referring to Example 5.11, complete and run the MATLAB script "em05p07.m" to plot the step response of the DT system $G_{Db}[z]$ (Eq. (P5.7.4)) (with sampling interval $T=0.1$[s]) as depicted in Fig. P5.7. Compare it with that of $G_D[z]$ (Eq. (E5.11.4)) in terms of how close they are to that of $G_A(s)=1/(s+1)$.

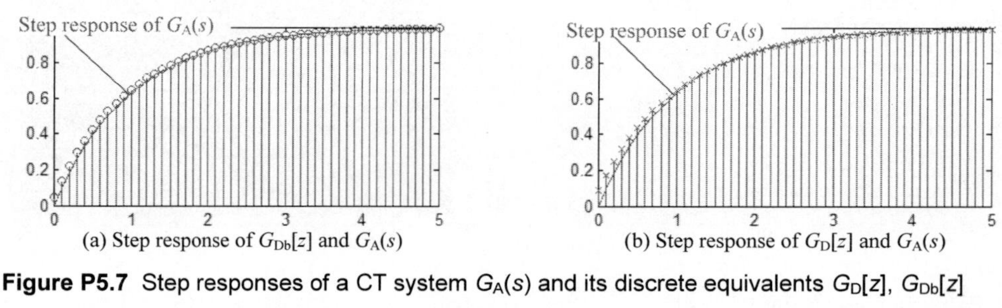

Figure P5.7 Step responses of a CT system $G_A(s)$ and its discrete equivalents $G_D[z]$, $G_{Db}[z]$

```
%em05p07.m
clear, clf
syms s z;
T=0.1; n=0:5/T; t=n*T; % Sampling interval and Sampling times
GAs=1/(s+1); % Analog system Eq.(E5.11.2)
yA=ilaplace(GAs*1/s), yAt=eval(yA); % Eq.(E5.11.5)
GDbz=subs(GAs,s,2/?*(?-1)/(z+?)) % Eq.(P5.7.4)
Gas=tf(1,[1 1]); % Transfer function object of the analog system
c2d(Gas,T,'tustin') % Discrete equivalent using Eq.(P5.7.3)
yDb=iztrans(????*z/(?-1)), yDbn=eval(yDb); % Eq.(5.5.6)
subplot(221), plot(t,yAt); hold on; stem(t,yDbn,'Markersize',5)
GDz=subs(GAs,s,(1-?^-1)/?); % Eq.(E5.11.4)
yD=iztrans(GDz*z/(?-1)), yDn=eval(yD); % Eq.(E5.11.6)
subplot(222), plot(t,yAt); hold on; stem(t,yDn,'x');
norm(yAt-yDbn), norm(yAt-yDn)
```

CHAPTER 6

Fourier Series and Fourier Transform

CHAPTER OUTLINE

6.1 CONTINUOUS-TIME FOURIER SERIES (CTFS)[Y-2]

6.1.1 Definition and Convergence Conditions

Let a function $x(t)$ be *periodic* with period P in t, that is,

$$x(t) = x(t+P) \quad \forall\, t \tag{6.1.1}$$

where P[sec] and $\omega_0=2\pi/P$[rad/s] are referred to as the *fundamental period* and *fundamental (angular) frequency*, respectively, if P is the smallest positive real number to satisfy Eq. (6.1.1) for periodicity. Suppose $x(t)$ satisfies at least one of the following conditions A and B:

< **Condition A** >

(**A1**) The periodic function $x(t)$ is square-integrable over the period P, i.e.,

$$\int_P |x(t)|^2\, dt < \infty \tag{6.1.2a}$$

where $\int_P$ means the integration over any interval of length P whatever its upper and lower limits are. This implies that the signal described by $x(t)$ has finite power.

< **Condition B**: *Dirichlet condition* >

(**B1**) The periodic function $x(t)$ has only a finite number of extrema and discontinuities in any one period.

(**B2**) These extrema are finite.
(**B3**) The periodic function $x(t)$ is absolutely integrable over the period P, i.e.,

$$\int_P |x(t)|\,dt < \infty \tag{6.1.2b}$$

Then the periodic function $x(t)$ can be represented by the following forms of *continuous-time Fourier series* (*CTFS*), each of which is called the *CTFS representation*:

<Trigonometric form>

$$x(t) = a_0 + \sum_{k=1}^{\infty} a_k \cos k\omega_0 t + \sum_{k=1}^{\infty} b_k \sin k\omega_0 t \quad \text{with } \omega_0 = \frac{2\pi}{P} \quad (P\text{: the period of } x(t)) \tag{6.1.3a}$$

where the Fourier coefficients a_0, a_k, and b_k are

$$\begin{aligned} a_0 &= \frac{1}{P}\int_P x(t)\,dt \quad \text{(the integral over one period } P\text{)} \\ a_k &= \frac{2}{P}\int_P x(t)\cos k\omega_0 t\, dt \\ b_k &= \frac{2}{P}\int_P x(t)\sin k\omega_0 t\, dt \end{aligned} \tag{6.1.3b}$$

<Magnitude-and-phase form>

$$x(t) = d_0 + \sum_{k=1}^{\infty} d_k \cos(k\omega_0 t + \phi_k) \tag{6.1.4a}$$

where the Fourier coefficients are

$$d_0 = a_0, \quad d_k = \sqrt{a_k^2 + b_k^2}, \quad \phi_k = \tan^{-1}\left(-\frac{b_k}{a_k}\right) \tag{6.1.4b}$$

<Sine-and-phase form>

$$x(t) = d_0' + \sum_{k=1}^{\infty} d_k' \sin(k\omega_0 t + \phi_k') \tag{6.1.5a}$$

where the Fourier coefficients are

$$d_0' = a_0 = d_0, \quad d_k' = \sqrt{a_k^2 + b_k^2} = d_k, \quad \phi_k' = \tan^{-1}\left(\frac{a_k}{b_k}\right) = \phi_k + \frac{\pi}{2} \tag{6.1.5b}$$

<Complex exponential form>

$$x(t) = \frac{1}{P}\sum_{k=-\infty}^{\infty} c_k\, e^{jk\omega_0 t} \tag{6.1.6a}$$

where the Fourier coefficients are

$$c_k = \int_P x(t) e^{-jk\omega_0 t}\, dt \quad \text{(the integral over one period } P\text{)} \tag{6.1.6b}$$

Here, the kth frequency $k\omega_0 (k>1)$ with the fundamental frequency $\omega_0=2\pi/P=2\pi f_0$[rad/s] (P: period) is referred to as the kth *harmonic*. The above four forms of Fourier representation are equivalent and their Fourier coefficients are related to each other as follows:

$$c_0 = \int_P x(t)dt = Pd_0 = Pa_0$$

$$c_k = \int_P x(t)e^{-jk\omega_0 t}dt = \int_P x(t)(\cos k\omega_0 t - j\sin k\omega_0 t)dt = \frac{P}{2}(a_k - jb_k) = \frac{P}{2}d_k\angle\phi_k$$

$$c_{-k} = \int_P x(t)e^{jk\omega_0 t}dt = \int_P x(t)(\cos k\omega_0 t + j\sin k\omega_0 t)dt = \frac{P}{2}(a_k + jb_k) = \frac{P}{2}d_k\angle-\phi_k = c_k^*$$

$$;\quad a_0 = \frac{c_0}{P},\ a_k = \frac{c_k+c_{-k}}{P} = \frac{2\,\mathrm{Re}\{c_k\}}{P},\ b_k = \frac{c_{-k}-c_k}{jP} = -\frac{2\,\mathrm{Im}\{c_k\}}{P}$$

The plot of the Fourier coefficients (6.1.4b), (6.1.5b), or (6.1.6b) versus the (discrete) frequency $k\omega_0$ is referred to as the *spectrum*, which can be used to describe the spectral contents of a signal, that is, depict what frequency components are contained in the signal and how they are distributed over the low/medium/high frequency range.

<Proof of the Complex Exponential Fourier Analysis Formula (6.1.6b)>

To show the validity of Eq. (6.1.6b), we substitute the Fourier synthesis formula (6.1.6a) with the (dummy) summation index k replaced by n into Eq. (6.1.6b) as

$$c_k \overset{?}{=} \int_P \frac{1}{P}\sum_{n=-\infty}^{\infty} c_n e^{jn\omega_0 t}e^{-jk\omega_0 t}dt = \sum_{n=-\infty}^{\infty} c_n \frac{1}{P}\int_P e^{j(n-k)\frac{2\pi}{P}t}dt \overset{\text{O.K.}}{=} c_k \tag{6.1.7}$$

This equality holds due to the *orthogonality* among the family of complex exponential functions $\{e^{jk\omega_0 t}\}$:

$$\frac{1}{P}\int_{-P/2}^{P/2} e^{j(n-k)\frac{2\pi}{P}t}dt = \begin{cases} \dfrac{1}{P\cdot j(n-k)2\pi/P}e^{j(n-k)\frac{2\pi}{P}t}\Big|_{-P/2}^{P/2} = 0 & \text{for } n\neq k \\ \dfrac{1}{P}\int_{-P/2}^{P/2} dt = 1 & \text{for } n=k \end{cases} = \delta[n-k] \tag{6.1.8}$$

where $\delta[n]$ is the *unit sample* or *Kronecker delta sequence* defined as

$$\delta[n] = \begin{cases} 1 & \text{for } n=0 \\ 0 & \text{elsewhere} \end{cases} \tag{6.1.9}$$

<Orthogonality>

With $f_1(t) = e^{jn\omega_0 t}$, $f_2(t) = e^{jk\omega_0 t}$, $a=-P/2$, $b=P/2$, and $n\neq k$, Eq. (6.1.8a) can be rewritten as

$$\int_a^b f_1(t)f_2^*(t)dt = 0 \tag{6.1.10}$$

which implies that the two functions $f_1(t)$ and $f_2(t)$ are orthogonal on the interval $[a, b]$. In Section 1.1, the word "orthogonal" was used to describe how two vectors are geometrically perpendicular. Why is the word used for describing a relation between two functions? It can be made possible by making an analogy between vectors and functions, and defining the inner (or dot) product of two functions as

$$f_1(t) \bullet f_2(t) = \int_a^b f_1(t) f_2^*(t)\, dt \tag{6.1.11}$$

<Effects of Vertical/Horizontal Translations of *x(t)* on the Fourier Coefficients>

- Translating $x(t)$ along the vertical axis by $\pm A$ (+: upward, −: downward) causes only the change of Fourier coefficient $d_0 = a_0$ for $k=0$ (DC component or average value) by $\pm A$.
- Translating $x(t)$ along the horizontal (time) axis by $\pm t_1$ (+: rightward, −: leftward) causes only the change of phases (ϕ_k's) by $\mp k\omega_0 t_1$, not affecting the magnitudes d_k or d_k' or $|c_k|$.

$$c_k' \overset{(6.1.6b)}{=} \int_P x(t-t_1)\, e^{-jk\omega_0 t} dt = \int_P x(t-t_1)\, e^{-jk\omega_0 (t-t_1+t_1)} dt$$

$$= e^{-jk\omega_0 t_1} \int_P x(t-t_1)\, e^{-jk\omega_0 (t-t_1)} dt \overset{(6.1.6b)}{=} c_k e^{-jk\omega_0 t_1} = |c_k| \angle(\phi_k - k\omega_0 t_1) \tag{6.1.12}$$

6.1.2 Examples of CTFS

If a periodic function $x(t)$ has one of the following three symmetries, we can make use of it to reduce the computation for computing the Fourier coefficients, as summarized in Table 6.1:

(1) Even symmetry: $x(t) = x(-t)$

(2) Odd symmetry: $x(t) = -x(-t)$

(3) Half-wave symmetry: $x(t) = -x(t + \frac{P}{2})$ (P: the period)

Note that an even (symmetric) periodic function can become an odd (symmetric) one just by translation and vice versa, as illustrated in Fig. 6.1, which shows rectangular and triangular waves.

Let us see some examples of finding the Fourier series coefficients.

Table 6.1 Reduced computation of CTFS coefficients exploiting symmetries of $x(t)$

Fourier coefficients	Symmetry of a periodic function $x(t)$ with period P		
	Even symmetry: $x(t)=x(-t)$	Odd symmetry: $x(t)=-x(-t)$	Half-wave symmetry: $x(t)=-x(t+P/2)$
a_0(DC term)	$a_0 = \frac{2}{P}\int_0^{P/2} x(t)\, dt$	0	0
a_k (k: odd)	$a_k = \frac{4}{P}\int_0^{P/2} x(t)\cos k\omega_0 t\, dt$	0	$a_k = \frac{4}{P}\int_0^{P/2} x(t)\cos k\omega_0 t\, dt$
a_k (k: even)	$a_k = \frac{4}{P}\int_0^{P/2} x(t)\cos k\omega_0 t\, dt$	0	0
b_k (k: odd)	0	$b_k = \frac{4}{P}\int_0^{P/2} x(t)\sin k\omega_0 t\, dt$	$b_k = \frac{4}{P}\int_0^{P/2} x(t)\sin k\omega_0 t\, dt$
b_k (k: even)	0	$b_k = \frac{4}{P}\int_0^{P/2} x(t)\sin k\omega_0 t\, dt$	0
c_k	$c_k = 2\int_0^{P/2} x(t)\cos k\omega_0 t\, dt$	$c_k = 2\int_0^{P/2} x(t)\sin k\omega_0 t\, dt$	$c_k = 2\int_0^{P/2} x(t)e^{-jk\omega_0 t} dt$ for odd k

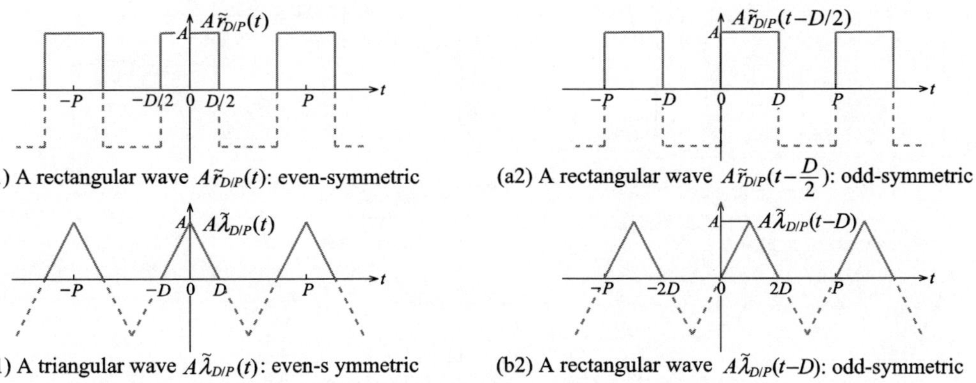

Figure 6.1 Examples of rectangular and triangular waves

(**Example 6.1**) Fourier Spectra of a Rectangular (Square) Wave and a Triangular Wave

(a) CTFS Spectrum of a Rectangular (Square) Wave (Fig. 6.1(a1))

Consider an even rectangular wave $x(t)$ with height A, duration D, and period P:

$$x(t) = A\tilde{r}_{D/P}(t) \text{ where } \tilde{r}_{D/P}(t) = \begin{cases} 1 & \text{for } |t-mP| \le D/2 \, (m\text{: an integer}) \\ 0 & \text{elsewhere} \end{cases} \tag{E6.1.1}$$

We use Eq. (6.1.3b)/(6.1.6b) to find the trigonometric/exponential Fourier coefficients as

$$a_0 \overset{(6.1.3b)}{=} \frac{1}{P}\int_{-P/2}^{P/2} A\tilde{r}_{D/P}(t)\,dt = \frac{A}{P}\int_{-D/2}^{D/2} 1\,dt = \frac{AD}{P} \tag{E6.1.2a}$$

$$a_k \underset{\text{even } x(t)}{\overset{(6.1.3b)}{=}} \frac{4}{P}\int_0^{D/2} Ar_{D/P}(t)\cos k\omega_0 t\,dt \overset{(C.36)}{=} \frac{4A}{P}\left.\frac{\sin k\omega_0 t}{k\omega_0}\right|_0^{D/2} = \frac{2AD}{P}\frac{\sin(k\pi D/P)}{k\pi D/P} \tag{E6.1.2b}$$

$$b_k \underset{\text{even } x(t)}{\overset{(6.1.3b)}{=}} \frac{2}{P}\int_{-P/2}^{P/2} A\tilde{r}_{D/P}(t)\sin k\omega_0 t\,dt \overset{\text{odd}}{=} 0 \quad \text{with } \omega_0 = \frac{2\pi}{P} \tag{E6.1.2c}$$

$$c_k \overset{(6.1.6b)}{=} \int_{-P/2}^{P/2} A\tilde{r}_{D/P}(t)e^{-jk\omega_0 t}\,dt = A\int_{-D/2}^{D/2} e^{-jk\omega_0 t}\,dt = \left.\frac{A}{-jk\omega_0}e^{-jk\omega_0 t}\right|_{-D/2}^{D/2}$$

$$= A\frac{e^{jk\omega_0 D/2} - e^{-jk\omega_0 D/2}}{jk\omega_0} = AD\frac{\sin(k\omega_0 D/2)}{k\omega_0 D/2} = AD\,\text{sinc}\left(k\frac{D}{P}\right) \text{ with } \omega_0 = \frac{2\pi}{P} \tag{E6.1.3}$$

Now we can use Eq. (6.1.3a)/(6.1.6a) to write the Fourier series representation of the rectangular wave $x(t)$ as

$$A\,\tilde{r}_{D/P}(t) \underset{(6.1.6a)\text{ with }(E6.1.3)}{\overset{(6.1.3a)\text{ with }(E6.1.2)}{=}} \begin{cases} \dfrac{AD}{P} + \displaystyle\sum_{k=1}^{\infty} \frac{2AD}{P}\frac{\sin(k\pi D/P)}{k\pi D/P}\cos k\omega_0 t \\ \dfrac{1}{P}\displaystyle\sum_{k=-\infty}^{\infty} AD\,\text{sinc}\left(k\frac{D}{P}\right)e^{jk\omega_0 t} \end{cases} \tag{E6.1.4}$$

In the case of $A=1$, $D=1$, and $P=2D=2$ as depicted in Fig. 6.2(a1), the magnitudes of c_k's, called the magnitude spectrum, are plotted for $k=-8\sim8$ in Fig. 6.2(b1).

(Q) What about the case of $P=D$, which corresponds to a constant DC (direct current) signal?

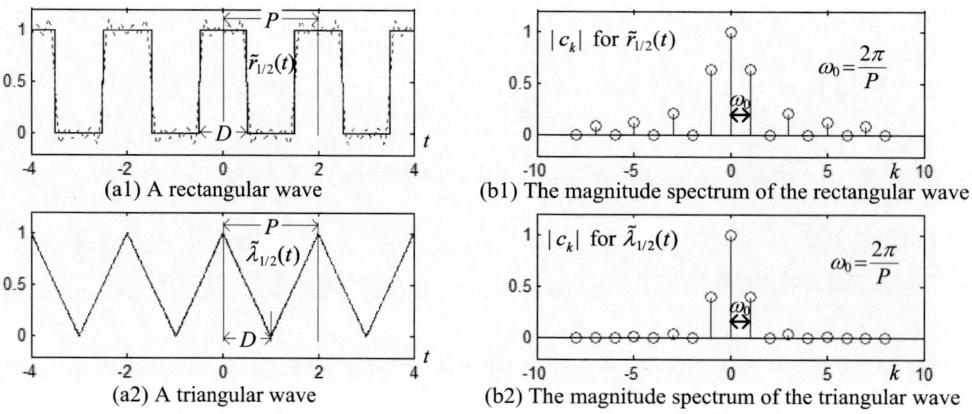

(a1) A rectangular wave (b1) The magnitude spectrum of the rectangular wave

(a2) A triangular wave (b2) The magnitude spectrum of the triangular wave

Figure 6.2 Rectangular/triangular waves and their CTFS spectra

(b) CTFS Spectrum of a Triangular Wave (Fig. 6.1(b1))

Consider an even triangular wave $x(t)$ with maximum height A, duration $[-D, D]$, and period P:

$$x(t) = A\tilde{\lambda}_{D/P}(t) \ \text{ where } \ \tilde{\lambda}_{D/P}(t) = \begin{cases} 1-|t|/D & \text{for } |t-mP| \le D \ (m\text{: an integer}) \\ 0 & \text{elsewhere} \end{cases} \tag{E6.1.5}$$

We use Eq. (6.1.3b)/(6.1.6b) to find the trigonometric/exponential Fourier coefficients as

$$a_0 = \frac{1}{P}\int_{-P/2}^{P/2} A\tilde{\lambda}_{D/P}(t)\,dt \overset{\text{even}}{=} \frac{2A}{P}\int_0^D (1-\frac{t}{D})\,dt \overset{(C.32)}{=} \frac{2A}{P}(t-\frac{t^2}{2D})\Big|_0^D = \frac{AD}{P} \tag{E6.1.6a}$$

$$a_k \overset{\text{even}}{=} \frac{4}{P}\int_0^D A(1-\frac{t}{D})\cos k\omega_0 t\,dt \ \text{ with } \omega_0 = \frac{2\pi}{P}$$

$$\overset{(C.36),(C.38)}{=} \frac{4A}{P}\left(\frac{\sin k\omega_0 t}{k\omega_0}\Big|_0^D - \frac{1}{D}\left(\frac{\sin k\omega_0 t}{k\omega_0}t\Big|_0^D - \int_0^D \frac{\sin k\omega_0 t}{k\omega_0}dt\right)\right)$$

$$\overset{(C.37)}{=} -\frac{4A}{PD}\frac{\cos k\omega_0 t}{(k\omega_0)^2}\Big|_0^D = \frac{4A}{PD}\frac{(1-\cos k\omega_0 D)}{(k\omega_0)^2} \overset{(C.14)}{=} \frac{2AD}{P}\frac{\sin^2(k\omega_0 D/2)}{(k\omega_0 D/2)^2} = \frac{2AD}{P}\frac{\sin^2(k\pi D/P)}{(k\pi D/P)^2} \tag{E6.1.6b}$$

$$b_k \underset{\text{even } x(t)}{\overset{(6.1.3b)}{=}} \frac{2}{P}\int_{-P/2}^{P/2} A\tilde{\lambda}_{D/P}(t)\sin k\omega_0 t\,dt \overset{\text{odd}}{=} 0 \tag{E6.1.6c}$$

$$c_k = \int_{-P/2}^{P/2} A(1-\frac{|t|}{D})e^{-jk\omega_0 t}\,dt = \int_{-D}^{D} A(1-\frac{|t|}{D})\cos(k\omega_0 t)\,dt$$

$$= 2\int_0^D A(1-\frac{t}{D})\cos(k\omega_0 t)\,dt = 2A(1-\frac{t}{D})\frac{1}{k\omega_0}\sin(k\omega_0 t)\Big|_0^D - \int_0^D 2A(-\frac{1}{D})\frac{1}{k\omega_0}\sin(k\omega_0 t)\,dt$$

$$= -2A\frac{1}{(k\omega_0)^2 D}\cos(k\omega_0 t)\Big|_0^D = 2AD\frac{1-\cos(k\omega_0 D)}{(k\omega_0 D)^2} = AD\frac{4\sin^2(k\omega_0 D/2)}{(k\omega_0 D/2)^2}$$

$$= AD\,\text{sinc}^2\left(k\frac{D}{P}\right) \ \text{ with } \omega_0 = \frac{2\pi}{P} \tag{E6.1.7}$$

Thus, we can write the Fourier series representation of the triangular wave $x(t)$ as

$$A\tilde{\lambda}_{D/P}(t) \underset{(6.1.6\text{a})\ \text{with}\,(\text{E}6.1.7)}{\overset{(6.1.3\text{a})\ \text{with}\,(\text{E}6.1.6)}{=}} \begin{cases} \dfrac{AD}{P}+\sum_{k=1}^{\infty}\dfrac{2AD}{P}\dfrac{\sin^2(k\pi D/P)}{(k\pi D/P)^2}\cos k\omega_0 t \\ \dfrac{1}{P}\sum_{k=-\infty}^{\infty} AD\,\text{sinc}^2\left(k\dfrac{D}{P}\right)e^{jk\omega_0 t} \end{cases} \tag{E6.1.8}$$

In the case of $A=1$, $D=1$, and $P=2D=2$ as depicted in Fig. 6.2(a2), the magnitudes of c_k's, called the magnitude spectrum, are plotted for $k=-8\sim 8$ in Fig. 6.2(b2).

(c) MATLAB program to get the Fourier spectra

```
%em06e01.m: plot Fig.6.2 (CTFS spectra of rectangular/triangular waves)
clear, clf
global P D
N=8; k=-N:N; % Range of frequency indices
for i=1:2
   if i==1, x='rect_wave'; P=2; D=1;
            a_true=D/P*[1 2*sinc([1:N]*D/P)]; % Eq.(E6.1.2)
            c_true=D*sinc(k*D/P); % Eq.(E6.1.3)
    else  x='tri_wave'; P=2; D=1;
           a_true=D/P*[1 2*sinc([1:N]*D/P).^2]; % Eq.(E6.1.6)
           c_true=D*sinc(k*D/P).^2; % Eq.(E6.1.7)
   end
   w0=2*pi/P; % Fundamental frequency
   tt=[-400:400]*P/200; xt=feval(x,tt); % Time interval and x(t)
   [a0,a,b]=CTFS_trigonometric(x,P,N);
   [c,kk]=CTFS_exponential(x,P,N);
   [a0 a; a_true], [c; c_true] % compare with true Fourier series coefs
   jkw0t=j*kk.'*w0*tt; xht = real(c/P*exp(jkw0t)); % Eq.(6.1.13)
   subplot(219+i*2), plot(tt,xt,'k-', tt,xht,'b:')
   axis([tt([1 end]) -0.2 1.2]), title('Periodic function x(t)')
   c_mag = abs(c); c_phase = angle(c);
   subplot(220+i*2), stem(kk,c_mag), title('CTFS Spectrum |X(k)|')
end
```

```
function y=rect_wave(t)
global P D
tmp=min(abs(mod(t,P)),abs(mod(-t,P)));   y=(tmp<=D/2);
```

```
function y=tri_wave(t)
global P D
tmp= min(abs(mod(t,P)),abs(mod(-t,P)));   y=(tmp<=D).*(1-tmp/D);
```

```
function [a0,a,b]=CTFS_trigonometric(x,P,N)
% Find the trigonometric Fourier coefficients a0,ak,bk for k=1:N
% x: A periodic function with period P
% P: Period, N: Maximum order of Fourier coefficients
w0=2*pi/P; % Fundamental frequency [rad/s]
xcoskw0t=[x '(t).*cos(k*w0*t)'];
x_coskw0t=inline(xcoskw0t,'t','k','w0');
xsinkw0t=[x '(t).*sin(k*w0*t)'];
x_sinkw0t=inline(xsinkw0t,'t','k','w0');
tol=1e-6; % Tolerance on numerical error
a0=quadl(x_coskw0t,-P/2,P/2,tol,[],0,w0)/P; % Eq.(6.1.3b)
for k=1:N
  a(k)=2/P*quadl(x_coskw0t,-P/2,P/2,tol,[],k,w0); % Eq.(6.1.3b)
  b(k)=2/P*quadl(x_sinkw0t,-P/2,P/2,tol,[],k,w0); % Eq.(6.1.3b)
end
```

```
function [c,kk]=CTFS_exponential(x,P,N)
% Find the complex exponential Fourier coefficients c(k) for k=-N:N
% x: A periodic function with period P
% P: Period, N: Maximum frequency index to specify the frequency range
w0=2*pi/P; % Fundamental frequency [rad/s]
xexp_jkw0t_= [x '(t).*exp(-j*k*w0*t)'];
xexp_jkw0t= inline(xexpjkw0t_,'t','k','w0');
kk=-N:N; % Frequency range
tol=1e-6; % Tolerance on numerical error
for k=kk
   c(k+N+1)=quadl(xexp_jkw0t,-P/2,P/2,tol,[],k,w0); % Eq.(6.1.6b)
end
```

Once you have defined and saved a periodic function (like '`rect_wave()`' or '`tri_wave()`' listed below) as an M-file, the MATLAB functions '`CTFS_trigonometric()`' and '`CTFS_exponential()`' can be used to find its trigonometric/exponential Fourier series coefficients. Interested readers are invited to run the above MATLAB script "em06e01.m" to get the CTFS coefficients and plot the spectra for the rectangular and triangular waves as in Fig. 6.2. The script performs two additional jobs for each of the two waves.

- It compares the hand-calculated results (Eqs. (E6.1.2)/(E6.1.3) and (E6.1.6)/(E6.1.7)) with the trigonometric/exponential Fourier series coefficients obtained using '`CTFS_trigonometric()`'/'`CTFS_exponential()`'.
- It plots the original signal $x(t)$ (as a solid line) together with its *Fourier reconstruction* $\hat{x}_N(t)$ (as a dotted line) consisting of frequency components up to the Nth harmonic:

$$\hat{x}_N(t) = \frac{1}{P}\sum_{k=-N}^{N} c_k\, e^{jk\omega_0 t} \tag{6.1.13}$$

At this point, you may wonder how a rectangular wave with discontinuities can be represented by the sum of trigonometric or complex exponential functions that are continuous for all t. To satisfy your curiosity, run the MATLAB script "em06e01_1.m," which uses the Fourier reconstruction formula (6.1.13) to reconstruct the rectangular wave $\tilde{x}_{D/P}(t)$ from its Fourier series coefficients $\{c_k, k=-N:N\}$, with increasing number of terms $N=1, 3, 9, 19,\ldots$ as in Fig. 6.3.

```
%em06e01_1.m : plot Fig.6.3 (CTFS reconstruction)
clear, clf
global P D
P=2; w0=2*pi/P; D=1; % Period, Fundamental frequency, and Duration
tt=[-400:400]*P/400; % Time interval of two periods
x = 'rect_wave';
xt = feval(x,tt); % Original signal
plot(tt,xt,'k:'), hold on
Ns= [1 3 9 19]; % Numbers of reconstruction terms
for N=Ns
   k=-N:N; %Frequency indices of the components to be summed in Eq.(6.1.13)
   jkw0t=j*k.'*w0*tt; % To construct the Fourier reconstruction terms
   c= D*sinc(k*D/P);
   xht = real(c/P*exp(jkw0t)); % Eq.(6.1.13)
   plot(tt,xht,'b')
   hold on, pause
end
axis([tt(1) tt(end) -0.2 1.2])
```

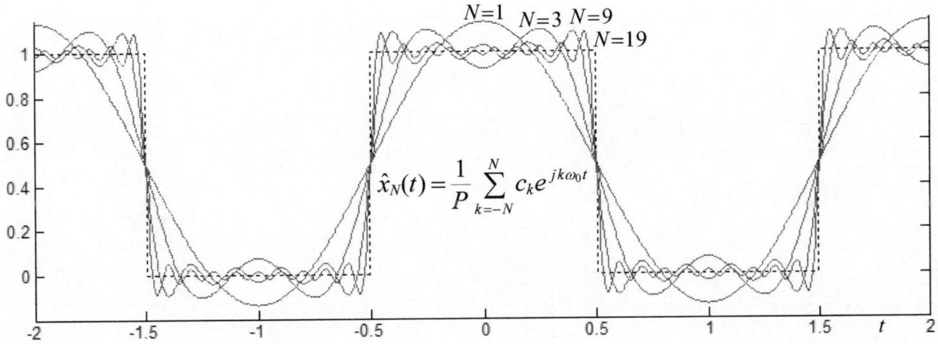

Figure 6.3 Fourier reconstruction of a rectangular wave with an increasing number of terms

[**Remark 6.1**] Convergence of Fourier Series Reconstruction

(1) The Fourier series convergence condition A stated in Section 6.1 guarantees that the Fourier coefficients are finite and the Fourier series reconstruction $\hat{x}_N(t)$ converges to the original time function $x(t)$ in the sense that

$$\int_P |\hat{x}_N(t)-x(t)|^2\, dt \rightarrow 0 \text{ as } N\rightarrow\infty$$

(2) The Fourier series convergence condition B stated in Section 6.1 guarantees the following:

- The Fourier coefficients are finite.
- The Fourier series reconstruction $\hat{x}_N(t)$ converges to the original time function $x(t)$ at every t, except the discontinuities of $x(t)$ and to the average value of the limits from the left/right at each discontinuity.

(3) Fig. 6.3 illustrates that $\hat{x}_N(t)$ has ripples around the discontinuities of $x(t)$, whose magnitude does not decrease as $N\to\infty$. This is called the *Gibbs phenomenon*.

(**Example 6.2**) Fourier Spectra of Rectified Cosine Waves

(a) Spectrum of a Cosine Wave (Fig. 6.4(a1))

Consider a cosine wave $x(t)=A\cos\omega_0 t$ with $\omega_0=2\pi/P$. In the case of single-tone periodic signal, the time function is the very Fourier series representation with a single nonzero coefficient

$$a_1 \overset{(6.1.3b)}{=} \frac{2A}{P}\int_{-P/2}^{P/2}\cos\omega_0 t\cos\omega_0 t\, dt \overset{(C.15)}{=} \frac{2A}{P}\int_{-P/2}^{P/2}\frac{1+\cos 2\omega_0 t}{2}\, dt = A \tag{E6.2.1}$$

The (magnitude) spectrum is depicted in Fig. 6.4(b1).

(b) Spectrum of a Half-Wave Rectified Cosine Wave (Fig. 6.4(a2))

Consider a half-wave rectified cosine wave $x_h(t)=\max(A\cos\omega_0 t, 0)$. We use Eq. (6.1.3b) to obtain the Fourier coefficients as

$$a_0 \overset{(6.1.3b)}{=} \frac{1}{P}\int_{-P/4}^{P/4} A\cos\omega_0 t\, dt \overset{\text{even}}{=} \frac{2A}{P}\int_0^{P/4}\cos\omega_0 t\, dt \overset{(C.36)}{=} \frac{2A}{\omega_0 P}\sin\omega_0 t\Big|_0^{P/4} = \frac{A}{\pi} \tag{E6.2.2}$$

$$a_k \underset{\text{even}}{\overset{(6.1.3b)}{=}} \frac{4}{P}\int_0^{P/4} A\cos\omega_0 t\cos k\omega_0 t\, dt \text{ with } \omega_0=\frac{2\pi}{P}$$

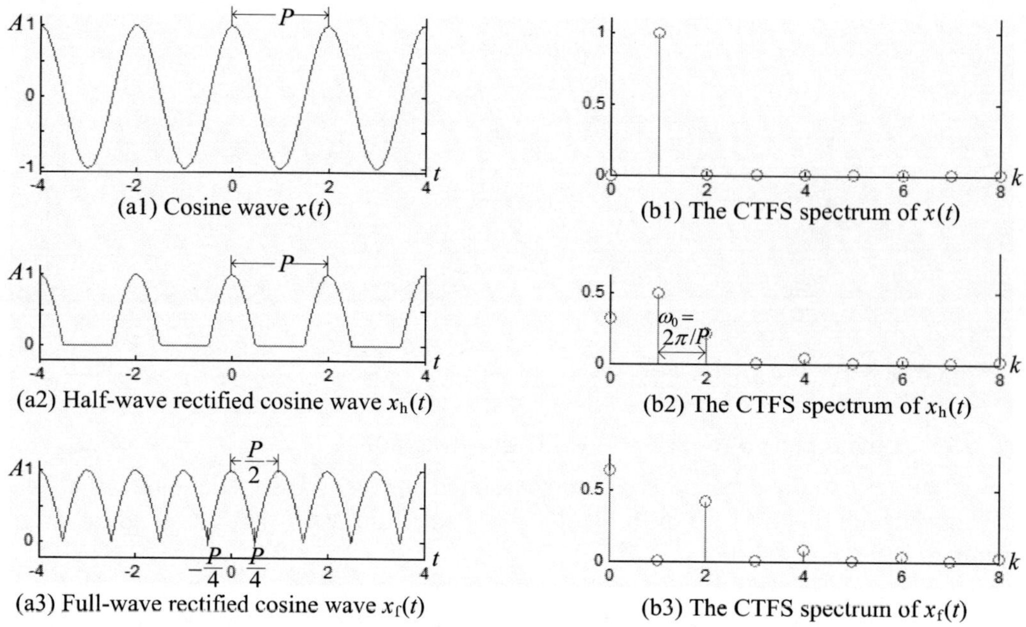

(a1) Cosine wave $x(t)$ (b1) The CTFS spectrum of $x(t)$

(a2) Half-wave rectified cosine wave $x_{\rm h}(t)$ (b2) The CTFS spectrum of $x_{\rm h}(t)$

(a3) Full-wave rectified cosine wave $x_{\rm f}(t)$ (b3) The CTFS spectrum of $x_{\rm f}(t)$

Figure 6.4 Half/full-wave rectified cosine waves and their CTFS spectra

$$\overset{\text{(C.11)}}{=} \frac{2A}{P}\int_0^{P/4}\left(\cos(k+1)\omega_0 t+\cos(k-1)\omega_0 t\right)dt$$

$$\overset{\text{(C.36)}}{=} \frac{2A}{(k+1)\omega_0 P}\sin(k+1)\omega_0 t\Big|_0^{P/4} + \frac{2A}{(k-1)\omega_0 P}\sin(k-1)\omega_0 t\Big|_0^{P/4}$$

$$\overset{k\neq 1}{=} \frac{A}{(k+1)\pi}\sin(k+1)\frac{\pi}{2} + \frac{A}{(k-1)\pi}\sin(k-1)\frac{\pi}{2}$$

$$\overset{k=2m(\text{even})}{=} (-1)^m\frac{A}{\pi}\left(\frac{1}{k+1}-\frac{1}{k-1}\right) = \frac{(-1)^{m+1}2A}{(k^2-1)\pi} \tag{E6.2.3}$$

$$a_1 \underset{\text{even}}{\overset{\text{(6.1.3b)}}{=}} \frac{4}{P}\int_0^{P/4}A\cos^2\omega_0 t\,dt \overset{\text{(C.15)}}{=} \frac{2A}{P}\int_0^{P/4}(1+\cos 2\omega_0 t)\,dt = \frac{A}{2} \tag{E6.2.4}$$

$$b_k \underset{\text{even}}{\overset{\text{(6.1.3b)}}{=}} \frac{2}{P}\int_{-P/4}^{P/4}A\cos\omega_0 t\,\sin k\omega_0 t\,dt \overset{\text{odd}}{=} 0 \tag{E6.2.5}$$

Thus, we can write the Fourier series representation of the half-wave rectified cosine wave as

$$x_{\rm h}(t) \underset{\text{(E6.2.2,3,4,5)}}{\overset{\text{(6.1.3a)}}{=}} \frac{A}{\pi}+\frac{A}{2}\cos\omega_0 t - \sum_{k=2m}^{\infty}\frac{(-1)^m 2A}{(k^2-1)\pi}\cos k\omega_0 t \tag{E6.2.6}$$

The corresponding (magnitude) spectrum is depicted in Fig. 6.4(b2).

(c) Spectrum of a Full-wave Rectified Cosine Wave (Fig. 6.4(a3))

Consider a full-wave rectified cosine wave $x_{\rm f}(t)=|A\cos\omega_0 t|$, which can be regarded as the sum of a half-wave rectified cosine wave $x_{\rm h}(t)$ and its $P/2$-shifted version $x_{\rm h}(t-P/2)$. Thus, instead of using Eq. (6.1.3b), we use Eq. (E6.2.6) to write its Fourier series representation as

$$x_f(t) = x_h(t) + x_h(t - \frac{P}{2}) \overset{(E6.2.6)}{=} \frac{A}{\pi} + \frac{A}{2}\cos\omega_0 t - \sum_{k=2m}^{\infty} \frac{(-1)^m 2A}{(k^2-1)\pi}\cos k\omega_0 t$$

$$+\frac{A}{\pi} + \frac{A}{2}\cos\omega_0(t - \frac{P}{2}) - \sum_{k=2m}^{\infty} \frac{(-1)^m 2A}{(k^2-1)\pi}\cos k\omega_0(t - \frac{P}{2}) \quad \text{with } \omega_0 P = 2\pi$$

$$\overset{(C.4)}{=} \frac{2A}{\pi} - \sum_{k=2m}^{\infty} \frac{(-1)^m 4A}{(k^2-1)\pi}\cos k\omega_0 t \overset{k=2m}{=} \frac{2A}{\pi} - \sum_{m=1}^{\infty} \frac{(-1)^m 4A}{(4m^2-1)\pi}\cos 2m\omega_0 t \tag{E6.2.7}$$

The corresponding (magnitude) spectrum is depicted in Fig. 6.4(b3).

(Note) In fact, the fundamental frequency of a full-wave rectified cosine wave is $2\omega_0 = 4\pi/P$, which is twice that of the original cosine wave.

(d) MATLAB program to get the Fourier spectra:

```
%em06e02.m : to plot Fig. 6.4
% plots the CTFS spectra of (half/full-rectified) cosine waves
clear, clf
global P
P=2; w0=2*pi/P; % Period and Fundamental frequency
N=8; kk=1:N; % Frequency indices
tt=[-400:400]*P/200; % Time interval of 4 periods
for i=1:3
   if i==1
     x ='cos_wave'; % Fourier coefficients from analytical results
     a0_t= 0; a_t= zeros(1,N); a_t(1)=1; b_t= zeros(1,N); % (E6.2.1)
    elseif i==2
     x ='cos_wave_half_rectified';
     a0_t= 1/pi; a_t= [1/2 zeros(1,N-1)]; b_t= zeros(1,N); % (E6.2.3)
     for m=1:floor(N/2), a_t(2*m)= -(-1)^m*2/(4*m^2-1)/pi; end
    else
     x ='cos_wave_full_rectified';
     a0_t= 2/pi; a_t= zeros(1,N); b_t= zeros(1,N); % (E6.2.7)
     for m=1:floor(N/2), a_t(2*m)= -(-1)^m*4/(4*m^2-1)/pi; end
   end
   xt = feval(x,tt); % Original signal
   [a0,a,b] = CTFS_trigonometric(x,P,N);
   discrepancy_between_numeric_and_analytic= [a0 a b]-[a0_t a_t b_t]
   xht = a0; % starting from the DC term
   for k=1:N % Fourier reconstruction (6.1.3a)
      xht = xht + a(k)*cos(k*w0*tt) + b(k)*sin(k*w0*tt);
   end
   subplot(319+i*2)
   plot(tt,xt,'k-', tt,xht,'r:')
   d1 = [a0 sqrt(a.^2+b.^2)]; phi1k = [0 atan2(a,b)]; % Eq.(6.1.3b)
   subplot(320+i*2)
   stem([0:length(d1)-1], d1)
end
```

```
function y=cos_wave(t)
global P, y=cos(2*pi/P*t);
```

```
function y=cos_wave_half_rectified(t)
global P, y=max(cos(2*pi/P*t),0);
```

```
function y=cos_wave_full_rectified(t)
global P, y=abs(cos(2*pi/P*t));
```

Just like the MATLAB script "em06e01.m" used to plot the spectra in Fig. 6.2, the above script "em06e02.m" can be run to yield Fig. 6.4, which shows the spectra of the above three signals that are defined and saved in M-files named "cos_wave.m," "cos_wave_half_rectified.m," and "cos_wave_full_rectified.m," respectively.

(Example 6.3) Fourier Spectrum of an Impulse Train

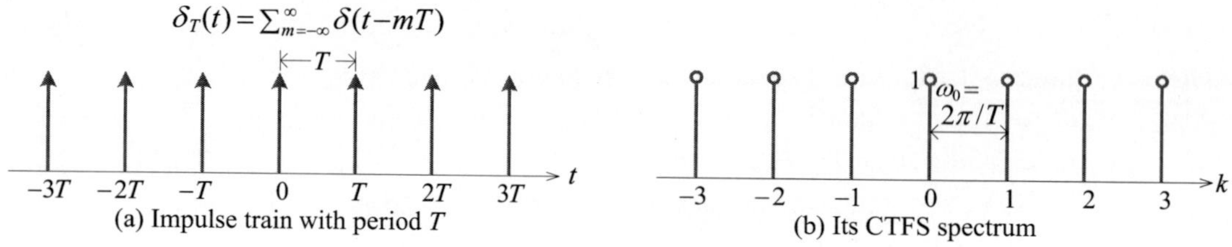

Figure 6.5 An impulse train and its CTFS spectrum

Consider an *impulse train* consisting of infinitely many shifted unit impulses that are equally spaced on the time axis:

$$\delta_T(t) = \sum_{m=-\infty}^{\infty} \delta(t-mT) \tag{E6.3.1}$$

We can use Eq. (6.1.6b) with $P=T$ and $\omega_0=2\pi/T$ to obtain the Fourier coefficients as

$$c_k = \int_{-T/2}^{T/2} \delta_T(t) e^{-jk\omega_0 t}\, dt \overset{\text{(E6.3.1)}}{=} \int_{-T/2}^{T/2} \sum_{m=-\infty}^{\infty} \delta(t-mT) e^{-jk\omega_0 t}\, dt$$

(since there is only one impulse $\delta(t)$ within the integration interval $[-T/2, T/2]$)

$$= \int_{-T/2}^{T/2} \delta(t) e^{-jk\omega_0 t}\, dt \underset{\text{with } t_1=0}{\overset{\text{(C.40)}}{=}} e^{-jk\omega_0 t}\Big|_{t=0} = 1 \;\forall\, k \tag{E6.3.2}$$

This means a flat spectrum that is uniformly distributed for every frequency index. Now we can use Eq. (6.1.6a) to write the Fourier series representation of the impulse train as

$$\delta_T(t) \underset{P=T}{\overset{\text{(6.1.6a)}}{=}} \frac{1}{T} \sum_{k=-\infty}^{\infty} c_k e^{jk\omega_0 t} \overset{\text{(E6.3.2)}}{=} \frac{1}{T} \sum_{k=-\infty}^{\infty} e^{jk\omega_0 t} \quad \text{with } \omega_0 = \frac{2\pi}{T} \tag{E6.3.3}$$

Figs. 6.5(a) and (b) show an impulse train and its spectrum, respectively.

6.2 CONTINUOUS-TIME FOURIER TRANSFORM (CTFT)[Y-2]

6.2.1 Definition and Convergence Conditions

In this section we will define the Fourier transform for aperiodic signals. Suppose we have an aperiodic signal $x(t)$ of finite duration $D>0$ and its *periodic extension* (*repetition*) $\tilde{x}_P(t)$ with period $P>D$ that is obtained by repeating $x(t)$ every P seconds.

Noting that as we choose the period P to be longer, $\tilde{x}_P(t)$ appears to be identical to $x(t)$ over a longer interval, we can think of $x(t)$ as the limit of $\tilde{x}_P(t)$ as $P\to\infty$. Since $\tilde{x}_P(t)$ is periodic with period P, it can be represented by the Fourier series of the form

$$\tilde{x}_P(t) = \frac{1}{P}\sum_{k=-\infty}^{\infty} c_k\, e^{jk\omega_0 t} = \frac{1}{2\pi}\sum_{k=-\infty}^{\infty} X(jk\omega_0)\, e^{jk\omega_0 t}\,\omega_0 \text{ with } \omega_0 = \frac{2\pi}{P}$$

where the CTFS coefficients are

$$X(jk\omega_0) = c_k = \int_P \tilde{x}_P(t)e^{-jk\omega_0 t}\,dt = \int_{-\infty}^{\infty} x(t)e^{-jk\omega_0 t}\,dt$$

Noting that $\tilde{x}_P(t) \to x(t)$ and $\omega_0=2\pi/P\to 0$ as $P\to\infty$, we let $\omega_0=d\omega$ & $k\omega_0=\omega$ and take the limits of the above two equations as $P\to\infty$ to write the *continuous-time Fourier transform* (*CTFT*) pair:

$$X(j\omega) = \mathcal{F}\{x(t)\} = \int_{-\infty}^{\infty} x(t)e^{-j\omega t}\,dt \;(\omega:\text{analog frequency})\;(\text{CTFT}) \tag{6.2.1a}$$

$$x(t) = \mathcal{F}^{-1}\{X(j\omega)\} = \frac{1}{2\pi}\int_{-\infty}^{\infty} X(j\omega)e^{j\omega t}\,d\omega \;(\text{ICTFT: Inverse CTFT}) \tag{6.2.1b}$$

where $X(j\omega)$, called the spectrum of $x(t)$, has values for a continuous variable ω (called the *analog frequency*) and is often written as $X(\omega)$ with the constant j omitted. Like the CTFS of periodic signals, the CTFT of aperiodic signals provides us with the information concerning the frequency contents of signals, while the concept of frequency describes how rapidly a signal changes rather than how fast it oscillates.

Note that the sufficient condition for the convergence of CTFT is obtained by replacing the square-integrability condition (6.1.2a) of Condition A with

$$\int_{-\infty}^{\infty} |x(t)|^2\,dt < \infty \tag{6.2.2a}$$

or by replacing the absolute-integrability condition (6.1.2b) of Condition B with

$$\int_{-\infty}^{\infty} |x(t)|\,dt < \infty \tag{6.2.2b}$$

[**Remark 6.2**] Fourier Transform and Laplace Transform

For any square-integrable or absolutely integrable causal function $x(t)$ such that $x(t)=0\ \forall\, t<0$, the Fourier transform can be obtained by substituting $s=j\omega$ into the Laplace transform:

$$X(j\omega) \overset{(6.2.1a)}{=} \int_{-\infty}^{\infty} x(t)e^{-j\omega t}\,dt \underset{\text{causal signal}}{\overset{x(t)=0 \text{ for } t<0}{=}} \int_{0}^{\infty} x(t)e^{-j\omega t}\,dt \overset{\text{Table A.7(0b)}}{=} X(s)\big|_{s=j\omega} \tag{6.2.3}$$

(**Example 6.4**) CTFT Spectra of a Rectangular (Square) Pulse and a Triangular Pulse

(a) CTFT Spectrum of a Rectangular (Square) Pulse (Fig. 6.6(a1))

Consider a single rectangular pulse with height A and duration D on the interval $[-D/2, D/2]$:

$$A\,r_D(t) = A\left(u_s\left(t+\frac{D}{2}\right)-u_s\left(t-\frac{D}{2}\right)\right) = \begin{cases} A & \text{for } -D/2 \le t < D/2 \\ 0 & \text{elsewhere} \end{cases} \tag{E6.4.1}$$

We use Eq. (6.2.1a) to obtain the CTFT as

$$AR_D(j\omega) \overset{(6.2.1a)}{=} \int_{-\infty}^{\infty} A r_D(t) e^{-j\omega t}\, dt = A\int_{-D/2}^{D/2} e^{-j\omega t}\, dt = \frac{A}{-j\omega} e^{-j\omega t}\Big|_{-D/2}^{D/2}$$

$$= A\frac{e^{j\omega D/2} - e^{-j\omega D/2}}{j\omega} = AD\frac{\sin(\omega D/2)}{\omega D/2} = AD\,\mathrm{sinc}\left(\frac{\omega D}{2\pi}\right) \tag{E6.4.2}$$

This CTFT spectrum is depicted in Fig. 6.6(b1).

(Note) As a by-product, we can apply the inverse CTFT formula (6.2.1b) for Eq. (E6.4.2) to get the integral of a sinc function:

$$r_D(t) \overset{(6.2.1b)}{=} \frac{1}{2\pi}\int_{-\infty}^{\infty} R_D(j\omega) e^{j\omega t}\, d\omega \overset{(E6.4.2)}{=} \frac{1}{2\pi}\int_{-\infty}^{\infty} D\frac{\sin(\omega D/2)}{\omega D/2} e^{j\omega t}\, d\omega$$

$$;\qquad u_s(t+\frac{D}{2}) - u_s(t-\frac{D}{2}) \overset{\omega=2w}{=} \frac{1}{\pi}\int_{-\infty}^{\infty} \frac{\sin(wD)}{w} e^{j2wt}\, dw$$

Substituting $t=0$ into this equation yields

$$\frac{1}{\pi}\int_{-\infty}^{\infty} \frac{\sin(wD)}{w}\, dw = u_s(\frac{D}{2}) - u_s(-\frac{D}{2}) = \mathrm{sign}(D) = \begin{cases} 1 & \text{for } D>0 \\ 0 & \text{for } D=0 \\ -1 & \text{for } D<0 \end{cases} \tag{E6.4.3}$$

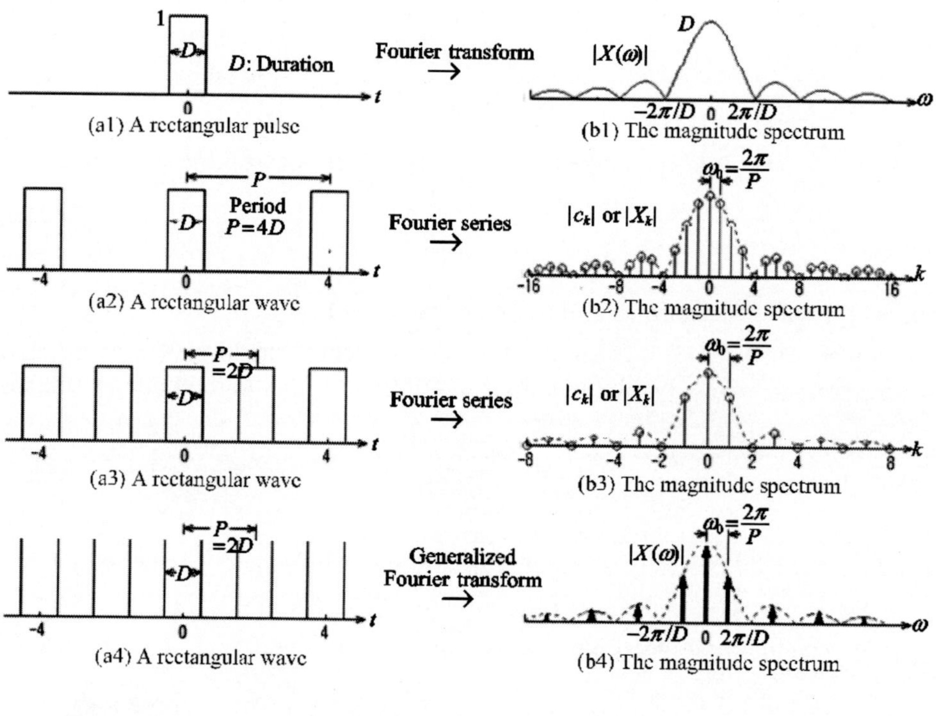

Figure 6.6 The CTFS or CTFT spectra of a rectangular pulse and waves

(b) CTFT Spectrum of a Triangular Pulse

Like Eq. (E6.1.7), which is the CTFS coefficient of a triangular wave, we can find the CTFT spectrum of a single triangular pulse $x(t)=A\lambda_D(t)$ as

$$X(j\omega) = A\Lambda_D(j\omega) = AD\frac{\sin^2(\omega D/2)}{(\omega D/2)^2} = AD\,\text{sinc}^2\left(\frac{\omega D}{2\pi}\right) \tag{E6.4.4}$$

Figs. 6.6(a1) and (b1) show a rectangular pulse and its CTFT spectrum, respectively, while Figs. 6.6(a2)/(a3) and (b2)/(b3) show two rectangular waves and their CTFS spectra, respectively. These figures present us with an observation about the relationship between the CTFT of a single pulse $x(t)$ and the CTFS of its periodic repetition $\tilde{x}_P(t)$ with period P, which is summarized in the following remark.

[**Remark 6.3**] Fourier Series and Fourier Transform

(1) We will mainly use the complex exponential Fourier coefficients, but seldom use the trigonometric or magnitude-and-phase form of the Fourier series. Thus, from now on, we denote the complex exponential Fourier coefficients of $x(t)$ by X_k instead of c_k, which has been used so far to distinguish it from the other Fourier coefficients a_k, b_k, or d_k.

(2) As can be seen from comparing Eqs. (E6.1.3) and (E6.4.2) or Eqs. (E6.1.7) and (E6.4.4), the relationship between the CTFT $X(j\omega)$ of $x(t)$ and the CTFS coefficient X_k of $\tilde{x}_P(t)$ (the periodic extension of $x(t)$ with period P) is as follows:

$$X(j\omega)\big|_{\omega=k\omega_0=2\pi k/P} = X(jk\omega_0) = X_k \tag{6.2.4}$$

As the period P gets longer so that the fundamental frequency or frequency spacing $\omega_0=2\pi/P$ decreases, the Fourier coefficients X_k's become more closely spaced samples of the CTFT $X(j\omega)$, implying that the set of CTFS coefficients approaches the CTFT as $P\to\infty$ (see Figs. 6.6(b3), (b2), and (b1)).

(3) Unlike the discrete frequency $k\omega_0$ of CTFS, the continuous frequency ω of CTFT describes how abruptly the signal changes rather than how often it oscillates.

(4) If the CTFT of a single pulse $x(t)$ and the CTFS of the periodic extension $\tilde{x}_P(t)$ were of different shapes in spite of the same shape of $x(t)$ and $\tilde{x}_P(t)$ over one period P, it would be so confusing for some one who wants the spectral information about a signal without knowing whether it is of finite duration or periodic. In this context, how lucky we are to have the same shape of spectrum (in the sense that CTFS are just samples of CTFT) whether we take the CTFT of $x(t)$ or the CTFS of $\tilde{x}_P(t)$! Furthermore, you will be happier to see that even the CTFT of $\tilde{x}_P(t)$ (Fig. 6.6(b4)) is also of the same shape as the CTFS of $\tilde{x}_P(t)$, because one might observe one period of $\tilde{x}_P(t)$ and mistake it for $x(t)$ so that he or she would happen to apply the CTFT for periodic signals. Are you puzzled at the CTFT of a periodic signal? Then rush into the next section.

6.2.2 (Generalized) CTFT of Periodic Signals

Since a periodic signal can satisfy neither the square-integrability condition (6.2.2a) nor the absolute-integrability condition (6.2.2b), the CTFTs of periodic signals are not only singular but also difficult to compute directly. For example, let us try to compute the CTFT of $x_k(t)=e^{jk\omega_0 t}$ by using Eq. (6.2.1a):

$$\mathcal{F}\{x_k(t)\} \overset{(6.2.1a)}{=} \int_{-\infty}^{\infty} e^{jk\omega_0 t} e^{-j\omega t}\,dt = \int_{-\infty}^{\infty} e^{-j(\omega-k\omega_0)t}\,dt \overset{(C.35)}{=} \frac{1}{-j(\omega-k\omega_0)} e^{-j(\omega-k\omega_0)t}\Big|_{-P}^{P}\Big|_{P\to\infty}$$

$$= \frac{1}{j(\omega-k\omega_0)}\left(e^{j(\omega-k\omega_0)P} - e^{-j(\omega-k\omega_0)P}\right)\Big|_{P=\infty} \overset{(C.22)}{=} \left(\frac{2\sin(\omega-k\omega_0)P}{(\omega-k\omega_0)}\right)\Big|_{P=\infty} = ? \qquad (6.2.5)$$

To get around this mathematical difficulty, let us find the ICTFT (inverse CTFT) of

$$X_k(j\omega) = 2\pi\delta(\omega-k\omega_0)$$

by applying the ICTFT formula (6.2.1b):

$$\mathcal{F}^{-1}\{X_k(j\omega)\} \overset{(6.2.1b)}{=} \frac{1}{2\pi}\int_{-\infty}^{\infty} 2\pi\,\delta(\omega-k\omega_0) e^{j\omega t}\,d\omega \overset{(C.40)}{=} e^{jk\omega_0 t}$$

This implies a (generalized) CTFT pair as

$$x_k(t) = e^{jk\omega_0 t} \overset{\mathcal{F}}{\leftrightarrow} X_k(j\omega) = 2\pi\,\delta(\omega-k\omega_0) \qquad (6.2.6)$$

Based on this relation, we get the CTFT of a periodic function $x(t)$ from its CTFS representation as follows:

$$x(t) \overset{(6.1.6a)}{=} \frac{1}{P}\sum_{k=-\infty}^{\infty} X_k e^{jk\omega_0 t} \text{ with } \omega_0 = \frac{2\pi}{P} \overset{\mathcal{F}}{\leftrightarrow} X(j\omega) \overset{(6.2.6)}{=} \frac{2\pi}{P}\sum_{k=-\infty}^{\infty} X_k\,\delta(\omega-k\omega_0) \qquad (6.2.7)$$

This implies that the CTFT of a periodic signal consists of a train of impulses on the frequency axis having the same shape of envelope as the CTFS spectrum. Fig. 6.6(b4) is an illustration of Eq. (6.2.7) as stated in Remark 6.3(4).

[**Remark 6.4**] Fourier Transform of a Periodic Signal

It would be cumbersome to directly find the CTFT of a periodic function. Thus, we had better find the CTFS coefficients first and then use Eq. (6.2.6) as illustrated in Eq. (6.2.7).

6.2.3 Examples of CTFT

(**Example 6.5**) CTFT of a Unilateral Exponential Function

For a unilateral (one-sided) decaying exponential function (Fig. 6.7(a)) with time constant $T>0$:

$$e_1(t) = \frac{1}{T}e^{-t/T}u_s(t) \text{ with } T>0, \qquad (E6.5.1)$$

we have the CTFT

$$E_1(j\omega) = \mathcal{F}\{e_1(t)\} \underset{(E6.5.1)}{\overset{(6.2.1a)}{=}} \int_{-\infty}^{\infty}\frac{1}{T}e^{-t/T}u_s(t)e^{-j\omega t}\,dt = \frac{1}{T}\int_0^{\infty} e^{-t/T}e^{-j\omega t}\,dt = \frac{1}{T}\int_0^{\infty} e^{-(1/T+j\omega)t}\,dt$$

$$\overset{(C.35)}{=} -\frac{1}{T(1/T+j\omega)}e^{-(1/T+j\omega)t}\Big|_0^{\infty} = \frac{1}{1+j\omega T} = \frac{1}{\sqrt{1+(\omega T)^2}}\angle -\tan^{-1}(\omega T) \qquad (E6.5.2)$$

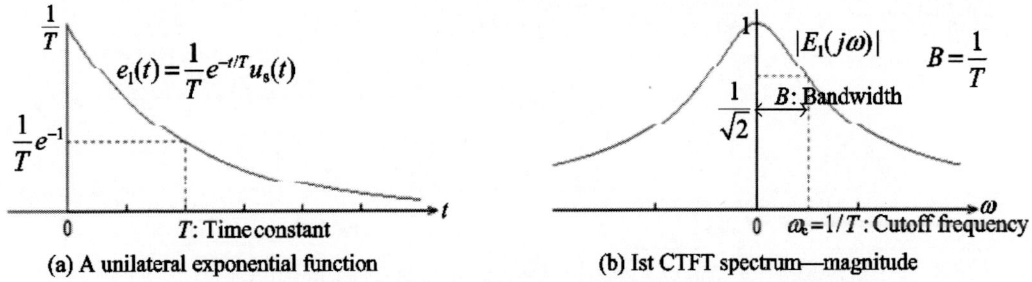

Figure 6.7 A unilateral exponential function and its CTFT spectrum

whose magnitude is depicted in Figure 6.7(b). From this magnitude spectrum, we see that $\omega_c=1/T$ is the half-power frequency at which $|E_1(j\omega)|$ is $1/\sqrt{2}$ times the maximum magnitude 1:

$$|E_1(j\omega)|=\frac{1}{\sqrt{1+(\omega T)^2}}=\frac{1}{\sqrt{2}};\ 1+(\omega T)^2=2;\ \omega T=1;\ \omega_c=\frac{1}{T} \tag{E6.5.3}$$

(Example 6.6) CTFT of an Even-Symmetric Exponential Function

For an even-symmetric exponential function (Fig. 6.8(a)) with time constant $T>0$:

$$e_2(t)=\frac{1}{2T}e^{-|t|/T}\ \text{ with } T>0 \tag{E6.6.1}$$

we have the CTFT

$$E_2(j\omega)=\mathcal{F}\{e_2(t)\}\underset{(E6.6.1)}{\overset{(6.2.1a)}{=}}\int_{-\infty}^{\infty}\frac{1}{2T}e^{-|t|/T}e^{-j\omega t}dt$$

$$=\frac{1}{2T}\left\{\int_{-\infty}^{0}e^{t/T}e^{-j\omega t}dt+\int_{0}^{\infty}e^{-t/T}e^{-j\omega t}dt\right\}$$

$$\overset{(C.35)}{=}\frac{1}{2T(1/T-j\omega)}e^{(1/T-j\omega)t}\Big|_{-\infty}^{0}-\frac{1}{2T(1/T+j\omega)}e^{-(1/T+j\omega)t}\Big|_{0}^{\infty}$$

$$=\frac{1/2}{1-j\omega T}+\frac{1/2}{1+j\omega T}=\frac{1}{1+(\omega T)^2} \tag{E6.6.2}$$

whose magnitude is depicted in Figure 6.8(b). We can use MATLAB

(Q) Why has the magnitude spectrum narrower bandwidth than that in Fig. 6.7(b)?

(A) It is because the signal in Fig. 6.8(a) is smoother than that in Fig. 6.7(a).

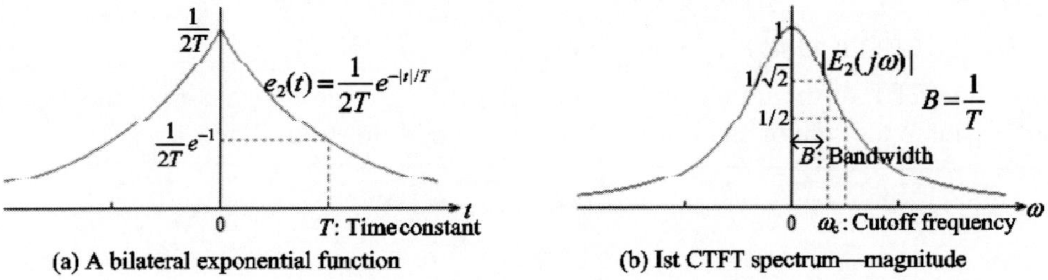

Figure 6.8 A bilateral exponential function and its CTFT spectrum

(Example 6.7) CTFT of the Unit Impulse ("Dirac Delta") Function

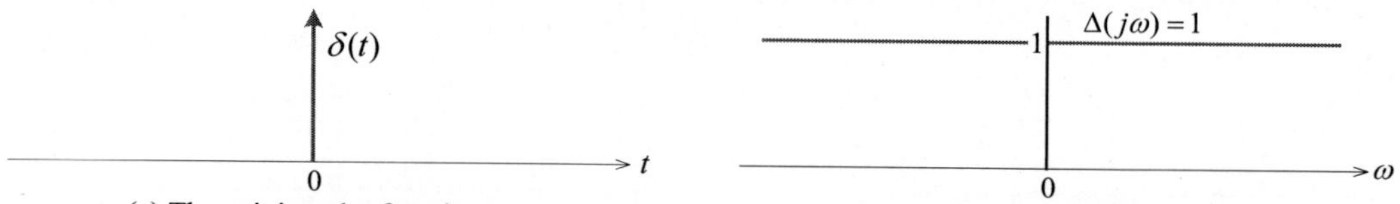

(a) The unit impulse function (b) Its CTFT spectrum

Figure 6.9 The unit impulse function and its CTFT spectrum

We can obtain the CTFT of the unit impulse function $\delta(t)$ as

$$\Delta(j\omega) = \mathcal{F}\{\delta(t)\} \overset{(6.2.1a)}{=} \int_{-\infty}^{\infty} \delta(t) e^{-j\omega t} dt \overset{(C.40)}{=} 1 \ \forall \ \omega \tag{E6.7.1}$$

This implies that an impulse signal has a flat or white spectrum, which is evenly distributed over all frequencies (see Fig. 6.9).

(Example 6.8) CTFT of a Constant Function

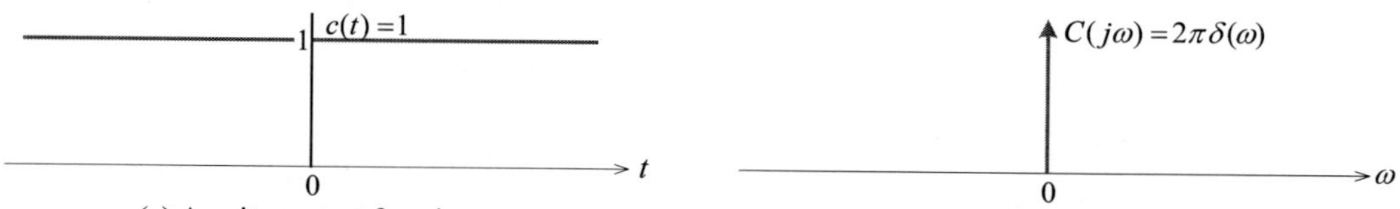

(a) A unit constant function (b) Its CTFT spectrum

Figure 6.10 A unit constant (DC) function and its CTFT spectrum

We can obtain the CTFT of the unit constant function $c(t)=1$ (Fig. 6.10(a)) as

$$C(j\omega) = \mathcal{F}\{1\} \overset{(6.2.1a)}{=} \int_{-\infty}^{\infty} 1 e^{-j\omega t} dt \underset{\text{with } k=0}{\overset{(6.2.6)}{=}} 2\pi\delta(\omega) \tag{E6.8.1}$$

This spectrum, which is depicted in Fig. 6.10(b), shows that a constant signal has only a DC component with zero frequency, i.e., has all of its (infinite) energy at $\omega=0$.

This can also be verified by using the inverse CTFT (6.2.1b) to show that

$$\mathcal{F}^{-1}\{2\pi\delta(\omega)\} \overset{(6.2.1b)}{=} \frac{1}{2\pi}\int_{-\infty}^{\infty} 2\pi\delta(\omega) e^{j\omega t} d\omega \overset{(C.40)}{=} 1 \tag{E6.8.2}$$

(Example 6.9) CTFT of the Unit Step Function

To compute the CTFT of the unit step function $u_s(t)$, let us write its even–odd decomposition, i.e., decompose $u_s(t)$ into the sum of an even function and an odd function (Fig. 6.11) as

$$u_s(t) = u_e(t) + u_o(t) \tag{E6.9.1}$$

where

$$u_e(t) = \frac{1}{2}(u_s(t) + u_s(-t)) = \begin{cases} 1/2 & \text{for } t\neq 0 \\ 1 & \text{for } t=0 \end{cases} \tag{E6.9.2}$$

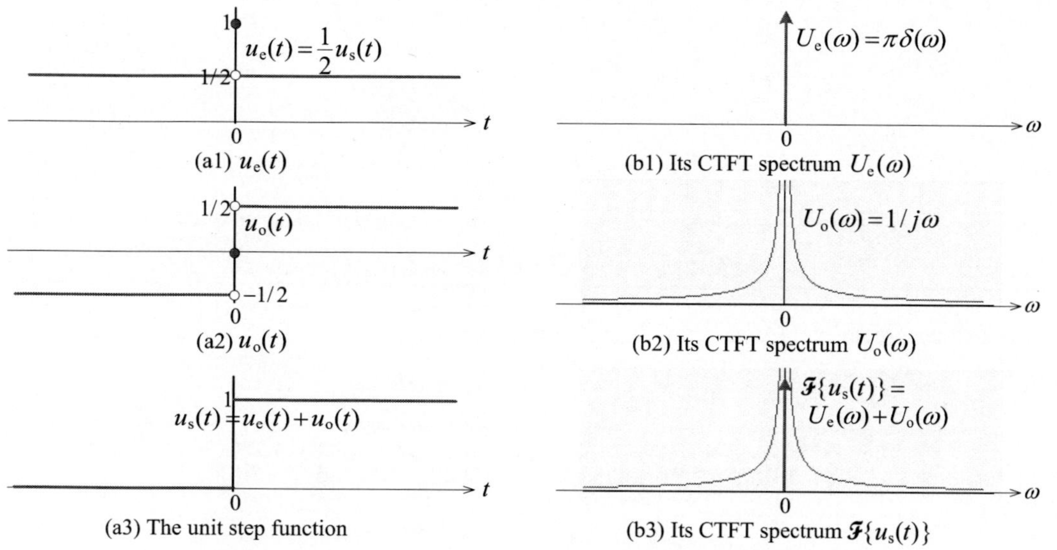

Figure 6.11 The unit step function and its CTFT spectrum

$$u_o(t)=\frac{1}{2}(u_s(t)-u_s(-t))=\frac{1}{2}\text{sign}(t)=\begin{cases}1/2 & \text{for } t>0\\ 0 & \text{for } t\neq 0\\ -1/2 & \text{for } t<0\end{cases} \tag{E6.9.3}$$

Then, noting that the even part $u_e(t)$ is a constant function of amplitude 1/2 except at t=0, we can use Eq. (E6.8.1) to write its CTFT as

$$U_e(j\omega)=\mathcal{F}\{u_e(t)\}=\frac{1}{2}\mathcal{F}\{1\}\overset{\text{(E6.8.1)}}{=}\pi\delta(\omega) \tag{E6.9.4}$$

On the other side, the CTFT of the odd part can be computed as

$$\begin{aligned}U_o(j\omega)&=\mathcal{F}\{u_o(t)\}\overset{\text{(6.2.1a)}}{=}\int_{-\infty}^{\infty}u_o(t)e^{-j\omega t}\,dt\\
&=\int_{-\infty}^{\infty}\overset{\text{odd}}{u_o(t)}(\overset{\text{even}}{\cos\omega t}-j\overset{\text{odd}}{\sin\omega t})\,dt\\
&=-j\int_{-\infty}^{\infty}\overset{\text{odd}\times\text{odd}=\text{even}}{u_o(t)\sin\omega t}\,dt=-j2\int_0^{\infty}\overset{\text{even}}{u_o(t)\sin\omega t}\,dt\\
&\overset{u_o(t)=1/2\text{ for }t>0}{=}-j\int_0^{\infty}\sin\omega t\,dt=-j\int_0^{\infty}\sin\omega t e^{-st}\,dt\Big|_{s=0}\\
&\overset{\text{Table A.7(0b)}}{=}-j\,\mathcal{L}\{\sin\omega t\}\Big|_{s=0}\overset{\text{Table A.8(9)}}{=}-j\frac{\omega}{s^2+\omega^2}\Big|_{s=0}=\frac{1}{j\omega}\end{aligned} \tag{E6.9.5}$$

where we have used the Laplace transform. Now we add these two results (E6.9.4) and (E6.9.5) to obtain the CTFT of the unit step function (Fig. 6.11) as

$$\mathcal{F}\{u_s(t)\}=\mathcal{F}\{u_e(t)+u_o(t)\}\overset{\text{(6.3.1)}}{=}U_e(j\omega)+U_o(j\omega)\overset{\text{(E6.9.4), (E6.9.5)}}{=}\pi\delta(\omega)+\frac{1}{j\omega} \tag{E6.9.6}$$

(Example 6.10) Inverse CTFT of an Ideal LPF (Lowpass Filter) Frequency Response

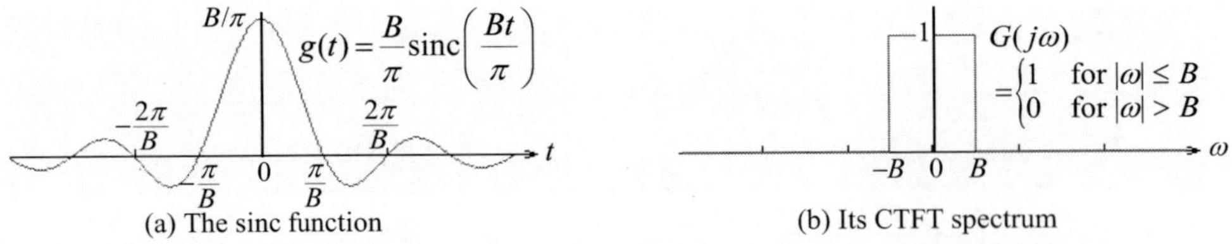

Figure 6.12 The sinc function and its CTFT spectrum

Let us consider a rectangular spectrum depicted in Fig. 6.12(b):

$$G(j\omega)=\begin{cases}1 & \text{for } |\omega|\le B\\ 0 & \text{for } |\omega|>B\end{cases} \tag{E6.10.1}$$

Taking the inverse CTFT of this rectangular spectrum yields the sinc function as

$$\begin{aligned} g(t)&=\mathcal{F}^{-1}\{G(j\omega)\}\overset{(6.2.1\text{b})}{=}\frac{1}{2\pi}\int_{-\infty}^{\infty}G(j\omega)e^{j\omega t}\,d\omega=\frac{1}{2\pi}\int_{-B}^{B}1e^{j\omega t}\,d\omega\\ &\overset{(\text{C.35})}{=}\frac{1}{2\pi jt}(e^{jBt}-e^{-jBt})=\frac{\sin Bt}{\pi t}=\frac{B}{\pi}\text{sinc}\left(\frac{Bt}{\pi}\right) \end{aligned} \tag{E6.10.2}$$

which is depicted in Fig. 6.12(a).

(Example 6.11) CTFT of an Impulse Train

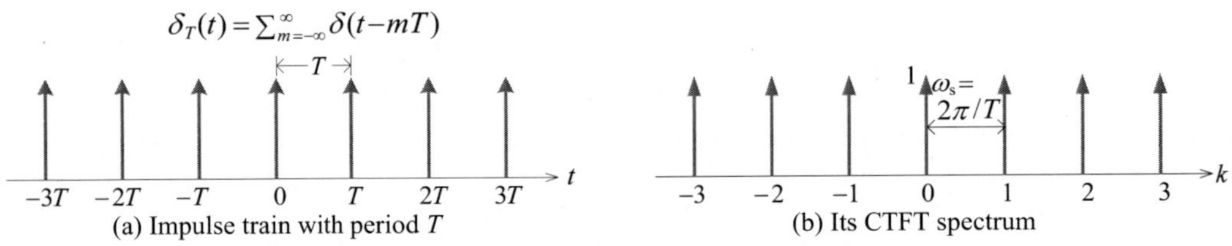

Figure 6.13 An impulse train and its CTFT spectrum

Let us consider an impulse train

$$\delta_T(t)=\sum_{m=-\infty}^{\infty}\delta(t-mT) \tag{E6.11.1}$$

Since this is a periodic function, we first write its CTFS representation from Eq. (E6.3.3) as

$$\delta_T(t)\overset{(\text{E6.3.3})}{=}\frac{1}{T}\sum_{k=-\infty}^{\infty}e^{jk\omega_s t}\ \text{ with } \omega_s=\frac{2\pi}{T} \tag{E6.11.2}$$

Then we use Eq. (6.2.6) with $\omega_0=\omega_s=2\pi/T$ to obtain the CTFT as

$$\Delta_T(j\omega)\underset{(6.2.6)}{\overset{(\text{E6.11.2})}{=}}\frac{2\pi}{T}\sum_{k=-\infty}^{\infty}\delta(\omega-k\omega_s)=\frac{2\pi}{T}\sum_{k=-\infty}^{\infty}\delta(\omega+k\omega_s)\ \text{ with } \omega_s=\frac{2\pi}{T} \tag{E6.11.3}$$

(Note) Applying Eq. (6.2.1b) to take the ICTFT of Eq. (E6.11.3) will produce Eq. (E6.11.2).

(Example 6.12) CTFT of Cosine/Sine Functions

(a) The CTFT of $x(t)=\sin(\omega_1 t)=(e^{j\omega_1 t}-e^{-j\omega_1 t})/j2$ can be obtained as

$$X(j\omega)=\mathcal{F}\{\sin(\omega_1 t)\}=\frac{1}{j2}\mathcal{F}\{e^{j\omega_1 t}-e^{-j\omega_1 t}\}\overset{(6.2.6)}{=}j\pi(\delta(\omega+\omega_1)-\delta(\omega-\omega_1)) \quad \text{(E6.12.1)}$$

(b) The CTFT of $x(t)=\cos(\omega_1 t)=(e^{j\omega_1 t}+e^{-j\omega_1 t})/2$ can be obtained as

$$X(j\omega)=\mathcal{F}\{\cos(\omega_1 t)\}=\frac{1}{2}\mathcal{F}\{e^{j\omega_1 t}+e^{-j\omega_1 t}\}\overset{(6.2.6)}{=}\pi(\delta(\omega+\omega_1)+\delta(\omega-\omega_1)) \quad \text{(E6.12.2)}$$

6.2.4 Properties of CTFT

In this section we discuss some basic properties of continuous-time Fourier transform (CTFT), which will provide us with an insight into the Fourier transform and the capability of taking easy ways to get the CTFTs or inverse CTFTs (ICTFTs).

(Note) From now on, we will use $X(\omega)$ instead of $X(j\omega)$ to denote the CTFT of $x(t)$.

6.2.4.1 Linearity

With $\mathcal{F}\{x(t)\}=X(\omega)$ and $\mathcal{F}\{y(t)\}=Y(\omega)$, we have

$$a\,x(t)+b\,y(t)\overset{\mathcal{F}}{\leftrightarrow}a\,X(\omega)+b\,Y(\omega) \quad (6.2.8)$$

which implies that the CTFT of a linear combination of many functions is the same linear combination of the individual transforms.

6.2.4.2 (Conjugate) Symmetry

In general, CTFT has the *time reversal property*:

$$\mathcal{F}\{x(-t)\}\overset{(6.2.1a)}{=}\int_{-\infty}^{\infty}x(-t)e^{-j\omega t}dt\overset{-t=\tau}{=}\int_{\infty}^{-\infty}x(\tau)e^{j\omega\tau}(-d\tau)\overset{\tau=t}{=}\int_{-\infty}^{\infty}x(t)e^{j\omega t}dt\underset{\omega\to-\omega}{\overset{(6.2.1a)}{=}}X(-\omega)$$

$$;\qquad x(-t)\overset{\mathcal{F}}{\leftrightarrow}X(-\omega) \quad (6.2.9)$$

In case $x(t)$ is a real-valued function, we have

$$X(-\omega)\underset{\omega\to-\omega}{\overset{(6.2.1a)}{=}}\int_{-\infty}^{\infty}x(t)e^{-j(-\omega)t}dt=\int_{-\infty}^{\infty}x(t)\,e^{-(-j)\omega t}dt\underset{j\to-j}{\overset{(6.2.1a)}{=}}X^*(\omega):\text{Complex conjugate of }X(\omega)$$

or equivalently,

$$\mathrm{Re}\{X(-\omega)\}+j\,\mathrm{Im}\{X(-\omega)\}=\mathrm{Re}\{X(\omega)\}-j\,\mathrm{Im}\{X(\omega)\}$$

$$;\qquad |X(-\omega)|\angle X(-\omega)=|X(\omega)|\angle -X(\omega) \quad (6.2.10)$$

This implies that the real/imaginary parts and magnitude/phase of the CTFT of a real-valued function are even/odd functions of frequency ω. Thus, when obtaining the CTFT of a real-valued time function, we need to compute it only for $\omega\ge 0$ since we can use the conjugate symmetry to generate the values for $\omega<0$ from those for $\omega>0$. In other words, for a real-valued time function, its magnitude and phase spectra are symmetrical about the vertical axis and the origin, respectively.

For an even and real-valued function $x_e(t)$ such that $x_e(-t)=x_e(t)$, its CTFT is also an even and real-valued function in frequency ω:

$$X_e(-\omega)\overset{(6.2.1a)}{\underset{\omega=-\omega}{=}}\int_{-\infty}^{\infty}x_e(t)e^{-j(-\omega)t}dt\overset{t=-\tau}{=}\int_{-\infty}^{\infty}x_e(-\tau)e^{-j\omega\tau}d\tau\overset{x_e(-\tau)=x_e(\tau)}{\underset{\text{even}}{=}}\int_{-\infty}^{\infty}x_e(\tau)e^{-j\omega\tau}d\tau\overset{(6.2.1a)}{=}X_e(\omega)\tag{6.2.11a}$$

Also for an odd and real-valued function $x_o(t)$ such that $x_o(-t)=-x_o(t)$, its Fourier transform is an odd and imaginary-valued function in frequency ω:

$$X_o(-\omega)\overset{(6.2.1a)}{\underset{\omega=-\omega}{=}}\int_{-\infty}^{\infty}x_o(t)e^{-j(-\omega)t}dt\overset{t=-\tau}{=}\int_{-\infty}^{\infty}x_o(-\tau)e^{-j\omega\tau}d\tau\overset{x_o(-\tau)=-x_o(\tau)}{\underset{\text{odd}}{=}}-\int_{-\infty}^{\infty}x_o(\tau)e^{-j\omega\tau}d\tau\overset{(6.2.1a)}{=}-X_o(\omega)\tag{6.2.11b}$$

Since any real-valued function $x(t)$ can be expressed as the sum of an even function and an odd one:

$$x(t)=x_e(t)+x_o(t)\ \text{ with }\ x_e(t)=\frac{1}{2}(x(t)+x(-t))\quad\text{and}\quad x_o(t)=\frac{1}{2}(x(t)-x(-t))$$

we can get the relations

$$X(\omega)=\mathcal{F}\{x_e(t)\}+\mathcal{F}\{x_o(t)\}=X_e(\omega)+X_o(\omega)$$
$$;\ \ \mathrm{Re}\{X(\omega)\}+j\,\mathrm{Im}\{X(\omega)\}=X_e(\omega)+X_o(\omega)$$

which implies

$$\text{even and real-valued } x_e(t)\ \overset{\mathcal{F}}{\leftrightarrow}\ \mathrm{Re}\{X(\omega)\}\ \text{ even and real-valued}\tag{6.2.12a}$$

$$\text{odd and real-valued } x_o(t)\ \overset{\mathcal{F}}{\leftrightarrow}\ j\,\mathrm{Im}\{X(\omega)\}\ \text{ odd and imaginary-valued}\tag{6.2.12b}$$

6.2.4.3 Conjugation

For the complex conjugate $x^*(t)$ of a time function $x(t)$, we have its CTFT

$$\mathcal{F}\{x^*(t)\}\overset{(6.2.1a)}{=}\int_{-\infty}^{\infty}x^*(t)e^{-j\omega t}dt=\left(\int_{-\infty}^{\infty}x(t)\,e^{j\omega t}dt\right)^*\overset{(6.2.1a)}{=}X^*(-\omega)$$

$$;\qquad x^*(t)\ \overset{\mathcal{F}}{\leftrightarrow}\ X^*(-\omega)\tag{6.2.13}$$

6.2.4.4 Time/Frequency Shifting (Real/Complex Translation)

For a shifted time function $x_1(t)=x(t-t_1)$, we have its CTFT as

$$\mathcal{F}\{x(t-t_1)\}\overset{(6.2.1a)}{=}\int_{-\infty}^{\infty}x(t-t_1)\,e^{-j\omega t}dt\overset{t-t_1\to t}{=}\int_{-\infty}^{\infty}x(t)\,e^{-j\omega(t+t_1)}dt\overset{(6.2.1a)}{=}\mathcal{F}\{x(t)\}e^{-j\omega t_1}$$

$$x(t-t_1)\ \overset{\mathcal{F}}{\leftrightarrow}\ X(\omega)\,e^{-j\omega t_1}\tag{6.2.14}$$

This implies that the real translation (time shifting) of $x(t)$ by t_1 along the t-axis will affect the CTFT on its phase by $-\omega t_1$, but not on its magnitude. This is similar to Eq. (6.1.12), which is the time shifting property of CTFS.

In duality with the time shifting property (6.2.14), the complex translation (frequency shifting) property holds:

$$x(t)e^{j\omega_1 t}\ \overset{\mathcal{F}}{\leftrightarrow}\ X(\omega-\omega_1)\tag{6.2.15}$$

6.2.4.5 Duality

As exhibited by some examples and properties, there exists a definite symmetry between Fourier transform pairs, stemming from a general property of *duality* between time and frequency. It can be proved by considering the following integral expression:

$$f(u) = \int_{-\infty}^{\infty} g(v)\, e^{-juv}\, dv \overset{v \to -v}{=} \int_{-\infty}^{\infty} g(-v)\, e^{juv}\, dv \tag{6.2.16}$$

This, with (ω,t) or $(\pm t, \mp\omega)$ for (u, v), yields the CTFT or ICTFT relation, respectively:

$$f(\omega) = \int_{-\infty}^{\infty} g(t)\, e^{-j\omega t}\, dt \overset{(6.2.1a)}{=} \mathcal{F}\{g(t)\}: \qquad g(t) \overset{\mathcal{F}}{\leftrightarrow} f(\omega) \tag{6.2.17a}$$

$$f(\pm t) = \int_{-\infty}^{\infty} g(\mp\omega)\, e^{j\omega t}\, d\omega \overset{(6.2.1b)}{=} 2\pi \mathcal{F}^{-1}\{g(\mp\omega)\}: \; f(\pm t) \overset{\mathcal{F}}{\leftrightarrow} 2\pi\, g(\mp\omega) \tag{6.2.17b}$$

It is implied that, if one CTFT relation holds, the substitution of $(\pm t, \mp\omega)$ for (ω,t) yields the other one, which is also valid. This property of duality can be used to derive/show a CTFT relation or property from another one. We have the following examples:

(Ex 0) $\quad g(t) = x(t) \underset{(6.2.17a)}{\overset{\mathcal{F}}{\leftrightarrow}} f(\omega) = X(\omega) \Leftrightarrow f(t) = X(t) \underset{(6.2.17b)}{\overset{\mathcal{F}}{\leftrightarrow}} 2\pi g(-\omega) = 2\pi x(-\omega)$

(Ex 1) $\quad x(t) = A\delta(t) \underset{(6.2.17a)}{\overset{\mathcal{F}}{\leftrightarrow}} X(\omega) \overset{(E6.7.1)}{=} A \Leftrightarrow X(t) = A \underset{(6.2.17b)}{\overset{\mathcal{F}}{\leftrightarrow}} 2\pi x(-\omega) = 2\pi A\delta(-\omega)$

(Ex 2) $\quad x(t) = \begin{cases} 1 & \text{for } |t| \le D \\ 0 & \text{for } |t| > D \end{cases} \underset{(6.2.17a)}{\overset{\mathcal{F}}{\leftrightarrow}} X(\omega) \overset{(E6.4.2)}{=} 2D\,\text{sinc}\left(\frac{D\omega}{\pi}\right)$ (See Example 6.4 and Fig. 6.14.)

$\Leftrightarrow \quad \frac{1}{2\pi} X(t) = \frac{B}{\pi}\text{sinc}\left(\frac{Bt}{\pi}\right) \underset{(6.2.17b)}{\overset{\mathcal{F}}{\leftrightarrow}} x(-\omega) = \begin{cases} 1 & \text{for } |\omega| \le B \\ 0 & \text{for } |\omega| > B \end{cases}$ (See Example 6.10)

(Ex 3) $\quad x(t) = A\delta(t - t_1) \underset{(6.2.17a)}{\overset{\mathcal{F}}{\leftrightarrow}} X(\omega) \overset{(E6.7.1)\&(6.2.14)}{=} Ae^{-j\omega t_1}$

$\Leftrightarrow X(-t) = Ae^{j\omega_1 t} \underset{(6.2.17b)}{\overset{\mathcal{F}}{\leftrightarrow}} 2\pi x(\omega) \overset{(6.2.6)}{=} 2\pi A\delta(\omega - \omega_1)$ (See Eq. (6.2.6).)

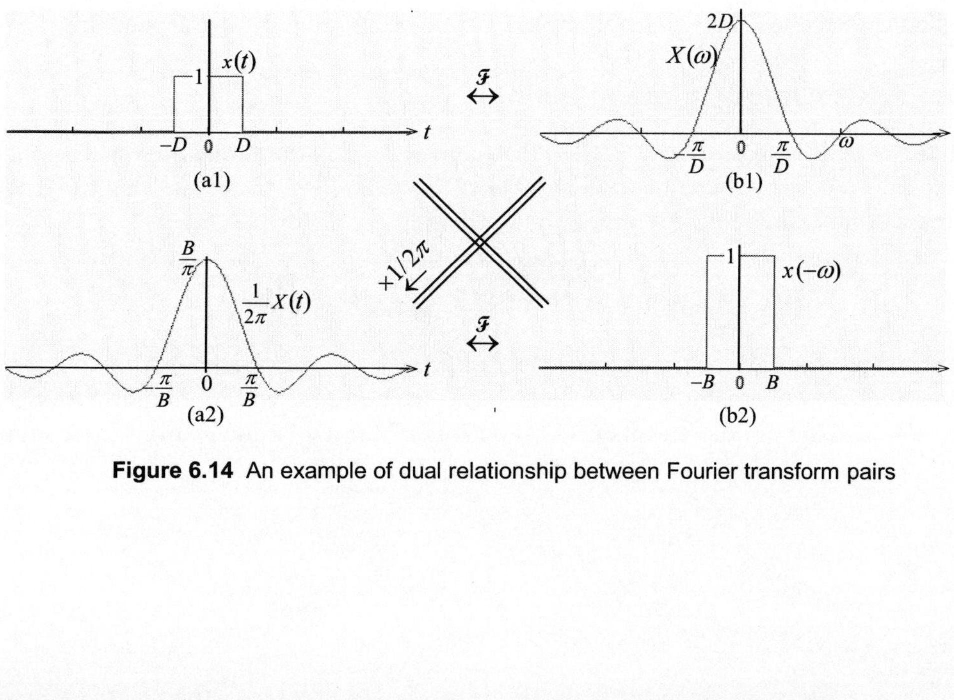

Figure 6.14 An example of dual relationship between Fourier transform pairs

6.2.4.6 Real Convolution

For the (linear) convolution of two functions $x(t)$ and $g(t)$

$$y(t) = x(t)*g(t) = \int_{-\infty}^{\infty} x(\tau)g(t-\tau)d\tau = \int_{-\infty}^{\infty} g(\tau)x(t-\tau)d\tau = g(t)*x(t) \qquad (6.2.18)$$

we can get its CTFT as

$$Y(\omega) = \mathcal{F}\{y(t)\} \overset{(6.2.1a)}{\underset{(6.2.18)}{=}} \int_{-\infty}^{\infty}\left\{\int_{-\infty}^{\infty} x(\tau)\, g(t-\tau)d\tau\right\} e^{-j\omega t}\,dt$$

$$= \int_{-\infty}^{\infty} x(\tau)\, e^{-j\omega\tau}\left\{\int_{-\infty}^{\infty} g(t-\tau)\, e^{-j\omega(t-\tau)}\,dt\right\} d\tau$$

$$\overset{t-\tau\to t}{\underset{(6.2.1a)}{=}} \int_{-\infty}^{\infty} x(\tau)\, e^{-j\omega\tau} G(\omega)\, d\tau$$

$$= G(\omega) \int_{-\infty}^{\infty} x(\tau)\, e^{-j\omega\tau} d\tau \overset{(6.2.1a)}{=} G(\omega)\, X(\omega)$$

$$;\quad y(t) = x(t) * g(t) \overset{\mathcal{F}}{\leftrightarrow} Y(\omega) = X(\omega)G(\omega) \qquad (6.2.19)$$

where $X(\omega)=\mathcal{F}\{x(t)\}$, $Y(\omega)=\mathcal{F}\{y(t)\}$, and $G(\omega)=\mathcal{F}\{g(t)\}$. This is the frequency-domain input–output relationship of a linear time-invariant (LTI) system with the input $x(t)$, the output $y(t)$, and the impulse response $g(t)$, where $G(\omega)$ is called the frequency response of the system.

On the other hand, if two functions $\tilde{x}_P(t)$ and $\tilde{g}_P(t)$ are periodic with common period P, their linear convolution does not converge. Accordingy, we need another definition of convolution, which is the *periodic* or *circular convolution* with the integration performed over one period:

$$\tilde{y}_P(t) = \tilde{x}_P(t) \underset{P}{*} \tilde{g}_P(t) = \int_P \tilde{x}_P(\tau)\tilde{g}_P(t-\tau)d\tau = \int_P \tilde{g}_P(\tau)\tilde{x}_P(t-\tau)d\tau = \tilde{g}_P(t) \underset{P}{*} \tilde{x}_P(t) \qquad (6.2.20)$$

where $\underset{P}{*}$ denotes the circular convolution with period P. Like the CTFT of a linear convolution, the CTFS coefficient of the periodic (circular) convolution turns out to be the multiplication of the CTFS coefficients of two periodic functions $\tilde{x}_P(t)$ and $\tilde{g}_P(t)$ (see Problem 6.3(a)):

$$\tilde{y}_P(t) = \tilde{x}_P(t) \underset{P}{*} \tilde{g}_P(t) \overset{\text{CTFS}}{\leftrightarrow} Y_k = X_k G_k \qquad (6.2.21)$$

where Y_k, X_k, and G_k are the CTFS coefficients of $\tilde{y}_P(t)$, $\tilde{x}_P(t)$, and $\tilde{g}_P(t)$, respectively.

6.2.4.7 Complex Convolution (Modulation)

In duality with the convolution property (6.2.19) that convolution in the time domain corresponds to multiplication in the frequency domain, we may expect the *modulation property* that multiplication in the time domain corresponds to convolution in the frequency domain:

$$y(t) = x(t)m(t) \overset{\mathcal{F}}{\leftrightarrow} Y(\omega) = \frac{1}{2\pi} X(\omega) * M(\omega) \qquad (6.2.22)$$

where $Y(\omega)=\mathcal{F}\{y(t)\}$, $X(\omega)=\mathcal{F}\{x(t)\}$, and $M(\omega)=\mathcal{F}\{m(t)\}$.

On the other hand, if two functions $\tilde{x}_P(t)$ and $\tilde{m}_P(t)$ are periodic with common period P, then their multiplication is also periodic and its CTFS coefficient can be obtained from the convolution sum (see Problem 6.3(b)):

$$\tilde{y}_P(t) = \tilde{x}_P(t)\tilde{m}_P(t) \overset{\text{CTFS}}{\leftrightarrow} Y_k = \frac{1}{P}\sum_{n=-\infty}^{\infty} X_n M_{k-n} \tag{6.2.23}$$

where Y_k, X_k, and M_k are the CTFS coefficients of $\tilde{y}_P(t)$, $\tilde{x}_P(t)$, and $\tilde{m}_P(t)$, respectively.

(**Example 6.13**) Sinusoidal Amplitude Modulation and Demodulation

(a) Sinusoidal Amplitude Modulation

Consider a sinusoidal amplitude-modulated (AM) signal

$$x_m(t) = x(t)m(t) = x(t)\cos(\omega_c t) \tag{E6.13.1}$$

Noting that the CTFT of the carrier signal $m(t)=\cos\omega_c t$ is

$$M(\omega) = \mathcal{F}\{m(t)\} = \mathcal{F}\{\cos(\omega_c t)\} \overset{\text{(E6.12.2)}}{=} \pi(\delta(\omega+\omega_c)+\delta(\omega-\omega_c)) \tag{E6.13.2}$$

we can use the modulation property (6.2.22) to get the CTFT of the AM signal as

$$X_m(\omega) = \mathcal{F}\{x_m(t)\} = \mathcal{F}\{x(t)\cos\omega_c t\} \overset{(6.2.22)}{=} \frac{1}{2\pi}X(\omega)*M(\omega)$$

$$\overset{\text{(E6.13.2)}}{=} \frac{1}{2\pi}X(\omega)*\pi(\delta(\omega+\omega_c)+\delta(\omega-\omega_c)) \overset{\text{(C.39)}}{=} \frac{1}{2}(X(\omega+\omega_c)+X(\omega-\omega_c)) \tag{E6.13.3}$$

This implies that the spectrum of the AM signal $x(t)\cos\omega_c t$ consists of the sum of two shifted and scaled versions of $X(\omega)=\mathcal{F}\{x(t)\}$. Note that this result can also be obtained by applying the frequency shifting property (6.2.15) to the following expression:

$$x_m(t) = x(t)\cos(\omega_c t) \overset{\text{(C.21)}}{=} \frac{1}{2}(x(t)e^{j\omega_c t} + x(t)e^{-j\omega_c t}) \tag{E6.13.4}$$

(b) Sinusoidal Amplitude Demodulation

In an AM communication system, the receiver demodulates the modulated signal $x_m(t)$ by multiplying the carrier signal as done for modulation at the transmitter:

$$x_d(t) = 2x_m(t)\cos(\omega_c t) \tag{E6.13.5}$$

We can use the modulation property (6.2.22) together with Eqs. (E6.13.3) and (E6.13.2) to express the CTFT of the demodulated signal in terms of the signal spectrum $X(\omega)$ as

$$X_d(\omega) = \mathcal{F}\{x_d(t)\} = \mathcal{F}\{x_m(t)2\cos\omega_c t\} \overset{(6.2.22)}{=} \frac{1}{2\pi}X_m(\omega)*2M(\omega)$$

$$\underset{\text{(E6.13.3)}}{\overset{\text{(E6.13.2)}}{=}} \frac{1}{2\pi}\frac{1}{2}\{X(\omega+\omega_c)+X(\omega-\omega_c)\}*2\pi\{\delta(\omega+\omega_c)+\delta(\omega-\omega_c)\}$$

$$\overset{\text{(C.39)}}{=} \frac{1}{2}\{X(\omega+2\omega_c)+X(\omega)+X(\omega)+X(\omega-2\omega_c)\}$$

$$= \frac{1}{2}X(\omega+2\omega_c)+X(\omega)+\frac{1}{2}X(\omega-2\omega_c) \tag{E6.13.6}$$

6.2.4.8 Time Differential/Integration — Frequency Multiplication/Division

By differentiating both sides of the inverse Fourier transform formula (6.2.1b) w.r.t. t, we obtain

$$\frac{dx(t)}{dt} \overset{(6.2.1b)}{=} \frac{1}{2\pi}\int_{-\infty}^{\infty} X(\omega)\left(\frac{d}{dt}e^{j\omega t}\right)d\omega = \frac{1}{2\pi}\int_{-\infty}^{\infty}(j\omega X(\omega))e^{j\omega t}\,d\omega \tag{6.2.24}$$

which yields the time-differentiation property of the CTFT as

$$\frac{dx(t)}{dt} \quad \overset{\mathcal{F}}{\leftrightarrow} \quad j\omega X(\omega) \tag{6.2.25}$$

This means that differentiation in the time domain results in multiplication by $j\omega$ in the frequency domain.

On the other hand, the time-integration property is obtained by expressing the integration of $x(t)$ in the form of the convolution of $x(t)$ and the unit step function $u_s(t)$, and then applying the convolution property (6.2.19) together with Eq. (E6.9.6) as

$$\begin{aligned}&\int_{-\infty}^{t} x(\tau)d\tau = \int_{-\infty}^{\infty} x(\tau)u_s(t-\tau)d\tau = x(t)*u_s(t)\\ &\underset{(6.2.19)}{\overset{\mathcal{F}}{\leftrightarrow}} \quad \mathcal{F}\{x(t)\}\mathcal{F}\{u_s(t)\} \overset{(E6.9.6)}{=} \pi X(\omega)\delta(\omega) + \frac{1}{j\omega}X(\omega) = \pi X(0)\delta(\omega) + \frac{1}{j\omega}X(\omega)\end{aligned} \tag{6.2.26}$$

where the additional impulse term $\pi X(0)\delta(\omega)$ on the RHS reflects the DC value resulting from the integration. This equation is slightly above our intuition that integration/differentiation in the time domain results in division/multiplication by $j\omega$ in the frequency domain.

6.2.4.9 Frequency Differentiation — Time Multiplication

By differentiating both sides of the CTFT formula (6.2.1a) w.r.t. ω, we obtain

$$\frac{dX(\omega)}{d\omega} \overset{(6.2.1a)}{=} \int_{-\infty}^{\infty} x(t)\left(\frac{d}{d\omega}e^{-j\omega t}\right)dt = -j\int_{-\infty}^{\infty}(t\,x(t))e^{-j\omega t}\,dt\,, \tag{6.2.27}$$

which yields the frequency-differentiation property of the CTFT as

$$t\,x(t) \quad \overset{\mathcal{F}}{\leftrightarrow} \quad j\frac{dX(\omega)}{d\omega} \tag{6.2.28}$$

This means that multiplication by t in the time domain results in differentiation w.r.t. ω and multiplication by j in the frequency domain.

6.2.4.10 Time and Frequency Scaling

The CTFT of $x(at)$ scaled along the time axis can be obtained as

$$\mathcal{F}\{x(at)\} = \int_{-\infty}^{\infty} x(at)e^{-j\omega t}\,dt \overset{t=\tau/a}{=} \frac{1}{|a|}\int_{-\infty}^{\infty} x(\tau)e^{-j\omega\tau/a}\,d\tau \overset{(6.2.1a)}{=} \frac{1}{|a|}X\left(\frac{\omega}{a}\right)$$

which yields the time and frequency scaling property of CTFT:

$$x(at) \quad \overset{\mathcal{F}}{\leftrightarrow} \quad \frac{1}{|a|}X\left(\frac{\omega}{a}\right) \tag{6.2.29}$$

This is another example of the dual (inverse) relationship between time and frequency. A common illustration of this property is the effect on frequency contents of playing back an audio tape at different speeds. If the playback speed is higher/slower than the recording speed, corresponding to compression($a>1$)/expansion($a<1$) in time, then the playback sounds get higher/lower, corresponding to expansion/compression in frequency. Also, you can see the expansion/compression effect of increasing T on $e_2(t)/E_2(\omega)$ along the t-/ω-axis, respectively.

6.2.4.11 Parseval's Relation (Rayleigh Theorem)

If $x(t)$ has finite energy and $\mathcal{F}\{x(t)\}=X(\omega)$, then we have

$$\int_{-\infty}^{\infty}|x(t)|^2\,dt = \frac{1}{2\pi}\int_{-\infty}^{\infty}|X(\omega)|^2\,d\omega \tag{6.2.30}$$

(Proof)

$$\int_{-\infty}^{\infty}|x(t)|^2\,dt = \int_{-\infty}^{\infty}x(t)x^*(t)\,dt \overset{(6.2.1b)}{=} \int_{-\infty}^{\infty}x(t)\left\{\frac{1}{2\pi}\int_{-\infty}^{\infty}X^*(\omega)e^{-j\omega t}\,d\omega\right\}dt$$

$$= \frac{1}{2\pi}\int_{-\infty}^{\infty}X^*(\omega)\left\{\int_{-\infty}^{\infty}x(t)e^{-j\omega t}\,dt\right\}d\omega \overset{(6.2.1a)}{=} \frac{1}{2\pi}\int_{-\infty}^{\infty}X^*(\omega)X(\omega)\,d\omega = \frac{1}{2\pi}\int_{-\infty}^{\infty}|X(\omega)|^2\,d\omega$$

This implies that the total energy in the signal $x(t)$ can be determined either by integrating $|x(t)|^2$ over all time or by integrating $|X(\omega)|^2/2\pi$ over all frequencies. For this reason, $|X(\omega)|^2$ is called the *energy-density spectrum* of the signal $x(t)$.

On the other hand, if $\tilde{x}_P(t)$ is periodic with period P and its CTFS coefficients are X_k's, then we have an analogous relation

$$\int_{-\infty}^{\infty}|\tilde{x}_P(t)|^2\,dt = \frac{1}{P}\sum_{k=-\infty}^{\infty}|X_k|^2 \tag{6.2.31}$$

where $|X_k|^2/P$ is called the *power-density spectrum* of the periodic signal $\tilde{x}_P(t)$.

6.3 DISCRETE-TIME FOURIER TRANSFORM (DTFT)[Y-2]

6.3.1 Definition and Convergence Conditions

As the discrete-time counterpart of the CTFT (continuous-time Fourier transform)

$$X(j\omega) = \mathcal{F}\{x(t)\} \overset{(6.2.1a)}{=} \int_{-\infty}^{\infty}x(t)e^{-j\omega t}\,dt$$

we define the Fourier transform of a discrete-time sequence $x[n]$ as

$$X(j\Omega) = X[e^{j\Omega}] = \mathcal{F}\{x[n]\} = \sum_{n=-\infty}^{\infty}x[n]e^{-j\Omega n} \tag{6.3.1}$$

which is called the *DTFT* (*discrete-time Fourier transform*). From now on, we will more often use $X(\Omega)$ than $X(j\Omega)$ or $X[e^{j\Omega}]$, for simplicity.

One striking difference between the DTFT and the CTFT is its periodicity (with period 2π) in the *digital frequency* variable Ω, which results from the fact that it is a function of $e^{j\Omega}$ periodic with period 2π in Ω, i.e., $e^{j(\Omega+2\pi n)}=e^{j\Omega}$. Based on the periodicity of the DTFT, we are going to use the CTFS (for periodic functions) to derive the *IDTFT* (*inverse discrete-time Fourier transform*) formula. To this end, we can use Eq. (6.1.6a) with $P=2\pi$, $\omega_0=2\pi/P=1$, $t=\Omega$, and $k=n$ to write the (continuous-frequency) Fourier series representation of $X(\Omega)$ as

$$X(\Omega) \overset{(6.1.6a)}{=} \frac{1}{2\pi} \sum_{n=-\infty}^{\infty} x_n e^{-j\Omega n} \tag{6.3.2a}$$

where the Fourier coefficients are

$$x_n \underset{P=2\pi, \omega_0=1, t=\Omega, k=-n}{\overset{(6.1.6b)}{=}} \int_{2\pi} X(\Omega) e^{j\Omega n} d\Omega \text{ (Integral over one period of length } 2\pi) \tag{6.3.2b}$$

Noting that Eq. (6.3.2a) is the same as Eq. (6.3.1) multiplied by a scaling factor $1/2\pi$, we can multiply Eq. (6.3.2b) by the same scaling factor $1/2\pi$ to write the IDTFT formula as

$$x[n] = \mathcal{F}^{-1}\{X(\Omega)\} = \frac{1}{2\pi} \int_{2\pi} X(\Omega) e^{j\Omega n} d\Omega \tag{6.3.3}$$

We call Eqs. (6.3.1) and (6.3.3) the DTFT pair, where Eq. (6.3.1) is the analysis equation and Eq. (6.3.3) is the synthesis equation.

Like the convergence conditions (6.2.2a) and (6.2.2b) for the CTFT, it can be stated that the DTFT will exist if the sequence $x[n]$ has finite energy, i.e.,

$$\sum_{n=-\infty}^{\infty} |x[n]|^2 < \infty \tag{6.3.4a}$$

or if it is absolutely summable, i.e.,

$$\sum_{n=-\infty}^{\infty} |x[n]| < \infty \tag{6.3.4b}$$

The DTFT formula can be cast into the following MATLAB function ‘DTFT()’:

```
function [X,ph]=DTFT(x,W,n0)
% Computes the DTFT of a given sequence x[n] for frequency vector W,
%  regarding the first sample as the n0-th one.
x=x(:).'; Nt=length(x); n=0:Nt-1;
if nargin<3, n0 = -floor(Nt/2); end
X= x*exp(-j*(n+n0)'*W); % Eq.(6.3.1)
if nargout==2, ph=angle(X); X=abs(X);  end
```

6.3.2 Examples of DTFT

(**Example 6.14**) DTFT of a Rectangular Pulse Sequence

For the rectangular pulse of duration $2M+1$ from $n=-M$ to M shown in Fig. 6.15(a1) and described by

$$r'_{2M+1}[n] = u_s[n+M] - u_s[n-M-1] \tag{E6.14.1}$$

we can apply Eq. (6.3.1) to get its DTFT as

$$R'_{2M+1}(\Omega) \overset{(6.3.1)}{=} \sum_{n=-\infty}^{\infty} r'_{2M+1}[n] e^{-j\Omega n} \overset{(E6.14.1)}{=} \sum_{n=-M}^{M} e^{-j\Omega n} \overset{(C.23)}{=} e^{j\Omega M} \frac{1-e^{-j\Omega(2M+1)}}{1-e^{-j\Omega}} \text{ (Dirichlet kernel)}$$

$$= e^{j\Omega M} \frac{e^{-j\Omega(2M+1)/2}(e^{j\Omega(2M+1)/2} - e^{-j\Omega(2M+1)/2})}{e^{-j\Omega/2}(e^{j\Omega/2} - e^{-j\Omega/2})} \overset{(C.22)}{=} \frac{\sin(\Omega(2M+1)/2)}{\sin(\Omega/2)} \tag{E6.14.2}$$

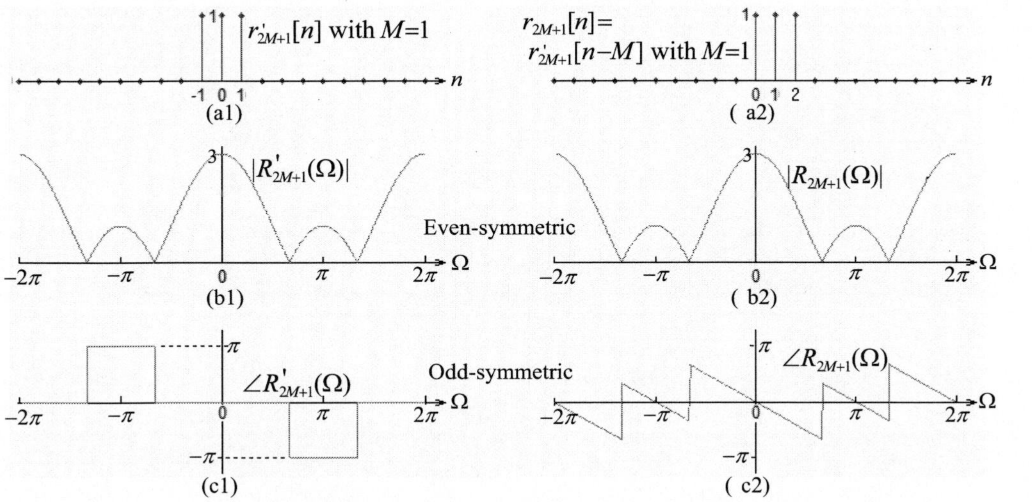

Figure 6.15 Two rectangular pulse sequences and their CTFT (magnitude and phase) spectra

whose magnitude and phase are depicted in Figs. 6.15(b1) and (c1), respectively. Likewise, for the rectangular pulse of duration $2M+1$ from $n=0$ to $2M$ shown in Fig. 6.15(a2) and described by

$$r_{2M+1}[n]=u_s[n]-u_s[n-2M-1]=r'_{2M+1}[n-M] \tag{E6.14.3}$$

we can apply Eq. (6.3.1) to get its DTFT as

$$R_{2M+1}(\Omega)\overset{(6.3.1)}{=}\sum_{n=-\infty}^{\infty} r_{2M+1}[n]e^{-j\Omega n}\overset{(E6.14.3)}{=}\sum_{n=0}^{2M} e^{-j\Omega n}\overset{(C.23)}{=}\frac{1-e^{-j\Omega(2M+1)}}{1-e^{-j\Omega}}\overset{(E6.14.2)}{=}R'_{2M+1}(\Omega)e^{-j\Omega M}$$

$$\overset{(E6.14.2)}{=}\frac{\sin(\Omega(2M+1)/2)}{\sin(\Omega/2)}e^{-j\Omega M}=\frac{\sin(\Omega(2M+1)/2)}{\sin(\Omega/2)}\angle -M\Omega \tag{E6.14.4}$$

whose magnitude and phase are depicted in Figs. 6.15(b2) and (c2), respectively.

(Example 6.15) DTFT of a Unilateral Exponential Sequence

For a unilateral (one-sided) decaying exponential sequence $e_1[n]=a^n u_s[n]$ with $|a|<1$ (Fig. 6.16(a1)), we can apply Eq. (6.3.1) to get its DTFT as

$$E_1(\Omega)\overset{(6.3.1)}{=}\sum_{n=-\infty}^{\infty} a^n u_s[n]e^{-j\Omega n}=\sum_{n=0}^{\infty}(ae^{-j\Omega})^n\overset{(C.23)}{=}\frac{1}{1-ae^{-j\Omega}}$$

$$=\frac{1}{1-ae^{-j\Omega}}\frac{1-ae^{j\Omega}}{1-ae^{j\Omega}}\overset{(C.20)}{=}\frac{1-a\cos\Omega-ja\sin\Omega}{1-2a\cos\Omega+a^2} \tag{E6.15.1}$$

whose magnitude and phase are

$$|E_1(\Omega)|=\frac{1}{\sqrt{(1-a\cos\Omega)^2+(a\sin\Omega)^2}} \quad\text{and}\quad \angle E_1(\Omega)=\tan^{-1}\frac{-a\sin\Omega}{1-a\cos\Omega} \tag{E6.15.2}$$

Figs. 6.16(a1–a2), (b1–b2), and (c2–c2) show the exponential sequences $e_1[n]=a^n u_s[n]$ with $a=\pm 0.5$ and ± 0.8 and their magnitude and phase spectra.

(Q) Among the four exponential sequences, which one has the most low-frequency contents and which one has the most high-frequency contents?

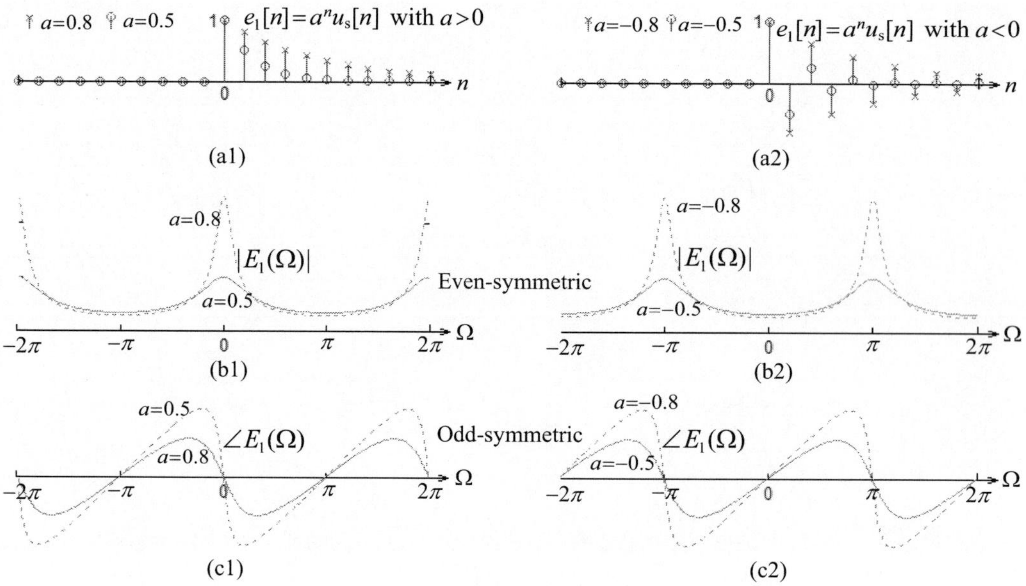

Figure 6.16 Four exponential sequences and their CTFT (magnitude and phase) spectra

(Example 6.16) DTFT of an Even-Symmetric Exponential Sequence

For the exponential sequence $e_2[n]=a^{|n|}$ with $|a|<1$, we can apply Eq. (6.3.1) to get its DTFT as

$$E_2(\Omega) \overset{(6.3.1)}{=} \sum_{n=-\infty}^{\infty} a^{|n|} e^{-j\Omega n} = \sum_{n=-\infty}^{-1} a^{-n} e^{-j\Omega n} + \sum_{n=0}^{\infty} a^{n} e^{-j\Omega n}$$

$$= \sum_{n=1}^{\infty} (ae^{j\Omega})^n + \sum_{n=0}^{\infty} (ae^{-j\Omega})^n$$

$$\overset{(C.23)}{=} \frac{ae^{j\Omega}}{1-ae^{j\Omega}} + \frac{1}{1-ae^{-j\Omega}} \overset{(C.20)}{=} \frac{1-a^2}{1-2a\cos\Omega+a^2} : \text{ real-valued} \qquad \text{(E6.16.1)}$$

whose magnitudes with $a=\pm 0.5$ and ± 0.8 are depicted in Fig. 6.17.

(Q) Among the four exponential sequences, which one has the most low-frequency contents and which one has the most high-frequency contents?

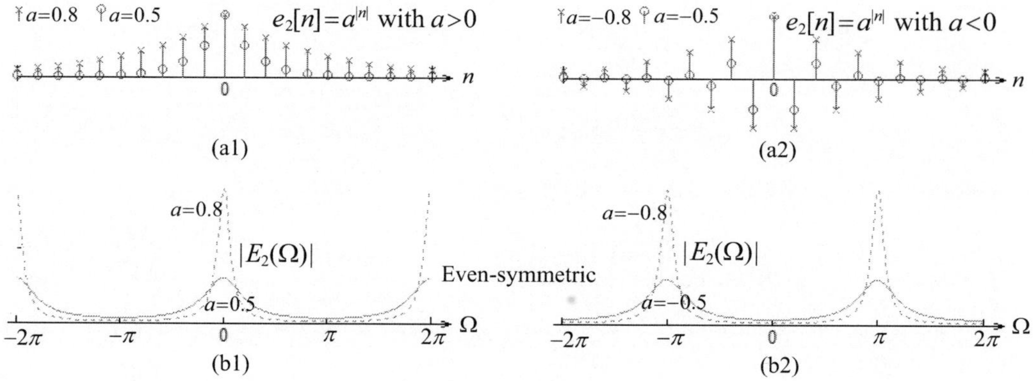

Figure 6.17 Four exponential sequences and their CTFT (magnitude) spectra

(Example 6.17) DTFT of the Unit Sample (Impulse) Sequence

For the impulse (unit sample) sequence $\delta[n]$, we can apply Eq. (6.3.1) to get its DTFT as

$$\Delta(\Omega) \overset{(6.3.1)}{=} \sum_{n=-\infty}^{\infty} \delta[n]\, e^{-j\Omega n} = \delta[n] e^{-j\Omega n}\Big|_{n=0} = 1 \tag{E6.17.1}$$

As with the continuous-time case (Eq. (E6.7.1) in Example 6.7), this implies that a discrete-time impulse signal has a flat or white spectrum, which is evenly distributed over all digital frequencies.

(Example 6.18) IDTFT of an Ideal LPF (Lowpass Filter) Frequency Response

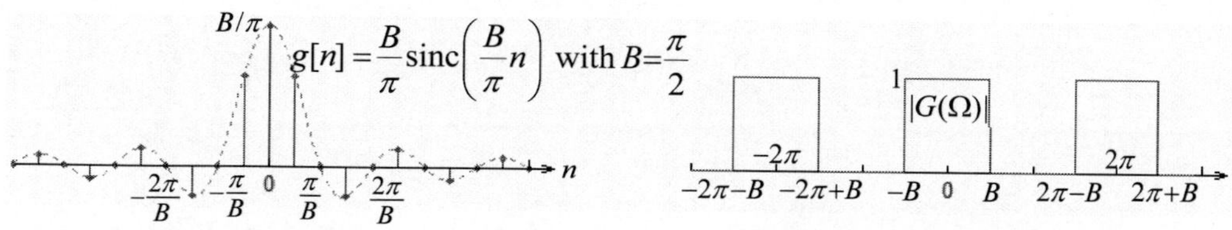

(a) A sinc sequence (b) The DTFT (magnitude) spectrum

Figure 6.18 A sinc sequence and its DTFT (magnitude) spectrum

Let us consider the frequency response of an ideal LPF depicted in Fig. 6.18(b):

$$G(\Omega) = \begin{cases} 1 & \text{for } |\Omega - 2m\pi| \le B \le \pi \ (m\text{: an integer}) \\ 0 & \text{elsewhere} \end{cases} \tag{E6.18.1}$$

We can take the IDTFT (inverse discrete-time Fourier transform) formula (6.3.3) to get the impulse response of the LPF, which is shown in Fig. 6.18(a):

$$g[n] = \mathcal{F}^{-1}\{G(\Omega)\} \overset{(6.3.3)}{=} \frac{1}{2\pi}\int_{-B}^{B} G(\Omega)\, e^{j\Omega n} d\Omega \underset{(C.35)}{\overset{(E6.18.1)}{=}} \frac{1}{2\pi j n}(e^{jBn} - e^{-jBn})$$

$$\overset{(C.22)}{=} \frac{\sin(Bn)}{\pi n} = \frac{B}{\pi}\,\text{sinc}\left(\frac{Bn}{\pi}\right) \tag{E6.18.2}$$

6.3.3 DTFT of Periodic Sequences

In analogy with the arguments in Section 6.2.2 for deriving Eq. (6.2.7) as the generalized CTFT of continuous-time periodic functions, we begin with applying the IDTFT formula (6.3.3) for an impulse train type spectrum

$$X(\Omega) = \sum_{i=-\infty}^{\infty} 2\pi\delta(\Omega - \Omega_0 - 2\pi i)$$

to get a complex sinusoidal sequence:

$$x[n] = \mathcal{F}^{-1}\{X(\Omega)\} \overset{(6.3.3)}{=} \frac{1}{2\pi}\int_{2\pi} X(\Omega)\, e^{j\Omega n} d\Omega = \int_{2\pi} \delta(\Omega - \Omega_0)\, e^{j\Omega n} d\Omega \overset{(C.40)}{=} e^{j\Omega_0 n}$$

This implies the following DTFT relation:

$$e^{j\Omega_0 n} \overset{\mathcal{F}}{\leftrightarrow} X(\Omega) = \sum_{i=-\infty}^{\infty} 2\pi\delta(\Omega - \Omega_0 - 2\pi i) \tag{6.3.5}$$

(Example 6.19) DTFT of a Constant Sequence

Let us consider a constant sequence $c[n]=1$. Noting that this is a kind of periodic function obtained by substituting $\Omega_0=0$ into the LHS of the DTFT relation (6.3.5), we can substitute $\Omega_0=0$ into the RHS of that DTFT relation to obtain the DTFT of $c[n]=1$ as

$$c[n]=1 \underset{(6.3.5)\text{ with }\Omega_0=0}{\overset{\mathcal{F}}{\leftrightarrow}} C(\Omega)=\sum_{i=-\infty}^{\infty} 2\pi\delta(\Omega-2\pi i) \tag{E6.19.1}$$

(Example 6.20) DTFT of a Sine/Cosine Sequence

(a) For $\sin(\Omega_0 n) \overset{(C.22)}{=} (e^{j\Omega_0 n}-e^{-j\Omega_0 n})/j2$, we use Eq. (6.3.5) to get its DTFT:

$$\sin(\Omega_0 n) \overset{\mathcal{F}}{\leftrightarrow} j\pi \sum_{i=-\infty}^{\infty}\{\delta(\Omega+\Omega_0-2\pi i)-\delta(\Omega-\Omega_0-2\pi i)\} \tag{E6.20.1}$$

(b) For $\cos(\Omega_0 n) \overset{(C.21)}{=} (e^{j\Omega_0 n}+e^{-j\Omega_0 n})/2$, we use Eq. (6.3.5) to get its DTFT:

$$\cos(\Omega_0 n) \overset{\mathcal{F}}{\leftrightarrow} \pi \sum_{i=-\infty}^{\infty}\{\delta(\Omega+\Omega_0-2\pi i)+\delta(\Omega-\Omega_0-2\pi i)\} \tag{E6.20.2}$$

(Example 6.21) DTFT of the Unit Step Sequence

As done for deriving the CTFT of the unit step function in Example 6.9, we first decompose the unit step sequence $u_s[n]$ into the sum of an even sequence and an odd sequence as

$$u_s[n]=u_e[n]+u_o[n] \tag{E6.21.1}$$

where

$$u_e[n]=\frac{1}{2}(u_s[n]+u_s[-n])=\frac{1}{2}(1+\delta[n]) \tag{E6.21.2}$$

$$u_o[n]=\frac{1}{2}(u_s[n]-u_s[-n])=\frac{1}{2}\text{sign}(n) \tag{E6.21.3}$$

Then we can take the DTFT of the even and odd parts as

$$U_e(\Omega)=\mathcal{F}\{u_e[n]\}=\frac{1}{2}\mathcal{F}\{\delta[n]\}+\mathcal{F}\{\frac{1}{2}\} \overset{(E6.17.1),(E6.19.1)}{=} \frac{1}{2}+\pi\sum_{i=-\infty}^{\infty}\delta(\Omega-2\pi i) \tag{E6.21.4}$$

$$U_o(\Omega)=\mathcal{F}\{u_o[n]\}=\frac{1}{2}(\mathcal{F}\{u_s[n]\}-\mathcal{F}\{u_s[-n]\})=\frac{1}{2}\left(\sum_{n=1}^{\infty}1e^{-j\Omega n}-\sum_{n=-\infty}^{-1}1e^{-j\Omega n}\right)$$

$$=\frac{1}{2}\sum_{n=1}^{\infty}(e^{-j\Omega n}-e^{j\Omega n}) \overset{(C.23)}{=} \frac{1}{2}\left(\frac{e^{-j\Omega}}{1-e^{-j\Omega}}-\frac{e^{j\Omega}}{1-e^{j\Omega}}\right) \overset{(C.20)}{=} \frac{-j\sin\Omega}{2(1-\cos\Omega)} \tag{E6.21.5}$$

Now we add these two results to obtain the DTFT of the unit step sequence as

$$\mathcal{F}\{u_s[n]\}=\frac{1}{2}\left(\frac{e^{-j\Omega}}{1-e^{-j\Omega}}+\frac{1}{1-e^{-j\Omega}}\right)+\frac{1}{2}+\pi\sum_{i=-\infty}^{\infty}\delta(\Omega-2\pi i)$$

$$=\frac{1}{1-e^{-j\Omega}}+\pi\sum_{i=-\infty}^{\infty}\delta(\Omega-2\pi i) \tag{E6.21.6}$$

6.3.4 Properties of DTFT

As with the CTFT (continuous-time Fourier transform), there are many properties of DTFT that provide us with further insight into the transform and can be used for reducing the computational complexity associated with it. Noting that there are striking similarities with the case of CTFT, we will simply state the properties unless their derivations and interpretations differ from those with the continuous-time case.

6.3.4.1 Periodicity

Since the DTFT $X(\Omega)$ defined by Eq. (6.3.1) is a function of $e^{j\Omega}$, it is always periodic with period 2π in Ω:

$$X(\Omega) = X[e^{j\Omega}] = X(\Omega + 2m\pi) \text{ for any integer } m \tag{6.3.6}$$

(Note) The periodicity lets us pay attention to the DTFT only for its one period, say, $-\pi \le \Omega \le \pi$.

6.3.4.2 Linearity

With $\mathcal{F}\{x[n]\} = X(\Omega)$ and $\mathcal{F}\{y[n]\} = Y(\Omega)$, we have

$$a\,x[n] + b\,y[n] \overset{\mathcal{F}}{\leftrightarrow} a\,X(\Omega) + b\,Y(\Omega) \tag{6.3.7}$$

which implies that the DTFT of a linear combination of many sequences is the same linear combination of the individual DTFTs.

6.3.4.3 (Conjugate) Symmetry

In general, the DTFT has the time reversal property:

$$\mathcal{F}\{x[-n]\} \overset{(6.3.1)}{=} \sum_{n=-\infty}^{\infty} x[-n]e^{-j\Omega n} \overset{-n=m}{=} \sum_{m=-\infty}^{\infty} x[m]e^{-j(-\Omega)m} \overset{(6.3.1)}{=} X(-\Omega)$$

$$; \qquad x[-n] \overset{\mathcal{F}}{\leftrightarrow} X(-\Omega) \tag{6.3.8}$$

In the where case $x[n]$ is a real-valued sequence, we have

$$X(-\Omega) \overset{(6.3.1)}{=} \sum_{n=-\infty}^{\infty} x[n]e^{-j(-\Omega)n} = \sum_{n=-\infty}^{\infty} x[n]e^{-(-j)\Omega n} \overset{(6.3.1)}{=} X^*(\Omega)\text{: complex conjugate of } X(\Omega)$$

or equivalently,

$$\mathrm{Re}\{X(-\Omega)\} + j\,\mathrm{Im}\{X(-\Omega)\} = \mathrm{Re}\{X(\Omega)\} - j\,\mathrm{Im}\{X(\Omega)\}$$

$$; \qquad |X(-\Omega)|\angle X(-\Omega) = |X(\Omega)|\angle -X(\Omega) \tag{6.3.9}$$

This implies that the magnitude/phase of the DTFT of a real-valued sequence is an even/odd function of frequency Ω.

Also in analogy with Eq. (6.2.12) for the CTFT, we have

$$\text{even and real-valued } x_e[n] \overset{\mathcal{F}}{\leftrightarrow} \mathrm{Re}\{X(\Omega)\} \text{ even and real-valued} \tag{6.3.10a}$$

$$\text{odd and real-valued } x_o[n] \overset{\mathcal{F}}{\leftrightarrow} j\,\mathrm{Im}\{X(\Omega)\} \text{ odd and imaginary-valued} \tag{6.3.10b}$$

6.3.4.4 Conjugation

$$x^*[n] \overset{\mathcal{F}}{\leftrightarrow} X^*(-\Omega) \text{ (like Eq. (6.2.13))} \tag{6.3.11}$$

6.3.4.5 Time/Frequency Shifting (Real/Complex Translation)

$$x[n-n_1] \overset{\mathcal{F}}{\leftrightarrow} X(\Omega)\, e^{-j\Omega n_1} = X(\Omega)\angle -n_1\Omega \text{ (like Eq. (6.2.14))} \tag{6.3.12}$$

$$x[n]e^{j\Omega_1 n} \overset{\mathcal{F}}{\leftrightarrow} X(\Omega-\Omega_1) \text{ (like Eq. (6.2.15))} \tag{6.3.13}$$

6.3.4.6 Real Convolution

The DTFT has the convolution property

$$y[n]=x[n]*g[n]=\sum_{m=-\infty}^{\infty} x[m]g[n-m]=\sum_{m=-\infty}^{\infty} x[n-m]g[m] \overset{\mathcal{F}}{\leftrightarrow} Y(\Omega)=X(\Omega)G(\Omega) \tag{6.3.14}$$

which can be derived in the same way with Eq. (6.2.19). This is very useful for describing the input-output relationship of a discrete-time LTI system with the input $x[n]$, the output $y[n]$, and the impulse response $g[n]$, where $G(\Omega)=\mathcal{F}\{g[n]\}$ is called the frequency response of the system.

6.3.4.7 Complex Convolution (Modulation/Windowing)

In analogy with Eq. (6.2.22) for the CTFT, the DTFT also has the complex convolution (modulation) property as

$$y[n]=x[n]\,m[n] \overset{\mathcal{F}}{\leftrightarrow} Y(\Omega)=\frac{1}{2\pi}X(\Omega)\underset{2\pi}{*}M(\Omega) \tag{6.3.15}$$

where $\underset{2\pi}{*}$ denotes a circular or periodic convolution with period 2π.

(**Example 6.22**) DTFT Spectrum of Rectangular Windowed Sinusoidal Sequence

We have the DTFT spectrum (Eq. (E6.20.2)) of a cosine sequence $x[n]=\cos(\Omega_0 n)=\cos(n\pi/2)$ as

$$X(\Omega)=\text{DTFT}\{\cos(\Omega_0 n)\}\overset{\text{(E6.20.2)}}{=}\pi\sum_{i=-\infty}^{\infty}\{\delta(\Omega+\Omega_0-2\pi i)+\delta(\Omega-\Omega_0-2\pi i)\}\,,\ \Omega_0=\frac{\pi}{2}\,. \tag{E6.22.1}$$

Truncating $x[n]$ to a finite-duration sequence of length 8 by multiplying the rectangular window

$$w[n]=\begin{cases}1 & \text{for } 0\le n\le 7\\ 0 & \text{elsewhere}\end{cases} \tag{E6.22.2}$$

whose DTFT is

$$R'_M(\Omega)\overset{\text{(E6.14.4)}}{=}\frac{\sin(M\Omega/2)}{\sin(\Omega/2)}e^{-j(M-1)\Omega/2} \text{ with } M=8 \tag{E6.22.3}$$

we obtain the windowed sinusoidal sequence as

$$x_w[n]=x[n]w[n] \tag{E6.22.4}$$

whose DTFT is

$$X_w(\Omega)\underset{\text{Complex convolution property}}{\overset{(6.3.15)}{=}}\frac{1}{2\pi}X(\Omega)\underset{2\pi}{*}R'_M(\Omega)\underset{\text{(C.39)}}{\overset{\text{(E6.22.1)}}{=}}\frac{1}{2}\left\{R'_M(\Omega+\Omega_0)+R'_M(\Omega-\Omega_0)\right\} \tag{E6.22.5}$$

```
%em06e22.m
%Fig. 6.19: Windowing Effect on DTFT spectrum of a Sinusoidal Sequence
clear, clf
n=[-4:19]; T=1; dt=0.02; t=[-4:dt:20]; % Time index vector
W0=pi/2; P=2*pi/W0; x=cos(W0*n); xt=cos(W0*t); % A sinusoidal Sequence
N=400; f=[-200:200]/N +1e-10; W=2*pi*f;  % Frequency vector
X=zeros(size(W)); X([101 301])=pi; pi2=pi*2; % Eq.(E6.22.1)/(E6.20.2)
subplot(321), stem(n,x,'.'), hold on, plot(t,xt,':')
subplot(322), plot(W,X)
% A rectangular window
wdw=zeros(size(n)); M=8; wdw(5+[0:M-1])=1; % Eq.(E6.22.2)
Wdw=DTFT(wdw,W,n(1)); % DTFT Wdw(W) of wdw[n] - Appendix E.3 for DTFT
RW=@(W,M)sin(W*M/2)./sin(W/2).*exp(-j*(M-1)/2*W);
Wdw_t=RW(W,M); % Eq.(E6.22.3)/(E6.14.4)
discrepancy_between_DTFT_and_E6_14_2 = norm(Wdw-Wdw_t)
subplot(323), stem(n,wdw,'.')
subplot(324), plot(W,abs(Wdw))
xw= x.*wdw; % Windowed sequence Eq.(E6.22.4)
Xw= DTFT(xw,W,n(1)); % DTFT spectrum Gw(W) of gw[n]
Xw1=(RW(W+W0,M)+RW(W-W0,M))/2; % Eq.(E6.22.5)
Xw2=conv_circular(X,Wdw,N)*N/pi2; % Numerical computation of Eq.(E6.22.5)
discrepancy2_between_circonv_and_DTFT=norm(abs(Xw(1:N))-abs(Xw2))
subplot(325), stem(n,xw,'.')
subplot(326), plot(W,abs(Xw), W,abs(Xw1),'r:')
```

```
function z=conv_circular(x,y,N)
% Circular convolution z(n)= (1/N) sum_m=0^N-1 x(m)*y(n-m)
if nargin<3, N=min(length(x),length(y)); end
x=x(1:N); y=y(1:N); y_circulated=fliplr(y); % Time reversal
for n=1:N
   y_circulated= [y_circulated(N) y_circulated(1:N-1)];
   z(n)= x*y_circulated'/N;
end
```

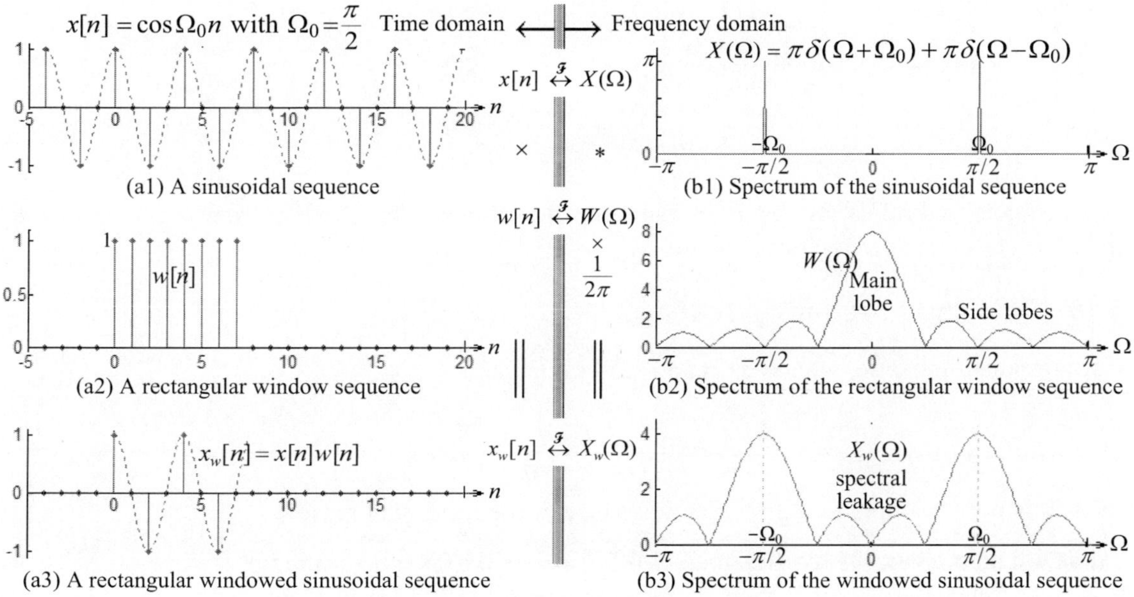

Figure 6.19 Effect of windowing on the DTFT spectrum

We can run the above MATLAB script "em06e22.m" to get Fig. 6.19, which shows the sinusoidal sequence $x[n]$, the rectangular window sequence $w[n]$, the rectangular windowed sinusoidal sequence $x_w[n]$, and their DTFT spectra.

There are a couple of observations to make about this. First, comparing $X(\Omega)$ (Fig. 6.19(b1)) and $X_w(\Omega)$ (Fig. 6.19(b3)) reveals that the windowing of a sequence results in *spectral leakage*, which makes the spectrum have many side lobe ripples (with low amplitudes) besides the two main peaks at $\pm\Omega_0$ (made smooth via convolution compared with the two spikes in Fig. 6.19(b1)). Second, the two DTFTs (plotted together in Fig. 6.19(b3)) of the windowed sinusoidal sequence $x_w[n]$ are the same where one has been obtained by using the DTFT formula (6.3.1) or the MATLAB function '`DTFT()`' and the other has been obtained by using the circular convolution (E6.22.5) or the MATLAB function '`conv_circular()`'.

(Q) How will the spectral leakage be affected by increasing the length M of the window?

6.3.4.8 Differencing and Accumulation in Time

As with the differentiation and integration property of the CTFT, the DTFT has the following differencing and accumulation properties:

$$x[n]-x[n-1] \;\overset{\mathcal{F}}{\leftrightarrow}\; (1-e^{-j\Omega})X(\Omega) \tag{6.3.16}$$

$$\sum_{m=-\infty}^{n} x[m] = x[n]*u_s[n] \;\overset{\mathcal{F}}{\leftrightarrow}\; \frac{1}{1-e^{-j\Omega}}X(\Omega) + \pi X(0)\sum_{i=-\infty}^{\infty}\delta(\Omega-2\pi i) \tag{6.3.17}$$

6.3.4.9 Frequency Differentiation

By differentiating both sides of the DTFT formula (6.3.1) w.r.t. Ω, we obtain

$$\frac{dX(\Omega)}{d\Omega} \overset{(6.3.1)}{=} -\sum_{n=-\infty}^{\infty} jnx[n]e^{-j\Omega n}$$

which yields the frequency-differentiation property of the DTFT as

$$nx[n] \;\overset{\mathcal{F}}{\leftrightarrow}\; j\frac{dX(\Omega)}{d\Omega} \tag{6.3.18}$$

This means that multiplication by n in the time domain results in differentiation w.r.t. Ω and multiplication by j in the frequency domain.

6.3.4.10 Time and Frequency Scaling

In the continuous-time case, we have Eq. (6.2.29) as

$$x(at) \;\overset{\mathcal{F}}{\leftrightarrow}\; \frac{1}{|a|}X\left(\frac{\omega}{a}\right)$$

However, it is not so simple to define $x[an]$ because of the following reasons:

- If a is not an integer, say, a=0.5, then $x[0.5n]|_{n=1}=x[0.5]$ is indeterminate.
- If a is an integer, say, a=2, then it does not merely speed up $x[n]$, but takes only the even-indexed samples of $x[n]$.

Thus we define a time-scaled version of $x[n]$ as

$$x_{(K)}[n] = \begin{cases} x[n/K] & \text{for } n{=}Kr \text{ (a multiple of } K\text{) with some integer } r \\ 0 & \text{elsewhere} \end{cases} \tag{6.3.19}$$

which is obtained by placing $(K-1)$ zeros between successive samples of $x[n]$. Then we apply the DTFT formula (6.3.1) to get

$$X_{(K)}(\Omega) \overset{(6.3.1)}{=} \sum_{n=-\infty}^{\infty} x_{(K)}[n]e^{-j\Omega n} \underset{\text{with } n=Kr}{\overset{(6.3.19)}{=}} \sum_{r=-\infty}^{\infty} x[r]e^{-j\Omega Kr} = X(K\Omega) \tag{6.3.20}$$

6.3.4.11 Parseval's Relation (Rayleigh Theorem)

If $x[n]$ has finite energy and the DTFT $X(\Omega)=\mathcal{F}\{x[n]\}$, then we have

$$\sum_{n=-\infty}^{\infty} |x[n]|^2 = \frac{1}{2\pi}\int_{2\pi} |X(\Omega)|^2 d\Omega \tag{6.3.21}$$

where $|X(\Omega)|^2$ is called the *energy-density spectrum* of the signal $x[n]$.

6.4 DISCRETE FOURIER TRANSFORM (DFT)[Y-2]

Before going into the DFT, let us consider the computational difference between the CTFS or CTFT for a periodic or an aperiodic continuous-time function $x(t)$ and the DTFT for the discrete-time sequence $x[n]$ consisting of the samples of $x(t)$. Since the CTFS (6.1.6b) and CTFT (6.2.1a) involving a continuous-time integral can be computed analytically for only $x(t)$ described by some basic functions, and require a numerical integration method for general finite-duration time functions, they are not practical when we need to find the spectra of real-world signals. The DTFT (6.3.1) is computationally advantageous over the CTFS or CTFT since it does not involve any continuous-time integration. However, it can be computed, say, by using the MATLAB function '`DTFT()`', only for finite-duration sequences unless the sequence is expressed by some basic sequences such as an exponential or sinc one. Therefore in practice, signals are truncated in various ways to obtain a set of finite number of samples [S-2]:

- A transient signal is assumed to be zero after it decays to a negligible amplitude.
- A periodic signal is sampled over an integral number of periods.
- A random signal is multiplied by a "window" of finite duration (short-time Fourier transform).

In any case, suppose we have a causal, finite-duration sequence $x[n]$ containing M samples. Then the DTFT formula (6.3.1) becomes

$$X(\Omega) = X[e^{j\Omega}] = \mathcal{F}\{x[n]\} = \sum_{n=0}^{N-1} x[n]e^{-j\Omega n} \tag{6.4.1}$$

where N is set to be greater than or equal to M (the number of samples). Noting that the frequency Ω takes continuous values despite its name "digital frequency", it may help improve the efficiency of computing Eq. (6.4.1) numerically to discretize Ω as

$$\Omega = k\Omega_0 = k\frac{2\pi}{N} \quad \text{for } 0 \le k \le N-1$$

This leads to another Fourier technique, called the N-point *DFT* (*discrete Fourier transform*), as follows:

$$X(k)=\text{DFT}_N\{x[n]\}=\sum_{n=0}^{N-1}x[n]e^{-j2\pi kn/N}=\sum_{n=0}^{N-1}x[n]W_N^{kn} \quad \text{for } k=0:N-1 \tag{6.4.2}$$

$$\text{with } W_N=e^{-j2\pi/N} : \textit{twiddle factor}$$

where N is called the *DFT size* and $\Omega_0=2\pi/N$ the *digital fundamental* or *resolution frequency*, just like $\omega_0=2\pi/P$ in the CTFS. Also, we can write the *IDFT* (*inverse discrete Fourier transform*) formula as

$$x[n]=\text{IDFT}_N\{X(k)\}=\frac{1}{N}\sum_{k=0}^{N-1}X(k)\,e^{j2\pi kn/N}=\frac{1}{N}\sum_{k=0}^{N-1}X(k)W_N^{-kn} \quad \text{for } n=0:N-1 \tag{6.4.3}$$

Here, by substituting Eq. (6.4.2) into Eq. (6.4.3), we are going to demonstrate the validity of the IDFT formula, which amounts to verifying that the set of DFT samples $\{X(k),\ k=0:N-1\}$ conveys all the information contained in the set of time-domain samples $\{x[n],\ n=0:N-1\}$, that is, $x[n]$ can be recovered perfectly from $X(k)$:

$$x[n]\underset{(6.4.3)}{\overset{?}{=}}\frac{1}{N}\sum_{k=0}^{N-1}X(k)W_N^{-kn}\overset{(6.4.2)}{=}\frac{1}{N}\sum_{k=0}^{N-1}\sum_{m=0}^{N-1}x[m]W_N^{km}W_N^{-kn}$$

$$=\sum_{m=0}^{N-1}x[m]\left(\frac{1}{N}\sum_{k=0}^{N-1}W_N^{k(m-n)}\right)=\sum_{m=0}^{N-1}x[m]\delta[(m-n)\bmod N]\overset{\text{O.K.}}{=}x[n]$$

('a mod b' or 'a modulo b'= the remainder after division of a by b)

where we have used the fact that

$$\frac{1}{N}\sum_{k=0}^{N-1}W_N^{k(m-n)}=\frac{1}{N}\sum_{k=0}^{N-1}e^{-j2\pi k(m-n)/N}$$
$$=\begin{cases}1 & \text{if } m-n=iN(i:\text{an integer})\\ 0 & \text{otherwise}\end{cases}=\delta[(m-n)\bmod N] \tag{6.4.4}$$

There are two observations about the DFT pair (6.4.2) and (6.4.3):

- The DFT sequence $X(k)$ is periodic in k with period N:

$$X(k+mN)\overset{(6.4.2)}{=}\sum_{n=0}^{N-1}x[n]e^{-j2\pi(k+mN)n/N}=\sum_{n=0}^{N-1}x[n]e^{-j2\pi kn/N}e^{-j2\pi mn}\overset{(6.4.2)}{=}X(k) \tag{6.4.5}$$

- The IDFT sequence $x[n]$ is periodic in n with period N:

$$x[n+mN]\overset{(6.4.3)}{=}\frac{1}{N}\sum_{k=0}^{N-1}X(k)e^{j2\pi k(n+mN)/N}=\frac{1}{N}\sum_{k=0}^{N-1}X(k)e^{j2\pi kn/N}e^{j2\pi km}\overset{(6.4.3)}{=}x[n] \tag{6.4.6}$$

The first observation (6.4.5) is no wonder since the DFT $X(k)$ inherits the periodicity of the DTFT $X(\Omega)$ as $X(k)$ originates from the samples of $X(\Omega)$ over its period. In contrast, the second observation is very surprising since there is no reason to believe that any sequence $x[n]$ is periodic and that with period equal to the DFT size N, which can be arbitrarily chosen as long as it covers the duration of $x[n]$. Because of this "illusory" periodicity, the shift involved with the time shifting property and the convolution related with the real convolution property of the DFT turn out to be not linear, but circular, as will be discussed subsequently. This headache is a cost that we should pay in return for the computational convenience of DFT, which is discrete in both time and frequency so that it needs only multiplication and summation instead of integration. If only the sequence $x[n]$ is really periodic with period N, the DFT is a perfect Fourier technique for analyzing the frequency characteristic, and thus it might well be referred to as the *discrete-time Fourier series* (*DTFS*). In this context, when we want to emphasize the periodicity of DFT, we call it *DFS* (*discrete Fourier series*) and write the DFS pair as

$$\tilde{X}(k) = \text{DFS}_N\{\tilde{x}[n]\} = \sum_{n=0}^{N-1} \tilde{x}[n]e^{-j2\pi kn/N} = \sum_{n=0}^{N-1} \tilde{x}[n]W_N^{kn} \tag{6.4.7}$$

$$\tilde{x}[n] = \text{IDFS}_N\{\tilde{X}(k)\} = \frac{1}{N}\sum_{k=0}^{N-1} \tilde{X}(k)e^{j2\pi kn/N} = \frac{1}{N}\sum_{k=0}^{N-1} \tilde{X}(k)W_N^{-kn} \tag{6.4.8}$$

The DFS formula can be cast into the MATLAB function '`DFS()`':

```
function [X,ph]=DFS(x,N,n0)
% N-point DFS/DFT of x[n] regarding the 1st sample as the n0-th one.
if nargin<3, n0=0; end
if nargin<2, N=length(x); end
n=n0+[0:length(x)-1]; k=[0:N-1];
X= x*exp(-j*2*pi*n.'*k/N); % Eq.(6.4.7)
if nargout==2, ph=angle(X); X=abs(X);  end
```

The DFT has properties similar to those of the CTFS, such as the real convolution property (6.2.19), since the DFT regards a time-domain sample sequence $x[n]$ as one period of its periodic extension (with period equal to the DFT size N) that can be described by

$$x[n] = \tilde{x}_N[n]\, r_N[n] \tag{6.4.9}$$

where $\tilde{x}_N[n]$ is the periodic extension (repetition) of $x[n]$ with period N and $r_N[n]$ is the rectangular pulse sequence with duration of n=0: $N-1$:

$$\tilde{x}[n] = \sum_{m=-\infty}^{\infty} x[n+mN] = x[n \bmod N] \tag{6.4.10a}$$

$$r_N[n] = u_s[n] - u_s[n-N] = \begin{cases} 1 & \text{for } 0 \le n \le N-1 \\ 0 & \text{elsewhere} \end{cases} \tag{6.4.10b}$$

On the other hand, the DFT sequence is born from the samples of the DTFT, and thus it inherits most of the DTFT properties, except for the fact that the real translation and convolution in the time domain are not linear but circular as with the CTFS. Therefore, we discuss only the two properties and summarize all the DFT properties in Table A.6 (Appendix A).

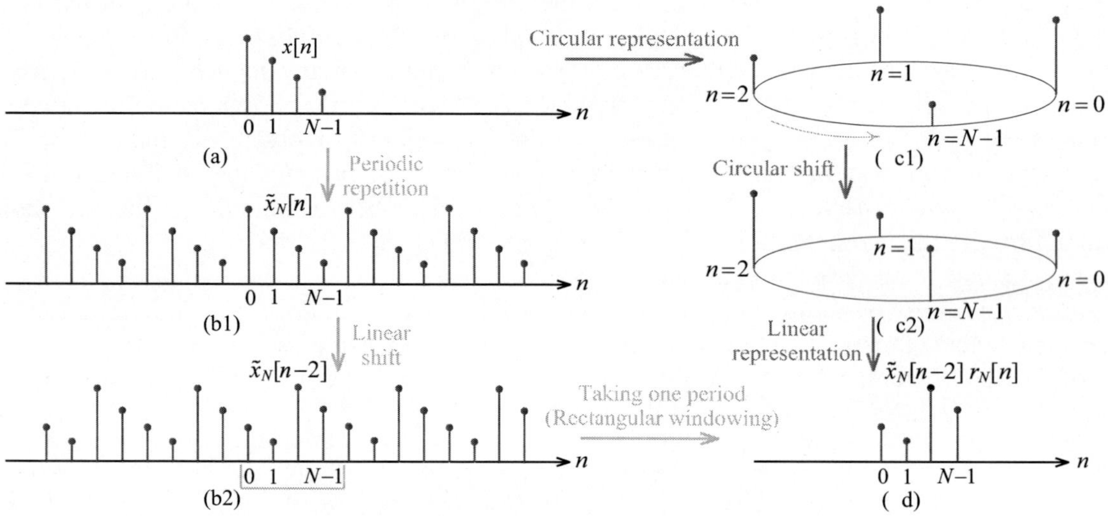

Figure 6.20 Circular shift of a sequence by 2 samples

<Real/Complex Translation — Circular Shift in Time/Frequency>

Figs. 6.20(a), (b1), (b2), and (d) show a sequence $x[n]$ of finite duration N=4, its periodic extension $\tilde{x}_N[n]$, its shifted version $\tilde{x}_N[n-M]\,(M=2)$, and its rectangular-windowed version $\tilde{x}_N[n-M]\,r_N[n]$, respectively. You can see more clearly from Figs. 6.20(a), (c1), (c2), and (d) that $\tilde{x}_N[n-M]\,r_N[n]$ is the circular-shifted (rotated) version of $x[n]$, where Figs. 6.20(c1) and (c2) show the circular representation of the finite-duration sequence $x[n]$ and its shifted or rotated version displayed around the circle with a circumference of N points. With this visual concept of *circular shift*, we can write the (circular) time-shifting property of DFS and DFT as

$$\tilde{x}_N[n-M]\,(\text{circular shift}) \overset{\text{DFS}}{\leftrightarrow} W_N^{Mk}\tilde{X}(k) \tag{6.4.11a}$$

$$\tilde{x}_N[n-M]\,r_N[n]\,(\text{one period of circular shift}) \overset{\text{DFT}}{\leftrightarrow} W_N^{Mk}X(k) \tag{6.4.11b}$$

We can also apply the duality between the time and frequency domains to write the (circular) frequency-shifting property of DFS/DFT:

$$W_N^{-Ln}\,\tilde{x}_N[n] \overset{\text{DFS}}{\leftrightarrow} \tilde{X}(k-L)\,(\text{circular shift}) \tag{6.4.12a}$$

$$W_N^{-Ln}\,x[n] \overset{\text{DFT}}{\leftrightarrow} \tilde{X}(k-L)\,r_N[k]\,(\text{one period of circular shift}) \tag{6.4.12b}$$

<Real/Complex Convolution — Circular Convolution in Time/Frequency>

Let us consider two sequences $x[n]$ and $y[n]$, both of duration $M \le N$ and each with DFT $X(k)$ and $Y(k)$, respectively. Also, let us denote the periodic extensions of the two sequences and their DFSs by $\tilde{x}[n]$, $\tilde{y}[n]$, $\tilde{X}(k)$, and $\tilde{Y}(k)$, respectively. Then we can write the product of the two DFSs

$$\tilde{X}(k) = \text{DFS}_N\{\tilde{x}[n]\} = \sum_{m=0}^{N-1}\tilde{x}[m]\,W_N^{km} \quad \text{and} \quad \tilde{Y}(k) = \text{DFS}_N\{\tilde{y}[n]\} = \sum_{r=0}^{N-1}\tilde{y}[r]\,W_N^{kr}$$

as

$$\tilde{X}(k)\tilde{Y}(k) = \sum_{m=0}^{N-1}\sum_{r=0}^{N-1} \tilde{x}[m]\,\tilde{y}[r]\,W_N^{k(m+r)} \tag{6.4.13}$$

and compute its IDFS (inverse discrete Fourier series) as

$$\mathrm{IDFS}_N\left\{\tilde{X}(k)\tilde{Y}(k)\right\} \overset{(6.4.8)}{=} \frac{1}{N}\sum_{k=0}^{N-1}\tilde{X}(k)\tilde{Y}(k)W_N^{-kn} \overset{(6.4.13)}{=} \sum_{m=0}^{N-1}\tilde{x}[m]\sum_{r=0}^{N-1}\tilde{y}[r]\left(\frac{1}{N}\sum_{k=0}^{N-1}W_N^{k(r-n+m)}\right)$$

$$\overset{(6.4.4)}{=} \sum_{m=0}^{N-1}\tilde{x}[m]\sum_{r=0}^{N-1}\tilde{y}[r]\,\delta[r-(n-m)] = \sum_{m=0}^{N-1}\tilde{x}[m]\,\tilde{y}[n-m] = \tilde{x}[n] \underset{N}{*} \tilde{y}[n] \tag{6.4.14}$$

where $\underset{N}{*}$ denotes the circular or periodic convolution sum with period N. This implies the real convolution property of DFS/DFT as

$$\tilde{x}[n] \underset{N}{*} \tilde{y}[n]\,(\text{circular convolution}) \overset{\text{DFS}}{\leftrightarrow} \tilde{X}(k)\,\tilde{Y}(k) \tag{6.4.15a}$$

$$(\tilde{x}[n] \underset{N}{*} \tilde{y}[n])\,r_N[n]\,(\text{one period of circular convolution}) \overset{\text{DFT}}{\leftrightarrow} X(k)\,Y(k) \tag{6.4.15b}$$

In duality with this, we can interchange the time and frequency indices to write the complex convolution property as

$$\tilde{x}[n]\,\tilde{y}[n] \overset{\text{DFS}}{\leftrightarrow} \frac{1}{N}\sum_{i=0}^{N-1}\tilde{X}(i)\,\tilde{Y}(k-i) = \frac{1}{N}\tilde{X}(k) \underset{N}{*} \tilde{Y}(k)\ (\text{circular convolution}) \tag{6.4.16a}$$

$$x[n]y[n] \overset{\text{DFT}}{\leftrightarrow} \frac{1}{N}\left(\tilde{X}(k) \underset{N}{*} \tilde{Y}(k)\right)r_N[k]\,(\text{one period of circular convolution}) \tag{6.4.16b}$$

Note that the shift of a periodic sequence along the time/frequency axis actually causes a rotation, making the sequence values appear to wrap around from the beginning of the sequence to the end.

6.5 FAST FOURIER TRANSFORM (FFT)

In this section we discuss the FFT algorithm that exploits the periodicity and symmetry of the discrete-time complex exponentials $e^{j2\pi nk/N}$ to significantly reduce the number of multiplications for the DFT computation. The FFT algorithm discussed here achieves its efficiency when N is a power of 2, i.e., $N=2^{LOG2N}$ for some integer $LOG2N$. This makes no practical problem since the length of $x[n]$ can be increased to the next power of 2 by zero-padding.

To get some understanding of the steps in the FFT algorithm, let us consider a sequence $x[n]$ for $0 \le n \le N-1$ with $N=2^{LOG2N}$. There are two approaches, each of which is based on the decimation process in the time and frequency domains, respectively.

6.5.1 Decimation-in-Time (DIT) FFT

In this approach, we break the N-point DFT into two $N/2$-point DFTs, one for even-indexed subsequence $x[2r]$, and the other for odd-indexed subsequence $x[2r+1]$, then break each $N/2$-point DFT into two $N/4$-point DFTs, and continue this process until 2-point DFTs appear. Specifically, we can write the N-point DFT of $x[n]$ as

$$X(k) \overset{(6.4.2)}{=} \sum_{n=0}^{N-1} x[n]W_N^{kn} = \sum_{n=2r(\text{even})} x[n]W_N^{kn} + \sum_{n=2r+1(\text{odd})} x[n]W_N^{kn}$$
$$= \sum_{r=0}^{N/2-1} x[2r]\,W_N^{2rk} + \sum_{r=0}^{N/2-1} x[2r+1]\,W_N^{(2r+1)k}$$
$$(W_N^{2rk} = e^{-j2\pi(2r)k/N} = e^{-j2\pi rk/(N/2)} = W_{N/2}^{rk})$$
$$= \sum_{r=0}^{N/2-1} x_e[r]W_{N/2}^{rk} + W_N^k \sum_{r=0}^{N/2-1} x_o[r]W_{N/2}^{rk}$$
$$\overset{(6.4.2)}{=} X_e(k) + W_N^k X_o(k) \text{ for } 0 \le k \le N-1 \tag{6.5.1}$$

so that

$$X(k) \overset{(6.5.1)}{=} X_e(k) + W_N^k X_o(k) \text{ for } 0 \le k \le N/2-1 \tag{6.5.2a}$$
$$X(k) \overset{(6.5.1)}{=} X_e(k) + W_N^k X_o(k) \text{ for } N/2 \le k \le N-1$$
$$;X(k+N/2) \overset{(6.5.1)}{=} X_e(k+N/2) + W_N^{k+N/2} X_o(k+N/2) = X_e(k) - W_N^k X_o(k) \text{ for } 0 \le k \le N/2-1 \tag{6.5.2b}$$

where $X_e(k)$ and $X_o(k)$ are $N/2$-point DFTs that are periodic with period $N/2$ in k. If $N/2$ is even, then we can again break each of $X_e(k)$ and $X_o(k)$ into two $N/4$-point DFTs in the same way:

$$X_e(k) \underset{\text{with } N\to N/2}{\overset{(6.5.1)}{=}} X_{ee}(k) + W_{N/2}^k X_{eo}(k) = X_{ee}(k) + W_N^{2k} X_{eo}(k) \text{ for } 0 \le k \le N/2-1 \tag{6.5.3a}$$
$$X_o(k) \underset{\text{with } N\to N/2}{\overset{(6.5.1)}{=}} X_{oe}(k) + W_{N/2}^k X_{oo}(k) = X_{oe}(k) + W_N^{2k} X_{oo}(k) \text{ for } 0 \le k \le N/2-1 \tag{6.5.3b}$$

If $N=2^{NLOG2}$, we repeat this procedure $NLOG2-1$ times to obtain $N/2$ 2-point DFTs, say, for $N=2^3$, as

$$X_{ee}(k) = \sum_{n=0}^{N/4-1} x_{ee}[n]W_{N/4}^{kn} = x[0] + (-1)^k x[4] \text{ with } x_{ee}[n] = x_e[2n] = x[2^2 n] \tag{6.5.4a}$$
$$X_{eo}(k) = \sum_{n=0}^{N/4-1} x_{eo}[n]W_{N/4}^{kn} = x[2] + (-1)^k x[6] \text{ with } x_{eo}[n] = x_e[2n+1] = x[2^2 n+2] \tag{6.5.4b}$$
$$X_{oe}(k) = \sum_{n=0}^{N/4-1} x_{oe}[n]W_{N/4}^{kn} = x[1] + (-1)^k x[5] \text{ with } x_{oe}[n] = x_o[2n] = x[2^2 n+1] \tag{6.5.4c}$$
$$X_{oo}(k) = \sum_{n=0}^{N/4-1} x_{oo}[n]W_{N/4}^{kn} = x[3] + (-1)^k x[7] \text{ with } x_{oo}[n] = x_o[2n+1] = x[2^2 n+2+1] \tag{6.5.4d}$$

Along with this procedure, we can draw the signal flow graph for an 8-point DFT computation as shown in Fig. 6.21(a). By counting the number of branches with a gain W_N^r (representing multiplications) and empty circles (denoting additions), we see that each stage needs N complex multiplications and N complex additions. Since there are $\log_2 N$ stages, we have a total of $N \log_2 N$ complex multiplications and additions. This is a substantial reduction of computation compared with the direct DFT computation, which requires N^2 complex multiplications and $N(N-1)$ complex additions since we must get N values of $X(k)$ for k=0:$N-1$, each $X(k)$ requiring N complex multiplications and $N-1$ complex additions.

Fig. 6.21(b) shows that a further reduction of multiplications is possible by making one complex multiplication $W_N^k X_o(k)$ shared for computing the two equations (6.5.2a) and (6.5.2b), called a *simplified butterfly computation*, so that the total number of complex multiplications can be halved to $(N/2)\log_2 N$, as depicted by the simplified signal flow graph in Fig. 6.21(c).

[Remark 6.5] Data Rearrangement in "Bit Reversed" Order

The signal flow graph in Fig. 6.21(a) or (c) shows that the input data $x[n]$ appear in the bit reversed order:

Position		Binary equivalent		Bit reversed		Sequence index
3	→	011	→	110	→	6
4	→	100	→	001	→	1

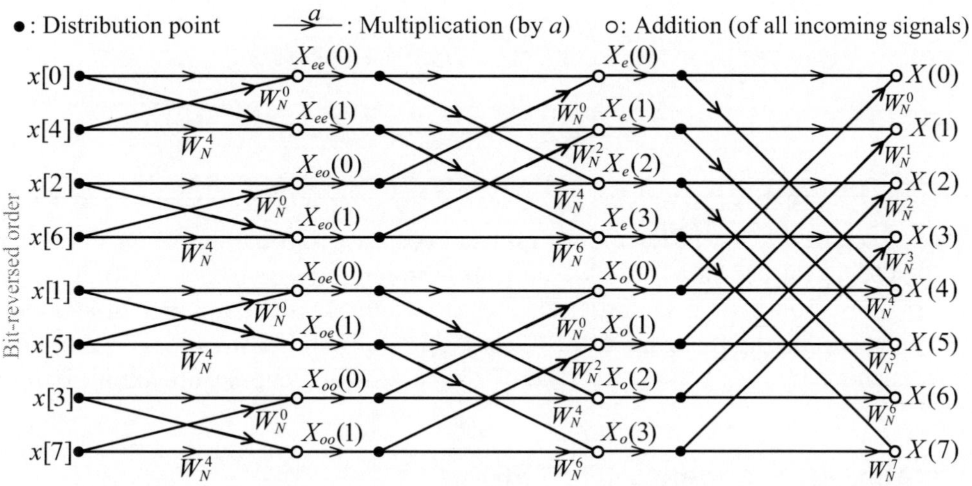

(a) Signal flow graph of DIT (decimation-in-time) FFT algorithm for computing an N=8-point DFT

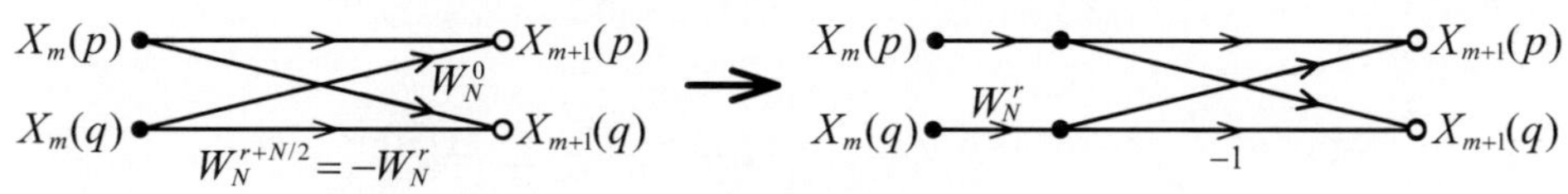

(b) Simplifying the butterfly computation

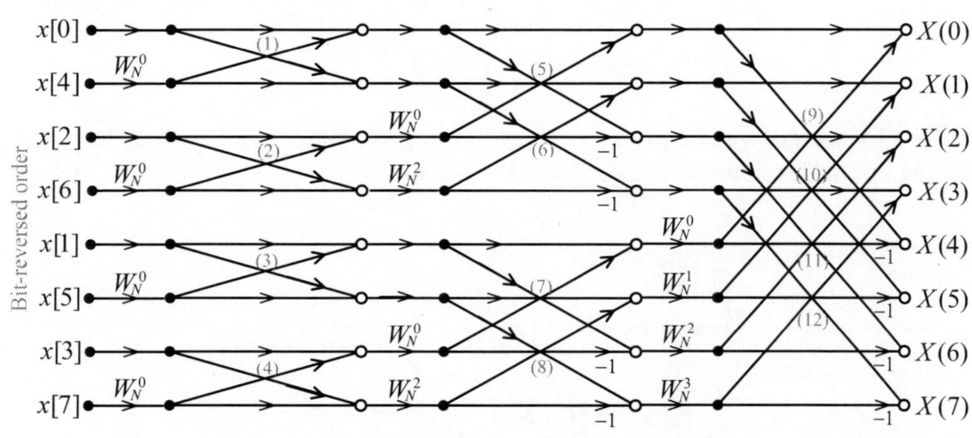

(c) Signal flow graph of simplified DIT FFT algorithm for computing an N=8-point DFT

Figure 6.21 Signal flow graph of DIT (decimation-in-time) FFT algorithm

```
function X=fft_DIT(x)
% Radix 2 Decimation in Time FFT algorithm
M=length(x); log2N=nextpow2(M); % Number of stages
N=2^log2N;  % Determine the DFT size N=2^(NLOG2)
X=[x(:); zeros(N-M,1)]; % Zero-padding so that length(x)=N.
X=bitrevorder(X); % Rearrange x[n] in bit-reversed order
for s=1:log2N    % For each stage
   Ns=2^s; NB=N/Ns; % DFT size, Number of DFT blocks at current stage
   XX=reshape(X,Ns,NB);
   for n=1:NB  % For each block of intermediate X(k)
      Xn=XX(:,n);
      W=exp(-j*2*pi*[0:Ns/2-1]/Ns).';
      tmp=Xn(Ns/2+[1:Ns/2]).*W;
      XX(:,n)=[Xn(1:Ns/2)+tmp; Xn(1:Ns/2)-tmp]; % Eqs.(6.5.2a,b)
   end
   X=XX(:);
end
if size(x,2)>1, X=X.'; end % To make the size of X equal to that of x
```

The DIT (decimation-in-time) FFT algorithm has been cast into the above MATLAB function '`fft_DIT()`'. Note that it uses a MATLAB built-in function '`nextpow2()`' to find the exponent log2N for the smallest power of 2 such that $2^{\log 2N} \geq M$, where M is the number of elements of the sequence $x[n]$ given as the input argument of '`fft_DIT()`'. Then it computes $N=2^{\log 2}$ and if M is smaller than N, appends $(N-M)$ zeros to the end of $x[n]$ so that the length of resulting $x[n]$ will be N – *zero padding*.

6.5.2 Decimation-in-Frequency (DIF) FFT

An alternative algorithm for computing the DFT results from breaking up or decimating the transform sequence $X(k)$. It begins by breaking up $X(k)$ as follows:

$$\begin{aligned} X(k) &\overset{(6.4.2)}{=} \sum_{n=0}^{N-1} x[n]\, W_N^{kn} = \sum_{n=0}^{N/2-1} x[n]\, W_N^{kn} + \sum_{n=N/2}^{N-1} x[n]\, W_N^{kn} \\ &= \sum_{n=0}^{N/2-1} x[n]\, W_N^{kn} + W_N^{kN/2} \sum_{n=0}^{N/2-1} x[n+N/2]\, W_N^{kn} \\ &= \sum_{n=0}^{N/2-1} \{x[n] + (-1)^k x[n+N/2]\} W_N^{kn} \end{aligned} \tag{6.5.6}$$

We separate this into two groups, i.e., one group of the even-indexed elements (with $k=2r$) and the other group of the odd-indexed elements (with $k=2r+1$) of $X(k)$:

$$\begin{aligned} X(2r) &= \sum_{n=0}^{N/2-1} \{x[n] + x[n+N/2]\} W_N^{2rn} \\ &= \sum_{n=0}^{N/2-1} \{x[n] + x[n+N/2]\} W_{N/2}^{rn} \quad \text{for } 0 \leq r \leq N/2-1 \end{aligned} \tag{6.5.7a}$$

$$\begin{aligned} X(2r+1) &= \sum_{n=0}^{N/2-1} \{x[n] - x[n+N/2]\}\, W_N^{(2r+1)n} \\ &= \sum_{n=0}^{N/2-1} \{x[n] - x[n+N/2]\}\, W_N^{n}\, W_{N/2}^{rn} \quad \text{for } 0 \leq r \leq N/2-1 \end{aligned} \tag{6.5.7b}$$

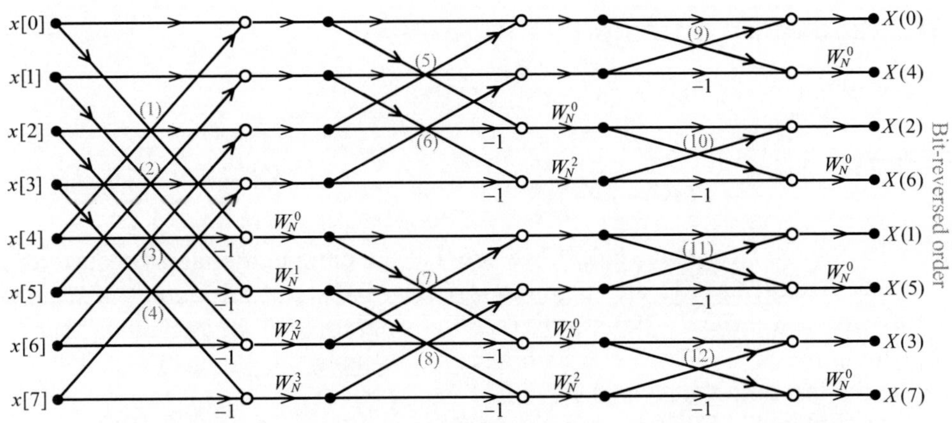

Figure 6.22 Signal flow graph of DIF (decimation-in-frequency) FFT algorithm (with simplified butterfly pattern) for computing an N=8-point DFT

```
function X=fft_DIF(x)
% Radix 2 Decimation in Frequency FFT algorithm
M=length(x); log2N=nextpow2(M); % Number of stages
N=2^log2N;   % Determine the DFT size N=2^(NLOG2)
X=[x(:); zeros(N-M,1)]; % Zero-padding so that length(x)=N.
for s=1:log2N    % For each stage
   Ns=2^(log2N+1-s); % DFT size at current stage
   NB=N/Ns; % Number of DFT blocks at current stage
   XX=reshape(X,Ns,NB);
   for n=1:NB  % For each block of intermediate Xk
      Xn=XX(:,n); W=exp(-j*2*pi*[0:Ns/2-1]/Ns).';
      XX(:,n)=[Xn(1:Ns/2)+Xn(Ns/2+[1:Ns/2]);     % Eq.(6.5.7a)
               (Xn(1:Ns/2)-Xn(Ns/2+[1:Ns/2])).*W]; % Eq.(6.5.7b)
   end
   X=XX(:);
end
X=bitrevorder(X); % Rearrange X(k) in bit-reversed order
if size(x,2)>1, X=X.'; end % To make the size of X equal to that of x
```

These are the $N/2$-point DFTs of the two sequences $(x[n]+x[n+N/2])$ and $(x[n]-x[n+N/2])W_N^n$, respectively. If $N=2^{NLOG2}$, we can proceed in the same way until it ends up with $N/2$ 2-point DFTs. The DIF FFT algorithm with simplified butterfly computation and the output data in the bit reversed order is illustrated for an 8-point DFT in Fig. 6.22.

The DIF FFT algorithm has been cast into the above MATLAB function '`fft_DIF()`'. Let us try the two MATLAB functions '`fft_DIT()`' and '`fft_DIF()`' for a 7-element sequence:

```
>>x=[0 1 0 2 0 1 0]; % A 7-element sequence
>>X1=fft_DIT(x), X2=fft_DIF(x), X3=fft(x)
>>log2N=nextpow2(length(x)); N=2^log2N; X4=fft(x,N)
```

Having run these MATLAB statements, you will see that X1 and X2 are the same as X4 (obtained by using the MATLAB built-in function as '`fft(x,N)`'), implying that '`fft_DIT()`' and '`fft_DIF()`' work fine, but X3 differs from them. Why is that? It is because '`fft(x)`' uses the length of x by default as the DFT size N unless N is given as its second input argument. Curious readers may follow the steps specified by the signal flow graphs in Figs. 6.21 and 6.22 to compute, say, $X(2)$ by hand.

6.5.3 Computation of IDFT Using FFT Algorithm

The inverse DFT is given by Eq. (6.4.3) as

$$x[n] = \mathrm{IDFT}_N\{X(k)\} = \frac{1}{N}\sum_{k=0}^{N-1} X(k)\, e^{j2\pi kn/N} = \frac{1}{N}\sum_{k=0}^{N-1} X(k) W_N^{-kn} \quad \text{for } n=0:N-1 \tag{6.5.8}$$

Comparing this with the DFT formula (6.4.2), we see that the computational procedure remains the same except that the twiddle factors are negative powers of W_N and the output is scaled by $1/N$. Therefore, an inverse fast Fourier transform (IFFT) program can be obtained from an FFT algorithm by replacing the input data $x[n]$'s with $X(k)$'s, changing the exponents of W_N to negative values, and scaling the last output data by $1/N$.

An alternative way to get the IDFT by using the FFT algorithm is derived by taking the complex conjugate of the IDFT formula (6.4.3) or (6.5.8) as follows:

$$x^*[n] = \frac{1}{N}\sum_{k=0}^{N-1} X^*(k) W_N^{kn} = \frac{1}{N}\mathrm{FFT}\{X^*(k)\}; \quad x[n] = \frac{1}{N}\left(\mathrm{FFT}\{X^*(k)\}\right)^* \tag{6.5.9}$$

It is implied that we can get $x[n]=\mathrm{IDFT}_N\{X(k)\}$ by taking the complex conjugate of $X(k)$, applying the FFT algorithm for $X^*(k)$, taking the complex conjugate of the output, and scaling by $1/N$.

The MATLAB functions '`fft(x,N)`'/'`ifft(X,N)`' apply the N-point FFT/IFFT algorithms for the data $x[n]/X(k)$ given as their first input argument if the second input argument N is given as a power of 2, or if N is not given and the length of the given data is a power of 2. If the length of the data sequence differs from the second input argument N, it will be zero-padded or truncated so that the resulting length will be N.

6.5.4 Interpretation of DFT Results

In general, the DFT takes a complex-valued sequence $\{x[n],\ n=0:N-1\}$ and produces another complex-valued sequence $\{X(k), k=0:N-1\}$. Then, what is the physical meaning of a DFT spectrum? To understand it, let us take some examples where we think of a discrete sequence $x[n]$ as the discrete-time version of an analog signal $x(t)$ (with sampling period T) such that $x[n]=x(nT)$.

(Example 6.23) DFT Spectrum of a Single-Tone Sinusoidal Wave

(a) Consider a 16-point discrete-time sequence

$$x_a[n] = \cos\left(\frac{15}{4}\pi t\right)\Big|_{t=nT=0.1n} = \cos\left(\frac{3}{8}\pi n\right), \quad n=0,1,\cdots,15 \tag{E6.23.1}$$

The DFT magnitude spectrum of this sequence, i.e., the magnitude of $X_a(k)=\mathrm{DFT}_{16}\{x_a[n]\}$, has been obtained by running the MATLAB script "em06e23.m" (listed below) and plotted in Fig. 6.23(a). It has a spike at the frequency index $k=3$ corresponding to the digital frequency as

$$\Omega_3 = 3\Omega_0 = 3\frac{2\pi}{N} = \frac{6\pi}{16} = \frac{3\pi}{8}\,[\text{rad/sample}] \tag{E6.23.2}$$

It has another spike at $k=13-N=-3$, but it is just like the mirror image of $k=3$. It can be assured by writing the discrete-time sequence (E6.23.1) as

$$x[n]=\frac{1}{2}\left(e^{j3\pi n/8}+e^{-j3\pi n/8}\right)\overset{\text{IDFT (6.4.3)}}{=}\frac{1}{N}\left(\frac{N}{2}e^{j2\pi 3n/N}+\frac{N}{2}e^{-j2\pi 3n/N}\right)\Bigg|_{N=16}$$

$$=\frac{1}{16}\left(8e^{j2\pi 3n/16}+8e^{j2\pi(13)n/16}\right)=\frac{1}{N}\sum_{k=0}^{N-1}X(k)\,e^{j2\pi kn/N}$$

$$\rightarrow \quad X(k)=8 \quad \text{for } k=3\ \&\ k=13 \tag{E6.23.3}$$

This explains why the magnitude of the DFT spectrum has two spikes of 8 at $k=3$ and $N-3=13$.

(b) Consider another discrete-time sequence

$$x_b[n]=\cos\left(\frac{15}{4}\pi t\right)\Bigg|_{t=nT=0.05n}=\cos\left(\frac{3}{16}\pi n\right),\quad n=0,1,\cdots,15 \tag{E6.23.4}$$

The magnitude spectrum of this sequence, i.e., the magnitude of $X_b(k)=\text{DFT}_{16}\{x_b[n]\}$, has also been obtained by running the MATLAB script "em06e23.m" and is plotted in Fig. 6.23(b), which looks quite different from Fig. 6.23(a). That is, so many nonzero values show up here and there in the magnitude spectrum, though the highest two values occur at $k=1$ and 2. It is because the *picket fence effect* (due to the frequency domain sampling of DTFT) hides the spectral peak, allowing the many side lobes (resulting from the *spectral leakage* due to the time domain windowing — see Example 6.22) to be seen. This is a contrast to the other cases where the picket fence effect coincidentally hides the side lobes except the main peak as if by magic. Still, we might fake a "far-fetched" story about the frequency corresponding to $k=1.5$ (between 1 and 2) as

$$\Omega_{1.5}=1.5\Omega_0=1.5\frac{2\pi}{N}=\frac{3\pi}{16}\ \text{[rad/sample]} \tag{E6.23.5}$$

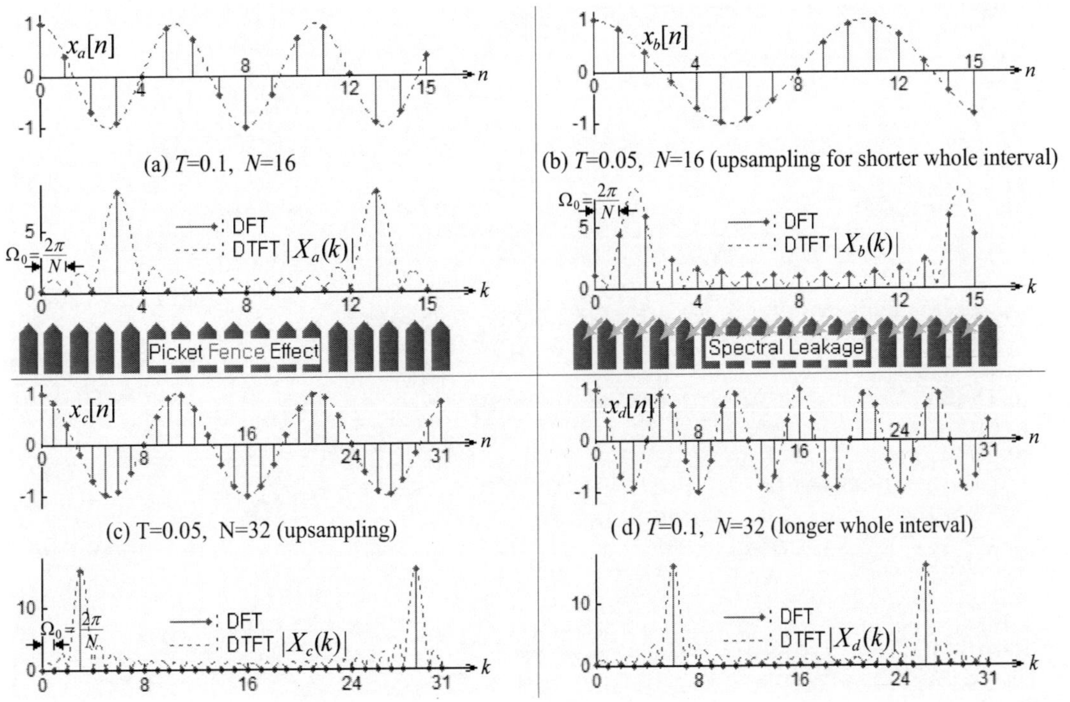

Figure 6.23 DFT spectra of a single-tone sequence depending on the sampling period and the whole interval

(c) Consider another discrete-time sequence

$$x_c[n] = \cos(\frac{15}{4}\pi t)\Big|_{t=nT=0.05n} = \cos(\frac{3}{16}\pi n), \quad n = 0, 1, \cdots, 31 \tag{E6.23.6}$$

The magnitude spectrum of this sequence, i.e., the magnitude of $X_c(k)=\text{DFT}_{32}\{x_c[n]\}$, has also been obtained by running the MATLAB script "em06e23.m" and is plotted in Fig. 6.23(c). Like Fig. 6.23(a), it has two spikes at k=3 and N−3=29 with only a difference in their amplitude (16), which is twice as large as those in (a) because of the increased number of samples (N=32). Thus, we can read the digital frequency as

$$\Omega_3 = 3\Omega_0 = 3\frac{2\pi}{N} = \frac{6\pi}{32} = \frac{3\pi}{16}\text{[rad/sample]} \tag{E6.23.7}$$

(d) Consider another discrete-time sequence

$$x_d[n] = \cos(\frac{15}{4}\pi t)\Big|_{t=nT=0.1n} = \cos(\frac{3}{8}\pi n), \quad n = 0, 1, \cdots, 31 \tag{E6.23.8}$$

The magnitude spectrum of this sequence, i.e., the magnitude of $X_d(k)=\text{DFT}_{32}\{x_d[n]\}$, has also been obtained by running the MATLAB script "em06e23.m" and is plotted in Fig. 6.23(d). It has two spikes at k=6 and N−6=26, which tells us about the existence of the digital frequency as

$$\Omega_6 = 6\Omega_0 = 6\frac{2\pi}{N} = \frac{12\pi}{32} = \frac{3\pi}{8}\text{[rad/sample]} \tag{E6.23.9}$$

```
%em06e23.m
% To plot Fig.6.23
clear, clf
w=15*pi/4; % The analog frequency of x(t) given by Eq.(6.23.1)
for i=1:4
   if i==1, T=0.1; N=16; wn1=421; wn2=423; Xmax=10;
            ts='cos(15pi*t/4) with T=0.1';
    elseif i==2, T=0.05; N=16; wn1=422; wn2=424; Xmax=10;
                 ts='cos(15pi*t/4) with T=0.05 (up-sampling/shorter)';
    elseif i==3, T=0.05; N=32; wn1=425; wn2=427; Xmax=20;
                 ts='cos(15pi*t/4) with T=0.05 (up-sampling)';
    else T=0.1; N=32; wn1=426; wn2=428; Xmax=20;
               ts='cos(15pi*t/4) with T=0.1';
   end
   n=[0:N-1]; xn=cos(w*n*T); X=fft(xn,N); k=[0:N-1];
   subplot(wn1)
   stem(n,xn,'.')
   title(ts), axis([n([1 end]) -1.2 1.2])
   nn=[0:(N-1)*20]/20; xnn=cos(w*nn*T);
   hold on, plot(nn,xnn,':')
   subplot(wn2)
   stem(k,abs(X),'.')
   axis([n([1 end]) 0 Xmax])
   kk=[k(1):0.01:k(end)]; W=kk*2*pi/N; X_DTFT=DTFT(xn,W,0);
   hold on, plot(kk,abs(X_DTFT),':')
end
```

6.6 FOURIER–BESSEL/LEGENDRE/CHEBYSHEV/COSINE/SINE SERIES

Fourier series is not the only orthogonal series expansion. In this section, two more orthogonal series expansions, Fourier–Bessel series and Fourier–Legendre series, are introduced, since they can be used to solve the PDEs (partial differential equations) in the next chapter.

6.6.1 Fourier–Bessel Series (see Section 12.6 of [Z-1])

Just as the cosine/sine or complex sinusoidal functions of frequency $k\omega_0 = 2k\pi/P$ are orthogonal to each other over the period P (see Eq. (6.1.10)), the Bessel function $J_n(x)$'s (Eq. (3.3.10)) are orthogonal w.r.t. the weight function x in the sense that

$$\int_0^{x_f} x J_n(\lambda_{n,k}x) J_n^*(\lambda_{n,m}x)\,dx = 0 \;\; \forall\, k \neq m \tag{6.6.1}$$

where $\lambda_{n,k}$ is obtained by dividing the kth positive root of a mixed type of BC (boundary condition)

$$b_1 J_n(x_f\lambda) + b_2 \lambda J_n'(x_f\lambda) = 0 \tag{6.6.2}$$

by x_f. A function $f(x)$ defined on $[0, x_f]$ can be expressed in terms of the orthogonal eigenfunctions $\{J_n(\lambda_{n,k}x),\ k = 1, 2, \cdots\}$ as

$$f(x) = c_0 + \sum_{k=1}^{\infty} c_k J_n(\lambda_{n,k}x) \tag{6.6.3}$$

where the coefficient c_k's are determined depending on the BC as follows:

(Case I) When the BC is $J_n(x_f\lambda) = 0$,

$$c_0 = 0 \text{ and } c_k = \frac{1}{x_f^2 J_{n+1}^2(x_f\lambda_{n,k})} \int_0^{x_f} x J_n(\lambda_{n,k}x) f(x)\,dx \text{ for } k \geq 1 \tag{6.6.4a}$$

(Case II) When the BC is $b_1 J_n(x_f\lambda) + b_2\lambda J_n'(x_f\lambda) = 0$,

$$c_0 = 0 \text{ and } c_k = \frac{2\lambda_{n,k}^2 \int_0^{x_f} x J_n(\lambda_{n,k}x) f(x)\,dx}{(x_f^2\lambda_{n,k}^2 - n^2 + b_1^2) J_n^2(\lambda_{n,k}x)} \text{ for } k \geq 1 \tag{6.6.4b}$$

(Case III) When the BC is $J_0'(x_f\lambda) = 0$,

$$c_0 = \frac{2}{x_f^2}\int_0^{x_f} x f(x)\,dx \text{ and } c_k = \frac{2\int_0^{x_f} x J_0(\lambda_{n,k}x) f(x)\,dx}{b_1^2 J_0^2(\lambda_{0,k}x_f)} \text{ for } k \geq 1 \tag{6.6.4c}$$

This generalized Fourier expansion (6.6.3) is called a *Fourier–Bessel series*. Note that Eqs. (P3.12.5) and (P3.12.3) with $v = 0$ can be used to rewrite the second and third BCs (each in Eqs. (6.6.4b/c)) as

$$b_1 J_n(x_f\lambda) + b_2\lambda J_n'(x_f\lambda) \overset{\text{(P3.12.5)}}{=} b_1 J_n(x_f\lambda) + b_2\lambda \frac{1}{2}\left(J_{n-1}(x_f\lambda) - J_{n+1}(x_f\lambda)\right) \tag{6.6.5}$$

$$J_n'(x_f\lambda) = 0 \overset{\text{(P3.12.3)}}{\rightarrow} J_{n+1}(x_f\lambda) = 0 \tag{6.6.6}$$

This Fourier–Bessel series expansion formula (6.6.3) with Eqs. (6.6.4a,b,c) has been cast into the MATLAB function 'Fourier_Bessel_series()'.

```
function [c,f_hat]=Fourier_Bessel_series(f,xf,b12,n,N)
% find the Fourier-Bessel series expansion of f(x) on [0,xf]
% with Bessel functions {Jn(lambda_k*x)} where the eigenvalue
%  lambda_k's are determined from the BC
%        b1*Jn(xf*lambda)+b2*lambda*Jn'(xf*lambda)=0
% n = Order of the Bessel functions involved in the series expansion
% N = Number of terms contained in the series expansion
%Copyleft: Won Y. Yang, wyyang53@hanmail.net, CAU for academic use
if nargin<5, N=20; end
if length(b12)==1, b1=b12; b2=0; else b1=b12(1); b2=b12(2); end
xs=[0:0.1:100]; xx=[0:0.01:xf]; EPS=1e-6;
if b2<EPS % Case I
  flambda = @(lambda)besselj(n,lambda); % J_n(lambda)=0
  % To find the initial guesses (lam0s) of zeros of J_n(lambda)=0
  fxs = 1./abs(flambda(xs)).^(1/10);
  [localmaxima,indms] = localmax(fxs,[],false); lam0s = xs(indms);
  f_hat = 0;
  for k=1:min(N,length(lam0s))
     lambda(k) = fsolve(flambda,lam0s(k))/xf; lamk=lambda(k);
     f_int = @(x)x.*besselj(n,lamk*x).*f(x);
     c(k)=2/xf^2/besselj(n+1,lamk*xf)^2*quad(f_int,0,xf); %Eq.(6.6.4a)
     f_hat = f_hat + c(k)*besselj(n,lamk*xx);  % Eq.(6.6.3)
  end
 elseif b1>EPS % Case II
  flambda = @(lambda)b1*besselj(n,lambda)+b2/xf*lambda/2.* ...
              (besselj(n-1,lambda)-besselj(n+1,lambda)); %Eq.(6.6.5)
  % To find the initial guesses (lam0s) of zeros of flambda=0
  fxs = 1./abs(flambda(xs)).^(1/10);
  [localmaxima,indms] = localmax(fxs,[],false); lam0s = xs(indms);
  f_hat = 0;
  for k=1:min(N,length(lam0s))
     lambda(k) = fsolve(flambda,lam0s(k))/xf; lamk=lambda(k);
     f_int = @(x)x.*besselj(n,lamk*x).*f(x);
     c(k) = 2*lamk^2/(xf^2*lamk^2-n^2 +(b1/b2*xf)^2) ...
                 /besselj(n,xf*lamk)^2*quad(f_int,0,xf); %Eq.(6.6.4b)
     f_hat = f_hat + c(k)*besselj(n,lamk*xx);  % Eq.(6.6.3)
  end
 else % Case III
  flambda = @(lambda)besselj(1,lambda); % Since Jo'(x)=J1(x) (6.6.6)
  % To find the initial guesses (lam0s) of zeros of flambda=0
  fxs = 1./abs(flambda(xs)).^(1/10);
  [localmaxima,indms] = localmax(fxs,[],false); lam0s = xs(indms);
  for k=1:length(lam0s), lambda(k)=fsolve(flambda,lam0s(k))/xf;  end
  f_int = @(x)x.*f(x);  f_hat = 2/xf^2*quad(f_int,0,xf);
  for k=1:min(N,length(lambda))
     lambda(k) = fsolve(flambda,lam0s(k))/xf;  lamk = lambda(k);
     f_int = @(x)x.*besselj(0,lamk*x).*f(x);
     c(k)=2/xf^2/besselj(0,xf*lamk)^2*quad(f_int,0,xf); %Eq.(6.6.4c)
     f_hat = f_hat + c(k)*besselj(0,lamk*xx);  % Eq.(6.6.3)
  end
end
```

(Example 6.24) Fourier–Bessel Series Expansions

Find the Fourier–Bessel series expansions of the following functions using Bessel functions of the same order as indicated in their BCs:

(a) $f_a(x)=x(3-x)$ for $0\le x\le x_f=3$ subject to the BC $J_1(x_f\lambda)=0$ (E6.24.1)

(b) $f_b(x)=\begin{cases} x & 0\le x\le 1.5 \\ 3-x & 1.5\le x\le x_f=3 \end{cases}$ subject to the BC $J_2(x_f\lambda)+\lambda J_2'(x_f\lambda)=0$ (E6.24.2)

We can run the following MATLAB script "em06e24.m" to find the Fourier–Bessel series expansions

$$f_a(x) \overset{(6.6.3)}{=} \sum_{k=1}^{N} c_k J_n(\lambda_{n,k} x) \text{ with } c_k = \overset{k=1}{4.0700}\ \overset{k=2}{-0.2837}\ \overset{k=3}{-0.2882}\ \overset{k=4}{-0.0692} \cdots \tag{E6.24.3}$$

$$f_b(x) \overset{(6.6.3)}{=} \sum_{k=1}^{N} c_k J_n(\lambda_{n,k} x) \tag{E6.24.4}$$

with $c_k = \overset{k=1}{1.7669}\ \overset{k=2}{1.6565}\ \overset{k=3}{-0.0022}\ 0.0678\ -0.0170\ 0.2872\ -0.0007\ 0.0168\ -0.0276\ 0.1264$

and plot the expansion (approximation) results with the number of terms N=2/4 and N=5/10, respectively, as in Figs. 6.24(a) and (b).

```
%em06e24.m
clear, clf
gss='brg';
for iter=1:2
   if iter<2, f=@(x)x.*(3-x);
                  xf=3; b1=1; b2=0; n=1; Ns=[2 4]; % Case I
    else   f=@(x)x.*(x<1.5)+(3-x).*(x>=1.5);
           xf=3; b1=1; b2=1; n=2; Ns=[5 10]; % Case II
   end
   subplot(2,2,iter)
   for iter1=1:length(Ns)
      N=Ns(iter1);
      [c,f_hat,xx]=Fourier_Bessel_series(f,xf,[b1 b2],n,N); c
      plot(xx,f_hat,gss(iter1)), hold on
   end
   plot(xx,f(xx),'k:'), xlim([0 xf])
end
```

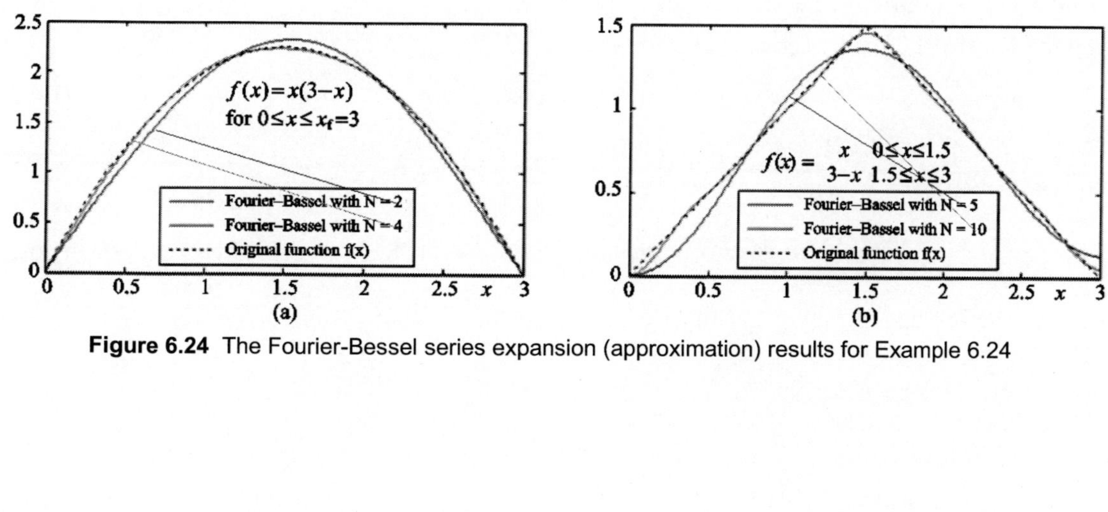

Figure 6.24 The Fourier-Bessel series expansion (approximation) results for Example 6.24

6.6.2 Fourier–Legendre Series (see Section 12.6 of [Z-1])

Legendre polynomial $P_m(x)$'s introduced in Section 3.3.2 are also orthogonal in the sense that

$$\int_0^{x_f} P_m(x)P_n^*(x)\,dx = \begin{cases} 2/(2m+1) & \text{if } m=n \\ 0 & \text{otherwise} \end{cases} \tag{6.6.7}$$

Therefore, a function $f(x)$ defined on $[-1,1]$ can be expressed in terms of the Legendre polynomials $\{P_m(x),\ m=0,1,2,\ldots\}$ as

$$f(x) = \sum_{k=0}^{\infty} c_k P_k(x) \tag{6.6.8a}$$

with

$$c_k = \frac{2k+1}{2}\int_{-1}^{1} f(x)P_k(x)\,dx \text{ for } k \ge 0 \tag{6.6.8b}$$

Note that with $x=\cos\theta$, Eq. (6.6.8b) becomes

$$c_k = \frac{2k+1}{2}\int_0^{\pi} f(\theta)P_k(\cos\theta)\sin\theta\,d\theta \text{ for } k \ge 0 \tag{6.6.9}$$

The Fourier–Legendre series expansion formula (6.6.8) has been cast into the following MATLAB function '`Fourier_Legendre_series()`'.

```
function [c,f_hat]=Fourier_Legendre_series(f,N)
% find the Fourier-Legendre series expansion of f(x) on [-1,1]
%  with Legendre polynomials {Pn(x)}
% N = Number of terms contained in the seires expansion
%Copyleft: Won Y. Yang, wyyang53@hanmail.net, CAU for academic use only
if nargin<2,  N=20;  end
f_hat=0; xx=[-1:0.01:1];
for k=1:N
   f_int = @(x)f(x).*mfun('P',k-1,x); % Integrand of Eq.(6.6.8b)
   c(k)  = (2*(k-1)+1)/2*quad(f_int,-1,1); % Eq.(6.6.8b)
   f_hat = f_hat + c(k)*mfun('P',k-1,xx); % Eq.(6.6.8a)
end
if nargout<1
  plot(xx,f(xx), xx,f_hat,'r:'); xlim([-1 1]);
  legend('Original function f(x)','Fourier-Legendre series expansion')
end
```

(Example 6.25) Fourier–Legendre Series Expansion

Find the Fourier–Legendre series expansion of

$$f(x) = \cos^{-1} x. \tag{E6.25.1}$$

We can run the MATLAB script "em06e25.m" to find a partial sum for the Fourier–Legendre series expansion as

$$f(x) \overset{(6.6.8)}{=} \sum_{k=0}^{N} c_k P_k(x) \text{ with } c_k = \overset{k=0}{1.5708}\ \overset{k=1}{-1.1781}\ \overset{k=2}{0}\ \overset{k=3}{-0.1718}\ \overset{k=4}{0}\ \cdots \tag{E6.25.2}$$

and plot the expansion (approximation) results with the number of terms $N=3$ and $N=5$ as in Fig. 6.25.

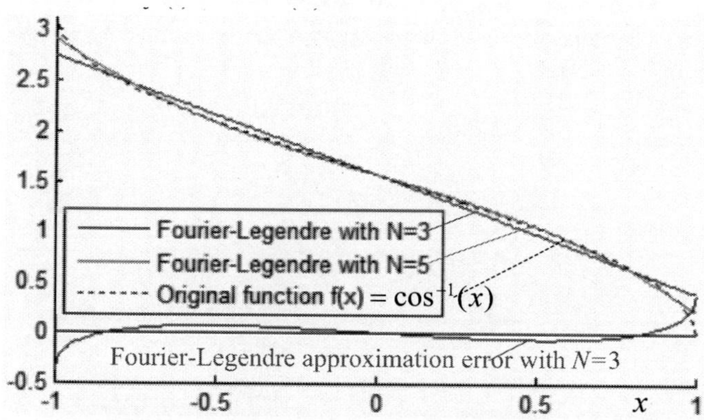

Figure 6.25 The Fourier-Legendre series expansion (approximation) results for Example 6.25

```
%em06e25.m
f=@(x)acos(x);
Ns=[3 5];  gss='brg';
for iter=1:length(Ns)
   N=Ns(iter1);
   [c,f_hat,xx] = Fourier_Legendre_series(f,N); c
   plot(xx,f_hat,gss(iter)), hold on
end
plot(xx,f(xx),'k:'), xlim([-1 1])
legend(['Fourier-Legendre with N=' num2str(Ns(1))],...
  ['Fourier-Legendre with N=' num2str(Ns(2))],'Original function f(x)')
```

6.6.3 Fourier–Chebyshev Series

Chebyshev polynomials $T_m(x)$'s introduced in Section 3.3.3 are orthogonal w.r.t. the weighting function $1/\sqrt{1-x^2}$ in the sense that

$$\int_{-1}^{1} \frac{T_m(x)T_n(x)}{\sqrt{1-x^2}}dx = \begin{cases} 0 & \text{for } m\neq n \\ \pi/2 & \text{for } m=n\neq 0 \\ \pi & \text{for } m=n=0 \end{cases}. \tag{6.6.10}$$

Therefore, a function $f(x)$ defined on [-1, 1] can be expressed in terms of the Chebyshev polynomials $\{T_m(x), m=0,1,2,\ldots\}$ as

$$f(x) = \sum_{k=0}^{\infty} c_k T_k(x) \tag{6.6.11a}$$

with

$$c_k = \begin{cases} \dfrac{1}{\pi}\displaystyle\int_{-1}^{1} f(x)\frac{T_k(x)}{\sqrt{1-x^2}}dx & \text{for } k=0 \\ \dfrac{2}{\pi}\displaystyle\int_{-1}^{1} f(x)\frac{T_k(x)}{\sqrt{1-x^2}}dx & \text{for } k>0 \end{cases} \tag{6.6.11b}$$

The Fourier–Chebyshev series expansion formula (6.6.11) has been cast into the MATLAB function '`Fourier_Chebyshev_series()`'.

```
function [c,f_hat]=Fourier_Chebyshev_series(f,N)
% find the Fourier-Chebyshev series expansion of f(x) on [-1,1]
%  with Chebyshev polynomials {Tn(x)}
% N = Number of terms contained in the seires expansion
%Copyleft: Won Y. Yang, wyyang53@hanmail.net, CAU for academic use only
f_hat=0; xx=[-1:0.01:1];
for k=1:N
   f_int=@(x)f(x).*mfun('T',k-1,x)./sqrt(1-x.^2);%Integrand of (6.6.11b)
   c(k) = ((k==1)+2*(k>1))/pi*quad(f_int,-1,1); % Eq.(6.6.11b)
   f_hat = f_hat + c(k)*mfun('T',k-1,xx); % Eq.(6.6.11a)
end
if nargout<1
  plot(xx,f(xx), xx,f_hat,'r:'); xlim([-1 1]);
  legend('Original function f(x)','Fourier-Chebyshev series expansion')
end
```

(Example 6.26) Fourier–Chebyshev Series Expansion

Find the Fourier–Chebyshev series expansion of

$$f(x)=\cos^{-1}x \tag{E6.26.1}$$

We can run the MATLAB script "em06e26.m" to find a partial sum for the Fourier–Chebyshev series expansion:

$$f(x)\overset{(6.6.11)}{=}\sum_{k=0}^{N} c_k T_k(x) \text{ with } c_k=\underset{k=0}{1.5708}\quad \underset{k=1}{-1.2732}\quad \underset{k=2}{0}\quad \underset{k=3}{-0.1415}\quad \underset{k=4}{0}\quad \cdots \tag{E6.26.2}$$

and plot the expansion (approximation) results with the number of terms N=3 and N=5 as in Fig. 6.26.

```
%em06e26.m
f=@(x)acos(x);
Ns=[3 5]; gss='brg';
for iter=1:length(Ns)
   N=Ns(iter1);
   [c,f_hat,xx] = Fourier_Chebyshev_series(f,N); c
   plot(xx,f_hat,gss(iter)), hold on
end
plot(xx,f(xx),'k:'), xlim([-1 1])
legend(['Fourier-Chebyshev with N=' num2str(Ns(1))],...
  ['Fourier-Chebyshev with N=' num2str(Ns(2))],'Original function f(x)')
```

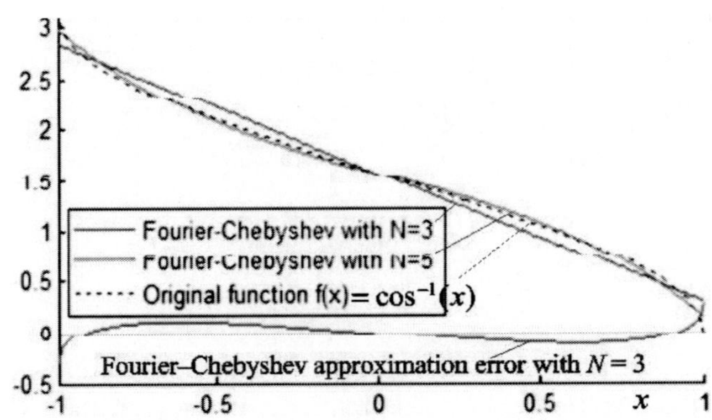

Figure 6.26 The Fourier–Legendre series expansion (approximation) results for Example 6.26

6.6.4 Fourier-Cosine/Sine Series

Cosine functions $\{\cos(k\omega_0 x), k=0,1,\ldots\}$ are orthogonal in the sense that

$$\int_{-P/2}^{P/2}\cos(m\omega_0 x)\cos(n\omega_0 x)\,dx=\begin{cases}0 & \text{for } m\neq n\\ P/2 & \text{for } m=n\neq 0\\ P & \text{for } m=n=0\end{cases}\quad \text{with } \omega_0=\frac{2\pi}{P} \tag{6.6.12}$$

Therefore, a function $f(x)$ defined on $[0, x_f]$ can be expressed in terms of the cosine functions as

$$f(x)=\sum_{k=0}^{\infty}c_k\cos(k\omega_0 x)\ \text{ with } \omega_0=\frac{2\pi}{2x_f}=\frac{\pi}{x_f}:\ \textit{Fourier-cosine series} \tag{6.6.13a}$$

with

$$c_k=\begin{cases}\dfrac{1}{x_f}\displaystyle\int_0^{x_f}f(x)\cos(k\omega_0 x)dx & \text{for } k=0\\[2ex] \dfrac{2}{x_f}\displaystyle\int_0^{x_f}f(x)\cos(k\omega_0 x)dx & \text{for } k>0\end{cases} \tag{6.6.13b}$$

where $f(x)$ is assumed to be an even periodic function with period $P=2x_f$. The Fourier-cosine series expansion formula (6.6.13) has been cast into the MATLAB function '`Fourier_cosine_series()`'.

```
function [c,f_hat,xx]=Fourier_cosine_series(f,xf,N)
% find the Fourier-cosine series expansion of f(x) on [0,xf]
%   with cosine functions {cos(k*w0*x), k=0,1,2,...} (w0=pi/xf)
%  assuming that it is an even (periodic) function on [-xf,xf].
% N = Number of terms contained in the seires expansion
%Copyleft: Won Y. Yang, wyyang53@hanmail.net, CAU for academic use only
if nargin<3,   N=20;   end
f_hat=0; xx=[0:0.01:xf];
w0=pi/xf; % Fundamental frequency 2*pi/P with P=2*xf
for k=1:N
   f_int = @(x)f(x).*cos((k-1)*w0*x);
   c(k)  = ((k==1)+2*(k>1))/xf*quad(f_int,0,xf); % Eq.(6.6.13b)
   f_hat = f_hat + c(k)*cos((k-1)*w0*xx); % Eq.(6.6.13a)
end
if nargout<1
  plot(xx,f(xx), xx,f_hat,'r:'); xlim([0 xf]);
  legend('Original function f(x)','Fourier-cosine series expansion')
end
```

(Example 6.27) Fourier-Cosine Series Expansion

Find the Fourier-cosine series expansions of

$$f_a(x)=(x-0.5)^2 \text{ on } [0,2] \tag{E6.27.1a}$$

and

$$f_b(x)=2.25-(x-1.5)^2 \text{ on } [0,3] \tag{E6.27.1b}$$

We can run the MATLAB script "em06e27.m" to find the partial sums for their Fourier-cosine series expansions:

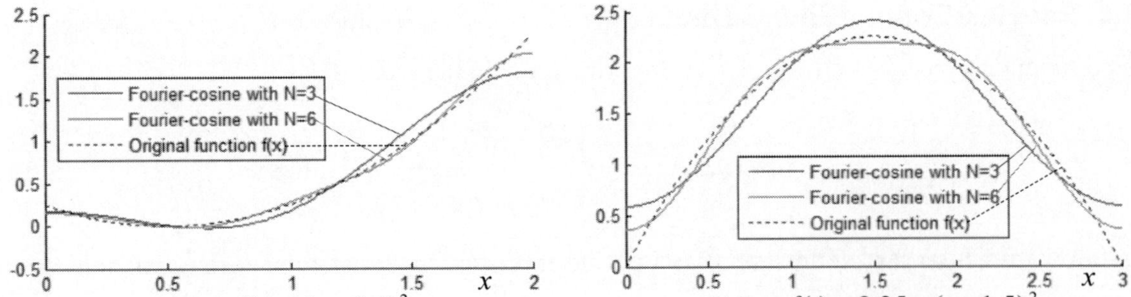

(a) For $f(x) = (x-0.5)^2$ (b) For $f(x) = 2.25-(x-1.5)^2$

Figure 6.27 The Fourier-cosine series expansion (approximation) results for Example 6.27

```
%em06e27.m
clear, clf
Ns=[3 6]; gss='brg';
for iter=1:2
   if iter==1, f=@(x)(x-0.5).^2; xf=2;
    else f=@(x)2.25-(x-1.5).^2; xf=3;
   end
   subplot(2,2,iter), hold on
   for iter1=1:length(Ns)
      N=Ns(iter1);
      [c,f_hat,xx] = Fourier_cosine_series(f,xf,N); c
      plot(xx,f_hat,gss(iter1))
   end
   plot(xx,f(xx),'k:'), xlim([0 xf])
   legend(['Fourier-cosine with N=' num2str(Ns(1))],...
     ['Fourier-cosine with N=' num2str(Ns(2))],'Original function f(x)')
end
```

$$f_a(x) \overset{(6.6.13)}{=} \sum_{k=0}^{N} c_k \cos(k\omega_0 x) \text{ with } \omega_0 = \frac{\pi}{2} \text{ and } c_k = \overset{k=0}{0.5833}\ \ \overset{k=1}{-0.8106}\ \ \overset{k=2}{0.4053}\ \cdots \quad \text{(E6.27.2a)}$$

$$f_b(x) \overset{(6.6.13)}{=} \sum_{k=0}^{N} c_k \cos(k\omega_0 x) \text{ with } \omega_0 = \frac{\pi}{3} \text{ and } c_k = \overset{k=0}{1.5}\ \ \overset{k=1}{0}\ \ \overset{k=2}{-0.9119}\ \ \overset{k=3}{0}\ \cdots \quad \text{(E6.27.2b)}$$

and plot the expansion (approximation) results with the number of terms N=3 and N=6 as in Fig. 6.27.

Sine functions $\{\sin(k\omega_0 x), k=1,2,\ldots\}$ are also orthogonal in the sense that

$$\int_{-P/2}^{P/2} \sin(m\omega_0 x)\sin(n\omega_0 x)\,dx = \begin{cases} 0 & \text{for } m \neq n \\ P/2 & \text{for } m=n\neq 0 \end{cases} \quad \text{with } \omega_0 = \frac{2\pi}{P} \tag{6.6.14}$$

Therefore, a function $f(x)$ defined on $[0, x_f]$ can be expressed in terms of the cosine functions as

$$f(x) = \sum_{k=1}^{\infty} c_k \sin(k\omega_0 x) \text{ with } \omega_0 = \frac{2\pi}{2x_f} = \frac{\pi}{x_f} : \textit{Fourier-sine series} \tag{6.6.15a}$$

with

$$c_k = \frac{2}{x_f}\int_0^{x_f} f(x)\sin(k\omega_0 x)dx \quad \text{for } k>0 \tag{6.6.15b}$$

```
function [c,f_hat,xx]=Fourier_sine_series(f,xf,N)
% find the Fourier-sine series expansion of f(x) on [0,xf]
%   with sine functions {sin(k*w0*x), k=1,2,...} (w0=pi/xf)
%  assuming that it is an odd (periodic) function on [-xf,xf].
% N = Number of terms contained in the seires expansion
%Copyleft: Won Y. Yang, wyyang53@hanmail.net, CAU for academic use only
if nargin<3,  N=20;  end
f_hat=0; xx=[0:0.01:xf];
w0=pi/xf; % Fundamental frequency 2*pi/P with P=2*xf
for k=1:N
   f_int = @(x)f(x).*sin(k*w0*x);
   c(k) = 2/xf*quad(f_int,0,xf); % Eq.(6.6.15b)
   f_hat = f_hat + c(k)*sin(k*w0*xx); % Eq.(6.6.15a)
end
if nargout<1
  plot(xx,f(xx), xx,f_hat,'r:'); xlim([0 xf]);
  legend('Original function f(x)','Fourier-sine series expansion')
end
```

where $f(x)$ is assumed to be an odd periodic function with period $P=2x_f$. The Fourier-sine series expansion formula (6.6.15) has been cast into the above MATLAB function 'Fourier_sine_series()'.

(**Example 6.28**) Fourier-Sine Series Expansion

Find the Fourier-sine series expansions of

$$f_a(x)=(x-0.5)^2 \text{ on } [0,2] \tag{E6.28.1a}$$

and

$$f_b(x)=2.25-(x-1.5)^2 \text{ on } [0,3] \tag{E6.28.1b}$$

We can run the MATLAB script "em06e28.m" to find the partial sums for their Fourier-sine series expansions:

$$f_a(x) \overset{(6.6.15)}{=} \sum_{k=0}^{N} c_k \sin(k\omega_0 x) \text{ with } \omega_0=\frac{\pi}{2} \text{ and } c_k=\overset{k=0}{0.5595}\ \ \overset{k=1}{-0.6366}\ \ \overset{k=2}{0.4923}\ \cdots \tag{E6.28.2a}$$

$$f_b(x) \overset{(6.6.15)}{=} \sum_{k=0}^{N} c_k \sin(k\omega_0 x) \text{ with } \omega_0=\frac{\pi}{3} \text{ and } c_k=\overset{k=0}{2.3221}\ \ \overset{k=1}{0}\ \ \overset{k=2}{0.086}\ \ \overset{k=3}{0}\ \cdots \tag{E6.28.2b}$$

and plot the expansion (approximation) results with the number of terms $N=3$ and $N=5$ as in Fig. 6.28.

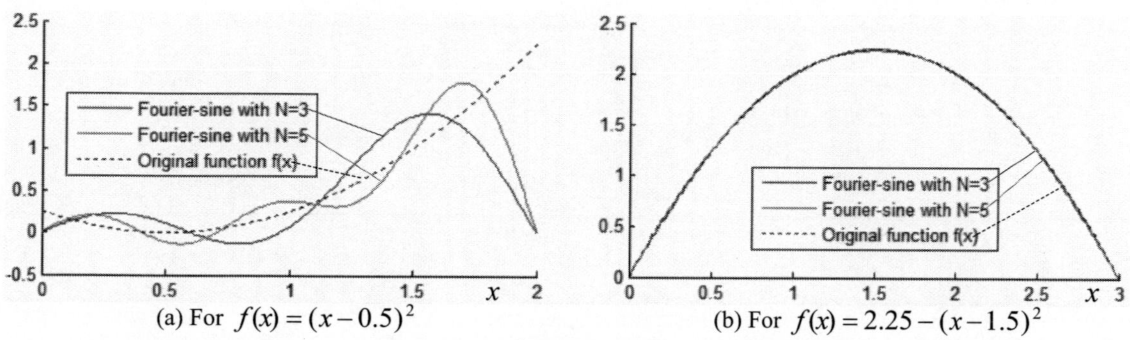

Figure 6.28 The Fourier-sine series expansion (approximation) results for Example 6.28

```
%em06e28.m
clear, clf
Ns=[3 5]; gss='brg';
for iter=1:2
   if iter==1, f=@(x)(x-0.5).^2; xf=2;
    else f=@(x)2.25-(x-1.5).^2; xf=3;
   end
   subplot(2,2,2+iter), hold on
   for iter1=1:length(Ns)
      N=Ns(iter1);
      [c,f_hat,xx] = Fourier_sine_series(f,xf,N); c
      plot(xx,f_hat,gss(iter1))
   end
   plot(xx,f(xx),'k:'), xlim([0 xf])
   legend(['Fourier-sine with N=' num2str(Ns(1))],...
     ['Fourier-sine with N=' num2str(Ns(2))],'Original function f(x)')
end
```

Comparing the Fourier-cosine and Fourier-sine series approximations based on Figs. 6.27(a)/(b) and 6.28(a)/(b), we might say that the former will be better for a function with nonzero boundary values, i.e., $f(0)\neq 0$ and $f(x_f)\neq 0$, while the latter will be better for a function with zero boundary values, i.e., $f(0)=0$ and $f(x_f)=0$.

Problems

6.1 Fourier Series Representations of Multi-Step and Sawtooth Waves

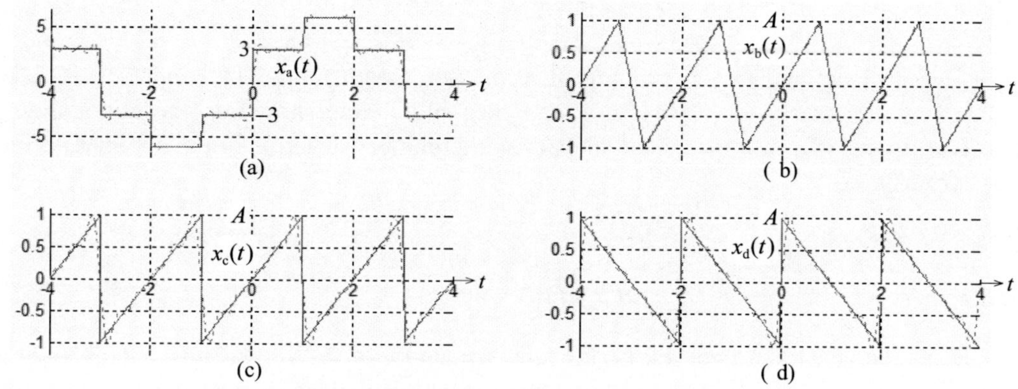

Figure P6.1 Periodic functions for Problem 6.1

(a) Noting that the multi-step wave shown as the solid line in Fig. P6.1(a) can be regarded as the sum of three scaled/time-shifted square waves with the common period 6[s] and different durations

$$x_a(t) = 6\,\tilde{r}_{3/6}(t-1.5) - 3 + 3\,\tilde{r}_{1/6}(t-1.5) - 3\,\tilde{r}_{1/6}(t-4.5) \qquad \text{(P6.1.1)}$$

use Eq. (E6.1.4) to find the CTFS representation, and then run the MATLAB script "em06p01a.m" to plot it (as shown as the dotted line in Fig. P6.1(a)) to see if it becomes close to $x_a(t)$ as the number of terms in the summation increases, say, from 5 to 10.

<Answer>

$$x_a(t) \underset{\text{(E6.1.4)}}{\overset{\text{(P6.1.1)}}{=}} 12\sum_{m=1}^{\infty}(-1)^{m+1}\frac{\sin((2m-1)\pi/2)+\sin((2m-1)\pi/6)}{(2m-1)\pi}\sin((2m-1)\omega_0 t) \qquad \text{(P6.1.2)}$$

```
%em06p01a.m
% to plot Fig.P6.1
clear, clf
P=6; w0=2*pi/P; % Period P, Fundamental frequency w0
Ts=0.01; % Sampling interval
x1=ones(1,1/Ts);
xt=[3*x1 -3*x1 -6*x1 -3*x1 0 3*x1 6*x1 3*x1 -3*x1]; % Original x(t)
tt=[-4:Ts:4]; % Whole time interval
N=10; % Number of terms in the following summation
xht=0; sgn=1;
for m=1:N % Restore from the CTFS representation up to N terms
   kpi=(2*m-1)*pi; kw0t=(2*m-1)*w0*tt;
   xht= xht + sgn*12/kpi*(sin(kpi/2) + sin(kpi/6))*sin(kw0t); %(P6.1.2)
   sgn= -sgn;
end
subplot(221), plot(tt,xt,'k', tt,xht,'r:')
```

(b) Noting that the saw-tooth wave shown as the solid line in Fig. P6.1(b) can be expressed as

$$x_b(t) = \begin{cases} \frac{A}{d/2}(t-kP) & \text{for } kP-\frac{d}{2} \le t < kP+\frac{d}{2} \\ \frac{A}{d/2}(P-2(t-kP)) & \text{for } kP+\frac{d}{2} \le t < (k+1)P-\frac{d}{2} \end{cases} \quad \text{(P6.1.3)}$$

use Eqs. (6.1.6a) and (6.1.6b) to find the complex exponential CTFS representation with the CTFS coefficients $c_k = (P/2)(a_k - jb_k)$ and then plot it (as shown as the dotted line in Fig. P6.1(b)) to see if it becomes close to $x_b(t)$ as the number of terms in the summation increases, say, from 3 to 6.

<Answer>

$$x_b(t) = \frac{8A}{d(P-d)} \sum_{k=1}^{\infty} \frac{\sin(k\omega_0 d/2)}{(k\omega_0)^2} \sin k\omega_0 t \quad \text{(P6.1.4)}$$

(c) Make use of (P6.1.4) to get the CTFS representations of the waveforms, each shown as solid lines in Figs. P6.1(c) and (d), and plot them to check if they become close to $x_c(t)$ and $x_d(t)$, respectively, as the number of terms in the summation increases, say, from 5 to 10.

6.2 Laplace Transform and Fourier Series of the Output of *RC* Circuit Excited by a Square Wave

For the output $v_o(t)$ of an *RC* (resistor-capacitor) circuit to a square wave input $v_i(t)$ (Fig. P6.2.1), we have two expressions:

$$v_{o,L}(t) = \begin{cases} V_m\left(1 - \frac{2}{1+e^{-D/RC}} e^{-(t-2kD)/RC}\right) & \text{for } 2kD \le t < (2k+1)D \\ V_m\left(-1 + \frac{2}{1+e^{-D/RC}} e^{-(t-(2k+1)D)/RC}\right) & \text{for } (2k+1)D \le t < (2k+2)D \end{cases} \quad \text{(P6.2.1)}$$

$$v_{o,F}(t) = \sum_{k=2m+1}^{\infty} \frac{4V_m}{k\pi\sqrt{1+(k\omega_0 RC)^2}} \sin(k\omega_0 t - \tan^{-1} k\omega_0 RC) \ \text{ with } \omega_0 = \frac{2\pi}{P} = \pi \quad \text{(P6.2.2)}$$

To show that these two expressions are equivalent, take the following steps to find the CTFS representation of Eq. (P6.2.1) and verify that it conforms with Eq. (P6.2.2).

(a) Check if the following equation is valid.

$$\int_0^{P/2} e^{-t/RC} e^{-jk\omega_0 t} dt = \frac{1}{-jk\omega_0 - 1/RC} e^{-(jk\omega_0 + 1/RC)t}\Big|_0^{P/2} = \frac{-RC}{1+jk\omega_0 RC}(e^{-jk\pi - P/2RC} - 1)$$

$$= \frac{RC(1-jk\omega_0 RC)}{1+(k\omega_0 RC)^2}(1-(-1)^k e^{-P/2RC}) \overset{\text{for } k\text{:odd}}{=} \frac{RC(1-jk\omega_0 RC)}{1+(k\omega_0 RC)^2}(1+e^{-P/2RC}) \quad \text{(P6.2.3)}$$

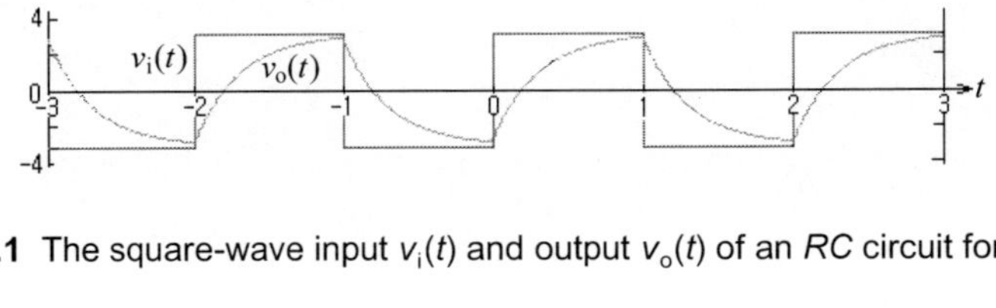

Figure P6.2.1 The square-wave input $v_i(t)$ and output $v_o(t)$ of an *RC* circuit for Problem 6.2

(b) Noting the half-wave symmetry of the output waveform $v_o(t)$ shown in Fig. P6.2.1, we see from Table 6.1 that the CTFS coefficients can be obtained by doubling the integration for a half period, and only for odd k's. Check if the exponential CTFS coefficients are as follows:

$$c_k \overset{(6.1.6b)\text{ for half-wave symmetry}}{=} 2\int_0^{P/2} v_{o,L}(t)\, e^{-jk\omega_0 t}\, dt \quad \text{for } k: \text{odd},\ P/2 = D$$

$$\overset{(P6.2.1a)}{=} 2\int_0^{D} V_m\left(1 - \frac{2}{1+e^{-D/RC}} e^{-t/RC}\right) e^{-jk\omega_0 t}\, dt$$

$$\underset{(C.35)}{\overset{(P6.2.3)}{=}} \frac{2V_m}{-jk\omega_0} e^{-jk\omega_0 t}\Big|_0^D - 4V_m \frac{RC(1-jk\omega_0 RC)}{1+(k\omega_0 RC)^2}$$

$$= \frac{2V_m}{-jk\omega_0}(e^{-jk\pi} - 1) - 4V_m \frac{RC(1-jk\omega_0 RC)}{1+(k\omega_0 RC)^2}$$

$$= -\frac{4V_m RC}{1+(k\omega_0 RC)^2} - j\frac{4V_m}{k\omega_0} + j\frac{4V_m k\omega_0 (RC)^2}{1+(k\omega_0 RC)^2}$$

$$= -\frac{4V_m RC}{1+(k\omega_0 RC)^2} - j\frac{4V_m}{k\omega_0 (1+(k\omega_0 RC)^2)} \tag{P6.2.4}$$

$$= \frac{P}{2}(a_k - jb_k) = D(a_k - jb_k)$$

(c) Using Eq. (6.1.6b) with Eq. (P6.2.4), find the CTFS representation of $v_o(t)$ in Fig. P6.2.1.

(d) Referring to the MATLAB script 'em06e01.m' for Example 6.1, make use of the MATLAB function '`CTFS_trigonometric()`' to find the CTFS coefficients $\{a_k$ and b_k, k=1, 3, 5, 7$\}$ and check if they agree with the corresponding ones in Eq. (P6.2.4). Plot the Fourier spectrum and the CTFS reconstruction (6.1.13) (with N=1 and 3) together with $v_{o,L}(t)$ (as depicted in Fig. P6.2.2) to see if they become close as the number of terms in the summation increases, say, from 2 to 4.

<Hint> You had better store the function defining Eq. (P6.2.1) as an M-file named, say, 'RC_output.m', complete the following program 'em06p02.m', and run it.

```
%em06p02.m
global P D Vm RC
P=2; w0=2*pi/P; D=1; Vm=pi; RC=0.3;
N=7; kk=0:N; % Frequency indices to analyze using Fourier analysis
tt=[-400:400]*P/200; % Time interval of 4 periods
x = 'RC_output'; % Name of the M-file function defining Eq.(P6.2.1)
[a0,a,b] = ?????????????????????????; % get the CTFS coefficients
k=1:2:N; kw0= k*w0; kw0RC21= 1+(kw0*RC).^2;
a_t(k)=-4*Vm*RC/D./kw0RC21; b_t(k)=???????????????????; %Eq.(P6.2.4)
a0_ak_bk=[a0 a b; 0 a_t b_t] % Numerical/analytical FS coefficients
xht = a0; % starting from the DC component
for k=1:N % Fourier series reconstruction (6.1.13)
   xht = xht + a(k)*cos(k*w0*tt) + b(k)*????????????;
end
subplot(221), xt = feval(x,tt); % Original signal
plot(tt,xt,'k-', tt,xht,'b:') % plot original/reconstructed signals
d1 = [a0 sqrt(?????????)]; % Magnitude of CTFS spectrum (6.1.4b)
subplot(222), stem(kk, d1) % plot the magnitude spectrum
```

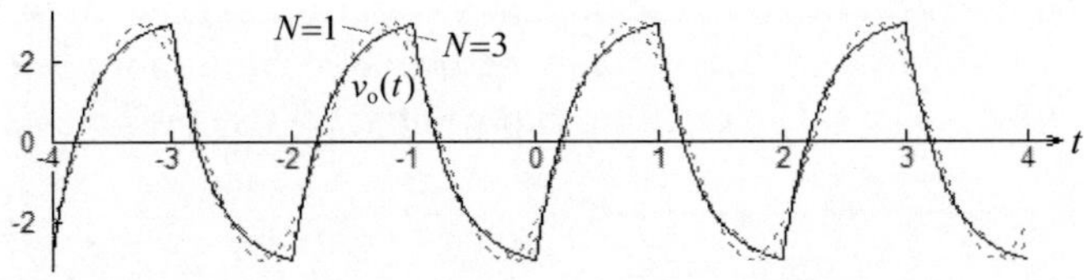

(a) The output of an *RC* filter to a square-wave input and its Fourier reconstruction

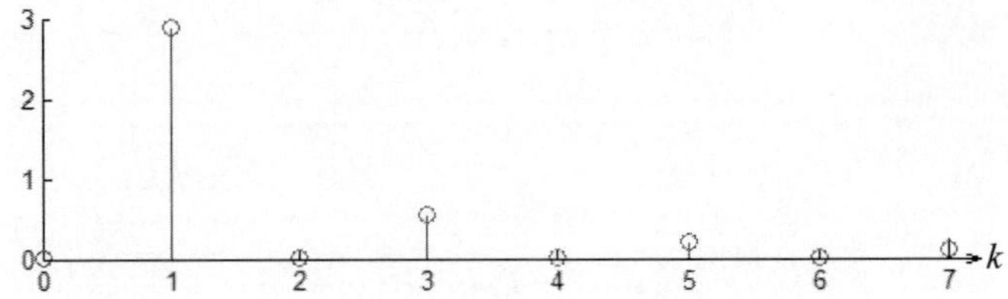

(b) The CTFS spectrum of the output of an *RC* filter to a square-wave input

Figure P6.2.2 The output, its Fourier reconstruction, and spectrum of an *RC* circuit for Problem 6.2

```
function y=RC_output(t)
global P D Vm RC
t= mod(t,P);
tmp= 2/(1+exp(-D/RC)); % Eq.(P6.2.1)
y= (t<=D)*Vm.*(1-tmp*exp(-t/RC)) + (t>D)*Vm.*(-1+tmp*exp(-(t-D)/RC));
```

6.3 Convolution/Modulation Properties of Fourier Series

Let $\tilde{x}_P(t)$ and $\tilde{y}_P(t)$ be periodic functions with common period P that have the following Fourier series representations, respectively:

$$\tilde{x}_P(t) = \frac{1}{P}\sum_{k=-\infty}^{\infty} X_k\, e^{jk\omega_0 t} \quad \text{and} \quad \tilde{y}_P(t) = \frac{1}{P}\sum_{k=-\infty}^{\infty} Y_k\, e^{jk\omega_0 t} \tag{P6.3.1}$$

(a) The periodic or circular convolution of two periodic functions is defined as

$$\tilde{z}_P(t) = \tilde{x}_P(t) * \tilde{y}_P(t) \overset{(6.2.20)}{=} \int_P \tilde{x}_P(\tau)\,\tilde{y}_P(t-\tau)\,d\tau = \int_P \tilde{y}_P(\tau)\tilde{x}_P(t-\tau)\,d\tau = \tilde{y}_P(t) * \tilde{x}_P(t) \tag{P6.3.2}$$

Show that the product of the two CTFS coefficients X_k and Y_k is the CTFS coefficient of this convolution (Eq. (6.2.20)) so that

$$\tilde{z}_P(t) = \frac{1}{P}\sum_{k=-\infty}^{\infty} Z_k\, e^{jk\omega_0 t} = \frac{1}{P}\sum_{k=-\infty}^{\infty} X_k Y_k\, e^{jk\omega_0 t} \tag{P6.3.3}$$

(b) Show that the CTFS coefficients of $\tilde{w}_P(t) = \tilde{x}_P(t)\tilde{y}_P(t)$ are

$$\tilde{w}_P(t) = \tilde{x}_P(t)\,\tilde{y}_P(t) \underset{(6.2.23)}{\overset{\text{CTFS}}{\leftrightarrow}} W_k = \frac{1}{P}\sum_{n=-\infty}^{\infty} X_n Y_{k-n} \tag{P6.3.4}$$

so that we have

$$\tilde{w}_P(t) = \frac{1}{P}\sum_{k=-\infty}^{\infty} W_k\, e^{jk\omega_0 t} \tag{P6.3.5}$$

6.4 Applications and Derivations of Several CTFT Properties

(a) Applying the frequency shifting property (6.2.15) of CTFT to Eq. (E6.8.1), derive Eq. (6.2.6).

(b) Applying the duality relation (6.2.17) to the time shifting property (6.2.14) of CTFT, derive the frequency shifting property (6.2.15).

(c) From the time-differentiation property (6.2.25) of CTFT that differentiation in the time domain corresponds to multiplication by $j\omega$ in the frequency domain, one might conjecture that integration in the time domain corresponds to division by $j\omega$ in the frequency domain. But Eq. (6.2.26) shows that the CTFT of $\int_{-\infty}^{t} x(\tau)d\tau$ has an additional impulse term reflecting the DC or average value of $x(t)$. Can you apply the time-differentiation property for Eq. (6.2.26) to derive the original Fourier relation $\mathcal{F}\{x(t)\} = X(\omega)$?

(d) Applying the duality relation (6.2.17) to the convolution property (6.2.19) of CTFT, derive the modulation property (6.2.22).

6.5 CTFT of a Triangular Pulse

Noting that as depicted in Fig. P6.5, a triangular pulse $D\lambda_D(t)$ of height D and duration $[-D, D]$ is the convolution of two identical rectangular pulses $r_D(t)$ of height 1 and duration $[-D/2, D/2]$, i.e.,

$$D\lambda_D(t) = r_D(t) * r_D(t) = \int_{-\infty}^{\infty} r_D(\tau) r_D(t-\tau)d\tau = \int_{-D/2}^{D/2} r_D(t-\tau)\,d\tau = \begin{cases} t+D & \text{for } -D \le t < 0 \\ -t+D & \text{for } 0 \le t < D \\ 0 & \text{elsewhere} \end{cases} \quad \text{(P6.5.1)}$$

apply the convolution property (6.2.19) to find the CTFT of a triangular pulse $D\lambda_D(t)$ as

$$D\lambda_D(t) \overset{\text{(P6.5.1)}}{=} r_D(t) * r_D(t) \underset{\text{convolution property}}{\overset{\mathcal{F}}{\leftrightarrow}} R_D(\omega) R_D(\omega) \overset{\text{(E6.4.2)}}{=} \boxed{} \quad \text{(P6.5.2)}$$

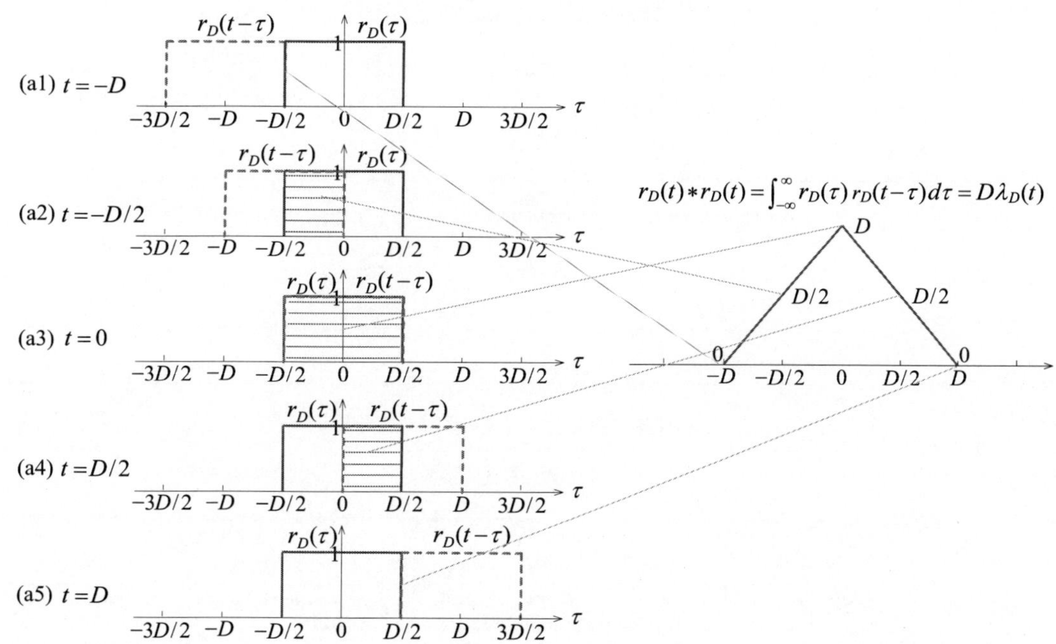

Figure P6.5 A triangular pulse as the convolution of two identical rectangular pulses for Problem 6.5

6.6 Spectrum Analysis of Amplitude Modulation (AM) Communication System

Consider the AM communication system of Fig. P6.6(a), and the CTFT spectrum $X(\omega)$ of the message signal $x(t)$ depicted in Fig. P6.6(b). Assume that the channel and LPF (lowpass filter) are LTI (linear time-invariant) systems so that their input–output relationships can be expressed as in Eq. (6.2.19), where the frequency responses of the channel and LPF are as follows:

$$H(\omega)=\begin{cases}1 & \text{for } |\omega-\omega_c|<B_C/2\\ 0 & \text{elsewhere}\end{cases} \quad \text{with } B_C\ge 2\omega_x \tag{P6.6.1}$$

$$G(\omega)=\begin{cases}1 & \text{for } |\omega|<B_R\\ 0 & \text{elsewhere}\end{cases} \quad \text{with } \omega_x<B_R<2\omega_c-\omega_x \tag{P6.6.2}$$

(a) Noting that the received and demodulated signals can be written as $r(t)=x_m(t)=x(t)\cos(\omega_c t)$ and $x_d(t)=2r(t)\cos(\omega_c t)$, respectively, express the CTFT $X_d(\omega)$ of $x_d(t)$ in terms of $X(\omega)$.

(b) Noting that the CTFT of the detected signal $d(t)$ can be written as $D(\omega)=G(\omega)X_d(\omega)$, sketch the spectra of $r(t)$, $x_d(t)$, and $d(t)$ on Figs. P6.6(c), (d), and (e).

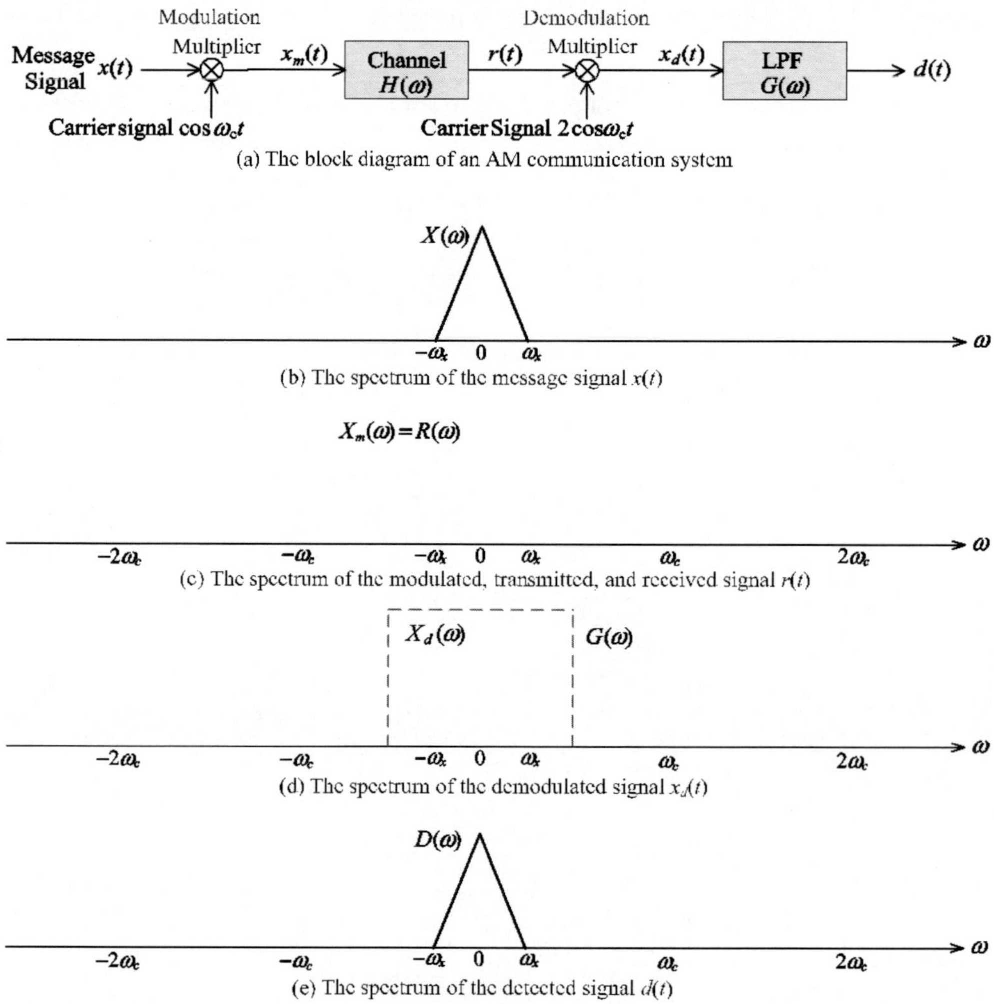

Figure P6.6 Amplitude-modulation (AM) communication for Problem 6.6

6.7 Hilbert Transform and Analytic Signal

(a) CTFT of Hilbert Transformed Signal

The *Hilbert transform* of a real-valued signal $x(t)$ is defined as the convolution of $x(t)$ and $1/\pi t$:

$$\hat{x}(t) = h(t) * x(t) = \frac{1}{\pi t} * x(t) = \frac{1}{\pi}\int_{-\infty}^{\infty} x(\tau)\frac{1}{t-\tau}d\tau \qquad \text{(P6.7.1)}$$

- First, find the CTFT of $h(t)=1/\pi t$ by applying the duality relation (6.2.17) for Eq. (E6.9.5), which states that

$$\frac{1}{2}\text{sign}(t) \underset{\text{(E6.9.5)}}{\overset{\mathcal{F}}{\leftrightarrow}} \frac{1}{j\omega} \underset{\omega\to t,t\to-\omega,\times 2\pi}{\overset{\text{duality}(6.2.17)}{\Leftrightarrow}} \frac{1}{jt} \overset{\mathcal{F}}{\leftrightarrow} \boxed{} \qquad \text{(P6.7.2)}$$

- Multiply both sides by j/π to get the CTFT of $h(t)=1/\pi t$ as

$$h(t) = \frac{1}{\pi t} \quad \overset{\mathcal{F}}{\leftrightarrow} \quad H(\omega) = \boxed{} \qquad \text{(P6.7.3)}$$

- Now, apply the convolution property to write the CTFT of $\hat{x}(t)$ in terms of $X(\omega)$ as

$$\hat{x}(t) = \frac{1}{\pi t} * x(t) \underset{\text{convolution property}}{\overset{\mathcal{F}}{\leftrightarrow}} \mathcal{F}\left(\frac{1}{\pi t}\right)\mathcal{F}(x(t)) = \boxed{} \qquad \text{(P6.7.4)}$$

(b) Inverse Hilbert Transform

Note that multiplying $-H(\omega)$ by the CTFT of the Hilbert transformed signal $\hat{x}(t)$ brings the spectrum back into that of the original signal $x(t)$:

$$-H(\omega)H(\omega)X(\omega) \overset{\text{(P6.7.3)}}{=} -(\boxed{})(\boxed{})X(\omega) = \boxed{} \qquad \text{(P6.7.5)}$$

This implies the *inverse Hilbert transform* as

$$(-h(t)) * \hat{x}(t) = \left(-\frac{1}{\boxed{}}\right) * \frac{1}{\boxed{}} * x(t) = \boxed{} \qquad \text{(P6.7.6)}$$

(c) Analytic Signal

The analytic signal, or analytic representation, of a real-valued signal $x(t)$ is defined as

$$x_a(t) = x(t) + j\hat{x}(t) \qquad \text{(P6.7.7)}$$

where $\hat{x}(t)$ is the Hilbert transform of $x(t)$ and j is the imaginary unit. Show that the CTFT of the analytic signal is

$$X_a(\omega) = \mathcal{F}\{x(t)\} + j\mathcal{F}\{\hat{x}(t)\} \cong 2u_s(\omega)X(\omega) \qquad \text{(P6.7.8)}$$

where $u_s(\omega)$ is the unit step function in frequency ω. This implies that $x_a(t)$ has only nonnegative frequency components of $x(t)$.

(d) Examples of Hilbert Transform and Analytic Signal
Show that the following Hilbert transform relations hold:

$$\cos(\omega_c t) \overset{\mathcal{H}}{\leftrightarrow} \sin(\omega_c t) \quad \text{(P6.7.9)}$$

$$\sin(\omega_c t) \overset{\mathcal{H}}{\leftrightarrow} -\cos(\omega_c t) \quad \text{(P6.7.10)}$$

In general, it would be difficult to get the Hilbert transform of a signal $x(t)$ directly from its definition (P6.7.1), and therefore, you had better take the CTFT $X(\omega)$ of $x(t)$, multiply $-j\sin(\omega)$, and then take the ICTFT (inverse CTFT) as

$$\hat{x}(t) = \mathcal{F}^{-1}\{\boxed{\qquad\qquad} X(\omega)\} \quad \text{(P6.7.11)}$$

Now, let us consider a narrowband signal

$$s(t) = m(t)\cos(\omega_c t + \phi) \text{ with } m(t)\text{: Bandpass signal} \quad \text{(P6.7.12)}$$

whose spectrum is (narrowly) bandlimited around the carrier frequency ω_c. Show that the Hilbert transform of $s(t)$ is

$$s(t) = m(t)\cos(\omega_c t + \phi) \overset{\mathcal{H}}{\leftrightarrow} \hat{s}(t) = m(t)\sin(\omega_c t + \phi) \quad \text{(P6.7.13)}$$

Also, show that the analytic signal for $s(t)$ can be expressed as

$$s_a(t) = s(t) + j\hat{s}(t) = m(t)e^{j(\omega_c t + \phi)} \quad \text{(P6.7.14)}$$

6.8 Ideal (Impulse or Instant) Sampler and Finite Pulsewidth Sampler

(a) Ideal Sampling

We can describe the output of the *ideal* (or impulse or instant) *sampler* to a given input $x(t)$ as

$$x_*(t) = x(t)\delta_T(t) \quad (\delta_T(t) = \textstyle\sum_{m=-\infty}^{\infty}\delta(t-mT)\text{: Impulse train}) \quad \text{(P6.8.1)}$$

This is illustrated in Fig. P6.8(a1). Here, the switching function has been modeled as an impulse train with period T and its CTFT is given by Eq. (E6.11.3) as

$$\Delta_T(\omega) \overset{\text{(E6.11.3)}}{=} \frac{2\pi}{T}\textstyle\sum_{k=-\infty}^{\infty}\delta(\omega + k\omega_s) \text{ with } \omega_s = \frac{2\pi}{T} \quad \text{(P6.8.2)}$$

which is shown in Fig. P6.8(c1). Apply the modulation property (6.2.22) together with Eq. (P6.8.2) to show that the CTFT of the ideal sampler output $x_*(t)$ can be expressed in terms of the input spectrum $X(\omega)$ as

$$X_*(\omega) = \frac{1}{T}\textstyle\sum_{k=-\infty}^{\infty}\boxed{\qquad\qquad} \quad \text{(P6.8.3)}$$

which is depicted in Fig. P6.8(d1).

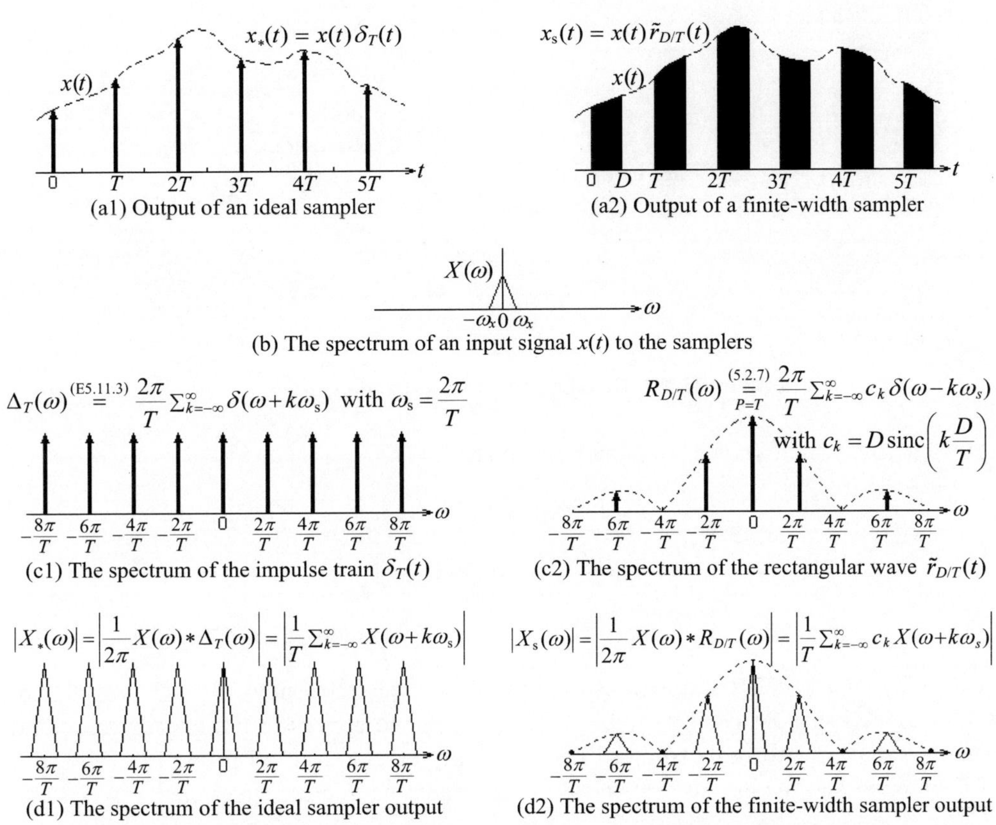

Figure P6.8 Ideal sampler and finite pulse-width sampler for Problem 6.8

(b) Finite Pulse-Width Sampling

We can describe the output of a *finite pulse-width sampler* with period T to a given input $x(t)$ as

$$x_s(t) = x(t)\tilde{r}_{D/T}(t) \; (\tilde{r}_{D/T}(t): \text{a rectangular wave with duration } D \text{ and period } T) \quad \text{(P6.8.4)}$$

This is illustrated in Fig. P6.8(a2). Here, the switching function has been modeled as a rectangular wave $\tilde{r}_{D/P}(t)$ with duration D and period T. Noting from Eq. (E6.1.4) that the CTFS representation of this rectangular wave is

$$\tilde{r}_{D/P}(t) \underset{A=1,\,P=T}{\overset{\text{(E6.1.4)}}{=}} \frac{1}{T}\sum_{k=-\infty}^{\infty} D_k \operatorname{sinc}\left(k\frac{D}{T}\right)e^{jk\omega_s t} \text{ with } \omega_s = \frac{2\pi}{T} \quad \text{(P6.8.5)}$$

we can use Eq. (6.2.7) to write its CTFT (Fig. P6.8(c2)) as

$$R_{D/P}(\omega) \underset{P=T}{\overset{(6.2.7)}{=}} \frac{2\pi}{T}\sum_{k=-\infty}^{\infty} c_k \boxed{\qquad\qquad} \text{ with } c_k = D\operatorname{sinc}\left(k\frac{D}{T}\right) \quad \text{(P6.8.6)}$$

Apply the modulation property (6.2.22) together with Eq. (P6.8.6) to express the CTFT of the finite pulse-width sampler output $x_s(t)$ in terms of the input spectrum $X(\omega)$ (Fig. P6.8(d2)) as

$$X_s(\omega) = \frac{1}{T}\sum_{k=-\infty}^{\infty} c_k \boxed{\qquad\qquad} \text{ with } c_k = D\operatorname{sinc}\left(k\frac{D}{T}\right) \quad \text{(P6.8.7)}$$

6.9 Continuous-Time Raised-Cosine Filter

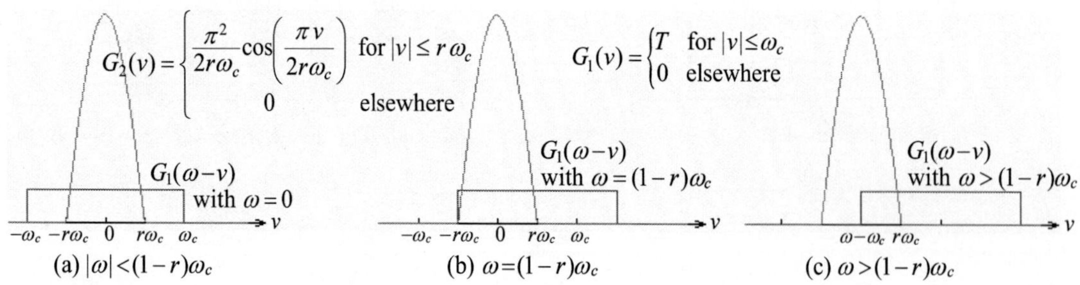

Figure P6.9.1 Convolution of $G_1(\omega)$ and $G_2(\omega)$

Let us consider the *raised-cosine* frequency response

$$G_{rc}(\omega)=\begin{cases}T=\pi/\omega_c, & |\omega|\le(1-r)\omega_c\\ \frac{T}{2}\left\{1+\cos\frac{\pi}{2r}\left(\frac{|\omega|}{\omega_c}-(1-r)\right)\right\}, & (1-r)\omega_c<|\omega|\le(1+r)\omega_c\\ 0 & (1+r)\omega_c<|\omega|\le 2\pi\end{cases} \quad \text{(P6.9.1)}$$

where $\omega_c=\pi/T$, r $(0\le r\le 1)$, and $r\omega_c$ are called the cutoff frequency, roll-off factor, and excess bandwidth, respectively. Proceed with the following steps to apply the ICTFT (inverse continuous-time Fourier transform) formula (6.2.1b) and the complex convolution property (Table A.2(9)) of CTFT for showing that the impulse response of the system having this frequency response is

$$g_{rc}(t)=\text{ICTFT}\{G_{rc}(\omega)\}=\frac{\sin(\omega_c t)}{\omega_c t}\frac{\cos(r\omega_c t)}{1-(2r\omega_c t)^2}=\text{sinc}\left(\frac{t}{T}\right)\frac{\cos(r\pi t/T)}{1-(2r\pi t/T)^2} \quad \text{(P6.9.2)}$$

(a) Referring to Fig. P6.9.1, show that the raised-cosine frequency response (P6.9.1) is the convolution of the following two frequency responses:

$$G_1(\omega)=\begin{cases}T & \text{for } |\omega|\le\omega_c\\ 0 & \text{elsewhere}\end{cases},\quad G_2(\omega)=\begin{cases}\frac{\pi^2}{2r\omega_c}\cos\left(\frac{\pi\omega}{2r\omega_c}\right) & \text{for } |\omega|\le r\omega_c\\ 0 & \text{elsewhere}\end{cases} \quad \text{(P6.9.3a,b)}$$

In other words, verify the following:

$$\frac{1}{2\pi}G_1(\omega)*G_2(\omega)=\frac{1}{2\pi}\int_{-\infty}^{\infty}G_1(\omega-v)G_2(v)\,dv$$

$$=\begin{cases}\frac{T\pi}{4r\omega_c}\int_{-r\omega_c}^{r\omega_c}\cos\left(\frac{\pi v}{2r\omega_c}\right)dv=\frac{T}{2}\sin\left(\frac{\pi v}{2r\omega_c}\right)\Big|_{-r\omega_c}^{r\omega_c}=T \quad \text{for } |\omega|\le(1-r)\omega_c & \text{(P6.9.4a)}\\ \frac{T\pi}{4r\omega_c}\int_{|\omega|-\omega_c}^{r\omega_c}\cos\left(\frac{\pi v}{2r\omega_c}\right)dv=\frac{T}{2}\sin\left(\frac{\pi v}{2r\omega_c}\right)\Big|_{|\omega|-\omega_c}^{r\omega_c}\\ =\frac{T}{2}\left\{\sin\left(\frac{\pi}{2}\right)-\sin\left(\frac{\pi(|\omega|-\omega_c)}{2r\omega_c}\right)\right\}=\frac{T}{2}\left\{1-\sin\left(\frac{\pi(|\omega|-(1-r)\omega_c)}{2r\omega_c}-\frac{\pi}{2}\right)\right\}\\ =\frac{T}{2}\left\{1+\cos\frac{\pi}{2r}\left(\frac{|\omega|}{\omega_c}-(1-r)\right)\right\} \quad \text{for } (1-r)\omega_c<|\omega|\le(1+r)\omega_c & \text{(P6.9.4b)}\end{cases}$$

(b) Verify that the ICTFT of (P6.9.3a) is

$$g_1(t) = \mathcal{F}^{-1}\{G_1(\omega)\} \underset{\text{(P6.9.3a)}}{\overset{\text{(6.2.1b)}}{=}} \frac{1}{2\pi}\int_{-\omega_c}^{\omega_c} Te^{j\omega t}\,d\omega = \frac{\pi}{\omega_c}\frac{\sin(\omega_c t)}{\pi t} = \text{sinc}\left(\frac{\omega_c t}{\pi}\right) = \text{sinc}\left(\frac{t}{T}\right) \quad \text{(P6.9.5)}$$

(c) Verify that the ICTFT of (P6.9.3b) is

$$g_2(t) = \mathcal{F}^{-1}\{G_2(\omega)\} \overset{\text{(6.2.1b)}}{=} \frac{1}{2\pi}\int_{-\infty}^{\infty} G_2(\omega)\,e^{j\omega t}\,d\omega$$

$$\overset{\text{(P6.9.3b)}}{=} \frac{1}{2\pi}\int_{-r\omega_c}^{r\omega_c} \frac{\pi^2}{2r\omega_c}\cos\left(\frac{\pi\omega}{2r\omega_c}\right)e^{j\omega t}\,d\omega = \frac{\pi}{2r\omega_c}\int_0^{r\omega_c}\cos\left(\frac{\pi\omega}{2r\omega_c}\right)\cos(\omega t)\,d\omega$$

$$\overset{\text{(C.11)}}{=} \frac{\pi}{4r\omega_c}\int_0^{r\omega_c}\left\{\cos\left(\frac{\pi\omega}{2r\omega_c}+\omega t\right)+\cos\left(\frac{\pi\omega}{2r\omega_c}-\omega t\right)\right\}d\omega$$

$$\overset{\text{(C.36)}}{=} \frac{\pi}{4r\omega_c}\left\{\frac{\boxed{\qquad\qquad}}{\pi/2r\,\omega_c+t}+\frac{\sin(\pi\,\omega/2r\,\omega_c-\omega t)}{\boxed{\qquad\qquad}}\right\}\Bigg|_0^{r\omega_c}$$

$$= \frac{\pi}{4r\omega_c}\left\{\frac{\sin(\pi/2+r\omega_c t)}{\pi/2r\,\omega_c+t}+\frac{\sin(\pi/2-r\omega_c t)}{\pi/2r\,\omega_c-t}\right\}$$

$$= \frac{1}{2}\left\{\frac{\cos(r\omega_c t)}{1+2r\omega_c t/\pi}+\frac{\cos(r\omega_c t)}{1-2r\omega_c t/\pi}\right\} = \frac{\cos(r\omega_c t)}{1-(2r\omega_c t/\pi)^2} = \frac{\cos(r\pi t/T)}{1-(2r\pi t/T)^2} \quad \text{(P6.9.6)}$$

(d) Use the complex convolution property of CTFT (Table A.2(9)) to verify that

$$g_1(t)g_2(t) \underset{\text{A.2(9)}}{\overset{\mathcal{F}}{\leftrightarrow}} \frac{1}{2\pi}G_1(\omega)*G_2(\omega) \underset{\text{(P6.9.1)}}{\overset{\text{(P6.9.4)}}{=}} G_{\text{rc}}(\omega) \quad \text{(P6.9.7)}$$

(e) Complete the following script, "plot_raised_cosine.m," to plot the impulse response of an RC (raised-cosine) filter with r=0.5 and $\omega_c=\pi/T=\pi/4$ for t=[−3T:3T]. If the roll-off factor r decreases to, say, zero, will the length of $g_{\text{rc}}(t)$ and the width of $G_{\text{rc}}(\omega)$ increase or decrease?

$$g_{\text{rc}}(t) = g_1(t)g_2(t) \underset{\text{(P6.9.6)}}{\overset{\text{(P6.9.5)}}{=}} \text{sinc}\left(\frac{\omega_c t}{\pi}\right)\frac{\cos(r\omega_c t)}{1-(2r\omega_c t/\pi)^2} = \text{sinc}\left(\frac{t}{T}\right)\frac{\cos(r\pi t/T)}{1-(2rt/T)^2} \quad \text{(P6.9.8)}$$

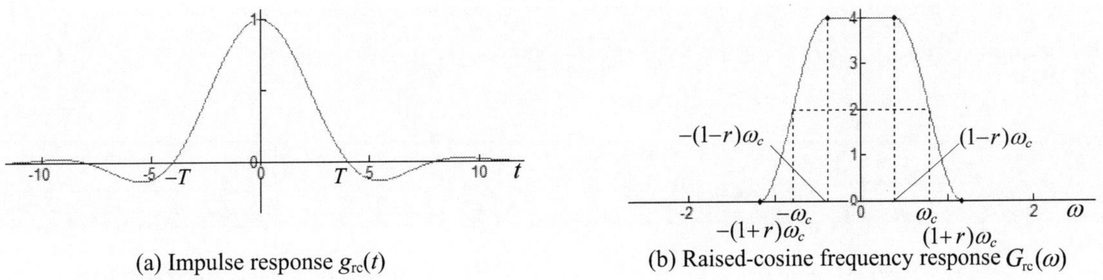

(a) Impulse response $g_{\text{rc}}(t)$ (b) Raised-cosine frequency response $G_{\text{rc}}(\omega)$

Figure P6.9.2 Impulse/frequency responses of an RC (raised-cosine) filter with r=0.5 and T=4

```
%plot_raised_cosine.m
% Raised-cosine impulse/frequency response in Fig. P6.9.1
clear, clf
r=0.5; pi2=pi*2; % Roll-off factor, etc
Ts=4; NTs=100; T=Ts/NTs; % Symbol period
N=3*NTs; t=[-N:N]*T+1e-6; N_nn=N*2+1; % Time range
wc=pi/Ts; % Cutoff frequency of raised-cosine filter
grc=sinc(t/Ts).*???(r*pi*t/??)./(1-(2*?*t/Ts).^2); % Eq.(P6.9.8)
subplot(221), plot(t,grc)
w=[-128:128]*(pi/128); % Frequency range
wawc=abs(w)/wc;
Grc=zeros(w);
Grc=Ts*(1-wawc>=r)+Ts/2*(1+cos(??/2/r*(wawc-1+r))).*(abs(1-wawc)<r);
subplot(222), plot(w,Grc)
```

6.10 CTFT Spectra of Lowpass Signals

Consider the two lowpass signals of Figs. P6.10(a1) and (a2) that can be described by

$$g_B(t) = \frac{B}{\pi}\text{sinc}\left(\frac{Bt}{\pi}\right) \quad \text{(P6.10.1)}$$

with $B=1$ and 1/2, respectively. Choose their spectra from Figs. P6.10(b1) and (b2).

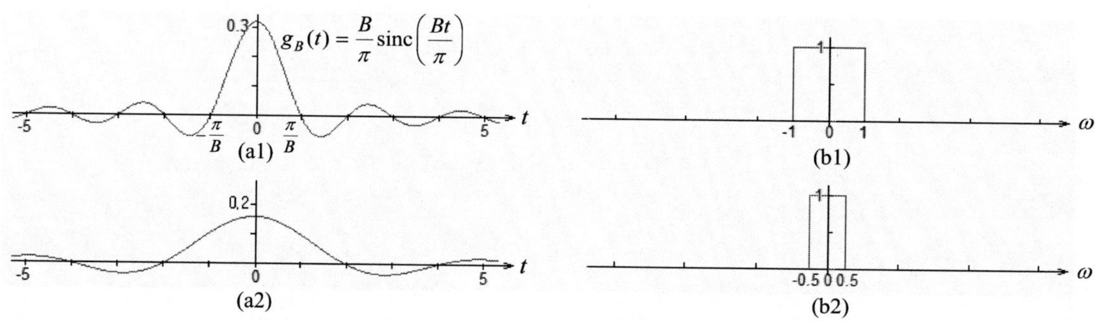

Figure P6.10 Sinc functions and their CTFT spectra

6.11 CTFT Spectrum of a Bandpass Signal

Consider the following bandpass signal described as a sinc function multiplied by a cosine function, which is depicted in Fig. P6.11(a):

$$x(t) = 2\frac{B}{\pi}\text{sinc}\left(\frac{B}{\pi}t\right)\cos\omega_p t \text{ with } B=\frac{\pi}{2} \text{ and } \omega_p=10B \quad \text{(P6.11.1)}$$

Referring to Fig. 6.12, plot the CTFT spectrum $X(\omega) = \mathcal{F}\{x(t)\}$.

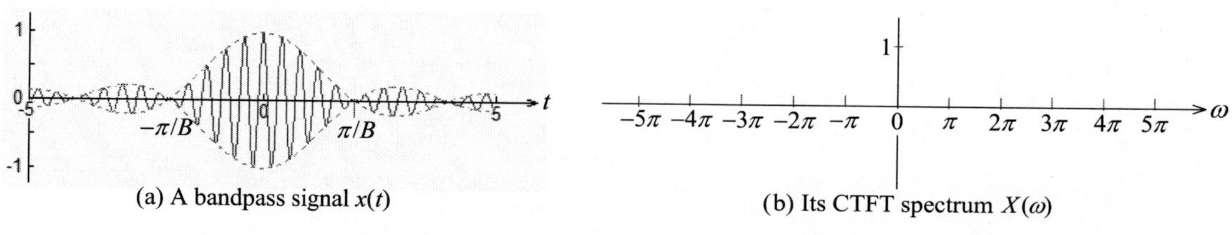

Figure P6.11 A bandpass signal and its CTFT spectrum

6.12 CTFT of a Gaussian Function

Show that the CTFT of the (normalized) *Gaussian function*

$$x(t) = \frac{1}{\sqrt{2\pi}\sigma} e^{-t^2/2\sigma^2} \text{ with } \int_{-\infty}^{\infty} x(t)\,dt = 1 \tag{P6.12.1}$$

is

$$X(\omega) = e^{-\omega^2\sigma^2/2} \tag{P6.12.2}$$

which is another Gaussian function. For this purpose, take the following steps:

1. Write the derivative of $x(t)$ w.r.t. t in terms of itself as

$$\frac{dx(t)}{dt} = \frac{-t}{\sigma^2} x(t) \tag{P6.12.3}$$

2. Use the time differentiation property (6.2.25) and frequency differentiation (time multiplication) property (6.2.28) to write the CTFT of both sides of Eq. (P6.12.3) as

$$\boxed{} = -\frac{j}{\sigma^2}\frac{d}{d\omega}X(\omega); \quad \frac{X'(\omega)}{X(\omega)} = \boxed{} \tag{P6.12.4}$$

3. Integrate both sides of Eq. (P6.12.4) for $[0, \omega]$ to write as

$$\int_0^{\omega} \frac{X'(\bar{\omega})}{X(\bar{\omega})} d\bar{\omega} = -\sigma^2 \int_0^{\omega} \bar{\omega}\, d\bar{\omega} \overset{(C.34)}{\rightarrow} \ln X(\bar{\omega})\big|_0^{\omega} = -\sigma^2 \frac{\bar{\omega}^2}{2}\bigg|_0^{\omega}$$

$$; \quad \ln X(\omega) - \ln X(0) = \ln \frac{X(\omega)}{X(0)} = -\sigma^2 \frac{\omega^2}{2} \tag{P6.12.5}$$

4. Take the exponential of both sides of Eq. (P6.12.5) to write as

$$\frac{X(\omega)}{X(0)} = \boxed{}; \quad X(\omega) = \boxed{} \tag{P6.12.6}$$

where $X(0)$, the DC component of $x(t)$, can be evaluated by using the double integral in Example 2.21:

$$X(0)^2 \underset{\omega=0}{\overset{(6.2.1a)}{=}} \left(\int_{-\infty}^{\infty} x(t)\,dt\right)\left(\int_{-\infty}^{\infty} x(\tau)\,d\tau\right) \overset{(P6.12.1)}{=} \frac{1}{2\pi\sigma^2}\iint e^{-(t^2+\tau^2)/2\sigma^2} dA$$
$$\overset{(2.6.7)}{=} \frac{1}{2\pi\sigma^2}\int_0^{2\pi}\int_0^{\infty} e^{-r^2/2\sigma^2} r\,dr\,d\theta = \frac{1}{2\pi}\int_0^{2\pi} e^{-r^2/2\sigma^2}\Big|_0^{\infty} d\theta = 1 \quad \rightarrow X(0) = 1 \tag{P6.12.7}$$

Note that this implies the *Gaussian integral identity*:

$$\int_{-\infty}^{\infty} e^{-r^2} dr = \boxed{} \tag{P6.12.8}$$

With $\alpha = 1/2\sigma^2$, we multiply Eqs. (P6.12.1) and (P6.12.2) by $\sqrt{\pi/\alpha} = \sqrt{2\pi}\sigma$ to write a CTFT pair:

$$e^{-\alpha t^2} \overset{\mathcal{F}}{\leftrightarrow} \boxed{} \tag{P6.12.9}$$

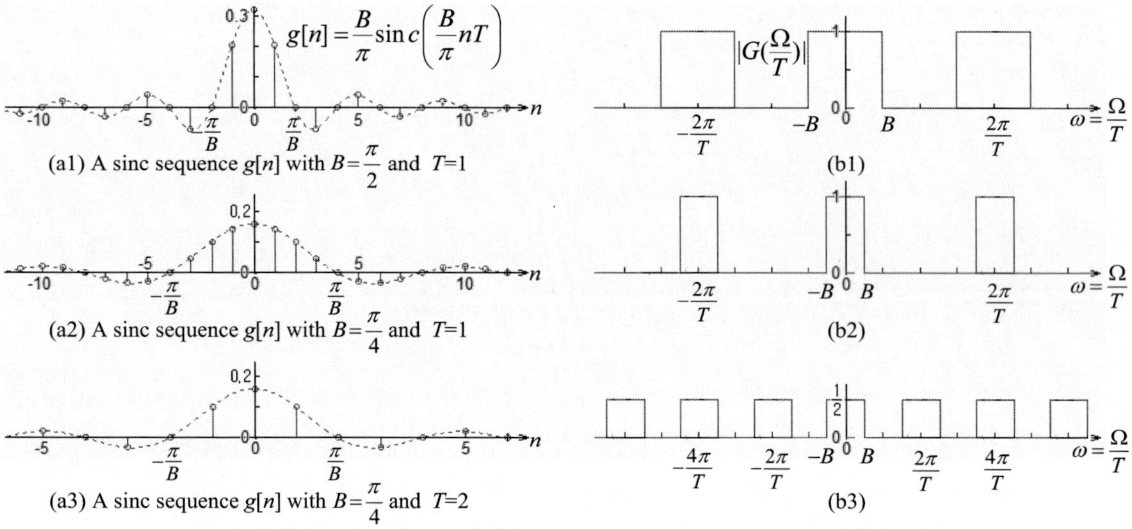

Figure P6.13 Sinc sequences and their DTFT (magnitude) spectra

6.13 DTFT Spectra of Lowpass Sequences

Consider the three lowpass sequences of Figs. P6.13(a1), (a2), and (a3) that can be described by

$$g[n]=\left.\frac{\sin(Bt)}{\pi t}\right|_{t=nT}=\frac{B}{\pi}\mathrm{sinc}\left(\frac{BnT}{\pi}\right) \tag{P6.13.1}$$

with different values of B and T as shown in the figure, where the analog frequency ω[rad/sec] is related to the digital frequency Ω[rad/sample] as

$$(\text{Analog frequency})\ \omega=\frac{\Omega\,(\text{Digital frequency})}{T\ (\text{Sampling period})} \tag{P6.13.2}$$

Choose their spectra from Figs. P6.13(b1), (b2), and (b3).

6.14 Properties of DTFT (Discrete-Time Fourier Transform)

(a) Prove the frequency-shifting property (6.3.13) by using the DTFT formula (6.3.1) or IDTFT formula (6.3.3).

$$x[n]e^{j\Omega_1 n}\overset{\mathcal{F}}{\leftrightarrow}X(\Omega-\Omega_1) \tag{P6.14.1}$$

(b) Prove the convolution property (6.3.14).

$$y[n]=x[n]*g[n]\overset{\mathcal{F}}{\leftrightarrow}Y(\Omega)=X(\Omega)G(\Omega) \tag{P6.14.2}$$

(c) Prove the modulation property (6.3.15).

$$y[n]=x[n]\,m[n]\overset{\mathcal{F}}{\leftrightarrow}Y(\Omega)=\frac{1}{2\pi}X(\Omega)*M(\Omega) \tag{P6.14.3}$$

(d) Prove the accumulation property (6.3.17) by using the convolution property (6.3.14) together with Eq. (E6.21.6).

$$\sum_{m=-\infty}^{n}x[m]=x[n]*u_s[n]\overset{\mathcal{F}}{\leftrightarrow}\frac{1}{1-e^{-j\Omega}}X(\Omega)+\pi X(0)\sum_{i=-\infty}^{n}\delta(\Omega-2\pi i) \tag{P6.14.4}$$

6.15 Discrete-Time Hilbert Transformer, Differentiator, and LPF (Lowpass Filter)

(a) $\begin{matrix}\text{Hilbert transformer}\\(-\pi/2\text{-phase shifter})\end{matrix}: H(\Omega)=\begin{cases}-j & \text{for } 0<\Omega<\pi\\ +j & \text{for } -\pi<\Omega<0\end{cases} \overset{\mathcal{F}^{-1}}{\leftrightarrow} h[n]=\begin{cases}2/n\pi & \text{for } n\text{: odd}\\ 0 & \text{for } n\text{: even}\end{cases}$ (P6.15a.1)

(a1) Prove the above DTFT pair.

(a2) To get the feel of the *Hilbert transformer* with frequency/impulse response $H(\Omega)/h[n]$, complete and run the MATLAB script "em06p15a.m" to get Fig. P6.15.1 where Figs. P6.15.1(a1) and (a2)/(a3) show the input $x[n]$ and its DTFT magnitude/phase $|X(\Omega)|/\angle X(\Omega)$, respectively, Figs. P6.15.1(b1) and (b2)/(b3) show the impulse response $h[n]$ and its DTFT (frequency response) magnitude/phase $|H(\Omega)|/\angle H(\Omega)$, respectively, and Figs. P6.15.1(c1) and (c2)/(c3) show the output $y[n]$ and its DTFT magnitude/phase $|Y(\Omega)|/\angle Y(\Omega)$, respectively. What is implied by the fact that a very small number close to zero is obtained as the value of 'Discrepancy' in the last statement of the script?

(a3) Noting that the time-domain input–output relationship of the Hilbert transformer with impulse response $h[n]$ is

$$\text{Time-domain input–output relationship: } y[n]=x[n]*h[n], \quad \text{(P6.15a.2)}$$

we can use the (   ) property (6.3.14) of the DTFT to write the frequency-domain input–output relationship as

$$Y(\Omega)=X(\Omega)H(\Omega)=|X(\Omega)|\,|H(\Omega)|\,e^{j\{\angle X(\Omega)+\angle H(\Omega)\}} \quad \text{(P6.15a.3)}$$

In line with Eq. (P6.15a.3), is $|Y(\Omega)|$ (Fig. P6.15.1(c2)) the product of $|X(\Omega)|$ (Fig. P6.15.1(a2)) and $|H(\Omega)|$ (Fig. P6.15.1(b2))? Also, is $\angle Y(\Omega)$ (Fig. P6.15.1(c3)) the sum of $\angle X(\Omega)$ (Fig. P6.15.1(a3)) and $\angle H(\Omega)$ (Fig. P6.15.1(b3))?

```
%em06p15a.m
% Hilbert transformer
clear, clf
M=50; n=-M:M; EPS=1.2; % Time range in samples
W1=0.1*pi; x=cos(W1*n); % A test input signal
W=[-50:-1 1:50]*(pi/50); % Digital frequency range
X=DTFT(x,W,n(1)); % Input and its DTFT spectrum
h=zeros(size(n));
h(2:2:end)=2/pi./[-(M-1):2:(M-1)]; % Impulse response (P6.15a.1)
H=DTFT(h,?,n(1)); % Frequency response
y=conv(x,?); ny=[-2*M:2*M]; y1=y(M+1+[0:2*M]); % Output
Y=DTFT(?,W,ny(1)); % DTFT spectrum of the output
subplot(331), stem(n,x,'.'); title('x[n]')
subplot(334), stem(n,h,'.'); title('h[n]')
subplot(337), stem(n,y1,'.'); title('y[n]')
Xmag=abs(X); Xph=angle(X).*(Xmag>EPS);
Hmag=abs(H); Hph=angle(H).*(Hmag>0.01);
Ymag=abs(Y); Yph=angle(Y).*(Ymag>EPS);
subplot(332), plot(W,Xmag); title('|X(W)')
subplot(333), plot(W,Xph); title('<X(W)')
subplot(335), plot(W,Hmag); title('|H(W)|')
subplot(336), plot(W,Hph); title('<H(W)')
subplot(338), plot(W,Ymag); title('|Y(W)|')
subplot(339), plot(W,Yph); title('<Y(W)')
Discrepancy = norm(H.*X-Y) % Y=H*X ?
```

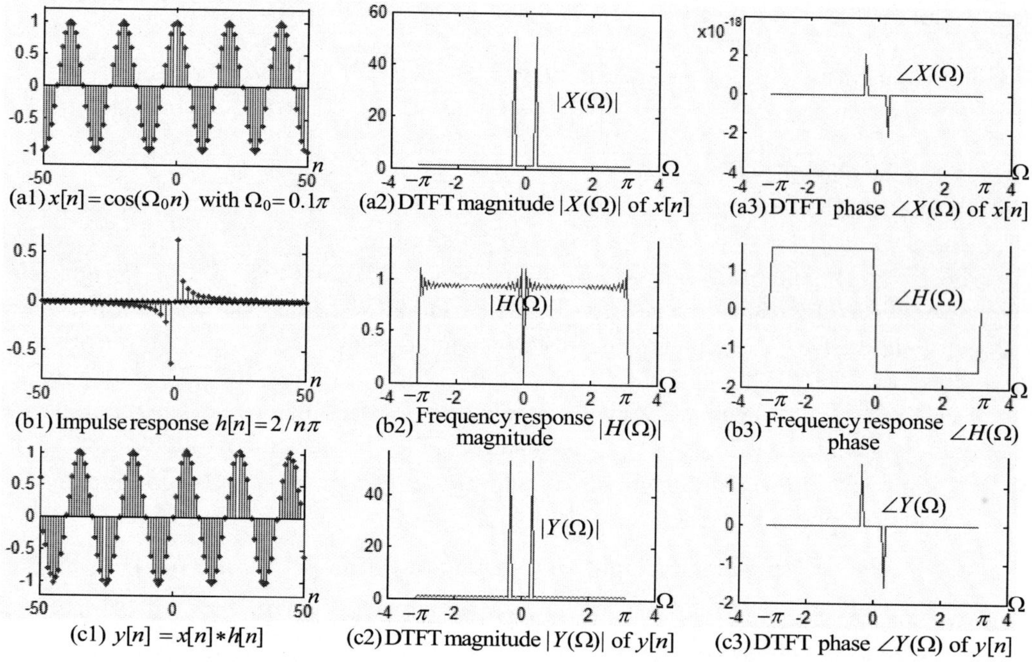

Figure P6.15.1 Input, impulse response, and output of a Hilbert transformer

(a4) The input (shown in Fig. P6.15.1(a1)) and its DTFT spectrum can be described by

$$x[n]=\cos(\Omega_0 n) \text{ with } \Omega_0=0.1\pi$$
$$\underset{\text{(E6.20.2)}}{\overset{\mathcal{F}}{\leftrightarrow}} X(\Omega)=\pi\sum_{i=-\infty}^{\infty}\{\delta(\Omega+\Omega_0-2\pi i)+\delta(\Omega-\Omega_0-2\pi i)\}$$
$$\overset{i=0}{=}\pi\{\delta(\Omega+\Omega_0)+\delta(\Omega-\Omega_0)\} \text{ for } -\pi\le\Omega\le\pi \qquad \text{(P6.15a.4)}$$

In light of this equation, do the DTFT magnitude/phase spectra $|X(\Omega)|/\angle X(\Omega)$ (Figs. P6.15.1(a2)/(a3)) look qualitatively fine? Eq. (P6.15a.3) with Eq. (P6.15a.1) can be used to find the output DTFT spectrum as

$$Y(\Omega)\overset{\text{(P6.15a.3)}}{=}X(\Omega)H(\Omega)\overset{\text{(P6.15a.1)}}{=}X(\Omega)(-j\,\text{sign}\,\Omega)$$
$$\overset{\text{(P6.15a.4)}}{=}\pi\{j\delta(\Omega+\Omega_0)-j\,\delta(\Omega-\Omega_0)\} \text{ for } -\pi\le\Omega\le\pi \qquad \text{(P6.15a.5)}$$
$$\overset{j=\exp(j\pi)}{=}\pi\{\delta(\Omega+\Omega_0)e^{j\pi/2}+\delta(\Omega-\Omega_0)e^{-j\pi/2}\}$$

In light of this equation, do the DTFT magnitude/phase spectra $|Y(\Omega)|/\angle Y(\Omega)$ (Figs. P6.15.1(c2)/(c3)) look qualitatively fine? Use Eq. (E6.20.1) to find the IDTFT $y[n]$ of $Y(\Omega)$. In light of the result, does $y[n]$ (Fig. P6.15.1(c1)) look fine? Comparing the output spectrum (P6.15a.5) with the input spectrum (P6.15a.4), do you think that the Hilbert transform may well be called a $-90°$-*phase shifter*?

(b) Differentiator: $H(\Omega)=j\Omega \overset{\mathcal{F}}{\leftrightarrow} h[n]=\dfrac{(-1)^n}{n}$ (P6.15b.1)

(b1) Prove the above DTFT pair.

(b2) To get the feel of *differentiator* with the frequency/impulse response $H(\Omega)/h[n]$, complete and run the MATLAB script "em06p15b.m" below to get Fig. P6.15.2.

(b3) In line with the general frequency-domain input–output relationship (P6.15a.3) or (6.3.14) of an LTI (linear time-invariant) system, is $|Y(\Omega)|$ (Fig. P6.15.2(c2)) the product of $|X(\Omega)|$ (Fig. P6.15.2(a2)) and $|H(\Omega)|$ (Fig. P6.15.2(b2))? Also, is $\angle Y(\Omega)$ (Fig. P6.15.2(c3)) the sum of $\angle X(\Omega)$ (Fig. P6.15.2(a3)) and $\angle H(\Omega)$ (Fig. P6.15.2(b3))?

(b4) The input (shown in Fig. P6.15.2(a1)) and its DTFT spectrum can be described by

$$x[n] = \sin(\Omega_0 n) \text{ with } \Omega_0 = 0.1\pi$$

$$\underset{(\text{E6.20.1})}{\overset{\mathcal{F}}{\leftrightarrow}} X(\Omega) = j\pi \sum_{i=-\infty}^{\infty} \{\delta(\Omega+\Omega_0-2\pi i)+\delta(\Omega-\Omega_0-2\pi i)\}$$

$$\overset{i=0}{=} j\pi\{\delta(\Omega+\Omega_0)-\delta(\Omega-\Omega_0)\} \text{ for } -\pi \le \Omega \le \pi \quad (\text{P6.15b.2})$$

$$\overset{j=\exp(j\pi)}{=} \pi\{(\Omega+\Omega_0)e^{j\pi/2} + \delta(\Omega-\Omega_0)e^{-j\pi/2}\}$$

In light of this equation, do the DTFT magnitude/phase spectra $|X(\Omega)|/\angle X(\Omega)$ (Figs. P6.15.2(a2)/(a3)) look qualitatively fine? Eq. (P6.15b.3) with Eq. (P6.15b.1) can be used to find the output DTFT spectrum as

$$Y(\Omega) \overset{(6.3.3)}{=} X(\Omega)H(\Omega) \overset{(\text{P6.15b.1})}{=} j\Omega X(\Omega)$$

$$\overset{(\text{P6.15b.2})}{=} j^2\pi\{-\Omega_0\,\delta(\Omega+\Omega_0) - \Omega_0\,\delta(\Omega-\Omega_0)\} \quad \text{for } -\pi \le \Omega \le \pi \quad (\text{P6.15b.3})$$

$$= \Omega_0\pi\{(\Omega+\Omega_0) + \delta(\Omega-\Omega_0)\}$$

In light of this equation, do the DTFT magnitude/phase spectra $|Y(\Omega)|/\angle Y(\Omega)$ (shown in Figs. P6.15.2(c2)/(c3)) look qualitatively fine? Use Eq. (E6.20.2) to find the IDTFT $y[n]$ of $Y(\Omega)$. In light of the result, does $y[n]$ (shown in Fig. P6.15.2(c1)) look fine? Comparing $y[n]$ with $x[n]$, do you think that the system with the frequency/impulse response $H(\Omega)/h[n]$ given by Eq. (P6.15b.1) may well be called a differentiator?

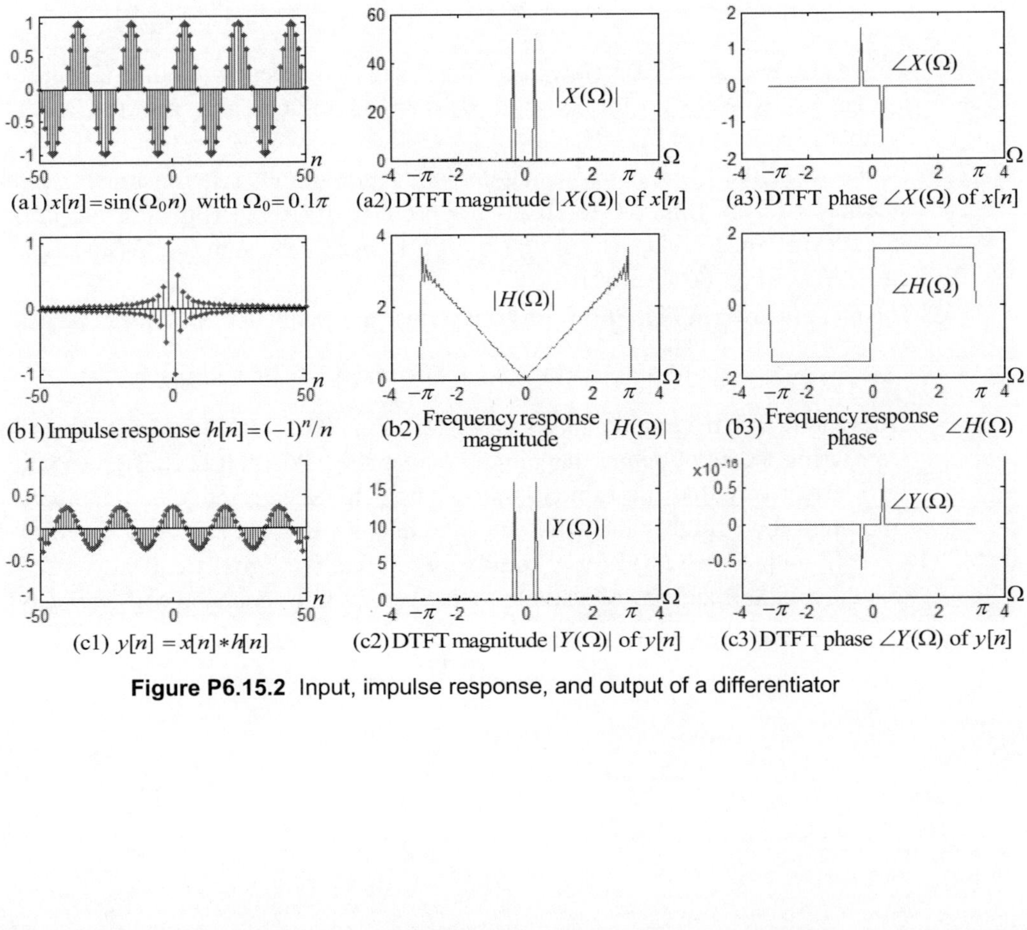

Figure P6.15.2 Input, impulse response, and output of a differentiator

```
%em06p15b.m
% Differentiator
clear, clf
M=50; n=-M:M; EPS=1e-2; % Time range in samples
W1=0.1*pi; x=sin(W1*n);  % A test input signal
W=[-50:-1 1:50]*(pi/50); % Digital frequency range
X=DTFT(x,W,n(1)); % Input and its DTFT spectrum
h=(-1).^n./n; h(M+1) = 0; % Impulse response (P6.15b.1)
H=DTFT(?,W,n(1)); % Frequency response
y=conv(?,h); ny=[-2*M:2*M]; y1=y(M+1+[0:2*M]); % Output
Y=DTFT(y,?,ny(1)); % DTFT spectrum of the output
subplot(331), stem(n,x,'.'); title('x[n]')
subplot(334), stem(n,h,'.'); title('h[n]')
subplot(337), stem(n,y1,'.'); title('y[n]')
Xmag=abs(X); Xph=angle(X).*(Xmag>EPS);
Hmag=abs(H); Hph=angle(H).*(Hmag>EPS);
Ymag=abs(Y); Yph=angle(Y).*(Ymag>EPS);
subplot(332), plot(W,Xmag); title('|X(W)')
subplot(333), plot(W,Xph); title('<X(W)')
subplot(335), plot(W,Hmag); title('|H(W)|')
subplot(336), plot(W,Hph); title('<H(W)')
subplot(338), plot(W,Ymag); title('|Y(W)|')
subplot(339), plot(W,Yph); title('<Y(W)')
Discrepancy = norm(H.*X-Y) % Y=H*X ?
```

Which property of DTFT is related to the fact that a very small number close to zero is obtained as the value of 'Discrepancy' in the last statement of the script?

(c) LPF with bandwidth or cutoff frequency B: $H(\Omega)=\begin{cases}1 & \text{for } |\Omega-2m\pi|\le B\le\pi\ (m\text{: an integer})\\ 0 & \text{elsewhere}\end{cases}$ (P6.15c.1)

$$\overset{\mathcal{F}}{\leftrightarrow}\quad h[n]=\frac{\sin(Bn)}{\pi n}=\frac{B}{\pi}\,\mathrm{sinc}\left(\frac{Bn}{\pi}\right)$$

(c1) To get the feel of an LPF (lowpass filter) whose frequency/impulse responses $H(\Omega)/h[n]$ are given by Eq. (P6.15c.1) with $B=0.3\pi$, complete and run the MATLAB script "em06p15c.m" to get Fig. P6.15.3.

(c2) In line with the general frequency-domain input–output relationship (6.3.14) of an LTI system, is $|Y(\Omega)|$ (Fig. P6.15.3(c2)) the product of $|X(\Omega)|$ (Fig. P6.15.3(a2)) and $|H(\Omega)|$ (Fig. P6.15.3(b2))? Also, is $\angle Y(\Omega)$ (Fig. P6.15.3(c3)) the sum of $\angle X(\Omega)$ (Fig. P6.15.3(a3)) and $\angle H(\Omega)$ (Fig. P6.15.3(b3))?

(c3) The input (shown in Fig. P6.15.3(a1)) is a two-tone sequence, which can be described by

$$x[n]=\sin(\Omega_1 n)+\cos(\Omega_2 n) \text{ with } \Omega_1=0.1\pi \text{ and } \Omega_2=0.5\pi \qquad \text{(P6.15c.2)}$$

Does the DTFT magnitude spectrum $|X(\Omega)|$ (Fig. P6.15.3(a2)) show the two frequencies? Comparing the input/output magnitude spectra (Fig. P6.15.3(a2)/(c2)), can you see that the LPF effectively has performed its function to keep/remove $\sin(\Omega_1 n)/\cos(\Omega_2 n)$ with frequency lower/higher than the cutoff frequency $B=0.3\pi$? Does Fig. P6.15.3(c1) show the LPF output $\sin(\Omega_1 n)$/? How could you tell that its digital frequency is $\Omega_1=0.1\pi$? Note that the digital frequency of a single-tone discrete-time sequence with period N[samples] is

$$\Omega=\frac{2\pi}{N} \qquad \text{(P6.15c.3)}$$

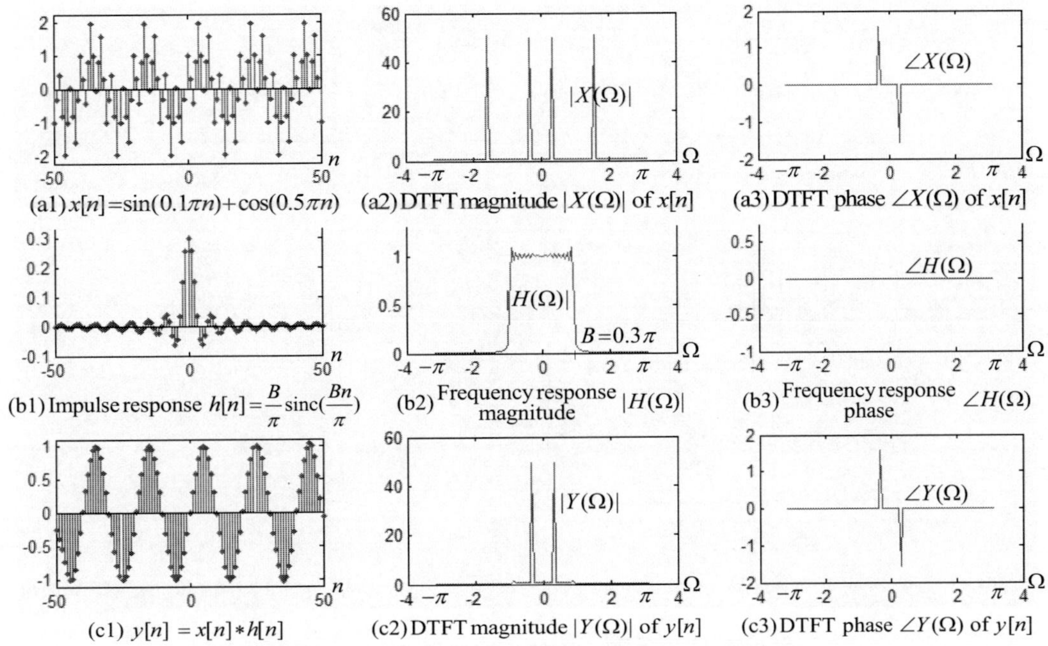

Figure P6.15.3 Input, impulse response, and output of an LPF (lowpass filter)

```
%em06p15c.m
% Lowpass Filter
clear, clf
M=50; n=-M:M; EPS=1.1; % Time range in samples
W1=0.1*pi; W2=0.5*pi; % Two digital frequencies
x = sin(W1*n) + cos(W2*n); % A test input signal
W = [-50:-1 1:50]*(pi/50); % Digital frequency range
X = DTFT(x,W,n(1)); % Input and its DTFT spectrum
B = 0.3*pi; h = B/pi*sinc(B*n/pi); % Impulse response (P6.15c.1)
W = [-50:-1 1:50]*(pi/50); % Digital frequency range
x = sin(W1*n);  X = DTFT(x,W,n(1)); % Input and its DTFT spectrum
h = (-1).^n./n; h(M+1) = 0; % Impulse response
H=DTFT(h,W,n(1)); % Frequency response
y=conv(x,h); ny=[-2*M:2*M]; y1=y(M+1+[0:2*M]); % Output
Y=DTFT(y,W,ny(1)); % DTFT spectrum of the output
% ... .. .. ..
Xmag=abs(X); Xph=angle(X).*(Xmag>EPS);
Hmag=abs(H); Hph=angle(H).*(Hmag>EPS);
Ymag=abs(Y); Yph=angle(Y).*(Ymag>EPS);
% ... .. .. ..
```

6.16 Effect of Rectangular Windowing for a Cosine Wave on Its DTFT

Noting from Eqs. (E6.20.2) and (E6.14.2) that the DTFTs of a cosine wave $x[n]=\cos(\Omega_0 n)$ and an even rectangular pulse sequence $r'_{2M+1}[n]$ of duration $2M+1$ are

$$X(\Omega)=\mathrm{DTFT}\{\cos(\Omega_0 n)\}\overset{(E6.20.2)}{=}\pi\sum\nolimits_{i=-\infty}^{\infty}(\delta(\Omega+\Omega_0-2\pi i)+\delta(\Omega-\Omega_0-2\pi i)) \quad (P6.16.1)$$

$$\text{and}\quad R'_{2M+1}(\Omega)=\mathrm{DTFT}\{r'_{2M+1}[n]\}\overset{(E6.14.2)}{=}\frac{\sin(\Omega(2M+1)/2)}{\sin(\Omega/2)} \quad (P6.16.2)$$

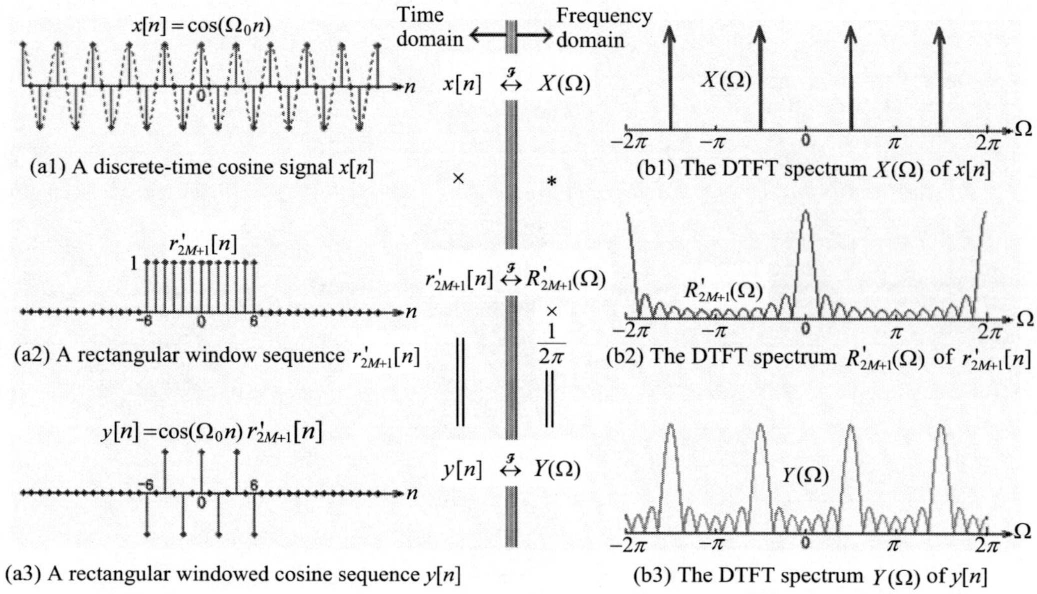

Figure P6.16 Effect of windowing on the DTFT spectrum

use the complex convolution property (6.3.15) to show that the DTFT of a rectangular-windowed cosine wave $y[n]=\cos(\Omega_0 n)r'_{2M+1}[n]$ is

$$Y(\Omega) = \text{DTFT}\{y[n]\} = \text{DTFT}\{\cos(\Omega_0 n)r'_{2M+1}[n]\} \overset{(6.3.15)}{=} \frac{1}{2\pi} X(\Omega) \underset{2\pi}{*} R'_{2M+1}(\Omega)$$

$$\underset{(C.39)}{\overset{(P6.16.1,2)}{=}} \frac{1}{2}\left(\frac{\sin((\Omega+\Omega_0)(2M+1)/2)}{\sin((\Omega+\Omega_0)/2)} + \frac{\sin((\Omega-\Omega_0)(2M+1)/2)}{\sin((\Omega-\Omega_0)/2)}\right) \tag{P6.16.3}$$

which is depicted together with $X(\Omega)$ (P6.16.1) and $R'_{2M+1}(\Omega)$ (P6.16.2) in Fig. P6.16. Compared with the spectrum $X(\Omega)$ of the cosine wave (Fig. P6.16(b1)), the spectrum $Y(\Omega)$ of the rectangular-windowed cosine wave (Fig. P6.16(b3)) has many side lobe ripples (with low amplitude) besides the two main peaks in $[-\pi, \pi]$, which is interpreted as the *spectral leakage* due to the rectangular windowing.

6.17 Discrete-Time Raised-Cosine Filter

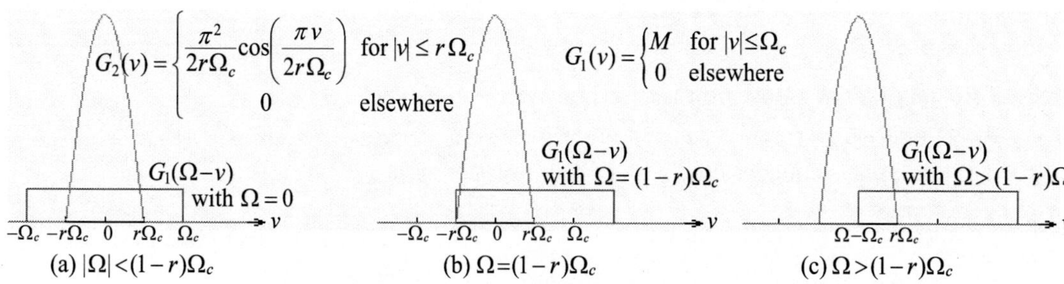

Figure P6.17.1 Convolution of $G_1(\Omega)$ and $G_2(\Omega)$

Let us consider the *raised-cosine* frequency response

$$G_{rc}(\Omega)=\begin{cases} M=\pi/\Omega_c, & |\Omega|\le(1-r)\Omega_c \\ \dfrac{M}{2}\left\{1+\cos\dfrac{\pi}{2r}\left(\dfrac{|\Omega|}{\Omega_c}-(1-r)\right)\right\}, & (1-r)\Omega_c<|\Omega|\le(1+r)\Omega_c \\ 0 & (1+r)\Omega_c<|\Omega|\le 2\pi \end{cases} \tag{P6.17.1}$$

where $\Omega_c=\pi/T$, r $(0\le r\le 1)$, and $r\Omega_c$ are called the cutoff frequency, the roll-off factor, and the excess bandwidth, respectively. Proceed with the following steps to apply the IDTFT (inverse discrete-time Fourier transform) formula (6.3.3) and the complex convolution property (Table A.4(9)) of DTFT for showing that the impulse response of the system having this frequency response is

$$g_{rc}[n]=\text{IDTFT}\{G_{rc}(\Omega)\}=\frac{\sin(\Omega_c n)}{\Omega_c n}\frac{\cos(r\Omega_c n)}{1-(2r\Omega_c n)^2}=\text{sinc}\left(\frac{\Omega_c n}{\pi}\right)\frac{\cos(r\Omega_c n)}{1-(2r\Omega_c n)^2} \tag{P6.17.2}$$

(a) Referring to Fig. P6.17.1, show that the raised-cosine frequency response (P6.17.1) is the periodic or circular convolution of the following two frequency responses:

$$G_1(\Omega)=\begin{cases} M \text{ for } |\Omega|\le\Omega_c \\ 0 \text{ elsewhere}\end{cases},\quad G_2(\Omega)=\begin{cases}\dfrac{\pi^2}{2r\Omega_c}\cos\left(\dfrac{\pi\Omega}{2r\Omega_c}\right) & \text{for } |\Omega|\le r\Omega_c \\ 0 & \text{elsewhere}\end{cases} \tag{P6.17.3a,b}$$

In other words, verify the following:

$$\frac{1}{2\pi}G_1(\Omega)\underset{2\pi}{\circledast}G_2(\Omega)=\frac{1}{2\pi}\int_{-\pi}^{\pi}G_1(\Omega-v)G_2(v)\,dv$$

$$=\begin{cases}\dfrac{M\pi}{4r\Omega_c}\displaystyle\int_{-r\Omega_c}^{r\Omega_c}\cos\left(\frac{\pi v}{2r\Omega_c}\right)dv=\frac{M}{2}\sin\left(\frac{\pi v}{2r\Omega_c}\right)\Bigg|_{-r\Omega_c}^{r\Omega_c}=M \quad \text{for } |\Omega|\le(1-r)\Omega_c & \text{(P6.17.4a)} \\[2ex] \dfrac{M\pi}{4r\Omega_c}\displaystyle\int_{|\Omega|-\Omega_c}^{r\Omega_c}\cos\left(\frac{\pi v}{2r\Omega_c}\right)dv=\frac{M}{2}\sin\left(\frac{\pi v}{2r\Omega_c}\right)\Bigg|_{|\Omega|-\Omega_c}^{r\Omega_c} \\ =\dfrac{M}{2}\left\{\sin\left(\dfrac{\pi}{2}\right)-\sin\left(\dfrac{\pi(|\Omega|-\Omega_c)}{2r\Omega_c}\right)\right\}=\dfrac{M}{2}\left\{1-\sin\left(\dfrac{\pi(|\Omega|-(1-r)\Omega_c)}{2r\Omega_c}-\dfrac{\pi}{2}\right)\right\} \\ =\dfrac{M}{2}\left\{1+\cos\dfrac{\pi}{2r}\left(\dfrac{|\Omega|}{\Omega_c}-(1-r)\right)\right\} \quad \text{for } (1-r)\Omega_c<|\Omega|\le(1+r)\Omega_c & \text{(P6.17.4b)}\end{cases}$$

(b) Verify that the IDTFT of $G_1(\Omega)$ (P6.17.3a) is

$$g_1[n]=\text{IDTFT}\{G_1(\Omega)\}\overset{(6.3.3)}{\underset{(P6.17.3a)}{=}}\frac{1}{2\pi}\int_{-\Omega_c}^{\Omega_c}Me^{j\Omega n}d\Omega$$

$$=\frac{\pi}{\Omega_c}\frac{\sin(\Omega_c n)}{\pi n}=\text{sinc}\left(\frac{\Omega_c n}{\pi}\right) \tag{P6.17.5}$$

(c) Verify that the IDTFT of $G_2(\Omega)$ (P6.17.3b) is

$$g_2[n] = \text{IDTFT}\{G_2(\Omega)\} \overset{(6.3.3)}{=} \frac{1}{2\pi}\int_{-\pi}^{\pi} G_2(\Omega)\, e^{j\Omega n} d\Omega$$

$$\overset{(\text{P6.17.3b})}{=} \frac{1}{2\pi}\int_{-r\Omega_c}^{r\Omega_c} \frac{\pi^2}{2r\Omega_c}\cos\left(\frac{\pi\Omega}{2r\Omega_c}\right) e^{j\Omega n} d\Omega = \frac{\pi}{2r\Omega_c}\int_0^{r\Omega_c} \cos\left(\frac{\pi\Omega}{2r\Omega_c}\right)\cos(\Omega n)\, d\Omega$$

$$\overset{(\text{C.11})}{=} \frac{\pi}{4r\Omega_c}\int_0^{r\Omega_c} \left\{\cos\left(\frac{\pi\Omega}{2r\Omega_c}+\Omega n\right)+\cos\left(\frac{\pi\Omega}{2r\Omega_c}-\Omega n\right)\right\} d\Omega$$

$$\overset{(\text{C.36})}{=} \frac{\pi}{4r\Omega_c}\left\{\frac{\sin(\pi\Omega/2r\Omega_c+n\Omega)}{\boxed{\qquad\qquad}}+\frac{\boxed{\qquad\qquad\qquad\qquad}}{\pi/2r\Omega_c-n}\right\}\Bigg|_0^{r\Omega_c}$$

$$= \frac{\pi}{4r\Omega_c}\left\{\frac{\sin(\pi/2+nr\Omega_c)}{\pi/2r\Omega_c+n}+\frac{\sin(\pi/2-nr\Omega_c)}{\pi/2r\Omega_c-n}\right\}$$

$$\overset{(\text{C.1})}{=} \frac{1}{2}\left\{\frac{\cos(nr\Omega_c)}{1+2nr\Omega_c/\pi}+\frac{\cos(nr\Omega_c)}{1-2nr\Omega_c/\pi}\right\} = \frac{\cos(nr\Omega_c)}{1-(2rn\Omega_c/\pi)^2} \qquad \text{(P6.17.6)}$$

(d) Use the complex convolution property of DTFT (Table A.4(9)) to verify that

$$g_1[n]g_2[n] \underset{\text{A.4(9)}}{\overset{\mathcal{F}}{\leftrightarrow}} \frac{1}{2\pi} G_1(\Omega) \underset{2\pi}{\circledast} G_2(\Omega) \underset{(\text{P6.17.1})}{\overset{(\text{P6.17.4})}{=}} G_{\text{rc}}(\Omega) \qquad \text{(P6.17.7)}$$

(e) Plot the impulse response of an RC (raised-cosine) filter with r=0.5 and $\Omega_c=\pi/M=\pi/4$:

$$g_{\text{rc}}[n] = g_1[n]g_2[n] \underset{(\text{P6.17.6})}{\overset{(\text{P6.17.5})}{=}} \boxed{\qquad\qquad\qquad} \frac{\cos(nr\Omega_c)}{1-(2rn\Omega_c/\pi)^2} = \text{sinc}\left(\frac{n}{M}\right)\frac{\cos(nr\pi/M)}{1-(2rn/M)^2} \qquad \text{(P6.17.8)}$$

for n=[−3M:3M]. Also, use the MATLAB function '`rcosflt()`' (with the 1st input argument x=1 (impulse input), the 2nd input argument Fd=1 (sampling frequency of filter input), the 3rd input argument Fs=Fd*M (sampling frequency of filter output), the 4th input argument filter_type='fir' or 'iir', and the 5th input argument r=0.5) to generate the impulse responses $g_{\text{rc1}}[n]$ and $g_{\text{rc2}}[n]$ of FIR/IIR raised-cosine filters, and plot them together with $g_{\text{rc}}[n]$. Also, find the DTFT of $g_{\text{rc}}[n]$ (by using '`DTFT()`') and plot it together with the raised-cosine frequency response (P6.17.1) against the digital frequency range Ω=[−32:32]×(π/32) to feel that the frequency response of an RC filter is just like Eq. (P6.17.1). Complete and run the MATLAB script "plot_raised_cosine_DTFT.m" to plot Fig. P6.17.2, where $g_{\text{rc}}[Mn]=\delta[m]$ can be seen from Fig. P6.17.2(a).

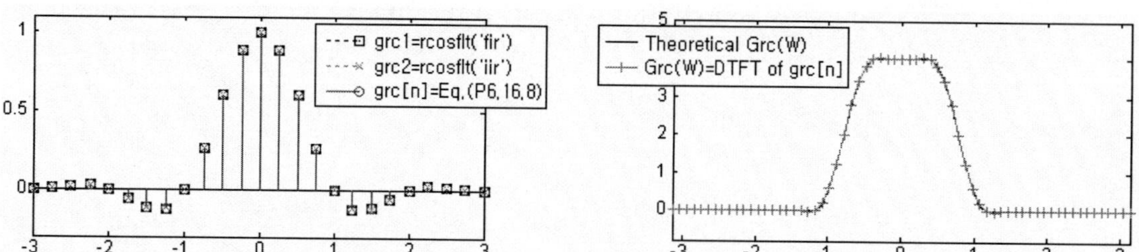

(a) Impulse response (b) Frequency response

Figure P6.17.2 Impulse/frequency responses of an RC (raised-cosine) filter with r=0.5 and M=4-times upsampling

```
%plot_raised_cosine_DTFT.m
% Raised-cosine impulse/frequency response in Fig.P6.17.1
clear, clf
r=0.5; pi2=pi*2; % Roll-off factor, etc
Ts=1; M=4; T=Ts/M; % Sampling interval (period) of filter output
Fd=1/Ts; Fs=Fd*M; % Sampling frequency of filter input/output
x=[1]; % Impulse input
% Design a raised cosine filter and implement it by using rcosflt()
grc1=rcosflt(x,Fd,Fs,'fir',r).'; % (length(x)+2*delay)*Fs/Fd
grc2=rcosflt(x,Fd,Fs,'iir',r).'; % with delay=3 by default.
[Max,im]=max(grc1); N=im-1; % Peak point
nn=-N:N; tt=nn*T; N_nn=N*2+1; % Time range
Wc=pi/M; % Cutoff frequency of raised-cosine filter to be designed
n=nn+1e-10; % 1e-10 has been added to avoid singular points
grc=sinc(Wc*n/pi).*cos(n*r*??)./(1-(2*?*n*Wc/??).^2); % Eq.(P6.17.8)
W=[-32:32]*(pi/32); Grc=DTFT(grc,W,-N); % DTFT of grc[n]
subplot(221)
stem(tt,grc1(1:N_nn),':ks'), hold on
stem(tt,grc2(1:N_nn),':rx'), stem(tt,grc)
r2=2*r; W1=(1-r)*Wc; W2=(1+r)*Wc;
for k=1:length(W)
   absWk=abs(W(k));
   if absWk<W1, Grc_theory(k)=M; % Eq.(P6.17.4a)
    elseif absWk<W2  % Eq.(P6.17.4b)
          Grc_theory(k)=M*(1+cos(pi/r2*(?????????????)))/2;
    else Grc_theory(k)=0; %1e-5 to take the logarithm for dB scale
   end
end
subplot(222), plot(W,Grc_theory,'k', W,real(Grc),'-r+')
```

6.18 DFT Spectra of Sampled Signals

With the sampling period $T = 0.1$[sec], we have sampled the following signals:

(1) $0.5\cos 3\pi t + \sin 2\pi t$

(2) $\cos 3\pi t + 0.5\sin 2\pi t$

(3) $0.5\cos 4\pi t + \sin \pi t$

(4) $\cos 4\pi t + 0.5\sin \pi t$

(5) $0.5\cos 23\pi t + 0.5\sin 2\pi t - 0.5\sin 18\pi t$

(6) $0.5\sin 21\pi t + 0.5\cos 24\pi t + 0.5\cos 16\pi t$

Noting that a sequence obtained by sampling a CT signal $x(t)$ every T[sec] can be expressed as $x[n]=x(nT)$, find the spectrum of each signal from the DFT spectra depicted in Fig. P6.18.

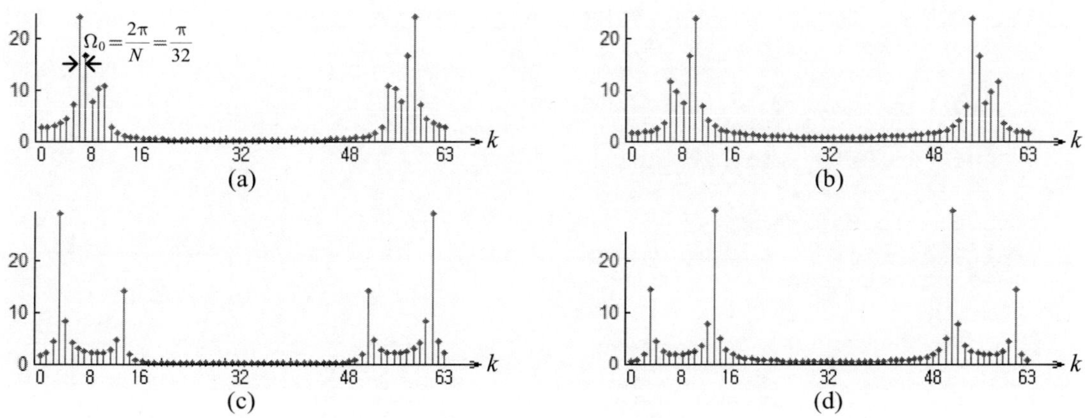

Figure P6.18 DFT spectra for Problem 6.18

6.19 To Use the z-Transform for Finding the Zeros of the DTFT of a Finite Sequence

Consider how to find the (approximate) zeros of the DTFT of a finite complex-valued sequence $\{q[n], n=0:N\}$:

$$Q[e^{j\Omega}] = \sum_{n=0}^{N} q_n e^{-j\Omega n} \text{ for } \Omega = 0:2\pi \tag{P6.19.1}$$

that is, the values of Ω such that $Q[e^{j\Omega}]=0$ where the sequence $q[n]$ is

q=[−0.0042 −0.0333+0.0026i −0.0348−0.1194i −0.0160+0.0195i 0.0726−0.0845i
−0.0071+0.0422i −0.0301+0.0561i 0.0487+0.0114i 0.5479 0.0487−0.0114i
−0.0301−0.0561i −0.0071−0.0422i 0.0726+0.0845i −0.0160−0.0195i −0.0348+0.1194i
−0.0333−0.0026i −0.0042]

We can plot the magnitude curve of $|Q[e^{j\Omega}]|$ as in Fig. P6.19(a). However, it is not that easy to find its zero(s) as local minima. An alternative, inspired by the fact that the DTFT is the evaluation of the z-transform $Q[z]$ on the unit circle $z=e^{j\Omega}$ (Remark 5.2), is to find the zeros of $Q[z]$ on or close to the unit circle. To try this method, complete and run the MATLAB script "find_roots_close_to_UC.m," where the number Nz of the zeros of $Q[e^{j\Omega}]$ is assumed to be known.

```
%find_roots_close_to_UC.m
clear, clf
% A complex-valued sequence
q=[-0.0042 -0.0333+0.0026i -0.0348-0.1194i -0.0160+0.0195i ...
   0.0726-0.0845i -0.0071+0.0422i -0.0301+0.0561i 0.0487+0.0114i ...
   0.5479 0.0487-0.0114i -0.0301-0.0561i -0.0071-0.0422i ...
   0.0726+0.0845i -0.0160-0.0195i -0.0348+0.1194i -0.0333-0.0026i ...
   -0.0042];
W=[-pi:0.01:pi]; % W=[0:0.01:2*pi]; % Frequency range
e_jW=exp(-j*W);
QejW=polyval(fliplr(q),e_jW); % Polynomial of decending order in z^-1
QejW1=freqz(q,1,W); % Alternatively, we can use freqz()
subplot(221) % To plot the magnitude curve
plot(W,abs(QejW), W,abs(QejW1),'r:'), hold on
dmin=0.5; Nz=3;
z=roots_close_to_UC(q,Nz,dmin,222);
Wz = angle(z); %ind = find(Wz<0); Wz(ind) = Wz(ind)+2*pi;
subplot(221) % To plot the results of using the z-transform
for i=1:Nz, plot(Wz(i)*[1 1],[0 1],'r:'); end
```

```
function z=roots_close_to_UC(c,Nz,dmin,graph_pane)
% Find Nz (distinct) roots of c(z)=0 closest to the UC (unit circle)
if nargin<3, dmin=0.1; end
roots_of_cz = roots(c); % Roots of c[z]=0
distinct_roots = roots_of_cz(1);
for i=2:length(c)-1
   root = roots_of_cz(i);
   if min(abs(root-distinct_roots))>dmin
     distinct_roots = [distinct_roots root];
   end
end
% To find the Nz closest roots to the unit circle
[sorted,inds] = sort(abs(abs(distinct_roots)-1));
z=distinct_roots(inds(1:Nz)); % Nz closest roots to the unit circle
```

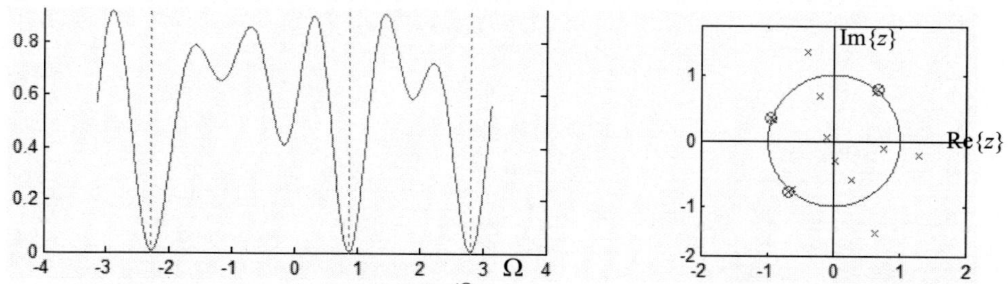

(a) Magnitude curve of the DTFT $Q[e^{j\Omega}]$ (b) The roots of $Q[z]=0$

Figure P6.19 Using the *z*-transform for finding the zeros of the DTFT of a finite sequence

6.20 Discrete-Time Convolution and Correlation, Periodogram

The convolution of two real-valued sequences $x[n]$ and $g[n]$ are defined as

$$y[n] = x[n]*g[n] = \sum_{m=-\infty}^{\infty} x[m]\,g[n-m] = \sum_{m=-\infty}^{\infty} x[n-m]\,g[m] \tag{P6.20.1}$$

and the *correlation* of two real-valued sequences $x[n]$ and $a[n]$ are defined as

$$\phi_{xa}[n] = \sum_{m=-\infty}^{\infty} x[m+n]\,a[m] \tag{P6.20.2}$$

To understand the computational relationship between correlation and convolution, which look quite similar, take the following steps.

(a) Show that, with $g[n]=a[-n]$, the convolution (P6.20.1) becomes equal to the correlation (P6.20.2):

$$y[n] = x[n]*a[-n] = \sum_{m=-\infty}^{\infty} x[n-m]\,a[-m] = \phi_{xa}[n] \tag{P6.20.3}$$

(b) To check the validity of Eq. (P6.20.3) experimentally, run the following MATLAB statements and see the results:

```
>>x=[1 -1]; a=[1 -1  2]; g=fliplr(a);
>>y=conv(x,g), [phi_xa,nlag]=xcorr(x,a)
  y       =   2        -3         2        -1
  phi_xa  =   2.0000   -3.0000    2.0000   -1.0000    0.0000
  nlag    =  -2        -1         0         1         2
```

These outputs mean that the convolution and correlation are

$$y[n] = [\ \overset{n=0}{2}\ \ \overset{1}{-3}\ \ \overset{2}{2}\ \ \overset{3}{-1}\ ] \tag{P6.20.4}$$

$$\phi_{xa}[n] = [\ \overset{n=-2}{2}\ \ \overset{-1}{-3}\ \ \overset{0}{2}\ \ \overset{1}{-1}\ \ \overset{2}{0}\ ] \tag{P6.20.5}$$

where nlag=[−2 −1 0 1 2] is the vector of sample indices for the correlation $\phi_{xa}[n]$. Let the lengths of $x[n]$ and $a[n]$ be denoted by N_x and N_a, respectively. From Fig. P6.20.1(a), we see that Eq. (P6.20.5) has been computed properly. However, curiously, $y[n]$ is not exactly the same as $\phi_{xa}[n]$ but is its 2-sample delayed version. What made the difference? The difference stems from that-to-right flipped vesion $g[n]$ of $a[n]$ (obtained using '`fliplr()`') is considered a causal sequence $a[2-n]$ ($D=N_a-1=2$-sample delayed version of $a[-n]$) rather than $a[-n]$ in '`conv()`'. That is why it is not $\phi_{xa}[n]=x[n]*a[-n]$ (Fig. P6.20.1(c)) but $\phi_{xa}[n-D]=x[n]*a[D-n]$ with $D=2$ (Fig. P6.20.1(b)) that has been computed by '`conv(x,g)`'. Therefore, to equalize $y[n]$ and $\phi_{xa}[n]$ as Eq. (P6.20.3), the starting point of $y[n]$ should be fixed as $n_0=-D$, like 'nlag', which has been returned as the second output argument of '`xcorr(x,a)`'.

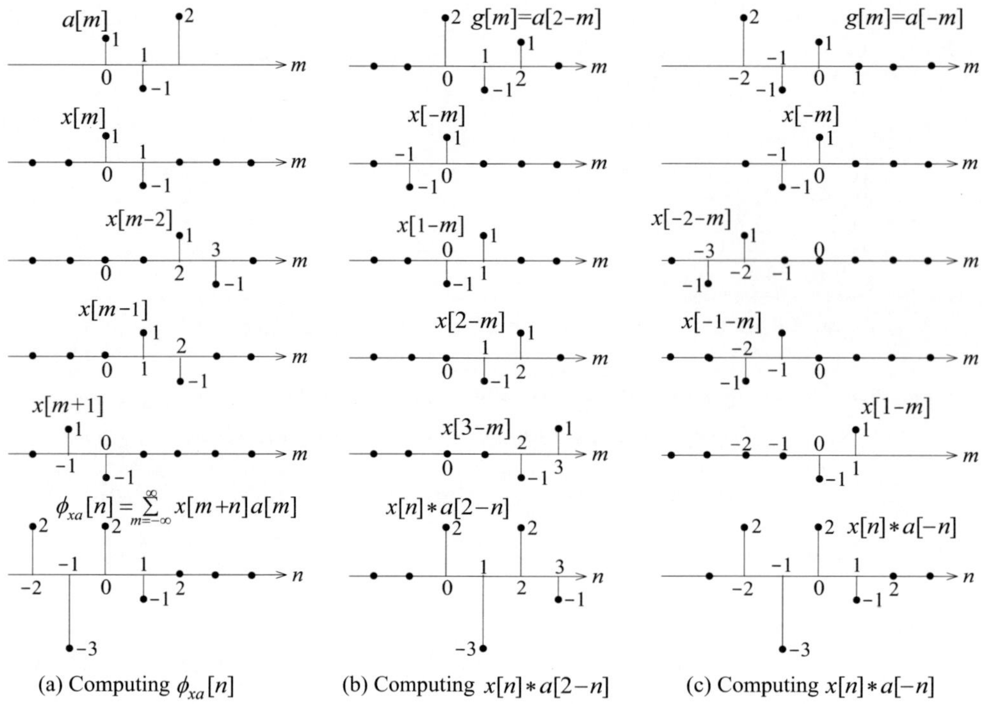

Figure P6.20.1 Computing the correlation between two sequences

How are the starting point n_0 and length L of the corelation $\phi_{xa}[n]$ determined from N_x or N_a?

(c) Use the real convolution property (6.3.14) and the conjuate symmetry (6.3.8) of DTFT to show that the DTFT of the cross-correlation (P6.20.3) between two different real-valued sequences $x[n]$ and $a[n]$ can be written as

$$\phi_{xa}[n] = x[n]*a[-n] \overset{\text{DTFT}}{\rightarrow} Y(\Omega) = X(\Omega)A(-\Omega) = X(\Omega)A^*(\Omega) \tag{P6.20.6}$$

(d) In the same context with Eq. (P6.20.6), show that the DTFT of the auto-correlation

$$\phi_{xx}[n] = x[n]*x[-n] = \sum_{m=-\infty}^{\infty} x[n+m]x[m] \tag{P6.20.7}$$

of a real-valued sequence $x[n]$ is

$$\Phi_{xx}(\Omega) = X(\Omega)X^*(\Omega) = |X(\Omega)|^2 \tag{P6.20.8}$$

(e) Noting that the N-point DFT (discrete Fourier transform) $X(k)$ of a finite-duration sequence $x[n]$ is an N-point discrete-frequency sequence obtained by sampling one period of $X(\Omega)$= DTFT$\{x[n]\}$ at $\Omega = k\Omega_0 = 2\pi k/N$ for $0 \le k \le N-1$, we can see from Eq. (P6.20.8) that the N-point DFT of the auto-correlation $\phi_{xx}[n] = x[n]*x[-n]$ of a finite-duration real-valued sequence $x[n]$ can be obtained from $X(k)$=DFT$_N\{x[n]\}$ as

$$\Phi_{xx}(k) = |X(k)|^2 \tag{P6.20.9}$$

Here, the DFT size N should be large enough to cover the length $2N_x-1$ of $\phi_{xx}[n]$ where N_x is the length of $x[n]$. To check the validity of Eq. (P6.20.9) experimentally, run the following MATLAB statements and see the results:

```
>>x=[1 -1  2]; Nx=length(x); % A sequence x and its length
>>phi_x=xcorr(x,x) % Auto-correlation of x
   phi_x =  2    -3     6    -3     2
>>N=2*Nx-1; % DFT size to cover the duration of phi_x
>>PHI_x=fft(phi_x,N) % N-point DFT of phi_x : LHS of Eq.(P6.20.9)
   PHI_x =  4.0000            -0.7361-0.5348i         3.7361+11.4984i
            3.7361-11.4984i   -0.7361+0.5348i
>>Xk2=abs(fft(x,N)).^2 % |X(k)|^2 : RHS of Eq.(P6.20.9)
   Xk2 =  4.0000   0.9098   12.0902   12.0902   0.9098
```

Does Eq. (P6.20.9) seem to be all right for this case? If not, try it with the DFT of the $-(N_x-1)$-sample circular-shift version of $\phi_{xx}[n-(N_x-1)]$:

```
>>phi_x1=circshift(phi_x,[0,-(Nx-1)])
  % To cicular-shift horizontally/vertically by 0/-(Nx-1)[samples]
   phi_x1 =  6    -3     2     2    -3
>>PHI_x1=fft(phi_x1,N)
   PHI_x1 =  4.0000   0.9098   12.0902   12.0902   0.9098
```

In light of the circular-shifting property (6.4.11b) of DFT:

$$\tilde{x}_N[n-M]r_N[n](\text{one period of circular shift}) \overset{\text{DFT}}{\leftrightarrow} W_N^{Mk}X(k)=e^{-j2\pi Mk/N}X(k) \quad \text{(P6.20.10)}$$

what should be multiplied by $\tilde{\Phi}_{xx}(k)=\text{DFT}_N\{\phi_{xx}[n-(N_x-1)]\}$ (instead of advancing the delayed correlation by (N_x-1) samples) to make up for the (N_x-1)-sample delay effect? Try the following MATLAB statement:

```
>>k=[0:N-1]; PHI_x2=PHI_x.*exp(j*2*pi*(Nx-1)*k/N)
```

(f) In connection with Eq. (P6.20.9), the MATLAB function '`periodogram()`' or '`cpsd()`' can be used to find/plot the PSD (power spectral density) estimate, called the *periodogram*, of a discrete-time signal $\{x[n], n=0:N-1\}$ that is defined as

$$P_{xx}(\Omega)=\frac{1}{2\pi N}\left|\sum_{n=0}^{N-1}x[n]w[n]e^{-j\Omega n}\right|^2 \text{ with } w[n]\text{: a window sequence} \quad \text{(P6.20.11a)}$$

$$P_{xx}(k)=P_{xx}(\Omega)\big|_{\Omega=\frac{2\pi k}{N}}=\frac{1}{2\pi N}\left|\text{DFT}_N\{x[n]w[n]\}\right|^2 \quad \text{(P6.20.11b)}$$

To get the feeling of the periodogram, find the (auto-)periodogram of a noisy 2-tone signal that is expressed as

$$x[n]=1.6\sin(\frac{2\pi}{64}10n)+1.2\cos(\frac{2\pi}{64}15n)+\sqrt{12\times 2}(u[n]-0.5) \quad \text{(P6.20.12)}$$

where $u[n]$ is a uniform random sequence (see Section 10.2) taking values on [0,1] that can be generated by using the MATLAB function '`rand()`'. To this end, complete and run the MATLAB script "em06p20.m" to plot a noisy 2-tone signal, its DFT magnitude spectrum, and its periodogram, as shown in Fig. P6.20.2, where the periodogram is computed three times, once by using Eq. (P6.20.11b) and twice by using the MATLAB functions '`periodogram()`' and '`cpsd()`', respectively. Which of these, the DFT spectrum and periodogram or the shows the spectral distribution of $x[n]$ more clearly (against the noise components)? What are the two major digital frequency components in radian/sample?

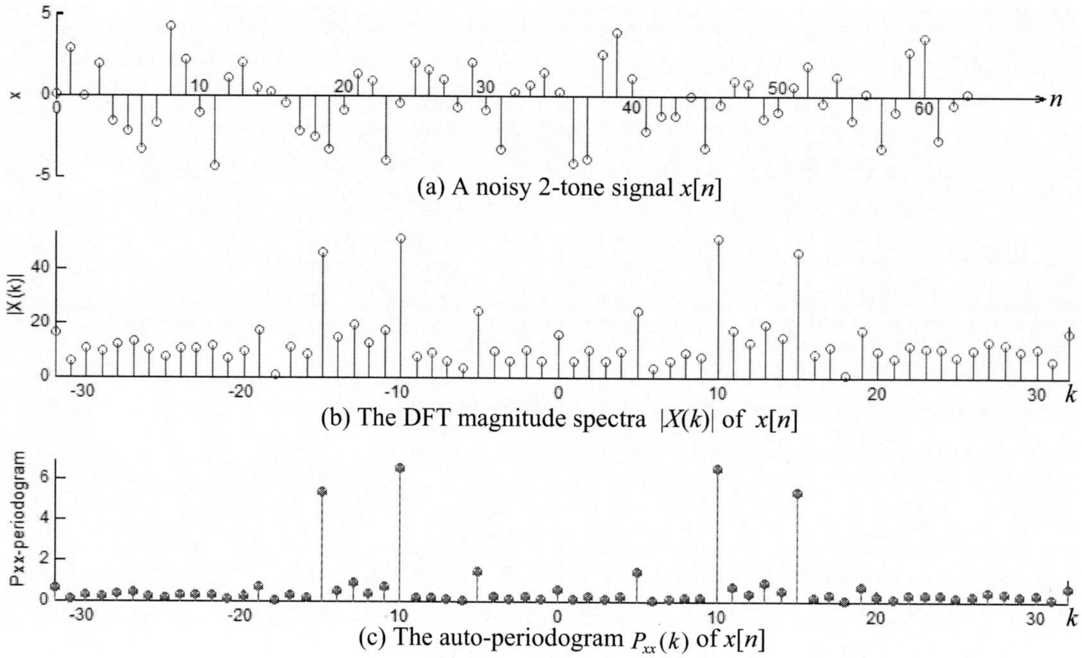

(a) A noisy 2-tone signal $x[n]$

(b) The DFT magnitude spectra $|X(k)|$ of $x[n]$

(c) The auto-periodogram $P_{xx}(k)$ of $x[n]$

Figure P6.20.2 The (auto-)periodogram of a noisy 2-tone signal

Note that the MATLAB command '`fftshift()`', swapping the left/upper and right/lower halves of a given vector or matrix, was used to rearrange the outputs of '`fft()`', '`periodogram()`', and '`cpsd()`' so that the spectra can have the zero-frequency components in the middle.

```
%em06p20.m : for Problem 6.20
% Auto-Periodogram of a noisy 2-tone signal
k1=10; k2=15; % Frequency indices of 2 tones
N=64; w1=2*pi*k1/N; w2=2*pi*k2/N; % 2 frequencies
P=2; Noise_Amp=sqrt(12*P);  % Noise power and amplitude
Nx=64;   n=0:Nx-1;   % Time index vector
% A noisy 2-tone signal
x=1.6*sin(w1*n)+1.2*cos(w2*n)+Noise_Amp*(rand(1,Nx)-0.5);%(P6.20.12)
subplot(411), stem(n,x), ylabel('x')
wn=ones(Nx,1); % A (rectangular) window vector as a column vector
kk=[-N/2:N/2]; X=fft(x,N); X=fftshift(X); X_=[X X(1)];
Pxx=abs(X_).^2/(2*?*??); % Auto-Periodogram of x by Eq.(P6.20.11b)
subplot(412), stem(kk,abs(X_)), ylabel('|X(k)|')
Noverlap = 0; % Number of overlap
[Pxx_per,W]=periodogram(x,wn,N,'twosided'); % Whole Periodogram of x
Pxx_per=fftshift(Pxx_per); Pxx_per_=[Pxx_per; Pxx_per(1)];
[Pxx_cpsd,W]=cpsd(x,x,wn,Noverlap,N,'twosided'); % Whole APSD of x
Pxx_cpsd=fftshift(Pxx_cpsd); Pxx_cpsd_=[Pxx_cpsd; Pxx_cpsd(1)];
subplot(413), stem(kk,Pxx), ylabel('Pxx-periodogram'), hold on,
stem(kk,Pxx_per_,'^') % Pxx obtained using periodogram()
stem(kk,Pxx_cpsd_,'r:.') % Pxx obtained using cpsd()
```

CHAPTER 7

Partial Differential Equations

CHAPTER OUTLINE

What is a partial differential equation (PDE)? It is a class of differential equations involving more than one independent variable. In this chapter, we consider a general second-order PDE in two independent variables x and y, which is written as

$$A(x,y)\frac{\partial^2 u}{\partial x^2} + B(x,y)\frac{\partial^2 u}{\partial x \partial y} + C(x,y)\frac{\partial^2 u}{\partial y^2} = f(x,y,u,\frac{\partial u}{\partial x},\frac{\partial u}{\partial y}) \tag{7.0.1}$$

$$\text{for } x_0 \le x \le x_f \text{ and } y_0 \le y \le y_f$$

subject to the boundary conditions given by

$$u(x,y_0) = b_{y0}(x),\ u(x,y_f) = b_{yf}(x),\ u(x_0,y) = b_{x0}(y), \text{ and } u(x_f,y) = b_{xf}(y) \tag{7.0.2}$$

These PDEs are classified into three groups:

elliptic PDE: if $B^2 - 4AC < 0$
parabolic PDE: if $B^2 - 4AC = 0$
hyperbolic PDE: if $B^2 - 4AC > 0$

These three types of PDEs are associated with equilibrium states, diffusion states, and oscillating systems, respectively. We will study some numerical methods for solving these PDEs, since their analytical solutions are usually difficult to find.

7.1 ELLIPTIC PDE[Y-1], Section 9.1

Let us consider a special type of elliptic equation called *Helmholtz's equation*

$$\nabla^2 u(x,y)+g(x,y)u(x,y)=\frac{\partial^2 u(x,y)}{\partial x^2}+\frac{\partial^2 u(x,y)}{\partial y^2}+g(x,y)u(x,y)=f(x,y) \tag{7.1.1}$$

over a domain $D=\{(x,y)\,|\,x_0\le x\le x_f,\ y_0\le y\le y_f\}$ with some *Dirichlet* (fixed) *boundary conditions* of

$$u(x_0,y)=b_{x0}(y),\ u(x_f,y)=b_{xf}(y),\ u(x,y_0)=b_{y0}(x),\ \text{and}\ u(x,y_f)=b_{yf}(x) \tag{7.1.2}$$

(Note) Eq. (7.1.1) is called *Poisson's equation* if $g(x,y)=0$ and it is called *Laplace's equation* if $g(x,y)=0$ and $f(x,y)=0$.

To apply the difference method as a numerical approach, we divide the domain into M_x sections, each of length $\Delta x=(x_f-x_0)/M_x$ along the x-axis, and into M_y sections, each of length $\Delta y=(y_f-y_0)/M_y$ along the y-axis, respectively, and then replace the second derivatives by the 3-point central difference approximation:

$$\left.\frac{\partial^2 u(x,y)}{\partial x^2}\right|_{x_j,y_i}\cong\frac{u_{i,j+1}-2u_{i,j}+u_{i,j-1}}{\Delta x^2}\quad\text{with } x_j=x_0+j\Delta x,\ y_i=x_0+i\Delta y \tag{7.1.3a}$$

$$\left.\frac{\partial^2 u(x,y)}{\partial y^2}\right|_{x_j,y_i}\cong\frac{u_{i+1,j}-2u_{i,j}+u_{i-1,j}}{\Delta y^2}\quad\text{with } u_{i,j}=u(x_j,y_i) \tag{7.1.3b}$$

so that, for every interior point $(x_j,\ y_i)$ with $1\le i\le M_y-1$ and $1\le j\le M_x-1$, we obtain the finite difference equation

$$\frac{u_{i,j+1}-2u_{i,j}+u_{i,j-1}}{\Delta x^2}+\frac{u_{i+1,j}-2u_{i,j}+u_{i-1,j}}{\Delta y^2}+g_{i,j}u_{i,j}=f_{i,j} \tag{7.1.4}$$

where

$$u_{i,j}=u(x_j,y_i),\ f_{i,j}=f(x_j,y_i),\ \text{and}\ g_{i,j}=g(x_j,y_i)$$

These equations can somehow be arranged into a system of simultaneous equations with the $(M_y-1)(M_x-1)$ variables $\{u_{1,1},u_{1,2},\dots,u_{1,M_x-1},u_{2,1},\dots,u_{2,M_x-1},\dots,u_{M_y-1,1},u_{M_y-1,2},\dots,u_{M_y-1,M_x-1}\}$. However, it seems to be messy to work with and we may be really in trouble as M_x and M_y become large. A simpler way is to use the iterative methods. To do so, we first need to shape the equations and the boundary conditions into the following form:

$$u_{i,j}=r_y(u_{i,j+1}+u_{i,j-1})+r_x(u_{i+1,j}+u_{i-1,j})+r_{xy}(g_{i,j}u_{i,j}-f_{i,j}) \tag{7.1.5a}$$

$$u_{i,0}=b_{x0}(y_i),\ u_{i,M_x}=b_{xf}(y_i),\ u_{0,j}=b_{y0}(x_j),\ \text{and}\ u_{M_y,j}=b_{yf}(x_j) \tag{7.1.5b}$$

where

$$\frac{\Delta y^2}{2(\Delta x^2+\Delta y^2)}=r_y,\ \frac{\Delta x^2}{2(\Delta x^2+\Delta y^2)}=r_x,\ \text{and}\ \frac{\Delta x^2\Delta y^2}{2(\Delta x^2+\Delta y^2)}=r_{xy} \tag{7.1.6}$$

How do we initialize this algorithm? If we have no prior knowledge about the solution, it is reasonable to take the average value of the boundary values as the initial values of $u_{i,j}$.

This process for solving Eq. (7.1.1) with BC (7.1.2) has been cast into the following MATLAB function 'pde_poisson()'.

```
function [u,x,y]=pde_poisson(f,g,bx0,bxf,by0,byf,D,Mx,My,tol,MaxIter)
% To solve u_xx +u_yy +g(x,y)u = f(x,y)
%  over the region D=[x0,xf,y0,yf]={(x,y)|x0<=x<=xf, y0<=y<=yf}
% subject to the BCs (boundary conditions):
%  u(x0,y) = bx0(y), u(xf,y) = bxf(y)
%  u(x,y0) = by0(x), u(x,yf) = byf(x)
% D = [x0 xf y0 yf] : Domain
%   = [x0 xf y0 yf; 0 1 1 0] : Dirichlet/Neumann/Neumann/Dirichlet BCs
% Mx = Number of subintervals along x-axis
% My = Number of subintervals along y-axis
% tol    : Error tolerance
% MaxIter: Maximum number of iterations
if nargin<11, MaxIter=500; end
if nargin<10, tol=1e-8; end
if nargin<9
  if nargin==8, My=Mx; else My=50; end
end
if nargin<8, Mx=50; end
x0=D(1); xf=D(2); y0=D(3); yf=D(4);
dx=(xf-x0)/Mx;   x=x0+[0:Mx]*dx;
dy=(yf-y0)/My;   y=y0+[0:My]'*dy;
Mx1=Mx+1; My1=My+1;
% Boundary conditions (7.1.5b)
for m=1:My1
   u(m,[1 Mx1])=[bx0(y(m)) bxf(y(m))]; % Left/Right side
end
for n=1:Mx1
   u([1 My1],n)=[by0(x(n)); byf(x(n))]; % Bottom/Top
end
% Initialize as the average of boundary values
sum_of_bv= sum(sum([u(2:My,[1 Mx1]) u([1 My1],2:Mx)']));
u(2:My,2:Mx)=sum_of_bv/(2*(Mx+My-2));
for i=1:My
  for j=1:Mx
     F(i,j)=f(x(j),y(i)); G(i,j)=g(x(j),y(i));
  end
end
dx2=dx*dx; dy2=dy*dy;  dxy2= 2*(dx2+dy2);
rx= dx2/dxy2; ry= dy2/dxy2;  rxy= rx*dy2;
for itr=1:MaxIter
  for j=2:Mx
    for i=2:My
      u(i,j)= ry*(u(i,j+1)+u(i,j-1)) +rx*(u(i+1,j)+u(i-1,j))...
               +rxy*(G(i,j)*u(i,j)-F(i,j)); % Eq.(7.1.5a)
    end
  end
  if itr>1& max(max(abs(u-u0)))<tol, break; end
  u0=u;
end
```

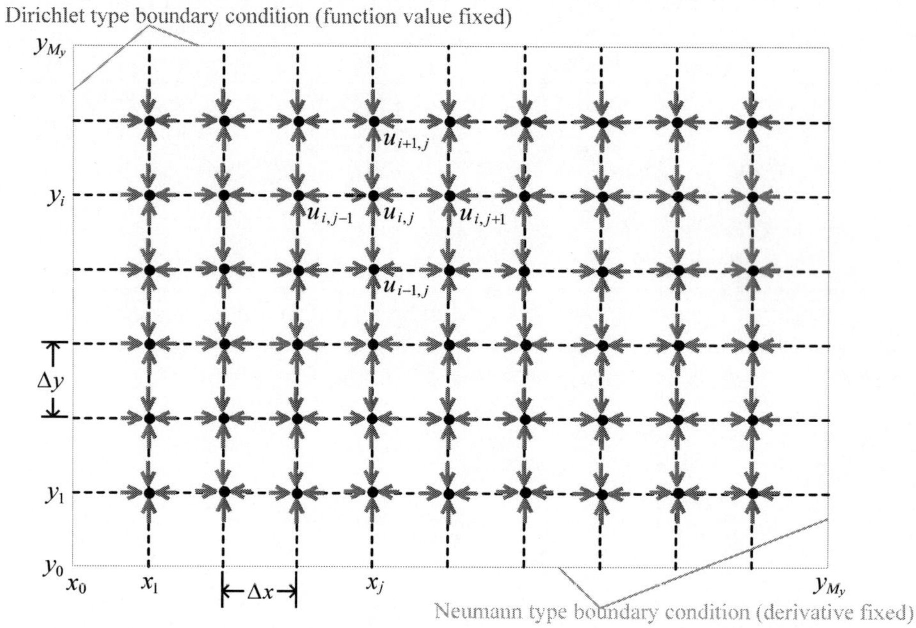

Figure 7.1 Grid for an elliptic PDE with Dirichlet/Neumann boundary conditions

Now, to see how an analytical approach can be taken to solve an elliptic equation, let us consider the following problem of solving Laplace's equation:

$$\frac{\partial^2 u(x,y)}{\partial x^2} + \frac{\partial^2 u(x,y)}{\partial y^2} = 0 \tag{7.1.7}$$

over a domain $D=\{(x,y)\,|\,0\le x\le x_f,\ 0\le y\le y_f\}$ subject to the Dirichlet BCs (boundary conditions):

$$u(0,y)=0,\ u(x_f,y)=0,\ u(x,0)=b_{y0}(x),\ \text{and}\ u(x,y_f)=0 \tag{7.1.8}$$

By the method of *separation of variables*, we substitute

$$u(x,y) = X(x)Y(y) \tag{7.1.9}$$

into the PDE (7.1.7) and divide by XY to write

$$X''Y + XY'' = 0\,;\ \frac{X''}{X} = -\frac{Y''}{Y} = -\lambda\,(\text{separation constant});\ \begin{cases} X''+\lambda X = 0 \\ Y''-\lambda Y = 0 \end{cases} \tag{7.1.10a,b}$$

Table 7.1 *Separable solutions* to Laplace's EQ. (7.1.7) ($\cosh x=(e^x+e^{-x})/2$, $\sinh x=(e^x-e^{-x})/2$)

λ	$X(x)$	$Y(y)$	$u(x,y)=X(x)Y(y)$
$\lambda=\omega^2>0$	$\cos\omega x,\ \sin\omega x$ if u is periodic in x.	$e^{\omega y},\ e^{-\omega y}$ $\cosh\omega y,\ \sinh\omega y$	$e^{\omega y}\cos\omega x,\ e^{\omega y}\sin\omega x$ $e^{-\omega y}\cos\omega x,\ e^{-\omega y}\sin\omega x$
$\lambda=0$	$1,\ x$	$1,\ y$	$1,\ x,\ y,\ xy$
$\lambda=-\omega^2<0$	$e^{\omega x},\ e^{-\omega x}$ $\cosh\omega x,\ \sinh\omega x$	$\cos\omega y,\ \sin\omega y$ if u is periodic in y	$e^{\omega x}\cos\omega y,\ e^{\omega x}\sin\omega y$ $e^{-\omega x}\cos\omega y,\ e^{-\omega x}\sin\omega y$

Depending on the sign of the separation constant λ, the separable solution to the Laplace's equation takes different forms as listed in Table 7.1. Also, we substitute Eq. (7.1.9) into the BC (boundary condition) (7.1.8) to write

$$u(0,y) = X(0)Y(y) = 0 \ \forall y \rightarrow X(0) = 0 \tag{7.1.11a}$$

$$u(x_f, y) = X(x_f)Y(y) = 0 \ \forall y \rightarrow X(x_f) = 0 \tag{7.1.11b}$$

$$u(x,0) = X(x)Y(0) = b_{y0}(x) \ \ \forall x \tag{7.1.11c}$$

$$u(x, y_f) = X(x)Y(y_f) = 0 \ \forall x \rightarrow Y(y_f) = 0 \tag{7.1.11d}$$

Referring to the second column of Table 7.1, the candidate for $X(x)$ that may satisfy Eq. (7.1.10a) together with the BC (7.1.11a) is

$$X(x) = \begin{cases} \sin\omega x & \text{if } \lambda=\omega^2>0 \\ x & \text{if } \lambda=0 \\ \sinh\omega x=(e^{\omega x}-e^{-\omega x})/2 & \text{if } \lambda=-\omega^2<0 \end{cases} \tag{7.1.12}$$

However, the second and third ones cannot satisfy the BC (7.1.11b) and therefore, the candidate for $X(x)$ is just the first one:

$$X(x) = \sin\omega x = \sin(\frac{n\pi}{x_f}x) \text{ with } \omega = \frac{n\pi}{x_f}(n = 1,2,3,\cdots) \text{ with } \lambda=\omega^2>0 \tag{7.1.13}$$

Now, referring to the third column of Table 7.1 for $\lambda=\omega^2=(n\pi/x_f)^2>0$, the candidate for $Y(y)$ satisfying the BC (7.1.11d) is

$$Y(y) = \sinh\omega(y_f - y) = \sinh(\frac{n\pi}{x_f}(y_f - y)) \text{ with } \omega = \frac{n\pi}{x_f}(n = 1,2,3,\cdots) \text{ for } \lambda=\omega^2>0 \tag{7.1.14}$$

Plugging Eqs. (7.1.13) and (7.1.14) into Eq. (7.1.9) yields the separable solutions:

$$u_n(x,y) = X(x)Y(y) = \sin(\frac{n\pi}{x_f}x)\sinh(\frac{n\pi}{x_f}(y_f - y)) \tag{7.1.15}$$

Hence, the general solution of the Laplace's equation can be written as

$$u(x,y) = \sum_{n=1}^{\infty} A(n)u_n(x,y) \overset{(7.1.15)}{=} \sum_{n=1}^{\infty} A(n)\sin(\frac{n\pi}{x_f}x)\sinh(\frac{n\pi}{x_f}(y_f - y)) \tag{7.1.16}$$

where the coefficient $A(n)$'s are to be fixed by the remaining BC (7.1.11c):

$$u(x,0) \overset{(7.1.16)}{=} \sum_{n=1}^{\infty} A(n)\sinh(\frac{n\pi}{x_f}y_f)\sin(\frac{n\pi}{x_f}x) \overset{(7.1.11c)}{=} b_{y0}(x) \tag{7.1.17}$$

Assuming that $b_{y0}(x)$ is an odd function so that it can be expressed as the sum of many sine functuions, we can use Eq. (6.6.15b) to find the Fourier-sine series coefficients as

$$b_n = A(n)\sinh(\frac{n\pi}{x_f}y_f) \underset{P=2x_f,\text{odd symmetry}}{\overset{(6.1.3b),\text{Table 6.1}}{=}} \frac{2}{x_f}\int_0^{x_f} b_{y0}(x)\sin(\frac{n\pi}{x_f}x)dx \tag{7.1.18}$$

$$\rightarrow A(n) = \frac{b_n}{\sinh(\frac{n\pi}{x_f}y_f)} = \frac{2}{x_f\sinh(\frac{n\pi}{x_f}y_f)}\int_0^{x_f} b_{y0}(x)\sin(\frac{n\pi}{x_f}x)dx \tag{7.1.19}$$

The solution to the Laplace's equation (7.1.7) with a more general Dirichlet BC

$$u(x_0,y)=b_{x0}(y),\ u(x_f,y)=b_{xf}(y),\ u(x,y_0)=b_{y0}(x),\ \text{and}\ u(x,y_f)=b_{yf}(x) \tag{7.1.20}$$

is (see <http://eqworld.ipmnet.ru/en/solutions/lpde/lpdetoc3.htm>)

$$\begin{aligned} u(x,y)=&\sum_{n=1}^{\infty} A_1(n)\sin\left(\frac{n\pi}{y_f}y\right)\sinh\left(\frac{n\pi}{y_f}(x_f-x)\right)+A_2(n)\sin\left(\frac{n\pi}{y_f}y\right)\sinh\left(\frac{n\pi}{y_f}x\right)\\ &+\sum_{n=1}^{\infty} A_3(n)\sin\left(\frac{n\pi}{x_f}x\right)\sinh\left(\frac{n\pi}{x_f}(y_f-y)\right)+A_4(n)\sin\left(\frac{n\pi}{x_f}x\right)\sinh\left(\frac{n\pi}{x_f}y\right)\end{aligned} \tag{7.1.21}$$

where the coefficients are

$$A_1(n)=\frac{2}{\lambda_n}\int_0^{y_f} b_{x0}(y)\sin(\frac{n\pi}{y_f}y)\,dy,\quad A_2(n)=\frac{2}{\lambda_n}\int_0^{y_f} b_{xf}(y)\sin(\frac{n\pi}{y_f}y)\,dy\ \text{ with }\lambda_n=y_f\sinh(\frac{n\pi}{y_f}x_f)$$

$$A_3(n)=\frac{2}{\mu_n}\int_0^{x_f} b_{y0}(x)\sin(\frac{n\pi}{x_f}x)\,dx,\quad A_4(n)=\frac{2}{\mu_n}\int_0^{x_f} b_{yf}(x)\sin(\frac{n\pi}{x_f}x)\,dx\ \text{ with }\mu_n=x_f\sinh(\frac{n\pi}{x_f}y_f)$$

(7.1.22a,b,c,d)

This solution process has been cast into the MATLAB function '`pde_Laplace_ana()`'.

```
function [u,A,xx,yy]=pde_Laplace_ana(xf,yf,bx0,bxf,by0,byf,Mx,My,Nterm)
% To solve the Laplace's equation u_xx +u_yy = 0
%  over the region D=[0,xf,0,yf]={(x,y)|0<=x<=xf, 0<=y<=yf}
% subject to the BCs:
%  u(0,y) = bx0(y), u(xf,y) = bxf(y)
%  u(x,0) = by0(x), u(x,yf) = byf(x)
% Mx = Number of subintervals along x-axis
% My = Number of subintervals along y-axis
% Nterm: Number of terms
if nargin<9, Nterm=10; end
if nargin<8, My=50; end
if nargin<7, Mx=50; end
x0=0; dx=(xf-x0)/Mx;  xx= x0+[0:Mx]*dx;
y0=0; dy=(yf-y0)/My;  yy= y0+[0:My]'*dy;
[X,Y]=meshgrid(xx,yy);
u=zeros(size(X));
syms x y
for n=1:Nterm
   npixf = n*pi/xf;  npiyf = n*pi/yf;
   lam(n)=yf*sinh(npiyf*xf);
   mu(n)=xf*sinh(npiyf*xf);
   A(1,n)=2/lam(n)*eval(int(bx0*sin(npiyf*y),0,yf)); % Eq.(7.1.22a)
   A(2,n)=2/lam(n)*eval(int(bxf*sin(npiyf*y),0,yf)); % Eq.(7.1.22b)
   A(3,n)=2/mu(n)*eval(int(by0*sin(npixf*x),0,xf)); % Eq.(7.1.22c)
   A(4,n)=2/mu(n)*eval(int(byf*sin(npixf*x),0,xf)); % Eq.(7.1.22d)
   u = u + A(1,n)*sin(npiyf*Y).*sinh(npiyf*(xf-X)) ...
         + A(2,n)*sin(npiyf*Y).*sinh(npiyf*X) ...
         + A(3,n)*sin(npixf*X).*sinh(npixf*(yf-Y)) ...
         + A(4,n)*sin(npixf*X).*sinh(npixf*Y); % Eq.(7.1.21)
end
```

(Example 7.1) Laplace's Equation — Steady-State Temperature Distribution over a Plate

Consider the Laplace's equation

$$\nabla^2 u(x,y)=\frac{\partial^2 u(x,y)}{\partial x^2}+\frac{\partial^2 u(x,y)}{\partial y^2}=0 \ \text{for}\ 0\le x\le\pi \ \text{and}\ 0\le y\le\pi \tag{E7.1.1}$$

subject to the boundary conditions (BCs)

$$u(0,y)=0,\ u(\pi,y)=0,\ u(x,0)=\sin(2x),\ \text{and}\ u(x,\pi)=0 \tag{E7.1.2a,b,c,d}$$

This corresponds to the problem of Eq. (7.1.7) subject to the BCs (7.1.8) where $b_{y0}(x)=\sin(2x)$. Thus, we can use Eq. (7.1.16) to write the solution

$$u(x,y)\overset{(7.1.16)}{=}\sum_{n=1}^{\infty}A(n)\sin(\frac{n\pi}{x_f}x)\sinh(\frac{n\pi}{x_f}(y_f-y))=\frac{1}{\sinh(2\pi)}\sin(2x)\sinh(2(\pi-y)) \tag{E7.1.3}$$

where the coefficients are obtained by using Eq. (7.1.19) as

$$A(n)\overset{(7.1.19)}{=}\frac{2}{x_f\sinh(n\pi y_f/x_f)}\int_0^{x_f}b_{y0}(x)\sin(\frac{n\pi}{x_f}x)\,dx=\begin{cases}1/\sinh(2\pi) & \text{for } n=2\\ 0 & \text{otherwise}\end{cases} \tag{E7.1.4}$$

The MATLAB script "em07e01.m," which solves the above Laplace's equation twice, once by using 'pde_poisson()' and once by using 'pde_Laplace_ana()', computes the discrepancy between the two solutions, and plots the solution graph as shown in Fig. 7.2.

```
%em07e01.m
clear, clf
f=@(x,y)0; g=@(x,y)0;   % Eq.(E7.1.1) w.r.t. Eq.(7.1.1)
% (Rectangular) Domain and BC types
x0=0; xf=pi; y0=0; yf=pi; D=[x0 xf y0 yf; 0 0 0 0];
% BCs
bx0=@(y)0; bxf=@(y)0;  by0=@(x)sin(2*x); byf=@(x)0; % Eq.(E7.1.2a,b,c,d)
% Numerical approach
% Numbers of grid points along x-/y-axes, Error tolerance,
Mx=50; My=50; tol=1e-5; MaxIter=600;
[un,xx,yy]= pde_poisson(f,g,bx0,bxf,by0,byf,D,Mx,My,tol,MaxIter);
% Analytical approach
K=10; [ua,A]=pde_Laplace_ana(xf,yf,bx0,bxf,by0,byf,Mx,My,K);
A, Discrepancy=norm(un-ua)/norm(ua)
mesh(xx,yy,un) % Plot the solution graph
```

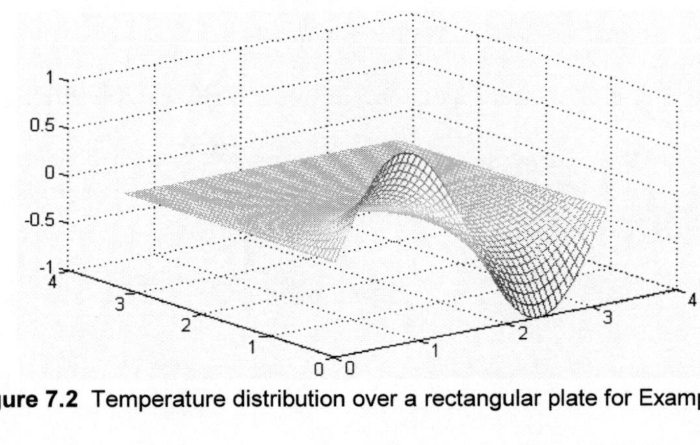

Figure 7.2 Temperature distribution over a rectangular plate for Example 7.1

7.2 PARABOLIC PDE[Y-1], Section 9.2

An example of a parabolic PDE is a one-dimensional (1D) *heat equation* describing the temperature distribution $u(x, t)$ (x: position, t: time) as

$$a^2 \frac{\partial^2 u(x,t)}{\partial x^2} = \frac{\partial u(x,t)}{\partial t} \quad \text{for } 0 \le x \le x_f,\ 0 \le t \le t_f \tag{7.2.1}$$

In order for this equation to be solvable, the BCs $u(0,t)=b_0(t)$ and $u(x_f,t)=b_{xf}(t)$ as well as the IC (initial condition) $u(x, 0)=i_0(x)$ should be provided.

To see how an analytical approach can be taken to solve a parabolic equation, let us consider the following problem of solving Eq. (7.2.1) subject to the boundary and initial conditions

$$u(0,t)=0 \text{ and } u(x_f,t)=0 \tag{7.2.2a,b}$$

$$u(x,0)=i_0(x) \tag{7.2.3}$$

By the method of separation of variables, we substitute

$$u(x, t) = X(x)\,T(t) \tag{7.2.4}$$

into the PDE (7.2.1) and divide by aXT to write

$$a^2 X''T = XT'; \quad \frac{X''}{X} = \frac{T'}{a^2 T} = -\lambda \text{ (separation constant)} \tag{7.2.5}$$

so that

$$X'' + \lambda X = 0 \text{ with } X(0) \underset{(7.2.4)}{\overset{(7.2.2a)}{\equiv}} 0 \text{ and } X(x_f) \underset{(7.2.4)}{\overset{(7.2.2b)}{\equiv}} 0 \tag{7.2.6}$$

$$T' + a^2 \lambda T = 0 \tag{7.2.7}$$

Referring to Table 7.1, the general solution to Eq. (7.2.6) is

$$X(x) = \begin{cases} c_1 \cos \omega x + c_2 \sin \omega x & \text{if } \lambda=\omega^2>0 \\ c_1 + c_2 x & \text{if } \lambda=0 \\ c_1 \cosh \omega x + c_2 \sinh \omega x & \text{if } \lambda=-\omega^2<0 \end{cases} \tag{7.2.8}$$

However, the second and third ones cannot satisfy the BC (7.2.2) without $c_1=c_2=0$ (unmeaningful) and therefore, when $\lambda=\omega^2>0$, the candidate for $X(x)$ satisfying the first BC $X(0)=0$ is just

$$X(x) = c_2 \sin \omega x \tag{7.2.9}$$

With $c_2 \ne 0$, we find the value of ω, not c_2, so that the second BC can be satisfied:

$$X(x_f) = c_2 \sin(\omega x_f) = 0 \ \rightarrow\ \omega = \frac{n\pi}{x_f},\quad n = 1, 2, \cdots \tag{7.2.10}$$

Substituting this into Eq. (7.2.9) yields

$$X_n(x) = c_2 \sin\left(\frac{n\pi}{x_f} x\right) \tag{7.2.11}$$

Also, noting that the values of the so-called eigenvalue $\lambda=\omega^2>0$ are

$$\lambda=\omega^2\overset{(7.2.10)}{=}\left(\frac{n\pi}{x_f}\right)^2 \tag{7.2.12}$$

we can write the general solution of Eq. (7.2.7), which is a first-order ODE, as

$$T_n(t)\overset{(3.2.5)}{=}c_3e^{-a^2\lambda t}\overset{(7.2.12)}{=}c_3e^{-a^2(n^2\pi^2/x_f^2)t} \tag{7.2.13}$$

Now, we substitute Eqs. (7.2.11) and (7.2.13) into Eq. (7.2.4) to write the general solution to Eq. (7.2.1) as

$$u(x,t)=\sum_{n=1}^{\infty}A(n)u_n(x,t)=\sum_{n=1}^{\infty}A(n)X_n(x)T_n(t)\underset{(7.2.13)}{\overset{(7.2.11)}{=}}\sum_{n=1}^{\infty}A(n)\sin(\frac{n\pi}{x_f}x)e^{-a^2(n^2\pi^2/x_f^2)t} \tag{7.2.14}$$

where the coefficient $A(n)$'s are to be fixed so that the initial condition (7.2.3) can be satisfied:

$$u(x,0)\overset{(7.2.14)}{=}\sum_{n=1}^{\infty}A(n)\sin(\frac{n\pi}{x_f}x)=i_0(x) \tag{7.2.15}$$

This implies that $A(n)$'s are the Fourier-sine series coefficients of $i_0(x)$ and accordingly, they can be computed as

$$A(n)\overset{(6.6.15b)}{=}\frac{2}{x_f}\int_0^{x_f}i_0(x)\sin(\frac{n\pi}{x_f}x)\,dx \tag{7.2.16}$$

This solution process has been cast into the MATLAB function '`pde_heat_ana()`'.

Now, let us consider some numerical methods for solving Eq. (7.2.1).

```
function [u,A,tt,xx]=pde_heat_ana(a,xf,tf,it0,M,N,Nterm)
% To solve a^2*u_xx = u_t
%  over the region D=[0,tf,0,xf]={(t,x)|0<=t<=tf, 0<=x<=xf}
% s.t. the BCs and IC:
%  u(t,0) = 0, u(t,xf) = 0, u(0,x) = it0(x)
% M = Number of subintervals along x-axis
% N = Number of subintervals along t-axis
% Nterm: Number of terms
if nargin<8, Nterm=10; end
if nargin<7, N=50; end
if nargin<6, M=50; end
x0=0; dx=(xf-x0)/M;  xx= x0+[0:M]*dx;
t0=0; dt=(tf-t0)/N;  tt= t0+[0:N]'*dt;
[T,X]=meshgrid(tt,xx);
u=zeros(size(X));
syms x
for n=1:Nterm
   npixf = n*pi/xf;
   A(n)=2/xf*eval(int(it0*sin(npixf*x),0,xf)); % Eq.(7.2.16)
   u = u + A(n)*sin(npixf*X).*exp(-a^2*npixf^2*T); % Eq.(7.2.14)
end
```

7.2.1 The Explicit Forward Euler Method

To apply the finite difference method, we divide the spatial domain $[0, x_f]$ into M sections, each of length $\Delta x = x_f/M$, and the time domain $[0, t_f]$ into N segments, each of duration $\Delta t = t_f/N$. Then we replace the second partial derivative on the LHS and the first partial derivative on the RHS of Eq. (7.2.1) by the central difference approximation and the forward difference approximation, respectively, so that we have

$$a^2 \frac{u_{i+1}^k - 2u_i^k + u_{i-1}^k}{\Delta x^2} = \frac{u_i^{k+1} - u_i^k}{\Delta t} \tag{7.2.17}$$

This can be cast into the following algorithm, called the *explicit forward Euler method*, which is to be solved iteratively:

$$u_i^{k+1} = r(u_{i+1}^k + u_{i-1}^k) + (1-2r)u_i^k \ \text{ with } r = a^2\frac{\Delta t}{\Delta x^2} \text{ for } i = 1, 2, ..., M-1 \tag{7.2.18}$$

To find the stability condition of this algorithm, we substitute a trial solution

$$u_i^k = \lambda^k e^{ji\pi/P} \ (P\text{: any nonzero integer}) \tag{7.2.19}$$

into Eq. (7.2.18) to get

$$\lambda = r(e^{j\pi/P} + e^{-j\pi/P}) + (1-2r) = 1 - 2r(1 - \cos(\frac{\pi}{P})) \tag{7.2.20}$$

Since we must have $|\lambda| \le 1$ for non-divergence, the stability condition turns out to be

$$r \overset{(7.2.18)}{=} a^2 \frac{\Delta t}{\Delta x^2} \le \frac{1}{2} \tag{7.2.21}$$

This implies that, as we decrease the spatial interval Δx for better accuracy, we must also decrease the time step Δt at the cost of more computations in order not to lose the stability.

The explicit forward Euler method has been cast into the MATLAB function 'pde_heat_exp()'.

```
function [u,x,t]=pde_heat_exp(a,xf,tf,it0,bx0,bxf,M,N)
% To solve a^2*u_xx = u_t  for 0<=x<=xf, 0<=t<=tf
%  IC : u(x,0)=it0(x)
%  BCs: u(0,t)=bx0(t), u(xf,t)=bxf(t)
% M = Number of subintervals along x-axis
% N = Number of subintervals along t-axis
dx=xf/M;  x=[0:M]'*dx;
dt=tf/N;  t=[0:N]*dt;
for i=1:M+1, u(i,1)= it0(x(i)); end
for n=1:N+1, u([1 M+1],n)=[bx0(t(n)); bxf(t(n))]; end
r=a^2*dt/dx/dx,  r1=1-2*r;
for k=1:N
   for i=2:M
      u(i,k+1)= r*(u(i+1,k)+u(i-1,k)) +r1*u(i,k); % Eq.(7.2.18)
   end
end
```

7.2.2 The Implicit Backward Euler Method

In this section, we consider another algorithm, called the implicit backward Euler method, which comes from substituting the backward difference approximation for the first partial derivative on the RHS of Eq. (7.2.1) as

$$a^2\frac{u_{i+1}^k-2u_i^k+u_{i-1}^k}{\Delta x^2}=\frac{u_i^k-u_i^{k-1}}{\Delta t} \tag{7.2.22}$$

$$;\ -ru_{i-1}^k+(1+2r)u_i^k-ru_{i+1}^k=u_i^{k-1}\ \text{with}\ r=a^2\frac{\Delta t}{\Delta x^2}\ \text{for}\ i=1,2,\ldots,M-1 \tag{7.2.23}$$

If the values of u_0^k and u_M^k at both end points are given from the Dirichlet (or fixed) BC, then the above equation can be cast into a system of simultaneous equations:

$$\begin{bmatrix} 1+2r & -r & 0 & \cdot & 0 & 0 \\ -r & 1+2r & -r & \cdot & 0 & 0 \\ 0 & -r & 1+2r & \cdot & 0 & 0 \\ \cdot & \cdot & \cdot & \cdot & \cdot & \cdot \\ 0 & 0 & 0 & \cdot & 1+2r & -r \\ 0 & 0 & 0 & \cdot & -r & 1+2r \end{bmatrix}\begin{bmatrix} u_1^k \\ u_2^k \\ u_3^k \\ \cdot \\ u_{M-2}^k \\ u_{M-1}^k \end{bmatrix}=\begin{bmatrix} u_1^{k-1}+ru_0^k \\ u_2^{k-1} \\ u_3^{k-1} \\ \cdot \\ u_{M-2}^{k-1} \\ u_{M-1}^{k-1}+ru_M^k \end{bmatrix} \tag{7.2.24}$$

How about the case where the values of $\partial u/\partial x|_{x=0}=b_0'(t)$ at one end are given? In that case, we approximate this *Neumann boundary condition* (with derivatives fixed) by

$$\frac{u_1^k-u_{-1}^k}{2\Delta x}=b_0'(k) \tag{7.2.25}$$

and mix it up with one more equation associated with the unknown variable u_0^k

$$-ru_{-1}^k+(1+2r)u_0^k-ru_1^k=u_0^{k-1} \tag{7.2.26}$$

to get

$$(1+2r)u_0^k-2ru_1^k=u_0^{k-1}-2rb_0'(k)\Delta x \tag{7.2.27}$$

We augment Eq. (7.2.24) with this to write

$$\begin{bmatrix} 1+2r & -2r & 0 & 0 & \cdot & 0 & 0 \\ -r & 1+2r & -r & 0 & \cdot & 0 & 0 \\ 0 & -r & 1+2r & -r & \cdot & 0 & 0 \\ 0 & 0 & -r & 1+2r & \cdot & 0 & 0 \\ \cdot & \cdot & \cdot & \cdot & \cdot & \cdot & \cdot \\ 0 & 0 & 0 & \cdot & \cdot & 1+2r & -r \\ 0 & 0 & 0 & \cdot & \cdot & -r & 1+2r \end{bmatrix}\begin{bmatrix} u_0^k \\ u_1^k \\ u_2^k \\ u_3^k \\ \cdot \\ u_{M-2}^k \\ u_{M-1}^k \end{bmatrix}=\begin{bmatrix} u_0^{k-1}-2rb_0'(k)\Delta x \\ u_1^{k-1} \\ u_2^{k-1} \\ u_3^{k-1} \\ \cdot \\ u_{M-2}^{k-1} \\ u_{M-1}^{k-1}+ru_M^k \end{bmatrix} \tag{7.2.28}$$

Equations such as Eq. (7.2.24) or (7.2.28) are really nice in the sense that they can be solved very efficiently by exploiting their tri-diagonal structures and are guaranteed to be stable owing to their diagonal dominancy. The unconditional stability of Eq. (7.2.24) can be shown by substituting Eq. (7.2.19) into Eq. (7.2.23):

$$-re^{-j\pi/P}+(1+2r)-re^{j\pi/P}=\frac{1}{\lambda};\ \lambda=\frac{1}{1+2r(1-\cos(\pi/P))};\ |\lambda|\le 1 \tag{7.2.29}$$

The MATLAB function '`pde_heat_imp()`' implements this algorithm to solve the PDE (7.2.1) subject to the ordinary (Dirichlet or fixed) BC via Eq. (7.2.24).

```
function [u,x,t]=pde_heat_imp(a,xf,tf,it0,bx0,bxf,M,N)
% To solve a^2*u_xx = u_t  for 0<=x<=xf, 0<=t<=tf
%  IC : u(x,0)=it0(x)
%  BCs: u(0,t)=bx0(t), u(xf,t)=bxf(t)
% M = Number of subintervals along x-axis
% N = Number of subintervals along t-axis
dx=xf/M;  x=[0:M]'*dx;   dt=tf/N;  t=[0:N]*dt;
for i=1:M+1, u(i,1)=it0(x(i)); end
for n=1:N+1, u([1 M+1],n)=[bx0(t(n)); bxf(t(n))]; end
r= a^2*dt/dx/dx;    r2= 1+2*r;
for i=1:M-1
   A(i,i)=r2; % Eq.(7.2.24)
   if i>1, A(i-1,i)= -r; A(i,i-1)= -r; end
end
for k=2:N+1
   b= [r*u(1,k); zeros(M-3,1); r*u(M+1,k)] +u(2:M,k-1); % Eq.(7.2.24)
   u(2:M,k)= trid(A,b);
end
```

7.2.3 The Crank–Nicholson Method

Here, let us go back to Eq. (7.2.22) and try to improve the implicit backward Euler method. The difference approximation on the LHS is taken at time point k, while the difference approximation on the RHS is taken at the midpoint between time k and $k-1$, if we regard it as the central difference approximation with time step $\Delta t/2$. Doesn't this seem to be inconsistent? How about taking the difference approximation of both sides at the same time point, say, the midpoint between k+1 and k, for balance? To do so, we take the average of the central difference approximations of the LHS at the midpoint between k+1 and k, yielding

$$\frac{a^2}{2}\left(\frac{u_{i+1}^{k+1}-2u_i^{k+1}+u_{i-1}^{k+1}}{\Delta x^2}+\frac{u_{i+1}^{k}-2u_i^{k}+u_{i-1}^{k}}{\Delta x^2}\right)=\frac{u_i^{k+1}-u_i^k}{\Delta t} \tag{7.2.30}$$

which leads to the so-called *Crank–Nicholson method*:

$$-ru_{i+1}^{k+1}+2(1+r)u_i^{k+1}-ru_{i-1}^{k+1}=ru_{i+1}^{k}+2(1-r)u_i^{k}+ru_{i-1}^{k}\ \text{with}\ r=a^2\frac{\Delta t}{\Delta x^2} \tag{7.2.31}$$

With the Dirichlet/Neumann BCs on x_0 and x_M, respectively, this can be cast into the following tri-diagonal system of equations:

$$\begin{bmatrix} 2(1+r) & -r & 0 & \cdot & 0 & 0 \\ -r & 2(1+r) & -r & \cdot & 0 & 0 \\ 0 & -r & 2(1+r) & \cdot & 0 & 0 \\ \cdot & \cdot & \cdot & \cdot & \cdot & \cdot \\ 0 & 0 & 0 & \cdot & 2(1+r) & -r \\ 0 & 0 & 0 & \cdot & -2r & 2(1+r) \end{bmatrix} \begin{bmatrix} u_1^{k+1} \\ u_2^{k+1} \\ u_3^{k+1} \\ \cdot \\ u_{M-1}^{k+1} \\ u_M^{k+1} \end{bmatrix}$$

$$= \begin{bmatrix} 2(1-r) & r & 0 & \cdot & 0 & 0 \\ r & 2(1-r) & r & \cdot & 0 & 0 \\ 0 & r & 2(1-r) & \cdot & 0 & 0 \\ \cdot & \cdot & \cdot & \cdot & \cdot & \cdot \\ 0 & 0 & 0 & \cdot & 2(1-r) & r \\ 0 & 0 & 0 & \cdot & 2r & 2(1-r) \end{bmatrix} \begin{bmatrix} u_1^{k} \\ u_2^{k} \\ u_3^{k} \\ \cdot \\ u_{M-1}^{k} \\ u_M^{k} \end{bmatrix} + \begin{bmatrix} r(u_0^{k+1}+u_0^k) \\ 0 \\ 0 \\ \cdot \\ 0 \\ 2r(b_M'(k+1)+b_M'(k)) \end{bmatrix} \tag{7.2.32}$$

This system of equations can also be solved very efficiently and its unconditional stability can be shown by substituting Eq. (7.2.19) into Eq. (7.2.31):

$$2\lambda(1+r(1-\cos(\pi/P))) = 2(1-r(1-\cos(\pi/P)));\ \lambda = \frac{1-r(1-\cos(\pi/P))}{1+r(1-\cos(\pi/P))};\ |\lambda| \le 1 \tag{7.2.33}$$

This solution process has been cast into the following MATLAB function '`pde_heat_CN()`'.

```
function [u,x,t]=pde_heat_CN(a,xf,tf,it0,bx0,bxf,M,N)
%solve a^2*u_xx = u_t  for 0<=x<=xf, 0<=t<=tf
% IC : u(x,0)=it0(x)
% BCs: u(0,t)=bx0(t), u(xf,t)=bxf(t)
% M = Number of subintervals along x-axis
% N = Number of subintervals along t-axis
dx=xf/M;   x=[0:M]'*dx;
dt=tf/N;   t=[0:N]*dt;
for i=1:M+1, u(i,1)= it0(x(i)); end
for n=1:N+1, u([1 M+1],n)= [bx0(t(n)); bxf(t(n))]; end
r=a^2*dt/dx/dx;   r1=2*(1-r); r2=2*(1+r);
for i=1:M-1
   A(i,i)=r1; % Eq.(7.2.32)
   if i>1, A(i-1,i)= -r; A(i,i-1)= -r; end
end
for k=2:N+1
   b=[r*u(1,k); zeros(M-3,1); r*u(M+1,k)] ...
          +r*(u(1:M-1,k-1)+u(3:M+1,k-1)) +r2*u(2:M,k-1);
   u(2:M,k)=trid(A,b); % Eq.(7.2.32)
end
```

7.2.4 Using the MATLAB Function 'pdepe()'

The MATLAB function '`pdepe()`' can be used to solve a system of 1D parabolic and elliptic PDEs subject to the IC (initial condition) and BCs (boundary conditions) given:

$$x^{-m}\frac{\partial}{\partial x}\left(x^m f\left(x,t,u,\frac{\partial u}{\partial x}\right)\right)+s\left(x,t,u,\frac{\partial u}{\partial x}\right)=c\left(x,t,u,\frac{\partial u}{\partial x}\right)\frac{\partial u}{\partial t} \quad \text{for } x_0\le x\le x_f,\ t_0\le t\le t_f \tag{7.2.34}$$

Note that m can be 0, 1, or 2, corresponding to slab, cylindrical, or spherical symmetry, and if $m>0$, the x_0 must be nonnegative. In order for this equation to be solvable, the BCs

$$\begin{aligned} p_0(x,t,u)+q_0(x,t)f(x,t,u,\partial u/\partial x)&=0 \ \ \forall t \ \text{ for } x=x_0\\ p_f(x,t,u)+q_f(x,t)f(x,t,u,\partial u/\partial x)&=0 \ \ \forall t \ \text{ for } x=x_f \end{aligned} \tag{7.2.35}$$

as well as the IC $u(x, t_0)=i_0(x)$ should be provided.

The syntax of '`pdepe()`' is

```
sol = pdepe(m,pdefun,icfun,bcfun,xmesh,tspan,options)
```

where its arguments are as follows:

m = Parameter to be set as 0/1/2 depending on the symmetry of slab/cylinder/sphere.

pdefun = A handle to a function returning the components c, f, and s of the PDE (7.2.34) with the syntax of

```
[c,f,s]=pdefun(x,t,u,dudx)
```

Note that c, f, and s are column vectors of the dimension of the system of PDEs where c stores the diagonal elements of the matrix $c(x,t,u,\partial u/\partial x)$.

icfun = A handle to a function $i_0(x)$ returning the column vector of ICs at x with the syntax of

```
u = icfun(x)
```

bcfun = A handle to a function that defines the BCs with the syntax of

```
[p0,q0,pf,qf]=bcfun(x0,u0,xf,uf,t)
```

xmesh = A vector [x0, x1, ..., xn] (of length greater than 2) specifying the points at which a numerical solution is requested for every value in tspan.

tspan = A vector [t0, t1, ..., tf] (of length greater than 2) specifying the points at which a solution is requested for every value in xmesh.

options = A structure of optional parameters that can be built as the output argument of '`odeset()`' as follows:

```
options = odeset('RelTol',1e-4,'AbsTol',[1e-4 1e-4]);
```

sol = A 3D array where sol(j,k,i) contains the ith component of the solution at (tspan(j),xspan(k)).

(Example 7.2) 1D (One-dimensional) Parabolic PDE — Heat Flow Equation

Consider the following parabolic PDE

$$a^2\frac{\partial^2 u(x,t)}{\partial x^2}=\frac{\partial u(x,t)}{\partial t} \quad \text{for } 0\le x\le 1 \text{ and } 0\le t\le 0.1 \tag{E7.2.1}$$

subject to the IC and BCs

$$u(x,0)=\sin\pi x,\ u(0,t)=0,\ \ u(1,t)=0 \tag{E7.2.2a,b,c}$$

Since this is the problem discussed in the beginning of Section 7.2, its solution is given by Eq. (7.2.14) as

$$u(x,t) \overset{(7.2.14)}{=} \sum_{n=1}^{\infty} A(n)\sin(\frac{n\pi}{x_f}x)e^{-a^2(n^2\pi^2/x_f^2)t} \underset{a=1,x_f=1}{\overset{(E7.2.4)}{=}} \sin(\pi x)e^{-a^2\pi^2 t} \tag{E7.2.3}$$

with the coefficient $A(n)$'s fixed by Eq. (7.2.16) as

$$A(n) \overset{(7.2.16)}{=} \frac{2}{x_f}\int_0^{x_f} i_0(x)\sin(\frac{n\pi}{x_f}x)dx \overset{x_f=1}{=} 2\int_0^1 \sin(\pi x)\sin(n\pi x)dx = \begin{cases} 1 & \text{for } n=1 \\ 0 & \text{for } n\neq 1 \end{cases} \tag{E7.2.4}$$

We composed the MATLAB script "em07e02.m" in order to use the functions 'pde_heat_ana()', 'pde_heat_exp()', 'pde_heat_imp()', 'pde_heat_CN()', and 'pdepe()' for solving this problem (with $a=1$), and we run the script to obtain the results shown in Fig. 7.3. Note that, with the spatial interval $\Delta x = x_f/M = 1/20$ and the time step $\Delta t = t_f/N = 0.1/100 = 0.001$, we have

$$r = a^2\frac{\Delta t}{\Delta x^2} = 1\frac{0.001}{(1/20)^2} = 0.4 \tag{E7.2.5}$$

```
function [u_pdepe,x,t]=em07e02
a=1; % The parameter of Eq.(E7.2.1)
it0=@(x)sin(pi*x); % IC (E7.2.2a)
bx0=@(t)0;  bxf=@(t)0; % BC (E7.2.2b,c)
xf=1; M=25;  tf=0.1; N=100; % r=0.625
[u_exp,x,t]=pde_heat_exp(a,xf,tf,it0,bx0,bxf,M,N); % Explicit method
[u_imp,x,t]=pde_heat_imp(a,xf,tf,it0,bx0,bxf,M,N); % Implicit method
[u_CN,x,t]=pde_heat_CN(a,xf,tf,it0,bx0,bxf,M,N); % Crank-Nicholson method
subplot(131), mesh(t,x,u_exp),
xlabel('t'), ylabel('x'), title('Numerical solution using pde heat exp()')
subplot(132), mesh(t,x,u_CN),
xlabel('t'), ylabel('x'), title('Numerical solution using pde heat CN()')
% Using pdepe() ------------------------------
m=0; sol=pdepe(m,@pdex1pde,@pdex1ic,@pdex1bc,x,t);
u_pdepe=sol(:,:,1); % Extract the first solution component as u.
% Analytical solution
uo=@(x,t)sin(pi*x)*exp(-pi*pi*t);
Uo=uo(x,t); aUo=abs(Uo)+eps; % Values of true analytical solution
Nterm=10; [u_an,A]=pde_heat_ana(a,xf,tf,it0,M,N,Nterm);
MN=M*N; err_an= norm((u_an-Uo)./aUo)/MN
err_exp= norm((u_exp-Uo)./aUo)/MN
err_imp= norm((u_imp-Uo)./aUo)/MN
err_CN=norm((u_CN-Uo)./aUo)/MN
err_pdepe=norm((u_pdepe.'-Uo)./aUo)/MN
err_an= norm((u_an-Uo)./aUo)/MN
subplot(133), mesh(t,x,u_pdepe.') % surf(t,x,u)
xlabel('t'), ylabel('x'), title('Numerical solution using pdepe()')
% -------------------------------------------------------------
function [c,f,s] = pdex1pde(x,t,u,DuDx)
% defines the PDE by setting the parameters
c = 1; f = DuDx;  s = 0; % c=1/a;
% -------------------------------------------------------------
function u0 = pdex1ic(x)
u0 = sin(pi*x); % Initial condition (E7.2.1)
% -------------------------------------------------------------
function [pl,ql,pr,qr] = pdex1bc(xl,ul,xr,ur,t)
pl = ul; ql = 0; % Left boundary condition
pr = ur; qr = 0; % Right boundary condition
```

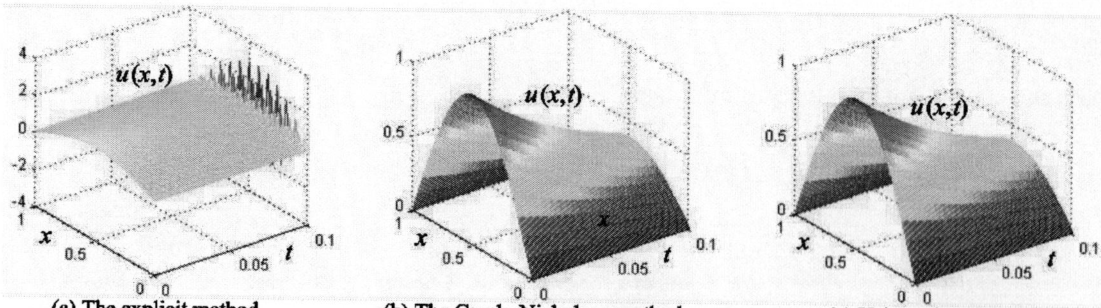

(a) The explicit method (b) The Crank–Nicholson method (c) Using pdepe()

Figure 7.3 Results of various algorithms for a 1D parabolic PDE — heat equation for Example 7.2

which satisfies the stability condition $r \le 1/2$ (Eq. (7.2.21)) and that all of the three methods lead to reasonably fair results with a relative error of about 0.013. But if we decrease the spatial interval to $\Delta x = 1/25$ for better resolution, we have $r=0.625$, violating the stability condition, and the explicit forward Euler method ('`pde_heat_exp()`') blows up because of instability, as shown in Fig. 7.3(a), while the implicit backward Euler method ('`pde_heat_imp()`') and the Crank–Nicholson method ('`pde_heat_CN()`') work quite well, as shown in Fig. 7.3(b),(c). Now, with the spatial interval $\Delta x = 1/25$ and the time step $\Delta t = 0.1/120$, the explicit method as well as the other ones works well with a relative error of less than 0.001 in return for somewhat (30%) more computations, despite the fact that $r=0.5208$ doesn't strictly satisfy the stability condition. This implies that the condition ($r \le 1/2$) for stability of the explicit forward Euler method is not a necessary one, but only a sufficient one. Besides, if it converges, its accuracy may be better than that of the implicit backward Euler method, but generally no better than that of the Crank–Nicholson method.

7.2.5 Two-Dimensional Parabolic PDEs

Another example of a parabolic PDE is a 2D *heat equation* describing the temperature distribution $u(x, y, t)$ ((x, y): position, t: time) as

$$a^2\left(\frac{\partial^2 u(x,y,t)}{\partial x^2}+\frac{\partial^2 u(x,y,t)}{\partial y^2}\right)=\frac{\partial u(x,y,t)}{\partial t} \quad \text{for } x_0 \le x \le x_f,\, y_0 \le y \le y_f,\, t_0 \le t \le t_f \quad (7.2.36)$$

In order for this equation to be solvable, we should be provided with the BCs

$$u(x_0, y, t) = b_{x0}(y,t),\ u(x_f, y, t) = b_{xf}(y,t),\ u(x, y_0, t) = b_{y0}(x,t), \text{ and } u(x, y_f, t) = b_{yf}(x,t)$$

as well as the IC $u(x, y, t_0) = i_{t0}(x, y)$.

We replace the first-order time derivative on the RHS by the 3-point central difference at the midpoint $(t_{k+1}+t_k)/2$, just as with the Crank–Nicholson method. We also replace one of the second-order derivatives, u_{xx} and u_{yy}, by the 3-point central difference approximation at time t_k and the other at time t_{k+1}, yielding

$$a^2\left(\frac{u_{i,j+1}^k-2u_{i,j}^k+u_{i,j-1}^k}{\Delta x^2}+\frac{u_{i+1,j}^{k+1}-2u_{i,j}^{k+1}+u_{i-1,j}^{k+1}}{\Delta y^2}\right)=\frac{u_{i,j}^{k+1}-u_{i,j}^k}{\Delta t} \quad (7.2.37a)$$

This seems to be attractive, since it can be formulated into a tri-diagonal system of equations in $u^{k+1}_{i+1,j}$, $u^{k+1}_{i,j}$ and $u^{k+1}_{i-1,j}$. But why do we treat u_{xx} and u_{yy} with discrimination, i.e., evaluate one at time t_k and the other at time t_{k+1} in a fixed manner? In an alternate manner, we write the difference equation for the next time point t_{k+1} as

$$a^2\left(\frac{u^{k+1}_{i,j+1}-2u^{k+1}_{i,j}+u^{k+1}_{i,j-1}}{\Delta x^2}+\frac{u^{k}_{i+1,j}-2u^{k}_{i,j}+u^{k}_{i-1,j}}{\Delta y^2}\right)=\frac{u^{k+2}_{i,j}-u^{k+1}_{i,j}}{\Delta t} \tag{7.2.37b}$$

```
function [u,x,y,t]=pde_heat2_ADI(a,D,tf,it0,bxyt,Mx,My,N)
%To solve a^2*(u_xx+u_yy)=u_t for D(1)<=x<=D(2), D(3)<=y<=D(4), 0<=t<=tf
%  IC: u(x,y,0) = it0(x,y)
%  BC: u(x,y,t) = bxyt(x,y,t) for (x,y)cB
% Mx/My = Number of subintervals along x/y-axis
% N = Number of subintervals along t-axis
dx=(D(2)-D(1))/Mx;   x=D(1)+[0:Mx]*dx;
dy=(D(4)-D(3))/My;   y=D(3)+[0:My]'*dy;
dt=tf/N;   t=[0:N]*dt;
% Initialization
for j=1:Mx+1,   for i=1:My+1, u(i,j)= it0(x(j),y(i));  end,  end
rx= a^2*dt/(dx*dx); rx1=1+2*rx; rx2=1-2*rx;
ry= a^2*dt/(dy*dy); ry1=1+2*ry; ry2=1-2*ry;
for j=1:Mx-1 % Eq.(7.2.39a)
   Ay(j,j)= ry1; if j>1, Ay(j-1,j)= -ry; Ay(j,j-1)= -ry; end
end
for i=1:My-1 % Eq.(7.2.39b)
   Ax(i,i)= rx1; if i>1, Ax(i-1,i)= -rx; Ax(i,i-1)= -rx; end
end
for k=1:N
  u_1=u;    t=k*dt;
  for i=1:My+1 % BCs
    u(i,[1 Mx+1])= feval(bxyt,x([1 Mx+1]),y(i),t);
  end
  for j=1:Mx+1
    u([1 My+1],j)= feval(bxyt,x(j),y([1;My+1]),t);
  end
  if mod(k,2)==0
    for i=2:My
      jj=2:Mx;
      bx= [ry*u(i,1) zeros(1,Mx-3) ry*u(i,My+1)] ...
                     +rx*(u_1(i-1,jj)+u_1(i+1,jj)) +rx2*u_1(i,jj);
      u(i,jj)= trid(Ay,bx')'; % Eq.(7.2.38a)
    end
   else
    for j=2:Mx
      ii=2:My;
      by= [rx*u(1,j); zeros(My-3,1); rx*u(Mx+1,j)] ...
                     +ry*(u_1(ii,j-1)+u_1(ii,j+1)) +ry2*u_1(ii,j);
      u(ii,j)= trid(Ax,by); % Eq.(7.2.38b)
    end
  end
end
```

This solution process, proposed by Peaceman and Rachford [P-1] and referred to as the *alternating direction implicit* (*ADI*) method, can be put into the following algorithm:

$$-r_y(u_{i-1,j}^{k+1}+u_{i+1,j}^{k+1})+(1+2r_y)u_{i,j}^{k+1}=r_x(u_{i,j-1}^{k}+u_{i,j+1}^{k})+(1-2r_x)u_{i,j}^{k}\ \text{ for } j=1,2,\ldots,M_x-1 \quad (7.2.38a)$$

$$-r_x(u_{i,j-1}^{k+2}+u_{i,j+1}^{k+2})+(1+2r_x)u_{i,j}^{k+2}=r_y(u_{i-1,j}^{k+1}+u_{i+1,j}^{k+1})+(1-2r_y)u_{i,j}^{k+1}\ \text{ for } i=1,2,\ldots,M_y-1 \quad (7.2.38b)$$

$$\text{with } r_x=a^2\Delta t/\Delta x^2,\ r_y=a^2\Delta t/\Delta y^2,\ \Delta x=(x_f-x_0)/M_x,\ \Delta y=(y_f-y_0)/M_y,\ \Delta t=(t_f-t_0)/N$$

This algorithm for solving a 2D heat equation (7.2.36) has been cast into the above MATLAB function, '`pde_heat2_ADI()`'.

(Example 7.3) A Parabolic PDE — 2D Temperature Diffusion over a Plate

Consider a 2D parabolic PDE

$$10^{-4}\left(\frac{\partial^2 u(x,y,t)}{\partial x^2}+\frac{\partial^2 u(x,y,t)}{\partial y^2}\right)=\frac{\partial u(x,y,t)}{\partial t}\ \text{ for } 0\le x\le 4,\ 0\le y\le 4, \text{ and } 0\le t\le 5000 \quad (E7.3.1)$$

subject to the IC and BCs:

$$u(x,y,0)=0 \ \text{ for all } (x,y) \quad (E7.3.2a)$$

$$u(x,y,t)=e^y\cos x-e^x\cos y \text{ for } x=0,\ x=4,\ y=0,\ y=4, \text{ and } \forall\, t \quad (E7.3.2b)$$

The following MATLAB script, "em07e03.m," which uses the function '`pde_heat2_ADI()`' to solve this equation, has been run to yield the solution graph at the final time as shown in Fig. 7.4.

```
%em07e03.m
clear, clf
a=1e-2;
it0=@(x,y)0; % IC (E7.3.2a)
bxyt=@(x,y,t)exp(y)*cos(x)-exp(x)*cos(y); % BC (E7.3.2b)
D=[0 4 0 4]; % Solution region
t0=0; tf=5000; Mx=40; My=40; N=50;
[u,x,y,t]=pde_heat2_ADI(a,D,tf,it0,bxyt,Mx,My,N);
mesh(x,y,u)
```

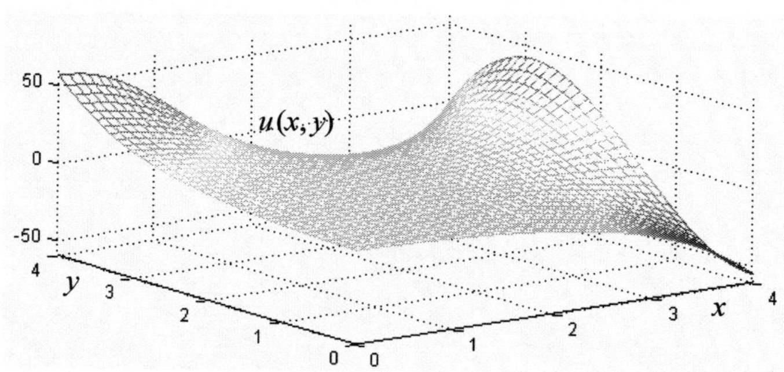

Figure 7.4 Temperature distribution over a plate for Example 7.3

7.3 HYPERBOLIC PDES[Y-1], Section 9.3

An example of a hyperbolic PDE is a 1D *wave equation* for the amplitude function $u(x, t)$ (x : position, t : time) as

$$a^2 \frac{\partial^2 u(x,t)}{\partial x^2} = \frac{\partial^2 u(x,t)}{\partial t^2} \quad \text{for } 0 \le x \le x_f \text{ and } 0 \le t \le t_f \tag{7.3.1}$$

In order for this equation to be solvable, the BCs $u(0,t)=b_{x0}(t)$ and $u(x_f, t)=b_{xf}(t)$ as well as the ICs $u(x,0)=i_0(x)$ and $\partial u/\partial t|_{t=0}(x,0)=i_0'(x)$ should be provided.

To feel how an analytical approach can be taken to solve a hyperbolic equation, let us consider the following problem of solving Eq. (7.3.1) subject to the BCs and ICs

$$u(0,t)=0 \text{ and } u(x_f, t)=0 \tag{7.3.2a,b}$$

$$u(x, 0)=i_0(x) \text{ and } u'(x,0)=i_0'(x) \tag{7.3.3a,b}$$

By the method of separation of variables, we substitute

$$u(x,t) = X(x)T(t) \tag{7.3.4}$$

into the PDE (7.3.1) and divide by a^2XT to write

$$a^2X''T = XT''; \quad \frac{X''}{X} = \frac{T''}{a^2T} = -\lambda \text{(separation constant)} \tag{7.3.5}$$

so that

$$X''+\lambda X = 0 \quad \text{with } X(0) \underset{(7.3.4)}{\overset{(7.3.2a)}{=}} 0 \text{ and } X(x_f) \underset{(7.3.4)}{\overset{(7.3.2b)}{=}} 0 \tag{7.3.6}$$

$$T''+a^2\lambda T = 0 \tag{7.3.7}$$

Like Eq. (7.2.11) for Eq. (7.2.6), the general solution to Eq. (7.3.6) can be written as

$$X_n(x) = c_2 \sin \omega x = c_2 \sin\left(\frac{n\pi}{x_f}x\right) \text{ with } \omega = \frac{n\pi}{x_f}, \; n = 1, 2, \cdots \tag{7.3.8}$$

Also, noting that the values of the so called eigenvalue $\lambda=\omega^2>0$ are

$$\lambda = \omega^2 \overset{(7.3.8)}{=} \left(\frac{n\pi}{x_f}\right)^2 \tag{7.3.9}$$

we can write the general solution of Eq. (7.3.7) as

$$T_n(t) \overset{(3.2.7)}{=} c_3 \cos(a\omega t) + c_4 \sin(a\omega t) \overset{(7.3.8)}{=} c_3 \cos(a\frac{n\pi}{x_f}t) + c_4 \sin(a\frac{n\pi}{x_f}t) \tag{7.3.10}$$

Now, we substitute Eqs. (7.3.8) and (7.3.10) into Eq. (7.3.4) to write the general solution to Eq. (7.3.1) as

$$u(x,t)=\sum_{n=1}^{\infty}A(n)u_n(x,t)=\sum_{n=1}^{\infty}A(n)X_n(x)T_n(t)$$

$$\overset{(7.3.8)}{\underset{(7.3.10)}{=}}\sum_{n=1}^{\infty}\sin(\frac{n\pi}{x_f}x)\left\{A_1(n)\cos(a\frac{n\pi}{x_f}t)+A_2(n)\sin(a\frac{n\pi}{x_f}t)\right\} \quad (7.3.11)$$

where the coefficients are to be fixed so that the ICs (7.3.3a,b) can be satisfied:

$$u(x,0)\overset{(7.3.11)}{=}\sum_{n=1}^{\infty}\sin(\frac{n\pi}{x_f}x)A_1(n)\overset{(7.3.3a)}{=}i_0(x)$$

$$\underset{\text{sine series}}{\overset{\text{Fourier}}{\rightarrow}} A_1(n)\overset{(6.6.15b)}{=}\frac{2}{x_f}\int_0^{x_f}i_0(x)\sin(\frac{n\pi}{x_f}x)dx \quad (7.3.12)$$

$$u'(x,0)\overset{(7.3.3b)}{=}i_0'(x)\overset{(7.3.11)}{\rightarrow}\sum_{n=1}^{\infty}\sin(\frac{n\pi}{x_f}x)\left\{-A_1(n)a\frac{n\pi}{x_f}\sin(a\frac{n\pi}{x_f}0)+A_2(n)a\frac{n\pi}{x_f}\cos(a\frac{n\pi}{x_f}0)\right\}$$

$$=\sum_{n=1}^{\infty}\sin(\frac{n\pi}{x_f}x)A_2(n)a\frac{n\pi}{x_f}\overset{(7.3.3b)}{=}i_0'(x)$$

$$\underset{\text{sine series}}{\overset{\text{Fourier}}{\rightarrow}} A_2(n)a\frac{n\pi}{x_f}\overset{(6.6.15b)}{=}\frac{2}{x_f}\int_0^{x_f}i_0'(x)\sin(\frac{n\pi}{x_f}x)dx\rightarrow A_2(n)=\frac{2}{an\pi}\int_0^{x_f}i_0'(x)\sin(\frac{n\pi}{x_f}x)dx \quad (7.3.13)$$

This solution process has been cast into the MATLAB function '`pde_wave_ana()`'.
Now, let us consider some numerical methods for solving Eq. (7.3.1).

```
function [u,A,tt,xx]=pde_wave_ana(a,xf,tf,it0,i1t0,M,N,Nterm)
% To solve a^2*u_xx = u_tt
%  over the region D=[0,tf,0,xf]={(t,x)|0<=t<=tf, 0<=x<=xf}
% s.t. the BCs/ICs:
%  u(t,0)=0, u(t,xf)=0, u(0,x)=it0(x), u'(0,x)=i1t0(x)
% M = Number of subintervals along x-axis
% N = Number of subintervals along t-axis
% Nterm: Number of terms
if nargin<8, Nterm=10; end
if nargin<7, N=50; end
if nargin<6, M=50; end
x0=0; dx=(xf-x0)/M;  xx= x0+[0:M]*dx;
t0=0; dt=(tf-t0)/N;  tt= t0+[0:N]'*dt;
[T,X]=meshgrid(tt,xx);
u=zeros(size(X));
syms x
for n=1:Nterm
   npixf=n*pi/xf; anpixfT=a*npixf*T;
   A(1,n)=2/xf*eval(int(it0(x).*sin(npixf*x),0,xf)); % Eq.(7.3.12)
   A(2,n)=2/a/n/pi*eval(int(i1t0(x).*sin(npixf*x),0,xf)); % Eq.(7.3.13)
   u = u + sin(npixf*X).*(A(1,n)*cos(anpixfT) ...
                   + A(2,n)*sin(anpixfT)); % Eq.(7.3.11)
end
```

7.3.1 The Explicit Central Difference Method

In the same way as with the parabolic PDEs, we replace the second derivatives on both sides of Eq. (7.3.1) by their 3-point central difference approximations as

$$a^2\frac{u_{i+1}^k-2u_i^k+u_{i-1}^k}{\Delta x^2}=\frac{u_i^{k+1}-2u_i^k+u_i^{k-1}}{\Delta t^2} \text{ with } \Delta x=\frac{x_f}{M} \text{ and } \Delta t=\frac{t_f}{N} \tag{7.3.14}$$

which leads to the explicit central difference method:

$$u_i^{k+1}=r(u_{i+1}^k+u_{i-1}^k)+2(1-r)u_i^k-u_i^{k-1} \text{ with } r=a^2\frac{\Delta t^2}{\Delta x^2} \tag{7.3.15}$$

Since $u_i^{-1}=u(x_i,-\Delta t)$ is not given, we cannot get u_i^1 directly from this formula (7.3.15) with $k=0$:

$$u_i^1=r(u_{i+1}^0+u_{i-1}^0)+2(1-r)u_i^0-u_i^{-1} \tag{7.3.16}$$

Therefore, we approximate the IC (7.3.3b) on the derivative by the central difference as

$$\frac{u_i^1-u_i^{-1}}{2\Delta t}=i_0'(x_i) \tag{7.3.17}$$

and make use of this to remove u_i^{-1} from Eq. (7.3.15):

$$u_i^1=r(u_{i+1}^0+u_{i-1}^0)+2(1-r)u_i^0-(u_i^1-2i_0'(x_i)\Delta t)$$

$$;\quad u_i^1=\frac{1}{2}r(u_{i+1}^0+u_{i-1}^0)+(1-r)u_i^0+i_0'(x_i)\Delta t \tag{7.3.18}$$

We use this equation (7.3.18) together with the ICs to get u_i^1 and then go on with Eq. (7.3.15) for $k=1,2,\ldots$. Here, note the following:

- We must have $r\le 1$ to guarantee the stability.
- The accuracy of the solution gets better as r becomes larger so that Δx decreases.

It is therefore reasonable to select $r=1$.

The stability condition can be obtained by substituting Eq. (7.2.19) into Eq. (7.3.15) and applying the Jury test [P-2]:

$$\lambda=2r\cos(\frac{\pi}{P})+2(1-r)-\lambda^{-1};\quad \lambda^2+2\{r(1-\cos(\frac{\pi}{P}))-1\}\lambda+1=0$$

We need the solution of this equation to be inside the unit circle for stability, which requires

$$r\le\frac{1}{1-\cos(\pi/P)};\quad r\overset{(7.3.15)}{=}a^2\frac{\Delta t^2}{\Delta x^2}\le 1 \tag{7.3.19}$$

This process to solve a 1D wave equation like Eq. (7.3.1) has been cast into the following MATLAB function '`pde_wave()`'.

```
function [u,x,t]=pde_wave(a,xf,tf,it0,i1t0,bx0,bxf,M,N)
% To solve a^2*u_xx = u_tt  for 0<=x<=xf, 0<=t<=tf
%   IC: u(x,0)=it0(x), u_t(x,0)=i1t0(x)
%   BC: u(0,t)=bx0(t), u(xf,t)=bxf(t)
% M = Number of subintervals along x-axis
% N = Number of subintervals along t-axis
dx=xf/M;   x=[0:M]'*dx;
dt=tf/N;   t=[0:N]*dt;
for i=1:M+1, u(i,1)= it0(x(i)); end
for k=1:N+1, u([1 M+1],k)=[bx0(t(k)); bxf(t(k))];  end
r=(a*dt/dx)^2;  r1=r/2;  r2=2*(1-r);
u(2:M,2)= r1*u(1:M-1,1) +(1-r)*u(2:M,1) +r1*u(3:M+1,1) ...
                 +dt*i1t0(x(2:M)); % Eq.(7.3.18)
for k=3:N+1
  u(2:M,k)= r*u(1:M-1,k-1) +r2*u(2:M,k-1) +r*u(3:M+1,k-1)...
                 -u(2:M,k-2); % Eq.(7.3.15)
end
```

(Example 7.4) A Hyperbolic PDE — 1D Wave (Vibration) on a String Fixed at Both Ends

Consider a 1D hyperbolic PDE

$$\frac{\partial^2 u(x,t)}{\partial x^2}=\frac{\partial^2 u(x,t)}{\partial t^2} \text{ for } 0 \le x \le 1 \text{ and } 0 \le t \le 2 \tag{E7.4.1}$$

subject to the BCs and ICs

$$u(0,t)=0, \quad u(1,t)=0 \text{ for all } t, \tag{E7.4.2a,b}$$

$$u(x,t)\big|_{t=0}=x(1-x), \quad \frac{\partial u}{\partial t}(x,t)\bigg|_{t=0}=0 \text{ for } t=0 \tag{E7.4.3a,b}$$

Noting that this problem exactly conforms with Eq. (7.3.1) subject to the BCs (7.3.2) and ICs (7.3.3), we can use Eqs. (7.3.12) and (7.3.13) to get the coefficients

$$A_1(n) \overset{(7.3.12)}{=} \frac{2}{x_f}\int_0^{x_f} i_0(x)\sin(\frac{n\pi}{x_f}x)\,dx \overset{(E7.4.3a)}{=} \frac{2}{x_f}\int_0^1 x(1-x)\sin(n\pi x)\,dx$$

$$= \frac{8\sin^2(n\pi/2)}{n^3\pi^3} = \begin{cases} 8/n^3\pi^3 & \text{for } k \text{ odd} \\ 0 & \text{for } k \text{ even} \end{cases} \tag{E7.4.4}$$

```
>>syms x n; int(x*(1-x)*sin(n*pi*x),0,1)
```

$$A_2(n) \overset{(7.3.13)}{=} \frac{2}{an\pi}\int_0^{x_f} i_0'(x)\sin(\frac{n\pi}{x_f}x)\,dx \underset{a=1}{\overset{(E7.4.3b)}{=}} \frac{2}{n\pi}\int_0^1 0\sin(\frac{n\pi}{x_f}x)\,dx = 0 \tag{E7.4.5}$$

and then use Eq. (7.3.11) to get the solution as

$$u(x,t) \underset{(E7.4.5)}{\overset{(7.3.11)}{=}} \sum_{n=1}^{\infty} A_1(n)\sin(\frac{n\pi}{x_f}x)\cos(a\frac{n\pi}{x_f}t)$$

$$\underset{a=1,x_f=1}{\overset{(E7.4.4)}{=}} \sum_{m=1}^{\infty} \frac{8}{(2m-1)^3\pi^3}\sin((2m-1)\pi x)\cos(a(2m-1)\pi t) \tag{E7.4.6}$$

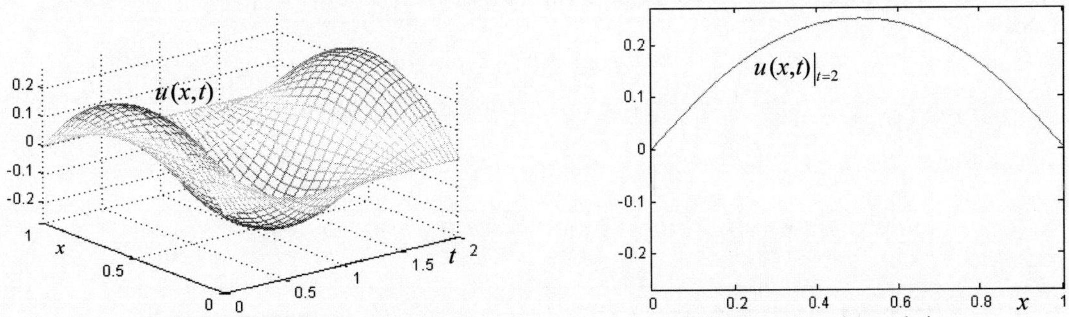

(a) The solution $u(x, t)$ (b) A snapshot of the solution at $t=t_f$

Figure 7.5 Solution to the 1D hyperbolic PDE for Example 7.4

We can compose the MATLAB script "em07e04.m," which solves this problem twice, once by using '`pde_wave_ana()`' and once by using '`pde_wave()`', and run it to get Fig. 7.5 and see a motion picture showing how $u(x, t)$ changes with time $t=n\Delta t$.

```
%em07e04.m
clear
a=1;
% BCs and ICs
bx0t=@(t)0;   bxft=@(t)0;        % Eq.(E7.4.2a,b)
it0=@(x)x.*(1-x); i1t0=@(x)0; % Eq.(E7.4.3a,b)
xf=1; M=20;   tf=2; N=50;
% Numerical approach
[un,x,t]=pde_wave(a,xf,tf,it0,i1t0,bx0t,bxft,M,N);
% Analytical approach
K=10; [ua,A]=pde_wave_ana(a,xf,tf,it0,i1t0,M,N,K); A
[T,X]=meshgrid(t,x);
ua1=0;
for m=1:floor(K/2) % Eq.(E7.4.6)
   nodd=2*m-1; ua1=ua1+8/(nodd*pi)^3*sin(nodd*pi*X).*cos(a*nodd*pi*T);
end
Discrepancy1=norm(un-ua)/norm(ua)
Discrepancy2=norm(ua-ua1)/norm(ua)
figure(1), clf
mesh(t,x,un)
figure(2), clf
for n=1:N % Motion picture
   plot(x,un(:,n)), axis([0 xf -0.3 0.3]), pause(0.2)
end
```

7.3.2 Two-Dimensional Hyperbolic PDEs

In this section, we consider a 2D *wave equation* for the amplitude function $u(x,y,t)$ ((x,y): position, t: time) as

$$a^2\left(\frac{\partial^2 u(x,y,t)}{\partial x^2}+\frac{\partial^2 u(x,y,t)}{\partial y^2}\right)=\frac{\partial^2 u(x,t)}{\partial t^2} \quad \text{for } 0\le x\le x_f,\ 0\le y\le y_f, \text{ and } 0\le t\le t_f \qquad (7.3.20)$$

In order for this equation to be solvable, we should be provided with the BCs

$$u(0,y,t)=b_{x0}(y,t),\ u(x_f,y,t)=b_{xf}(y,t),\\ u(x,0,t)=b_{y0}(x,t),\ \text{and}\ u(x,y_f,t)=b_{yf}(x,t) \quad (7.3.21\text{a,b,c,d})$$

as well as the ICs

$$u(x,y,0)=i_0(x,y)\ \text{and}\ \left.\frac{\partial u}{\partial t}\right|_{t=0}(x,y,0)=i_0'(x,y) \quad (7.3.22)$$

```
function [u,x,y,t]=pde_wave2(a,D,tf,it0,i1t0,bxyt,Mx,My,N)
%To solve a^2*(u_xx+u_yy)=u_tt for D(1)<=x<=D(2),D(3)<=y<=D(4),0<=t<=tf
%  IC: u(x,y,0) = it0(x,y), u_t(x,y,0) = i1t0(x,y)
%  BC: u(x,y,t) = bxyt(x,y,t) for (x,y)cBoundary
% Mx/My = Number of subintervals along x/y-axis
% N = Number of subintervals along the t-axis
dx=(D(2)-D(1))/Mx; x=D(1)+[0:Mx]*dx;
dy=(D(4)-D(3))/My; y=D(3)+[0:My]'*dy;
dt=tf/N;  t=[0:N]*dt;
% Initialization
u=zeros(My+1,Mx+1);  ut=zeros(My+1,Mx+1);
for j=2:Mx
  for i=2:My,  u(i,j)=it0(x(j),y(i)); ut(i,j)=i1t0(x(j),y(i));  end
end
adt2= a^2*dt*dt; rx= adt2/(dx*dx); ry=adt2/(dy*dy);
rxy1= 1-rx-ry; rxy2= rxy1*2;
u_1=u;
for k=0:N
  t=k*dt;
  for i=1:My+1 % BC
     u(i,[1 Mx+1])= [bxyt(x(1),y(i),t) bxyt(x(Mx+1),y(i),t)];
  end
  for j=1:Mx+1
     u([1 My+1],j)= [bxyt(x(j),y(1),t); bxyt(x(j),y(My+1),t)];
  end
  if k==0
    for i=2:My
      for j=2:Mx  % Eq.(7.3.27)
        u(i,j)=0.5*(rx*(u_1(i,j-1)+u_1(i,j+1))...
            +ry*(u_1(i-1,j)+u_1(i+1,j)))+rxy1*u(i,j)+dt*ut(i,j);
      end
    end
   else
    for i=2:My
      for j=2:Mx  % Eq.(7.3.24)
        u(i,j)= rx*(u_1(i,j-1)+u_1(i,j+1))...
           +ry*(u_1(i-1,j)+u_1(i+1,j)) +rxy2*u(i,j) -u_2(i,j);
      end
    end
  end
  u_2=u_1;  u_1=u; % update the buffer memory
  mesh(x,y,u), axis([0 2 0 2 -0.1 0.1]), pause
end
```

In the same way as with the 1D case, we replace the second derivatives on both sides by their 3-point central difference approximation as

$$a^2\left(\frac{u_{i,j+1}^k-2u_{i,j}^k+u_{i,j-1}^k}{\Delta x^2}+\frac{u_{i+1,j}^k-2u_{i,j}^k+u_{i-1,j}^k}{\Delta y^2}\right)=\frac{u_{i,j}^{k+1}-2u_{i,j}^k+u_{i,j}^{k-1}}{\Delta t^2} \tag{7.3.23}$$

$$\text{with } \Delta x=\frac{x_f}{M_x},\ \Delta y=\frac{y_f}{N_y},\text{ and } \Delta t=\frac{T}{N}$$

which leads to the explicit central difference method:

$$u_{i,j}^{k+1}=r_x(u_{i,j+1}^k+u_{i,j-1}^k)+2(1-r_x-r_y)u_{i,j}^k+r_y(u_{i+1,j}^k+u_{i-1,j}^k)-u_{i,j}^{k-1} \tag{7.3.24}$$

$$\text{with } r_x=a^2\frac{\Delta t^2}{\Delta x^2} \text{ and } r_y=a^2\frac{\Delta t^2}{\Delta y^2}$$

Since $u_{i,j}^{-1}=u(x_j,y_i,-\Delta t)$ is not given, we cannot get $u_{i,j}^1$ directly from this formula (7.3.24) with k=0:

$$u_{i,j}^1=r_x(u_{i,j+1}^0+u_{i,j-1}^0)+2(1-r_x-r_y)u_{i,j}^0+r_y(u_{i+1,j}^0+u_{i-1,j}^0)-u_{i,j}^{-1} \tag{7.3.25}$$

Therefore, we approximate the IC on the derivative by the central difference as

$$\frac{u_{i,j}^1-u_{i,j}^{-1}}{2\Delta t}=i_0'(x_j,y_i) \tag{7.3.26}$$

and make use of this to remove $u_{i,j}^{-1}$ from Eq. (7.3.25) to have

$$u_{i,j}^1=\frac{1}{2}\left\{r_x(u_{i,j+1}^0+u_{i,j-1}^0)+r_y(u_{i+1,j}^0+u_{i-1,j}^0)\right\}+(1-r_x-r_y)u_{i,j}^0+i_0'(x_j,y_i)\Delta t \tag{7.3.27}$$

We use this equation together with the ICs to get $u_{i,j}^1$ and then go on using Eq. (7.3.24) for k=1,2,…. A sufficient condition for stability [S-1, Sec. 9.6] is

$$r=\frac{4a^2\Delta t^2}{\Delta x^2+\Delta y^2}\le 1 \tag{7.3.28}$$

This process to solve the 2D wave equation (7.3.20) has been cast into the above MATLAB function '`pde_wave2()`'.

(**Example 7.5**) A Hyperbolic PDE — 2D Wave (Vibration) over a Square Membrane

Consider a 2D hyperbolic PDE

$$\frac{1}{4}\left(\frac{\partial^2 u(x,y,t)}{\partial x^2}+\frac{\partial^2 u(x,y,t)}{\partial y^2}\right)=\frac{\partial^2 u(x,y,t)}{\partial t^2} \text{ for } 0\le x\le 2,\ 0\le y\le 2, \text{ and } 0\le t\le 2 \tag{E7.5.1}$$

subject to the zero BCs and the ICs:

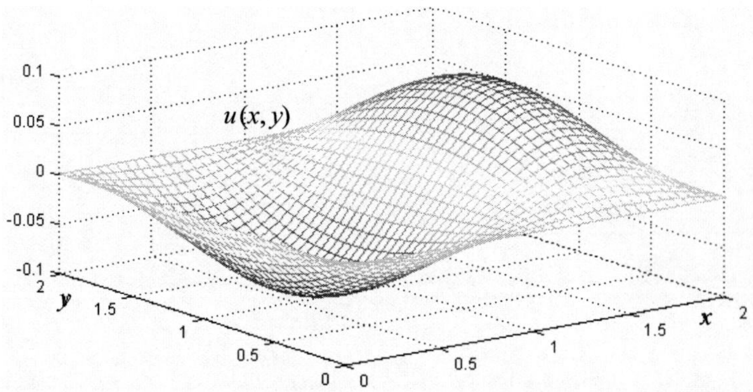

Figure 7.6 Solution to the 2D hyperbolic PDE — vibration of a square membrane (Example 7.5)

$$u(0, y, t) = 0,\ u(2, y, t) = 0,\ u(x, 0, t) = 0,\ u(x, 2, t) = 0, \tag{E7.5.2a,b,c,d}$$

$$u(x,y,t)\big|_{t=0} = 0.1\sin(\pi x)\sin(\pi y/2),\quad \frac{\partial u}{\partial t}(x,y,t)\bigg|_{t=0} = 0 \ \text{ for } t=0 \tag{E7.5.3a,b}$$

We have composed the following MATLAB script,"em07e05.m," to use the function 'pde_wave2()' for solving the PDE, and we run it to get Fig. 7.6 and see a motion picture showing how $u(x, t)$ changes with time $t = n\Delta t$. Note that we can be sure of stability since we have

$$r \overset{(7.3.28)}{=} \frac{4a^2\Delta t^2}{\Delta x^2 + \Delta y^2} = \frac{4(1/4)(2/20)^2}{(2/20)^2 + (2/20)^2} = \frac{1}{2} \le 1 \tag{E7.5.4}$$

```
%em07e05.m
clear, clf
bxyt=@(x,y,t)0; % BC (E7.5.2)
it0=@(x,y)0.1*sin(pi*x)*sin(pi*y/2); % IC (E7.5.3a)
i1t0=@(x,y)0; % IC (E7.5.3b)
a=1/2;
D=[0 2 0 2]; tf=2; Mx=40; My=40; N=40;
[u,x,y,t]=pde_wave2(a,D,tf,it0,i1t0,bxyt,Mx,My,N);
```

7.4 PDES IN OTHER COORDINATE SYSTEMS

7.4.1 PDE in Polar/Cylindrical Coordinates

Let us consider a simple elliptic PDE — Laplace's equation in cylindrical coordinates:

$$\frac{1}{r}\frac{\partial}{\partial r}\left(r\frac{\partial u(r,\phi,z)}{\partial r}\right) + \frac{1}{r^2}\frac{\partial^2 u(r,\phi,z)}{\partial \phi^2} + \frac{\partial^2 u(r,\phi,z)}{\partial z^2} = \frac{1}{r}\frac{\partial u}{\partial r} + \frac{\partial^2 u}{\partial^2 r} + \frac{1}{r^2}\frac{\partial^2 u}{\partial \phi^2} + \frac{\partial^2 u}{\partial z^2} = 0 \tag{7.4.1}$$

Here, referring to the Laplacian operator (2.4.6b) in the cylindrical coordinate system, the radian coordinate is denoted by r instead of ρ.

For further simplification, we deal with a *Laplace's equation with cylindrical symmetry* (i.e., no variation along the z-direction so that $\partial u/\partial z = 0$) for the circular domain $\{(r,\phi),\ 0 \le r \le R = 1,\ 0 \le \phi \le 2\pi\}$ of radius $R=1$:

$$\frac{1}{r}\frac{\partial}{\partial r}\left(r\frac{\partial u(r,\phi)}{\partial r}\right)+\frac{1}{r^2}\frac{\partial^2 u(r,\phi)}{\partial \phi^2}=\frac{1}{r}\frac{\partial u(r,\phi)}{\partial r}+\frac{\partial^2 u(r,\phi)}{\partial^2 r}+\frac{1}{r^2}\frac{\partial^2 u(r,\phi)}{\partial \phi^2}=0 \tag{7.4.2}$$

subject to the BC:

$$u(R,\phi)=b_R(\phi)=\sin 3\phi \ \text{ for } R=1 \text{ and } 0\le\phi\le 2\pi \tag{7.4.3}$$

By the method of separation of variables, we substitute

$$u(r,\phi)=\Upsilon(r)\Phi(\phi) \tag{7.4.4}$$

into the PDE (7.4.2) and multiply both sides by $r^2/\Upsilon(r)\Phi(\phi)$ to write

$$\frac{r}{\Upsilon(r)}\frac{\partial}{\partial r}\left(r\frac{\partial \Upsilon(r)}{\partial r}\right)+\frac{1}{\Phi(\phi)}\frac{\partial^2\Phi(\phi)}{\partial\phi^2}=0\ ;\quad \frac{r}{\Upsilon(r)}\frac{\partial}{\partial r}\left(r\frac{\partial \Upsilon(r)}{\partial r}\right)=-\frac{1}{\Phi(\phi)}\frac{\partial^2\Phi(\phi)}{\partial\phi^2}=-\lambda$$

$$;\quad r\frac{\partial}{\partial r}\left(r\frac{\partial \Upsilon(r)}{\partial r}\right)+\lambda\Upsilon(r)=0 \quad (7.4.5\text{a}) \qquad \frac{\partial^2\Phi(\phi)}{\partial\phi^2}-\lambda\Phi(\phi)=0 \quad (7.4.5\text{b})$$

Noting that $u(r,\phi)$ is periodic with period $2\pi/k$ (k: an integer) in ϕ and so is $\Phi(\phi)$, the candidates for $\Phi(\phi)$ are chosen from the third column of Table 7.1 with the separation constant $\lambda=-\omega^2=-k^2$ as

$$\Phi_k(\phi)=C(k)\cos k\phi+D(k)\sin k\phi : \textit{Azimuthal solution.} \tag{7.4.6}$$

Eq. (7.4.5a) with $\lambda=-\omega^2=-k^2$ can be arranged into the form of Cauchy–Euler equation (3.2.23) as

$$r^2\Upsilon''(r)+r\Upsilon'(r)-k^2\Upsilon(r)=0 \tag{7.4.7}$$

Since the auxiliary (indicial) equation of this Cauchy–Euler equation and its roots are

$$m(m-1)+m-k^2=0;\ m^2-k^2=0;\ m=\pm k \tag{7.4.8}$$

the solution of Eq. (7.4.7) can be found from Eq. (3.2.26) as

$$\Upsilon_k(r)=A(n)r^k+B(n)r^{-k} : \textit{Radial solution} \tag{7.4.9}$$

Plugging Eqs. (7.4.6) and (7.4.9) into Eq. (7.4.4) yields the separable solutions:

$$u_k(r,\phi)\overset{(7.4.4)}{=}\Upsilon_k(r)\Phi_k(\phi)\underset{(7.4.6)}{\overset{(7.4.9)}{=}}\left(A(k)r^k+B(k)r^{-k}\right)\left(C(k)\cos k\phi+D(n)\sin k\phi\right)$$

$$=r^k\left(A_k\cos k\phi+B_k\sin k\phi\right)+r^{-k}\left(C_k\cos k\phi+D_k\sin k\phi\right) \tag{7.4.10}$$

Hence, the general solution of the Laplace's equation can be written as

$$u(r,\phi)=A_0+\sum_{k=1}^{\infty}r^k\left(A_k\cos k\phi+B_k\sin k\phi\right)+r^{-k}\left(\cancel{C}_k\cos k\phi+\cancel{D}_k\sin k\phi\right) \tag{7.4.11}$$

where in most cases, the coefficient C_k's and D_k's are set to zero in order to keep the solution bounded at the center (with $r=0$) of the circular domain.

```
function [u,rs,phs,A0,A,B] = pde_Laplace_circle_ana(R,bR,Mr,Mph,Nterm)
%solve u_rr +1/r*u_r + (1/r^2)*u_phph = 0 for 0<=r<=R, 0<=ph<=2pi
% BC: u(R,ph)=bR(ph)
% Mr  = # of subintervals along the r(radial)-axis
% Mph = # of subintervals along the ph(circular)-axis
% rs, phs : Radial/Circular grids
if nargin<5, Nterm=20; end % Number of terms to add in the partial sum
syms ph   % Declare ph as a symbolic variable
dr = R/Mr; rs = [0:Mr]*dr;  % Radial grid points
dph = 2*pi/Mph;  phs = [0:Mph]*dph; % Circular grid points
[Ths,Rs]=meshgrid(phs,rs); % Grid
A0 = eval(int(bR,ph,0,2*pi))/2/pi; u=A0;    % Eq.(7.4.13a)
for k=1:Nterm
   A(k) = eval(int(bR*cos(k*ph),ph,0,2*pi))/pi/R^k; % Eq.(7.4.13b)
   B(k) = eval(int(bR*sin(k*ph),ph,0,2*pi))/pi/R^k; % Eq.(7.4.13c)
   u = u + Rs.^k.*(A(k)*cos(k*Ths)+B(k)*sin(k*Ths)); % Eq.(7.4.11)
end
```

Now, to find the coefficients of the solution, we substitute Eq. (7.4.11) into the BC (7.4.3):

$$u_k(R,\phi) \overset{(7.4.11)}{=} A_0+\sum_{k=1}^{\infty} R^k\left(A_k\cos k\phi + B_k\sin k\phi\right)+R^{-k}\left(\cancel{\cdot}\right) \overset{(7.4.3)}{=} \sin 3\phi \tag{7.4.12}$$

and, regarding this as a Fourier series representation (6.1.3a) of the BC function $\sin 3\phi$, we use Eq. (6.1.3b) with $P=2\pi$ (due to the assumed periodicity of $u(r,\phi)$ in ϕ) to compute the coefficients as

$$\begin{cases} A_0=\dfrac{1}{2\pi}\displaystyle\int_0^{2\pi} b_R(\phi)\,d\phi \\ A_k=\dfrac{1}{R^k\pi}\displaystyle\int_0^{2\pi} b_R(\phi)\cos k\phi\, d\phi \quad \text{for } k\ge 1. \\ B_k=\dfrac{1}{R^k\pi}\displaystyle\int_0^{2\pi} b_R(\phi)\sin k\phi\, d\phi \quad \text{for } k\ge 1 \end{cases} \tag{7.4.13a,b,c}$$

This analytical solution process has been cast into the above MATLAB function, '`pde_Laplace_circle_ana()`'.

On the other hand, we can use the central difference approximations of the derivatives (like in Eq. (7.2.30)) to discretize the PDE (7.4.2) and the BC (7.4.3) as

$$\frac{1}{r}\frac{\partial u(r,\phi)}{\partial r}+\frac{\partial^2 u(r,\phi)}{\partial^2 r}+\frac{1}{r^2}\frac{\partial^2 u(r,\phi)}{\partial \phi^2} \overset{(7.4.2)}{=} 0 \;\rightarrow\; \frac{1}{r}\frac{u(m+1,n)-u(m-1,n)}{2\Delta r}$$
$$+\frac{u(m+1,n)-2u(m,n)+u(m-1,n)}{(\Delta r)^2}+\frac{1}{r^2}\frac{u(m,n+1)-2u(m,n)+u(m,n-1)}{(\Delta\phi)^2}=0$$

$$;\; u(m,n)=\frac{1}{a}\left\{\begin{array}{l}\dfrac{1}{r(m)}\dfrac{u(m+1,n)-u(m-1,n)}{2\Delta r}+\dfrac{u(m+1,n)+u(m-1,n)}{(\Delta r)^2} \\ \qquad +\dfrac{1}{r(m)^2}\dfrac{u(m,n+1)+u(m,n-1)}{(\Delta\phi)^2}\end{array}\right\} \text{ with } a=\frac{2}{(\Delta r)^2}+\frac{2}{r^2(\Delta\phi)^2} \tag{7.4.14a}$$

$$\text{s.t. the BC:}\quad u(M_r+1,n)=b_R(\phi(n)) \overset{(7.4.3)}{=} \sin(3\phi(n)) \text{ for } 1\le n\le M_\phi+1 \tag{7.4.14b}$$

```
function [u,rs,ths,iter] = pde_Laplace_circle(R,bR,Mr,Mph,tol)
%solve u_rr +1/r*u_r + (1/r^2)*u_phph = 0 for 0<=r<=R, 0<=ph<=2pi
% BC: u(R,ph) = bR(ph)
% Mr = # of subintervals along the r(radial)-axis
% Nph = # of subintervals along the ph(circular)-axis
% rs, phs : Radial/Circular grids
dr = R/Mr; rs = [0:Mr]*dr;  dr2 = dr^2;
dph = 2*pi/Mph; phs = [0:Mph]*dph;  dph2 = dph^2;
if nargin<5, tol = 1e-4; end % Error tolerance
u = zeros(Mr+1,Mph+1);
if isnumeric(bR),  u(Mr+1,:) = bR;
 else u(Mr+1,:)=feval(bR,phs); % BC on the circle
end
for iter=1:500
   u0 = u;
   for m=2:Mr
      rm = rs(m); rm2 = rm^2;
      a = 2/dr2 + 2/rm^2/dph2; % Eq.(7.4.14)
      for n=1:Mph
         nm1=(n-1)*(n>1)+Mph*(n<=1); np1=(n+1)*(n<Mph)+1*(n>=Mph);
         u(m,n)=((u(m-1,n)+u(m+1,n))/dr2+(u(m+1,n)-u(m-1,n))/2/rm/dr
                   + (u(m,nm1)+u(m,np1))/rm2/dph2)/a; % Eq.(7.4.14)
      end
   end
   if norm(u-u0)/norm(u0)<tol, break, end
end
```

where the circular domain is divided into $M_r \times M_\phi$ segments as illustrated in Fig. 7.7(a). This numerical solution process has been cast into the above MATLAB function, '`pde_Laplace_circle()`'.

The following MATLAB script "em07_04_1a.m" can be run to graph the analytical and numerical solutions of the PDE (7.4.2) with the BC (7.4.3) for the circular domain divided into 10×72 segments (Fig. 7.7(a)) as shown in Fig. 7.7(b). It also gives the coefficients computed by Eq. (7.4.13a,b,c) as

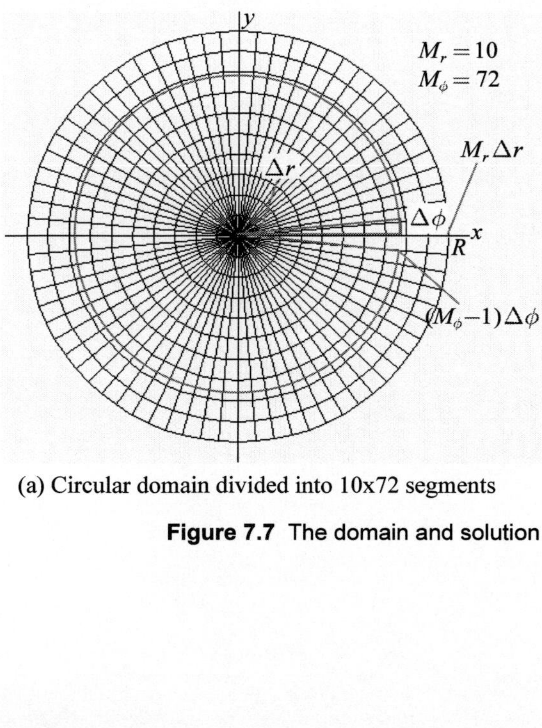

(a) Circular domain divided into 10x72 segments

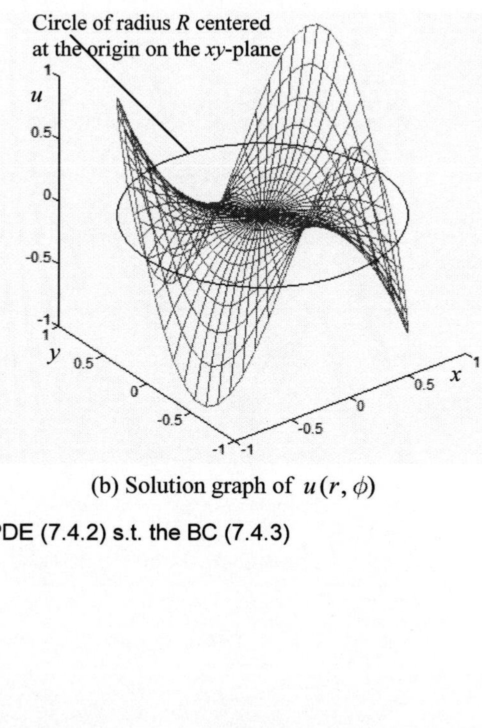

(b) Solution graph of $u(r, \phi)$

Figure 7.7 The domain and solution of the PDE (7.4.2) s.t. the BC (7.4.3)

```
%em07_04_1a.m
% solves the PDE (7.4.2) s.t. the BC (7.4.3) on a circular domain
clear, clf
R = 1; % Radius of the circular region
bR=@(ph)sin(3*ph); % BC b(R,ph) (7.4.3) on the circle
Mr=10; Mph=72; % Numbers of segments along the radial/circular axes
K=10; % The number of terms to be added in Eq.(7.4.11)
[ua,ras,phas,A0,A,B]=pde_Laplace_circle_ana(R,bR,Mr,Mph,K); % Analytical sol
A0,A,B % gives the coefficients (7.4.13a,b,c)
[un,rs,phs,Niter]=pde_Laplace_circle(R,bR,Mr,Mph); % Numerical sol
Discrepancy = norm(un-ua)/norm(ua)
subplot(221), mesh_polar(ras,phas,ua), axis('equal')
Ph=-37.5; Th=30; view([Ph Th])
title('Analytical solution u')
subplot(222), mesh_polar(rs,phs,un), axis('equal')
title('Numerical solution u')
```

```
function mesh_polar(rs,phs,u)
% Polar mesh of u on (rs x ph)
[M,N]=size(u); lr=length(rs); lt=length(phs);
if lr~=M|lt~=N, error('Incompatible dimensions in mesh_polar!'); end
for m=1:lr, p=rs(m)*exp(j*phs); plot3(real(p),imag(p),u(m,:)); hold on; end
for m=1:lt, p=rs*exp(j*phs(m)); plot3(real(p),imag(p),u(:,m));   end
```

```
>>em07_04_01a
  A0 =    0
  A  =    0    0    0    0    0    0    0    0    0    0
  B  =    0    0    1    0    0    0    0    0    0    0
```

This means the analytical solution (7.4.11) is

$$u(r,\phi)\overset{(7.4.11)}{=}A_0+\sum_{n=1}^{\infty}r^n\left(A_n\cos n\phi+B_n\sin n\phi\right)=r^3\sin 3\phi \tag{7.4.15}$$

Now, we deal with a *Laplace's equation with azimuthal or radial symmetry* (i.e., rotational invariance or independence on the azimuth angle ϕ so that $\partial u/\partial\phi=0$)

$$\frac{1}{r}\frac{\partial}{\partial r}\left(r\frac{\partial u(r,z)}{\partial r}\right)+\frac{\partial^2 u(r,z)}{\partial z^2}=\frac{1}{r}\frac{\partial u(r,z)}{\partial r}+\frac{\partial^2 u(r,z)}{\partial^2 r}+\frac{\partial^2 u(r,z)}{\partial z^2}=0 \tag{7.4.16}$$

for the cylindrical domain $\{(r,z), 0\le r\le R, 0\le z\le z_f\}$ of radius R subject to the BC:

$$u(R,z)=0 \text{ for } 0\le z\le z_f,\ u(r,0)=b_0(r), \text{ and } u(r,z_f)=b_f(r) \text{ for } 0\le r\le R \tag{7.4.17a,b,c}$$

By the method of separation of variables, we substitute

$$u(r,z)=\Upsilon(r)Z(z) \tag{7.4.18}$$

into the PDE (7.4.16) and multiply the both sides by $1/\Upsilon(r)Z(z)$ to write

$$\frac{1}{\Upsilon(r)}\frac{1}{r}\frac{\partial}{\partial r}\left(r\frac{\partial \Upsilon(r)}{\partial r}\right)+\frac{1}{Z(z)}\frac{\partial^2 Z(z)}{\partial z^2}=0;\quad \frac{1}{\Upsilon(r)}\frac{1}{r}\frac{\partial}{\partial r}\left(r\frac{\partial \Upsilon(r)}{\partial r}\right)=-\frac{1}{Z(z)}\frac{\partial^2 Z(z)}{\partial z^2}=-\lambda$$

$$; \quad \frac{\partial}{\partial r}\left(r\frac{\partial \Upsilon(r)}{\partial r}\right)+\lambda r\Upsilon(r)=0 \quad (7.4.19a) \qquad \frac{\partial^2 Z(z)}{\partial z^2}-\lambda Z(z)=0 \quad (7.4.19b)$$

Assuming that $u(r,z)$ is not periodic in z and neither is $Z(z)$, the candidate for $Z(z)$ is chosen from the third column of Table 7.1 with the separation constant $\lambda=\omega^2$ as

$$Z(z)=C(\omega)\cosh\omega z+D(\omega)\sinh\omega z : \textit{Axial solution.} \tag{7.4.20}$$

Eq. (7.4.19a) with $\lambda=\omega^2$ can be multiplied by r and arranged into the form of parametric Bessel equation (3.3.21) (with $\lambda=\omega$ and $v=0$) of order 0 as

$$r^2\Upsilon''(r)+r\Upsilon'(r)+\omega^2r^2\Upsilon(r)=0 \tag{7.4.21}$$

whose general solution can be written from Eq. (3.3.22) as

$$\Upsilon(r) \underset{\lambda=\omega,\ v=0}{\overset{(3.3.22)}{=}} A(\omega)J_0(\omega r)+B(\omega)Y_0(\omega r) : \textit{Radial solution} \tag{7.4.22}$$

where the coefficient $B(\omega)$ of the second term containing $Y_0(x)$ (such that $Y_0(0)=-\infty$ as depicted in Fig. 3.15(b)) must be zero to keep $\Upsilon(r)$ from being unbounded at the center ($r=0$).

Plugging Eqs. (7.4.22) and (7.4.20) into Eq. (7.4.18) yields the separable solutions:

$$u(r,z)\overset{(7.4.18)}{=}\Upsilon(r)Z(z)\underset{(7.4.20)}{\overset{(7.4.22)}{=}}(C(\omega)\cosh\omega z+D(\omega)\sinh\omega z)J_0(\omega r) \tag{7.4.23}$$

where the coefficient $A(\omega)$ has been absorbed into $C(\omega)$ and $D(\omega)$. We can substitute this general solution into the BC (7.4.17a) to write

$$u(R,z)\overset{(7.4.23)}{=}(C(\omega)\cosh\omega z+D(\omega)\sinh\omega z)J_0(R\omega)\overset{(7.4.17a)}{=}0 \text{ for } 0\le z\le 2$$

$$\rightarrow \qquad J_0(R\omega)=0 \tag{7.4.24}$$

This implies that the values of the eigenvalue ω can be determined by dividing the (infinitely many) zeros of the Bessel function of order 0 by R:

$$\omega_k=\frac{x_k}{R} \text{ where } x_k\text{'s are the roots of } J_0(x)=0 \tag{7.4.25}$$

Once the values of the eigenvalue ω are determined, we write the general solution as

$$u(r,z)=\sum_{k=1}^{\infty}(C_k\cosh\omega_k z+D_k\sinh\omega_k z)J_0(r\omega_k) \tag{7.4.26}$$

Substituting Eq. (7.4.26) into the BCs (7.4.17b) and (7.4.17c) yields

$$u(r,0)\overset{(7.4.23)}{=}\sum_{k=1}^{\infty}(C_k\cosh\omega_k 0+D_k\sinh\omega_k 0)J_0(r\omega_k)\overset{(7.4.17b)}{=}b_0(r) \text{ for } 0\le r\le R \tag{7.4.27a}$$

$$u(r,z_f)\overset{(7.4.23)}{=}\sum_{k=1}^{\infty}(C_k\cosh\omega_k z_f+D_k\sinh\omega_k z_f)J_0(r\omega_k)\overset{(7.4.17c)}{=}b_f(r) \text{ for } 0\le r\le R \tag{7.4.27b}$$

Regarding each of these equations as the Fourier–Bessel series expansions (Eq. (6.6.3)) of $b_0(r)$ and $b_f(r)$, respectively, subject to the BC (7.4.24) (corresponding to Case I), we use Eq. (6.6.4a) with $n=0$ to determine their Fourier–Bessel coefficients, and consequently, C_k and D_k as

```
function [u,rs,zs,C,D,w]=pde_Laplace_cylinder_ana(R,zf,b0,bf,Mr,Mz,K)
%solve u_rr +1/r*u_r + u_zz = 0 for 0<=r<=R, 0<=z<=zf
% BCs: u(R,z) = 0, u(r,z0) = b0(r), u(r,zf) = bf(r)
% Mr = # of subintervals along the r(radial)-axis
% Mz = # of subintervals along the z-axis
% K  = Number of coefficients/terms to be found/added in partial sum
% rs, zs : Radial/Vertical grids
%Copyleft: Won Y. Yang, wyyang53@hanmail.net, CAU for academic use only
if nargin<7, K=20; end;
syms r
dr = R/Mr; rs = [0:Mr]*dr;  % Radial grid points
dz = zf/Mz;  zs = [0:Mz]'*dz; % Grid points along the z-axis
J0x=@(x)besselj(0,x); % Bessel function of the first kind of order 0
% To find the initial guesses (x0s) of zeros of J_0(x)=0
xs=[0:0.1:100];
fxs = 1./abs(J0x(xs)).^(1/10);
[localmaxima,indms] = localmax(fxs,[],false); x0s = xs(indms);
u = 0; options=optimset('Display','off');
C=zeros(1,K); D=zeros(1,K);
for k=1:K
   x(k) = fsolve(J0x,x0s(k),options); wk=x(k)/R; w(k)=wk; % Eq.(7.4.25)
   rJ0b0 = @(r)r.*J0x(wk*r).*feval(b0,r);
   rJ0bf = @(r)r.*J0x(wk*r).*feval(bf,r);
   Rb1x2 = R^2*besselj(1,x(k))^2; wkzf = wk*zf;
   C(k)=2/Rb1x2*quad(rJ0b0,0,R); % Eq.(7.4.28a)
   D(k)=(2/Rb1x2*quad(rJ0bf,0,R)-C(k)*cosh(wkzf))/sinh(wkzf);%(7.4.28b)
   u = u + (C(k)*cosh(wk*zs)+D(k)*sinh(wk*zs))*besselj(0,wk*rs);
                                                          % Eq.(7.4.26)
end
if nargout<1
   [Rs,Zs]=meshgrid(rs,zs); pcolor(Rs,Zs,u), hold on,
   [Rs,Zs]=meshgrid(-rs,zs); pcolor(Rs,Zs,fliplr(u)), colorbar
  axis('equal')
end
```

$$C_k \underset{(7.4.27a)}{\overset{(6.6.4a)}{=}} \frac{2}{R^2 J_1^2(R\omega_k)} \int_0^R r J_0(\omega_k r) b_0(r) dr \tag{7.4.28a}$$

$$D_k \underset{(7.4.27b)}{\overset{(6.6.4a)}{=}} \frac{1}{\sinh(\omega_k z_f)} \left\{ \frac{2}{R^2 J_1^2(R\omega_k)} \int_0^R r J_0(\omega_k r) b_f(r) dr - C_k \cosh(\omega_k z_f) \right\} \tag{7.4.28b}$$

This analytical solution process has been cast into the above MATLAB function '`pde_Laplace_cylinder_ana()`'.

On the other hand, we can use the central difference approximations of the derivatives (like in Eq. (7.2.30)) to discretize the PDE (7.4.16) and the BCs (7.4.17a,b,c) as

$$\frac{1}{r}\frac{\partial u(r,z)}{\partial r} + \frac{\partial^2 u(r,z)}{\partial^2 r} + \frac{\partial^2 u(r,z)}{\partial z^2} \overset{(7.4.16)}{=} 0 \rightarrow \frac{1}{r}\frac{u(m+1,n)-u(m-1,n)}{2\Delta r}$$
$$+\frac{u(m+1,n)-2u(m,n)+u(m-1,n)}{(\Delta r)^2} + \frac{u(m,n+1)-2u(m,n)+u(m,n-1)}{(\Delta z)^2} = 0$$

$$; \; u(m,n) = \frac{1}{a}\left\{ \begin{matrix} \dfrac{1}{r(m)} \dfrac{u(m+1,n)-u(m-1,n)}{2\Delta r} \\ + \dfrac{u(m+1,n)+u(m-1,n)}{(\Delta r)^2} + \dfrac{u(m,n+1)+u(m,n-1)}{(\Delta z)^2} \end{matrix} \right\} \text{ with } a = \frac{2}{(\Delta r)^2} + \frac{2}{(\Delta z)^2} \quad (7.4.29)$$

s.t. the BC:

$$u(M_r+1,n) \overset{(7.4.17a)}{=} 0 \text{ for } 1 \le n \le M_z+1 \quad (7.4.30a)$$

$$u(m,1) \overset{(7.4.17b)}{=} b_0(r(m)) \text{ and } u(m,M_z+1) \overset{(7.4.17c)}{=} b_f(r(m)) \text{ for } 1 \le m \le M_r+1 \quad (7.4.30b,c)$$

where a half of the cross-section of the cylindrical domain is divided into $M_r \times M_z$ segments. This numerical solution process has been cast into the MATLAB function 'pde_Laplace_cylinder()'.

```
function [u,rs,zs,iter]=pde_Laplace_cylinder(R,zf,bR,b0,bf,Mr,Mz,tol)
%solve u_rr +1/r*u_r + u_zz = 0 for 0<=r<=R, 0<=z<=zf
% BCs: u(R,z) = bR, u(r,0) = b0(r), u(r,zf) = bf(r)
% Mr = # of subintervals along the r(radial)-axis
% Mz = # of subintervals along the z-axis
% tol = Error tolerance
% rs, zs : Radial/Vertical grids
% iter   : Number of iterations performed
%Copyleft: Won Y. Yang, wyyang53@hanmail.net, CAU for academic use only
if nargin<8, tol = 1e-4; end  % Error tolerance
dr = R/Mr;  rs = [0:Mr]*dr;   % Radial grid points
dz = zf/Mz;  zs = [0:Mz]*dz;  % Grid points along the z-axis
[Rs,Zs]=meshgrid(rs,zs); % Grid
u = zeros(Mz+1,Mr+1);  dr2=dr^2; dz2=dz^2;
if isnumeric(bR),  u(:,Mr+1)=bR;
 else u(:,Mr+1)=feval(bR,zs); % BC (7.4.30a) on the cylindrical surface
end
if isnumeric(b0),  u(1,:)=b0;
 else u(1,:)=feval(b0,rs); % BC (7.4.30b) on the lower disk
end
if isnumeric(bf),  u(Mz+1,:)=bf;
 else u(Mz+1,:)=feval(bf,rs); % BC (7.4.30c) on the upper disk
end
a = 2/dr2 + 2/dz2;
for iter=1:2000
   u0 = u;
   for m=2:Mr
      rm = rs(m);
      for n=Mz:-1:2
         u(n,m)=((u(n,m+1)+u(n,m-1))/dr2
                  +(u(n,m+1)-u(n,m-1))/2/rm/dr
                  + (u(n+1,m)+u(n-1,m))/dz2)/a; % Eq.(7.4.29)
      end
   end
   u(:,1)=u(:,2); % Set u at the center to their neighbors.
   if norm(u-u0)/norm(u0)<tol, break, end
end
```

The following MATLAB script, "em07_04_1b.m," can be run to graph the analytical and numerical solutions of the PDE (7.4.16) s.t. the BCs:

$$u(R,z)=0 \ \text{for} \ 0 \le z \le z_f=2 \tag{7.4.31a}$$

$$u(r,0)=b_0(r)=0, \text{ and } u(r,z_f)=b_f(r)=10 \ \text{ for } \ 0 \le r \le R=3 \tag{7.4.31b,c}$$

for the circular domain divided into $M_r \times M_z = 30 \times 20$ segments as shown in Fig. 7.8. It also gives the coefficients of the analytical solution (7.4.26) that are computed by Eqs. (7.4.28a,b) through the MATLAB function 'pde_Laplace_cylinder_ana()' as

```
>>em07_04_1b; C, D, w
   C =      0        0        0        0        0        0        0
   D = 6.7201  -0.5375   0.0532  -0.0056   0.0006  -0.0001   0.0000
   w = 0.8016   1.8400   2.8846   3.9305   4.9770   6.0237   7.0705
```

This means the analytical solution (7.4.26) is

$$\begin{aligned} u(r,z) &= \sum_{k=1}^{\infty}\left(C_k \cosh\omega_k z + D_k \sinh\omega_k z\right)J_0(r\omega_k) = 6.7201\sinh(0.8016z)J_0(0.8016r) \\ &\quad -0.5375\sinh(1.84z)J_0(1.84r) + 0.0532\sinh(2.8846z)J_0(2.8846r) - 0.0056\cdots \end{aligned} \tag{7.4.32}$$

```
%em07_04_1b.m
% solves the PDE (7.4.16) s.t. the BC (7.4.17) on a cylindrical domain
R=3; zf=2; % Radius and Height of the cylindrical domain
bR=@(z)0; % BC b(R,z) (7.4.31a) on  the cylindrical surface
b0=@(r)0;   bf=@(r)10; % BCs b(r,0) and b(r,zf) (7.4.31b,c)
Mr=30; Mz=20; % Numbers of segments along the radial and z-axes
K = 7; % Number of coefficients/terms to be found/added
[ua,ras,zas,C,D,w]=pde_Laplace_cylinder_ana(R,zf,b0,bf,Mr,Mz,K);
tol = 1e-4; % Error tolerance
[un,rs,zs,Niter] = pde_Laplace_cylinder(R,zf,bR,b0,bf,Mr,Mz,tol);
% Discrepancy between the analytical and numerical solutions
Discrepancy = norm(un-ua)/norm(ua)
% Plot the solution graph
[Rs,Zs]=meshgrid(rs,zs); pcolor(Rs,Zs,un), hold on,
[Rs,Zs]=meshgrid(-rs,zs); pcolor(Rs,Zs,un), colorbar
axis('equal'), axis([-R-0.2,R+0.2 -0.5 zf+0.5]), axis off
```

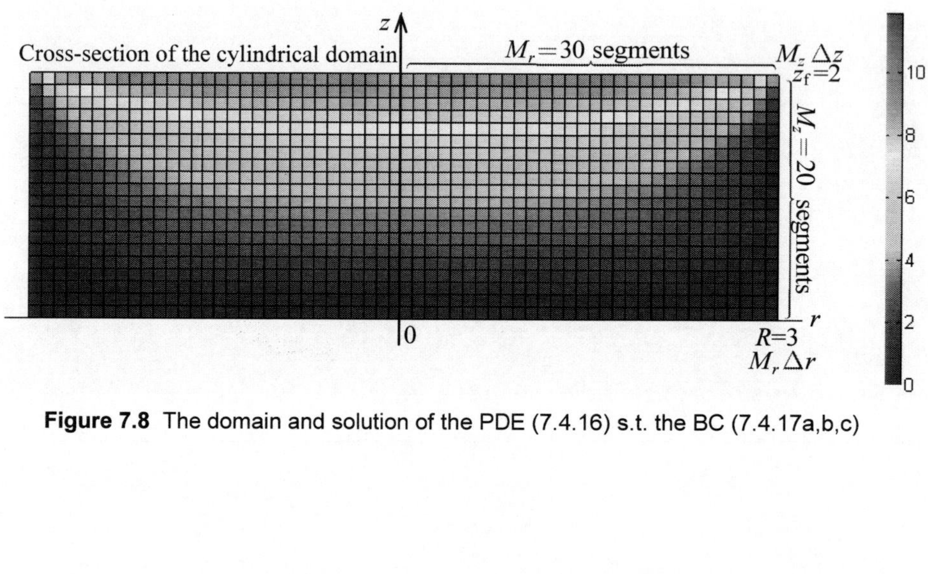

Figure 7.8 The domain and solution of the PDE (7.4.16) s.t. the BC (7.4.17a,b,c)

Now, let us consider the following hyperbolic PDE — a 1D wave equation with rotational invariance (i.e., independence of the azimuth angle ϕ or azimuthal symmetry) in the cylindrical coordinate system:

$$a^2\left(\frac{\partial^2 u(r,t)}{\partial^2 r}+\frac{1}{r}\frac{\partial u(r,t)}{\partial r}\right)=\frac{\partial^2 u(r,t)}{\partial t^2} \tag{7.4.33}$$

with the BC and ICs on a circular domain of radius R as

$$u(R,t)=0\ \forall\, t\geq 0,\quad u(r,0)=i_0(r),\ \text{and}\ \left.\frac{\partial u}{\partial t}\right|_{t=0}=i_{10}(r)\ \text{for}\ 0<r<R \quad (7.4.34\text{a,b,c})$$

Noting that this PDE can model the behavior of a vibrating membrane or drum, we will show that the general solution of this PDE is

$$u(r,t)=\sum_{k=1}^{\infty}u_k(t)=\sum_{k=1}^{\infty}\{A_k\cos(a\omega_k t)+B_k\sin(a\omega_k t)\}J_0(\omega_k r) \tag{7.4.35}$$

where $J_0(\cdot)$ is the Bessel function of the first kind, or cylinder function, or cylindrical harmonics of order $v=0$ (see Section 3.3.1), and $R\omega_k$'s are the (infinitely many) positive roots of $J_0(\cdot)=0$. Each term $u_k(t)$ of the solution (P7.8.3) is called a *standing wave* with radial vibration frequency $a\omega_k$.

By the method of separation of variables, we substitute

$$u(r,t)=\Upsilon(r)T(t) \tag{7.4.36}$$

into the PDE (7.4.33) and divide by $a^2\,\Upsilon(r)T(t)$ to write

$$\frac{\Upsilon''+\Upsilon'/r}{\Upsilon}=\frac{T''}{a^2T}=-\lambda=-\omega^2\rightarrow\begin{cases}r\Upsilon''+\Upsilon'+\omega^2r\Upsilon=0 & (7.4.37\text{a})\\ T''+a^2\omega^2T=0 & (7.4.37\text{b})\end{cases}$$

where the sign of the separation constant $\lambda=\omega^2$ is made negative so that $T(t)$ can be periodic in t, and accordingly, so is $u(r,t)$. Thus, the solution of Eq. (7.4.37b) can be written as

$$T(t)=C(\omega)\cos(a\omega t)+D(\omega)\sin(a\omega t) \tag{7.4.38}$$

Comparing Eq. (7.4.37a) and Eq. (3.3.21) (with order $v=0$), we see that Eq. (7.4.37a) is nothing but the parametric Bessel equation of order 0 that has the general solution (3.3.22) with $v=0$, $\lambda=\omega$, and $x=r$:

$$\Gamma(r)\underset{v=0,\lambda=\omega,x=r}{\overset{(3.3.22)}{=}}A(\omega)J_0(\omega r)+B(\omega)Y_0(\omega r)=A(\omega)J_0(\omega r) \tag{7.4.39}$$

where the coefficient $B(\omega)$ of the second term containing $Y_0(x)$ (such that $Y_0(0)=-\infty$ as depicted in Fig. 3.15(b)) must be zero to keep $\Upsilon(r)$ from being unbounded at the center ($r=0$). Thus, from the BC (7.4.34a), we get an equation to be satisfied by the parameter ω as

$$u(r,t)\overset{(7.4.36)}{=}\Upsilon(r)T(t)\underset{(7.4.38)}{\overset{(7.4.39)}{=}}A(\omega)J_0(\omega r)\{C(\omega)\cos(a\omega t)+D(\omega)\sin(a\omega t)\}\underset{(7.4.34\text{a})}{\overset{r=R}{=}}0\ \forall\, t\geq 0$$

$$\rightarrow J_0(\omega R)=0 \tag{7.4.40}$$

That is why Eq. (7.4.35) is the solution of the 1D wave equation (7.4.33).

```
function [u,rs,ts,A,B,w]=pde_wave_circle_ana(a,R,tf,i0,i10,bR,Mr,Mt,K)
%solve a^2*(u_rr +1/r*u_r) = u_tt for 0<=r<=R, 0<=t<=tf
%  ICs : u(r,0) = i0(r), u_t(r,0) = i10(r)
%  BCs : u(R,t) = bR  for r=R and all t>=0
% Mr = # of subintervals along the r(radial)-axis
% Mt = Number of subintervals along the t-axis
% K  = Number of coefficients/terms to be found/added
% rs, ts : Radial/Temporal grids
% A, B   : The Fourier-Bessel coefficients
%Copyleft: Won Y. Yang, wyyang53@hanmail.net, CAU for academic use only
if nargin<9, N = 10;  end
dr = R/Mr; rs = [0:Mr]'*dr;  % Radial grid points
dt = tf/Mt;  ts = [0:Mt]*dt; % Time points
J0x=@(x)besselj(0,x)-bR; % Bessel function of the first kind of order 0
xs=[0:0.1:100];   fxs = 1./abs(J0x(xs)).^(1/10);
[localmaxima,indms] = localmax(fxs,[],false); x0s = xs(indms);
u = 0; options=optimset('Display','off');
for k=1:K
   x(k) = fsolve(J0x,x0s(k),options); w(k)=x(k)/R;
   rJ0i0 = @(r)r.*J0x(w(k)*r).*feval(i0,r);
   A(k) = 2/R^2/besselj(1,x(k))^2*quad(rJ0i0,0,R); % Eq.(7.4.42a)
   rJ0i10 = @(r)r.*J0x(w(k)*r).*feval(i10,r);
   B(k) = 2/a/w(k)/R^2/besselj(1,x(k))^2  ...
             *quad(rJ0i1t0,0,R);                   % Eq.(7.4.42b)
   awt = a*w(k)*ts;
   u = u+besselj(0,w(k)*rs)*(A(k)*cos(awt)+B(k)*sin(awt)); %Eq.(7.4.35)
end
```

Substituting Eq. (7.4.35) into the ICs (7.4.34b) and (7.4.34c) yields

$$u(r,0) \overset{(7.4.35)}{=} \sum_{k=1}^{\infty} \{A_k \cos(a\omega_k 0) + B_k \sin(a\omega_k 0)\} J_0(\omega_k r) \overset{(7.4.34b)}{=} i_0(r) \tag{7.4.41a}$$

$$u'(r,0) \overset{(7.4.35)}{=} \sum_{k=1}^{\infty} \{-A_k a\omega_k \sin(a\omega_k 0) + B_k a\omega_k \cos(a\omega_k 0)\} J_0(\omega_k r) \overset{(7.4.34c)}{=} i_{10}(r) \tag{7.4.41b}$$

Regarding each of these equations as the Fourier–Bessel series expansions (Eq. (6.6.3) with $n=0$) of $i(r)$ and $i_1(r)$, respectively, subject to the BC (7.4.40) ($J_0(R\omega)=0$) corresponding to Case I, we use Eq. (6.6.4a) with $n=0$ to determine their Fourier–Bessel coefficients, and consequently, A_k and B_k as

$$\begin{cases} A_k \overset{(6.6.4a)}{=} \dfrac{2}{R^2 J_1^2(R\omega_k)} \displaystyle\int_0^R r J_0(R\omega_k) i_0(r)\,dr & \text{for } k \ge 1 \\ B_k \overset{(6.6.4a)}{=} \dfrac{2}{a\omega_k R^2 J_1^2(R\omega_k)} \displaystyle\int_0^R r J_0(R\omega_k) i_{10}(r)\,dr & \text{for } k \ge 1 \end{cases} \tag{7.4.42a,b}$$

where ω_k's are obtained by dividing the positive root x_k's of $J_0(x)=0$ by R.

This process to get the analytical solution (7.4.35) for $\{r(m)=mR/M_r, t(n)=nt_f/M_t$ for $m=0:M_r$ and $n=0:M_t\}$ has been cast into the above MATLAB function 'pde_wave_circle_ana()'. To use it for finding the solution $u(r,t)$ of the wave equation (7.4.33) with $a=1$ (for the circular domain of radius $R=1$ divided into $M=50$ segments and $0 \le t \le t_f=2$[s] divided into $N=100$ subintervals) subject to the following BC and ICs:

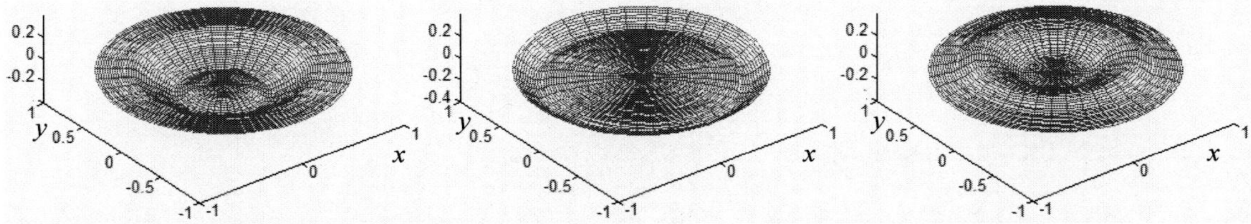

(a) For t=0.5 (b) For t=1 (c) For t=1.5

Figure 7.9 Solution graphs for the wave equation (7.4.33) subject to the BC and IC (7.4.43a,b,c)

$$u(R,t)=0 \text{ for } t\ge 0,\; u(r,0)=0 \text{ for } 0<r<0.2,\text{ and } \left.\frac{\partial u}{\partial t}\right|_{t=0}=3 \text{ for } 0<r<0.2, \tag{7.4.43a,b,c}$$

we run the MATLAB script "em07_04_1c.m" to get the plots of the vibrating drum at t=0.5, 1, and 1.5[s] as shown in Fig. 7.9.

```
%em07_04_1c.m
% solves the PDE (7.4.33) s.t. the BC and IC (7.4.43a,b,c)
clear, clf
a=1; R=1; tf=2; % Parameters of the PDE
i0=@(r)0; % IC (7.4.43b)
v0=3; r0=0.2; i10 = @(r)-v0*(r<r0); % IC (7.4.43c) only for 0<=r<=r0=0.2;
bR = 0; % BC (7.4.43a) on the boundary circle with r=R
Mr=50; Mt=100; % Numbers of segments along the radial and time axes
[ua,ras,tas] = pde_wave_circle_ana(a,R,tf,i0,i10,bR,Mr,Mt);
phs = [0:10:360]*(pi/180); Mph = length(phs);
for n=1:Mt  % Animation
   mesh_polar(ras,phs,repmat(ua(:,n),1,Mph));
   axis('equal'); axis([-1 1 -1 1 -0.4 0.4]), view([-37.5 30])
   if mod(n,25)==1, pause, end
   pause(0.1), if n<Mt, cla; end
end
```

To get the numerical solution of the PDE (7.4.33), we use the central difference approximations of derivatives to discretize the PDE as

$$a^2\left(\frac{u_{m+1}^k-2u_m^k+u_{m-1}^k}{\Delta r^2}+\frac{u_{m+1}^k-u_{m-1}^k}{r2\Delta r}\right)=\frac{u_m^{k+1}-2u_m^k+u_m^{k-1}}{\Delta t^2} \tag{7.4.44}$$

This leads to the following difference equation, like Eq. (7.3.24):

$$u_m^{k+1}=c_1(r_m)u_{m+1}^k+c_2u_m^k+c_3(r_m)u_{m-1}^k-u_m^{k-1} \tag{7.4.45}$$

where

$$c_1(r_m)=a^2\frac{\Delta t^2}{\Delta r^2}\left(1+\frac{\Delta r}{2r_m}\right),\; c_2=2\left(1-a^2\frac{\Delta t^2}{\Delta r^2}\right),\text{ and } c_3(r_m)=a^2\frac{\Delta t^2}{\Delta r^2}\left(1-\frac{\Delta r}{2r_m}\right) \tag{7.4.46a,b,c}$$

Since $u_m^{-1}=u(r_m,-\Delta t)$ is not given, u_m^1 cannot be obtained from this formula (7.4.45) with k=0:

$$u_m^1=c_1(r_m)u_{m+1}^0+c_2u_m^0+c_3(r_m)u_{m-1}^0-u_m^{-1} \tag{7.4.47}$$

Therefore, we approximate the IC (7.4.43c) on the derivative by the central difference as

$$\frac{u_m^1 - u_m^{-1}}{2\Delta t} = u_0'(r_m) = i_{10}(r_m) \tag{7.4.48}$$

and make use of this to remove u_m^{-1} from Eq. (7.4.47) to write

$$\begin{aligned} u_m^1 &= \frac{1}{2}\left(c_1(r_m)u_{m+1}^0 + c_2 u_m^0 + c_3(r_m)u_{m-1}^0\right) + \Delta t\, u_0'(r_m) \\ &= \frac{1}{2}\left(c_1(r_m)u_0(r_{m+1}) + c_2 u_0(r_m) + c_3(r_m)u_0(r_{m-1})\right) + \Delta t\, u_0'(r_m) \end{aligned} \tag{7.4.49}$$

We use this equation together with the ICs to get u_m^1 and then go on using Eq. (7.4.45) for k=1,2,.... This numerical solution process has been cast into the MATLAB function '`pde_wave_circle()`'. Interested readers are encouraged to use it for finding the numerical solution $u(r,t)$ of the wave equation (7.4.33) with a=1 (for the circular domain of radius R=1 divided into M_r=50 segments and time range $0 \le t \le t_f$=2[s] divided into M_t=100 sub-intervals) subject to the BC and ICs (7.4.43a,b,c), and see if it is close to the analytical solution depicted in Fig. 7.9.

```
function [u,rs,ts,iter] = pde_wave_circle(a,R,tf,i0,i10,bR,Mr,Mt)
%solve a^2*(u_rr +1/r*u_r) = u_tt for 0<=r<=R, 0<=t<=tf
% ICs : u(r,0) = i0(r), u_t(r,0) = i10(r)
% BCs : u(R,t) = bR for r=R and all t>=0
% Mr = # of subintervals along the r(radial)-axis
% Mt = Number of subintervals along the t-axis
% u  : Solution u(r,t)
% rs, ts : Radial/Temporal grids
%Copyleft: Won Y. Yang, wyyang53@hanmail.net, CAU for academic use
dr=R/Mr; rs=[0:Mr]'*dr;  % Radial grid points
dt=tf/Mt;   ts=[0:Mt]*dt; % Time points
% Note that dt should be small enough to make a*dt<dr for stability.
% Initialization
u=zeros(Mr+1,Mt+1);  u_t=zeros(Mr+1,Mt+1);
if isnumeric(it0), u(:,1)=it0;  else u(:,1)=feval(i0,rs); end
if isnumeric(i1t0), u_t(:,1)=i1t0; else u_t=feval(i10,rs); end
if isnumeric(bR), u(Mr+1,:)=bR;  else u(Mr+1,:)=feval(bR,ts); end
adtdr2=(a*dt/dr)^2;  rs1=rs(2:end); % Eq.(7.4.46)
c1=adtdr2*(1+dr/2./rs1); c2=2*(1-adtdr2), c3=adtdr2*(1Dr/2./rs1);
m=2:Mr; MaxIter=50;
for iter=1:MaxIter
   u0 = u;
   u(m,2)=0.5*(c1(m).*u(m+1,1)+c2*u(m,1)+c3(m).*u(m-1,1)) ...
            +dt*u_t(m);  % Eq.(7.4.49)
   for k=3:Mt+1
      u(m,k) = c1(m).*u(m+1,k-1) +c2*u(m,k-1) ...
            +c3(m).*u(m-1,k-1) -u(m,k-2);  % Eq.(7.4.45)
   end
   u(1,:) = u(2,:); % Approximation for around the center
   difference=norm(u-u0)/norm(u0);
   if iter>5&difference<0.001, break; end
end
```

7.4.2 PDEs in Spherical Coordinates

Let us consider the Laplace's equation with azimuthal symmetry in the spherical coordinate system:

$$\frac{\partial^2 u(r,\theta)}{\partial r^2}+\frac{2}{r}\frac{\partial u(r,\theta)}{\partial r}+\frac{1}{r^2}\frac{\partial^2 u(r,\theta)}{\partial \theta^2}+\frac{\cot\theta}{r^2}\frac{\partial u(r,\theta)}{\partial \theta}=0 \text{ for } 0<r\le R,\ 0\le\theta\le\pi \quad (7.4.50)$$

subject to the BC

$$u(R,\theta)=b(\theta) \quad \text{for } 0\le\theta\le\pi \quad (7.4.51)$$

It will be shown that the general solution of this PDE over a spherical region is

$$u(r,\theta)=\sum_{n=0}^{\infty}A_n r^n P_n(\cos\theta),\ P_n(x)\text{: Legendre polynomial of degree } n \text{ (Section 3.3.2)} \quad (7.4.52)$$

with

$$A_n=\frac{2n+1}{2R^n}\int_0^{\pi} b(\theta)P_n(\cos\theta)\sin\theta\, d\theta \quad (7.4.53)$$

By the method of separation of variables, we substitute

$$u(r,\theta)=\Upsilon(r)\Theta(\theta) \quad (7.4.54)$$

into the PDE (7.4.50) and multiply by $r^2/\Upsilon(r)\Theta(\theta)$ to write

$$\frac{r^2\Upsilon''(r)+2r\Upsilon'(r)}{\Upsilon(r)}=-\frac{\Theta''(\theta)+(\cot\theta)\Theta'(\theta)}{\Theta(\theta)}=\lambda=\omega^2$$

$$;\ \begin{cases} r^2\Upsilon''(r)+2r\Upsilon'(r)-\omega^2\Upsilon(r)=0 & (7.4.55a) \\ (\sin\theta)\dfrac{d^2\Theta}{d\theta^2}+(\cos\theta)\dfrac{d\Theta}{d\theta}+\omega^2(\sin\theta)\Theta(\theta)=0 & (7.4.55b) \end{cases}$$

where the sign of the separation constant ω^2 has been set to be positive to disallow any oscillation of $u(r,\theta)$ or $\Upsilon(r)$ along the r-axis. Here, with $x=\cos\theta$, we can substitute

$$\frac{d\Theta(x)}{d\theta}=\frac{dx}{d\theta}\frac{d\Theta}{dx}=-\sin\theta\frac{d\Theta}{dx} \quad (7.4.56a)$$

$$\frac{d^2\Theta(x)}{d\theta^2}=\frac{d}{d\theta}\frac{d\Theta(x)}{d\theta}\overset{(7.4.56a)}{=}\frac{d}{d\theta}(-\sin\theta)\frac{d\Theta}{dx}+(-\sin\theta)\frac{dx}{d\theta}\frac{d^2\Theta}{dx^2}=-\cos\theta\frac{d\Theta}{dx}+\sin^2\theta\frac{d^2\Theta}{dx^2} \quad (7.4.56b)$$

into Eq. (7.4.55b) and divide both sides by $\sin\theta$ to write

$$(1-x^2)\frac{d^2\Theta}{dx^2}-2x\frac{d\Theta}{dx}+\omega^2\Theta(x)=0 \quad (7.4.57)$$

Amazingly, this coincides with the Legendre's equation (3.3.29):

$$(1-x^2)y''-2xy'+n(n+1)y=0 \text{ for } n>0 \text{ and } |x=\cos\theta|<1 \quad (7.4.58)$$

```
function [u,rs,ths,A]=pde_Laplace_sphere_ana(R,bR,Mr,Mth,N)
%solve u_rr +2/r*u_r +(1/r^2)*u_thth +cot(th)/r^2*u_th = 0
%   for 0<=r<=R, 0<=th<=pi
% BC: u(R,th) = bR(th)   for 0<=th<=pi
% Mr = # of subintervals along the r(radial)-direction
% Mth = # of subintervals along the th(circular)-direction
% N : Number of coefficients/terms found/added
% rs, ths : Radial/Circular grids
%Copyleft: Won Y. Yang, wyyang53@hanmail.net, CAU for academic use
syms th
dr = R/Mr; rs = [0:Mr]*dr;  % Radial grid points
dth = pi/Mth;  ths = [0:Mth]*dth; % Circular grid points
[Ths,Rs]=meshgrid(ths,rs); % Grid
if nargin<5, N=5; end;
f_int=@(th,n)feval(bR,th).*mfun('P',n,cos(th)).*sin(th); % Eq.(7.4.53)
u = 0;
for n=0:N
   A(n+1)=(2*n+1)/2/R^n*quad(f_int,0,pi,1e-4,[],n); % Eq.(7.4.53)
   u = u + A(n+1)*Rs.^n.*mfun('P',n,cos(Ths)); % Eq.(7.4.52)
end
```

That is why we let the separation constant be $\omega^2=n(n+1)$ and write the solution of Eq. (7.4.55b) as a Legendre polynomial of degree n:

$$\Theta_n(x) = P_n(x) \overset{x=\cos\theta}{\rightarrow} \Theta_n(\theta) = P_n(\cos\theta) \tag{7.4.59}$$

On the other hand, Eq. (7.4.55a) with $\omega^2=n(n+1)$ is a second-order Cauchy–Euler equation (3.2.23). Since its indicial equation, with $\omega^2=n(n+1)$, has the roots:

$$m(m-1)+2m-n(n+1) = 0;\ (m+n+1)(m-n)=0;\ m=n,-(n+1) \tag{7.4.60}$$

the solution of Eq. (7.4.55a) can be found from Eq. (3.2.26) as

$$\Upsilon_n(r) \underset{(7.4.60)}{\overset{(3.2.26)}{=}} A_n r^n + \cancel{B_n} r^{-(n+1)} \tag{7.4.61}$$

However, in order for $u(r,\theta)=\Upsilon(r)\Theta(\theta)$ to be bounded at the center ($r=0$), the coefficient B_n of the second term must be zero. Now, plugging Eqs. (7.4.59) and (7.4.61) into Eq. (7.4.54) yields the separable solutions:

$$u_n(r,\theta) \overset{(7.4.54)}{=} \Upsilon_n(r)\Theta_n(\theta) \underset{(7.4.59)}{\overset{(7.4.61)}{=}} A_n r^n P_n(\cos\theta) \tag{7.4.62}$$

Hence, the general solution of the Laplace's equation (7.4.50) can be written as Eq. (7.4.52), where the coefficient A_n's are determined by substituting the general solution into the BC (7.4.51):

$$u(R,\theta) \overset{(7.4.52)}{=} \sum_{n=0}^{\infty} A_n R^n P_n(\cos\theta) \overset{(7.4.51)}{=} b(\theta) \quad \text{for } 0\le\theta\le\pi \tag{7.4.63}$$

and using Eq. (6.6.9) (for Fourier–Legendre series coefficients), which leads to Eq. (7.4.53). This solution process has been cast into the above MATLAB function '`pde_Laplace_sphere_ana()`'. We can use it to find the solution $u(r,\theta)$ of the Laplace's equation (7.4.50) (for the cross-section of the spherical domain: $D=\{(r,\theta), 0\le r\le R=1, 0\le\theta\le\pi\}$ of radius $R=1$ subject to the BC:

$$u(R,\theta) = b_R(\theta) = 100(1-\frac{\theta}{\pi}) \quad \text{for } R=1 \text{ and } 0\le\theta\le\pi \tag{7.4.64}$$

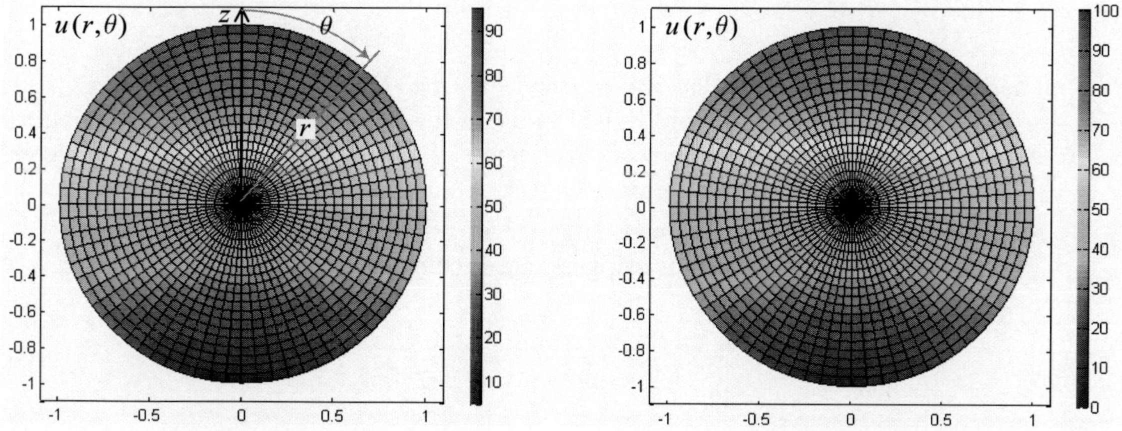

(a) Solution graph obtained by an analytical approach (b) Solution graph obtained by a numerical method

Figure 7.10 The domain and solution of the PDE (7.4.50) s.t. the BC (7.4.67)

To this end, the MATLAB script "em07_04_2.m" has been run to yield the solution graph for the domain divided into $M_r \times M_\theta = 20 \times 36$ segments as shown in Fig. 7.10(a).

```
%em07_04_2.m
clear, clf
R=1; % Radius of the sphere
bR = @(th)100*(1-th/pi); % BC (7.4.64)
Mr=20; Mth=36; % Numbers of segments along the r-/th-directions
N=5; % Number of coefficients/terms to be added for Eq.(7.4.63)
[ua,rs,ths,A]=pde_Laplace_sphere_ana(R,bR,Mr,Mth,N); A
[Ths,Rs]=meshgrid(ths,rs); % Grid
pcolor(Rs.*sin(Ths),Rs.*cos(Ths),ua), hold on
pcolor(-Rs.*sin(Ths),Rs.*cos(Ths),ua), axis('equal'), colorbar
```

As a numerical approach, we can use the central difference approximations of derivatives (like in Eq. (7.2.30)) to discretize the PDE (7.4.50) and the BC (7.4.64) as

$$\frac{\partial^2 u(r,\theta)}{\partial r^2}+\frac{2}{r}\frac{\partial u(r,\theta)}{\partial r}+\frac{1}{r^2}\frac{\partial^2 u(r,\theta)}{\partial \theta^2}+\frac{\cot\theta}{r^2}\frac{\partial u(r,\theta)}{\partial \theta}\overset{(7.4.49)}{=}0$$

$$\begin{aligned}\rightarrow&\frac{u(m+1,n)-2u(m,n)+u(m-1,n)}{(\Delta r)^2}+\frac{2}{r}\frac{u(m+1,n)-u(m-1,n)}{2\Delta r}\\&+\frac{1}{r^2}\frac{u(m,n+1)-2u(m,n)+u(m,n-1)}{(\Delta\theta)^2}+\frac{\cot\theta}{r^2}\frac{u(m,n+1)-u(m,n-1)}{2\Delta\theta}=0\end{aligned} \quad (7.4.65)$$

$$;\quad u(m,n)=\frac{1}{a}\left\{\begin{aligned}&\frac{u(m+1,n)+u(m-1,n)}{(\Delta r)^2}+\frac{1}{r(m)}\frac{u(m+1,n)-u(m-1,n)}{\Delta r}\\&+\frac{1}{r(m)^2}\frac{u(m,n+1)+u(m,n-1)}{(\Delta\theta)^2}+\frac{\cot\theta(n)}{r(m)^2}\frac{u(m,n+1)-u(m,n-1)}{2\Delta\theta}\end{aligned}\right\} \quad (7.4.66a)$$

$$\text{with } a=\frac{2}{(\Delta r)^2}+\frac{2}{r(m)^2(\Delta\theta)^2} \quad (7.4.66b)$$

$$\text{s.t. the BC: } u(M_r+1,n) \overset{(7.4.51)}{=} b_R(\theta(n)) \overset{(7.4.64)}{=} 100\left(1-\frac{\theta(n)}{\pi}\right) \text{ for } 1\le n\le M_\theta+1 \tag{7.4.67}$$

This numerical solution process has been cast into the MATLAB function '`pde_Laplace_sphere()`'. Interested readers are encouraged to use it for finding the numerical solution $u(r,\theta)$ of Eq. (7.4.50) (for the cross-section of the spherical domain of radius R=1 divided into $M_r\times M_\theta$=20×36 segments) subject to the BC (7.4.67), as shown in Fig. 7.10(b).

(Q) It seems that Eq. (7.4.66a) becomes singular at the north/south poles: $\theta=0^\circ$ ($n=1$) and $=180^\circ$ ($n=M_\theta+1$). How could those singular cases be dealt with?

```
function [u,rs,ths,k]=pde_Laplace_sphere(R,bR,Mr,Mth,tol)
%solve u_rr +2/r*u_r +(1/r^2)*u_thth +cot(th)/r^2*u_th = 0
%   for 0<=r<=R, 0<=th<=pi
% BC: u(R,th) = bR(th)   for 0<=th<=pi
% Mr = # of subintervals along the r(radial)-axis
% Mth = # of subintervals along the th(circular)-axis
% tol = Error tolerance
% rs, ths : Radial/Circular grids
% k : Number of iterations performed
%Copyleft: Won Y. Yang, wyyang53@hanmail.net, CAU for academic use
if nargin<5, tol = 1e-4; end  % Error tolerance
dr = R/Mr; rs = [0:Mr]*dr;   % Radial grid points
dth = pi/Mth;  ths = [0:Mth]*dth; % Circular grid points
[Ths,Rs]=meshgrid(ths,rs); % Grid
u = zeros(Mr+1,Mth+1);  dr2 = dr^2; dth2 = dth^2;
if isnumeric(bR),  u(Mr+1,:) = bR;  % Eq.(7.4.51)
 else u(Mr+1,:) = feval(bR,ths); % BC on the sphere
end
for k=1:1000
   u0 = u;
   for m=2:Mr
      rm = rs(m); rm2 = rm^2;
      for n=2:Mth
         a = 2/dr2 + 2/rm2/dth2;     % Eq.(7.4.66b)
         nm1=(n-1)*(n>1)+2*(n<=1); np1=(n+1)*(n<=Mth)+Mth*(n>Mth);
         u(m,n)=((u(m+1,n)+u(m-1,n))/dr2 +(u(m+1,n)-u(m-1,n))/rm/dr
           + (u(m,np1)+u(m,nm1))/rm2/dth2 ...
           + cot(ths(n))*(u(m,np1)-u(m,nm1))/rm2/2/dth)/a; % Eq.(7.4.66a)
      end
   end
   % Set the value of u at the center to mean of those around it.
   u(1,:) = mean(u(2,:));
   % Set the values of u at th=0 and pi to their adjacent ones.
   u(1:Mr,1) = u(1:Mr,2); u(1:Mr,Mth+1) = u(1:Mr,Mth);
   if norm(u-u0)/norm(u0)<tol, break, end
end
if nargout<1
  [Ths,Rs]=meshgrid(ths,rs); % Grid
  pcolor(Rs.*sin(Ths),Rs.*cos(Ths),u), hold on
  pcolor(-Rs.*sin(Ths),Rs.*cos(Ths),u), axis('equal'), colorbar
end
```

7.5 LAPLACE/FOURIER TRANSFORMS FOR SOLVING PDES

7.5.1 Using the Laplace Transform for PDEs

This section will show that the Laplace transform (LT) can be used to deal with linear constant-coefficient PDEs (as well as ODEs), especially when the spatial variable ranges over the entire line or half-line. A premise on which to apply the LT for solving PDEs in two independent variables, say, x and t, is that differentiating w.r.t a variable, say x, and taking the LT w.r.t. another variable t are interchangeable in the sense that

$$\underbrace{\mathcal{L}_t\left\{\frac{\partial f(x,t)}{\partial x}\right\}}_{\text{Differentiation w.r.t. }x\text{-Laplace transform w.r.t. }t} \overset{(4.1.1)}{=} \int_{0-}^{\infty}\frac{\partial f(x,t)}{\partial x}e^{-st}dt = \underbrace{\frac{\partial}{\partial x}\int_{0-}^{\infty} f(x,t)\,e^{-st}dt}_{\text{Laplace transform w.r.t. }t\text{-Differentiation w.r.t. }x} = \frac{\partial}{\partial x}\underbrace{\mathcal{L}_t\{f(x,t)\}}_{\text{function of }s\,\&\,x} = \frac{d}{dx}F(x,s) \quad (7.5.1)$$

where $\mathcal{L}_t$ denotes the Laplace transform w.r.t. t and the partial derivative in the second to last term becomes the total derivative because $F(x,s)=\mathcal{L}_t\{f(x,t)\}$ is considered as a function of only one variable x (with s as a complex parameter). See Section 11.13 of [K-2] or Section 11.8 of [W-6] for details.

Like the process of using the LT for solving ODEs, the process of using the LT to get the solution $u(x,t)$ of a PDE begins with taking the LT of the PDE w.r.t., say, t to make an ODE in a variable x (with a parameter s), followed by solving the ODE for the transformed solution $U(x,s)$, determining the coefficients (in terms of s) from the (Laplace-transformed) BC(s), and then by taking the ILT (inverse Laplace transform). See the following examples.

(**Example 7.6**) Laplace Transform for a Parabolic PDE — 1D Heat Equation over a Line

Consider the following 1D heat equation (dealt with in Example 7.2):

$$a^2\frac{\partial^2 u(x,t)}{\partial x^2} = \frac{\partial u(x,t)}{\partial t} \quad \text{for } 0\le x\le 1 \text{ and } 0\le t\le 1 \qquad \text{(E7.6.1)}$$

subject to the IC and BCs

$$u(x,0)=\sin\pi x\ \ \forall x,\ \ u(0,t)=0\ \forall t,\ \ u(1,t)=0\ \forall t. \qquad \text{(E7.6.2a,b,c)}$$

Taking the s-transform (LT w.r.t. t) of Eq. (E7.6.1) with the IC (E7.6.2a) yields

$$\mathcal{L}_t\left\{\overset{\text{LHS}}{a^2\frac{\partial^2 u(x,t)}{\partial x^2}}\right\} \overset{(7.5.1)}{=} a^2\frac{d^2}{dx^2}U(x,s) = \mathcal{L}_t\left\{\overset{\text{RHS}}{\frac{\partial u(x,t)}{\partial t}}\right\} \underset{\text{or (4.2.2)}}{\overset{\text{Table A.7(5)}}{=}} sU(x,s)-u(x,0) \overset{(E7.6.2a)}{=} sU(x,s)-\sin\pi x$$

$$; \qquad a^2\frac{d^2}{dx^2}U(x,s) - sU(x,s) = -\sin\pi x \qquad \text{(E7.6.3)}$$

Since this is a second-order ODE (in x with s as a constant), we use Eq. (4.2.3) with $n=2$ and Table A.8(9) to take its $\tilde{s}$-transform (LT w.r.t. x with $\tilde{s}$ as a complex parameter) to write

$$\tilde{s}^2\tilde{U}(\tilde{s}) - \tilde{s}U(0,s) - \left.\frac{d}{dx}U(x,s)\right|_{x=0}^{?} - \frac{s}{a^2}\tilde{U}(\tilde{s}) = \frac{-\pi/a^2}{\tilde{s}^2+\pi^2} \overset{(E7.6.2b)}{\to} (\tilde{s}^2-\frac{s}{a^2})\tilde{U}(\tilde{s}) = \frac{-\pi/a^2}{\tilde{s}^2+\pi^2} + ? \quad \text{(E7.6.4)}$$

where

$$\tilde{U}(\tilde{s}) = \mathcal{L}_x\{U(x,s)\}\ \text{(LT w.r.t. } x) \qquad \text{(E7.6.5)}$$

Since $dU(x,s)/dx|_{x=0}$ is not yet clear, we solve this equation provisionally just for the particular solution part of $\tilde{U}(\tilde{s})$ (stemming from $u(x,0)=\sin\pi x$) and get its PFE (partial fraction expansion) as

$$\tilde{U}(\tilde{s}) \overset{\text{(E7.6.4)}}{=} \frac{1}{\tilde{s}^2-s/a^2}\frac{-\pi/a^2}{\tilde{s}^2+\pi^2} = \frac{A}{\tilde{s}-\sqrt{s}/a}+\frac{B}{\tilde{s}+\sqrt{s}/a}+\frac{C\tilde{s}+D\pi}{\tilde{s}^2+\pi^2} \tag{E7.6.6}$$

$$= \frac{(A+B+C)\tilde{s}^3+(A\sqrt{s}/a-B\sqrt{s}/a+D\pi)\tilde{s}^2+((A+B)\pi^2-sC)\tilde{s}+(A\sqrt{s}\pi/a-B\sqrt{s}\pi/a-Ds/a^2)\pi}{(\tilde{s}^2-s/a^2)(\tilde{s}^2+\pi^2)}$$

where

$$\left.\begin{array}{c} A+B+C=0 \rightarrow C=-(A+B) \\ A\sqrt{s}/a-B\sqrt{s}/a+D\pi=0 \rightarrow A\sqrt{s}/a-B\sqrt{s}/a=-D\pi \\ (A+B)\pi^2-sC=0 \rightarrow -C(\pi^2+s)=0;\ C=0 \\ (A\sqrt{s}\pi/a-B\sqrt{s}\pi/a-Ds/a^2)\pi=-\pi/a^2 \rightarrow -D\pi^2-Ds/a^2=-1/a^2 \end{array}\right\} \rightarrow \begin{array}{c} C=0,\ A=-B \\ D=\dfrac{1}{s+a^2\pi^2} \end{array} \tag{E7.6.7}$$

Taking the inverse $\tilde{s}$-transform (ILT w.r.t. x with s as a constant) of Eq. (E7.6.6) yields

$$U(x,s)=\mathcal{L}_x^{-1}\{\tilde{U}(s)\} \underset{\text{A.8(6),(9)}}{\overset{\text{(E7.6.6,7)}}{=}} Ae^{\sqrt{s}x/a}+Be^{-\sqrt{s}x/a}+\frac{1}{s+a^2\pi^2}\sin\pi x \tag{E7.6.8}$$

where A and B are determined from the (s-transformed) BCs

$$\left.\begin{array}{l} u(0,t) \overset{\text{(E7.6.2b)}}{=} 0 \overset{\mathcal{L}_t}{\rightarrow} U(0,s)=0 \overset{\text{(E7.6.8)}}{\rightarrow} Ae^{\sqrt{s}0}+Be^{-\sqrt{s}0}+\dfrac{1}{s+a^2\pi^2}\sin\pi 0=0 \\ u(1,t) \overset{\text{(E7.6.2c)}}{=} 0 \overset{\mathcal{L}_t}{\rightarrow} U(1,s)=0 \overset{\text{(E7.6.8)}}{\rightarrow} Ae^{\sqrt{s}1/a}+Be^{-\sqrt{s}1/a}+\dfrac{1}{s+a^2\pi^2}\sin\pi 1=0 \end{array}\right\} \rightarrow A=B=0 \tag{E7.6.9}$$

Thus, Eq. (E7.6.8) can be written as

$$U(x,s) \underset{\text{(E7.6.9)}}{\overset{\text{(E7.6.8)}}{=}} \frac{1}{s+a^2\pi^2}\sin\pi x \tag{E7.6.10}$$

We can take the inverse s-transform (ILT w.r.t. t) of this equation to get

$$u(x,t)=\mathcal{L}_t^{-1}\{U(x,s)\} \underset{\text{A.8(6)}}{\overset{\text{(E7.6.10)}}{=}} e^{-a^2\pi^2 t}\sin(\pi x) \tag{E7.6.11}$$

This conforms with Eq. (E7.2.3).

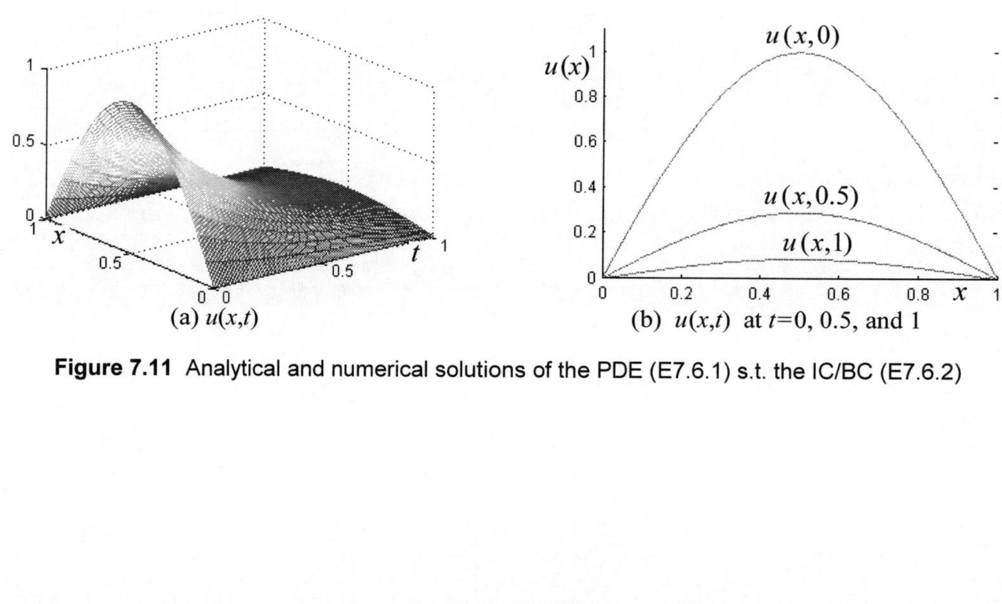

Figure 7.11 Analytical and numerical solutions of the PDE (E7.6.1) s.t. the IC/BC (E7.6.2)

The analytical solution (E7.6.11) for the PDE (E7.6.1) with a=0.5 (for $0<x<5$) subject to the IC and BCs (E7.6.2) and its snapshots are plotted in Figs. 7.11(a) and (b), which have been obtained by running the MATLAB script "em07e06.m."

```
%em07e06.m
a=0.5; w=pi;
% BCs and ICs
it0=@(x)sin(w*x); bx0t=@(t)0; bxft=@(t)0;   % Eq.(E7.6.2a,b,c)
xf=1; M=20;  tf=1; N=100;
% Numerical approach
[un,x,t]=pde_heat_CN(a,xf,tf,it0,bx0t,bxft,M,N);
% Analytical solution using the Laplace transform
[T,X]=meshgrid(t,x);
ua = sin(w*X).*exp(-a^2*w^2*T); % Eq.(E7.6.11)
Discrepancy=norm(ua-un)/norm(ua)
subplot(121), mesh(t,x,ua), axis([0 tf 0 xf 0 1])
title('Analytical solution')
subplot(122)
for n=1:N % motion picture (animation)
   plot(x,u(:,n), x,un(:,n),'r:'), axis([0 xf 0 1.2]), pause(0.1)
end
```

(**Example 7.7**) Laplace Transform for Hyperbolic PDE — 1D Wave Equation on a String

Consider the 1D wave equation (dealt with in Example 7.4):

$$a^2\frac{\partial^2 u(x,t)}{\partial x^2}=\frac{\partial^2 u(x,t)}{\partial t^2} \quad \text{for } 0\le x\le 1 \text{ and } 0\le t\le 2 \tag{E7.7.1}$$

subject to the BCs and ICs

$$u(0,t)=0 \ \forall\, t, \quad u(1,t)=0 \ \forall\, t \tag{E7.7.2a,b}$$

$$u(x,t)\big|_{t=0}=\sin\pi x \ \forall\, x, \quad \left.\frac{\partial u(x,t)}{\partial t}\right|_{t=0}=0 \ \forall\, x \tag{E7.7.3a,b}$$

Taking the s-transform of Eq. (E7.7.1) with the IC (E7.7.3a,b) yields

$$\mathcal{L}_t\left\{\overbrace{a^2\frac{\partial^2 u(x,t)}{\partial x^2}}^{\text{LHS}}\right\}\overset{(7.5.1)}{=}a^2\frac{d^2}{dx^2}U(x,s)$$

$$=\mathcal{L}_t\left\{\overbrace{\frac{\partial^2 u(x,t)}{\partial t^2}}^{\text{RHS}}\right\}\overset{(4.2.3)}{=}s^2U(x,s)-s\,u(x,0)-\left.\frac{\partial u(x,t)}{\partial t}\right|_{t=0}\overset{(E7.7.3a,b)}{=}s^2U(x,s)-s\sin\pi x$$

$$;\quad \frac{d^2}{dx^2}U(x,s)-(s/a)^2U(x,s)=-(s/a^2)\sin\pi x \tag{E7.7.4}$$

Since this is a second-order ODE (in x with s as a constant), we use Eq. (4.2.3) with n=2 and Table A.8(9) to take its $\tilde{s}$-transform (LT w.r.t. x with $\tilde{s}$ as a complex parameter) to write

$$\tilde{s}^2\tilde{U}(\tilde{s})-\tilde{s}U(0,s)-\left.\frac{d}{dx}U(x,s)\right|_{x=0}-\frac{s^2}{a^2}\tilde{U}(\tilde{s})=\frac{-s\pi/a^2}{\tilde{s}^2+\pi^2}\overset{(E7.7.2a)}{\rightarrow}(\tilde{s}^2-\frac{s^2}{a^2})\tilde{U}(\tilde{s})=\frac{-s\pi/a^2}{\tilde{s}^2+\pi^2}+? \tag{E7.7.5}$$

where

$$\tilde{U}(\tilde{s}) = \mathcal{L}_x\{U(x,s)\} \text{ (Laplace transform w.r.t. } x) \tag{E7.7.6}$$

Since $dU(x,s)/dx|_{x=0}$ is not yet clear, we solve this equation provisionally just for the particular solution part of $\tilde{U}(\tilde{s})$ (stemming from $u(x,0)=\sin \pi x$) and get its PFE (partial fraction expansion) as

$$\tilde{U}(\tilde{s}) \overset{(E7.7.5)}{=} \frac{1}{\tilde{s}^2-(s/a)^2}\frac{-s\pi/a^2}{\tilde{s}^2+\pi^2} = \frac{A}{\tilde{s}-s/a}+\frac{B}{\tilde{s}+s/a}+\frac{C\tilde{s}+D\pi}{\tilde{s}^2+\pi^2} \tag{E7.7.7}$$

$$= \frac{(A+B+C)\tilde{s}^3+(As-Bs+D\pi)\tilde{s}^2+((A+B)\pi^2-s^2C)\tilde{s}+(As\pi-Bs\pi-Ds^2)\pi}{(\tilde{s}^2-s^2/a^2)(\tilde{s}^2+\pi^2)}$$

where

$$\left.\begin{array}{c} A+B+C=0 \rightarrow C=-(A+B) \\ As-Bs+D\pi=0 \rightarrow As-Bs=-D\pi \\ (A+B)\pi^2-s^2C=0 \rightarrow -C(\pi^2+s^2)=0;\ C=0 \\ (As\pi-Bs\pi-Ds^2/a^2)\pi=-s\pi/a^2 \rightarrow -Da^2\pi^2-Ds^2=-s \end{array}\right\} \rightarrow \begin{array}{c} C=0 \\ D=\dfrac{s}{s^2+a^2\pi^2} \end{array} \tag{E7.7.8}$$

Taking the inverse $\tilde{s}$-transform (ILT w.r.t. x with s as a constant) of Eq. (E7.7.7) yields

$$U(x,s) = \mathcal{L}_x^{-1}\{\tilde{U}(s)\} \underset{A.8(6),(9)}{\overset{(E7.7.7,8)}{=}} Ae^{sx/a}+Be^{-sx/a}+\frac{s}{s^2+a^2\pi^2}\sin \pi x \tag{E7.7.9}$$

where A and B are determined from the (s-transformed) BCs

$$\left.\begin{array}{l} u(0,t) \overset{(E7.7.2a)}{=} 0 \overset{\mathcal{L}_t}{\rightarrow} U(0,s)=0 \overset{(E7.7.9)}{\rightarrow} Ae^{s0}+Be^{-s0}+\dfrac{s}{s+a^2\pi^2}\sin \pi 0=0 \\ u(1,t) \overset{(E7.7.2b)}{=} 0 \overset{\mathcal{L}_t}{\rightarrow} U(1,s)=0 \overset{(E7.7.9)}{\rightarrow} Ae^{s1/a}+Be^{-s1/a}+\dfrac{s}{s+a^2\pi^2}\sin \pi 1=0 \end{array}\right\} \rightarrow \begin{array}{c} A=0 \\ B=0 \end{array} \tag{E7.7.10}$$

Thus, Eq. (E7.7.9) can be written as

$$U(x,s) \underset{(E7.7.10)}{\overset{(E7.7.9)}{=}} \frac{s}{s^2+a^2\pi^2}\sin \pi x \tag{E7.7.11}$$

We can take the inverse s-transform (ILT w.r.t. t) of this equation to get

$$u(x,t) = \mathcal{L}_t^{-1}\{U(x,s)\} \underset{A.8(9)}{\overset{(E7.7.11)}{=}} \cos(a\pi t)\sin(\pi x) \tag{E7.7.12}$$

(Q) Can you get the same solution by substituting $a=1$ and $i_0(x)=\sin \pi x$ in Example 7.4?

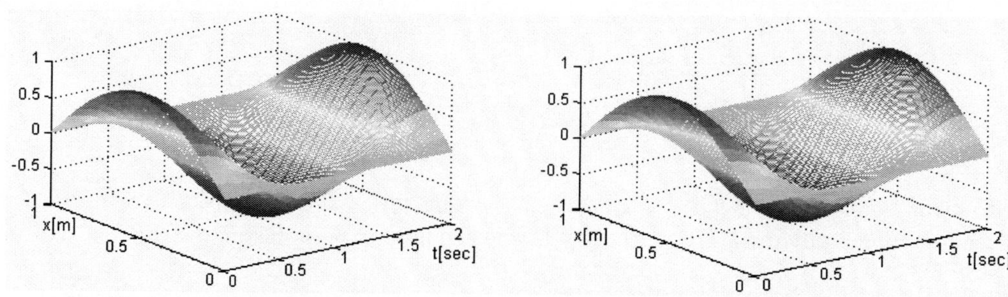

(a) Numerical solution $u(x,t)$ (b) Analytical solution $u(x,t)$

Figure 7.12 Analytical and numerical solutions of the PDE (E7.7.1) s.t. the BC/IC (E7.7.2)/(E7.7.3)

```
%em07e07.m
clear, clf
a=1; w=pi; % The parameter of Eq.(E7.7.1)
bx0=@(t)0; bxf=@(t)0; % BCs (E7.7.2a,b)
it0=@(x)sin(w*x); i1t0=@(x)0; % ICs (E7.7.3a,b)
xf=1; M=20; dx=xf/M; tf=2; N=200; dt=tf/N;  % r=0.04
% Numerical solution
[un,x,t]=pde_wave(a,xf,tf,it0,i1t0,bx0,bxf,M,N);
subplot(121), mesh(t,x,un)
xlabel('t'), ylabel('x')
title('Numerical solution using pde_wave()')
% Analytical solution
[T,X]=meshgrid(t,x);
ua = sin(w*X).*cos(a*w*T); % Eq.(E7.7.12)
Discrepancy=norm(ua-un)/norm(ua)
subplot(122), mesh(t,x,ua)
xlabel('t'), ylabel('x'), title('Analytical solution')
```

The analytical solution (E7.7.12) and a numerical solution (using '`pde_wave()`') for the PDE (E7.7.1) with $a=1$ subject to the BCs (E7.7.2a,b) and ICs (E7.7.3a,b) are shown in Fig. 7.12, which has been obtained by running the above MATLAB script, "em07e07.m."

(Example 7.8) Laplace Transform for Hyperbolic PDE — Wave Equation on a String

Consider the 1D wave equation

$$a^2\frac{\partial^2 u(x,t)}{\partial x^2}=\frac{\partial^2 u(x,t)}{\partial t^2}\ \text{ for } 0\le x\le 2 \text{ and } 0\le t\le 2 \qquad \text{(E7.8.1)}$$

subject to the BCs and ICs

$$u(0,t)=f(t) \text{ for } 0\le t<1,\ \lim_{x\to\infty} u(x,t)=0\ \forall\, t \qquad \text{(E7.8.2a,b)}$$

$$u(x,t)\big|_{t=0}=0\ \forall\, x,\quad \left.\frac{\partial u(x,t)}{\partial t}\right|_{t=0}=0\ \forall\, x \qquad \text{(E7.8.3a,b)}$$

Taking the s-transform of Eq. (E7.8.1) with the IC (E7.8.3a,b) yields

$$\mathcal{L}_t\left\{\overset{\text{LHS}}{a^2\frac{\partial^2 u(x,t)}{\partial x^2}}\right\}\overset{(7.5.1)}{=}a^2\frac{d^2}{dx^2}U(x,s)$$

$$=\mathcal{L}_t\left\{\overset{\text{RHS}}{\frac{\partial^2 u(x,t)}{\partial t^2}}\right\}\overset{(4.2.3)}{=}s^2U(x,s)-su(x,0)-\left.\frac{\partial u(x,t)}{\partial t}\right|_{t=0}\overset{(\text{E7.8.3a,b})}{=}s^2U(x,s)$$

$$;\quad a^2\frac{d^2}{dx^2}U(x,s)-s^2U(x,s)=0 \qquad \text{(E7.8.4)}$$

Since this is a second-order ODE (in x with s as a constant), we use Eq. (4.2.3) with n=2 and Table A.8(9) to take its $\tilde{s}$-transform (LT w.r.t. x with $\tilde{s}$ as a complex parameter) to write

$$a^2\left\{\tilde{s}^2\tilde{U}(\tilde{s})-\tilde{s}U(0,s)-\left.\frac{d}{dx}U(x,s)\right|_{x=0}\right\}-s^2\tilde{U}(\tilde{s})=0\overset{(\text{E7.8.2a})}{\rightarrow}\left(\tilde{s}^2-\frac{s^2}{a^2}\right)\tilde{U}(\tilde{s})=\tilde{s}F(s)+? \quad \text{(E7.8.5)}$$

where

$$\tilde{U}(\tilde{s}) = \mathcal{L}_x\{U(x,s)\} \text{ (LT w.r.t. } x\text{) and } F(s) = \mathcal{L}_t\{u(0,t) = f(t)\} \tag{E7.8.6}$$

Since $dU(x,s)/dx|_{x=0}$ is not yet clear, we solve this equation provisionally just for the particular solution part of $\tilde{U}(\tilde{s})$ (stemming from $u(0,t)=f(t)$) and get its PFE (partial fraction expansion) as

$$\tilde{U}(\tilde{s}) \overset{\text{(E7.8.5)}}{=} \frac{\tilde{s}}{\tilde{s}^2-(s/a)^2}F(s) = \frac{A}{\tilde{s}-s/a}F(s) + \frac{B}{\tilde{s}+s/a}F(s) \tag{E7.8.7}$$

Taking the inverse $\tilde{s}$-transform (ILT w.r.t. x with s as a constant) of Eq. (E7.8.7) yields

$$U(x,s) = \mathcal{L}_x^{-1}\{\tilde{U}(s)\} \underset{\text{A.8(6)}}{\overset{\text{(E7.8.7)}}{=}} AF(s)e^{sx/a} + BF(s)e^{-sx/a} \tag{E7.8.8}$$

Here, taking the s-transform of the second BC (E7.8.2b), we get

$$\mathcal{L}_t\left\{\lim_{x\to\infty} u(x,t)\right\} = \int_0^\infty e^{-st} \lim_{x\to\infty} u(x,t)dt = \lim_{x\to\infty}\int_0^\infty e^{-st}u(x,t)dt = \lim_{x\to\infty} U(x,s) \overset{\text{(E7.8.2b)}}{=} 0 \tag{E7.8.9}$$

In light of Eq. (E7.8.9), the first term of Eq. (E7.8.8) should be zero so that we can write the inverse s-transform (in t) of this equation as

$$u(x,t) = \mathcal{L}_t^{-1}\{U(x,s)\} \underset{\text{A.7(2)}}{\overset{\text{(E7.8.8)}}{=}} Bf\left(t-\frac{x}{a}\right) \underset{B=1}{\overset{\text{(E7.8.2a)}}{=}} f\left(t-\frac{x}{a}\right) \text{ for } \frac{x}{a} \le t < \frac{x}{a}+1. \tag{E7.8.10}$$

```
%em07e08.m
a=1; w=2*pi;
% BCs and ICs
bx0t=@(t)sin(w*t).*(t<=1); bxft=@(t)0; % Eq.(E7.8.2a,b)
it0=@(x)0; i1t0=@(x)0;                 % Eq.(E7.8.3a,b)
xf=2; M=40;  tf=2; N=40;
% Numerical approach
[un,x,t]=pde_wave(a,xf,tf,it0,i1t0,bx0t,bxft,M,N);
% Analytical approach using the Laplace transform
[T,X]=meshgrid(t,x);
ua = bx0t(T-X/a).*(T>=X/a).*(T<X/a+1); % Eq.(E7.8.10)
Discrepancy=norm(ua-un)/norm(ua)
subplot(221), mesh(t,x,ua), title('Analytical solution')
subplot(222)
for n=1:N % motion picture
   plot(x,ua(:,n), x,un(:,n),'r:'), axis([0 xf -1.1 1.1]), pause(0.1)
end
title('Animation')
```

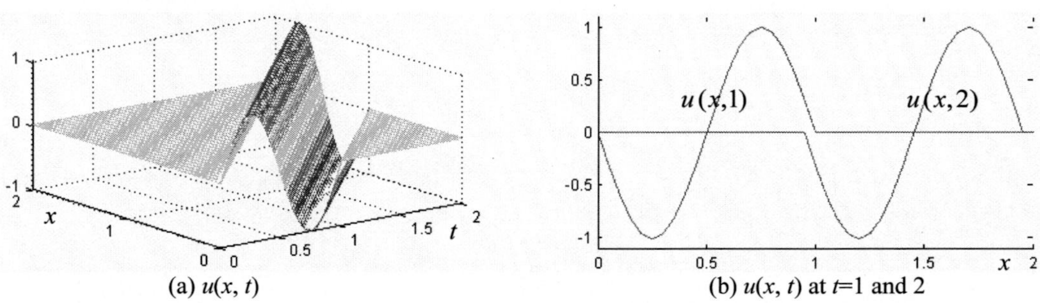

(a) $u(x, t)$ (b) $u(x, t)$ at t=1 and 2

Figure 7.13 Animated plots of the solution of the PDE (E7.8.1) s.t. the BC/IC (E7.8.2)/(E7.8.3)

The analytical solution (E7.8.10) and a numerical solution (using '`pde_wave()`') for the PDE (E7.8.1) with a=1 and $f(t)$=sin $2\pi t$ subject to the BCs (E7.8.2a,b) and ICs (E7.8.3a,b) are shown in Fig. 7.13, which has been obtained by running the above MATLAB script, "em07e08.m."

(**Example 7.9**) Laplace Transform for Hyperbolic PDE — 1D Wave Equation

Consider the 1D wave equation

$$a^2\frac{\partial^2 u(x,t)}{\partial x^2}+\sin\omega x=\frac{\partial^2 u(x,t)}{\partial t^2}\quad\text{for } 0\le x\le x_f=1 \text{ and } 0\le t\le 1 \tag{E7.9.1}$$

subject to the BCs and ICs

$$u(0,t)=0,\ \ u(x_f,t)=0\ \ \forall t \tag{E7.9.2a,b}$$

$$u(x,t)\big|_{t=0}=0,\qquad \frac{\partial u(x,t)}{\partial t}\bigg|_{t=0}=0\ \ \forall x \tag{E7.9.3a,b}$$

Taking the s-transform of Eq. (E7.9.1) with the IC (E7.9.3a,b) (w.r.t. t) yields

$$\mathcal{L}_t\left\{\overset{\text{LHS}}{a^2\frac{\partial^2 u(x,t)}{\partial x^2}}\right\}\overset{(7.5.1)}{\underset{\text{A.8(3)}}{=}}a^2\frac{d^2}{dx^2}U(x,s)+\frac{1}{s}\sin\omega x$$

$$=\mathcal{L}_t\left\{\overset{\text{RHS}}{\frac{\partial^2 u(x,t)}{\partial t^2}}\right\}\overset{(4.2.3)}{=}s^2U(x,s)-su(x,0)-\frac{\partial u(x,t)}{\partial t}\bigg|_{t=0}\overset{(\text{E7.9.3a,b})}{=}s^2U(x,s)$$

$$;\qquad a^2\frac{d^2}{dx^2}U(x,s)-s^2U(x,s)=-\frac{1}{s}\sin\omega x. \tag{E7.9.4}$$

We use Eq. (4.2.3) with n=2 and Table A.8(9) to take its $\tilde{s}$-transform (LT w.r.t. x with $\tilde{s}$ as a complex parameter) to write

$$a^2\left\{\tilde{s}^2\tilde{U}(\tilde{s})-\tilde{s}U(0,s)-\frac{d}{dx}U(x,s)\bigg|_{x=0}\right\}-s^2\tilde{U}(\tilde{s})=-\frac{\omega}{s(\tilde{s}^2+\omega^2)}$$
$$\overset{(\text{E7.9.2a})}{\longrightarrow}\left(\tilde{s}^2-\frac{s^2}{a^2}\right)\tilde{U}(\tilde{s})=-\frac{\omega}{a^2s(\tilde{s}^2+\omega^2)}+? \tag{E7.9.5}$$

Since $dU(x,s)/dx\big|_{x=0}$ is not yet clear, we solve this equation provisionally just for the particular solution part of $\tilde{U}(\tilde{s})$ (stemming from $\sin x$) and get its PFE (partial fraction expansion) as

$$\tilde{U}(\tilde{s})\overset{(\text{E7.9.5})}{=}\frac{-1}{\tilde{s}^2-(s/a)^2}\frac{\omega/a^2s}{\tilde{s}^2+\omega^2}=\frac{A}{\tilde{s}-s/a}+\frac{B}{\tilde{s}+s/a}+\frac{C\tilde{s}+D\omega}{\tilde{s}^2+\omega^2} \tag{E7.9.6}$$
$$=\frac{-\omega/a^2s}{(2s/a)\{(s/a)^2+\omega^2\}}\left(\frac{1}{\tilde{s}-s/a}-\frac{1}{\tilde{s}+s/a}\right)+\frac{C\tilde{s}+D\omega}{\tilde{s}^2+\omega^2}$$
$$=\frac{(A+B+C)\tilde{s}^3+(As/a-Bs/a+D\omega)\tilde{s}^2+(A\omega^2+B\omega^2-Cs^2/a^2)\tilde{s}+A\omega^2s/a-B\omega^2s/a-D\omega s^2/a^2}{\{\tilde{s}^2-(s/a)^2\}(\tilde{s}^2+\omega^2)}$$

where

$$A=-B=\frac{-\omega/a^2s}{(2s/a)\{(s/a)^2+\omega^2\}},\ C=-(A+B)=0$$
$$D=\frac{(B-A)s/a}{\omega}=\frac{1/a^2s}{(s/a)^2+\omega^2}=\frac{1}{s\{s^2+(a\omega)^2\}} \tag{E7.9.7}$$

```
%em07e09.m
a=1; w=2*pi; g=@(x)sin(w*x);
% BCs and ICs
bx0t=@(t)0; bxft=@(t)0;  % Eq.(E7.9.2a,b)
it0=@(x)0;  i1t0=@(x)0;  % Eq.(E7.9.3a,b)
xf=1; M=20;  tf=1; N=100;
% Numerical approach
[un,x,t]=pde_wave1(a,g,xf,tf,it0,i1t0,bx0t,bxft,M,N);
% Analytical approach using the Laplace transform
[T,X]=meshgrid(t,x);
ua = sin(w*X).*(1-cos(a*w*T))/(a*w)^2; % Eq.(E7.9.9)
Discrepancy=norm(ua-un)/norm(ua)
subplot(121), mesh(t,x,ua), title('Analytical solution')
subplot(122)
for n=1:N % motion picture
   plot(x,ua(:,n), x,un(:,n),'r:'), pause(0.1)
end
```

```
function [u,x,t]=pde_wave1(a,g,xf,tf,it0,i1t0,bx0,bxf,M,N)
%solve a^2*u_xx + g(x) = u_tt   for 0<=x<=xf, 0<=t<=tf
% Initial Condition: u(x,0) = it0(x), u_t(x,0) = i1t0(x)
% Boundary Condition: u(0,t) = bx0(t), u(xf,t)=bxf(t)
% M = Number of subintervals along x-axis
% N = Number of subintervals along t-axis
dx=xf/M;  x=[0:M]'*dx;
dt=tf/N;  t=[0:N]*dt;  dt2=dt^2;
if isnumeric(g), gx=repmat(g,M+1,1)*dt2; else gx=feval(g,x)*dt2; end
for i=1:M+1, u(i,1)= it0(x(i)); end
for k=1:N+1, u([1 M+1],k)= [bx0(t(k)); bxf(t(k))];  end
r= (a*dt/dx)^2,  r1= r/2;  r2= 2*(1-r);
u(2:M,2)= r1*u(1:M-1,1) +(1-r)*u(2:M,1) +r1*u(3:M+1,1) +dt*i1t0(x(2:M))
                +gx(2:M);
for k=3:N+1
   u(2:M,k)= r*u(1:M-1,k-1) +r2*u(2:M,k-1) +r*u(3:M+1,k-1) -u(2:M,k-2)
                +gx(2:M);
end
```

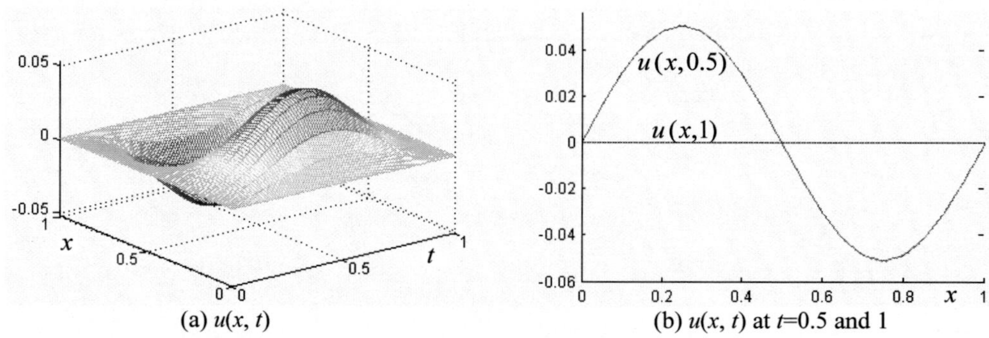

(a) $u(x, t)$ (b) $u(x, t)$ at t=0.5 and 1

Figure 7.14 Animated plots of the solution of the PDE (E7.9.1) s.t. the BC/IC (E7.9.2)/(E7.9.3)

Taking the inverse $\tilde{s}$-transform of Eq. (E7.9.6) with Eq. (E7.9.7) yields

$$U(x,s) = \mathcal{L}_x^{-1}\{\tilde{U}(s)\} \underset{\text{A.8(9)}}{\overset{\text{(E7.9.6)}}{=}} \frac{\sin\omega x}{s\{s^2+(a\omega)^2\}} = \frac{\sin\omega x}{(a\omega)^2}\left\{\frac{1}{s} - \frac{s}{s^2+(a\omega)^2}\right\} \tag{E7.9.8}$$

We can get the inverse s-transform (ILT w.r.t. t) of this equation as

$$u(x,t) = \mathcal{L}_t^{-1}\{U(x,s)\} \underset{\text{A.8(3),(11)}}{\overset{\text{(E7.9.7)}}{=}} \frac{1}{(a\omega)^2}\sin(\omega x)(1-\cos a\omega t) \tag{E7.9.9}$$

The analytical solution (E7.9.9) and a numerical solution (using '`pde_wave1()`') for the PDE (E7.9.1) with a=1 and ω=2π subject to the BCs (E7.9.2a,b) and ICs (E7.9.3a,b) are shown in Fig. 7.14, which has been obtained by running the above MATLAB script, "em07e09.m."

7.5.2 Using the Fourier Transform for PDEs

Like the Laplace transform (LT), the Fourier transform (FT) can also be used to deal with linear constant-coefficient PDEs when the spatial variable ranges over the entire line or half-line. A premise on which to use the FT for solving PDEs in two independent variables, say, x and t, is that differentiating w.r.t a variable and taking the FT w.r.t. another variable are interchangeable:

$$\underbrace{\mathcal{F}_t\left\{\frac{\partial f(x,t)}{\partial x}\right\} \overset{(6.2.1a)}{=} \int_{-\infty}^{\infty}\frac{\partial f(x,t)}{\partial x}e^{-j\omega t}dt}_{\text{Differentiation}-\text{Fourier transform}} = \underbrace{\frac{\partial}{\partial x}\int_{-\infty}^{\infty}f(x,t)e^{-j\omega t}dt = \frac{\partial}{\partial x}\mathcal{F}_t\{f(x,t)\}}_{\text{Fourier transform}-\text{Differentiation}} \overset{\text{function of }\omega\,\&\,x}{=} \frac{d}{dx}F(x,\omega) \tag{7.5.2}$$

$$\mathcal{F}_t\left\{\frac{\partial f(x,t)}{\partial t}\right\} \overset{(6.2.25)}{=} j\omega\mathcal{F}_t\{f(x,t)\} = j\omega F(x,\omega) \tag{7.5.3}$$

$$\underbrace{\mathcal{F}_x\left\{\frac{\partial f(x,t)}{\partial t}\right\} \overset{(6.2.1a)}{=} \int_{-\infty}^{\infty}\frac{\partial f(x,t)}{\partial t}e^{-j\nu x}dx}_{\text{Differentiation}-\text{Fourier transform}} = \underbrace{\frac{\partial}{\partial t}\int_{-\infty}^{\infty}f(x,t)e^{-j\nu x}dx = \frac{\partial}{\partial t}\mathcal{F}_x\{f(x,t)\}}_{\text{Fourier transform}-\text{Differentiation}} \overset{\text{function of }\nu\,\&\,t}{=} \frac{d}{dt}F(\nu,t) \tag{7.5.4}$$

$$\mathcal{F}_x\left\{\frac{\partial f(x,t)}{\partial x}\right\} \overset{(6.2.25)}{=} j\nu\mathcal{F}_x\{f(x,t)\} = j\nu F(\nu,t) \tag{7.5.5}$$

See the following examples.

(**Example 7.10**) Fourier Transform for Elliptic PDE — Laplace's Equation for Upper Half Plane

Consider the Laplace's equation

$$\frac{\partial^2 u(x,y)}{\partial x^2} + \frac{\partial^2 u(x,y)}{\partial y^2} = 0 \ \text{ for } -\infty < x < \infty \text{ and } y > 0 \tag{E7.10.1}$$

subject to the BC

$$u(x,0) = b_{y0}(x) \ \text{ for all } x \text{ and } y = 0 \tag{E7.10.2}$$

Taking the ν-transform (FT w.r.t. x) of Eq. (E7.10.1) with the BC (E7.10.2) yields

$$\mathcal{F}_x\left\{\frac{\partial^2 u(x,y)}{\partial x^2}\right\} + \mathcal{F}_x\left\{\frac{\partial^2 u(x,y)}{\partial y^2}\right\} \underset{(7.5.4)}{\overset{(7.5.5)}{=}} (j\nu)^2 U(\nu,y) + \frac{d^2}{dy^2}U(\nu,y) = 0$$

$$;\ \frac{d^2}{dy^2}U(\nu,y) - \nu^2 U(\nu,y) = 0 \ \text{ with } U(\nu,0) = \mathcal{F}_x\{u(x,0)\} = \mathcal{F}_x\{b_{y0}(x)\} = B_{y0}(\nu)\,. \tag{E7.10.3}$$

This is a second-order ODE (in y with v as a constant) with the IC $U(v,0)=B_{y0}(v)$. Noting that its characteristic equation (Eq. (3.2.6)) is $s^2-v^2=0$ with roots of $s=\pm|v|$, its solution can be written as

$$U(v,y)=K_1(v)e^{-|v|y}+K_2(v)e^{|v|y} \tag{E7.10.4}$$

where $K_2(v)=0$ to keep $U(v,y)$ bounded$K_1(v)$ can be determined from the BC (E7.10.2):

$$U(v,0)=K_1(v)e^{-|v|y}\overset{y=0}{=}K_1(v)=B_{y0}(v) \tag{E7.10.5}$$

Substituting these coefficients into Eq. (E7.10.4) yields

$$U(v,y)=B_{y0}(v)e^{-|v|y} \tag{E7.10.6}$$

Regarding this as the product of two FT expressions, $B_{y0}(v)$ and $H(v)=e^{-|v|y}$, we can use the real convolution property A.2(8) of the FT to write the inverse v-transform (ICTFT w.r.t. x) of Eq. (E7.10.6) as

$$u(x,y)=\mathcal{F}_x^{-1}\{U(v,y)\}=\mathcal{F}_x^{-1}\{B_{y0}(v)\}*\mathcal{F}_x^{-1}\{e^{-|v|y}\}$$
$$\overset{A.3(16)}{=}b_{y0}(x)*\frac{1}{\pi}\frac{y}{x^2+y^2}\overset{(4.2.8)}{=}\frac{1}{\pi}\int_{-\infty}^{\infty}b_{y0}(t)\frac{y}{(x-t)^2+y^2}dt \tag{E7.10.7}$$

```
%em07e10.m
clear, clf
f=@(x,y)0; g=@(x,y)0;   % Eq.(E7.10.1) w.r.t. (7.1.1)
% (Rectangular) Domain and BC types
x0=-5; xf=5; y0=1e-6; yf=10; D=[x0 xf y0 yf; 0 0 0 0];
bx0=@(y)0; bxf=@(y)0;                % BC
by0=@(x)1*(abs(x)<1); byf=@(x)0; % BC (E7.10.10)
% Numbers of grid points along x-/y-axes,
Mx=50; My=50;  dx=(xf-x0)/Mx; dy=(yf-y0)/My;
xx=x0+dx*[0:Mx]; yy=y0+dy*[0:My];
% Analytical approach
h=@(x,y)y./(x.^2+y.^2)/pi; % Eq.(E7.10.8)
ua(1,:) = by0(xx);
Me=ceil(2/dx); xe=[[-Me:-1]*dx+xx(1) xx xx(end)+[1:Me]*dx];
for n=2:length(yy)
   ue = conv(by0(xx),h(xe,yy(n)))*dx; % Eq.(E7.10.7)
   ua(n,:) = ue(ceil((max(-x0,0)+2)/dx)+[1:Mx+1]);
end
subplot(121), mesh(xx,yy,ua)
xlabel('x'), ylabel('y')
title('Analytical solution')
% Numerical approach
% BCs are taylored to the analytical solution since D is finite here.
bx0=@(y)by0(xe)*h(x0-xe,y)'*dx; % Eq.(E7.10.7) at x=x0
bxf=@(y)by0(xe)*h(xf-xe,y)'*dx; % Eq.(E7.10.7) at x=xf
byf=@(x)by0(xe)*h(x-xe,yy(My+1))'*dx; % Eq.(E7.10.7) at y=yf
tol=1e-5; MaxIter=600; % Error tolerance and Maximum iteration number
[un,x,y]=pde_poisson(f,g,bx0,bxf,by0,byf,D,Mx,My,tol,MaxIter);
Discrepancy=norm(ua-un)/norm(ua)
% Plot the solution graph
subplot(122), mesh(xx,yy,un)
xlabel('x'), ylabel('y')
title('Numerical solution using pde poisson()')
```

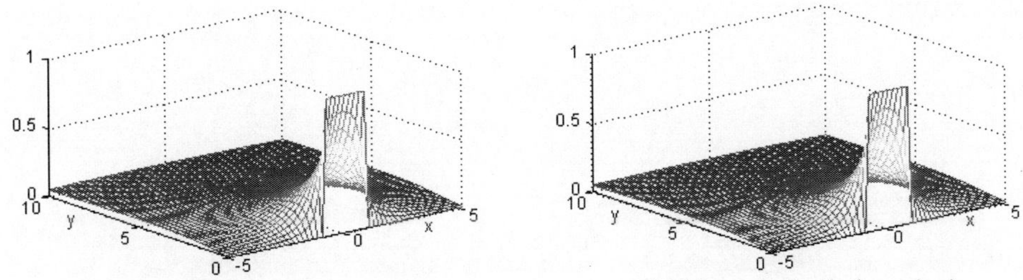

(a) Analytical solution $u(x,y)$ (b) Numerical solution $u(x,y)$

Figure 7.15 Analytical and numerical solutions of the PDE (E7.10.1) s.t. the BC (E7.10.10)

This is the convolution (w.r.t. x) of the two inverse v-transforms

$$\mathcal{F}_x^{-1}\{B_{y0}(v)\} = b_{y0}(x) \text{ and } h(x,y) = \mathcal{F}_x^{-1}\{e^{-|v|y}\} \overset{\text{A.3(16)}}{=} \frac{1}{\pi}\frac{y}{x^2+y^2} \tag{E7.10.8}$$

where the second inverse v-transform has been obtained by applying the duality (Table A.2(7)) to the CTFT pair (E6.6.1)–(E6.6.2) with y in place of $1/T$:

$$\overset{\text{(E6.6.1)}}{\frac{y}{2}e^{-y|t|}} \overset{\mathcal{F}_t}{\leftrightarrow} \overset{\text{(E6.6.2)}}{\frac{1}{1+(\omega/y)^2}} \overset{\text{(6.2.16)}}{\Leftrightarrow} \frac{1}{1+(t/y)^2} \overset{\mathcal{F}_t}{\leftrightarrow} y\pi e^{-y|\omega|} \overset{\times 1/y\pi}{\rightarrow} \frac{1}{\pi}\frac{y}{t^2+y^2} \overset{\mathcal{F}_t}{\leftrightarrow} e^{-y|\omega|} \tag{E7.10.9}$$

The analytical solution (E7.10.9) and a numerical solution (using '`pde_poisson()`') for the PDE (E7.10.1) subject to the BCs

$$u(x,0) = b_{y0}(x) = \begin{cases} 1 & \text{for } |x|<1 \\ 0 & \text{elsewhere} \end{cases} \tag{E7.10.10}$$

are shown in Fig. 7.15, which has been obtained by running the above MATLAB script, "em07e10.m."

(Example 7.11) Fourier Transform for Elliptic PDE — Laplace's Equation for Right Quarter Plane

Consider the Laplace's equation

$$\frac{\partial^2 u(x,y)}{\partial x^2} + \frac{\partial^2 u(x,y)}{\partial y^2} = 0 \text{ for } x>0 \text{ and } y>0 \tag{E7.11.1}$$

subject to the BC

$$u(x,0) = b_{y0}(x) \quad \text{for } x>0 \text{ and } y=0 \tag{E7.11.2a}$$

$$u(0,y) = 0 \quad \text{for } x=0 \text{ and } y>0 \tag{E7.11.2b}$$

[O-1] presents a very clever idea to use the solution (E7.10.7) for the upper half plane (in the previous example) assuming that another BC has been imposed for $x<0$ (on the negative x-axis) as

$$u(x,0) = -b_{y0}(-x) \quad \text{for } x<0 \text{ and } y=0 \tag{E7.11.3}$$

This is an odd extension of the BC (E7.11.2a) to the entire x-axis, which is expected to result in $u(0,y)=0$ for all y so that the BC (E7.11.2b) can naturally be satisfied. With these two BCs (E7.11.2a) and (E7.11.3), Eq. (E7.10.7) becomes

$$u(x,y) = \frac{1}{\pi}\left\{\int_{-\infty}^{0} -b_{y0}(-t)\frac{y}{(x-t)^2+y^2}dt + \int_0^{\infty} b_{y0}(t)\frac{y}{(x-t)^2+y^2}dt\right\}$$

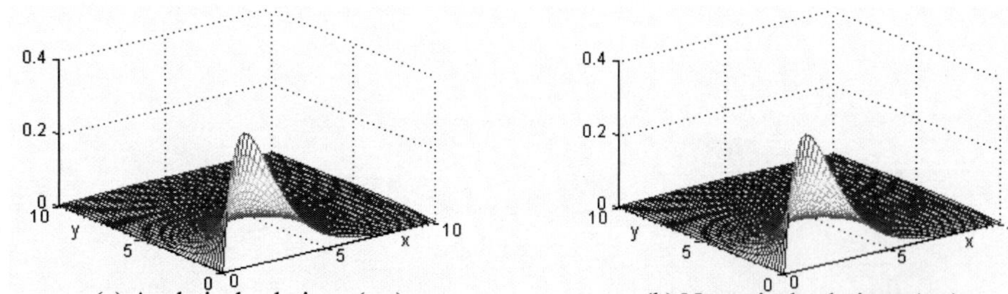

(a) Analytical solution $u(x,y)$ (b) Numerical solution $u(x,y)$

Figure 7.16 Analytical and numerical solutions of the PDE (E7.11.1) s.t. the BC (E7.11.5)

$$\overset{t\to -t}{\underset{dt\to -dt}{=}} \frac{y}{\pi}\left\{\int_0^\infty -b_{y0}(t)\frac{1}{(x+t)^2+y^2}dt+\int_0^\infty b_{y0}(t)\frac{1}{(x-t)^2+y^2}dt\right\}$$

$$=\frac{y}{\pi}\int_0^\infty\left\{\frac{1}{(x-t)^2+y^2}-\frac{1}{(x+t)^2+y^2}\right\}b_{y0}(t)\,dt \tag{E7.11.4}$$

The analytical solution (E7.11.4) and a numerical solution (using '`pde_poisson()`') for the PDE (E7.11.1) subject to the BCs

$$u(x,0)=b_{y0}(x)=xe^{-x} \text{ for } x\ge 0 \tag{E7.11.5}$$

are shown in Fig. 7.16, which has been obtained by running the MATLAB script "em07e11.m."

```
%em07e11.m
f=@(x,y)0;  g=@(x,y)0;  % Eq.(E7.11.1) w.r.t. (7.1.1)
x0=1e-6; xf=10; y0=1e-6; yf=10; D=[x0 xf y0 yf;0 0 0 0]; %Domain, BC types
bx0=@(y)0; bxf=@(y)0; by0=@(x)x.*exp(-x); byf=@(x)0; % BCs (E7.11.2,5)
% Numbers of grid points along x-/y-axes,
Mx=50; My=50;  dx=(xf-x0)/Mx; dy=(yf-y0)/My;
xx=x0+dx*[0:Mx]; yy=y0+dy*[0:My].';
% Analytical approach
h=@(t,x,y)(y./((t-x).^2+y.^2)-y./((t+x).^2+y.^2)).*by0(t)/pi; %(E7.11.4)
ua(1,:) = by0(xx); ua(:,1) = bx0(yy);
for n=2:length(yy) % Eq.(E7.11.4)
   for m=2:length(xx), ua(n,m)=quad(h,0,10,1e-5,[],xx(m),yy(n)); end
end
subplot(121), mesh(xx,yy,ua), title('Analytical solution')
% Numerical approach
% BCs are taylored to the analytical solution since D is finite here.
bx0=@(y)quad(h,0,10,1e-4,[],x0,y);  bxf=@(y)quad(h,0,10,1e-4,[],xf,y);
byf=@(x)quad(h,0,10,1e-4,[],x,yf);
tol=1e-5; MaxIter=600; % Error tolerance and Maximum iteration number
[un,x,y]= pde_poisson(f,g,bx0,bxf,by0,byf,D,Mx,My,tol,MaxIter);
Discrepancy=norm(ua-un)/norm(ua)
% Plot the solution graph
subplot(122), mesh(xx,yy,un), title('Numerical solution')
```

(Example 7.12) Fourier Transform for a Parabolic PDE — 1D Heat Equation over an Entire Line

Consider the following 1D heat equation:

$$a^2\frac{\partial^2 u(x,t)}{\partial x^2}=\frac{\partial u(x,t)}{\partial t} \quad \text{for } -\infty<x<\infty \text{ and } t>0 \tag{E7.12.1}$$

subject to the IC

$$u(x,0)=i_0(x)\ \forall x. \tag{E7.12.2}$$

Taking the v-transform (FT w.r.t. x) of Eq. (E7.12.1) yields

$$\mathcal{F}_x\left\{\overbrace{a^2\frac{\partial^2 u(x,t)}{\partial x^2}}^{\text{LHS}}\right\}\overset{(7.5.5)}{=}a^2(jv)^2\tilde{U}(v,t)=-(av)^2\tilde{U}(v,t)=\mathcal{F}_x\left\{\overbrace{\frac{\partial u(x,t)}{\partial t}}^{\text{RHS}}\right\}\overset{(7.5.4)}{=}\frac{d}{dt}\tilde{U}(v,t)$$

$$;\qquad \frac{d}{dt}\tilde{U}(v,t)+(av)^2\tilde{U}(v,t)=0 \tag{E7.12.3}$$

where

$$\tilde{U}(v,t)=\mathcal{F}_x\{u(x,t)\}\ \text{(FT w.r.t. } x) \tag{E7.12.4}$$

$$\text{IC:}\quad \tilde{U}(v,0)=\mathcal{F}_x\{u(x,0)\}=\mathcal{F}_x\{i_0(x)\} \tag{E7.12.5}$$

This is a first-order ODE (in t with v as a constant) with the IC (E7.12.5), which can be solved to yield

$$\tilde{U}(v,t)=\tilde{U}(v,0)e^{-(av)^2t}\overset{(E7.12.5)}{=}\mathcal{F}_x\{i_0(x)\}e^{-(av)^2t}\ \text{for } t>0 \tag{E7.12.6}$$

This is the product of two v-transform expressions, and the inverse v-transform of the latter can be found from the FT pair (P6.12.1)–(P6.12.2) or A.3(17), with $\omega=v$ and $\sigma^2/2=a^2t$ as

$$\mathcal{F}_x^{-1}\{e^{-(av)^2t}\}=\frac{1}{\sqrt{4\pi ta}}e^{-x^2/4a^2t}\ :\ \textit{Fundamental solution}\text{ to the heat equation (E7.12.1).} \tag{E7.12.7}$$

Then, using the real convolution property A.2(8) of FT, we write the inverse v-transform of Eq. (E7.12.6) as the x-domain convolution of two inverse v-transform expressions, $i_0(x)$ and $\mathcal{F}_x^{-1}\{e^{-(av)^2t}\}$

$$u(x,t)\overset{(E7.12.4)}{=}\mathcal{F}_x^{-1}\{\tilde{U}(v,t)\}\underset{A.2(8)}{\overset{(E7.12.6)}{=}}i_0(x)*\mathcal{F}_x^{-1}\{e^{-(av)^2t}\}\overset{(E7.12.7)}{=}i_0(x)*\frac{1}{\sqrt{4\pi ta}}e^{-x^2/4a^2t} \tag{E7.12.8}$$

The analytical solution (E7.12.8) and a numerical solution (using '`pde_heat_CN()`') for the PDE (E7.12.1) with $a=0.5$ subject to the IC and BCs

$$u(x,0)=\begin{cases}\sin(\pi x) & \text{for } 2\le x\le 3\\ 0 & \text{elsewhere}\end{cases} \tag{E7.12.9}$$

$$u(0,t)=0\ \forall t,\ u(5,t)=0\ \forall t \tag{E7.12.10a,b}$$

are shown in Fig. 7.17, which has been obtained by running the MATLAB script "em07e12.m."

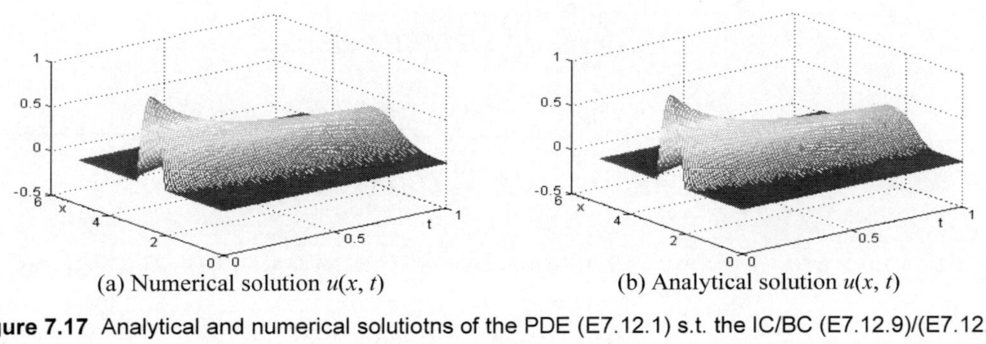

(a) Numerical solution $u(x, t)$ (b) Analytical solution $u(x, t)$

Figure 7.17 Analytical and numerical solutiotns of the PDE (E7.12.1) s.t. the IC/BC (E7.12.9)/(E7.12.10)

```
%em07e12.m
clear, clf
a=0.5; % The parameter of Eq.(E7.12.1)
it0=@(x)sin(pi*x).*(x>=2).*(x<=3); % IC (E7.12.9)
bx0=@(t)0;   bxf=@(t)0; % BCs (E7.12.10)
xf=5; M=100; dx=xf/M; tf=1; N=100; dt=tf/N;   % r=1
% Numerical solution
[u_CN,x,t]=pde_heat_CN(a,xf,tf,it0,bx0,bxf,M,N);
subplot(121), mesh(t,x,u_CN)
xlabel('t'), ylabel('x'), title('Numerical solution using pde_heat_CN()')
% Analytical solution
g=@(x,t)exp(-x.^2/(4*a^2*t))/sqrt(4*pi*t)/a; % Eq.(E7.12.8)
ua(:,1) = it0(x); % From IC
% Extend the x-interval of g(x,t) for [-1 0] to get a proper convolution
Me=ceil(1/dx); xe=[[-Me:-1].'*dx+x(1); x];
for n=2:length(t)
   ue = conv(it0(x),g(xe,t(n)))*dx; % Eq.(E7.12.8)
   ua(:,n) = ue(Me+[1:M+1]);
end
Discrepancy=norm(u_CN-ua)/norm(ua)
subplot(122), mesh(t,x,ua)
xlabel('t'), ylabel('x'), title('Analytical solution')
```

(Example 7.13) Fourier Transform for a Hyperbolic PDE — 1D Wave Equation over an Entire Line

Consider the following 1D wave equation:

$$a^2\frac{\partial^2 u(x,t)}{\partial x^2}=\frac{\partial^2 u(x,t)}{\partial t^2} \text{ for } -\infty<x<\infty \text{ and } t>0 \tag{E7.13.1}$$

subject to the IC

$$u(x,t)\big|_{t=0}=g(x) \text{ and } \frac{\partial}{\partial t}u(x,t)\Big|_{t=0}=h(x) \left(\text{with } \int_{-\infty}^{\infty}h(x)\,dx=0\right)\ \forall x \tag{E7.13.2}$$

Taking the v-transform (FT w.r.t. x) of Eq. (E7.13.1) yields

$$\mathcal{F}_x\left\{\overset{\text{LHS}}{a^2\frac{\partial^2 u(x,t)}{\partial x^2}}\right\}\overset{(7.5.5)}{=}a^2(jv)^2\tilde{U}(v,t)=-(av)^2U(v,t)=\mathcal{F}_x\left\{\overset{\text{RHS}}{\frac{\partial^2 u(x,t)}{\partial t^2}}\right\}\overset{(7.5.4)}{=}\frac{d^2}{dt^2}\tilde{U}(v,t)$$

$$;\quad \frac{d^2}{dt^2}\tilde{U}(v,t)+(av)^2\tilde{U}(v,t)=0 \tag{E7.13.3}$$

where

$$\tilde{U}(v,t)=\mathcal{F}_x\{u(x,t)\} \text{ (FT w.r.t. } x) \tag{E7.13.4}$$

$$\text{IC1: } \tilde{U}(v,t)\big|_{t=0}=\mathcal{F}_x\{u(x,0)\}=\mathcal{F}_x\{g(x)\}=G(v) \tag{E7.13.5a}$$

$$\text{IC2: } \frac{d}{dt}\tilde{U}(v,t)\Big|_{t=0}=\mathcal{F}_x\left\{\frac{d}{dt}u(x,0)\right\}=\mathcal{F}_x\{h(x)\}=H(v) \tag{E7.13.5b}$$

This is a second-order ODE (in t with v as a constant) with the ICs (E7.13.5), which can be solved to yield

$$\tilde{U}(v,t)=A(v)\cos(avt)+B(v)\sin(avt) \tag{E7.13.6}$$

where the coefficients $A(v)$ and $B(v)$ can be determined from the ICs (E7.13.5a,b):

$$\tilde{U}(v,0) = A(v)\cos(av0) \overset{\text{(E7.13.5a)}}{=} G(v) \rightarrow A(v) = G(v) \tag{E7.13.7a}$$

$$\frac{d}{dt}\tilde{U}(v,t)\Big|_{t=0} = -avA(v)\sin(av0) + avB(v)\cos(av0) \overset{\text{(E7.13.5b)}}{=} H(v) \rightarrow B(v) = \frac{H(v)}{av} \tag{E7.13.7b}$$

Substituting Eqs. (E7.13.7a,b) into Eq. (E7.13.6) yields

$$\tilde{U}(v,t) \underset{\text{(E7.13.7a,b)}}{\overset{\text{(E7.13.6)}}{=}} G(v)\cos(avt) + \frac{H(v)}{av}\sin(avt)$$

$$\underset{\text{(C.22)}}{\overset{\text{(C.21)}}{=}} G(v)\frac{1}{2}(e^{jvat} + e^{-jvat}) + \frac{H(v)}{ajv}\frac{1}{2}(e^{jvat} - e^{-jvat}) \tag{E7.13.8}$$

Using the time(t)/space(x)-shifting property A.2(5) of the Fourier transform

$$x(t-t_1) \underset{\text{A.2(5)}}{\overset{\mathcal{F}_t}{\leftrightarrow}} X(\omega)\,e^{-j\omega t_1} \Leftrightarrow g(x-x_1) \overset{\mathcal{F}_x}{\leftrightarrow} G(v)e^{-jvx_1} \tag{E7.13.9}$$

the inverse v-transform of the first term can be found as

$$\mathcal{F}_x^{-1}\left\{G(v)\frac{1}{2}(e^{jvat}+e^{-jvat})\right\} \underset{x_1=\mp at}{\overset{\text{(E7.13.9)}}{=}} \frac{1}{2}\{g(x+at)+g(x-at)\} \tag{E7.13.10}$$

Noting that the time(t)/space(x)-integration property A.2(11) of the Fourier transform yields

$$\mathcal{F}_x^{-1}\left\{\frac{H(v)}{ajv}\right\} = \frac{1}{a}\int_{-\infty}^{x} h(\xi)d\xi \ \text{ with } H(0) = \int_{-\infty}^{\infty} h(\xi)d\xi = 0 \tag{E7.13.11}$$

we again use the space(x)-shifting property A.2(5) of the v-transform to get the inverse v-transform of the second term of Eq. (E7.13.8) as

$$\mathcal{F}_x^{-1}\left\{\frac{H(v)}{ajv}\frac{1}{2}(e^{jvat}-e^{-jvat})\right\} \underset{\text{A.2(5)}}{\overset{\text{(E7.13.11)}}{=}} \frac{1}{2a}\left\{\int_{-\infty}^{x+at} h(\xi)d\xi - \int_{-\infty}^{x-at} h(\xi)d\xi\right\} = \frac{1}{2a}\int_{x-at}^{x+at} h(\xi)d\xi \tag{E7.13.12}$$

Thus, we can add Eqs. (E7.13.10) and (E7.13.12) to write the inverse v-transform of Eq. (E7.13.8) as

$$u(x,t) = \mathcal{F}_x^{-1}\{\tilde{U}(v,t)\} = \frac{1}{2}\{g(x+at)+g(x-at)\} + \frac{1}{2a}\int_{x-at}^{x+at} h(\xi)d\xi \tag{E7.13.13}$$

This is the general solution to the 1D wave equation like Eq. (E7.13.1) (with propagation speed a), which is called the *d'Alembert formula*.

The analytical solution (E7.13.13) and a numerical solution (using '`pde_wave()`') for the PDE (E7.13.1) with $a=1$ subject to the IC and BCs

$$u(0,t)=0 \ \forall\, t, \ \ u(6,t)=0 \ \forall\, t \tag{E7.13.14a,b}$$

$$u(x,t)\big|_{t=0} = \begin{cases}(x-2)(4-x) & \text{for } 2\le x\le 4 \\ 0 & \text{elsewhere}\end{cases}, \quad \frac{\partial u(x,t)}{\partial t}\Big|_{t=0} = \begin{cases}\cos(\pi(x-3)) & \text{for } 2\le x\le 4 \\ 0 & \text{elsewhere}\end{cases} \tag{E7.13.15a,b}$$

are shown in Fig. 7.18, which has been obtained by running the MATLAB script "em07e13.m."

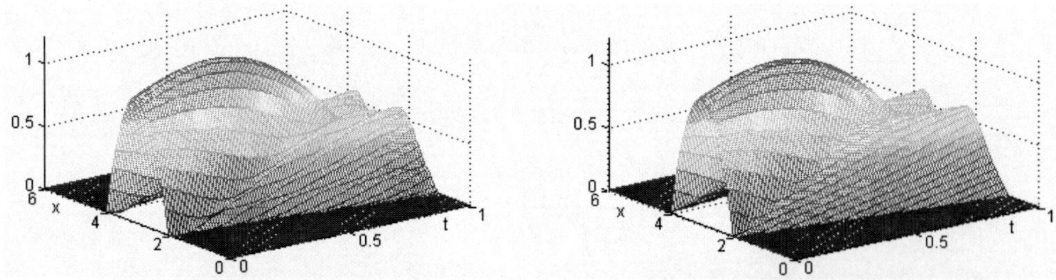

(a) Numerical solution $u(x, t)$ (b) Analytical solution $u(x, t)$

Figure 7.18 Analytical and numerical solutions of the PDE (E7.13.1) s.t. the BC/IC (E7.13.14)/(E7.13.15)

```
%em07e13.m
clear, clf
a=1; % The parameter of Eq.(E7.13.1)
bx0=@(t)0;  bxf=@(t)0; % BCs (E7.14.14a,b)
it0=@(x)(x-2).*(4-x).*(x>=2).*(x<=4); % IC (E7.13.15a)
i1t0=@(x)cos(pi*(x-3)).*(x>=2).*(x<=4); % IC (E7.13.15b)
xf=6; M=60; dx=xf/M; tf=1; N=100; dt=tf/N;  % r=0.01
% Numerical solution
[un,x,t]=pde_wave(a,xf,tf,it0,i1t0,bx0,bxf,M,N);
subplot(121), mesh(t,x,un)
xlabel('t'), ylabel('x')
title('Numerical solution using pde_wave()')
% Analytical solution
ua(:,1) = it0(x); % From IC
for n=2:length(t)
   for m=1:M+1
      x1=x(m)-a*t(n);  x2=x(m)+a*t(n);
      ua(m,n)=(it0(x1)+it0(x2))/2+quad(i1t0,x1,x2)/a/2; % Eq.(E7.13.13)
   end
end
Discrepancy=norm(un-ua)/norm(ua)
subplot(122), mesh(t,x,ua)
xlabel('t'), ylabel('x'), title('Analytical solution')
```

Problems

7.1 Elliptic PDEs — Poisson Equations

Use the MATLAB function 'pde_poisson()' (in Section 7.1) to solve the following PDEs, and graph the solutions by using the MATLAB command 'mesh()' as shown in Fig. P7.1.

(a)
$$\nabla^2 u(x,y)=\frac{\partial^2 u(x,y)}{\partial x^2}+\frac{\partial^2 u(x,y)}{\partial y^2}=0 \text{ for } 0\le x\le 1 \text{ and } 0\le y\le 1 \quad \text{(P7.1.1)}$$

subject to the BCs

$$u(0,y)=0,\ u(1,y)=0,\ u(x,0)=\sin\pi x,\ u(x,1)=0 \quad \text{(P7.1.2)}$$

Divide the solution region into $M_x\times M_y$=20×20 sections.

Also, use Eq. (7.1.16) and/or the MATLAB function 'pde_Laplace_ana()' (in Section 7.1) to find the analytical solution.

(b)
$$\nabla^2 u(x,y)=\frac{\partial^2 u(x,y)}{\partial x^2}+\frac{\partial^2 u(x,y)}{\partial y^2}=0 \text{ for } 0\le x\le 1 \text{ and } 0\le y\le \frac{\pi}{2} \quad \text{(P7.1.3)}$$

subject to the BCs

$$u(0,y)=4\cos(3y),\ u(1,y)=4e^{-3}\cos(3y),\ u(x,0)=4e^{-3x},\ u(x,\pi/2)=0 \quad \text{(P7.1.4)}$$

Divide the solution region into $M_x\times M_y$=20×20 sections.

(c)
$$\frac{\partial^2 u(x,y)}{\partial x^2}+\frac{\partial^2 u(x,y)}{\partial y^2}=10e^{2x+y} \text{ for } 0\le x\le 1 \text{ and } 0\le y\le 2 \quad \text{(P7.1.5)}$$

subject to the BCs

$$u(0,y)=2e^{y},\ u(1,y)=2e^{2x+y},\ u(x,0)=2e^{2x},\ u(x,2)=2e^{2x+2} \quad \text{(P7.1.6)}$$

Divide the solution region into $M_x\times M_y$=20×40 sections.

(d)
$$\frac{\partial^2 u(x,y)}{\partial x^2}+\frac{\partial^2 u(x,y)}{\partial y^2}=x+y \text{ for } 0\le x\le 1 \text{ and } 0\le y\le 1 \quad \text{(P7.1.7)}$$

subject to the BCs

$$u(0,y)=y^2,\ u(1,y)=1,\ u(x,0)=x^2,\ u(x,1)=1 \quad \text{(P7.1.8)}$$

Divide the solution region into $M_x\times M_y$=5×10 sections.

(e)
$$\frac{\partial^2 u(x,y)}{\partial x^2}+\frac{\partial^2 u(x,y)}{\partial y^2}-12.5\pi^2 u(x,y)=-25\pi^2\cos(\frac{5\pi}{2}x)\cos(\frac{5\pi}{2}y) \text{ for } 0\le x,y\le 0.4 \quad \text{(P7.1.9)}$$

subject to the BCs

$$u(0,y)=\cos(\frac{5\pi}{2}y),\ u(0.4,y)=-\cos(\frac{5\pi}{2}y) \quad \text{(P7.1.10a,b)}$$

$$u(x,0)=\cos(\frac{5\pi}{2}x),\ u(x,0.4)=-\cos(\frac{5\pi}{2}x) \quad \text{(P7.1.10c,d)}$$

Divide the solution region into $M_x\times M_y$=40×40 sections.

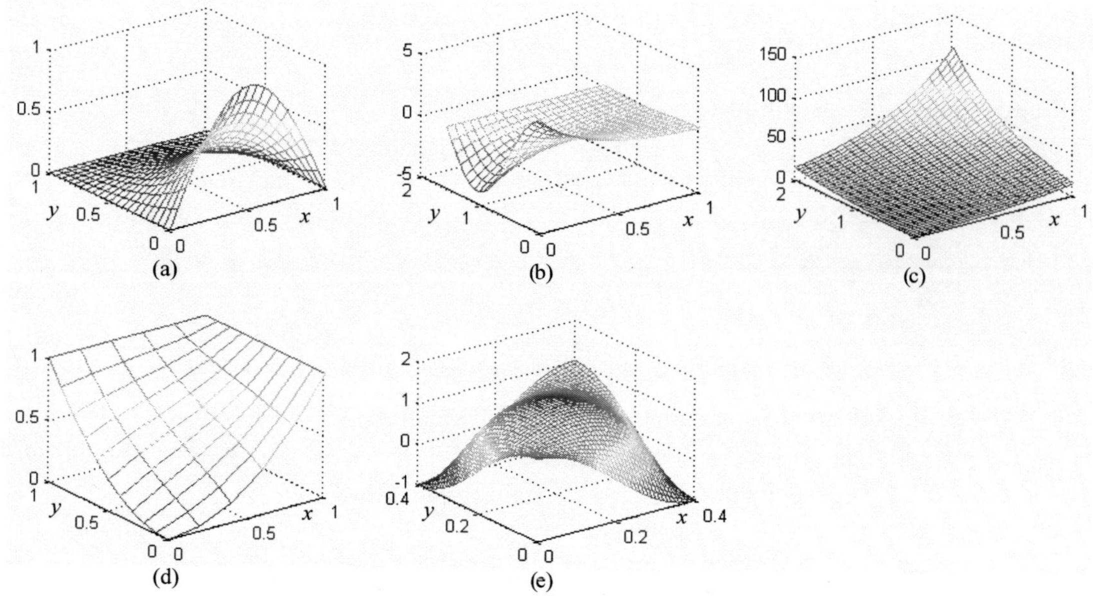

Figure P7.1 Solution graphs for Problem 7.1

7.2 Elliptic PDE — Laplace's Equation with Dirichlet Boundary Conditions

(a) Consider the following PDE having unity coefficients:

$$\frac{\partial^2 u(x,y)}{\partial x^2}+\frac{\partial^2 u(x,y)}{\partial y^2}=0 \tag{P7.2a.1}$$

over a domain $D=\{(x,y)\,|\,0\le x\le x_f,\ 0\le y\le y_f\}$ subject to the Dirichlet BCs:

$$u(0,y)=0,\ u(x_f,y)=0,\ u(x,0)=0,\ \text{and } u(x,y_f)=b_{yf}(x) \tag{P7.2a.2}$$

By the method of separation of variables, we substitute

$$u(x,y) = X(x)Y(y) \tag{P7.2a.3}$$

into the PDE (P7.2a.1) and divide by XY to write

$$X''Y+XY''=0;\quad \frac{X''}{X}=-\frac{Y''}{Y}=-\lambda;\quad \begin{cases} X''+\lambda X=0 \\ Y''-\lambda Y=0 \end{cases} \tag{P7.2a.4}$$

Also, we substitute Eq. (P7.2a.3) into the BC (P7.2a.2) to write

$$u(0,y)=X(0)Y(y)=0 \rightarrow X(0)=0 \tag{P7.2a.5a}$$

$$u(x_f,y)=X(x_f)Y(y)=0 \rightarrow X(x_f)=0 \tag{P7.2a.5b}$$

$$u(x,0)=X(x)Y(0)=0 \rightarrow Y(0)=0 \tag{P7.2a.5c}$$

$$u(x,y_f)=X(x)Y(y_f)=b_{yf}(x) \tag{P7.2a.5d}$$

Referring to the second column of Table 7.1, the candidates for $X(x)$ satisfying Eq. (P7.2a.4) together with the BC (P7.2a.5a) are

$$X(x)=\begin{cases}\sin\omega x & \text{if } \lambda=\omega^2>0\\ x & \text{if } \lambda=0\\ \sinh\omega x=(e^{\omega x}-e^{-\omega x})/2 & \text{if } \lambda=-\omega^2<0\end{cases} \tag{P7.2a.6}$$

However, the second and third ones cannot satisfy the BC (P7.2a.5b) and therefore, the candidate for $X(x)$ is just

$$X(x)=\boxed{} \text{ with } \omega=\frac{n\pi}{x_f}(n=1,2,3,\cdots) \text{ with } \lambda=\omega^2>0 \tag{P7.2a.7}$$

Now, referring to the third column of Table 7.1 for $\lambda=\omega^2=(n\pi/x_f)^2>0$, the candidate for $Y(y)$ satisfying the BC (P7.2a.5c) is

$$Y(y)=\sinh\omega y=\sinh(\frac{n\pi}{x_f}y) \text{ with } \omega=\frac{n\pi}{x_f}(n=1,2,3,\cdots) \text{ for } \lambda=\omega^2>0. \tag{P7.2a.8}$$

Plugging Eqs. (P7.2a.7) and (P7.2a.8) into Eq. (P7.2a.3) yields the separable solutions:

$$u_n(x,y)=X(x)Y(y)=\sin(\frac{n\pi}{x_f}x)\sinh(\frac{n\pi}{x_f}y) \tag{P7.2a.9}$$

Hence, the general solution of the Laplace's equation can be written as

$$u(x,y)=\sum_{n=1}^{\infty}A(n)u_n(x,y)\overset{(P7.2a.9)}{=}\sum_{n=1}^{\infty}A(n)\sin(\frac{n\pi}{x_f}x)\sinh(\frac{n\pi}{x_f}y) \tag{P7.2a.10}$$

where the coefficient $A(n)$'s are to be fixed by the remaining BC (P7.2a.5d):

$$u(x,y_f)\overset{(P7.2a.10)}{=}\sum_{n=1}^{\infty}A(n)\sinh(\frac{n\pi}{x_f}y_f)\sin(\frac{n\pi}{x_f}x)\overset{(P7.2a.5d)}{=}\boxed{} \tag{P7.2a.11}$$

Regarding this as the Fourier sine series for $b_{yf}(x)$, the coefficients can be obtained as

$$b_n=A(n)\sinh(\frac{n\pi}{x_f}y_f)\underset{P=2x_f\text{, odd symmetry}}{\overset{(6.1.3b),\text{ Table 6.1}}{=}}\frac{2}{x_f}\int_0^{x_f}\boxed{}\,dx \tag{P7.2a.12}$$

$$\rightarrow A(n)=\frac{b_n}{\sinh(\frac{n\pi}{x_f}y_f)}=\frac{2}{x_f\sinh(\frac{n\pi}{x_f}y_f)}\int_0^{x_f}\boxed{}\,dx \tag{P7.2a.13}$$

Choose the variable name together with its indices corresponding to these coefficients $A(n)$ from the MATLAB function '`pde_Laplace_ana()`' (in Section 7.1). To solve the PDE (P7.2a.1) over a domain $D=\{(x,y)\,|\,0\le x\le 1,\ 0\le y\le 1\}$ subject to the Dirichlet BCs:

$$u(0,y)=0,\ u(1,y)=0,\ u(x,0)=0, \text{ and } u(x,1)=\sin(2\pi x) \tag{P7.2a.14}$$

complete the MATLAB script "em07p02a.m" and run it to get the solution graph as shown in Fig. P7.2(a).

```
%em07p02a.m
clear, clf
f=@(x,y)0;    g=@(x,y)0;
% (Rectangular) Domain and BC types
x0=0; xf=1; y0=0; yf=1; D=[x0 xf y0 yf; 0 0 0 0];
% BCs
bx0=@(y)0; bxf=@(y)0; by0=@(x)0; byf=@(x)sin(2*??*x); % Eq.(P7.2a.14)
% Numerical approach
% Numbers of grid points along x-/y-axes, Error tolerance,
Mx=50; My=50; tol=1e-4; MaxIter=500;
[un,xx,yy]= pde_poisson(f,g,bx0,bxf,by0,byf,D,Mx,My,tol,MaxIter);
% Analytical approach
K=10; [ua,A]=pde_Laplace_ana(xf,yf,bx0,bxf,by0,byf,Mx,My,K);
Discrepancy1=norm(ua-un)/norm(ua)
syms x;
ua1=0;   [X,Y]=meshgrid(xx,yy);
for n=1:N
   npixf=n*pi/xf;
   Atmp=2/xf/sinh(npixf*??)*int(???*sin(npixf*x),0,xf);%Eq.(P7.2a.13)
   A1(n)=eval(Atmp); % Symbolic(output of int())-to-Numeric conversion
   ua1 = ua1 + A1(n)*sin(npixf*X).*sinh(npixf*?); % Eq.(P7.2a.11)
end
[A; A1] % For comparison (check)
Discrepancy2=norm(ua-ua1)/norm(ua)
% Plot the solution graphs
subplot(121), mesh(xx,yy,un), view([-37.5 10])
subplot(122), mesh(xx,yy,ua), view([-37.5 10])
```

(b) Consider the same PDE

$$\frac{\partial^2 u(x,y)}{\partial x^2}+\frac{\partial^2 u(x,y)}{\partial y^2}=0 \tag{P7.2b.1}$$

over a domain $D=\{(x,y)\,|\,0\le x\le x_f=\infty,\ 0\le y\le y_f\}$ subject to the Dirichlet BCs:

$$u(0,y)=u_0\sin(\pi y/y_f),\ u(\infty,y)=0,\ u(x,0)=0,\ \text{and}\ u(x,y_f)=0 \tag{P7.2b.2}$$

By the method of separation of variables, we substitute

$$u(x,y) = X(x)Y(y) \tag{P7.2b.3}$$

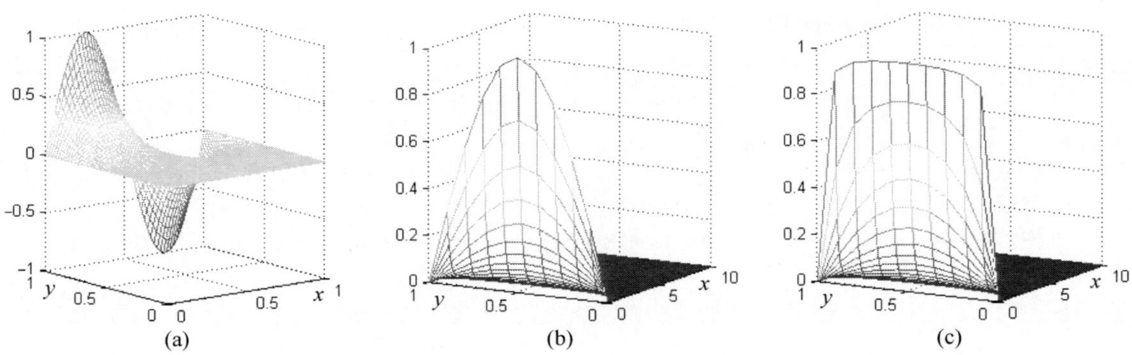

Figure P7.2 Solution graphs for Problem 7.2

into the PDE (P7.2b.1) and divide by XY to write

$$X''Y + XY'' = 0\,; \quad \frac{X''}{X} = -\frac{Y''}{Y} = -\lambda; \quad \begin{cases} X''+\lambda X = 0 \\ Y''-\lambda Y = 0 \end{cases} \tag{P7.2b.4}$$

Also, we substitute Eq. (P7.2b.3) into the BC (P7.2b.2) to write

$$u(0,y) = X(0)Y(y) = u_0 \sin(\pi y / y_f) \tag{P7.2b.5a}$$

$$u(\infty,y) = X(\infty)Y(y) = 0 \;\rightarrow X(\infty) = 0 \tag{P7.2b.5b}$$

$$u(x,0) = X(x)Y(0) = 0 \;\rightarrow Y(0) = 0 \tag{P7.2b.5c}$$

$$u(x,y_f) = X(x)Y(y_f) = 0 \;\rightarrow Y(y_f) = 0 \tag{P7.2b.5d}$$

Referring to the second column of Table 7.1, the candidate for $X(x)$ that may satisfy Eq. (P7.2b.4) together with the BC (P7.2b.5b) is

$$X(x) = \boxed{} \text{ with } \lambda = -\omega^2 < 0 \tag{P7.2b.6}$$

Now, referring to the third column of Table 7.1 for $\lambda = -\omega^2 = (n\pi/y_f)^2 > 0$, the candidate for $Y(y)$ satisfying the BC (P7.2b.5c,d) is

$$Y(y) = \boxed{} \text{with } \omega = \frac{n\pi}{y_f}(n = 1,2,3,\cdots) \text{ for } \lambda = -\omega^2 < 0 \tag{P7.2b.7}$$

Plugging Eqs. (P7.2b.6) and (P7.2b.7) into Eq. (P7.2b.3) yields the separable solutions:

$$u_n(x,y) = X(x)Y(y) = e^{-n\pi x/y_f} \sin(\frac{n\pi}{y_f} y) \tag{P7.2b.8}$$

Hence, the general solution of the Laplace's equation can be written as

$$u(x,y) = \sum_{n=1}^{\infty} A(n)\, u_n(x,y) \overset{\text{(P7.2b.8)}}{=} \sum_{n=1}^{\infty} A(n) e^{-n\pi x/y_f} \sin(\frac{n\pi}{y_f} y) \tag{P7.2b.9}$$

```
%em07p02b.m
f=@(x,y)0; u0=1;
if ~exist('g'), g=@(x,y)0; end
% (Rectangular) Domain and BC types
x0=0; xf=10; y0=0; yf=1; D=[x0 xf y0 yf; 0 0 0 0];
% BCs
bx0=@(y)sin(????); bxf=@(y)0; by0=@(x)0; byf=@(x)0; % Eq.(P7.2b.11)
% Numerical approach
% Numbers of grid points along x-/y-axes, Error tolerance,
Mx=100; My=10; tol=1e-5; MaxIter=500;
[un,xx,yy]=pde_poisson(f,g,bx0,bxf,by0,byf,D,Mx,My,tol,MaxIter);
% Analytical approach
N=20; [ua,A]=pde_Laplace_ana(xf,yf,bx0,bxf,by0,byf,Mx,My,N);
Discrepancy1=norm(un-ua)/norm(ua)
% Plot the solution graph
subplot(132), mesh(xx,yy,un), view([-60 10])
[X,Y]=meshgrid(xx,yy);
A1(1)=u0; N=length(A1); ua1=0;  % Eqs.(P7.2b.9) with (P7.2b.10)
for n=1:N, ua1=ua1+A1(n)*???(-n*pi/yf*X).*sin(n*pi/??*Y); end
Discrepancy2=norm(ua-ua1)/norm(ua)
```

where the coefficient $A(n)$'s are to be fixed by the remaining BC (P7.2b.5a):

$$u(0,y_f) \overset{\text{(P7.2b.9)}}{=} \sum_{n=1}^{\infty} A(n) e^{-n\pi 0/y_f} \sin(\frac{n\pi}{y_f}y) \overset{\text{(P7.2b.5a)}}{=} u_0 \sin(\frac{\pi}{y_f}y) \rightarrow A(n) = \begin{cases} \boxed{} & \text{for } n=1 \\ 0 & \text{otherwise} \end{cases} \quad \text{(P7.2b.10)}$$

To solve the PDE (P7.2b.1) over a domain $D=\{(x,y)\,|\,0 \le x \le x_f=10,\ 0 \le y \le y_f=1\}$, subject to the Dirichlet BCs:

$$u(0,y) = \sin \pi y,\ u(10,y) \approx 0,\ u(x,0)=0, \text{ and } u(x,1)=0 \quad \text{(P7.2b.11)}$$

complete the above MATLAB script, "em07p02b.m," and run it to get the solution graph as shown in Fig. P7.2(b).

(c) Consider the same PDE as Eq. (P7.2b.1)

$$\frac{\partial^2 u(x,y)}{\partial x^2} + \frac{\partial^2 u(x,y)}{\partial y^2} = 0 \quad \text{(P7.2c.1)}$$

over a domain $D=\{(x,y)\,|\,0 \le x \le x_f=10,\ 0 \le y \le y_f=1\}$, but with slightly different Dirichlet BCs:

$$u(0,y) = u_0,\ u(10,y) \approx 0,\ u(x,0)=0, \text{ and } u(x,1)=0 \quad \text{(P7.2c.2)}$$

Referring to the second column of Table 7.1, the candidate for $X(x)$ that may satisfy Eq. (P7.2c.2b) not exactly but closely is

$$X(x) = \boxed{} \quad \text{with } \lambda = -\omega^2 < 0 \quad \text{(P7.2c.3)}$$

Referring to the third column of Table 7.1 for $\lambda = -\omega^2 = (n\pi/y_f)^2 > 0$, the candidate for $Y(y)$ satisfying the BC (P7.2c.2c,d) is

$$Y(y) = \boxed{} \quad \text{with } \omega = \frac{n\pi}{y_f} (n=1,2,3,\cdots) \text{ for } \lambda = -\omega^2 < 0 \quad \text{(P7.2c.4)}$$

Plugging Eqs. (P7.2c.3) and (P7.2c.4) into Eq. (P7.2b.3) yields the separable solutions:

$$u_n(x,y) = X(x)Y(y) = e^{-n\pi x/y_f} \sin(\frac{n\pi}{y_f}y) \quad \text{(P7.2c.5)}$$

Hence, the general solution of the Laplace's equation can be written as

$$u(x,y) = \sum_{n=1}^{\infty} C(n)\, u_n(x,y) \overset{\text{(P7.2c.5)}}{=} \sum_{n=1}^{\infty} C(n) e^{-n\pi x/y_f} \sin(\frac{n\pi}{y_f}y) \quad \text{(P7.2c.6)}$$

where the coefficient $C(n)$'s are to be fixed by the remaining BC (P7.2c.2a):

$$u(0,y_f) \overset{\text{(P7.2c.6)}}{=} \sum_{n=1}^{\infty} C(n) e^{-n\pi 0/y_f} \sin(\frac{n\pi}{y_f}y) = \sum_{n=1}^{\infty} C(n) \sin(\frac{n\pi}{y_f}y) \overset{\text{(P7.2c.2a)}}{=} u_0 \quad \text{(P7.2c.7)}$$

$$\rightarrow C(n) \underset{P=2y_f\text{, odd symmetry}}{\overset{\text{(6.1.3b), Table 6.1}}{=}} \frac{2}{y_f} \int_0^{y_f} \boxed{}\, dy$$

$$= -\frac{2u_0}{y_f} \frac{y_f}{n\pi} \cos(\frac{n\pi}{y_f}y) \Big|_0^{y_f} = \boxed{} \text{ for odd } n \quad \text{(P7.2c.8)}$$

To solve the PDE (P7.2c.1) with u_0=1, complete the MATLAB script "em07p02c.m" (including the corresponding equation number) to get the solution graph as shown in Fig. P7.2(c).

```
%em07p02c.m
clear
f=@(x,y)0; u0=1;
g=@(x,y)0;
% (Rectangular) Domain and BC types
x0=0; xf=10; y0=0; yf=1; D=[x0 xf y0 yf; 0 0 0 0];
% BCs
bx0=@(y)??;  bxf=@(y)0; by0=@(x)0;  byf=@(x)0; % Eq.(???????)
% Numerical approach
% Numbers of grid points along x-/y-axes, Error tolerance,
Mx=100; My=10; tol=1e-5; MaxIter=500;
[un,xx,yy,itr]=pde_poisson(f,g,bx0,bxf,by0,byf,D,Mx,My,tol,MaxIter);
% Analytical approach
K=20;
[ua,A]=pde_Laplace_ana(xf,yf,bx0,bxf,by0,byf,Mx,My,K);
Discrepancy1=norm(un-ua)/norm(ua)
% Plot the solution graph
subplot(133), mesh(xx,yy,un), view([-60 10])
[X,Y]=meshgrid(xx,yy);
% In case where the BC at x=x0 and 0<=y<=yf is bx0=@(y)u0
ua1 = 0;
for n=1:2:K
   C(n)=4*u0/n/pi; % Eq.(???????)
   ua1 = ua1 + C(n)*exp(-n*pi/yf*?).*sin(n*pi/yf*Y); % Eq.(???????)
end
Discrepancy2=norm(ua1-ua)/norm(ua)
```

7.3 Elliptic PDE — Laplace's Equation with Mixed Dirichlet/Neumann Boundary Conditions

Consider the following PDE having unity coefficients:

$$\frac{\partial^2 u(x,y)}{\partial x^2}+\frac{\partial^2 u(x,y)}{\partial y^2}=0 \tag{P7.3.1}$$

over a domain $D=\{(x,y)\,|\,0\le x\le x_f,\ 0\le y\le y_f\}$ subject to the mixed BCs:

$$\left.\frac{\partial u}{\partial x}\right|_{x=0}=0,\ \left.\frac{\partial u}{\partial x}\right|_{x=x_f}=0,\ u(x,0)=0,\ \text{and}\ u(x,y_f)=b_{yf}(x) \tag{P7.3.2a,b,c,d}$$

By the method of separation of variables, we substitute

$$u(x,y)=X(x)Y(y) \tag{P7.3.3}$$

into the PDE (P7.3.1) and divide by XY to write

$$X''Y+XY''=0;\quad \frac{X''}{X}=-\frac{Y''}{Y}=-\lambda;\quad \begin{cases}X''+\lambda X=0\\ Y''-\lambda Y=0\end{cases} \tag{P7.3.4a,b}$$

Also, we substitute Eq. (P7.3.3) into the BC (P7.3.2) to write

$$\partial u/\partial x\big|_{x=0}=0\,;\; X'(0)Y(y)=0 \rightarrow X'(0)=0, \tag{P7.3.5a}$$

$$\partial u/\partial x\big|_{x=x_f}=0\,;\; X'(x_f)Y(y)=0 \rightarrow X'(x_f)=0 \tag{P7.3.5b}$$

$$u(x,0)=X(x)Y(0)=0 \rightarrow Y(0)=0 \tag{P7.3.5c}$$

$$u(x,y_f)=X(x)Y(y_f)=b_{yf}(x) \tag{P7.3.5d}$$

Referring to the second column of Table 7.1, the candidate for $X(x)$ that may satisfy Eq. (P7.3.4a) together with the BC (P7.3.5a) is

$$X(x)=\begin{cases}\cos\omega x & \text{if } \lambda=\omega^2>0\\ 1 & \text{if } \lambda=0\\ \text{None} & \text{if } \lambda=-\omega^2<0\end{cases} \tag{P7.3.6}$$

In order for the first one ($X(x)=\cos\omega x$) to satisfy the BC (P7.3.5b), the value of ω should be

$$X'(x_f)=-\omega\sin(\omega x_f)=0\,;\; \omega=\boxed{\quad}\,(n=1,2,3,\cdots) \text{ with } \lambda=\omega^2>0 \tag{P7.3.7}$$

Now, referring to the third column of Table 7.1 for $\lambda=\omega^2=(n\pi/x_f)^2>0$, the candidate for $Y(y)$ satisfying the BC (P7.3.5c) is

$$Y(y)=\boxed{\qquad\qquad} \text{ with } \omega=\frac{n\pi}{x_f}(n=1,2,3,\cdots) \text{ for } \lambda=\omega^2>0 \tag{P7.3.8}$$

Plugging Eqs. (P7.3.6) and (P7.3.8) with Eq. (P7.3.7) into Eq. (P7.3.3) yields the separable solutions:

$$u_n(x,y)=X(x)Y(y)\overset{(P7.3.6)}{\underset{(P7.3.8)}{=}}\cos\left(\frac{n\pi}{x_f}x\right)\sinh\left(\frac{n\pi}{x_f}y\right). \tag{P7.3.9}$$

Also, corresponding to the second one ($X(x)=1$) with $\lambda=0$, the candidate for $Y(y)$ satisfying Eq. (P7.3.4b) and the BC (P7.3.5c) can be obtained from the third column of Table 7.1 as

$$Y(y)=\boxed{\quad} \tag{P7.3.10}$$

Hence, the general solution of Eq. (P7.3.1) subject to the BCs (P7.3.2) can be written as

$$u(x,y)=A(0)\boxed{\quad}+\sum_{n=1}^{\infty}A(n)u_n(x,y)\overset{(P7.3.9)}{=}A(0)y+\sum_{n=1}^{\infty}A(n)\sinh\left(\frac{n\pi}{x_f}y\right)\cos\left(\frac{n\pi}{x_f}x\right) \tag{P7.3.11}$$

where the coefficient $A(n)$'s are to be fixed by the remaining BC (P7.3.5d):

$$u(x,y_f)\overset{(P7.3.11)}{=}A(0)y_f+\sum_{n=1}^{\infty}A(n)\sinh\left(\frac{n\pi}{x_f}y_f\right)\cos\left(\frac{n\pi}{x_f}x\right)\overset{(P7.3.5d)}{=}b_{yf}(x) \tag{P7.3.12}$$

Regarding this as the Fourier cosine series for $b_{yf}(x)$, the coefficients can be obtained as

$$a_0=A(0)y_f\overset{(6.1.3b)}{=}\frac{1}{x_f}\int_0^{x_f}b_{yf}(x)\,dx \tag{P7.3.13a}$$

$$\rightarrow A(0)=\frac{a_0}{y_f}=\frac{1}{x_f y_f}\int_0^{x_f}b_{yf}(x)\,dx \tag{P7.3.14a}$$

$$a_n = A(n)\sinh(\frac{n\pi}{x_f}y_f) \underset{P=2x_f,\ \text{even symmetry}}{\overset{(6.1.3b),\text{Table 6.1}}{=}} \frac{2}{x_f}\int_0^{x_f} b_{yf}(x) \qquad dx \tag{P7.3.13b}$$

$$\to A(n) = \frac{a_n}{\sinh(\frac{n\pi}{x_f}y_f)} = \frac{2}{x_f\sinh(\frac{n\pi}{x_f}y_f)}\int_0^{x_f} \qquad \cos(\frac{n\pi}{x_f}x)\,dx \tag{P7.3.14b}$$

This algorithm can be cast into the MATLAB function 'pde_Laplace_mixed_BC_ana()'. Find the MATLAB statements using Eq. (P7.3.14) to compute $A(n)$'s.

```
function [u,A,xx,yy]=pde_Laplace_mixed_BC_ana(xf,yf,byf,Mx,My,Nterm)
% To solve the Laplace's equation u_xx +u_yy = 0
%  over the region D=[0,xf,0,yf]={(x,y)|0<=x<=xf, 0<=y<=yf}
% s.t. the BCs:
%  u'(0,y)=0, u'(xf,y)=0,  u(x,0)=0, u(x,yf)=byf(x)
% Mx = Number of subintervals along the x-axis
% My = Number of subintervals along the y-axis
% Nterm: Number of terms
if nargin<6, N=10; end
if nargin<5, My=50; end
if nargin<4, Mx=50; end
x0=0; dx=(xf-x0)/Mx;  xx= x0+[0:Mx]*dx;
y0=0; dy=(yf-y0)/My;  yy= y0+[0:My]'*dy;
[X,Y]=meshgrid(xx,yy);
syms x y
A0=eval(int(byf+0*x,0,xf))/xf/yf*Y; %Eq.(P7.3.???)
u=A0*Y; % the 1st term of Eq.(P7.3.??)
for n=1:Nterm
  npixf = n*pi/xf;
  A(n)=2/xf/sinh(npixf*yf)*eval(int(byf*cos(npixf*x),0,xf)); %(P7.3.???)
  u = u + A(n)*sinh(npixf*Y).*cos(npixf*X); %Eq.(P7.3.??)
end
```

To use the MATLAB functions 'pde_poisson()' (Section 7.1) and 'pde_Laplace_mixed_BC_ana()' for solving the above problem with $b_{yf}(x)=\cos(2\pi x)$, complete and run the MATLAB script "em07p03.m" to get the solution graph as in Fig. P7.3. Note that the seventh input argument D of 'pde_poisson()' is a 2×4 matrix whose second row consists of 0 or 1 depending on whether the corresponding BC is of Dirichlet or Neumann type.

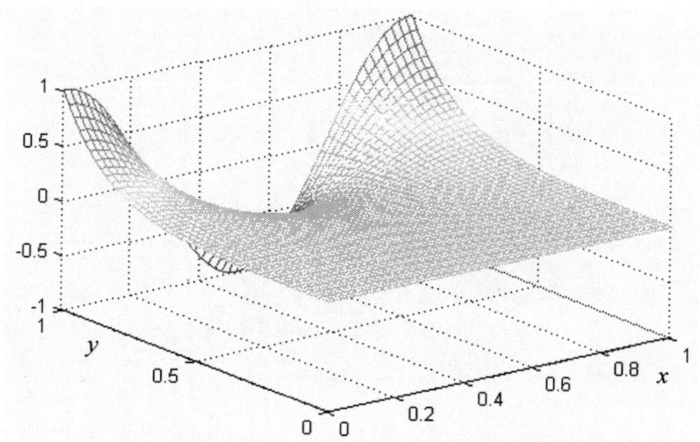

Figure P7.3 Solution graph for Problem 7.3

```
%em07p03.m
clear, clf
f=@(x,y)0; % Eq.(P7.3.1) w.r.t. Eq.(7.1.1)
g=@(x,y)0;
% (Rectangular) Domain and BC types
x0=0; xf=1; y0=0; yf=1;   D=[x0 xf y0 yf; 1 ? 0 ?];
% BCs
bx0=@(y)0;    bxf=@(y)0; % Eq.(P7.3.2a,b)
by0=@(x)0;    byf=@(x)???????????; % Eq.(P7.3.2c,d)
% Numerical approach
Mx=50; My=50; % Numbers of grid points along x-/y-axes
tol=1e-4;         % Error tolerance
MaxIter=600;    % Maximum number of iterations
[un,xx,yy]= pde_poisson(f,g,bx0,bxf,by0,byf,D,Mx,My,tol,MaxIter);
% Analytical approach
K=10; [ua,A]=pde_Laplace_mixed_BC_ana(xf,yf,byf,Mx,My,K); A
% Plot the solution graph
mesh(xx,yy,un)
Discrepancy1=norm(un-ua)/norm(ua)
```

7.4 Elliptic PDE — Laplace's Equation in Cylindrical Coordinates

Consider the following PDE in the cylindrical coordinate system:

$$\frac{1}{r}\frac{\partial}{\partial r}\left(r\frac{\partial u(r,\phi)}{\partial r}\right)+\frac{1}{r^2}\frac{\partial^2 u(r,\phi)}{\partial \phi^2}=\frac{1}{r}\frac{\partial R(r)}{\partial r}+\frac{\partial^2 R(r)}{\partial^2 r}+\frac{1}{r^2}\frac{\partial^2 u(r,\phi)}{\partial \phi^2}=0 \qquad \text{(P7.4.1)}$$

for the semicircular domain $\{(r,\phi), 0\le r\le R=1, 0\le\phi\le\pi\}$ (Fig. P7.4(a)) subject to the BCs:

$$u(R,\phi)=b_R(\phi)=\sin 2\phi \text{ for } R=1 \text{ and } 0\le\theta\le\pi,\ u(r,0)=0, \text{ and } u(r,\pi)=0 \quad \text{(P7.4.2a,b,c)}$$

First, as an analytical approach, modify the MATLAB function 'pde_Laplace_circle_ana()' (Section 7.4.1) and use it to get the analytical solution. Second, as a numerical approach based on Eq. (7.4.14), use the MATLAB function 'pde_Laplace_semicircle()' to get the numerical solution. To this end, complete and run the script "em07p04.m" to graph the analytical and numerical solutions for the semicircular domain divided into $M_r\times M_\phi$=10×36 segments, as shown in Fig. P7.4(b).

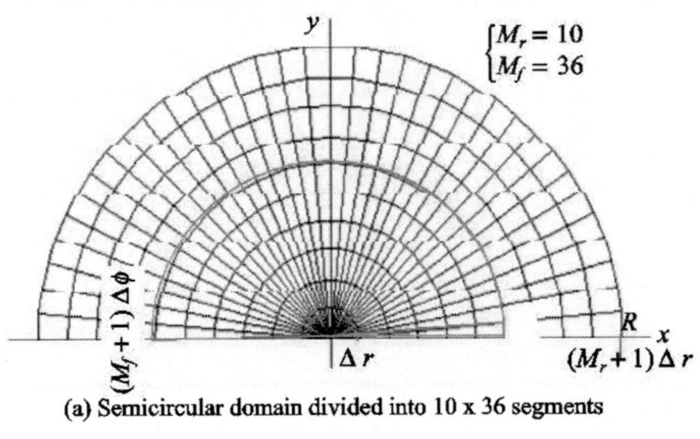

(a) Semicircular domain divided into 10 x 36 segments

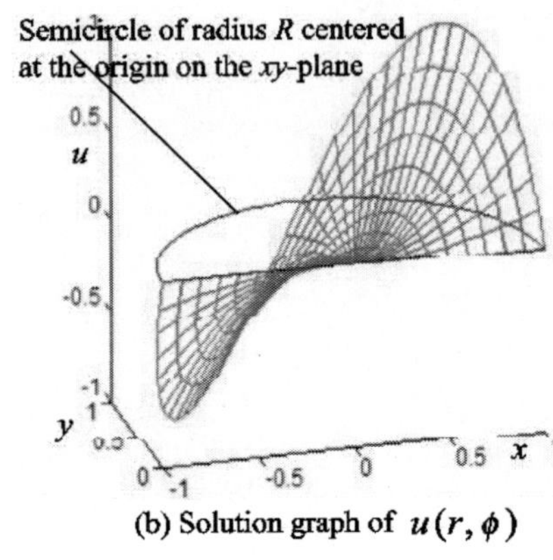

(b) Solution graph of $u(r,\phi)$

Figure P7.4 The domain and solution of the PDE (P7.4.1) s.t. the BC (P7.4.2)

```
function [u,rs,phs,k]=pde_Laplace_semicircle(R,bR,b0,bpi,Mr,Mph,tol)
%solve u_rr +1/r*u_r +(1/r^2)*u_phph = 0 for 0<=r<=R, 0<=ph<=pi
% BCs: u(R,ph) = bR(ph), u(r,0) = b0(r), u(r,pi) = bpi(r)
% Mr  = # of subintervals along the r(radial)-axis
% Mph = # of subintervals along the ph(circular)-axis
% tol = Error tolerance
% k   = Number of iterations performed
if nargin<7, tol = 1e-4; end   % Error tolerance
dr = R/Mr;  rs = [0:Mr]*dr;  dr2 = dr^2;
dph = pi/Mph;  phs = [0:Mph]*dph;  dph2 = dph^2;
u = zeros(Mr+1,Mph+1);
if isnumeric(b0),  u(:,1)=b0;
 else u(:,1)=feval(b0,r); % BC on the radial line with ph=ph0
end
if isnumeric(bpi),  u(:,Mph+1)=bpi;
 else u(:,Mph+1)=feval(bpi,rs); % BC on the radial line with ph=phN
end
if isnumeric(bR),  u(Mr+1,:) = bR;
 else u(Mr+1,:) = feval(bR,phs); % BC on the semicircle
end
u(1,:) = (u(1,1)+u(1,Mph+1))/2;
for k=1:500
   u0 = u;
   for m=2:Mr
      rm = rs(m); rm2 = rm^2;
      for n=2:Mph
         a = 2/dr2 + 2/rm^2/dph2;
         u(m,n) = ((u(m-1,n)+u(m+1,n))/dr2   ...
                  + (u(m+1,n)-u(m-1,n))/2/rm/dr ...
                  + (u(m,n-1)+u(m,n+1))/rm2/dph2)/a;
      end
   end
   if norm(u-u0)/norm(u0)<tol, break, end
end
```

```
%em07p04.m
% solves the PDE (P7.4.1) s.t. the BC (P7.4.2)
% on a semicircular boundary
clear, clf
R = 1; % Radius of the semicircular region
bR = @(ph)sin(2*??); % BC b(R,ph) (P7.4.2a) on the semicircle
b0 = 0;  bpi = 0;  % BCs b(r,0) and b(r,pi) on the two radial axes
Mr=10; Mph=36; % Numbers of segments along the radial/circular axes
% Analytical solution
[ua,ras,phas,A0,A,B] = ...
     pde_Laplace_semicircle_ana(R,b?,Mr,Mph);
A0, A, B
% Numerical solution
[un,rs,phs,K] = pde_Laplace_semicircle(R,bR,b0,b??,Mr,Mph);
Discrepancy = norm(un-ua)/norm(ua)
subplot(223)
mesh_polar(rs,phs,ua), Ph=-10; Th=10; view([Ph Th])
title('Analytical solution ua')
subplot(224)
mesh_polar(rs,phs,un), view([Ph Th])
title('Numerical solution u')
```

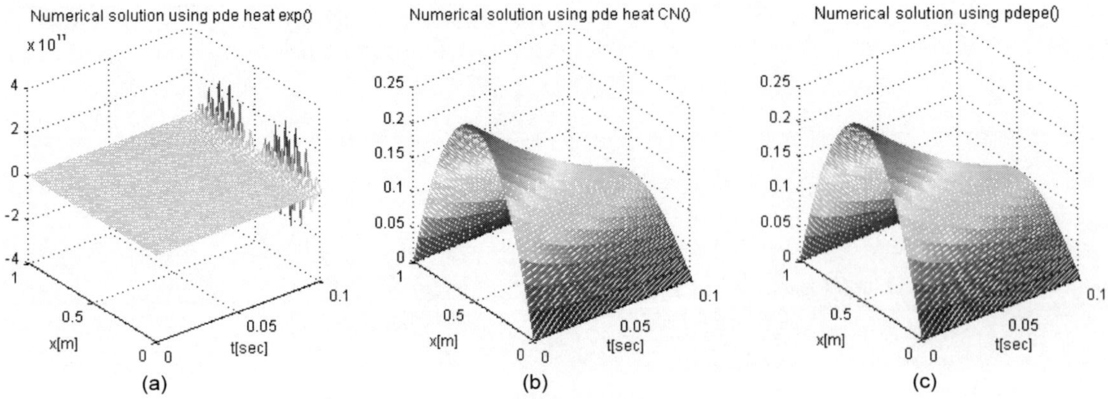

Figure P7.5 Solution graphs for Problem 7.5

7.5 Parabolic PDEs — Heat Equations

Modify the script "em07e02.m" (in Sec. 7.2.4), run it to solve the following heat equation by using the MATLAB functions '`pde_heat_exp()`', '`pde_heat_imp()`', '`pde_heat_CN()`', '`pdepe()`', and '`pde_heat_ana()`', and then graph the solutions as shown in Fig. P7.5.

$$a^2\frac{\partial^2 u(x,t)}{\partial x^2}=\frac{\partial u(x,t)}{\partial t}\ (a=1)\ \text{ for } 0\le x\le 1 \text{ and } 0\le t\le 0.1 \tag{P7.5.1}$$

subject to the IC and BCs:

$$u(x,0)=x(1-x),\ u(0,t)=0,\ u(1,t)=0 \tag{P7.5.2}$$

7.6 Parabolic PDEs — Heat Equations

Modify the script "em07e02.m" (in Sec. 7.2.4), run it to solve the following heat equations by using the MATLAB functions '`pde_heat_exp()`', '`pde_heat_imp()`', and '`pde_heat_CN()`', and then graph the solutions as shown in Fig. P7.6.

(a)

$$a^2\frac{\partial^2 u(x,t)}{\partial x^2}=\frac{\partial u(x,t)}{\partial t}\ (a=1)\ \text{ for } 0\le x\le 1 \text{ and } 0\le t\le 0.1 \tag{P7.6.1}$$

subject to the IC and BCs

$$u(x,0)=x^4,\ u(0,t)=0,\ u(1,t)=1 \tag{P7.6.2a,b,c}$$

(i) With the solution region divided into $M\times N=10\times 20$ sections, does the explicit forward Euler method converge? What is the value of $r\overset{(7.2.21)}{=}a\Delta t/(\Delta x)^2$?

(ii) If you increase M and N to make $M\times N=20\times 40$ for better accuracy, does the explicit forward Euler method still converge? What is the value of $r=a\Delta t/(\Delta x)^2$?

(iii) What is the number N of sub-intervals along the t-axis that we should choose in order to keep the same value of r for $M=20$? With that value of r, does the explicit forward Euler method converge?

```
%em07p06a.m: solve_heat
a=1; xf=1; tf=0.1; M=10; N=20; %r=0.5
it0=@(x)x.^4;  bx0=@(y)0;  bxf=@(y)1; % Eq.(P7.6.2a,b,c)
[u1,x,t]=pde_heat_exp(a,xf,tf,it0,bx0,bxf,M,N);
[u2,x,t]=pde_heat_imp(a,xf,tf,it0,bx0,bxf,M,N);
[u3,x,t]=pde_heat_CN(a,xf,tf,it0,bx0,bxf,M,N);
mesh(t,x,u3)
```

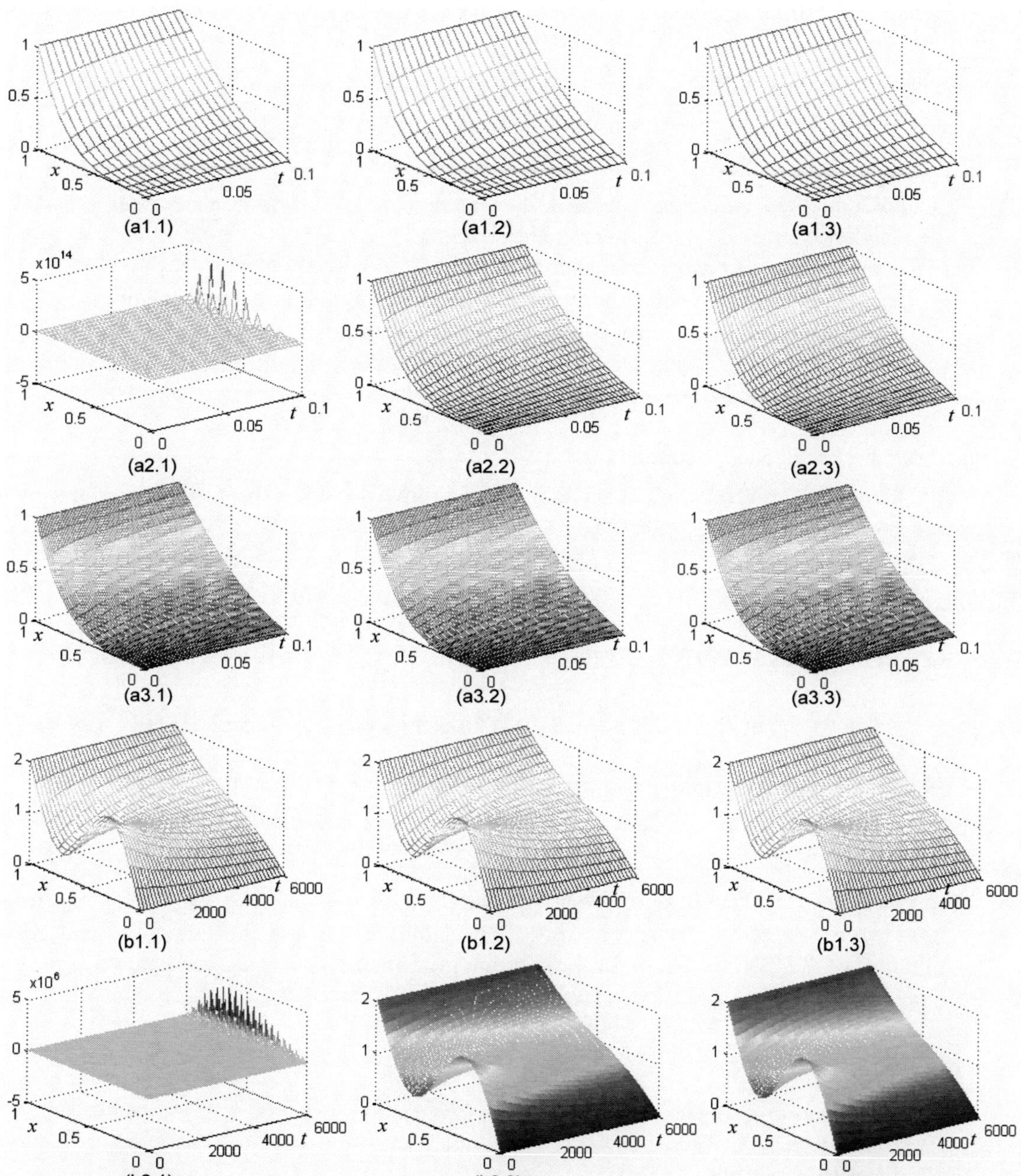

Figure P7.6 Solution graphs for Problem 7.6

(b) $$10^{-5}\frac{\partial^2 u(x,t)}{\partial x^2}=\frac{\partial u(x,t)}{\partial t} \text{ for } 0\le x\le 1 \text{ and } u(x,0)=x^4,\ u(0,t)=0 \quad \text{(P7.6.3)}$$

subject to the IC and BCs

$$u(x,0)=2x+\sin(2\pi x),\ u(0,t)=0,\ u(1,t)=2 \quad \text{(P7.6.4)}$$

(i) With the solution region divided into $M\times N$=20×40 sections, does the explicit forward Euler method converge? What is the value of $r\overset{(7.2.21)}{=}a\Delta t/(\Delta x)^2$? Does the numerical stability condition (7.2.21) seem to be so demanding?

(ii) If you increase M and N to make the solution grid size $M\times N$=40×160 for better accuracy, does the explicit forward Euler method still converge? What is the value of $r=a\Delta t/(\Delta x)^2$? Does the numerical stability condition (7.2.21) seem to be so demanding?

(iii) With the solution grid size of $M\times N$=40×200, does the explicit forward Euler method converge? What is the value of $r=a\Delta t/(\Delta x)^2$?

7.7 Hyperbolic PDEs — Wave Equations

Modify the script "em07e04.m" (in Sec. 7.3) and run it to solve the following PDEs by using '`pde_wave()`' and/or '`pde_wave_ana()`' if possible.

(a) $$2^2\frac{\partial^2 u(x,t)}{\partial x^2}=\frac{\partial^2 u(x,t)}{\partial t^2} \text{ for } 0\le x\le 1 \text{ and } 0\le t\le 1 \quad \text{(P7.7a.1)}$$

subject to the ICs and BCs

$$u(x,0)=0,\quad \left.\frac{\partial u(x,t)}{\partial t}\right|_{t=0}=5\sin(\pi x),\ u(0,t)=0,\ u(1,t)=0 \quad \text{(P7.7a.2,3,4,5)}$$

Noting that the true analytical solution is

$$u_o(x,t)=\frac{2.5}{\pi}\sin(\pi x)\sin(2\pi t) \quad \text{(P7.7a.6)}$$

complete the MATLAB script "em07p07a.m" so that it can use the MATLAB functions '`pde_wave()`' and '`pde_wave_ana()`' (Section 7.3) to get the numerical and analytical solutions of the PDE (P7.7a.1) for the solution grid of size $M\times N$ as given below.

```
%em07p07a.m
% To solve wave equation (1D Hyperbolic PDE) (P7.7a.1)
clear, clf
a=2; % Propagation speed parameter in Eq. (P7.7a.1)
xf=1; tf=1; % End points of the solution region
it0=@(x)0; i1t0=@(x)5*???(pi*x); % ICs (P7.7a.2,3)
bx0t=@(t)0; bxft=@(t)0; % BCs (P7.7a.4,5)
M=20; N=50; % Solution grid sizes
% Numerical solution
[un,x,t]=pde_wave(a,xf,tf,???,i1t0,bx0t,bxft,M,N);
% Analytical solution
Nterm=10; [ua,A]=pde_wave_ana(a,xf,tf,it0,????,M,N,Nterm); A
Discrepancy=norm(un-ua)/norm(ua)
% True analytical solution
uo=2.5/pi*sin(pi*x)*sin(2*pi*t); % Eq.(P7.7a.6)
err1=norm(un-uo)/norm(uo) % Error
% To plot the numerical solution
mesh(t,x,un), xlabel('t'), ylabel('x')
```

(i) With the solution region divided into $M{\times}N$=20×50 sections, what is the value of $r \overset{(7.3.19)}{=} a(\Delta t)^2/(\Delta x)^2$ where M/N are the numbers of sub-intervals along the x/t-axis, respectively? Find the relative error $||u_n - u_o||/||u_o||$ of the numerical solution u_n (obtained by using the function 'pde_wave()') w.r.t. the true solution u_o (Eq. (P7.7a.6)). To get the answers to this and the following questions, run the above MATLAB script, "em07p07a.m," after making appropriate modifications as needed.

(ii) With the solution grid size of $M{\times}N$=40×100, find the value of r and the relative error of the numerical solution.

(iii) With $M{\times}N$=50×100, find the value of r and the relative error of the numerical solution to check and see whether has been improved.

(iv) With $M{\times}N$=51×100, find the value of r and the relative error of the numerical solution to check and see whether has been improved. Discuss the result in connection with the stability condition (7.3.19). What do you think the best value of r is?

(v) Use Eq. (7.3.11) with the values of the coefficients $\{A_1(n), A_2(n)\}$ obtained by running the 11th line of the script "em07p07a.m" to write the analytical solution (of Eq. (P7.7a.1)) and compare it with Eq. (P7.7a.6).

(b)
$$2.5^2 \frac{\partial^2 u(x,t)}{\partial x^2} = \frac{\partial^2 u(x,t)}{\partial t^2} \quad \text{for } 0 \le x \le \pi \text{ and } 0 \le t \le 0.4\pi \tag{P7.7b.1}$$

subject to the IC and BCs

$$u(x,0) = \sin(2x), \quad \left.\frac{\partial u(x,t)}{\partial t}\right|_{t=0} = 0, \quad u(0,t) = 0, \text{ and } u(\pi,t) = 0 \quad \text{(P7.7b.2,3,4,5)}$$

Noting that the true analytical solution is

$$u_o(x, t) = \sin(2x)\cos(5t) \tag{P7.7b.6}$$

complete the MATLAB script "em07p07b.m" so that it can use the MATLAB functions 'pde_wave()'/'pde_wave_ana()' to get the numerical/analytical solutions of the PDE (P7.7b.1) for the given size $M{\times}N$ of the solution grid.

```
%em07p07b.m
% To solve the wave equation (1D Hyperbolic PDE) Eq.(P7.7b.1)
clear, clf
a=2.5; % Propagation speed parameter in Eq.(P7.7b.1)
xf=pi;  tf=0.4*pi;
it0=@(x)???(2*x); i1t0=@(x)0; % ICs (P7.7a.2,3)
bx0t=@(t)0;  bxft=@(t)0; % BCs (P7.7a.4,5)
M=20; N=40; % Solution grid size
% Numerical solution
[un,x,t]=pde_wave(a,xf,tf,it0,????,bx0t,bxft,M,N);
% Analytical solution
Nterm=10; [ua,A]=pde_wave_ana(a,xf,tf,???,i1t0,M,N,Nterm);
A
Discrepancy=norm(un-ua)/norm(ua)
% True anaytical solution
uo=sin(2*x)*cos(5*t); % Eq.(P7.7b.6)
err1=norm(un-uo)/norm(uo) % Error
mesh(t,x,un)
xlabel('t'), ylabel('x')
```

(i) With the solution grid size of $M \times N = 20 \times 50$, what is the value of $r \overset{(7.3.19)}{=} a(\Delta t)^2/(\Delta x)^2$? Find the relative error $\|u_n - u_o\|/\|u_o\|$ of the numerical solution u_n (obtained by using the function '`pde_wave()`') w.r.t. the true solution u_o (Eq. (P7.7b.6)). To get the answers to this and the following questions, run the above MATLAB script, "em07p07b.m," after making appropriate modifications as needed.

(ii) With $M \times N = 40 \times 50$, find the value of r and the relative error of the numerical solution.

(iii) With N (the number of sub-intervals along the t-axis) increased to 45 for better accuracy so that $M \times N = 45 \times 50$, find the value of r and discuss its effect on the relative error of the numerical solution. What do you think the best value of r is?

(iv) Use Eq. (7.3.11) with the values of the coefficients $\{A_1(n), A_2(n)\}$ obtained by running the 12th line of the script "em07p07b.m" to write the analytical solution (of Eq. (P7.7b.1)) and compare it with Eq. (P7.7b.6).

(c)
$$a^2 \frac{\partial^2 u(x,t)}{\partial x^2} = \frac{\partial^2 u(x,t)}{\partial t^2} \text{ for } 0 \le x \le x_f \text{ and } 0 \le t \le t_f \; (a=1, x_f=10, t_f=10) \quad \text{(P7.7c.1)}$$

subject to the ICs and BCs

$$u(x,0) = \begin{cases} i_{t0}(x) = (x-2)(3-x) & \text{for } 2 \le x \le 3 \\ 0 & \text{elsewhere} \end{cases} \quad \text{(P7.7c.2)}$$

$$\left.\frac{\partial u(x,t)}{\partial t}\right|_{t=0} = h(x) = 0, \; u(0,t) = 0, \text{ and } u(x_f, t) = 0 \quad \text{(P7.7c.3,4,5)}$$

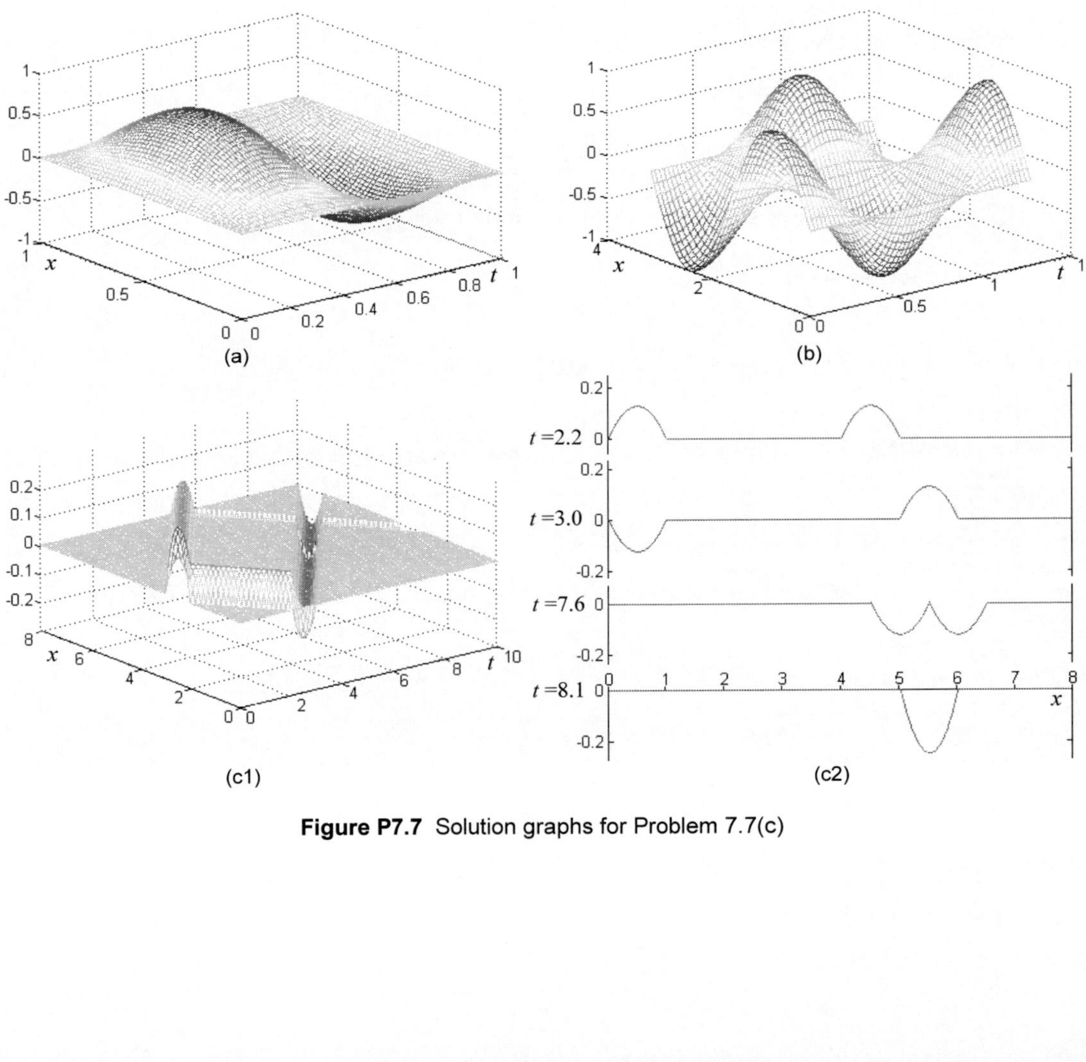

Figure P7.7 Solution graphs for Problem 7.7(c)

```
%em07p07c.m
% To solve the wave equation (1D Hyperbolic PDE) (P7.7c.1)
clear, clf
a=1; % The parameter of Eq.(P7.7c.1)
xf=10; M=100;  tf=10; N=100; dt=tf/N;
it0=@(x)(x-2).*(3-x).*(2<x&x<3);  i1t0=@(x)0; % ICs (P7.7c.2,3)
it0_s='(x-2).*(3-x).*(2<x&x<3)'; % u(x,0) defined as a string
bx0t=@(t)0;  bxft=@(t)0;  % BCs (P7.7c.4,5)
[un,x,t]=pde_wave(a,xf,tf,it0,i1t0,bx0t,bxft,M,N);
subplot(223), mesh(t,x,un), axis([0 tf 0 xf -0.3 0.3])
xlabel('t'), ylabel('x')
% Analytical approach
K=51; [ua,A]=pde_wave_anal(a,xf,tf,it0_s,i1t0,M,N,K); A
% Discrepancy between the numerical and analytical solutions
Discrepancy1=norm(un-ua)/norm(un)
% Analytical solution 2
ua1(:,1) = it0(x); % From IC
f=@(x)it0(x).*(x<=??)-it0(2*??-x).*(x>xf); % Eq.(P7.7c.8)
% To make g(x) odd and periodic with period 2*xf in x
f_tilde=@(x)f(mod(x,2*xf));
ua1(:,1)=it0(x);
for n=2:length(t)
   for m=1:M+1
      x1=x(m)?a*t(n);  x2=x(m)+a*t(n);
      ua1(m,n)=(f_tilde(??)+f_?????(x2))/2; % Eq.(P7.7c.7)
      %ua1(m,n)=(it0(x1)+it0(x2))/2;
   end
end
Discrepancy2=norm(un-ua1)/norm(un)
subplot(221), mesh(t,x,un), xlabel('t'), ylabel('x')
subplot(222)
for n=1:N
   plot(x,un(:,n)), text(xf/2,-0.5,['t=' num2str(t(n))]);
   axis([0 xf -0.6 0.6]), pause(0.1)
end
```

```
function [u,A,tt,xx]=pde_wave_anal(a,xf,tf,it0,i1t0,M,N,Nterm)
% The same as 'pde_wave_ana()' (Sec. 7.3) except for the for loop.
tol=1e-4;
for n=1:Nterm
  npixf = n*pi/xf; anpixfT=a*npixf*T;
  if isa(it0,'function_handle')
   A(1,n)=2/xf*eval(int(it0*sin(npixf*x),0,xf)); %Eq.(7.3.12)
  else % if it0 has been given as a string
   it0_sin = inline([it0 '.*sin(n*pi/xf*x)'],'x','n','xf');
   A(1,n)=2/xf*quadl(it0_sin,0,xf,tol,[],n,xf); % Eq.(7.3.12)
  end
  if isa(i1t0,'function_handle')
   A(2,n)=2/a/n/pi*eval(int(i1t0*sin(npixf*x),0,xf));%Eq.(7.3.13)
   else % if i1t0 has been given as a string
   i1t0_sin = inline([i1t0 '.*sin(n*pi/xf*x)'],'x','n','xf');
   A(2,n)=2/xf*quadl(i1t0_sin,0,xf,tol,[],n,xf); % Eq.(7.3.13)
  end
  u = u + sin(npixf*X).*(A(1,n)*cos(anpixfT)+ ...
                        A(2,n)*sin(anpixfT)); % Eq.(7.3.11)
end
```

If the spatial solution region (let us call this "string") were $-\infty<x<\infty$, the alalytical solution to the PDE (P7.7c.1) could be obtained from Eq. (E7.14.13), called the d'Alembert formula, as

$$u(x,t)=\frac{1}{2}\{i_{t0}(x+at)+i_{t0}(x)(x-at)\}+\frac{1}{2a}\int_{x-at}^{x+at}\overset{0}{i'_{t0}}(\xi)\,d\xi \tag{P7.7c.6}$$

However, since the string is finite as $[0, x_f=10]$, let us claim that the solution will be [W-5]

$$u(x,t)=\frac{1}{2}\{\tilde{f}(x+at)+\tilde{f}(x-at)\} \tag{P7.7c.7}$$

where $\tilde{f}(x)$ is the repetition (with period $2x_f$) of

$$f(x)=\begin{cases} i_{t0}(x) & \text{for } 0\le x<x_f \\ -i_{t0}(2x_f-x) & \text{for } x_f\le x<2x_f \end{cases} \tag{P7.7c.8}$$

Note that $\tilde{f}(x)$ is not only periodic but also odd, i.e., anti-symmetric w.r.t. $x=0$ because

$$\tilde{f}(-x)\underset{\text{with period } 2x_f}{\overset{\text{periodicity}}{=}}\tilde{f}(2x_f-x)\overset{\text{(P7.7c.8)}}{=}-i_{t0}(x)=-\tilde{f}(x) \tag{P7.7c.9}$$

Eq. (P7.7c.7) can easily be shown to satisfy the PDE (P7.7c.1):

$$\text{LHS: } a^2\frac{\partial^2 u(x,t)}{\partial x^2}\overset{\text{(P7.7c.7)}}{=}\frac{a^2}{2}\left\{\frac{\partial^2}{\partial x^2}\tilde{f}(x+at)+\frac{\partial^2}{\partial x^2}\tilde{f}(x-at)\right\} \tag{P7.7c.10a}$$

$$\text{RHS: }\frac{\partial^2 u(x,t)}{\partial t^2}\overset{\text{(P7.7c.7)}}{=}\frac{1}{2}\left\{a^2\frac{\partial^2}{\partial (at)^2}\tilde{f}(x+at)+(-a)^2\frac{\partial^2}{\partial (-at)^2}\tilde{f}(x-at)\right\} \tag{P7.7c.10b}$$

where

$$\frac{\partial^2}{\partial x^2}\tilde{f}(x+at)=\frac{\partial^2}{\partial (at)^2}\tilde{f}(x+at) \tag{P7.7c.11a}$$

$$\frac{\partial^2}{\partial x^2}\tilde{f}(x-at)=\frac{\partial^2}{\partial (-at)^2}\tilde{f}(x-at) \tag{P7.7c.11b}$$

Also, Eq. (P7.7c.7) can easily be shown to satisfy the ICs (P7.7c.2,3) and BCs PDE (P7.7c.4,5):

$$u(x,0)\underset{\text{(P7.7c.8)}}{\overset{\text{(P7.7c.7)}}{=}}\frac{1}{2}\{i_{t0}(x+at)+i_{t0}(x-at)\}\Big|_{t=0}=i_{t0}(x)\text{: (P7.7c.2) OK}$$

$$\left.\frac{\partial u(x,t)}{\partial t}\right|_{t=0}\underset{\text{(P7.7c.8)}}{\overset{\text{(P7.7c.7)}}{=}}\frac{1}{2}\frac{\partial}{\partial t}\{i_{t0}(x+at)+i_{t0}(x-at)\}\Big|_{t=0}=\frac{1}{2}\left\{a\frac{\partial}{\partial x}i_{t0}(x+at)-a\frac{\partial}{\partial x}i_{t0}(x-at)\right\}\Big|_{t=0}=0$$

$$\text{: (P7.7c.3) OK}$$

$$u(0,t)\overset{\text{(P7.7c.7)}}{=}\frac{1}{2}\{\tilde{f}(x+at)+\tilde{f}(x-at)\}\Big|_{x=0}=\frac{1}{2}\{\tilde{f}(at)+\tilde{f}(-at)\}\overset{\text{odd}}{=}0\text{: (P7.7c.4) OK}$$

$$u(x_f,t)\overset{\text{(P7.7c.7)}}{=}\frac{1}{2}\{\tilde{f}(x+at)+\tilde{f}(x-at)\}\Big|_{x=x_f}=\frac{1}{2}\{\tilde{f}(x_f+at)+\tilde{f}(x_f-at)\}\underset{\text{odd}}{\overset{\text{periodicity}}{=}}0\text{ :(P7.7c.5) OK}$$

Therefore, Eq. (P7.7c.7) has been shown to be an analytic solution to the PDE (P7.7c.1) subject to the ICs (P7.7c.2,3) and BCs PDE (P7.7c.4,5).

It is noteworthy that Eq. (7.3.11) can also be used to find the solution to the PDE (P7.7c.1), but the MATLAB function '`pde_wave_ana()`' should be modified as the above MATLAB function '`pde_wave_ana1()`' to be applicable when the functions describing the ICs contain logical expressions (like Eq. (P7.7c.2)) so that they cannot be symbolic-integrable. Such ICs involving non-symbolic-integrable functions should be expressed by a string (a sequence of characters), like 'it0_s' in the 7th line of the above MATLAB script "em07p07c.m", to be given as input arguments of '`pde_wave_ana1()`'.

(i) Complete the MATLAB script "em07p07c.m" so that it can use the MATLAB functions '`pde_wave()`'/Eq. (7.3.11)('`pde_wave_ana1()`')/Eq. (P7.7c.7) to get one numerical solution u_n and two analytical solutions u_a and u_{a1} of the PDE (P7.7c.1) for the given size $M \times N$ of the solution grid.

(ii) With $M \times N = 100 \times 100$ resulting in $r = a(\Delta t)^2/(\Delta x)^2 = 1$, find the discrepancies $\|u_n - u_a\|/\|u_n\|$ and $\|u_n - u_{a1}\|/\|u_n\|$. Which of these u_a and u_{a1} or, do you think is better? If you increase/decrease the number K of terms in Eq. (7.3.11) in the 13th line of the MATLAB script "em07p07c.m," does the discrepancy $\|u_n - u_a\|/\|u_n\|$ decrease/increase?

(iii) Referring to the motion picture with the current time printed, which is made by the last 'for' loop in the MATLAB script "em07p07c.m," estimate the time when one of the two separated pulses propagating leftwards is reflected and reversed. How about the time when the two separated pulses are reunited?

7.8 PDE with Cylindrical Symmetry in Cylindrical Coordinates

Consider the Laplace's equation (7.4.16) with rotational invariance:

$$\frac{1}{r}\frac{\partial}{\partial r}\left(r\frac{\partial u(r,z)}{\partial r}\right) + \frac{\partial^2 u(r,z)}{\partial z^2} = \frac{1}{r}\frac{\partial u(r,z)}{\partial r} + \frac{\partial^2 u(r,z)}{\partial^2 r} + \frac{\partial^2 u(r,z)}{\partial z^2} = 0 \tag{P7.8.1}$$

over a cylindrical domain of radius $R=2$ and height 2 (divided into $M_r \times M_z = 20 \times 40$ segments) subject to the following BCs

$$u(R,z) = 0 \text{ for } 0 \le z \le z_f = 2 \tag{P7.8.2a}$$

$$u(r,0) = 0, \text{ and } u(r,z_f) = \begin{cases} 10 & \text{for } 0 \le r < 1 \\ 0 & \text{elsewhere} \end{cases} \text{ for } 0 \le r \le R = 2 \tag{P7.8.2b,c}$$

(i) Using Eq. (7.4.26), write the analytical solution $u(r,z)$ as a partial sum of the first few significant terms.

(ii) Graph the analytical and numerical solutions as shown in Fig. P7.8.

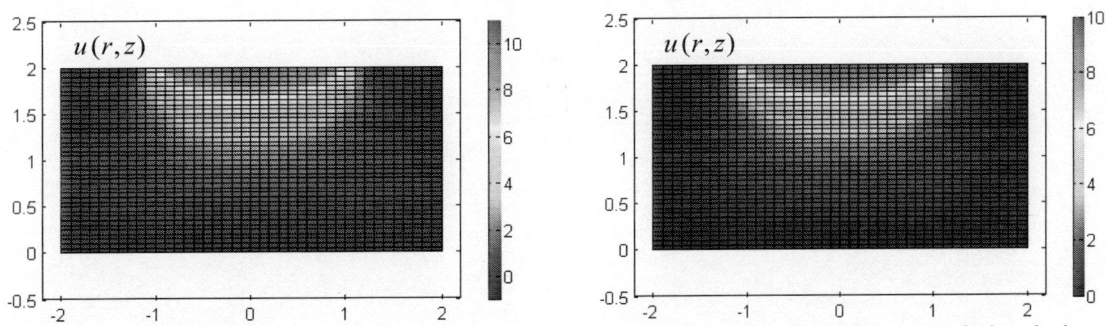

(a) Solution graph obtained by an analytical approach (b) Solution graph obtained by a numerical method

Figure P7.8 Analytical/numerical solutions for the PDE (P7.8a.1) s.t. the BC (P7.8a.2)

7.9 PDE with Azimuthal Symmetry in Cylindrical Coordinates

Consider the wave equation (7.4.33):

$$2^2\left(\frac{\partial^2 u(r,t)}{\partial^2 r}+\frac{1}{r}\frac{\partial u(r,t)}{\partial r}\right)=\frac{\partial^2 u(r,t)}{\partial t^2} \tag{P7.9.1}$$

for the circular domain of radius R=1 (divided into M=50 segments) and $0 \le t \le t_f$=2[s] (divided into N=200 subintervals) subject to the following BC and ICs:

$$u(R,t)=0 \ \text{ for } t\ge 0 \tag{P7.9.2a}$$

$$u(r,0)=i_0(r)=0.5(1-\frac{r}{R}), \text{ and } \left.\frac{\partial u}{\partial t}\right|_{t=0}=0 \ \text{ for } 0<r<R \tag{P7.9.2b,c}$$

```
%em07p09.m
% solves the PDE (7.4.33) s.t. the BC and IC (P7.9.2a,b,c)
a=2; R=1; tf=2; % Parameters of the PDE
% ICs and BCs
i0 = @(r)0.5*(1-r/R);
v0=0; r0=0.2; i10 = @(r)-v0*(r<r0); % Only for 0<=r<=r0;
bR = 0; % BC on the boundary circle with r=R
Mr=50; Mt=199; % Numbers of segments along the radial and time axes
%Mr=50; Mt=200; % Numbers of segments along the radial and time axes
[ua,ras,tas,A,B,w] = pde_wave_circle_???(a,R,tf,i0,i10,bR,Mr,Mt);
[A;B;w]
[u,rs,ts,iter] = pde_wave_??????(a,R,tf,i0,i10,bR,Mr,Mt); iter
Discrepancy = norm(u(2:Mr,1:Mt)-ua(2:Mr,1:Mt))/norm(ua)
animation = 1;
if animation>0
  subplot(121), hold on
  phs=[0:10:360]*(pi/180); Mph=length(phs);  dt=tf/Mt;
  for n=1:Mt
     mesh_polar(ras,phs,repmat(ua(:,n),1,Mph)); axis('equal');
     axis([-1 1 -1 1 -0.4 0.4]), view([-37.5 30])
     text(-1,-0.5,-0.2,['t=' num2str(tas(n))])
     if abs(tas(n)-0.5)<dt|abs(tas(n)-1.5)<dt, pause, end
     pause(0.1), if n<Mt, cla; end
  end
  ps = R*exp(j*phs);
  plot3(real(ps),imag(ps),zeros(size(ps)),'k')
end
```

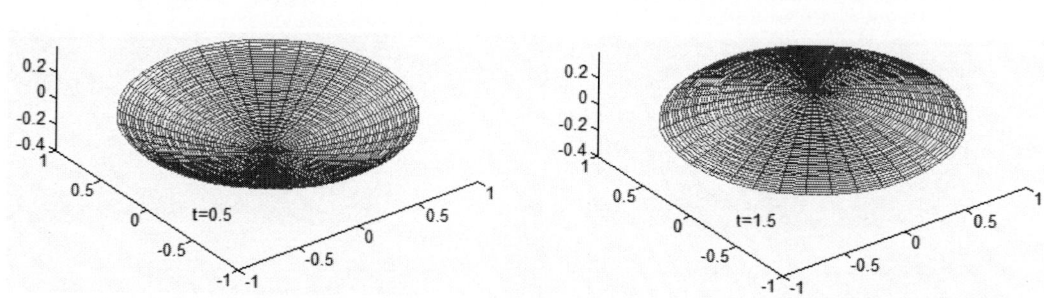

(a) Analytical solution graph at t=0.5 (b) Numerical solution graph at t=1.5

Figure P7.9 Analytical/numerical solutions for the PDE (P7.9.1) s.t. the BC (P7.9.2)

(i) Referring to Eq. (7.4.35), write the analytical solution $u(r,t)$ as a partial sum of the first three terms where the coefficients can be obtained by completing and running the above MATLAB script, "em07p09.m." Plot it at t=0.5[s] and 1.5[s] as shown in Figs. P7.9(a) and (b).

(ii) Evaluate how close the numerical solutions obtained with the time interval divided into N=199 subintervals is to the analytical solution. What is the value of c_2 defined by Eq. (7.4.46b)? If there is something wrong with the numerical solution, do the same with a larger number (N=200) of subintervals along the time. What is the value of c_2? Can you make any conjecture about the numerical stability condition on the value of c_2?

7.10 PDE in Spherical Coordinates

Consider the Laplace's equation (7.4.49) with rotational invariance:

$$\frac{\partial^2 u(r,\theta)}{\partial r^2}+\frac{2}{r}\frac{\partial u(r,\theta)}{\partial r}+\frac{1}{r^2}\frac{\partial^2 u(r,\theta)}{\partial \theta^2}+\frac{\cot\theta}{r^2}\frac{\partial u(r,\theta)}{\partial \theta}=0 \tag{P7.10.1}$$

for a spherical domain $\{0<r\le R, 0\le\theta\le\pi\}$ of radius R=1 (divided into $M_r\times M_\theta$ =20×36 segments) subject to the BC

$$u(R,\theta)=b(\theta)=100\cos(\theta/2) \quad \text{for } 0\le\theta\le\pi \tag{P7.10.2}$$

(i) Referring to Eq. (7.4.52), write the solution $u(r,\theta)$ as a partial sum of the first three terms.

(ii) Graph the analytical and numerical solutions as shown in Fig. P7.10.

```
%em07p10.m
R=1; % Radius of the sphere
bR=@(th)100*cos(th/2); % Boundary condition (P7.10.2)
Mr=20; Mth=36; % Numbers of segments along the r-/th-directions
N=10; % Number of coefficients/terms to be added for Eq.(7.4.62)
[ua,ras,thas,A]=pde_Laplace_sphere_ana(R,bR,Mr,Mth,N); A
tol=1e-4; [un,rs,ths,k]=pde_Laplace_sphere(R,bR,Mr,Mth,tol); k
Discrepancy = norm(ua-un)/norm(ua)
[Ths,Rs]=meshgrid(ths,rs); % Grid
subplot(121)
pcolor(Rs.*sin(Ths),Rs.*cos(Ths),ua), hold on
pcolor(-Rs.*sin(Ths),Rs.*cos(Ths),ua), axis('equal'), colorbar
subplot(122)
pcolor(Rs.*sin(Ths),Rs.*cos(Ths),un), hold on
pcolor(-Rs.*sin(Ths),Rs.*cos(Ths),un), axis('equal'), colorbar
```

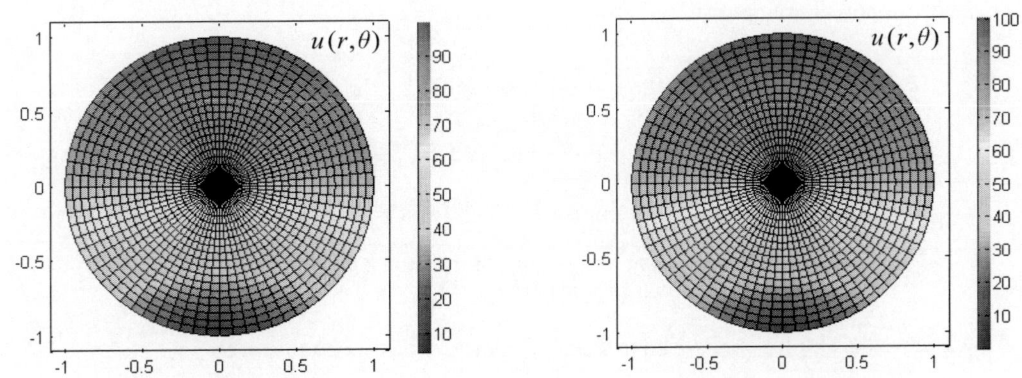

(a) Analytical solution graph

(b) Numerical solution graph

Figure P7.10 Analytical/numerical solutions for the PDE (P7.10.1) s.t. the BC (P7.10.2)

7.11 Laplace Transform for a 1D Heat Equation on a Line

Consider the following 1D heat equation:

$$a^2\frac{\partial^2 u(x,t)}{\partial x^2}=\frac{\partial u(x,t)}{\partial t}\quad \text{for } x>0 \text{ and } t>0 \tag{P7.11.1}$$

subject to the IC and BCs

$$u(x,0)=0\ \forall x,\quad u(0,t)=u_0,\quad u(\infty,t)=0\ \forall t \tag{P7.11.2a,b,c}$$

(a) To solve this PDE by using the Laplace transform (LT), proceed as follows:

Referring to Example 7.7, take the s-transform (LT w.r.t. t) of Eq. (P7.11.1) with the IC (E7.11.2a) to write

$$\mathcal{L}_t\left\{\overset{\text{LHS}}{a^2\frac{\partial^2 u(x,t)}{\partial x^2}}\right\}\overset{(7.5.1)}{=}a^2\frac{d^2}{dx^2}U(x,s)=\mathcal{L}_t\left\{\overset{\text{RHS}}{\frac{\partial u(x,t)}{\partial t}}\right\}\overset{\text{A.7(5)}}{=}sU(x,s)-u(x,0)\overset{\text{(P7.11.2a)}}{=}sU(x,s)$$

$$;\quad a^2\frac{d^2}{dx^2}U(x,s)-sU(x,s)=0\ \text{ with } U(x,s)=\mathcal{L}_t\{u(x,t)\} \tag{P7.11.3}$$

Since this is a second-order ODE (in x with s as a constant), we use Table A.7(5) twice to take its LT (w.r.t. x with $\tilde{s}$ as a complex parameter) to write

$$\tilde{s}^2\tilde{U}(\tilde{s})-\tilde{s}U(0,s)-\frac{d}{dx}U(x,s)\Big|_{x=0}-\frac{s}{a^2}\tilde{U}(\tilde{s})=0\ \text{ with } \tilde{U}(\tilde{s})=\mathcal{L}_x\{U(x,s)\} \tag{P7.11.4}$$

Since $dU(x,s)/dx|_{x=0}$ is not yet clear, we solve this equation provisionally just for the solution part of $\tilde{U}(\tilde{s})$ (stemming from $u(x,0)=u_0$) and get its PFE (partial fraction expansion) as

$$\tilde{U}(\tilde{s})\overset{\text{(E7.6.4)}}{=}\frac{\tilde{s}u_0/s+\cdots}{\tilde{s}^2-s/a^2}=\frac{A}{\tilde{s}-\sqrt{s}/a}+\frac{B}{\tilde{s}\boxed{}}\ \text{ with } U(0,s)=\mathcal{L}_t\{u(0,t)\}=\frac{u_0}{s}. \tag{P7.11.5}$$

Taking the inverse $\tilde{s}$ -transform (ILT w.r.t. x with s as a constant) of Eq. (E7.6.6) yields

$$U(x,s)=\mathcal{L}_x^{-1}\{\tilde{U}(s)\}\underset{\text{A.8(6)}}{\overset{\text{(P7.11.5)}}{=}}Ae^{\sqrt{s}x/a}+B\boxed{} \tag{P7.11.6}$$

where A and B are determined from the (s-transformed) BCs

$$\left.\begin{array}{l} u(0,t)\overset{\text{(P7.11.2b)}}{=}u_0\overset{\mathcal{L}_t}{\rightarrow}U(0,s)=\dfrac{u_0}{s}\overset{\text{(P7.11.6)}}{\rightarrow}Ae^{\sqrt{s}0}+Be^{-\sqrt{s}0}=\dfrac{u_0}{s}\\ u(\infty,t)\overset{\text{(P7.11.2c)}}{=}0\overset{\mathcal{L}_t}{\rightarrow}U(\infty,s)=0\overset{\text{(P7.11.6)}}{\rightarrow}Ae^{\sqrt{s}\infty/a}+Be^{-\sqrt{s}\infty/a}=0 \end{array}\right\}\rightarrow A=0,\ B=\boxed{} \tag{P7.11.7}$$

Thus, Eq. (P7.11.6) can be written as

$$U(x,s)\underset{\text{(P7.11.7)}}{\overset{\text{(P7.11.6)}}{=}}\boxed{} \tag{P7.11.8}$$

We can take the inverse s-transform (ILT w.r.t. t) of this equation to get

$$u(x,t)=\mathcal{L}_t^{-1}\{U(x,s)\}\underset{\text{A.8(16) with } a\to x/a}{\overset{\text{(E7.6.10)}}{=}}\boxed{} \tag{P7.11.9}$$

```
%em07p11.m
a=5; u0=1; xf=10; tf=0.2;
% BCs and ICs
it0=@(x)0;     % Eq.(P7.11.2a)
bx0t=@(t)u0; bxft=@(t)0;   % Eq.(P7.11.2b,c)
M=40; N=200; MN=M*N;
% Numerical approach
[un,x,t]=pde_heat_CN(a,xf,tf,it0,????,bxft,M,N);
% Analytical solution using the Laplace transform
[T,X]=meshgrid(t,x); T=T+eps;
ua = u0*erfc(?/2/a./sqrt(?)); % Eq.(P7.11.9)
Discrepancy=norm(ua-un)/norm(ua)
subplot(121), mesh(t,x,un), view([-110 10])
subplot(122), mesh(t,x,ua), view([-110 10])
```

(b) To check the validity of the analytical solution (P7.11.9) to the PDE (P7.11.1) with $a=5$ and $u_0=1$ for a finite line [0, 10] and for a finite time interval [0, 0.2], complete and run the above MATLAB script, "em07p11.m," to see that the analytical solution is close to a numerical solution.

7.12 Bilateral Laplace Transform for a 1D Heat Equation on the Entire Line

Consider the following 1D heat equation:

$$a^2\frac{\partial^2 u(x,t)}{\partial x^2}=\frac{\partial u(x,t)}{\partial t} \quad \text{for } -\infty<x<\infty \text{ and } t>0 \tag{P7.12.1}$$

subject to the IC

$$u(x,0)=i_0(x)\ \forall x \tag{P7.12.2}$$

While this problem was dealt with by using the Fourier transform in Example 7.12, let us use the Laplace transform (LT) to solve it here.

Referring to Example 7.7, we take the s-transform (LT w.r.t. t) of Eq. (P7.12.1) with the IC (P7.12.2) to write

$$\mathcal{L}_t\left\{\overbrace{a^2\frac{\partial^2 u(x,t)}{\partial x^2}}^{\text{LHS}}\right\}\overset{(7.5.1)}{=}a^2\frac{d^2}{dx^2}U(x,s)=\mathcal{L}_t\left\{\overbrace{\frac{\partial u(x,t)}{\partial t}}^{\text{RHS}}\right\}\overset{\text{A.7(5)}}{=}sU(x,s)-u(x,0)\overset{(\text{P7.12.2})}{=}sU(x,s)-i_0(x)$$

$$;\qquad a^2\frac{d^2}{dx^2}U(x,s)-sU(x,s)=-i_0(x) \text{ with } U(x,s)=\mathcal{L}_t\{u(x,t)\}. \tag{P7.12.3}$$

Since this is a second-order ODE (in x with s as a constant), we use Table A.7(5) twice to take its bilateral LT (w.r.t. x with $\tilde{s}$ as a complex parameter) to write

$$\tilde{s}^2\tilde{\mathbb{U}}(\tilde{s})-\frac{s}{a^2}\tilde{\mathbb{U}}(\tilde{s})=-\frac{1}{a^2}\tilde{\mathbb{I}}_0(\tilde{s}) \tag{P7.12.4}$$

with

$$\tilde{\mathbb{U}}(\tilde{s})=L_x\{U(x,s)\} \text{ and } \tilde{\mathbb{I}}_0(\tilde{s})=L_x\{i_0(x)\} \text{ (bilateral Laplace transform)} \tag{P7.12.5}$$

Eq. (P7.12.4) can be solved for $\tilde{\mathbb{U}}(\tilde{s})$ to yield

$$\tilde{\mathbb{U}}(\tilde{s})=\frac{-1}{\tilde{s}^2-s/a^2}\frac{1}{a^2}\tilde{\mathbb{I}}_0(\tilde{s}) \tag{P7.12.6}$$

This is the product of two bilateral $\tilde{s}$-transforms, whose inverse bilateral $\tilde{s}$-transforms can be found from Table A.10(7) with $a \to \sqrt{s}/a$ and written from Eq. (P7.12.5b), respectively, as

$$L_x^{-1}\left\{\frac{-1}{\tilde{s}^2 - s/a^2}\right\} \underset{a\to\sqrt{s}/a}{\overset{\text{A.10(7)}}{=}} \boxed{}, \quad L_x^{-1}\left\{\frac{1}{a^2}\tilde{\mathbb{I}}_0(\tilde{s})\right\} \overset{\text{(P7.12.5b)}}{=} \frac{1}{a^2} i_0(x) \tag{P7.12.7}$$

Thus, the convolution property A.7(4) of the bilateral LT can be used to write the inverse bilateral LT of Eq. (P7.12.6) as the x-domain convolution of these two inverse bilateral LT expressions, as

$$U(x,s) = \frac{1}{2a\sqrt{s}} e^{-(\sqrt{s}/a)|x|} * i_0(x). \tag{P7.12.8}$$

Lastly, we can use Table A.8(14) with $|x|/a$ in place of a to take the inverse s-transform (LT w.r.t. t) of Eq. (P7.12.8):

$$u(x,t) = \mathcal{L}_t^{-1}\{U(x,s)\} \overset{\text{(P7.12.8)}}{=} i_0(x) * \mathcal{L}_t^{-1}\left\{\frac{1}{2a\sqrt{s}} e^{-(\sqrt{s}/a)|x|}\right\} \underset{a\to|x|/a}{\overset{\text{A.10(16)}}{=}} i_0(x) * \boxed{} \tag{P7.12.9}$$

7.13 Fourier Transform for Elliptic PDE – Laplace's Equation

Consider the Laplace's equation

$$\frac{\partial^2 u(x,y)}{\partial x^2} + \frac{\partial^2 u(x,y)}{\partial y^2} = 0 \text{ for } -\infty < x < \infty \text{ and } y > 0 \tag{P7.13.1}$$

subject to the boundary conditions (BCs)

$$u(x,0) = b_{y0}(x) = \sin(2x) \text{ for all } x \text{ and } y = 0 \tag{P7.13.2}$$

If we use Eq. (E7.10.7), which was derived by using the Fourier transform in Example 7.10, we will get

$$u(x,y) \overset{\text{(E7.10.7)}}{=} b_{y0}(x) * \frac{1}{\pi}\frac{y}{x^2+y^2} \underset{\text{(P7.13.2)}}{\overset{\text{(4.2.8)}}{=}} \frac{1}{\pi}\int_{-\infty}^{\infty} \sin(2\pi t)\frac{y}{(x-t)^2+y^2} dt. \tag{P7.13.3}$$

As an alternative, but still referring to Example 7.10, use Eq. (E7.10.4) to write the v-transformed solution as

$$U(v,y) \overset{\text{(E7.10.4)}}{=} K_1(v) e^{-|v|y} \tag{P7.13.4}$$

where $K_1(v)$ can be determined from the BC (E7.13.2):

$$U(v,0) \underset{y=0}{\overset{\text{(P7.13.4)}}{=}} K_1(v) = B_{y0}(v) = \mathcal{F}_x\{b_{y0}(x) \overset{\text{(P7.13.2)}}{=} \sin(2x)\} \overset{\text{(E6.12.1)}}{=} j\pi\{\delta(v+2) - \delta(v-2)\} \tag{P7.13.5}$$

Substituting these coefficients into Eq. (P7.13.4) yields

$$U(v,y) = \boxed{} \tag{P7.13.6}$$

Now take the inverse v-transform (ICTFT w.r.t. x) of Eq. (E7.10.6) to find the solution as

$$u(x,y) = \mathcal{F}_x^{-1}\{U(v,y)\} \overset{\text{(6.2.1b)}}{=} \frac{1}{2\pi}\int_{-\infty}^{\infty} \boxed{} dv$$

$$\overset{\text{(C.40)}}{=} \frac{j}{2} \boxed{} = \boxed{} \tag{P7.13.7}$$

Finally, substitute this solution into Eq. (P7.13.1) to check its validity.

7.14 Using '`pdepe()`' to Solve a 1D Heat Equation on a Circular Domain

Consider the following 1D heat equation in the cylindrical coordinate system:

$$a^2\left(\frac{\partial^2 u(r,t)}{\partial^2 r}+\frac{1}{r}\frac{\partial u(r,t)}{\partial r}\right)=\frac{\partial u(r,t)}{\partial t} \tag{P7.14.1}$$

s.t. the IC and BC on a circular domain of radius R as

$$u(r,0)=i_0(r)=1-r^2 \text{ for } 0<r<R,\quad u(R,t)=0 \text{ for } 0\le t\le t_f \tag{P7.14.2,3}$$

To get the numerical solution of this PDE, we use the central/forward difference approximations of derivatives in LHS/RHS to discretize the PDE as

$$a^2\left(\frac{u_{m+1}^k-2u_m^k+u_{m-1}^k}{\Delta r^2}+\frac{u_{m+1}^k-u_{m-1}^k}{r2\Delta r}\right)=\frac{u_m^{k+1}-u_m^k}{\Delta t}. \tag{P7.14.4}$$

This leads to the following difference equation, like Eq. (7.3.24):

$$u_m^{k+1}=c_1(r_m)u_{m+1}^k+c_2u_m^k+c_3(r_m)u_{m-1}^k \tag{P7.14.5}$$

where

$$c_1(r_m)=a^2\frac{\Delta t}{\Delta r^2}\left(1+\frac{\Delta r}{2r_m}\right),\ c_2=1-2a^2\frac{\Delta t}{\Delta r^2},\ \text{and } c_3(r_m)=a^2\frac{\Delta t}{\Delta r^2}\left(1-\frac{\Delta r}{2r_m}\right) \tag{P7.14.6a,b,c}$$

This numerical solution process has been cast into the above MATLAB function, '`pde_heat_circle()`'. See [W-5] for the analytical solutions of many 2D PDEs on a circular region.

Another way to find a numerical solution to Eq. (P7.14.1) is to use the MATLAB function '`pdepe()`', which is supposed to solve initial-boundary problems for 1D parabolic-elliptic PDEs:

$$x^{-m}\frac{\partial u}{\partial x}\left(x^m f\left(x,t,u,\frac{\partial u}{\partial x}\right)\right)+s\left(x,t,u,\frac{\partial u}{\partial x}\right)=c\left(x,t,u,\frac{\partial u}{\partial x}\right)\frac{\partial u}{\partial t} \tag{P7.14.7}$$

with IC (for $x_0\le x\le x_f$) and BCs (for $0\le t\le t_f$) given by

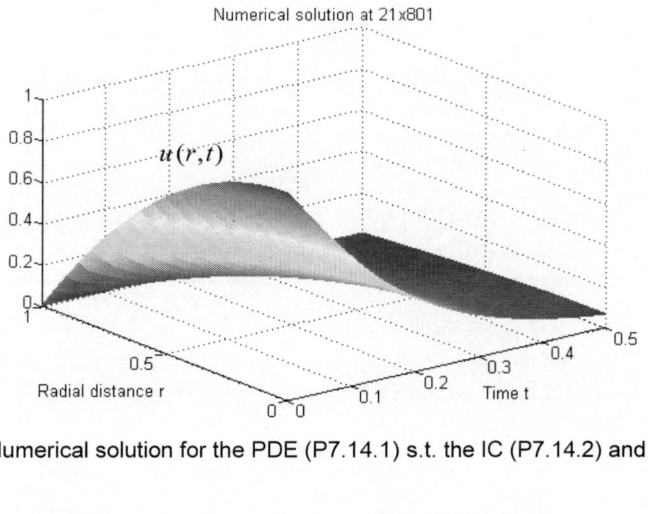

Figure P7.14 Numerical solution for the PDE (P7.14.1) s.t. the IC (P7.14.2) and BC (P7.14.3)

$$u(x,t_0) = i_0(x) \tag{P7.14.8}$$

$$p_0(x,t,u) + q_0(x,t)f\left(x,t,u,\frac{\partial u}{\partial x}\right)\Big|_{x=x_0} = 0, \quad p_f(x,t,u) + q_f(x,t)f\left(x,t,u,\frac{\partial u}{\partial x}\right)\Big|_{x=x_f} = 0 \tag{P7.14.9a,b}$$

```
function [u,rs,ts,iter]=pde_heat_circle(a,R,tf,i0,bR,Mr,Mt)
%solve a^2*(u_rr +1/r*u_r) = u_t for 0<=r<=R, 0<=t<=tf
% ICs : u(r,0)=i0(r)
% BCs : u(R,t)=bR for r=R and all t>=0
% Mr = # of subintervals along the r(radial)-axis
% Mt = Number of subintervals along the t-axis
% u  : Solution u(r,t)
% rs, ts : Radial/Temporal grids
%Copyleft: Won Y. Yang, wyyang53@hanmail.net, CAU for academic use
dr=R/Mr; rs=[0:Mr]'*dr; % Radial grid points
dt=tf/Mt; ts=[0:Mt]*dt; % Time points
% dt must be small enough to make 2*(a/dr)^2*dt<1 for stability.
u=zeros(Mr+1,Mt+1); %Initialization
if isnumeric(it0), u(:,1)=it0;  else u(:,1)=feval(i0,rs); end
if isnumeric(bR), u(Mr+1,:)=bR;  else u(Mr+1,:)=feval(bR,ts); end
% Eq.(P7.14.5a,b,c)
adtdr2=dt*(a/dr)^2;  rs1=rs(2:end);
c1=adtdr2*(1+dr/2./rs1); c2=1-2*adtdr2; c3=adtdr2*(1Dr/2./rs1);
m=2:Mr; MaxIter=50;
for iter=1:MaxIter
   u0 = u;
   for k=2:Mt+1  % Eq.(P7.14.4)
      u(m,k) = c1(m).*u(m+?,k-?) +c2*u(m,k-?) +c3(m).*u(m-?,k-?);
   end
   u(1,:) = u(2,:); % Approximation for around the center
   difference=norm(u-u0)/norm(u0);
   if iter>5&difference<0.001, break; end
end
```

```
function [u,r,t]=pde_heat_circle_pdepe(a,R,tf,Mr,Mt)
% This uses pdepe() to solve 1D heat eq on the circular region.
% Usage 1: pde_heat_circle_pdepe(a,R,tf,Mr,Mt)
% Usage 2: pde_heat_circle_pdepe(a,rs,ts)
m = 1; % Cylindrical coordinate system
if length(R)>1, rs=R; else Nr=Mr+1; rs=linspace(0,R,Nr); end
if length(tf)>1, ts=tf; else Nt=Mt+1; ts=linspace(0,tf,Nt); end
sol = pdepe(m,@pdex1pde,@pdex1ic,@pdex1bc,rs,ts);
u = sol(:,:,1); % Extract the first solution component as u.
% ---------------------------------------------------------------
function [c,f,s]=pdex1pde(x,t,u,DuDx) % describing the PDE
a=1; c = 1/?^2;  f = DuDx; s = 0;
% ---------------------------------------------------------------
function u0=pdex1ic(x) % Initial condition
u0 = 1-?^2; % Eq.(P7.14.2)
% ---------------------------------------------------------------
function [p0,q0,pf,qf]=pdex1bc(x0,u0,xf,uf,t) % Boundary conditions
p0 = 0; q0 = 1;  % Virtually, no BC at x0
pf = uf; qf = ?; % u(xf,t)=0 (Eq.(P7.14.3)) at xf
```

Noting that substituting $x=r$, $m=1$, $f=\partial u/\partial r$, $s=0$, and $c=1/a^2$ into Eq. (P7.14.7) yields Eq. (P7.14.1) and referring to pertinent MATLAB documentation, complete the above MATLAB function, 'pde_heat_circle_pdepe()', which uses 'pdepe()' to solve Eq. (P7.14.1). Then complete the above MATLAB function, 'pde_heat_circle()', so that it can be used to find a numerical solution for the PDE (P7.14.1) subject to the IC (P7.14.2) and BC (P7.14.3). Then complete and run the following MATLAB script, "em07p14.m," to find two numerical solutions for the PDE (P7.14.1) with $a=1$ (for the circular domain of radius $R=1$ divided into $M_r=20$ segments and time range for $0\le t\le t_f=0.5$[s] divided into $M_t=800$ sub-intervals), one using 'pde_heat_circle()' and one using 'pde_heat_circle_pdepe()', and graph the solution as in P7.14 from the second numerical solution. Check the validity of the two numerical solutions based on the difference between them.

```
%em07p14.m
clear, clf
a=1; % PDE parameter
R=1; tf=0.5; % Boundaries of the solution region
i0=@(r)1-?.^2; bR=@(th)0; % IC (P7.14.2) and BC (P7.14.3)
Mr=20; Nr=Mr+1; %r=linspace(0,R,Nr);
Mt=800; Nt=Mt+1; %t=linspace(0,tf,Nt);
[un,rs,ts,iter] = pde_heat_circle(a,R,tf,i0,bR,Mr,Mt);
un_pdepe = pde_heat_circle_pdepe(a,rs,ts);
Discrepancy = norm(un-un_pdepe.')/norm(un_pdepe)
mesh(ts,rs,un_pdepe.') % Plot the solution graph
title(['Numerical solution on ' num2str(Nr) 'x' num2str(Nt) ' grid'])
xlabel('Time t'), ylabel('Radial distance r')
```

CHAPTER 8

Complex Analysis

CHAPTER OUTLINE

Complex analysis, i.e., the theory of functions of a complex variable, is the branch of mathematical analysis that investigates functions of complex numbers. It is useful in many branches of mathematics and physics, as well as hydrodynamics, thermodynamics, and electrical engineering.

8.1 FUNCTIONS OF A COMPLEX VARIABLE

8.1.1 Complex Numbers and Their Powers/Roots

A *complex number* can be written in the rectangular (Cartesian), polar, and exponential forms as

$$z = \overset{\text{Rectangular form}}{x + jy} = \overset{\text{Polar form}}{r\angle\theta} = \overset{\text{Exponential form}}{re^{j\theta}} \overset{(C.20)}{=} r(\cos\theta + j\sin\theta) \text{ with } \overset{\text{Absolute value or modulus}}{r = \sqrt{x^2+y^2}} \text{ and } \overset{\text{Phase or argument}}{\theta = \tan^{-1}(y/x)} \quad (8.1.1)$$

where $x = r\cos\theta / y = r\sin\theta$ are real numbers, each called the *real/imaginary parts,* represented by $\mathrm{Re}\{z\}/\mathrm{Im}\{z\}$, respectively, and j is the *imaginary unit* $\sqrt{-1}$. $r = |z| = \sqrt{x^2+y^2}$ and $\theta = \tan^{-1}(y/x)$ are called the *absolute value* or *modulus* and *phase* (*angle*) or *argument* of z, respectively.

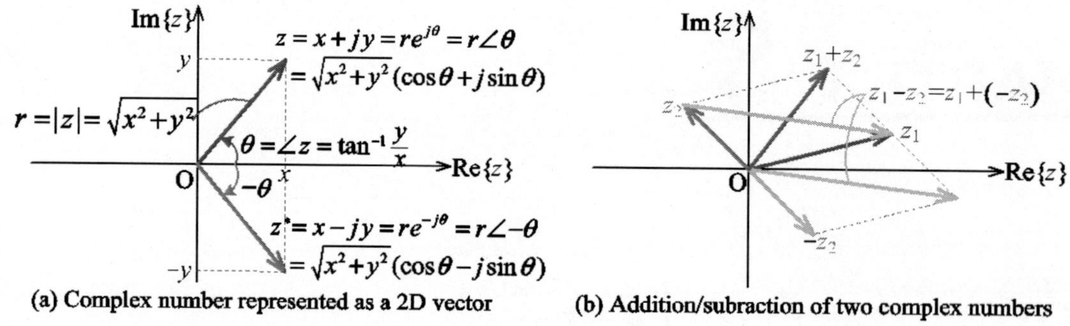

Figure 8.1 Graphical representation of a complex number and the addition/subtraction of complex numbers

A complex number is sometimes written as an ordered pair $z=(x,y)$ (in component form) so that it can be represented by a point and interpreted as a 2D vector in a coordinate plane (called the complex plane), as depicted in Fig. 8.1(a). Fig. 8.1(a) also shows the (complex) *conjugate* z^* of z (with the sign of the imaginary part reversed) as

$$z^* = x - jy \tag{8.1.2}$$

A complex number with zero real part and nonzero imaginary part is called a *pure imaginary number*. Two complex numbers, $z_1 = x_1 + jy_1$ and $z_2 = x_2 + jy_2$, are said to be *equal* if their real/imaginary parts are equal, i.e.,

$$z_1 = z_2 \Leftrightarrow \text{Re}\{z_1\} = \text{Re}\{z_2\} \text{ and } \text{Im}\{z_1\} = \text{Im}\{z_2\} \tag{8.1.3}$$

The addition and subtraction of complex numbers are defined as (see Fig. 8.1(b))

$$z_1 \pm z_2 = (x_1 + jy_1) \pm (x_2 + jy_2) = (x_1 \pm x_2) + j(y_1 \pm y_2) \tag{8.1.4}$$

The multiplication of complex numbers can be done in rectangular/polar forms as

$$\text{Rectangular form:} \quad (x_1 + jy_1)(x_2 + jy_2) = (x_1x_2 - y_1y_2) + j(x_1y_2 + x_2y_1) \tag{8.1.5a}$$

$$\text{Polar form:} \quad (r_1\angle\theta_1)(r_2\angle\theta_2) = r_1 e^{j\theta_1} r_2 e^{j\theta_2} = r_1r_2e^{j(\theta_1+\theta_2)} = r_1r_2\angle(\theta_1+\theta_2) \tag{8.1.5b}$$

The division of complex numbers can be done in rectangular/polar forms as

$$\text{Rectangular form:} \quad \frac{x_1+jy_1}{x_2+jy_2} = \frac{x_1+jy_1}{x_2+jy_2}\frac{x_2-jy_2}{x_2-jy_2} = \frac{x_1x_2+y_1y_2}{x_2^2+y_2^2} + j\frac{y_1x_2-x_1y_2}{x_2^2+y_2^2} \tag{8.1.6a}$$

$$\text{Polar form:} \quad \frac{r_1\angle\theta_1}{r_2\angle\theta_2} = \frac{r_1e^{j\theta_1}}{r_2e^{j\theta_2}} = \frac{r_1}{r_2}e^{j(\theta_1-\theta_2)} = \frac{r_1}{r_2}\angle(\theta_1-\theta_2) \tag{8.1.6b}$$

Note that the sum, difference, and product of a complex number z and its conjugate z^* yield

$$z + z^* = (x+jy) + (x-jy) = 2x = 2\text{Re}\{z\} \rightarrow \text{Re}\{z\} = \frac{z+z^*}{2} \tag{8.1.7a}$$

$$z - z^* = (x+jy) - (x-jy) = j2y = j2\,\text{Im}\{z\} \rightarrow \text{Im}\{z\} = \frac{z-z^*}{j2} \tag{8.1.7b}$$

$$zz^* = (x+jy)(x-jy) = x^2+y^2 = |z|^2 = r^2 \tag{8.1.7c}$$

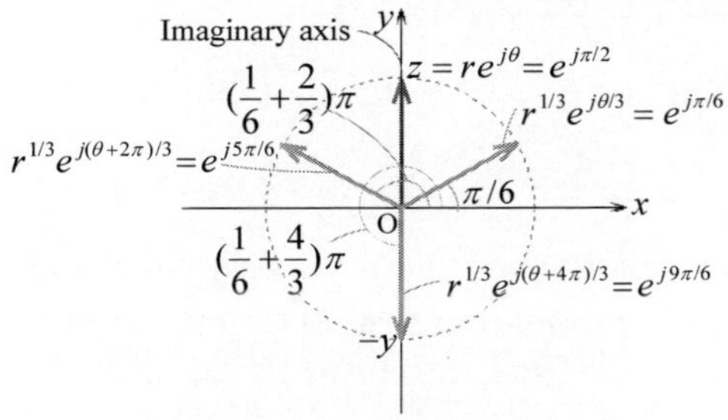

Figure 8.2 3rd roots of z=*j*

Integer powers and *N*th roots of a complex number can easily be computed in polar form:

$$z^N=(re^{j\theta})^N=r^Ne^{jN\theta}=r^N(\cos N\theta+j\sin N\theta) \text{ with an integer } N \tag{8.1.8}$$

$$z_n^{1/N}=r^{1/N}e^{j(\theta+2\pi n)/N}=r^{1/N}\left(\cos\frac{\theta+2\pi n}{N}+j\sin\frac{\theta+2\pi n}{N}\right) \text{ for } n=0,1,\cdots,N-1 \tag{8.1.9}$$

where $z_n^{1/N}=r^{1/N}(\cos(\theta/N)+j\sin(\theta/N))$ is called the *principal Nth root* of z. Fig. 8.2 shows the three 3rd roots of $z=j=re^{j\pi/2}$ (with r=1), all of which are equally spaced on a circle of radius $r^{1/3}=1$ centered at the origin, where the angle difference of two adjacent *N*th roots is $2\pi/N$:

$$z_n^{1/N}\overset{(8.1.9)}{\underset{N=3}{=}}r^{1/3}e^{j(\theta+2\pi n)/3}\overset{r=1}{\underset{\theta=\pi/2}{=}}\cos\frac{(3+4n)\pi}{6}+j\sin\frac{(3+4n)\pi}{6} \text{ for } n=0,1,2 \tag{8.1.10}$$

Note that in MATLAB, if `i` and `j` are not used for any other purpose, they represent the imaginary unit $j=\sqrt{-1}$ by default. Try typing the following statements at the MATLAB prompt and see what they yield:

```
>>z1=1+2i; z2=3-4i; % Two arbitrary complex numbers z1 and z2
>>z1*z2  % Multiplying z1 by z2
>>z1/z2  % Dividing z1 by z2
>>r1=abs(z1) % Absolute value or modulus of z1, i.e., |z1|
  r1 =  2.2361
>>sqrt(real(z1)^2+imag(z1)^2) % Also, |z1|
>>th1=angle(z1), atan2(imag(z1),real(z1)) % Phase/Argument of z1 in rad
  th1 =  1.1071
>>th1*180/pi, rad2deg(th1) % Radian-to-Degree conversion
  ans =  63.4349
>>r1*exp(j*th1) % Polar-to-Rectangular conversion
   ans = 1.0000 + 2.0000i
>>C=[1+i  1-2i; -1+3i -1-4i] % A complex matrix
   C1 =  1.0000 + 1.0000i   1.0000 - 2.0000i
        -1.0000 + 3.0000i  -1.0000 - 4.0000i
>>C1=C' % Conjugate transpose of C
   C1 =  1.0000 - 1.0000i  -1.0000 - 3.0000i
         1.0000 + 2.0000i  -1.0000 + 4.0000i
```

8.1.2 Functions of a Complex Variable

A complex function, i.e., a function of a complex variable $z=x+jy$, can be written as

$$w=f(z)=u(x,y)+j\,v(x,y) \tag{8.1.11}$$

where u and v are real-valued functions of two real variables x and y.

If a function $f(z)$ is defined in a neighborhood of z_0 and for any $\varepsilon>0$, there exists a $\delta>0$ such that $|f(z)-L|<\varepsilon$ if $0<|z-z_0|<\delta$, and it is said to have a limit at z_0, written as

$$\lim_{z\to z_0} f(z)=L \tag{8.1.12}$$

Just like the derivative (Eq. (2.1.1)) of a real function, the (*complex*) *derivative* of a complex function $w=f(z)$ is defined as

$$\frac{dw}{dz}=\frac{d}{dz}f(z)=\lim_{\Delta z\to 0}\frac{f(z+\Delta z)-f(z)}{\Delta z} \tag{8.1.13}$$

The function is said to be (*complex*) *differentiable* at a point $z=z_0$ if the derivative exists at that point in the sense that Eq. (8.1.12) approaches the same limit from any direction, which is more demanding than the differentiability of a real function. For example, consider a complex function

$$w=f(z)=\frac{1-z}{1+z} \tag{8.1.14}$$

Using Eq. (8.1.13) or the quotient rule (C.31), we can find the derivative of this function

$$\frac{dw}{dz}\overset{(8.1.13)}{\underset{(8.1.14)}{=}}\lim_{\Delta z\to 0}\frac{\dfrac{1-(z+\Delta z)}{1+(z+\Delta z)}-\dfrac{1-z}{1+z}}{\Delta z}=\lim_{\Delta z\to 0}\frac{\dfrac{2z-2(z+\Delta z)}{(1+z)(1+z+\Delta z)}}{\Delta z}=\frac{-2}{(1+z)^2} \tag{8.1.15a}$$

$$\frac{dw}{dz}\overset{(C.31)}{\underset{(8.1.13)}{=}}\frac{(-1)(1+z)-(1-z)\cdot 1}{(1+z)^2}=\frac{-2}{(1+z)^2} \tag{8.1.15b}$$

which turns out to exist everywhere except at $z=-1$. For another example, consider a complex function of z as

$$w=\mathrm{Re}(z)=x=u(x,y)+j\,v(x,y)\ \text{ with }\ u(x,y)=x\ \text{ and }\ v(x,y)=0 \tag{8.1.16}$$

We can use Eq. (8.1.13) to find the derivative of this function

$$\frac{dw}{dz}\overset{(8.1.13)}{\underset{(8.1.16)}{=}}\lim_{\Delta z\to 0}\frac{\mathrm{Re}(z+\Delta z)-\mathrm{Re}(z)}{\Delta z}=\lim_{\Delta x,\Delta y\to 0}\frac{x+\Delta x-x}{\Delta x+j\Delta y}$$

$$=\begin{cases}0 & \text{if } z \text{ is approached along a vertical line (parallel to the } y\text{-axis) with } \Delta x=0\\ 1 & \text{if } z \text{ is approached along a horizontal line (parallel to the } x\text{-axis) with } \Delta y=0\end{cases} \tag{8.1.17}$$

which exists nowhere, implying that the complex function (8.1.16) is not differentiable.

A complex function $w=f(z)$ is said to be *analytic* (or *holomorphic*) at a point $z=z_0$ if it is differentiable in some neighborhood of z_0 as well as at z_0. For example, consider a complex function

$$w=f(z)=|z|^2=x^2+y^2 \tag{8.1.18}$$

We can use Eq. (8.1.12) to find the derivative of this function

$$\frac{dw}{dz}\overset{(8.1.12)}{=}\lim_{\Delta z\to 0}\frac{f(z+\Delta z)-f(z)}{\Delta z}\overset{(8.1.17)}{=}\lim_{\Delta x,\Delta y\to 0}\frac{(x+\Delta x)^2+(y+\Delta y)^2-(x^2+y^2)}{\Delta x+j\Delta y}$$
$$=\lim_{\Delta x,\Delta y\to 0}\frac{2x\Delta x+2y\Delta y+(\Delta x^2+\Delta y^2)}{\Delta x+j\Delta y}=\lim_{\Delta x,\Delta y\to 0}2\frac{x\Delta x+y\Delta y}{\Delta x+j\Delta y}$$
$$=\begin{cases}-j2y & \text{if } z \text{ is approached vertically with } \Delta x=0\\ 2x & \text{if } z \text{ is approached horizontally with } \Delta y=0\end{cases} \tag{8.1.19}$$

which exists nowhere except at the origin, implying that the complex function (8.1.18) is analytic nowhere. For another example, consider a complex function

$$w=f(z)=z^2=x^2-y^2+j2xy \tag{8.1.20}$$

Using Eq. (8.1.13) or directly using (C.26), we can find the derivative of this function

$$\frac{dw}{dz}\underset{(8.1.20)}{\overset{(8.1.12)\text{ or }(C.26)}{=}}2z \tag{8.1.21}$$

which exists everywhere, implying that the complex function (8.1.20) is analytic everywhere so that it may be called an *entire function*.

8.1.3 Cauchy–Riemann Equations

The Cauchy–Riemann differential equations, named after Augustin Cauchy and Bernhard Riemann, refer to the system of two PDEs (partial differential equations) that must be satisfied by complex differentiable functions for analyticity.

[Theorem 8.1.1a: Cauchy–Riemann (CR) Equations]

Let a complex function $f(z)=u(x,y)+j\,v(x,y)$ be differentiable at a point $z=x+jy$. Then at z, the first-order partial derivatives of u and v exist and satisfy the Cauchy–Riemann equations:

$$\frac{\partial u}{\partial x}=\frac{\partial v}{\partial y}\quad\text{and}\quad\frac{\partial u}{\partial y}=-\frac{\partial v}{\partial x} \tag{8.1.22a,b}$$

so that the complex derivative of $f(z)$ can be written as

$$f'(z)=\frac{\partial u}{\partial x}+j\frac{\partial v}{\partial x}=\frac{\partial v}{\partial y}-j\frac{\partial u}{\partial y} \tag{8.1.23}$$

(Proof)
We use Eq. (8.1.13) to write the complex derivative of the function $f(z)=u(x,y)+j\,v(x,y)$ as

$$f'(z)\overset{(8.1.13)}{=}\lim_{\Delta z\to 0}\frac{f(z+\Delta z)-f(z)}{\Delta z}=\lim_{\Delta x,\Delta y\to 0}\frac{u(x+\Delta x,y+\Delta y)+jv(x+\Delta x,y+\Delta y)-u(x,y)-jv(x,y)}{\Delta x+j\Delta y}$$
$$=\begin{cases}\lim\limits_{\Delta x\to 0,\ \Delta y=0}\dfrac{u(x+\Delta x,y)-u(x,y)+j(v(x+\Delta x,y)-v(x,y))}{\Delta x}=\dfrac{\partial u}{\partial x}+j\dfrac{\partial v}{\partial x}: \text{Horizontal limit} & (8.1.24a)\\ \lim\limits_{\Delta x=0,\ \Delta y\to 0}\dfrac{u(x,y+\Delta y)-u(x,y)+j(v(x,y+\Delta y)-v(x,y))}{j\Delta y}=\dfrac{\partial v}{\partial y}-j\dfrac{\partial u}{\partial y}: \text{Vertical limit} & (8.1.24b)\end{cases}$$

The complex differentiability of $f(z)$ forces the horizontal and vertical limits to be equal so that the real/imaginary parts of Eqs. (8.1.24a) and (8.1.24b) can be equated pairwise, as expressed by Eq, (8.1.22). This completes the proof.

Note that if the continuity of u, v, and their partial derivatives are added to the precondition of the above theorem for a domain D, then the complex function $f(z)=u(x,y)+jv(x,y)$ is guaranteed to be analytic in D and as a consequence, each of $u(x,y)$ and $v(x,y)$ is *harmonic* in the sense that they have continuous second-order partial derivatives in D and satisfy *Laplace's equation*:

$$\begin{cases} \dfrac{\partial u}{\partial x} \overset{(8.1.22a)}{=} \dfrac{\partial v}{\partial y} \rightarrow \dfrac{\partial^2 u}{\partial x^2} = \dfrac{\partial^2 v}{\partial x \partial y} \\ \dfrac{\partial u}{\partial y} \overset{(8.1.22b)}{=} -\dfrac{\partial v}{\partial x} \rightarrow \dfrac{\partial^2 u}{\partial y^2} = -\dfrac{\partial^2 v}{\partial y \partial x} \end{cases} \rightarrow \nabla^2 u = \dfrac{\partial^2 u}{\partial x^2} + \dfrac{\partial^2 u}{\partial y^2} = 0$$

$$\begin{cases} \dfrac{\partial u}{\partial x} \overset{(8.1.22a)}{=} \dfrac{\partial v}{\partial y} \rightarrow \dfrac{\partial^2 u}{\partial y \partial x} = \dfrac{\partial^2 v}{\partial y^2} \\ \dfrac{\partial u}{\partial y} \overset{(8.1.22b)}{=} -\dfrac{\partial v}{\partial x} \rightarrow \dfrac{\partial^2 u}{\partial x \partial y} = -\dfrac{\partial^2 v}{\partial x^2} \end{cases} \rightarrow \nabla^2 v = \dfrac{\partial^2 v}{\partial x^2} + \dfrac{\partial^2 v}{\partial y^2} = 0 \qquad (8.1.25)$$

where $v(x,y)$ is called the *conjugate harmonic function* of $u(x,y)$.

[Theorem 8.1.1b: Cauchy–Riemann (CR) Equations in Polar Form]

Let a complex function $f(z)=u(r,\theta)+j\,v(r,\theta)$ be differentiable at a point $z=re^{j\theta}$. Then at z, the first-order partial derivatives of u and v exist and satisfy the CR equations:

$$\frac{\partial u}{\partial r} = \frac{1}{r}\frac{\partial v}{\partial \theta} \quad \text{and} \quad \frac{\partial u}{\partial \theta} = -r\frac{\partial v}{\partial r} \qquad (8.1.26a,b)$$

so that the complex derivative of $f(z)$ can be written as

$$f'(z) = e^{-j\theta}\left(\frac{\partial u}{\partial r} + j\frac{\partial v}{\partial r}\right) = \frac{1}{r}e^{-j\theta}\left(\frac{\partial v}{\partial \theta} - j\frac{\partial u}{\partial \theta}\right). \qquad (8.1.27)$$

(Proof)

We use Eq. (8.1.13) to write the complex derivative of the function $f(z)=u(r,\theta)+j\,v(r,\theta)$ as

$$f'(z) \overset{(8.1.13)}{=} \lim_{\Delta z \to 0} \frac{f(z+\Delta z)-f(z)}{\Delta z} = \lim_{\Delta r, \Delta\theta \to 0} \frac{u(r+\Delta r,\theta+\Delta\theta)-u(r,\theta)+j(v(r+\Delta r,\theta+\Delta\theta)-v(r,\theta))}{(r+\Delta r)e^{j(\theta+\Delta\theta)}-re^{j\theta}=e^{j\theta}\Delta r}$$

$$= \begin{cases} \lim\limits_{\substack{\Delta r \to 0 \\ \Delta\theta = 0}} \dfrac{u(r+\Delta r,\theta)-u(r,\theta)+j(v(r+\Delta r,\theta)-v(r,\theta))}{(r+\Delta r)e^{j\theta}-re^{j\theta}=e^{j\theta}\Delta r} = \left(\dfrac{\partial u}{\partial r}+j\dfrac{\partial v}{\partial r}\right)e^{-j\theta} : \text{Radial limit} & (8.1.28a) \\ \lim\limits_{\substack{\Delta r \to 0 \\ \Delta\theta \to 0}} \dfrac{u(r,\theta+\Delta\theta)-u(r,\theta)+j(v(r,\theta+\Delta\theta)-v(r,\theta))}{re^{j(\theta+\Delta\theta)}-re^{j\theta} \approx jre^{j\theta}\Delta\theta} = \left(\dfrac{\partial v}{\partial \theta}-j\dfrac{\partial u}{\partial \theta}\right)\dfrac{e^{-j\theta}}{r} : \text{Circular limit} & (8.1.28b) \end{cases}$$

The complex differentiability of $f(z)$ forces the radial and circular limits to be equal so that the real/imaginary parts of Eqs. (8.1.28a) and (8.1.28b) can be equated pairwise, as expressed by Eq, (8.1.26). This completes the proof.

It is interesting that if $u(x,y)$ and $v(x,y)$ are the harmonic functions satisfying the Cauchy–Riemann equations, their level curves $f_1(x,y)=u(x,y)-c_1=0$ and $f_2(x,y)=v(x,y)-c_2=0$ are orthogonal because the inner product of their gradient vectors is zero:

$$\nabla f_1 = \frac{\partial f_1}{\partial x}\mathbf{u}_x + \frac{\partial f_1}{\partial y}\mathbf{u}_y = \frac{\partial u}{\partial x}\mathbf{u}_x + \frac{\partial u}{\partial y}\mathbf{u}_y \tag{8.1.29a}$$

$$\nabla f_2 = \frac{\partial f_2}{\partial x}\mathbf{u}_x + \frac{\partial f_2}{\partial y}\mathbf{u}_y = \frac{\partial v}{\partial x}\mathbf{u}_x + \frac{\partial v}{\partial y}\mathbf{u}_y \tag{8.1.29b}$$

$$\begin{aligned}\nabla f_1 \bullet \nabla f_2 &= \left(\frac{\partial u}{\partial x}\mathbf{u}_x + \frac{\partial u}{\partial y}\mathbf{u}_y\right)\bullet\left(\frac{\partial v}{\partial x}\mathbf{u}_x + \frac{\partial v}{\partial y}\mathbf{u}_y\right) = \frac{\partial u}{\partial x}\frac{\partial v}{\partial x} + \frac{\partial u}{\partial y}\frac{\partial v}{\partial y}\\ &\overset{(8.1.22)}{=} \frac{\partial v}{\partial y}\left(-\frac{\partial u}{\partial y}\right) + \frac{\partial u}{\partial y}\frac{\partial v}{\partial y} = 0\end{aligned} \tag{8.1.30}$$

(Example 8.1) Cauchy–Riemann (CR) Equations

Check if the following complex functions satisfy the CR equations.

(a) $$f(z) = u(x,y) + j\,v(x,y) = \frac{x^2-y^2}{(x^2+y^2)^2} - j\frac{2xy}{(x^2+y^2)^2} \tag{E8.1.1}$$

This satisfies the CR equations (8.1.22a,b):

$$\frac{\partial u}{\partial x} = \frac{2x(x^2+y^2)^2 - 4x(x^2-y^2)(x^2+y^2)}{(x^2+y^2)^4} = -\frac{2x(x^2+y^2)^2 - 8xy^2(x^2+y^2)}{(x^2+y^2)^4} = \frac{\partial v}{\partial y} \tag{E8.1.2a}$$

$$\frac{\partial u}{\partial y} = \frac{-2y(x^2+y^2)^2 - 4y(x^2-y^2)(x^2+y^2)}{(x^2+y^2)^4} = \frac{2y(x^2+y^2)^2 - 8x^2y(x^2+y^2)}{(x^2+y^2)^4} = -\frac{\partial v}{\partial x} \tag{E8.1.2b}$$

It may be convenient to use the MATLAB command '`diff()`' for symbolic differentiation:

```
>>syms x y; u=(x^2-y^2)/(x^2+y^2)^2; v=-2*x*y/(x^2+y^2)^2;
>>simplify(diff(u,x)-diff(v,y)) % Eq.(E8.1.2a)
>>simplify(diff(u,y)+diff(v,x)) % Eq.(E8.1.2b)
   ans = 0,   ans = 0
```

It might have been easier to deal with the CR equations in polar form when u and v can be expressed in polar coordinates, as in this case:

$$f(z) = u + jv \underset{\substack{x=r\cos\theta\\ y=r\sin\theta}}{\overset{(E8.1.1)}{=}} \frac{\cos^2\theta - \sin^2\theta}{r^2} - j\frac{2\cos\theta\sin\theta}{r^2} \underset{(C.16)}{\overset{(C.17)}{=}} \frac{\cos 2\theta}{r^2} - j\frac{\sin 2\theta}{r^2} \overset{(C.20)}{=} \frac{1}{r^2}e^{-j2\theta} \tag{E8.1.3}$$

Then we can use Eq. (8.1.26) to check if this satisfies the CR equations:

$$\frac{\partial u}{\partial r} = -\frac{2}{r^3}\cos 2\theta \overset{\text{OK}}{=} \frac{1}{r}\frac{\partial}{\partial \theta}\left(-\frac{1}{r^2}\sin 2\theta\right) = \frac{1}{r}\frac{\partial v}{\partial \theta} \tag{E8.1.4a}$$

$$\frac{\partial u}{\partial \theta} = -\frac{2}{r^2}\sin 2\theta \overset{\text{OK}}{=} -r\frac{\partial}{\partial r}\left(-\frac{1}{r^2}\sin 2\theta\right) = -r\frac{\partial v}{\partial r} \tag{E8.1.4b}$$

In fact, we can replace $re^{j\theta}$ in Eq. (E8.1.3) with z to write the complex function in compact form as

$$f(z)=\frac{1}{z^2} \tag{E8.1.5}$$

According to Theorem 8.1.1, the differentiability of this function $f(z)=1/z^2$ is enough to guarantee that the CR equations are satisfied by the real part u and imaginary part v of $f(z)$.

(b) $f(z)=u(x,y)+jv(x,y)=x^3+xy^2+j(x^2y+y^3)$ (E8.1.6)

This fails to satisfy the CR equations (8.1.22a,b):

$$\frac{\partial u}{\partial x}=3x^2+y^2\neq x^2+3y^2=\frac{\partial v}{\partial y} \text{ unless } x=\pm y \tag{E8.1.7a}$$

$$\frac{\partial u}{\partial y}=2xy\neq -2xy=-\frac{\partial v}{\partial x} \text{ unless } xy=0 \tag{E8.1.7b}$$

(Example 8.2) Harmonic Function and Conjugate Harmonic Function

Show that the following function is harmonic and find its conjugate harmonic function:

$$u(x,y)=x^3-3xy^2-2y \tag{E8.2.1}$$

First, it is harmonic since it satisfies Laplace's equation (8.1.24):

$$\nabla^2 u=\frac{\partial^2 u}{\partial x^2}+\frac{\partial^2 u}{\partial y^2}=6x-6x\overset{\text{OK}}{=}0 \tag{E8.2.2}$$

Second, to find the conjugate harmonic function $v(x, y)$, we use the CR equations:

$$\frac{\partial u}{\partial x}=3x^2-3y^2=\frac{\partial v}{\partial y}\rightarrow v(x,y)=\int(3x^2-3y^2)dy=3x^2y-y^3+h(x)$$

$$\text{and } \frac{\partial u}{\partial y}=-6xy-2=-\frac{\partial v}{\partial x}\rightarrow 6xy+2=\frac{\partial v}{\partial x}=6xy+h'(x)\rightarrow h(x)=2x+C$$

$$;\quad v(x,y)=3x^2y-y^3+h(x)=3x^2y-y^3+2x+C \tag{E8.2.3}$$

Thus, an analytic function with u/v as its real/imaginary parts can be formed as

$$f(z)=x^3-3xy^2-2y+j(3x^2y-y^3+2x+C)=z^3+j(2z+C) \tag{E8.2.4}$$

The orthogonality of level curves $u=c_1$ and $v=c_2$ can be shown by checking to see whether inner product is zero:

$$\nabla u\bullet\nabla v=\{(3x^2-3y^2)\mathbf{u}_x+(-6xy-2)\mathbf{u}_y\}\bullet\{(6xy+2)\mathbf{u}_x+(3x^2-3y^2)\mathbf{u}_y\}=0 \tag{E8.2.5}$$

[Theorem 8.1.2: Vector Fields and Analyticity]

Let a complex function $f(z)=u(x,y)-j\,v(x,y)$ be analytic in a domain D so that u and v can satisfy the CR equations. Then the 2D vector field $\mathbf{F}=u(x,y)\,\mathbf{u}_x+v(x,y)\mathbf{u}_y=u(x,y)+j\,v(x,y)$ (in complex form) is incompressible (not divergent) and irrotational (not circulant):

$$\text{div}\,\mathbf{F}=\nabla\bullet\mathbf{F}=0: \text{incompressible (not divergent)} \tag{8.1.31}$$

$$\text{curl}\,\mathbf{F}=\nabla\times\mathbf{F}=\mathbf{0}: \text{irrotational (not circulant)} \tag{8.1.32}$$

The converse is also true.

(Proof)

$$\nabla \cdot \mathbf{F} \overset{(2.4.4a)}{=} \frac{\partial F_x}{\partial x} + \frac{\partial F_y}{\partial y} = \frac{\partial u}{\partial x} + \frac{\partial v}{\partial y} \underset{-v \text{ in place of } v}{\overset{(8.1.22a)}{=}} 0 : \text{incompressible (not divergent)}$$

$$\nabla \times \mathbf{F} \overset{(2.4.5a)}{=} \left(\frac{\partial F_y}{\partial x} - \frac{\partial F_x}{\partial y} \right) \mathbf{u}_z = \left(\frac{\partial v}{\partial x} - \frac{\partial u}{\partial y} \right) \mathbf{u}_z \underset{-v \text{ in place of } v}{\overset{(8.1.22b)}{=}} \mathbf{0} : \text{irrotational (not circulant) Q.E.D.}$$

In the situation where Theorem 8.1.2 holds, the vector field **F** has a *complex potential*

$$g(z) = \phi(x,y) + j\psi(x,y) \text{ with } \frac{\partial \phi}{\partial x} = u = \frac{\partial \psi}{\partial y} \text{ and } \frac{\partial \phi}{\partial y} = v = -\frac{\partial \psi}{\partial x} \tag{8.1.33}$$

which has the *complex velocity*

$$\frac{d}{dz} g(z) = f(z) = u(x,y) - jv(x,y) = \mathbf{F}^* \tag{8.1.34}$$

This can be verified by writing

$$dg = \frac{\partial g}{\partial x}dx + \frac{\partial g}{\partial y}dy = \left(\frac{\partial \phi}{\partial x} + j\frac{\partial \psi}{\partial x} \right) dx + \left(\frac{\partial \phi}{\partial y} + j\frac{\partial \psi}{\partial y} \right) dy = u(dx + jdy) - jv(dx + jdy) = (u - jv)dz$$

Note that $\phi(x,y)$ and $\psi(x,y)$, called a real (*velocity*) *potential function* and a *stream function*, respectively, also satisfy the CR equations, and consequently, are harmonic functions satisfying Laplace's equation:

$$\nabla^2 \phi = \frac{\partial^2 \phi}{\partial x^2} + \frac{\partial^2 \phi}{\partial y^2} = 0, \qquad \nabla^2 \psi = \frac{\partial^2 \psi}{\partial x^2} + \frac{\partial^2 \psi}{\partial y^2} = 0 \tag{8.1.35a,b}$$

To have a physical interpretation of the complex potential function (8.1.33), let us see the following example.

(**Example 8.3**) Complex Potential for a Flow around a Corner

Consider a particle whose path in a fluid is parameterized by $z(t) = x(t) + jy(t)$ and whose horizontal/vertical velocities are

$$\begin{aligned} \frac{dx}{dt} &= u(x,y) = 2x \\ \frac{dy}{dt} &= v(x,y) = -2y \end{aligned} \tag{E8.3.1}$$

Noting that u and $-v$ satisfy the CR equations

$$\frac{\partial u}{\partial x} = 2 = \frac{\partial(-v)}{\partial y} \quad \text{and} \quad \frac{\partial u}{\partial y} = 0 = -\frac{\partial(-v)}{\partial x} \tag{E8.3.2}$$

we can find the complex potential function (8.1.33) consisting of the potential/stream functions as

$$\begin{aligned} g(z) &\overset{(8.1.33)}{=} \phi(x,y) + j\psi(x,y) \text{ with } \frac{\partial \phi}{\partial x} = u = 2x = \frac{\partial \psi}{\partial y} \text{ and } \frac{\partial \phi}{\partial y} = v = -2y = -\frac{\partial \psi}{\partial x} \\ &= x^2 - y^2 + j(2xy) = (x + jy)^2 = z^2 \end{aligned} \tag{E8.3.3}$$

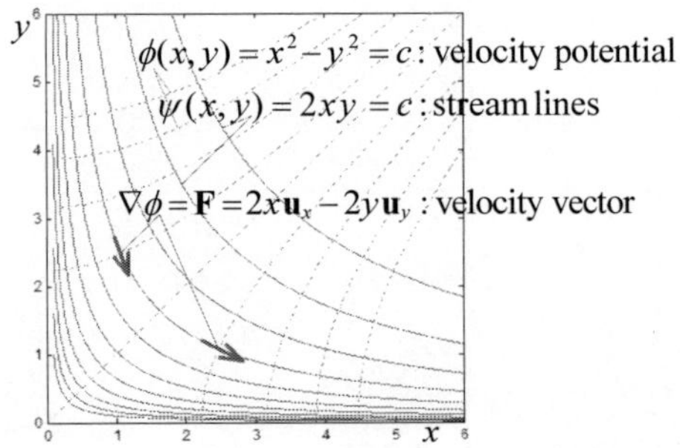

Figure 8.3 Stream lines, equipotential lines, and velocity vectors describing a flow around a corner

Why is $\phi(x,y)$ called a velocity potential? Because its derivatives w.r.t. x and y, i.e., $\partial\phi/\partial x = u = 2x$ and $\partial\phi/\partial y = v = -2y$, are the velocities along the x- and y-axes at (x, y), respectively. Why is $\psi(x,y)$ called a stream? Because its level curves $\psi(x,y) = 2xy = c$ (depicted in Fig. 8.3) are *stream lines* (for the flow of the particle) to which the velocity vector field $\mathbf{F} = u(x,y)\,\mathbf{u}_x + v(x,y)\,\mathbf{u}_y = 2x\,\mathbf{u}_x - 2y\,\mathbf{u}_y$ is tangential where

$$\frac{d\psi}{dt} = \frac{\partial\psi}{\partial x}\frac{dx}{dt} + \frac{\partial\psi}{\partial y}\frac{dy}{dt} = (-v)u + uv = 0 \tag{E8.3.4}$$

Note that the solution of the differential equation (E8.3.1) with initial values $(x(0), y(0))$ is

$$x(t) = x(0)e^{2t} \quad \text{and} \quad y(t) = y(0)e^{-2t} \tag{E8.3.5}$$

The trace of solution points on the path determined by these equations belongs to the stream lines.

From the viewpoint of the relationship between x and y (with time t hidden), the differential equation (E8.3.1) can be treated as an exact DE since it satisfies Eq. (3.1.9):

$$\frac{dy}{dx} = \frac{v}{u} = \frac{-2y}{2x};\ 2y\,dx + 2x\,dy = 0 \ \left(\text{exact DE since } \frac{\partial M}{\partial y} = 2 = \frac{\partial N}{\partial x}\right) \tag{E8.3.6}$$

Just like Eq. (3.1.11), this yields the solution (Eq. (3.1.12)):

$$xy = c_1 \tag{E8.3.7}$$

which describes the stream lines $\psi(x,y) = c$. Fig. 8.3 also shows the *equipotential lines* (as dotted lines) corresponding to

$$\phi(x,y) = x^2 - y^2 = c \tag{E8.3.8}$$

such that

$$\nabla\phi \overset{(2.1.7)}{=} \mathbf{u}_x\frac{\partial\phi}{\partial x} + \mathbf{u}_y\frac{\partial\phi}{\partial y} = 2x\,\mathbf{u}_x - 2y\,\mathbf{u}_y = u\,\mathbf{u}_x + v\,\mathbf{u}_y = \mathbf{F}: \textit{velocity vector field} \tag{E8.3.9}$$

Note that the velocity vector field $\mathbf{F}$ is a tangent vector to the path $\mathbf{z}(t) = x(t)\mathbf{u}_x + y(t)\mathbf{u}_y = x(0)e^{2t}\mathbf{u}_x + y(0)e^{-2t}\mathbf{u}_y$, that is,

$$\mathbf{z}'(t) = x'(t)\mathbf{u}_x + y'(t)\mathbf{u}_y \overset{(E8.3.5)}{=} 2x(0)e^{2t}\mathbf{u}_x - 2y(0)e^{-2t}\mathbf{u}_y \overset{(E8.3.5)}{=} 2x(t)\mathbf{u}_x - 2y(t)\,\mathbf{u}_y = \mathbf{F} \tag{E8.3.10}$$

Here, it is interesting and useful to see that *we may more easily find the complex potential function* (E8.3.3) *by constructing the complex velocity function* (8.1.34) *and integrating it w.r.t. z* as

$$\frac{d}{dz}g(z) \overset{(8.1.34)}{=} f(z) = u - jv \overset{(E8.3.1)}{=} 2x + j2y = 2z \ \rightarrow\ g(z) = \int f(z)\,dz = \int 2z\,dz = z^2. \tag{E8.3.11}$$

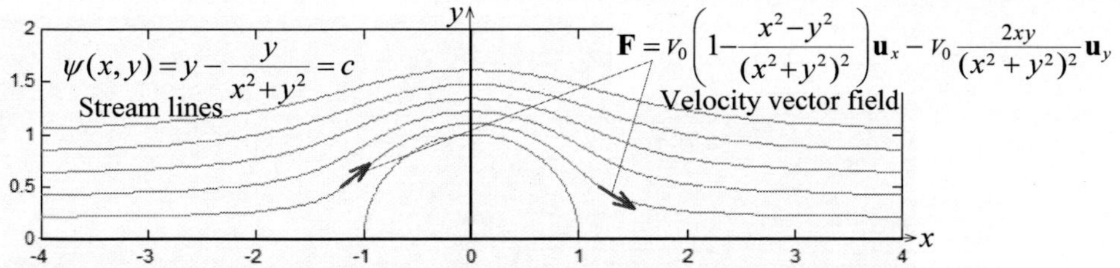

Figure 8.4 Stream lines and velocity vectors describing a flow around a cylinder

(Example 8.4) Complex Potential for a Flow around a Cylinder (dealt with in Example 3.13)

Consider a particle whose path in a fluid is parameterized by $z(t)=x(t)+jy(t)$ and whose horizontal/vertical velocities are

$$\frac{dx}{dt}=u(x,y)=V_0\left(1-\frac{x^2-y^2}{(x^2+y^2)^2}\right)$$
$$\frac{dy}{dt}=v(x,y)=V_0\frac{-2xy}{(x^2+y^2)^2} \tag{E8.4.1}$$

Noting that u and $-v$ satisfy the CR equations (8.1.22):

$$\frac{\partial u}{\partial x}=-V_0\frac{2x(x^2+y^2)^2-4x(x^2-y^2)(x^2+y^2)}{(x^2+y^2)^4}=V_0\frac{2x(x^2+y^2)^2-8xy^2(x^2+y^2)}{(x^2+y^2)^4}=\frac{\partial(-v)}{\partial y} \tag{E8.4.2a}$$

$$\frac{\partial u}{\partial y}=V_0\frac{2y(x^2+y^2)^2+4y(x^2-y^2)(x^2+y^2)}{(x^2+y^2)^4}=-V_0\frac{2y(x^2+y^2)^2-8x^2y(x^2+y^2)}{(x^2+y^2)^4}=-\frac{\partial(-v)}{\partial x} \tag{E8.4.2b}$$

we can find the complex potential function with these horizontal/vertical velocities as

$$g(z)\overset{(8.1.33)}{=}\phi(x,y)+j\psi(x,y)=V_0\left(x+\frac{x}{x^2+y^2}\right)+jV_0\left(y-\frac{y}{x^2+y^2}\right) \tag{E8.4.3}$$

where

$$\frac{\partial\phi}{\partial x}=u=V_0\left(1-\frac{x^2-y^2}{(x^2+y^2)^2}\right)=\frac{\partial\psi}{\partial y}\quad\text{and}\quad\frac{\partial\phi}{\partial y}=v=V_0\frac{-2xy}{(x^2+y^2)^2}=-\frac{\partial\psi}{\partial x}$$

It will be easier to find this complex potential function (E8.4.3) by constructing the complex velocity function $f(z)=u(x,y)-jv(x,y)$ (Eq. (8.1.34)) and integrating it w.r.t. z:

$$\frac{d}{dz}g(z)\overset{(8.1.34)}{=}u-jv=V_0\left(1-\frac{x^2-y^2}{(x^2+y^2)^2}\right)-jV_0\frac{-2xy}{(x^2+y^2)^2}=V_0-V_0\frac{(x-jy)^2}{(x+jy)^2(x-jy)^2}=V_0-\frac{V_0}{z^2}$$

$$\overset{\text{Integrate}}{\rightarrow}g(z)=\int V_0(1-\frac{1}{z^2})\,dz=V_0\left(z+\frac{1}{z}\right)=V_0\left(x+jy+\frac{1}{x+jy}\right)=V_0\left(x+\frac{x}{x^2+y^2}\right)+jV_0\left(y-\frac{y}{x^2+y^2}\right) \tag{E8.4.4}$$

Equating the imaginary part of this complex potential function to a constant yields the equation describing the stream lines depicted in Fig. 8.4:

$$y-\frac{y}{x^2+y^2}=\frac{c}{V_0} \tag{E8.4.5}$$

This suggests a novel approach (unlike that explained in Example 3.5) to solving exact DEs.

8.1.4 Exponential and Logarithmic Functions

The *complex exponential function* is defined as

$$e^z = e^{x+jy} = e^x e^{jy} \overset{\text{Euler's formula}}{\underset{\text{(C.20)}}{=}} e^x(\cos y + j\sin y) \tag{8.1.36}$$

Note the following properties of the complex exponential function e^z:

- It is analytic in the entire z-plane so that it can be called an entire function, because the real and imaginary parts of e^z, i.e., $u(x,y)=e^x\cos y$ and $v(x,y)=e^x\sin y$ and their first-order partial derivatives, are continuous and satisfy the CR equations in the entire z-plane:

$$\frac{\partial u}{\partial x} = e^x \cos y = \frac{\partial v}{\partial y} \text{ and } \frac{\partial u}{\partial y} = -e^x \sin y = -\frac{\partial v}{\partial x} \tag{8.1.37}$$

  Therefore, just like the real exponential function, it can be expanded as the Maclaurin series as

$$e^z = \sum_{n=0}^{\infty} \frac{z^n}{n!} = 1 + z + \frac{z^2}{2!} + \frac{z^3}{3!} + \cdots \tag{8.1.38}$$

  This can be another definition of the complex exponential function.
- Unlike the real exponential function, it is periodic with the complex period $j2\pi$ in the sense that

$$e^{z+j2\pi n} = e^z e^{j2\pi n} = e^z \tag{8.1.39}$$

The *complex logarithmic function* is defined as the inverse of the complex exponential function:

$$w = \ln z = u + jv = \ln z + j\arg(z) \text{ for } z = e^w = e^{u+jv} = x + jy \neq 0 \tag{8.1.40}$$

The relationship between the real/imaginary parts of $w=u+jv$ and $z=x+jy$ can be derived as

$$x \overset{(8.1.36)}{=} e^u \cos v \text{ and } y \overset{(8.1.36)}{=} e^u \sin v \rightarrow |z| = \sqrt{x^2+y^2} = e^u \text{ and } \frac{y}{x} = \tan v$$

$$\rightarrow u = \ln|z| \text{ and } v = \arg(z) = \text{Arg}(z) + 2n\pi \text{ with } -\pi < \text{Arg}(z) = \tan^{-1}\frac{y}{x} \leq \pi \tag{8.1.41a,b}$$

Here, Arg(z) denotes the principal argument of z within the interval $(-\pi, \pi]$, i.e., $\{\theta \mid -\pi < \theta \leq \pi\}$. To remove the ambiguity of the multi-valued complex logarithmic function, the *principal logarithmic function*, denoted by Ln z, is defined as the principal value of ln z with $n=0$:

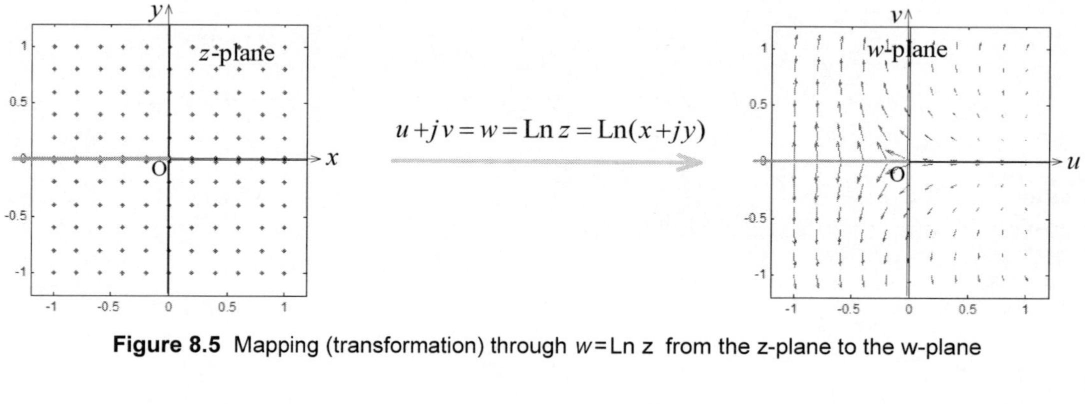

Figure 8.5 Mapping (transformation) through w=Ln z from the z-plane to the w-plane

$$w = \mathrm{Ln}\, z = u + jv \text{ with } u = \ln|z| = \frac{1}{2}\ln(x^2+y^2) \text{ and } v = \mathrm{Arg}(z) = \tan^{-1}\frac{y}{x} \tag{8.1.42}$$

Note that the complex logarithmic function $f(z)=\mathrm{Ln}\, z$ is non-analytic on the nonpositive real axis including $z=0$ (see Fig. 8.5). Why? First, $f(0)=\mathrm{Ln}\, 0$ is not even defined. Second, for any $z<0$ on the negative real axis, Arg(z) is π or $-\pi$, depending on whether z is approached from the upper or lower half-plane. Still, it is analytic in the complex plane, except for the nonpositive real axis.

The *complex power* of a complex number is defined as

$$z^w = e^{w \ln z} \text{ for } z \neq 0 \tag{8.1.43}$$

Some properties of the real logarithmic function hold for the complex one:

$$\ln(z_1 z_2) = \ln z_1 + \ln z_2 \text{ and } \ln\left(\frac{z_1}{z_2}\right) = \ln z_1 - \ln z_2 \tag{8.1.44a,b}$$

However, these properties do not hold for Ln z. The derivative of the complex logarithmic function is like that of the real one:

$$\frac{d}{dz}\ln z = \frac{1}{z} \text{ and } \frac{d}{dz}\mathrm{Ln}\, z = \frac{1}{z} \tag{8.1.45a,b}$$

8.1.5 Trigonometric and Hyperbolic Functions

Just like the real trigonometric functions, the *complex trigonometric functions* of a complex variable $z=x+jy$ are defined as

$$\cos z = \frac{e^{jz}+e^{-jz}}{2}, \qquad \sin z = \frac{e^{jz}-e^{-jz}}{j2} \tag{8.1.46a,b}$$

$$\tan z = \frac{\sin z}{\cos z}, \quad \cot z = \frac{1}{\tan z}, \quad \sec z = \frac{1}{\cos z}, \quad \csc z = \frac{1}{\sin z} \tag{8.1.47a,b,c,d}$$

The forms of the derivatives of the complex trigonometric functions are the same as the real trigonometric functions:

$$\frac{d}{dz}\sin z = \cos z, \quad \frac{d}{dz}\cos z = -\sin z, \quad \frac{d}{dz}\tan z = \sec^2 z, \quad \frac{d}{dz}\cot z = -\csc^2 z, \cdots \tag{8.1.48a,b,c,d}$$

Note that cos z and sin z can be expressed in terms of the real trigonometric/hyperbolic functions as

$$\sin z = \frac{e^{j(x+jy)} - e^{-j(x+jy)}}{j2} = \sin x \cosh y + j \cos x \sinh y \tag{8.1.49a}$$

$$\cos z = \frac{e^{j(x+jy)} + e^{-j(x+jy)}}{2} = \cos x \cosh y - j \sin x \sinh y \tag{8.1.49b}$$

Note also that the trigonometric formulas for the real trigonometric functions apply for the complex trigonometric functions, too:

$$\sin(-z) = -\sin z, \qquad \cos(-z) = \cos z \tag{8.1.50a,b}$$

$$\cos^2 z + \sin^2 z = 1, \quad \cos 2z = \cos^2 z - \sin^2 z, \quad \sin 2z = 2\sin z \cos z \tag{8.1.51a,b,c}$$

$$\sin(z_1 \pm z_2) = \sin z_1 \cos z_2 \pm \cos z_1 \sin z_2 \tag{8.1.52a}$$

$$\cos(z_1 \pm z_2) = \cos z_1 \cos z_2 \mp \sin z_1 \sin z_2 \tag{8.1.52b}$$

Just like the real hyperbolic functions, the *complex hyperbolic functions* of a complex variable $z=x+jy$ are defined as

$$\cosh z = \frac{e^z+e^{-z}}{2}, \qquad \sinh z = \frac{e^z-e^{-z}}{2}, \tag{8.1.53a,b}$$

$$\tanh z = \frac{\sinh z}{\cosh z}, \quad \coth z = \frac{1}{\tanh z}, \quad \mathrm{sech}\, z = \frac{1}{\cosh z}, \quad \mathrm{csch}\, z = \frac{1}{\sinh z} \tag{8.1.54a,b,c,d}$$

The forms of the derivatives of the complex hyperbolic functions are the same as the real hyperbolic functions:

$$\frac{d}{dz}\sinh z = \cosh z, \quad \frac{d}{dz}\cosh z = \sinh z, \quad \frac{d}{dz}\tanh z = \mathrm{sech}^2 z, \quad \frac{d}{dz}\coth z = -\mathrm{csch}^2 z, \cdots \tag{8.1.55a,b,c,d}$$

The MATLAB command '`diff()`' can be used to find these derivatives:

```
>>syms z, dcoshz=diff(coth(z)), simplify(dcoshz+csch(z)^2)
   dcoshz = 1 - coth(z)^2,    0
```

Note also that, unlike the real trigonometric and hyperbolic functions, the complex trigonometric and hyperbolic functions are related to each other:

$$\sin z = -j\sinh(jz), \qquad \cos z = \cosh(jz) \tag{8.1.56a,b}$$

$$\sinh z = -j\sin(jz), \qquad \cosh z = \cos(jz) \tag{8.1.57a,b}$$

8.1.6 Inverse Trigonometric/Hyperbolic Functions

The inverse sine function is defined as

$$w = \sin^{-1} z \text{ (called 'arcsin } z\text{')} \quad \text{for } z = \sin w \tag{8.1.58a}$$

Another expression of $\sin^{-1} z$ can be obtained as follows:

$$z = \sin w \overset{(8.1.46b)}{=} \frac{e^{jw}-e^{-jw}}{j2}; \quad e^{j2w} - j2ze^{jw} - 1 = 0 \overset{\text{Quadratic formula}}{\rightarrow} e^{jw} = jz + (1-z^2)^{1/2}$$

$$\overset{\text{Logarithm}}{\rightarrow} jw = \ln\{jz + (1-z^2)^{1/2}\}; \quad \sin^{-1} z = w = -j\ln\{jz + (1-z^2)^{1/2}\} \tag{8.1.58b}$$

Likewise, the inverse cosine and tangent functions can be expressed as

$$w = \cos^{-1} z \text{ (called 'arccos } z\text{')} \text{ for } z = \cos w \rightarrow w = -j\ln\{z + j(1-z^2)^{1/2}\} \tag{8.1.59}$$

$$w = \tan^{-1} z \text{ (called 'arctan } z\text{')} \text{ for } z = \tan w \rightarrow w = \frac{j}{2}\ln\frac{j+z}{j-z} \tag{8.1.60}$$

The derivatives of the inverse sine, cosine, and tangent functions can be found as

$$z=\sin w;\ dz=\cos w\,dw;\ \frac{d}{dz}(\sin^{-1}z)=\frac{dw}{dz}=\frac{1}{\cos w}=\frac{1}{(1-\sin^2 w)^{1/2}}=\frac{1}{(1-z^2)^{1/2}} \tag{8.1.61}$$

$$z=\cos w;\ dz=-\sin w\,dw;\ \frac{d}{dz}(\cos^{-1}z)=\frac{dw}{dz}=\frac{-1}{\sin w}=\frac{-1}{(1-\cos^2 w)^{1/2}}=-\frac{1}{(1-z^2)^{1/2}} \tag{8.1.62}$$

$$z=\tan w;\ dz=\sec^2 w\,dw;\ \frac{d}{dz}(\tan^{-1}z)=\frac{dw}{dz}=\frac{1}{\sec^2 w}=\frac{1}{1+\tan^2 w}=\frac{1}{1+z^2} \tag{8.1.63}$$

The inverse hyperbolic sine function is defined as

$$w=\sinh^{-1}z \quad \text{for} \quad z=\sinh w \tag{8.1.64a}$$

Another expression of $\sinh^{-1}z$ can be obtained as follows:

$$z=\sinh w \overset{(8.1.53b)}{=} \frac{e^w-e^{-w}}{2};\ e^{2w}-2ze^w-1=0 \overset{\text{Quadratic formula}}{\rightarrow} e^w=z+(1+z^2)^{1/2}$$

$$;\qquad w=\sinh^{-1}z=\ln\{z+(1+z^2)^{1/2}\} \tag{8.1.64b}$$

Likewise, the inverse hyperbolic cosine and tan functions can be expressed as

$$w=\cosh^{-1}z \ \text{ for } \ z=\cosh w \rightarrow w=\ln\{z+(z^2-1)^{1/2}\} \tag{8.1.65}$$

$$w=\tanh^{-1}z \ \text{ for } \ z=\tanh w \rightarrow w=\frac{1}{2}\ln\frac{1+z}{1-z} \tag{8.1.66}$$

The derivatives of the inverse hyperbolic sine, cosine, and tangent functions can be found as

$$z=\sinh w;\ dz=\cosh w\,dw;\ \frac{d}{dz}(\sinh^{-1}z)=\frac{dw}{dz}=\frac{1}{\cosh w}=\frac{1}{(1+\sin^2 w)^{1/2}}=\frac{1}{(z^2+1)^{1/2}} \tag{8.1.67}$$

$$z=\cosh w;\ dz=\sinh w\,dw;\ \frac{d}{dz}(\cosh^{-1}z)=\frac{dw}{dz}=\frac{1}{\sinh w}=\frac{1}{(\cosh^2 w-1)^{1/2}}=\frac{1}{(z^2-1)^{1/2}} \tag{8.1.68}$$

$$z=\tanh w;\ dz=\text{sech}^2 w\,dw;\ \frac{d}{dz}(\tanh^{-1}z)=\frac{dw}{dz}=\frac{1}{\text{sech}^2 w}=\frac{1}{1-\tanh^2 w}=\frac{1}{1-z^2} \tag{8.1.69}$$

8.2 CONFORMAL MAPPING

8.2.1 Conformal Mappings

A complex function of a complex variable $z=x+jy$, written as

$$w=f(z)=u(x,y)+j\,v(x,y) \tag{8.2.1}$$

can be interpreted as a (planar) mapping or transformation of a curve or region $z=a$ in the z-plane to the *image* $b=f(a)$ of a in the w-plane. Illustrating a mapping requires two graphs as in Fig. 8.6, one for curves/regions in the z-plane (xy-plane) and one for their images in the w-plane (uv-plane).

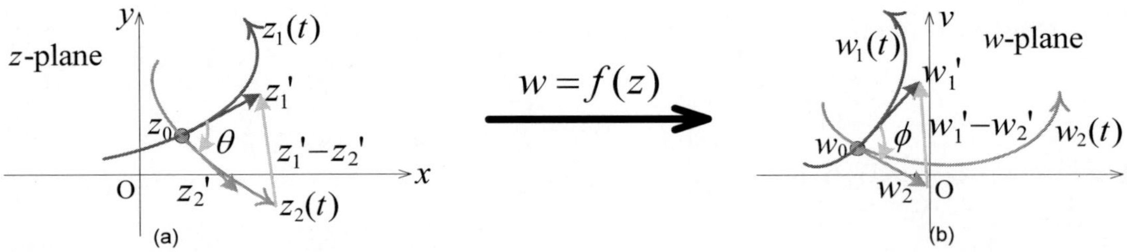

Figure 8.6 Angle between 2 curves in the *z*-plane and angle between their image curves in the *w*-plane

A complex mapping $w=f(z)$ defined on a domain D is said to be *conformal* if it preserves the angles between any two curves in D, except at (critical) points where $f'(z)=0$. What does "*angle-preserving*" mean? To figure it out, suppose there are two curves $z_1(t)$ and $z_2(t)$ (like Fig. 8.6(a)) that intersect at $z=z_0$ in the z-plane. The angle θ between the two curves are the angle between their tangent vectors $z_1'(t)$ and $z_2'(t)$ at z_0, which can be found by applying Eq. (1.1.8) (the law of cosines) to the triangle constructed by $z_1'(t)$ and $z_2'(t)$ (Fig. 8.6(a)) as

$$\theta = \cos^{-1}\left(\frac{|z_1'|^2+|z_2'|^2-|z_1'-z_2'|^2}{2|z_1'||z_2'|}\right) \tag{8.2.2}$$

Likewise, the angle c between the two corresponding image curves (Fig. 8.6(b)) can be found as

$$\phi = \cos^{-1}\left(\frac{|w_1'|^2+|w_2'|^2-|w_1'-w_2'|^2}{2|w_1'||w_2'|}\right) = \cos^{-1}\left(\frac{|f'(z_0)z_1'|^2+|f'(z_0)z_2'|^2-|f'(z_0)z_1'-f'(z_0)z_2'|^2}{2|f'(z_0)z_1'||f'(z_0)z_2'|}\right) \tag{8.2.3}$$

where we have used

$$w'(t) = \frac{dw}{dt} \underset{(2.1.4)}{\overset{\text{chain rule}}{=}} \frac{dw}{dz}\frac{dz}{dt} = f'(z)z'(t) \tag{8.2.4}$$

In fact, if only $f'(z_0)\neq 0$, the angle ϕ (Eq. (8.2.3) (with $|f'(z_0)|^2$ factored out from its numerator/denominator) is the same as θ (Eq. (8.2.2)), which is referred to as "angle-preserving." That is why it can be stated that if $f(z)$ is analytic in the domain D (containing z_0) and $f'(z_0)\neq 0$, then it is conformal at $z=z_0$.

(Q) Why do we use $z_1(t)$, $z_1'(t)$, … (instead of vector notations in bold characters) to denote the parameterized curves or their tangent vectors?

(A) Because the 2D vectors are represented by complex expressions. That is why the absolute values of complex expressions have been taken instead of the norms of vector expressions.

(**Example 8.5**) The Mapping $w=f(z)=z+1/z$

Consider the mapping described by the complex function

$$w=f(z)=z+1/z \tag{E8.5.1}$$

Note that this mapping is conformal at all values of z except $z=\pm 1$ (*critical* or *stationary points* at which $f'(z)=1-1/z^2=0$) and $z=0$ (at which $f'(z)$ does not exist). To find the images of the radial lines (with constant phase θ) and circular lines (with constant modulus r) in the z-plane mapped to the w-plane, let us substitute $z=re^{j\theta}$ into Eq. (E8.5.1) to write

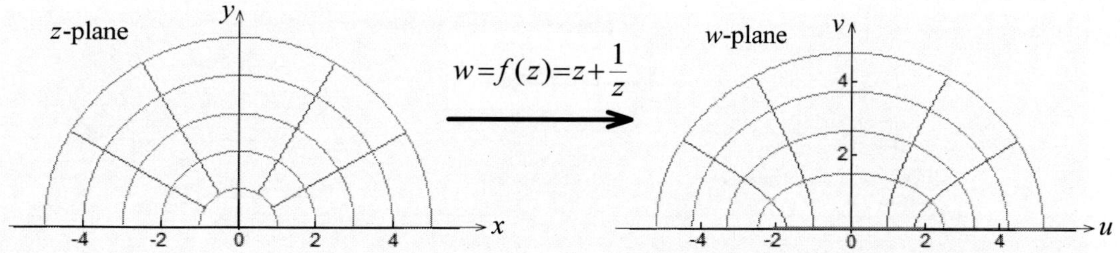

Figure 8.7 Mapping (transformation) of the radial/circular lines from the z-plane to the w-plane under w=z+1/z

$$u+jv=w=z+\frac{1}{z}=re^{j\theta}+\frac{1}{r}e^{-j\theta}=(r+\frac{1}{r})\cos\theta+j(r-\frac{1}{r})\sin\theta \qquad \text{(E8.5.2)}$$

$$;\begin{cases}\frac{u^2}{\cos^2\theta}-\frac{v^2}{\sin^2\theta}=(r+\frac{1}{r})^2-(r-\frac{1}{r})^2=4 & \text{for the images of radial lines with angle } \theta \\ \frac{u^2}{(r+1/r)^2}+\frac{v^2}{(r-1/r)^2}=1 & \text{for the images of circular lines with radius } r\end{cases} \qquad \text{(E8.5.3a,b)}$$

Fig. 8.7 shows the radial/circular lines in the z-plane and their images in the w-plane under the mapping $w=z+1/z$.

Now, to show that the images of circular lines and radial lines are perpendicular to each other in the w-plane, as the circular lines and radial lines are perpendicular to each other in the z-plane, we use Eq. (2.4.3b) to find their gradient vectors and see if their inner product is zero:

$$f(u,v)=\frac{u^2}{\cos^2\theta}-\frac{v^2}{\sin^2\theta}-4=0; \qquad \nabla f=\frac{\partial f}{\partial u}\mathbf{u}_u+\frac{\partial f}{\partial v}\mathbf{u}_v=\frac{2u}{\cos^2\theta}\mathbf{u}_u-\frac{2v}{\sin^2\theta}\mathbf{u}_v \qquad \text{(E8.5.4a)}$$

$$g(u,v)=\frac{u^2}{(r+1/r)^2}+\frac{v^2}{(r-1/r)^2}-1=0; \ \nabla g=\frac{\partial g}{\partial u}\mathbf{u}_u+\frac{\partial g}{\partial v}\mathbf{u}_v=\frac{2u}{(r+1/r)^2}\mathbf{u}_u+\frac{2v}{(r-1/r)^2}\mathbf{u}_v \qquad \text{(E8.5.4b)}$$

$$;\qquad \nabla f\bullet\nabla g \overset{\text{(E8.5.4a)}}{\underset{\text{(E8.5.4b)}}{=}} \frac{2u}{\cos^2\theta}\frac{2u}{(r+1/r)^2}-\frac{2v}{\sin^2\theta}\frac{2v}{(r-1/r)^2} \overset{\text{(E8.5.2)}}{=} \frac{4u^2}{u^2}-\frac{4v^2}{v^2}=4-4=0 \qquad \text{(E8.5.5)}$$

(Example 8.6) The Mapping $w=f(z)=e^z$

Consider the mapping described by the complex function

$$w=f(z)=e^z \qquad \text{(E8.6.1)}$$

To find the images of the radial lines (with constant phase θ) and circular lines (with constant modulus r) in the z-plane mapped to the w-plane, let us substitute $z=re^{j\theta}$ into Eq. (E8.6.1) to write

$$\rho e^{j\phi}=w=e^z=e^{re^{j\theta}}=e^{r\cos\theta}e^{jr\sin\theta}; \quad \rho=e^{r\cos\theta} \text{ and } \phi=r\sin\theta \qquad \text{(E8.6.2)}$$

$$;\begin{cases}\sqrt{\phi^2+\ln^2\rho}=r & \text{for the images of circular lines with radius } r \\ \frac{\phi}{\ln\rho}=\frac{r\sin\theta}{r\cos\theta}=\tan\theta & \text{for the images of radial lines with angle } \theta\end{cases} \qquad \text{(E8.6.3a,b)}$$

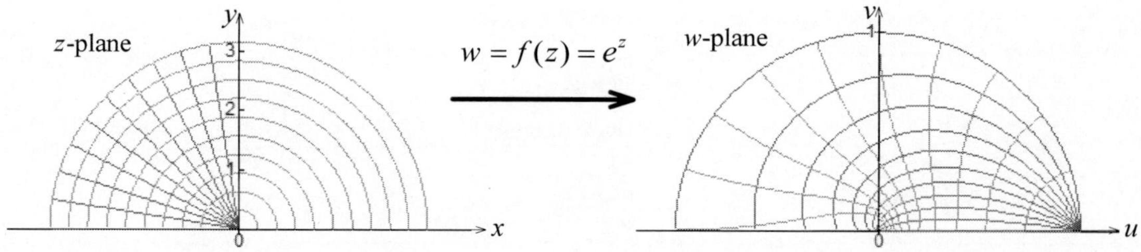

Figure 8.8 Mapping (transformation) of the radial/circular lines from the z-plane to the w-plane under $w = e^z$

The MATLAB script "em08f08.m" can be run to yield Fig. 8.8, which shows the radial/ circular lines in the z-plane and their images in the w-plane under the mapping $w{=}e^z$.

Now, to show that the images of circular/radial lines are perpendicular to each other in the w-plane, as the circular lines/radial lines are perpendicular to each other in the z-plane, we use Eq. (2.4.3b) to find their gradient vectors and see if their inner product is zero:

$$f(\rho,\phi)=\sqrt{\phi^2+\ln^2\rho}-r=0;\ \nabla f=\frac{\partial f}{\partial\rho}\mathbf{u}_\rho+\frac{1}{\rho}\frac{\partial f}{\partial\phi}\mathbf{u}_\phi=\frac{(\ln\rho)/\rho}{\sqrt{\phi^2+\ln^2\rho}}\mathbf{u}_\rho+\frac{1}{\rho}\frac{\phi}{\sqrt{\phi^2+\ln^2\rho}}\mathbf{u}_\phi \quad \text{(E8.6.4a)}$$

$$g(\rho,\phi)=\frac{\phi}{\ln\rho}-\tan\theta=0;\ \nabla g=\frac{\partial g}{\partial\rho}\mathbf{u}_\rho+\frac{1}{\rho}\frac{\partial g}{\partial\phi}\mathbf{u}_\phi=-\frac{\phi}{\ln^2\rho}\frac{1}{\rho}\mathbf{u}_\rho+\frac{1}{\rho}\frac{1}{\ln\rho}\mathbf{u}_\phi \quad \text{(E8.6.4b)}$$

$$;\ \nabla f\cdot\nabla g\overset{\text{(E8.6.4a)}}{\underset{\text{(E8.6.4b)}}{=}}\left(\frac{(\ln\rho)/\rho}{\sqrt{\phi^2+\ln^2\rho}}\mathbf{u}_\rho+\frac{1}{\rho}\frac{\phi}{\sqrt{\phi^2+\ln^2\rho}}\mathbf{u}_\phi\right)\cdot\left(-\frac{\phi}{\ln^2\rho}\frac{1}{\rho}\mathbf{u}_\rho+\frac{1}{\rho}\frac{1}{\ln\rho}\mathbf{u}_\phi\right)\overset{\text{Table1.1}}{=}0 \quad \text{(E8.6.5)}$$

```
%em08f08.m
clear, clf
ths=pi-[0:0.1:1]*(pi/2); % Several values of angle in the 2nd quadrant
r=[0.01:0.01:1]*pi; % Many values of radial distance on a radial line
for n=1:length(ths) % Radial lines with th=const
   th_=ths(n)*ones(size(r)); % Each value of angle
   [x,y]=pol2cart(th_,r); w=exp(x+j*y)+j*eps;
   subplot(121), polar(th_,r); hold on
   subplot(122), plot(w); hold on
end
rs=[0.1:0.1:1]*pi; % Values of radius for several circular lines
th=0:0.01:pi; % Many values of angle on a circular line
for n=1:length(rs) % Circular lines with r=const
   r_=rs(n)*ones(size(th)); % Each value of radius
   [x,y]=pol2cart(th,r_); w=exp(x+j*y)+j*eps;
   subplot(121), polar(th,r_,'r'); hold on
   subplot(122), plot(w,'r'); hold on
end
axis('equal'), axis([-1.2 1.2 -1.2 1.2])
subplot(121), axis('equal'), axis([-4 4 -4 4])
```

8.2.2 Linear Fractional Transformations

Linear fractional transformation (*LFT*), also called *Mobius transformation*, is a class of conformal mappings that are described by

$$w = f(z) = \frac{az+b}{cz+d} \text{ with } ad-bc \neq 0 \tag{8.2.5}$$

where *a*, *b*, *c*, and *d* are complex constants such that $ad-bc \neq 0$. Since its derivative is

$$f'(z) \underset{(C.31)}{\overset{(8.2.5)}{=}} \frac{a(cz+d)-(az+b)c}{(cz+d)^2} = \frac{ad-bc}{(cz+d)^2} \tag{8.2.6}$$

an LFT is conformal as long as $ad-bc \neq 0$ and $z \neq -d/c$.

Note that an LFT can be regarded as a sequence of four simpler transformations:

$$w = f(z) = f_4(f_3(f_2(f_1(z)))) \tag{8.2.7}$$

where

(1) $f_1(z) = z + \frac{d}{c}$ (translation by d/c)

(2) $f_2(z) = \frac{1}{z}$ (inversion and reflection w.r.t. the real axis)

(3) $f_3(z) = \frac{bc-ad}{c^2}z$ (dilation and rotation)

(4) $f_4(z) = z + \frac{a}{c}$ (translation by a/c).

The *inverse LFT* can easily be derived by the compositions of the inverse functions:

$$z = f_1^{-1}(f_2^{-1}(f_3^{-1}(f_4^{-1}(w)))) = \frac{bc-ad}{c^2(w-a/c)} - \frac{d}{c} = \frac{dw-b}{-cw+a} \tag{8.2.8}$$

It is interesting that associating an LFT and its inverse with the matrix *A*, consisting of its coefficients and its adjoint matrix, respectively, as

$$A = \begin{bmatrix} a & b \\ c & d \end{bmatrix} \text{ and } \mathrm{adj}(A) = \begin{bmatrix} d & -b \\ -c & a \end{bmatrix} \tag{8.2.9a,b}$$

makes it easy to deal with the inversion and composition of LFTs.

(**Example 8.7**) Composition of Two LFTs Using Matrix

We can find the composite mapping $w = f_2^{-1}(f_1(z))$ with

$$f_1(z) = \frac{2z+j}{jz-2} \text{ and } f_2(z) = \frac{2z+j}{jz-1} \tag{E8.7.1}$$

as

$$\left(\mathrm{adj}\begin{bmatrix} 2 & j \\ j & -1 \end{bmatrix}\right)\begin{bmatrix} 2 & j \\ j & -2 \end{bmatrix} = \begin{bmatrix} -1 & -j \\ -j & 2 \end{bmatrix}\begin{bmatrix} 2 & j \\ j & -2 \end{bmatrix} = \begin{bmatrix} -1 & j \\ 0 & -3 \end{bmatrix} \rightarrow f_2^{-1}(f_1(z)) = \frac{-z+j}{0z-3} \tag{E8.7.2}$$

A feature of LFTs is to map circles and lines to circles and lines where a line can also be considered as a (generalized) circle of infinitely long radius centered at infinity. Another feature is that it can be specified by a triple (z_1, z_2, z_3) in the *z*-plane and its corresponding triple (w_1, w_2, w_3) in the *w*-plane as

$$g^{-1}(f(z)) \tag{8.2.10}$$

with

$$f(z)=\frac{(z-z_1)(z_2-z_3)}{(z-z_3)(z_2-z_1)}=\frac{(z_2-z_3)z-(z_2-z_3)z_1}{(z_2-z_1)z-(z_2-z_1)z_3} \tag{8.2.11a}$$

$$g(w)=\frac{(w-w_1)(w_2-w_3)}{(w-w_3)(w_2-w_1)}=\frac{(w_2-w_3)w-(w_2-w_3)w_1}{(w_2-w_1)w-(w_2-w_1)w_3} \tag{8.2.11b}$$

where $(z-z_1)(z_2-z_3)/(z-z_3)(z_2-z_1)$ is called the *cross-ratio* of the complex numbers z, z_1, z_2, and z_3, which is invariant under any LFT. Note that if one of the points is ∞, the quotient of the differences containing the point should be zero. In view of Eq. (8.2.9), using Eq. (8.2.10) to find the LFT mapping a triple (z_1, z_2, z_3) in the z-plane onto another triple (w_1, w_2, w_3) in the w-plane can be cast into the MATLAB function '`LFT()`'.

```
function A = LFT(z,w)
% LFT mapping z=[z1,z2,z3] onto w=[w1,w2,w3].
if w(1)==inf,   W = [0  w(3)-w(2); -1  w(3)]; % Eq.(8.2.11b) to (8.2.9)
 elseif w(2)==inf,   W = [1 -w(1); 1 -w(3)];
 elseif w(3)==inf,   W = [-1 w(1); 0  w(1)-w(2)];
 else W=[w(2)-w(3)  (w(3)-w(2))*w(1); w(2)-w(1)  (w(1)-w(2))*w(3)];
end
if z(1)==inf,   Z=[0 z(3)-z(2); -1  z(3)]; % Eq.(8.2.11a) to (8.2.9)
 elseif z(2)==inf,   Z = [1 -z(1); 1  -z(3)];
 elseif z(3)==inf,   Z = [-1  z(1); 0  z(1)-z(2)];
 else Z=[z(2)-z(3)  (z(3)-z(2))*z(1); z(2)-z(1)  (z(1)-z(2))*z(3)];
end
A=det(W)*inv(W)*Z; A=A/A(1,1); % Eq.(8.2.10) in the form of Eq.(8.2.9)
syms z;
pretty((A(1,1)*z+A(1,2))/(A(2,1)*z+A(2,2))) % in the form of Eq.(8.2.9)
```

(**Example 8.8**) Constructing an LFT

We can construct an LFT mapping a triple $(-j, 0, j)$ to $(j, 1, -j)$ by using Eq. (8.2.10) with Eq. (8.2.11a,b) in matrix notation as

$$\left(\mathrm{adj}\begin{bmatrix} w_2-w_3 & -(w_2-w_3)w_1 \\ w_2-w_1 & -(w_2-w_1)w_3\end{bmatrix}\right)\begin{bmatrix} z_2-z_3 & -(z_2-z_3)z_1 \\ z_2-z_1 & -(z_2-z_1)z_3\end{bmatrix}$$

$$=\left(\mathrm{adj}\begin{bmatrix} 1+j & -(1+j)j \\ 1-j & (1-j)j\end{bmatrix}\right)\begin{bmatrix} 0-j & (0-j)j \\ 0+j & -(0+j)j\end{bmatrix}=\begin{bmatrix} 1+j & -1+j \\ -1+j & 1+j\end{bmatrix}\begin{bmatrix} -j & 1 \\ j & 1\end{bmatrix}=\begin{bmatrix} -j2 & j2 \\ j2 & j2\end{bmatrix}$$

$$\to g^{-1}(f(z))=\frac{-j2z+j2}{j2z+j2}=\frac{-z+1}{z+1} \text{ (See Example 8.10 and Fig. 8.10)} \tag{E8.8.1}$$

The above MATLAB function, '`LFT()`', can also be used to find this LFT:

```
>>LFT([-j  0  j],[j  1  -j])
      z - 1
   - -----
      z + 1
```

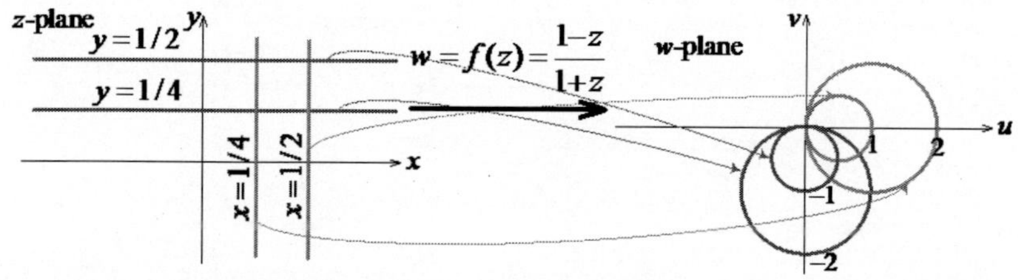

Figure 8.9 Mapping of the real/imaginary axes from the z-plane to the w-plane under $w=f(z)=1/z$

(Example 8.9) $w=f(z)=1/z$ as an LFT

Consider an LFT described by the complex function

$$w=f(z)=\frac{1}{z}=\frac{1}{x+jy}=\frac{x}{x^2+y^2}-j\frac{y}{x^2+y^2}=u(x,y)+j\,v(x,y) \tag{E8.9.1}$$

(a) Find the image of a horizontal line $y=c_1$ (with $c_1>0$) in the z-plane mapped to the w-plane.

To generate out the image, we substitute $z=x+jc_1$ into Eq. (E8.9.1) to get

$$w=\left.\frac{1}{z}\right|_{z=x+jc_1}=\frac{1}{x+jc_1}=\frac{x}{x^2+c_1^2}-j\frac{c_1}{x^2+c_1^2}=u+j\,v \tag{E8.9.2}$$

The image can be determined from the relationship between $u=\mathrm{Re}(w)$ and $v=\mathrm{Im}(w)$ as

$$u^2+v^2=\left(\frac{x}{x^2+c_1^2}\right)^2+\left(\frac{-c_1}{x^2+c_1^2}\right)^2=\frac{1}{x^2+c_1^2}=-\frac{v}{c_1};\ u^2+\left(v+\frac{1}{2c_1}\right)^2=\left(\frac{1}{2c_1}\right)^2 \tag{E8.9.3}$$

This describes a circle of radius $1/2c_1$ centered at $(0,-1/2c_1)$ in the w-plane (see Fig. 8.9).

(b) Find the image of the vertical line $x=c_2$ (with $c_2>0$) in the z-plane mapped to the w-plane.

To find out the image, we substitute $z=c_2+jy$ into Eq. (E8.9.1) to get

$$w=\left.\frac{1}{z}\right|_{z=c+jy}=\frac{1}{c+jy}=\frac{c}{c^2+y^2}-j\frac{y}{c^2+y^2}=u+j\,v \tag{E8.9.4}$$

The image can be determined from the relationship between $u=\mathrm{Re}(w)$ and $v=\mathrm{Im}(w)$ as

$$u^2+v^2=\left(\frac{c_2}{c_2^2+y^2}\right)^2+\left(\frac{-y}{c_2^2+y^2}\right)^2=\frac{1}{c_2^2+y^2}=\frac{u}{c_2};\ \left(u-\frac{1}{2c_2}\right)^2+v^2=\left(\frac{1}{2c_2}\right)^2 \tag{E8.9.5}$$

This describes a circle of radius $1/2c_2$ centered at $(1/2c_2, 0)$ in the w-plane (see Fig. 8.9).

(c) Check if the two images described by Eqs. (E8.9.3) and (E8.9.5) are perpendicular to each other in the w-plane, as the horizontal and vertical lines are perpendicular to each other in the z-plane.

We can use Eq. (2.4.3a) to find their gradient vectors and see if their inner product is zero:

$$f(u,v)=u^2+\left(v+\frac{1}{2c_1}\right)^2-\left(\frac{1}{2c_1}\right)^2=0;\ \nabla f=\frac{\partial f}{\partial u}\mathbf{u}_u+\frac{\partial f}{\partial v}\mathbf{u}_v=2u\mathbf{u}_u+2\left(v+\frac{1}{2c_1}\right)\mathbf{u}_v \tag{E8.9.6a}$$

$$g(u,v)=\left(u-\frac{1}{2c_2}\right)^2+v^2-\left(\frac{1}{2c_2}\right)^2=0;\ \nabla g=\frac{\partial f}{\partial u}\mathbf{u}_u+\frac{\partial f}{\partial v}\mathbf{u}_v=2\left(u-\frac{1}{2c_2}\right)\mathbf{u}_u+2v\,\mathbf{u}_v \tag{E8.9.6b}$$

$$;\ \nabla f\bullet\nabla g\underset{\text{(E8.9.6b)}}{\overset{\text{(E8.9.6a)}}{=}}4u\left(u-\frac{1}{2c_2}\right)+4v\left(v+\frac{1}{2c_1}\right)=2u\left(u-\frac{1}{c_2}\right)+2v^2+2u^2+2v\left(v+\frac{1}{c_1}\right)\underset{\text{(E8.9.5)}}{\overset{\text{(E8.9.3)}}{=}}0 \tag{E8.9.7}$$

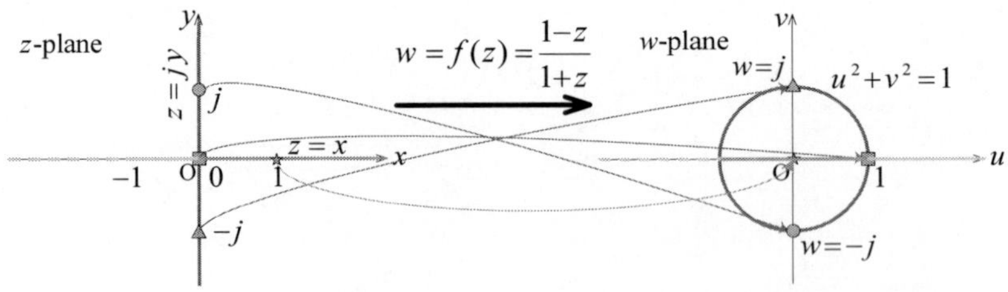

Figure 8.10 Mapping of the real/imaginary axes from the z-plane to the w-plane under $w=\frac{1-z}{1+z}$

(**Example 8.10**) Bilinear (or Tustin) Transform $w=f(z)=(1-z)/(1+z)$ as an LFT

Consider the BLT (Bilinear Transform) described by the complex function

$$w=f(z)=\frac{1-z}{1+z}=\frac{1-(x+jy)}{1+(x+jy)}=\frac{1-x^2-y^2}{(1+x)^2+y^2}-j\frac{2y}{(1+x)^2+y^2}=u(x,y)+j\,v(x,y) \quad \text{(E8.10.1)}$$

(a) Find the image of the imaginary axis $z=jy$ (including $z=j$, 0, and $-j$) in the z-plane mapped to the w-plane.

To find out the image, we substitute $z=jy$ into Eq. (E8.10.1) to get

$$w=\left.\frac{1-z}{1+z}\right|_{z=jy}=\frac{1-jy}{1+jy}=\frac{1^2-y^2}{1^2+y^2}-j\frac{2y}{1^2+y^2}=u+j\,v=-j,1,\text{ and } j \;\text{ for } z=j,0,\text{ and } -j \quad \text{(E8.10.2)}$$

The image can be determined from the relationship between $u=\text{Re}(w)$ and $v=\text{Im}(w)$ as

$$u^2+v^2=\left(\frac{1-y^2}{1+y^2}\right)^2+\left(\frac{2y}{1+y^2}\right)^2=\frac{(1-2y^2+y^4)+4y^2}{(1+y^2)^2}=1 \quad \text{(E8.10.3)}$$

This describes the unit circle centered at the origin in the w-plane (see Fig. 8.10).

(b) Find the image of the real axis $z=x$ (including $z=1$) in the z-plane mapped to the w-plane.

To find out the image, we substitute $z=x$ into Eq. (E8.10.1) to get

$$w=\left.\frac{1-z}{1+z}\right|_{z=x}=\frac{1-x}{1+x}=0 \text{ for } x=1. \quad \text{(E8.10.4)}$$

The image is also the real axis (see Fig. 8.10).

(c) The intersection angle of the real/imaginary axes is 90° and it is preserved under the BLT.

8.3 INTEGRATION OF COMPLEX FUNCTIONS

8.3.1 Line Integrals and Contour Integrals

Like the real definite integral, the (complex) *line integral* of a continuous complex function $f(z)$ (of a complex variable $z=x+jy$) along a piecewise smooth, oriented (directed) curve C (called the *path of integration*) in the complex plane (like the one depicted in Fig. 8.11) is defined as

$$\int_C f(z)\,dz=\lim_{\max|\Delta z_k|\to 0}\sum_{k=1}^{n} f(z_k^{\text{o}})\Delta z_k \quad \text{with } \Delta z_k=z_k-z_{k-1} \quad (8.3.1)$$

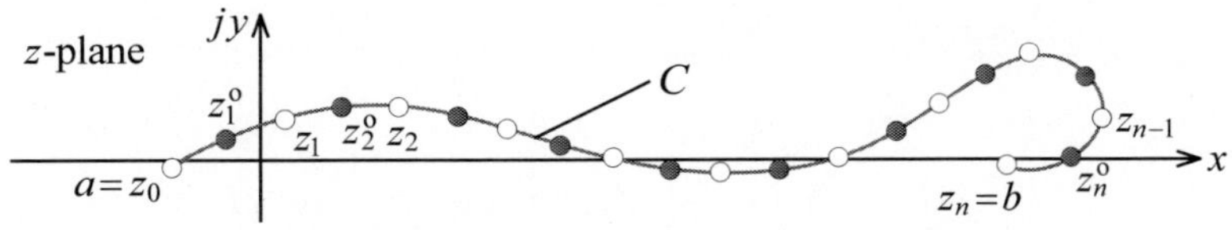

Figure 8.11 A curve C partitioned into many subarcs

where the curve C is described by a complex-valued function of a real variable t (called the parameter) as $\{z(t)=x(t)+jy(t),\ a\le t\le b\}$, and can be divided into n subarcs (each denoted by $\Delta z_k = z_k - z_{k-1} = z(t_k) - z(t_{k-1})$) according to the partition $\{a=t_0<t_1<\cdots<t_n=b\}$ of $[a, b]$, and z_k^o is a sample point on the kth subarc Δz_k.

If the curve is closed, then the line integral is referred to as a *contour integral*. The line integral (8.3.1) can be evaluated by

$$\int_C f(z)\,dz = \int_a^b f(z(t))z'(t)\,dt = \int_a^b (u+jv)(x'+jv')\,dt \tag{8.3.2}$$

where $z'(t) = dz(t)/dt = x'(t) + jy'(t)$. It has the following properties:

$$\int_{C_1\text{-}C_2} f(z)dz = \int_{C_1} f(z)dz + \int_{C_2} f(z)dz \tag{8.3.3}$$

$$\int_C f(z)\,dz = -\int_{C^-} f(z)\,dz \tag{8.3.4}$$

where the directions of curves C_1 and C_2 are supposed to be consistent and C^- denotes the same curve as C, but with reversed direction.

It is interesting that the line integral of a 2D vector field $\mathbf{f}(x,y) = u(x,y)\mathbf{u}_x + v(x,y)\mathbf{u}_y$ along a curve C described by $\mathbf{l}(t)=x(t)\mathbf{u}_x+y(t)\mathbf{u}_y$ is equal to the real part of the line integral of the conjugate of the corresponding complex function $f(z) = f(x,y) = u(x,y) + jv(x,y)$:

$$\int_C f^*(z)dz = \int_C (u-jv)(dx+jdy) = \int_C (u\,dx+v\,dy) + j\int_C (u\,dy-v\,dx) = \int_C \mathbf{f}\bullet d\mathbf{l} + j\int_C (u\,dy-v\,dx)$$

$$;\qquad \mathrm{Re}\left\{\int_C f^*(z)\,dz\right\} = \int_C \mathbf{f}\bullet d\mathbf{l} \tag{8.3.5}$$

(Example 8.11) Line Integral of an Analytic Function — Path Independent

Consider the complex function

$$f(z) = z^2. \tag{E8.11.1}$$

Find the line integral of $f(z)$ along the two oriented curves C_1 and C_2 depicted in Fig. 8.12.

$$\text{(a)}\ \int_{C_1} f(z)dz \underset{\text{(E8.11.1)}}{\overset{\text{(8.3.2)}}{=}} \int_0^{\pi/2} z^2(t)\,z'(t)dt \overset{z=e^{jt}}{=} \int_0^{\pi/2} e^{j2t}\,je^{jt}dt = \frac{1}{3}e^{j3t}\Big|_0^{\pi/2} = \frac{1}{3}(-j-1) \tag{E8.11.2}$$

$$\text{(b)}\ \int_{C_2} f(z)dz \underset{\text{(E8.11.1)}}{\overset{\text{(8.3.2)}}{=}} \int_0^1 z^2(t)\,z'(t)dt \overset{z=1-t+jt}{=} \int_0^1 (1-t+jt)^2(-1+j)\,dt \tag{E8.11.3}$$

$$= \int_0^1 \{2t^2-1+j(1-4t+2t^2)\}dt = \{\frac{2}{3}t^3 - t + j(t-2t^2+\frac{2}{3}t^3)\}\Big|_0^1 = -\frac{1}{3} - j\frac{1}{3}$$

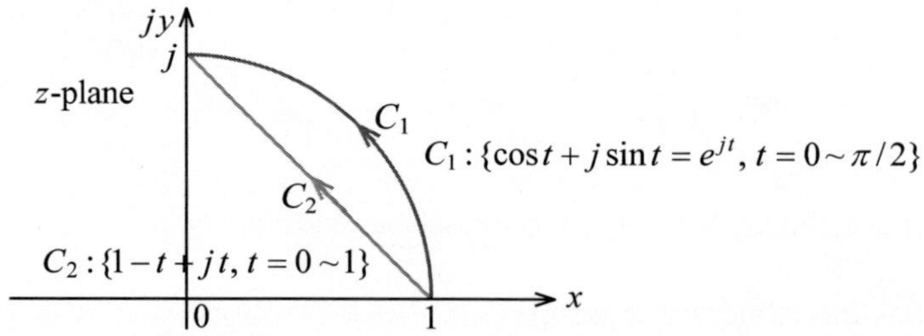

Figure 8.12 Paths of integration for the line integral of Example 8.11

Note that these two integral values are equal, which implies that the line integral is path independent so that it depends only on the endpoints z_a and z_b of the curve. In fact, since the function $f(z)=z^2$ is an entire function, being analytic everywhere in the complex plane, so that it has an antiderivative (also called a primitive or an indefinite integral) like $F(z)=z^3/3$ such that $dF(z)/dz = f(z)$, its line integral can, directly and independently of the path of integration, be evaluated as

$$\text{(c)} \int_{C_1} f(z)\,dz \overset{\text{(E8.11.1)}}{=} \frac{1}{3}z^3\Big|_{z_a=1}^{z_b=j} = \frac{1}{3}(-j-1) \tag{E8.11.4}$$

(Example 8.12) Line Integral of a Nonanalytic Function — Path Dependent

Consider the complex function

$$f(z) = \text{Re}(z) = x\,. \tag{E8.12.1}$$

Find the line integral of $f(z)$ along the two oriented curves C_1 and C_2 depicted in Fig. 8.12.

$$\text{(a)} \int_{C_1} f(z)\,dz \underset{\text{(E8.12.1)}}{\overset{\text{(8.3.2)}}{=}} \int_0^{\pi/2} \text{Re}\{z(t)\}\,z'(t)\,dt \overset{z=e^{jt}}{=} \int_0^{\pi/2} (\cos t) je^{jt}\,dt = \int_0^{\pi/2} (-\cos t \sin t + j\cos^2 t)\,dt$$

$$\underset{\text{(C.15)}}{\overset{\text{(C.16)}}{=}} \frac{1}{2}\int_0^{\pi/2} \{-\sin 2t + j(1+\cos 2t)\}\,dt = \frac{1}{4}\{\cos 2t + j(2t+\sin 2t)\}\Big|_0^{\pi/2} = \frac{1}{4}(-2+j\pi) \tag{E8.12.2}$$

$$\text{(b)} \int_{C_2} f(z)\,dz \underset{\text{(E8.12.1)}}{\overset{\text{(8.3.2)}}{=}} \int_0^1 \text{Re}\{z(t)\}\,z'(t)\,dt \overset{z=1-t+jt}{=} \int_0^1 (1-t)(-1+j)\,dt$$

$$= \int_0^1 \{-1+t+j(1-t)\}\,dt \overset{\text{(C.34)}}{=} \{-t+\frac{1}{2}t^2 + j(t-\frac{1}{2}t^2)\}\Big|_0^1 = -\frac{1}{2} + j\frac{1}{2} \tag{E8.12.3}$$

Note that these two integral values differ from each other, which implies that the line integral is path dependent due to the non-analyticity of the complex integrand function.

(Q) What decides the independence or dependence of path for the line integrals in the above two examples?

(A) It depends on whether or not the integrand function is analytic. In other words, the analyticity of the integrand function guarantees the path independence of the line integral, as stated in the Cauchy–Goursat Theorem (Section 8.3.2).

(Example 8.13) Contour Integral of $f(z) = (z - z_0)^m$

Find the contour integral of the complex function

$$f(z) = (z - z_0)^m \text{ with a constant } z_0 \text{ and an integer } m \tag{E8.13.1}$$

along the counterclockwise oriented circle C of radius R centered at z_0.

Noting that the circle is described by $\{z=z_0+Re^{jt},\ t=0\sim 2\pi\}$, we can compute the contour integral as

$$\oint_C f(z)\,dz \overset{\text{(E8.13.1)}}{=} \oint_C (z-z_0)^m dz \overset{z=z_0+Re^{jt}}{=} \int_0^{2\pi} R^m e^{jmt} jRe^{jt} dt = jR^{m+1}\int_0^{2\pi} e^{j(m+1)t} dt$$

$$= \begin{cases} j2\pi & \text{if } m=-1 \\ 0 & \text{otherwise} \end{cases} \tag{E8.13.2}$$

where the radius R of the circular path of integration does not matter. Especially for $m=-1$, the anti-derivative $\text{Ln}\,z$ of $1/z$ can be used to compute the contour integral:

$$\oint_C f(z)\,dz \overset{\text{(E8.13.1)}}{=} \oint_C (z-z_0)^{-1} dz = \text{Ln}\,(z-z_0)\Big|_{z=z_0+Re^{-j\pi^-}}^{z=z_0+Re^{j\pi^-}}$$

$$\overset{\text{(8.1.42)}}{=} (\ln R + j\pi^-) - (\ln R - j\pi^-)$$

$$= j2\pi \tag{E8.13.3}$$

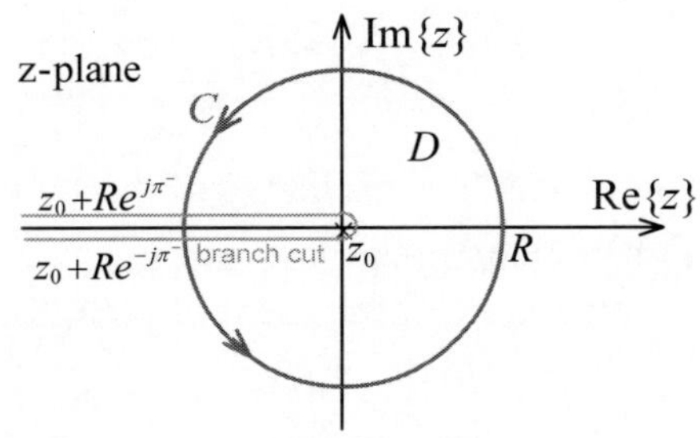

Figure 8.13 Contour of integration for Example 8.13

where $\pi^-=\pi-\varepsilon$ (ε: a positive infinitesimal) and the path of integration is approximated by $\{Re^{-j\pi^-}\sim 0\sim Re^{j\pi^-}\}$ so that the half infinite line left to $z=z_0$ (on which Ln z fails to be analytic) can be excluded from the region D.

8.3.2 Cauchy–Goursat Theorem

Here is the Cauchy's Intergral Theorem, which was discovered by the French mathematician Augustin-Louis Cauchy (1789–1857).

[Theorem 8.3.1: Cauchy's Integral Theorem]

If a complex function $f(z)=u(x,y)+j\,v(x,y)$ is analytic (or holomorphic) in a simply connected domain D, and $f'(z)$ is continuous in D, then for every simple contour C in D,

$$\oint_C f(z)\,dz = 0 \tag{8.3.6}$$

where a domain is said to be *simply connected* if any simple closed curve can be shrunk to a point continuously within the domain. Note that a curve is said to be *simple* if it does not cross itself.

(Proof)

We can use Green's theorem (2.7.1) and the Cauchy–Riemann (CR) equations (8.1.22) to prove this theorem:

$$\oint_C f(z)\,dz = \oint_C (u(x,y) + j\,v(x,y))(dx + j\,dy)$$

$$= \oint_C (u(x,y)dx - v(x,y)dy) + j\oint_C (v(x,y)dx + u(x,y)dy)$$

$$\overset{\text{Green's theorem}}{=} \iint_R \left(-\frac{\partial v}{\partial x} - \frac{\partial u}{\partial y}\right) dA + j\iint_R \left(\frac{\partial u}{\partial x} - \frac{\partial v}{\partial y}\right) dA \overset{\text{C-R eqs}}{=} 0$$

since we have the C-R equations $\dfrac{\partial u}{\partial x} = \dfrac{\partial v}{\partial y}$ and $\dfrac{\partial u}{\partial y} = -\dfrac{\partial v}{\partial x}$.

Another French mathematician, Édouard Jean-Baptiste Goursat (1858–1936), released the second precondition ("continuity of $f'(z)$") of the Cauchy's integral theorem to make the following theorem:

[Theorem 8.3.2: Cauchy–Goursat Theorem]

If a complex function $f(z)$ is analytic (or holomorphic) in a simply connected domain D, then for every simple contour (closed path) C in D,

$$\oint_C f(z)\,dz = 0 \tag{8.3.7}$$

[Remark 8.1] Corollaries of the Cauchy–Goursat Theorem

(1) If the complex integrand function $f(z)$ is analytic in a simply connected domain D, then its line integral is independent of path. More specifically, for any two paths C and C_1 in D, we have

$$\int_C f(z)dz = \int_{C_1} f(z)dz \;:\; \textit{path independence} \tag{8.3.8}$$

which can be proved, referring to Fig. 8.14, as follows:

$$0 \overset{\text{Cauchy's theorem}}{\underset{\text{Fig. 8.14(b)}}{=}} \int_{C\text{-}C_1^-} f(z)dz \overset{(8.3.3)}{=} \int_C f(z)dz + \int_{C_1^-} f(z)\,dz \overset{(8.3.4)}{\underset{\text{Fig. 8.14(a)}}{=}} \int_C f(z)dz - \int_{C_1} f(z)dz$$

In this case, the line integral along a curve with initial/terminal points z_0/z_1 can be computed by using an antiderivative $F(z)$ such that $F'(z)=f(z)$:

$$\int_C f(z)dz = F(z_1) - F(z_0) \tag{8.3.9}$$

as long as the paths are all contained in D.

(2) Consider two closed contours C and C_1 surrounding a hole (of nonanalyticity) in a doubly connected region D, as illustrated in Fig. 8.15(a). Note from Fig. 8.15(b) that, with a line (cut) L connecting any two points each on C and C_1, respectively, the dotted domain enclosed by the set of curves $C\text{-}C_{01}\text{-}C_1^-\text{-}C_{10}$ is simply connected, so that the Cauchy's theorem gives

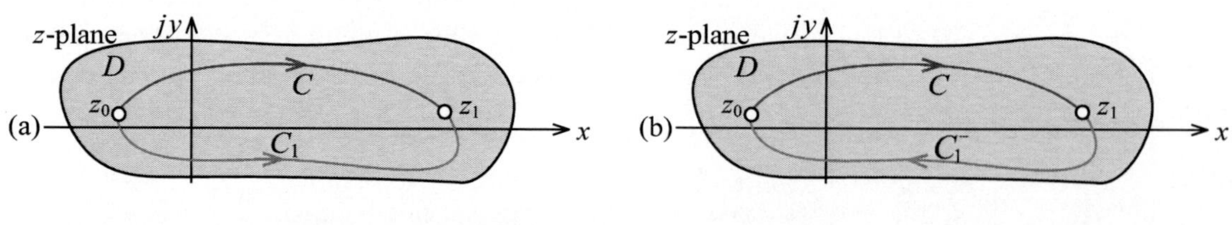

Figure 8.14 Paths of integration — path independence

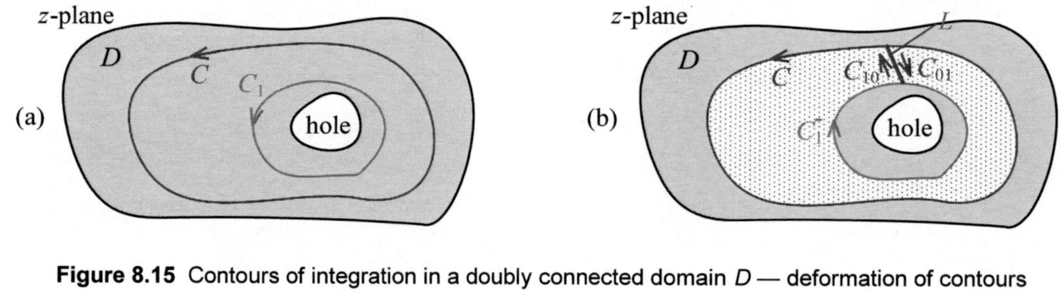

Figure 8.15 Contours of integration in a doubly connected domain D — deformation of contours

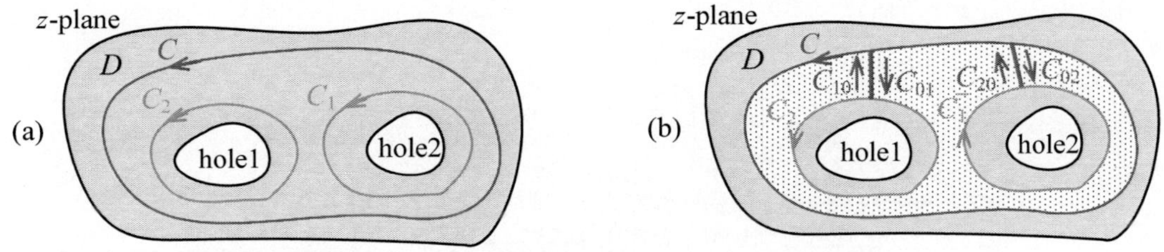

Figure 8.16 Contours of integration in a multiply connected domain *D*

$$0 \underset{\text{Fig. 8.15(b)}}{\overset{\text{Cauchy's theorem}}{=}} \oint_{C-C_{01}-\bar{C_1}-C_{10}} f(z)dz \overset{(8.3.3)}{=} \oint_C f(z)dz + \int_{C_{01}} f(z)dz + \oint_{\bar{C_1}} f(z)dz + \int_{C_{10}} f(z)dz$$

$$\underset{\text{Fig. 8.15(a)}}{\overset{(8.3.4)}{=}} \oint_C f(z)dz - \oint_{C_1} f(z)dz$$

$$; \qquad \oint_C f(z)dz = \oint_{C_1} f(z) : \textit{Deformation of contours} \tag{8.3.10}$$

(3) Eqs. (8.3.8) (independence of path) and (8.3.10) (deformation of contours) imply that the path/contour of integration can be continuously deformed without changing the value of the line/contour integral.

(4) Referring to Fig. 8.16, Eq. (8.3.10) can be generalized into the case of multiply connected domain (with n holes of nonanalyticity) as

$$\oint_C f(z)\,dz = \sum_{k=1}^{n} \oint_{C_k} f(z)\,dz \tag{8.3.11}$$

(**Example 8.14**) Contour Integral of $f(z)=z^m$

Note from the result of Example 8.13 with $z_0=0$ that we have

$$\oint_C z^m dz \overset{(E8.13.2)}{=} \begin{cases} j2\pi & \text{if } m=-1 \\ 0 & \text{otherwise} \end{cases} \tag{E8.14.1}$$

where the path C of integration is the counterclockwise oriented circle C of radius R centered at z_0.

In view of the Cauchy-Goursat theorem (Eq. (8.3.7)), it is natural that

$$\oint_C z^m dz = 0 \ \text{ for } m \neq -1$$

However, it is questionable why the value of the contour integral is not zero, which differs from the conclusion of the theorem. Can we justify it? Yes, because the contour C contains the pole $z=z_0$ so that C cannot be thought of as being in a simply connected domain D, as required by the precondition of the theorem.

(Q) How can the value of the contour integral be zero for $m<-1$ even if the precondition is still not satisfied?

(**Example 8.15**) Contour Integral Using PFE (Partial Fraction Expansion)

Find the contour integral of the complex function

$$f(z) = \frac{z^3-3z^2-3z-3}{z^4+2z^3+2z^2+2z+1} = \frac{z-1}{(z+1)^2} - \frac{2}{z^2+1} \overset{\text{PFE}}{=} \frac{1}{z+1} - \frac{2}{(z+1)^2} + \frac{j}{z-j} - \frac{j}{z+j} \tag{E8.15.1}$$

along the counterclockwise oriented circle C of radius $R=2$ centered at $z_0=0$, as shown in Fig. 8.17.

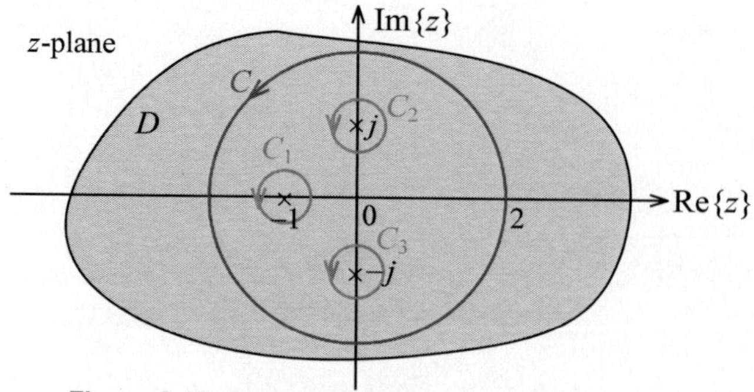

Figure 8.17 Contour of integration for Example 8.15

Noting that the complex integrand function $f(z)$ is not analytic at its poles $z=-1, +j$, and $-j$, each of which is enclosed by circular contours C_1, C_2, and C_3 in Fig. 8.17, we can use Eq. (8.3.11) to write the contour as

$$\oint_C f(z)\,dz \overset{(8.3.11)}{\underset{(E8.15.1)}{=}} \oint_{C_1}\left(\frac{1}{z+1}-\frac{2}{(z+1)^2}+\cancel{\frac{j}{z-j}}-\cancel{\frac{j}{z+j}}\right)dz + \oint_{C_2}\left(\cancel{\frac{1}{z+1}}-\cancel{\frac{2}{(z+1)^2}}+\frac{j}{z-j}-\cancel{\frac{j}{z+j}}\right)dz$$
$$+\oint_{C_3}\left(\cancel{\frac{1}{z+1}}-\cancel{\frac{2}{(z+1)^2}}+\cancel{\frac{j}{z-j}}-\frac{j}{z+j}\right)dz \tag{E8.15.2}$$

Since $1/(z+1)$ and $-2/(z+1)^2$ are analytic on/inside C_2 and C_3, $j/(z-j)$ is analytic on/inside C_1 and C_3, and $-j/(z+j)$ is analytic on/inside C_1 and C_2, their corresponding integrals are zero (by Cauchy's theorem, i.e., Eq. (8.3.7)) so that Eq. (E8.15.2) becomes

$$\oint_C f(z)\,dz \overset{(E8.15.2)}{\underset{(8.3.7)}{=}} \oint_{C_1}\left(\frac{1}{z+1}-\frac{2}{(z+1)^2}\right)dz + \oint_{C_2}\frac{j}{z-j}dz - \oint_{C_3}\frac{j}{z+j}dz$$
$$\overset{(E8.13.2)}{=} j2\pi + j(j2\pi) - j(j2\pi) = j2\pi \tag{E8.15.3}$$

8.3.3 Cauchy's Integral Formula

[Theorem 8.3.3: Cauchy's Integral Formula]

If a complex function $f(z)$ is analytic (or holomorphic) at all points except its single simple pole $z=z_0$ in a simply connected domain D, then it can be expressed in terms of another complex function $g(z)$, which is analytic everywhere in D, as

$$f(z) = \frac{g(z)}{z-z_0}$$

and for any simple closed contour C enclosing z_0 in D,

$$\oint_C f(z)dz = \oint_C \frac{g(z)}{z-z_0}dz = j2\pi g(z_0); \quad g(z_0) = \frac{1}{j2\pi}\oint_C \frac{g(z)}{z-z_0}dz \tag{8.3.12}$$

where $g(z_0)$ is called the *residue* of $f(z)$ at $z=z_0$ and denoted by

$$g(z_0) = (z-z_0)f(z)\big|_{z=z_0} = \operatorname*{Res}_{z=z_0} f(z)$$

[Theorem 8.3.4: Cauchy's Integral Formula for Derivatives]

If a complex function $g(z)$ is analytic (or holomorphic) in a simply connected domain D, then for any point z_0 in D and any simple closed contour C enclosing z_0 in D, its nth derivative is

$$g^{(n)}(z_0) = \left.\frac{d^n g(z)}{dz^n}\right|_{z=z_0} = \frac{n!}{j2\pi}\oint_C \frac{g(z)}{(z-z_0)^{n+1}}dz \tag{8.3.13}$$

(**Example 8.16**) Contour Integral Using Cauchy's Integral Formula

Find the contour integral of the complex function

$$f(z) = \frac{2z}{z^2-1} \overset{\text{PFE}}{=} \frac{1}{z-1} + \frac{1}{z+1} \tag{E8.16.1}$$

along the counterclockwise-oriented circle C of radius $R=2$ centered at $z_0=0$, as depicted in Fig. 8.18.

$$\text{(a)}\quad \oint_{C_1} f(z)dz \overset{\text{(E8.16.1)}}{=} \oint_{C_1}\frac{2z}{z^2-1}dz = \oint_{C_1}\frac{2z/(z+1)}{z-1}dz \overset{(8.3.12)}{=} j2\pi\left.\frac{2z}{z+1}\right|_{z=1} = j2\pi \tag{E8.16.2}$$

$$\text{(b)}\quad \oint_{C_2} f(z)dz \overset{\text{(E8.16.1)}}{=} \oint_{C_1}\frac{2z}{z^2-1}dz = \oint_{C_1}\frac{g(z)=2z/(z-1)}{z+1}dz \overset{(8.3.12)}{=} j2\pi\left.\frac{2z}{z-1}\right|_{z=-1} = j2\pi \tag{E8.16.3}$$

$$\text{(c)}\quad \oint_{C_3} f(z)dz \overset{\text{(E8.16.1)}}{=} \oint_{C_3}\frac{2z}{z^2-1}dz \underset{f(z)\text{ is analytic in }C_3}{\overset{(8.3.7)}{=}} 0 \tag{E8.16.4}$$

$$\text{(d)}\quad \oint_{C} f(z)dz \overset{(8.3.11)}{=} \oint_{C_1}\frac{2z}{z^2-1}dz + \oint_{C_2}\frac{2z}{z^2-1}dz + \oint_{C_3}\frac{2z}{z^2-1}dz \overset{\text{(E8.16.2,3,4)}}{=} j2\pi + j2\pi + 0 = j4\pi \tag{E8.16.5}$$

Eq. (8.3.11) can be used directly to get the same result:

$$\begin{aligned}\oint_C f(z)dz &\underset{\text{(E8.16.1)}}{\overset{(8.3.11)}{=}} \oint_{C_1}\left(\frac{1}{z-1}+\frac{1}{z+1}\right)dz + \oint_{C_2}\left(\frac{1}{z-1}+\frac{1}{z+1}\right)dz \\ &\overset{(8.3.7)}{=} \oint_{C_1}\frac{1}{z-1}dz + \oint_{C_2}\frac{1}{z+1}dz \overset{\text{(E8.13.2)}}{=} j2\pi + j2\pi = j4\pi\end{aligned} \tag{E8.16.6}$$

Here, we have used the fact that, since $1/(z+1)$ is analytic on/inside C_1 and $1/(z-1)$ is analytic on/inside C_2, their corresponding integrals are zero according to Cauchy's theorem, i.e., Eq. (8.3.7).

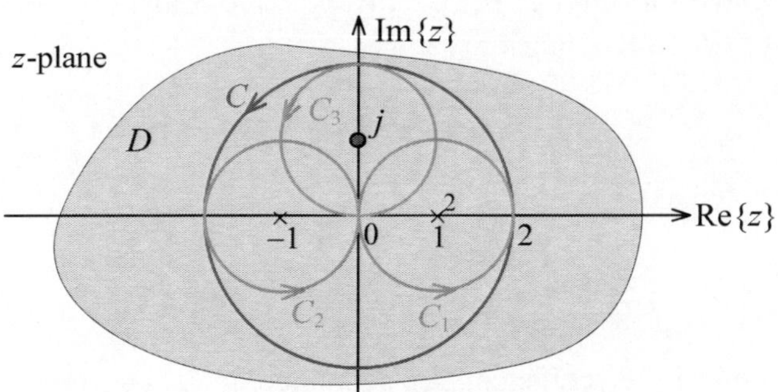

Figure 8.18 Contours of integration for Example 8.16

(Example 8.17) Contour Integral Using Cauchy's Integral Formula

Consider the contour integral of the following complex function:

$$f(z) = \frac{2(z+1)}{(z-1)^2(z^2+1)} \tag{E8.17.1}$$

(a) Find the contour integral of $f(z)$ along a counterclockwise-oriented circle enclosing $z=1$, but neither $z=j$ nor $z=-j$, like the circle C_1 of radius $R=1$ centered at $z_0=1$ that is depicted in Fig. 8.18.

The Cauchy's integral formula can be used to evaluate the contour integral:

$$\oint_{C_1} f(z)\,dz \overset{\text{(E8.17.1)}}{=} \oint_{C_1} \frac{2(z+1)}{(z-1)^2(z^2+1)} dz = \oint_{C_1} \frac{g(z)=2(z+1)/(z^2+1)}{(z-1)^2} dz$$

$$\underset{n=1}{\overset{\text{(8.3.13)}}{=}} j2\pi \frac{d}{dz}\frac{2(z+1)}{z^2+1}\bigg|_{z=1} = j2\pi \frac{2(z^2+1)-2(z+1)2z}{(z^2+1)^2}\bigg|_{z=1} = -j2\pi \tag{E8.17.2}$$

(b) Find the contour integral of $f(z)$ along a counterclockwise-oriented circle enclosing $z=j$, but neither $z=1$ nor $z=-j$, like the circle C_3 of radius $R=1$ centered at $z_0=j$ that is depicted in Fig. 8.18.

The Cauchy's integral formula can be used to evaluate the contour integral:

$$\oint_{C_3} f(z)\,dz \overset{\text{(E8.17.1)}}{=} \oint_{C_3} \frac{2(z+1)}{(z-1)^2(z^2+1)} dz = \oint_{C_3} \frac{g(z)=2(z+1)/(z-1)^2(z+j)}{z-j} dz$$

$$\overset{\text{(8.3.12)}}{=} j2\pi \frac{2(z+1)}{(z-1)^2(z+j)}\bigg|_{z=j} = j2\pi \frac{2(j+1)}{(j-1)^2 j2} = (-1+j)\pi \tag{E8.17.3}$$

8.4 SERIES AND RESIDUES

8.4.1 Sequences and Series

Referring to Fig. 8.19, an infinite sequence $\{z_1, z_2, \ldots, z_n, \ldots\}$ is said to *converge* or to be *convergent* to L or to have the limit L if, for any positive real number ε (however small but not zero), there is an integer $N(\varepsilon)$ such that

$$|z_n - L| < \varepsilon \;\; \forall\, n < N \tag{8.4.1}$$

The convergence is denoted as

$$\lim_{n\to\infty} z_n = L \quad \text{or} \quad z_n \to L \text{ as } n \to \infty$$

where L is called the *limit* of the sequence. If a sequence does not converge, it is said to *diverge* or to be *divergent*.

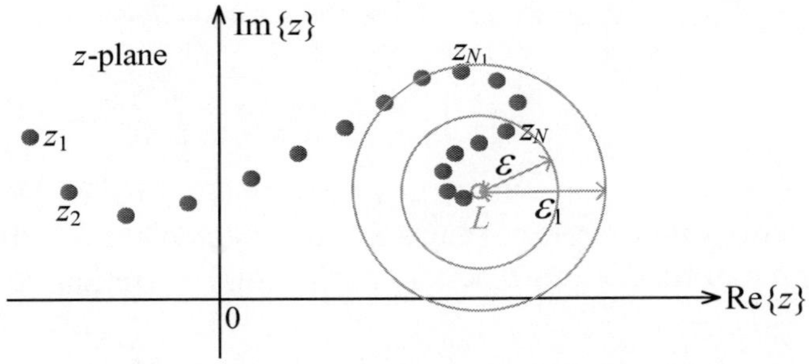

Figure 8.19 Convergent complex sequence

An infinite sequence $\{z_1, z_2, \ldots, z_n, \ldots\}$ is said to be a *Cauchy sequence* if, for any positive real number ε (however small but not zero), there is an integer $N(\varepsilon)$ such that

$$|z_n - z_m| < \varepsilon \quad \forall\, n, m > N \tag{8.4.2}$$

A sequence is convergent iff (if and only if) it is a Cauchy sequence.

Let $\{z_1, z_2, \ldots, z_n, \ldots\}$ be an infinite sequence of real or complex numbers. Then the sum of the numbers (terms) belonging to the sequence is called the *(infinite) series*:

$$\sum_{n=1}^{\infty} z_n = z_1 + z_2 + z_3 + \cdots \tag{8.4.3}$$

The sum (S_N) of the first N terms belonging to the sequence and that (R_N) of the remaining terms

$$S_N = \sum_{n=1}^{N} z_n = z_1 + z_2 + \cdots + z_N \tag{8.4.4}$$

$$R_N = \sum_{n=N+1}^{\infty} z_n = z_{N+1} + z_{N+2} + z_{N+3} + \cdots \tag{8.4.5}$$

are called the Nth *partial sum* and the *remainder* of the series after the Nth term, respectively.

In particular, for a geometric series with common ratio z/a:

$$\sum_{n=1}^{\infty} (z/a)^{n-1} = 1 + (z/a) + (z/a)^2 + \cdots + (z/a)^{n-1} + \cdots \tag{8.4.6}$$

the Nth partial sum is

$$S_N = \sum_{n=1}^{N} (z/a)^{n-1} = 1 + (z/a) + (z/a)^2 + \cdots + (z/a)^{N-1} = \frac{1-(z/a)^N}{1-z/a} \tag{8.4.7}$$

because

$$S_N \overset{(8.4.7)}{=} 1 + (z/a) + (z/a)^2 + \cdots + (z/a)^{N-2} + (z/a)^{N-1}$$

$$(z/a) S_N = (z/a) + (z/a)^2 + (z/a)^3 + \cdots + (z/a)^{N-1} + (z/a)^N$$

$$S_N - (z/a) S_N = 1 - (z/a)^N; \quad (1 - z/a) S_N = 1 - (z/a)^N; \quad S_N = \frac{1-(z/a)^N}{1-z/a}$$

For $|z/a|<1$, we have $(z/a)^n \to 0$ as $n \to \infty$ so that the geometric series (8.4.6) converges to

$$S_\infty = \sum_{n=0}^{\infty} (z/a)^n = \frac{1}{1-z/a} = -\frac{a}{z-a} \tag{8.4.8}$$

Here $\{z \,|\, |z|<|a|\}$, the area for which the series converges, is called the *region of convergence* (*ROC*) and it is the interior of the circle of radius $R=|a|$ centered at $z=0$ (the origin).

Let us consider another form of geometric series with common ratio $(z/b)^{-1}$:

$$\sum_{n=1}^{\infty} (z/b)^{-n} = (z/b)^{-1} + (z/b)^{-2} + \cdots + (z/b)^{-n} + \cdots = \frac{(z/b)^{-1}}{1-(z/b)^{-1}} = \frac{bz^{-1}}{1-bz^{-1}} = \frac{b}{z-b} \tag{8.4.9}$$

This series converges for $|(z/b)^{-1}|<1$, and therefore the ROC of this series is $\{z \,|\, |z|>|b|\}$, which is the exterior of the circle of radius $R=|b|$ centered at $z=0$.

Eqs. (8.4.8) and (8.4.9) can be used for Taylor/Laurent series expansions:

$$\frac{1}{1-z/a} = -\frac{a}{z-a} (\text{with ROC } |z|<|a|) = \sum_{n=1}^{\infty} (z/a)^{n-1} = 1 + a^{-1}z + a^{-2}z^2 + \cdots \tag{8.4.10}$$

$$\frac{bz^{-1}}{1-bz^{-1}} = \frac{b}{z-b} (\text{with ROC } |z|>|b|) = \sum_{n=1}^{\infty} (bz^{-1})^n = bz^{-1} + b^2 z^{-2} + \cdots \tag{8.4.11}$$

Just substituting $(z-c)$ for z into these equations yields

$$\frac{1}{1-(z-c)/a} = -\frac{a}{z-(a+c)} (\text{with ROC } |z-c|<|a|)$$
$$= \sum_{n=1}^{\infty} ((z-c)/a)^{n-1} = 1 + a^{-1}(z-c) + a^{-2}(z-c)^2 + \cdots \tag{8.4.12}$$

$$\frac{b/(z-c)}{1-b/(z-c)} = \frac{b}{z-(b+c)} (\text{with ROC } |z-c|>|b|)$$
$$= \sum_{n=1}^{\infty} (b/(z-c))^n = b(z-c)^{-1} + b^2(z-c)^{-2} + \cdots. \tag{8.4.13}$$

8.4.2 Taylor Series

An infinite series of the form

$$\sum_{n=0}^{\infty} a_n (z-c)^n = a_0 + a_1(z-c) + a_2(z-c)^2 + \cdots \tag{8.4.14}$$

with a variable z (called the *expansion variable*), and constants $\{a_0, a_1, a_2, \ldots\}$ (called the *coefficients*) and c (called the *center* or *expansion point*) is called the *power series* of z about c. A power series has a circle of convergence region $D=\{z \,|\, r<|z-z_0|<R\}$ such that it converges/diverges for z inside/outside the circle.

If a function $f(z)$ is analytic in a neighborhood (open disk) D of a point $z=z_0$, then it can be uniquely represented by a power series called the *Taylor series* with center at z_0:

$$f(z) = \sum_{n=0}^{\infty} a_n (z-z_0)^n \tag{8.4.15}$$

with the coefficients

$$a_n = \frac{f^{(n)}(z_0)}{n!} \overset{(8.3.13)}{=} \frac{1}{j2\pi}\oint_C \frac{f(z)}{(z-z_0)^{n+1}}dz \text{ (}C\text{: any simple closed curve in the open disk)} \quad (8.4.16)$$

The coefficient a_n's satisfy the inequality [K-2]

$$|a_n| \le \frac{M}{r^n} \text{ (}M\text{: the maximum of } |f(z)| \text{ on the circle } |z-a|=r) \quad (8.4.17)$$

implying that a_n's tend to be decreasing as the order n increases and accordingly, the partial sum of the Taylor series becomes closer as the number of terms increases. The Taylor series with center at $z=0$ is called the *Maclaurin series*.

We can find (finite) Taylor series expansions of elementary functions by using the MATLAB function '`taylor()`', whose usage is summarized as follows:

- `taylor(f)` gives the 5th-order Maclaurin series expansion of `f`. w.r.t. a default variable closest to x.
- `taylor(f,y)` with a symbolic variable (y) gives the 5th-order Maclaurin series expansion of `f(y)`.
- `taylor(f,x,c)` gives the 5th-order Taylor series expansion of `f(x)` about expansion point `x=c`.
- `taylor(f,z,c,'order',N+1)` gives the `N`th-order Taylor series expansion of `f(z)` about `z=c`.

(cf) Before using the function '`taylor()`', one should declare the arguments of the function as symbols by running a statement like "`syms x t.`"

(cf) In case the function has several arguments, it is good practice to put the expansion variable as the second input argument of '`taylor()`', where '`taylor()`' takes the one closest (alphabetically) to '`x`' as the expansion variable by default if it has been declared as a symbolic variable and is contained as an input argument of the function `f`.

(cf) The MATLAB function '`sym2poly()`' can be used to extract the coefficients from the Taylor series expansion obtained as a symbolic expression.

For example, we can find the 3rd-order Taylor series expansion of $f(z)=e^z$ about $z=0$ by running the following MATLAB statements:

```
>>syms z; zo=0; N=3;
>>T=taylor(exp(-z),z,zo,'order',N+1); pretty(T), Tp=sym2poly(T)
      3    2
     z    z
   - -- + -- - z + 1              Tp = -0.1667    0.5000   -1.0000    1.0000
     6    2
```

Note that the elementary functions such as exponential/trigonometric/logarithmic functions have the following Maclaurin series representations:

$$e^z = 1 + \frac{z}{1!} + \frac{z^2}{2!} + \cdots = \sum_{n=0}^{\infty} \frac{z^n}{n!} \quad (8.4.18)$$

$$\sin z = z - \frac{z^3}{3!} + \frac{z^5}{5!} - \cdots = \sum_{n=0}^{\infty} (-1)^n \frac{z^{2n+1}}{(2n+1)!} \quad (8.4.19)$$

$$\cos z = 1 - \frac{z^2}{2!} + \frac{z^4}{4!} - \cdots = \sum_{n=0}^{\infty} (-1)^n \frac{z^{2n}}{(2n)!} \quad (8.4.20)$$

$$\sinh z = z + \frac{z^3}{3!} + \frac{z^5}{5!} - \cdots = \sum_{n=0}^{\infty} \frac{z^{2n+1}}{(2n+1)!} \quad (8.4.21)$$

$$\cosh z = 1 + \frac{z^2}{2!} + \frac{z^4}{4!} + \cdots = \sum_{n=0}^{\infty} \frac{z^{2n}}{(2n)!} \tag{8.4.22}$$

$$\mathrm{Ln}(1 \pm z) = \pm z - \frac{z^2}{2} \pm \frac{z^3}{3} - \cdots = \sum_{n=1}^{\infty} (-1)^{n+1} \frac{(\pm z)^n}{n} \tag{8.4.23}$$

$$\mathrm{Ln}\frac{1+z}{1-z} = 2\left(z + \frac{z^3}{3} + \frac{z^5}{5} + \cdots\right) = \sum_{n=1}^{\infty} \frac{2z^{2n-1}}{2n-1} \tag{8.4.24}$$

(Example 8.18) Numerical/Symbolic Methods for Taylor Series Expansion

Let us find the 3rd-order Taylor series expansion of $f(z)=e^{-z}$ about center $z_0=1$ and plot it together with $f(z)=e^{-z}$ for $z=[0, 2]$. Eqs. (8.4.15) and (8.4.16) can be used to find the expansion as

$$\begin{aligned} &f(z_0) + f'(z_0)(z - z_0) + \frac{1}{2!} f''(z_0)(z - z_0)^2 + \frac{1}{3!} f^{(3)}(z_0)(z - z_0)^3 \\ &= e^{-1} - e^{-1}(z-1) + \frac{1}{2} e^{-1}(z-1)^2 - \frac{1}{6} e^{-1}(z-1)^3 = -e^{-1}\left(\frac{1}{6}z^3 - z^2 + \frac{5}{2}z - \frac{8}{3}\right) \end{aligned} \tag{E8.18.1}$$

We run the following MATLAB script "em08e18.m" to get Fig. 8.20, which shows that the 3rd-order Taylor series expansion is close to e^{-z} around the center $z=1$, supporting the validity of the Taylor series expansion result.

```
%em08e18.m
syms z; f=@(z)exp(-z); % Original function
N=3; zo=1; % To find 3rd-order Taylor series expansion of e^-z about zo
Ts=taylor(exp(-z),z,zo,'Order',N+1), Tsp=sym2poly(Ts)
Ta=-exp(-1)*[1/6  -1  5/2  -8/3] % Eq.(E8.18.1)
zz=[0:0.01:2]; plot(zz,f(zz), zz,polyval(Tsp,zz),'r:')
```

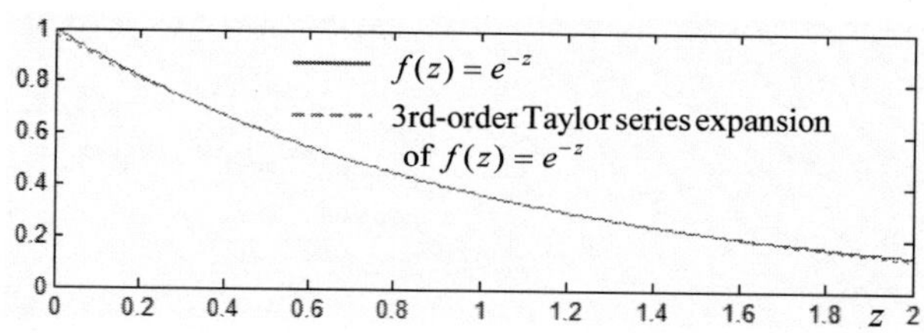

Figure 8.20 3rd-order Taylor series expansion of $f(z)$=exp($-z$)

8.4.3 Laurent Series

If a function $f(z)$ is analytic in an annular region $D=\{z|r<|z-z_0|<R\}$ around a point $z=z_0$, then it can be uniquely represented by a power series called the *Laurent series* with center at z_0:

$$f(z) = \sum_{n=-\infty}^{\infty} a_n(z - z_0)^n = \overbrace{\sum_{n=1}^{\infty} a_{-n}(z - z_0)^{-n}}^{\text{Principal part}} + \sum_{n=0}^{\infty} a_n(z - z_0)^n = \Sigma^- + \Sigma^+ \tag{8.4.25}$$

with the coefficients

$$a_n = \frac{1}{j2\pi} \oint_C \frac{f(z)}{(z-z_0)^{n+1}} dz \text{ for } n= 0, \pm 1, \cdots \text{ (}C\text{: any simple closed curve in } D\text{)} \tag{8.4.26}$$

where $\Sigma^- = \sum_{n=1}^{\infty} a_{-n}(z - z_0)^{-n}$ and $\Sigma^+ = \sum_{n=0}^{\infty} a_n(z - z_0)^n$ converge for $|z-z_0|>r$ and $|z-z_0|<R$, respectively.

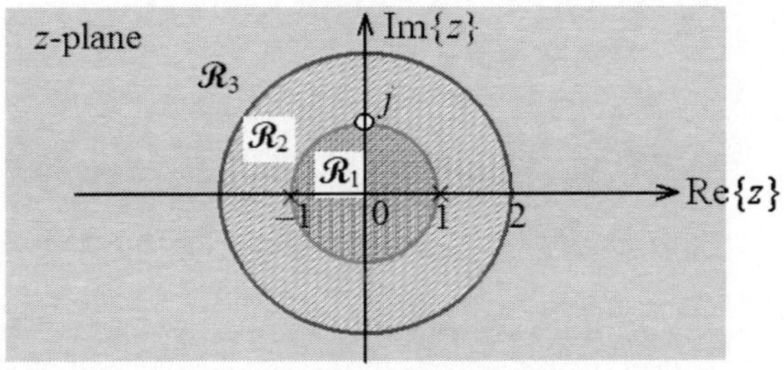

Figure 8.21 Region of convergences (ROCs) for Example 8.19

(**Example 8.19**) Laurent Series Expansion

Find the Laurent series expansions of

$$f(z)=\frac{1}{z^2-3z+2}=\frac{1}{(z-2)(z-1)}\overset{\text{PFE}}{=}\frac{1}{1-z}-\frac{1}{2-z}=f_1(z)+f_2(z) \tag{E8.19.1}$$

each of which is valid for (a) $\mathcal{R}_1$: $|z|<1$, (b) $\mathcal{R}_2$: $1<|z|<2$, and (c) $\mathcal{R}_3$: $|z|>2$ (see Fig. 8.21).

(a) For $\mathcal{R}_1$: $|z|<1$, we use Eq. (8.4.10) with $a=1$ and $a=2$ to find the Laurent series expansions of $f_1(z)$ and $f_2(z)$:

$$f(z)=\frac{1}{1-z}-\frac{1/2}{1-z/2}\underset{\text{with }a=1,2}{\overset{(8.4.10)}{=}}(1+z+z^2+\cdots)-\frac{1}{2}\left\{1+2^{-1}+2^{-2}z+\cdots\right\}=\sum_{n=0}^{\infty}(1-2^{-(n+1)})\,z^n \tag{E8.19.2}$$

Note that both $f_1(z)$ (with pole $z=1$) and $f_2(z)$ (with pole $z=2$), having their poles outside the boundary circle of the ROC $\mathcal{R}_1$, have been expanded into nonnegative powers of z.

(b) For $\mathcal{R}_2$: $1<|z|<2$, we use Eq. (8.4.11) with $b=1$ to find the Laurent series expansion of $f_1(z)$ and use the part of Eq. (E8.19.2) corresponding to $f_2(z)$ in Eq. (E8.19.2):

$$f(z)=\frac{-z^{-1}}{1-z^{-1}}-\frac{1/2}{1-z/2}\underset{\text{with }b=1}{\overset{(8.4.11)}{=}}-(z^{-1}+z^{-2}+\cdots)-(2^{-1}+2^{-2}z+2^{-3}z^2+\cdots) \tag{E8.19.3}$$

Note that $f_1(z)$ (with pole $z=1$ inside the inner boundary circle of ROC $\mathcal{R}_2$) and $f_2(z)$ (having pole $z=2$ outside the outer boundary circle of ROC $\mathcal{R}_2$) have been expanded into negative and nonnegative powers of z, respectively (see Remarks 5.1 and 5.4).

(c) For $\mathcal{R}_3$: $|z|>2$, we use the part of Eq. (E8.19.3) corresponding to $f_1(z)$ in Eq. (E8.19.3) and use Eq. (8.4.11) with $b=2$ to find the Laurent series expansion of $f_2(z)$:

$$f(z)=\frac{-z^{-1}}{1-z^{-1}}+\frac{z^{-1}}{1-2z^{-1}}\underset{\text{with }b=1,2}{\overset{(8.4.11)}{=}}-(z^{-1}+z^{-2}+\cdots)+(z^{-1}+2z^{-2}+2^2z^{-3}+\cdots) \tag{E8.19.4}$$

Note that both $f_1(z)$ (with pole $z=1$) and $f_2(z)$ (with pole $z=2$), having their poles inside the boundary circle of the ROC $\mathcal{R}_3$, have been expanded into negative powers of z.

(**Example 8.20**) Laurent Series Expansion

Find the Laurent series expansions of

$$f(z)=\frac{1}{z(1-z)}\overset{\text{PFE}}{=}\frac{1}{z}+\frac{1}{1-z}=f_1(z)+f_2(z) \tag{E8.20.1}$$

for each of the following ROCs.

(a) For $0<|z|<1$, we use Eq. (8.4.10) with $a=1$ to find the Laurent series expansion of $f(z)$:

$$f(z)=\frac{1}{z}+\frac{1}{1-z}\underset{\text{with }a=1}{\overset{(8.4.10)}{=}}z^{-1}+1+z+z^2+\cdots \tag{E8.20.2}$$

(b) For $|z|>1$, we use Eq. (8.4.11) with $b=1$ to find the Laurent series expansion of $f(z)$:

$$f(z)=\frac{1}{z}-\frac{1}{z-1}\overset{(8.4.11)}{\underset{\text{with } b=1}{=}}z^{-1}-z^{-1}-z^{-2}-z^{-3}-\cdots=-z^{-2}-z^{-3}-\cdots \tag{E8.20.3}$$

(c) For $0<|z-1|<1$, we use Eq. (8.4.12) with $a=-1$ and $c=1$ to find the Laurent expansion of $f(z)$:

$$f(z)=-\frac{1}{z-1}+\frac{1}{1-(z-1)/(-1)}\overset{(8.4.12)}{\underset{a=-1,c=1}{=}}-\frac{1}{z-1}+1-(z-1)+(z-1)^2-(z-1)^3+\cdots \tag{E8.20.4}$$

(d) For $|z-1|>1$, we use Eq. (8.4.13) with $b=-1$ and $c=1$ to find the Laurent expansion of $f(z)$:

$$f(z)=-\frac{1}{z-1}+\frac{1/(z-1)}{1-(-1)/(z-1)}\overset{(8.4.13)}{\underset{b=-1,c=1}{=}}-\frac{1}{z-1}+(z-1)^{-1}-(z-1)^{-2}+(z-1)^{-3}-\cdots \tag{E8.20.5}$$

(e) For $|z-2|<1$, we use Eq. (8.4.12) to find the Laurent expansions of $f_1(z)$ and $f_2(z)$:

$$f_1(z)=\frac{1}{z}=\frac{1}{(z-2)+2}\overset{|z-2|<1}{=}\frac{1/2}{1-(z-2)/(-2)}\overset{(8.4.12)}{\underset{a=-2,c=2}{=}}\frac{1}{2}-\frac{1}{2^2}(z-2)+\frac{1}{2^3}(z-2)^2+\cdots$$

$$f_2(z)=\frac{1}{1-z}=\frac{1}{-1-(z-2)}\overset{|z-2|<1}{=}\frac{-1}{1-(z-2)/(-1)}\overset{(8.4.12)}{\underset{a=-1,c=2}{=}}-1+(z-2)-(z-2)^{-2}+(z-2)^{-3}+\cdots$$

(E8.20.6)

(f) For $1<|z-2|<2$, we use Eqs. (8.4.12)/(8.4.13) to find the Laurent expansions of $f_1(z)/f_2(z)$:

$$f_1(z)=\frac{1}{z}=\frac{1}{(z-2)+2}\overset{|z-2|<2}{=}\frac{1/2}{1-(z-2)/(-2)}\overset{(8.4.12)}{\underset{a=-2,c=2}{=}}\frac{1}{2}-\frac{1}{2^2}(z-2)+\frac{1}{2^3}(z-2)^2+\cdots$$

$$f_2(z)=\frac{1}{1-z}=\frac{1}{-1-(z-2)}\overset{|z-2|>1}{=}\frac{-1/(z-2)}{1-(-1)/(z-2)}\overset{(8.4.13)}{\underset{b=-1,c=2}{=}}-(z-2)^{-1}+(z-2)^{-2}-(z-2)^{-3}+\cdots$$

(E8.20.7)

(g) For $|z-2|>2$, we use Eq. (8.4.13) to find the Laurent expansions of $f_1(z)/f_2(z)$:

$$f_1(z)=\frac{1}{z}=\frac{1}{(z-2)+2}\overset{|z-2|>2}{=}\frac{1/(z-2)}{1-(-2)/(z-2)}\overset{(8.4.13)}{\underset{b=-2,c=2}{=}}(z-2)^{-1}-2(z-2)^{-2}+4(z-2)^{-3}+\cdots$$

$$f_2(z)=\frac{1}{1-z}=\frac{1}{-1-(z-2)}\overset{|z-2|>2}{=}\frac{-1/(z-2)}{1-(-1)/(z-2)}\overset{(8.4.13)}{\underset{b=-1,c=2}{=}}-(z-2)^{-1}+(z-2)^{-2}-(z-2)^{-3}+\cdots$$

(E8.20.8)

8.4.4 Residues and Residue Theorem

Let a complex function $f(z)$ have an *isolated singularity* or *singular point* at $z=z_p$ where $f(z)$ fails to be analytic, i.e., not differentiable, while being differentiable in $D=\{z|0<|z-z_p|<r\}$ (the neighborhood of $z=z_p$). Then its Laurent expansion about the singularity z_p

$$f(z)=\sum_{n=-\infty}^{\infty}a_n(z-z_p)^n=\overbrace{\sum_{n=1}^{\infty}a_{-n}(z-z_p)^{-n}}^{\text{Principal part}}+\sum_{n=0}^{\infty}a_n(z-z_p)^n \tag{8.4.27}$$

is valid around the center z_p. The isolated singular point z_p of $f(z)$ is classified according to the number M of nonzero terms contained in the principal part (involving negative powers of $(z-z_p)$) of the Laurent expansion of $f(z)$ about z_p:

(i) If $M=0$, it is called a *removable singularity*.
(ii) If $0<M<\infty$, it is called a *pole* of *order* (or *multiplicity*) N (the highest degree of $(z-z_p)^{-1}$) and a *simple pole* when $N=1$.
(iii) If $M=\infty$, it is called an *essential singularity*.

On the other hand, if $f(z_0)=0$, z_0 is called a *zero* of the function $f(z)$. Also, it is called a *zero* of *order* N if

$$f(z_0)=0,\ f'(z_0)=0,\ \cdots,\ f^{(N-1)}(z_0)=0,\ \text{but } f^{(N)}(z_0)\neq 0 \tag{8.4.28}$$

so that the Taylor expansion of the function $f(z)$ can be written in the form of

$$f(z)=(z-z_0)^N\left\{a+a_{N+1}(z-z_0)+a_{N+2}(z-z_0)^2+\cdots\right\} \tag{8.4.29}$$

[Remark 8.2] Residues at a Simple Pole or a Pole of Order N

The coefficient a_{-1} of $(z-z_p)^{-1}$ in the Laurent series is called the *residue* of f at z_p and is denoted by $\text{Res}\{f(z), z_p\}$. Note the following about the residue:

(1) If z_p is a simple pole, the residue can be found as

$$\text{Res}\{f(z),z_p\}=\lim_{z\to z_p}(z-z_p)f(z) \tag{8.4.30a}$$

(2) If $f(z)=p(z)/q(z)$, $p(z_p)\neq 0$, $q(z_p)=0$, and $q'(z_p)\neq 0$, the residue can be found as

$$\text{Res}\{f(z),z_p\}=\frac{p(z_p)}{q'(z_p)}. \tag{8.4.30b}$$

(3) If z_p is a pole of order N, the residue can be found as

$$\text{Res}\{f(z),z_p\}=\frac{1}{(N-1)!}\lim_{z\to z_p}\left\{\frac{d^{N-1}}{dz^{N-1}}(z-z_p)^N f(z)\right\} \tag{8.4.30c}$$

[Theorem 8.4.1: Cauchy's Residue Theorem]

Let a complex function f be analytic on and within a simple closed contour C, except at a finite number of poles $\{z_1, z_2, \ldots, z_N\}$, included in a simply connected domain D. Then the contour integral of f in the counterclockwise direction around C can be computed by multiplying $j2\pi$ with the sum of the residues of f at the poles (enclosed by C):

$$\oint_C f(z)dz = j2\pi\sum_{n=1}^{N}\text{Res}\{f(z),z_n\} \tag{8.4.31}$$

(Example 8.21) Contour Integral Using Cauchy's Residue Theorem

Consider the contour integral of the following complex function that was dealt with in Example 8.17:

$$f(z)=\frac{2(z+1)}{(z-1)^2(z^2+1)} \tag{E8.21.1}$$

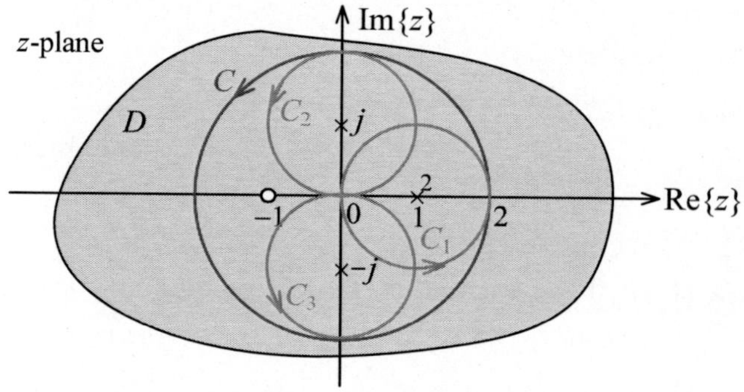

Figure 8.22 Contours of integration for Example 8.21

(a) Find the contour integral of $f(z)$ along a counterclockwise-oriented circle (like C_1 in Fig. 8.22) enclosing $z=1$, but neither $z=j$ nor $z=-j$.

We can use the Cauchy's residue theorem to evaluate the contour integral:

$$\oint_{C_1} f(z)\,dz \overset{(8.4.31)}{=} j2\pi\,\mathrm{Res}\{f(z),\ z_P=1\} \overset{(8.4.30c)}{=} j2\pi \lim_{z\to 1}\left\{\frac{d}{dz}(z-1)^2 f(z)\right\}$$

$$\overset{(E8.21.1)}{=} j2\pi \left.\frac{d}{dz}\frac{2(z+1)}{z^2+1}\right|_{z=1} = j2\pi \left.\frac{2(z^2+1)-2(z+1)2z}{(z^2+1)^2}\right|_{z=1} = -j2\pi \qquad \text{(E8.21.2)}$$

(b) Find the contour integral of $f(z)$ along a counterclockwise-oriented circle (like C_2/C_3 in Fig. 8.22) enclosing $z=+j/-j$, but neither $z=1$ nor $z=-j/+j$.

We can use the Cauchy's residue theorem to evaluate the contour integral:

$$\oint_{C_2 \text{ or } C_3} f(z)\,dz \overset{(8.4.31)}{=} j2\pi\,\mathrm{Res}\{f(z),\ z_P=\pm j\} \overset{(8.4.30a)}{=} j2\pi \lim_{z\to\pm j}\{(z\mp j)f(z)\}$$

$$\overset{(E8.21.1)}{=} j2\pi \left.\frac{2(z+1)}{(z-1)^2(z\pm j)}\right|_{z=\pm j} = j2\pi\frac{2(\pm j+1)}{(\pm j-1)^2(\pm j2)} = (-1\pm j)\pi \qquad \text{(E8.21.3)}$$

(c) Find the contour integral of $f(z)$ along a counterclockwise-oriented circle (like C in Fig. 8.22) enclosing all the three poles $z=1$ and $z=\pm j$.

We can use the Cauchy's residue theorem to evaluate the contour integral:

$$\oint_C f(z)\,dz \overset{(8.4.31)}{=} j2\pi\sum_{n=1}^{N}\mathrm{Res}\{f(z), z_n\} \overset{(E8.21.2\&3)}{=} -j2\pi+(-1+j)\pi+(-1-j)\pi = -2(1+j)\pi \qquad \text{(E8.21.4)}$$

8.4.5 Real Intergrals Using Residue Theorem

The residue theorem can be used to evaluate certain classes of real integrals involving a rational function $f(\cdot)$.

(1) $\int_0^{2\pi} f(\cos a\theta, \sin b\theta)\, d\theta$

By the change of variables $z=e^{j\theta}$, we can convert this form of integral into a *contour integral around the unit circle*, which can be evaluated by using the residue theorem:

$$z=e^{j\theta}=\cos\theta+j\sin\theta;\qquad dz=je^{j\theta}d\theta=jz\,d\theta$$

$$\Rightarrow d\theta=\frac{dz}{jz},\quad \cos a\theta \overset{(C.21)}{=} \frac{1}{2}(z^{a}+z^{-a}),\quad \sin b\theta \overset{(C.22)}{=} \frac{1}{j2}(z^{b}-z^{-b}) \tag{8.4.32}$$

$$;\ \int_0^{2\pi} f(\cos a\theta,\sin b\theta)\,d\theta \;\rightarrow\; \oint_C f\left(\frac{1}{2}(z^{a}+z^{-a}),\frac{1}{j2}(z^{b}-z^{-b})\right)\frac{dz}{jz} \tag{8.4.33}$$

(Example 8.22) Real Integral Using Residue Theorem

Evaluate the integral

$$I=\int_0^{2\pi}\frac{d\theta}{\sqrt{2}-\cos\theta} \tag{E8.22.1}$$

By the change of variables (8.4.32), it can be evaluated as

$$I=\int_0^{2\pi}\frac{d\theta}{\sqrt{2}-\cos\theta} \underset{a=1}{\overset{(8.4.32)}{=}} \oint_C\frac{dz/jz}{\sqrt{2}-(z+z^{-1})/2}=\oint_C\frac{j2dz}{z^2-2\sqrt{2}\,z+1}=\oint_C\frac{j2dz}{(z-\sqrt{2}+1)(z-\sqrt{2}-1)}$$

$$\overset{(8.4.31)}{=} j2j2\pi\,\mathrm{Res}(f(z),\, \underset{\substack{\text{pole inside}\\ \text{the unit circle}}}{z=\sqrt{2}-1}) \overset{(8.4.30a)}{=} -4\pi\lim_{z\to\pm j}\left\{(z-\sqrt{2}+1)f(z)\right\}$$

$$=-4\pi\left.\frac{1}{z-\sqrt{2}-1}\right|_{\sqrt{2}-1}=2\pi \tag{E8.22.2}$$

(2) Improper integral $\int_{-\infty}^{\infty} f(x)\,dx$ where the denominator polynomial of $f(x)$ is of degree at least two higher than that of the numerator polynomial.

Noting that adding an infinite semicircular path to the integration path from $x=-\infty$ to ∞ yields a closed curve, we can convert this form of integral w.r.t. a real variable x into a part of the *contour integral (w.r.t. a complex variable z in place of x) around the infinite upper semicircle C detouring the poles on the real axis* (as illustrated in Fig. 8.23), which can be evaluated by using the residue theorem:

$$\oint_C f(z)\,dz=\oint_{C_R\text{-}C_{l1}\text{-}\{C_r^-\}\text{-}C_{l2}} f(z)\,dz=\oint_{C_{l1}\text{-}C_{l2}} f(z)\,dz+\oint_{\{C_r^-\}} f(z)\,dz+\oint_{C_R} f(z)\,dz$$

$$\underset{(8.3.4)}{\overset{z=Re^{j\theta}}{=}} \oint_{C_{l1}\text{-}C_{l2}} f(z)\,dz-\oint_{\{C_r\}} f(z)\,dz+\lim_{R\to\infty} jR\int_0^{\pi} f(Re^{j\theta})e^{j\theta}\,d\theta$$

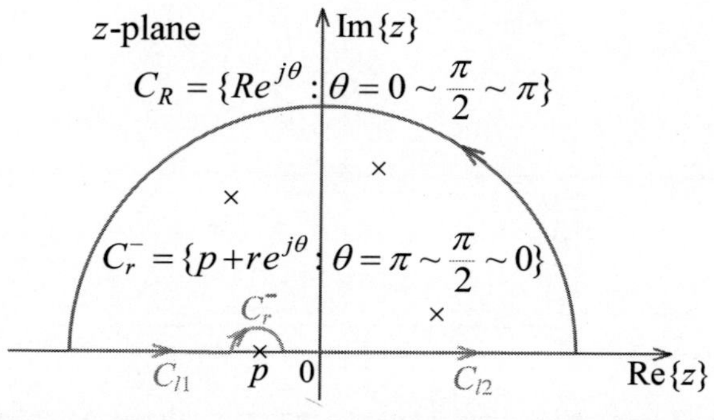

Figure 8.23 Infinite semicircular contour of integration detouring the poles on the real axis

$$; \int_{-\infty}^{\infty} f(x)\,dx = \oint_{C_{l1}\text{-}C_{l2}} f(z)\,dz = \oint_C f(z)\,dz + \oint_{\{C_r\}} f(z)\,dz$$

$$\overset{(8.4.31)}{=} j2\pi \sum_{\text{at poles inside } C} \text{Res}\{f(z), z_n\} + j\pi \sum_{\text{at poles on real axis}} \text{Res}\{f(z), z_n\} \quad (8.4.34)$$

(3) Fourier integrals $\int_{-\infty}^{\infty} f(x)\cos\omega x\,dx$ and $\int_{-\infty}^{\infty} f(x)\sin\omega x\,dx$ ($\omega>0$: a positive number) where the degree of the denominator polynomial of $f(x)$ is one or more higher than that of the numerator polynomial.

In the same context with the previous integral (2), we can use Eq. (8.4.34) to find the Fourier integrals:

$$\int_{-\infty}^{\infty} f(x)e^{j\omega x}dx = j2\pi \sum\nolimits_C \text{Res} + j\pi \sum\nolimits_{\text{real}} \text{Res} \quad (8.4.35)$$

with

$$\begin{cases} \sum_C \text{Res} = \sum_{\text{at poles inside } C} \text{Res}\{f(z)e^{j\omega z}, z_n\} \\ \sum_{\text{real}} \text{Res} = \sum_{\text{at poles on real axis}} \text{Res}\{f(z)e^{j\omega z}, z_n\} \end{cases}$$

This equation can be rewritten as

$$\int_{-\infty}^{\infty} f(x)(\cos\omega x + j\sin\omega x)dx$$
$$= -2\pi\,\text{Im}\left\{\sum\nolimits_C \text{Res}\right\} - \pi\,\text{Im}\left\{\sum\nolimits_{\text{real}} \text{Res}\right\} + j2\pi\,\text{Re}\left\{\sum\nolimits_C \text{Res}\right\} + j\pi\,\text{Re}\left\{\sum\nolimits_{\text{real}} \text{Res}\right\}$$

Equating the real/imaginary parts of both sides yields

$$\int_{-\infty}^{\infty} f(x)\cos\omega x\,dx = -2\pi\,\text{Im}\left\{\sum\nolimits_C \text{Res}\right\} - \pi\,\text{Im}\left\{\sum\nolimits_{\text{real}} \text{Res}\right\} \quad (8.4.36a)$$

$$\int_{-\infty}^{\infty} f(x)\sin\omega x\,dx = 2\pi\,\text{Re}\left\{\sum\nolimits_C \text{Res}\right\} + \pi\,\text{Re}\left\{\sum\nolimits_{\text{real}} \text{Res}\right\} \quad (8.4.36b)$$

(**Example 8.23**) Improper Integral Using Residue Theorem

Evaluate the integral

$$I = \int_{-\infty}^{\infty} \frac{1+x}{x(1+x^2)}dx = \int_{-\infty}^{\infty} \frac{x+1}{x(x+j)(x-j)}dx \quad (E8.23.1)$$

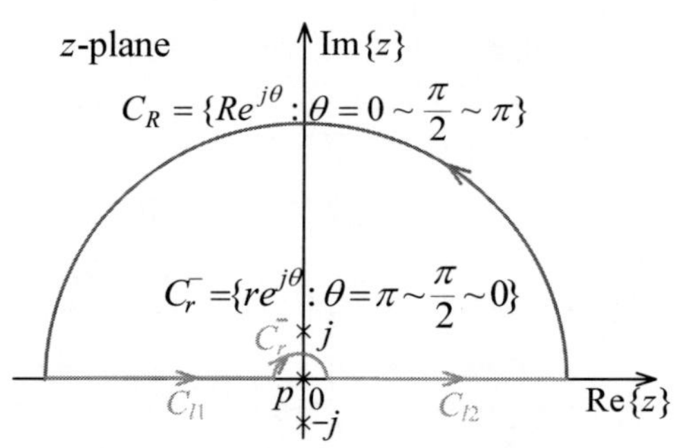

Figure 8.24 Infinite semicircular contour of integration detouring the poles on the real axis

Noting that the denominator of the integrand is of higher degree than the numerator by 2 and referring to Fig. 8.24 (showing the integration contour and poles of the integrand), we can use Eq. (8.3.34) to evaluate this integral as

$$\int_{-\infty}^{\infty} f(x)\,dx \overset{(8.4.34)}{=} j2\pi \sum_{\text{at poles inside } C} \text{Res}\{f(z), z_n\} + j\pi \sum_{\text{at poles on real axis}} \text{Res}\{f(z), z_n\}$$

$$\overset{(\text{E8.23.1})}{=} j2\pi\,\text{Res}\left\{\frac{z+1}{z(z+j)(z-j)}, j\right\} + j\pi\,\text{Res}\left\{\frac{z+1}{z(z+j)(z-j)}, 0\right\}$$

$$\overset{(8.4.30\text{a})}{=} j2\pi \left.\frac{z+1}{z(z+j)}\right|_{z=j} + j\pi \left.\frac{z+1}{z^2+1}\right|_{z=0} = j2\pi \frac{j+1}{j(j2)} + j\pi = (1-j)\pi + j\pi = \pi \quad (\text{E8.23.2})$$

(Example 8.24) Improper Integral Using Residue Theorem

Evaluate the integral

$$I = \int_{-\infty}^{\infty} \frac{\cos\omega x}{x^2+a^2} dx = \int_{-\infty}^{\infty} \frac{\cos\omega x}{(x+ja)(x-ja)} dx = \int_{-\infty}^{\infty} f(x)\cos\omega x\,dx \text{ with } f(x) = \frac{1}{x^2+a^2} \quad (\text{E8.24.1})$$

Noting that $f(z)e^{j\omega z} = e^{j\omega z}/(z^2+a^2)$ has only one pole ($z = ja$) in the infinite upper semicircle and no pole on the real axis (see Fig. 8.25), we can use Eq. (8.4.36a) to evaluate this integral as

$$\int_{-\infty}^{\infty} \frac{\cos\omega x}{x^2+a^2} dx \overset{(8.4.36\text{a})}{=} -2\pi\,\text{Im}\left\{\sum_C \text{Res}\right\} = -2\pi\,\text{Im}\left\{\left.\frac{e^{j\omega z}}{(z+ja)}\right|_{z=ja}\right\} = -2\pi\,\text{Im}\left\{-j\frac{e^{-a\omega}}{2a}\right\} = \frac{e^{-a\omega}\pi}{a}$$

(E8.24.2)

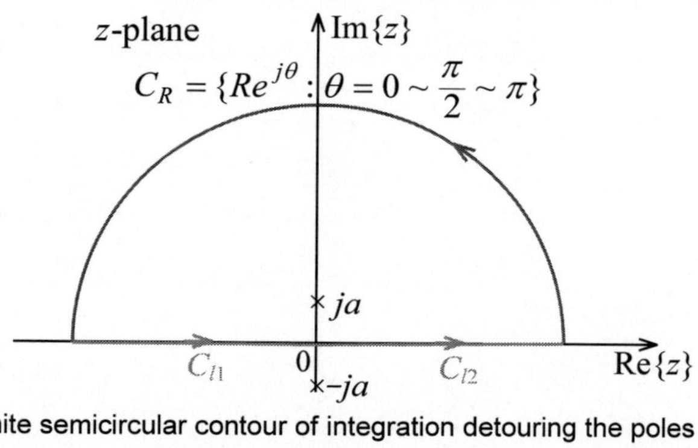

Figure 8.25 Infinite semicircular contour of integration detouring the poles on the real axis

(Example 8.25) Improper Integral Using Residue Theorem

Evaluate the integral

$$I = \int_{-\infty}^{\infty} \frac{\sin x}{x} dx = \int_{-\infty}^{\infty} \frac{1}{x}\sin x\,dx = \int_{-\infty}^{\infty} f(x)\sin x\,dx \text{ with } f(x) = \frac{1}{x} \quad (\text{E8.25.1})$$

Noting that $f(z)e^{jz} = e^{jz}/z$ has no pole in the infinite upper semicircle and only one pole ($z=0$) on the real axis (see Fig. 8.26), we can use Eq. (8.3.36b) to evaluate this integral as

$$\int_{-\infty}^{\infty} \frac{\sin x}{x} dx \overset{(8.4.36\text{b})}{=} \pi\,\text{Re}\left\{\sum_{\text{real}} \text{Res}\right\} = \pi\,\text{Re}\left\{z\left.\frac{e^{jz}}{z}\right|_{z=0}\right\} = \pi \quad (\text{E8.25.2})$$

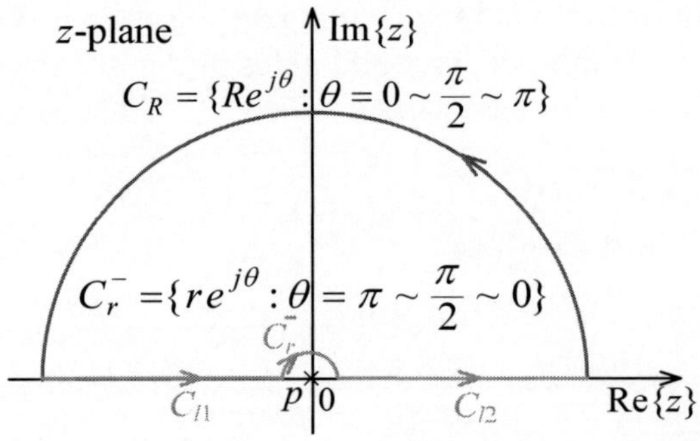

Figure 8.26 Infinite semicircular contour of integration detouring the poles on the real axis

The MATLAB function '`int()`' can be used for evaluating this integral:

```
>>syms x;  int(sin(x)/x,-inf,inf)
   ans =  pi
```

Problems

8.1 Complex Number Computations

(a) For $z_1=26-j39$ and $z_2=2+j3$, compute z_1z_2 and z_1/z_2.

(b) Find the polar expressions for $z_1=1+j\sqrt{3}$ and $z_2=\sqrt{2}-j\sqrt{2}$.

(c) Find the rectangular expressions for $z_1=2e^{j\pi/3}$ and $z_2=2\angle-\pi/4$.

(d) Compute $z_1=(1+j\sqrt{3})^6$ and $z_2=(\sqrt{2}-j\sqrt{2})^8$.

(e) Find the four 4th roots of $z_1=1$ and the three cubic roots of $z_2=-1$.

8.2 Cauchy–Riemann (CR) Equations

In connection with the complex differentiability of the following complex functions w.r.t. $z=x+jy$, check if their real/imaginary parts satisfy the CR equations.

(a) $f(z)=|z|^2+jz=x^2+y^2+j(x+jy)=x^2+y^2-y+jx$

(b) $f(z)=e^z=e^x\cos y+je^x\sin y$

(c) $f(z)=z^*=x-jy$

(d) $f(z)=\dfrac{1}{z}=\dfrac{x}{x^2+y^2}-j\dfrac{y}{x^2+y^2}=\dfrac{\rho\cos\phi}{\rho^2}-j\dfrac{\rho\sin\phi}{\rho^2}=\dfrac{\cos\phi}{\rho}-j\dfrac{\sin\phi}{\rho}$

(e) $f(z)=zz^*+z-z^*=x^2+y^2+j2y$

8.3 Harmonic Function

Show that the following functions are harmonic and find their (conjugate) harmonic functions. Check if the level curves of u and v are orthogonal to each other. You can get more credits on (b) and (b) if you solve them both in Cartesian and polar coordinates.

(a) Find the conjugate harmonic function for

$$u(x,y)=4xy^3-4x^3y \tag{P8.3a.1}$$

(b) Find the conjugate harmonic function for

$$u(x,y)=\ln\sqrt{x^2+y^2}=\frac{1}{2}\ln(x^2+y^2)\ \text{ or }\ u(\rho,\phi)=\ln\rho \tag{P8.3b.1}$$

(c) Find the harmonic function for

$$v(x,y)=y-\frac{y}{x^2+y^2}\ \text{ or }\ v(\rho,\phi)=(\rho-\frac{1}{\rho})\sin\phi \tag{P8.3c.1}$$

8.4 Complex Potential for a Flow

Consider the particles whose paths are parameterized by $z(t)=x(t)+jy(t)$ and whose horizontal/vertical velocities are as follows:

(a)
$$\frac{dx}{dt}=u(x,y)=1, \quad \frac{dy}{dt}=v(x,y)=0 \tag{P8.4a.1}$$

Find the complex potential function. Plot the stream lines and velocity vectors at (0,1) and (0, −1) in Fig. P8.4.1(a).

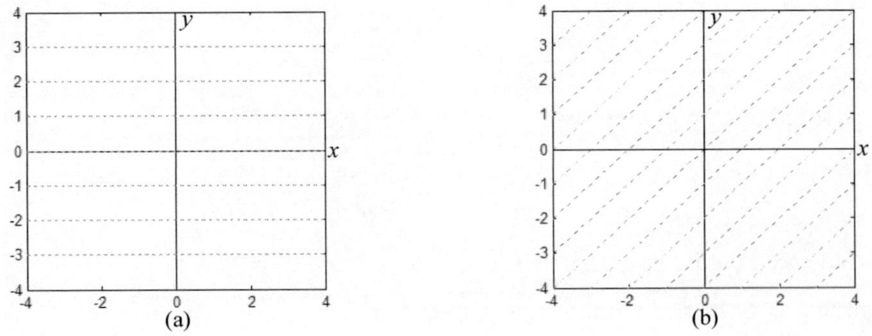

Figure P8.4.1 Stream lines and velocity vectors describing the uniform flows for Problem 8.4(a) and (b)

(b) $$\frac{dx}{dt} = u(x,y) = C\cos\alpha \qquad \text{with } C = 1 \text{ and } \alpha = \frac{\pi}{4} \qquad \text{(P8.4b.1)}$$
$$\frac{dy}{dt} = v(x,y) = C\sin\alpha$$

Find the complex potential function. Plot the stream lines and velocity vectors at (0,1) and (0, −1) in Fig. P8.4.1(b). Compare the complex potential function and stream lines with those in (a).

(c) $$\frac{dx}{dt} = u(x,y) = 2y \qquad \text{(P8.4c.1)}$$
$$\frac{dy}{dt} = v(x,y) = -2x$$

Find the complex potential function. Complete the MATLAB script "em08p04c.m" and run it to plot the stream lines and velocity vectors at (1,0), (2,0), (0,1), and (0,2) as shown in Fig. 8.4.2(a). Compare the complex potential function and stream lines with those in Example 8.3.

```
%em08p04c.m
cs=0.1*2.^[3:8]; cs=[-fliplr(cs) 0 cs]; % Level values
for n=1:length(cs) % To plot the stream lines
   h=ezplot(['?^2-?^2-' num2str(cs(n))]), hold on
   set(h,'color','r')
end
ps=[1 0; 2 0; 0 1; 0 2];
for n=1:size(ps,1)
   x=ps(n,1); y=ps(n,2); vx=2*y; vy=-2*x; % Eq.(P8.4b.2)
   quiver(x,y,vx,vy,'b','Linewidth',2)
end
```

(d) $$\frac{dx}{dt} = u(x,y) = x^2 - y^2 \qquad \text{(P8.4d.1)}$$
$$\frac{dy}{dt} = v(x,y) = -2xy$$

Find the complex potential function. Modify the above MATLAB script, "em08p04c.m," and run it to plot the stream lines and velocity vectors at (1, 1) and (1, 2) as shown in Fig. 8.4.2(b). Compare the complex potential function and stream lines with those in Example 8.3.

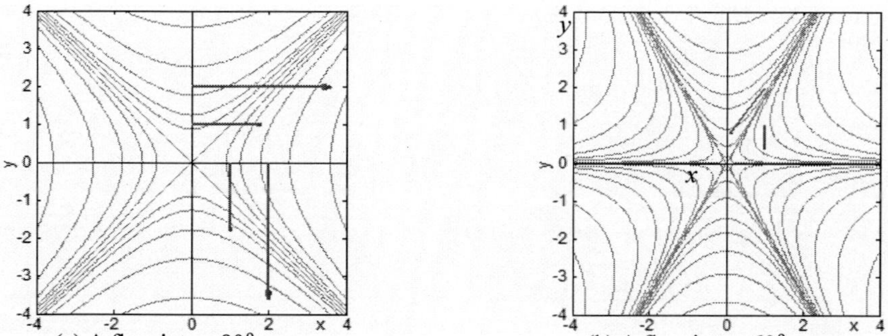

(a) A flow into a 90° corner (b) A flow into a 60° corner

Figure P8.4.2 Stream lines and velocity vectors describing the flows for Problem 8.4(c) and (d)

8.5 Complex Exponential/Logarithmic/Power Functions

Fill in the blanks:

(a) $e^{2(x+jy)} \overset{(8.1.36)}{=} e^{(\quad)}(\cos(\quad)+j(\qquad))$

(b) $\ln(1+j\sqrt{3}) \underset{(8.1.41a,b)}{\overset{(8.1.40)}{=}} \ln(\quad)+j(\qquad+\qquad)$

```
>>log(1+j*sqrt(3))
```

(c) $\ln(-1) \underset{(8.1.41a,b)}{\overset{(8.1.40)}{=}} \ln(\quad)+j(\qquad+\qquad)$

(d) $(\sqrt{3}+j)^{1+j} \overset{(8.1.43)}{=} e^{(\qquad)\ln(\qquad)} = e^{(\qquad)(\ln 2+j(\qquad))} = e^{\ln 2-(\qquad)+j(\qquad)}$

$= e^{\ln 2-(\qquad)}(\cos(\ln 2\qquad)+j\sin(\qquad+\pi/6+2n\pi)) = e^{-2n\pi}(0.4108+j1.1113)$

```
>>(sqrt(3)+j)^(1+j), exp(log(2)-pi/6)*exp(j*(log(2)+pi/6))
```

8.6 Complex Trigonometric/Hyperbolic Functions

(a) Verify Eqs. (8.1.49a) and (8.1.49b).

(b) Verify Eqs. (8.1.55a,b) and (8.1.56a,b).

(c) Verify the following equalities:

$$|\cos z|^2 = \cos^2 x + \sinh^2 y$$

$$|\sin z|^2 = \sin^2 x + \sinh^2 y$$

(d) Find the values of $\sin(\pi+j)$ and $\cos(\pi+j)$.

(e) Solve $\sin z = \sqrt{2}$ for z.

```
>>asin(sqrt(2)+[j -j]*eps)  % For reference
```

(f) Solve $\cos z = \sqrt{2}$ for z.

```
>>acos(sqrt(2)+[j -j]*eps)  % For reference
```

(g) Solve $\sinh z = \sqrt{3}$ for z.

```
>>asinh(sqrt(3))
```

(h) Solve $\cosh z = \sqrt{2}$.

```
>>acosh(sqrt(2))
```

8.7 Inverse Complex Trigonometric/Hyperbolic Functions

(a) Find $\sin^{-1}\sqrt{2}$. `>>asin(sqrt(2))`

(b) Find $\cos^{-1}\sqrt{2}$. `>>acos(sqrt(2))`

(c) Find $\tan^{-1}(2+j)$. `>>atan(2+j)`

(d) Find $\sinh^{-1} j$. `>>asinh(j)`

(e) Find $\cosh^{-1}\sqrt{2}$. `>>acosh(sqrt(2))`

(f) Find $\tanh^{-1} j$. `>>atanh(j)`

8.8 Conformal Mapping

Consider the mapping described by the complex function

$$w = f(z) = \sin z. \tag{P8.8.1}$$

Note that this mapping is conformal at all values of z except $z=(2n+1)\pi/2$ (at which $f'(z) = \cos z = 0$). To find the images of the vertical lines (with constant x) and horizontal lines (with constant y) in the z-plane mapped to the w-plane, let us substitute $z=x+jy$ into Eq. (P8.8.1) to write

$$u + jv = w = \sin z = \frac{e^{jz}-e^{-jz}}{j2} = \frac{e^{j(x+jy)}-e^{-j(x+jy)}}{j2} = \frac{e^{-y}(\cos x+j\sin x)-e^{y}(\cos x-j\sin x)}{j2}$$

$$= \begin{cases} \dfrac{(e^{-y}+e^{y})\sin c_1+j(e^{y}-e^{-y})\cos c_1}{2} = u+jv & \text{for } x=c_1 \\ \dfrac{(e^{-c_2}+e^{c_2})\sin x+j(e^{c_2}-e^{-c_2})\cos x}{2} = u+jv & \text{for } y=c_2 \end{cases} \tag{P8.8.2}$$

Find the equations describing the images and verify that they are orthogonal. Complete the MATLAB script "em08p08.m" and run it to get Fig. P8.8, which shows the vertical/horizontal lines in the z-plane and their images as the hyperbolas/ellipses in the w-plane under the mapping $w = \sin z$.

```
%em08p08.m
clear, clf
cs=[-7:7]*(pi/8); % Constant
for n=1:length(cs) % Vertical lines with x=const
   y=[-pi:0.01:pi]; x=cs(n)*ones(size(y));
   w=???(x+j*y)+j*eps;
   if n<4, gs='r'; elseif n==8, gs='g'; else gs='b';  end
   subplot(121), plot(x,y,gs); hold on
   subplot(122), plot(w,gs); hold on
end
for n=1:length(cs) % Horizontal lines with y=const
   x=[-pi:0.01:pi]; y=cs(n)*ones(size(x));
   w=sin(?????)+j*eps;
   if n<4, gs='r'; elseif n==8, gs='g'; else gs='b';  end
   subplot(121), plot(x,y,gs); hold on
   subplot(122), plot(w,gs); hold on
end
axis('equal')
subplot(121), axis('equal'), axis([-3.5 3.5 -3.5 3.5])
```

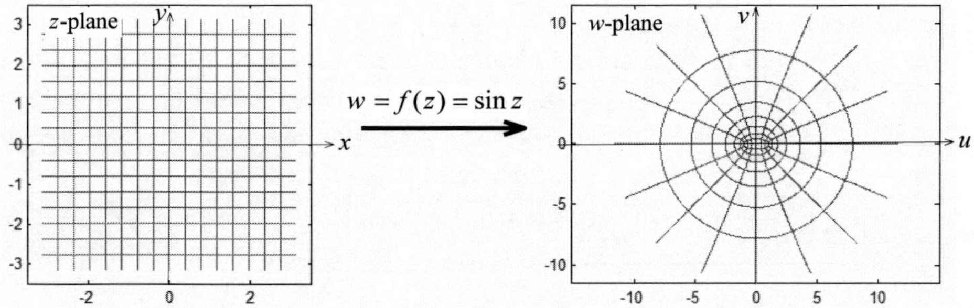

Figure P8.8 Mapping (transformation) of the vertical/horizontal lines from the z-plane to the w-plane under $w=\sin z$

8.9 Complex Potential

Consider a 2D Laplace's equation in the Cartesian coordinate system and its solution that is harmonic:

$$\nabla^2\phi = \frac{\partial^2\phi}{\partial x^2} + \frac{\partial^2\phi}{\partial y^2} = 0: \ \ \phi(x,y) = ax+b \tag{P8.9.1}$$

(a) Show that the conjugate harmonic function ψ of ϕ is

$$\psi(x,y) = ay+c \tag{P8.9.2}$$

by verifying that ϕ and ψ satisfy the CR equations (8.1.22a,b).

Here, we could derive such a conjugate harmonic function ψ that satisfies the CR equations. However, if we have noticed that $\phi(x,y)$ is mainly the real part of a function of $z=x+jy$, like Eq. (P8.9.3), ψ can easily be determined as just the imaginary part of such a function of z.

(b) Construct the complex potential having ϕ as its real part (called the real potential).

$$g(z) \underset{z=x+jy}{\overset{(8.1.33)}{=}} \phi(x,y) + j\psi(x,y) = \boxed{} \quad \text{where } z=x+jy \tag{P8.9.3}$$

(c) Do the lines of $\psi(x,y)=$const intersect with the equipotential lines $\phi(x,y)=$const at right angles? Note that if $\phi(x,y)=$const denotes an electric potential, then $\psi(x,y)=$const describes lines of electric force (field).

8.10 Complex Potential and Complex Velocity

Consider a 2D Laplace's equation in the polar coordinate system and its solution that is harmonic:

$$\nabla^2\phi = \frac{\partial^2\phi}{\partial r^2} + \frac{1}{r}\frac{\partial\phi}{\partial r} + \frac{1}{r^2}\frac{\partial^2\phi}{\partial\theta^2} = 0: \ \ \phi(r,\theta) = a\ln r + b \tag{P8.10.1}$$

(a) Show that the conjugate harmonic function ψ of ϕ is

$$\psi(r,\theta) = a\theta + c \tag{P8.10.2}$$

by verifying that ϕ and ψ satisfy the CR equations (8.1.26a,b) in polar form.

(b) Construct the complex potential having ϕ as its real part (called the real potential).

$$g(z) \underset{z=re^{j\theta}}{\overset{(8.1.33)}{=}} \phi(r,\theta) + j\psi(r,\theta) = \boxed{} \quad \text{where } z=re^{j\theta} \tag{P8.10.4}$$

(c) Do the lines of $\psi(r,\theta)$=const intersect with the equipotential lines $\phi(r,\theta)$=const at right angles? Note that if $\phi(r,\theta)$=const denotes an electric potential, then $\psi(r,\theta)$=const describes lines of electric force (field).

(d) Use Eq. (8.1.34) to find the complex velocity $f(z)$ of the flow described by complex potential (P8.10.4). Then take the conjugate of $f(z)$ to find the vector field $\mathbf{F}(x,y)=u(x,y)+jv(x,y)$.

(e) Is the vector field **F** obtained in (d) equal to the gradient $\nabla\phi$ of the real potential ϕ?

$$\nabla\phi \overset{(2.4.3b)}{\underset{(P8.10.1)}{=}} \frac{\partial\phi}{\partial r}\mathbf{u}_r + \frac{1}{r}\frac{\partial\phi}{\partial\theta}\mathbf{u}_\theta = \boxed{}\,\mathbf{u}_r \overset{(1.1.58a)}{=} \frac{a}{\boxed{}}\left(\frac{\overset{\cos\theta}{\boxed{}}}{\sqrt{x^2+y^2}}\mathbf{u}_x + \frac{\overset{\sin\theta}{\boxed{}}}{\sqrt{x^2+y^2}}\mathbf{u}_y\right) \quad \text{(P8.10.5)}$$
$$= \boxed{}$$

(f) To show that $\mathbf{F}=\nabla\phi$ holds if ϕ and ψ have continuous second partial derivatives so that they satisfy the CR equations, use Eq. (8.1.24a) to write the derivative of the complex potential $g(z,y)=\phi(x,y)+j\psi(x,y)$ and see if its conjugate conforms with the gradient of ϕ:

$$f(z) \overset{(8.1.34)}{=} \frac{d}{dz}g(z) \overset{(8.1.24a)}{\underset{(8.1.33)}{=}} \frac{\partial\phi}{\partial x} + j\boxed{} \overset{(8.1.22b)}{\underset{u=\phi,\,v=\Psi}{=}} \frac{\partial\phi}{\partial x} - j\boxed{} \overset{\text{conjugate}}{\rightarrow} \frac{\partial\phi}{\partial x} + j\boxed{} = \boxed{} \quad \text{(P8.10.6)}$$

8.11 Complex Potential and Complex Velocity

Consider the electric potential $\phi(z)$ at $z=x+jy$ from the origin $(0+j0)$, which is formed due to two oppositely (positively/negatively) and uniformly charged lines of infinite length, each located at $z=1+j0$ and $z=-1+j0$ on the z-plane so that it can be expressed as

$$\phi(z) = c\,\mathrm{Ln}\frac{|z-1|}{|z-(-1)|} = c\,\mathrm{Ln}\frac{|x+jy-1|}{|x+jy+1|} \quad (c: \text{a constant}) \quad \text{(P8.11.1)}$$

(a) Noting that this is a (real) potential function corresponding to the real part of the complex logarithmic function (see Eqs. (8.1.40) and (8.1.41a)) called the complex potential:

$$g(z) = c\ \mathrm{Ln}\frac{z-1}{z+1} = \phi(z) + j\psi(z) \quad \text{(P8.11.2)}$$

use Eq. (8.1.42) to write the imaginary part of Eq. (P8.11.1) as the stream function (describing the electric field):

$$\psi(z) = \mathrm{Im}\left(\mathrm{Ln}\frac{z-1}{z-(-1)}\right) \overset{(8.1.42)}{=} \mathrm{Arg}\left(\boxed{}\right) = \tan^{-1}\boxed{} - \tan^{-1}\boxed{} \quad \text{(P8.11.3)}$$

(b) To find the potential lines as the images of the vertical lines in the $g(z)$-plane (corresponding to $\phi(z)$=const) mapped to the z-plane, rewrite Eq. (P8.11.1) as

$$\frac{(x-1)^2+y^2}{(x+1)^2+y^2} = e^{2\phi/c};\quad (e^{2\phi/c}-1)(x^2+y^2+1) + 2(e^{2\phi/c}+1)x = 0$$

$$;\ \begin{cases} \left(x+\boxed{}\right)^2 + y^2 = \left(\dfrac{e^{2\phi/c}+1}{e^{2\phi/c}-1}\right)^2 - 1 = \dfrac{4e^{2\phi/c}}{(e^{2\phi/c}-1)^2} & \text{for } \phi\neq 0 \\ x=0 & \text{for } \phi=0 \end{cases} \quad \text{(P8.11.4)}$$

(c) To find the stream lines or electric field lines as the images of the horizontal lines in the $g(z)$-plane (corresponding to $\psi(z)$=const) mapped to the z-plane, rewrite Eq. (P8.11.3) as

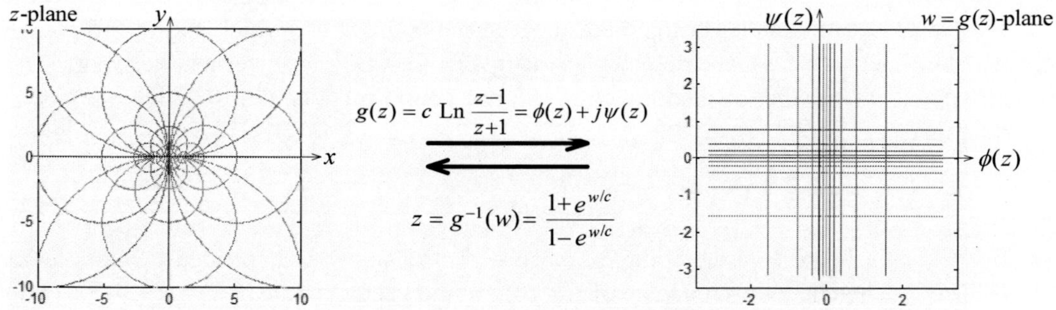

Figure P8.11 Mapping of the vertical/horizontal lines from the *w*-plane to the z-plane under $w = \text{Ln}\,\frac{z-1}{z+1}$

$$\tan\psi \underset{(C.7)}{\overset{(P8.11.3)}{=}} \frac{\dfrac{y}{x-1} - \dfrac{y}{x+1}}{1 + \dfrac{y}{x-1}\dfrac{y}{x+1}} = \frac{2y}{x^2+y^2-1}; \quad \begin{cases} x^2 + \left(y - \boxed{}\right)^2 = 1 + \left(\dfrac{1}{\tan\psi}\right)^2 & \text{for } \psi \neq 0 \\ y = 0 & \text{for } \psi = 0 \end{cases} \quad \text{(P8.11.5)}$$

```
%em08p11.m
cs=[-4 -2 -1 -0.5 -0.25 0 0.25 0.5 1 2 4]*(pi/8); c=1; % Constants
psis=[-pi:0.01:pi];
for n=1:length(cs) % Vertical lines with phi=const
   phi=cs(n); phis=phi*ones(size(psis)); w=phis+j*psis;
   z=-(exp(w/c)+1)./(exp(???)-1); % Eq.(P8.11.6)
   xs=real(z); ys=imag(?);
   if phi==0, yy=[-6:0.1:6]; xx=zeros(size(yy)); gs='g';
    else
      if phi<0,  gs='r';  else gs='b';  end
      ephi=exp(phi/c); e2phi=ephi^2;
      x1=(e2phi+1)/(?????-1);
      r=2*????/abs(e2phi-1);
      xx=-x1+[-r:0.1:r]; yy=sqrt(max(r^2-(xx+x1).^2,0)); %Eq.(P8.11.4)
   end
   subplot(121), plot(xs,ys,gs); hold on
   plot(xx,yy,'m:', xx,-yy,'m:'), axis('equal'), axis([-10 10 -10 10])
   subplot(122), plot(phis,psis,gs); hold on
end
axis('equal'), axis([-3.5 3.5 -3.5 3.5])
phis=[-pi:0.01:pi];
for n=1:length(cs) % Horizontal lines with psi=const
   psi=cs(n); psis=psi*ones(size(phis)); w=phis+j*psis;
   z=-(exp(???)+1)./(exp(w/c)-1); % Eq.(P8.11.6)
   xs=real(?); ys=imag(z);
   if psi==0, xx=[-6:0.1:6]; yy=zeros(size(xx));
    else
      if psi<0,  gs='r';  else gs='b';  end
      tanpsi=tan(psi); y1=1/tanpsi;
      r=sqrt(1+1/??????^2);
      yy=y1+[-r:0.1:r]; xx=sqrt(max(r^2-(yy-y1).^2,0)); %Eq.(P8.11.5)
   end
   subplot(121), plot(xs,ys,gs); hold on
   plot(xx,yy,'m:', -xx,yy,'m:'),
   subplot(122), plot(phis,psis,gs); hold on
end
```

(d) To check the validity of these derivations about the complex potential, complete the above MATLAB script, "em08p11.m," and run it to get Fig. P8.11, which shows the potential/field (stream) lines in the z-plane obtained as the inverse images of the vertical/horizontal lines in the $g(z)$-plane where the inverse function of the complex potential function $g(z)$ is used for reference:

$$w = g(z) = c\ \mathrm{Ln}\frac{z-1}{z+1} \rightarrow z = g^{-1}(w) = \frac{1+e^{w/c}}{1-e^{w/c}} \tag{P8.11.6}$$

(e) Use Eq. (8.1.34) to find the complex velocity $f(z)$ of the flow described by complex potential (P8.11.2). Then take the conjugate of $f(z)$ to find the vector field $\mathbf{F}(x,y)=u(x,y)+jv(x,y)$.

$$f(z) \overset{(8.1.34)}{=} \frac{d}{dz}g(z) \overset{(P8.11.2)}{=} \frac{d}{dz}\left(c\ \mathrm{Ln}\frac{z-1}{z+1}\right) \underset{(8.1.45b)}{\overset{(2.1.4)}{=}} \boxed{}$$

$$\overset{z=x+jy}{=} \boxed{} \tag{P8.11.7}$$

$$;\ \mathbf{F} = f^*(z) = u + jv = \boxed{} \tag{P8.11.8}$$

(f) Check if the vector field **F** obtained in (e) equals the gradient $\nabla\phi$ of the real potential ϕ. Instead of doing hand calculations, you can complete and run the MATLAB script "em08p11e.m."

```
%em08p11e.m
syms x y
phi=log(((x-1)^2+y^2)/((x+1)^2+y^2))/?; % Eq.(P8.11.1)
u = 2*(x^2-?^2-1)/((x^2-y^2-1)^2+(2*?*y)^2); % Eq.(P8.11.7)
v = 4*x*y/((?^2-y^2-1)^2+(2*x*?)^2);  % Eq.(P8.11.7)
F = u+i*?; % Vector field (P8.11.8)
grad_phi=gradienta(???); % The gradient of phi
del_phi=grad_phi(?)+?*grad_phi(2); % in complex form
discrepancy_between_del_phi_and_F=simplify(del_phi-F)
```

8.12 Complex Potential and Complex Velocity

(a) Consider a vector field, called a source flow, with the horizontal/vertical velocities as

$$u = \frac{x}{x^2+y^2} \underset{y=r\sin\theta}{\overset{x=r\cos\theta}{=}} \frac{\cos\theta}{r} \quad \text{and} \quad v = \frac{y}{x^2+y^2} \underset{y=r\sin\theta}{\overset{x=r\cos\theta}{=}} \frac{\sin\theta}{r} \tag{P8.12a.1}$$

(i) Find the complex potential $g(z)=\phi(z)+j\psi(z)$ by integrating the complex velocity function $f(z)=u(x,y)-jv(x,y)$ where $z=x+jy=re^{j\theta}$.

(ii) Conversely, use Eqs. (8.1.22a,b) to find the horizontal/vertical velocities u/v by differentiating $\phi(z)$ or $\psi(z)$ w.r.t. x or y:

$$u(x,y) \underset{\text{with } u=\phi}{\overset{(8.1.22a)}{=}} \frac{\partial\phi}{\partial x} =$$

```
>>syms x y; phi=log(sqrt(x^2+y^2)); diff(phi)
```

$$v(x,y) \underset{\text{with } u=\phi}{\overset{(8.1.22b)}{=}} \frac{\partial\phi}{\partial y} =$$

```
>>diff(phi,y)
```

(iii) Also, use Eqs. (8.1.26a,b) to find the radial/tangential (azimuthal) velocities v_r/v_θ by differentiating $\phi(z)$ or $\psi(z)$ w.r.t. r or θ:

$$v_r(r,\theta) \underset{\text{with } u=\phi}{\overset{(8.1.26a)}{=}} \frac{\partial\phi(r,\theta)}{\partial r} =$$

$$v_\theta(r,\theta) \underset{\text{with } u=\phi}{\overset{(8.1.26b)}{=}} \frac{1}{r}\frac{\partial\phi(r,\theta)}{\partial\theta} =$$

(iv) Show that the radial/tangential (azimuthal) velocities v_r/v_θ can be obtained by using Eq. (1.1.58b) to make a Cartesian-to-polar conversion of the horizontal/vertical velocities u/v.

$$[\mathbf{u}_x \;\; \mathbf{u}_y]\begin{bmatrix}(\cos\theta)/r\\(\sin\theta)/r\end{bmatrix} \overset{(1.1.58b)}{=} [\mathbf{u}_r \;\; \mathbf{u}_\theta]\begin{bmatrix}\cos\theta & \sin\theta\\-\sin\theta & \cos\theta\end{bmatrix}\begin{bmatrix}(\cos\theta)/r\\(\sin\theta)/r\end{bmatrix} = [\mathbf{u}_r \;\; \mathbf{u}_\theta]\begin{bmatrix} \\ \end{bmatrix}$$

(v) Plot the equipotential lines, the stream lines, and the velocity vectors at (1,0) and (0,1) by completing and running the MATLAB script "em08p12a.m."

```
%em08p12a.m
syms x y r th
u = x/(?^2+y^2);   v = ?/(x^2+y^2); % Vector field (P8.12a.1)
frth=subs(u-j*v,{x,y},{r*cos(th),r*sin(th)});
pretty(simplify(frth))
% Level values of r for equipotential lines
rs1=[1:4]; ths1=[0:180]*(pi/90);
% Level values of theta for stream lines
rs2=[0:0.02:5]; ths2=[0:45:360]*(pi/180);
subplot(231), hold on
for n=1:length(rs1), polar(ths1,rs1(n)*ones(size(ths1))); end
for n=1:length(ths2), polar(ths2(n)*ones(size(rs2)),rs2,'r'); end
ps=[1 0; 0 1]; % Points at which to plot the velocity vector
for n=1:size(ps,1)
   xn=ps(n,1); yn=ps(n,2);
   vx=subs(u,{x,y},{xn,yn}); vy=subs(v,{x,y},{xn,yn});
   quiver(xn,yn,vx,vy,'k','Linewidth',2)
end
axis('equal'); axis('off')
```

(b) Consider a vector field, called a source flow, with the horizontal/vertical velocities as

$$u = \frac{-y}{x^2+y^2} \underset{y=r\sin\theta}{\overset{x=r\cos\theta}{=}} \frac{-\sin\theta}{r} \quad\text{and}\quad v = \frac{x}{x^2+y^2} \underset{y=r\sin\theta}{\overset{x=r\cos\theta}{=}} \frac{\cos\theta}{r} \tag{P8.12b.1}$$

(i) Find the complex potential $g(z)=\phi(z)+j\psi(z)$ by integrating the complex velocity function $f(z)=u(x,y)-jv(x,y)$ where $z=x+jy=re^{j\theta}$.

(ii) Conversely, use Eqs. (8.1.22a,b) to find the horizontal/vertical velocities u/v by differentiating $\phi(z)$ or $\psi(z)$ w.r.t. x or y:

$$u(x,y) \underset{\text{with } u=\phi}{\overset{(8.1.22a)}{=}} \frac{\partial\phi}{\partial x} =$$

```
>>phi=atan(y/x); diff(phi)
```

$$v(x,y) \underset{\text{with } u=\phi}{\overset{(8.1.22b)}{=}} \frac{\partial\phi}{\partial y} =$$

```
>>diff(phi,y)
```

(iii) Also, use Eqs. (8.1.26a,b) to find the radial/tangential (azimuthal) velocities v_r/v_θ by differentiating $\phi(z)$ or $\psi(z)$ w.r.t. r or θ:

$$v_r(r,\theta) \underset{\text{with}\,u=\phi}{\overset{(8.1.26a)}{=}} \frac{\partial\phi(r,\theta)}{\partial r} =$$

$$v_\theta(r,\theta) \underset{\text{with}\,u=\phi}{\overset{(8.1.26b)}{=}} \frac{1}{r}\frac{\partial\phi(r,\theta)}{\partial\theta} =$$

(iv) Show that the radial/tangential (azimuthal) velocities v_r/v_θ can be obtained by using Eq. (1.1.58b) to make a Cartesian-to-polar conversion of the horizontal/vertical velocities u/v.

$$[\mathbf{u}_x\ \mathbf{u}_y]\begin{bmatrix}(-\sin\theta)/r\\(\cos\theta)/r\end{bmatrix} \overset{(1.1.58b)}{=} [\mathbf{u}_r\ \mathbf{u}_\theta]\begin{bmatrix}\cos\theta & \sin\theta\\-\sin\theta & \cos\theta\end{bmatrix}\begin{bmatrix}(-\sin\theta)/r\\(\cos\theta)/r\end{bmatrix} = [\mathbf{u}_r\ \mathbf{u}_\theta]\begin{bmatrix} \\ \end{bmatrix}$$

(v) Plot the equipotential lines, the stream lines, and the velocity vectors at (1,0) and (0,1) by completing and running the MATLAB script "em08p12b.m."

```
%em08p12b.m
syms x y r th
u = -?/(x^2+y^2); v = x/(x^2+?^2); % Vector field (P8.12b.1)
frth = subs(u-j*v,{x,y},{r*cos(th),r*sin(th)});
pretty(simplify(frth))
rs1=[1:4]; ths1=[0:180]*(pi/90);
rs2=[0:0.02:5]; ths2=[0:45:360]*(pi/180);
subplot(232), hold on
for n=1:length(ths2), polar(ths2(n)*ones(size(rs2)),rs2); end
for n=1:length(rs1), polar(ths1,rs1(n)*ones(size(ths1)),'r'); end
ps=[1 0; 0 1]; % Points at which to plot the velocity vector
for n=1:size(ps,1)
   xn=ps(n,1); yn=ps(n,2);
   vx=subs(u,{x,y},{xn,yn}); vy=subs(v,{x,y},{xn,yn});
   quiver(xn,yn,vx,vy,'k','Linewidth',2)
end
axis('equal'); axis('off')
```

```
%em08p12c.m
syms x y r th
u=2*(x^2-y^2)/((x^2-y^2)^2+4*x^2*y^2); % Vector field (P8.12c.1)
v=4*x*y/((x^2-y^2)^2+4*x^2*y^2);
frth = subs(u-j*v,{x,y},{r*cos(th),r*sin(th)});
pretty(simplify(frth))
subplot(233), hold on
cs=1*0.5.^[0:3]; cs=[-fliplr(cs) 0 cs];
for n=1:length(cs)
   h=ezplot(['-x/(x^2+y^2)-' num2str(cs(n))]); set(h,'color','b')
   h=ezplot(['y/(x^2+y^2)-' num2str(cs(n))]); set(h,'color','r')
end
plot([-5 5],[0 0],'k'), plot([0 0],[-5 5],'k')
% Points at which to plot the velocity vector
ps=[1 1;0 2; -1 1; -1 -1;0 -2; 1 -1];
for n=1:size(ps,1)
   xn=ps(n,1); yn=ps(n,2);
   vx=subs(u,{x,y},{xn,yn}); vy=subs(v,{x,y},{xn,yn});
   quiver(xn,yn,vx,vy,'k','Linewidth',2)
end
grid on, axis('equal'), axis([-5 5 -5 5]), axis('off')
```

(c) Consider a vector field, called a source flow, with the horizontal/vertical velocities as

$$u=\frac{2(x^2-y^2)}{(x^2+y^2)^2}\underset{y=r\sin\theta}{\overset{x=r\cos\theta}{=}}\frac{2\cos 2\theta}{r^2} \quad \text{and} \quad v=\frac{4xy}{(x^2+y^2)^2}\underset{y=r\sin\theta}{\overset{x=r\cos\theta}{=}}\frac{2\sin 2\theta}{r^2} \tag{P8.12c.1}$$

(i) Find the complex potential $g(z)=\phi(z)+j\psi(z)$ by integrating the complex velocity function $f(z)=u(x,y)-jv(x,y)$ where $z=x+jy=re^{j\theta}$.

(ii) Plot the equipotential lines, the stream lines, and the velocity vectors at (1,1), (0,2), (−1,1), (−1,−1), (0,−2), and (1,−1) by completing and running the above MATLAB script "em08p12c.m."

8.13 Linear Fractional Transformation (LFT) — Smith Chart

Consider the mapping described by the complex function

$$w=f(z)=\frac{z-1}{z+1} \tag{P8.13.1}$$

```
%em08p13.m
syms u v
x=(1-u^2-v^2)/((1-u)^2+v^2);   y=2*v/((1-u)^2+v^2);
simplify((u-x/(x+1))^2+v^2-1/(x+1)^2), simplify((u-1)^2+(v-1/y)^2-1/y^2)
xs=[-2:0.5:2]; ys=xs; % Constants
for n=1:length(xs) % Vertical lines with x=const
   y_=-8:0.01:8; x=xs(n); x_=x*ones(size(y_)); z=x_+j*??;
   subplot(121), plot(z,gs); hold on
   subplot(122), w=(???)./(z+1+j*eps); plot(w); hold on
   if x==-1, plot([1 1],[-5 5],'m:')
    else uo=x/(?+1); r=1/(?+1);
         u=??+[-r:0.01:r]; v=sqrt(r^2-(u-uo).^2); % Eq.(8.13.4)
         plot(u,v,'m:', u,-v,'m:')
   end
end
for n=1:length(ys) % Radial lines with th=const
   x_=-8:0.01:8; y=ys(n); y_=y*ones(size(x_)); z=??+j*y_;
   subplot(121), plot(z,gs); axis('equal'), axis([-3 3 -3 3]); hold on
   subplot(122), w=(z-1)./(???+j*eps); plot(w); hold on
   if y==0, plot([-4 6],[0 0],'m:')
    else uo=?; vo=1/?; r=1/y;
         u=uo+[-r:0.01:r]; v=vo+sqrt(r^2-(u-??).^2); % Eq.(8.13.5)
         plot(u,v,'m:', u,-v,'m:');
         axis('equal'), axis([-4 6 -5 5])
   end
end
```

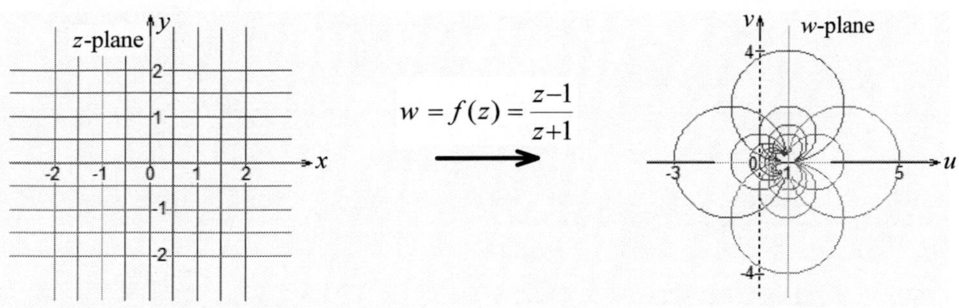

Figure P8.13 Mapping of the horizontal/vertical lines from the z-plane to the w-plane under $w=(z-1)/(z+1)$

To find the images of the vertical/horizontal lines in the z-plane mapped to the w-plane, let us write the inverse mapping function

$$z = g(w) = \frac{1+w}{1-w} \tag{P8.13.2}$$

and substitute $w = u + jv$ into Eq. (P8.13.2) to write

$$x + jy = \frac{1+u+jv}{1-u-jv} = \frac{(1+u+jv)(1-u+jv)}{(1-u)^2+v^2} = \frac{1-u^2-v^2+j2v}{(1-u)^2+v^2} \tag{P8.13.3}$$

Find the equations for the images:

$$(u-\square)^2 + v^2 = \left(\frac{1}{x+1}\right)^2 \text{ for } x = const \neq -1 \text{ and } u = 1 \text{ for } x = -1 \tag{P8.13.4}$$

$$(u-1)^2 + (v-\square)^2 = \left(\frac{1}{y}\right)^2 \text{ for } y = const \neq 0 \text{ and } v = 0 \text{ for } y = 0 \tag{P8.13.5}$$

Complete the above MATLAB script, "em08p13.m," and run it to get Fig. P8.13, which shows the vertical/horizontal lines in the z-plane and their images as the circles/circles in the w-plane under the mapping $w=(z-1)/(z+1)$. Is the orthogonality of vertical/horizontal lines in the z-plane inherited to their images in the w-plane, as implied by the conformality of the LFT?

8.14 Linear Fractional Transformation (LFT) — Nichols Chart

Consider the mapping described by the complex function

$$w = f(z) = \frac{z}{1+z} \tag{P8.14.1}$$

To find the images of the radial/circular lines in the w-plane mapped to the z-plane, let us write the inverse mapping function

$$z = g(w) = \frac{w}{1-w} \tag{P8.14.2}$$

and substitute $w = re^{j\theta}$ into Eq. (P8.14.2) to write

$$x + jy = \frac{w}{1-w} = \frac{re^{j\theta}}{1-re^{j\theta}} = \frac{re^{j\theta}(1-re^{-j\theta})}{(1-re^{j\theta})(1-re^{-j\theta})} = \frac{r(e^{j\theta}-r)}{1-2r\cos\theta+r^2} = \frac{r(\cos\theta-r)+jr\sin\theta}{1-2r\cos\theta+r^2}$$

Find the equations for the images:

$$\left(x-\square\right)^2 + y^2 = \left(\frac{r}{1-r^2}\right)^2 \text{ for } r \neq 1 \text{ and } (x,y) = \left(-\frac{1}{2}, \frac{\sin\theta}{4\sin^2(\theta/2)}\right) \text{ for } r = 1 \tag{P8.14.3}$$

$$\left(x+\frac{1}{2}\right)^2 + \left(y-\square\right)^2 = \left(\frac{1}{2\sin\theta}\right)^2 \text{ for } \theta \neq n\pi \text{ and } (x,y) = \left(-\frac{r}{r\mp 1}, 0\right) \text{ for } \theta = n\pi \tag{P8.14.4}$$

This implies

$$\frac{r^2}{1-r^2} + \frac{r}{1-r^2}\cos\phi + j\frac{r}{1-r^2}\sin\phi = x + jy \qquad \text{for } r = c_1 \tag{P8.14.5}$$

$$-\frac{1}{2}+\frac{1}{2\sin\theta}\cos\phi+j\left(\frac{1}{2\tan\theta}+\frac{1}{2\sin\theta}\sin\phi\right)=x+jy \quad \text{for } \theta=c_2 \tag{P8.14.6}$$

Complete the MATLAB script "em08p14.m" and run it to get Fig. P8.14, which shows the radial/circular lines in the w-plane and their images as the circles/circles in the z-plane under the mapping $z=w/(1-w)$. Is the orthogonality of radial/circular lines in the w-plane inherited to their images in the z-plane, as implied by the conformality of the LFT?

```
%em08p14.m
clear, clf
syms r th
x=r*(cos(th)-r)/(1-2*r*cos(th)+r^2);
y=r*sin(th)/(1-2*r*cos(th)+r^2);
pretty(simplify((x-r^2/(1-r^2))^2+y^2)) % r^2/(1-r^2)^2
simplify((x+1/2)^2+(y-1/2/tan(th))^2) % 1/(4*sin(th)^2)
rs=10.^([-12:3:12]/20+1e-8);
ths=deg2rad([-175 -150:30:150 175]);
for n=1:length(rs) % Circular lines with r=const
   th_=-pi:0.01:pi; r=rs(n);
   w=r*exp(j*th_); z=w./(???); % Eq.(P8.14.2)
   subplot(121), plot(z); hold on
   if r==1, plot([-1/2 -1/2],[-5 5],'m:') % u=-1/2
    else xo=r^2/(1-r^2); R=r/(1-r^2);
          x=xo+[-R:0.01:R]; y=sqrt(R^2-(x-xo).^2); % Eq.(P8.14.3)
          plot(x,y,'m:', x,-y,'m:')
   end
   subplot(122), plot(w); hold on
end
for n=1:length(ths) % Radial lines with th=const
   r_=logspace(-2,1,500); th=ths(n);
   w=r_*exp(j*th); z=?./(1-w); % Eq.(P8.14.1)
   subplot(121), plot(z); hold on
   if th==0, plot([-6 6],[0 0],'m:') % u=1
    else xo=-1/2; yo=1/2/tan(th); R=1/2/sin(th);
          x=xo+[-R:0.01:R]; y=yo+sqrt(R^2-(x-xo).^2); % Eq.(P8.14.4)
          plot(x,y,'m:', x,-y,'m:')
   end
   subplot(122), plot(w); hold on
end
axis('equal'), axis([-6 6 -6 6])
subplot(121), axis('equal'), axis([-4 3 -3 3])
```

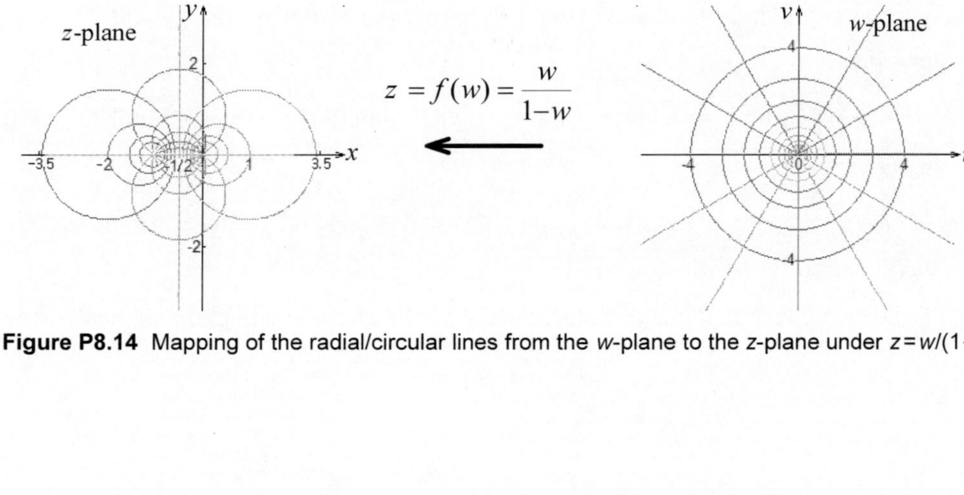

Figure P8.14 Mapping of the radial/circular lines from the w-plane to the z-plane under $z=w/(1-w)$

8.15 Linear Fractional Transformation

Construct an LFT mapping a triple (1, $-j$, -1) in the z-plane to (-1, 0, 1) in the w-plane. You can complete and run the MATLAB script "em08p15.m" to see the vertical/horizontal lines in the z-plane and their images as the circles/circles in the w-plane under the LFT.

```
%em08p15.m
clear, clf
LFT([1 -j  -1],[-1  0  1])
xs=[-2:1:2]; ys=xs; % Constants
y_=-8:0.01:8; % Whole range of y
for n=1:length(xs) % Vertical lines with x=const
   x=xs(n); z=x+j*y_;
   w=-(z+?)./(?*z+1);
   subplot(121), plot(z,'b'); hold on
   subplot(122), plot(w,'b'); hold on
end
axis('equal'), axis([-2.5 2.5 -1.5 3.5])
subplot(121), axis('equal'), axis([-3 3 -3 3])
x_=-8:0.01:8; % Whole range of x
for n=1:length(ys) % Horizontal lines with y=const
   y=ys(n); z=x_+j*(y+eps);
   w=-(?+j)./(j*?+1);
   subplot(121), plot(z,'r'); hold on
   subplot(122), plot(w,'r'); hold on
end
```

8.16 Complex Integration

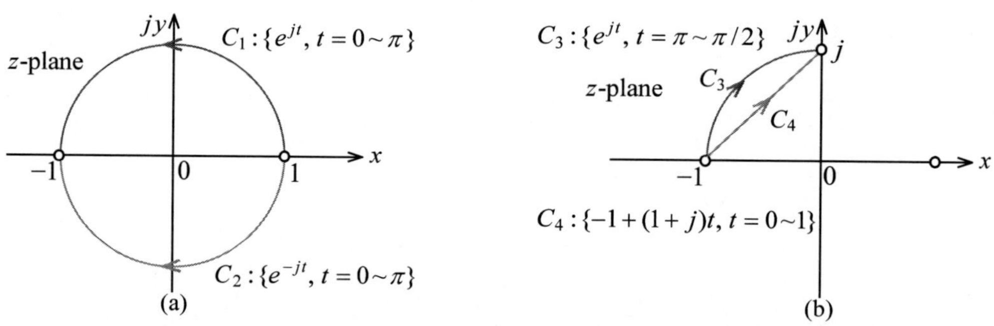

Figure P8.16 Paths of integration for the line integral of Problem 8.16

(a) For $f(z)=z^2$, find $\int_{C_1} f(z)dz$ and $\int_{C_2} f(z)dz$ where the curves C_1 and C_2 are shown in Fig. P8.16(a).

(b) For $f(z)=z^{-1}$, find $\int_{C_1} f(z)dz$ and $\int_{C_2} f(z)dz$ where the curves C_1 and C_2 are shown in Fig. P8.16(a).

(c) For $f(z)=|z|^2=x^2+y^2$, find $\int_{C_3} f(z)dz$ and $\int_{C_4} f(z)dz$ where the curves C_3 and C_4 are shown in Fig. P8.16(b).

(d) For the above complex integration results, explain why they are path independent or dependent.

8.17 Taylor Series

To check the validity of the following Taylor series expansions for $\sin^{-1}x$ and $\tan^{-1}x$, complete and run the following MATLAB functions to get their values for an arbitrary value of x, say, x=0.3, where the coefficients of the Taylor series are found in two ways, once by using the corresponding equation (P8.17.1) or (P8.17.2), and the other by using the MATLAB function '`taylor()`'.

$$\text{(a) } \sin^{-1}x = \sum_{n=0}^{\infty} \frac{(2n)!}{4^n (n!)^2 (2n+1)} x^{2n+1} = x + \frac{(2\cdot1)!}{4(1)^2}\frac{1}{3}x^3 + \frac{(2\cdot2)!}{4^2(1\cdot2)^2}\frac{1}{5}x^5 + \cdots \text{ for } |x| \le 1 \qquad \text{(P8.17.1)}$$

```
%em08p17a.m
x0=0.3; tol=1e-3; [y,e]=asin_my(x0,tol)
syms x; T=taylor(asin(x),'order',6);
Tp=fliplr(sym2poly(T))
```

```
function [y,e]=asin_my(x,tol)
if nargin<2, tol=1e-3;  end
an=1; a(2)=an; x2=x^2;
xn2=x; y=a(2)*xn2;
for n=1:100
   n2=n*2; xn2=xn2*x2;
   an=an*n2*(n2-1)/4/n^2; % To generate the coefficients recursively
   a(2*n+2)=an/(n2+1);
   dy=a(n2+2)*xn2;
   y = y+dy;
   if abs(dy)<tol, break; end
end
e = asin(x)-y; % Difference from the exact value
```

$$\text{(b) } \tan^{-1}x = \sum_{n=0}^{\infty} \frac{(-1)^n}{2n+1} x^{2n+1} = x - \frac{1}{3}x^3 + \frac{1}{5}x^5 - \cdots \quad \text{for } |x| \le 1,\ x \ne \pm j \qquad \text{(P8.17.2)}$$

```
%em08p17b.m
x0=0.3; tol=1e-3; [y,e]=atan_my(x0,tol)
syms x; T=taylor(asin(x),'order',6);
Tp=fliplr(sym2poly(T))
```

```
function [y,e]=atan_my(x,tol)
if nargin<2, tol = 1e-3;  end
a(2)=1; x2=x^2;
xn2=x; sgn=1; y=a(2)*xn2;
for n=1:100
   sgn=-sgn; n2=n*2; xn2=xn2*x2;
   a(n2+2)=sgn/(n2+1);
   dy=a(n2+2)*xn2;
   y = y+dy;
   if abs(dy)<tol, break; end
end
e = atan(x)-y; % Difference from the exact value
```

CHAPTER 9

Optimization

CHAPTER OUTLINE

Optimization involves finding the minimum/maximum of an objective function $f(x)$ subject to some constraint $x \in S$. If there is no constraint for x to satisfy, or equivalently, S is the universe, it is called an unconstrained optimization, and otherwise, it is a constrained optimization. In this chapter, we will cover several unconstrained optimization techniques such as the golden search method, the quadratic approximation method, Nelder–Mead method, the steepest descent method, and Newton's method. As for constrained optimization, we will only introduce the MATLAB built-in functions together with the functions for unconstrained optimization. Note that we don't have to distinguish maximization and minimization, because maximizing $f(x)$ is equivalent to minimizing $-f(x)$, and so, without loss of generality, we deal only with the minimization problems.

9.1 UNCONSTRAINED OPTIMIZATION[Y-1], Section 7.1

9.1.1 Golden Search Method

This method is applicable to an unconstrained minimization problem such that the solution interval $[a, b]$ is known and the objective function $f(x)$ is *uni-modal* within the interval, i.e., the sign of its derivative $f'(x)$ changes at most once in $[a, b]$, so that $f(x)$ decreases/increases monotonically for $[a, x^o]/[x^o, b]$, where x^o is the solution that we are looking for. The so-called golden search procedure is summarized below and has been cast into the function '`opt_gs()`'. The MATLAB script "test_opt_gs.m" below has been composed to use this function for finding the minimum point of the objective function

$$f(x) = (x^2 - 4)^2/8 - 1 \tag{9.1.1}$$

Figure 9.1 shows how the function '`opt_gs()`' proceeds towards the minimum point step by step.

<Golden Search Procedure>

(Step 1) Pick up the two points $c=a+(1-r)h$ and $d=a+rh$ inside the interval $[a, b]$, where $(r=(\sqrt{5}-1)/2)$ and $h=b-a$.

(Step 2) If the values of $f(x)$ at the two points are almost equal, i.e., $f(a)\approx f(b)$ and the width h of the interval is sufficiently small, i.e., $h\approx 0$, then stop the iteration to exit the loop and declare $x^o=c$ or $x^o=d$ depending on whether $f(c)<f(d)$ or not. Otherwise, go to Step 3.

(Step 3) If $f(c)<f(d)$, let d be the new upper bound of the interval: $b\leftarrow d$; otherwise, let c be the new lower bound of the interval: $a\leftarrow c$. Then, go to Step 1.

```
function [xo,fo]=opt_gs(f,a,b,r,TolX,TolFun,k)
h=b-a; rh=r*h; c=b-rh; d=a+rh; % Step 1
fc=feval(f,c); fd=feval(f,d);
if k<=0|(abs(h)<TolX&abs(fc-fd)<TolFun)
  if fc<=fd,   xo=c; fo=fc;
   else  xo=d; fo=fd;       % Step 2
  end
  if k==0, fprintf('Just the best in given # of iterations'), end
else
  if fc<fd, [xo,fo]=opt_gs(f,a,d,r,TolX,TolFun,k-1);
   else   [xo,fo]=opt_gs(f,c,b,r,TolX,TolFun,k-1); % Step 3
  end
end
```

```
%test_opt_gs.m
% performs the golden search method
f=@(x)(x.*x-4).^2/8-1; % Objective function (9.1.1)
a=0; b=3;    % Boundary points of the solution interval
r=(sqrt(5)-1)/2;   % Golden ratio (9.1.3)
TolX=1e-4; TolFun=1e-4; MaxIter=100;
[xo,fo]=opt_gs(f,a,b,r,TolX,TolFun,MaxIter)
```

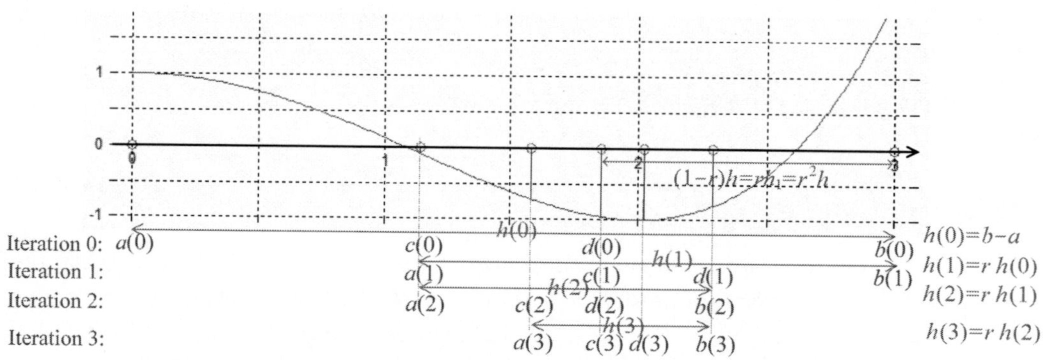

Figure 9.1 Illustration of the golden search method

Note the following about the golden search procedure:

- At every iteration, the new interval width is

$$b-c=b-\{a+(1-r)(b-a)\}=rh \quad \text{or} \quad d-a=a+rh-a=rh \tag{9.1.2}$$

so that it becomes r times the old interval width $(b-a=h)$.

- The *golden ratio* r is fixed so that a point $c_1=b_1-rh_1=b-r^2h$ in the new interval $[c, b]$ conforms with $d=a+rh=b-(1-r)h$, i.e.,

$$r^2=1-r; \quad r^2+r-1=0; \quad r=\frac{-1+\sqrt{5}}{2} \tag{9.1.3}$$

9.1.2 Quadratic Approximation Method

The idea of this method is to approximate the objective function $f(x)$ by a quadratic function $p(x)$ crossing the previous three (estimated solution) points, and to keep updating the three points by replacing one of them with the minimum point of $p(x)$. More specifically, for the three points

$$\{(x_0,f_0),(x_1,f_1),(x_2,f_2)\} \quad \text{with } x_0<x_1<x_2$$

we find the interpolation polynomial $p(x)$ of degree 2 to fit them as, say, the Lagrange polynomial:

$$p_2(x)=f_0\frac{(x-x_1)(x-x_2)}{(x_0-x_1)(x_0-x_2)}+f_1\frac{(x-x_0)(x-x_2)}{(x_1-x_0)(x_1-x_2)}+f_2\frac{(x-x_0)(x-x_1)}{(x_2-x_0)(x_2-x_1)} \tag{9.1.4}$$

and replace one of them with the zero of the derivative, i.e., the root of $dp(x)/dx=0$:

$$x_3=\frac{f_0(x_1^2-x_2^2)+f_1(x_2^2-x_0^2)+f_2(x_0^2-x_1^2)}{2\{f_0(x_1-x_2)+f_1(x_2-x_0)+f_2(x_0-x_1)\}}=\frac{(f_0-f_2)x_1^2+(f_1-f_0)x_2^2+(f_2-f_1)x_0^2}{2\{(f_0-f_2)x_1+(f_1-f_0)x_2+(f_2-f_1)x_0\}} \tag{9.1.5}$$

This can be found by running the following MATLAB statements:

```
>>syms x x0 x1 x2 x3 f0 f1 f2 f3
>>p=f0*(x-x1)*(x-x2)/(x0-x1)/(x0-x2)+... % Lagrange polynomial Eq.(9.1.4)
     f1*(x-x2)*(x-x0)/(x1-x2)/(x1-x0)+f2*(x-x0)*(x-x1)/(x2-x0)/(x2-x1)
>>x3=solve(diff(p)); pretty(simplify(x3)) % solve dp(x)/dx=0 to find x3

            2          2          2          2          2          2
     f0 x1  - f1 x0  - f0 x2  + f2 x0  + f1 x2  - f2 x1
   -------------------------------------------------------------
   2 f0 x1 - 2 f1 x0 - 2 f0 x2 + 2 f2 x0 + 2 f1 x2 - 2 f2 x1
```

We keep updating the three points this way until $|x_2-x_0|\approx 0$ and/or $|f(x_2)-f(x_0)|\approx 0$, when we stop the iteration and declare x_3 as the minimum point. The rule for updating the three points is as follows:

1. In case $x_0<x_3<x_1$, we take $\{x_0,x_3,x_1\}$ or $\{x_3,x_1,x_2\}$ as the new set of three points, depending on whether $f(x_3)<f(x_1)$ or not.
2. In case $x_1<x_3<x_2$, we take $\{x_1,x_3,x_2\}$ or $\{x_0,x_1,x_3\}$ as the new set of three points, depending on whether $f(x_3)\le f(x_1)$ or not.

This procedure, called the quadratic approximation method, has been cast into the MATLAB function '`opt_quad()`', which has the recursive (self-calling) structure. The MATLAB script "test_opt_quad.m" has been composed to use this function and the MATLAB built-in function '`fminbnd()`' for finding the minimum point of the objective function (9.1.1). Figure 9.2 shows how the function '`opt_quad()`' proceeds towards the minimum point step by step.

```
function [xo,fo]=opt_quad(f,x0,TolX,TolFun,MaxIter)
% search for the minimum of f(x) by quadratic approximation method
if length(x0)>2, x012=x0(1:3);
 else
   if length(x0)==2, a=x0(1); b=x0(2);  else a=x0-10; b=x0+10;  end
   x012= [a (a+b)/2 b];
end
f012= f(x012);
[xo,fo]=opt_quad0(f,x012,f012,TolX,TolFun,MaxIter);
% ---------------------------------------------------
function [xo,fo]=opt_quad0(f,x012,f012,TolX,TolFun,k)
x0= x012(1); x1= x012(2); x2= x012(3);
f0= f012(1); f1= f012(2); f2= f012(3);
nd= [f0-f2 f1-f0 f2-f1]*[x1*x1 x2*x2 x0*x0; x1 x2 x0].';
x3= nd(1)/2/nd(2); f3=feval(f,x3); % Eq.(9.1.5)
if k<=0|abs(x3-x1)<TolX|abs(f3-f1)<TolFun
  xo=x3;  fo=f3;
  if k==0, fprintf('Just the best in given # of iterations'), end
 else
  if x3<x1
    if f3<f1, x012=[x0 x3 x1]; f012= [f0 f3 f1];
     else  x012=[x3 x1 x2]; f012= [f3 f1 f2];
    end
   else
    if f3<=f1, x012=[x1 x3 x2]; f012= [f1 f3 f2];
     else  x012=[x0 x1 x3]; f012= [f0 f1 f3];
    end
  end
  [xo,fo]=opt_quad0(f,x012,f012,TolX,TolFun,k-1);
end
```

```
%test_opt_quad.m  to perform the quadratic approximation method
f=@(x) (x.*x-4).^2/8-1;
a=0; b=3; TolX=1e-5; TolFun=1e-8; MaxIter=100;
[xoq,foq]=opt_quad(f,[a b],TolX,TolFun,MaxIter)
% Minimum point and its function value
[xob,fob]=fminbnd(f,a,b) % MATLAB built-in function
```

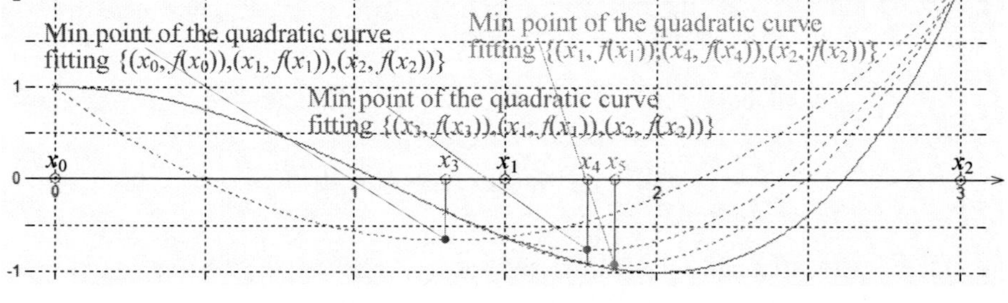

Figure 9.2 Illustration of the quadratic approximation method

9.1.3 Nelder–Mead Method

The Nelder–Mead method is applicable to the minimization of a multi-variable objective function, for which neither the golden search method nor the quadratic approximation method can be applied. The algorithm of the Nelder–Mead method summarized in the box below has been cast into the MATLAB function '`Nelder0()`'. Note that with N-dimensional case ($N>2$), this algorithm should be repeated for each subplane as implemented in the outer function '`opt_Nelder()`'.

<Nelder–Mead Algorithm>

(Step 1) Let the initial or previous three estimated solution points be a, b, and c where

$$f(a)<f(b)<f(c).$$

(Step 2) If the three points or their function values are sufficiently close to each other, then declare a to be the minimum and terminate the procedure.

(Step 3) Otherwise, expecting that the minimum we are looking for may be at the opposite side of the worst point c over the line $\overline{ab}$ (see Fig. 9.3(a)), take

(i) $$e=m+2(m-c) \text{ where } m=(a+b)/2, \tag{9.1.6a}$$

and if $f(e)<f(b)$, take e as the new c; otherwise, take

(ii) $$r=(m+e)/2=2m-c, \tag{9.1.6b}$$

and if $f(r)<f(c)$, take r as the new c; if $f(r)\geq f(b)$, take

(iii) $$s=(c+m)/2, \tag{9.1.6c}$$

and if $f(s)<f(c)$, take s as the new c; otherwise, give up the two points b and c, and take m and $c_1=(a+c)/2$ as the new b and c, reflecting our expectation that the minimum would be around a.

(Step 4) Go back to Step 1.

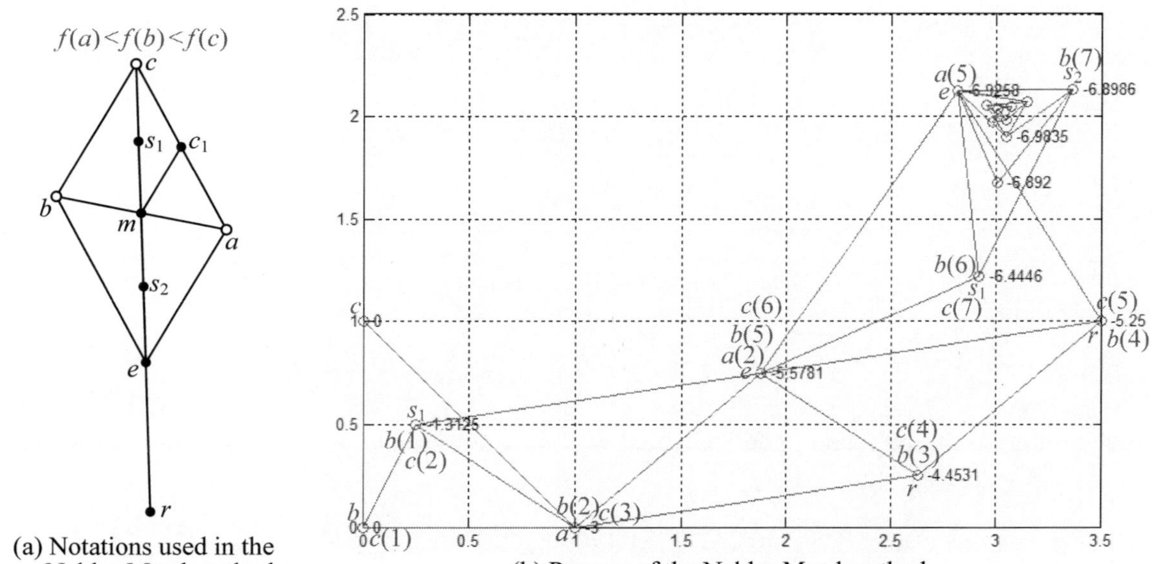

(a) Notations used in the Nelder-Mead method

(b) Process of the Nelder-Mead method

Figure 9.3 Illustration of the Nelder–Mead method

```
function [xo,fo]=opt_Nelder(f,x0,TolX,TolFun,MaxIter)
if numel(x0)==1 % for 1-dimensional case
  [xo,fo]=opt_quad(f,x0,TolX,TolFun); return
end
N=size(x0,2); S=eye(N);
for i=1:N % repeat the procedure for each directional subplane
   Nr=size(x0,1); i1=mod(i,N)+1;
   if Nr==3, abc=x0;  else abc=[x0; x0+S(i,:); x0+S(i1,:)];  end
   abc=abc(:,rotate_l([1:N],i-1));
   fabc=[feval(f,abc(1,:)); feval(f,abc(2,:)); feval(f,abc(3,:))];
   [x0,fo]=Nelder0(f,abc,fabc,TolX,TolFun,MaxIter);
   x0=x0(:,rotate_r([1:N],i-1));
   if N<3, break; end % No repetition needed for a 2D case
end
xo=x0;
%-------------------------------------------------
function [xo,fo]=Nelder0(f,abc,fabc,TolX,TolFun,k)
%Copyleft: Won Y. Yang, wyyang53@hanmail.net, CAU for academic use only
[fabc,I]=sort(fabc); a=abc(I(1),:); b=abc(I(2),:); c=abc(I(3),:);
fa=fabc(1); fb=fabc(2); fc=fabc(3);  fba=fb-fa; fcb=fc-fb;
if k<=0|abs(fba)+abs(fcb)<TolFun|abs(b-a)+abs(c-b)<TolX
  xo=a; fo=fa;
  if k==0, fprintf('Just best in given # of iterations'), end
 else
  m=(a+b)/2;  e=3*m-2*c;  fe=feval(f,e); % Step 3(i)
  if fe<fb, c=e;  fc=fe;
   else
     r=(m+e)/2;  fr=feval(f,r);
     if fr<fc,  c=r;  fc=fr;  end
     if fr>=fb
       s=(c+m)/2;  fs=feval(f,s);
       if fs<fc,  c=s;  fc=fs;          % Step 3(iii)
        else b=m; c=(a+c)/2; fb=feval(f,b); fc=feval(f,c); % Step 3(ii)
       end
     end
  end
  [xo,fo]= Nelder0(f,[a;b;c],[fa fb fc],TolX,TolFun,k-1);
end
```

The MATLAB script "test_opt_Nelder.m" has been composed to minimize a two-variable objective function

$$f(x_1,x_2)=x_1^2-x_1x_2-4x_1+x_2^2-x_2 \tag{9.1.7}$$

whose minimum can be found in an analytical way, i.e., by setting the partial derivatives of $f(x_1,x_2)$ w.r.t. x_1 and x_2 to zero as

$$\begin{matrix}\frac{\partial}{\partial x_1}f(x_1,x_2)=2x_1-x_2-4=0\\ \frac{\partial}{\partial x_2}f(x_1,x_2)=2x_2-x_1-1=0\end{matrix}\;;\;\begin{bmatrix}2&-1\\-1&2\end{bmatrix}\begin{bmatrix}x_1\\x_2\end{bmatrix}=\begin{bmatrix}4\\1\end{bmatrix}\;;\;\mathbf{x}^{\mathrm{o}}=\begin{bmatrix}x_1\\x_2\end{bmatrix}=\frac{1}{3}\begin{bmatrix}2&1\\1&2\end{bmatrix}\begin{bmatrix}4\\1\end{bmatrix}=\begin{bmatrix}3\\2\end{bmatrix} \tag{9.1.8}$$

```
%test_opt_Nelder.m
clear, clf
f=@(x)x(1)*(x(1)-4-x(2))+x(2)*(x(2)-1); % Objective function
x0=[0 0], TolX=1e-4; TolFun=1e-9; MaxIter=100;
[xon,fon]=opt_Nelder(f,x0,TolX,TolFun,MaxIter)
% Minimum point and its function value
[xos,fos]=fminsearch(f,x0) % use the MATLAB built-in function
```

This MATLAB script also uses the MATLAB function '`fminsearch()`' to minimize the same objective function for crosscheck. The minimization process is illustrated in Fig. 9.3(b).

9.1.4 Steepest Descent Method

```
function [xo,fo]=opt_steep(f,x0, TolX,TolFun,alpha0,MaxIter)
% minimize the ftn f by the steepest descent method.
% Input:  f = Objective ftn to be given as a string 'f'
%         x0= Initial guess of the solution
% Output: xo= Minimum point reached
%         fo= f(x(o))
%Copyleft: Won Y. Yang, wyyang53@hanmail.net, CAU for academic use only
if nargin<6, MaxIter=100; end % Maximum # of iteration
if nargin<5, alpha0=10; end % Initial step size
if nargin<4, TolFun=1e-8; end %|f(x)|<TolFun wanted
if nargin<3, TolX=1e-6; end %|x(k)-x(k-1)|<TolX wanted
x=x0;  fx0= feval(f,x0); fx=fx0;
alpha=alpha0;  kmax1=25;
warning=0; % Number of vain wanderings to find the optimum step size
for k=1:MaxIter
   g = grad(f,x); g = g/norm(g); % Gradient as a row vector
   alpha = alpha*2; % for trial move in negative gradient direction
   fx1 = feval(f,x-alpha*2*g);
   for k1=1:kmax1 % To find the optimum step size(alpha) by line search
      fx2 = fx1; fx1 = feval(f,x-alpha*g);
      if fx0>fx1+TolFun & fx1<fx2-TolFun %fx0>fx1<fx2
         den = 4*fx1-2*fx0-2*fx2;
         num = den-fx0+fx2; % Eq.(9.1.5)
         alpha = alpha*num/den; % Eq.(9.1.10)
         x = x-alpha*g;
         fx = feval(f,x); % Eq.(9.1.11)
         break;
       else  alpha= alpha/2;
      end
   end
   if k1>=kmax1, warning=warning+1; % failed to find optimum stepsize
     else  warning= 0;
   end
   if warning>=2|(norm(x-x0)<TolX&abs(fx-fx0)<TolFun), break;  end
   x0=x;  fx0=fx;
end
xo=x;  fo=fx;
if k==MaxIter, fprintf('Just best in %d iterations',MaxIter); end
```

This method finds the minimum of an N-dimensional objective function in the direction of a negative gradient

$$-\mathbf{g}(\mathbf{x}) = -\nabla f(\mathbf{x}) = -\left[\frac{\partial f(\mathbf{x})}{\partial x_1} \ \frac{\partial f(\mathbf{x})}{\partial x_2} \ \cdots \ \frac{\partial f(\mathbf{x})}{\partial x_N}\right]^T \tag{9.1.9}$$

with the step size α_k (at iteration k) adjusted so that the objective function value is minimized along that direction by a (one-dimensional) line search technique like the quadratic approximation method. The algorithm of the steepest descent method is summarized in the following box and has been cast into the above MATLAB function, '`opt_steep()`'.

<Steepest Descent Algorithm>

(Step 0) With the iteration number $k=0$, find the function value $f_0=f(\mathbf{x}_0)$ for the initial point $\mathbf{x}_0$.

(Step 1) Increment the iteration number k by one and find the step size α_{k-1} along the direction of the negative gradient $-\mathbf{g}_{k-1}$ by a (one-dimensional) line search like the quadratic approximation method.

$$\alpha_{k-1} = \underset{\alpha}{\text{ArgMin}}\, f(\mathbf{x}_{k-1} - \alpha \mathbf{g}_{k-1}/\|\mathbf{g}_{k-1}\|) \tag{9.1.10}$$

(Step 2) Move the approximate minimum by the step size α_{k-1} along the direction of the negative gradient $-\mathbf{g}_{k-1}$ to get the next point:

$$\mathbf{x}_k = \mathbf{x}_{k-1} - \alpha_{k-1}\mathbf{g}_{k-1}/\|\mathbf{g}_{k-1}\| \tag{9.1.11}$$

(Step 3) If $\mathbf{x}_k \approx \mathbf{x}_{k-1}$ and $f(\mathbf{x}_k) \approx f(\mathbf{x}_{k-1})$, then declare $\mathbf{x}_k$ as the minimum and terminate the procedure. Otherwise, go back to Step 1.

Fig. 9.4 shows that the steepest descent method proceeds in the negative gradient direction (perpendicular to the contours) of the objective function to find the minimum point in the line and repeats such an iteration till the minimum point is reached. The MATLAB script "test_opt_steep.m" has been composed to minimize the objective function (9.1.7) by using the steepest descent method.

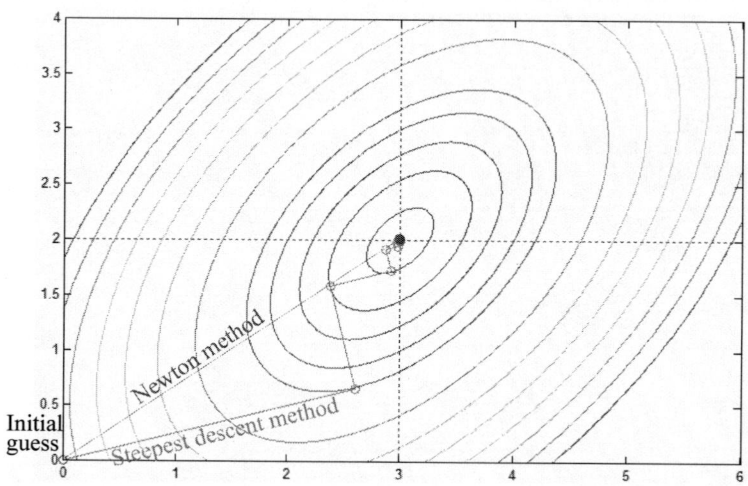

Figure 9.4 Illustration of the Newton's method and steepest descent method

```
%test_opt_steep.m
f=@(x)x(1)*(x(1)-4-x(2))+x(2)*(x(2)-1); % Objective function
x0=[0 0]; TolX=1e-4; TolFun=1e-9; alpha0=1; MaxIter=100;
[xo,fo]=opt_steep(f,x0,TolX,TolFun,alpha0,MaxIter)
```

9.1.5 Newton's Method

Like the steepest descent method, Newton's method also uses the gradient to find the minimum point of an objective function. Such gradient-based optimization methods are supposed to reach a point at which the gradient is (close to) zero. In this context, the optimization of an objective function $f(\mathbf{x})$ is equivalent to finding a zero of its gradient $\mathbf{g}(\mathbf{x})$, which in general is a vector-valued function of a vector-valued independent variable $\mathbf{x}$. Therefore, if we have the gradient function $\mathbf{g}(\mathbf{x})$ of the objective function $f(\mathbf{x})$, we can solve the system of nonlinear equations $\mathbf{g}(\mathbf{x})=\mathbf{0}$ to get the minimum of $f(\mathbf{x})$ by using Newton's method.

The backgrounds of Newton's method as well as the steepest descent method can be shown by taking the Taylor series of, say, a two-variable objective function $f(x_1, x_2)$:

$$f(x_1,x_2) \approx f(x_{1k},x_{2k}) + \left[\frac{\partial f}{\partial x_1} \;\; \frac{\partial f}{\partial x_2}\right]\Bigg|_{\substack{x_1=x_{1k}\\x_2=x_{2k}}} \begin{bmatrix} x_1-x_{1k} \\ x_2-x_{2k}\end{bmatrix}$$

$$+\frac{1}{2}[x_1-x_{1k} \;\; x_2-x_{2k}]^{\mathrm{T}} \begin{bmatrix} \partial^2 f/\partial x_1^2 & \partial^2 f/\partial x_1\partial x_2 \\ \partial^2 f/\partial x_2\partial x_1 & \partial^2 f/\partial x_2^2 \end{bmatrix}\Bigg|_{\substack{x_1=x_{1k}\\x_2=x_{2k}}} \begin{bmatrix} x_1-x_{1k} \\ x_2-x_{2k}\end{bmatrix}$$

$$;\; f(\mathbf{x}) \approx f(\mathbf{x}_k) + \nabla f(\mathbf{x})^{\mathrm{T}}\Big|_{\mathbf{x}_k}[\mathbf{x}-\mathbf{x}_k] + \frac{1}{2}[\mathbf{x}-\mathbf{x}_k]^{\mathrm{T}} J(\nabla f(\mathbf{x}))\Big|_{\mathbf{x}_k}[\mathbf{x}-\mathbf{x}_k]$$

$$; \qquad f(\mathbf{x}) \cong f(\mathbf{x}_k) + \mathbf{g}_k^T[\mathbf{x}-\mathbf{x}_k] + \frac{1}{2}[\mathbf{x}-\mathbf{x}_k]^T H_k[\mathbf{x}-\mathbf{x}_k] \tag{9.1.12}$$

where $\mathbf{g}_k$ and H_k are the gradient vector and Hessian matrix of $f(\mathbf{x})$ at $\mathbf{x}=\mathbf{x}_k$, each defined as

$$\mathbf{g}_k = \nabla f(\mathbf{x})\big|_{\mathbf{x}=\mathbf{x}_k} = \begin{bmatrix} \partial f/\partial x_1 \\ \vdots \\ \partial f/\partial x_N \end{bmatrix}\Bigg|_{x_i=x_{ik}} \tag{9.1.13}$$

$$H_k = J(\mathbf{g})\big|_{\mathbf{x}_k} \text{(Jacobian of the gradient)} = \begin{bmatrix} \frac{\partial g_1}{\partial x_1} & \cdots & \frac{\partial g_1}{\partial x_N} \\ \vdots & \ddots & \vdots \\ \frac{\partial g_N}{\partial x_1} & \cdots & \frac{\partial g_N}{\partial x_N} \end{bmatrix} = \begin{bmatrix} \frac{\partial^2 f}{\partial x_1^2} & \cdots & \frac{\partial^2 f}{\partial x_1\partial x_N} \\ \vdots & \ddots & \vdots \\ \frac{\partial^2 f}{\partial x_N\partial x_1} & \cdots & \frac{\partial^2 f}{\partial x_N^2} \end{bmatrix} \tag{9.1.14}$$

In the light of this equation, we can see that the value of the objective function at point $\mathbf{x}_{k+1}$ updated by the steepest descent algorithm (9.1.11):

$$\mathbf{x}_{k+1} \overset{(9.1.11)}{=} \mathbf{x}_k - \alpha_k\, \mathbf{g}_k/\|\mathbf{g}_k\|$$

is most likely smaller than that at the old point $\mathbf{x}_k$, with the third term in Eq. (9.1.12) neglected:

$$f(\mathbf{x}_{k+1}) \overset{(9.1.12)}{\approx} f(\mathbf{x}_k) + \mathbf{g}_k^T[\mathbf{x}_{k+1} - \mathbf{x}_k] \overset{(9.1.11)}{=} f(\mathbf{x}_k) - \alpha \mathbf{g}_k^T \mathbf{g}_k / \| \mathbf{g}_k \|$$

$$; \qquad f(\mathbf{x}_{k+1}) - f(\mathbf{x}_k) \approx -\alpha \mathbf{g}_k^T \mathbf{g}_k / \| \mathbf{g}_k \| \le 0$$

$$; \qquad f(\mathbf{x}_{k+1}) \le f(\mathbf{x}_k) \tag{9.1.15}$$

```
function J=jacob(g,x,h,varargin)
% Jacobian of g(x)
% Copyleft: Won Y. Yang, wyyang53@hanmail.net, CAU for academic use only
if nargin<3,  h=.0001; end
N= length(x); h2=2*h; %h12=12*h;
x=x(:).'; Ih=i*h*eye(N); %Ih2=Ih*2;
for n=1:N
  g1=feval(g,x+Ih(n,:),varargin{:});
  g2=feval(g,x-Ih(n,:),varargin{:});
  g12=imag(g1-g2)/h2;
  J(:,n)=g12(:); % Central difference approximation of Eq.(9.1.14)
end
if (size(g,1)==size(g,2))&sum(sum(isnan(g)))==0&rank(g)<N
  fprintf('At x=%12.6e, Jacobian singular with J=',x);  disp(g);
end
```

```
function [x,gx,xx]=Newtons(g,x0,TolX,MaxIter,varargin)
% to solve a set of nonlinear eqs  g1(x)=0, g2(x)=0,..
%Input:  g       = 1st-order vector ftn equivalent to a set of eqns
%        x0      = Initial guess of the solution
%        TolX    = Upper limit of |x(k)-x(k-1)|
%        MaxIter = Maximum # of iteration
%output: x       = Point which the algorithm has reached
%        gx      = g(x(last))
%        xx      = History of x
% Copyleft: Won Y. Yang, wyyang53@hanmail.net, CAU for academic use only
h=1e-4; TolFun=1e-8; EPS=1e-6;
gx=feval(g,x0,varargin{:});
Ng=length(gx); Nx=length(x0);
if Ng~=Nx, error('Incompatible dimensions of f and x0!'); end
if nargin<4, MaxIter=100; end
if nargin<3, TolX=EPS; end
xx(1,:)=x0(:).';  gx0= norm(gx);
for k=1:MaxIter
  J=jacob(g,xx(k,:),h,varargin{:}); % Jacobian/Hessian
  if rank(J)<Nx
    k=k-1;
    fprintf('Jacobian singular! with det(J)=%12.6e\n',det(J));
    break;
  else
     dx= -J\gx(:); %-[dfdx]^-1*gx; Eq.(9.1.17)
  end
  xx(k+1,:)= xx(k,:)+dx.'; % Eq.(9.1.17)
  fx=feval(g,xx(k+1,:),varargin{:}); gxn=norm(gx);
  if fxn<TolFun|norm(dx)<TolX, break; end
end
x= xx(k+1,:);
```

Slightly different from this strategy of the steepest descent algorithm, Newton's method tries to go straight to the zero of the gradient of the approximate objective function (9.1.12):

$$\frac{df(\mathbf{x})}{d\mathbf{x}} \overset{(9.1.12)}{\underset{(B.2),(B.6)}{=}} \mathbf{g}_k + H_k[\mathbf{x}-\mathbf{x}_k] = \mathbf{0}\,; \quad \mathbf{x} = \mathbf{x}_k - H_k^{-1}\mathbf{g}_k \tag{9.1.16}$$

by the updating rule

$$\mathbf{x}_{k+1} \overset{(9.1.16)}{=} \mathbf{x}_k - H_k^{-1}\mathbf{g}_k \tag{9.1.17}$$

In fact, Eq. (9.1.16) is essentially Newton's method[Y-1](Section 4.6), trying to find a zero of the gradient vector function $\mathbf{g}(\mathbf{x})=\nabla f(\mathbf{x})$ (of the objective function $f(\mathbf{x})$) based on its Jacobian, which corresponds to the Hessian of $f(\mathbf{x})$ (see Appendix B). Therefore, it can be implemented by using any nonlinear (vector) equation solver like the MATLAB function '`fsolve()`' or '`Newtons()`' (listed above). Thus, we will just define the gradient function $\mathbf{g}(\mathbf{x})$ and let '`fsolve()`' solve $\mathbf{g}(\mathbf{x})=\mathbf{0}$ to find an extremum point of $f(\mathbf{x})$.

Now, to see the performance of Newton's method, we compose the MATLAB script "test_opt_Newton.m," which actually solves $\mathbf{g}(\mathbf{x})=\mathbf{0}$ for the gradient function

$$\mathbf{g}(\mathbf{x}) = \nabla f(\mathbf{x}) = \begin{bmatrix} \partial f/\partial x_1 \\ \partial f/\partial x_2 \end{bmatrix} = \begin{bmatrix} 2x_1 - x_2 - 4 \\ 2x_2 - x_1 - 1 \end{bmatrix} \tag{9.1.18}$$

of the objective function (9.1.7):

$$f(\mathbf{x}) = f(x_1, x_2) = x_1^2 - x_1x_2 - 4x_1 + x_2^2 - x_2$$

Running the script yields a part of Fig. 9.4, which illustrates that Newton's method proceeds directly toward the minimum point. Note that Newton's method usually reaches the minimum in fewer iterations than the steepest descent method, and especially in the case of a quadratic objective function $f(\mathbf{x})$ like Eq. (9.1.7), just a single iteration is enough.

```
%test_opt_Newton.m
% minimizes an objective ftn f(x) by the Newton's method.
f=@(x)x(1).^2-4*x(1)-x(1).*x(2)+x(2).^2-x(2); % Eq.(9.1.7)
g=@(x)[2*x(1)-x(2)-4  2*x(2)-x(1)-1];          % Eq.(9.1.18)
x0=[0 0]; % Initial guess
xo=fsolve(g,x0) % An extremum point reached
f(xo) % the corresponding value of the objective function
% Alternatively, Newtons() can be used.
TolX=1e-4; MaxIter=50;
xo_Newtons=Newtons(g,x0,TolX,MaxIter)
```

```
>>test_opt_Newton
  xo = [3.0000  2.0000], ans= -7
```

[Remark 9.1] A Weak Point of Newton's Method

Newton's method is usually more efficient than the steepest descent method if it works as illustrated above. However, regrettably, it is not guaranteed to reach a (local) minimum point. Depending on the initial guess and the objective function, it may end up with one of the extrema having zero gradient, which is not necessarily a minimum, but possibly a saddle point (i.e., a local maximum w.r.t. some variable(s) and a local minimum w.r.t. other variable(s)) or even a maximum, while the steepest descent method reaches a minimum most of the time, but usually in more iterations.

9.2 CONSTRAINED OPTIMIZATION[Y-1], Section 7.2

In this section, only the concept of constrained optimization is introduced. The explanation for the usage of the corresponding MATLAB functions is postponed until the next section.

9.2.1 Lagrange Multiplier Method

A class of common optimization problems subject to equality constraints may be nicely handled by the Lagrange multiplier method. Consider an optimization problem with M equality constraints:

$$\text{Min } f(\mathbf{x}) \tag{9.2.1a}$$

$$\text{s.t. (subject to) } \mathbf{h}(\mathbf{x}) = \begin{bmatrix} h_1(\mathbf{x}) \\ h_2(\mathbf{x}) \\ \vdots \\ h_M(\mathbf{x}) \end{bmatrix} = \mathbf{0} \tag{9.2.1b}$$

According to the Lagrange multiplier method, this problem can be converted to the following unconstrained optimization problem:

$$\text{Min } l(\mathbf{x}, \boldsymbol{\lambda}) = f(\mathbf{x}) + \boldsymbol{\lambda}^{\mathrm{T}} \mathbf{h}(\mathbf{x}) = f(\mathbf{x}) + \sum_{m=1}^{M} \lambda_m h_m(\mathbf{x}) \text{ with } \boldsymbol{\lambda}^{\mathrm{T}} = [\lambda_1 \; \lambda_2 \; \dots \; \lambda_M] \tag{9.2.2}$$

The solution of this problem, if it exists, can be obtained by setting the derivatives of this new objective function $l(\mathbf{x}, \boldsymbol{\lambda})$ w.r.t. $\mathbf{x}$ and $\boldsymbol{\lambda}$ to zero:

$$\frac{\partial}{\partial \mathbf{x}} l(\mathbf{x}, \boldsymbol{\lambda}) \overset{(9.2.2)}{=} \frac{\partial}{\partial \mathbf{x}} f(\mathbf{x}) + \boldsymbol{\lambda}^{\mathrm{T}} \frac{\partial}{\partial \mathbf{x}} \mathbf{h}(\mathbf{x}) = \nabla f(\mathbf{x}) + \sum_{m=1}^{M} \lambda_m \nabla h_m(\mathbf{x}) = \mathbf{0} \tag{9.2.3a}$$

$$\frac{\partial}{\partial \boldsymbol{\lambda}} l(\mathbf{x}, \boldsymbol{\lambda}) \overset{(9.2.2)}{=} \mathbf{h}(\mathbf{x}) = \mathbf{0} \tag{9.2.3b}$$

Note that the solution for this system of equations is the extremum of the objective function. We may determine if it is a minimum or maximum, from the positive/negative-definiteness of the second derivative (Hessian matrix) of $l(\mathbf{x}, \boldsymbol{\lambda})$ w.r.t. $\mathbf{x}$ at that solution point. Let us see the following examples.

[Remark 9.2] Inequality Constraints with the Lagrange Multiplier Method

Even if the optimization problem may involve inequality constraints like $g_n(\mathbf{x}) \le 0$, we can convert them to equality constraints by introducing the (nonnegative) slack variables y_j^2 as

$$g_n(\mathbf{x}) + y_n^2 = 0 \tag{9.2.4}$$

Then the Lagrange multiplier method can be used to handle the optimization problem like an equality-constrained problem with the additional slack variables.

(Example 9.1) Minimization Using the Lagrange Multiplier Method

Consider the following minimization problem subject to a single equality constraint:

$$\text{Min } f(\mathbf{x}) = x_1^2 + x_2^2 \tag{E9.1.1a}$$

$$\text{s.t. } h(\mathbf{x}) = x_1 + x_2 - 2 = 0 \tag{E9.1.1b}$$

We can substitute the equality constraint $x_2 = 2 - x_1$ into the objective function (E9.1.1a) so that this problem becomes an unconstrained optimization problem, as

$$\text{Min } f(x_1) = x_1^2 + (2 - x_1)^2 = 2x_1^2 - 4x_1 + 4 \tag{E9.1.2}$$

which can be easily solved by setting the derivative of this new objective function w.r.t. x_1 to zero as

$$\frac{\partial}{\partial x_1} f(x_1) = 4x_1 - 4 = 0\,;\;\; x_1 = 1 \;\text{ and }\; x_2 \overset{\text{(E9.1.1b)}}{=} 2 - x_1 = 1 \tag{E9.1.3}$$

Alternatively, we can apply the Lagrange multiplier method:

$$\text{Min } l(\mathbf{x}, \lambda) \overset{(9.2.2)}{=} f(\mathbf{x}) + \lambda h(\mathbf{x}) \overset{\text{(E9.1.1)}}{=} x_1^2 + x_2^2 + \lambda(x_1 + x_2 - 2) \tag{E9.1.4}$$

$$\frac{\partial}{\partial x_1} l(\mathbf{x}, \boldsymbol{\lambda}) \underset{\text{(E9.1.4)}}{\overset{(9.2.3a)}{=}} 2x_1 + \lambda = 0;\;\; x_1 = -\frac{\lambda}{2} \tag{E9.1.5a}$$

$$\frac{\partial}{\partial x_2} l(\mathbf{x}, \boldsymbol{\lambda}) \underset{\text{(E9.1.4)}}{\overset{(9.2.3a)}{=}} 2x_2 + \lambda = 0;\;\; x_2 = -\frac{\lambda}{2} \tag{E9.1.5b}$$

$$\frac{\partial}{\partial \lambda} l(\mathbf{x}, \boldsymbol{\lambda}) \underset{\text{(E9.1.4)}}{\overset{(9.2.3b)}{=}} x_1 + x_2 - 2 = 0 \overset{\text{(E9.1.5a,b)}}{\rightarrow} -\frac{\lambda}{2} - \frac{\lambda}{2} - 2 = 0\,;\; \lambda = -2 \tag{E9.1.5c}$$

$$;\, x_1 \overset{\text{(E9.1.5a)}}{=} -\frac{\lambda}{2} \overset{\text{(E9.1.5c)}}{=} 1 \;\text{ and }\; x_2 \overset{\text{(E9.1.5b)}}{=} -\frac{\lambda}{2} \overset{\text{(E9.1.5c)}}{=} 1 \text{ (See Fig. 9.5)} \tag{E9.1.6}$$

In this example, the substitution of (linear) equality constraints is more convenient than the Lagrange multiplier method. However, it is not always the case, as illustrated by the next example.

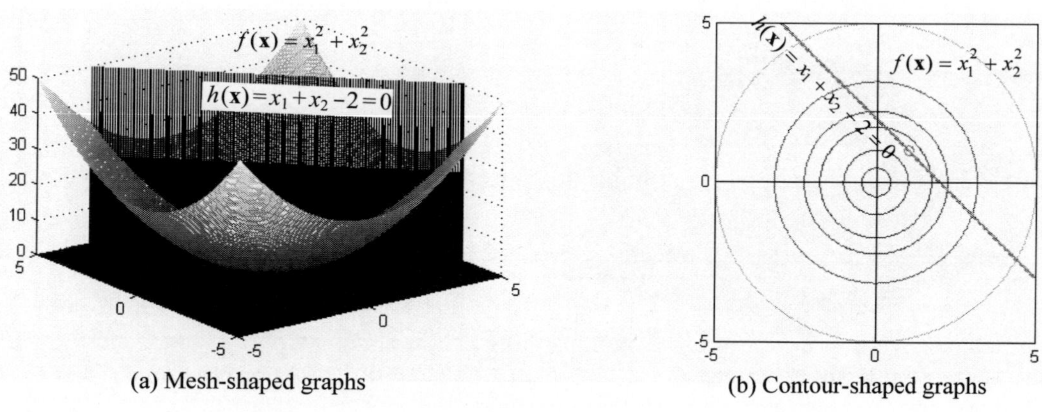

Figure 9.5 The objective and constraint functions for Example 9.1

(**Example 9.2**) Optimization Using the Lagrange Multiplier Method

Consider the following optimization problem subject to a single nonlinear equality constraint:

$$\text{Min } f(\mathbf{x}) = x_1 + x_2 \tag{E9.2.1a}$$

$$\text{s.t. } h(\mathbf{x}) = x_1^2 + x_2^2 - 2 = 0 \tag{E9.2.1b}$$

Noting that it is absurd to substitute the equality constraint (E9.2.1b) into the objective function (E9.2.1a), we apply the Lagrange multiplier method:

$$\text{Min } l(\mathbf{x}, \lambda) \overset{(9.2.2)}{=} f(\mathbf{x}) + \lambda h(\mathbf{x}) \overset{(\text{E9.2.1})}{=} x_1 + x_2 + \lambda(x_1^2 + x_2^2 - 2) \tag{E9.2.2}$$

$$\frac{\partial}{\partial x_1} l(\mathbf{x}, \boldsymbol{\lambda}) \underset{(\text{E9.2.2})}{\overset{(9.2.3\text{a})}{=}} 1 + 2\lambda x_1 = 0; \ x_1 = -\frac{1}{2\lambda} \tag{E9.2.3a}$$

$$\frac{\partial}{\partial x_2} l(\mathbf{x}, \boldsymbol{\lambda}) \underset{(\text{E9.2.2})}{\overset{(9.2.3\text{a})}{=}} 1 + 2\lambda x_2 = 0; \ x_2 = -\frac{1}{2\lambda} \tag{E9.2.3b}$$

$$\frac{\partial}{\partial \lambda} l(\mathbf{x}, \boldsymbol{\lambda}) \underset{(\text{E9.2.2})}{\overset{(9.2.3\text{b})}{=}} x_1^2 + x_2^2 - 2 = 0 \overset{(\text{E9.2.3a,b})}{\rightarrow} \left(-\frac{1}{2\lambda}\right)^2 + \left(-\frac{1}{2\lambda}\right)^2 = 2; \ \lambda = \pm\frac{1}{2} \tag{E9.2.3c}$$

$$; x_1 \overset{(\text{E9.2.3a})}{=} -\frac{1}{2\lambda} \overset{(\text{E9.2.3c})}{=} \mp 1 \ \text{ and } \ x_2 \overset{(\text{E9.2.3b})}{=} -\frac{1}{2\lambda} \overset{(\text{E9.2.3c})}{=} \mp 1 \ \text{(See Fig. 9.6)} \tag{E9.2.4}$$

Now, to tell whether each of these is a (local) minimum or maximum, we should determine the positive/negative-definiteness of the second derivative (Hessian matrix) of $l(\mathbf{x}, \lambda)$ w.r.t. $\mathbf{x}$:

$$H = \frac{\partial^2}{\partial \mathbf{x}^2} l(\mathbf{x}, \lambda) = \begin{bmatrix} \partial^2 l/\partial x_1^2 & \partial^2 l/\partial x_1 \partial x_2 \\ \partial^2 l/\partial x_2 \partial x_1 & \partial^2 l/\partial x_2^2 \end{bmatrix} = \begin{bmatrix} 2\lambda & 0 \\ 0 & 2\lambda \end{bmatrix} \tag{E9.2.5}$$

This matrix is positive/negative-definite if the sign of λ is positive/negative. Therefore, the solution $(x_1, x_2)=(-1,-1)$ corresponding to $\lambda=1/2$ is a minimum, while the solution $(x_1, x_2)=(1,1)$ corresponding to $\lambda=-1/2$ is a maximum (see Fig. 9.6).

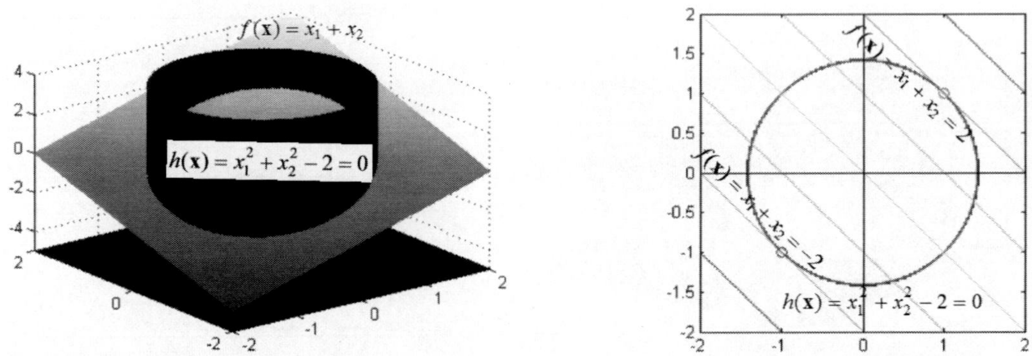

(a) Mesh-shaped graphs (b) Contour-shaped graphs

Figure 9.6 The objective and constraint functions for Example 9.2

(Example 9.3) Optimization Using the Lagrange Multiplier Method

Consider the following optimization problem subject to a single nonlinear equality constraint:

$$\text{Min or Max } f(\mathbf{x}) = \mathbf{x}^T R\mathbf{x} = [x_1 \;\; x_2]\begin{bmatrix} 5 & 3 \\ 3 & 5 \end{bmatrix}\begin{bmatrix} x_1 \\ x_2 \end{bmatrix} \tag{E9.3.1a}$$

$$\text{s.t. } h(\mathbf{x}) = 1 - x_1^2 + x_2^2 - 1 = 1 - \mathbf{x}^T\mathbf{x} = 0 \tag{E9.3.1b}$$

We apply the Lagrange multiplier method:

$$\text{Min or Max } l(\mathbf{x},\lambda) \overset{(9.2.2)}{=} f(\mathbf{x}) + \lambda h(\mathbf{x}) \overset{(E9.3.1)}{=} \mathbf{x}^T R\mathbf{x} + \lambda(1 - \mathbf{x}^T\mathbf{x}) \tag{E9.3.2}$$

$$\frac{\partial}{\partial \mathbf{x}} l(\mathbf{x},\lambda) \underset{(E9.3.2),(B.6)}{\overset{(9.2.3a)}{=}} 2R\mathbf{x} - 2\lambda\mathbf{x} = 0; \;\; (R - \lambda I)\mathbf{x} = \mathbf{0} \tag{E9.3.3a}$$

$$\frac{\partial}{\partial \lambda} l(\mathbf{x},\lambda) \underset{(E9.3.2)}{\overset{(9.2.3b)}{=}} 1 - \mathbf{x}^T\mathbf{x} = 0 \tag{E9.3.3b}$$

Comparing Eq. (E9.3.3a) with Eq. (1.2.12), we see that the solution $\mathbf{x}$ and λ to Eqs. (9.3.3a,b) are nothing but the unit eigenvector(s) and eigenvalue(s) of the coefficient matrix R, which can be obtained by using the built-in MATLAB function '`eig()`' as

```
>>R=[5 3; 3 5]; [V,L]=eig(R)
    V =  -0.7071     0.7071        L =    2         0
          0.7071     0.7071               0         8
```

$$\mathbf{x}_{o,a} = \begin{bmatrix} -1/\sqrt{2} \\ 1/\sqrt{2} \end{bmatrix}, \; \lambda_a = 2 \text{ and } \mathbf{x}_{o,b} = \begin{bmatrix} 1/\sqrt{2} \\ 1/\sqrt{2} \end{bmatrix}, \; \lambda_b = 8 \tag{E9.3.4a,b}$$

Now, to tell whether each of these is a (local) minimum or maximum, we should determine the positive/negative-definiteness of the second derivative (Hessian matrix) of $l(\mathbf{x},\lambda)$ w.r.t. $\mathbf{x}$:

$$H = \frac{\partial^2}{\partial \mathbf{x}^2} l(\mathbf{x},\lambda) \overset{(E9.3.3a)}{=} 2(R - \lambda I) = 2\begin{bmatrix} 5-\lambda & 3 \\ 3 & 5-\lambda \end{bmatrix} = \begin{bmatrix} 6 & 6 \\ 6 & 6 \end{bmatrix} \text{ or } \begin{bmatrix} -6 & 6 \\ 6 & -6 \end{bmatrix} \text{ at } \lambda = 2 \text{ or } 8 \tag{E9.2.5}$$

This matrix (corresponding to the second derivative) is positive/negative-semidefinite at λ=2/8 and therefore, the solution $\mathbf{x}_a = (-1/\sqrt{2}, 1/\sqrt{2})$ corresponding to λ=2 is a (local) minimum, while the solution $\mathbf{x}_b = (1/\sqrt{2}, 1/\sqrt{2})$ corresponding to λ=8 is a (local) maximum (see Fig. 9.7).

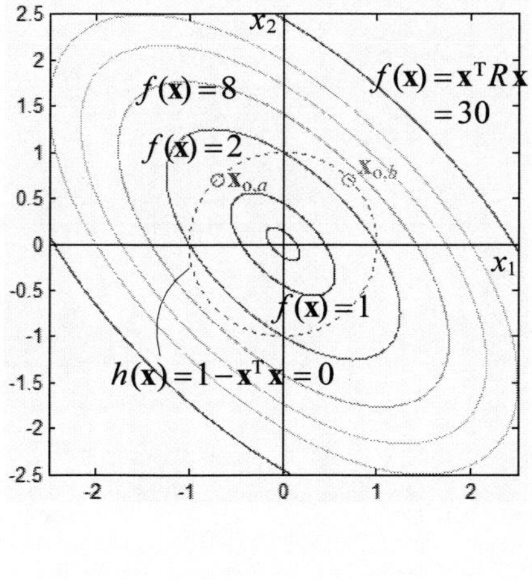

Figure 9.7 Contours of the objective function $f(\mathbf{x}) = \mathbf{x}^T R\mathbf{x}$ and the constraint curve for Example 9.3

(**Example 9.4**) Optimization Using the Lagrange Multiplier Method

Consider the following optimization problem subject to linear equality constraints:

$$\text{Min } f(\mathbf{x}) = \frac{1}{2}\mathbf{x}^T R\,\mathbf{x} \ (R > 0 : \text{positive definite}) \tag{E9.4.1a}$$

$$\text{s.t.} \quad \mathbf{h}(\mathbf{x}) = A\mathbf{x} - \mathbf{b} = \mathbf{0} \tag{E9.4.1b}$$

We apply the Lagrange multiplier method:

$$\text{Min } l(\mathbf{x},\boldsymbol{\lambda}) \overset{(9.2.2)}{=} f(\mathbf{x}) + \boldsymbol{\lambda}^T\mathbf{h}(\mathbf{x}) \overset{(E9.4.1)}{=} \frac{1}{2}\mathbf{x}^T R\,\mathbf{x} + \boldsymbol{\lambda}^T(A\mathbf{x} - \mathbf{b}) \tag{E9.4.2}$$

$$\frac{\partial}{\partial \mathbf{x}} l(\mathbf{x},\boldsymbol{\lambda}) \underset{(E9.4.2),(B.6),(B.2)}{\overset{(9.2.3a)}{\equiv}} R\mathbf{x} + A^T\boldsymbol{\lambda} = \mathbf{0}\,; \quad \mathbf{x} = -R^{-1}A^T\boldsymbol{\lambda} \tag{E9.4.3a}$$

$$\frac{\partial}{\partial \boldsymbol{\lambda}} l(\mathbf{x},\boldsymbol{\lambda}) \underset{(E9.4.2),(B.6)}{\overset{(9.2.3b)}{\equiv}} A\mathbf{x} - \mathbf{b} = 0 \overset{(E9.4.3a)}{\rightarrow} -AR^{-1}A^T\boldsymbol{\lambda} - \mathbf{b} = 0\,; \ \boldsymbol{\lambda} = -[AR^{-1}A^T]^{-1}\mathbf{b} \tag{E9.4.3b}$$

Substituting Eq. (E9.4.3b) into Eq. (E9.4.3a) yields the optimal solution:

$$\mathbf{x}_o \overset{(E9.4.3a)}{=} -R^{-1}A^T\boldsymbol{\lambda} \overset{(E9.4.3b)}{=} R^{-1}A^T[AR^{-1}A^T]^{-1}\mathbf{b} \tag{E9.4.4}$$

Now, to check the validity of this solution, let us use this result to find the solution of the following minimization problem:

$$\text{Min } f(\mathbf{x}) = \frac{1}{2}\mathbf{x}^T \begin{bmatrix} 5 & 3 \\ 3 & 5 \end{bmatrix} \mathbf{x} \text{ with } R = \begin{bmatrix} 5 & 3 \\ 3 & 5 \end{bmatrix} > 0 : \text{positive definite} \tag{E9.4.5a}$$

$$\text{s.t. } h(\mathbf{x}) = [-1 \quad 1] - 1 = 0 \tag{E9.4.5b}$$

Eq. (E9.4.4) yields the solution

$$\mathbf{x}^o \overset{(E9.4.4)}{=} R^{-1}A^T[AR^{-1}A^T]^{-1}\mathbf{b} \overset{(E9.4.5)}{=} \begin{bmatrix} 5 & 3 \\ 3 & 5 \end{bmatrix}^{-1} \begin{bmatrix} -1 \\ 1 \end{bmatrix} \left([-1 \ \ 1] \begin{bmatrix} 5 & 3 \\ 3 & 5 \end{bmatrix}^{-1} \begin{bmatrix} -1 \\ 1 \end{bmatrix} \right)^{-1} 1 = \begin{bmatrix} -0.5 \\ 0.5 \end{bmatrix} \tag{E9.4.6}$$

as depicted in Fig. 9.8. This can also be found by using the general constrained minimization function '`fmincon()`' (Section 9.3.2):

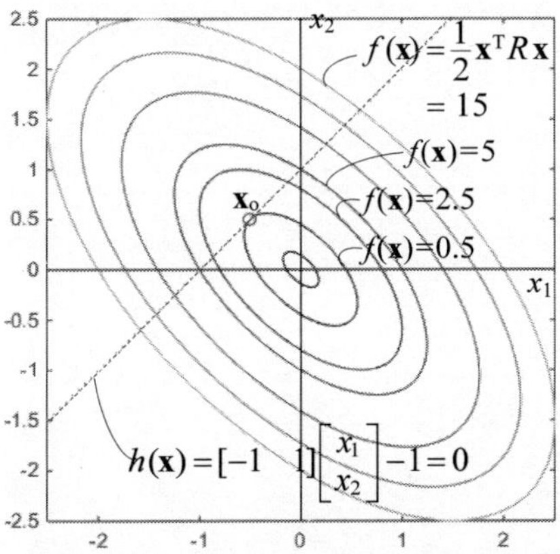


```
R=[5 3; 3 5];
fx=@(x)x'*R*x/2; % Objective function
Aeq=[-1 1]; b=1; Beq=[b];
RI=R^-1; xo=RI*Aeq.'*(Aeq*RI*Aeq.')^-1*Beq
x0 = [0; 0]; % Initial value
A=[]; B=[];
[xo_con,fo_con]=fmincon(fx,x0,A,B,Aeq,Beq)
```

Figure 9.8 Contours of the objective function $f(\mathbf{x}) = \mathbf{x}^T R\,\mathbf{x}$ and the constraint curve for Example 9.4

9.2.2 Penalty Function Method

This method is practically very useful for dealing with general constrained optimization problems involving equality/inequality constraints. It is really attractive for optimization problems with fuzzy or loose constraints that are not so strict with zero tolerance.

Consider the following problem:

$$\text{Min } f(\mathbf{x}) \tag{9.2.5a}$$

$$\text{s.t. } \mathbf{h}(\mathbf{x}) = \begin{bmatrix} h_1(\mathbf{x}) \\ \vdots \\ h_M(\mathbf{x}) \end{bmatrix} = \mathbf{0} \text{ and } \mathbf{g}(\mathbf{x}) = \begin{bmatrix} g_1(\mathbf{x}) \\ \vdots \\ g_L(\mathbf{x}) \end{bmatrix} \le \mathbf{0} \tag{9.2.5b}$$

The penalty function method consists of two steps. The first step is to construct a new objective function

$$\text{Min } l(\mathbf{x}) = f(\mathbf{x}) + \sum_{m=1}^{M} w_m h_m^2(\mathbf{x}) + \sum_{l=1}^{L} v_l \psi_l(g_l(\mathbf{x})) \tag{9.2.6}$$

by including the constraint terms in such a way that violating the constraints would be penalized through the large value of the constraint terms in the objective function (to be minimized), while satisfying the constraints would not affect the objective function. The second step is to minimize the new objective function with no constraints by using a non-gradient-based method that is applicable to unconstrained optimization problems, like the Nelder–Mead method (presented in Section 9.1.3). Why don't we use a gradient-based optimization method? Because the inequality constraint terms $v_l \psi_l(g_l(\mathbf{x}))$ attached to the objective function are often zero as long as $\mathbf{x}$ stays inside the (permissible) region satisfying the corresponding constraint ($g_l(\mathbf{x}) \le 0$), and increase very steeply (like $\psi_l(g_l(\mathbf{x})) = \exp(e_l g_l(\mathbf{x}))$ as $\mathbf{x}$ goes out of the region; consequently, the gradient of the new objective function may not carry useful information about the direction along which the value of the objective function decreases.

From an application point of view, it might be a good feature of this method that we can make the weighting coefficient (w_m, v_m, and e_m) on each penalizing constraint term large/small depending on how strictly it should be satisfied. Let us see the following example.

(**Example 9.5**) Minimization by the Penalty Function Method

Consider the following minimization problem subject to several nonlinear inequality constraints:

$$\text{Min } f(\mathbf{x}) = \{(x_1+1.5)^2+5(x_2-1.7)^2\}\{(x_1-1.4)^2+0.6(x_2-0.5)^2\} \tag{E9.5.1a}$$

$$\text{s.t. } g(\mathbf{x}) = \begin{bmatrix} -x_1 \\ -x_2 \\ 3x_1-x_1x_2+4x_2-7 \\ 2x_1+x_2-3 \\ 3x_1-4x_2^2-4x_2 \end{bmatrix} \le \begin{bmatrix} 0 \\ 0 \\ 0 \\ 0 \\ 0 \end{bmatrix} \tag{E9.5.1b}$$

According to the penalty function method, we construct a new objective function (9.2.6) as

$$\text{Min } l(\mathbf{x}) = \{(x_1+1.5)^2+5(x_2-1.7)^2\}\{(x_1-1.4)^2+0.6(x_2-0.5)^2\} + \sum_{l=1}^{5} v_l \psi_l(g_l(\mathbf{x})) \tag{E9.5.2a}$$

where

$$v_l = 1,\ e_l = 1\ \ \forall\, l = 1, \cdots, 5 \tag{E9.5.2b}$$

$$\psi\left(g_l(\mathbf{x})\right) = \begin{cases} 0 & \text{if } g_l(\mathbf{x}) \le 0 \text{ (constraint satisfied)} \\ \exp(e_l g_l(\mathbf{x})) & \text{if } g_l(\mathbf{x}) > 0 \text{ (constraint violated)} \end{cases} \tag{E9.5.2c}$$

```
%em09e05.m for Example 9.5
% to solve a constrained optimization problem by penalty ftn method.
f='em09e05_f'; % (Penalized) Objective function
x0=[0.4 0.5]; % Initial guess
TolX=1e-4; TolFun=1e-9; alpha0=1;
[xo_N,fo_N]=opt_Nelder(f,x0) % Nelder method
[fc_N,fo_N,co_N]=em09e05_f(xo_N) % its results with constraint margin
[xo_s,fo_s]=fminsearch(f,x0) % MATLAB built-in fminsearch()
[fc_s,fo_s,co_s]=em09e05_f(xo_s) % its results with constraint margin
[xo_u,fo_u]=fminunc(f,x0) % MATLAB built-in fminunc()
[fc_u,fo_u,co_u]=em09e05_f(xo_u) % its results with constraint margin
plot(xo_N(1),xo_N(2),'ro', xo_s(1),xo_s(2),'rs'), hold on
plot(xo_u(1),xo_u(2),'rx')
```

```
function [fc,f,c]=em09e05_f(x)
% The objective function and constraint vector
f=((x(1)+1.5)^2+5*(x(2)-1.7)^2)*((x(1)-1.4)^2+.6*(x(2)-.5)^2); (E9.5.1a)
c=[-x(1); -x(2); 3*x(1)-x(1)*x(2)+4*x(2)-7;
   2*x(1)+x(2)-3; 3*x(1)-4*x(2)^2-4*x(2)]; % Constraint vector (E9.5.1b)
v=[1 1 1 1 1]; e=[1 1 1 1 1].'; % Weighting coefficient vector
fc= f + v*((c>0).*exp(e.*c)); % New objective function (E9.5.2a)
```

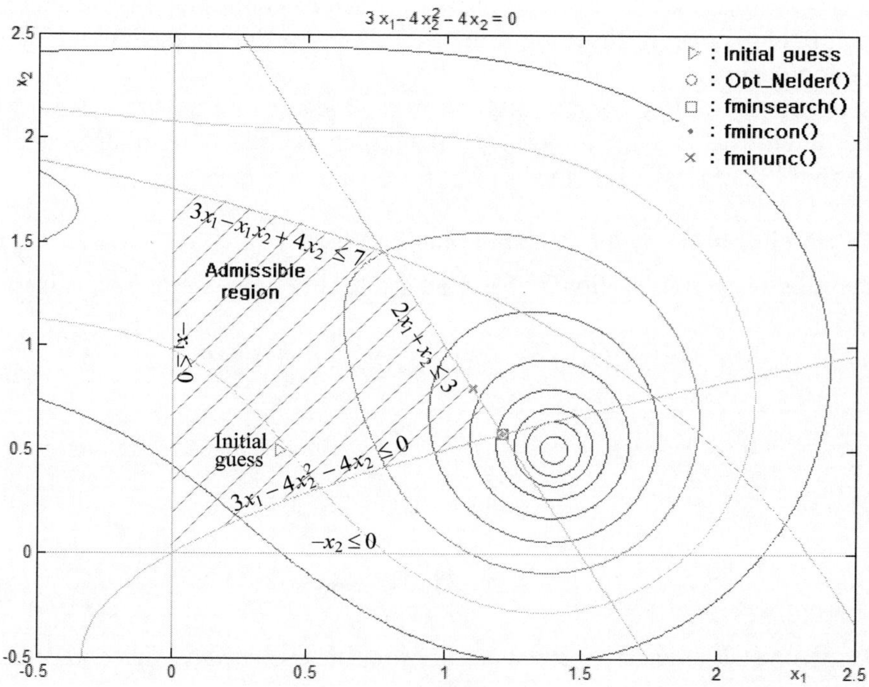

Figure 9.9 Minimum points in the admissible region (satisfying the constraints) for Example 9.5

Note that the shape of the penalty function as well as the values of the weighting coefficients is set by the users to cope with their own problems. Then, we apply an unconstrained optimization technique like the Nelder–Mead method, which is not a gradient-based approach. Here, we make an M-file containing the MATLAB function '`em09e5_f()`' to find the values of the objective function (E9.5.1a), the constraint function (E9.5.1b), and the new objective function (E9.5.2a) with penalized constrained terms. Also, we compose and run the MATLAB script "em09e05.m," which uses the function '`opt_Nelder()`' and also the MATLAB built-in functions '`fminsearch()`' and '`fminunc()`' to minimize the penalized objective function (E9.5.2a).

```
>>em09e05
xo_N =  1.2118   0.5765 %Min point
fo_N =  0.5322 % Min value
co_N = -1.2118 % Constraint margin
       -0.5765
       -1.7573 % High margin
       -0.0000 % No margin
       -0.0000 % No margin

xo_s =  1.2118   0.5765 %Min point
fo_s =  0.5322 % Min value
```

```
xo_u =  1.1024   0.7951 %Min point
fo_u =  1.5300 % Not a minimum
co_u = -1.1024 % Constraint margin
       -0.7951
       -1.3888
       -0.0000
       -2.4020
Solver stopped prematurely.

fminunc stopped because it exceeded
   the function evaluation limit,
options.MaxFunEvals = 200 (the
                  default value).
```

Fig. 9.9 shows the minimum points in the admissible region (satisfying the constraints) reached by each function, where the gradient-based optimization function '`fminunc()`' had difficulty with the penalized objective function, while the non-gradient-based functions '`opt_Nelder()`' and '`fminsearch()`' had no problem finding the constrained minimum, which is on the intersection of the two boundary curves corresponding to the 4th and 5th constraints of (E9.5.1b).

9.3 MATLAB BUILT-IN FUNCTIONS FOR OPTIMIZATION[Y-1], Section 7.3

In this section, we will apply several MATLAB built-in unconstrained optimization functions including '`fminsearch()`' and '`fminunc()`' for the same problem, expecting that their nuances will be clarified. Our intention is not to compare or evaluate the performances of the sophisticated functions, but rather to let the readers get a feel for their functional differences. We also introduce some other MATLAB functions like '`linprog()`'/'`intlinprog()`' (each implementing a regular/integer Linear Programming (LP) scheme) and '`fmincon()`', designed for solving constrained optimization problems. Interested readers should run the tutorial function '`tutdemo`', which demonstrate the usages and performances of the two nonlinear optimization solvers, i.e., '`fminunc()`' and '`fmincon()`'.

9.3.1 Unconstrained Optimization

To try applying the unconstrained optimization functions introduced in Section 9.1 and see how they work, we compose the following MATLAB script "test_opt_unc.m," which uses those functions for solving the problem

$$\text{Min } f(\mathbf{x}) = (x_1 - 0.5)^2 (x_1 + 1)^2 + (x_2 + 1)^2 (x_2 - 1)^2 \tag{9.3.1}$$

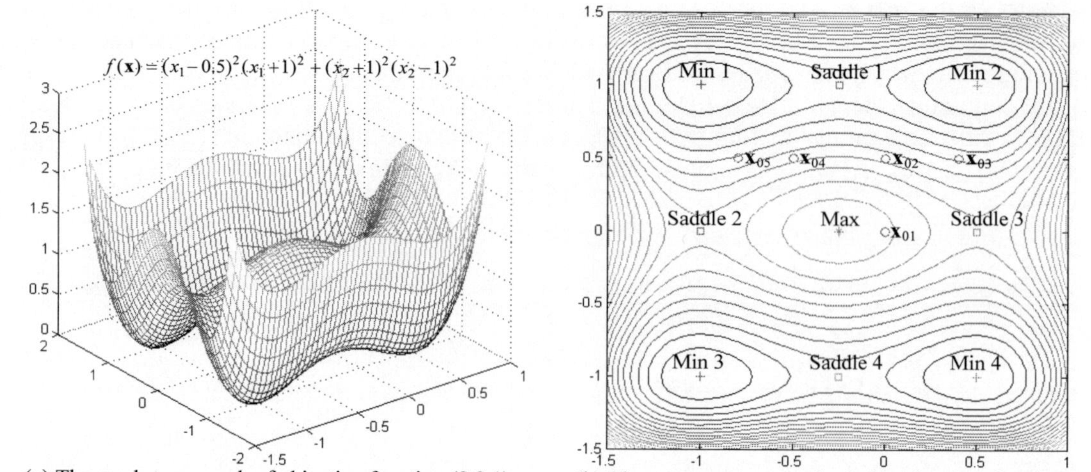

(a) The mesh-type graph of objective function (9.3.1) (b) The contour-type graph of objective function (9.3.1)

Figure 9.10 The contours, local minima/maxima, and saddle points of the objective function (9.3.1)

where the (local) maximum/minimum/saddle points as well as the contours of this objective function are depicted in Fig. 9.10.

```
%test_opt_unc.m
f=@(x) (x(1)-0.5).^2.*(x(1)+1).^2+(x(2)+1).^2.*(x(2)-1).^2; % Eq.(9.3.1)
g=@(x) [2*(x(1)-0.5)*(x(1)+1)*(2*x(1)+0.5) 4*(x(2)^2-1).*x(2)];
x0=[0 0.5] % Initial guess
[xon,fon]=opt_Nelder(f,x0) % Min point, its ftn value by opt_Nelder
[xos,fos]=fminsearch(f,x0) % Min point, its ftn value by fminsearch()
[xost,fost]=opt_steep(f,x0) % Min point, its ftn value by opt_steep()
TolX=1e-4; MaxIter=100; xont=Newtons(g,x0,TolX,MaxIter)
[xou,fou]=fminunc(f,x0) % Min point, its ftn value by fminunc()
```

Table 9.1 Results of running unconstrained optimization functions with various initial values

Initial guess $\mathbf{x}_0$	`opt_Nelder`	`fminsearch`	`opt_steep`	`fminunc`	`Newtons`
$\mathbf{x}_{01}$=[0 , 0]	[-1, 1] (Min 1)	[0.5 , 1] (Min 2)	[0.5, 0] (Saddle 3)	[0.5 , 0] (Saddle 3)	[-0.25, 0] (Max)
$\mathbf{x}_{02}$=[0 , 0.5]	[0.5, 1] (Min 2)	[0.02, 1] (lost)	[0.5 , 1] (Min 2)	[0.5 , 1] (Min 2)	[-0.25, -1] (Saddle 4)
$\mathbf{x}_{03}$=[0.4, 0.5]	[0.5, 1] (Min 2)	[0.5 , 1] (Min 2)	[0.5 , 1] (Min 2)	[0.5 , 1] (Min 2)	[0.5, -1] (Min 4)
$\mathbf{x}_{04}$=[-0.5, 0.5]	[0.5, 1] (Min 2)	[-1 , 1] (Min 1)	[-1 , 1] (Min 1)	[-1, 1] (Min 1)	[-0.25, -1] (Saddle 4)
$\mathbf{x}_{05}$=[-0.8, 0.5]	[-1, 1] (Min 1)	[-1 , 1] (Min 1)	[-1 , 1] (Min 1)	[-1, 1] (Min 1)	[-1 , -1] (Min 3)

To show that whether each function succeeds in finding a minimum point depends mainly on the initial value $\mathbf{x}_0$, the results of running those MATLAB functions with various initial values are summarized in Table 9.1. It can be seen from this table that the gradient-based optimization functions like '`opt_steep()`', '`fminunc()`', and '`Newtons()`' sometimes get to a saddle point or even a maximum point (Remark 9.1), and that the functions do not always approach the extremum that is closest to the initial point. It is interesting to note that even the non-gradient-based MATLAB built-in function '`fminsearch()`' may get lost, while our function '`opt_Nelder()`'

works well for this case. We cannot, however, conclude that this function is better than one based on only one trial, because there may be many problems for which the MATLAB built-in function works well, but our function does not. What can be stated over this happening is that no human work is free from defect.

Now we try using a MATLAB built-in function '`lsqnonlin(f,a0,..)`', which presents a nonlinear least squares (NLLS) solution to the following minimization problem:

$$\underset{\mathbf{a}}{\text{Min}} \sum_{n=1}^{N} f_n^2(\mathbf{a}) \tag{9.3.2}$$

The function needs at least a vector or matrix function $\mathbf{f}(\mathbf{a})$ and an initial guess $\mathbf{a}_0$ (of the minimum point) as its 1st and 2nd input arguments, respectively, where the components of $\mathbf{f}(\mathbf{a})=[f_1(\mathbf{a})\ f_2(\mathbf{a})\ \dots\ f_N(\mathbf{a})]^T$ are squared, summed, and then minimized over $\mathbf{x}$. To learn the usage and functions of '`lsqnonlin()`', we compose the following MATLAB script "test_opt_lsq.m," which uses it to find a second-degree polynomial approximating (or fitting) the following function at $x=\{-2,-1,0,1,1\}$ as shown in Fig. 9.11:

$$y=g(x)=\frac{1}{1+8x^2} \tag{9.3.3}$$

Two other MATLAB functions, '`polyfit(x,y,M)`' and '`lsqcurvefit(gax,a0,x,y)`' are used to find the solutions to the following curve fitting problems, respectively:

$$\underset{\mathbf{a}}{\text{Min}} \sum\nolimits_{n=1}^{N} \left\{ \sum\nolimits_{m=1}^{M} a_m x_n^{M-m} - y_n \right\}^2 \text{ with } y_n=g(x_n) \tag{9.3.4}$$

$$\underset{\mathbf{a}}{\text{Min}} \sum\nolimits_{n=1}^{N} \{\hat{g}(\mathbf{a},x_n)-y_n\}^2 \text{ with } y_n=g(x_n) \tag{9.3.5}$$

where the curve-fitting functions with '`polyfit()`' are restricted to only polynomials.

```
%test_opt_lsq.m
% try using lsqnonlin() for a vector-valued objective ftn F(x)
g=@(x)1./(1+8*x.^2); % True function to approximate by a polynomial ftn
x=-2+[0:4]; % The vector of points where curve fitting is evaluated.
y=g(x); % True function values at x
f=@(a)polyval(a,x)-y; % Error vector ftn whose squared sum to minimize
M=2; a0=zeros(1,M+1); % Initial guess of polynomial coefficient vector
a_lsq=lsqnonlin(f,a0) % Parameter estimate using lsqnonlin()
a_poly=polyfit(x,y,M) % Parameter estimate using polyfit()
gax=@(a,x)polyval(a,x); % Curve-fitting function for g(x)
a_lsqcf=lsqcurvefit(gax,a0,x,y) % LS curve-fitting by ga(x)
xx=-2+[0:400]/100; gxx=g(xx); % To plot the curve fitting result
plot(xx,polyval(a_lsq,xx),'r:', xx,gxx,'b', x,y,'x')
```

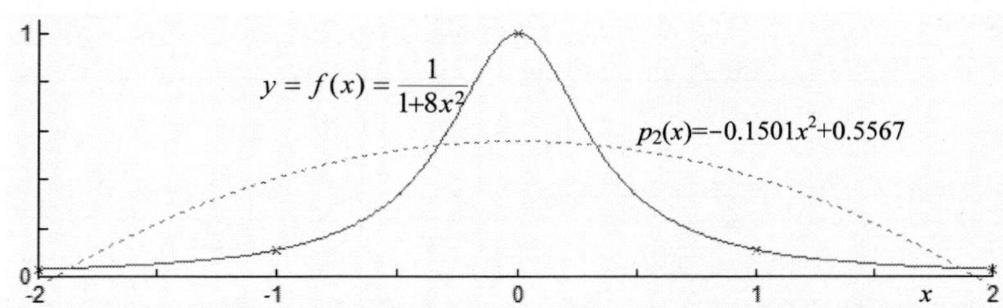

Figure 9.11 The result of using the MATLAB built-in function 'lsqnonlin()' for polynomial approximation (curve fitting)

9.3.2 Constrained Optimization

Generally, constrained optimization problems are very complicated and difficult to deal with. Thus we will not cover the topic in detail and instead, we will just introduce a MATLAB built-in function '`fmincon()`', which can be widely used to solve optimization problems subject to some constraints (not involved with integers):

$$\text{Min } f(\mathbf{x}) \tag{9.3.6}$$

$$\text{s.t. } A\mathbf{x} \le \mathbf{b},\ A_{\rm eq}\mathbf{x} \le \mathbf{b}_{\rm eq},\ \mathbf{c}(\mathbf{x}) \le 0,\ A_{\rm eq}\mathbf{x} \le \mathbf{b}_{\rm eq},\ \text{and } \mathbf{l} \le \mathbf{x} \le \mathbf{u} \tag{9.3.7}$$

Its usage is summarized in the following box, where the objective function f (the first input argument) and the constraint function fcon (the 9th input argument) are supposed to be both continuous and have continuous first derivatives, since '`fmincon()`' is a gradient-based method.

<Usage of '`fmincon()`' for constrained optimization>

`[xo,fo,.]=`**`fmincon`**`(f,x0,A,b,Aeq,beq,l,u,fcon,options,p1,p2,.)`

- Input arguments (At least four input arguments, `'ftn'`, `x0`, `A`, and `B, are` required)
  - `f` : an objective function $f(\mathbf{x})$ to be minimized, defined as an inline function, a function handle, or in an M-file whose file name enclosed by single quotes (`''`) should be put as the first input argument f.
  - `x0` : an initial guess $\mathbf{x}_0$ of the solution
  - `A,b` : linear inequality *constraints* $A\mathbf{x} \le \mathbf{b}$; to be given as `[]` if not applied.
  - `Aeq,beq`: linear equality *constraints* $A_{eq}\mathbf{x} = \mathbf{b}_{eq}$; to be given as `[]` if not applied.
  - `l,u` : lower/upper bound vectors such that $\mathbf{l} \le \mathbf{x} \le \mathbf{u}$; to be given as `[]` if not applied, set `l(i)=-inf/u(i)=inf` if `x(i)` is not bounded below/above.
  - `fcon` : a nonlinear constraint function, defined as an inline function, a function handle, or in an M-file, and supposed to return two output arguments for a given $\mathbf{x}$; the 1st one being the LHS (vector) of inequality constraints $\mathbf{c}(\mathbf{x}) \le \mathbf{0}$ and the 2nd one being the LHS (vector) of equality constraints $\mathbf{c}_{\rm eq}(\mathbf{x}) = \mathbf{0}$; to be given as `[]` if not applied.
  - `options`: used for setting the display parameter, the tolerances for $\mathbf{x}_0$ & $f(\mathbf{x}_0)$, etc; to be given as `[]` if not applied. For details, type '`help optimoptions`' or '`help optimset`' at the MATLAB prompt.
  - `p1,p2,.`: the problem-dependent parameters to be passed to the objective function f and also to the constraint function fcon if it is given.
- Output arguments
  - `xo` : the minimum point ($\mathbf{x}_0$) reached in the permissible region satisfying the constraints
  - `fo` : the minimized function value $f(\mathbf{x}_0)$

```
%test_fmincon.m
% use fmincon() to solve the constrained optimization problem (E9.5.1)
fo=@(x)((x(1)+1.5)^2+5*(x(2)-1.7)^2)*((x(1)-1.4)^2+.6*(x(2)-0.5)^2);
x0=[0 0.5] % Initial guess
A=[]; B=[]; Aeq=[]; Beq=[]; % No linear constraints
l=-inf*ones(size(x0)); u=inf*ones(size(x0)); % No lower/upperbound
options=optimset('LargeScale','off'); % Just [] is OK.
[xo_con,fo_con]=fmincon(fo,x0,A,B,Aeq,Beq,l,u,'fcon_em09e05',options)
[co,ceqo]=fcon_em09e05(xo_con) % to see how the constraints are.
```

```
function [c,ceq]=fcon_em09e05(x)
c=[-x(1); -x(2); 3*x(1)-x(1)*x(2)+4*x(2)-7; % Eq.(E9.5.1b)
   2*x(1)+x(2)-3; 3*x(1)-4*x(2)^2-4*x(2)]; % Inequality constraints
ceq=[]; % Equality constraints
```

To practice using the MATLAB function '`fmincon()`', we compose the above MATLAB script "test_fmincon.m," which uses '`fmincon()`' to solve the problem presented in Example 9.5. Interested readers are welcome to run it and observe the result to check if it agrees with that of Example 9.5.

9.3.3 Linear Programming (LP)

An LP problem is a constrained minimization problem with linear objective and constraint functions such as

$$\text{Min } f(\mathbf{x}) = \mathbf{f}^T\mathbf{x} \tag{9.3.8a}$$

$$\text{s.t. } A\mathbf{x} \le \mathbf{b},\ A_{eq}\mathbf{x} \le \mathbf{b}_{eq},\ \text{and } \mathbf{l} \le \mathbf{x} \le \mathbf{u} \tag{9.3.8b}$$

LP problems can be solved by using the MATLAB built-in function '`linprog()`' as

'`[xo,fo]=linprog(f,A,b,Aeq,Beq,l,u,x0,options)`'

This MATLAB function produces the solution $\mathbf{x}_o$ (in column vector form if it is a vector) and the minimized value $f(\mathbf{x}_o)$ of the objective function as its 1st and 2nd output arguments `xo` and `fo`, where the objective function and constraints excluding the constant term are linear w.r.t. the independent (decision) variables. It works for such linear optimization problems as Eq. (9.3.8) more efficiently than the general constrained optimization function '`fmincon()`'.

The usage of '`linprog()`' is exemplified by the MATLAB script "test_LP.m," which uses the function for solving an LP problem, described as

$$\text{Min } f(\mathbf{x}) = \mathbf{f}^T\mathbf{x} = [-3\ \ -2]\begin{bmatrix} x_1 \\ x_2 \end{bmatrix} = -3x_1 - 2x_2 \tag{9.3.9a}$$

$$\text{s.t. } A\mathbf{x} = \begin{bmatrix} 3 & 4 \\ 2 & 1 \\ -3 & 2 \end{bmatrix}\begin{bmatrix} x_1 \\ x_2 \end{bmatrix} \begin{matrix} \le \\ \le \\ = \end{matrix} \begin{bmatrix} 7 \\ 3 \\ 2 \end{bmatrix} = \mathbf{b} \text{ and } \mathbf{l} = \begin{bmatrix} 0 \\ 0 \end{bmatrix} \le \mathbf{x} = \begin{bmatrix} x_1 \\ x_2 \end{bmatrix} \le \begin{bmatrix} 10 \\ 10 \end{bmatrix} = \mathbf{u} \tag{9.3.9b}$$

```
%test_LP.m
% to solve the Linear Programming problem (9.3.9).
% Min f*x=-3*x(1)-2*x(2) s.t. A*x<=b, Aeq*x=beq, and l<=x<=u
f=[-3 -2]; % Coefficient vector of the objective function
A=[3 4; 2 1]; b=[7; 3]; % Inequality constraint A*x<=b
Aeq=[-3 2]; beq=2; % Equality constraint Aeq*x=beq
l=[0 0]; u=[10 10]; % Lower/upper bound l<=x<=u
[xo_lp,fo_lp]=linprog(f,A,b,Aeq,beq,l,u)
cons_satisfied=[A; Aeq]*xo_lp-[b; beq] % How constraints are satisfied
% To apply fmincon() for the LP problem,
f_090309a=@(x)-3*x(1)-2*x(2); % The objective function
x0=[0 0]; % Initial point
[xo_con,fo_con]=fmincon(f_090309a,x0,A,b,Aeq,beq,l,u)
```

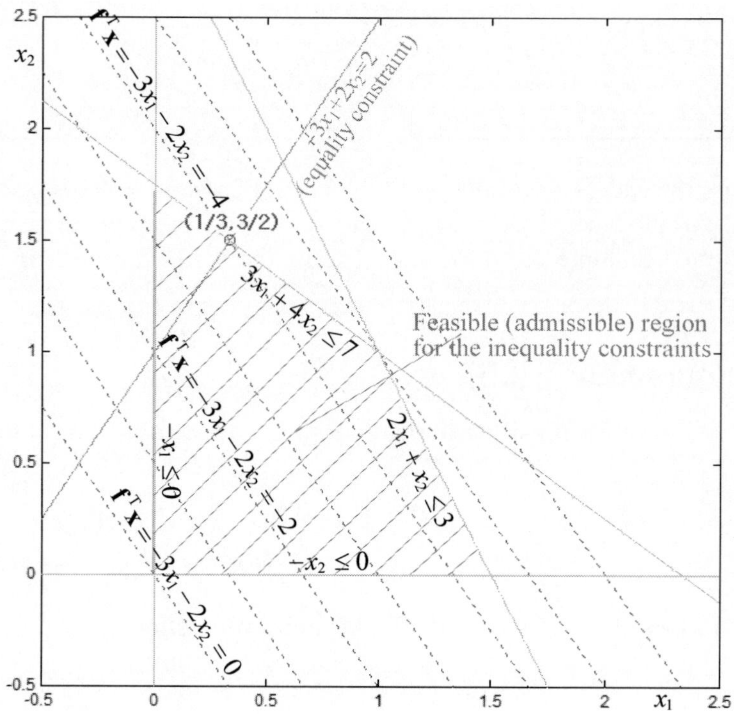

Figure 9.12 The objective function, constraints, and solution of an LP problem

The script also applies the general constrained minimization function '`fmincon()`' to solve the same problem for crosscheck. Running the script yields

```
>>test_LP
    xo_lp= 0.3333       % Solution using 'linprog()'
           1.5000
    fo_lp=-4.0000       % Minimum value of the objective function f(x)
    cons_satisfied=  -0.0000 %<=0(inequality)
                     -0.8333 %<=0(inequality)
                     -0.0000 %=0(equality)
    xo_con=[0.3333   1.5000] % Solution using 'fmincon()'
```

In this result, the solutions obtained by using the two functions '`linprog()`' and '`fmincon()`' agree with each other, satisfying the inequality/equality constraints, as can be seen in Fig. 9.12.

Here, the *simplex method*, which is the standard algorithm (developed by George Dantzig in 1946) for solving LP problems, is introduced. It begins with constructing the *simplex tableau* from the coefficients of the linear inequality/equality constraints and objective function as

$$\begin{array}{r|rrrrrrc|l} & x_1 & x_2 & s_1 & s_2 & s_3 & s_4 & \mathbf{b} & \text{Basic variables} \\ A & 3 & 4 & 1 & 0 & 0 & 0 & 7 & s_1 \\ & 2 & 1 & 0 & 1 & 0 & 0 & 3 & s_2 \\ A_{eq} & -3 & 2 & 0 & 0 & 1 & 0 & 2(\mathbf{b}_{eq}) & s_3 \\ -A_{eq} & 3 & -2 & 0 & 0 & 0 & 1 & -2(-\mathbf{b}_{eq}) & s_4 \\ \hline \mathbf{f}^T(\text{price}) & -3 & -2 & 0 & 0 & 0 & 0 & \textcircled{0} & \text{Minus current cost (objective function value)} \end{array} \tag{9.3.10}$$

where the bound constraints are not included. Note that the equality constraint

$$A_{eq}\mathbf{x}=\mathbf{b}_{eq}$$

has been split into two opposite inequality constraints:

$$A_{eq}\mathbf{x}\le\mathbf{b}_{eq} \tag{9.3.11a}$$

$$A_{eq}\mathbf{x}\ge\mathbf{b}_{eq},\ \text{or equivalently,}\ -A_{eq}\mathbf{x}\le-\mathbf{b}_{eq} \tag{9.3.11b}$$

This simplex tableau amounts to writing the constraints as

$$\begin{array}{rrrrrrcr} 3x_1 & +4x_2 & +s_1 & & & & = & 7\\ 2x_1 & +x_2 & & +s_2 & & & = & 3\\ -3x_1 & +2x_2 & & & +s_3 & & = & 2\\ 3x_1 & -2x_2 & & & & +s_4 & = & -2 \end{array} \tag{9.3.12}$$

where $\{s_1, \ldots, s_4\}$ are called the *slack variables*. Currently, this set of linear equations has the basic solution

$$\begin{array}{c} \text{Nonbasic variables}\quad\text{Basic variables}\\ [\ x_1\ \ x_2\quad s_1\ \ s_2\ \ s_3\ \ s_4\ ]=[\ 0\ \ 0\ \ 7\ \ 3\ \ 2\ \ -2\ ] \end{array} \tag{9.3.13}$$

where the variables $\{s_1, \ldots, s_4\}$ (corresponding to single columns) and others $\{x_1, x_2\}$ are said to be *basic variables* and *nonbasic variables*, respectively, and the corresponding value of objective function (with opposite sign) can be found as zero at the lower-right corner of the simplex tableau.

The simplex method continues as follows where the simplex tableau (9.3.10) is denoted by A_{bf}:

(Step1) Find the minimum negative entry $A_{bf,\min}=\text{Min}\{A_{bf}(\text{end},1{:}N)<0\}$ among the first N (the number of variables $\{x_j\text{'s}\}$) of the last (objective function) row. The column having that entry is chosen as the *pivot column* since the (nonbasic) variable corresponding to that column j_p (having the negative highest price) is expected to decrease the objective function value the most by being a new basic variable (what we call the *entering* (*incoming*) *variable*) – the *steepest descent rule*. If all $A_{bf}(\text{end},1{:}N)$'s are nonnegative, it implies that there is no candidate for entering variable that can contribute to decreasing the objective value, and consequently, go to Step 4 to find the optimal solution in the current tableau.

(Step2) Find the pivot element a_{i_p,j_p} with minimum ratio $\text{Min}\{b_i/a_{i,j_p}$ such that $a_{i_p,j_p}>0\}$ in the pivot column — the *minimum ratio rule*. If there is no positive a_{i,j_p} in the pivot column, the process should be stopped and declared an *unbounded case* where the objective function value can be decreased no more, without violating the constraints.

(Step 3) Do the pivot operation on the pivot element a_{i_p,j_p}, i.e., make elementary row operations to make $a_{i_p,j_p}=1$ so that it can be only one nonzero entry in the pivot column, as is done with Gauss elimination (Section 1.3.4). This operation amounts to selecting the basic variable corresponding to the pivot row i_p as a *departing* (*leaving*) *variable* and replacing it by the entering variable (corresponding to the pivot column j_p determined in Step 1) in the new set of basic variables. Then go to Step 1.

(Step 4) Set the values of basic variables (corresponding to single columns) to the value (corresponding to the single one) in the last (RHS) column. Also, set the optimal value of the objective function to the opposite-signed value of the entry at the lower-right corner of the tableau. Then the process is terminated.

Now, let us use the simplex method for the simplex tableau (9.3.10):

Basic variables	x_1	x_2	s_1	s_2	s_3	s_4	b_i	b_i/a_{i1}
s_1	3	4	1	0	0	0	7	7/3
s_2	2	1	0	1	0	0	3	3/2
s_3	−3	2	0	0	1	0	2	×
s_4	3	−2	0	0	0	1	−2	−2/3
	−3	−2	0	0	0	0	0	

(9.3.14-1)

Choosing the 1st column (having the minimum negative entry among the first N=2 entries in the last row) as the pivot column (by the steepest descent rule) and the 4th row (having the minimum ratio) as the pivot row, do the pivot operation on a_{41} of the tableau (9.3.14-1) to write

Basic variables	x_1	x_2	s_1	s_2	s_3	s_4	b_i	b_i/a_{i1}
s_1	0	6	1	0	0	−1	9	9/6
s_2	0	7/3	0	1	0	−2/3	13/3	13/7
s_3	0	0	0	0	1	1	0	×
x_1	1	−2/3	0	0	0	1/3	−2/3	×
	0	−4	0	0	0	1	−2	−{−3(−2/3)}

(9.3.14-2)

Now, choosing the 2nd column (having the minimum negative entry among the first N=2 entries in the last row) as the pivot column (by the steepest descent rule) and the 1st row (having the minimum ratio) as the pivot row, do the pivot operation on a_{12} of the tableau (9.3.14-2) to write

Basic variables	x_1	x_2	s_1	s_2	s_3	s_4	b_i	
x_2	0	1	1/6	0	0	−1/6	3/2	
s_2	0	0	−7/18	1	0	−5/18	5/6	
s_3	0	0	0	0	1	1	0	
x_1	1	0	1/9	0	0	2/9	1/3	
	0	0	2/3	0	0	1/3	4	−{−3·1/3−2·3/2}

(9.3.14-3)

It is interesting that the entry at the lower-right corner of the tableau is the negative objective function value. Now, we can see no nonzero entry among the first N=2 entries in the last row, which implies that the following optimal result has been reached:

$$\overbrace{[x_1\ \ x_2}^{\text{basic}}\ \ \overbrace{s_1}^{\text{nonbasic}}\ \ \overbrace{s_2\ \ s_3}^{\text{basic}}\ \ \overbrace{s_4}^{\text{nonbasic}}] = [1/3\ \ 3/2\ \ 0\ \ 5/6\ \ 0\ \ 0] \tag{9.3.15}$$

where the (current) basic variables $\{x_1, x_2, s_2, s_3\}$ corresponding to single columns are set to their corresponding RHS values (in the last column) and those of other (nonbasic) variables $\{s_1, s_4\}$ are set to zero. The value of the objective function at the optimal solution

$$\mathbf{x}^o = \begin{bmatrix} x_1^o \\ x_2^o \end{bmatrix} = \begin{bmatrix} 1/3 \\ 3/2 \end{bmatrix} \tag{9.3.16}$$

is set to the opposite-signed value of the lower-right corner of the last tableau (9.3.14-3):

$$f(\mathbf{x}^o) = -4 \tag{9.3.17}$$

as conformed with the solution obtained by '`linprog()`' above.

(Q) What is implied by the values of slack variables that are $[s_1\ s_2\ s_3\ s_4] = [0\ \ 5/6\ \ 0\ \ 0]$?

(A) It implies that the optimal solution (9.3.16) satisfies the constraints (9.3.12) tightly except for the 2nd one, which is satisfied by a margin of 5/6.

```
function [xo,fmin,Ab] = simplex(f,A,b,Aeq,beq)
% Uses the simplex method to solve an LP (linear programming) problem:
%   Min f(x) = f.'*x  s.t. A*x <= b and Aeq*x = beq
[M,N] = size(A); % Size of the coefficient matrix for inequality
[Meq,Neq] = size(Aeq); % Size of the coefficient matrix for equality
[M,N] = size(A); % Size of coefficient matrix for inequality
[Meq,Neq] = size(Aeq); % Size of the coefficient matrix for equality
MMeq2=M+Meq*2;  EPS=1e-6;
% Construct the simplex tableau like Eq.(9.3.10)
if Meq==0
  A1=A;  b1=b(:); % No equality constraint
 else
  A1=[A; Aeq; -Aeq]; b1=[b(:); beq; -beq];
  if N~=Neq, error('N must be equal to Neq in simplex'); end
end
Ab = [A1 eye(MMeq2) b1; f.' zeros(1,MMeq2) 0]
M1=MMeq2+1; NM1=N+M1;  % Row/Column sizes of the simplex tableau
% Choose the pivot column with minimum entry in the last row
[AbM1_min,jp]=min(Ab(M1,1:NM)); % Index of the most negative entry
while AbM1_min<-EPS
   indp = find(Ab(1:MMeq2,jp)>0);
   if isempty(indp), disp('Unbounded case!');  end
   ratios=Ab(indp,NM1)./Ab(indp,jp); %indp=indnz(find(ratios>0));
   [Abmin,imin] = min(ratios);
   ip = indp(imin); % Pivot row ip by the minimum ratio rule
   % Pivoting
   Ab(ip,:) = Ab(ip,:)/Ab(ip,jp);
   for i=1:M1
      if i~=ip, Ab(i,:) = Ab(i,:)-Ab(i,jp)*Ab(ip,:); end
   end
   [AbM1_min,jp]=min(Ab(M1,1:NM)); % Index of the most negative entry
end
% Last step to determine the optimal solution
xo = zeros(N,1);
if sum(Ab(1:MMeq2,NM1)<-EPS)>0, disp('Infeasible case!'); end
for j=1:N
   ind1=find(abs(Ab(1:MMeq2,j))>EPS);
    if length(ind1)==1, xo(j,1)=Ab(ind1,NM1);  end
end
fmin = -Ab(M1,NM1); % Optimal objective function value
```

```
%test_simplex.m
f=[-3; -2]; % Objective function coefficients to be minimized
A=[3 4; 2 1]; b=[7;3]; % Inequality constraint coefficients and RHS
Aeq=[-3 2]; beq=[2]; % Equality constraint coefficients and RHS
[xo,fmin,Ab_final] = simplex(f,A,b,Aeq,beq)
```

This simplex method has been cast into the above MATLAB function 'simplex()'. To solve the above LP problem, you can run the above MATLAB script "test_simplex.m," which uses 'simplex()'.

```
>>test_simplex
   xo =    1/3            fo =   -4  % Optimal solution
           3/2
```

9.3.4 Mixed Integer Linear Programming (MILP)

An MILP problem is a constrained minimization problem with linear objective and constraint functions such as

$$\text{Min } f(\mathbf{x}) = \mathbf{f}^{\mathrm{T}}\mathbf{x} \tag{9.3.18a}$$

$$\text{s.t. } A\mathbf{x} \le \mathbf{b},\ A_{\text{eq}}\mathbf{x} \le \mathbf{b}_{\text{eq}},\ \text{and } \mathbf{l} \le \mathbf{x} \le \mathbf{u} \tag{9.3.18b}$$

where some variables x(intcon) are restricted to be integer-valued. MILP problems can be solved by using the MATLAB built-in function '`intlinprog()`' as

'`[xo,fo]=intlinprog(f,intcon,A,b,Aeq,Beq,l,u,options)`'.

where '`intcon`' is a set of the indices of integer variables. This MATLAB function produces the solution $\mathbf{x}_o$ (in column vector form) and the minimized value of the objective function $f(\mathbf{x}_o)$ as its 1st and 2nd output arguments `xo` and `fo`.

The usage of '`intlinprog()`' is exemplified by the MATLAB script "test_ILP.m," which uses the function for solving an MILP problem described as

$$\text{Min } f(\mathbf{x}) = \mathbf{f}^{\mathrm{T}}\mathbf{x} = [-2\ \ -1]\begin{bmatrix} x_1 \\ x_2 \end{bmatrix} = -2x_1 - x_2 \tag{9.3.19a}$$

$$\text{s.t. } A\mathbf{x} = \begin{bmatrix} 3 & -1 \\ 1 & 5 \end{bmatrix}\begin{bmatrix} x_1 \\ x_2 \end{bmatrix} \begin{matrix} \le \\ \le \end{matrix} \begin{bmatrix} 6 \\ 16 \end{bmatrix} = \mathbf{b} \text{ and } \mathbf{l} = \begin{bmatrix} 0 \\ 0 \end{bmatrix} \le \mathbf{x} = \begin{bmatrix} x_1 \\ x_2 \end{bmatrix} \le \begin{bmatrix} 10 \\ 10 \end{bmatrix} = \mathbf{u} \tag{9.3.19b}$$

where x_1 and x_2 must be integer-valued (by the integral constraints).

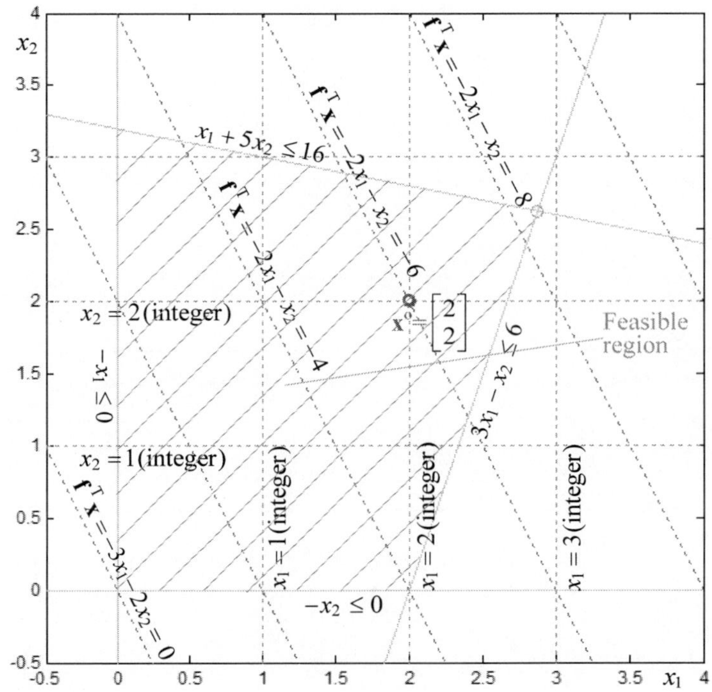

Figure 9.13 The objective function, constraints, and solution of an ILP problem

```
%test_ILP.m
% to solve the MILP problem (9.3.19).
% Min f*x=-3*x(1)-2*x(2) s.t. Ax<=b, Aeq=beq, and l<=x<=u
clear, clf
x0=[0 0]; % Initial point
f=[-2 -1]; % Coefficient vector of the objective function
intcon=[1 2]; % x(intcon)=x([1 2]) must be integer-valued
A=[3 -1; 1 5]; b=[6; 16]; % Inequality constraint Ax<=b
Aeq=[]; beq=[]; % No equality constraint Aeq*x=beq
l=[0 0]; u=[10 10]; % Lower/upper bound l<=x<=u
[xo,fo]=intlinprog(f,intcon,A,b,Aeq,beq,l,u)
```

Running the script "test_ILP.m" yields the following results, as shown in Fig. 9.13:

```
>>test_ILP
    xo =    2.0000            fo =   -6
            2.0000
```

(Q) Referring to Fig. 9.13, can you get the ILP solution simply by rounding up the LP solution?

There are two techniques that can be used to deal with ILP problems, the *branch-and-bound (BB) method* and the cutting plane method. First, to get the feel of the *BB method*, let us use it to solve the above ILP problem:

0. Relax (ignore temporarily) the integral constraints to turn the ILP (integer linear programming) problem into an LP one (*LP relaxation*):

$$\text{Min } f(\mathbf{x}) = \mathbf{f}^T\mathbf{x} = [-2 \;\; -1]\begin{bmatrix} x_1 \\ x_2 \end{bmatrix} = -2x_1 - x_2 \tag{9.3.20a}$$

$$\text{s.t. } A\mathbf{x} = \begin{bmatrix} 3 & -1 \\ 1 & 5 \end{bmatrix}\begin{bmatrix} x_1 \\ x_2 \end{bmatrix} \begin{matrix} \le \\ \le \end{matrix} \begin{bmatrix} 6 \\ 16 \end{bmatrix} = \mathbf{b} \text{ and } \mathbf{l} = \begin{bmatrix} 0 \\ 0 \end{bmatrix} \le \mathbf{x} = \begin{bmatrix} x_1 \\ x_2 \end{bmatrix} \le \begin{bmatrix} 10 \\ 10 \end{bmatrix} = \mathbf{u} \tag{9.3.20b}$$

This can be solved by using the LP solver '`linprog()`' as

```
f=[-2 -1]; % Coefficient vector of the objective function
A=[3 -1; 1 5]; b=[6; 16]; Aeq=[]; beq=[]; %Inequality/Equality constraint
l=[0 0]; u=[10 10]; % Lower/upper bound l<=x<=u
[xo1,fo1]=linprog(f,A,b,Aeq,beq,l,u) % -> [23/8; 21/8] with f=-67/8
```

to yield the solution

$$\mathbf{x}_1^o = \begin{bmatrix} 23/8 \\ 21/8 \end{bmatrix} \text{ with } f(\mathbf{x}_1^o) = -67/8 \tag{9.3.21}$$

1. First, to trim x_2=21/8 (which is more fractional than x_1 in the sense that its fractional part is closer to 0.5 than that of x_1=23/8) of $\mathbf{x}_1^o$ into an integer, we branch the relaxed LP problem into two subproblems (that we call SP11 and SP12, respectively), each with additional constraints $x_2 \le 2 = \text{floor}(2.625)$ and $x_2 \ge 3 = \text{ceil}(2.625)$, and solve them:

```
[xo11,fo11]=linprog(f,A,b,[],[],[0;0],[10;2]) %->[8/3;2], f=-22/3
[xo12,fo12]=linprog(f,A,b,[],[],[0;3],[10;10]) %-> [1;3], f=-5
```

to get

$$\mathbf{x}_{11}^o = \begin{bmatrix} 8/3 \\ 2 \end{bmatrix} \text{ with } f(\mathbf{x}_{11}^o) = -\frac{22}{3}, \quad \mathbf{x}_{12}^o = \begin{bmatrix} 1 \\ 3 \end{bmatrix} \text{ with } f(\mathbf{x}_{12}^o) = -5 \tag{9.3.22-1}$$

Here, $\mathbf{x}_{12}^{o} = [1 \quad 3]^T$ is registered as a candidate for the optimal solution, since it satisfies the integral constraints. But $\mathbf{x}_{11}^{o} = [8/3 \quad 2]^T$ does not satisfy the integral constraint on x_1, which forces us to branch on x_1.

1-1. To trim $x_1 = 8/3$ of $\mathbf{x}_{11}^{o}$ into an integer, we branch on x_1 to make two subproblems (that we call SP111 and SP112, respectively), each with additional constraints $x_1 \le 2 = \text{floor}(8/3)$ and $x_1 \ge 3 = \text{ceil}(8/3)$, and solve them:

```
[xo111,fo111]=linprog(f,A,b,Aeq,beq,[0;0],[2;2]) %-> [2;2], f=-6
[xo112,fo112]=linprog(f,A,b,Aeq,beq,[3;0],[10;2]) %->[3;400/167], f=-8.39
```

to get

$$\mathbf{x}_{111}^{o} = \begin{bmatrix} 2 \\ 2 \end{bmatrix} \text{ with } f(\mathbf{x}_{111}^{o}) = -6, \quad \mathbf{x}_{112}^{o} = \begin{bmatrix} 3 \\ 400/167 \end{bmatrix} \text{ with } f(\mathbf{x}_{112}^{o}) = -\frac{1402}{167} \tag{9.3.22-2}$$

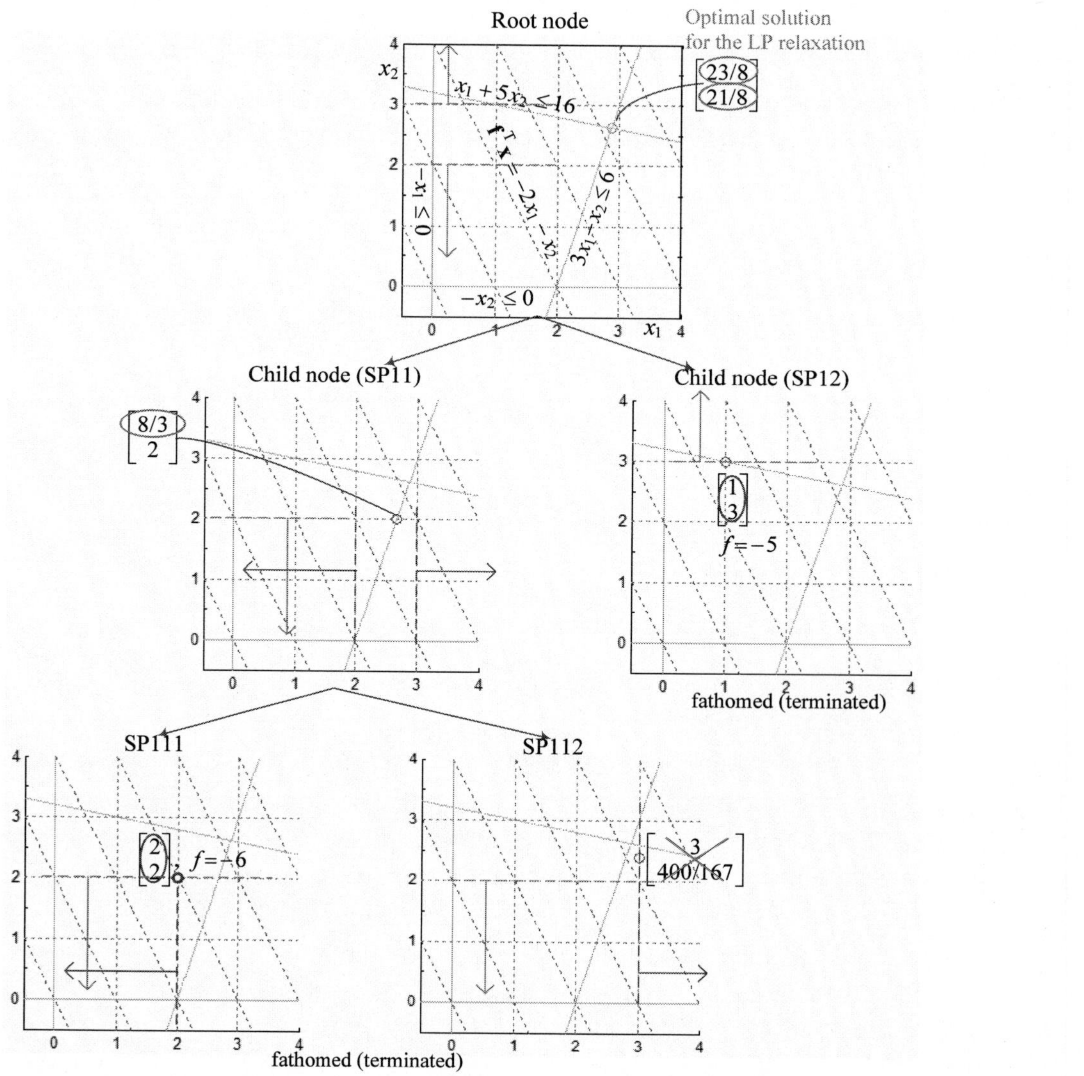

Figure 9.14 Branch-and-bound (BB) search (enumeration) tree for finding the solution of ILP problem (9.3.19)

Here, $\mathbf{x}^o_{111}=[2\ \ 2]^T$ is registered as a candidate for the optimal solution, since it satisfies the integral constraints, while $\mathbf{x}^o_{112}=[3\ \ 400/167]^T$, violating the (first) inequality constraint (9.3.20b), should be terminated due to infeasibility.

2. Now that all the branches are terminated with feasible or infeasible solution nodes, we choose the best feasible solution (with the least objective function value), which is

$$\mathbf{x}^o_{111}=\begin{bmatrix}2\\2\end{bmatrix} \text{ with } f(\mathbf{x}^o_{111})=-6 \tag{9.3.22-3}$$

This process of the BB (branch-and-bound) method is depicted by the search (enumeration) tree in Fig. 9.14 and has been cast into the MATLAB function 'branch_and_bound()'. To solve the above ILP problem, run the MATLAB script "test_BB.m," which uses 'branch_and_bound()':

```
>>test_BB
   xo =     2              fo =    -6  % Optimal solution
            2
```

```
function [xo,fo] = branch_and_bound(f,A,b,Aeq,beq,l,u)
% Uses the simplex method to solve an ILP (linear programming) problem:
%    Min f(x) = f.'*x   s.t. A*x <= b and Aeq*x = beq
% for the optimal solution xo with the optimized objective ftn value fo
Nx=length(f); % Number of objective function coefficients and variables
if nargin<6, l=zeros(Nx,1); u=1e10*ones(Nx,1); end
[xo,fo,icase] = branch_and_bound0(f,A,b,Aeq,beq,l,u);
```

```
function [xo,fo,icase] = branch_and_bound0(f,A,b,Aeq,beq,l,u)
%Copyleft: Won Y. Yang, wyyang53@hanmail.net, CAU for academic use only
[xo,fo,icase] = linprog(f,A,b,Aeq,beq,l,u);
% Find the index of the most fractional variable of LP optimal solution
if icase<1, xo=NaN; fo=1e10; return; end
[rmax,imf] = max(abs(xo-round(xo))); % Which variable to branch on?
EPS = 1e-6;
if rmax>EPS
  l1=l; u1=u; u1(imf)=floor(xo(imf));
  l2=l; u2=u; l2(imf)=ceil(xo(imf));
  [xo1,fos(1),icases(1)]=branch_and_bound0(f,A,b,Aeq,beq,l1,u1);
  [xo2,fos(2),icases(2)]=branch_and_bound0(f,A,b,Aeq,beq,l2,u2);
  icase = max(icases);
  if icases(1)==1
    if icases(2)==1, [fo,imin] = min(fos);
                     if imin==1, xo=xo1;  else xo = xo2; end
      else  xo = xo1; fo = fos(1);
    end
   else
    if icases(2)==1, xo=xo2; fo=fos(2);  else xo=NaN; fo=1e10; end
  end
end
```

```
%test_BB.m
f=[-2; -1]; % Objective function coefficients to be minimized
A=[3 -1; 1 5]; b=[6; 16]; % Inequality constraint Ax<=b
[xo,fmin]=branch_and_bound(f,A,b,[],[])
```

Now, to get the feel of the *(Gomory's) cutting plane method*, let us use it to solve the above ILP problem:

0. Like the BB method, the cutting plane method also begins with the LP relaxation (without caring about integral constraints), which yields the optimal simplex tableau as follows:

```
f=[-2 -1]; % Coefficient vector of the objective function
A=[3 -1; 1 5]; b=[6; 16]; % Inequality constraint Ax<=b
Aeq=[]; beq=[]; % No equality constraint Aeq*x=beq
[xo1,fo1,Ab]=simplex(f,A,b,Aeq,beq) % -> [23/8; 21/8] with f=-67/8
```

$$\begin{array}{ccccc} x_1 & x_2 & s_1 & s_2 & b_i \\ \end{array} \qquad\qquad \begin{array}{ccccc} x_1 & x_2 & s_1 & s_2 & b_i \end{array}$$

$$\begin{bmatrix} 3 & -1 & 1 & 0 & 6 \\ 1 & 5 & 0 & 1 & 16 \\ \hdashline -2 & -1 & 0 & 0 & 0 \end{bmatrix} \xrightarrow{\text{Simplex method}} \begin{bmatrix} 1 & 0 & 5/16 & 1/16 & 23/8 \\ 0 & 1 & -1/16 & 3/16 & 21/8 \\ \hdashline 0 & 0 & 9/16 & 5/16 & 67/8 \end{bmatrix} \tag{9.3.23}$$

This implies that the LP solution, i.e., the optimal solution for the LP relaxation, is

$$\mathbf{x}_1^{\mathrm{o}} = \begin{bmatrix} x_{1,1}^{\mathrm{o}} \\ x_{2,1}^{\mathrm{o}} \end{bmatrix} = \begin{bmatrix} 23/8 \\ 21/8 \end{bmatrix} \text{ with } f(\mathbf{x}_1^{\mathrm{o}}) = -\frac{67}{8} \tag{9.3.24}$$

1. Let us impose the integral constraint on x_1 and x_2 one by one. Which one first? Since $x_{2,1}^{\mathrm{o}} = 21/8$ is more fractional than $x_{1,1}^{\mathrm{o}} = 23/8$ in the sense that the distance of 21/8 from 20/8 (the closest midpoints between integers) is shorter than that of 23/8 from 20/8, we first try to fix x_2 by taking the positive fractional part of the constraint involving x_2 (in the 2nd row of the simplex tableau (9.3.23), called a *source row*,) as

From the 2nd row of simplex tableau (9.3.23)

$$0x_1 + 1x_2 - \frac{1}{16}s_1 + \frac{3}{16}s_2 = \frac{21}{8} \rightarrow x_2 + \left(-1 + \frac{15}{16}\right)s_1 + \frac{3}{16}s_2 = 2 + \frac{5}{8}$$

$$\rightarrow \underbrace{x_2 - s_1 - 2}_{\text{Integer}} = \frac{5}{8} - \frac{15}{16}s_1 - \frac{3}{16}s_2 \le 0 \rightarrow -\frac{15}{16}s_1 - \frac{3}{16}s_2 \le -\frac{5}{8} \tag{9.3.25a}$$

writing this inequality (called a *Gomory cut*) with a new slack variable s_3 as

$$-\frac{15}{16}s_1 - \frac{3}{16}s_2 + s_3 = -\frac{5}{8} \tag{9.3.25b}$$

and adding it to the simplex tableau as

$$\begin{array}{c} \begin{array}{ccccc} x_1 & x_2 & s_1 & s_2 & b_i \end{array} \\ \begin{bmatrix} 1 & 0 & 5/16 & 1/16 & 23/8 \\ 0 & 1 & -1/16 & 3/16 & 21/8 \\ \hdashline 0 & 0 & 9/16 & 5/16 & 67/8 \end{bmatrix} \end{array} \xrightarrow[\text{(9.3.25b)}]{\text{add Gomory cut}} \begin{array}{c} \begin{array}{cccccc} x_1 & x_2 & s_1 & s_2 & s_3 & b_i \end{array} \\ \begin{bmatrix} 1 & 0 & 5/16 & 1/16 & 0 & 23/8 \\ 0 & 1 & -1/16 & 3/16 & 0 & 21/8 \\ 0 & 0 & -15/16 & -3/16 & 1 & -5/8 \\ \hdashline 0 & 0 & 9/16 & 5/16 & 0 & 67/8 \end{bmatrix} \end{array} \tag{9.3.26}$$

$$\text{Ratio} \quad \left|\frac{9/16}{-15/16} = \frac{9}{-15}\right| < \left|\frac{5/16}{-3/16} = \frac{5}{-3}\right|$$

2. From the added constraint row M=3 (determined as the pivot row), choose the entry $A(M, j) = -15/16$ with the minimum absolute ratio $\text{Min}\{|A(M+1,j)/A(M,j)|, j=1:N\}(N=5)$ as the pivot element by the *minimum ratio rule*. Pivot the tableau on the pivot element, which yields

$$\begin{array}{cccccc} x_1 & x_2 & s_1 & s_2 & s_3 & b_i \\ \left[\begin{array}{cccccc} 1 & 0 & 5/16 & 1/16 & 0 & 23/8 \\ 0 & 1 & -1/16 & 3/16 & 0 & 21/8 \\ 0 & 0 & -15/16 & -3/16 & 1 & -5/8 \\ \hline 0 & 0 & 9/16 & 5/16 & 0 & 67/8 \end{array}\right] \end{array} \xrightarrow{\text{Pivoting}} \begin{array}{cccccc} x_1 & x_2 & s_1 & s_2 & s_3 & b_i \\ \left[\begin{array}{cccccc} 1 & 0 & 0 & 0 & 1/3 & 8/3 \\ 0 & 1 & 0 & 1/5 & -1/15 & 8/3 \\ 0 & 0 & 1 & 1/5 & -16/5 & 2/3 \\ \hline 0 & 0 & 0 & 1/5 & 3/5 & 8 \end{array}\right] \end{array} \quad (9.3.27)$$

(Q) Has the objective function value been improved by adding the new constraint (called a Gomory cut)? Why is that?

3. If all the RHS entries (in the rightest column) are nonnegative integers except for $A(M+1,N+1)$ (which is the negative objective function value), go to the final step where the current basic variables (corresponding the single-one columns) are set to the corresponding RHS values as the optimal feasible solution, and stop. If there is any negative entry in the RHS column, pivot the tableau on the pivot element which is in the same row with that entry and the column determined by the minimum ratio rule. Repeat the pivot operations until no negative entry appears in the RHS column. Since all the RHS entries are nonnegative (where $A(M+1,N+1)$ at the lower-right corner does not matter), we go to the next step, which is to impose the integral constraint on another noninteger basic variable as done in Step 1.
4. Which is the most fractional among the $M=3$ entries in the RHS column? They are all equal in the closeness to the midpoints between integers, and hence, we choose the first one as the source row to make a new Gomory cut:

$$\underbrace{1x_1+\frac{1}{3}s_3=\frac{8}{3}}_{\text{From the 1st row of simplex tableau (9.3.27)}}=2+\frac{2}{3} \;\rightarrow\; \overbrace{x_1-2}^{\text{Integer}}=\frac{2}{3}-\frac{1}{3}s_3\le 0 \rightarrow -\frac{1}{3}s_3\le-\frac{2}{3}$$

$$\rightarrow \boxed{-\frac{1}{3}s_3+\overset{\text{A new slack variable}}{s_4}=-\frac{2}{3}} \quad (9.3.28)$$

then add it to the simplex tableau, choose the pivot element by the minimum ratio rule, and pivot the tableau as

$$\begin{array}{ccccccc} x_1 & x_2 & s_1 & s_2 & s_3 & s_4 & b_i \\ \left[\begin{array}{ccccccc} 1 & 0 & 0 & 0 & 1/3 & 0 & 8/3 \\ 0 & 1 & 0 & 1/5 & -1/15 & 0 & 8/3 \\ 0 & 0 & 1 & 1/5 & -16/15 & 0 & 2/3 \\ 0 & 0 & 0 & 0 & -1/3 & 1 & -2/3 \\ \hline 0 & 0 & 0 & 1/5 & 3/5 & 0 & 8 \end{array}\right] \end{array} \xrightarrow{\text{Pivoting}} \begin{array}{ccccccc} x_1 & x_2 & s_1 & s_2 & s_3 & s_4 & b_i \\ \left[\begin{array}{ccccccc} 1 & 0 & 0 & 0 & 0 & 1 & 2 \\ 0 & 1 & 0 & 1/5 & 0 & -1/5 & 14/5 \\ 0 & 0 & 1 & 1/5 & 0 & -16/5 & 14/5 \\ 0 & 0 & 0 & 0 & 1 & -3 & 2 \\ \hline 0 & 0 & 0 & 1/5 & 0 & 9/5 & 34/5 \end{array}\right] \end{array} \quad (9.3.29)$$

$$\text{Ratio} \quad \left|\frac{3/5}{-1/3}\right|=\frac{9}{5} \text{ (Only one nonzero ratio)}$$

where this pivoting operation can be performed by using a MATLAB function 'pivoting' as follows:

```
>>A=[1 0 0 0 1/3 0 8/3; 0 1 0 1/5 -1/15 0 8/3; 0 0 1 1/5 -16/15 0 2/3;
   0 0 0 0 -1/3 1 -2/3; 0 0 0 1/5 3/5 0 8];  pivoting(A,4,5)

  ans =  1       0       0       0       0       1       2
         0       1       0     1/5       0    -1/5    14/5
         0       0       1     1/5       0   -16/5    14/5
         0       0       0       0       1      -3       2
         0       0       0     1/5       0     9/5    34/5
```

```
function Ap = pivoting(A,ip,jp)
[M,N] = size(A); % Size of coefficient matrix for inequality
Ap(ip,:) = A(ip,:)/A(ip,jp);
for i=1:M
   if i~=ip, Ap(i,:) = A(i,:)-A(i,jp)*Ap(ip,:); end
end
```

5. All the RHS entries are nonnegative (except for $A(M+1,N+1)$ with M=4), but some of them are still fractional. That is why we need to keep imposing the integral constraint on them. Taking the second row as the source row, make a new Gomory cut:

From the 2nd row of simplex tableau (9.3.29)

$$1\cdot x_2+\frac{1}{5}s_2+\left(-1+\frac{4}{5}\right)s_4=2+\frac{4}{5}\rightarrow \underbrace{x_2-s_4-2}_{\text{Integer}}=\frac{4}{5}-\frac{1}{5}s_2-\frac{4}{5}s_4\le 0$$

$$\rightarrow -\frac{1}{5}s_2-\frac{4}{5}s_4+\underbrace{s_5}_{\text{A new slack variable}}=-\frac{4}{5} \tag{9.3.30}$$

then add it to the simplex tableau, choose the pivot element by the minimum ratio rule, and pivot the tableau as

$$\begin{array}{c} \begin{array}{cccccccc} x_1 & x_2 & s_1 & s_2 & s_3 & s_4 & s_5 & b_i \end{array} \\ \left[\begin{array}{cccccccc} 1 & 0 & 0 & 0 & 0 & 1 & 0 & 2 \\ 0 & 1 & 0 & 1/5 & 0 & -1/5 & 0 & 14/5 \\ 0 & 0 & 1 & 1/5 & 0 & -16/5 & 0 & 14/5 \\ 0 & 0 & 0 & 0 & 1 & -3 & 0 & 2 \\ 0 & 0 & 0 & -1/5 & 0 & -4/5 & 1 & -4/5 \\ \hline 0 & 0 & 0 & 1/5 & 0 & 9/5 & 0 & 34/5 \end{array}\right] \end{array} \xrightarrow{\text{Pivoting}} \begin{array}{c} \begin{array}{cccccccc} x_1 & x_2 & s_1 & s_2 & s_3 & s_4 & s_5 & b_i \end{array} \\ \left[\begin{array}{cccccccc} 1 & 0 & 0 & 0 & 0 & 1 & 0 & 2 \\ 0 & 1 & 0 & 0 & 0 & -1 & 1 & 2 \\ 0 & 0 & 1 & 0 & 0 & -4 & 1 & 2 \\ 0 & 0 & 0 & 0 & 1 & -3 & 0 & 2 \\ 0 & 0 & 0 & 1 & 0 & 4 & -5 & 4 \\ \hline 0 & 0 & 0 & 0 & 0 & 1 & 0 & 6 \end{array}\right] \end{array} \tag{9.3.31}$$

$$\text{Ratio}\quad \left|\frac{1/5}{-1/5}\right|=1<\left|\frac{9/5}{-4/5}\right|=\frac{9}{5}$$

Now, all the RHS entries except for $A(M+1,N+1)$ are nonnegative integral numbers so that we can go to the final step.

6. We set the current basic variables (corresponding to the single-one columns) to their corresponding RHS values as the optimal feasible solution:

$$[\underbrace{x_1\ \ x_2\ \ s_1\ \ s_2\ \ s_3}_{\text{Basic variables}}\ \ \underbrace{s_4\ \ s_5}_{\text{Nonbasic variables}}]=[2\ \ 2\ \ 2\ \ 4\ \ 2\ \ 0\ \ 0] \tag{9.3.32}$$

with the optimized (minimized) objective function value $f_o=-6$ (the opposite-signed entry at the lower-right corner). This conforms to the solution (9.3.22-3) that was obtained by using the BB method, and is depicted in Fig. 9.13.

This cutting plane method has been cast into the MATLAB function '`cutting_plane()`' listed below. To use '`cutting_plane()`' for solving the above ILP problem, run the MATLAB script "test_cutting_plane.m":

```
%test_cutting_plane.m
% Use 'cutting_plane()' to solve the ILP problem (9.3.19).
f=[-2; -1]; % Coefficient vector of the objective function to minimize
A=[3 -1; 1 5]; b=[6; 16]; % Inequality constraint Ax<=b
[xo,fmin,Ab] = cutting_plane(f,A,b,[],[])
```

```
function [xo,fmin,Ab] = cutting_plane(f,A,b,Aeq,beq)
% Uses the cutting plane method to solve an ILP problem:
%   Min f(x) = f.'*x  s.t. A*x <= b and Aeq*x = beq
% for the optimal solution xo with optimized objective ftn value fo
%Copyleft: Won Y. Yang, wyyang53@hanmail.net, CAU for academic use only
Nx=length(f); % Number of objective function coefficients and variables
[xo,fmin,Ab] = simplex(f,A,b,Aeq,beq);
[M1,N1]=size(Ab); M=M1-1; N=N1-1; EPS=1e-6;
indfs=find(sum(abs(Ab(1:M,1:N)-round(Ab(1:M,1:N))),2)>EPS);
[rmax,im] = max(abs(Ab(indfs,N1)-round(Ab(indfs,N1))));
is=indfs(im); % Index of the most fractional RHS entry for source row
while ~isempty(is)&rmax>EPS
  Ab1=Ab;
  Ab1(:,end+1)=Ab(:,end); Ab1(:,end-1)=zeros(size(Ab,1),1);
  Ab1(end+1,:)=[Ab(end,1:end-1) 0 Ab(end,end)];
  cut_constraint = floor(Ab(is,:))-Ab(is,:);
  Ab1(end-1,:)=[cut_constraint(1,1:end-1) 1 cut_constraint(1,end)];
  Ab=Ab1  % Tableau with a new Gomory cut added
  [M1,N1]=size(Ab); M=M1-1; N=N1-1;
  ip=M; % Pivot row
  indnz = find(abs(Ab(ip,1:N))>EPS&abs(Ab(M1,1:N))>EPS);
  if isempty(indnz), break; end
  [ratio_min,imin] = min(abs(Ab(M1,indnz)./Ab(ip,indnz))); %
  jp=indnz(imin); % Pivot column
  Ab=pivoting(Ab,M,jp) % Tableau updated by pivoting
  indfs=find(sum(abs(Ab(1:M,1:N)-round(Ab(1:M,1:N))),2)>EPS);
  [rmax,im] = max(abs(Ab(indfs,N1)-round(Ab(indfs,N1))));
  is=indfs(im); % Index of the most fractional RHS entry for source row
end
% Last step to determine the optimal solution
for j=1:Nx
   ind1=find(abs(Ab(1:M,j))>EPS);
   if length(ind1)==1, xo(j,1)=Ab(ind1,N1); end
end
fmin = -Ab(M1,N1); % Optimal objective function value
```

```
>>test_cutting_plane
   xo =    2             fo =   -6  % Optimal solution
           2
```

Before ending this section, let us see a BILP (binary integer linear programming) problem, which is a constrained minimization problem with linear objective and constraint functions such as

$$\text{Min } f(\mathbf{x}) = \mathbf{f}^{\mathrm{T}}\mathbf{x} \tag{9.3.33a}$$

$$\text{s.t. } A\mathbf{x} \le \mathbf{b},\ A_{\mathrm{eq}}\mathbf{x} \le \mathbf{b}_{\mathrm{eq}},\ \text{and } \mathbf{l} \le \mathbf{x} \le \mathbf{u} \tag{9.3.33b}$$

where some variables x(intcon) are restricted to be binary-valued, i.e., either 0 or 1. BILP problems can be solved by using the MATLAB built-in function '`intlinprog()`' with the binary variables bounded in [0, 1]. For example, let us consider a BILP problem described as

$$\text{Min } f(\mathbf{x}) = \mathbf{f}^{\mathrm{T}}\mathbf{x} = [-3 \quad -2]\begin{bmatrix} x_1 \\ x_2 \end{bmatrix} = -3x_1 - 2x_2 \tag{9.3.34a}$$

$$\text{s.t. } A\mathbf{x} = \begin{bmatrix} 1 & 4 \\ 4 & 1 \end{bmatrix} \begin{bmatrix} x_1 \\ x_2 \end{bmatrix} \begin{matrix} \le \\ \le \end{matrix} \begin{bmatrix} 12 \\ 13 \end{bmatrix} = \mathbf{b} \text{ and } \mathbf{l} = \begin{bmatrix} 0 \\ 0 \end{bmatrix} \le \mathbf{x} = \begin{bmatrix} x_1 \\ x_2 \end{bmatrix} \le \begin{bmatrix} 10 \\ 10 \end{bmatrix} = \mathbf{u} \tag{9.3.34b}$$

where x_1 must be binary-valued and x_2 can take any (real) value in [0, 10].

```
%test_BILP.m
% to solve the BILP problem (9.3.34).
f=[-3 -2]; % Coefficient vector of the objective function (9.3.34a)
A=[1 4; 4 1]; b=[12; 13]; % Inequality constraint Ax<=b  (9.3.34b)
Aeq=[]; beq=[]; % No equality constraint Aeq*x=beq
intcon=[1]; % x(1) must be binary-valued
l=[0 0]; u=[1 10]; % Lower/upper bound l<=x<=u
[xo,fo]=intlinprog(f,intcon,A,b,Aeq,beq,l,u)
```

Running the above script "test_BILP.m" yields the following result:

```
>>test_BILP
    xo =     1.0000            fo =   -8.5000
             2.7500
```

Problems

9.1 Nelder–Mead, Steepest Descent, Newton, `fminunc()`, `fminsearch()`

Consider a two-variable objective function

$$f(\mathbf{x}) = x_1^4 - 12x_1^2 - 4x_1 + x_2^4 - 16x_2^2 - 5x_2 - 20\cos(x_1 - 2.5)\cos(x_2 - 2.9) \tag{P9.1.1}$$

with gradient vector function

$$\mathbf{g}(\mathbf{x}) = \nabla f(\mathbf{x}) = \begin{bmatrix} 4x_1^3 - 24x_1 - 4 + 20\sin(x_1 - 2.5)\cos(x_2 - 2.9) \\ 4x_2^3 - 32x_2 - 5 + 20\cos(x_1 - 2.5)\sin(x_2 - 2.9) \end{bmatrix} \tag{P9.1.2}$$

which has nine zeros, as listed in Table P9.1.1.

Table P9.1.1 Extrema (maxima/minima) and saddle points of the function (P9.1.1)

Points	Signs of $(\partial^2 f/\partial x_1^2, \partial^2 f/\partial x_2^2)$		Points	Signs of $(\partial^2 f/\partial x_1^2, \partial^2 f/\partial x_2^2)$	
(1) [0.6965 -0.1423]	−, −	M	(6) [−1.6926 −0.1183]		
(2) [2.5463 -0.1896]	+, −	S	(7) [−2.6573 −2.8219]		
(3) [2.5209 2.9027]	+, +	m	(8) [−0.3227 −2.4257]	−, +	S
(4) [−0.3865 2.9049]			(9) [2.5216 −2.8946]		
(5) [−2.6964 2.9031]					

(Note) M: maximum, m: minimum, S: saddle point (a local maximum w.r.t. one variable and a local minimum w.r.t. the other variable.)

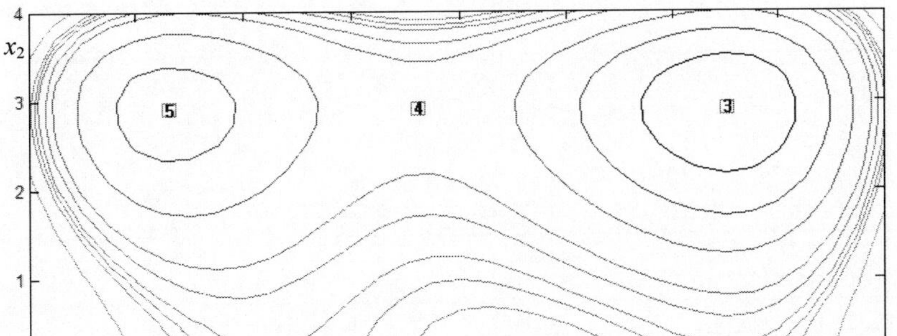

Figure P9.1 The contour, extrema, and saddle points of objective function (P9.1.1)

```
%em09p01.m
% to minimize an objective ftn f(x) given by Eq.(P9.1.1)
clear
f=@(x)x(1)^4-12*x(1)^2-4*x(1)+x(2)^4-16*x(2)^2-5*x(2)
      -20*cos(x(1)-2.5)*cos(x(2)-2.9);
g='em09p01_g';
l=[-4 -4]; u=[4 4]; % Lower/upper bounds
x1=l(1):0.1:u(1); x2=l(2):0.1:u(2); [X1,X2]=meshgrid(x1,x2);
for m=1:length(x1)
   for n=1:length(x2), F(n,m)=feval(f,[x1(m) x2(n)]); end
end
figure(1), clf, mesh(X1,X2,F), figure(2), clf,
contour(x1,x2,F,[-125 -100 -75 -50 -40 -30 -25 -20 0 50])
x0s=[0 0; 3 0; 3 3; 0 3; -3 3; -3 0; -3 -3; 0 -3; 3 -3];
for i=1:size(x0s,1) % find the local extrema by using fsolve()
   x=fsolve(g,x0s(i,:)); [g1,g2]=feval(g,x); xg(i,:)=[x f(x) g1 g2];
end
TolX=1e-4; TolFun=1e-8; MaxIter=100;
x0s=[0 0; 1 0; 1 1; 0 1; -1 1; -1 0; -1 -1; 0 -1; 1 -1; 2 2;-2 -2];
for iter=1:size(x0s,1)
  x0=x0s(iter,:)
  [xo_nd,fo_nd]=opt_Nelder(f,x0,TolX,TolFun,MaxIter);
  fprintf('xo_nd=[%10.4f %10.4f], fo_nd=%10.2f\n',xo_nd,fo_nd)
  [xo_st,fo_st]=opt_steep(f,x0);
  fprintf('xo_st=[%10.4f %10.4f], fo_st=%10.2f\n',xo_st,fo_st)
  [xo_nt,go_nt,xx]=Newtons(g,x0,TolFun,MaxIter);
  fprintf('xo_nt=[%10.4f %10.4f], fo_nt=%10.2f\n',xo_nt,f(xo_nt))
  [xo_uc,fo_uc]=fminunc(f,x0);
  fprintf('xo_uc=[%10.4f %10.4f], fo_uc=%10.2f\n',xo_uc,fo_uc)
  [xo_sh,fo_sh]=fminsearch(f,x0);
  fprintf('xo_sh=[%10.4f %10.4f], fo_sh=%10.2f\n',xo_sh,fo_sh)
end
```

```
function [df,d2f]=em09p01_g(x) % the 1st/2nd derivatives
df(1)=4*x(1)^3-24*x(1)-4+20*sin(x(1)-2.5)*cos(x(2)-2.9); %(P9.1.2)
df(2)=4*x(2)^3-32*x(2)-5+20*cos(x(1)-2.5)*sin(x(2)-2.9); %(P9.1.2)
d2f(1)=12*x(1)^2-24+20*cos(x(1)-2.5)*cos(x(2)-2.9);       %(P9.1.3)
d2f(2)=12*x(2)^2-32+20*cos(x(1)-2.5)*cos(x(2)-2.9);       %(P9.1.3)
```

Table P9.1.2 Points reached by the several optimization functions

Initial point $\mathbf{x}_0$	Points reached by each function				
	opt_Nelder( )	opt_steep( )	opt_Newton( )	fminunc( )	fminsearch( )
(0, 0)	(5)/m				
(1, 0)		(3)/m			
(1, 1)			(9)/m		
(0, 1)				(3)/m	
(-1, 1)			(1)/M		(5)/m
(-1, 0)					
(-1, -1)			(6)/S		
(0, -1)	(9)/m				
(1, -1)		(9)/m			
(2, 2)			(3)/m		
(-2, -2)				(7)/m	

(a) From the contour plots that you get by running the above MATLAB script, "em09p01.m," determine the characteristic of each of the nine points, i.e., whether it is a maximum (M), a minimum (m), or a saddle point (S) that is a minimum w.r.t. one variable and a maximum w.r.t. another variable. Support your judgement by identifying the signs of the second derivatives of $f(\mathbf{x})$ w.r.t. x_1 and x_2. Fill in Table P9.1.1 with the results.

$$\begin{aligned}\partial^2 f/\partial x_1^2 &= 12x_1^2 - 24 + 20\cos(x_1 - 2.5)\cos(x_2 - 2.9)\\ \partial^2 f/\partial x_2^2 &= 12x_1^2 - 32 + 20\cos(x_1 - 2.5)\cos(x_2 - 2.9)\end{aligned} \quad \text{(P9.1.3)}$$

(b) Use the MATLAB functions '`opt_Nelder()`', '`opt_steep()`', '`opt_Newton()`', '`fminunc()`', and '`fminsearch()`' to minimize the objective function (P9.1.1) and fill in Table P9.1.2 with the number and character of the point reached by each method.

(c) Overall, the point reached by each routine depends on the starting point, i.e., the initial value of the independent variable as well as the chatacteristic of its algorithm. Fill in the blanks in the following sentences. If the routines start from the initial point (1,0), (1,1), (0,1), or (2,2), they mostly succeed to reach the minimum ( ). If launched from (−1,1) or (−1,0), they mostly go to the closest local minimum ( ) and if launched from (−1,−1) or (−2,−2), they mostly go to the closest local minimum ( ). If launched from (0,−1), they mostly go to one of the two closest local minima ( ) and ( ). If launched from (1,−1), they mostly go to the closest local minimum ( ).

9.2 Constrained Optimization and *Principal Component Analysis* (*PCA*)

PCA is a technique that can be used to simplify/compress data consisting of many values of several variables by reducing the dimensionality of the data, i.e., by finding a smaller set of (new) variables, called PCs (principal components), that retains most of the information about the data. Note that the information of the data about a variable can be measured by the variance, which can be regarded as the uncertainty of the variable when its value is unknown.

How can the variance of data be computed? If the data about a variable x_1 is given as an $N\times 1$ vector $\mathbf{x}_1$ (N: the number of the data), its (sample) variance is

$$\sigma_{x_1}^2 = \frac{1}{N}\sum_{n=1}^{N} \underline{x}_{1n}^2 = \frac{1}{N}\underline{\mathbf{x}}_1^{\mathrm{T}}\underline{\mathbf{x}}_1 = \frac{1}{N}(\mathbf{x}_1 - m_{\mathbf{x}_1})^{\mathrm{T}}(\mathbf{x}_1 - m_{\mathbf{x}_1}) \text{ with } m_{\mathbf{x}_1} = \frac{1}{N}\sum_{n=1}^{N} x_{1n}: \text{mean} \quad \text{(P9.2.1)}$$

where $\underline{\mathbf{x}}_1 = \mathbf{x}_1 - m_{\mathbf{x}_1}$ is called the demeaned or detrended data in the sense that it is data with its mean or trend removed, say, by using the MATLAB function '`detrend()`'. If the detrended data about M variables (or properties) $\{x_1,\dots,x_M\}$ or the corresponding vector $\mathbf{x}=[x_1\ \dots\ x_M]^{\mathrm{T}}$ are given as an $N\times M$ matrix $\underline{X}$, their (sample) covariance matrix is defined as

$$C = \frac{1}{N-1}\underline{X}^{\mathrm{T}}\underline{X} \text{ } (N: \text{the number of the data, } M: \text{the number of variables}) \quad \text{(P9.2.2)}$$

Note that this matrix is symmetric, and also from Section 1.2.4 or 1.2.7.1, note that the eigenvectors of a symmetric matrix C (to its distinct eigenvalues) are orthogonal.

To find the first PC $\mathbf{z}_1=\underline{X}\mathbf{w}_1$ (which accounts for the largest variance of the data) or its weighting (or loading), vector $\mathbf{w}_1$ can be formulated as a constrained maximization problem like (E9.3.1):

$$\text{Max } f(\mathbf{w}_1) = \text{var}\{\mathbf{z}_1\} = \mathbf{z}_1^{\mathrm{T}}\mathbf{z}_1 = \mathbf{w}_1^{\mathrm{T}}\underline{X}^{\mathrm{T}}\underline{X}\mathbf{w}_1 = \mathbf{w}_1^{\mathrm{T}}C\mathbf{w}_1 \quad \text{(P9.2.3a)}$$

$$\text{s.t. } h(\mathbf{w}_1) = \mathbf{w}_1^{\mathrm{T}}\mathbf{w}_1 - 1 = 0 \quad \text{(P9.2.3b)}$$

This can be solved by applying the Lagrange multiplier method:

$$\text{Max } l(\mathbf{w}_1,\lambda_1) \overset{(9.2.2)}{=} f(\mathbf{w}_1) + \lambda_1 h(\mathbf{w}_1) \overset{(\text{P9.2.3})}{=} \mathbf{w}_1^T C \mathbf{w}_1 + \lambda_1(1-\mathbf{w}_1^T\mathbf{w}_1) \tag{P9.2.4}$$

$$\frac{\partial}{\partial \mathbf{w}_1} l(\mathbf{w}_1,\lambda_1) \underset{(\text{P9.2.4}),(\text{B.6})}{\overset{(9.2.3\text{a})}{=}} 2C\mathbf{w}_1 - 2\lambda_1\mathbf{w}_1 = \mathbf{0}\,;\;\; (C-\lambda_1 I)\mathbf{w}_1 = \mathbf{0} \tag{P9.2.5a}$$

$$\frac{\partial}{\partial \lambda_1} l(\mathbf{w}_1,\lambda_1) \underset{(\text{P9.2.4}),(\text{B.6})}{\overset{(9.2.3\text{b})}{=}} 1-\mathbf{w}_1^T\mathbf{w}_1 = 0 \tag{P9.2.5b}$$

Matching Eq. (P9.2.5a) with Eq. (1.2.12), we see that the solution $\mathbf{w}_1$ and λ_1 to Eqs. (P9.2.5a,b) are nothing but the unit eigenvector(s) and eigenvalue(s) of the covariance matrix C. Besides, since premultiplying Eq. (P9.2.5a) with $\mathbf{w}_1^T$ yields

$$\mathbf{w}_1^T C\mathbf{w}_1 = \lambda_1 \mathbf{w}_1^T\mathbf{w}_1 \overset{(\text{P9.2.5b})}{=} \lambda_1 \tag{P9.2.6}$$

and this is the (original) objective function to maximize, the eigenvalue should be the maximum eigenvalue. Now, to find the second PC $\mathbf{z}_2 = \underline{X}\,\mathbf{w}_2$ (which accounts for the second largest variance of the data) or its weighting vector $\mathbf{w}_2$, we solve another constrained maximization problem:

$$\text{Max } f(\mathbf{w}_2) = \mathbf{w}_2^T C \mathbf{w}_2 \tag{P9.2.7a}$$

$$\text{s.t. } h_1(\mathbf{w}_2) = \mathbf{w}_2^T\mathbf{w}_2 - 1 = 0 \quad \text{and} \quad h_2(\mathbf{w}_2) = \mathbf{w}_2^T\mathbf{w}_1 = 0 \tag{P9.2.7b}$$

This amounts to finding the second maximum eigenvalue λ_2 and the corresponding eigenvector $\mathbf{w}_2$ of the covariance matrix C, where the second constraint (of orthogonality) can be ignored, since the eigenvectors of a symmetric matrix C (corresponding to its distinct eigenvalues) are already orthogonal. Therefore, to reduce the dimension of the data X from M to $p<M$, we postmultiply the data matrix $\underline{X}$ by an $M\times p$ (loading) matrix W_{1p} (called the principal component coefficient or feature space matrix), which has the p (normalized) eigenvectors $\{\mathbf{w}_1, \mathbf{w}_2, \ldots, \mathbf{w}_p\}$ of C corresponding to the p largest eigenvalues, to get the compressed data Y_{1p} (called the score matrix):

$$Y_{1p} = \underline{X} W_{1p} \;\text{ with }\; W_{1p} = [\mathbf{w}_1 \; \mathbf{w}_2 \cdots \mathbf{w}_p] \tag{P9.2.8}$$

To restore the original data, we postmultiply this (compressed) data matrix Y_{1p} by W_{1p}^T:

$$\hat{\underline{X}} = Y_{1p} W_{1p}^T \tag{P9.2.9}$$

```
%test_PCA.m
clear
N=1000; % Data size
X(:,[1 3])=randn(N,2); X(:,2)=X(:,1)/2+0.1*randn(N,1);
Xb = X - repmat(mean(X),N,1); % demeaning data
C=Xb'*Xb/(N-1); % Eq.(P9.2.2)
[V,L]=eig(C); L=diag(L);
[L,ind]=sort(L,'descend'); % Sorting eigenvalues in descending order
V=V(:,ind); % Sorting eigenvectors in the order of eigenvalues
[U,S,W]=svd(Xb) % SVD
[Wp,Yp,lamdas]=princomp(Xb); % Using the MATLAB function princomp()
[Wp V W], [L lamdas]
p=2; W1p=W(:,1:p); % Transformation matrix (P9.2.8)
Y1p=Xb*???; % Data compression (P9.2.8)
Xbh=???*W1p'; % Data restoration (P9.2.9)
Discrepancy=norm(Xbh-Xb)/norm(Xb) % Has the data been restored well?
```

The PCA problem can be solved by using the ED (eigendecomposition) (Section 1.2.11) of $\underline{X}^T\underline{X}$ or SVD (singular value decomposition) (Section 1.2.10.4) of the detrended data matrix $\underline{X}$, as illustrated in the above MATLAB script "test_PCA.m," where the MATLAB function '`princomp()`' has also been used for practice and crosscheck. Complete and run the script twice, once with $p=1$ and once with $p=2$, to compare the two cases in terms of the relative discrepancy (between the restored data $\hat{\underline{X}}$ and the original detrended data $\underline{X}$) and the approximate compression ratio $3/p$.

9.3 Constrained Optimization and Penalty Function Method

Consider the problem of minimizing a nonlinear objective function

$$\underset{\mathbf{x}}{\text{Min}}\, f(\mathbf{x}) = -3x_1 - 2x_2 + M(3x_1 - 2x_2 + 2)^2 \ (M\text{: a large positive number}) \qquad \text{(P9.3.1a)}$$

subject to the constraints

$$\begin{bmatrix} 3 & 4 \\ -2 & -1 \end{bmatrix}\begin{bmatrix} x_1 \\ x_2 \end{bmatrix} \begin{matrix} \le 7 \\ \ge -3 \end{matrix} \quad \text{and} \quad \mathbf{l} = \begin{bmatrix} 0 \\ 0 \end{bmatrix} \le \mathbf{x} = \begin{bmatrix} x_1 \\ x_2 \end{bmatrix} \le \begin{bmatrix} 10 \\ 10 \end{bmatrix} = \mathbf{u} \qquad \text{(P9.3.1b)}$$

(a) With the two values of the weighting factor $M=20$ and 10000 in the objective function (P9.3.1a), use the MATLAB built-in function '`fmincon()`' to find the solutions to the above constrained minimization problem. To do this job, you might have to make the variable parameter M pass to the objective function either through '`fmincon()`' or directly by declaring the parameter as global both in the main program and in the M-file defining (P9.3.1a). In case you are going to have the parameter passed through '`fmincon()`' to the objective function, you should have the parameter included in the input argument list of the objective function as

```
>>em09p03_fM=@(x,M)-3*x(1)-2*x(2)+M*(3*x(1)-2*x(2)+2)^2;
```

Additionally, you should give empty matrices (`[]`) as the 9th input argument (for a nonlinear inequality/equality constraint function '`fcon`') as well as the 10th one (for '`options`'), and the value of M as the 11th one for the function '`fmincon()`':

```
>>xo=fmincon(em09p03_fM,x0,A,b,[],[],l,u,[],[],M)
```

You can complete the MATLAB script "em09p03.m" and run it.

(b) Noting that the 3rd (squared) term of the objective function (P9.3.1a) has its minimum value of zero for $3x_1-2x_2+2=0$ and thus, it actually represents the penalty (Section 9.2.2) imposed for not satisfying the equality constraint

$$3x_1 - 2x_2 + 2 = 0 \qquad \text{(P9.3.2)}$$

tell which of the solutions obtained in (a) is more likely to satisfy this constraint, and support your answer by comparing the values of the left-hand side of this equality for the two solutions.

(c) Removing the 3rd term from the objective function and splitting the equality constraint into two reversed inequality constraints, we can modify the problem as follows:

$$\underset{\mathbf{x}}{\text{Min}}\, f(\mathbf{x}) = -3x_1 - 2x_2 \qquad \text{(P9.3.3a)}$$

subject to the constraints

$$\begin{bmatrix} 3 & 4 \\ -2 & -1 \\ 3 & -2 \\ 3 & -2 \end{bmatrix}\begin{bmatrix} x_1 \\ x_2 \end{bmatrix} \begin{matrix} \le 7 \\ \ge -3 \\ \le -2 \\ \ge -2 \end{matrix} \quad \text{and} \quad \mathbf{l}=\begin{bmatrix} 0 \\ 0 \end{bmatrix} \le \mathbf{x}=\begin{bmatrix} x_1 \\ x_2 \end{bmatrix} \le \begin{bmatrix} 10 \\ 10 \end{bmatrix} = \mathbf{u} \tag{P9.3.3b}$$

Noting that this fits the linear programming, use the MATLAB function '`linprog()`' to solve this problem.

(d) Treating the equality constraint separately from the inequality constraints, we can modify the problem as follows:

$$\underset{\mathbf{x}}{\text{Min}}\, f(\mathbf{x}) = -3x_1 - 2x_2 \tag{P9.3.4a}$$

subject to the constraints

$$\begin{bmatrix} 3 & -2 \\ 3 & 4 \\ -2 & -1 \end{bmatrix}\begin{bmatrix} x_1 \\ x_2 \end{bmatrix} \begin{matrix} = -2 \\ \le 7 \\ \ge -3 \end{matrix} \quad \text{and} \quad \mathbf{l}=\begin{bmatrix} 0 \\ 0 \end{bmatrix} \le \mathbf{x}=\begin{bmatrix} x_1 \\ x_2 \end{bmatrix} \le \begin{bmatrix} 10 \\ 10 \end{bmatrix} = \mathbf{u} \tag{P9.3.4b}$$

Solve this problem twice, once by using '`linprog()`' and once by using '`fmincon()`', and see if the solutions agree with the solution obtained in (c). Which takes less time?

```
%em09p03.m - constrained optimization
em09p03_f=@(x)-?*x(1)-2*x(?); % Objective function (P9.3.1)
em09p03_fM=@(x,M)-3*x(1)-2*x(2)+?*(3*x(1)-2*x(2)+2)^2; % Penalized
A=[3 4;2 1]; b=[7; 3]; Aeq=[3 -2]; beq=-2;
l=[0 0]; u=[10 10]; % nonlcon=[]; options=[];
disp('(a) using fmincon() with the equality represented by penalty')
x0=[0 0]; M=20; xo1=fmincon(em09p03_fM,x0,A,b,[],[],l,u,[],[],M)
M=10000; xo2=fmincon(em09p03_??,x0,A,b,[],[],l,u,[],[],?)
disp('(b) how well the constraints are satisfied')
[A; Aeq]*[xo1' xo2']-[b; beq]*[1 1] %<=0 & =0?
disp('(c) LP with two symmetric inequalities for one equality')
f=[-3 -2]; A=[3 4;2 1;3 -2;-3 2]; b=[7; 3; -2; 2];
xo_lp1=linprog(f,A,b,[],[],l,u)
disp('(d1) using LP with equality as it is')
f=[-3 -2]; A=[3 4;2 1]; b=[7; 3];
tic, xo_lp2=linprog(?,A,?,Aeq,???,l,?), time_lp=toc
disp('(d2) using fmincon() with the equality as it is')
tic, xo_con=fmincon(em09p03_?,x0,A,b,???,beq,?,u), time_con=toc
```

9.4 Constrained Optimization and Penalty Function Method

Consider the following constrained minimization problem:

$$\text{Min}\, f(\mathbf{x}) = \{(x_1+1.5)^2+5(x_2-1.7)^2\}\{(x_1-1.4)^2+0.6(x_2-0.5)^2\} \tag{P9.4.1a}$$

$$\text{s.t.}\quad \mathbf{g}(\mathbf{x}) = \begin{bmatrix} -x_1 \\ -x_2 \\ 3x_1 - x_1x_2 + 4x_2 - 7 \\ 2x_1 + x_2 - 2.9 \\ 3x_1 - 4x_2^2 - 4x_2 \end{bmatrix} \le \begin{bmatrix} 0 \\ 0 \\ 0 \\ 0 \\ 0 \end{bmatrix} \tag{P9.4.1b}$$

(a) Noting that this is the same problem that was dealt with in Example 9.5, except for the 4th constraint being changed slightly, modify the MATLAB script file "em09e05.m" to make a new one named "em09p04.m" and run it to solve this problem. Do the solution points reached by the MATLAB routines '`opt_Nelder()`'/ '`fminsearch()`'/'`fminunc()`' satisfy all the constraints so that they are in the admissible region? If not, tell which one of the five constraints is the most violated.

(b) Suppose the fourth constraint was not satisfied by one or more of the MATLAB routines in (a). Then, how would you modify the weighting coefficient vector v (in the penalized objective function saved as an M-file named "em09e05_f.m") so that the violated constraint can be paid more respect? Choose one of the following two weighting coefficient vectors:

(i) **v**= `[1 1 1 1/3 1]` (ii) **v**=`[1 1 1 3 1]`

and modify the function file "em09e05_f.m" with the coefficient vector that you have chosen. Then run the script "em09p04.m" twice, once with the initial guess $\mathbf{x}_0$=[0.4 0.5] and once with $\mathbf{x}_0$=[0.2 4], fill in the blanks of Table P9.4 with the results, and see if the fourth constraint is still violated by the points reached by the optimization routines.

(c) Rather than the penalty method, use the constrained optimization routine '`fmincon()`' to solve the problem and fill in Table P9.4 with the results as to the point reached and the corresponding values of the objective/constraint functions.

(d) Based on the results listed in Table P9.4, circle the right word in each of the parentheses in the following sentences:

- For penalty methods, the non-gradient-based minimization functions like '`Nelder()`'/ '`fminsearch()`' may work (better, worse) than the gradient-based minimization function '`fminunc()`'.
- If some constraint is violated, you had better (increase, decrease) the corresponding weight coefficient.

Table P9.4 The results of penalty methods depending on the initial guess and weighting factor

v		Starting point $\mathbf{x}_0$=[0.4 0.5]				$\mathbf{x}_0$=[0.2 4]			
		Nelder	fminsearch	fminunc	fmincon	Nelder	fminsearch	fminunc	fmincon
[1 1 1 1 1]	$\mathbf{x}^o$	1.17 0.56		1.23 0.58	1.17 0.56		1.17 0.56		
	f^o	0.76		0.45	0.76		0.76		
	$\mathbf{c}^o$	-1.17 -0.56 -1.90 -0.00 -0.00		-1.23 -0.58 -1.70 0.14 -0.00	-1.17 -0.56 -1.90 -0.00 -0.00		-1.17 -0.56 -1.90 -0.00 -0.00		
[1 1 1 3 1]	$\mathbf{x}^o$		1.17 0.56		-	1.17 0.56		0.82 0.43	-
	f^o		0.76		-	0.76		4.55	-
	$\mathbf{c}^o$		-1.17 -0.56 -1.90 -0.00 -0.00		-	-1.17 -0.56 -1.90 -0.00 -0.00		-0.82 -0.43 -3.17 -0.83 -0.00	-

9.5 Nonnegative Constrained LS and Constrained Optimization

Consider the problem of minimizing a nonlinear objective function

$$\text{Min } \|C\mathbf{x}-\mathbf{d}\|^2 = [C\mathbf{x}-\mathbf{d}]^T[C\mathbf{x}-\mathbf{d}] \tag{P9.5.1a}$$

subject to the constraints

$$x = \begin{bmatrix} x_1 \\ x_2 \end{bmatrix} \geq \begin{bmatrix} 0 \\ 0 \end{bmatrix} = \mathbf{l} \tag{P9.5.1b}$$

where

$$C = \begin{bmatrix} 1 & 2 \\ 3 & 4 \\ 5 & 1 \end{bmatrix}, \quad \mathbf{d} = \begin{bmatrix} 5.1 \\ 10.8 \\ 6.8 \end{bmatrix} \tag{P9.5.1c}$$

(a) Noting that this problem has no other constraints than the lower bound, apply the constrained linear least squares function '`lsqlin()`' to find the solution.
(b) Noting that the lower bounds for all the variables are zeros, apply the MATLAB built-in function '`lsqnonneg()`' to find the solution.
(c) Use the general purpose constrained optimization routine '`fmincon()`' to find the solution.

```
%em09p05.m
% use lsqlin(constrained LS),lsqnonneg(nonnegative LS) & fmincon
C=[1 2; 3 4; ? ?]; d=[??? 10.8 6.8].'; l=[0 0]; u=[]; % inf*[1 1];
% To use lsqlin(),
xo_ll=lsqlin(C,d,[],[],[],[],l) % To solve Cx≈d with no constraint Ax<=b
% To use lsqnonneg(),
xo_nl=lsqnonneg(C,d) % Nonnegative LS solution to Cx≈d
% To use fmincon(),
x0=zeros(1,2);  %x(1,:)=[.2 .8]; %Initial point
em09p05_f=@(x)(C*x(:)-?).'*(?*x(:)-d);
xo_con=fmincon(em09p05_f,x0,[],[],[],[],l)
```

9.6 Constrained Optimization Problems

Solve the following constrained optimization problems by using the MATLAB built-in function '`fmincon()`' and fill in Table P9.6 with the results.

(a)

$$\text{Min}\, f(\mathbf{x}) = x_1^3 - 5x_1^2 + 6x_1 + x_2^2 - 2x_2 + x_3 \tag{P9.6.1a}$$

subject to the constraints

$$\begin{aligned} x_1^2 + x_2^2 - x_3 &\leq 0 \\ x_1^2 + x_2^2 + x_3^2 &\geq 6 \quad \text{and} \quad \mathbf{x} = \begin{bmatrix} x_1 \\ x_2 \\ x_3 \end{bmatrix} \geq \begin{bmatrix} 0 \\ 0 \\ 0 \end{bmatrix} = \mathbf{l} \\ x_3 &\leq 5 \end{aligned} \tag{P9.6.1b}$$

Try the function '`fmincon()`' with the initial guesses listed in Table P9.6.

```
function [C,Ceq]=em09p06a_c(x)
C=[x(1).^2+x(2).^2-x(3); 6-x(1).^2-x(2).^2-x(3).^2; x(3)-5];
Ceq=[]; % No equality constraint
```

```
%em09p06a.m
fa=@(x)x(1)^3-5*x(1)^2+6*x(1)+x(2)^2-2*x(2)+x(3);
x0s=[0 0 0;1 1 5;0 0 5;1 0 2];
A=[]; b=[]; Aeq=[]; beq=[]; l=zeros(size(x0s(1,:))); u=[];
for i=1:size(x0s,1)
  x0=x0s(i,:)
  [xo,fo]=fmincon(fa,x0,A,b,Aeq,beq,l,u,'em09p06a_c')
end
```

(b1) $$\text{Max } f(\mathbf{x}) = x_1 x_2 x_3 \quad \text{(P9.6.2a)}$$

subject to the constraints

$$x_1x_2 + x_2x_3 + x_1x_3 = 3 \ \text{ and } \ \mathbf{x} = \begin{bmatrix} x_1 \\ x_2 \\ x_3 \end{bmatrix} \geq \begin{bmatrix} 0 \\ 0 \\ 0 \end{bmatrix} \quad \text{(P9.6.2b)}$$

Try the function '`fmincon()`' with the initial guesses listed in Table P9.6.

(b2) $$\text{Min } f(\mathbf{x}) = x_1 x_2 x_3 \quad \text{(P9.6.3a)}$$

subject to the constraints (P9.6.2b).
Try the function '`fmincon()`' with the initial guesses listed in Table P9.6.

(c1) $$\text{Max } f(\mathbf{x}) = x_1x_2 + x_2x_3 + x_1x_3 \quad \text{(P9.6.4a)}$$

subject to the constraints

$$x_1 + x_2 + x_3 = 3 \ \text{ and } \ \mathbf{x} = \begin{bmatrix} x_1 \\ x_2 \\ x_3 \end{bmatrix} \geq \begin{bmatrix} 0 \\ 0 \\ 0 \end{bmatrix} \quad \text{(P9.6.4b)}$$

Try the function '`fmincon()`' with the initial guesses listed in Table P9.6.

(c2) $$\text{Min } f(\mathbf{x}) = x_1x_2 + x_2x_3 + x_1x_3 \quad \text{(P9.6.5a)}$$

subject to the constraints (P9.6.4b).
Try the function '`fmincon()`' with the initial guesses listed in Table P9.6.

(d) $$\text{Min } f(\mathbf{x}) = \frac{10000}{x_1 x_2^2} \quad \text{(P9.6.6a)}$$

Table P9.6 The results of applying '`fmincon()`' with different initial guess

	Initial guess $\mathbf{x}_0$	Lower bound	$\mathbf{x}^o$, $f(\mathbf{x}^o)$	Remark (warning)
(a)	[0 0 0]	0		
	[1 1 5]	0	[1.29 0.57 2.00], 2.74	a local minimum
(b1)	[0 0 0]	0	[1.00 1.00 1.00], −1	the maximum with f^o=-(−1)=1 O
	[10 10 10]	0		
(b2)	[0 0 0]	0		
	[2 2 2]	0	[1.00 1.00 1.00], 1	not a minimum, but the maximum X
	[0.1 0.1 3]	0		
(c1)	[0 0 0]	0		
	[2 2 2]	0	[1 1 1], −3	the maximum with f^o=-(−3)=3 O
	[0 1 2]	0		
(c2)	[0 0 0]	0		
	[2 2 2]	0		
	[0 1 2]	0	[0 0 3], 0	one of many minima O
(d)	[1.0 0.5]	0		
	[2 5]	0	[5.77 8.17], 25.98	a local minimum
(e)	[0 0 0]	None	[1.00 2.50 2.50], 0.5	the global minimum
	[1 1 1]	None		

subject to the constraints

$$x_1^2 + x_2^2 = 100 \quad \text{and} \quad \mathbf{x} = \begin{bmatrix} x_1 \\ x_2 \end{bmatrix} > \begin{bmatrix} 0 \\ 0 \end{bmatrix} \tag{P9.6.6b}$$

Try the function '`fmincon()`' with the initial guesses listed in Table P9.6.

(e)
$$\text{Min}\, f(\mathbf{x}) = (x_1-1)^2 + (x_2-3)^2 + (x_3-2)^2 \tag{P9.6.7a}$$

subject to the constraint

$$x_1 \le x_2 \le x_3 \tag{P9.6.7b}$$

Express this constraint in the form of

$$\begin{bmatrix} \quad & & \quad \end{bmatrix} \begin{bmatrix} x_1 \\ x_2 \\ x_3 \end{bmatrix} \le \begin{bmatrix} 0 \\ 0 \end{bmatrix} \tag{P9.6.6b}$$

and try the function '`fmincon()`' with the initial guesses listed in Table P9.6.

(Note) To use '`fmincon()`' for solving a maximization problem, we must reverse the sign of the objective function. Note also that the objective functions (P9.6.3a) and (P9.6.5a) have infinitely many minima with the value $f(\mathbf{x})=0$ in the admissible region satisfying the constraints.

9.7 Constrained Optimization on Location

A company has three factories that are located at the points (−16, 4), (6, 5), and (3,−9), respectively, in the x_1x_2-plane, and the numbers of deliveries to those factories are 5, 6, and 10 per month, respectively. The company has a plan to build a new warehouse in its site bounded by

$$|x_1 - 1| + |x_2 - 1| \le 2 \tag{P9.7.1}$$

and wants to minimize the monthly mileage of delivery trucks in determining the location of a new warehouse on the assumption that the distance between two points represents the driving distance.

(a) What is the objective function that must be defined in the following script, "em09p07.m"?
(b) Find the statement defining the inequality constraint (P9.7.1) in the MATLAB codes below.
(c) Complete and run the MATLAB script "em09p07.m" to get the optimum location of the new warehouse.

```
%em09p07.m
% to solve the warehouse location problem
f=@(x)sqrt(sum((repmat(x,1,3)-[-16 ? 3; 4 5 -9]).^2))*[5;6;??];
x0=[1;1]; A=[]; b=[]; Aeq=[]; beq=[]; l=[]; u=[];
xo=fmincon(f,x0,A,b,Aeq,beq,l,u,'em09p07_c')
```

```
function [C,Ceq]=em09p07_c(x)
C=sum(abs(x-[1;?]))-2;  Ceq=[];
```

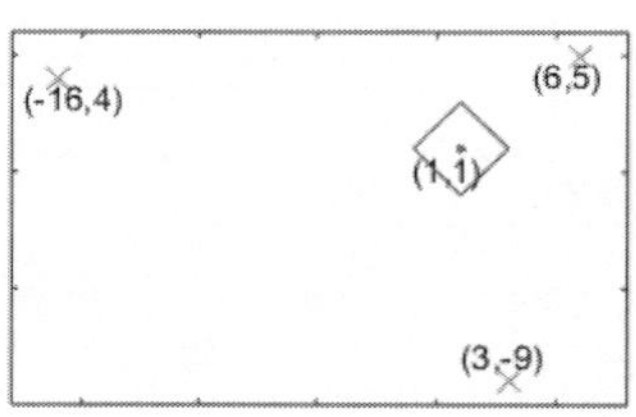

Figure P9.7 Location of a warehouse to be fixed

9.8 Linear Programming (LP)

Consider the problem of maximizing a linear objective function

$$\text{Max}\ f(\mathbf{x}) = \mathbf{f}^T\mathbf{x} = [3 \quad 2 \ -1]\begin{bmatrix} x_1 \\ x_2 \\ x_3 \end{bmatrix} \tag{P9.8.1a}$$

subject to the constraints

$$A\mathbf{x} = \begin{bmatrix} 3 & -2 & 0 \\ -3 & -4 & 0 \\ -2 & -1 & 0 \end{bmatrix}\begin{bmatrix} x_1 \\ x_2 \\ x_3 \end{bmatrix} \begin{matrix} = \\ \geq \\ \geq \end{matrix} \begin{bmatrix} -2 \\ -7 \\ -3 \end{bmatrix} = \mathbf{b} \ \text{and} \ \mathbf{l} = \begin{bmatrix} 0 \\ 0 \\ 0 \end{bmatrix} \leq \mathbf{x} = \begin{bmatrix} x_1 \\ x_2 \\ x_3 \end{bmatrix} \leq \begin{bmatrix} 10 \\ 10 \\ 10 \end{bmatrix} = \mathbf{u} \tag{P9.8.1b}$$

Complete the script "em09p08.m" and run it to use the MATLAB built-in functions '`linprog()`' and '`fmincon()`' for solving this problem.

```
%em09p08.m - maximization with Linear Programming
disp('Max [3 2 -1][x1 x2 x3]'' using LP(Linear Programming)')
f=-[3 2 ??]; % Coefficients of LP objective ftn to be minimized
A=[3 ? 0; ? 1 0]; b=[7;?]; % Inequality(<=) constraints
Aeq=[3 -2 ?]; beq=-2; % Equality(=) constraints
l=[0 0 0]; u=[10 10 10]; %lower/upperbound of the variables
tic, [xo_lp,fo_lp]=linprog(f,A,b,Aeq,beq,l,u), toc
fc=@(x)-?*x(1)-2*x(?)+x(3); % Objective ftn of fmincon()
x0=zeros(3,1); % Starting point
tic, [xo_con,fo_con]=fmincon(fc,x0,A,b,Aeq,beq,l,u), toc
format rat, xo_lp, -fo_lp, xo_con, -fo_con, format short
```

9.9 Linear Programming (LP) Problem Formulation

A company uses two machines, A and B, to make two products, P and Q, (each yielding profits of \$25 and \$30, respectively). A, available for 40 hours, needs the processing times of 50 and 24 minutes to make P and Q, respectively, and B, available for 35 hours, needs the processing times of 30 and 33 minutes to make P and Q, respectively. The company is supposed to determine the numbers of units of P and Q to produce so that the total profit can be maximized. Formulate this problem as an LP one.

9.10 Dual Simplex Method

Consider the problem of minimizing a linear objective function

$$\text{Min}\ f(\mathbf{x}) = \mathbf{f}^T\mathbf{x} = [2 \quad 1]\begin{bmatrix} x_1 \\ x_2 \end{bmatrix} = 2x_1 + x_2 \tag{P9.10.1a}$$

$$\text{s.t.}\ \ A\mathbf{x} = \begin{bmatrix} -3 & -1 \\ -1 & -2 \end{bmatrix}\begin{bmatrix} x_1 \\ x_2 \end{bmatrix} \begin{matrix} \leq \\ \leq \end{matrix} \begin{bmatrix} -8 \\ -6 \end{bmatrix} = \mathbf{b} \ \text{and} \ \mathbf{x} \geq \mathbf{0} \tag{P9.10.1b}$$

To apply the simplex method, we construct the simplex tableau as

$$\begin{matrix} x_1 & x_2 & s_1 & s_2 & b_i \end{matrix}$$
$$\begin{bmatrix} -3 & -1 & 1 & 0 & -8 \\ -1 & -2 & 0 & 1 & -6 \\ 2 & 1 & 0 & 0 & 0 \end{bmatrix} \tag{P9.10.2b}$$

But we can no longer continue the simplex process, since there is no negative entry in the last row, implying that no (entering) variable can contribute to decreasing the objective function value. Besides, the current solution

$$\overset{\text{Nonbasic variables}}{[\;x_1\;\;x_2} \quad \overset{\text{Basic variables}}{s_1\;\;s_2\;]} = [\,0 \quad 0 \;\; -8 \;\; -6\,] \tag{P9.10.3}$$

is not feasible because it does not satisfy the constraints, as revealed by the negative entries in the RHS column. Is there nothing we can do about this puzzling situation? If we plot the objective function together with the constraints as in Fig. P9.10(a), we can see the optimal solution $\mathbf{x}^o=[2 \;\; 2]^T$ with $f(\mathbf{x}^o)=6$ at a vertex of the polygon-type feasible region, implying that there must be a systematic way to get it. A breakthrough would be to convert the original (primal) LP problem into its dual LP problem as

$$\text{Max } f(\mathbf{y}) = -\mathbf{b}^T\mathbf{y} = [8 \;\; 6]\begin{bmatrix} y_1 \\ y_2 \end{bmatrix} \qquad\qquad \text{Min } f(\mathbf{y}) = \mathbf{b}^T\mathbf{y} = [-8 \;\; -6]\begin{bmatrix} y_1 \\ y_2 \end{bmatrix}$$

$$\text{s.t. } A^T\mathbf{y} = \begin{bmatrix} -3 & -1 \\ -1 & -2 \end{bmatrix}\begin{bmatrix} y_1 \\ y_2 \end{bmatrix} \begin{matrix} \ge \\ \ge \end{matrix} \begin{bmatrix} -2 \\ -1 \end{bmatrix} = -\mathbf{f} \text{ or equivalently, s.t. } -A^T\mathbf{y} = \begin{bmatrix} 3 & 1 \\ 1 & 2 \end{bmatrix}\begin{bmatrix} y_1 \\ y_2 \end{bmatrix} \begin{matrix} \le \\ \le \end{matrix} \begin{bmatrix} 2 \\ 1 \end{bmatrix} = \mathbf{f} \tag{P9.10.4}$$

$$\mathbf{y} \ge 0 \qquad\qquad\qquad\qquad \mathbf{y} \ge 0$$

where the objective function coefficient vector **f** and the RHS vector **b** are switched, and the constraint coefficient matrix A is transposed and opposite-signed. Then we can construct the simplex tableau and continue with the simplex process as

$$\begin{matrix} x_1 & x_2 & s_1 & s_2 & b_i \end{matrix}\qquad\qquad\qquad\qquad\qquad\qquad \begin{matrix} x_1 & x_2 & s_1 & s_2 & b_i \end{matrix}$$

$$\begin{bmatrix} \textcircled{3} & 1 & 1 & 0 & 2 \\ 1 & 2 & 0 & 1 & 1 \\ -8 & -6 & 0 & 0 & 0 \end{bmatrix} \rightarrow \begin{bmatrix} 1 & 1/3 & 1/3 & 0 & 2/3 \\ 0 & \textcircled{5/3} & -1/3 & 1 & 1/3 \\ 0 & -10/3 & 8/3 & 0 & 16/3 \end{bmatrix} \rightarrow \begin{bmatrix} 1 & 0 & 2/5 & -1/5 & 3/5 \\ 0 & 1 & -1/5 & 3/5 & 1/5 \\ 0 & 0 & 2 & 2 & 6 \end{bmatrix} \tag{P9.10.5}$$

which yields the optimal solution $\mathbf{y}^o=[3/5 \;\; 1/5]^T$ with $f(\mathbf{y}^o)=6$. Still, we are not relieved, because it is not clear how to find $\mathbf{x}^o$ even if the objective function values of the primal and dual LP problems are the same: $f(\mathbf{y}^o)=f(\mathbf{x}^o)=6$. (In fact, the values of $\mathbf{x}^o=[x_1^o \;\; x_2^o]$ can be found from the last row in the columns corresponding to the slack variables s_1 and s_2, respectively.) What is an alternative? It is to modify the simplex method into the dual simplex method as follows.

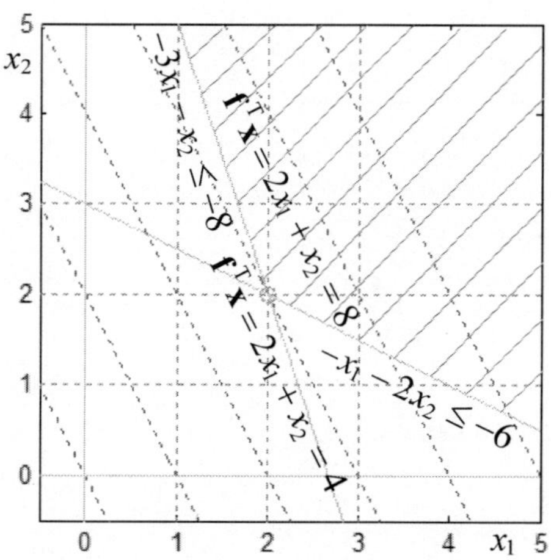

(a) For the primal LP problem (P9.10.1)

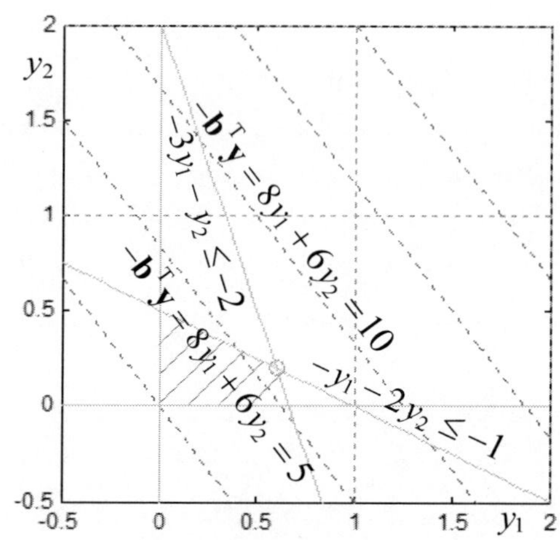

(b) For the dual LP problem (P9.10.4)

Figure P9.10 Plot of the objective function and constraints of the primal/dual LP problems

```
function [xo,fmin,Ab] = simplex_dual(f,A,b,Aeq,beq)
% Uses the dual simplex method to solve an LP problem:
%   Min f(x) = f.'*x  s.t. A*x <= b and Aeq*x = beq
% for the optimal solution xo with optimized objective function value
[M,N] = size(A); % Size of coefficient matrix for inequality
fT=f(:).';
[Meq,Neq] = size(Aeq);  % Size of coefficient matrix for equality
MMeq2=M+Meq*2; EPS=1e-6;
% Construct the simplex tableau like Eq.(9.3.8)
if Meq==0,  A1=A;  b1=b(:);  % No equality constraint
 else A1=[A; Aeq; -Aeq]; b1=[b(:); beq; -beq];
      if N~=Neq, error('N must be equal to Neq in simplex_dual()'); end
end
Ab = [A1 eye(MMeq2) b1; fT zeros(1,MMeq2) 0]
M1=MMeq2+1; NM=N+MMeq2; NM1=NM+1; % Row/Column sizes of the tableau
% Choose the pivot row with the minimum entry in the last row
[AbN1_min,ip]=min(Ab(1:MMeq2,NM1)); % Index of the most negative entry
while AbN1_min<-EPS
  indp = find(AbN1_min*Ab(ip,1:NM)>EPS&abs(Ab(M1,1:NM))>EPS);
  if isempty(indp), disp('Infeasible case!'); break; end
  ratios=abs(Ab(M1,indp)./Ab(ip,indp));
  [Abmin,jmin] = min(ratios);
  jp = indp(jmin); % Pivot column jp by the minimum ratio rule
  % Pivoting
  Ab(ip,:) = Ab(ip,:)/Ab(ip,jp);
  for i=1:M1
     if i~=ip, Ab(i,:) = Ab(i,:)-Ab(i,jp)*Ab(ip,:); end
  end
  [AbN1_min,ip]=min(Ab(1:MMeq2,NM1)); %Index of the most negative entry
end
% Last step to determine the optimal solution
xo = zeros(N,1);
if sum(Ab(1:MMeq2,NM1)<-EPS)>0, disp('Infeasible case!'); end
for j=1:N
   ind1=find(abs(Ab(1:MMeq2,j))>EPS);
   if length(ind1)==1, xo(j,1)=Ab(ind1,NM1);  end
end
fmin = -Ab(M1,NM1); % Optimal objective function value
```

Primal Simplex Method

(Step 0) Construct the simplex tableau (9.3.8) as an $(M+1)\times(M+N+1)$ matrix A from the M constraints and objective function in N variables.

(Step 1) If there is no negative entry in the last row, go to Step 3. Otherwise, choose the j_pth column containing the most negative entry as the pivot column.

(Step 2) If there is no positive entry in the pivot column, declare "an unbounded case" and stop. Otherwise, choose the pivot element $a_{i_p,j_p}>0$ with minimum ratio $\text{Min}\{b_i/a_{i,j_p}\}$ in the pivot column, do the pivot operation on that element, and go to Step 1.

(Step 3) Set the values of basic variables (corresponding to single-one columns) to the value (corresponding to the single one) in the last (RHS) column. Also, set the optimal value of the objective function to the opposite-signed value of the entry at the lower-right corner of the tableau. Note that if there is any negative entry in the RHS column, the solution is infeasible. Then the process is terminated.

Dual Simplex Method

(Step 0) The same as that of the primal simplex method.

(Step 1) If there is no negative entry in the last (RHS) column, go to Step 3. Otherwise, choose the i_pth row containing the most negative entry as the pivot row.

(Step 2) If there is no negative entry in the pivot row, declare "an infeasible case" and stop. Otherwise, choose the pivot element $a_{i_p,j_p}<0$ with minimum ratio Min{ $a_{M+1,j}/a_{i_p,j}$ } in the pivot row, do the pivot operation on that element, and go to Step 1.

(Step 3) The same as that of the primal simplex method.

The dual simplex method has been cast into the above MATLAB function, '`simplex_dual()`'. Try the following four ways to solve the above LP problem (P9.10.1):

(i) Try the simplex method (implemented by '`simplex()`').

```
>>f=[2 1]; A=[-3 -1; -1 -2]; b=[-8; -6];
>>[xo_simplex,fo_simplex,Ab_simplex]=simplex(f,A,b,[],[])
```

(ii) Convert the LP problem into its dual form (P9.10.4) and use the simplex method (implemented by '`simplex()`'). Read the value of $\mathbf{x}^o=[x_1^o \ \ x_2^o]$ from the resulting simplex tableau.

```
>>[yo_dual,fo_dual,Ab_dual] = simplex(b,????,f,[],[])
```

(iii) Use the dual simplex method (implemented by '`simplex_dual()`') directly.

```
>>[xo_dual,fo_dual1,Ab_dual1] = simplex_????(f,A,b,[],[])
```

(iv) Use the MATLAB built-in function '`linprog()`'.

```
>>[xo_linprog,fo_linprog] = linprog(f,A,b,[],[])
```

9.11 Mixed Integer Linear Programming (MILP)

Consider the problem of maximizing a linear objective function

$$\text{Max } f(\mathbf{x})=\mathbf{f}^T\mathbf{x}=[3 \ \ 2]\begin{bmatrix}x_1\\x_2\end{bmatrix}=3x_1+2x_2 \tag{P9.11.1a}$$

subject to the constraints

$$A\mathbf{x}=\begin{bmatrix}1&4\\4&1\end{bmatrix}\begin{bmatrix}x_1\\x_2\end{bmatrix}\begin{matrix}\le\\\le\end{matrix}\begin{bmatrix}12\\13\end{bmatrix}=\mathbf{b} \text{ and } \mathbf{l}=\begin{bmatrix}0\\0\end{bmatrix}\le\mathbf{x}=\begin{bmatrix}x_1\\x_2\end{bmatrix}\le\begin{bmatrix}5\\5\end{bmatrix}=\mathbf{u} \tag{P9.11.1b}$$

where x_2 must be integer-valued.

Complete the MATLAB script "em09p11.mm," run it to solve this MILP problem, and plot the solution point as in Fig. 9.13 to check the validity of the solution by inspection.

```
%em09p11.m
f=[-3 ??]; % Coefficient vector of the objective function
A=[1  ?; 4  1]; b=[??; 13]; % Inequality constraint Ax<=b
Aeq=[]; beq=[]; % No equality constraint Aeq*x=beq
l=[0  0]; u=[5  5]; % Lower/upper bound l<=x<=u
for iter=1:3
   if iter==1,  intcon=[1]; % x(1) must be integer-valued
    elseif iter==2,  intcon=[2]; % x(2) must be integer-valued
    else intcon=[1 2]; % x([1 2]) must be integer-valued
   end
   [xo,fo]=intlinprog(f,intcon,A,b,Aeq,beq,l,u)
   subplot(2,2,iter), plot_LP2(f,A,b,Aeq,beq,xo,l,u,[0:2:10])
end
```

```
function [xo_lp,fo_lp,icase] = plot_LP2(f,A,b,Aeq,beq,x1,l,u,fvs)
% Plot the objective ftn and constraints of LP problem with 2 vars
if nargin<9, fvs=[-20:2:20]; end % Objective ftn values for contour
if nargin<7, l=[0 0]; u=[10 10]; end  % Lower/upper bound 0<=x<=u
if nargin<6, x1=[]; end  % Some special point
xleft=-0.5; xright=u(1); ybottom=-0.5; ytop=u(2);
xx=[xleft:0.01:xright]; yy=[ybottom:0.01:ytop];
[M,N]=size(A);   [Meq,Neq]=size(Aeq);
if N==2
  for m=1:length(fvs)   % Objective function
     yy=(fvs(m)-f(1)*xx)/f(2); plot(xx,yy,':'), hold on
  end
  for i=1:M, yy=(b(i)-A(i,1)*xx)/A(i,2); plot(xx,yy,'g');  end
  for i=1:Meq, yy=(beq(i)-Aeq(i,1)*xx)/Aeq(i,2); plot(xx,yy,'r');  end
  plot([xleft xright],[0 0],'g', [0 0],[ybottom ytop],'g')
  xis=[1:3]; yis=[1:3];
  for xi=xis, plot([xi xi],[ybottom ytop],'r:');  end
  for yi=yis, plot([xleft xright],[yi yi],'r:');  end
  [xo_lp,fo_lp,icase]=linprog(f,A,b,Aeq,beq,l,u);
  if nargin>5&length(x1)==2, plot(x1(1),x1(2),'bo');
   elseif icase==1,  plot(xo_lp(1),xo_lp(2),'mo');
  end
  axis('equal'), axis([xleft xright ybottom ytop])
end
```

9.12 Branch-and-Bound Method and Cutting Plane Method for Integer Linear Programming (ILP)

Consider the following ILP:

$$\text{Min } f(\mathbf{x}) = \mathbf{f}^{\mathrm{T}}\mathbf{x} = [-2 \quad -1]\begin{bmatrix} x_1 \\ x_2 \end{bmatrix} = -2x_1 - x_2 \tag{P9.12.1a}$$

$$\text{s.t. } A\mathbf{x} = \begin{bmatrix} 3 & -1 \\ -1 & 7 \end{bmatrix}\begin{bmatrix} x_1 \\ x_2 \end{bmatrix} \begin{matrix} \le \\ \le \end{matrix} \begin{bmatrix} 6 \\ 14 \end{bmatrix} = \mathbf{b} \text{ and } \mathbf{l} = \begin{bmatrix} 0 \\ 0 \end{bmatrix} \le \mathbf{x} = \begin{bmatrix} x_1 \\ x_2 \end{bmatrix} \le \begin{bmatrix} 10 \\ 10 \end{bmatrix} = \mathbf{u} \tag{P9.12.1b}$$

where x_1 and x_2 must be integer-valued.

(a) Find the solution for the LP relaxation without the integral constraints.

```
f=[-2 -1]; % Coefficient vector of the objective function
A=[3 -1; -1 7]; b=[6; 14]; % Inequality constraint Ax<=b
l=[0 0]; u=[10 10]; % Lower/upper bound l<=x<=u
[xo_LP,fo_LP,A_LP]=simplex(f,A,b,[],[])
```

(b) Use the BB (branch-and-bound) method to get the optimal feasible solution of the ILP.

```
[xo_bb,fo_bb]=branch_and_?????(f,A,b,[],[])
```

(c) Use the cutting plane method to get the optimal feasible solution of the ILP.

```
[xo_cut,fo_cut,Ab_cut]=cutting_?????(f,A,b,[],[])
```

CHAPTER 10

Probability and Random Variables

CHAPTER OUTLINE

As a means of quantifying uncertainty, the importance of probability and statistics is increasing for many engineering purposes, such as simulating the performance of products to be designed and analyzing experimental data. This chapter presents an introduction to probability and random variables.

10.1 PROBABILITY

10.1.1 Definition of Probability

Consider an experiment (trial), like shooting a die, which has one of several *outcomes* (elements) like 1 or 2 or ... or 6. With the *sample space* (universe) defined as the set S of all possible outcomes, like $\{1,2,3,4,5,6\}$, an *event* A is a set of outcomes of a random phenomenon and can be expressed as a subset of S. The *probability* is a mapping of every event into a real number satisfying the following axioms:

(P1) $P(A) \geq 0$ where $P(A)$ is the probability that the experiment's outcome is a member of $A \subseteq S$.

(P2) $P(S) = 1$ (certain) since any event belongs to the sample space S.

(P3) If the events $A_1, A_2, \ldots$ are (*mutually*) *exclusive*, i.e., $A_i \cap A_j = \phi$(disjoint) $\forall\, i \neq j$, the following relations hold:

$$P(A_1 \cup A_2 \cup \cdots) = P(A_1) + P(A_2) + \cdots\,; \quad P(\bigcup_i A_i) = \sum_i P(A_i) \tag{10.1.1}$$

$$P(A) \leq 1,\;\; P(\phi) = 0 \text{ (impossible)}, \text{ and } P(\bar{A}) = 1 - P(A) \tag{10.1.2}$$

where $\bar{A} = A^c = S \setminus A$ denotes the complement of A and ϕ denotes the empty or null set.

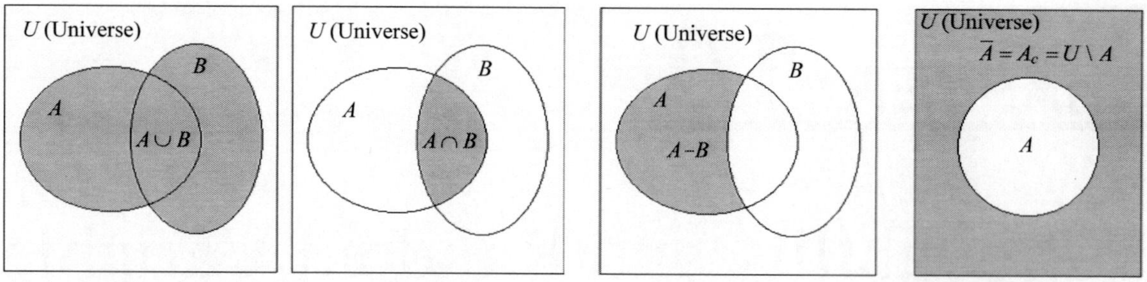

(a) $A \cup B$:Union of A and B (b) $A \cap B$:Intersection of A and B (c) $A - B$: Set difference of A and B (d) $\overline{A}$:Complement of A

Figure 10.1 Illustration of union, intersection, and set difference of two sets *A* and *B*, and complement

There are some consequences of the axioms:

$$\text{If } A_1 \subseteq A_2, \text{ then } P(A_1) \le P(A_2) \text{ and } P(A_2 \setminus A_1) = P(A_2) - P(A_1) \tag{10.1.3}$$

$$\text{For any events } A_1 \text{ and } A_2, \ P(A_1 \cup A_2) = P(A_1) + P(A_2) - P(A_1 \cap A_2) : \text{Addition rule} \tag{10.1.4}$$

Fig. 10.1 illustrates the union, intersection, and set difference of two sets *A* and *B*, and the complement of a set *A*. It is expected to be helpful in understanding the close relationship between the probability theory and the set theory.

10.1.2 Permutations and Combinations

A *permutation* of a finite number of elements refers to their rearrangement or reordering. How many permutations of *N* different elements are there? Referring to Fig. 10.2, the number of permutations of 4 different elements can be computed as

$$4 \cdot 3 \cdot 2 \cdot 1 = 4! \ (4 \text{ factorial})$$

This can be extended into the number of permutations of *N* different elements as

$$_N\mathsf{P}_N = N(N-1)\cdots 1 = N! \ (N \text{ factorial}) \tag{10.1.5}$$

where $N!$ can be computed by using the MATLAB command '`factorial(N)`'. A good approximation of $N!$ for large N is *Stirling's formula* (named after the Scottish mathematician James Stirling) [W-4]:

$$N! \approx \sqrt{2\pi N}\left(\frac{N}{e}\right)^N \tag{10.1.6}$$

<table>
<tr><td colspan="6">A</td><td colspan="6">B</td><td colspan="6">C</td><td colspan="6">D</td><td>4 choices for the 1st</td></tr>
<tr><td colspan="2">B</td><td colspan="2">C</td><td colspan="2">D</td><td colspan="2">A</td><td colspan="2">C</td><td colspan="2">D</td><td colspan="2">A</td><td colspan="2">B</td><td colspan="2">D</td><td colspan="2">A</td><td colspan="2">B</td><td colspan="2">C</td><td>3 choices for the 2nd</td></tr>
<tr><td>C</td><td>D</td><td>B</td><td>D</td><td>B</td><td>C</td><td>C</td><td>D</td><td>A</td><td>D</td><td>A</td><td>C</td><td>B</td><td>D</td><td>A</td><td>D</td><td>A</td><td>B</td><td>B</td><td>C</td><td>A</td><td>C</td><td>A</td><td>B</td><td>2 choices for the 3rd</td></tr>
<tr><td>D</td><td>C</td><td>D</td><td>B</td><td>C</td><td>B</td><td>D</td><td>C</td><td>D</td><td>A</td><td>C</td><td>A</td><td>D</td><td>B</td><td>D</td><td>A</td><td>B</td><td>A</td><td>C</td><td>B</td><td>C</td><td>A</td><td>B</td><td>A</td><td>No choices for the last</td></tr>
</table>

4·3·2·1=4! (non repetitive) permutations

Figure 10.2 Counting the number of nonrepetitive permutations of 4 elements

The MATLAB command '`randperm(N,k)`' returns a row vector containing k (unique) numbers selected randomly from the integers {1: N}.

If N elements can be divided into M groups, each of which consists of N_m (m=1,...,M) indistinguishable elements, the number of permutations is

$$\frac{N!}{N_1!N_2!\cdots N_M!} \tag{10.1.7}$$

A *combination* of a finite number of elements means to pick (select) a given number of elements regardless of their order. The number of different k-combinations (groups of k elements) from N elements, with no repetition, is

$$\mathsf{C}(N,k) = {}_N\mathsf{C}_k(\text{"N choose k"}) = \binom{N}{k} = \frac{N!}{k!(N-k)!} = \frac{N\cdot(N-1)\cdot\cdots\cdot(N-k+1)}{k!} \tag{10.1.8}$$

This can be computed by using the MATLAB command '`nchoosek(N,k)`'. Note that

$$\binom{N}{k} = \binom{N}{N-k} \quad\text{and}\quad \binom{N}{0} = \binom{N}{N} = 1 \tag{10.1.9a,b}$$

since choosing k ones among N elements is equivalent to discarding $(N-k)$ ones. Also, the number of different k-combinations among N elements, with repetitions allowed,

$$\left(\!\binom{N}{k}\!\right)(\text{"N multichoose k"}) = \binom{N+k-1}{k} = \frac{(N+k-1)!}{k!(N-1)!} = \frac{(N+k-1)\cdot(N+k-2)\cdot\cdots\cdot N}{k!} \tag{10.1.10}$$

The number of "k-permutations of N", i.e., permutations of k ones (selected with no repetition) among N elements can be obtained by multiplying Eq. (10.1.8) with $k!$ as

$$\mathsf{P}(N,k) = {}_N\mathsf{P}_k(\text{"N permutation k"}) = \binom{N}{k}k! \overset{(10.1.8)}{=} \frac{N!}{(N-k)!} \tag{10.1.11}$$

Also, referring to Fig. 10.3, showing that the number of different permutations of 2-combinations (selected with repetitions) among 3 elements is 3^2, the number of k-permutations (selected with repetitions) among N elements is

$$N^k \tag{10.1.12}$$

Note that the MATLAB command '`perms(v)`' (with a vector v of length N as its input argument) returns an $N!\times N$ matrix containing all the N-permutations of the N elements of v as its rows. Also, the MATLAB command '`combnk(v,k)`' (with a vector v of length N as its first input argument) returns an ${}_N\mathsf{C}_k\times k$ matrix containing all the k-combinations among the N elements of v:

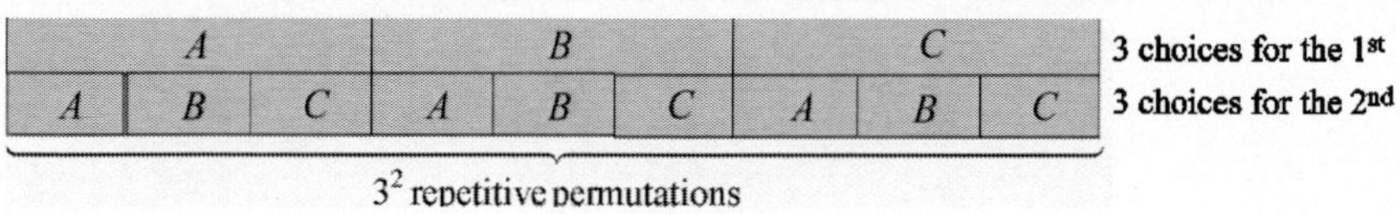

Figure 10.3 Counting the number of repetitive permutations of 2 ones from 3 elements with repetitions

```
>> perms([1 2 3])

   ans =
          3     2     1
          3     1     2
          2     3     1
          2     1     3
          1     2     3
          1     3     2
```

```
>>combnk({'A','B','C','D'},2)

   ans =
          'C'     'D'
          'B'     'D'
          'B'     'C'
          'A'     'D'
          'A'     'C'
          'A'     'B'
```

10.1.3 Joint Probability, Conditional Probability, and Bayes' Rule

The *joint probability* $P(A \cap B)$ that two events A and B occur together is related to the *conditional probability* $P(B|A)$ (read "the probability of B, given A") that the event B occurs, given the occurrence of the event A as follows:

$$P(B\,|\,A) = \frac{P(A \cap B)}{P(A)} \text{ with } P(A) > 0 \tag{10.1.13}$$

If the probability of one event is not affected by the occurrence of the other, they are said to be *independent* of each other, and

$$P(B\,|\,A) = P(B) = \frac{P(A \cap B)}{P(A)} \rightarrow P(A \cap B) = P(A)P(B) \tag{10.1.14}$$

On the other hand, if the possible cause of the event B is one of the mutually exclusive events $\{A_1, A_2, \ldots\}$, the *posteriori probability* $P(A_i|B)$ that A_i turns out to be the cause of the event B after B is known to have occurred is written in terms of the *priori probability* $P(A_i)$, that we guess A_i to have occurred before or without knowing about the occurrence of B, as follows:

$$P(A_i\,|\,B) \overset{(10.1.13)}{=} \frac{P(A_i \cap B)}{P(B)} \overset{(10.1.13)}{=} \frac{P(B|A_i)P(A_i)}{\sum_j P(B|A_j)P(A_j)} \text{ with } P(B) > 0 \tag{10.1.15}$$

where $P(B|A_i)$ is the likelihood that B will occur after A_i occurs and

$$\cup_i A_i = S \text{ (sample space)}, \quad A_i \cap A_j = \phi \text{ (mutually exclusive) } \forall\, i \neq j$$

This relation is called the *Bayes' rule*, which is often used to determine probabilities of (disjoint and exhaustive) events by incorporating information about subsequent events.

[**Remark 10.1**] Addition Rule, Multiplication Rule, and Complementation Rule

Note the following rules about probability, where A and B are events in a sample space S:

Addition rule: $$P(A \cup B) \overset{(10.1.4)}{=} P(A) + P(B) - P(A \cap B) \underset{\text{mutually exclusive}}{\overset{\text{If } A \text{ and } B \text{ are}}{=}} P(A) + P(B) \tag{10.1.16}$$

Multiplication rule: $$P(A \cap B) \overset{(10.1.13)}{=} P(A)P(B\,|\,A) = P(B)P(A\,|\,B) \underset{\text{independent}}{\overset{\text{If } A \text{ and } B \text{ are}}{=}} P(A)P(B) \tag{10.1.17}$$

Complementation rule: $$P(\bar{A}) = 1 - P(A) \tag{10.1.18}$$

(**Example 10.1**) Sum and Product of Probabilities

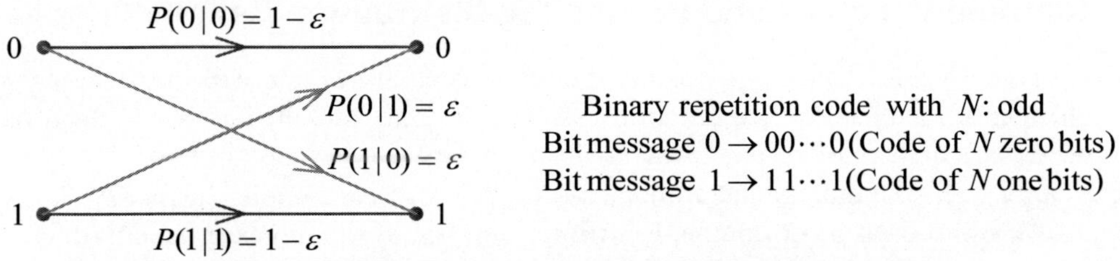

Figure 10.4 Binary symmetric channel with crossover probability ε

Consider the binary symmetric channel (BSC) depicted in Fig. 10.4, which transmits one of two binary values 0 and 1 every bit time interval with crossover (error) probability ε. Suppose the bit messages 0 and 1 are coded into N-zero and N-one sequences (N: odd), respectively, and transmitted through the BSC to a receiver, which decodes each N-bit sequence into 0 or 1 by the majority rule, i.e., depending on which of 0 or 1 is more than the other one. Noting that the crossover probability of each bit of the transmitted N-bit codes is constantly ε irrespective of whether or not other bits become erroneous, i.e., the event of each bit transmission is independent of that of other bits' transmission, the probability that more than $(N-1)/2$ bits out of the N bits of each code become erroneous can be written as

$$P_e = \sum_{k=(N+1)/2}^{N} {}_N\mathrm{C}_k \, \varepsilon^k (1-\varepsilon)^{N-k} \text{ where } {}_N\mathrm{C}_k = \frac{N!}{k!(N-k)!} \text{ and } k! = k(k-1)\cdots 2\cdot 1 \quad \text{(E10.1.1)}$$

(Q1) Why does the product of several ε's and $(1-\varepsilon)$'s constitute each term in this equation?
(Q2) Why is each $\varepsilon^k(1-\varepsilon)^{N-k}$ multiplied by ${}_N\mathrm{C}_k$ and then summed up to make P_e?

(**Example 10.2**) Application of Bayes' Rule

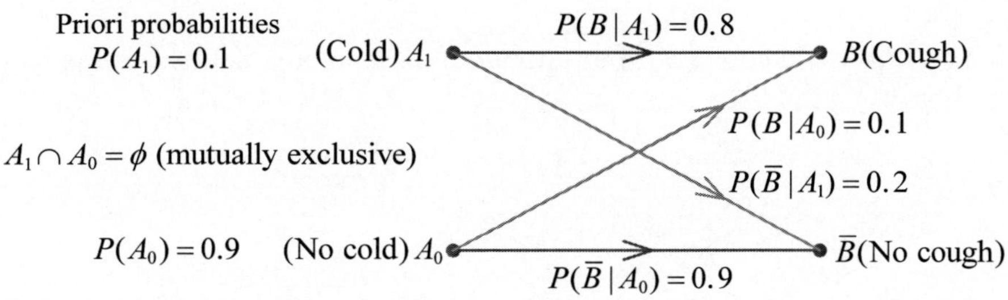

Figure 10.5 Transition probabilities from cold to cough

Suppose that the priori probability of a person catching a cold is $P(A_1)=0.1$ based on his living condition, health condition, and medical data including his disease history (before a diagnosis of his cough). The transition probabilities from cold (A) to cough (B) are depicted in Fig. 10.5.

We can use Eq. (10.1.15) to find the posteriori probabilities of his having caught a cold after being diagnosed as coughing or no coughing as

$$P(A_1|B) \overset{(10.1.15)}{=} \frac{P(B|A_1)P(A_1)}{P(B|A_0)P(A_0)+P(B|A_1)P(A_1)} = \frac{0.8\times0.1}{0.1\times0.9+0.8\times0.1} = \frac{8}{17} \quad \text{(E10.2.1)}$$

$$P(A_1|\bar{B}) \overset{(10.1.15)}{=} \frac{P(\bar{B}|A_1)P(A_1)}{P(\bar{B}|A_0)P(A_0)+P(\bar{B}|A_1)P(A_1)} = \frac{0.2\times0.1}{0.9\times0.9+0.2\times0.1} = \frac{2}{83} \quad \text{(E10.2.2)}$$

10.2 RANDOM VARIABLES

10.2.1 Random Variables and Probability Distribution/Density Functions

A *random* or *stochastic variable* (possibly being a vector), which will be denoted by a bold character like **x** in this chapter, is a function assigning a (real-valued) number to each outcome (in the sample space) of a random experiment. There are two types of random variables. One is a *discrete random variable* that can take on values from a discrete sample space, say $S_{\mathbf{x}}=\{x_0, x_1, \ldots\}$, consisting of a finite or at most countably infinite number of outcomes (elements). The other is a *continuous random variable* that can take on any real-valued number within an interval or a collection of intervals.

For a discrete random variable, the *probability mass function* (*pmf*) is defined as the set of probabilities

$$p_{\mathbf{x}}(x_i) = P\{\mathbf{x} = x_i\} \tag{10.2.1}$$

of the outcomes in its sample space $S_{\mathbf{x}}=\{x_0, x_1, \ldots\}$.

The *probability distribution function* (*PDF*) or *cumulative distribution function* (*CDF*) of a random variable **x** is defined as

$$F_{\mathbf{x}}(x) = P\{\mathbf{x} \le x\} \tag{10.2.2}$$

Note that the PDF is monotonically increasing from 0 to 1 as its argument value x increases and is continuous from the right in the sense that

$$F_{\mathbf{x}}(a^+) = \lim_{h\to 0} F_{\mathbf{x}}(a+h) = F_{\mathbf{x}}(a) \tag{10.2.3}$$

The PDF of a discrete random variable **x** can be written as

$$F_{\mathbf{x}}(x) = \sum\nolimits_k p_{\mathbf{x}}(x_k) u_s(x-x_k) \text{ with the unit step function } u_s(x) = \begin{cases} 1 & \text{for } x \ge 0 \\ 0 & \text{for } x < 0 \end{cases} \tag{10.2.4}$$

The *probability density function* (*pdf*) of a random variable **x** is defined as the derivative of the PDF:

$$f_{\mathbf{x}}(x) = \frac{d}{dx} F_{\mathbf{x}}(x) \tag{10.2.5}$$

The pdf of a discrete random variable **x** can be written as

$$f_{\mathbf{x}}(x) \overset{(10.2.5)}{=} \frac{d}{dx} F_{\mathbf{x}}(x) \overset{(10.2.4)}{=} \sum\nolimits_k p_{\mathbf{x}}(x_k) \delta(x-x_k) \text{ with } \begin{matrix}\text{the unit impulse function} \\ \delta(x) = du_s(x)/dx\end{matrix} \tag{10.2.6}$$

where the delta or unit impulse function $\delta(x)$ satisfies

$$\begin{aligned} &\delta(x) = \frac{d}{dx} u_s(x) \;\rightarrow\; \int_{-\infty}^{x} \delta(t)\,dt = u_s(x); \\ &\int_{-\infty}^{\infty} \delta(x)\,dx = \int_{-\Delta}^{\Delta} \delta(x)\,dx \overset{(4.1.7)}{=} 1 \; (\Delta\text{: an infinitesimally small positive number}) \end{aligned} \tag{10.2.7}$$

According to these definitions, the probability of **x** taking a real number in an interval (a, b) can be expressed as

$$P(a < \mathbf{x} \le b) = P(\mathbf{x} \le b) - P(\mathbf{x} \le a) = F_{\mathbf{x}}(b) - F_{\mathbf{x}}(a) = \int_a^b f_{\mathbf{x}}(x)\,dx \tag{10.2.8}$$

If the pdf of a random variable **x** is given as $f_{\mathbf{x}}(x)$, its PDF $F_{\mathbf{x}}(x)$ can be obtained by integrating the pdf:

$$F_{\mathbf{x}}(x) = \int_{-\infty}^{x} f_{\mathbf{x}}(t)\,dt \tag{10.2.9}$$

(**Example 10.3**) PDF/pdf of a Binomial Random Variable and a Uniform Random Variable

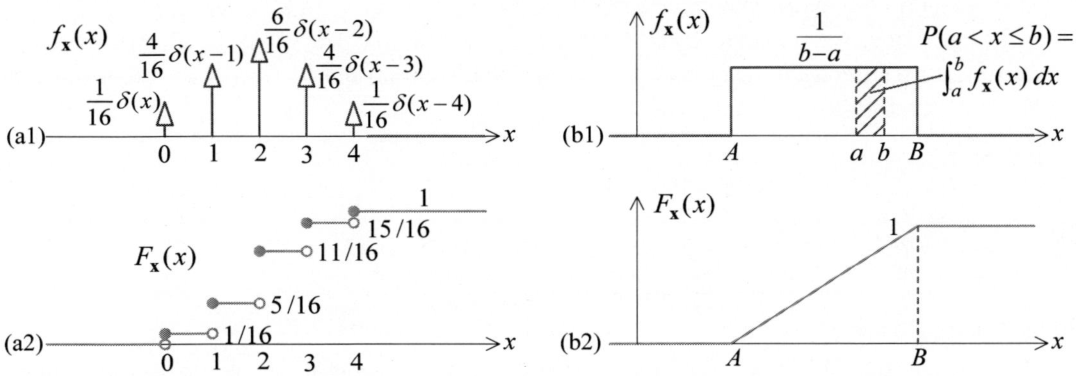

Figure 10.6 PDF/pdf of a binomial random variable and a uniform random variable

(a) Consider a binomial random variable **x**, which is the number of heads in N=4 tosses of a coin. Its pmf is

$$p_{\mathbf{x}}(k) = P\{\mathbf{x}=k\} = {}_4\mathrm{C}_k\, p^k (1-p)^{4-k} \text{ with } p=\frac{1}{2} \text{ and } k \in S_{\mathbf{x}} = \{0,1,2,3,4\} \tag{E10.3.1}$$

$$= \begin{bmatrix} \frac{1}{16} & \frac{4}{16} & \frac{6}{16} & \frac{4}{16} & \frac{1}{16} \end{bmatrix}$$

where the values can be obtained by running the following MATLAB statements:

```
>>N=4; p=0.5; for k=0:N, px(k+1)=nchoosek(4,k)*p^k*(1-p)^(N-k); end
>>px
   px =   0.0625   0.2500   0.3750   0.2500   0.0625   % [1 4 6 4 1]/16
```

As depicted in Fig. 10.6(a2) and (a1), its PDF and pdf are

$$F_{\mathbf{x}}(x) \overset{(10.2.4)}{=} \sum_k p_{\mathbf{x}}(k) u_s(x-k) \overset{(E10.3.1)}{=} \sum_{k=0}^{4} {}_4\mathrm{C}_k\, 0.5^k (1-0.5)^{4-k} u_s(x-k) \tag{E10.3.2}$$

$$f_{\mathbf{x}}(x) \overset{(10.2.6)}{=} \sum_k p_{\mathbf{x}}(k) \delta(x-k) \overset{(E10.3.1)}{=} \sum_{k=0}^{4} {}_4\mathrm{C}_k\, 0.5^k (1-0.5)^{4-k} \delta(x-k) \tag{E10.3.3}$$

(b) Consider a uniform random variable **y** that takes on values equally likely from the interval $[A, B]$. As depicted in Fig. 10.6(b1) and (b2), its pdf and PDF are

$$f_{\mathbf{x}}(x) = \begin{cases} \frac{1}{B-A} & \text{for } A \le x \le B \\ 0 & \text{elsewhere} \end{cases} \quad (\textit{uniform distribution}) \tag{E10.3.4}$$

$$F_{\mathbf{x}}(x) \overset{(10.2.9)}{=} \int_{-\infty}^{x} f_{\mathbf{x}}(t)\,dt \overset{(E10.3.4)}{=} \int_{A}^{x} f_{\mathbf{x}}(t)\,dt = \begin{cases} 0 & \text{for } x<A \\ \frac{x-A}{B-A} & \text{for } A \le x \le B \\ 1 & \text{for } x>B \end{cases} \tag{E10.3.5}$$

(**Example 10.4**) Uniform Number and Gaussian Number Generated by MATLAB

(a) Uniform Number

We can use the MATLAB function '`rand(M,N)`' to generate an $M\times N$ matrix composed of numbers having the standard uniform distribution $U(0,1)$ on [0, 1], and transform this matrix by

$$y=(b-a)x+a \tag{E10.4.1}$$

to make a random matrix having a general uniform distribution $U(a, b)$ on $[a, b]$.

For practice, we run the MATLAB script "em10e04a.m" to generate a 1×10^5 vector having uniform distribution $U(-1,1)$, and then draw the histogram that is similar to the uniform pdf (Eq. (10.2.36)), as shown in Fig. 10.7(a).

```
%em10e04a.m
N=1e5; Nb=40; b=1; a=-1; % Number of samples and bins
u_noise=rand(1,N); % An 1xN random vector with U(0,1)
u_noise1=(b-a)*u_noise+a; % Eq.(E10.4.1) for U(a,b)
subplot(221), hist(u_noise1,Nb) % Histogram
```

(b) Gaussian (Normal) Number

We can use the MATLAB function '`randn(M,N)`' to generate an $M\times N$ matrix composed of numbers having the standard normal (Gaussian) distribution $N(0,1)$ with zero mean and unit variance, and transform this matrix by

$$y=\sigma x+m \tag{E10.4.2}$$

to make a random $M\times N$ matrix having a general normal (Gaussian) distribution $N(m,\sigma^2)$.

For practice, we run the MATLAB script "em10e04b.m" to generate a 1×10^5 vector having normal distribution $N(0,1/4)$ with mean m=1 and variance $\sigma^2=1/4$, and then draw the histogram that is similar to the Gaussian pdf (Eq. (10.2.33)), as shown in Fig. 10.7(b).

```
%em10e04b.m
N=1e5; Nb=40; % Number of samples and bins
g_noise=randn(1,N); % An 1xN random vector with N(0,1)
m=1; sigma=1/2; % Average and sigma=sqrt(deviation)
g_noise1=sigma*g_noise+m; % Eq.(E10.4.2) for N(m,sigma^2)
subplot(222), hist(g_noise1,Nb) % Histogram
[ns,cs]=hist(g_noise1,Nb); % ns: Frequency counts, cs: Bin locations
dx=cs(2)-cs(1); % Bin width
x=m+[-500:500]*(sigma/100); % The range on the x-axis
f=exp(-(x-m).^2/(2*sigma^2))/sqrt(2*pi)/sigma; % Eq.(10.2.33)
hold on, plot(x,N*dx*f,'r')
sum(g_noise1)/N-m, sum((g_noise1-m).^2)/N-sigma^2
```

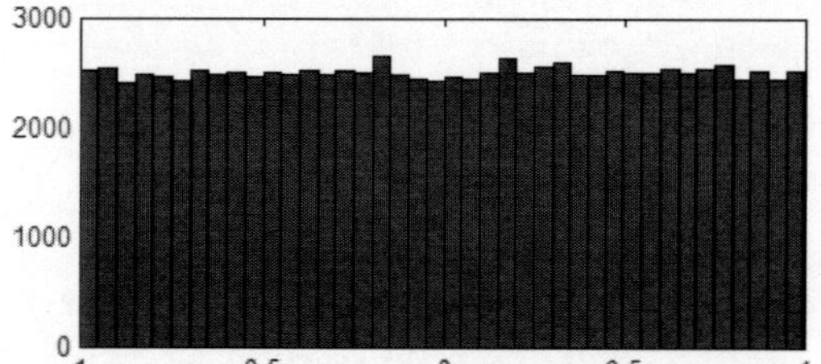

(a) Histogram of 1×10^5 uniform random vector

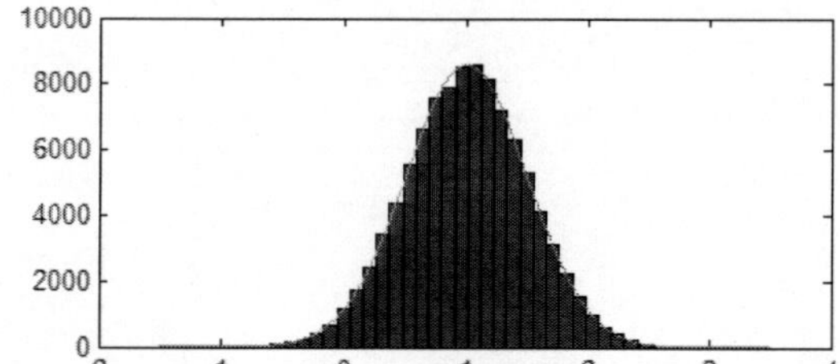

(b) Histogram of 1×10^5 Gaussian random vector

Figure 10.7 Histograms of uniform and Gaussian random sequences

10.2.2 Joint Probability Density Function

For two random variables **x** and **y**, their joint PDF and pdf are defined as

$$F_{\mathbf{x},\mathbf{y}}(x,y) = P(\mathbf{x} \le x,\, \mathbf{y} \le y) \tag{10.2.10}$$

$$f_{\mathbf{x},\mathbf{y}}(x,y) = \frac{\partial^2}{\partial x \partial y} F_{\mathbf{x},\mathbf{y}}(x,y) \tag{10.2.11}$$

If the joint pdf of **x** and **y** is given as $f_{\mathbf{x},\mathbf{y}}(x,y)$, their joint PDF $F_{\mathbf{x},\mathbf{y}}(x,y)$ can be obtained by integrating the pdf:

$$F_{\mathbf{x},\mathbf{y}}(x,y) = \int_{-\infty}^{y}\int_{-\infty}^{x} f_{\mathbf{x},\mathbf{y}}(x',y')\,dx'dy' \tag{10.2.12}$$

From the *joint pdf*, we can obtain the *marginal pdf* of each random variable **x** and **y** as follows:

$$f_{\mathbf{x}}(x) = \int_{-\infty}^{\infty} f_{\mathbf{x},\mathbf{y}}(x,y)\,dy, \qquad f_{\mathbf{y}}(y) = \int_{-\infty}^{\infty} f_{\mathbf{x},\mathbf{y}}(x,y)\,dx \tag{10.2.13a,b}$$

(**Example 10.5**) Joint pdf of Two Jointly Gaussian Random Variables

Figs. 10.8(a) and (b) show the joint pdfs of two jointly Gaussian random variables with correlation coefficients $\rho = 0.9$ and $\rho = 0$, respectively, where the joint Gaussian pdf is

$$f_{\mathbf{x},\mathbf{y}}(x,y) = \frac{1}{2\pi\sqrt{1-\rho^2}} e^{-(x^2-2\rho xy+y^2)/2(1-\rho^2)} \text{ for } -\infty < x, y < \infty. \tag{E10.5.1}$$

The marginal pdf of one random variable, say **x**, can be obtained by using Eq. (10.2.13a) as

$$f_{\mathbf{x}}(x) \overset{(10.2.13a)}{=} \int_{-\infty}^{\infty} f_{\mathbf{x},\mathbf{y}}(x,y)\,dy \overset{(E10.5.1)}{=} \frac{e^{-x^2/2(1-\rho^2)}}{2\pi\sqrt{1-\rho^2}} \int_{-\infty}^{\infty} e^{-(y^2-2\rho xy)/2(1-\rho^2)}\,dy$$

$$= \frac{e^{-x^2/2(1-\rho^2)} e^{\rho^2 x^2/2(1-\rho^2)}}{\sqrt{2\pi}} \int_{-\infty}^{\infty} \frac{e^{-(y-\rho x)^2/2(1-\rho^2)}}{\sqrt{2\pi(1-\rho^2)}}\,dy = \frac{e^{-x^2/2}}{\sqrt{2\pi}} \tag{E10.5.2}$$

where the last integral is 1 since its integrand is a Gaussian pdf with mean ρx and variance $1-\rho^2$.

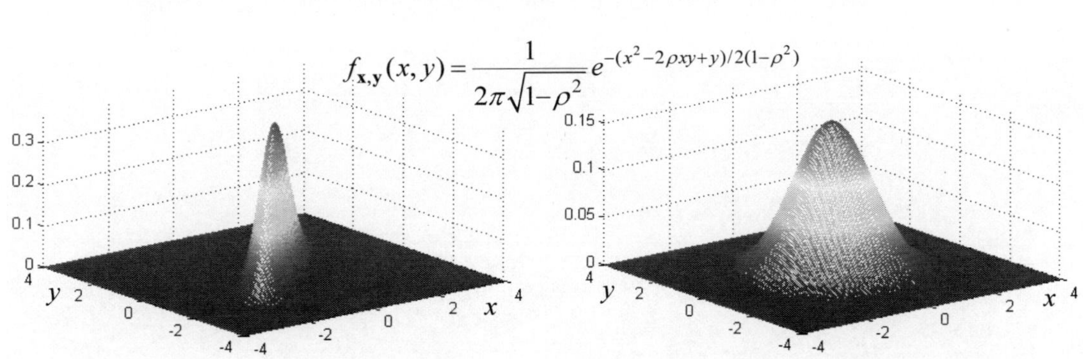

(a) Correlation coefficient $\rho = 0.9$ (b) Correlation coefficient $\rho = 0$

Figure 10.8 Joint pdfs of two jointly Gaussian random variables

10.2.3 Conditional Probability Density Function

For two random variables **x** and **y**, the *conditional pdf* of **x** given **y**=y and that of **y** given **x**=x are defined as

$$f_{\mathbf{x}|\mathbf{y}}(x\,|\,y)=\frac{f_{\mathbf{x},\mathbf{y}}(x,y)}{f_{\mathbf{y}}(y)} \quad \text{and} \quad f_{\mathbf{y}|\mathbf{x}}(y\,|\,x)=\frac{f_{\mathbf{x},\mathbf{y}}(x,y)}{f_{\mathbf{x}}(x)} \tag{10.2.14a,b}$$

respectively. Thus the joint pdf of **x** and **y** can be written as

$$f_{\mathbf{x},\mathbf{y}}(x,y)=f_{\mathbf{x}|\mathbf{y}}(x\,|\,y)f_{\mathbf{y}}(y)=f_{\mathbf{y}|\mathbf{x}}(y\,|\,x)f_{\mathbf{x}}(x) \tag{10.2.15}$$

The following examples will help you realize the concepts of the joint and conditional pdfs.

(Example 10.6) Joint/Marginal/Conditional pdfs

Let us consider two random variables **x** and **y**, whose joint pdf is

$$f_{\mathbf{x},\mathbf{y}}(x,y)=\begin{cases}2 & \text{for } 0\le y\le x\le 1\\ 0 & \text{elsewhere}\end{cases} \tag{E10.6.1}$$

as shown in Fig. 10.9(a). We can use Eq. (10.2.13) to obtain the marginal pdfs of **x** and **y** from the joint pdf (E10.6.1) as

$$f_{\mathbf{x}}(x)\overset{(10.2.13a)}{=}\int_{-\infty}^{\infty}f_{\mathbf{x},\mathbf{y}}(x,y)\,dy\overset{(E10.6.1)}{=}\int_0^x 2\,dy=2x \quad \text{for } 0\le x\le 1 \tag{E10.6.2a}$$

$$f_{\mathbf{y}}(y)\overset{(10.2.13b)}{=}\int_{-\infty}^{\infty}f_{\mathbf{x},\mathbf{y}}(x,y)\,dx\overset{(E10.6.1)}{=}\int_y^1 2\,dx=2(1-y) \quad \text{for } 0\le y\le 1 \tag{E10.6.2b}$$

From these joint and marginal pdfs, we can use Eq. (10.2.14) to obtain the conditional pdfs of **x** given **y**=y and **y** given **x**=x as

$$f_{\mathbf{x}|\mathbf{y}}(x\,|\,y)\overset{(10.2.14a)}{=}\frac{f_{\mathbf{x},\mathbf{y}}(x,y)}{f_{\mathbf{y}}(y)}\underset{(E10.6.2b)}{\overset{(E10.6.1)}{=}}\frac{2}{2(1-y)}=\frac{1}{1-y} \quad \text{for } 0\le y\le x\le 1 \tag{E10.6.3a}$$

$$f_{\mathbf{y}|\mathbf{x}}(y\,|\,x)\overset{(10.2.14b)}{=}\frac{f_{\mathbf{x},\mathbf{y}}(x,y)}{f_{\mathbf{x}}(x)}\underset{(E10.6.2a)}{\overset{(E10.6.1)}{=}}\frac{2}{2x}=\frac{1}{x} \quad \text{for } 0\le y\le x\le 1 \tag{E10.6.3b}$$

that are shown in Figs. 10.9(b) and (c), respectively. Fig. 10.9(b) together with Eq. (E10.6.3a) shows that the larger the value y of **y** is, the more concentrated the value x of **x** is to $[y, 1]$. Fig. 10.9(c) together with Eq. (E10.6.3b) shows that the smaller the value x of **x** is, the more concentrated the value y of **y** is to $[0, x]$.

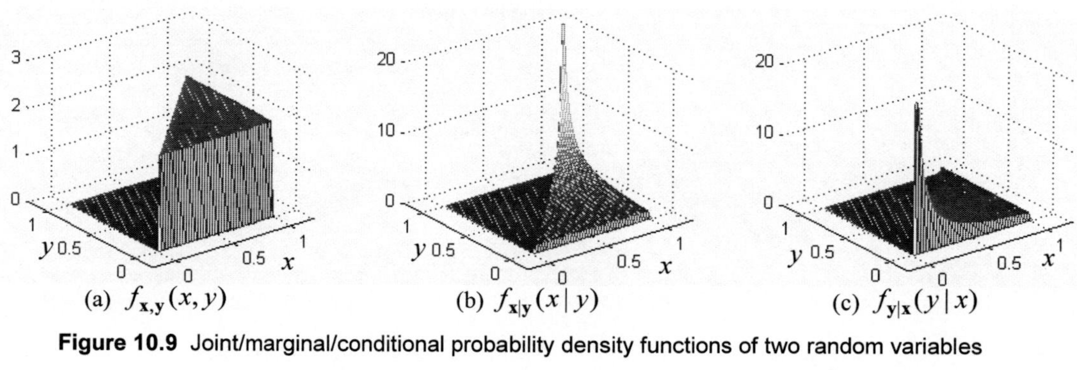

(a) $f_{\mathbf{x},\mathbf{y}}(x,y)$ (b) $f_{\mathbf{x}|\mathbf{y}}(x\,|\,y)$ (c) $f_{\mathbf{y}|\mathbf{x}}(y\,|\,x)$

Figure 10.9 Joint/marginal/conditional probability density functions of two random variables

(Example 10.7) Marginal/Conditional/Joint pdfs

Let the measured value **y** of a random variable **x** (whose value can be in [−1, 1]) be the sum of **x** and a measurement noise **n** (another random variable):

$$\mathbf{y} = \mathbf{x} + \mathbf{n} \tag{E10.7.1}$$

where the marginal pdfs of **x** and **n** are as follows (Fig. 10.10(a)):

$$f_{\mathbf{x}}(x) = \begin{cases} 1/2 & \text{for } -1 \le x \le 1 \\ 0 & \text{elsewhere} \end{cases} \quad \text{and} \quad f_{\mathbf{n}}(n) = \begin{cases} 2(1-|2n|) & \text{for } -0.5 \le n \le 0.5 \\ 0 & \text{elsewhere} \end{cases} \tag{E10.7.2a,b}$$

The conditional pdf of **y** given **x**=x can be written by substituting $n=y-x$ into Eq. (E10.7.2b) as

$$f_{\mathbf{y}|\mathbf{x}}(y\,|\,x) = \begin{cases} 2(1-|2(y-x)|) & \text{for } x-0.5 \le y \le x+0.5 \\ 0 & \text{elsewhere} \end{cases} \tag{E10.7.3}$$

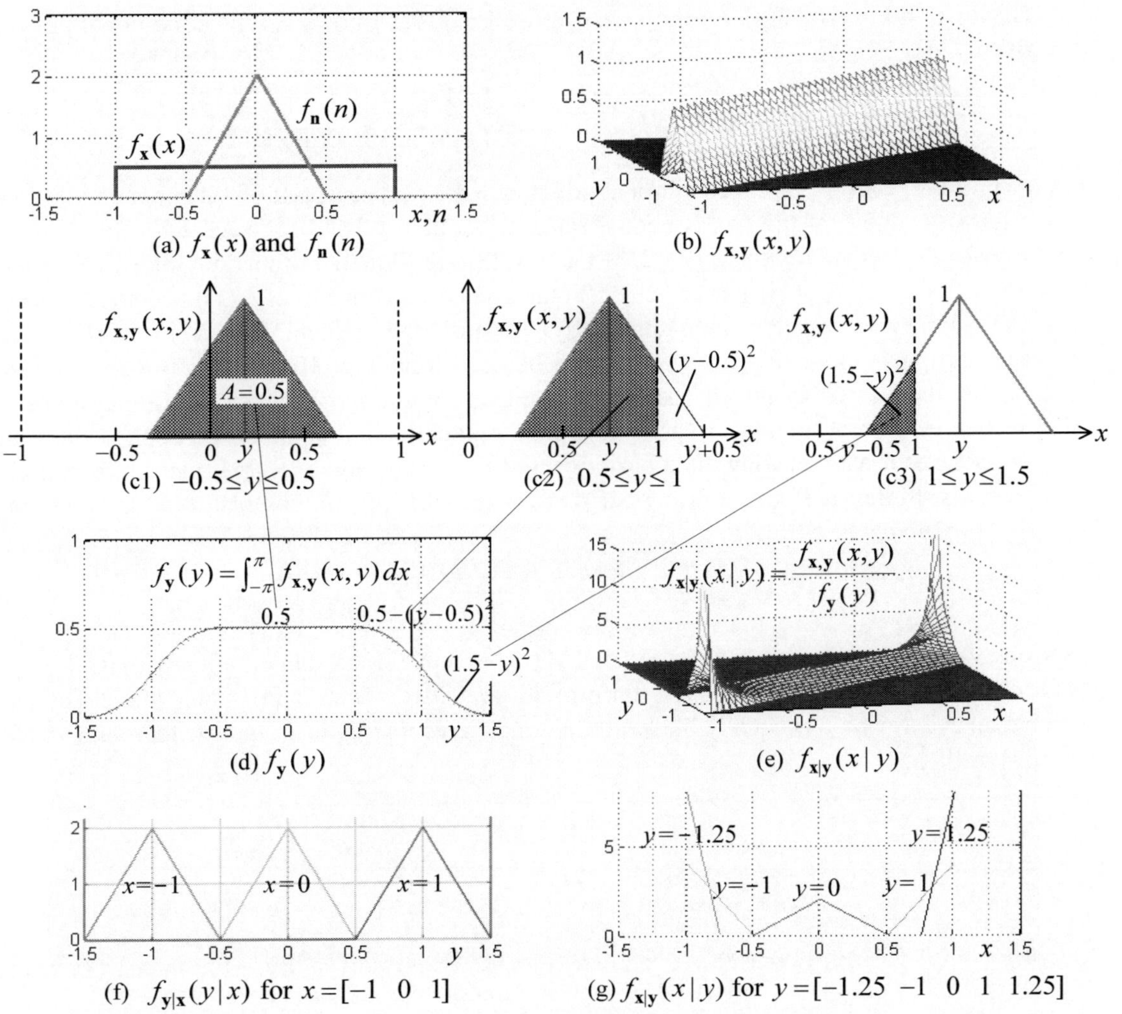

Figure 10.10 Marginal/joint/conditional probability density functions of two random variables

We can use Eq. (10.2.15) together with Eqs. (E10.7.3) and (E10.7.2a) to get the joint pdf of **x** and **y** as

$$f_{\mathbf{x},\mathbf{y}}(x,y) \overset{(10.2.15)}{=} f_{\mathbf{y}|\mathbf{x}}(y\,|\,x) f_{\mathbf{x}}(x) \underset{(\text{E}10.7.2\text{a})}{\overset{(\text{E}10.7.3)}{=}} \begin{cases} 1-|2(y-x)| & \text{for } x-0.5\le y\le x+0.5 \text{ and } -1\le x\le 1 \\ 0 & \text{elsewhere} \end{cases} \tag{E10.7.4}$$

This is shown in Fig. 10.10(b). Also, referring to Figs. 10.10(c1)|(c2)|c3), we can use Eq. (10.2.13b) together with Eq. (E10.7.4) to get the marginal pdf of **y** as

$$f_{\mathbf{y}}(y) \overset{(10.2.13\text{b})}{=} \int_{-\infty}^{\infty} f_{\mathbf{x},\mathbf{y}}(x,y)\,dx \overset{(\text{E}10.7.4)}{=} \int_{\max(-1,y-0.5)}^{\min(1,y+0.5)} (1-|2(y-x)|)\,dx$$

$$\underset{(\text{c}1,\text{c}2,\text{c}3)}{\overset{\text{Fig.}10.10}{=}} \begin{cases} (1.5+y)^2 & \text{for } -1.5\le y\le -1 \\ 0.5-(y+0.5)^2 & \text{for } -1\le y\le -0.5 \\ 0.5 & \text{for } -0.5\le y\le 0.5 \\ 0.5-(y-0.5)^2 & \text{for } 0.5\le y\le 1 \\ (1.5-y)^2 & \text{for } 1\le y\le 1.5 \\ 0 & \text{elsewhere} \end{cases} \tag{E10.7.5}$$

This is shown in Fig. 10.10(d). Further, we can use Eq. (10.2.14a) to get the conditional pdf $f_{\mathbf{x}|\mathbf{y}}(x|y)$ of **x** given $\mathbf{y}=y$:

$$f_{\mathbf{x}|\mathbf{y}}(x\,|\,y) \overset{(10.2.14\text{a})}{=} \frac{f_{\mathbf{x},\mathbf{y}}(x,y)\,(\text{Fig.}10.10(\text{b}))}{f_{\mathbf{y}}(y)\ (\text{Fig.}10.10(\text{d}))} \tag{E10.7.6}$$

as shown in Fig. 10.10(e). This conditional pdf is called a posteriori pdf of **x** since it can be evaluated only after the value y of the consequential random variable **y** has been revealed. The values of the conditional pdf $f_{\mathbf{x}|\mathbf{y}}(x|y)$ with $y=[-1.25\ -1\ 0\ 1\ 1.25]$ are plotted as functions of x in Fig. 10.10(g). Why do the values of $f_{\mathbf{x}|\mathbf{y}}(x|y)$ at $x=-1/+1$ (boundaries of the domain of $f_{\mathbf{x}}(x)$) become large as y gets close to $-1.5/+1.5$, respectively? Apparently, because the denominator $f_{\mathbf{y}}(y)$ of Eq. (E10.7.6) decreases as y gets close to $-1.5/+1.5$, as can be seen from Fig. 10.10(d). It is very interesting to observe that the less likely the outcome (y) of an event **y**, the more valuable information (about the other related event **x**) it contains. Even apart from the likelihood of y, the observation of the value of **y** itself contributes toward making the a posteriori pdf $f_{\mathbf{x}|\mathbf{y}}(x|y)$ (Fig. 10.10(g)) more informative about **x** compared with the (uniform) a-priori pdf $f_{\mathbf{x}}(x)$ (Fig. 10.10(a)). In contrast, the the conditional pdf $f_{\mathbf{y}|\mathbf{x}}(y|x)$ of **y** given $\mathbf{x}=x$

$$f_{\mathbf{y}|\mathbf{x}}(y\,|\,x) \overset{(10.2.14\text{a})}{=} \frac{f_{\mathbf{x},\mathbf{y}}(x,y)\,(\text{Fig.}10.10(\text{b}))}{f_{\mathbf{x}}(x)\ (\text{Fig.}10.10(\text{a}))} \tag{E10.7.7}$$

does not vary with x as depicted in Fig. 10.10(f), except for its range. Why is that? Because the denominator $f_{\mathbf{x}}(x)$ of Eq. (E10.7.7) is uniform all over its domain $x=[-1,\ 1]$. This is similar to the fact that an event, which is equally likely all the time, gives no implication other than time about other events.

```
%em10e07_jcpdf.m
% Plot f(x,y), f(y), and f(x|y)  Figs.10.10(b),(d), and (e).
f_xy=@(x,y)max((1-abs(2*(y-x))),0); % Eq.(E10.7.4)
tol=1e-4; h=0.05; xx=[-1:h:1]; yy=[-1.5:h:1.5]; Ny=length(yy);
for m=1:Ny
  y=yy(m); fXY(m,:)=feval(f_xy,xx,y); fy(m)=quad(f_xy,-1,1,tol,[],y);
  fx_given_y(m,:) = fXY(m,:)/fy(m); % Eqs.(E10.7.4-6)
end
[X,Y]=meshgrid(xx,yy); subplot(322), mesh(X,Y,fXY), title('f(x,y)')
subplot(323), plot(yy,fy); subplot(324), mesh(X,Y,fx_given_y)
```

(Example 10.8) Error Probability based on Conditional pdfs

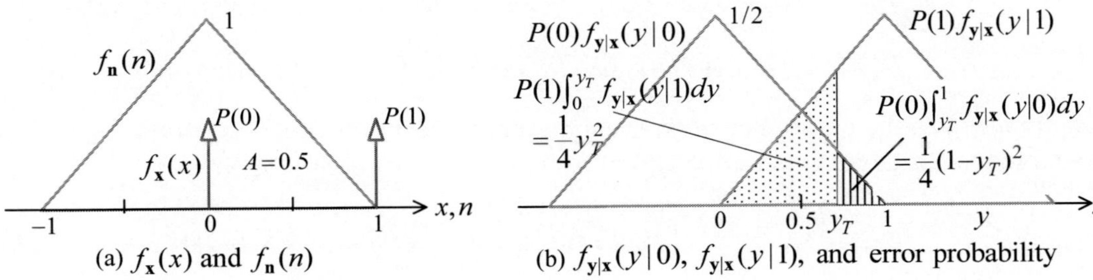

Figure 10.11 Conditional probability density function and probability of error

Suppose a random binary message **x** (whose value is equally likely 0 or 1) is transmitted through a channel to a receiver where a noise **n** (with the pdf $f_{\mathbf{n}}(n)$ as depicted in Fig. 10.11(a)) is added to **x** in the channel. Then the received signal **y** can be modeled as the sum of **x** and the additive noise **n**:

$$\mathbf{y} = \mathbf{x} + \mathbf{n} \tag{E10.8.1}$$

where the pdfs of **x** and **n** (Fig. 10.11(a)) can be written as

$$f_{\mathbf{x}}(x) = P(0)\delta(x) + P(1)\delta(x-1) \quad \text{and} \quad f_{\mathbf{n}}(n) = \begin{cases} (1-|n|) & \text{for } -1\le n\le 1 \\ 0 & \text{elsewhere} \end{cases} \tag{E10.8.2a,b}$$

The conditional pdf of **y** given **x**=0 or 1 can be written by substituting $n=y$ or $n=y-1$ into Eq. (E10.8.2b) as

$$f_{\mathbf{y}|\mathbf{x}}(y|x=0) = 1-|y| \quad \text{and} \quad f_{\mathbf{y}|\mathbf{x}}(y|x=1) = 1-|y-1| \tag{E10.8.3}$$

This is depicted in Fig. 10.11(b). Suppose the receiver tries to detect the received signal by judging it as 0 or 1, depending on whether it is smaller or larger than some threshold value y_T. We can find the probability of error that **x**=0|1 is to be misdetected as 1|0 as follows:

$$P_e(y_T) = P(1)\int_0^{y_T} f_{\mathbf{y}|\mathbf{x}}(y|1)\,dy + P(0)\int_{y_T}^1 f_{\mathbf{y}|\mathbf{x}}(y|0)\,dy \overset{\text{Fig.10.11(b)}}{=} \frac{1}{4}y_T^2 + \frac{1}{4}(1-y_T)^2 \tag{E10.8.4}$$

The optimal value of the threshold for the estimation (detection) of **x** can be obtained from the minimum of $P_e(y_T)$, which can be found by setting the derivative of Eq. (E10.8.4) w.r.t. y_T to zero:

$$\frac{d}{dy_T}P_e(y_T) = y_T - \frac{1}{2} = 0; \quad y_T^o = \frac{1}{2} \tag{E10.8.5}$$

```
%em10e08_error_probability.m
N=1e6; x=randi([0 1],1,N); % Number of binary message data
n=rand(1,N)+rand(1,N)-1; % What pdf will n have? (Refer to Eq.(10.2.21).)
hist(n,100); % See the shape of this histogram if you are curious.
y=x+n; % Received signal
yTs=[1/4 1/2 3/4]; % Threshold values
% To choose one of these yT's yielding the lowest probability of error
for i=1:length(yTs)
   detected=(y>yTs(i)); % Detected result
   Pes(i)=sum(detected~=x)/N; % Simulated error probability
end
[Pe_min,imin]=min(Pes); yT_best=yTs(imin), Pe_min % Optimal threshold
yTo=1/2; Peo=(2*yTo^2-2*yTo+1)/4 % Error probability (E10.8.4)/(E10.8.5)
```

10.2.4 Independence

Two events A and B with $P(A)>0$ and $P(B)>0$ are said to be *independent* of each other if

$$P(A\cap B)=P(A)P(B) \quad \text{or} \quad P(A|B)=P(A) \quad \text{or} \quad P(B|A)=P(B) \qquad (10.2.16\text{a,b,c})$$

which implies that the occurrence of an event does not affect the occurrence of the other event.

Two random variables $\mathbf{x}$ and $\mathbf{y}$ are said to be *independent* of each other if

$$F_{\mathbf{x},\mathbf{y}}(x,y)=F_{\mathbf{x}}(x)F_{\mathbf{y}}(y) \quad \text{or} \quad f_{\mathbf{x},\mathbf{y}}(x,y)=f_{\mathbf{x}}(x)f_{\mathbf{y}}(x) \qquad (10.2.17)$$

10.2.5 Function of a Random Variable

For a random variable $\mathbf{x}$, a differentiable function $\mathbf{u}=g(\mathbf{x})$ of $\mathbf{x}$ is another random variable, and its pdf is obtained from the pdf of $\mathbf{x}$ as

$$f_{\mathbf{u}}(u)=\sum_i \left.\frac{f_{\mathbf{x}}(x)}{|dg/dx|}\right|_{x=x_i} \quad \text{where } x_i\text{'s are the roots of } g(x)=u \qquad (10.2.18)$$

so that the probability of $\mathbf{u}$ being between u and $u+du$ can be expressed as (see Fig. 10.12)

$$P(u\le \mathbf{u}<u+du)=f_{\mathbf{u}}(u)du=\sum_i \left.\frac{f_{\mathbf{x}}(x)}{|dg/dx|}\right|_{x=x_i} du=\sum_i f_{\mathbf{x}}(x_i)\,dx_i\,.$$

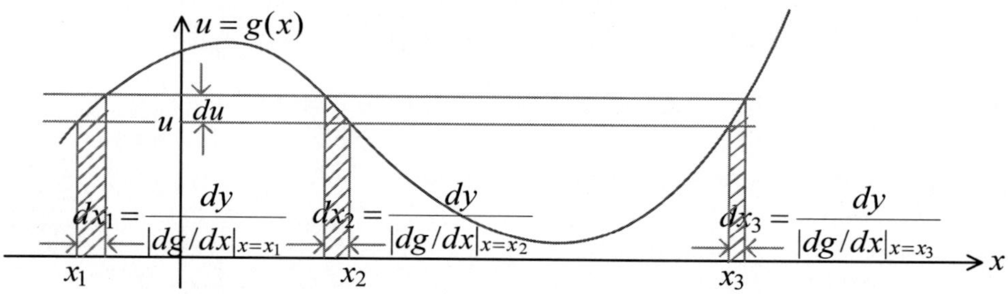

Figure 10.12 The pdf of a random variable as a differentiable function of another random variable

(Example 10.9) The pdf of a Random Variable That is a Function of Another Random Variable

Let us find the pdf of $\mathbf{u}=\sin(\mathbf{x})$ where $\mathbf{x}$ is a uniform random variable distributed in $(0, 2\pi]$. We can use Eq. (10.2.18) to get it as

$$f_{\mathbf{u}}(u)\overset{(10.2.18)}{=}\sum_i \left.\frac{f_{\mathbf{x}}(x)}{|dg/dx|}\right|_{x=x_i}=\frac{1/2\pi}{|\cos(x_1)|}+\frac{1/2\pi}{|\cos(x_2)|}=\frac{1/2\pi}{\sqrt{1-u^2}}+\frac{1/2\pi}{|-\sqrt{1-u^2}|}=\frac{1}{\pi\sqrt{1-u^2}} \qquad (E10.9.1)$$

Run the MATLAB script "em10e09.m" to check the validity of this result.

```
%em10e09.m
N=1e7; x=2*pi*rand(1,N);  u=sin(x);
Nb=50; [nu,cu]=hist(u,Nb); du=cu(2)-cu(1);
fu=1/pi./sqrt(1-cu.^2); % Eq.(E10.9.1)
plot(cu,nu/N, cu,fu*du,'r:');
```

Especially if the function $u=g(x)$ is a one-to-one correspondence mapping so that its inverse function $x=h(u)=g^{-1}(u)$ exists and is differentiable, the above equation for the pdf of $\mathbf{u}=g(\mathbf{x})$ becomes

$$f_{\mathbf{u}}(u)=f_{\mathbf{x}}(h(u))\frac{dh(u)}{du} \tag{10.2.19}$$

For two random variables $\mathbf{x}$ and $\mathbf{y}$, a differentiable function $\mathbf{u}=g(\mathbf{x},\mathbf{y})$ is also another random variable and its PDF can be obtained from the joint pdf of $\mathbf{x}$ and $\mathbf{y}$ as follows:

$$F_{\mathbf{u}}(u)=P((\mathbf{x},\mathbf{y})\in D_u)=\iint_{D_u} f_{\mathbf{x},\mathbf{y}}(x,y)\,dx\,dy \text{ where } D_u=\{(x,y)\,|\,g(x,y)\le u\} \tag{10.2.20}$$

Especially if $\mathbf{u}=g(\mathbf{x},\mathbf{y})=\mathbf{x}\pm\mathbf{y}$ where $\mathbf{x}$ and $\mathbf{y}$ are independent of each other, then the pdf of $\mathbf{u}$ is the convolution of the two pdfs of $\mathbf{x}$ and $\mathbf{y}$ as

$$f_{\mathbf{u}}(u)=f_{\mathbf{x}}(u)*f_{\pm\mathbf{y}}(u)=\int_{-\infty}^{\infty} f_{\mathbf{x}}(u\mp y)f_{\mathbf{y}}(\pm y)dy=\int_{-\infty}^{\infty} f_{\mathbf{x}}(x)f_{\mathbf{y}}(\pm(u-x))dx \tag{10.2.21}$$

(Example 10.10) The pdf of the Sum of Two Random Variables

Consider the two independent random variables $\mathbf{x}$ and $\mathbf{y}$, each having the uniform pdfs

$$f_{\mathbf{x}}(x)=\begin{cases}1/2 & \text{for } -1\le x<1\\ 0 & 0\end{cases} \quad\text{and}\quad f_{\mathbf{y}}(y)=\begin{cases}1 & \text{for } 0\le x<1\\ 0 & 0\end{cases} \tag{E10.10.1}$$

respectively, as shown in Fig. 10.13(a).

(a) Find the pdf of $\mathbf{u}=\mathbf{x}+\mathbf{y}$.

We can use Eq. (10.2.21) to find the pdf of $\mathbf{u}$:

$$f_{\mathbf{u}}(u)=f_{\mathbf{x}}(u)*f_{\mathbf{y}}(u)=\int_{-\infty}^{\infty} f_{\mathbf{x}}(x)f_{\mathbf{y}}(u-x)dx \tag{E10.10.2}$$

As illustrated in Example 4.6, the Laplace transform can be used to compute this convolution as

$$\mathcal{L}\{f_{\mathbf{x}}(u)\}=\frac{1/2}{s}(e^{s}-e^{-s}) \text{ and } \mathcal{L}\{f_{\mathbf{y}}(u)\}=\frac{1}{s}(1-e^{-s})$$

$$;\ f_{\mathbf{u}}(u)=\mathcal{L}^{-1}\left\{\frac{1/2}{s^2}(e^{s}-1-e^{-s}+e^{-2s})\right\}=\frac{1}{2}\{(t+1)u_s(t+1)-tu_s(t)-(t-1)u_s(t-1)+(t-2)u_s(t-2)\} \tag{E10.10.3}$$

This pdf is depicted in Fig. 10.13(b).

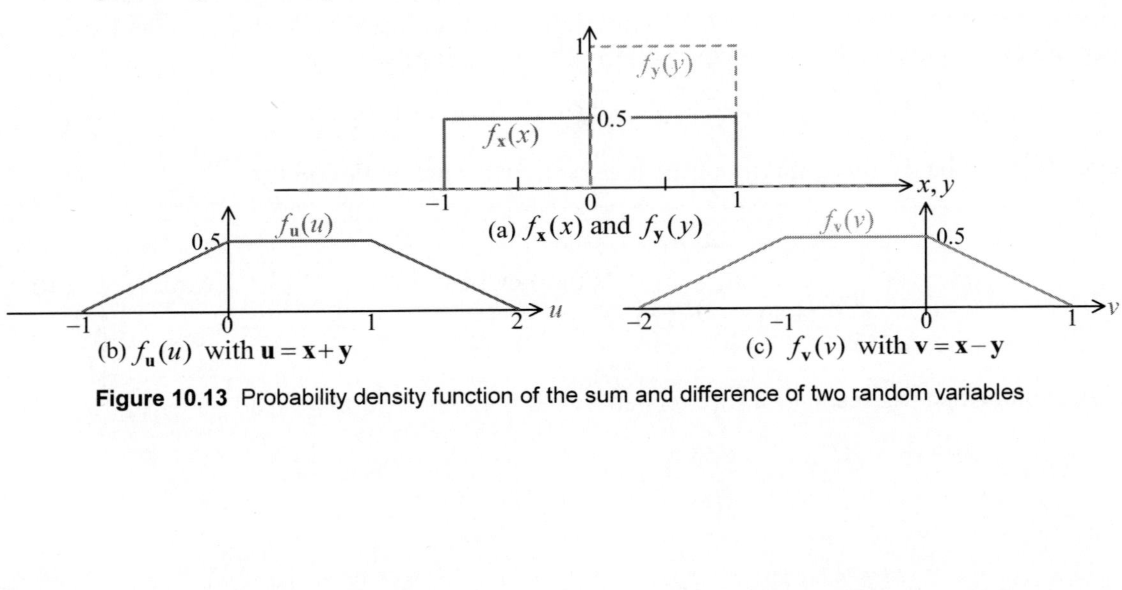

Figure 10.13 Probability density function of the sum and difference of two random variables

```
%em10e10.m
% The pdf of the sum and difference of two rvs.
x0=-1; xf=1; fx=@(x)(x>=x0&x<xf)/(xf-x0); % pdf of x: Eq.(E10.10.1a)
y0=0; yf=1; fy=@(y)(y>=y0&y<yf)/(yf-y0); % pdf of y: Eq.(E10.10.1b)
y_0=-yf; y_f=-y0; fy_=@(y)fy(-y); % pdf of y_: mirror version of f(y)
h=0.01; x=[x0:h:xf]; y=[y0:h:yf]; y_=[y_0:h:y_f];
subplot(221)
fu=conv(fx(x),fy(y))*h; % pdf of u=x+y: Eq.(E10.10.2)
Nfu=length(fu); u0=x0+y0; u=u0+[0:Nfu-1]*h; % Range of the convolution
plot(u,fu,'r'); xlabel('u'), title('f_u(u)')
subplot(222)
fv=conv(fx(x),fy_(y_))*h; % pdf of v=x-y: Eq.(E10.10.4)
Nfv=length(fv); v0=x0+y_0; v=v0+[0:Nfv-1]*h; % Range of the convolution
plot(v,fv,'r'); xlabel('v'), title('f_v(v)')
```

(b) Find the pdf of $\mathbf{v}=\mathbf{x}-\mathbf{y}$.

We can use Eq. (10.2.21) to find the pdf of $\mathbf{v}$:

$$f_{\mathbf{u}}(u)=f_{\mathbf{x}}(u)*f_{\mathbf{y}}(-u)=\int_{-\infty}^{\infty} f_{\mathbf{x}}(x)f_{\mathbf{y}}(x-u)dx \tag{E10.10.4}$$

Here, to use the MATLAB function '`conv()`' (rather than the Laplace transform) for computing this convolution, we compose and run the above script, "em10e10.m," to get Figs. 10.13(b) and (c).

Additionally, the joint pdf of two functions $\mathbf{u}=g_1(\mathbf{x},\mathbf{y})$ and $\mathbf{v}=g_2(\mathbf{x},\mathbf{y})$ can be obtained from that of $\mathbf{x}$ and $\mathbf{y}$ as

$$f_{\mathbf{u},\mathbf{v}}(u,v)=\sum_i \left.\frac{f_{\mathbf{x},\mathbf{y}}(x,y)}{|J(x,y)|}\right|_{(x,y)=(x_i,y_i)} \tag{10.2.22}$$

where (x_i, y_i)'s are the roots of two simultaneous equations $\{g_1(x,y)=u$ and $g_2(x,y)=v\}$, and $|J(x,y)|$ is the absolute value of the determinant of the *Jacobian* matrix, defined as

$$J(x,y)=\begin{bmatrix}\frac{\partial}{\partial x}g_1(x,y) & \frac{\partial}{\partial y}g_1(x,y)\\ \frac{\partial}{\partial x}g_2(x,y) & \frac{\partial}{\partial y}g_2(x,y)\end{bmatrix}: \text{the Jacobian of } \mathbf{g}=\begin{bmatrix}g_1(x,y)\\ g_2(x,y)\end{bmatrix} \text{ w.r.t. } \begin{bmatrix}x\\ y\end{bmatrix} \tag{10.2.23}$$

Especially if the relation between the two pairs (x, y) and (u, v) through g_1 and g_2 is a one-to-one correspondence mapping so that the inverse function is defined by $x=h_1(u,v)$ and $y=h_2(u,v)$, which are differentiable, the above equation for the joint pdf of $\mathbf{u}$ and $\mathbf{v}$ becomes

$$f_{\mathbf{u},\mathbf{v}}(u,v)=f_{\mathbf{x},\mathbf{y}}(h_1(u,v),h_2(u,v))|J(u,v)| \tag{10.2.24}$$

where $J(u,v)$ is the *Jacobian* matrix of the inverse transformation defined as

$$J(u,v)=\begin{bmatrix}\frac{\partial}{\partial u}h_1(u,v) & \frac{\partial}{\partial v}h_1(u,v)\\ \frac{\partial}{\partial u}h_2(u,v) & \frac{\partial}{\partial v}h_2(u,v)\end{bmatrix}: \text{the Jacobian of } \mathbf{h}=\begin{bmatrix}h_1(u,v)\\ h_2(u,v)\end{bmatrix} \text{ w.r.t. } \begin{bmatrix}u\\ v\end{bmatrix} \tag{10.2.25}$$

For example, when (u, v) and (x, y) are related by the linear transformation as

$$\begin{bmatrix} u \\ v \end{bmatrix} = A \begin{bmatrix} x \\ y \end{bmatrix} = \begin{bmatrix} a_{11} & a_{12} \\ a_{21} & a_{22} \end{bmatrix} \begin{bmatrix} x \\ y \end{bmatrix} \quad \text{with } A = \begin{bmatrix} a_{11} & a_{12} \\ a_{21} & a_{22} \end{bmatrix} \text{: nonsingular} \tag{10.2.26}$$

the joint pdf of two functions **u** and **v** can be obtained from that of **x** and **y** as

$$f_{[\mathbf{u}\ \mathbf{v}]}([u\ v]^{\mathrm{T}}) \overset{(10.2.22)}{=} \sum_i \left. \frac{f_{[\mathbf{x}\ \mathbf{y}]}([x\ y]^{\mathrm{T}})}{|J(x,y)|} \right|_{(x,y)=(x_i,y_i)} = \frac{f_{[\mathbf{x}\ \mathbf{y}]}(A^{-1}[u\ v]^{\mathrm{T}})}{|A|} \tag{10.2.27}$$

(Example 10.11) Joint pdf of Two Random Variables Dependent on Two Other Random Variables

Suppose **x** and **y** are two independent zero-mean, unit-variance jointly Gaussian random variables, whose joint pdf is

$$f_{\mathbf{x},\mathbf{y}}(x,y) = \frac{1}{2\pi} e^{-(x^2+y^2)/2} \tag{E10.11.1}$$

Let two random variables **R** and **θ** be functions of **x** and **y** as

$$\begin{bmatrix} \mathbf{R}=\mathbf{r}^2 \\ \boldsymbol{\theta} \end{bmatrix} = \begin{bmatrix} g_1(\mathbf{x},\mathbf{y}) = \mathbf{x}^2+\mathbf{y}^2 \\ g_2(\mathbf{x},\mathbf{y}) = \tan^{-1}(\mathbf{y}/\mathbf{x}) \end{bmatrix} \tag{E10.11.2}$$

Noting that the determinant of the Jacobian matrix of Eq. (E10.11.2) w.r.t. $[\mathbf{x}\ \ \mathbf{y}]^{\mathrm{T}}$ is

$$J(x,y) \overset{(10.2.23)}{=} \begin{bmatrix} \frac{\partial}{\partial x} g_1(x,y) & \frac{\partial}{\partial y} g_1(x,y) \\ \frac{\partial}{\partial x} g_2(x,y) & \frac{\partial}{\partial y} g_2(x,y) \end{bmatrix} \underset{(C.32,33)}{\overset{(E10.11.2)}{=}} \begin{bmatrix} 2x & 2y \\ \frac{-y}{x^2+y^2} & \frac{x}{x^2+y^2} \end{bmatrix}$$

$$; \qquad |J| = \frac{2(x^2+y^2)}{x^2+y^2} = 2 \tag{E10.11.3}$$

we can use Eq. (10.2.22) to find the joint pdf of $\mathbf{R}=\mathbf{r}^2$ and **θ** as

```
%em10e11.m
fx=@(x,m,sigma)exp(-(x-m).^2/2/sigma^2)/sqrt(2*pi)/sigma;
fxy=@(x,y,sigma)exp(-(x.^2+y.^2)/2/sigma^2)/2/pi/sigma^2;
N=1e6; [x,y]=Gaussian_rv(N); % Sample size, Two Gaussian random variables
B=4; sigma=1; % To take only values with absolute values less than B=4
x=x(find(abs(x)<B)); Nx=length(x);
y=y(find(abs(y)<B)); Ny=length(y);
Nb=50; [nx,cx]=hist(x,Nb); [ny,cy]=hist(y,Nb); % Histograms
dx=cx(2)-cx(1); dy=cy(2)-cy(1);
mx=0; fcx=fx(cx,mx,sigma); % Theoretical pdf of x
my=0; fcy=fx(cy,my,sigma); % Theoretical pdf of y
subplot(221), plot(cx,nx/Nx, cx,fcx*dx,'r:'); %Simulated/theoretical fx(x)
subplot(222), plot(cx,ny/Ny, cy,fcy*dy,'r:'); %Simulated/theoretical fy(y)
[X,Y]=meshgrid(cx,cy); [nX,nY]=meshgrid(nx,ny);
Jpdf=(nX.*nY)/(Nx*Ny); % Simulated jpdf of {x,y} generated by Gaussian_v()
subplot(223), mesh(X,Y,Jpdf); axis([-4 4 -4 4 0 max(max(Jpdf))])
Jpdf1 = fxy(X,Y,sigma)*dx*dy; % Theoretical jpdf of {x,y}
Discrepancy = norm(Jpdf-Jpdf1)/norm(Jpdf)
phi_xy=xcorr(x-mean(x),y-mean(y)); % Cross correlation between x and y
rxy=mean(phi_xy)/sqrt(var(x)*var(y)) % correlation coefficient
```

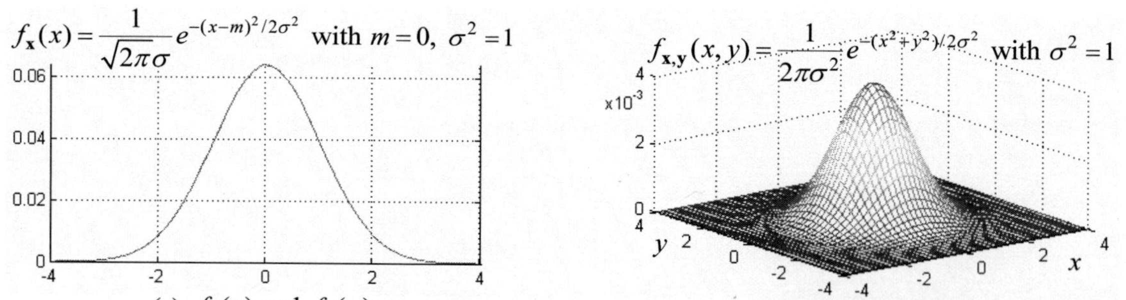

(a) $f_{\mathbf{x}}(x)$ and $f_{\mathbf{y}}(y)$ (b) Joint pdf $f_{\mathbf{x},\mathbf{y}}(x,y)$

Figure 10.14 Marginal/joint pdfs of two random variables generated

$$f_{\mathbf{R},\boldsymbol{\theta}}(R,\theta)\overset{(10.2.22)}{=}\sum_i \left.\frac{f_{\mathbf{x},\mathbf{y}}(x,y)}{|J(x,y)|}\right|_{(x,y)=(x_i,y_i)}\underset{(E10.11.3)}{\overset{(E10.11.1)}{=}}\left.\frac{1}{4\pi}e^{-(x^2+y^2)/2}\right|_{\substack{x=r\cos\theta\\ y=r\sin\theta}}\overset{r^2=R}{=}\frac{1}{4\pi}e^{-R/2}$$

$$=\frac{1}{4\pi}e^{-R/2}=\frac{1}{2}e^{-R/2}\frac{1}{2\pi}=f_{\mathbf{R}}(R)f_{\boldsymbol{\theta}}(\theta) \tag{E10.11.4}$$

This joint pdf turns out to be the product of the marginal pdfs of an exponential random variable $\mathbf{R}=\mathbf{r}^2$ and a uniform random variable $\boldsymbol{\theta}$ where

$$f_{\mathbf{R}}(R)=\frac{1}{2}e^{-R/2}\rightarrow F_{\mathbf{R}}(R)\underset{R>0}{\overset{(10.2.9)}{=}}\int_{-\infty}^{R}f_{\mathbf{R}}(t)\,dt\underset{R>0}{=}\frac{1}{2}\int_0^R e^{-t/2}\,dt=1-e^{-R/2}\ \text{ for } R>0 \tag{E10.11.5a}$$

$$f_{\boldsymbol{\theta}}(\theta)=\frac{1}{2\pi}\ \text{ for } 0\le\theta<2\pi \tag{E10.11.5b}$$

This implies that $\mathbf{R}=\mathbf{r}^2$ and $\boldsymbol{\theta}$ are independent, as **x** and **y** are.

A practical implication of Example 10.11 is a method of generating two independent zero-mean, unit-variance jointly Gaussian random variables **x** and **y** with the joint pdf of Eq. (E10.11.1) as follows:

(Step 1) Generate two independent (standard) uniform random variables $\mathbf{u}_1$ and $\mathbf{u}_2$ with uniform distribution $U(0,1)$.

(Step 2) Referring to Problem 10.3(a), make two random variables $\mathbf{R}=\mathbf{r}^2$ and $\boldsymbol{\theta}$ from $\mathbf{u}_1$ and $\mathbf{u}_2$:

$$\begin{bmatrix}\mathbf{R}=\mathbf{r}^2\\ \boldsymbol{\theta}\end{bmatrix}=\begin{bmatrix}-2\ln\mathbf{u}_1\\ 2\pi\mathbf{u}_2\end{bmatrix}\left(u_1=F_{\mathbf{R}}(R)=1-e^{-R/2}\ \underset{\substack{(P10.3.2)\\ \text{with }\lambda=1/2}}{\overset{\text{Inverse}}{\longrightarrow}}\ R=F_{\mathbf{R}}^{-1}(u_1)=-2\ln(1-u_1)\right) \tag{10.2.28}$$

(Step 3) From $\mathbf{R}=\mathbf{r}^2$ and $\boldsymbol{\theta}$, generate two random variables **x** and **y** as

$$\begin{bmatrix}\mathbf{x}\\ \mathbf{y}\end{bmatrix}=\begin{bmatrix}\sqrt{\mathbf{R}}\cos\boldsymbol{\theta}\\ \sqrt{\mathbf{R}}\sin\boldsymbol{\theta}\end{bmatrix}\ \text{(Inverse of the 2-variable function (E10.11.2))} \tag{10.2.29}$$

This algorithm has been cast into the MATLAB function '`Gaussian_rv()`'. To appreciate how well it works, run the above MATLAB script, "em10e11.m," to get the marginal/joint pdfs of **x** and **y** as depicted in Fig. 10.14.

```
function [x,y]=Gaussian_rv(N)
u1=rand(1,N);  u2=rand(1,N); % Step 1
r=sqrt(-2*log(u1));  th=2*pi*u2; % Step 2 (Eq.(10.2.28))
x=r.*cos(th);  y=r.*sin(th); % Step 3 (Eq.(10.2.29))
```

10.2.6 Expectation, Variance, and Correlation

For a continuous/discrete random variable **x** with pdf/pmf $f_{\mathbf{x}}(x)/p_{\mathbf{x}}(x_k)$, its *expectation* or *mean* (expected value) and *variance* or *deviation* are defined as

$$m_{\mathbf{x}} = \mathcal{E}(\mathbf{x}) = \int_{-\infty}^{\infty} x\, f_{\mathbf{x}}(x)\, dx \ \text{ for a continuous random variable with pdf } f_{\mathbf{x}}(x) \tag{10.2.30a}$$

$$m_{\mathbf{x}} = \mathcal{E}(\mathbf{x}) = \textstyle\sum_k x_k\, p_{\mathbf{x}}(x_k) \ \text{ for a discrete random variable with pmf } p_{\mathbf{x}}(x_k) \tag{10.2.30b}$$

$$\mathrm{var}(\mathbf{x}) = \sigma_{\mathbf{x}}^2 = \mathcal{E}\{(\mathbf{x}-m_{\mathbf{x}})^2\} = \mathcal{E}\{\mathbf{x}^2\} - 2m_{\mathbf{x}}\mathcal{E}\{\mathbf{x}\} + m_{\mathbf{x}}^2 = \mathcal{E}\{\mathbf{x}^2\} - m_{\mathbf{x}}^2 \tag{10.2.31}$$

With these definitions, the tail probability that the value of the random variable **x** goes away from its mean value $m_{\mathbf{x}}$ by further than t is upper-bounded as follows:

$$\textit{Chebyshev inequality}:\ P\{|x-m_{\mathbf{x}}|>t\} \le \frac{\sigma_{\mathbf{x}}^2}{t^2} \tag{10.2.32}$$

Especially if **x** has a *normal* (*Gaussian*) *distribution* with pdf as

$$f_{\mathbf{x}}(x) = \frac{1}{\sqrt{2\pi}\sigma_{\mathbf{x}}} e^{-(x-m_{\mathbf{x}})^2/2\sigma_{\mathbf{x}}^2} \tag{10.2.33}$$

its mean and variance are $m_{\mathbf{x}}$ and $\sigma_{\mathbf{x}}^2$, respectively, and this normal distribution is denoted by

$$\mathbf{x} \sim N(m_{\mathbf{x}}, \sigma_{\mathbf{x}}^2).$$

The probability that the value x of a Gaussian random variable **x** goes away from its mean value $m_{\mathbf{x}}$ by further than t is expressed in terms of the *error function* erf(•) or the *complementary error* (*co-error*) *function* erfc(•), as

$$Q(t) = \frac{1}{\sqrt{2\pi}} \int_t^{\infty} e^{-x^2/2} dx = \frac{1}{2}\,\mathrm{erfc}\left(\frac{t}{\sqrt{2}}\right) \tag{10.2.34}$$

where

$$\mathrm{erfc}(t) = \frac{2}{\sqrt{\pi}} \int_t^{\infty} e^{-x^2} dx = 1 - \underbrace{\frac{2}{\sqrt{\pi}} \int_0^{t} e^{-x^2} dt}_{\text{Error function}} = 1 - \mathrm{erf}(t) \tag{10.2.35}$$

If **x** has a *uniform distribution* on the interval $[a, b]$ with pdf

$$f_{\mathbf{x}}(x) = \frac{1}{b-a}\big(u_s(x-a) - u_s(x-b)\big) = \begin{cases} 1/(b-a) & \text{for } a \le x \le b \\ 0 & \text{elsewhere} \end{cases} \tag{10.2.36}$$

as denoted by

$$\mathbf{x} \sim U(a, b)$$

its mean and variance are as follows:

$$m_{\mathbf{x}} \overset{(10.2.30a)}{=} \int_{-\infty}^{\infty} x f_{\mathbf{x}}(x)\, dx \overset{(10.2.36)}{=} \int_a^b \frac{x}{b-a} dx \overset{(C.34)}{=} \left.\frac{x^2}{2(b-a)}\right|_a^b = \frac{1}{2}(a+b) \tag{10.2.37}$$

$$\sigma_{\mathbf{x}}^2 \overset{(10.2.31)}{=} \int_{-\infty}^{\infty} (x-m_{\mathbf{x}})^2 f_{\mathbf{x}}(x)\, dx \underset{(10.2.37)}{\overset{(10.2.36)}{=}} \int_a^b \left(x - \frac{a+b}{2}\right)^2 \frac{1}{b-a} dx \overset{(C.34)}{=} \left.\frac{1}{3(b-a)}\left(x-\frac{a+b}{2}\right)^3\right|_a^b = \frac{1}{12}(b-a)^2 \tag{10.2.38}$$

Additionally, the expectation or mean value of a function $\mathbf{u}=g(\mathbf{x})$ of a random variable $\mathbf{x}$ can be written as

$$\mathcal{E}\{g(\mathbf{x})\} = \int_{-\infty}^{\infty} g(x) f_{\mathbf{x}}(x)\,dx \tag{10.2.39}$$

Now we define the *covariance* $\sigma_{\mathbf{xy}}^2$, the *correlation coefficient* $\rho_{\mathbf{xy}}$, and the *correlation* $\phi_{\mathbf{xy}}$ for two real-valued random variables $\mathbf{x}$ and $\mathbf{y}$ as follows:

$$\sigma_{\mathbf{xy}}^2 = \mathrm{cov}(\mathbf{x},\mathbf{y}) = \mathcal{E}\{(\mathbf{x}-m_{\mathbf{x}})(\mathbf{y}-m_{\mathbf{y}})\} = \mathcal{E}\{\mathbf{xy}\} - m_{\mathbf{x}} m_{\mathbf{y}} \tag{10.2.40}$$

$$\rho_{\mathbf{xy}} = \frac{\mathcal{E}\{(\mathbf{x}-m_{\mathbf{x}})(\mathbf{y}-m_{\mathbf{y}})\}}{\sqrt{\mathcal{E}\{(\mathbf{x}-m_{\mathbf{x}})^2\}\mathcal{E}\{(\mathbf{y}-m_{\mathbf{y}})^2\}}} = \frac{\mathrm{cov}(\mathbf{x},\mathbf{y})}{\sqrt{\mathrm{var}(\mathbf{x})\,\mathrm{var}(\mathbf{y})}} = \frac{\sigma_{\mathbf{xy}}^2}{\sigma_{\mathbf{x}}\sigma_{\mathbf{y}}} \tag{10.2.41}$$

$$\phi_{\mathbf{xy}} = \mathrm{cor}(\mathbf{x},\mathbf{y}) = \mathcal{E}\{\mathbf{xy}\} \tag{10.2.42}$$

Especially if two random variables $\mathbf{x}$ and $\mathbf{y}$ have *jointly Gaussian* (*normal*) *distribution* with pdf

$$f_{\mathbf{x},\mathbf{y}}(x,y) = \frac{1}{2\pi\sigma_{\mathbf{x}}\sigma_{\mathbf{y}}\sqrt{1-\rho_{\mathbf{xy}}^2}} \exp\left\{\frac{-1}{2(1-\rho_{\mathbf{xy}}^2)}\left(\frac{(x-m_{\mathbf{x}})^2}{\sigma_{\mathbf{x}}^2} - 2\rho_{\mathbf{xy}}\frac{(x-m_{\mathbf{x}})(y-m_{\mathbf{y}})}{\sigma_{\mathbf{x}}\sigma_{\mathbf{y}}} + \frac{(y-m_{\mathbf{y}})^2}{\sigma_{\mathbf{y}}^2}\right)\right\} \tag{10.2.43}$$

their covariance becomes

$$\sigma_{\mathbf{xy}}^2 = \mathrm{cov}(\mathbf{x},\mathbf{y}) \overset{(10.2.41)}{=} \rho_{\mathbf{xy}}\sigma_{\mathbf{x}}\sigma_{\mathbf{y}} \tag{10.2.44}$$

Note that the jointly Gaussian (normal) pdf (10.2.43) can be written in vector form as

$$\begin{aligned} f_{[\mathbf{x}\,\mathbf{y}]}([x\ y]^T) &= \frac{1}{2\pi|\Sigma_{[\mathbf{x}\,\mathbf{y}]}|^{1/2}} \exp\left\{-\frac{1}{2}\begin{bmatrix} x-m_{\mathbf{x}} \\ y-m_{\mathbf{y}} \end{bmatrix}^T \Sigma_{[\mathbf{x}\,\mathbf{y}]}^{-1} \begin{bmatrix} x-m_{\mathbf{x}} \\ y-m_{\mathbf{y}} \end{bmatrix}\right\} \\ &= \frac{1}{2\pi\sigma_{\mathbf{x}}\sigma_{\mathbf{y}}\sqrt{1-\rho_{\mathbf{xy}}^2}} \exp\left\{\frac{-1}{2\sigma_{\mathbf{x}}^2\sigma_{\mathbf{y}}^2(1-\rho_{\mathbf{xy}}^2)}[x-m_{\mathbf{x}}\ \ y-m_{\mathbf{y}}]\begin{bmatrix} \sigma_{\mathbf{y}}^2 & -\rho_{\mathbf{xy}}\sigma_{\mathbf{x}}\sigma_{\mathbf{y}} \\ -\rho_{\mathbf{xy}}\sigma_{\mathbf{x}}\sigma_{\mathbf{y}} & \sigma_{\mathbf{x}}^2 \end{bmatrix}\begin{bmatrix} x-m_{\mathbf{x}} \\ y-m_{\mathbf{y}} \end{bmatrix}\right\} \end{aligned} \tag{10.2.45}$$

where the matrix $\Sigma_{[\mathbf{x}\,\mathbf{y}]}$, called the *covariance matrix*, of the random vector $[\mathbf{x}\ \ \mathbf{y}]^T$ is

$$\Sigma_{[\mathbf{x}\,\mathbf{y}]} = \mathcal{E}\left\{\begin{bmatrix} \mathbf{x}-m_{\mathbf{x}} \\ \mathbf{y}-m_{\mathbf{y}} \end{bmatrix}\begin{bmatrix} \mathbf{x}-m_{\mathbf{x}} \\ \mathbf{y}-m_{\mathbf{y}} \end{bmatrix}^T\right\} = \mathcal{E}\left\{\begin{bmatrix} (\mathbf{x}-m_{\mathbf{x}})^2 & (\mathbf{x}-m_{\mathbf{x}})(\mathbf{y}-m_{\mathbf{y}}) \\ (\mathbf{y}-m_{\mathbf{y}})(\mathbf{x}-m_{\mathbf{x}}) & (\mathbf{y}-m_{\mathbf{y}})^2 \end{bmatrix}\right\} = \begin{bmatrix} \sigma_{\mathbf{x}}^2 & \sigma_{\mathbf{xy}}^2 \\ \sigma_{\mathbf{yx}}^2 & \sigma_{\mathbf{y}}^2 \end{bmatrix}. \tag{10.2.46}$$

If two random variables $\mathbf{x}$ and $\mathbf{y}$ are *uncorrelated* with each other, i.e. $\mathrm{cov}(\mathbf{x},\mathbf{y})=0$, we have

$$\rho_{\mathbf{xy}} = 0;\quad \mathcal{E}\{\mathbf{xy}\} = \mathcal{E}\{\mathbf{x}\}E\{\mathbf{y}\} = m_{\mathbf{x}}m_{\mathbf{y}} \tag{10.2.47}$$

$$\mathrm{var}(\mathbf{x}+\mathbf{y}) = \mathrm{var}(\mathbf{x}) + \mathrm{var}(\mathbf{y}) \tag{10.2.48}$$

Two random variables $\mathbf{x}$ and $\mathbf{y}$ are said to be *orthogonal* to each other if

$$\mathcal{E}\{\mathbf{xy}\} = 0 \tag{10.2.49}$$

[Remark 10.2] Independence and Correlation

If two random variables **x** and **y** are independent of each other, they are also uncorrelated with each other; but the converse is not true unless they have jointly Gaussian (normal) distribution. In other words, even if they are uncorrelated with each other, they may be dependent on each other.

(Q) Is it possible for non-Gaussian random variables to be uncorrelated but not independent?
(A) Yes. See Example 10.13.

(Example 10.12) Mean and Variance

(a) The pmf of a *geometric random variable* **k**, which can model the number of trials before the first success in a series of repeated independent *Bernoulli trial*s with the success/failure probability p/q, is

$$p_{\mathbf{k}}(k) = p(1-p)^k \text{ for } k \text{ :nonnegative integers} \tag{E10.12a.1}$$

Its mean can be found by using Eq. (10.2.30b) as

$$m_{\mathbf{k}} = \mathcal{E}\{\mathbf{k}\} \overset{(10.2.30b)}{=} \sum_k k\, p_{\mathbf{k}}(k) \underset{k>0}{\overset{(E10.12a.1)}{=}} \sum_{k=1}^{\infty} k\,p(1-p)^k \overset{1-p=q}{=} \sum_{k=0}^{\infty} k\,pq^k$$

$$\overset{(C.26)}{=} \sum_{k=1}^{\infty} pq\frac{d}{dq}q^k = pq\frac{d}{dq}\sum_{k=1}^{\infty} q^k \overset{(C.23)}{=} pq\frac{d}{dq}\frac{q}{1-q} = pq\frac{1-q+q}{(1-q)^2} \overset{1-p=q}{=} \frac{q}{p} \tag{E10.12a.2}$$

Its variance can be found by using Eq. (10.2.31) as

$$\mathrm{var}(\mathbf{k}) \overset{(10.2.31)}{=} \mathcal{E}\{\mathbf{k}^2\} - m_{\mathbf{k}}^2 \overset{(10.2.30b)}{=} \sum_k k^2 p_{\mathbf{k}}(k) - m_{\mathbf{k}}^2 \underset{k>0}{\overset{(E10.12a.1)}{=}} \sum_{k=1}^{\infty} k^2 p(1-p)^k - m_{\mathbf{k}}^2$$

$$\underset{(E10.12a.2)}{\overset{1-p=q}{=}} pq^2\sum_{k=1}^{\infty} k^2 q^{k-2} - \frac{q^2}{p^2} = pq^2\sum_{k=1}^{\infty}\big(k(k-1)+k-1+1\big)q^{k-2} - \frac{q^2}{p^2}$$

$$\overset{(C.26)}{=} pq^2\left(\frac{d^2}{dq^2}\sum_{k=1}^{\infty} q^k + \frac{d}{dq}\sum_{k=1}^{\infty} q^{k-1} + \sum_{k=1}^{\infty} q^{k-2}\right) - \frac{q^2}{p^2}$$

$$\overset{(C.23)}{=} pq^2\left(\frac{d^2}{dq^2}\frac{q}{1-q} + \frac{d}{dq}\frac{1}{1-q} + \frac{1}{q(1-q)}\right) - \frac{q^2}{p^2} = pq^2\left(\frac{2}{(1-q)^3} + \frac{1}{(1-q)^2} + \frac{1}{q(1-q)}\right) - \frac{q^2}{p^2}$$

$$\overset{1-p=q}{=} \frac{2q^2}{p^2} + \frac{q^2}{p} + q - \frac{q^2}{p^2} = q\frac{2q+pq+p^2-q}{p^2} \overset{1-p=q}{=} \frac{q}{p^2} \tag{E10.12a.3}$$

```
%em10e12a.m: % Geometric random variable
p=0.2; q=1-p; Ns=1e5; M=30; % Parameters/Length of a geometric rv
for i=1:Ns
   pfs=rand(1,M)>(1-p); % A series of 1/0 with success probability p
   inds=find(pfs==1);   % Find the indices of 1's (successes)
   if isempty(inds), kk(i)=M; % More than M failures before success
    else  kk(i) = inds(1)-1; % Number of failures before first success
   end
end
%kk=random('geo',p,[1 Ns]); % kk=geornd(p,1,Ns);
m=mean(kk), q/p      % Mean, Eq.(E10.12a.2)
sgm2=var(kk), q/p^2 % Variance, Eq.(E10.12a.3)
Nb=max(kk)+1; cs=[0:Nb]; hist(kk,cs); hold on
pmf_geo = p*(1-p).^cs*Ns; % Eq.(E10.12a.1)
plot(cs,pmf_geo,'r'), title('Histogram/pdf of a geometric random seq')
```

Noting that the MATLAB function 'geornd()' or 'random()' can be used to generate geometric random numbers, we can run the above script, "em10e12a.m," to get the histogram and (scaled) pmf of a geometric random sequence, as shown in Fig. 10.15(a).

(b) The pmf of a *binomial random variable* **k**, which can model the number of successes in n independent *Bernoulli trial*s with the probability of success/failure p/q, is

$$p_{\mathbf{k}}(k \text{ successes in } n \text{ trials}) = \binom{n}{k} p^k (1-p)^{n-k} \tag{E10.12b.1}$$

While its mean/variance may be found by using Eqs. (10.2.30b)/(10.2.31), we deduce them. That is, the expected number of success is p for one trial so that the mean of **k** for n trials must be

$$m_{\mathbf{k}} = \mathcal{E}\{\mathbf{k}\} = np \tag{E10.12b.2}$$

The variance of **k** for one trial (n=1) is

$$\text{var}_1(\mathbf{k}) \underset{(10.2.30b)}{\overset{(10.2.31)}{=}} \sum_k^{n=1} k^2 p_{\mathbf{k}}(k) - m_{\mathbf{k}}^2 \underset{(E10.12b.2)}{\overset{(E10.12b.1)}{=}} 0^2(1-p) + 1^2 p - p^2 = p(1-p)$$

so that the variance of **k** for n trials must be

$$\text{var}(\mathbf{k}) \overset{(10.2.48)}{=} n\,\text{var}_1(\mathbf{k}) = np(1-p) = npq \tag{E10.12b.3}$$

We can run the script "em10e12b.m" to get the histogram and (scaled) pmf of a binomial random sequence, as shown in Fig. 10.15(b).

```
%em10e12b.m
% Binomial random variable
p=0.4; q=1-p; Ns=1e5; M=30; % Parameters/Length of a binomial rv
for i=1:Ns
   pfs = rand(1,M)<p; % A series of 1/0 with success probability p
   kk(i) = sum(pfs); % Number of successes
end
%kk=random('bino',M,p,[1 Ns]);
m=mean(kk), M*p       % Mean, Eq.(E10.12b.2)
sgm2=var(kk), M*p*q   % Variance, Eq.(E10.12b.3)
Nb=max(kk)+1; cs=[0:Nb]; % Number of bins and bin centers
hist(kk,cs); hold on
for k=0:Nb
   pmf_bin(k+1) = nchoosek(M,k)*p^k*(1-p)^(M-k); % Eq.(E10.12b.1)
end
plot(cs,pmf_bin*Ns,'r')
title('Histogram/pdf of a binomial random sequence')
```

(c) The pdf of an *exponential random variable* **t**, which can model the (waiting) time t until some event occurs, is

$$f_{\mathbf{t}}(t) = \frac{1}{\mu} e^{-t/\mu} \quad \text{for } t > 0 \tag{E10.12c.1}$$

Its mean can be found by using Eq. (10.2.30a) as

$$m_{\mathbf{t}} = \mathcal{E}\{\mathbf{t}\} \overset{(10.2.30a)}{=} \int_{-\infty}^{\infty} t f_{\mathbf{t}}(t)\,dt \overset{(E10.12c.1)}{=} \int_0^{\infty} t \frac{1}{\mu} e^{-t/\mu} dt \underset{(C.35)}{\overset{(C.38)}{=}} \cancel{-te^{-t/\mu}\Big|_0^{\infty}} + \int_0^{\infty} e^{-t/\mu} dt \overset{(C.35)}{=} -\mu e^{-t/\mu}\Big|_0^{\infty} = \mu \tag{E10.12c.2}$$

```
%em10e12c.m
% Exponential random variable
clear, %clf
mu=2; Ns=1e5; % Parameter/Length of an exponential rv
tt=random('exp',mu,[1 Ns]); % generate an exponential random sequence
m=mean(tt), mu       % Mean, Eq.(E10.12c.2)
sgm2=var(tt), mu^2   % Variance, Eq.(E10.12c.3)
% Dump too large values (bigger than 6*mu)
B=mu*6; ttB=tt(find(tt<B)); NsB=length(ttB);
subplot(223)
Nb=50; cs=[0.5:Nb]*(B/Nb); dt=cs(2)-cs(1);
hist(ttB,cs); hold on
pdf_exp = exp(-cs/mu)/mu; % Eq.(E10.12c.1)
plot(cs,pdf_exp*dt*Ns,'r')
xlim([-1 cs(end)+1])
title('Histogram/pdf of an exponential random sequence')
```

Its variance can be found by using Eq. (10.2.31) as

$$\begin{aligned}\mathrm{var}(\mathbf{t}) &\overset{(10.2.31)}{=} \mathcal{E}\{\mathbf{t}^2\} - m_{\mathbf{t}}^2 \underset{(\mathrm{E}10.12\mathrm{c}.2)}{\overset{(10.2.30\mathrm{a})}{=}} \int_{-\infty}^{\infty} t^2 f_{\mathbf{t}}(t)\, dt - \mu^2 \overset{(\mathrm{E}10.12\mathrm{c}.1)}{=} \int_0^{\infty} t^2 \frac{1}{\mu} e^{-t/\mu} dt - \mu^2 \\ &\underset{(\mathrm{C}.35)}{\overset{(\mathrm{C}.38)}{=}} \cancel{-t^2 e^{-t/\mu}\Big|_0^{\infty}} + 2\int_0^{\infty} t\, e^{-t/\mu} dt - \mu^2 \underset{(\mathrm{C}.35)}{\overset{(\mathrm{C}.38)}{=}} \cancel{-2\mu t e^{-t/\mu}\Big|_0^{\infty}} + 2\mu \int_0^{\infty} e^{-t/\mu} dt - \mu^2 \\ &\overset{(\mathrm{C}.35)}{=} 2\mu(-\mu) e^{-t/\mu}\Big|_0^{\infty} - \mu^2 = \mu^2 . \end{aligned} \qquad (\mathrm{E}10.12\mathrm{c}.3)$$

We can run the above script, "em10e12c.m," to get the histogram and (scaled) pmf of an exponential random sequence, as shown in Fig. 10.15(c).

(d) The pmf of a *Poisson random variable* **k**, which can model the number of occurrences in an interval, is

$$p_{\mathbf{k}}(k) = \frac{\lambda^k}{k!} e^{-\lambda} \quad \text{for } k \text{ :nonnegative integers} \qquad (\mathrm{E}10.12\mathrm{d}.1)$$

Eq. (10.2.30b) can be used to find the mean as

$$m_{\mathbf{k}} \underset{(\mathrm{E}10.12\mathrm{d}.1)}{\overset{(10.2.30\mathrm{b})}{=}} \sum_{k=0}^{\infty} k \frac{\lambda^k}{k!} e^{-\lambda} = \sum_{k=1}^{\infty} \frac{\lambda^k}{(k-1)!} e^{-\lambda} \overset{k \to k+1}{=} \sum_{k=0}^{\infty} \frac{\lambda^{k+1}}{k!} e^{-\lambda} = \lambda e^{-\lambda} \sum_{k=0}^{\infty} \frac{\lambda^k}{k!} \overset{(\mathrm{C}.25)}{=} \lambda \qquad (\mathrm{E}10.12\mathrm{d}.2)$$

```
%em10e12d.m
% Poisson random variable
clear, %clf
lam=10; Ns=1e5; % Parameters/Length of a Poisson rv
kk=random('poiss',lam,[1 Ns]); % generate a Poisson random sequence
m=mean(kk), lam     % Mean, Eq.(E10.12d.2)
sgm2=var(kk), lam   % Variance, Eq.(E10.12d.3)
subplot(224)
Nb=max(kk)+1; cs=[0:Nb];
hist(kk,cs); hold on
pmf_poi = lam.^cs./factorial(cs)*exp(-lam); % Eq.(E10.12d.1)
plot(cs,pmf_poi*Ns,'r')
title('Histogram/pdf of a Poisson random sequence')
```

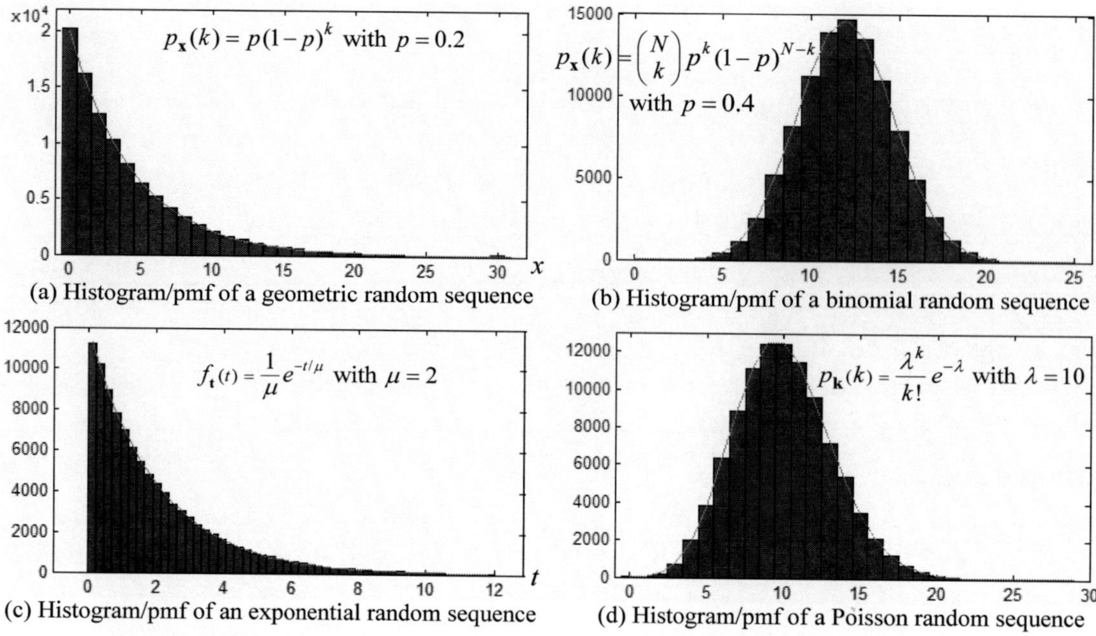

Figure 10.15 Histograms and pdfs of the four random sequences in Example 10.12(a)–(d)

Eq. (10.2.31) can be used to find the variance as

$$\sigma_{\mathbf{k}}^2 \underset{(10.2.30b)}{\overset{(10.2.31)}{=}} \sum_{k=0}^{\infty} k^2 p(k) - m_{\mathbf{k}}^2 \overset{(E10.12d.1)}{=} \sum_{k=0}^{\infty} k^2 \frac{\lambda^k}{k!} e^{-\lambda} - m_{\mathbf{k}}^2 = \lambda \sum_{k=1}^{\infty} k \frac{\lambda^{k-1}}{(k-1)!} e^{-\lambda} - \lambda^2$$

$$\overset{k \to k+1}{=} \lambda \sum_{k=0}^{\infty} (k+1) \frac{\lambda^k}{k!} e^{-\lambda} - \lambda^2 \underset{(C.25)}{\overset{(E10.12d.2)}{=}} \lambda(\lambda+1) - \lambda^2 = \lambda \quad \text{(E10.12d.3)}$$

We can run the above script, "em10e12d.m," to get the histogram and (scaled) pmf of a Poisson random sequence, as shown in Fig. 10.15(d).

(e) To figure out the relationship between the Poisson and exponential distributions, suppose the number of occurrences of an event for t units of time can be modeled as a Poisson random integer $\mathbf{k}$ whose pmf is as

$$p_{\mathbf{k}}(k) = \frac{(\lambda t)^k}{k!} e^{-\lambda t} \text{ for } k \text{:nonnegative integers} \quad \text{(E10.12e.1)}$$

where the average number of occurrences per unit time is λ. Then, what meaning can be given to $p_{\mathbf{k}}(0) = e^{-\lambda t}$? It can be thought of as the probability of zero (no) occurrence for t units of time. Viewing $p_{\mathbf{k}}(0) = e^{-\lambda t}$ as a function of (continuous) time t, it can be regarded as the probability that the waiting time $\mathbf{t}$ from the last occurrence to the next occurrence will be longer than t:

$$P(\mathbf{t} > t) = e^{-\lambda t} \quad \text{(E10.12e.2)}$$

Then the probability of the waiting time $\mathbf{t}$ (between successive occurrences) shorter than or equal to t is

$$F_{\mathbf{t}}(t) = 1 - P(\mathbf{t} > t) \overset{(E10.12e.2)}{=} 1 - e^{-\lambda t} \quad \text{(E10.12e.3)}$$

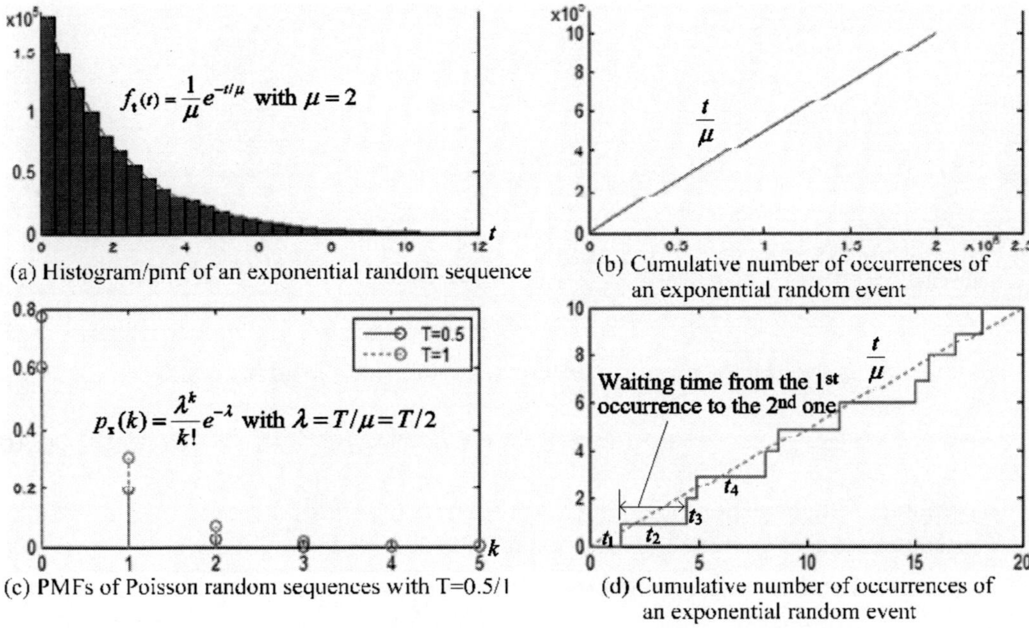

Figure 10.16 Exponential and Poisson random sequences in Example 10.12(e)

```
%Poisson_exponential.m
mu=2; % mu: Mean inter-occurrence time
T=0.5; lambda=T/mu; % lambda: Average number of occurrences per T
Ns=5e5; tt=random('exp',mu,[1 Ns]); % generate Poisson random times
kk=0:5;  time(kk+1)=0;
for n=1:Ns
   if tt(n)>T  % Only for tt(n)>T
     time(1) = time(1)+tt(n)-T; % Time of no occurrence (k=0) during T
     time(2) = time(2)+T; % Time of one occurrence (k=1) during T
   end
end
p_sim(1:2) = time(1:2)./sum(tt);
kk1=kk(3:end);
for k=kk1
   for n=2:Ns-k
      tmp = sum(tt(n+[0:k-2]));
      if tmp<=T & tmp+tt(n+k-1)>T % k occurrences during T
        time(k+1)=time(k+1)+tt(n+k-2); % Time of k occurrences during T
      end
   end
   % Probability of k occurrences during T
   p_sim(k+1) = time(k+1)./sum(tt(2:Ns-k));
end
p_Poisson(kk+1)=lambda.^kk./factorial(kk)*exp(-lambda); %Eq.(E10.12c.1)
%p_Poisson(kk+1) = poisspdf(kk,lambda);
[p_sim; p_Poisson] % Simulated and Theoretical pmfs
B=mu*6; ttB=tt(find(tt<B)); NsB=length(ttB);
Nb=30; [ns,cs]=hist(ttB,Nb); dt=cs(2)-cs(1);
pdf_exp = exp(-cs/mu)/mu; % Eq.(E10.12c.1)
subplot(221), hist(ttB,Nb), hold on, plot(cs,pdf_exp*dt*Ns,'r')
title('Histogram/pdf of the exponential random sequence')
```

```
subplot(223)
stem(kk,p_Poisson(1,:))
legend(['T=' num2str(T)])
title('pmf of the Poisson random variable')
subplot(222)
t=[0 cumsum(tt)];
stairs(t,[0:Ns]), hold on, plot(t,t/mu,'r-.')
title('Cumulative number of occurrences and t/mu vs. time')
subplot(224)
t=[0 cumsum(tt)];
stairs(t,[0:Ns]), hold on, plot(t,t/mu,'r-.'); axis([0 20*mu 0 20])
title('Cumulative number of occurrences and t/mu vs. time')
```

Regarding this as the PDF of **t**, we can differentiate it to find the pdf as

$$f_{\mathbf{t}}(t) = \frac{d}{dt} F_{\mathbf{t}}(t) \overset{\text{(E10.12e.3)}}{=} \lambda e^{-\lambda t} \tag{E10.12e.4}$$

This conforms with the exponential pdf (E10.12c.1) with $\mu=1/\lambda$.

We can run the above script, "Poisson_exponential.m," to do the following:

- To generate a random sequence {tt(n), n=1:Ns} with an exponential pdf

$$f_{\mathbf{t}}(t) = \frac{1}{\mu} e^{-t/\mu} \text{ with } \mu = 2 \text{ for } t > 0 \tag{E10.12e.5}$$

  and plot their histogram together with the theoretical exponential pdf, as shown in Fig. 10.16(a).
- Suppose that {tt(n), n=1:Ns} are the intervals between successive occurrences of an event, i.e., an event occurs successively at

$$t = \text{tt}(1),\ \text{tt}(1)+\text{tt}(2),\ \text{tt}(1)+\text{tt}(2)+\text{tt}(3), \cdots \tag{E10.12e.6}$$

  and plot the accumulated number of occurrences vs. t as shown in Fig. 10.16(b), and partially (for horizontal zoom-in) in Fig. 10.16(d).
- To find the sum 'time(k)' of times through which the number of occurrences for T=0.5[s] is k, divide it by the whole time to get the probability density $p_d(k)$ of the number of occurrences for T[s] being k for each k=0, 1, 2,.., and plot $p_d(k)$ together with the theoretical Poisson pmf

$$p_{\mathbf{k}}(k) = \frac{\lambda^k}{k!} e^{-\lambda} \text{ with } \lambda = \frac{T}{\mu} \text{ for } k \text{ :nonnegative integers} \tag{E10.12e.6}$$

  as shown in Fig. 10.16(c). Note that μ and λ can be interpreted as the mean inter-occurrence (arrival) time and the mean occurrence rate, also called the event rate; that is, the average number of occurrences per unit time), respectively.

(f) The pdf of a *Rayleigh random variable* **x**, which can model the envelope of a narrowband Gaussian noise or the lifetime of an object, is

$$p_{\mathbf{x}}(x) = \frac{x}{b^2} e^{-x^2/2b^2} \text{ for } x \geq 0 \tag{E10.12f.1}$$

Eq. (10.2.30b) can be used to find the mean as

$$m_{\mathbf{x}} \underset{\text{(E10.12f.1)}}{\overset{\text{(10.2.30a)}}{=}} \int_0^\infty x \frac{x}{b^2} e^{-x^2/2b^2} dx \overset{\text{(C.38)}}{=} -x\, e^{-x^2/2b^2}\Big|_0^\infty + \int_0^\infty e^{-x^2/2b^2} dx$$

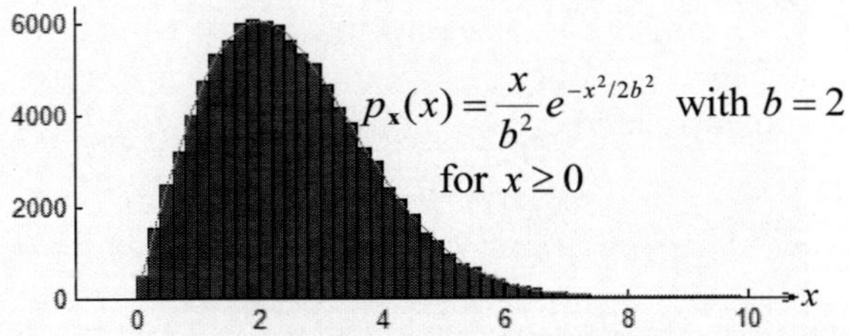

Figure 10.17 Histogram and pdf of the Rayleigh random sequence in Example 10.12(f)

$$m_{\mathbf{x}}^2 = \left(\int_0^\infty e^{-x^2/2b^2} dx\right)\left(\int_0^\infty e^{-y^2/2b^2} dy\right) = \int_0^\infty \int_0^\infty e^{-(x^2+y^2)/2b^2} dx\,dy$$

$$; \underset{x^2+y^2 \to \rho^2}{\overset{\text{rect-to-polar}}{=}} \int_0^{\pi/2} \int_0^\infty e^{-\rho^2/2b^2} \rho\, d\rho\, d\phi = \int_0^{\pi/2} -b^2 e^{-\rho^2/2b^2}\Big|_0^\infty d\phi = \frac{\pi}{2} b^2$$

$$; \quad m_{\mathbf{x}} = \sqrt{m_{\mathbf{x}}^2 = \left(\frac{\pi}{2} b^2\right)} = \sqrt{\frac{\pi}{2}}\, b \qquad \text{(E10.12f.2)}$$

Eq. (10.2.31) can be used to find the variance as

$$\sigma_{\mathbf{x}}^2 \underset{(10.2.30a)}{\overset{(10.2.31)}{=}} \int_0^\infty x^2 \frac{x}{b^2} e^{-x^2/2b^2} dx - m_{\mathbf{x}}^2 \overset{(C.38)}{=} -x^2 e^{-x^2/2b^2}\Big|_0^\infty + \int_0^\infty 2x e^{-x^2/2b^2} dx - m_{\mathbf{x}}^2 = \frac{4-\pi}{2} b^2 \qquad \text{(E10.12f.3)}$$

We can run the script "em10e12f.m" to get the histogram and (scaled) pmf of a Rayleigh random sequence, as shown in Fig. 10.17.

```
%em10e12f.m
% Rayleigh random variable
clear, clf
b=2; Ns=1e5; % Parameter/Length of a Rayleigh random variable
xx=random('Rayl',b,[1 Ns]); % generate a Rayleigh random sequence
m=mean(xx), sqrt(pi/2)*b    % Mean, Eq.(E10.12f.2)
sgm2=var(xx), (4-pi)/2*b^2 % Variance, Eq.(E10.12f.3)
% Dump too large values (bigger than B=5*b)
B=b*5; xxB=xx(find(xx<B)); NsB=length(xxB);
subplot(221)
Nb=50; cs=[0.5:Nb]*(B/Nb); dt=cs(2)-cs(1);
pdf_Ray = cs/b^2.*exp(-cs.^2/(2*b^2)); % Eq.(E10.12f.1)
hist(xxB,cs); hold on, plot(cs,pdf_Ray*dt*Ns,'r')
title('Histogram/pdf of a Rayleigh random sequence')
```

(Example 10.13) "Independent" and "Uncorrelated"

Consider two random variables **x** and **y** that are both dependent on a uniform random variable **θ** as

$$\begin{bmatrix} \mathbf{x} \\ \mathbf{y} \end{bmatrix} = \begin{bmatrix} \cos\boldsymbol{\theta} \\ \sin\boldsymbol{\theta} \end{bmatrix} \qquad \text{(E10.13.1)}$$

where the pdf of **θ** is

$$f_{\boldsymbol\theta}(\theta)=\begin{cases}1/2\pi & \text{for } -\pi<\theta<\pi\\ 0 & \text{elsewhere}\end{cases} \tag{E10.13.2}$$

To decide whether they are uncorrelated or not, let us see if Eq. (10.2.47) holds:

$$\begin{aligned}\mathcal{E}\{\mathbf{x}\mathbf{y}\}&\overset{(10.2.39)}{=}\int_{-\infty}^{\infty}g(\theta)f_{\boldsymbol\theta}(\theta)\,d\theta\underset{(\text{E}10.13.1)}{\overset{(\text{E}10.13.2)}{=}}\frac{1}{2\pi}\int_{-\pi}^{\pi}\cos\theta\sin\theta\,d\theta\\ &\overset{(\text{C}.16)}{=}\frac{1}{4\pi}\int_{-\pi}^{\pi}\sin 2\theta\,d\theta=0=m_{\mathbf{x}}m_{\mathbf{y}}\ \text{ since } m_{\mathbf{x}}=0 \text{ and } m_{\mathbf{y}}=0\end{aligned} \tag{E10.13.3}$$

Thus, Eq. (10.2.47) is satisfied, implying that **x** and **y** are uncorrelated, while they are not independent since they are related as

$$x^2+y^2=\cos^2\theta+\sin^2\theta=1 \tag{E10.13.4}$$

(Example 10.14) Mean and Covariance of Two Jointly Gaussian Random Variables

Consider how the covariance matrix K of two jointly Gaussian random variables **x** and **y** is changed by a linear transformation A, where the jointly Gaussian (normal) pdf is given by Eq. (10.2.45) as

$$f_{[\mathbf{x}\,\mathbf{y}]}([x\ y]^T)\overset{(10.2.45)}{=}\frac{1}{2\pi|\Sigma_{[\mathbf{x}\,\mathbf{y}]}|^{1/2}}\exp\left\{-\frac{1}{2}\begin{bmatrix}x-m_{\mathbf{x}}\\ y-m_{\mathbf{y}}\end{bmatrix}^T\Sigma_{[\mathbf{x}\,\mathbf{y}]}^{-1}\begin{bmatrix}x-m_{\mathbf{x}}\\ y-m_{\mathbf{y}}\end{bmatrix}\right\}\ \text{with}\ \Sigma_{[\mathbf{x}\,\mathbf{y}]}=\begin{bmatrix}\sigma_{\mathbf{x}}^2 & \rho_{\mathbf{xy}}\sigma_{\mathbf{x}}\sigma_{\mathbf{y}}\\ \rho_{\mathbf{xy}}\sigma_{\mathbf{x}}\sigma_{\mathbf{y}} & \sigma_{\mathbf{y}}^2\end{bmatrix} \tag{E10.14.1}$$

The joint pdf of two random variables **u** and **v** generated from **x** and **y** by a linear transformation

$$\begin{bmatrix}u\\ v\end{bmatrix}=A\begin{bmatrix}x\\ y\end{bmatrix} \tag{E10.14.2}$$

can be obtained by using Eq. (10.2.27) together with Eq. (E10.14.1) as

$$\begin{aligned}f_{[\mathbf{u}\,\mathbf{v}]}([u\ v]^T)&\overset{(10.2.27)}{=}\frac{f_{[\mathbf{x}\,\mathbf{y}]}(A^{-1}[u\ v]^T)}{|A|}\\ &\overset{(\text{E}10.14.1)}{=}\frac{1}{2\pi|A\Sigma_{[\mathbf{x}\,\mathbf{y}]}A^T|^{1/2}}\exp\left\{-\frac{1}{2}\left[\begin{bmatrix}u\\ v\end{bmatrix}-A\begin{bmatrix}m_{\mathbf{x}}\\ m_{\mathbf{y}}\end{bmatrix}\right]^T A^{-T}\Sigma_{[\mathbf{x}\,\mathbf{y}]}^{-1}A^{-1}\left[\begin{bmatrix}u\\ v\end{bmatrix}-A\begin{bmatrix}m_{\mathbf{x}}\\ m_{\mathbf{y}}\end{bmatrix}\right]\right\}\end{aligned} \tag{E10.14.3}$$

Compared to Eq. (E10.14.1), this implies that the mean and covariance matrix of the jointly Gaussian random vector $[\mathbf{u}\ \ \mathbf{v}]^T$ are

$$\begin{bmatrix}m_{\mathbf{u}}\\ m_{\mathbf{v}}\end{bmatrix}=\mathcal{E}\left\{\begin{bmatrix}\mathbf{u}\\ \mathbf{v}\end{bmatrix}\right\}=A\begin{bmatrix}m_{\mathbf{x}}\\ m_{\mathbf{y}}\end{bmatrix}\quad\text{and}\quad\Sigma_{[\mathbf{u}\,\mathbf{v}]}=\text{cov}\left\{\begin{bmatrix}\mathbf{u}\\ \mathbf{v}\end{bmatrix}\right\}=A\Sigma_{[\mathbf{x}\,\mathbf{y}]}A^T \tag{E10.14.4}$$

This example suggests that a jointly Gaussian vector with a given covariance matrix can be generated from two uncorrelated Gaussian random variables.

(Example 10.15) LMMSE (Linear Mimimum Mean Square Error) Estimator

Consider the problem of estimating a random variable **x** based on the value of another random variable (observation) **y** by a linear estimator $\hat{\mathbf{x}}=g(\mathbf{y})=a\mathbf{y}+\text{b}$, where the coefficients a and b are so determined as to minimize the objective function

$$J(a,b)=\mathcal{E}\left\{(\mathbf{x}-\hat{\mathbf{x}})^2\right\}=\mathcal{E}\left\{(\mathbf{x}-(a\mathbf{y}+b))^2\right\} \tag{E10.15.1}$$

which is called the *MSE* (*mean square error*).

First, the coefficient b is determined by setting the derivative of $J(a,b)$ (with a=constant) w.r.t. b to zero:

$$\frac{d}{db}J(a,b)\overset{(E10.15.1)}{=}\frac{d}{db}\mathcal{E}\left\{(\mathbf{x}-(a\mathbf{y}+b))^2\right\}=\frac{d}{db}\mathcal{E}\left\{\mathbf{x}^2-2\mathbf{x}(a\mathbf{y}+b)+(a\mathbf{y}+b)^2\right\}$$

$$=\mathcal{E}\left\{\frac{d}{db}\left(-2\mathbf{x}b+(a\mathbf{y}+b)^2\right)\right\}=-2\mathcal{E}\{\mathbf{x}\}+2\left(a\mathcal{E}\{\mathbf{y}\}+b\right)=0$$

$$;\qquad b^{\mathrm{o}}=\mathcal{E}\{\mathbf{x}\}-a\mathcal{E}\{\mathbf{y}\}=\mathbf{m_x}-a\mathbf{m_y} \tag{E10.15.2}$$

Second, the coefficient a can be determined by setting the derivative of $J(a, b)$ (with b determined as above) w.r.t. a to zero:

$$\frac{d}{da}J(a,b)\overset{(E10.15.1)}{=}\frac{d}{da}\mathcal{E}\left\{(\mathbf{x}-(a\mathbf{y}+b))^2\right\}\overset{(E10.15.2)}{=}\frac{d}{da}\mathcal{E}\left\{\left((\mathbf{x}-m_\mathbf{x})-a(\mathbf{y}-m_\mathbf{y})\right)^2\right\}$$

$$=-2\mathcal{E}\left\{(\mathbf{x}-m_\mathbf{x})(\mathbf{y}-m_\mathbf{y})\right\}+2a\mathcal{E}\left\{(\mathbf{y}-m_\mathbf{y})^2\right\}\underset{(10.2.31)}{\overset{(10.2.40)}{=}}-2\,\mathrm{cov}(\mathbf{x},\mathbf{y})+2a\,\mathrm{var}(\mathbf{y})=0$$

$$;\qquad a^{\mathrm{o}}=\frac{\mathrm{cov}(\mathbf{x},\mathbf{y})=\sigma_{\mathbf{xy}}^2}{\mathrm{var}(\mathbf{y})=\sigma_\mathbf{y}^2}\overset{(10.2.44)}{=}\frac{\rho_{\mathbf{xy}}\sigma_\mathbf{x}\sigma_\mathbf{y}}{\sigma_\mathbf{y}^2}=\rho_{\mathbf{xy}}\frac{\sigma_\mathbf{x}}{\sigma_\mathbf{y}} \tag{E10.15.3}$$

Thus, we have got the linear *MMSE* (*minimum mean square error*) estimator of $\mathbf{x}$ based on the measurement (observation) of $\mathbf{y}$ as

$$\hat{\mathbf{x}}_{\mathrm{MMSE}}=a^{\mathrm{o}}\mathbf{y}+b^{\mathrm{o}}\overset{(E10.15.2)}{=}a^{\mathrm{o}}\mathbf{y}+m_\mathbf{x}-a^{\mathrm{o}}m_\mathbf{y}\overset{(E10.15.3)}{=}\rho_{\mathbf{xy}}\frac{\sigma_\mathbf{x}}{\sigma_\mathbf{y}}(\mathbf{y}-m_\mathbf{y})+m_\mathbf{x} \tag{E10.15.4}$$

Here are two observations:

(i) Consider the gain $\rho_{\mathbf{xy}}(\sigma_{\mathbf{xy}}/\sigma_\mathbf{y})$, which determines how heavily the information contained in $\mathbf{y}$, i.e., the deviation $(\mathbf{y}-m_\mathbf{y})$ of measurement of $\mathbf{y}$ from its mean value $m_\mathbf{y}$, should be reflected in estimating $\mathbf{x}$. It will be large if the correlation $\rho_{\mathbf{xy}}$ (between $\mathbf{x}$ and $\mathbf{y}$) is high and/or the ratio of $\sigma_\mathbf{y}$ over $\sigma_\mathbf{x}$ is small, meaning higher relative reliability of the observation $\mathbf{y}$ over that of $\mathbf{x}$. Doesn't it agree with our intuition?

(ii) The gain $\rho_{\mathbf{xy}}(\sigma_{\mathbf{xy}}/\sigma_\mathbf{y})$ equals the ratio of the (1,2) entry to the (2,2) entry of $\Sigma_{[\mathbf{x}\,\mathbf{y}]}$ (Eq. (10.2.46)).

(iii) The error variance of this MMSE estimator turns out to be

$$\mathcal{E}\left\{(\hat{\mathbf{x}}_{\mathrm{MMSE}}-\mathbf{x})^2\right\}\underset{(10.2.41)}{\overset{(E10.15.4)}{=}}\mathcal{E}\left\{\left\{\frac{\sigma_{\mathbf{xy}}^2}{\sigma_\mathbf{y}^2}(\mathbf{y}-m_\mathbf{y})+m_\mathbf{x}-\mathbf{x}\right\}^2\right\}\underset{(10.2.31)}{\overset{(10.2.40)}{=}}\frac{\sigma_{\mathbf{xy}}^4}{\sigma_\mathbf{y}^4}\sigma_\mathbf{y}^2+\sigma_\mathbf{x}^2-2\frac{\sigma_{\mathbf{xy}}^2}{\sigma_\mathbf{y}^2}\sigma_{\mathbf{xy}}^2$$

$$\overset{(10.2.44)}{=}\left(1-\rho_{\mathbf{xy}}^2\right)\sigma_\mathbf{x}^2<\sigma_\mathbf{x}^2=\mathcal{E}\left\{(\mathbf{x}-m_\mathbf{x})^2\right\} \tag{E10.15.5}$$

This implies that the (posteriori) error variance of the MMSE estimator is always smaller than the (prior) error variance of $\mathbf{x}$ as long as $\rho_{\mathbf{xy}}^2>0$. Doesn't it support the raison d'être of MMSE?

Also, it can be shown that the error of the linear MMSE (LMMSE) estimator is orthogonal to the observation $\mathbf{y}$, which is called the *orthogonality principle*:

$$\mathcal{E}\{(\hat{\mathbf{x}}_{\text{MMSE}}-\mathbf{x})\mathbf{y}\} \overset{(\text{E10.15.4})}{\underset{(10.2.41)}{=}} \mathcal{E}\left\{\left(\frac{\sigma_{\mathbf{xy}}^2}{\sigma_{\mathbf{y}}^2}(\mathbf{y}-m_{\mathbf{y}})+m_{\mathbf{x}}-\mathbf{x}\right)(\mathbf{y}-m_{\mathbf{y}}+m_{\mathbf{y}})\right\}$$

$$=\frac{\sigma_{\mathbf{xy}}^2}{\sigma_{\mathbf{y}}^2}\mathcal{E}\{(\mathbf{y}-m_{\mathbf{y}})(\mathbf{y}-m_{\mathbf{y}})\}-\mathcal{E}\{(\mathbf{x}-m_{\mathbf{x}})(\mathbf{y}-m_{\mathbf{y}})\}+\frac{\sigma_{\mathbf{xy}}^2}{\sigma_{\mathbf{y}}^2}\mathcal{E}\{(\mathbf{y}-m_{\mathbf{y}})\}m_{\mathbf{y}}+\mathcal{E}\{\mathbf{x}-m_{\mathbf{x}}\}m_{\mathbf{y}}$$

$$\overset{(10.2.40)}{\underset{(10.2.31)}{=}}\frac{\sigma_{\mathbf{xy}}^2}{\sigma_{\mathbf{y}}^2}\sigma_{\mathbf{y}}^2-\sigma_{\mathbf{xy}}^2=0$$

$$; \qquad \mathcal{E}\{(\hat{\mathbf{x}}_{\text{MMSE}}-\mathbf{x})\mathbf{y}\}=0 \tag{E10.15.6}$$

Since the MMSE estimator $\hat{\mathbf{x}}_{\text{MMSE}}$ is a linear function of $\mathbf{y}$ (with an additive constant) and $\mathcal{E}\{\hat{\mathbf{x}}_{\text{MMSE}}-\mathbf{x}\}=0$, it is also orthogonal to the estimation error:

$$\mathcal{E}\{(\hat{\mathbf{x}}_{\text{MMSE}}-\mathbf{x})\hat{\mathbf{x}}_{\text{MMSE}}\}=0 \tag{E10.15.7}$$

10.2.7 Conditional Expectation

The conditional expectations (means) of a random variable $\mathbf{x}$ and its function $g(\mathbf{x})$ given the occurrence of an event A are defined as follows:

$$\mathcal{E}\{\mathbf{x}\,|\,A\}=\int_{-\infty}^{\infty}x\,f_{\mathbf{x}|A}(x\,|\,A)\,dx \tag{10.2.50}$$

$$\mathcal{E}\{g(\mathbf{x})\,|\,A\}=\int_{-\infty}^{\infty}g(x)\,f_{\mathbf{x}|A}(x\,|\,A)\,dx \tag{10.2.51}$$

If the event A is that the value of another random variable $\mathbf{y}$ is y, then these equations become

$$\mathcal{E}\{\mathbf{x}\,|\,\mathbf{y}=y\}=\int_{-\infty}^{\infty}x\,f_{\mathbf{x}|\mathbf{y}}(x\,|\,\mathbf{y}=y)\,dx \tag{10.2.52}$$

$$\mathcal{E}\{g(\mathbf{x})\,|\,\mathbf{y}=y\}=\int_{-\infty}^{\infty}g(x)\,f_{\mathbf{x}|\mathbf{y}}(x\,|\,\mathbf{y}=y)\,dx \tag{10.2.53}$$

(Example 10.16) Conditional Expectation — MMSE (Minimum Mean Square Error) Estimate

Let the joint pdf of two Gaussian random variables $\mathbf{x}$ and $\mathbf{y}$ be

$$f_{\mathbf{x},\mathbf{y}}(x,y)=\frac{1}{(2\pi)^{1/2}|\Sigma|^{1/2}}\exp\left\{-\frac{1}{2}[x-m_{\mathbf{x}}\;\; y-m_{\mathbf{y}}]\,\Sigma^{-1}\begin{bmatrix}x-m_{\mathbf{x}}\\ y-m_{\mathbf{y}}\end{bmatrix}\right\} \text{ with } \Sigma=\begin{bmatrix}\sigma_{\mathbf{x}}^2 & \rho_{\mathbf{xy}}\sigma_{\mathbf{x}}\sigma_{\mathbf{y}}\\ \rho_{\mathbf{xy}}\sigma_{\mathbf{x}}\sigma_{\mathbf{y}} & \sigma_{\mathbf{y}}^2\end{bmatrix}$$

$$\overset{(10.2.43)}{=}\frac{1}{2\pi\sigma_{\mathbf{x}}\sigma_{\mathbf{y}}\sqrt{1-\rho_{\mathbf{xy}}^2}}\exp\left\{\frac{-1}{2(1-\rho_{\mathbf{xy}}^2)}\left(\frac{(x-m_{\mathbf{x}})^2}{\sigma_{\mathbf{x}}^2}-2\rho_{\mathbf{xy}}\frac{(x-m_{\mathbf{x}})(y-m_{\mathbf{y}})}{\sigma_{\mathbf{x}}\sigma_{\mathbf{y}}}+\frac{(y-m_{\mathbf{y}})^2}{\sigma_{\mathbf{y}}^2}\right)\right\}$$

$$=\frac{1}{2\pi\sigma_{\mathbf{x}}\sigma_{\mathbf{y}}\sqrt{1-\rho_{\mathbf{xy}}^2}}\exp\left\{\frac{-1}{2(1-\rho_{\mathbf{xy}}^2)}\left(\begin{matrix}\frac{(x-m_{\mathbf{x}})^2}{\sigma_{\mathbf{x}}^2}-2\rho_{\mathbf{xy}}\frac{(x-m_{\mathbf{x}})(y-m_{\mathbf{y}})}{\sigma_{\mathbf{x}}\sigma_{\mathbf{y}}}\\ +\rho_{\mathbf{xy}}^2\frac{(y-m_{\mathbf{y}})^2}{\sigma_{\mathbf{y}}^2}+(1-\rho_{\mathbf{xy}}^2)\frac{(y-m_{\mathbf{y}})^2}{\sigma_{\mathbf{y}}^2}\end{matrix}\right)\right\} \tag{E10.16.1}$$

We can use Eq. (10.2.14a) to find the conditional pdf of **x** given **y**=y as

$$f_{\mathbf{x}|\mathbf{y}}(x\,|\,y) = \frac{f_{\mathbf{x},\mathbf{y}}(x,y)}{f_{\mathbf{y}}(y) = \frac{1}{\sqrt{2\pi}\sigma_{\mathbf{y}}} e^{-(y-m_{\mathbf{y}})^2/2\sigma_{\mathbf{y}}^2}}$$

$$\overset{\text{(E10.16.1)}}{=} \frac{1}{\sqrt{2\pi\sigma_{\mathbf{x}}^2(1-\rho_{\mathbf{xy}}^2)}} \exp\left\{\frac{-1}{2(1-\rho_{\mathbf{xy}}^2)\sigma_{\mathbf{x}}^2}\left(x-m_{\mathbf{x}}-\rho_{\mathbf{xy}}\frac{\sigma_{\mathbf{x}}}{\sigma_{\mathbf{y}}}(y-m_{\mathbf{y}})\right)^2\right\} \quad \text{(E10.16.2)}$$

From this conditional pdf, we can find the conditional expectation of **x** based on the value of **y**=y as

$$\mathcal{E}\{\mathbf{x}\,|\,\mathbf{y}=y\} \overset{(10.2.52)}{=} \int_{-\infty}^{\infty} x f_{\mathbf{x}|\mathbf{y}}(x\,|\,y)\,dx$$

$$\overset{\text{(E10.16.2)}}{=} \frac{1}{\sqrt{2\pi\sigma_{\mathbf{x}}(1-\rho_{\mathbf{xy}}^2)}} \int_{-\infty}^{\infty} x \exp\left\{\frac{-1}{2(1-\rho_{\mathbf{xy}}^2)\sigma_{\mathbf{x}}^2}\left(x-m_{\mathbf{x}}-\rho_{\mathbf{xy}}\frac{\sigma_{\mathbf{x}}}{\sigma_{\mathbf{y}}}(y-m_{\mathbf{y}})\right)^2\right\} dx$$

$$\overset{x-m_{\mathbf{x}}-\rho\sigma_{\mathbf{x}}(y-m_{\mathbf{y}})/\sigma_{\mathbf{y}} \to x'}{=} \frac{1}{\sqrt{2\pi\sigma_{\mathbf{x}}(1-\rho_{\mathbf{xy}}^2)}} \int_{-\infty}^{\infty} \left(\underset{\text{Odd function}}{x} + m_{\mathbf{x}} + \rho_{\mathbf{xy}}\frac{\sigma_{\mathbf{x}}}{\sigma_{\mathbf{y}}}(y-m_{\mathbf{y}})\right) \exp\left\{\frac{-1}{2(1-\rho_{\mathbf{xy}}^2)\sigma_{\mathbf{x}}^2} x^2\right\} dx$$

$$;\quad \mathcal{E}\{\mathbf{x}\,|\,\mathbf{y}=y\} = \left(m_{\mathbf{x}} + \rho_{\mathbf{xy}}\frac{\sigma_{\mathbf{x}}}{\sigma_{\mathbf{y}}}(y-m_{\mathbf{y}})\right) \frac{1}{\sqrt{2\pi}\sigma} \int_{-\infty}^{\infty} e^{-x/2\sigma^2} dx = m_{\mathbf{x}} + \rho_{\mathbf{xy}}\frac{\sigma_{\mathbf{x}}}{\sigma_{\mathbf{y}}}(y-m_{\mathbf{y}}) \quad \text{(E10.16.3)}$$

This happens to be identical with the *MMSE* (*minimum mean square error*) estimate of **x** (based on the value of **y**), that was obtained as Eq. (E10.15.4). Also, it may be regarded as the *MAP* (*maximum a posteriori*) estimate of **x** given **y**=y, because it maximizes the conditional (a posteriori) pdf (E10.16.2) over x where **y** is supposed to have resulted from **x**.

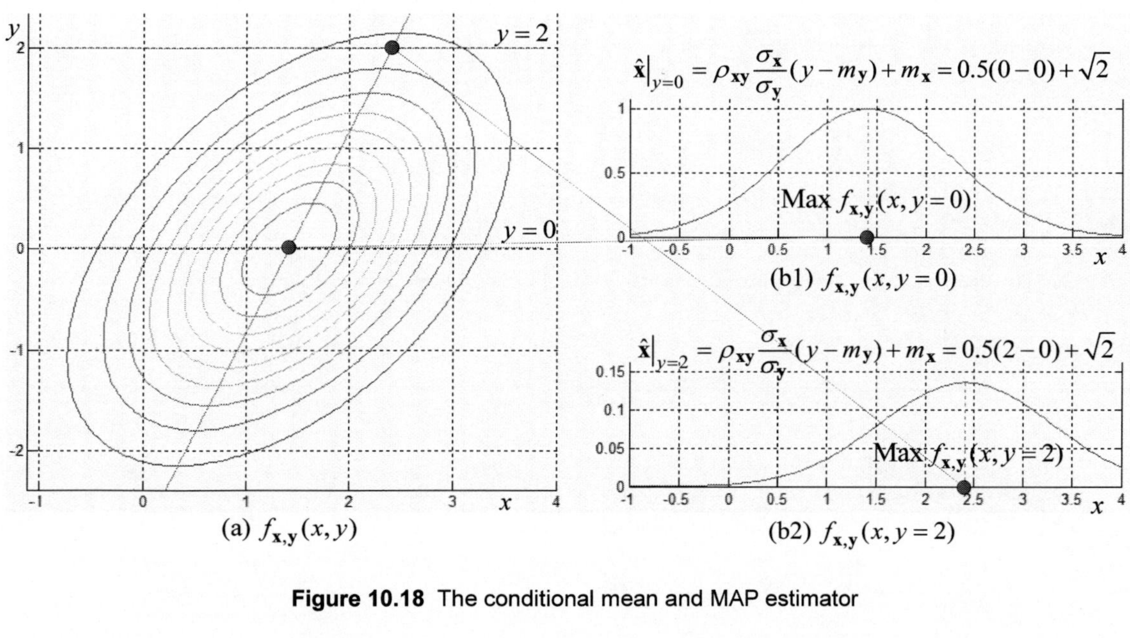

Figure 10.18 The conditional mean and MAP estimator

To get a deeper understanding of conditional mean, it is good to see Fig. 10.18, which shows the joint pdf $f_{\mathbf{x},\mathbf{y}}(x,y)$ (Eq. (E10.16.1) with $m_\mathbf{x}=\sqrt{2}$, $m_\mathbf{y}=0$, $\sigma_\mathbf{x}=1$, $\sigma_\mathbf{y}=1$, and $\rho_\mathbf{xy}=0.5$) of **x** and **y** (Fig. 10.18(a)), and $f_{\mathbf{x},\mathbf{y}}(x,y)$'s given y=0/2 (Fig. 10.18(b1)/(b2)) where the maxima of $f_{\mathbf{x},\mathbf{y}}(x,y=0)$ and $f_{\mathbf{x},\mathbf{y}}(x,y=2)$ conform with the MAP estimates obtained from Eq. (E10.16.3):

$$\hat{\mathbf{x}}_{\text{MAP}}\Big|_{y=0} \overset{\text{(E10.16.3)}}{=} m_\mathbf{x}+\rho_\mathbf{xy}\frac{\sigma_\mathbf{x}}{\sigma_\mathbf{y}}(y-m_\mathbf{y})=\sqrt{2}+0.5(0-0)=\sqrt{2} \quad \text{(E10.16.4a)}$$

$$\hat{\mathbf{x}}_{\text{MAP}}\Big|_{y=2} \overset{\text{(E10.16.3)}}{=} m_\mathbf{x}+\rho_\mathbf{xy}\frac{\sigma_\mathbf{x}}{\sigma_\mathbf{y}}(y-m_\mathbf{y})=\sqrt{2}+0.5(2-0)=\sqrt{2}+1 \quad \text{(E10.16.4b)}$$

Further, to be sure of that in this case, $\hat{\mathbf{x}}=\sqrt{2}+0.5\mathbf{y}$ is the best linear estimator of **x** given the measured value of **y** in terms of MSE, let us compare its MSE and the MSEs of many other arbitrary linear estimators $\hat{\mathbf{x}}=b+a\mathbf{y}$ (with a around 0.5 and b around $\sqrt{2}$). It can be done by running the MATLAB script "em10e16.m" to get the MSEs for several pairs of $\{a, b\}$, as listed in Table 10.1, which shows that the pair $\{a=0.5, b=\sqrt{2}\}$ results in the minimum MSE.

Table 10.1 MSEs of several linear estimators $\hat{\mathbf{x}}=b+a\mathbf{y}$ with different values of $\{a, b\}$

a \ b	$\sqrt{2}-0.2$	$\sqrt{2}-0.1$	$\sqrt{2}$	$\sqrt{2}+0.1$	$\sqrt{2}+0.2$
0.3	0.8301	0.7987	0.7874	0.7960	0.8247
0.4	0.7994	0.7680	0.7567	0.7653	0.7940
0.5	0.7886	0.7572	0.7459	0.7545	0.7832
0.6	0.7976	0.7663	0.7549	0.7636	0.7923
0.7	0.8265	0.7952	0.7839	0.7926	0.8212

```
%em10e16.m
N=1e5; rho=0.5;
mx=1; my=0; sgm2x=1; sgm2y=1; sgmx=sqrt(sgm2x); sgmy=sqrt(sgm2y);
% Desired covariance matrix
Sgm2_xy_desired=[sgm2x rho*sgmx*sgmy; rho*sgmx*sgmy sgm2y]
mu=1; mv=1; sgm2u=(1+rho)*sgm2x; sgm2v=(1-rho)*sgm2x;
sgmu=sqrt(sgm2u); sgmv=sqrt(sgm2v);
% To generate a pair of two uncorrelated Gaussian rvs
%  with covariance matrix Sgm2_uv=[sgm2u 0; 0 sgm2v]
u = randn(1,N)*sgmu + mu;     v = randn(1,N)*sgmv + mv;
% To transform this pair of rvs into a pair of two correlated rvs
V=[1 1; 1 -1]/sqrt(2); % Transformation matrix
xy=V*[u; v]; x=xy(1,:); y=xy(2,:);
% To check if the resulting rvs have the wanted covariance matrix.
mxy=V*[mu; mv]; mx=mxy(1), my=mxy(2)
sgm2x_measured = var(x); sgm2y_measured = var(y);
[sgm2x_measured sgm2x], [sgm2y_measured sgm2y]
rho_xy_measured=mean((x-mx).*(y-my)); [rho_xy_measured rho]
% Coefficients of MAP estimator Eq.(E10.16.3)
ao=rho*sgmx/sgmy; bo=mx;
% To check if the MAP estimate (conditional mean) has minimum MSE.
da=0.1; as = ao+[-2:2]*da;   db=0.1; bs = bo+[-2:2]*db;
for na=1:length(as)
   for nb=1:length(bs)
      MSEs(na,nb) = mean((x-as(na)*(y-my)-bs(nb)).^2);
   end
end
[0 bs; as.' MSEs] % Above table showing that MSE is minimized by x=ao*y+bo
```

Note that in the script "em10e16.m," to generate a pair of two correlated random sequences ($x[n]$, $y[n]$), represented by random vector $[\mathbf{x}\ \mathbf{y}]^T$ with covariance matrix

$$\Sigma_{[\mathbf{x}\,\mathbf{y}]} = \begin{bmatrix} \sigma_{\mathbf{x}}^2 & \rho_{\mathbf{xy}}\sigma_{\mathbf{x}}\sigma_{\mathbf{y}} \\ \rho_{\mathbf{xy}}\sigma_{\mathbf{x}}\sigma_{\mathbf{y}} & \sigma_{\mathbf{y}}^2 \end{bmatrix} = \begin{bmatrix} 1 & 0.5 \\ 0.5 & 1 \end{bmatrix} \tag{E10.16.5}$$

two uncorrelated Gaussian sequences $\{(u[n], v[n]), n=1:10^5\}$ have been generated as a realization of random vector $[\mathbf{u}\ \mathbf{v}]^T$ with mean vector and covariance matrix as

$$\begin{bmatrix} m_{\mathbf{u}} \\ m_{\mathbf{v}} \end{bmatrix} = \begin{bmatrix} 1 \\ 1 \end{bmatrix} \text{ and } \Sigma_{[\mathbf{u}\,\mathbf{v}]} = \text{cov}\left\{\begin{bmatrix} \mathbf{u} \\ \mathbf{v} \end{bmatrix}\right\} = \begin{bmatrix} \sigma_{\mathbf{u}}^2 & \rho_{\mathbf{uv}}\sigma_{\mathbf{u}}\sigma_{\mathbf{v}} \\ \rho_{\mathbf{uv}}\sigma_{\mathbf{u}}\sigma_{\mathbf{v}} & \sigma_{\mathbf{v}}^2 \end{bmatrix} \underset{\rho_{\mathbf{uv}}=0}{\overset{\sigma_{\mathbf{u}}^2=1.5,\ \sigma_{\mathbf{v}}^2=0.5}{=}} \begin{bmatrix} 1.5 & 0 \\ 0 & 0.5 \end{bmatrix} \tag{E10.16.6}$$

and then multiplied by a transformation matrix

$$V = \frac{1}{\sqrt{2}}\begin{bmatrix} 1 & 1 \\ 1 & -1 \end{bmatrix} \tag{E10.16.7}$$

which is a modal matrix of $\Sigma_{[\mathbf{x}\,\mathbf{y}]}$. Then the mean vector and covariance matrix of the resulting pair of two random sequences ($x[n]$, $y[n]$) will be

$$\begin{bmatrix} m_{\mathbf{x}} \\ m_{\mathbf{y}} \end{bmatrix} \overset{\text{(E10.14.4a)}}{=} V\begin{bmatrix} m_{\mathbf{u}} \\ m_{\mathbf{v}} \end{bmatrix} \underset{\text{(E10.16.6a)}}{\overset{\text{(E10.16.7)}}{=}} \frac{1}{\sqrt{2}}\begin{bmatrix} 1 & 1 \\ 1 & -1 \end{bmatrix}\begin{bmatrix} 1 \\ 1 \end{bmatrix} = \begin{bmatrix} \sqrt{2} \\ 0 \end{bmatrix} \tag{E10.16.8a}$$

$$\Sigma_{[\mathbf{x}\,\mathbf{y}]} \overset{\text{(E10.14.4b)}}{=} V\Sigma_{[\mathbf{u}\,\mathbf{v}]}V^T \underset{\text{(E10.16.6b)}}{\overset{\text{(E10.16.7)}}{=}} \frac{1}{\sqrt{2}}\begin{bmatrix} 1 & 1 \\ 1 & -1 \end{bmatrix}\begin{bmatrix} 1.5 & 0 \\ 0 & 0.5 \end{bmatrix}\frac{1}{\sqrt{2}}\begin{bmatrix} 1 & 1 \\ 1 & -1 \end{bmatrix}^T = \begin{bmatrix} 1 & 0.5 \\ 0.5 & 1 \end{bmatrix} \tag{E10.16.8b}$$

(Q) Why is the modal matrix V of $\Sigma_{[\mathbf{x}\,\mathbf{y}]}$ used for the transformation here?

(A) Because Eq. (E10.14.4b) can be matched with the eigendecomposition $A=V\Lambda V^{-1}$ (Eq. (1.2.25)) with $\Lambda=\Sigma_{[\mathbf{u}\,\mathbf{v}]}$ (a diagonal matrix), $A=\Sigma_{[\mathbf{x}\,\mathbf{y}]}$ (a symmetric matrix), and V (a modal matrix of $\Sigma_{[\mathbf{x}\,\mathbf{y}]}$), and besides, we can find an orthonormal modal matrix V such that $V^{-1}=V^T$ since $\Sigma_{[\mathbf{x}\,\mathbf{y}]}$ is real and symmetric (see Sections 1.2.4 and 1.2.7.2).

10.2.8 Central Limit Theorem — Normal Convergence Theorem

The central limit theorem (CLT) is the basis of the wide and diverse applicability of Gaussian random variables:

[Central Limit Theorem (CLT)]

Let $\{\mathbf{x}_1, \mathbf{x}_2, \ldots, \mathbf{x}_L\}$ be *i.i.d.* (*independent identically distributed*) random variables with finite mean m and variance σ^2. Then, irrespective of the distribution of the individual variables, their average, called the *sample mean*,

$$\mathbf{y}_L = \frac{1}{L}\sum_{i=1}^{L}\mathbf{x}_i \tag{10.2.54}$$

has an *asymptotically normal* (Gaussian) distribution $N(m,\sigma^2/L)$ with mean m and variance σ^2/L in the sense that, as the number L of the random variables increases, its pdf converges to

$$\lim_{L\to\infty} f_{\mathbf{y}_L}(y_L) = \frac{1}{\sqrt{2\pi\sigma^2/L}}\exp\left\{-\frac{(y_L-m)^2}{2\sigma^2/L}\right\} \tag{10.2.55}$$

where

$$m_{\mathbf{y}} = \mathcal{E}\{\mathbf{y}_L\} \overset{(10.2.54)}{=} \mathcal{E}\left\{\frac{1}{L}\sum_{i=1}^{L}\mathbf{x}_i\right\} = \frac{1}{L}\sum_{i=1}^{L}\mathcal{E}\{\mathbf{x}_i\} = \frac{1}{L}Lm = m \tag{10.2.56}$$

$$\sigma_{\mathbf{y}}^2 = \mathrm{var}(\mathbf{y}_L) \overset{(10.2.31)}{=} \mathcal{E}\{(\mathbf{y}_L - m_{\mathbf{y}})^2\} \overset{(10.2.54)}{=} \mathcal{E}\left\{\left(\frac{1}{L}\sum_{i=1}^{L}\mathbf{x}_i - \frac{1}{L}\sum_{i=1}^{L}m\right)^2\right\}$$

$$= \frac{1}{L^2}\mathcal{E}\left\{\left(\sum_{i=1}^{L}(\mathbf{x}_i - m)\right)^2\right\} \underset{\mathcal{E}\{(\mathbf{x}_i-m)(\mathbf{x}_j-m)\}=0}{\overset{\text{independence}}{=}} \frac{1}{L^2}\sum_{i=1}^{L}\mathcal{E}\left\{(\mathbf{x}_i-m)^2\right\} \overset{(10.2.31)}{=} \frac{1}{L}\sigma^2 \tag{10.2.57}$$

Eqs. (10.2.55) and (10.2.57) imply that the Gaussian approximation steadily improves and the variance of the sample mean gets closer to zero as the number L of random variables increases.

The CLT explains why the outcomes of the events appearing in many applications are often assumed to be Gaussian. To realize the physical meaning and check the validity of the CLT, see the following example.

(Example 10.17) Central Limit Theorem (CLT)

Let us plot the histogram of 10^6 sample means of L uniform random numbers for L=2 and 10 by running the MATLAB script "em10e17.m." This yields Fig. 10.19, which shows that the histogram gets closer to a Gaussian pdf with a smaller variance as the number L of samples for computing the sample mean increases from 2 to 10.

```
%em10e17.m
% Plot Fig. 10.19 to check the validity of the CLT
Ns=1e6; Nb=50; % Numbers of samples and bins
mx=1/2; % Mean of standard uniform noise: Eq.(10.2.37) (a=0, b=1)
sgmx2=1/12; % Variance of standard uniform noise with U(0,1): Eq.(10.2.38)
sgmx=sqrt(sgmx2);
Ls=[2 10]; % For two simulations, each with 2 and 10 random variables
for i=1:length(Ls)
   L=Ls(i); % Number of random variables
   y = sum(rand(L,Ns))/L; % Eq.(10.2.54)
   subplot(2,2,i); hist(y,Nb), hold on
   [ns,cs]=hist(y,Nb); % ns: Frequency counts, cs: Bin locations
   dy=cs(2)-cs(1); % Bin width
   my=mx; sgmy2=sgmx2/L; % Average/variance by Eq.(10.2.56/57)
   y=my+[-500:500]*(my/500); % Range on the y-axis
   fy=exp(-(y-my).^2/2/sgmy2)/sqrt(2*pi*sgmy2); % Gaussian pdf
   plot(y,Ns*dy*fy,'r')
end
```

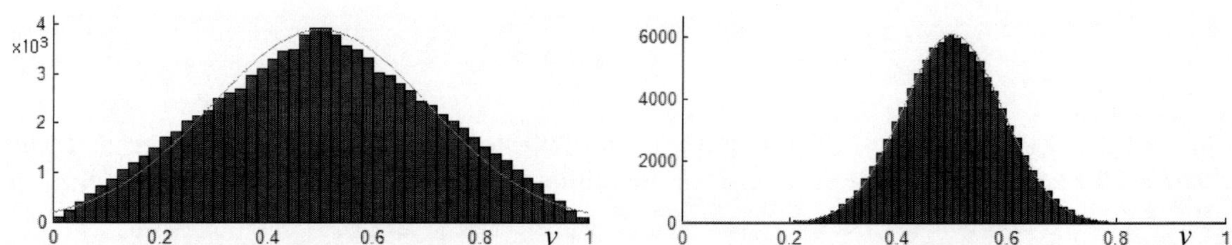

(a) The histogram of 10^6 samples of $\mathbf{y} = \frac{1}{2}\sum_i \mathbf{x}_i$ (b) The histogram of 10^6 samples of $\mathbf{y} = \frac{1}{10}\sum_i \mathbf{x}_i$

Figure 10.19 Central limit theorem (CLT) — the distributions of sample means

10.3 ML ESTIMATOR AND MAP ESTIMATOR

Let **y** be a random variable whose value depends on the value of a (possibly random) parameter **x** and a random noise **n** as

$$\mathbf{y} = h(\mathbf{x}) + \mathbf{n} \tag{10.3.1}$$

where the posterior pdf $f_{\mathbf{x}|\mathbf{y}}(x|y)$ (given **y**=y) and priori pdf $f_{\mathbf{x}}(x)$ of **x** are related by the Bayes' rule as

$$f_{\mathbf{x}|\mathbf{y}}(x\,|\,y) \overset{(10.2.14a)}{=} \frac{f_{\mathbf{x},\mathbf{y}}(x,y)}{f_{\mathbf{y}}(y)} \underset{(10.2.13b)}{\overset{(10.2.15)}{=}} \frac{f_{\mathbf{y}|\mathbf{x}}(y|x)f_{\mathbf{x}}(x)}{\int f_{\mathbf{y}|\mathbf{x}}(y|x)f_{\mathbf{x}}(x)\,dx} \quad \text{with } f_{\mathbf{y}}(y)>0 \tag{10.3.2}$$

The *ML* (*maximum likelihood*) *estimator* to find the value of x, which "most likely" might have caused y, is defined as

$$\hat{x}_{\mathrm{ML}}(y) = \operatorname*{Arg\,Max}_{x} L(x\,|\,y) \ \text{ or } \ \operatorname*{Arg\,Max}_{x} f_{\mathbf{y}|\mathbf{x}}(y|x) \tag{10.3.3a,b}$$

where $L(x|y)$ is the *log-likelihood* (*function*) defined as

$$L(x\,|\,y) = \ln f_{\mathbf{y}|\mathbf{x}}(y|x) \tag{10.3.4}$$

The *MAP* (*maximum a posteriori*) *estimator* is defined as

$$\hat{x}_{\mathrm{MAP}}(y) = \operatorname*{Arg\,Max}_{x} f_{\mathbf{x},\mathbf{y}}(x,y) = \operatorname*{Arg\,Max}_{x} f_{\mathbf{y}|\mathbf{x}}(y|x)\,f_{\mathbf{x}}(x) \tag{10.3.5}$$

where the priori probability $f_{\mathbf{x}}(x)$ is additionally considered compared with the ML estimator.

To figure out the difference between these two estimators, suppose there are two sensors that measure the position x of an object, each with different accuracy, and their outputs (measurements) $\mathbf{y}_1$ and $\mathbf{y}_2$ are modeled by independent Gaussian random variables, each having the pdfs as

$$\mathbf{y}_1 = x+\mathbf{n}_1 :\ f_{\mathbf{y}_1|\mathbf{x}}(y|\,x) = \frac{1}{\sqrt{2\pi}\sigma_{\mathbf{y}_1}}\exp\left\{\frac{-(y-x)^2}{\sigma_{\mathbf{y}_1}^2}\right\} :\ N(x,\sigma_{\mathbf{y}_1}^2) \tag{10.3.6a}$$

$$\mathbf{y}_2 = x+\mathbf{n}_2 :\ f_{\mathbf{y}_2|\mathbf{x}}(y|\,x) = \frac{1}{\sqrt{2\pi}\sigma_{\mathbf{y}_2}}\exp\left\{\frac{-(y-x)^2}{\sigma_{\mathbf{y}_2}^2}\right\} :\ N(x,\sigma_{\mathbf{y}_2}^2) \tag{10.3.6b}$$

Then the conditional joint pdf of the two sensor outputs $\mathbf{y}_1$ and $\mathbf{y}_2$ given x is

$$f_{\mathbf{y}_1,\mathbf{y}_2|\mathbf{x}}(y_1,y_2|\,x) = f_{\mathbf{y}_1|\mathbf{x}}(y|\,x)f_{\mathbf{y}_2|\mathbf{x}}(y|\,x) \overset{(10.3.6)}{=} \frac{1}{2\pi\sigma_{\mathbf{y}_1}\sigma_{\mathbf{y}_2}}\exp\left\{-\frac{(y_1-x)^2}{2\sigma_{\mathbf{y}_1}^2}-\frac{(y_2-x)^2}{2\sigma_{\mathbf{y}_2}^2}\right\} \tag{10.3.7}$$

The ML estimator of x given the two sensor outputs y_1 and y_2 is obtained by using Eq. (10.3.3), i.e., in such a way that Eq. (10.3.7) is maximized as

$$\hat{x}_{\mathrm{ML}}(y_1,y_2) \overset{(10.3.3a)}{=} \operatorname*{Arg\,Max}_{x} \ln\{f_{\mathbf{y}_1,\mathbf{y}_2|\mathbf{x}}(y_1,y_2|\,x)\}$$

$$; \frac{d}{dx}\ln\{f_{\mathbf{y}_1,\mathbf{y}_2|\mathbf{x}}(y_1,y_2|x)\} \overset{(10.3.7)}{=} \frac{d}{dx}\left\{-\frac{(y_1-x)^2}{2\sigma_{\mathbf{y}_1}^2}-\frac{(y_2-x)^2}{2\sigma_{\mathbf{y}_2}^2}\right\} = \frac{y_1-x}{\sigma_{\mathbf{y}_1}^2}+\frac{y_2-x}{\sigma_{\mathbf{y}_2}^2}=0$$

$$; \qquad \hat{x}_{ML}(y_1,y_2) = \frac{\sigma_{\mathbf{y}_1}^{-2}y_1+\sigma_{\mathbf{y}_2}^{-2}y_2}{\sigma_{\mathbf{y}_1}^{-2}+\sigma_{\mathbf{y}_2}^{-2}} \tag{10.3.8}$$

This is nothing but a weighted average of the two sensor outputs y_1 and y_2, each having *the reciprocal of its variance* ($\sigma_{\mathbf{y}_i}^{-2}$) as the weighting factor, where $\sigma_{\mathbf{y}_i}^{-2}$ can be regarded as a measure of reliability of an observed information. (Q) Can you rely on data with a large variance?

On the other hand, if the prior pdf of **x** is available as

$$f_{\mathbf{x}}(x) = \frac{1}{\sqrt{2\pi\sigma_{\mathbf{x}}^2}}\exp\left\{-\frac{(x-m_{\mathbf{x}})^2}{2\sigma_{\mathbf{x}}^2}\right\}:\ N(m_{\mathbf{x}},\sigma_{\mathbf{x}}^2) \tag{10.3.9}$$

the MAP estimator of x given the two sensor outputs y_1 and y_2 is obtained by using Eq. (10.3.5) as

$$\hat{x}_{MAP}(y) \overset{(10.3.5)}{=} \operatorname*{Arg\,Max}_x f_{\mathbf{x},\mathbf{y}}(x,y) = \operatorname*{Arg\,Max}_x f_{\mathbf{y}|\mathbf{x}}(y|x)\,f_{\mathbf{x}}(x)$$

$$; \frac{d}{dx}\ln\{f_{\mathbf{y}_1,\mathbf{y}_2|\mathbf{x}}(y_1,y_2|x)\,f_{\mathbf{x}}(x)\} \underset{(10.3.9)}{\overset{(10.3.7)}{=}} \frac{d}{dx}\left\{-\frac{(y_1-x)^2}{2\sigma_{\mathbf{y}_1}^2}-\frac{(y_2-x)^2}{2\sigma_{\mathbf{y}_2}^2}-\frac{(x-m_{\mathbf{x}})^2}{2\sigma_{\mathbf{x}}^2}\right\} = \frac{y_1-x}{\sigma_{\mathbf{y}_1}^2}+\frac{y_2-x}{\sigma_{\mathbf{y}_2}^2}-\frac{x-m_{\mathbf{x}}}{\sigma_{\mathbf{x}}^2}=0$$

$$; \qquad \hat{x}_{MAP}(y_1,y_2) = \frac{\sigma_{\mathbf{y}_1}^{-2}y_1+\sigma_{\mathbf{y}_2}^{-2}y_2+\sigma_{\mathbf{x}}^{-2}m_{\mathbf{x}}}{\sigma_{\mathbf{y}_1}^{-2}+\sigma_{\mathbf{y}_2}^{-2}+\sigma_{\mathbf{x}}^{-2}} \tag{10.3.10}$$

This seems to be another weighted average of the two sensor outputs y_1 and y_2 together with the prior information $m_{\mathbf{x}}$ about **x** as if it were an additional sensor output.

[Remark 10.3] Fisher Information

The *score function* for estimating a parameter x is defined as the gradient (the partial derivative w.r.t. x) of the log-likelihood function (natural logarithm of the likelihood function), which is the sensitivity of the likelihood function w.r.t. the parameter x:

$$\frac{\partial}{\partial x}L(x|y) \overset{(10.3.4)}{=} \frac{\partial}{\partial x}\ln f_{\mathbf{y}|\mathbf{x}}(y|x) \tag{10.3.11}$$

where the score function has zero mean:

$$\mathcal{E}\left\{\frac{\partial}{\partial x}\ln f_{\mathbf{y}|\mathbf{x}}(y|x)\right\} \overset{(10.2.53)}{=} \int_{-\infty}^{\infty}\left(\frac{\partial}{\partial x}\ln f_{\mathbf{y}|\mathbf{x}}(y|x)\right)f_{\mathbf{y}|\mathbf{x}}(y|x)\,dy$$

$$= \int_{-\infty}^{\infty}\frac{\partial f_{\mathbf{y}|\mathbf{x}}(y|x)/\partial x}{f_{\mathbf{y}|\mathbf{x}}(y|x)}f_{\mathbf{y}|\mathbf{x}}(y|x)\,dy = \frac{\partial}{\partial x}\int_{-\infty}^{\infty}f_{\mathbf{y}|\mathbf{x}}(y|x)\,dy = \frac{\partial}{\partial x}1 = 0 \tag{10.3.12}$$

The variance (second moment) of the score function is called the *Fisher information*:

$$I(x) = \mathrm{var}\left(\frac{\partial}{\partial x}\ln f_{\mathbf{y}|\mathbf{x}}(y|x)\right) \underset{(10.3.12)}{\overset{(10.2.31)}{=}} \mathcal{E}\left\{\left(\frac{\partial}{\partial x}\ln f_{\mathbf{y}|\mathbf{x}}(y|x)\right)^2\right\} = \int_{-\infty}^{\infty}\left(\frac{\partial}{\partial x}\ln f_{\mathbf{y}|\mathbf{x}}(y|x)\right)^2 f_{\mathbf{y}|\mathbf{x}}(y|x)\,dy \tag{10.3.13}$$

If $\ln f_{\mathbf{y}|\mathbf{x}}(y|x)$ is twice differentiable w.r.t. x, the Fisher information can be written as

$$I(x) = -\mathcal{E}\left\{\frac{\partial^2}{\partial x^2}\ln f_{\mathbf{y}|\mathbf{x}}(y|x)\right\} \tag{10.3.14}$$

since

$$\frac{\partial^2}{\partial x^2}\ln f_{\mathbf{y}|\mathbf{x}}(y|x) = \frac{\partial}{\partial x}\frac{\frac{\partial}{\partial x}f_{\mathbf{y}|\mathbf{x}}(y|x)}{f_{\mathbf{y}|\mathbf{x}}(y|x)} \overset{(C.31)}{=} \frac{\left(\frac{\partial^2}{\partial x^2}f_{\mathbf{y}|\mathbf{x}}(y|x)\right)f_{\mathbf{y}|\mathbf{x}}(y|x)}{\left(f_{\mathbf{y}|\mathbf{x}}(y|x)\right)^2} - \frac{\frac{\partial}{\partial x}f_{\mathbf{y}|\mathbf{x}}(y|x)\frac{\partial}{\partial x}f_{\mathbf{y}|\mathbf{x}}(y|x)}{\left(f_{\mathbf{y}|\mathbf{x}}(y|x)\right)^2}$$

$$= \frac{\frac{\partial^2}{\partial x^2}f_{\mathbf{y}|\mathbf{x}}(y|x)}{f_{\mathbf{y}|\mathbf{x}}(y|x)} - \left(\frac{\frac{\partial}{\partial x}f_{\mathbf{y}|\mathbf{x}}(y|x)}{f_{\mathbf{y}|\mathbf{x}}(y|x)}\right)^2 = \frac{\frac{\partial^2}{\partial x^2}f_{\mathbf{y}|\mathbf{x}}(y|x)}{f_{\mathbf{y}|\mathbf{x}}(y|x)} - \left(\frac{\partial}{\partial x}\ln f_{\mathbf{y}|\mathbf{x}}(y|x)\right)^2$$

so that

$$\overset{-(10.3.14)}{\mathcal{E}\left\{\frac{\partial^2}{\partial x^2}\ln f_{\mathbf{y}|\mathbf{x}}(y|x)\right\}} = \int_{-\infty}^{\infty}\frac{\frac{\partial^2}{\partial x^2}f_{\mathbf{y}|\mathbf{x}}(y|x)}{f_{\mathbf{y}|\mathbf{x}}(y|x)}f_{\mathbf{y}|\mathbf{x}}(y|x)\,dy - \mathcal{E}\left\{\left(\frac{\partial}{\partial x}\ln f_{\mathbf{y}|\mathbf{x}}(y|x)\right)^2\right\}$$

$$= \frac{\partial^2}{\partial x^2}\overbrace{\int_{-\infty}^{\infty}f_{\mathbf{y}|\mathbf{x}}(y|x)\,dy}^{1} - \mathcal{E}\left\{\left(\frac{\partial}{\partial x}\ln f_{\mathbf{y}|\mathbf{x}}(y|x)\right)^2\right\} = -\overset{-(10.3.13)}{\mathcal{E}\left\{\left(\frac{\partial}{\partial x}\ln f_{\mathbf{y}|\mathbf{x}}(y|x)\right)^2\right\}} \tag{10.3.15}$$

(Example 10.18) MLE (ML Estimator) of Parameters for Gaussian Distribution

Given some observations $\{x_1, x_2, \ldots, x_N\}$ of a Gaussian random variable $\mathbf{x}$, the likelihood function of those observations can be written as

$$f_{\mathbf{x}_1,\cdots,\mathbf{x}_N|m,\sigma}(x_1,\cdots,x_N|m,\sigma^2) = \prod_{n=1}^{N}\frac{1}{\sqrt{2\pi}\sigma}\exp\left\{-\frac{(x_n-m)^2}{2\sigma^2}\right\} \tag{E10.18.1}$$

We can take the logarithm of this to write the log-likelihood function as

$$L(m,\sigma^2|x_1,\cdots,x_N) \overset{(10.3.4)}{=} \ln f(x_1,\cdots,x_N|m,\sigma^2) \overset{(E10.18.1)}{=} -\frac{1}{2}N\ln\left(2\pi\sigma^2\right) - \sum_{n=1}^{N}\frac{(x_n-m)^2}{2\sigma^2} \tag{E10.18.2}$$

To find the MLE of m that maximizes this, we set its derivative w.r.t. m to zero:

$$\frac{\partial L}{\partial m} \overset{(E10.18.2)}{=} \sum_{n=1}^{N}\frac{(x_n-m)}{\sigma^2} = 0; \qquad \hat{m} = \frac{1}{N}\sum_{n=1}^{N}x_n \tag{E10.18.3}$$

Likewise, setting the derivative of Eq. (10.18.2) (with $m = \hat{m}$) w.r.t. σ^2 to zero yields the MLE of σ^2:

$$\frac{\partial L}{\partial(\sigma^2)} \overset{(E10.18.2)}{=} -\frac{N}{2\sigma^2} + \sum_{n=1}^{N}\frac{(x_n-\hat{m})^2}{2(\sigma^2)^2} = 0; \quad \hat{\sigma}^2 = \frac{1}{N}\sum_{n=1}^{N}(x_n-\hat{m})^2 \tag{E10.18.4}$$

Note that the MLEs of mean and variance of a Gaussian variable are equal to the sample mean and sample variance, respectively. If the set x of observations is given as numeric values, the MATLAB function '`mle(x,'distribution','Normal')`' can be used to find the MLEs of m and σ^2.

(**Example 10.19**) Fisher Information

Let the likelihood function of a parameter x given observed data $\{y_1, y_2\}$ be

$$f_{\mathbf{y}_1,\mathbf{y}_2|\mathbf{x}}(y_1, y_2|x) = \frac{1}{2\pi\sigma_{\mathbf{y}_1}\sigma_{\mathbf{y}_2}}\exp\left\{-\frac{(y_1-x)^2}{2\sigma_{\mathbf{y}_1}^2}-\frac{(y_2-x)^2}{2\sigma_{\mathbf{y}_2}^2}\right\} \tag{E10.19.1}$$

Then the log-likelihood function is

$$L(x|y) \overset{(10.3.4)}{=} \ln f_{\mathbf{y}|\mathbf{x}}(y|x) \overset{(E10.19.1)}{=} -\ln\left(2\pi\sigma_{\mathbf{y}_1}\sigma_{\mathbf{y}_2}\right)-\frac{(y_1-x)^2}{2\sigma_{\mathbf{y}_1}^2}-\frac{(y_2-x)^2}{2\sigma_{\mathbf{y}_2}^2} \tag{E10.19.2}$$

and the Fisher information is

$$I(x) \overset{(10.3.14)}{=} -\mathcal{E}\left\{\frac{\partial^2}{\partial x^2}L(x|y)\right\} \overset{(E10.19.2)}{=} \mathcal{E}\left\{\frac{\partial^2}{\partial x^2}\left\{\frac{(y_1-x)^2}{2\sigma_{\mathbf{y}_1}^2}+\frac{(y_2-x)^2}{2\sigma_{\mathbf{y}_2}^2}\right\}\right\}$$
$$= -\mathcal{E}\left\{\frac{\partial}{\partial x}\left(\frac{y_1-x}{\sigma_{\mathbf{y}_1}^2}+\frac{y_2-x}{\sigma_{\mathbf{y}_2}^2}\right)\right\} = \mathcal{E}\left\{\frac{1}{\sigma_{\mathbf{y}_1}^2}+\frac{1}{\sigma_{\mathbf{y}_2}^2}\right\} = \frac{1}{\sigma_{\mathbf{y}_1}^2}+\frac{1}{\sigma_{\mathbf{y}_2}^2} \tag{E10.19.3}$$

This may have an interpretation that the Fisher information is additive in the sense that the information (about the parameter) contributed by multiple independent measurements (experiments) is the sum of the information obtained from each measurement.

[**Remark 10.4**] Unbiased Estimator

The *bias (function)* of an estimator for a parameter is the difference between the estimator's expected value and the true value of the parameter:

$$\text{Bias}(\hat{x}) = \mathcal{E}\{\hat{x}\} - x = \mathcal{E}\{\hat{x}-x\} \tag{10.3.16}$$

An estimator is said to be *unbiased* if its bias is zero:

$$\text{Bias}(\hat{x}) = 0;\ \ \mathcal{E}\{\hat{x}\} = x \tag{10.3.17}$$

(**Example 10.20**) Unbiased Estimator

Suppose we have a set of N observed values, $\{x_1, x_2, \ldots, x_N\}$, of a random variable $\mathbf{x}$ with mean $m_\mathbf{x}$ and variance $\sigma_\mathbf{x}^2$. Then the *sample mean*, *population variance*, and *sample variance* are defined as

$$\bar{x} = \frac{1}{N}\sum_{i=1}^{N}x_i\ ,\quad s^2 = \frac{1}{N}\sum_{i=1}^{N}(x_i-\bar{x})^2\ ,\quad s_\mathrm{u}^2 = \frac{1}{N-1}\sum_{i=1}^{N}(x_i-\bar{x})^2 \tag{E10.20.1,2,3}$$

respectively. First, regarding the N observations $\{x_1, x_2, \ldots, x_N\}$ as N i.i.d. (independent and identically distributed) random variables with mean $m_\mathbf{x}=\mathcal{E}\{\mathbf{x}\}$ and variance $\sigma_\mathbf{x}^2$, let us see if the sample mean $\bar{x}$ is an unbiased estimator of the mean $m_\mathbf{x}$:

$$\mathcal{E}\{\bar{x}\} \overset{(E10.20.1)}{=} \mathcal{E}\left\{\frac{1}{N}\sum_{i=1}^{N}x_i\right\} = \frac{1}{N}\sum_{i=1}^{N}\mathcal{E}\{x_i\} = \frac{1}{N}Nm_\mathbf{x} = m_\mathbf{x} \tag{E10.20.4}$$

This implies that the sample mean $\bar{x}$ is an unbiased estimator of the mean $m_\mathbf{x}$.

Now, let us see if the population variance s^2 is an unbiased estimator of the variance $\sigma_{\mathbf{x}}^2 = \mathcal{E}\{(\mathbf{x}-m_{\mathbf{x}})^2\}$:

$$\mathcal{E}\{s^2\} \overset{(\text{E10.20.2})}{=} \mathcal{E}\left\{\frac{1}{N}\sum_{i=1}^{N}(x_i-\bar{x})^2\right\} = \frac{1}{N}\mathcal{E}\left\{\sum_{i=1}^{N}(x_i-m_{\mathbf{x}}+m_{\mathbf{x}}-\bar{x})^2\right\}$$

$$= \frac{1}{N}\mathcal{E}\left\{\sum_{i=1}^{N}\left((x_i-m_{\mathbf{x}})^2+2(x_i-m_{\mathbf{x}})(m_{\mathbf{x}}-\bar{x})+(m_{\mathbf{x}}-\bar{x})^2\right)\right\}$$

$$\overset{(10.2.31)}{=} \frac{1}{N}\left\{N\sigma_{\mathbf{x}}^2-2N\mathcal{E}\left\{(m_{\mathbf{x}}-\bar{x})^2\right\}+N\mathcal{E}\left\{(m_{\mathbf{x}}-\bar{x})^2\right\}\right\} \overset{(\text{E10.20.1})}{=} \sigma_{\mathbf{x}}^2 - \mathcal{E}\left\{\left(\frac{1}{N}\sum_{i=1}^{N}m_{\mathbf{x}}-\frac{1}{N}\sum_{i=1}^{N}x_i\right)^2\right\}$$

$$= \sigma_{\mathbf{x}}^2 - \frac{1}{N^2}\mathcal{E}\left\{\left(\sum_{i=1}^{N}(m_{\mathbf{x}}-x_i)\right)^2\right\} = \sigma_{\mathbf{x}}^2 - \frac{1}{N^2}\mathcal{E}\left\{\sum_{i=1}^{N}(m_{\mathbf{x}}-x_i)^2+2\sum_{i=1}^{N}\sum_{j\neq i}^{N}(m_{\mathbf{x}}-x_i)(m_{\mathbf{x}}-x_j)\right\}$$

$$\underset{\text{among } x_i\text{'s}}{\overset{\text{Independence}}{=}} \sigma_{\mathbf{x}}^2 - \frac{1}{N}\sigma_{\mathbf{x}}^2 = \frac{N-1}{N}\sigma_{\mathbf{x}}^2 \neq \sigma_{\mathbf{x}}^2 \tag{E10.20.5}$$

This implies that the population variance s^2 is not an unbiased estimator, but a biased estimator of the variance $\sigma_{\mathbf{x}}^2$. In contrast, the sample variance is a biased estimator of the variance:

$$\mathcal{E}\{s_{\mathrm{u}}^2\} \overset{(\text{E10.20.3})}{=} \mathcal{E}\left\{\frac{1}{N-1}\sum_{i=1}^{N}(x_i-\bar{x})^2\right\} = \frac{N}{N-1}\mathcal{E}\left\{\frac{1}{N}\sum_{i=1}^{N}(x_i-\bar{x})^2\right\} = \frac{N}{N-1}\mathcal{E}\{s^2\} \overset{(\text{E10.20.5})}{=} \frac{N}{N-1}\frac{N-1}{N}\sigma_{\mathbf{x}}^2 = \sigma_{\mathbf{x}}^2 \tag{E10.20.6}$$

[Remark 10.5] Cramer–Rao (Lower) Bound

The variance or MSE (mean square error) of any unbiased estimator $\hat{x}$ of an unknown deterministic parameter x is lower-bounded by the *Cramer–Rao bound* (*CRB*), which is defined as the reciprocal of the Fisher information:

$$\text{var}\{\hat{x}\} \geq \frac{1}{I(x)}: \textit{ Cramer–Rao Inequality} \tag{10.3.18}$$

(Example 10.21) Cramer–Rao (Lower) Bound

Suppose the mean $m_{\mathbf{x}}$ of a Gaussian random variable x for which a set of independently observed data is given as $\{x_1, x_2, \ldots, x_N\}$ has the likelihood function as

$$f_{\mathbf{x}_1,\mathbf{x}_2,\cdots,\mathbf{x}_N|m_{\mathbf{x}}}(x_1,x_2,\cdots,x_N \mid m_{\mathbf{x}}) = \frac{1}{(\sqrt{2\pi}\sigma_{\mathbf{x}})^N}\exp\left\{-\sum_{i=1}^{N}\frac{(x_i-m_{\mathbf{x}})^2}{2\sigma_{\mathbf{x}}^2}\right\} \tag{E10.21.1}$$

Then the log-likelihood function is

$$L(m_{\mathbf{x}} \mid x_1,x_2,\cdots,x_N) \overset{(10.3.4)}{=} \ln f_{\mathbf{x}_1,\mathbf{x}_2,\cdots,\mathbf{x}_N|m_{\mathbf{x}}}(x_1,x_2,\cdots,x_N \mid m_{\mathbf{x}}) \overset{(\text{E10.21.1})}{=} -N\ln\left(\sqrt{2\pi}\sigma_{\mathbf{x}}\right)-\sum_{i=1}^{N}\frac{(x_i-m_{\mathbf{x}})^2}{2\sigma_{\mathbf{x}}^2} \tag{E10.21.2}$$

and the Fisher information is

$$I(m_{\mathbf{x}}) \overset{(10.3.14)}{=} -\mathcal{E}\left\{\frac{\partial^2}{\partial m_{\mathbf{x}}^2} L(m_{\mathbf{x}}|x_1,x_2,\cdots,x_N)\right\} \overset{(E10.21.2)}{=} \sum_{i=1}^{N}\frac{1}{\sigma_{\mathbf{x}}^2} = \frac{N}{\sigma_{\mathbf{x}}^2} \tag{E10.21.3}$$

Like in Example 10.19, it is implied that each observation x_i with the same pdf makes an equal contribution to the growth of information.

Substituting this Fisher information into the RHS of Eq. (10.3.18) yields the Cramer–Rao inequality for the sample mean as the estimator of the true mean $m_{\mathbf{x}}$ as

$$\begin{aligned}
\text{var}\{\hat{m}_{\mathbf{x}} = \bar{x}\} &\underset{(E10.20.1)}{\overset{(10.2.31)}{=}} \mathcal{E}\left\{\left(\frac{1}{N}\sum_{i=1}^{N}x_i - \mathcal{E}\left\{\frac{1}{N}\sum_{i=1}^{N}x_i\right\}\right)^2\right\} = \mathcal{E}\left\{\left(\frac{1}{N}\sum_{i=1}^{N}(x_i - m_{\mathbf{x}})\right)^2\right\} \\
&= \frac{1}{N^2}\mathcal{E}\left\{\sum_{i=1}^{N}(m_{\mathbf{x}}-x_i)^2 + 2\sum_{i=1}^{N}\sum_{j\neq i}^{N}(m_{\mathbf{x}}-x_i)(m_{\mathbf{x}}-x_j)\right\} \\
&\underset{(10.2.31)}{\overset{\text{independence}}{=}} \frac{1}{N}\sigma_{\mathbf{x}}^2 \overset{(10.3.18)}{\geq} \frac{1}{I(m_{\mathbf{x}}) = N/\sigma_{\mathbf{x}}^2}
\end{aligned} \tag{E10.21.4}$$

In this case, the equality holds, implying that the sample mean is the best MMSE estimator of the mean $m_{\mathbf{x}}$.

Problems

10.1 Bayes' Rule

Table P10.1 shows a diagnostic data of illness based on a blood test for a population of 10,000 people.

Table P10.1 A diagnostic data of illness based on a blood test

	Good test result	Bad test result	
Healthy people	9,120	480	9,600
Ill people	8	392	400
	9,128	872	10,000

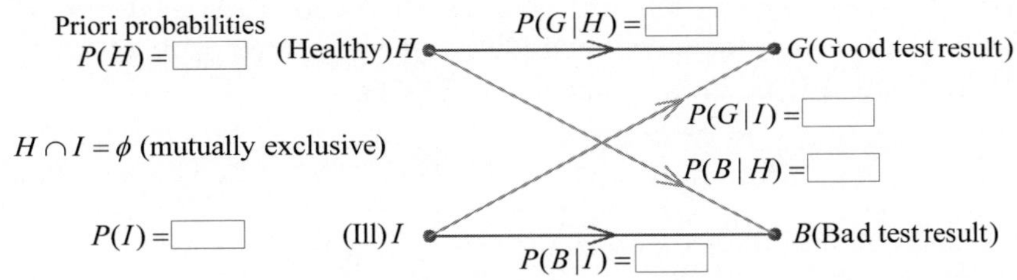

Figure P10.1 Transition probabilities from health condition to blood test result

(a) Assuming the sample size of 10,000 people is large enough to show a sufficient statistic information about the sensisivity and specificity of the blood test w.r.t. the disease, find the probabilities $P(H)$ (of being healthy), $P(I)$ (of being ill), $P(G|H)$ (of a healthy person to have a good test result), $P(B|H)$ (of a healthy person to have a bad test result), $P(G|I)$ (of an ill person to have a good test result), and $P(B|I)$ (of an ill person to have a bad test result) from the data listed in Table 10.1. Based on this information, complete the above transition probability diagram (in Fig. P10.1) from health condition to blood test result, as in Fig. 10.5.

(b) Use the Bayes' rule (10.1.15) to determine the posteriori probabilities of a person to be diagnosed with illness, $P(I|G)$ and $P(I|B)$, depending on whether his/her blood test result is good or not. Check the validity of your answers based on Table P10.1.

10.2 PDF (Probability Distribution Function or CDF) and pdf (Probability Densify Function)

The normal (Gaussian) pdf with mean m and variance σ^2 is

$$f_{\mathbf{x}}(x)=\frac{1}{\sqrt{2\pi}}e^{-x^2/2} \tag{P10.2.1}$$

Show that the normal (Gaussian) PDF can be written in terms of the error function as

$$F_{\mathbf{x}}(x)=\int_{-\infty}^{x}f_{\mathbf{x}}(t)\,dt=\int_{-\infty}^{0}f_{\mathbf{x}}(t)\,dt+\int_{0}^{x}f_{\mathbf{x}}(t)\,dt=\frac{1}{2}\left(1+\int_{-x}^{x}f_{\mathbf{x}}(t)\,dt\right)$$

$$\overset{\text{(P10.2.1)}}{=}\frac{1}{2}\left(1+\frac{1}{\sqrt{2\pi}}\int_{-x}^{x}e^{-t^2/2}dt\right)\overset{t/\sqrt{2}\to t}{=}\frac{1}{2}\left(1+\boxed{}\right)$$

$$\overset{\text{(10.2.35)}}{=}\frac{1}{2}\left(1+\text{erf}\left(\boxed{}\right)\right) \tag{P10.2.2}$$

10.3 Generating Random Variables with a Given PDF/pdf (Sec. 3.11 of [L-1])

(a) *Transformation method* to generate a random variable $\mathbf{x}$ with a given PDF $F_\mathbf{x}(\mathbf{x})$

(Step 1) Generate a standard uniform random number $\mathbf{u}$ distributed in [0, 1] (by using the MATLAB function '`rand()`').

(Step 2) Generate another random variable $\mathbf{x} = F_\mathbf{x}^{-1}(\mathbf{u})$.

How can this new random variable $\mathbf{x}$ have the given PDF $F_\mathbf{x}(\mathbf{x})$?

$$P(\mathbf{x} \le x) = P(F_\mathbf{x}^{-1}(\mathbf{u}) \le x) = P(\mathbf{u} \le F_\mathbf{x}(x)) = \boxed{} \quad \text{where } 0 \le F_\mathbf{x}(x) \le 1 \qquad \text{(P10.3.1)}$$

Note that this method works when we can explicitly find the inverse function of the given PDF. To check the validity of this method of generating a random variable $\mathbf{x}$ with a given PDF, say, $F_\mathbf{x}(x) = 1 - e^{-\lambda x}$ for $x \ge 0$ ($\lambda = 0.8$), complete and run the MATLAB script "em10p03a1.m" to plot the pdf and PDF of the generated random variable $\mathbf{x}$ as shown in Fig. P10.3.1. Note that the inverse function of $F_\mathbf{x}(\mathbf{x})$ is

$$u = F_\mathbf{x}(x) = 1 - e^{-\lambda x} \text{ for } x \ge 0 \rightarrow x = F_\mathbf{x}^{-1}(u) = -\frac{1}{\lambda}\ln(1-u) \qquad \text{(P10.3.2)}$$

Note also that the pdf of $\mathbf{x}$ must be the first derivative of $F_\mathbf{x}(\mathbf{x})$ as

$$f_\mathbf{x}(x) = \frac{d}{dx}F_\mathbf{x}(x) = \lambda e^{-\lambda x} \text{ for } x \ge 0 \qquad \text{(P10.3.3)}$$

```
%em10p03a1.m : to try the transformation method
clear
N=1e6; Nb=100; % Numbers of random data and bins for histogram
u=rand(1,N); lambda=0.8;
x=-log(?)/lambda; % Eq.(P10.3.2)
% Note that (1-u) is also a standard uniform random variable as u.
B=5; Nb=100; % Right bound, Number of bins for histogram
x_below_B = x(find(x<B));
[ns,cs]=hist(x_below_B,Nb);
dx=cs(2)-cs(1); cs=cs+dx/2;
xx=[0:Nb]*dx; % Range of x on which to plot the pde/PDF graphs
fx=lambda*exp(-??????*xx); % pdf by Eq.(P10.3.3)
subplot(221), plot(cs,ns/N/dx, xx,fx,'r-.')
Fx=1-exp(-lambda*xx); % PDF by Eq.(P10.3.2)
subplot(222), plot(cs,cumsum(ns)/N, xx,Fx,'r-.')
```

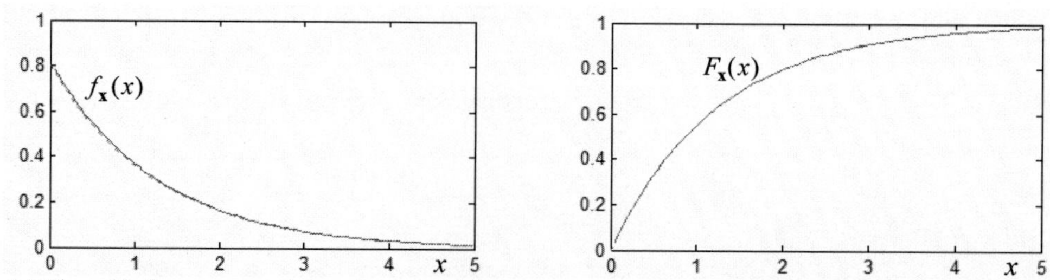

(a) The pdf of an exponential random variable $\mathbf{x}$ (b) The PDF of $\mathbf{x}$

Figure P10.3.1 pdf/PDF of an exponential random sequence generated by the transformation method

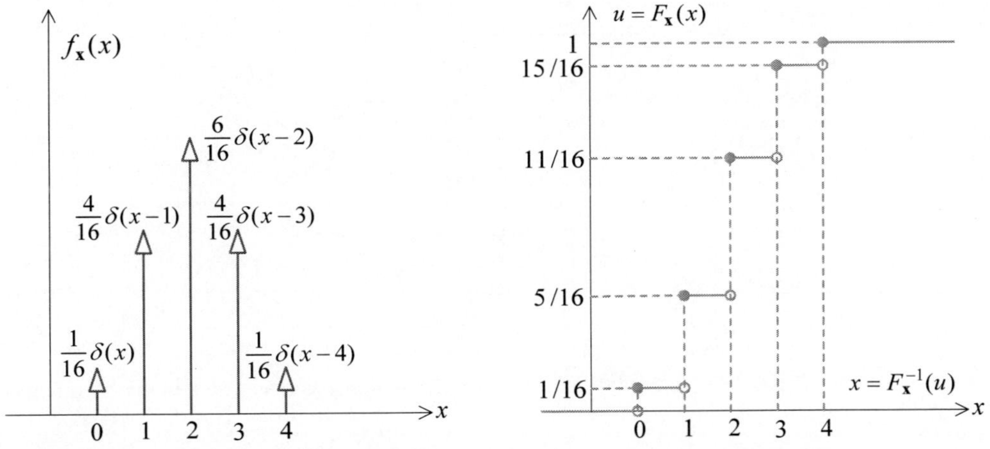

(a) The pdf of a binomial random variable **x** (b) The PDF of **x**

Figure P10.3.2 pdf/PDF of a binomial random variable

As another example, let us use the transformation method to generate the binomial random variable whose pdf/PDF are depicted in Fig. 10.6(a1)/(a2) or Fig. P10.3.2(a)/(b), respectively. Referring to the graph of $u=F_{\mathbf{x}}(x)$ (Fig. P10.3.2(b)), a new random variable **x** can be generated from the standard uniform random variable **u** by the following rule:

$$x = F_{\mathbf{x}}^{-1}(u) = \begin{cases} 0 & \text{for } 0 \le u < 1/16 \\ 1 & \text{for } 1/16 \le u < 5/16 \\ 2 & \text{for } 5/16 \le u < 11/16 \\ 3 & \text{for } 11/16 \le u < 15/16 \\ 4 & \text{for } 15/16 \le u \le 1 \end{cases} \tag{P10.3.4}$$

To check if the new random variable **x** really has the pdf shown in Fig. P10.3.2(a), complete and run the MATLAB script "em10p03a2.m" to plot the pmf and PDF of the discrete random variable **x** as shown in Fig. P10.3.3.

```
%em10p03a2.m : to try the transformation method
N=1e6; Nb=100; % Numbers of random data and bins for histogram
u=rand(1,N);
x=(1/16<=u&u<5/16)+2*(5/16<=u&u<?????)+3*(?????<=u&u<15/16)...
     +4*(15/16<=u); % Eq.(P10.3.4)
Nb=100; [ns,cs]=hist(x,Nb); dx=cs(2)-cs(1); cs=cs+dx/2; % Histogram
subplot(223), plot(cs,ns/N), axis([0 5 0 1])
subplot(224), plot(cs,cumsum(ns)/N), axis([0 5 0 1])
```

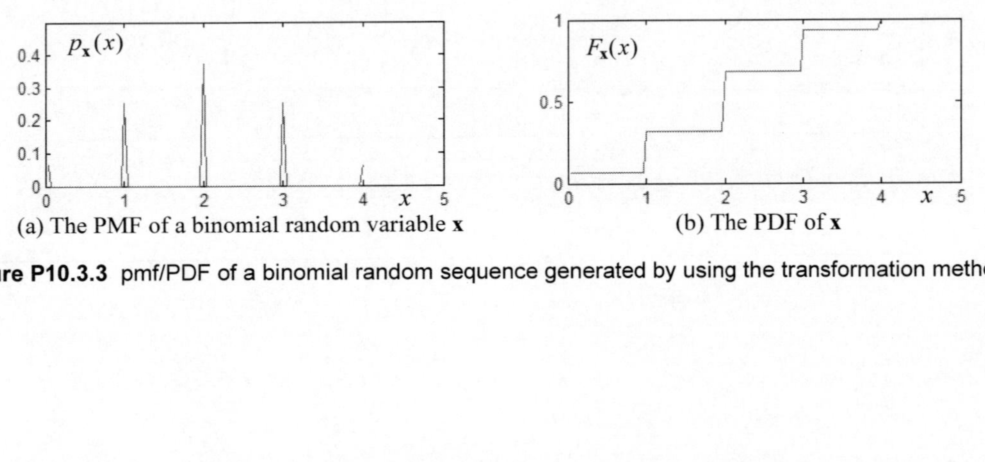

(a) The PMF of a binomial random variable **x** (b) The PDF of **x**

Figure P10.3.3 pmf/PDF of a binomial random sequence generated by using the transformation method

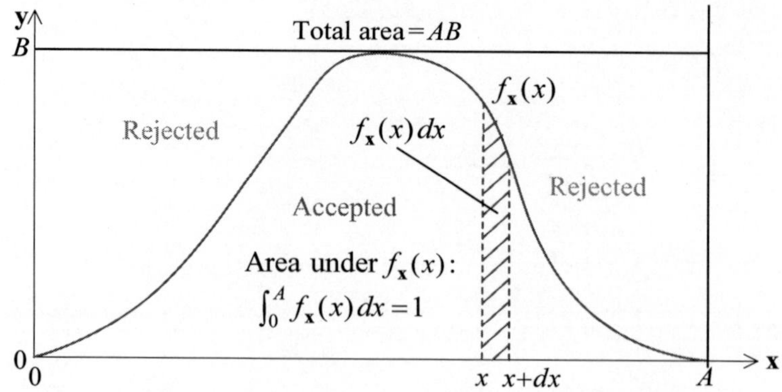

Figure P10.3.4 pdf/PDF of a binomial random variable and a uniform random variable

(b) *Rejection method* to generate a random variable **x** with a given pdf $f_{\mathbf{x}}(\mathbf{x})$ taking on values in a range [0, B] for a domain [0, A]. Referring to Fig. P10.3.4, we take the following steps:

(Step 1) Generate a uniform random number **x** distributed in [0, A].
(Step 2) Generate another uniform random variable **y** distributed in [0, B].
(Step 3) Accept only **x** satisfying $y \le f_{\mathbf{x}}(\mathbf{x})$, rejecting the other ones.

How can these accepted random numbers **x** have the given pdf $f_{\mathbf{x}}(\mathbf{x})$?

$$P(x < \mathbf{x} \le x + dx \mid \mathbf{x} \text{ accepted}) = \frac{P(\{x<\mathbf{x}\le x+dx\} \cap \{\mathbf{x} \text{ accepted}\})}{P(\mathbf{x} \text{ accepted}) = \dfrac{\text{Area of the region below } f_{\mathbf{x}}(x)}{\text{Total area the retangle}}}$$

$$= \frac{\square x/A) \times (\square\,/B)}{\int_0^B f_{\mathbf{x}}(x)dx/AB} = \square\, dx \quad \text{since } \int_0^B f_{\mathbf{x}}(x)dx = 1 \qquad \text{(P10.3.5)}$$

To check the validity of this method of generating a random variable **x** with a given pdf, say $f_{\mathbf{x}}(x) = \lambda e^{-\lambda x}$ for $x \ge 0$ ($\lambda = 0.8$), complete and run the MATLAB script "em10p03b.m" to plot the pdf and PDF of the generated random variable **x** as shown in Fig. P10.3.1.

```
%em10p03b.m : to try the rejection method
N=1e6; Nb=100; % Numbers of random data and bins for histogram
A=5; B=0.8; % Lengths of the domain/range of desired pdf
f=@(x,lambda)lambda*exp(-lambda*x); % Eq.(P10.3.3)
lambda=B;
[x,Nx]=rejection_??????(f,A,B,N,??????);
Nb=100; % Number of bins for histogram
[ns,cs]=hist(x,Nb); %ns:Frequency counts, cs:Bin locations
dx=cs(2)-cs(1); cs=cs+dx/2;
xx=[0:Nb]*dx; % Range of x on which to plot the pde/PDF graphs
fx=f(xx,??????); % pdf by Eq.(P10.3.3)
subplot(221), plot(cs,ns/Nx/dx, xx,fx,'r-.');
Fx=1-exp(-lambda*xx); % PDF by Eq.(P10.3.2)
subplot(222), plot(cs,cumsum(ns)/Nx, xx,Fx,'r-.');
```

```
function [x,Nx]=rejection_method(f,A,B,N,varargin)
x=A*rand(1,N); y=B*rand(1,N); % Step 1 and Step 2
x=x(find(y<=feval(f,x,varargin{:}))); % Step 3
x_below_A = x(find(x<A)); Nx=length(x);
```

10.4 Joint, Marginal, and Conditional pdfs

Consider two random variables **x** and **y**, whose joint pdf is

$$f_{\mathbf{x},\mathbf{y}}(x,y)=\begin{cases}2e^{-x}e^{-y} & \text{for } 0\le y\le x<\infty\\ 0 & \text{elsewhere}\end{cases} \tag{P10.4.1}$$

(a) Find the marginal pdf of **x**.
(b) Find the marginal pdf of **y**.
(c) Find the conditional pdf of **x** given **y**=y.
(d) Find the conditional pdf of **y** given **x**=x.

10.5 Covariance Matrix of Jointly Gaussian Random Vector

Referring to Example 10.14, we can make a zero-mean random vector $[\mathbf{x}\ \ \mathbf{y}]^{\mathrm{T}}$ consisting of two uncorrelated jointly Gaussian random variables **x** and **y**, whose covariance matrix is

$$\Sigma_{\mathbf{xy}}=\operatorname{cov}\left\{\begin{bmatrix}\mathbf{x}\\ \mathbf{y}\end{bmatrix}\right\}=\begin{bmatrix}(\sigma_{\mathbf{u}}^2+\sigma_{\mathbf{v}}^2+\sqrt{(\sigma_{\mathbf{u}}^2-\sigma_{\mathbf{v}}^2)^2+4\sigma_{\mathbf{uv}}^4})/2 & 0\\ 0 & (\sigma_{\mathbf{u}}^2+\sigma_{\mathbf{v}}^2-\sqrt{(\sigma_{\mathbf{u}}^2-\sigma_{\mathbf{v}}^2)^2+4\sigma_{\mathbf{uv}}^4})/2\end{bmatrix} \tag{P10.5.1}$$

and multiply it with

$$A=\begin{bmatrix}\cos\theta & -\sin\theta\\ \sin\theta & \cos\theta\end{bmatrix}\ \text{ with }\ \theta=\frac{1}{2}\tan^{-1}\frac{2\sigma_{\mathbf{uv}}^2}{\sigma_{\mathbf{u}}^2-\sigma_{\mathbf{v}}^2} \tag{P10.5.2}$$

to create another zero-mean Gaussian random vector $[\mathbf{u}\ \ \mathbf{v}]^{\mathrm{T}}$ whose covariance matrix is

$$\Sigma_{\mathbf{uv}}=\operatorname{cov}\left\{\begin{bmatrix}\mathbf{u}\\ \mathbf{v}\end{bmatrix}\right\}\overset{(E10.14.4)}{=}A\,\Sigma_{\mathbf{xy}}A^{\mathrm{T}}=\begin{bmatrix}\sigma_{\mathbf{u}}^2 & \sigma_{\mathbf{uv}}^2\\ \sigma_{\mathbf{uv}}^2 & \sigma_{\mathbf{v}}^2\end{bmatrix}=\begin{bmatrix}2 & 1\\ 1 & 1\end{bmatrix} \tag{P10.5.3}$$

To this end, complete the MATLAB script "em10p05.m," run it, and explain the simulation results.

```
%em10p05.m
sgmu2=2; sgmv2=1; sgmuv2=1; % Eq.(P10.5.3)
tmp1=sgmu2+sgmv2; tmp2=sqrt((sgmu2-sgmv2)^2+4*sgmuv2^2);
sgmx2=(tmp1+tmp2)/2;  sgmx=sqrt(sgmx2);
sgmy2=(tmp1-tmp2)/2;  sgmy=sqrt(sgmy2);
Sigma_xy=diag([sgmx2 ?????]); % Eq.(P10.5.1)
th=atan2(2*sgmuv2,sgmu2-sgmv2)/2;
A=[cos(th) ????????; sin(th) cos(th)] % Eq.(P10.5.2)
Sigma_uv=A*????????*A.'; % Eq.(P10.5.3)
[Sigma_uv [sgmu2 sgmuv2; sgmuv2 sgmv2]] % Eqs.(P10.5.3) and (10.2.46)
N=5e6; x = sgmx*randn(1,N);  y = ????*randn(1,N);
uv = ?*[x; y]; % A realization of new random vector [u; v]
uv1 = Gaussian_rv_with_cov(Sigma_uv,N); % Alternative
m_uv=mean(uv,2); Sigma_uv_sim = uv*uv'/N - m_uv*m_uv'; % Eq.(10.2.40)
[Sigma_uv  Sigma_uv_sim] % Eq.(10.2.46)
[V,L]=eig(Sigma_uv); [L Sigma_xy], [V A] % Eigendecomposition of Sigma
```

```
function [u,Sigmau_sim]=Gaussian_rv_with_cov(Sigmau,N)
M=size(Sigmau,1); x=randn(M,N); % N Realozations of M-dim Gaussian rv
A=chol(Sigmau).'; % Cholesky factorization
u=A*x; mu=mean(u,2); mus=repmat(mu,1,N);
Sigma_u_sim=(u-mus)*(u-mus)'/N % Or, u*u'/N - mu*mu'; %Eq.(10.2.40)
```

10.6 Marginal/Conditional/Joint pdfs of Random Variables

Consider the measured value **y** of a random variable **x** (whose value can be in $[-\pi, \pi]$) as the sum of **x** and a measurement noise **n** (another random variable independent of **x**):

$$\mathbf{y} = \mathbf{x} + \mathbf{n} \tag{P10.6.1}$$

where the marginal pdfs of **x** and **n** are

$$f_{\mathbf{x}}(x) = \begin{cases} 1/2\pi & \text{for } -\pi \le x \le \pi \\ 0 & \text{elsewhere} \end{cases} \quad \text{and} \quad f_{\mathbf{n}}(n) = \frac{1}{\sqrt{2\pi}} e^{-n^2/2} \tag{P10.6.2a,b}$$

The conditional pdf of **y** given **x**=x can be written by substituting $n=y-x$ into Eq. (P10.6.2b) as

$$f_{\mathbf{y}|\mathbf{x}}(y\,|\,x) = \frac{1}{\sqrt{2\pi}} e^{-(y-x)^2/2} \text{ for } -\pi \le x \le \pi \tag{P10.6.3}$$

We can use Eq. (10.2.15) together with Eqs. (P10.6.3) and (P10.6.2a) to get the joint pdf of **x** and **y** as

$$f_{\mathbf{x},\mathbf{y}}(x,y) \overset{(10.2.15)}{=} f_{\mathbf{y}|\mathbf{x}}(y\,|\,x) f_{\mathbf{x}}(x) \underset{(\text{P10.6.2a})}{\overset{(\text{P10.6.3})}{=}} \frac{1}{2\pi^{3/2}} e^{-(y-x)^2/2} \text{ for } \begin{matrix} -\pi \le x \le \pi \\ -\infty \le y \le \infty \end{matrix} \tag{P10.6.4}$$

Also, we can use Eq. (10.2.13b) together with Eq. (P10.6.4) to get the marginal pdf of **y** as

$$f_{\mathbf{y}}(y) \overset{(10.2.13\text{b})}{=} \int_{-\infty}^{\infty} f_{\mathbf{x},\mathbf{y}}(x,y)\,dx \overset{(\text{P10.6.4})}{=} \frac{1}{2\pi^{3/2}} \int_{-\pi}^{\pi} e^{-(y-x)^2/2} dx \tag{P10.6.5}$$

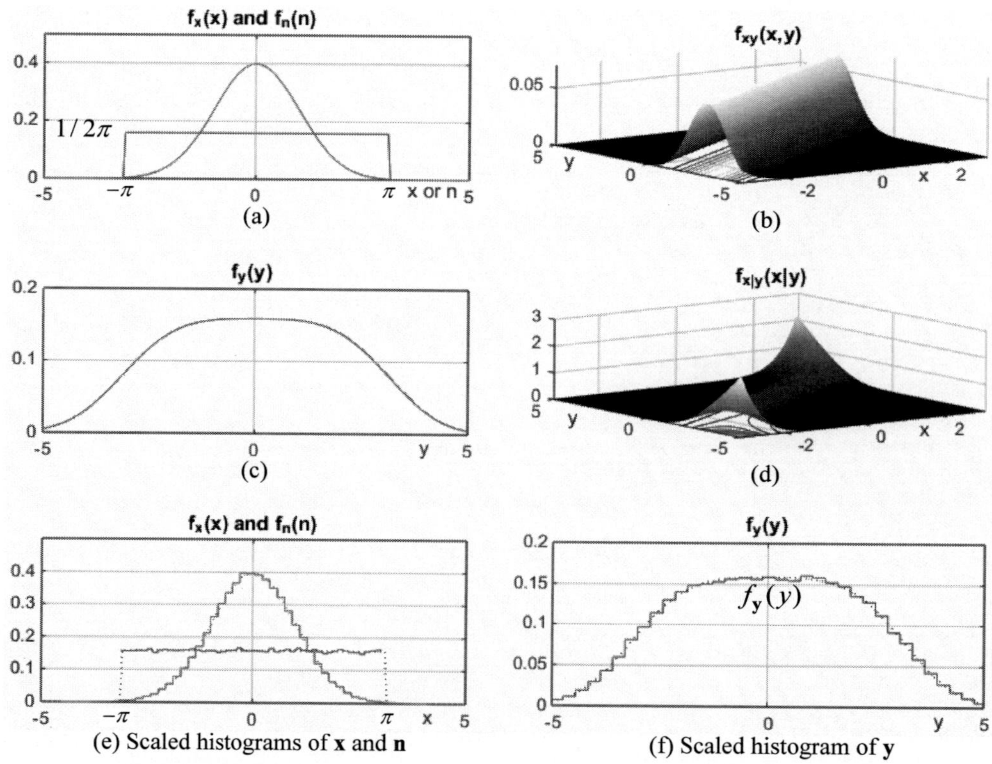

Figure P10.6 Marginal/joint/conditional probability density functions of two random variables

Further, we can use Eq. (10.2.14a) to get the conditional pdf $f_{\mathbf{x}|\mathbf{y}}(x|y)$ of **x** given **y**=y:

$$f_{\mathbf{x}|\mathbf{y}}(x \mid y) \overset{(10.2.14\text{a})}{=} \frac{f_{\mathbf{x},\mathbf{y}}(x,y)}{f_{\mathbf{y}}(y)} \tag{P10.6.6}$$

(a) Complete and run the MATLAB script "em10p06_jcpdf.m" to plot $f_{\mathbf{x}}(x)$, $f_{\mathbf{x},\mathbf{y}}(x,y)$, $f_{\mathbf{y}}(y)$, and $f_{\mathbf{x}|\mathbf{y}}(x|y)$, as in Figs. P10.6(a)–(d).

```
%em10p06_jcpdf.m
sigma=1; tol=1e-4;
fx=@(x)(abs(x)<pi)/2/pi; % fx(x) Eq.(P10.6.2a)
fn=@(n)exp(-n.^2/2/sigma^2)/sqrt(2*pi)/sigma; % fn(n) Eq.(P10.6.2b)
fy_x=@(y,x)fn(y-?); % fy|x(y|x) Eq.(P10.6.3)
fxy=@(x,y)f???(y,x).*f?(x); % jpdf: fxy(x,y) Eq.(P10.6.4)
h=0.05; xx=[-pi:h:pi]; xx1=[-5:h:5]; yy=[-5:h:5]; Ny=length(yy);
subplot(321)
plot(xx1,fx(xx1), xx1,fn(xx1),'r'); title('f_x(x) and f_n(n)')
for m=1:Ny
   y=yy(m);
   fXY(m,:)=feval(fxy,xx,y); % fxy(x,ym)
   fy(m)=quad(f??,-pi,pi,tol,[],y); % fy(ym) Eq.(P10.6.5)
   fx_y(m,:)=fXY(m,:)/f?(m); % fx|y(x|ym)  Eq.(P10.6.6)
end
subplot(323)
fy1=conv(fx(xx),fn(xx))*h;
Nfy1=length(fy1); Nfy=length(fy); Nd=(Nfy1-Nfy)/2;
fy1=fy1(Nd+[1:Nfy]); % take the middle half corresponding to yy
plot(yy,fy, yy,fy1,'r-.'); title('f_y(y)')
subplot(322)
[X,Y]=meshgrid(xx,yy); meshc(X,Y,fXY); title('f_{xy}(x,y)')
subplot(324), meshc(X,Y,fx_y); title('f_{x|y}(x|y)')
```

(b) Complete and run the MATLAB script "em10p06b.m" to generate Ns=10^5 random numbers corresponding to Eq. (P10.6.1) and plot the histograms/pdfs of **x**, **n**, and **y**=**x**+**n** as shown in Figs. P10.6(e) and (f).

```
%em10p06b.m
Ns=1e5; Nb=50; % Numbers of samples and bins for histogram (pdf)
x = rand(1,Ns)*2*pi-??; % Uniform numbers with U[-pi,+pi]
n = randn(1,Ns); % Gaussian random numbers with N(0,1)
n(find(n>5))=5; n(find(n<-5))=-5;
y = ? + ?; % Eq.(P10.6.1)
% Histogram data and Bin widths for x and n
[nxs,cxs]=hist(x,Nb); dx=cxs(2)-cxs(1);
[nns,cns]=hist(n,Nb); dn=cns(2)-cns(1);
subplot(321)
stairs(cxs-dx/2,nxs/Ns/dx); hold on % plot the scaled histogram of x
stairs(cns-dn/2,nns/Ns/dn,'r') % plot the scaled histogram of y
plot(xx1,fx(xx1),'b:', xx1,fn(xx1),'r:') % Theoretical pdfs
axis([-5 5 0 0.5]), grid on
xlabel('x'), title('f_x(x) and f_n(n)')
% Histogram data and Bin width for y
[nys,cys]=hist(y,Nb); dy=cys(2)-cys(1);
subplot(322), stairs(cys-dy/2,nys/Ns/??); % scaled histogram of y
hold on, plot(yy,fy,'b:') % Theoretical pdf
axis([-5 5 0 0.2]), grid on, xlabel('y'), title('f_y (y)')
```

```
%em10p06c.m
% To find the theoretical value of correlation coefficient rho
sgm2x_t=(pi-(-??))^2/??; % Eq.(P10.6.1)
sgm2n_t=1; % Variance of a standard Gaussian rv
sgm2y_t=sgm2?_t+sgm2?_t; % Eq.(P10.6.2)
sgm2xy_t=sgm2?_t; % Eq.(P10.6.3)
% Theoretical value correlation coefficient rho
rho=sgm2??_t/sqrt(sgm2?_t*sgm2?_t) % Eq.(P10.6.4)
% To compute the correlation coefficient from the simulation results
mx=mean(x); my=mean(y); % Means of x and y
sgm2x=var(x); sgm2y=var(y); % Variances of x and y
sgm2xy=mean((x-m?).*(?-my)); % Covariance of x and y by Eq.(10.2.40)
% Simulated value correlation coefficient rho
rho_h=sgm2??/sqrt(sgm2?*sgm2?) % Eq.(10.2.41)
```

(c) Referring to Section 10.2.6, find the variances $\sigma_{\mathbf{x}}^2$, $\sigma_{\mathbf{y}}^2$, $\sigma_{\mathbf{xy}}^2$ (covariance), and correlation coefficient $\rho_{\mathbf{xy}}$ of **x** and **y**=**x**+**n** as follows:

$$\sigma_{\mathbf{x}}^2 \overset{(10.2.38)}{=} (b-a)^2/12 = \boxed{} \tag{P10.6.1}$$

$$\sigma_{\mathbf{y}}^2 \overset{(10.2.31)}{=} \mathcal{E}\{\mathbf{y}^2\} - m_{\mathbf{y}}^2 = \mathcal{E}\{(\mathbf{x}+\mathbf{n})(\mathbf{x}+\mathbf{n})\} \overset{\text{Independence}}{=} \boxed{} \tag{P10.6.2}$$

$$\sigma_{\mathbf{xy}}^2 \overset{(10.2.40)}{=} \mathcal{E}\{\mathbf{xy}\} - m_{\mathbf{x}}m_{\mathbf{y}} = \mathcal{E}\{\mathbf{x}(\mathbf{x}+\mathbf{n})\} \overset{\text{Independence}}{=} \boxed{} \tag{P10.6.3}$$

$$; \qquad \rho_{\mathbf{xy}} \overset{(10.2.41)}{=} \frac{\sigma_{\mathbf{xy}}^2}{\sigma_{\mathbf{x}}\sigma_{\mathbf{y}}} \overset{(\text{P10.6.1}\sim3)}{=} \boxed{} \tag{P10.6.4}$$

Complete and run the above MATLAB script, "em10p06c.m," to check the validity of your answers.

10.7 Probability Density Function (pdf) Changed by a Transformation of Random Variable

Let the pdf of a random variable **x** be $f_{\mathbf{x}}(x)$.

(a) Linear Transformation of a Random Variable

Show that the pdf of another random variable $\mathbf{y}=g(\mathbf{x})=a\mathbf{x}+b$ can be obtained from $f_{\mathbf{x}}(x)$ as

$$f_{\mathbf{y}}(y) = \frac{1}{|a|} f_{\mathbf{x}}\left(g^{-1}(y)\right) = \frac{1}{|a|} f_{\mathbf{x}}\left(\frac{y-b}{a}\right) \tag{P10.7.1}$$

(b) Squaring of a Random Variable

Show that the pdf of another random variable $\mathbf{y}=g(\mathbf{x})=\mathbf{x}^2$ can be obtained from $f_{\mathbf{x}}(x)$ as

$$f_{\mathbf{y}}(y) = \begin{cases} \dfrac{1}{2\sqrt{y}}\left\{f_{\mathbf{x}}\left(\sqrt{y}\right) + f_{\mathbf{x}}\left(-\sqrt{y}\right)\right\} & \text{for } y \ge 0 \\ 0 & \text{for } y < 0 \end{cases} \tag{P10.7.2}$$

To make this feel real, complete and run the MATLAB script "em10p07b.m" to generate a set **x** of $N_s=10^5$ Gaussian random numbers with mean $m=0$ and variance $\sigma^2=1$, make a set of $\mathbf{y}=\mathbf{x}^2$, and plot their histograms together with their analytical pdfs, as Fig. P10.7.

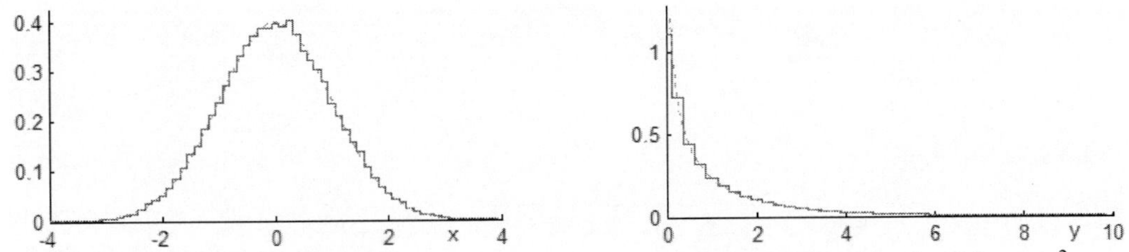

(a) Histogram and pdf for **x** (b) Histogram and pdf for **y**=**x**^{2}

Figure P10.7 Histograms and pdfs for **x** and **y**=**x**^{2}

```
%em10p07b.m
fx=@(x,m,sgm)exp(-(x-m).^2/2/sgm^2)/sqrt(2*pi)/sgm; % Gaussian pdf
fy=@(y,m,sgm)(fx(sqrt(y),m,sgm)+ ...
     fx(?????(y),m,sgm))/2./sqrt(y).*(y>0); % Eq.(P10.7.2)
% Generate Gaussian random numbers of size 1xNs
Ns=1e5; m=0; sgm=1;
xx=random('norm',m,sgm,[1,Ns]);
xmin=m-4*sgm; xmax=m+4*sgm; x=[xmin:0.01:xmax];
fx_x=fx(x,m,sgm); % Numeric values of the pdf of x
fxmax=max(fx_x); dx=sgm/8; cxs=xmin:dx:xmax; % Bin (center) locations
nxs=hist(xx,cxs); % Histogram data of x
subplot(221)
stairs(cxs-dx/2,nxs/Ns/dx); hold on  % plot the histogram of x
plot(x,fx_x,'r:'), axis([xmin xmax 0 fxmax*1.1]) % plot the pdf of x
xlabel('x'), title('Histogram and pdf of x')
yy=xx.^?; % To create a set of y=x^2
ymin=0.1; ymax=m^2+10*sgm^2; y=[ymin:0.02:ymax];
fy_y=fy(y,m,sgm); % Numeric values of the pdf of y
fymax=max(fy_y); dy=sgm^2/4; cys=0:dy:ymax; % Bin (center) locations
nys=hist(yy,cys); % Histogram data of y
subplot(222)
stairs(cys-dy/2,nys/Ns/dy); hold on  % plot the histogram of y
plot(y,fy_y,'r:'), axis([0 ymax 0 fymax*1.1]) % plot the pdf of y
xlabel('y'), title('Histogram and pdf of y=x^2')
```

10.8 Sum and Difference of Two Random Variables

Suppose there are two independent random variables **x** and **y**, each with uniform distribution $U[0, 1]$. Noting that the convolution of two identical rectangular pulses $r_D(t-D/2)$ of height 1 and duration $[0, D]$ is a triangular pulse of duration $[0, 2D]$, let us use Eq. (10.2.21) to find the pdfs of the sum and difference of **x** and **y**, as shown in Figs. P10.8(a) and (b), respectively.

(a) Show that the pdf of $\mathbf{z}=\mathbf{x}+\mathbf{y}$ is

$$f_{\mathbf{z}}(z) \overset{(10.2.21)}{=} f_{\mathbf{x}}(z)*f_{\mathbf{y}}(z)=\int_{-\infty}^{\infty} f_{\mathbf{x}}(z-y)f_{\mathbf{y}}(y)dy=\begin{cases} z & \text{for } 0\le z<1 \\ 2-z & \text{for } 1\le z<2 \\ 0 & \text{elsewhere}\end{cases} \quad \text{(P10.8.1)}$$

(b) Show that the pdf of $\mathbf{w}=\mathbf{x}-\mathbf{y}$ is

$$f_{\mathbf{w}}(w) \overset{(10.2.21)}{=} f_{\mathbf{x}}(w)*f_{\mathbf{y}}(-w)=\int_{-\infty}^{\infty} f_{\mathbf{x}}(w-y)f_{\mathbf{y}}(-y)dy=\int_{-\infty}^{\infty} f_{\mathbf{x}}(w+y)f_{\mathbf{y}}(y)dy=\begin{cases} w+1 & \text{for } -1\le w<0 \\ 1-w & \text{for } 0\le w<1 \\ 0 & \text{elsewhere}\end{cases}$$

(P10.8.2)

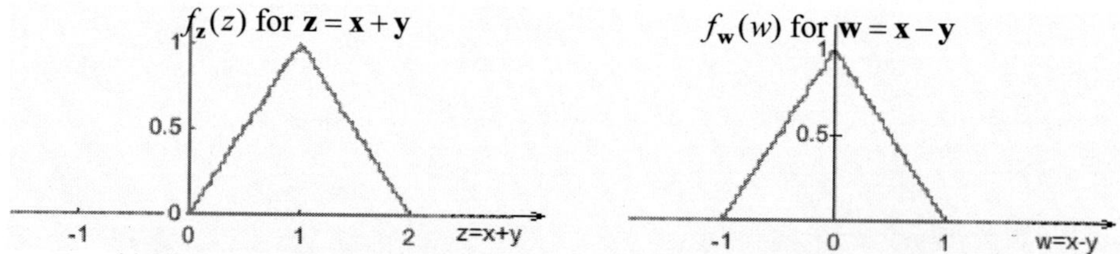

(a) Histogram and pdf for **z**=**x**+**y** (b) Histogram and pdf for **w**=**x**−**y**

Figure P10.8 Histograms and pdfs for **z=x+y** and **w=x−y**

(c) Complete and run the MATLAB script "sum_diff_of_random_numbers.m" to generate two vectors **x** and **y**, each of 10^5 uniform random numbers, **z**=**x**+**y**, and **w**=**x**−**y**, and plot the histograms and pdfs of **z** and **w**, as shown in Figs. P10.8(a) and (b).

```
%sum_diff_of_random_numbers.m
fz=@(z)(z>0&z<=1).*z+(z>1&z<=2).*(2-z); % pdf of z=x+y Eq.(P10.8.1)
fw=@(w)(w>-1&w<=0).*(w+1)+(w>0&w<=1).*(1-w); % pdf of w Eq.(P10.8.2)
Ns=1e5; Nb=50; % Numbers of samples and bins
x=rand(1,Ns); y=rand(1,Ns);
z=x?y; w=x?y;
[nxs,cxs]=hist(x,Nb); dx=cxs(2)-cxs(1);
[nys,cys]=hist(y,Nb); dy=cys(2)-cys(1);
[nzs,czs]=hist(z,Nb); dz=czs(2)-czs(1);
[nws,cws]=hist(w,Nb); dw=cws(2)-cws(1);
h=0.005; zz=[-2:h:3]; ww=[-2:h:3]; % Ranges to plot the graphs on
subplot(321)
stairs(czs-dz/2,nzs/Ns/dz); % plot the historgram of z
hold on, plot(zz,fz(zz),'r:') % plot the pdf of z
axis([-2 3 0 1.2]), xlabel('z=x+y'), title('f_z (z)')
subplot(322)
stairs(cws-dw/2,nws/Ns/dw); % plot the historgram of w
hold on, plot(ww,fw(ww),'r:') % plot the pdf of w
axis([-2 3 0 1.2]), xlabel('w=x-y'), title('f_w (w)')
```

10.9 Difference between Two Jointly Gaussian Random Variables

Consider two independent Gaussian random variables **x** and **y** whose pdfs are

$$f_{\mathbf{x}}(x)=\frac{1}{\sqrt{2\pi}\sigma}e^{-(x-m)^2/2\sigma^2}\sim N(m,\sigma^2)\ \text{and}\ f_{\mathbf{y}}(y)=\frac{1}{\sqrt{2\pi}\sigma}e^{-y^2/2\sigma^2}\sim N(0,\sigma^2)\quad\text{(P10.9.1a,b)}$$

Complete the following derivation of the pdf of **z**=**x**−**y**:

$$f_{\mathbf{z}}(z)\overset{(10.2.21)}{=}\int_{-\infty}^{\infty}f_{\mathbf{x}}(z+y)f_{\mathbf{y}}(y)dy\overset{(\text{P}10.9.1)}{=}\frac{1}{\sqrt{2\pi}\sigma}\frac{1}{\sqrt{2\pi}\sigma}\int_{-\infty}^{\infty}e^{-(z+y-m)^2/2\sigma^2}e^{-y^2/2\sigma^2}dy$$

$$=\frac{1}{\sqrt{2\pi}\Box}\int_{-\infty}^{\infty}e^{-\{\Box\}^2/2(\sigma/\sqrt{2})^2}dy\,\frac{1}{\sqrt{2\pi}\sqrt{2}\sigma}e^{-(z-m)^2/4\sigma^2}$$

$$=\frac{1}{\sqrt{2\pi}\sqrt{2}\sigma}e^{-(z-m)^2/2\Box}\quad\text{(P10.9.2)}$$

Also, complete the MATLAB script "em10p09.m" and run it to plot the histogram and pdf of **z** as shown in Fig. P10.9. Will there be any difference for **z**=**x**+**y**?

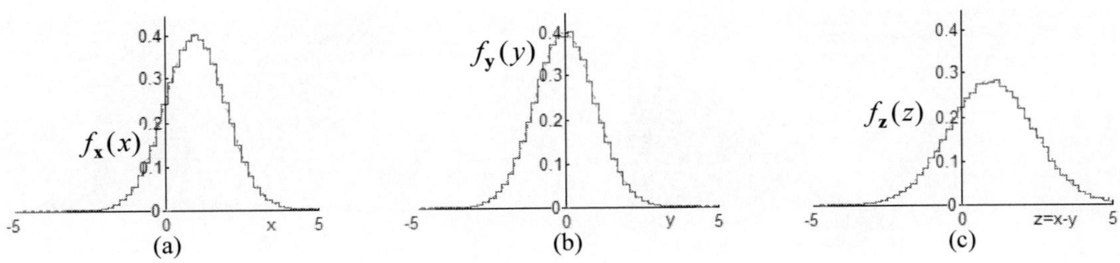

Figure P10.9

```
%em10p09.m
Ns=1e5; m=1; sgm=1; % Number of samples
x=m+sgm*randn(1,Ns); % pdf of x with Gaussian distribution N(m,sgm^2)
y=randn(1,Ns);  % pdf of y with N(0,sgm^2)
z=x?y; % Difference between x and y
fz=@(z)exp(-(z-m).^2/(4*sgm^2))/(sqrt(4*pi)*sgm); % Eq.(P10.9.2)
db=0.2; bin=[-4:db:4];
h=0.01; xx=[-4:h:4]; yy=[-4:h:4]; zz=[-4:h:4];
nxs=hist(x,bin); nys=hist(y,bin); nzs=hist(z,bin);
stairs(bin-db/2,nzs/Ns/db) % plot the historgram of z
hold on, plot(zz,fz(zz),'r:') % plot the pdf of z
```

10.10 Maximum and Minimum of Two Random Variables

(a) Consider the maximum $\mathbf{z}$ of two random variables $\mathbf{x}$ and $\mathbf{y}$:

$$\mathbf{z} = \max(\mathbf{x}, \mathbf{y}) \tag{P10.10a.1}$$

Since $\mathbf{z} \le z$ if and only if $\mathbf{x} \le z$ and $\mathbf{y} \le z$, the PDF of $\mathbf{z}$ can be written as

$$F_{\mathbf{z}}(z) = P(\mathbf{z} \le z) = P(\mathbf{x} \le z, \mathbf{y} \le z) = F_{\mathbf{xy}}(z,z) = \int_{-\infty}^{z}\int_{-\infty}^{\square} f_{\mathbf{xy}}(x,y)\,dx\,dy \tag{P10.10a.2}$$

and the pdf of $\mathbf{z}$ can be obtained by differentiating $F_{\mathbf{z}}(z)$ w.r.t. z as

$$f_{\mathbf{z}}(z) = \frac{d}{dz}F_{\mathbf{z}}(z) \tag{P10.10a.3}$$

Complete and run the MATLAB script "test_pdf_max.m" (including '`pdf_of_max()`') to plot the pdfs of $\mathbf{z}=\max(\mathbf{x}, \mathbf{y})$ for the following two cases as shown in Fig. P10.10.1:

(i) $$f_{\mathbf{xy}}(x,y) = \begin{cases} 1/4 & \text{for } -1 \le x < 1 \text{ and } -1 \le y < 1 \\ 0 & \text{elsewhere} \end{cases} \tag{P10.10a.4}$$

(ii) $$f_{\mathbf{xy}}(x,y) = \frac{1}{2\pi\sigma_{\mathbf{x}}\sigma_{\mathbf{y}}} e^{-(x-m_{\mathbf{x}})^2/2\sigma_{\mathbf{x}}^2 - (y-m_{\mathbf{y}})^2/2\sigma_{\mathbf{y}}^2} \tag{P10.10a.5}$$

```
function fz=pdf_of_max(fxy,z,xmin,ymin)
for i=1:length(z)
   Fz(i)=dblquad(f??,?min,z(i),ymin,?(i)); % Eq.(P10.10a.2)
end
fz=[??ff(Fz)./diff(z) 0]; % Eq.(P10.10a.3)
```

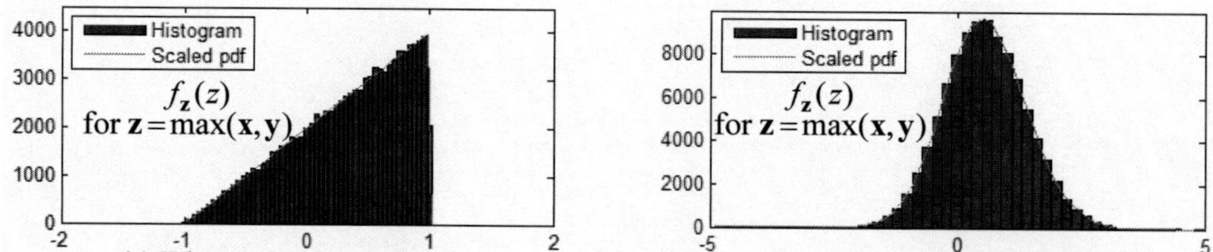

(a) When $\mathbf{x}\sim U(-1,1)$, $\mathbf{y}\sim U(-1,1)$ (b) When $\mathbf{x}\sim N(0,1)$, $\mathbf{y}\sim N(0,1)$

Figure P10.10.1

```
%test_pdf_max.m
clear, clf
disp('(a) x~U(-1,1), y~U(-1,1)')
subplot(221)
% (i) Jointly uniform
Ns=1e5; Nb=50; ax=-1; bx=1; ay=-1; by=1; B=max(bx,by);
xx = ax+(bx-ax)*rand(1,Ns);
yy = ay+(by-ay)*rand(1,Ns);
zz = ???(xx,yy); % Eq.(P10.10a.1)
zmin=max(ax,ay); zmax=max(bx,by);
cs=zmin+(zmax-zmin)/Nb*[0:Nb];
hist(zz,cs); hold on
zz = cs; dz = zz(2)-zz(1);
fxy=@(x,y)double((x>=ax)&(x<=bx)&(y>=ay)&(y<=by))/((bx-ax)*(by-ay));
fzz = pdf_of_???(fxy,zz,ax,ay)*dz*Ns;
plot(zz+dz/2,fzz,'r')
disp('(b) x~N(0,1), y~N(0,1)')
subplot(222)
% (ii) Jointly Gaussian
Ns=1e5; Nb=50; mx=0; sgmx=1; my=0; sgmy=1;
xmin=mx-5*sgmx; ymin=my-5*sgmy; xmax=mx+5*sgmx; ymax=my+5*sgmy;
xx = mx+sgmx*randn(1,Ns);
yy = my+sgmy*randn(1,Ns);
zz = ???(xx,yy); % Eq.(P10.10a.1)
zmin=max(xmin,ymin); zmax=max(xmax,ymax);
zz = zz(find(zz<zmax)); zz = zz(find(zz>zmin)); Nz=length(zz);
cs=zmin+(zmax-zmin)/Nb*[0:Nb];
hist(zz,cs); hold on
zz = cs; dz = zz(2)-zz(1);
fxy=@(x,y)exp(-(x-mx).^2/(2*sgmx^2)).*exp(-(y-my).^2/(2*sgmy^2))...
    /(2*pi*sgmx*sgmy); % Gaussian jpdf
fzz = pdf_of_max(fxy,zz,xmin,ymin)*dz*Nz;
plot(zz+dz/2,fzz,'r')
```

(b) Consider the minimum $\mathbf{z}$ of two random variables $\mathbf{x}$ and $\mathbf{y}$:

$$\mathbf{z}=\min(\mathbf{x},\mathbf{y}) \tag{P10.10b.1}$$

Since $\mathbf{z}\le z$ if and only if $\mathbf{x}\le z$ or $\mathbf{y}\le z$, the PDF of $\mathbf{z}$ can be written as

$$F_{\mathbf{z}}(z)=P(\mathbf{z}\le z)=P(\mathbf{x}\le z \text{ or } \mathbf{y}\le z)=\square-P(\mathbf{x}>z\,\square\,\mathbf{y}>z)$$

$$=1-\int_{\square}^{\infty}\int_{z}^{\infty} f_{\mathbf{xy}}(x,y)\,dx\,dy \tag{P10.10b.2}$$

and the pdf of **z** can be obtained by differentiating $F_{\mathbf{z}}(z)$ w.r.t. z as Eq. (P10.10a.3). Complete and run the MATLAB script "test_pdf_min.m" (including 'pdf_of_min()') to plot the pdfs of **z**=min(**x**, **y**) for the above two cases (i) and (ii), as shown in Fig. P10.10.2.

```
%test_pdf_min.m
disp('(a) x~U(-1,1), y~U(-1,1)')
% (i) Jointly uniform
Ns=1e5; Nb=50; ax=-1; bx=1; ay=-1; by=1; B=max(bx,by);
xx = ax+(bx-ax)*rand(1,Ns);
yy = ay+(by-ay)*rand(1,Ns);
zz = min(xx,yy);  % Eq.(P10.10b.1)
zmin=min(ax,ay); zmax=min(bx,by);
cs=zmin+(zmax-zmin)/Nb*[0:Nb];
subplot(221)
hist(zz,cs); hold on
zz = cs; dz = zz(2)-zz(1);
% Jointly uniform pdf
fxy=@(x,y)double((x>=ax)&(x<=bx)&(y>=ay)&(y<=by))/((bx-ax)*(by-ay));
fzz = pdf_of_???(fxy,zz,bx,by)*dz*Ns;
plot(zz+dz/2,fzz,'r')
disp('(b) x~N(0,1), y~N(0,1)')
% (ii) Jointly Gaussian
Ns=1e5; Nb=50; B=4;  mx=0; sgmx=1; my=0; sgmy=1;
xmin=mx-5*sgmx; ymin=my-5*sgmy; xmax=mx+5*sgmx; ymax=my+5*sgmy;
xx = mx+sgmx*randn(1,Ns);
yy = my+sgmy*randn(1,Ns);
zz = ???(xx,yy);  % Eq.(P10.10b.1)
zmin=min(xmin,ymin); zmax=min(xmax,ymax);
zz = zz(find(zz<B)); zz = zz(find(zz>-B)); Nz=length(zz);
[ns,cs] = hist(zz,Nb);
subplot(222)
hist(zz,Nb); hold on
zz = cs; dz = zz(2)-zz(1);
% Jointly Gaussian pdf
fxy=@(x,y)exp(-(x-mx).^2/(2*sgmx^2)).*exp(-(y-my).^2/(2*sgmy^2))...
         /(2*pi*sgmx*sgmy);
fzz = pdf_of_min(fxy,zz,xmax,ymax)*dz*Nz;
plot(zz+dz/2,fzz,'r')
```

```
function fz=pdf_of_min(fxy,z,xmax,ymax)
for i=1:length(z)
   Fz(i) = 1-dblquad(fxy,?(i),xmax,z(i),?max); % Eq.(P10.10b.2)
end
fz=[diff(Fz)./di??(z) 0]; % Eq.(P10.10a.3)
```

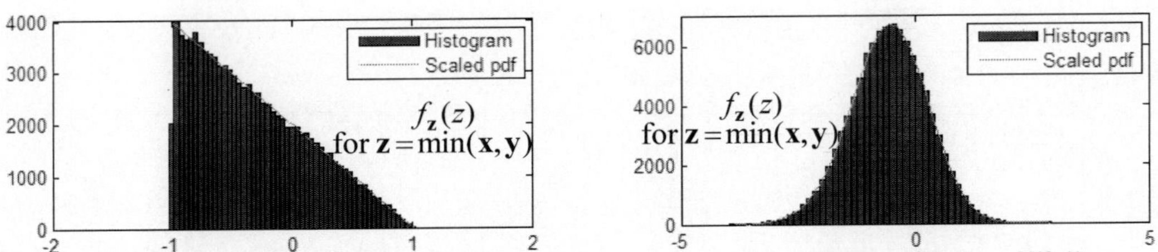

(a) When **x**~$U(-1,1)$, **y**~$U(-1,1)$ (b) When **x**~$N(0,1)$, **y**~$N(0,1)$

Figure P10.10.2

10.11 Product and Quotient of Two Random Variables

(a) Consider the product **z** of two random variables **x** and **y**:

$$\mathbf{z}=\mathbf{x}\,\mathbf{y} \tag{P10.11a.1}$$

The conditional pdf of **z** given $\mathbf{y}=y$, i.e., $g(\mathbf{x})=y\mathbf{x}$, is obtained by using Eq. (10.2.18) as

$$f_{\mathbf{z}|\mathbf{y}}(z\,|\,y)\overset{(10.2.18)}{=}\sum_i \frac{f_{\mathbf{x}}(x|y)}{|dg/dx|}\bigg|_{x=x_i} \quad \text{where } x_i\text{'s are the roots of } g(x)=yx=z$$

$$=\frac{f_{\mathbf{x}}(x|y)}{|y|}\bigg|_{x=z/y}=\frac{f_{\mathbf{x}}((z/y)|y)}{|y|} \tag{P10.11a.2}$$

and the marginal pdf of **z** is obtained by using Eq. (10.2.13a) as

$$f_{\mathbf{z}}(z)\overset{(10.2.13a)}{=}\int_{-\infty}^{\infty} f_{\mathbf{z},\mathbf{y}}(z,y)\,dy\overset{(10.2.15)}{=}\int_{-\infty}^{\infty} f_{\mathbf{z}|\mathbf{y}}(z|y)\,f_{\mathbf{y}}(y)dy\overset{(P10.11a.2)}{=}\int_{-\infty}^{\infty}\frac{f_{\mathbf{x}}((z/y)|y)}{|y|}f_{\mathbf{y}}(y)\,dy$$

$$\overset{(10.2.15)}{=}\int_{-\infty}^{\infty}\frac{f_{\mathbf{xy}}(z/y,y)}{|y|}dy \tag{P10.11a.3}$$

```
%test_pdf_xy.m
clear, clf
disp('(a) x~U(-1,1), y~U(-1,1)')
% (i) Jointly uniform
Ns=1e5; Nb=50;  ax=-1; bx=1; ay=-1; by=1; B=max(bx,by);
xx = ax+(bx-ax)*rand(1,Ns);
yy = ay+(by-ay)*rand(1,Ns);
zz = xx.*yy; % Eq.(P10.11a.1)
zmin=min(zz); zmax=max(zz);
cs=zmin+(zmax-zmin)/Nb*[0:Nb];
subplot(221), hist(zz,cs); hold on
zz = cs; dz = zz(2)-zz(1);
% Uniform jpdf
fxy=@(x,y)double(x>=ax&x<=bx&y>=ay&y<=by)/((bx-ax)*(by-ay));
tol=1e-4;
fzz = pdf_of_??(fxy,zz,ay,by,tol)*dz*Ns;
plot(zz,fzz,'r'), legend('Histogram','Scaled pdf')
disp('(b) x~N(0,1), y~N(0,1)')
% (ii) Jointly Gaussian
Ns=1e5; Nb=50;  mx=0; sgmx=1; my=0; sgmy=1;
xmin=mx-5*sgmx; ymin=my-5*sgmy; xmax=mx+5*sgmx; ymax=my+5*sgmy;
xx = mx+sgmx*randn(1,Ns);  yy = my+sgmy*randn(1,Ns);
zz = xx.?yy;  % Eq.(P10.11a.1)
zmin=-4; zmax=4;
zz = zz(find(zz<zmax)); zz = zz(find(zz>zmin)); Nz=length(zz);
cs=zmin+(zmax-zmin)/Nb*[0:Nb];
subplot(222),  hist(zz,cs); hold on
zz = cs; dz = zz(2)-zz(1); %Nz=Ns;
fxy=@(x,y)exp(-(x-mx).^2/(2*sgmx^2)).*exp(-(y-my).^2/(2*sgmy^2))...
    /(2*pi*sgmx*sgmy);  % Gaussian jpdf
fzz = pdf_of_??(fxy,zz,ymin,ymax,tol)*dz*Nz;
plot(zz,fzz,'r'), legend('Histogram','Scaled pdf')
```

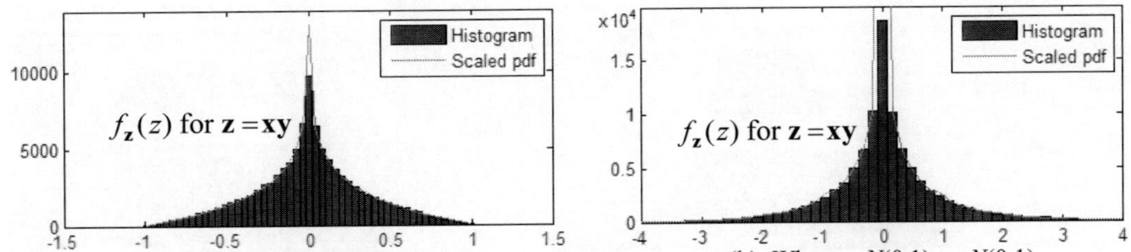

(a) When $x \sim U(-1,1)$, $y \sim U(-1,1)$ (b) When $x \sim N(0,1)$, $y \sim N(0,1)$

Figure P10.11.1

```
function fz=pdf_of_xy(fxy,z,ymin,ymax,tol)
if nargin<5,   tol=1e-4;   end
for i=1:length(z)
   fy = @(y)fxy(?(i)./(y+eps),y)./(abs(?)+eps); % Eq.(P10.11a.3)
   fz(i) = quad(fy,ymin,ymax,tol);
end
```

Complete and run the above MATLAB script, "test_pdf_xy.m," (including '`pdf_of_xy()`') to plot the pdfs of $\mathbf{z}=\mathbf{xy}$ for the following two cases, as shown in Fig. P10.11.1:

$$\text{(i)}\quad f_{\mathbf{xy}}(x,y) = \begin{cases} 1/4 & \text{for } -1 \le x < 1 \text{ and } -1 \le y < 1 \\ 0 & \text{elsewhere} \end{cases} \tag{P10.11a.4}$$

$$\text{(ii)}\quad f_{\mathbf{xy}}(x,y) = \frac{1}{2\pi\sigma_{\mathbf{x}}\sigma_{\mathbf{y}}} e^{-(x-m_{\mathbf{x}})^2/2\sigma_{\mathbf{x}}^2 - (y-m_{\mathbf{y}})^2/2\sigma_{\mathbf{y}}^2} \tag{P10.11a.5}$$

(b) Consider the quotient $\mathbf{z}$ of two random variables $\mathbf{x}$ and $\mathbf{y}$:

$$\mathbf{z} = \mathbf{x}/\mathbf{y} \tag{P10.11b.1}$$

The conditional pdf of $\mathbf{z}$ given $\mathbf{y}=y$, i.e., $g(\mathbf{x})=\mathbf{x}/y$, is obtained by using Eq. (10.2.18) as

$$f_{\mathbf{z}|\mathbf{y}}(z\,|\,y) \overset{(10.2.18)}{=} \sum_i \left.\frac{f_{\mathbf{x}}(x|y)}{|dg/dx|}\right|_{x=x_i} \quad \text{where } x_i\text{'s are the roots of } g(x) = x/y = z$$

$$= \left.\frac{f_{\mathbf{x}}(x|y)}{|1/y|}\right|_{x=yz} = |y|\, f_{\mathbf{x}}(yz\,|\,y) \tag{P10.11b.2}$$

and the marginal pdf of $\mathbf{z}$ is obtained by using Eq. (10.2.13a) as

$$f_{\mathbf{z}}(z) \overset{(10.2.13a)}{=} \int_{-\infty}^{\infty} f_{\mathbf{z},\mathbf{y}}(z,y)\,dy \overset{(10.2.15)}{=} \int_{-\infty}^{\infty} f_{\mathbf{z}|\mathbf{y}}(z|y)\, f_{\mathbf{y}}(y)\,dy \overset{(P10.11b.2)}{=} \int_{-\infty}^{\infty} |y|\, f_{\mathbf{x}}(yz|y)\, f_{\mathbf{y}}(y)\,dy$$

$$\overset{(10.2.15)}{=} \int_{-\infty}^{\infty} |y|\, f_{\mathbf{xy}}(yz,y)\,dy \tag{P10.11b.3}$$

```
function fz=pdf_of_x_div_y(fxy,z,ymin,ymax,tol)
if nargin<5,   tol=1e-4;   end
for i=1:length(z)
   fy = @(y)fxy(?(i)*y,y).*abs(?); % Eq.(P10.11b.3)
   fz(i) = quad(fy,ymin,ymax,tol);
end
```

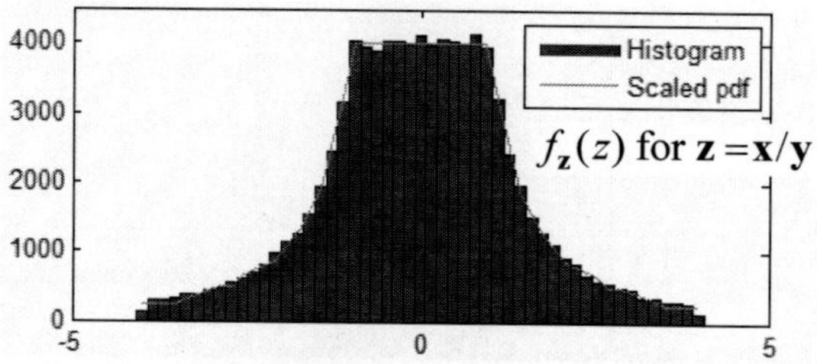

(a) When $\mathbf{x}\sim U(-1,1)$, $\mathbf{y}\sim U(-1,1)$

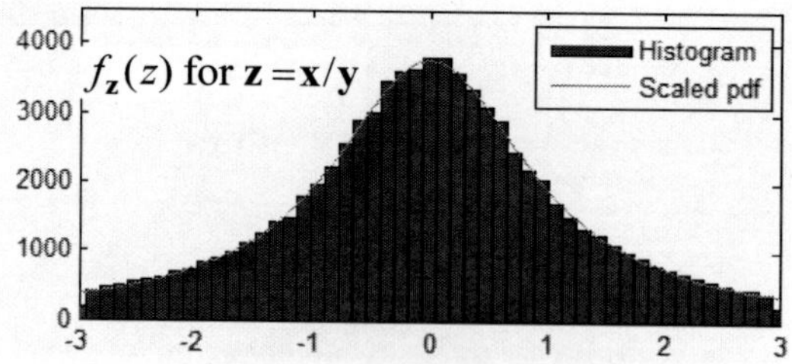

(b) When $\mathbf{x}\sim N(0,1)$, $\mathbf{y}\sim N(0,1)$

Figure P10.11.2

Complete and run the MATLAB script "test_pdf_x_div_y.m" (including 'pdf_of_x_div_y()') to plot the pdfs of $\mathbf{z}=\mathbf{x}/\mathbf{y}$ for the above two cases (i) and (ii), as shown in Fig. P10.11.2.

```
%test_pdf_x_div_y.m
disp('(a) x~U(-1,1), y~U(-1,1)')
% (i) Jointly uniform
Ns=1e5; Nb=50;  ax=-1; bx=1; ay=-1; by=1; B=5;
xx = ax+(bx-ax)*rand(1,Ns);
yy = ay+(by-ay)*rand(1,Ns);
zz = xx./yy; % Eq.(P10.11b.1)
zmin=-4; zmax=4;
zz=zz(find(zz<zmax)); zz=zz(find(zz>zmin));
Nz=length(zz);
cs=zmin+(zmax-zmin)/Nb*[0:Nb];
subplot(221)
hist(zz,cs); hold on
zz = cs; dz = zz(2)-zz(1);
% Uniform jpdf
fxy=@(x,y)double(x>=ax&x<=bx&y>=ay&y<=by)/((bx-ax)*(by-ay));
tol=1e-4;
fzz = pdf_of_x_???_y(fxy,zz,ay,by,tol)*dz*Ns;
plot(zz,fzz,'r'), axis([zmin-1 zmax+1 0 4500])
legend('Histogram','Scaled pdf')
disp('(b) x~N(0,1), y~N(0,1)')
% (ii) Jointly Gaussian
Ns=1e5; Nb=50;  mx=0; sgmx=1; my=0; sgmy=1;
xmin=mx-3*sgmx; ymin=my-3*sgmy; xmax=mx+3*sgmx; ymax=my+3*sgmy;
xx = mx+sgmx*randn(1,Ns);
yy = my+sgmy*randn(1,Ns);
zz = xx.?yy;  % Eq.(P10.11b.1)
%zmin=min(zz); zmax=max(zz);
zmin=-3; zmax=3;
zz = zz(find(zz<zmax)); zz = zz(find(zz>zmin)); Nz=length(zz);
cs=zmin+(zmax-zmin)/Nb*[0:Nb];
subplot(222)
hist(zz,cs); hold on
zz = cs; dz = zz(2)-zz(1);
fxy=@(x,y)exp(-(x-mx).^2/(2*sgmx^2)).*exp(-(y-my).^2/(2*sgmy^2))...
    /(2*pi*sgmx*sgmy);  % Gaussian jpdf
fzz = pdf_of_x_???_y(fxy,zz,ymin,ymax,tol)*dz*Ns;
plot(zz,fzz,'r'), axis([zmin zmax 0 4500])
legend('Histogram','Scaled pdf')
```

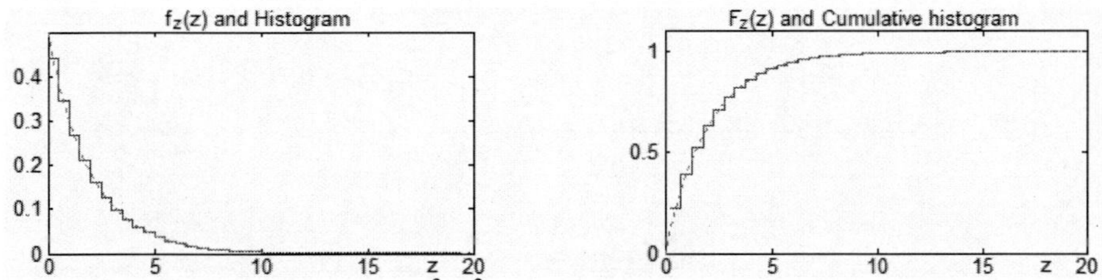

(a) pdf $f_z(z)$ and histogram of $\mathbf{z}=\mathbf{x}^2+\mathbf{y}^2$ (b) PDF $F_z(z)$ and cumulative histogram of $\mathbf{z}=\mathbf{x}^2+\mathbf{y}^2$

Figure P10.12

10.12 Sum of Two Squared Random Variables

Consider two independent Gaussian random variables $\mathbf{x}$ and $\mathbf{y}$ whose joint pdf is

$$f_{\mathbf{xy}}(x,y)=\frac{1}{2\pi\sigma^2}e^{-(x^2+y^2)/2\sigma^2} \tag{P10.12.1}$$

(a) Derive the joint PDF and pdf of $\mathbf{z}=\mathbf{x}^2+\mathbf{y}^2$ as follows:

$$F_{\mathbf{z}}(z)\overset{(10.2.12)}{=}\iint_{x^2+y^2\le z} f_{\mathbf{xy}}(x,y)\,dx\,dy=\frac{1}{2\pi\sigma^2}\iint_{x^2+y^2\le z} e^{-(x^2+y^2)/2\sigma^2}\,dx\,dy$$

$$\underset{(2.6.7)}{\overset{\text{Cartesian-to-polar}}{=}}\frac{1}{2\pi\sigma^2}\int_0^{2\pi}\int_0^{\sqrt{z}} e^{-r^2/2\sigma^2}\,r\,dr\,d\theta=\frac{1}{2\pi}\int_0^{2\pi}d\theta\left(-e^{-r^2/2\sigma^2}\right)\Big|_0^{\sqrt{z}}$$

$$=(\boxed{})\,u_s(z) \tag{P10.12.2}$$

$$f_{\mathbf{z}}(z)=\frac{d}{dz}F_{\mathbf{z}}(z)=\boxed{}\;u_s(z) \tag{P10.12.3}$$

(b) Complete and run the MATLAB script "em10p12.m" to plot the pdf and PDF of $\mathbf{z}=\mathbf{x}^2+\mathbf{y}^2$, together with the histogram and cumulative histogram for 10^6 realizations of $\mathbf{z}$, as shown in Figs. P10.12(a) and (b).

```
%em10p12.m
clear, clf
Ns=1e6; m=0; sgm=1; % Number of samples, Mean, and Standard deviation
x=m+sgm*randn(1,Ns); % Gaussian random numbers with m=0 and sigma=1
y=m+sgm*randn(1,Ns); % Gaussian random numbers with m=0 and sigma=1
z=?.^2+?.^2;
Fz=@(z)(1-exp(-?/2/sgm^2)).*(z>=0); % PDF of z by Eq.(P10.12.2)
fz=@(z)???(-z/2/???^2)/2/sgm^2; % pdf of z by Eq.(P10.12.3)
zmax=20; db=0.5; bin=[0:db:zmax]; % Bins
dz=0.05; zz=[0:dz:zmax]; % Range of z
subplot(222)
nzs=hist(z,bin); stairs(bin-db/2,nzs/Ns/db), hold on
plot(zz,fz(zz),'r:'), title('f_z(z) and Histogram')
axis([0 zmax 0 0.5]), xlabel('z')
subplot(222)
cumulative_sum_of_nzs=cumsum(nzs);
stairs(bin,cumulative_sum_of_nzs/Ns), hold on
plot(zz,Fz(zz),'r:'), title('F_z(z) and Cumulative histogram')
axis([0 zmax 0 1.1]), xlabel('z')
```

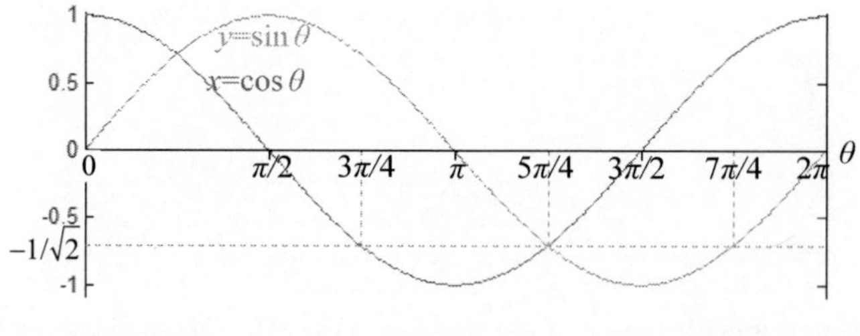

Figure P10.13 Probability of error based on the conditional pdfs

10.13 Independence between Two Random Variables

In Example 10.13, two random variables **x** and **y** that are both dependent on a random variable $\boldsymbol{\theta}$ with uniform distribution $U(0,2\pi)$ as

$$\begin{bmatrix}\mathbf{x}\\ \mathbf{y}\end{bmatrix}=\begin{bmatrix}u(\boldsymbol{\theta})=\cos\boldsymbol{\theta}\\ v(\boldsymbol{\theta})=\sin\boldsymbol{\theta}\end{bmatrix} \tag{P10.13.1}$$

were shown to be uncorrelated and claimed to be not independent. Show that they do not satisfy the independence condition (10.2.17) through the following steps:

- Noting that the pdf of $\mathbf{y}=\sin\boldsymbol{\theta}$ was derived (using Eq. (10.2.18)) as Eq. (E10.9.1):

$$f_{\mathbf{y}}(y)\overset{(10.2.18)}{=}\sum_i \frac{f_{\boldsymbol{\theta}}(\theta)}{|dv/d\theta|}\bigg|_{\theta=\theta_i}\overset{\text{Fig.P10.13}}{=}\frac{1/2\pi}{|\cos(\theta_1)|}+\frac{1/2\pi}{|\cos(\theta_2)|}=\frac{1}{\pi\sqrt{1-y^2}} \tag{P10.13.2}$$

derive the pdf of $\mathbf{x}=\cos\boldsymbol{\theta}$.

- Integrate $f_{\mathbf{x}}(x)$ and $f_{\mathbf{y}}(y)$ w.r.t. x and, y, respectively, to find the PDFs of **x** and **y**.
- Find $F_{\mathbf{x}}(-1/\sqrt{2})=P(x\le -1/\sqrt{2})$, $F_{\mathbf{y}}(-1/\sqrt{2})=P(y\le -1/\sqrt{2})$, and $F_{\mathbf{xy}}(-1/\sqrt{2},-1/\sqrt{2})=P(x\le -1/\sqrt{2}, y\le -1/\sqrt{2})$ (referring to Fig. P10.13), and check if

$$F_{\mathbf{xy}}\left(-\frac{1}{\sqrt{2}},-\frac{1}{\sqrt{2}}\right)\neq F_{\mathbf{x}}\left(-\frac{1}{\sqrt{2}}\right)F_{\mathbf{y}}\left(-\frac{1}{\sqrt{2}}\right) \tag{P10.13.3}$$

10.14 Error Probability Based on Conditional pdfs

Suppose a random binary message **x** (whose value is equally likely to be a_0 or a_1) is transmitted through a channel to a receiver where a zero-mean Gaussian random noise **n** is added to **x** in the channel. Then the received signal **y** can be modeled as the sum of **x** and the additive noise **n**:

$$\mathbf{y}=\mathbf{x}+\mathbf{n} \tag{P10.14.1}$$

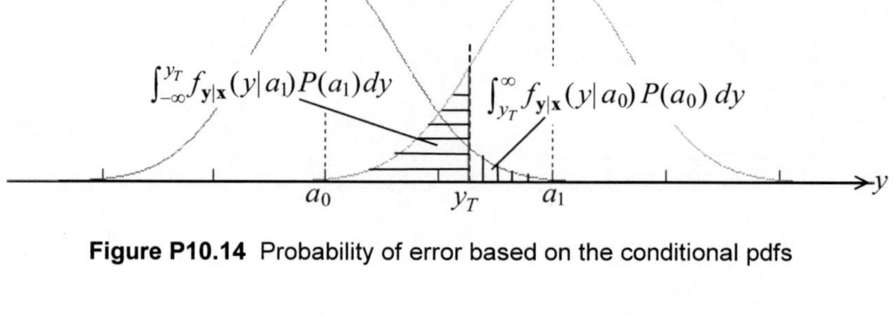

Figure P10.14 Probability of error based on the conditional pdfs where the pdfs of **x** and **n** can be written as

$$f_{\mathbf{x}}(x)=P(a_0)\delta(x-a_0)+P(a_0)\delta(x-a_1)=\frac{1}{2}\delta(x-a_0)+\frac{1}{2}\delta(x-a_1) \quad \text{(P10.14.2a)}$$

$$f_{\mathbf{n}}(n)=\frac{1}{\sqrt{2\pi}\sigma}e^{-(n-m)^2/2\sigma^2}\overset{m=0}{=}\frac{1}{\sqrt{2\pi}\sigma}e^{-n^2/2\sigma^2} \quad \text{(P10.14.2b)}$$

Let the receiver try to detect the transmitted message signal **x** by judging the received signal **y** as a_0 or a_1 depending on whether it is smaller or larger than some threshold value y_T.

(a) Find the conditional pdfs of **y** given $\mathbf{x}=a_0$ or a_1, i.e., $f_{\mathbf{y}|\mathbf{x}}(y|x=a_0)$ and $f_{\mathbf{y}|\mathbf{x}}(y|x=a_1)$.

(b) Show that the probability of error that $\mathbf{x}=a_0|a_1$ is misdetected as $a_1|a_0$ is

$$P_e(y_T)\overset{\text{Fig.P10.14}}{=}P(a_1)\int_{-\infty}^{y_T}f_{\mathbf{y}|\mathbf{x}}(y|a_1)\,dy+P(a_0)\int_{y_T}^{\infty}f_{\mathbf{y}|\mathbf{x}}(y|a_0)\,dy$$

$$\overset{P(a_0)=P(a_1)=1/2}{=}\frac{1}{2\sqrt{2\pi}\sigma}\int_{-\infty}^{y_T}e^{-(y-a_1)^2/2\sigma^2}dy+\frac{1}{2\sqrt{2\pi}\sigma}\int_{y_T}^{\infty}e^{-(y-a_0)^2/2\sigma^2}dy \quad \text{(P10.14.3)}$$

(c) Noting that the derivative of Eq. (P10.14.3) w.r.t. y_T is

$$\frac{d}{dy_T}P_e(y_T)=\frac{1}{2\sqrt{2\pi}\sigma}\left\{e^{-(y_T-a_1)^2/2\sigma^2}-e^{-(y_T-a_0)^2/2\sigma^2}\right\} \quad \text{(P10.14.4)}$$

show that the optimal value of the threshold y_T such that $P_e(y_T)$ is minimized is

$$y_{To}=(a_0+a_1)/2 \quad \text{(P10.14.5)}$$

and with this threshold, the probability of error is

$$P_e(y_{To})\overset{\text{(P10.14.3)}}{\underset{\text{(P10.14.5)}}{=}}\frac{1}{\sqrt{2\pi}\sigma}\int_{(a_0+a_1)/2}^{\infty}e^{-(y-a_0)^2/2\sigma^2}dy\overset{(y-a_0)/\sigma\to t}{=}\frac{1}{\sqrt{2\pi}}\boxed{}$$

$$\overset{(10.2.34)}{=}Q\left(\frac{\boxed{}}{\sigma}\right)=Q\left(\frac{\text{half the distance between } a_1 \text{ and } a_0}{\text{square root of noise variance}}\right) \quad \text{(P10.14.6)}$$

(d) Run the MATLAB script "em10p14_error_probability.m" to check the validity of the error probability $P_e(y_T)$ given by Eq. (P10.14.6). With $y_T=[-1/2\ -1/4\ 0\ 1/4\ 1/2]$, compare the simulated error probabilities.

```
%em10p14_error_probability.m
N=1e7; % Number of binary message data
a=[-1 1]; % Two equally likely numbers
x=a(randi([1 2],1,N)); % 1xN Message data consisting of -1 or 1
sigma=1/2; n=sigma*randn(1,N); y=x+n; % signal received with noise
yT=(a(1)+a(2))/2; % The optimal threshold value (P10.14.5)
detected=a(1)*(y<=yT)+a(2)*(y>yT); % Detected result
Pe=sum(detected~=x)/N  % Simulated error probability
Q((a(2)-a(1))/2/sigma) % Theoretical error probability Eq.(P10.14.6)
```

10.15 Joint pdf of Functions of Random Variables[L-1]

(a) Consider the two uncorrelated random variables **r** and **θ** whose pdfs are

$$f_{\mathbf{r}}(r)=\frac{r}{\sigma^2}e^{-r^2/2\sigma^2} \text{ for } r>0 \text{ (Rayleigh pdf)} \quad \text{(P10.15a.1)}$$

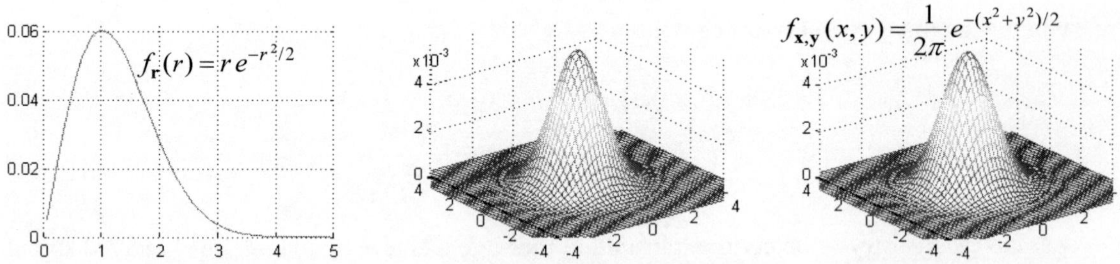

(a) Histogram of $r[n]$ and pdf of $\mathbf{r}$ (b) Histogram of $\{x[n], y[n]\}$ (c) Joint pdf of $\{\mathbf{x}, \mathbf{y}\}$

Figure P10.15.1 Simulated histograms and theoretical pdfs for Problem 10.15(a)

$$f_{\boldsymbol{\theta}}(\theta) = \frac{1}{2\pi} \text{ for } 0 \le \theta < 2\pi \tag{P10.15a.2}$$

so that their joint pdf is

$$f_{\mathbf{r},\boldsymbol{\theta}}(r,\theta) = f_{\mathbf{r}}(r) f_{\boldsymbol{\theta}}(\theta) = \begin{cases} \dfrac{r}{2\pi\sigma^2} e^{-r^2/2\sigma^2} & \text{for } r>0 \text{ and } 0\le\theta<2\pi \\ 0 & \text{elsewhere} \end{cases} \tag{P10.15a.3}$$

Show that the joint pdf of two random variables $\mathbf{x}$ and $\mathbf{y}$ that are functions of $\mathbf{r}$ and $\boldsymbol{\theta}$ as

$$\begin{bmatrix} \mathbf{x} \\ \mathbf{y} \end{bmatrix} = \begin{bmatrix} \mathbf{r}\cos\boldsymbol{\theta} \\ \mathbf{r}\sin\boldsymbol{\theta} \end{bmatrix} \tag{P10.15a.4}$$

is

$$f_{\mathbf{x,y}}(x,y) = \frac{1}{2\pi\sigma^2} e^{-(x^2+y^2)/2\sigma^2} \text{ for } -\infty < x, y < \infty \tag{P10.15a.5}$$

```
%em10p15a.m
clear, clf
fr=@(r,sgm)r/sgm^2.*exp(-r.^2/2/sgm^2); %Eq.(P10.15a.1) Rayleigh pdf
sgm=1; N=1e6;
r=random('Rayleigh',1,1,N); % A Rayleigh rv
th=random('Uniform',0,2*pi,1,N); % A uniform rv
B=5; r=r(find(r<B)); th=th(find(r<B)); Nr=length(r);
Nb=50; [nr,cr]=hist(r,Nb); % Histogram data of r
dr=cr(2)-cr(1); fcr=fr(cr,sgm); % Theoretical Rayleigh pdf of r
subplot(231)
plot(cr,nr/Nr, cr,fcr*dr,'r:'); grid on
% Random variable transformation
x=?.*cos(th); y=r.*???(th); % Eq.(P10.15a.4)
fxy=@(x,y)exp(-(x.^2+?.^2)/2)/?/pi; % Eq.(P10.15a.5)
x=x(find(abs(x)<B)); Nx=length(x);
y=y(find(abs(y)<B)); Ny=length(y);
Nb=50; [nx,cx]=hist(x,Nb); [ny,cy]=hist(y,Nb); % Histograms of x & y
dx=cx(2)-cx(1); dy=cy(2)-cy(1);
[X,Y]=meshgrid(cx,cy); [nX,nY]=meshgrid(nx,ny);
Jpdf = (nX.*nY)/(Nx*Ny); % Simulated joint pdf of x and y
subplot(232), mesh(X,Y,Jpdf);
Jpdf1 = fxy(X,Y)*dx*dy; % Theoretical joint pdf of x and y
subplot(233), mesh(X,Y,Jpdf1);
```

Note that a Rayleigh random sequence $\{r[n], 1{:}N\}$ with the pdf (P10.15a.1) can be generated by using the MATLAB function '`random()`' as

```
>>r=random('Rayleigh',sgm,1,N);
```

Then complete and run the above MATLAB script, "em10p15a.m," to simulate this situation and plot the simulated histograms and theoretical pdfs of $\{r[n]$ for $1{:}N\}$ and $\{(x[n], y[n])$ for $1{:}N\}$ ($N=10^6$), as shown in Fig. P10.15.1.

(b) Consider a pair of two uncorrelated Gaussian random variables **x** and **y** whose joint pdf is

$$f_{\mathbf{x},\mathbf{y}}(x,y)=\frac{1}{2\pi\sigma^2}e^{-(x^2+y^2)/2\sigma^2}\ \text{ for } -\infty<x,y<\infty \tag{P10.15b.1}$$

and another pair of two random variables **r** and **θ** that are functions of **x** and **y** as

$$\begin{bmatrix}\mathbf{z}\\ \boldsymbol{\theta}\end{bmatrix}=\begin{bmatrix}g_1(\mathbf{x},\mathbf{y})=\sqrt{(A\cos\theta_0+\mathbf{x})^2+(A\sin\theta_0+\mathbf{y})^2}\\ g_2(\mathbf{x},\mathbf{y})=\tan^{-1}\{(A\sin\theta_0+\mathbf{y})/(A\cos\theta_0+\mathbf{x})\}\end{bmatrix}\ (A,\ \theta_0\text{: constants}) \tag{P10.15b.2}$$

Noting that the inverse of this transformation is

$$\begin{bmatrix}\mathbf{z}\cos\boldsymbol{\theta}\\ \mathbf{z}\sin\boldsymbol{\theta}\end{bmatrix}=\begin{bmatrix}\mathbf{x}+A\cos\theta_0\\ \mathbf{y}+A\sin\theta_0\end{bmatrix}\rightarrow\begin{bmatrix}\mathbf{x}\\ \mathbf{y}\end{bmatrix}=\begin{bmatrix}h_1(\mathbf{z},\boldsymbol{\theta})=\mathbf{z}\cos\boldsymbol{\theta}-A\cos\theta_0\\ h_2(\mathbf{z},\boldsymbol{\theta})=\mathbf{z}\sin\boldsymbol{\theta}-A\sin\theta_0\end{bmatrix} \tag{P10.15b.3}$$

show that the joint pdf of **r** and **θ** is

$$f_{\mathbf{z},\boldsymbol{\theta}}(z,\theta)=\begin{cases}\dfrac{z}{2\pi\sigma^2}e^{-\{(z\cos\theta-A\cos\theta_0)^2+(z\sin\theta-A\sin\theta_0)^2\}^2/2\sigma^2}\\ \quad=\dfrac{z}{2\pi\sigma^2}e^{-(z^2-2zA\cos(\theta-\theta_0)+A^2\}/2\sigma^2} & \text{for } z>0 \text{ and } 0\le\theta<2\pi\\ \quad 0 & \text{elsewhere}\end{cases} \tag{P10.15b.4}$$

Also, show that the marginal pdf of **z** can be written as

$$f_{\mathbf{z}}(z)\overset{(10.2.13)}{=}\int_0^{2\pi}f_{\mathbf{z},\boldsymbol{\theta}}(z,\theta)\,d\theta\overset{(\text{P10.15b.4})}{=}\frac{z\,u_s(z)}{2\pi\sigma^2}\int_0^{2\pi}e^{-(z^2-2zA\cos(\theta-\theta_0)+A^2)/2\sigma^2}\,d\theta$$

$$=\frac{z\,u_s(z)}{\sigma^2}e^{-(z^2+A^2)/2\sigma^2}I_0\left(\frac{zA}{\sigma^2}\right)\text{: Rice pdf} \tag{P10.15b.5}$$

$$\text{with } I_0(\beta)\underset{(3.3.15)}{\overset{(3.3.25)}{=}}\frac{1}{2\pi}\int_0^{2\pi}e^{-\beta\cos(\theta-\theta_0)}\,d\theta\text{: the 0th-order modified Bessel function of the first kind} \tag{P10.15b.6}$$

where the unit step function $u_s(z)$ has been multiplied to imply that $f_{\mathbf{z}}(z)$ is zero for $z<0$.

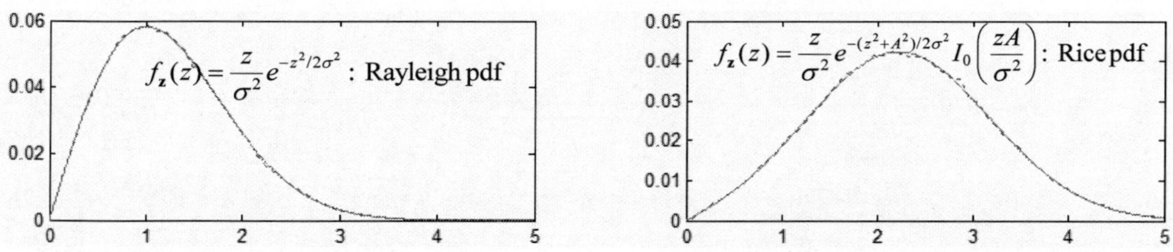

(a) Histogram/pdf of $\{z[n], n=1{:}N_s\}$ for $A=0$ (b) Histogram/pdf of $\{z[n], n=1{:}N_s$ for $A=2$

Figure P10.15.2 Simulated histograms and theoretical pdfs for Problem 10.15(b)

```
%em10p15b.m
clear, clf
f=@(z,A,sgm)(z>=0).*z/???^2.*exp(-(?.^2+A^2)/2/sgm^2).* ...
          besseli(0,z*?/sgm^2); % Eq.(P10.15b.5)
sgm=1; th0=pi/4;  Ns=1e4;
B=5; zz=0:0.01:B; Nb=50; As=[0 2];
for m=1:length(As)
   A=As(m);
   x = randn(1,Ns);
   y = randn(1,Ns);
   zr = A*cos(th0) + x;
   zi = A*sin(th0) + y;
   z = sqrt(zr.^2+??.^2); % Eq.(P10.15b.2a)
   z_below_B = z(find(z<B));
   Nz = length(z);
   subplot(210+m)
   [ns,cs]=hist(z_below_B,Nb); % Histogram data of z
   dz=cs(2)-cs(1);
   fzz = f(zz,A,sgm); % Numeric values of the pdf of z
   plot(zz,fzz*dz,'r',cs,ns/Nz,':')
   title(['Histogram for z=sqrt(A*cos(th)+x)^2+(A*sin(th)+y)^2) and
       Rice pdf with A=' num2str(A)])
end
```

Then complete and run the above MATLAB script, "em10p15b.m," to simulate this situation and plot the simulated histograms and theoretical pdfs of $\{z[n]$ for $1{:}N\}$ ($N=10^4$) for $A=0$ and $A=2$, as shown in Fig. P10.15.2.

10.16 Joint pdf of Functions of Random Variables

Consider two random variables **x** and **y** that are generated from two independent uniform random variables $\mathbf{u}_1$ and $\mathbf{u}_2$ (distributed in [−1, 1]), as

$$\begin{bmatrix}\mathbf{x}_1\\ \mathbf{x}_2\end{bmatrix}=\begin{bmatrix}g_1(\mathbf{u}_1,\mathbf{u}_2)=\mathbf{u}_1\sqrt{-2(\ln\mathbf{R})/\mathbf{R}}\\ g_2(\mathbf{u}_1,\mathbf{u}_2)=\mathbf{u}_2\sqrt{-2(\ln\mathbf{R})/\mathbf{R}}\end{bmatrix}\quad\text{with }\ \mathbf{R}=\mathbf{u}_1^2+\mathbf{u}_2^2 \tag{P10.16.1}$$

Note the following:

- The determinant of the Jacobian matrix of Eq. (P10.16.1) w.r.t. $[\mathbf{u}_1\ \ \mathbf{u}_2]^T$ is

$$J(u_1,u_2)\overset{(10.2.23)}{=}\begin{vmatrix}\dfrac{\partial}{\partial u_1}g_1(u_1,u_2) & \dfrac{\partial}{\partial u_2}g_1(u_1,u_2)\\ \dfrac{\partial}{\partial u_1}g_2(u_1,u_2) & \dfrac{\partial}{\partial u_2}g_2(u_1,u_2)\end{vmatrix} \tag{P10.16.2}$$

$$\overset{(P10.16.1)}{=}\begin{vmatrix}\sqrt{\dfrac{-2\ln R}{R}}+u_1\left(\dfrac{-2\ln R}{R}\right)^{-1/2}\dfrac{-2+2\ln R}{2R^2}\dfrac{\partial R}{\partial u_1} & u_1\left(\dfrac{-2\ln R}{R}\right)^{-1/2}\dfrac{-2+2\ln R}{2R^2}\dfrac{\partial R}{\partial u_2}\\ u_2\left(\dfrac{-2\ln R}{R}\right)^{-1/2}\dfrac{-2+2\ln R}{2R^2}\Box & \sqrt{\dfrac{-2\ln R}{R}}+u_2\left(\dfrac{-2\ln R}{R}\right)^{-1/2}\dfrac{-2+2\ln R}{2R^2}\Box\end{vmatrix}$$

$$;|J|=\left\|\begin{bmatrix} \sqrt{\frac{-2\ln R}{R}}+u_1\left(\frac{-2\ln R}{R}\right)^{-1/2}\frac{-2+2\ln R}{2R^2}2u_1 & u_1\left(\frac{-2\ln R}{R}\right)^{-1/2}\frac{-2+2\ln R}{2R^2}2u_2 \\ u_2\left(\frac{-2\ln R}{R}\right)^{-1/2}\frac{-2+2\ln R}{2R^2}2u_1 & \sqrt{\frac{-2\ln R}{R}}+u_2\left(\frac{-2\ln R}{R}\right)^{-1/2}\frac{-2+2\ln R}{2R^2}2u_2 \end{bmatrix}\right\|$$

$$=\frac{-2\ln R}{R}+\frac{-2+2\ln R}{R^2}(u_1^2+u_2^2)=\boxed{} \quad \text{(P10.16.3)}$$

- $f_{\mathbf{u}_1,\mathbf{u}_2}(u_1,u_2)=\boxed{}$ for $-1<u_1,u_2<1$ (P10.16.4)

- $x_1^2+x_2^2 \overset{\text{(P10.16.1)}}{=} (u_1^2+u_2^2)\frac{-2\ln R}{R}=R\frac{-2\ln R}{R}=-2\ln R\,;\; R=e^{-(x_1^2+x_2^2)/2}$ (P10.16.5)

Show that the joint pdf of $\mathbf{x}_1$ and $\mathbf{x}_2$ is

$$f_{\mathbf{x}_1,\mathbf{x}_2}(x_1,x_2)\overset{(10.2.22)}{=}\sum_i \frac{f_{\mathbf{u}_1,\mathbf{u}_2}(u_1,u_2)}{|J(u_1,u_2)|}\Bigg|_{(u_1,u_2)=(u_1^i,u_2^i)} \overset{\text{(P10.16.3,4)}}{=} \boxed{}$$

$$\overset{\text{(P10.16.5)}}{=}\boxed{} \overset{\times 4/\pi}{\rightarrow} \frac{1}{2\pi}e^{-(x_1^2+x_2^2)/2}=\frac{1}{\sqrt{2\pi}}e^{-x_1^2/2}\frac{1}{\sqrt{2\pi}}e^{-x_2^2/2}=f_{\mathbf{x}_1}(x_1)f_{\mathbf{x}_2}(x_2) \quad \text{(P10.16.6)}$$

This implies that Eq. (P10.16.1) can be used to generate two independent zero-mean, unit-variance jointly Gaussian random variables $\mathbf{x}_1$ and $\mathbf{x}_2$, whose joint pdf is Eq. (P10.16.6). This idea has been cast into the MATLAB function '`Gaussian_rv1()`'. Use this function in place of '`Gaussian_rv()`' in the program "em10e11.m" to see if it works well.

```
function [x1,x2]=Gaussian_rv1(N)
u1=2*rand(1,N)-1;  u2=2*rand(1,N)-1; % Step 1
R=u1.^2+u2.^2; ind=find(R<1); % to scale the jpdf by 2^2/pi
u1=u1(ind); u2=u2(ind); R=R(ind);
tmp=sqrt(-2*log(R)./R);
x1=u1.*tmp; x2=u2.*tmp; % Eq.(P10.16.1)
```

10.17 Markov Property — Lack of Memory

Consider a (possibly infinite) series of independent *Bernoulli trials* that succeeds and fails with probability p and $q=1-p$, respectively. Then the probability that the first success occurs at $(k+1)$th trial so that the waiting time to the first success is k trials can be expressed as

$$p_{\mathbf{x}}(k)=p(1-p)^k \text{ for } k\text{ :nonnegative integers} \quad \text{(P10.17.1)}$$

This implies that the waiting time in a Bernoulli process can be modeled as a geometric random variable $\mathbf{x}$, which has a pdf like Eq. (P10.17.1).

(a) Verify that the probability $P(\mathbf{x}\geq k)$ that the waiting time to the first success is at least k is

$$P(\mathbf{x}\geq k)=(1-p)^k=q^k \quad \text{(P10.17.2)}$$

This is nothing but the probability that the first k trials all fail.

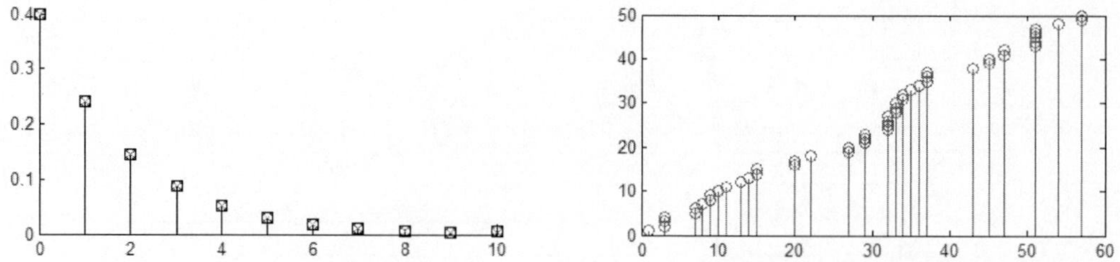

(a) PMF and histograms of waiting times (b) Accumulated number of successes vs. time

Figure P10.17

(b) Verify the conditional probability that the first success will occur at the $(N+k)$th trial after the first N trials failed is

$$P(\mathbf{x}=N+k \mid \mathbf{x}\geq N)=\frac{P(\mathbf{x}=N+k)\ \text{(P10.17.1)}}{P(\mathbf{x}\geq N)\ \text{(P10.17.2)}} = \boxed{} = P(\mathbf{x}=k) \quad \text{(P10.17.3)}$$

This implies that the geometric random variable has the *Markov property* (having no memory) that the probability of observing k failures before a success is independent of past history, since each Bernoulli trial succeeds or fails with the same probability independently of the number of preceding trials.

(c) Complete the MATLAB script "em10p17c.m" and run it to get Fig. P10.17, which implies that the waiting time to the first success in a Bernoulli process (i.e., a sequence of binary random variables taking values from {1(success/head), 0(failure/tail)}) with the probability p of success has the geometric distribution with parameter p.

```
%em10p17c.m
N_exp=1e5; N_trial=11; % # of experiments and trials per experiment
p=0.4; q=1-p; % Probabilities of success/failure
for i_e=1:N_exp
   % Bernoulli process with success probability p
   x_bern=randsrc(1,N_trial,[1 0; p 1-p]);
   indices=find(x_bern==?); % To get the first success time
   if length(indices)>0, waiting_times(i_e)=indices(1)-1;
    else waiting_times(i_e)=11; % Actually more than 10;
   end
end
kk=[0:10]; pmfk=hist(waiting_times,kk); % Histogram
subplot(221), stem(kk,pmfk/?????), hold on
stem(kk,p*q.^??,'rx') % Eq.(P10.17.1)
% Another way to generate the geometric random variable
waiting_times1=random('geo',p,1,N_exp);
pmfk1=hist(waiting_times1,kk); stem(kk,pmfk1/N_exp,'ks')
event_times=cumsum(waiting_times1(1:50));
number_of_events_occurred=1:length(event_times);
subplot(212), stem(event_times,number_of_events_happened)
[mean(waiting_times1) q/p] % Mean of the geometric random variable
[var(waiting_times1) q/p^2] % Var of the geometric random variable
```

(d) Noting the mean number of successes in k independent Bernoulli trials is kp, suppose k Bernoulli trials are performed in t[sec] so that the mean number of successes per unit time is

$$\lambda=\frac{kp}{t} \quad \text{(P10.17.4)}$$

If we let $k \to \infty$ and $p = \lambda t / k \to 0$ with λt fixed, the probability that the waiting time is at least t[sec] (corresponding to k trials) is

$$P(\mathbf{t} \ge t) = \lim_{k\to\infty} P(\mathbf{x} \ge k) \overset{\text{(P10.17.2)}}{=} \lim_{k\to\infty} (1-p)^k \overset{p=\lambda t/k}{=} \lim_{k\to\infty}\left(1 - \frac{\lambda t}{k}\right)^k \overset{\text{(C.25)}}{=} \boxed{} \quad \text{(P10.17.5)}$$

This conforms with Eq. (E10.12e.2), which yields a continuous exponential random variable **t** with pdf

$$f_{\mathbf{t}}(t) \overset{\text{(E10.12e.4)}}{=} \lambda e^{-\lambda t} \quad \text{(P10.17.6)}$$

Note that this random variable has the *Markov property*, since

$$f_{\mathbf{t}}(\mathbf{t} = T + t \mid \mathbf{t} \ge T) = \frac{f_{\mathbf{t}}(\mathbf{t} = T + t)}{P(\mathbf{t} \ge T)} \underset{\text{(P10.17.5)}}{\overset{\text{(P10.17.6)}}{=}} \frac{\lambda e^{-\lambda(T+t)}}{e^{-\lambda T}} = \lambda e^{-\lambda t} \overset{\text{(P10.17.6)}}{=} f_{\mathbf{t}}(\mathbf{t} = t) \quad \text{(P10.17.7)}$$

10.18 M/M/1 Queueing System (Section 9.3 of [L-1])

The M/M/1 queueing system, shown in Fig. P10.18.1(a), is a queueing system (with an infinite capacity) in which customers arrive with exponential interarrival times of mean $1/\lambda$ (arrival rate λ) and are served by a single server with exponential service times of mean $1/\mu$ (service rate μ) (see Example 10.12(c)–(e) for exponential and Poisson random variables) where

$$\text{Arrival rate} = \lambda < \mu = \text{Service rate} \quad \text{(P10.18.1)}$$

Fig. P10.18.1(b) shows the flow (transition rate) diagram for the M/M/1 queueing system.

(a) Let p_k denote the probability of the number N_q of customers in the queueing system (waiting in the queue or being served by the server) being k. Referring to Fig. P10.18.1(b), note that the limiting or equilibrium or steady-state probabilities, p_k's, satisfy

$$\left.\begin{array}{l} -\lambda p_0 + \mu p_1 = 0 \\ \lambda p_{k-1} - (\lambda + \mu) p_k + \mu p_{k+1} = 0 \;\text{ for } k = 1, 2, \cdots \end{array}\right\}$$

$$\overset{\rho=\lambda/\mu}{\rightarrow} \left.\begin{array}{l} p_1 = (\lambda/\mu) p_0 = \rho p_0 \\ p_{k+1} - \rho p_k = p_k - \rho p_{k-1} = 0 \;\text{ for } k = 1, 2, \cdots \end{array}\right\} \rightarrow p_k = \rho^k p_0 \quad \text{(P10.18.2)}$$

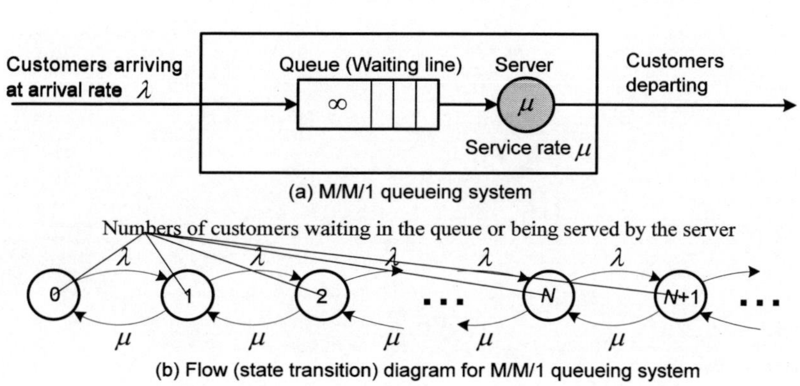

Figure P10.18.1 M/M/1 queueing system

Since the sum of p_k's for all k=0,1,2,... must be 1, we get

$$\sum_{k=0}^{\infty} p_k \overset{\text{(P10.18.2)}}{=} \sum_{k=0}^{\infty} \rho^k p_0 \overset{\text{(C.23)}}{=} \frac{1}{1-\rho} p_0 = 1\,;\;\; p_0 = \boxed{} \tag{P10.18.3}$$

so that we can rewrite the pmf (P10.18.2) as

$$p_k = \rho^k p_0 = \rho^k \boxed{} \quad \text{for } k = 1, 2, \cdots \tag{P10.18.4}$$

(b) Noting that Eq. (P10.18.4) is the pmf of a geometric random variable (Eq. (E10.12a.1)), find the mean number of customers in the queueing system:

$$\mathcal{E}\{N_q\} \overset{\text{(10.2.30b)}}{=} \sum_{k=0}^{\infty} k p_k \underset{p=1-\rho}{\overset{\text{(E10.12a.2)}}{=}} \boxed{} \tag{P10.18.5}$$

(c) Using *Little's law* (*formula*) that the mean number $\mathcal{E}\{N_q\}$ of customers in the queueing system is the product of the (customer) arrival rate λ and the mean (total) time $\mathcal{E}\{T_q\}$ of a customer being in the queueing system:

$$\text{Little's law (formula)}: \; \mathcal{E}\{N_q\} = \lambda \mathcal{E}\{T_q\} \tag{P10.18.6}$$

find $\mathcal{E}\{T_q\}$:

$$\mathcal{E}\{T_q\} \overset{\text{(P10.18.6)}}{=} \frac{\mathcal{E}\{N_q\}}{\lambda} \overset{\text{(P10.18.5)}}{=} \boxed{} \tag{P10.18.7}$$

Note that the mean waiting time in the queue can be obtained by subtracting the mean service time $\mathcal{E}\{T_s\}=1/\mu$ from $\mathcal{E}\{T_q\}$:

$$\mathcal{E}\{T_w\} = \mathcal{E}\{T_q\} - \mathcal{E}\{T_s\} \overset{\text{(P10.18.7)}}{=} \frac{1}{\mu-\lambda} - \frac{1}{\mu} = \frac{\lambda/\mu}{\mu-\lambda} \overset{\rho=\lambda/\mu}{=} \frac{\rho}{1-\rho}\mathcal{E}\{T_s\} = \rho\,\mathcal{E}\{T_q\} \tag{P10.18.8}$$

(d) Find the *server utilization* by subtracting the steady state probability p_0 of the queueing system being empty (idle) from one:

$$1 - p_0 \overset{\text{(P10.18.3)}}{=} \boxed{} \tag{P10.18.9}$$

(e) Complete and run the above MATLAB script, "test_queue_MM1.m," into get the three figures–Figs. P10.18.2(a)-(c). From Fig. P10.18.2(a), see how the number of customers in the queueing system varies for t=0~8[s] (by typing 'Nqs(1:8)' at the MATLAB prompt) and tell what the sequence means in connection with Figs. P10.18.2(a) and (c).

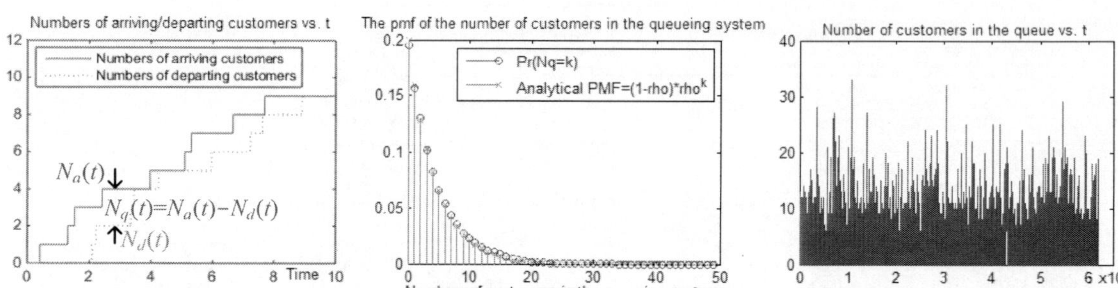

(a) Cumulative numbers of arriving/departing customers
(b) PMF for the number of customers in the system
(c) Number of customers in the queueing system vs. t

Figure P10.18.2 M/M/1 queueing system

```
%test_queue_MM1.m
clear, figure(1), clf
lambda=0.8; mu=1; % Arrival rate and Service rate
Ne=1e5;  % Number of events (customer arrival or departure)
Cq=50; % Virtual maximum capacity of the queue
t_Nqs=zeros(1,Cq); % Times for each number of customers in the system
tA_q=zeros(1,Cq); % Arrival times for the customers waiting in queue
tt=zeros(1,Ne+1);
Nq=0; % Number of customers in the system
At_i=random('exp',1/lambda);
tA=At_i; tAs=[0 tA]; tDs=[tA];
A_or_D=1; Nq=Nq+A_or_D; Nqs(1)=Nq;
t_A_or_D = tAs(end); tA_q(Nq)=t_A_or_D;
At_i=random('exp',1/lambda); % An inter-arrival time
St=random('exp',1/mu); % A service time
Na=1; Nd=0; % Numbers of arriving/departing customers
for ne=1:Ne
   tA = tAs(end)+At_i;  tD = max(tA_q(1),tDs(end))+St;
   if Nq<1|tA<tD
     A_or_D=1; Nq = Nq+A_or_D; t_A_or_D=tA; tAs=[tAs tA];
     tA_q(Nq)=tA; % Arrival time of the last customer in the system
     Na=Na+1; % Accumulative number of arriving customers
     At_i=random('exp',1/lambda); % An inter-arrival time
   elseif Nq>0
     A_or_D=-1; Nq = Nq+A_or_D;
     Nd=Nd+1; % Accumulative number of departing customers
     t_A_or_D = tD; tDs=[tDs tD];
     Tq(Nd) = tD - tA_q(1);
     tA_q(1:end-1) = tA_q(2:end);
     St=random('exp',1/mu); % A service time
   end
   Nq0 = Nqs(ne); Nqs(ne+1) = Nq;
   tt(ne+1)=t_A_or_D; t_Nqs(Nq0+1) = t_Nqs(Nq0+1) + tt(ne+1)-tt(ne);
end
rho=lambda/mu;
subplot(221)
stairs(tAs,[0:Na]), hold on, stairs(tDs,[0:Nd],'r:')
title('Numbers of arriving/departing customers vs. t')
legend('Number of arriving customers','Number of departing customers')
Nqs(1:20) % First 10 numbers of customers in the queueing system
Tq_mean=mean(Tq)
Tq_mean1=mean(tDs-tAs(1:length(tDs)))
1/(1-rho)*(1/??) % Eq.(P10.18.7)
subplot(222)
kk=[0:Cq-1]; stem(kk,t_Nqs/tt(end)), hold on,
stem(kk,(1-rho)*rho.^??,'r') % Eq.(P10.18.4)
legend('Pr(Nq=k)','Analytical PMF=(1-rho)*rho^k')
subplot(223)
stairs(tt,Nqs)
title('Number of customers in the queue vs. t')
mean_number_of_customers_in_queue=t_Nqs*[0:Cq-1].'/sum(t_Nqs)
rho/(1-???) % Eq.(P10.18.9)
% Little's formula
discrepancy=abs(mean_number_of_customers_in_queue-Tq_mean*lambda)...
      /mean_number_of_customers_in_queue  % Eq.(P10.18.6)
%Type sedemo_mm1 to run a Simulink model for M/M/1 queue
```

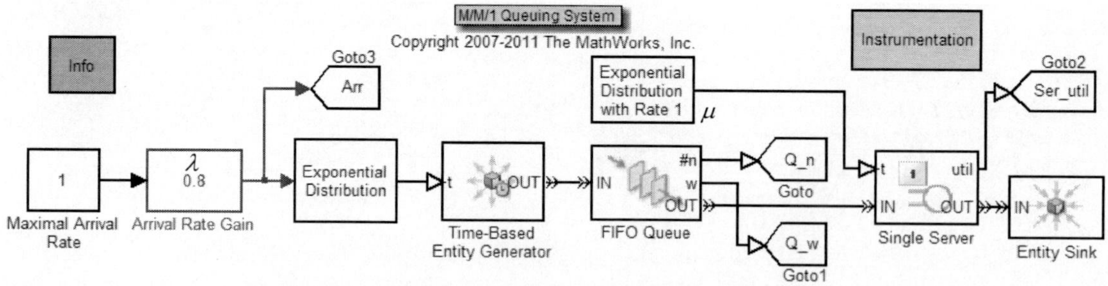

(a) Simulink model "sedemo_mm1.mdl" to simulate an $M/M/1$ queueing system

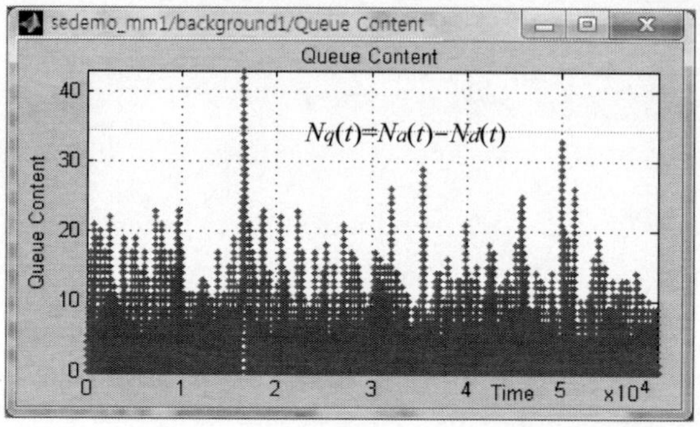

(b) Queue content showing $Nq(t)$ (the number of customers waiting in the queue)

Figure P10.18.3 Simulation of an $M/M/1$ queueing system with a Simulink model "sedemo_mm1.mdl"

(Note) You can type 'sedemo_mm1' at the MATLAB prompt to run the Simulink model 'sedemo_mm1.mdl' (Fig. P10.18.3(a)) and see the queue content $N_q(t)$ (Fig. P10.18.3(b)).

10.19 LMMSE (Linear Mimimum Mean Square Error) Estimator

To see the performance of the LMMSE estimator (introduced in Example 10.15), do the following:

(a) Referring to Example 10.14, premultiply $[\mathbf{x}\ \mathbf{y}]^T$, composed of two jointly uncorrelated Gaussain random variables $\mathbf{x}$ and $\mathbf{y}$ with mean $[m_\mathbf{x}\ m_\mathbf{y}]^T=[1\ \ 1]^T$ and covariance matrix

$$\Sigma_{[\mathbf{x}\ \mathbf{y}]}=\begin{bmatrix}1+\rho & 0\\ 0 & 1-\rho\end{bmatrix}\ \text{with}\ \rho=0.8 \tag{P10.19.1}$$

by A to generate two jointly Gaussain (correlated) random variables $\mathbf{u}$ and $\mathbf{v}$ with covariance matrix

$$\Sigma_{[\mathbf{u}\ \mathbf{v}]}\overset{\text{(E10.14.4b)}}{=}A\,\Sigma_{[\mathbf{x}\ \mathbf{y}]}A^T=\begin{bmatrix}1/\sqrt{2} & 1/\sqrt{2}\\ 1/\sqrt{2} & -1/\sqrt{2}\end{bmatrix}\begin{bmatrix}1+\rho & 0\\ 0 & 1-\rho\end{bmatrix}\begin{bmatrix}1/\sqrt{2} & 1/\sqrt{2}\\ 1/\sqrt{2} & -1/\sqrt{2}\end{bmatrix}=\boxed{} \tag{P10.19.2}$$

(b) Use the MMSE estimator (E10.15.4)

$$\hat{\mathbf{u}}_{\text{MMSE}}\overset{\text{(E10.15.4)}}{=}\rho\frac{\sigma_\mathbf{u}}{\sigma_\mathbf{v}}(\mathbf{v}-m_\mathbf{v})+m_\mathbf{u}=0.8\,\mathbf{v}+\sqrt{2} \tag{P10.19.3}$$

to estimate the value of $\mathbf{u}$ from the measured value of $\mathbf{v}$ for N=10^4 samples and compare the resulting MSE with those obtained with other arbitrary estimators like $\hat{\mathbf{u}}=(\rho+0.1)\mathbf{v}+\sqrt{2}$ or $\hat{\mathbf{u}}=\rho\mathbf{v}+1.3$. If you like, complete the MATLAB script "em10p19b.m" and run it to get the MSE values.

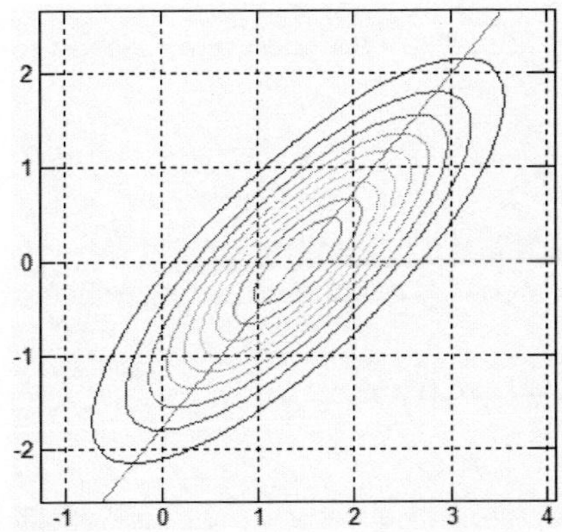

Figure P10.19 MMSE estimator in Problem 10.19

(c) Complete and run the MATLAB script "em10p19c.m" to plot the contour for the joint pdf of **u** and **v** and the MMSE estimator (P10.19.3), as shown in Fig. P10.19 (like Fig. 10.18(a)). Discuss the implication of the figure for the estimator.

```
%em10p19b.m
clear
N=1e4; rho=0.8;
mx=1; my=1; sgm2u=1; sgm2v=1;
sgmu=sqrt(sgm2u);  sgmv=sqrt(sgm2u);
sgm2x=(1+rho)*sgm2u;  sgm2y=(1-rho)*sgm2u;
sgmx=sqrt(sgm2x);  sgmy=sqrt(sgm2y);
S = [sgm2x 0; 0 sgm2y];
x = normrnd(mx,sgmx); % randn(1,N)*sgmx + mx;
y = normrnd(my,sgmy); % randn(1,N)*sgmy + my;
A=[1 1; 1 -1]/sqrt(2);
uv=A*[x; y]; u=uv(1,:); v=uv(2,:);
muv=A*[mx; my]; mu=muv(1); mv=muv(2);
ao = rho*sgmu/sgmv;
uh = ??*(v-mv)+mu; % Eq.(P10.19.3)
MSE=mean((uh-u).^2)
uh1 = (rho+0.1)*sgmu/sgmv*(v-mv) + mu;  MSE1=mean((uh?-u).^2)
uh2 = ao*(v-mv) + 1.3;  MSE2=mean((uh?-u).^2)
```

```
%em10p19c.m
clf
uu = mu+[-3:0.05:3];  vv = mv+[-3:0.05:3];
ASAT=A*?*?'; % The covariance matrix of [u;v]=A*[x;y]
[V,L] = eig(ASAT);
[V A] % Is A similar to the modal matrix V of AKAT?
for iu=1:length(uu)
  for iv=1:length(vv)
    ui=uu(iu);  vi=vv(iv);
    uvi=[ui-mu; vi-mv];
    fuv(iv,iu)=exp(-uvi'*(????\uvi)/2); %Joint pdf Eq.(10.2.45)
  end
end
contour(uu,vv,fuv), hold on
uh3 = ???*sgmu/????*(vv-mv) + mu; % Eq.(E10.15.4)
plot(uh3,vv,'r'), grid on, axis('equal')
```

10.20 LMMSE (Linear Mimimum Mean Square Error) Estimator

Suppose there is a random signal $\mathbf{x}$ with Gaussian distribution $N(m_\mathbf{x}, \sigma_\mathbf{x}^2)$, and we have a linear measurement process

$$\mathbf{y} = a\mathbf{x} + \mathbf{v} \text{ with } \mathbf{v}: N(0, \sigma_\mathbf{v}^2) \tag{P10.20.1}$$

where a random noise $\mathbf{v}$ having Gaussian distribution $N(0, \sigma_\mathbf{v}^2)$ uncorrelated with $\mathbf{x}$ is added.

(a) Verify that the autovariance of $\mathbf{y}$ and the crosscovariance of $\mathbf{x}$ and $\mathbf{y}$ are

$$\sigma_\mathbf{y}^2 = \mathcal{E}\{(\mathbf{y} - m_\mathbf{y})^2\} = \mathcal{E}\left\{(a\mathbf{x}+\mathbf{v}-am_\mathbf{x})^2\right\} = \mathcal{E}\left\{a^2(\mathbf{x}-m_\mathbf{x})^2 + 2(\mathbf{x}-m_\mathbf{x})\mathbf{v}+\mathbf{v}^2\right\}$$

$$= a^2\sigma_\mathbf{x}^2 + \boxed{} \tag{P10.20.2}$$

$$\sigma_\mathbf{xy}^2 = \mathcal{E}\left\{(\mathbf{x}-m_\mathbf{x})(\mathbf{y}-m_\mathbf{y})\right\} = \mathcal{E}\{(\mathbf{x}-m_\mathbf{x})(a\mathbf{x}+\mathbf{v}-am_\mathbf{x})\} = \mathcal{E}\left\{a(\mathbf{x}-m_\mathbf{x})^2 + (\mathbf{x}-m_\mathbf{x})\mathbf{v}\right\}$$

$$= a\boxed{} \tag{P10.20.3}$$

so that the covariance matrix of $[\mathbf{x}\ \ \mathbf{y}]^T$ is

$$\Sigma_{[\mathbf{x}\,\mathbf{y}]} = \mathcal{E}\left\{\begin{bmatrix}\mathbf{x}-m_\mathbf{x}\\ \mathbf{y}-m_\mathbf{y}\end{bmatrix}\begin{bmatrix}\mathbf{x}-m_\mathbf{x}\\ \mathbf{y}-m_\mathbf{y}\end{bmatrix}^T\right\} = \mathcal{E}\left\{\begin{bmatrix}(\mathbf{x}-m_\mathbf{x})^2 & (\mathbf{x}-m_\mathbf{x})(\mathbf{y}-m_\mathbf{y})\\ (\mathbf{y}-m_\mathbf{y})(\mathbf{x}-m_\mathbf{x}) & (\mathbf{y}-m_\mathbf{y})^2\end{bmatrix}\right\} = \begin{bmatrix}\boxed{}\end{bmatrix} \tag{P10.20.4}$$

(b) Referring to Eq. (E10.15.4) and the second observation in Example 10.15, find the LMMSE estimator of $\mathbf{x}$ based on the measurement process $\mathbf{y}$

$$\hat{\mathbf{x}}_{\text{MMSE}} \overset{\text{(E10.15.4)}}{=} g_{\text{MMSE}}(\mathbf{y} - a\,m_\mathbf{x}) + m_\mathbf{x} \tag{P10.20.5}$$

with

$$\text{MMSE estimator gain: } g_{\text{MMSE}} \overset{\text{(E10.15.3)}}{=} \frac{\text{cov}(\mathbf{x},\mathbf{y})}{\text{cov}(\mathbf{y})} \overset{\text{(P10.20.4)}}{=} \boxed{} \tag{P10.20.6}$$

(c) Find the MSE of this estimator:

$$\mathcal{E}\left\{(\hat{\mathbf{x}}_{\text{MMSE}} - \mathbf{x})^2\right\} \overset{\text{(P10.20.5)}}{=} \mathcal{E}\left\{\left(g(\mathbf{y}-a\,m_\mathbf{x})+m_\mathbf{x}-\mathbf{x}\right)^2\right\} \overset{\text{(P10.20.1)}}{=} \mathcal{E}\left\{\left(g(a\mathbf{x}+\mathbf{v}-a\,m_\mathbf{x})+m_\mathbf{x}-\mathbf{x}\right)^2\right\}$$

$$= \mathcal{E}\left\{\left((ga-1)(\mathbf{x}-m_\mathbf{x})+g\mathbf{v}\right)^2\right\} = (ga-1)^2\mathcal{E}\left\{(\mathbf{x}-m_\mathbf{x})^2\right\} + g^2\mathcal{E}\{\mathbf{v}^2\}$$

$$\overset{\text{(P10.20.6)}}{=} \left(\frac{a^2\sigma_\mathbf{x}^2}{a^2\sigma_\mathbf{x}^2+\sigma_\mathbf{v}^2} - 1\right)^2\sigma_\mathbf{x}^2 + \left(\frac{a\sigma_\mathbf{x}^2}{a^2\sigma_\mathbf{x}^2+\sigma_\mathbf{v}^2}\right)^2\sigma_\mathbf{v}^2 = \frac{\sigma_\mathbf{x}^2}{\boxed{}+1} = \frac{\sigma_\mathbf{x}^2}{\text{SNR}+1} \tag{P10.20.7}$$

where the SNR (signal-to-noise ratio) has been defined as

$$\text{SNR} = \frac{\text{Signal power}}{\text{Noise power}} = \frac{a^2\sigma_\mathbf{x}^2}{\boxed{}} \tag{P10.20.8}$$

(d) To take advantage of the orthogonality principle (E10.15.6):

$$\mathcal{E}\left\{(\hat{\mathbf{x}}_{\text{MMSE}} - \mathbf{x})\mathbf{y}\right\} = 0 \tag{P10.20.9}$$

we define the inner product between random variables **x** and **y** as the expectation of their product:

$$\mathbf{x} \bullet \mathbf{y} = \mathcal{E}\{\mathbf{xy}\} \tag{P10.20.10a}$$

and the norm of a random variable **x** as

$$\|\mathbf{x}\| = \sqrt{\mathbf{x}\bullet\mathbf{x}} = \sqrt{\mathcal{E}\{\mathbf{x}^2\}} \tag{P10.20.10b}$$

Then we can use Eq. (1.1.10) to find the projection of $\mathbf{x} - m_\mathbf{x}$ on $\mathbf{y} - m_\mathbf{y}$ as

$$\text{proj}_{\mathbf{y}-m_\mathbf{y}}(\mathbf{x}-m_\mathbf{x}) \overset{(1.1.10)}{=} \{(\mathbf{x}-m_\mathbf{x})\bullet(\mathbf{y}-m_\mathbf{y})\}\frac{\mathbf{y}-m_\mathbf{y}}{\|\mathbf{y}-m_\mathbf{y}\|^2} = \mathcal{E}\{(\mathbf{x}-m_\mathbf{x})(\mathbf{y}-m_\mathbf{y})\}\frac{\mathbf{y}-m_\mathbf{y}}{\mathcal{E}\{(\mathbf{y}-m_\mathbf{y})^2\}}$$

$$\underset{\text{uncorrelated}}{\overset{(\text{P10.20.1})}{=}} \mathcal{E}\{(\mathbf{x}-m_\mathbf{x})a(\mathbf{x}-m_\mathbf{x})\}\frac{\mathbf{y}-m_\mathbf{y}}{\mathcal{E}\{(a(\mathbf{x}-m_\mathbf{x})+\mathbf{v})^2\}} = \boxed{}(\mathbf{y}-a m_\mathbf{x}) \tag{P10.20.11}$$

This implies that the projection of **x** on $\mathbf{y} - m_\mathbf{y}$ will be Eq. (P10.20.5).

(e) To understand the LMMSE estimator better, perform two soft experiments using Eq. (P10.20.5) to estimate a random signal **x** with Gaussian distribution $N(m_\mathbf{x}=1, \sigma_\mathbf{x}^2=3^2)$, each based on one of the following two linear measurements of **x**:

$$\mathbf{y}_1 = 2\mathbf{x} + \mathbf{v} \text{ with } \mathbf{v}: N(0, 2^2) \tag{P10.20.12a}$$

$$\mathbf{y}_2 = -\mathbf{x} + \mathbf{v} \text{ with } \mathbf{v}: N(0, 2^2) \tag{P10.20.12b}$$

```
%em10p20.m
N=1e4; % Number of samples
mx=1; sgmx=3; mv=0; sgmv=2; a=2;
for n=1:N
   x(n)=random('normal',mx,sgmx);
   v(n)=random('normal',mv,sgmv);
   y(n) = ?*x(n) + ?(n); % Eq.(P10.20.1)
end
sgmx2=var(x); discrepancy_var_x=abs(sgmx2-sgmx^2)
sgmv2=var(v); discrepancy_var_v=abs(sgmv2-sgmv^2)
cov_y=a^2*sgmx^2+sgmv^2; cov_xy=a*sgmx^22; % (P10.20.2), (P10.20.3)
rho = cov_??/sgmx/sqrt(cov_?)  % Correlation coefficient (10.2.41)
g_MMSE = cov_??/???_y; % MMSE estimator gain (P10.20.6)
gs = [0.5:0.1:1.5];  Ngs=length(gs);
for i=1:Ngs  % For several estimator gains
   xh = gs(i)*g_MMSE*(y-a*mx) + mx; % Estimate of x by Eq.(P10.20.5)
   MSEs(i) = mean((xh-x).^?); % MSE Eq.(10.2.31)
   if gs(i)==1
     xh_MMSE=xh; MSE_MMSE=MSEs(i); % MMSE estimate and its MSE
   end
   orthognonalities(i) = abs(mean((xh-x).*y)); % Orthogonal?
end
disp('MSE')
[MSE_MMSE  sgmx2*sgmv2/(a^2*sgmx2+sgmv2)] % Eq.(P10.20.7)
subplot(221)
plot(gs,MSEs,'.-'); title('MSEs with different gains')
subplot(222), plot(gs,orthognonalities,'.-')
title('Orthogonalities with different gains')
subplot(223)
plot(x,y,'.', xh_MMSE,y,'r'), title('y vs. x')
```

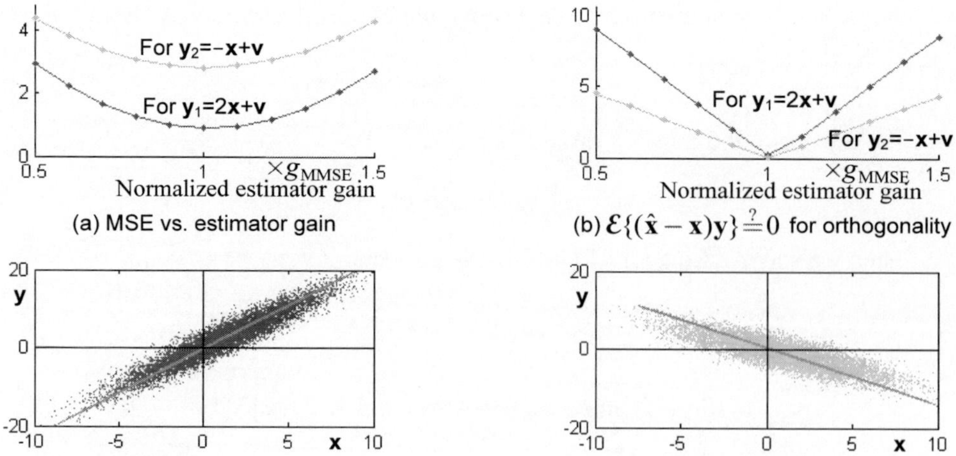

(a) MSE vs. estimator gain (b) $\mathcal{E}\{(\hat{\mathbf{x}}-\mathbf{x})\mathbf{y}\}\overset{?}{=}0$ for orthogonality

(c1) **x** and its estimator vs. $\mathbf{y}_1$=2**x**+**v** (c2) **x** and its estimator vs. $\mathbf{y}_2$=−**x**+**v**

Figure P10.20.1

Table P10.20 MSEs of the MMSE estimator based on two linear measurements

Measurement	Correlation $\rho_{\mathbf{xy}}=\sigma_{\mathbf{xy}}^2/\sigma_{\mathbf{x}}\sigma_{\mathbf{y}}$	SNR= $a^2\sigma_{\mathbf{x}}^2/\sigma_{\mathbf{v}}^2$	MMSE estimator $\hat{\mathbf{x}}=g(\mathbf{y}-am_{\mathbf{x}})+m_{\mathbf{x}}$	Theoretial/Simulated MSEs
$\mathbf{y}_1=2\mathbf{x}+\mathbf{v}$ with $\mathbf{v}\sim N(0,2^2)$	0.949		$\hat{\mathbf{x}}=0.45(\mathbf{y}-2\times 1)+1$	/ 0.
$\mathbf{y}_2=-\mathbf{x}+\mathbf{v}$ with $\mathbf{v}\sim N(0,2^2)$		2.25		2.769 /

and plot the MSEs of several estimators with different gains around the MMSE gain g_{MMSE}. To this end, complete/modify and run the above MATLAB script, "em10p20.m," to get Figs. P10.20.1(a), (b), and (c1)/(c2). What implications do Figs. P10.20.1(a) and (b) have on the MSE and orthogonality? Do the MSEs (obtained from your simulation) conform with those obtained from Eq. (P10.20.7)?

$$\text{MSE}_1 \overset{\text{(P10.20.7)}}{=} \frac{\sigma_{\mathbf{x}}^2\sigma_{\mathbf{v}}^2}{a^2\sigma_{\mathbf{x}}^2+\sigma_{\mathbf{v}}^2} = \quad , \quad \text{MSE}_2 \overset{\text{(P10.20.7)}}{=} \frac{3^2 2^2}{(-1)^2 3^2+2^2} = 2.769 \quad \text{(P10.20.13a,b)}$$

Fill up Table P10.20 with the pertinent computation and simulation results. It implies that if the MSE (increases, decreases) as the SNR and the (magnitude of) correlation coefficient $\rho_{\mathbf{xy}}$ decreases so that **y** becomes (more, less) informative about **x**.

(f) Referring to Fig. P10.20.2, use the orthogonality principle (P10.20.9) to find the LMMSE estimator for the case of measurement vector:

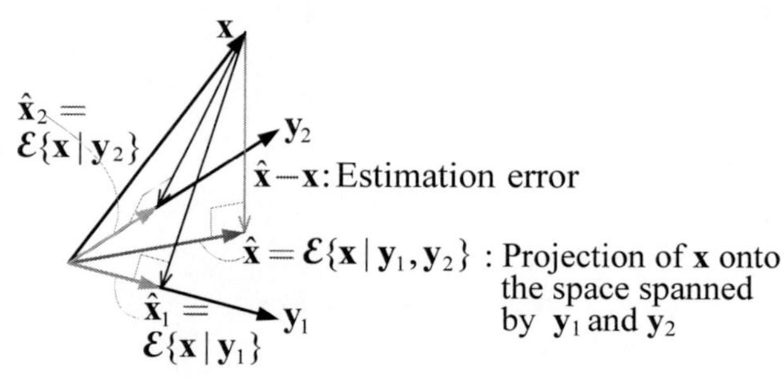

Figure P10.20.2 Orthogonality principle

$$\mathbf{y}=A\mathbf{x}+\mathbf{v}=\begin{bmatrix}2\\-1\end{bmatrix}\mathbf{x}+\begin{bmatrix}\mathbf{v}_1\\\mathbf{v}_2\end{bmatrix} \text{ with } \mathbf{x}:\mathcal{N}(m_{\mathbf{x}}=1,\ \sigma_{\mathbf{x}}^2=3^2),\ \mathbf{v}_1,\mathbf{v}_2:\mathcal{N}(0,\ 2^2) \quad \text{(P10.20.14)}$$

To this end, substitute $\hat{\mathbf{x}}\overset{\text{(P10.20.5)}}{=}G(\mathbf{y}-Am_{\mathbf{x}})+m_{\mathbf{x}}$ and Eq. (P10.20.14) to write the LHS of Eq. (P10.20.9) as

$$\begin{aligned}\mathcal{E}\{(\hat{\mathbf{x}}-\mathbf{x})\mathbf{y}^{\mathrm{T}}\} &\underset{\text{(P10.20.14)}}{\overset{\text{(P10.20.5)}}{=}} \mathcal{E}\{(G(\mathbf{y}-Am_{\mathbf{x}})+m_{\mathbf{x}}-\mathbf{x})(A\mathbf{x}+\mathbf{v})^{\mathrm{T}}\}\\ &\overset{\text{(P10.20.14)}}{=} \mathcal{E}\{(GA(\mathbf{x}-m_{\mathbf{x}})+G\mathbf{v}+m_{\mathbf{x}}-\mathbf{x})(A(\mathbf{x}-m_{\mathbf{x}})+Am_{\mathbf{x}}+\mathbf{v})^{\mathrm{T}}\}\\ &=G\mathcal{E}\{A(\mathbf{x}-m_{\mathbf{x}})(\mathbf{x}-m_{\mathbf{x}})^{\mathrm{T}}A^{\mathrm{T}}+\mathbf{v}\mathbf{v}^{\mathrm{T}}\}-\mathcal{E}\{(\mathbf{x}-m_{\mathbf{x}})(\mathbf{x}-m_{\mathbf{x}})^{\mathrm{T}}A^{\mathrm{T}}\}\\ &=G\,\mathrm{cov}(\mathbf{y})-\mathrm{cov}(\mathbf{x},\mathbf{y})=G\Sigma_{\mathbf{y}}-\Sigma_{\mathbf{xy}} \end{aligned} \quad \text{(P10.20.15)}$$

where

$$\mathrm{cov}(\mathbf{y})=\Sigma_{\mathbf{y}}\overset{(10.2.46)}{=}\mathcal{E}\{(\mathbf{y}-m_{\mathbf{y}})(\mathbf{y}-m_{\mathbf{y}})^{\mathrm{T}}\}\underset{m_{\mathbf{y}}=Am_{\mathbf{x}}}{\overset{\text{(P10.20.14)}}{=}}\mathcal{E}\{(A(\mathbf{x}-m_{\mathbf{x}})+\mathbf{v})(A(\mathbf{x}-m_{\mathbf{x}})+\mathbf{v})^{\mathrm{T}}\}$$

$$\overset{\text{uncorrelated}}{=}\mathcal{E}\{A(\mathbf{x}-m_{\mathbf{x}})(\mathbf{x}-m_{\mathbf{x}})^{\mathrm{T}}A^{\mathrm{T}}+\mathbf{v}\mathbf{v}^{\mathrm{T}}\}=A\Sigma_{\mathbf{x}}A^{\mathrm{T}}+\Sigma_{\mathbf{v}}=A\sigma_{\mathbf{x}}^2A^{\mathrm{T}}+\Sigma_{\mathbf{v}} \quad \text{(P10.20.16a)}$$

$$\mathrm{cov}(\mathbf{x},\mathbf{y})=\Sigma_{\mathbf{xy}}=\mathcal{E}\{(\mathbf{x}-m_{\mathbf{x}})(\mathbf{y}-m_{\mathbf{y}})^{\mathrm{T}}\}\underset{m_{\mathbf{y}}=Am_{\mathbf{x}}}{\overset{\text{(P10.20.14)}}{=}}\mathcal{E}\{(\mathbf{x}-m_{\mathbf{x}})(A(\mathbf{x}-m_{\mathbf{x}})+\mathbf{v})^{\mathrm{T}}\}\overset{\text{uncorrelated}}{=}\Sigma_{\mathbf{x}}A^{\mathrm{T}} \quad \text{(P10.20.16b)}$$

Note that, compared with Eq. (P10.20.9) or (E10.15.6), the transpose of **y** has been taken because it is not a scalar but a (column) vector (see Eq. (10.2.46) for the definition of the covariance matrix.) Then, according to the principle of orthogonality (Eq. (P10.20.9)), set Eq. (P10.20.15) equal to zero to determine the MMSE estimator as

$$G=\boxed{}\Sigma_{\mathbf{y}}^{-1}=\sigma_{\mathbf{x}}^2A^{\mathrm{T}}\left[A\sigma_{\mathbf{x}}^2A^{\mathrm{T}}+\begin{bmatrix}\sigma_{\mathbf{v}_1}^2&0\\0&\sigma_{\mathbf{v}_2}^2\end{bmatrix}\right]^{-1}$$

$$\overset{\text{(P10.20.12)}}{=}3^2[2\ \ -1]\left(\begin{bmatrix}4&-2\\-2&1\end{bmatrix}\cdot 3^2+\begin{bmatrix}2^2&0\\0&2^2\end{bmatrix}\right)^{-1}=[\overset{g_1}{18/49}\ \ \overset{g_2}{-9/49}] \quad \text{(P10.20.17)}$$

Why do you think g_1 is larger than g_2?

The MSE of the LMMSE estimator is obtained as

$$\begin{aligned}\mathrm{MSE}&=\mathcal{E}\{(\hat{\mathbf{x}}_{\mathrm{MMSE}}-\mathbf{x})^2\}\overset{\text{(P10.20.5)}}{=}\mathcal{E}\{[G(\mathbf{y}-Am_{\mathbf{x}})+m_{\mathbf{x}}-\mathbf{x}][G(\mathbf{y}-Am_{\mathbf{x}})+m_{\mathbf{x}}-\mathbf{x}]^{\mathrm{T}}\}\\ &=G\Sigma_{\mathbf{y}}G^{\mathrm{T}}-G\Sigma_{\mathbf{yx}}-\Sigma_{\mathbf{xy}}G^{\mathrm{T}}+\Sigma_{\mathbf{x}}\overset{\text{(P10.20.16)}}{=}\Sigma_{\mathbf{xy}}\Sigma_{\mathbf{y}}^{-1}\Sigma_{\mathbf{y}}\Sigma_{\mathbf{y}}^{-1}\Sigma_{\mathbf{yx}}-2\Sigma_{\mathbf{xy}}\Sigma_{\mathbf{y}}^{-1}\Sigma_{\mathbf{yx}}+\Sigma_{\mathbf{x}}\\ &=\Sigma_{\mathbf{x}}-\Sigma_{\mathbf{xy}}\Sigma_{\mathbf{y}}^{-1}\boxed{}\overset{\text{(P10.20.16b)}}{=}\Sigma_{\mathbf{x}}-\Sigma_{\mathbf{x}}A^{\mathrm{T}}\Sigma_{\mathbf{y}}^{-1}A^{\mathrm{T}}\Sigma_{\mathbf{x}}\\ &=\sigma_{\mathbf{x}}^2-\sigma_{\mathbf{x}}^2A^{\mathrm{T}}\left[\boxed{}+\begin{bmatrix}\sigma_{\mathbf{v}_1}^2&0\\0&\sigma_{\mathbf{v}_2}^2\end{bmatrix}\right]^{-1}A\sigma_{\mathbf{x}}^2\\ &=3^2-3^2[2\ \ -1]\left[\begin{bmatrix}2\\-1\end{bmatrix}3^2[2\ \ -1]+\begin{bmatrix}4&0\\0&4\end{bmatrix}\right]^{-1}\begin{bmatrix}2\\-1\end{bmatrix}3^2=\boxed{}\end{aligned} \quad \text{(P10.20.18)}$$

where

$$\Sigma_{\mathbf{xy}}=\mathrm{cov}(\mathbf{x},\mathbf{y})=\mathcal{E}\{[\mathbf{x}-m_{\mathbf{x}}][\mathbf{y}-m_{\mathbf{y}}]^{\mathrm{T}}\}=\Sigma_{\mathbf{yx}}^{\mathrm{T}} \quad \text{(P10.20.19)}$$

Why do you think that the MSE has become less than Eq. (P10.20.13a,b)? To simulate this estimator, complete and run the MATLAB script "em10p20f.m" and check if the MSE is close to the theoretical value obtained from Eq. (P10.20.18).

```
%em10p20f.m
N=1e4; % Number of samples
A=[2; -1]; mx=1; sgmx=3; mv=0; sgmv1=2; sgmv2=2;
Cov_x=sgmx^2;
Cov_y=A*sgmx^2*A' + diag([sgmv1^2 sgmv2^2]); % Eq.(P10.20.16a)
Cov_xy=Cov_x*A'; % Eq.(P10.20.16b)
G_MMSE=Cov_??/Cov_? % MMSE estimator gain by Eq.(P10.20.17)
for n=1:N
   x(n) = random('normal',mx,sgmx);
   v = [random('normal',mv,sgmv1); random('normal',mv,sgmv)];
   y = A*x(n) + v;   % Eq.(P10.20.14)
   xh(n)=G_MMSE*(?-A*??)+??; % MMSE estimate of x: %Eq.(P10.20.5)
end
MSE = [mean((x?-x).^?) Cov_?-Cov_xy/Cov_?*Cov_xy'] %Eq.(P10.20.18)
```

(g) To get the feel of the physical meaning of the covariance or correlation coefficient through Figs. P10.20.1(c1)/(c2), consider the covariance between $\mathbf{x}$ and $\mathbf{y}_1$ or $\mathbf{y}_2$:

$$\text{cov}(\mathbf{x},\mathbf{y}_1) \overset{(10.2.40)}{=} \mathcal{E}\{(\mathbf{x}-m_\mathbf{x})(\mathbf{y}_1-m_\mathbf{y})\} \overset{(P10.20.12a)}{=} \mathcal{E}\{2(\mathbf{x}-m_\mathbf{x})^2+(\mathbf{x}-m_\mathbf{x})\mathbf{v}\} = 2\sigma_\mathbf{x}^2 \quad (P10.20.20a)$$

$$\text{cov}(\mathbf{x},\mathbf{y}_2) \overset{(10.2.40)}{=} \mathcal{E}\{(\mathbf{x}-m_\mathbf{x})(\mathbf{y}_2-m_\mathbf{y})\} \overset{(P10.20.12b)}{=} \mathcal{E}\{-(\mathbf{x}-m_\mathbf{x})^2+(\mathbf{x}-m_\mathbf{x})\mathbf{v}\} = -\sigma_\mathbf{x}^2 \quad (P10.20.20b)$$

where the correlation coefficient is defined by Eq. (10.2.41):

$$\rho_{\mathbf{xy}} \overset{(10.2.41)}{=} \frac{\text{cov}(\mathbf{x},\mathbf{y})}{\sqrt{\text{var}(\mathbf{x})\,\text{var}(\mathbf{y})}} = \frac{a\sigma_\mathbf{x}^2}{\sigma_\mathbf{x}\sqrt{a^2\sigma_\mathbf{x}^2+\sigma_\mathbf{v}^2}} = \frac{\text{sign}(a)}{\sqrt{1+\sigma_\mathbf{v}^2/a^2\sigma_\mathbf{x}^2}} \quad (P10.20.21)$$

Noting that Figs. P10.20.1(c1)/(c2) depict how $\mathbf{x}$ and $\mathbf{y}_1$ or $\mathbf{y}_2$ affect each other's variation around their mean values, we may realize that a positive/negative covariance between two random variables implies that they vary in the (same/opposite, opposite/same) direction, i.e., if one increases from its mean, the other increases/decreases from its mean. The larger/smaller the magnitude of correlation coefficnent is, the (more/less, less/more) dependent they are on each other.

10.21 Fisher Information and Cramer–Rao (Lower) Bound

Suppose the parameter λ of a Poisson random variable $\mathbf{x}$ (with pmf $p(x)=\lambda^x e^{-\lambda}/x!$ for any nonnegative integer x) for which the observed data is given as $\{x_1, x_2, \ldots, x_N\}$ has the likelihood function as

$$p_{\mathbf{x}_1,\mathbf{x}_2,\cdots,\mathbf{x}_N|\lambda}(x_1,x_2,\cdots,x_N\,|\,\lambda) = \prod_{i=1}^{N}\frac{\lambda^{x_i}}{x_i\,!}e^{-\lambda} \quad (P10.21.1)$$

(a) Show that the log-likelihood function of the parameter λ is

$$L(\lambda|x_1,x_2,\cdots,x_N) \overset{(10.3.4)}{=} \ln p_{\mathbf{x}_1,\mathbf{x}_2,\cdots,\mathbf{x}_N|\lambda}(x_1,x_2,\cdots,x_N\,|\,\lambda) \overset{(P10.21.1)}{=} \sum_{i=1}^{N}x_i\ln\lambda - \sum_{i=1}^{N}\ln x_i\,! - N\lambda \quad (P10.21.2)$$

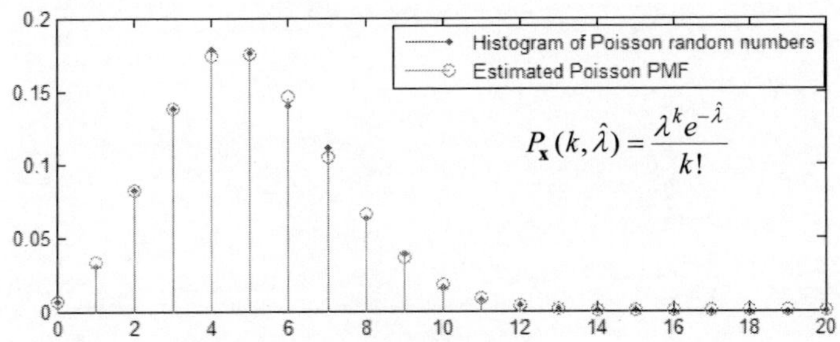

Figure P10.21

(b) Show that the ML (maximum likelihood) estimator of the parameter λ is equal to the sample mean:

$$\hat{\lambda}_{\text{ML}} = \frac{1}{N}\sum_{i=1}^{N} x_i = \bar{x}\,(\text{sample mean}) \tag{P10.21.3}$$

(c) Find the Fisher information:

$$I(\lambda) \overset{(10.3.14)}{=} -\mathcal{E}\left\{\frac{\partial^2}{\partial \lambda^2} L(\lambda|x_1,x_2,\cdots,x_N)\right\} \overset{(\text{P10.21.2})}{=} \sum_{i=1}^{N} \mathcal{E}\{x_i \boxed{} = \boxed{} \tag{P10.21.4}$$

(d) Write the Cramer–Rao inequality for the ML estimator λ_{ML} of λ:

$$\text{var}\{\hat{\lambda}_{\text{ML}} = \bar{x}\} \overset{(10.2.31)}{=} \mathcal{E}\left\{\left(\frac{1}{N}\sum_{i=1}^{N} x_i - \mathcal{E}\left\{\frac{1}{N}\sum_{i=1}^{N} x_i\right\}\right)^2\right\} = \mathcal{E}\left\{\left(\frac{1}{N}\sum_{i=1}^{N}(x_i - m_{\mathbf{x}})\right)^2\right\}$$

$$\underset{\text{among } x_i\text{'s}}{\overset{\text{uncorrelated}}{=}} \frac{1}{N}\sigma_{\mathbf{x}}^2 \overset{(\text{E10.12d.3})}{=} \frac{1}{N}\boxed{} \geq \frac{1}{I(\lambda) = \boxed{}} \tag{P10.21.5}$$

In this case, does the equality hold? What is implied by that?

(e) Perform a soft experiment of estimating the parameter λ of Poisson distribution by taking the following steps:

1. Use '`poissrnd()`' to generate N=10^4 Poisson random numbers and use '`hist()`'/'`stem()`' to get/plot the histogram of the numbers.
2. Use Eq. (P10.21.3) or '`mle(x,'distribution','Poisson')`' to find the ML estimate of λ and use '`poisspdf(x,lambda)`'/'`stem()`' to get/plot the Poisson pmf with the estimated λ.

```
%em10p21e.m
N=1e4; lambda=5;
% To generate N Poisson random numbers
x=poissrnd(lambda,1,N);
dx=1; bin=[0:dx:50]; % Bin locations
ns=hist(x,bin); % Frequency counts
stem(bin,ns/N/dx,'.'), hold on, axis([0 20 0 0.2])
% To find the MLE of lambda of Poisson pdf for x
lambdah=mle(x,'distribution','Poisson')
mean(x) % Eq.(P10.21.3): Sample mean of the Poisson random numbers
% To plot the Poisson pdf with the estimate of lambda for [0:20]
xs=[0:20]; fxh=poisspdf(xs,???????); stem(xs,fxh,'ro')
legend('Histogram of Poisson random nums','Estimated Poisson pmf')
```

To this end, complete and run the above MATLAB script, "em10p21e.m," to get Fig. P10.21.

Appendix A: Tables of Various Transforms

Table A.1 Definition and properties of the CTFS (continuous-time Fourier series)

(0) Definition	Synthesis: $\tilde{x}_P(t) \overset{(6.1.6a)}{=} \frac{1}{P}\sum_{k=-\infty}^{\infty} X_k e^{jk\omega_0 t}$, P : period of $\tilde{x}_P(t)$ Analysis: $X_k \overset{(6.1.6b)}{=} \int_P \tilde{x}_P(t) e^{-jk\omega_0 t}\, dt$, $\omega_0 = \frac{2\pi}{P}$
(1) Linearity	$a\tilde{x}_P(t) + b\tilde{y}_P(t) \overset{\mathcal{F}}{\leftrightarrow} aX_k + bY_k$
(2) Time reversal	$\tilde{x}_P(-t) \overset{\mathcal{F}}{\leftrightarrow} X_{-k}$
(3) Symmetry for real-valued functions	Real-valued $\tilde{x}_P(t) = \tilde{x}_{P,\mathrm{e}}(t) + \tilde{x}_{P,\mathrm{o}}(t) \overset{\mathcal{F}}{\leftrightarrow} X_k = X_{-k}^*$ Real-valued and even $\tilde{x}_{P,\mathrm{e}}(t) \overset{\mathcal{F}}{\leftrightarrow} X_{k,\mathrm{e}} = \mathrm{Re}\{X_k\}$ Real-valued and odd $\tilde{x}_{P,\mathrm{o}}(t) \overset{\mathcal{F}}{\leftrightarrow} X_{k,\mathrm{o}} = j\,\mathrm{Im}\{X_k\}$
(4) Conjugation	$\tilde{x}_P^*(t) \overset{\mathrm{CTFS}}{\leftrightarrow} X_{-k}^*$
(5) Time shifting (Real translation)	$\tilde{x}_P(t-t_1) \leftrightarrow X_k e^{-jk\omega_0 t_1} = \|X_k\| \angle(\phi_k - k\omega_0 t_1)$
(6) Frequency shifting (Complex translation)	$\tilde{x}_P(t) e^{jM\omega_0 t} \overset{\mathrm{CTFS}}{\leftrightarrow} X_{k-M}$
(7) Real convolution (periodic/circular)	$\tilde{x}_P(t) * \tilde{y}_P(t) = \int_P \tilde{x}(\tau)\tilde{y}(t-\tau)\, d\tau \overset{\mathrm{CTFS}}{\leftrightarrow} X_k Y_k$
(8) Complex convolution	$\tilde{x}_P(t)\tilde{y}_P(t) \overset{\mathrm{CTFS}}{\leftrightarrow} \frac{1}{P}\sum_{m=-\infty}^{\infty} X_m Y_{k-m}$
(9) Time differentiation	$\frac{dx(t)}{dt} \overset{\mathrm{CTFS}}{\leftrightarrow} jk\omega_0 X_k$
(10) Time integration	$\int_{-\infty}^{t} x(\tau)\, d\tau \begin{pmatrix}\text{finite-valued and periodic}\\ \text{only if } X_0 = 0\end{pmatrix} \overset{\mathrm{CTFS}}{\leftrightarrow} \frac{1}{jk\omega_0} X_k$
(11) Scaling	$x(at)$, $a>0$ (periodic with period P/a) $\overset{\mathrm{CTFS}}{\leftrightarrow}$ X_k (with fundamental frequency $a\omega_0$)
(12) Parseval's relation	$\int_{-\infty}^{\infty} \|\tilde{x}_P(t)\|^2\, dt = \frac{1}{P}\sum_{k=-\infty}^{\infty} \|X_k\|^2$

Table A.2 Definition and properties of the CTFT (continuous-time Fourier transform)

(0) Definition	$X(j\omega) = \mathcal{F}\{x(t)\} \overset{(6.2.1a)}{=} \int_{-\infty}^{\infty} x(t)e^{-j\omega t}\,dt$
(1) Linearity	$a\,x(t) + b\,y(t) \overset{\mathcal{F}}{\leftrightarrow} a\,X(\omega) + b\,Y(\omega)$
(2) Time reversal	$x(-t) \underset{(6.2.9)}{\overset{\mathcal{F}}{\leftrightarrow}} X(-\omega)$
(3) Symmetry for real-valued functions	Real-valued $x(t) = x_e(t) + x_o(t) \overset{\mathcal{F}}{\leftrightarrow} X(\omega) = X^*(-\omega)$ Real-valued and even $x_e(t) \underset{(6.2.12a)}{\overset{\mathcal{F}}{\leftrightarrow}} X_e(\omega) = \mathrm{Re}\{X(\omega)\}$ Real-valued and odd $x_o(t) \underset{(6.2.12b)}{\overset{\mathcal{F}}{\leftrightarrow}} X_o(\omega) = j\,\mathrm{Im}\{X(\omega)\}$
(4) Conjugation	$x^*(t) \underset{(6.2.13)}{\overset{\mathcal{F}}{\leftrightarrow}} X^*(-\omega)$
(5) Time shifting (Real translation)	$x(t - t_1) \underset{(6.2.14)}{\overset{\mathcal{F}}{\leftrightarrow}} X(\omega)\,e^{-j\omega t_1} = X(\omega)\angle -t_1\omega$
(6) Frequency shifting (Complex translation)	$x(t)e^{j\omega_1 t} \underset{(6.2.15)}{\overset{\mathcal{F}}{\leftrightarrow}} X(\omega - \omega_1)$
(7) Duality	$g(t) \overset{\mathcal{F}}{\leftrightarrow} f(\omega) \underset{(6.2.17)}{\Leftrightarrow} f(t) \overset{\mathcal{F}}{\leftrightarrow} 2\pi g(-\omega)$
(8) Real convolution	$y(t) = x(t) * g(t) \underset{(6.2.19)}{\overset{\mathcal{F}}{\leftrightarrow}} Y(\omega) = X(\omega)G(\omega)$
(9) Complex convolution	$y(t) = x(t)m(t) \underset{(6.2.22)}{\overset{\mathcal{F}}{\leftrightarrow}} Y(\omega) = \frac{1}{2\pi} X(\omega) * M(\omega)$
(10) Time differentiation	$\frac{dx(t)}{dt} \underset{(6.2.25)}{\overset{\mathcal{F}}{\leftrightarrow}} j\omega X(\omega)$
(11) Time integration	$\int_{-\infty}^{t} x(\tau)\,d\tau \underset{(6.2.26)}{\overset{\mathcal{F}}{\leftrightarrow}} \pi X(0)\delta(\omega) + \frac{1}{j\omega} X(\omega)$
(12) Scaling	$x(at) \underset{(6.2.29)}{\overset{\mathcal{F}}{\leftrightarrow}} \frac{1}{\lvert a\rvert} X\left(\frac{\omega}{a}\right)$
(13) Time multiplication Frequency differentiation	$t\,x(t) \underset{(6.2.28)}{\overset{\mathcal{F}}{\leftrightarrow}} j\frac{dX(\omega)}{d\omega}$
(14) Parseval's relation	$\int_{-\infty}^{\infty} \lvert x(t)\rvert^2 dt \underset{(6.2.30)}{=} \frac{1}{2\pi} \int_{-\infty}^{\infty} \lvert X(\omega)\rvert^2 d\omega$

Table A.3 CTFS/CTFT for basic functions

function $x(t)$	CTFS X_k (for periodic function)	CTFT $X(\omega)$ (for periodic/aperiodic function)
(1) $\frac{1}{P}\sum_{k=-\infty}^{\infty} X_k e^{jk\omega_0 t},\ \omega_0=\frac{2\pi}{P}$	X_k	$\overset{(6.2.7)}{\rightarrow} \frac{2\pi}{P}\sum_{k=-\infty}^{\infty} X_k\delta(\omega-k\omega_0)$
(2) $e^{j\omega_0 t},\ \omega_0=\frac{2\pi}{P}$	$X_k=\begin{cases}P & \text{for } k=1\\ 0 & \text{elsewhere}\end{cases}$	$2\pi\delta(\omega-\omega_0)$
(3) $\cos\omega_0 t,\ \omega_0=\frac{2\pi}{P}$	$X_k=\begin{cases}P/2 & \text{for } k=\pm 1,\cdots\\ 0 & \text{elsewhere}\end{cases}$	$\overset{(E6.12.2)}{\rightarrow} \pi(\delta(\omega+\omega_0)+\delta(\omega-\omega_0))$
(4) $\sin\omega_0 t,\ \omega_0=\frac{2\pi}{P}$	$X_k=\begin{cases}-jP/2 & \text{for } k=1\\ jP/2 & \text{for } k=-1\\ 0 & \text{elsewhere}\end{cases}$	$\overset{(E6.12.1)}{\rightarrow} j\pi(\delta(\omega+\omega_0)-\delta(\omega-\omega_0))$
(5) Constant function $x(t)=1$	$X_k=\begin{cases}P & \text{for } k=0\\ 0 & \text{elsewhere}\end{cases}$	$\overset{(E6.8.1)}{\rightarrow} 2\pi\delta(\omega)$
(6) Impulse train $\sum_{i=-\infty}^{\infty}\delta(t-iP)$	$X_k=1\ \forall\ k$	$\frac{2\pi}{P}\sum_{k=-\infty}^{\infty}\delta(\omega-k\omega_0)\ ,\ \omega_0=\frac{2\pi}{P}$
(7) Rectangular wave $\sum_{i=-\infty}^{\infty} r_D(t-iP), D\le P$	$X_k=D\,\mathrm{sinc}\left(\frac{k\omega_0 D}{2\pi}\right),\ \omega_0=\frac{2\pi}{P}$	$\frac{2\pi}{P}\sum_{k=-\infty}^{\infty} X_k\delta(\omega-k\omega_0)$
(8) Rectangular pulse $r_D(t)=u_s(t+\frac{D}{2})-u_s(t-\frac{D}{2})$	N/A for an aperiodic function	$D\,\mathrm{sinc}(\frac{D}{2\pi}\omega)\overset{(E6.4.2)}{=}\frac{\sin(D\omega/2)}{\omega/2}$
(9) $\frac{\sin(Bt)}{\pi t}=\frac{B}{\pi}\mathrm{sinc}\left(\frac{Bt}{\pi}\right)$	N/A for an aperiodic function	$\overset{(E6.10.1)}{\rightarrow}\begin{cases}1 & \text{for } \lvert\omega\rvert\le B\\ 0 & \text{elsewhere}\end{cases}$
(10) $\delta(t)$	N/A for an aperiodic function	1
(11) $u_s(t)$	N/A for an aperiodic function	$\overset{(E6.9.6)}{\rightarrow}\frac{1}{j\omega}+\pi\delta(\omega)$
(12) $e^{-at}u_s(t)$, $\mathrm{Re}\{a\}>0$	N/A for an aperiodic function	$\frac{1}{a+j\omega}$
(13) $te^{-at}u_s(t)$, $\mathrm{Re}\{a\}>0$	N/A for an aperiodic function	$\frac{1}{(a+j\omega)^2}$
(14) $\frac{t^{M-1}}{(M-1)!}e^{-at}u_s(t)$, $\mathrm{Re}\{a\}>0$	N/A for an aperiodic function	$\frac{1}{(a+j\omega)^M}$
(15) $e^{-a\lvert t\rvert}$, $\mathrm{Re}\{a\}>0$	N/A for an aperiodic function	$\frac{2a}{a^2+\omega^2}$
(16) $\frac{a}{a^2+t^2}$, $\mathrm{Re}\{a\}>0$	N/A for an aperiodic function	$\pi e^{-a\lvert\omega\rvert}$
(17) $x(t)=\frac{1}{\sqrt{2\pi}\sigma}e^{-t^2/2\sigma^2}$	N/A for an aperiodic function	$e^{-\omega^2\sigma^2/2}$

Table A.4 Definition and properties of the DTFT (discrete-time Fourier transform)

(0) Definition	DTFT: $X(\Omega)=X[e^{j\Omega}]=\mathcal{F}\{x[n]\}\overset{(6.3.1)}{=}\sum_{n=-\infty}^{\infty}x[n]e^{-j\Omega n}$ IDTFT: $x[n]=\mathcal{F}^{-1}\{X(\Omega)\}\overset{(6.3.3)}{=}\frac{1}{2\pi}\int_{2\pi}X(\Omega)e^{j\Omega n}d\Omega$
(1) Periodicity	$X(\Omega)=X[e^{j\Omega}]=X(\Omega+2m\pi)$ for any integer m
(2) Linearity	$ax[n]+by[n]\overset{\mathcal{F}}{\leftrightarrow}aX(\Omega)+bY(\Omega)$
(3) Time reversal	$x[-n]\underset{(6.3.8)}{\overset{\mathcal{F}}{\leftrightarrow}}X(-\Omega)$
(4) Symmetry for real-valued sequences	Real-valued $x[n]=x_e[n]+x_o[n]\overset{\mathcal{F}}{\leftrightarrow}X(\Omega)=X^*(-\Omega)$ Real-valued and even $x_e[n]\underset{(6.3.10a)}{\overset{\mathcal{F}}{\leftrightarrow}}X_e(\Omega)=\mathrm{Re}\{X(\Omega)\}$ Real-valued and odd $x_o[n]\underset{(6.3.10b)}{\overset{\mathcal{F}}{\leftrightarrow}}X_o(\Omega)=j\,\mathrm{Im}\{X(\Omega)\}$
(5) Conjugation	$x^*[n]\underset{(6.3.11)}{\overset{\mathcal{F}}{\leftrightarrow}}X^*(-\Omega)$
(6) Time shifting (Real translation)	$x[n-n_1]\underset{(6.3.12)}{\overset{\mathcal{F}}{\leftrightarrow}}X(\Omega)e^{-j\Omega n_1}=X(\Omega)\angle-n_1\Omega$
(7) Frequency shifting (Complex translation)	$x[n]e^{j\Omega_1 n}\underset{(6.3.13)}{\overset{\mathcal{F}}{\leftrightarrow}}X(\Omega-\Omega_1)$
(8) Real convolution	$x[n]*g[n]\underset{(6.3.14)}{\overset{\mathcal{F}}{\leftrightarrow}}X(\Omega)G(\Omega)$
(9) Complex convolution (circular)	$x[n]\,m[n]\underset{(6.3.15)}{\overset{\mathcal{F}}{\leftrightarrow}}\frac{1}{2\pi}X(\Omega)\underset{2\pi}{*}M(\Omega)$ where $\underset{2\pi}{*}$ denotes the circular convolution with period 2π
(10) Differencing in time	$x[n]-x[n-1]\underset{(6.3.16)}{\overset{\mathcal{F}}{\leftrightarrow}}(1-e^{-j\Omega})X(\Omega)$
(11) Summation in time	$\sum_{m=-\infty}^{n}x[m]=x[n]*u_s[n]\underset{(6.3.17)}{\overset{\mathcal{F}}{\leftrightarrow}}\frac{1}{1-e^{-j\Omega}}X(\Omega)+\pi X(0)\sum_{i=-\infty}^{\infty}\delta(\Omega-2\pi i)$
(12) Scaling	$x_{(K)}[n]=\begin{cases}x[n/K]=x[r] & \text{for } n=rK\\ 0 & \text{elsewhere}\end{cases}\underset{(6.3.20)}{\overset{\mathcal{F}}{\leftrightarrow}}X(K\Omega)$
(13) Time multiplication Frequency differentiation	$n\,x[n]\underset{(6.3.18)}{\overset{\mathcal{F}}{\leftrightarrow}}j\frac{dX(\Omega)}{d\Omega}$
(14) Parseval's relation	$\sum_{n=-\infty}^{\infty}\lvert x[n]\rvert^2\overset{(6.3.21)}{=}\frac{1}{2\pi}\int_{2\pi}\lvert X(\Omega)\rvert^2d\Omega$

Table A.5 Definition and properties of the DFT/DFS (discrete-time Fourier series)

(0) Definition	$X(k) = \text{DFT}_N\{x[n]\} \overset{(6.4.2)}{=} \sum_{n=0}^{N-1} x[n]\, e^{-j2\pi kn/N} = \sum_{n=0}^{N-1} x[n]\, W_N^{kn}$ $x[n] = \text{IDFT}_N\{X(k)\} \overset{(6.4.3)}{=} \frac{1}{N}\sum_{k=0}^{N-1} X(k)\, e^{j2\pi kn/N} = \frac{1}{N}\sum_{k=0}^{N-1} X(k)\, W_N^{-kn}$
(1) Periodicity	$X(k) = X(k+mN)$ for any integer m
(2) Linearity	$a\,x[n] + b\,y[n] \overset{\text{DFT}}{\leftrightarrow} a\,X(k) + b\,Y(k)$
(3) Time reversal	$\tilde{x}[-n] r_N[n] \overset{\text{DFT}}{\leftrightarrow} \tilde{X}_N(-k)\, r_N[k]$ $(\, r_N[k] = u_s[k] - u_s[k-N])$ where $\tilde{x}_N[n]$ is the periodic repetition of $x[n]$ with period N
(4) Symmetry for real-valued sequences	Real-valued $x[n] = x_e[n] + x_o[n] \overset{\text{DFT}}{\leftrightarrow} X(k) = \tilde{X}_N^*(-k)\, r_N[k]$ Real-valued and even $x_e[n] \overset{\text{DFT}}{\leftrightarrow} X_e(k) = \text{Re}\{X(k)\}$ Real-valued and odd $x_o[n] \overset{\text{DFT}}{\leftrightarrow} X_o(k) = j\,\text{Im}\{X(k)\}$
(5) Conjugation	$x^*[n] \overset{\text{DFT}}{\leftrightarrow} \tilde{X}_N^*(-k)\, r_N[k]$
(6) Duality	$g[n] \overset{\text{DFT}}{\leftrightarrow} f(k) \Leftrightarrow f[n] \overset{\text{DFT}}{\leftrightarrow} N\,\tilde{g}_N(-k) r_N[k]$
(7) Time shifting (circular)	$\tilde{x}_N[n-M] r_N[n]$(one period of circular shift) $\overset{\text{DFT}}{\leftrightarrow} W_N^{Mk} X(k)$
(8) Frequency shifting (circular) (Complex translation)	$W_N^{-Ln}\, x[n] \overset{\text{DFT}}{\leftrightarrow} \tilde{X}_N(k-L)\, r_N[k]$ (one period of circular shift)
(9) Real convolution	$\left(\sum_{i=0}^{N-1} \tilde{x}(i)\, \tilde{y}(n-i)\right) r_N[n] \overset{\text{DFT}}{\leftrightarrow} X(k) Y(k)$
(10) Complex (circular) convolution	$x[n]\, y[n] \overset{\text{DFT}}{\leftrightarrow} \frac{1}{N}\left(\sum_{i=0}^{N-1} \tilde{X}_N(i)\, \tilde{Y}_N(k-i)\right) r_N[k]$
(11) Scaling	$x_{(K)}[n] = \begin{cases} x[n/K] & \text{for } n = rK \\ 0 & \text{elsewhere} \end{cases} \overset{\text{DFT}}{\leftrightarrow} \tilde{X}_N(Kk)\, r_N[k]$
(12) Parseval's relation	$\sum_{n=<N>} \lvert x[n]\rvert^2 = \frac{1}{N} \sum_{k=<N>} \lvert X(k)\rvert^2$ where $<N> = \{0, 1, \cdots, N-1\}$

Table A.6 DFS/DFT and DTFT for basic sequences

Sequence $x[n]$	DFS $\tilde{X}(k)$ /DFT $X(k)$ (for periodic/aperiodic and finite-duration sequences)	DTFT $X(\Omega)$
(1) $\frac{1}{P}\sum_{k=<N>} X_k e^{j2\pi kn/P}, P=NT$	$\tilde{X}(k)=\frac{1}{T}X_k$	$\frac{2\pi}{P}\sum_{k=-\infty}^{\infty} X_k\delta(\Omega-k\frac{2\pi}{N})$
(2) $e^{jK\Omega_0 n},\ \Omega_0=\frac{2\pi}{N}$	$\tilde{X}(k)=\begin{cases}N & \text{for } k=K, K\pm N, K\pm 2N,\cdots\\ 0 & \text{elsewhere}\end{cases}$	$\overset{(6.3.5)}{\rightarrow} 2\pi\sum_{k=-\infty}^{\infty}\delta(\Omega-K\Omega_0-2\pi k)$
(3) $\cos K\Omega_0 n,\ \Omega_0=\frac{2\pi}{N}$	$\tilde{X}(k)=\begin{cases}N/2 & \text{for } k=\pm K, \pm K\pm N,\cdots\\ 0 & \text{elsewhere}\end{cases}$	$\overset{(E6.20.2)}{\rightarrow} \pi\sum_{k=-\infty}^{\infty}\begin{matrix}\delta(\Omega+K\Omega_0-2\pi k)\\ +\delta(\Omega-K\Omega_0-2\pi k)\end{matrix}$
(4) $\sin K\Omega_0 n,\ \Omega_0=\frac{2\pi}{N}$	$\tilde{X}(k)=\begin{cases}-jN/2 & \text{for } k=K, K\pm N,\cdots\\ jN/2 & \text{for } k=-K, -K\pm N,\cdots\\ 0 & \text{elsewhere}\end{cases}$	$\overset{(E6.20.1)}{\rightarrow} j\pi\sum_{k=-\infty}^{\infty}\begin{matrix}\delta(\Omega+K\Omega_0-2\pi k)\\ -\delta(\Omega-K\Omega_0-2\pi k)\end{matrix}$
(5) Constant sequence $x[n]=1$	$\tilde{X}(k)=\begin{cases}N & \text{for } k=0, \pm N, \pm 2N,\cdots\\ 0 & \text{elsewhere}\end{cases}$	$\overset{(E6.19.1)}{\rightarrow} 2\pi\sum_{k=-\infty}^{\infty}\delta(\Omega-2\pi k)$
(6) $\sum_{i=-\infty}^{\infty}\delta[n-iN]$	$\tilde{X}(k)=1\ \forall\ k$	$\frac{2\pi}{N}\sum_{k=-\infty}^{\infty}\delta(\Omega-k\Omega_0),\ \Omega_0=\frac{2\pi}{N}$
(7) Rectangular wave $\sum_{i=-\infty}^{\infty} r_{2M+1}[n-iM],\ 2M+1\le N$	$\tilde{X}(k)=\frac{\sin(k\Omega_0(2M+1)/2)}{\sin(k\Omega_0/2)},\ \Omega_0=\frac{2\pi}{N}$	$\frac{2\pi}{N}\sum_{k=-\infty}^{\infty}\tilde{X}(k)\delta(\Omega-k\Omega_0)$
(8) Rectangular pulse $r_{2M+1}[n]$ $=u_s[n+M]-u_s[n-M-1]$	$X(k)=\frac{\sin(k\Omega_0(2M+1)/2)}{\sin(k\Omega_0/2)},\ \Omega_0=\frac{2\pi}{N}$	$\overset{(E6.14.2)}{\rightarrow} \frac{\sin(\Omega(2M+1)/2)}{\sin(\Omega/2)}$
(9) $\frac{\sin(Bn)}{\pi n}=\frac{B}{\pi}\text{sinc}\left(\frac{Bn}{\pi}\right)$	N/A for an infinite-duration aperiodic sequence	$\overset{(E6.18.1)}{\rightarrow}\begin{cases}1 & \text{for } \lvert\Omega-2m\pi\rvert\le B\le\pi\\ 0 & \text{elsewhere}\end{cases}$
(10) $\delta[n]$	$X(k)=1\ \forall\ 0\le k\le N-1$	$\overset{(E6.17.1)}{\rightarrow} 1$
(11) $u_s[n]$	N/A for an infinite-duration aperiodic sequence	$\overset{(E6.21.6)}{\rightarrow} \frac{1}{1-e^{-j\Omega n}}+\pi\sum_{i=-\infty}^{\infty}\delta(\Omega-2\pi i)$
(12) $a^n u_s[n]$ with $\lvert a\rvert<1$	N/A for an infinite-duration aperiodic sequence	$\overset{(E6.15.1)}{\rightarrow} \frac{1}{1-ae^{-j\Omega}}$
(13) $(n+1)a^n u_s[n]$ with $\lvert a\rvert<1$	N/A for an infinite-duration aperiodic sequence	$\frac{1}{(1-ae^{-j\Omega})^2}$
(14) $a^{\lvert n\rvert}$ with $\lvert a\rvert<1$	N/A for an infinite-duration aperiodic sequence	$\overset{(E6.16.1)}{\rightarrow} \frac{1-a^2}{1-2a\cos\Omega+a^2}$

Table A.7 Definitions and properties of the Laplace transform and z-transform

Laplace transform		z-transform	
(0a) $\mathbb{X}(s)=L\{x(t)\}=\int_{-\infty}^{\infty}x(t)e^{-st}dt$		Bilateral (two-sided)	$\mathbb{X}[z]=Z\{x[n]\}=\sum_{n=-\infty}^{\infty}x[n]z^{-n}$
(0b) $X(s)=\mathcal{L}\{x(t)\}=\int_{0}^{\infty}x(t)e^{-st}dt$		Unilateral (one-sided)	$X[z]=\mathcal{Z}\{x[n]\}=\sum_{n=0}^{\infty}x[n]z^{-n}$
(1) Linearity			
$aX(s)+bY(s)$	$ax(t)+by(t)$	$ax[n]+by[n]$	$aX[z]+bY[z]$
(2) Time shifting (Real translation)			
$e^{-st_1}\mathbb{X}(s)$	$x(t-t_1)$	$x[n-n_1]$	$z^{-n_1}\mathbb{X}[z]$
$e^{-st_1}\left\{X(s)+\int_{-t_1}^{0}x(\tau)e^{-s\tau}d\tau\right\}$	$x(t-t_1),\ t_1>0$	$x[n-n_1],\ n_1>0$	$z^{-n_1}\left\{X[z]+\sum_{m=-n_1}^{-1}x[m]z^{-m}\right\}$
$e^{st_1}\left\{X(s)-\int_{0}^{t_1}x(\tau)e^{-s\tau}d\tau\right\}$	$x(t+t_1),\ t_1>0$	$x[n+n_1],\ n_1>0$	$z^{n_1}\left\{X[z]-\sum_{m=0}^{n_1-1}x[m]z^{-m}\right\}$
(3) Frequency shifting (Complex translation)			
$X(s-s_1)$	$e^{s_1t}x(t)$	$z_1^n\,x[n]$	$X[z/z_1]$
(4) Real convolution			
$G(s)X(s)$	$g(t)*x(t)$	$g[n]*x[n]$	$G[z]X[z]$
(5) Time derivative (Differentiation property)		Differencing	
<Bilateral> $s\mathbb{X}(s)$ <Unilateral> $sX(s)-x(0)$	$\frac{d}{dt}x(t)$	$x[n]-x[n-1]$	<Bilateral> $(1-z^{-1})\mathbb{X}[z]$ <Unilateral> $(1-z^{-1})X[z]-x[-1]$
(6) Time integral (Integration property)		Summation	
<Bilateral> $\frac{1}{s}\mathbb{X}(s)$ <Unilateral> $\frac{1}{s}\left(X(s)+\int_{-\infty}^{t}x(\tau)\,d\tau\right)$	$\int_{-\infty}^{t}x(\tau)d\tau$	$\sum_{m=-\infty}^{n}x[m]$	<Bilateral> $\frac{1}{1-z^{-1}}\mathbb{X}[z]$ <Unilateral> $\frac{1}{1-z^{-1}}(X[z]+\sum_{m=-\infty}^{-1}x[m])$
(7) Complex derivative			
$\frac{d}{ds}X(s)$	$-t\,x(t)$	$-n\,x[n]$	$z\frac{d}{dz}X[z]$
(8) Complex convolution			
$\frac{1}{2\pi j}\int_{\sigma_0-\infty}^{\sigma_0+\infty}X(v)Y(s-v)\,dv$	$x(t)y(t)$	$x[n]y[n]$	$\frac{1}{2\pi j}\oint_c X[z/v]Y[v]\,dv$
(9) Time scaling			
$\lvert a\rvert X(as)$	$x(t/a)$	$x[n/M]$ (M: a positive integer)	$X[z^M]$
(10) Time reversal			
$\mathbb{X}(-s)$	$x(-t)$	$x[-n]$	$\mathbb{X}[z^{-1}]$
(11) Initial value theorem			
$x(0)=\lim_{s\to\infty}sX(s)$		$x[0]=\lim_{z\to\infty}X[z]$	
(12) Final value theorem			
$x(\infty)=\lim_{s\to 0}sX(s)$		$x[\infty]=\lim_{z\to 1}(1-z^{-1})X[z]$	

Table A.8 Laplace transforms and z-transforms of basic functions/sequences

$X(s)=\mathcal{L}\{x(t)\}$	$x(t)$		$x[n]$	$X[z]=\mathcal{Z}\{x[n]\}$
1	$\delta(t)$	(1)	$\frac{1}{T}\delta[n]$	$\frac{1}{T}$
$e^{-t_1 s}$	$\delta(t-t_1)$	(2)	$\frac{1}{T}\delta[n-n_1]$	$\frac{1}{T}z^{-n_1}$
$\frac{1}{s}$	$u_s(t)$	(3)	$u_s[n]$	$\frac{z}{z-1}=\frac{1}{1-z^{-1}}$
$\frac{1}{s^2}$	$t\,u_s(t)$	(4)	$nT\,u_s[n]$	$\frac{Tz}{(z-1)^2}$
$\frac{M!}{s^{M+1}}$	$t^M u_s(t)$	(5)	$(nT)^M u_s[n]$	$\lim_{a\to 0}(-1)^M \frac{\partial^M}{\partial a^M}\left(\frac{z}{z-e^{-aT}}\right)$
$\frac{1}{s+a}$	$e^{-at}u_s(t)$	(6)	$e^{-aTn}u_s[n]=b^n u_s[n]$	$\frac{z}{z-e^{-aT}}=\frac{z}{z-b}\ (b=e^{-aT})$
$\frac{1}{(s+a)^2}$	$t\,e^{-at}u_s(t)$	(7)	$nTe^{-aTn}u_s[n]$	$\frac{Tze^{-aT}}{(z-e^{-aT})^2}$
$\frac{M!}{(s+a)^{M+1}}$	$t^M e^{-at}u_s(t)$	(8)	$(nT)^M e^{-aTn}u_s[n]$	$(-1)^M \frac{\partial^M}{\partial a^M}\left(\frac{z}{z-e^{-aT}}\right)$
$\frac{\omega}{s^2+\omega^2}$	$\sin\omega t\,u_s(t)$	(9)	$\sin(\omega Tn)u_s[n]$	$\frac{z\sin\omega T}{z^2-2z\cos\omega T+1}$
$\frac{\omega}{(s+\sigma)^2+\omega^2}$	$e^{-\sigma t}\sin\omega t\,u_s(t)$	(10)	$e^{-\sigma Tn}\sin(\omega Tn)u_s[n]$	$\frac{ze^{-\sigma T}\sin\omega T}{(z-e^{-\sigma T}\cos\omega T)^2+e^{-2\sigma T}\sin^2\omega T}$
$\frac{s}{s^2+\omega^2}$	$\cos\omega t\,u_s(t)$	(11)	$\cos(\omega Tn)u_s[n]$	$\frac{z(z-\cos\omega T)}{z^2-2z\cos\omega T+1}$
$\frac{s+\sigma}{(s+\sigma)^2+\omega^2}$	$e^{-\sigma t}\cos\omega t\,u_s(t)$	(12)	$e^{-\sigma Tn}\cos(\omega Tn)u_s[n]$	$\frac{z(z-e^{-\sigma T}\cos\omega T)}{(z-e^{-\sigma T}\cos\omega T)^2+e^{-2\sigma T}\sin^2\omega T}$
$\frac{1}{s}e^{-s^2/4}\mathrm{erfc}(\frac{s}{2})$	$\mathrm{erf}(t)$	(13)		
$\frac{e^{-a\sqrt{s}}}{\sqrt{s}}$	$\frac{e^{-a^2/4t}}{\sqrt{\pi t}}$	(14)		
$e^{-a\sqrt{s}}$	$\frac{ae^{-a^2/4t}}{2\sqrt{\pi t^3}}$	(15)		
$\frac{e^{-a\sqrt{s}}}{s}$	$\mathrm{erfc}\left(\frac{a}{2\sqrt{t}}\right)$	(16)		
Transfer function $G(s)=\frac{Y(s)}{U(s)}=C[sI-A]^{-1}B+D$	State equation $\mathbf{x}'(t)=A\mathbf{x}(t)+Bu(t)$ $\mathbf{y}(t)=C\mathbf{x}(t)+Du(t)$	(17)	State equation $\mathbf{x}[n+1]=A\mathbf{x}[n]+Bu[n]$ $\mathbf{y}[n]=C\mathbf{x}[n]+Du[n]$	Transfer function $G[z]=\frac{Y[z]}{U[z]}\overset{(5.5.5)}{=}C[zI-A]^{-1}B+D$

(Note) See <http://www.math.ttu.edu/~gilliam/ttu/s10/m3351_s10/c15_laplace_trans_pdes.pdf> for (13–16).

Table A.9 Bilateral z-transforms of basic sequences

	$x[n]$	$\mathbb{X}[z] = \mathcal{Z}\{x[n]\} = \sum_{n=-\infty}^{\infty} x[n]z^{-n}$ with ROC
(1)	$\delta[n]$	1 All z
(2)	$\delta[n-n_1]$	z^{-n_1} All z except 0 (for $n_1>0$) or ∞ (for $n_1<0$)
(3)	$u_s[n]$	$\frac{z}{z-1} = \frac{1}{1-z^{-1}}$ $\quad |z|>1$
(4)	$-u_s[-n-1]$	$\frac{z}{z-1} = \frac{1}{1-z^{-1}}$ $\quad |z|<1$
(5)	$e^{-\sigma T n}u_s[n] = a^n u_s[n]$	$\frac{z}{z-a} = \frac{1}{1-az^{-1}}\ (a = e^{-\sigma T})$ $\quad |z|>|a|$
(6)	$-b^n u_s[-n-1]$	$\frac{z}{z-b} = \frac{1}{1-bz^{-1}}$ $\quad |z|<|b|$
(7)	$e^{-\sigma T|n|} = a^{|n|}$	$\frac{z}{z-a} - \frac{z}{z-1/a}\ (a = e^{-\sigma T})$ $\quad |a|<|z|<\left|\frac{1}{a}\right|$
(8)	$n\, u_s[n]$	$\frac{z}{(z-1)^2}$ $\quad |z|>1$
(9)	$-\, n\, u_s[-n-1]$	$\frac{z}{(z-1)^2}$ $\quad |z|<1$
(10)	$na^n u_s[n]$	$\frac{az}{(z-a)^2}$ $\quad |z|>|a|$
(11)	$-nb^n u_s[-n-1]$	$\frac{bz}{(z-b)^2}$ $\quad |z|<|b|$
(12)	$n^2 a^n u_s[n]$	$\frac{az(z+a)}{(z-a)^3}$ $\quad |z|>|a|$
(13)	$-n^2 b^n u_s[-n-1]$	$\frac{bz(z+b)}{(z-b)^3}$ $\quad |z|<|b|$
(14)	$\frac{n(n-1)\cdots(n-K+2)}{(K-1)!} a^{n-(K-1)} u_s[n]$	$\frac{z}{(z-a)^K}$ $\quad |z|>|a|$
(15)	$\sin(\Omega_1 n) u_s[n]$	$\frac{z\sin\Omega_1}{(z-\cos\Omega_1)^2+\sin^2\Omega_1} = \frac{z\sin\Omega_1}{z^2-2z\cos\Omega_1+1}$ $\quad |z|>1$
(16)	$\cos(\Omega_1 n) u_s[n]$	$\frac{z(z-\cos\Omega_1)}{(z-\cos\Omega_1)^2+\sin^2\Omega_1} = \frac{z(z-\cos\Omega_1)}{z^2-2z\cos\Omega_1+1}$ $\quad |z|>1$
(17)	$r^n \sin(\Omega_1 n) u_s[n]$	$\frac{zr\sin\Omega_1}{(z-r\cos\Omega_1)^2+r^2\sin^2\Omega_1} = \frac{zr\sin\Omega_1}{z^2-2zr\cos\Omega_1+r^2}$ $\quad |z|>|r|$
(18)	$r^n \cos(\Omega_1 n) u_s[n]$	$\frac{z(z-r\cos\Omega_1)}{(z-r\cos\Omega_1)^2+r^2\sin^2\Omega_1} = \frac{z(z-r\cos\Omega_1)}{z^2-2zr\cos\Omega_1+r^2}$ $\quad |z|>|r|$

Table A.10 Bilateral Laplace transforms of basic functions

	$x(t)$	$\mathbb{X}(s)=L\{x(t)\}=\int_{-\infty}^{\infty}x(t)e^{-st}$ with ROC	
(1)	$\delta(t)$	1	All s
(2)	$\delta(t-t_1)$	e^{-st_1}	All s except $-\infty$ (for $t_1>0$) or ∞ (for $t_1<0$)
(3)	$u_s(t)$	$\frac{1}{s}$	$\mathrm{Re}\{s\}>0$
(4)	$-u_s(-t)$	$\frac{1}{s}$	$\mathrm{Re}\{s\}<0$
(5)	$e^{-at}u_s(t)$ with $a\geq 0$	$\frac{1}{s+a}$	$\mathrm{Re}\{s\}>-a$
(6)	$-e^{-bt}u_s(-t)$ with $b\leq 0$	$\frac{1}{s+b}$	$\mathrm{Re}\{s\}<-b$
(7)	$e^{-a\vert t\vert}$ with $a>0$	$\frac{1}{s+a}-\frac{1}{s-a}=-\frac{2a}{s^2-a^2}$	$\vert\mathrm{Re}\{s\}\vert<a$
(8)	$\frac{t^K}{K!}e^{-at}u_s(t)$ with $a\geq 0$	$\frac{1}{(s+a)^K}$	$\mathrm{Re}\{s\}>-a$
(9)	$-\frac{t^K}{K!}e^{-bt}u_s(-t)$ with $b\leq 0$	$\frac{1}{(s+a)^K}$	$\mathrm{Re}\{s\}<-b$
(10)	$e^{-\sigma t}\cos(\omega t)u_s(t)$ with $\sigma\geq 0$	$\frac{s+\sigma}{(s+\sigma)^2+\omega^2}$	$\mathrm{Re}\{s\}>-\sigma$
(11)	$e^{-\sigma t}\sin(\omega t)u_s(t)$ with $\sigma\geq 0$	$\frac{\omega}{(s+\sigma)^2+\omega^2}$	$\mathrm{Re}\{s\}>-\sigma$
(12)	$-e^{-\sigma t}\cos(\omega t)u_s(-t)$ with $\sigma\leq 0$	$\frac{s+\sigma}{(s+\sigma)^2+\omega^2}$	$\mathrm{Re}\{s\}<-\sigma$
(13)	$-e^{-\sigma t}\sin(\omega t)u_s(-t)$ with $\sigma\leq 0$	$\frac{\omega}{(s+\sigma)^2+\omega^2}$	$\mathrm{Re}\{s\}<-\sigma$
(14)	$\mathrm{erf}(t)$	$\frac{1}{s}e^{-s^2/4}\mathrm{erfc}(\frac{s}{2})$	
(15)	$\frac{1}{\sqrt{\pi t}}$	$\frac{1}{\sqrt{s}}$	
(16)	$\frac{e^{-a^2/4t}}{\sqrt{\pi t}}$	$\frac{e^{-a\sqrt{s}}}{\sqrt{s}}$	
(17)	$\frac{ae^{-a^2/4t}}{2\sqrt{\pi t^3}}$	$e^{-a\sqrt{s}}$	
(18)	$\mathrm{erfc}\left(\frac{a}{2\sqrt{t}}\right)$	$\frac{e^{-a\sqrt{s}}}{s}$	

Appendix B: Differentiation w.r.t. Vector

The first derivative of a scalar-valued function $f(\mathbf{x})$ w.r.t. the vector $\mathbf{x}=[x_1\ \ x_2]^T$ is called the *gradient* of $f(\mathbf{x})$ and is defined as

$$\nabla f(\mathbf{x}) = \frac{d}{d\mathbf{x}} f(\mathbf{x}) = \begin{bmatrix} \partial f/\partial x_1 \\ \partial f/\partial x_2 \end{bmatrix} \tag{B.1}$$

Based on this definition, we can write the following equation:

$$\frac{\partial}{\partial \mathbf{x}} \mathbf{x}^T \mathbf{y} = \frac{\partial}{\partial \mathbf{x}} \mathbf{y}^T \mathbf{x} = \frac{\partial}{\partial \mathbf{x}} (x_1 y_1 + x_2 y_2) \overset{(B.1)}{=} \begin{bmatrix} y_1 \\ y_2 \end{bmatrix} = \mathbf{y} \tag{B.2}$$

$$\frac{\partial}{\partial \mathbf{x}} \mathbf{x}^T \mathbf{x} = \frac{\partial}{\partial \mathbf{x}} (x_1^2 + x_2^2) = 2 \begin{bmatrix} x_1 \\ x_2 \end{bmatrix} = 2\mathbf{x} \tag{B.3}$$

Also with an $M \times N$ matrix $A=[a_{mn}]$, we have

$$\frac{\partial}{\partial \mathbf{x}} \mathbf{x}^T A \mathbf{y} = \frac{\partial}{\partial \mathbf{x}} \mathbf{y}^T A^T \mathbf{x} = A\mathbf{y} \tag{B.4a}$$

$$\frac{\partial}{\partial \mathbf{x}} \mathbf{y}^T A \mathbf{x} = \frac{\partial}{\partial \mathbf{x}} \mathbf{x}^T A^T \mathbf{y} = A^T \mathbf{y} \tag{B.4b}$$

where

$$\mathbf{x}^T A \mathbf{y} = \sum_{m=1}^{M} \sum_{n=1}^{N} a_{mn} x_m y_n \overset{\text{if } M=N=2}{=} a_{11} x_1 y_1 + a_{12} x_1 y_2 + a_{21} x_2 y_1 + a_{22} x_2 y_2 \tag{B.5}$$

Especially for a symmetric (and square) matrix A (with $M=N$), we have

$$\frac{\partial}{\partial \mathbf{x}} \mathbf{x}^T A \mathbf{x} = (A + A^T)\mathbf{x} \overset{\text{if } A \text{ is symmetric}}{\rightarrow} 2A\mathbf{x} \tag{B.6}$$

The second derivative of a scalar function $f(\mathbf{x})$ w.r.t. the vector $\mathbf{x}=[x_1\ \ x_2]^T$ is called the *Hessian* (*matrix*) of $f(\mathbf{x})$ and is defined as

$$H(\mathbf{x}) = \nabla^2 f(\mathbf{x}) = \frac{d^2}{d\mathbf{x}^2} f(\mathbf{x}) = \begin{bmatrix} \partial^2 f/\partial x_1^2 & \partial^2 f/\partial x_1 \partial x_2 \\ \partial^2 f/\partial x_2 \partial x_1 & \partial^2 f/\partial x_2^2 \end{bmatrix} \tag{B.7}$$

Based on this definition, we can write the following equation:

$$\frac{d^2}{d\mathbf{x}^2} \mathbf{x}^T A \mathbf{x} = A + A^T \overset{\text{if } A \text{ is symmetric}}{\rightarrow} 2A \tag{B.8}$$

The first derivative of a vector-valued function $\mathbf{f}(\mathbf{x})$ w.r.t. the vector $\mathbf{x}=[x_1\ \ x_2]^T$ is called the *Jacobian* (*matrix*) of $\mathbf{f}(\mathbf{x})$ and is defined as

$$J(\mathbf{x}) = \frac{d}{d\mathbf{x}} \mathbf{f}(\mathbf{x}) = \begin{bmatrix} \partial f_1/\partial x_1 & \partial f_1/\partial x_2 \\ \partial f_2/\partial x_1 & \partial f_2/\partial x_2 \end{bmatrix} \tag{B.9}$$

Note that the Hessian of a scalar function $f(\mathbf{x})$ of a vector $\mathbf{x}$ is equal to the *Jacobian* of the gradient of $f(\mathbf{x})$:

$$H\{f(\mathbf{x})\} = J\{\nabla f(\mathbf{x})\} \tag{B.10}$$

Appendix C: Useful Formulas

$\sin(A\pm\pi/2) = \pm\cos A$	(C.1)	$\cos(A\pm\pi/2) = \mp\sin A$	(C.2)
$\sin(A\pm\pi) = -\sin A$	(C.3)	$\cos(A\pm\pi) = -\cos A$	(C.4)
$\sin(A\pm B) = \sin A\cos B \pm \cos A\sin B$	(C.5)	$\tan(A\pm B) = \dfrac{\tan A \pm \tan B}{1\mp \tan A\tan B}$	(C.7)
$\cos(A\pm B) = \cos A\cos B \mp \sin A\sin B$	(C.6)		
$\sin A\sin B = \frac{1}{2}\{\cos(A-B) - \cos(A+B)\}$	(C.8)	$\sin A\cos B = \frac{1}{2}\{\sin(A+B) + \sin(A-B)\}$	(C.9)
$\cos A\sin B = \frac{1}{2}\{\sin(A+B) - \sin(A-B)\}$	(C.10)	$\cos A\cos B = \frac{1}{2}\{\cos(A+B) + \cos(A-B)\}$	(C.11)
$\sin A + \sin B = 2\sin\left(\frac{A+B}{2}\right)\cos\left(\frac{A-B}{2}\right)$	(C.12)	$\cos A + \cos B = 2\cos\left(\frac{A+B}{2}\right)\cos\left(\frac{A-B}{2}\right)$	(C.13)
$\sin^2 A = \frac{1}{2}(1-\cos 2A)$	(C.14)	$\cos^2 A = \frac{1}{2}(1+\cos 2A)$	(C.15)
$\sin 2A = 2\sin A\cos A$	(C.16)	$\cos 2A = \cos^2 A - \sin^2 A = 1-2\sin^2 A = 2\cos^2 A - 1$	(C.17)
$a\cos A - b\sin A = \sqrt{a^2+b^2}\cos(A+\theta)$ with $\theta = \tan^{-1}\frac{b}{a}$			(C.18)
$a\sin A + b\cos A = \sqrt{a^2+b^2}\sin(A+\theta)$ with $\theta = \tan^{-1}\frac{b}{a}$			(C.19)
Euler identity: $e^{\pm j\theta} = \cos\theta \pm j\sin\theta$	(C.20)	$e^{j\theta} + e^{-j\theta} = 2\cos\theta$	(C.21)
		$e^{j\theta} - e^{-j\theta} = j2\sin\theta$	(C.22)
Sum of (infinite) geometric series: $\sum_{n=0}^{N} r^n = \frac{1-r^{N+1}}{1-r} \xrightarrow{N\to\infty} \sum_{n=0}^{\infty} r^n = \frac{1}{1-r},\ \lvert r\rvert < 1$			(C.23)
Binomial expansion: $(a+b)^N = \sum_{n=0}^{N} {}_N C_n a^{N-n} b^n$ with ${}_N C_n = {}_N C_{N-n} = \frac{{}_N P_n}{n!} = \frac{N!}{(N-n)!n!}$			(C.24)
$e^x = \lim_{t\to\infty}\left(1+\frac{x}{t}\right)^t = \sum_{n=0}^{\infty}\frac{1}{n!}x^n = 1+\frac{1}{1!}x+\frac{1}{2!}x^2+\frac{1}{3!}x^3+\cdots \xrightarrow{x\to 0} e^x = 1+x$			(C.25)
$\frac{d}{dt}t^n = nt^{n-1}$	(C.26)	$\int t^n\,dt = \frac{1}{n+1}t^{n+1}$ for $n\neq -1$, $\int t^{-1}dt = \ln\lvert t\rvert$	(C.34)
$\frac{d}{dt}e^{at} = ae^{at}$	(C.27)	$\int e^{at}dt = \frac{1}{a}e^{at}$	(C.35)
$\frac{d}{dt}\cos\omega t = -\omega\sin\omega t$	(C.28)	$\int \cos\omega t\,dt = \frac{1}{\omega}\sin\omega t$	(C.36)
$\frac{d}{dt}\sin\omega t = \omega\cos\omega t$	(C.29)	$\int \sin\omega t\,dt = -\frac{1}{\omega}\cos\omega t$	(C.37)
$\frac{d}{dt}(uv) = u\frac{d}{dt}v + v\frac{d}{dt}u$	(C.30)	$\int u\frac{dv}{dt}dt = uv - \int v\frac{du}{dt}dt$ (Partial integral)	(C.38)
$\frac{d}{dt}\left(\frac{v}{u}\right) = \frac{u(dv/dt) - v(du/dt)}{u^2}$	(C.31)	$x(t) * \delta(t-t_1) = x(t-t_1)$	(C.39)
$\frac{d}{dx}\left(\tan^{-1}\frac{y}{x}\right) = \frac{-y}{x^2+y^2}$	(C.32)	$\int_{-\infty}^{+\infty} x(t)\,\delta(t-t_1)\,dt = x(t_1)$	(C.40)
$\frac{d}{dy}\left(\tan^{-1}\frac{y}{x}\right) = \frac{x}{x^2+y^2}$	(C.33)	$\int_{-\infty}^{+\infty} e^{j\omega t}dt = 2\pi\,\delta(\omega)$	(C.41)

Appendix D: Gamma Function

For any complex number x with a positive real part, the *gamma function* is defined as

$$\Gamma(x) = \int_0^\infty t^{x-1} e^{-t} dt \tag{D.1}$$

where $\Gamma(1) = \int_0^\infty e^{-t} dt = -e^{-t}\Big|_0^\infty = 1$. We can integrate this by parts

$$\Gamma(x) \overset{(D.1)}{=} \int_0^\infty t^{x-1} e^{-t} dt \overset{(C.38)}{=} -t^{x-1} e^{-t}\Big|_0^\infty + (x-1)\int_0^\infty t^{x-2} e^{-t} dt \overset{(D.1)}{=} (x-1)\Gamma(x-1)$$

to get

$$\Gamma(x) = (x-1)\Gamma(x-1) \tag{D.2}$$

If x is a positive integer n, then it reduces to the factorial:

$$\Gamma(n) = (n-1)\Gamma(n-1) = (n-1)(n-2)\Gamma(n-2) = (n-1)(n-2)\cdots 1 = (n-1)! \tag{D.3}$$

That is why the gamma function is called the generalized factorial function.

Note that

$$\Gamma(\frac{1}{2}) \overset{(D.1)}{=} \int_0^\infty t^{1/2-1} e^{-t} dt \underset{dt=2u\,du}{\overset{t=u^2}{=}} 2\int_0^\infty e^{-u^2} du = \sqrt{\pi} \tag{D.4}$$

because

$$\left(\Gamma(\frac{1}{2})\right)^2 = \left(2\int_0^\infty e^{-u^2} du\right)\left(2\int_0^\infty e^{-v^2} dv\right) = 4\int_0^\infty \int_0^\infty e^{-(u^2+v^2)} du\, dv$$

$$\underset{u=r\cos\theta,\, v=r\sin\theta}{\overset{(2.6.7)}{=}} 4\int_0^{\pi/2} \int_0^\infty e^{-r^2} r\, dr\, d\theta = -2\int_0^{\pi/2} (e^{-r^2})\Big|_0^\infty d\theta = 2\int_0^{\pi/2} 1 d\theta = \pi \tag{D.5}$$

Also, for half-integers $x = n + 1/2$, we have

$$\Gamma(\frac{1}{2} + n) = \frac{1\cdot 3\cdot 5\cdots (2n-1)}{2^n}\sqrt{\pi} \tag{D.6}$$

$$\Gamma(\frac{1}{2} - n) = \frac{(-2)^n}{1\cdot 3\cdot 5\cdots (2n-1)}\sqrt{\pi} \tag{D.7}$$

(Note) Visit the web page <http://mathworld.wolfram.com/GammaFunction.html>.

Appendix E: MATLAB

On a Windows-based computer with MATLAB software installed, you can double-click on the MATLAB icon (like the one on the right side) to start MATLAB. Once MATLAB has been started, you will see the MATLAB desktop as in Fig. E.1, which has the MATLAB Command window in the lower middle, where a blinking cursor appears to the right of the MATLAB prompt '>>', waiting for you to type in MATLAB commands/statements. It also contains other windows labeled Current Folder, Editor, and Workspace (showing the contents of MATLAB memory).

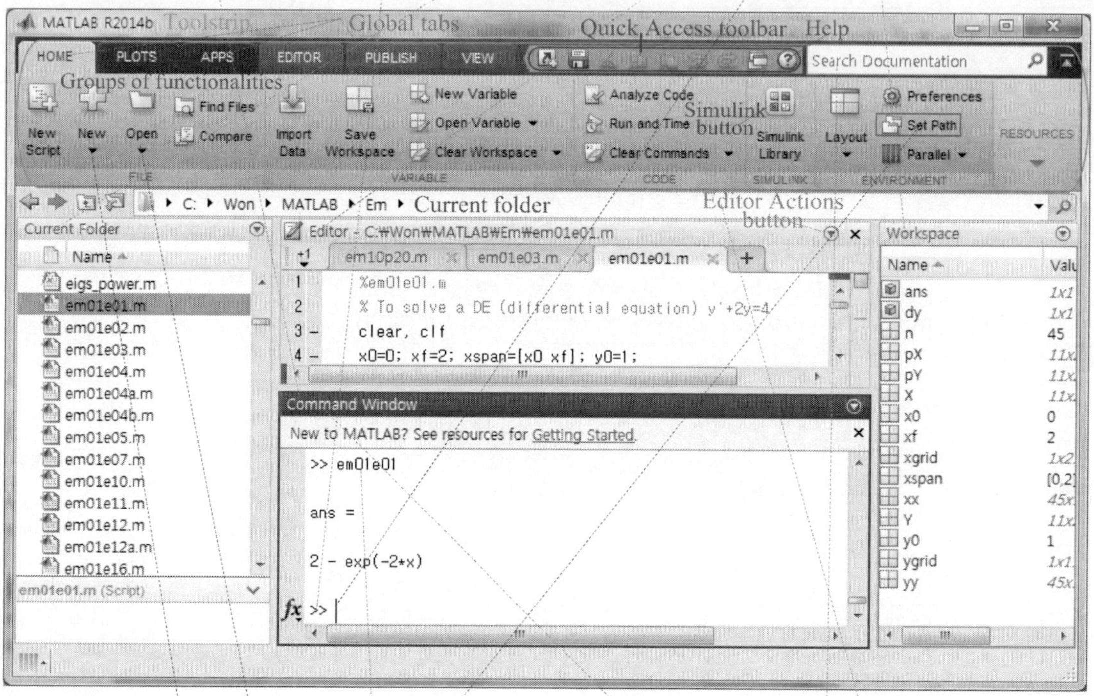

Figure E.1 The MATLAB desktop with Command/Editor/Current folder/Workspace windows

How do we work with the MATLAB Command window?

- By clicking on "New"/"Open" in the groups of functionalities under the HOME tab and then selecting an item in their drop-down menus, you can create/edit any file with the MATLAB editor.
- By typing any MATLAB commands/statements in the MATLAB Command Window, you can use various powerful mathematic/graphic functions of MATLAB.
- If you have an M-file that contains a series of commands/statements for performing a target procedure, you can type in the filename (without the extension '.m') to run it.
- By clicking on "Set Path" in the right part of the toolstrip, you can make the MATLAB search path list include or exclude the paths containing the files you want to do not want to be run.
- By typing a search term like a keyword or a function name into the Search Documentation box at top right, you can search help documentation for what you want to know about.
- To undock, i.e., separate the Editor window from the MATLAB desktop, click on the Editor Actions button (down-arrow shaped) in the top-right part of the Editor window and select "Undock" in its drop-down menu. To dock the Editor window back to the desktop, click on the Editor Actions button and select "Dock Editor" in its drop-down menu.

Some mathematical functions and special reserved constants/variables defined in MATLAB are listed in Table E.1. Table E.2 shows the graphic line specifications used in the plot( ) command.

Table E.1 Functions and Variables in MATLAB

function	Remark	function	Remark
`cos(x)`		`exp(x)`	exponential function
`sin(x)`		`log(x)`	natural logarithm
`tan(x)`		`log10(x)`	common logarithm
`acos(x)`	$\cos^{-1}(x)$	`abs(x)`	absolute value
`asin(x)`	$\sin^{-1}(x)$	`angle(x)`	phase of a complex number [rad]
`atan(x)`	$-\pi/2 \le \tan^{-1}(x) \le \pi/2$	`sqrt(x)`	square root
`atan2(y,x)`	$-\pi \le \tan^{-1}(y,x) \le \pi$	`real(x)`	real part
`cosh(x)`	$(e^x + e^{-x})/2$	`imag(x)`	imaginary part
`sinh(x)`	$(e^x - e^{-x})/2$	`conj(x)`	complex conjugate
`tanh(x)`	$(e^x - e^{-x})/(e^x + e^{-x})$	`round(x)`	the nearest integer (round-off)
`acosh(x)`	$\cosh^{-1}(x)$	`fix(x)`	the nearest integer towards 0
`asinh(x)`	$\sinh^{-1}(x)$	`floor(x)`	the greatest integer $\le x$
`atanh(x)`	$\tanh^{-1}(x)$	`ceil(x)`	the smallest integer $\ge x$
`max`	maximum and its index	`sign(x)`	1(positive)/0/−1(negative)
`min`	minimum and its index	`mod(y,x)`	remainder of y/x
`sum`	sum	`rem(y,x)`	remainder of y/x
`prod`	product	`eval(f)`	evaluate an expression
`norm`	norm	`feval(f,a)`	function evaluation
`sort`	sort in the ascending order	`polyval`	value of a polynomial function
`clock`	present time	`poly`	polynomial with given roots
`find`	index of element(s)	`roots`	roots of polynomial
`tic`	start a stopwatch timer	`toc`	read the stopwatch timer (elapsed time from tic)
`date`	present date		
reserved variables with special meaning			
`i, j`	$\sqrt{-1}$	`pi`	π
`eps`	machine epsilon	`Inf, inf`	largest number (∞)
realmax,realmin	largest/smallest positive number	`NaN`	Not_a_Number (undetermined)
`end`	the end of for-loop or if, while,	`break`	exit while/for loop
`nargin`	# of input arguments	`nargout`	# of output arguments
`varargin`	variable input argument list	`varargout`	variable output argument list

Table E.2 Graphic line specifications used in the `plot()` command

Line type	*Point type (Marker symbol)*			*Color*	
- solid line	.(dot)	+(plus)	*(asterisk)	r: red	m: magenta
: dotted line	^: Δ	>: >	o(circle)	g: green	y: yellow
-- dashed line	p: ☆	v: ∇	x: x-mark	b: blue	c: cyan (sky blue)
-. dash-dot	d: ◊	<: <	s: square(□)	k: black	

E.1 Convolution and Correlation

`C=conv(A,B)`

- For two sequences $A=[a_1\ a_2\ \cdots\ a_N]$ and $B=[b_1\ b_2\ \cdots\ b_M]$, this function computes their (linear) convolution $C=[c_1\ c_2\ \cdots\ c_{N+M-1}]$ as

$$c_{n+1}=\sum_{m=\max(0,n-N+1)}^{\min(n,M-1)} a_{n-m+1}b_{m+1} \tag{E.1}$$

- It can be used to multiply two polynomials whose coefficients are A and B, respectively:

$$(a_1x^{N-1}+\cdots+a_{N-1}x+a_N)(b_1x^{M-1}+\cdots+b_{M-1}x+b_M)=c_1x^{N+M-2}+\cdots+c_{N+M-2}x+c_{N+M-1}$$

`C=xcorr(A,B)` returns the correlation of two sequences A and B:

$$c_{n+1}=\sum_{m=\max(0,-n)}^{\min(N-1-n,M-1)} a_{n+m+1}b_{m+1} \tag{E.2}$$

```
function z=conv_circular(x,y,N)
% Circular convolution z(n)= (1/N) sum_m=0^N-1 x(m)*y(n-m)
if nargin<3, N=min(length(x),length(y)); end
x=x(1:N); y=y(1:N); y_crc= fliplr(y);
for n=1:N
   y_crc = [y_crc(N) y_crc(1:N-1)];    z(n) = x*y_crc.'/N;
end
```

E.2 CTFS (Continuous-Time Fourier Series)

```
function [c,kk]=CTFS_exponential(x,P,N)
% Find the complex exponential Fourier coefficients c(k) for k=-N:N
% x: A periodic function with period P
% P: Period,  N: Maximum frequency index to specify the frequency range
w0=2*pi/P; % the fundamental frequency [rad/s]
xexp_jkw0t_= [x '(t).*exp(-j*k*w0*t)'];
xexp_jkw0t= inline(xexp_jkw0t_,'t','k','w0');
kk=-N:N; tol=1e-6; % the frequency range tolerance on numerical error
for k=kk
   c(k+N+1)= quadl(xexp_jkw0t,-P/2,P/2,tol,[],k,w0); % Eq.(6.1.6b)
end
```

E.3 DTFT (Discrete-Time Fourier Transform)

```
function [X,ph]=DTFT(x,W,n0)
% DTFT of x[n] for frequency W regarding the 1st sample as the n0-th one.
Nt =length(x); n=0:Nt-1;
if nargin<3, n0 = -floor(Nt/2); end
X= x*exp(-j*(n+n0)'*W); % Eq.(6.3.1)
if nargout==2,  ph=angle(X); X=abs(X);  end
```

E.4 DFS/DFT (Discrete Fourier Series/Transform)

```
function [X,ph]=DFS(x,N,n0)
% N-point DFS/DFT of x[n] regarding the 1st sample as the n0-th one.
if nargin<3, n0 = 0; end
n=n0+[0:length(x)-1]; k=[0:N-1];
X= x*exp(-j*2*pi*n'*k/N); % Eq.(6.4.7)
if nargout==2,  ph=angle(X); X=abs(X);  end
```

E.5 FFT (Fast Fourier Transform)

<table>
<tr><td>

`X= fft(x,N)` or `fft(x)`

- This function computes the N-point DFT X(k) of a given sequence x[n], where the sequence will be zero-padded or truncated so that its resulting length will be N. If the 2nd input argument is not given, the DFT size will be set to the length of the sequence. In a case where the 1st input argument x is a matrix, X will also be a matrix, each column of which is the FFT of the corresponding column of x.
- Note that, to exploit the computational efficiency of FFT, the DFT size N should be chosen as a power of 2 greater than or equal to the length of the sequence.

</td></tr>
<tr><td>

`x= ifft(X,N)` or `ifft(X)`

- This function computes the N-point IDFT x[n] of a given sequence X(k), where the sequence will be zero-padded or truncated so that its resulting length will be N. If the 2nd input argument is not given, the IDFT size will be set to the length of the sequence. In the case where the 1st input argument x is a matrix, x will also be a matrix, each column of which is the IDFT of the corresponding column of X.
- Note that, to exploit the computational efficiency of FFT, the IDFT size N should be chosen as a power of 2 greater than or equal to the length of the sequence.

</td></tr>
<tr><td>

`y= fftshift(x)`

- This function swaps the first and second halves of the input vector x if x is a vector, and swaps quadrants one and three with quadrants two and four if x is a matrix. It is used to rearrange the FFT sequence so that the zero (DC) frequency component is located at the center with the negative/positive frequency ones on the left/right side.

</td></tr>
</table>

E.6 Input/Output through File

The MATLAB script "input_output_data.m" illustrates how the MATLAB commands "save"/"load" are used to write/read data, respectively, into/from a MAT-file or an ASCII file. The MATLAB command "clear" is used to delete some or all of the contents from the MATLAB meneory (workspace).

```
% input_output_data.m
clear
x=[1 2]; y=[3 4 5];
save sig x y % save x and y in a MATLAB data file 'sig.mat'
clear('y')
display('After y has been cleared, does y exist?')
if exist('y')~=0, disp('Yes'), y
 else  disp('No')
end
load sig y % read y from the MATLAB data file 'sig.mat'
disp('After y has been loaded the file sig.mat, does y exist?')
if isempty('y'), disp('No')
 else disp('Yes'), y
end
fprintf('x(2)=%5.2f \n', x(2))
save y.dat y /ascii % save y into the ASCII data file 'y.dat'
% The name of the ASCII data file must be the same as the variable name.
load y.dat % read y from the ASCII data file 'y.dat'
str='prod(y)'; % ready to compute the produce of the elements of y
eval(str) % evaluate the string expression 'prod(y)'
```

E.7 Symbolic Coordinate Conversion

```
function v_cyl = cart2cyl_v(v_cart)
% Cartesian-to-cylindrical coordinate conversion of a 3D vector
%Copyleft: Won Y. Yang, wyyang53@hanmail.net, CAU for academic use only
syms x y r ph z
sinph=sin(ph); cosph=cos(ph);
v_cart1=subs(v_cart,{x,y},{r*cosph,r*sinph});
v_cyl = [cosph sinph 0; -sinph cosph 0; 0 0 1]*v_cart1(:);% Eq.(1.1.58b)
```

```
function v_cart = cyl2cart_v(v_cyl)
% Cylindrical-to-Cartesian coordinate conversion of a 3D vector
%Copyleft: Won Y. Yang, wyyang53@hanmail.net, CAU for academic use only
syms x y z
r=sqrt(x^2+y^2); ph=atan(y/x); sinph=sin(ph); cosph=cos(ph);
v_cyl1 = subs(v_cyl,{r,ph},{r,ph});
v_cart = [cosph -sinph 0; sinph cosph 0; 0 0 1]*v_cyl1(:); % Eq.(1.1.58a)
```

```
function v_sph = cart2sph_v(v_cart)
% Cartesian-to-spherical coordinate conversion of a 3D vector
%Copyleft: Won Y. Yang, wyyang53@hanmail.net, CAU for academic use only
syms x y z r th ph
sinth=sin(th); costh=cos(th); sinph=sin(ph); cosph=cos(ph);
v_cart1=subs(v_cart,{x,y,z},{r*sinth*cosph,r*sinth*sinph,r*costh});
v_sph = [sinth*cosph sinth*sinph costh; costh*cosph costh*sinph -sinth;
         -sinph        cosph       0]*v_cart1(:); % Eq.(1.1.59b)
```

```
function v_cart = sph2cart_v(v_sph)
% Spherical-to-Cartesian coordinate conversion of a 3D vector
%Copyleft: Won Y. Yang, wyyang53@hanmail.net, CAU for academic use only
syms x y z r th ph
r=sqrt(x^2+y^2+z^2); th=atan(sqrt(x^2+y^2)/z); ph=atan(y/x);
sinth=sin(th); costh=cos(th); sinph=sin(ph); cosph=cos(ph);
v_sph1 = subs(v_sph,{r,th,ph},{r,th,ph});
v_cart = [sinth*cosph costh*cosph -sinph; sinth*sinph costh*sinph cosph;
           costh       -sinth       0]*v_sph1(:); % Eq.(1.1.59a)
% Example of usage:
% v_cart=sph2cart_v([r 0 0]); pretty(simplify(v_cart))
```

The above MATLAB functions '`cart2cyl_v()`', '`cyl2cart_v()`', '`cart2sph_v()`', and '`sph2cart_v()`' are used to convert a literal expression of 3D-vector w.r.t. the Cartesian coordinate system into that w.r.t. the cylindrcal/spherical coordinate system and vice versa. See the following examples:

```
>> syms x y z; v_cyl=cart2cyl_v([x y z]); pretty(simplify(v_cyl))
    +-   -+
    |  r  |
    |     |
    |  0  |
    |     |
    |  z  |
    +-   -+
>> v_sph=cart2sph_v([x y z]); pretty(simplify(v_sph))
    +-   -+
    |  r  |
    |     |
    |  0  |
    |     |
    |  0  |
    +-   -+
>> syms r; v_cart=sph2cart_v([r 0 0]); pretty(simplify(v_cart))
```

E.8 Symbolic Line/Surface Integrals

```
function I = int_line(f,lt,t0,tf)
% Input:
%   f  = Vector or Scalar integrand function in x, y, and z
%   lt = Integration path vector parameterized by t
% Output:
%   I  = Line integral of f over dlt from t=t0 to tf
%Copyleft: Won Y. Yang, wyyang53@hanmail.net, CAU for academic use only
syms x y z t
dlt = diff(lt);
if length(f)<2 % Scalar integrand function
  if length(dlt)==1, ft = subs(f,x,lt(1));
   elseif length(dlt)==2,
    ft = subs(f,{x,y},{lt(1),lt(2)})*sqrt(dlt(1)^2+dlt(2)^2);
   elseif length(dlt)==3,
    ft=subs(f,{x,y,z},{lt(1),lt(2),lt(3)})*sqrt(dlt(1)^2+dlt(2)^2+dlt(3)^2);
  end
  I = int(ft,t0,tf);
 else
  ft = subs(f,{x,y,z},{lt(1),lt(2),lt(3)});
  I = int(dot(ft,dlt),t0,tf);
end
%Usage:
% syms x y z t; int_line(2*[x y z]/(x^2+y^2+z^2)^(3/2),[1 0 1-t],0,1/2)
```

```
function I = int_surf(G,f,x0f,y0f,order)
% Input:
%   G = Vector or Scalar integrand function in x, y, and z
%   f = Surface function f(x,y)
%   order = 0/1 to determine which one of x and y to integrate over
% Output:
%   I = Surface integral of G over R for x=x0f(1)~x0f(2), y=y0f(1)~y0f(2)
%Copyleft: Won Y. Yang, wyyang53@hanmail.net, CAU for academic use only
if nargin<5, order=0; end
syms x y z v w
if length(f)==1 %
  if length(G)<2 % Scalar integrand function
    integrand = G*sqrt(1+diff(f,x)^2+diff(f,y)^2);
   else
    n = [diff(f,x) diff(f,y) diff(f,z)];
    un = n/sqrt(n(1)^2+n(2)^2+n(3)^2);
    integrand = un*G(:);
  end
  if order==0, I = int(int(integrand,x,x0f(1),x0f(2)),y,y0f(1),y0f(2));
   else  I = int(int(integrand,y,y0f(1),y0f(2)),x,x0f(1),x0f(2));
  end
 else
  Gvw = simplify(subs(G,{x,y,z},{f(1),f(2),f(3)}));
  dsdvw = cross(diff(f,v),diff(f,w));
  if length(G)<2 % Scalar integrand function
    integrand = Gvw*sqrt(dsdvw(1)^2+dsdvw(2)^2+dsdvw(3)^2);
   else    integrand = Gvw(:).'*dsdvw(:);
  end
  if order==0, I = int(int(integrand,v,x0f(1),x0f(2)),w,y0f(1),y0f(2));
   else  I = int(int(integrand,w,y0f(1),y0f(2)),v,x0f(1),x0f(2));
  end
end
%Usage:
% I=int_surf([x y 0]/(x^2+y^2)^(3/2),[cos(v) sin(v) w],[0 2*pi],[0 1],0)
```

E.9 Inverse Laplace transform

```
function x=ilaplace_my(B,A)
% To find the inverse Laplace transform of B(s)/A(s) using residue()
% Copyleft: Won Y. Yang, wyyang53@hanmail.net, CAU for academic use only
if ~isnumeric(B)
  [B,A]=numden(simplify(B)); B=sym2poly(B); A=sym2poly(A);
end
[r,p,k]= residue(B,A); EPS = 1e-4;
N= length(r); x=[]; n=1;
while n<=N
  if n<N & abs(imag(p(n)))>EPS & abs(sum(imag(p([n n+1]))))<EPS
    % For complex poles p(n) and p(n+1)
    sigma=real(p(n)); w=imag(p(n)); Kc=2*real(r(n)); Ks=-2*imag(r(n));
    sigma_=num2str(sigma); w_=num2str(w); Kc_=num2str(Kc); Ks_=num2str(Ks);
    x = [x '+'];
    if abs(sigma)>EPS, x = [x 'exp(' sigma_ '*t).*']; end
    if abs(Kc)>EPS&abs(Ks)>EPS
      x = [x '(' Kc_ '*cos(' w_ '*t)'];
      if Ks>EPS, x = [x '+']; end
      x = [x  Ks_ '*sin(' w_ '*t))'];
     elseif abs(Kc)>EPS, x = [x  '(' Kc_ '*cos(' w_ '*t))'];
     else    x = [x '(' Ks_ '*sin(' w_ '*t))'];
    end
    n = n + 2;
   elseif n<=N & abs(imag(r(n)))<EPS
    % For a real pole p(n)
    if r(n)>0, x = [x ' + ']; end
    x = [x  num2str(r(n))];
    if abs(p(n))>EPS,  x = [x  '*exp(' num2str(p(n)) '*t)'];  end
    n = n+1;
  end
end
if ~isempty(k)
  Kd = num2str(k(end));
  if k(end)>0, x = [x ' + ']; end
  x = [x Kd '*dirac(t)'];
end
%Usage 1: B=[4 0 2], A=[6 4 4 1]; ilaplace_my(B,A)
%Usage 2: syms s; ilaplace_my((4*s^2+2)/(6*s^3+4*s^2+4*s+1))
```

To obtain the inverse Laplace transform of a given function of s, say,

$$X(s) = \frac{10}{s(s^2+2s+5)} = \frac{2}{s} - \frac{2(s+1)+2}{(s+1)^2+2^2} \tag{E.3}$$

we can use the MATLAB function "`ilaplace()`" as

```
>>syms s;  Xs=10/s/(s^2+2*s+5);
>>xt=ilaplace(Xs)
   xt =  2 - 2*exp(-t)*(cos(2*t) + sin(2*t)/2)
```

However, for some functions of s, say,

$$X(s) = \frac{4s^2+2}{6s^3+4s^2+4s+1} \tag{E.4}$$

the MATLAB function "`ilaplace()`" does not work properly:

```
>>syms s;  Xs=(4*s^2+2)/(6*s^3+4*s^2+4*s+1);
>>xt=ilaplace(Xs); pretty(xt)
       ---                                        ---              2
       \              exp(r12 t)                  \            r12  exp(r12 t)
   2   /      ---------------------- + 4   /      ----------------------
       ---                2                       ---              2
    r12 in #1 2 (9 r12  + 4 r12 + 2)    r12 in #1 2 (9 r12  + 4 r12 + 2)
  where
                     /                2                 \
                     |     3    2 s12      2 s12        |
        #1 == RootOf| s12   + ------ + ----- + 1/6, s12 |
                     \              3        3            /
```

In this case, the private MATLAB function "`ilaplace_my()`" may be an alternative on the premise that $X(s)$ has no multiple poles:

```
>>xt=ilaplace_my(Xs)
>>B=[4 0 2]; A=[6 4 4 1]; xt=ilaplace_my(B,A) % An alternative usage
   xt =
      +exp(-0.18364*t).*(-0.066184*cos(0.72318*t)-0.328*sin(0.72318*t))
      + 0.73285*exp(-0.29938*t)
```

This has been obtained from the PFE (partial fraction expansion) of $X(s)$, which can be found by using "`residue()`" as

```
>>[r,p,k]=residue(B,A); k, [r  p]
   k =  []
   ans =   -0.0331 + 0.1640i    -0.1836 + 0.7232i
           -0.0331 - 0.1640i    -0.1836 - 0.7232i
            0.7329              -0.2994
```

This implies that the PFE form of Eq. (E.4) is

$$X(s)=\frac{4s^2+2}{6s^3+4s^2+4s+1}=\frac{-0.0331+j0.164}{s+0.1836-j0.7232}+\frac{-0.0331-j0.164}{s+0.1836+j0.7232}+\frac{0.7329}{s+0.2994} \tag{E.5}$$

If our final goal is to plot the inverse Laplace transform $x(t)$ over a time period, we can try the following MATLAB statements on the premise that $X(s)$ has no multiple poles:

```
>>t=0:0.01:10; % Time vector
>>xt=r.'*exp(p*t); % x(t) = Σ_n r_n e^{jp_n t}
>>xt1 = exp(-0.18364*t).*(-0.066184*cos(0.72318*t)-0.328*sin(0.72318*t))
        + 0.73285*exp(-0.29938*t);
>>plot(t,xt, t,xt1,'r:')
```

E.10 Examples of Script File

There are two kinds of M-files saved with a filename having an ".m" extension, functions and scripts:

- Functions, which in general accept input arguments and return output arguments where internal variables are local to the functions as long as they are not declared as global variables.

  As illustrated by the above function "`ilaplace_my().`" the first line of a function file is like

  ```
  function [output1,output2,…]=my_fun(input1,input2,…)
  ```

- Scripts, which operate on data in the workspace without accepting/returning any input/output arguments.

The following M-file, "em01f04.m," used to plot Fig. 1.4 (in Section 1.1.2.4), is a script file. To understand it in detail, let us run the program line by line and see what is happening.

```
>>x=[-2:0.1:2]; y=[-2:0.1:2]; [X,Y]=meshgrid(x,y); % x- and y-grids

  X = -2.0000  -1.9000  …   1.9000   2.0000 % 41 x 41 grid matrix
      -2.0000  -1.9000  …   1.9000   2.0000
       . . . . . . . . . . . . . . . . . .
      -2.0000  -1.9000  …   1.9000   2.0000

  Y = -2.0000  -2.0000  …  -2.0000  -2.0000 % 41 x 41 grid matrix
      -1.9000  -1.9000  …  -1.9000  -1.9000
       . . . . . . . . . . . . . . . . . .
       2.0000   2.0000  …   2.0000   2.0000
>>a=1; b=-1; c=9; d=2;  Z=(-d-a*X-b*Y)/c;
```

% Make a matrix Z for the grid matrix (X,Y) by the plane equation (1.1.12) $ax+by+cz+d=0$ with $a=1, b=-1, c=9$, and $d=2$.

```
>>n=[a b c].';
```
% Make a normal vector $\mathbf{n}=[a\ b\ c]^T$ to the plane (1.1.12)

```
>>mesh(X,Y,Z), hold on
```

% Plot the plane as a mesh-type graph and hold it so that it can be overlapped with the next graph.

```
>>p1=[-0.5 0.5 2].';
```
% Take an arbitrary point $\mathbf{p}_1$ not on the plane.

```
>>p0=zeros(3,1); p0(1)=p1(1)+1; p0(2)=p1(2)-1;
  p0(3)=(-d-a*p0(1)-b*p0(2))/c;
```

% Take an arbitrary point $\mathbf{p}_0$ on the plane satisfying Eq.(1.1.12).

```
>>p01=p0-p1;
```

% Make the vector $\mathbf{p}_{01}=\mathbf{p}_0-\mathbf{p}_1$ from $\mathbf{p}_1$ to $\mathbf{p}_0$.

```
>>un=n/norm(n);
```
% The unit normal vector $\mathbf{u}_n=\mathbf{n}/||\mathbf{n}||$

```
>>p21=dot(p01,un)*un % Projection p21 of p01 on n
```

% $\mathbf{p}_{21}=\text{proj}_{\mathbf{n}}\mathbf{p}_{01} \overset{(1.1.10)}{=} (\text{comp}_{\mathbf{n}}\mathbf{p}_{01})\frac{\mathbf{n}}{||\mathbf{n}||} \overset{(1.1.9)}{=} (\mathbf{p}_{01}\bullet\frac{\mathbf{n}}{||\mathbf{n}||})\frac{\mathbf{n}}{||\mathbf{n}||} = (\mathbf{p}_{01}\bullet\mathbf{u}_n)\,\mathbf{u}_n$

```
>>p2=p1+p21;
```

```
% Projection p2 = p1 + p21 of p1 onto the plane: Eq.(1.1.17)
```

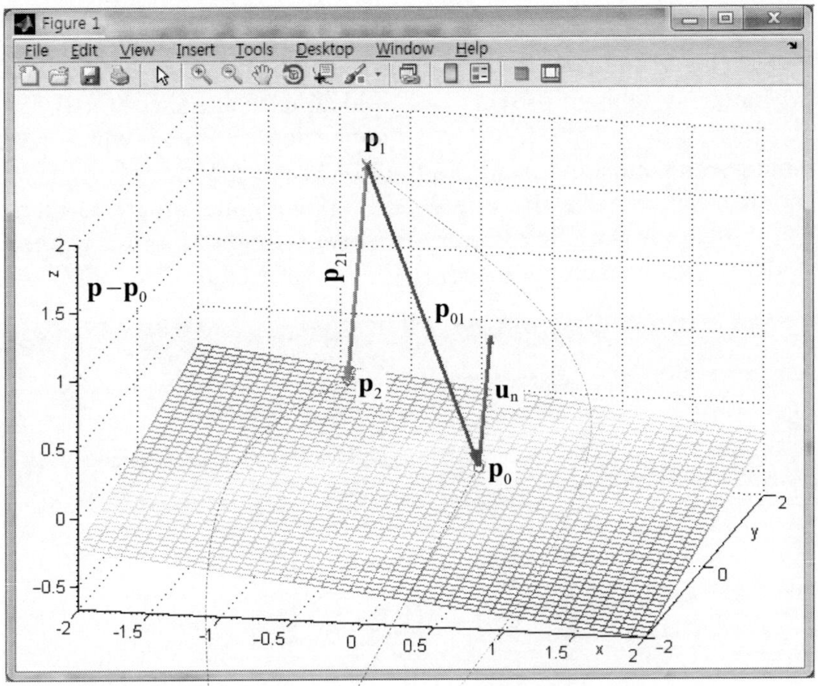

Figure E.2 Distance from a point to a plane (Fig. 1.4)

To plot the three points $\mathbf{p}_0$, $\mathbf{p}_1$, and $\mathbf{p}_2$, and the three vectors $\mathbf{u}_n$ (at $\mathbf{p}_0$), $\mathbf{p}_{01}$ (at $\mathbf{p}_1$), and $\mathbf{p}_{21}$ (at $\mathbf{p}_1$) as depicted in Fig. E.2 (or Fig. 1.4), the following statements have been executed:

```
plot3(p0(1),p0(2),p0(3),'o') % Plot p0 with a blue 'o' mark
plot3(p1(1),p1(2),p1(3),'rx') % Plot p1 with a red 'x' mark
plot3(p2(1),p2(2),p2(3),'v') % Plot p2 with a blue 'v' mark
quiver3(p0(1),p0(2),p0(3),un(1),un(2),un(3),0) % un from p0
quiver3(p1(1),p1(2),p1(3),p01(1),p01(2),p01(3),0) % p01 from p1
quiver3(p1(1),p1(2),p1(3),p21(1),p21(2),p21(3),0) % p21 from p1
```

where the last input argument of '`plot3()`' has been set to zero for no scaling.

The M-file "em01f04.m" containing all these statements can be run to execute them at the same time for getting Fig. E.2 by typing just the filename without the extension name ".m" at the MATLAB prompt:

```
>>em01f04
```

Appendix F: Simulink

Simulink, integrated with MATLAB, is a platform for multi-domain simulation and model-based design of dynamic systems. It provides a graphical user interface (GUI) for building a model as a block diagram (a collection of blocks connected by lines), which enables you to draw models as you would with pencil and paper. In the MATLAB Command window, you can double-click on the Simulink button (like the one in the upper-right corner) to start Simulink. Once Simulink has been started, the Simulink Library Browser window (Fig. F.1) appears. You can click on File>New>Model to create an empty model (initially named "untitled.mdl") as shown in Fig. F.1.

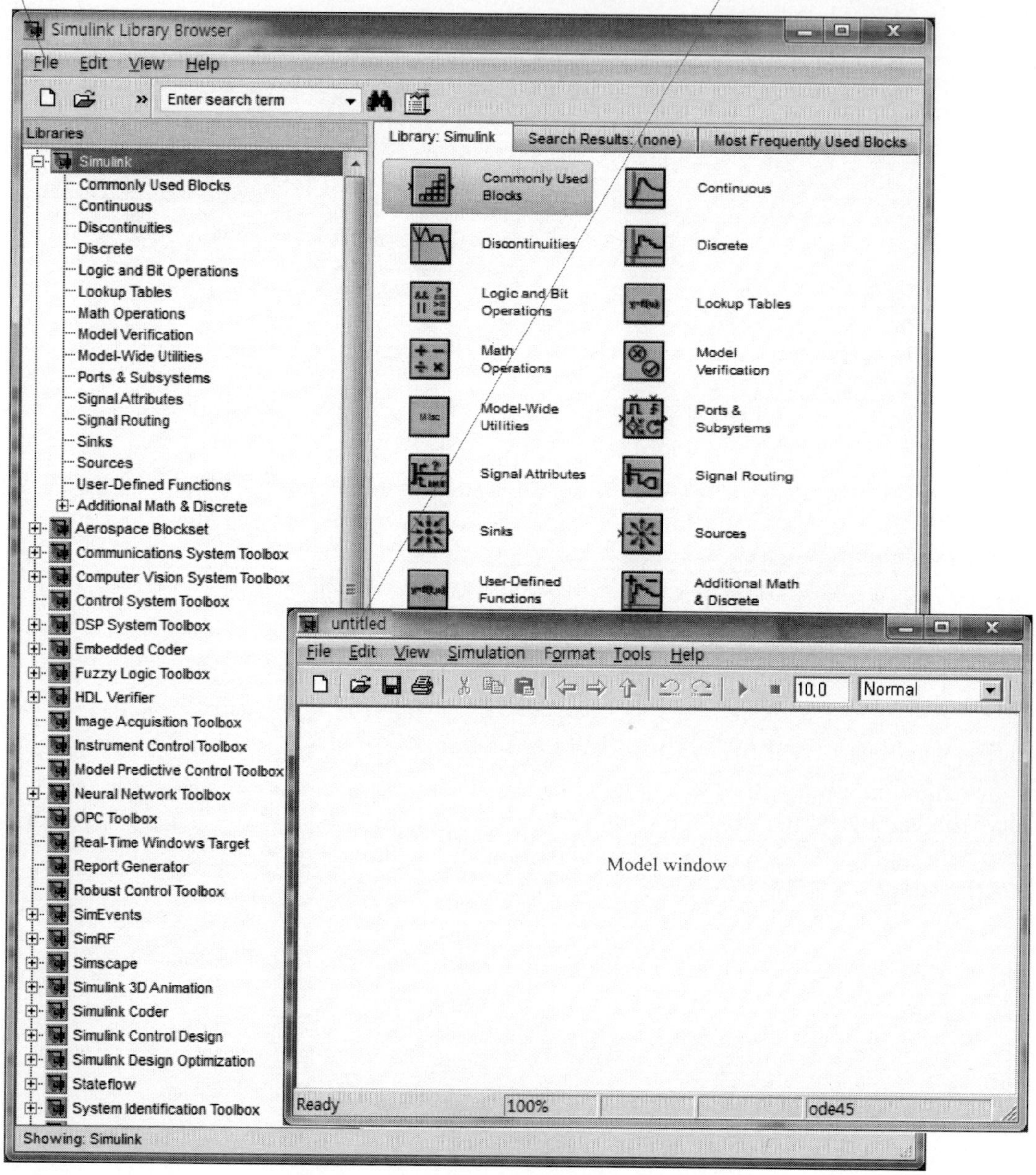

Figure F.1 The Simulink Library Browser window

The empty model can be completed by clicking and dragging the blocks (which you need) from the libraries (blocksets/toolboxes) in the Simulink Library Browser window (into the model window) and interconnecting them, where you may have to double-click on some blocks (with parameters) to open the parameter configuration dialog box and specify the parameter values.

F.1 An Example of Simulation with Simulink

For example, consider the Simulink model (named "em04e09_sim.mdl" and shown in Figs. 4.4(a) and F.2(a)), which was composed in Example 4.9 to simulate the step response of a continuous-time second-order system with transfer function

$$G(s)=\frac{Y(s)}{U(s)}=\frac{1}{s^2+3s+2} \tag{F.1}$$

where the excitation (input) block "Step," the continuous-time system block "Transfer Fcn," and the measurement (output) block "Scope" have been dragged from the three libraries, i.e., "Sources", "Continuous", and "Sinks", respectively. Fig. F.2(a) also shows the parameter configuration dialog boxes of the two blocks, "Step" and "Transfer Fcn", where the values of the parameters are specified.

To run the model, click on Simulation>Configuration_Parameters (in the top menu bar) to open the Configuration Parameters dialog box, set the simulation parameters (Fig. F.2(b)), and click OK.

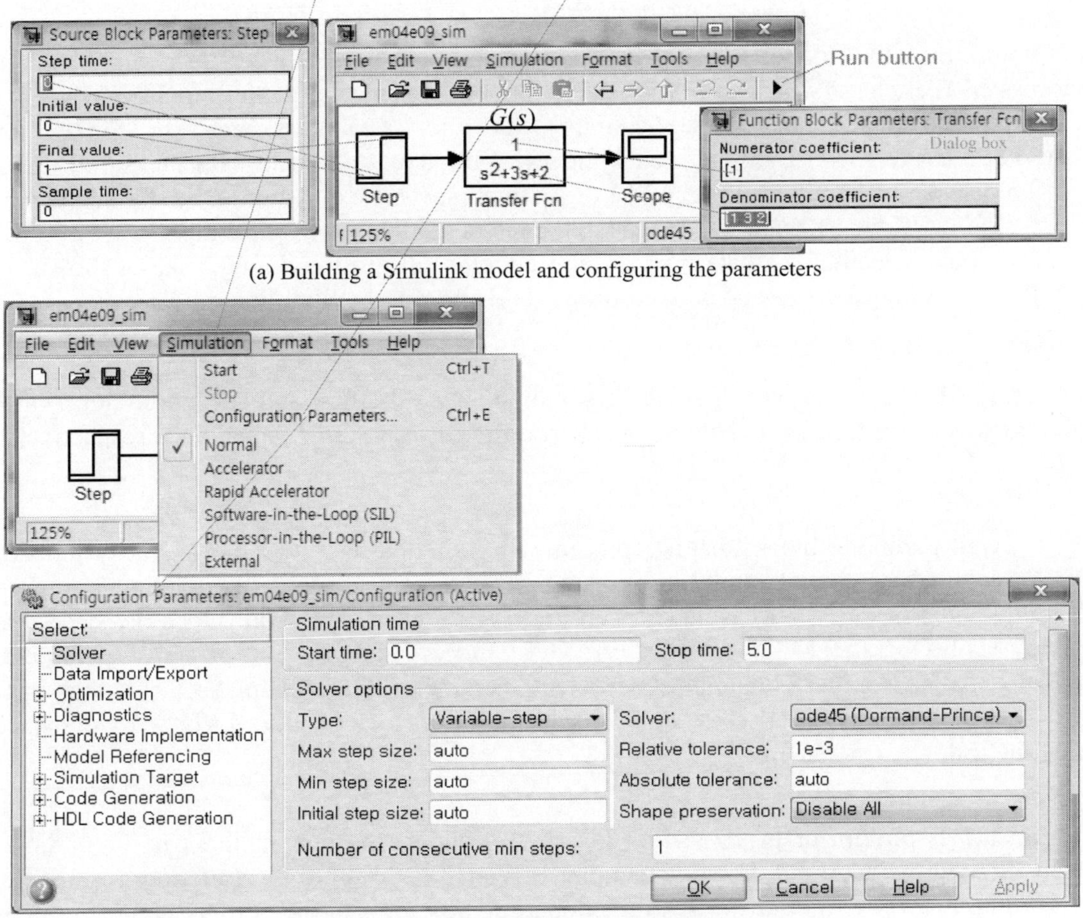

(a) Building a Simulink model and configuring the parameters

(b) Setting the (simulation) configuration parameters

Figure F.2 An example of building a Simulink model and configuring its parameters

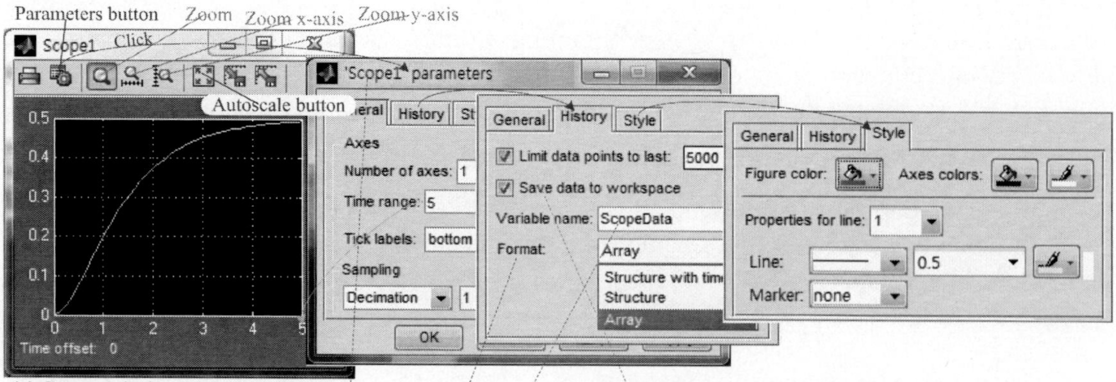

(a) Scope window showing simulation results (b) Scope parameters dialog box depending on which tab you click on

Figure F.3 The Scope window and its parameters dialog box

Once the Simulink model is composed and its parameters are specified, click on the Run (▶) button to run it and double-click on the "Scope" block to see the simulation result as shown in Fig. F.3(a). If you have saved the model as an MDL-file named, say, "em04e09_sim.mdl", you can also run it by using the MATLAB command "`sim()`" in the Command window or an M-file as follows:

```
>>Simulation_time=5; sim('em04e09_sim',Simulation_time)
```

Note that a Simulink model not only refers to all the (MATLAB) workspace variables but also can access M-files, MAT-files, and workspace variables by means of the blocks, such as "From File." "From Workspace" (in "Sources" library), "To File," "To Workspace" (in "Sinks" library), and "MATLAB Fcn" (in "User-Defined Functions" library). Also, the data measured through a "Scope" block can be exported into the MATLAB workspace by clicking on the Parameters button in the Scope window to open the Scope parameters dialog box, checking the "Save data to workspace" box, selecting the data Format, and clicking OK in the dialog box, as shown in Fig. F.3(b). If you have chosen "Array" in the Format field and run the Simulink model, you can type

```
>>plot(ScopeData(:,1),ScopeData(:,2))
```

at the MATLAB prompt to plot the data versus time, since the first and second columns of the 2-column array `ScopeData` have the time vector and the corresponding signal data, respectively.

F.2 Subsystem

A *subsystem* is a group of blocks that is replaced with a single Subsystem block. Subsystems can be used to create a hierarchical model comprising many layers. To practice making a subsystem, modify the model in Fig. F.2(a) as depicted in Fig. F.4(a) where the 'Add' block has been dragged from the 'Math Operations' library (to the model window) and its "List of signs" parameter has been set to "+−." Also, the "Numerator coefficient" parameter of the "Transfer Fcn" block has been set to K, whose value is supposed to be set arbitrarily, say, to 1 (in the MATLAB workspace) by running the following statement:

```
>>K=2;
```

at the MATLAB prompt or in an M-file. Fig. F.4(b) shows the rectangular part (bounding box) taken by clicking point A and dragging to point B, which is going to be made into a subsystem as shown in Fig. F.4(c) by right-clicking on the bounding box to open the pop-up menu (Fig. F.4(b1)) and selecting "Create Subsystem."

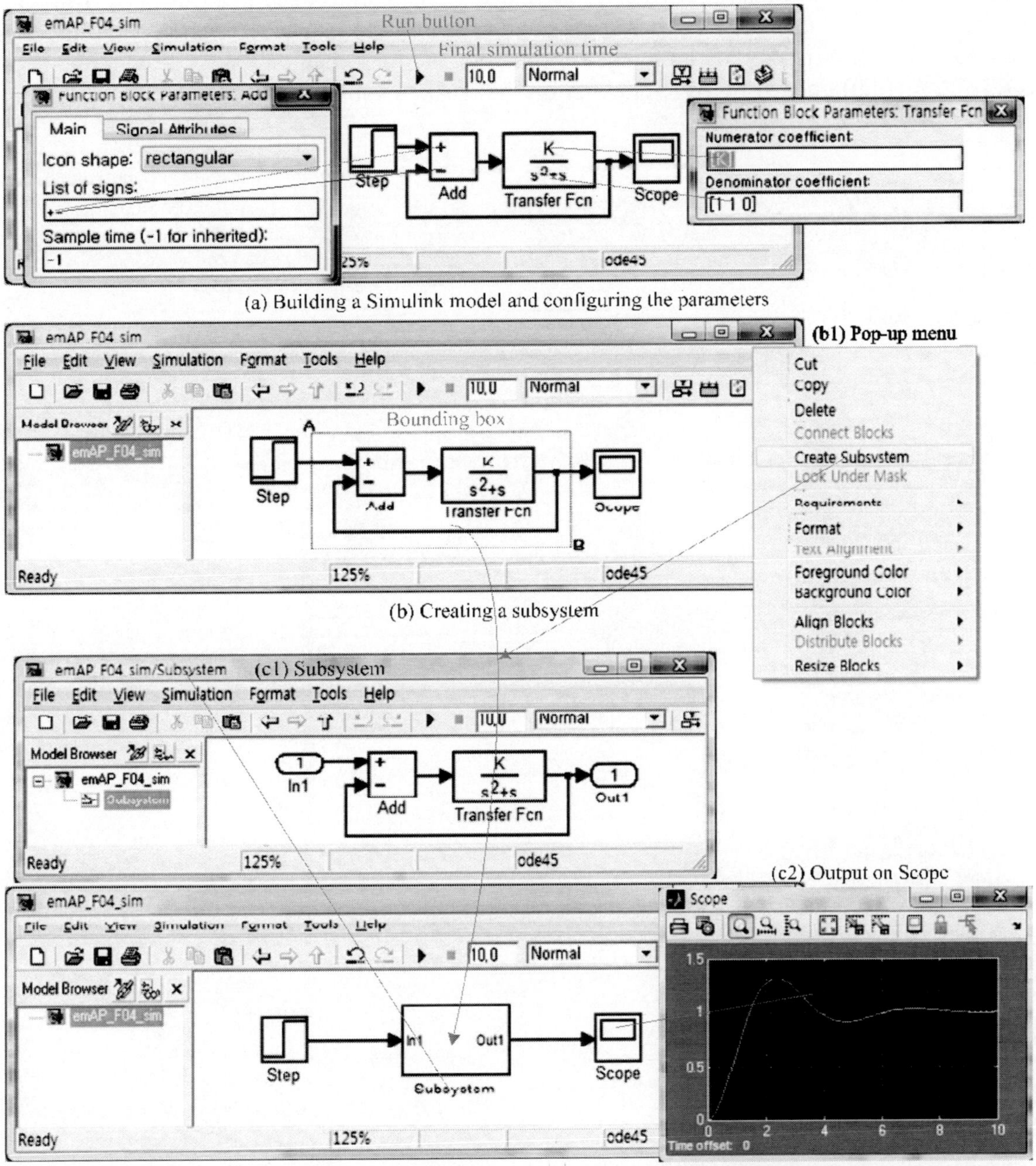

(a) Building a Simulink model and configuring the parameters

(b) Creating a subsystem

(c) Simulink model with a subsystem

Figure F.4 Making a part of Simulink model a subsystem ("emAP_F04_sim.mdl")

Once a subsystem has been made as illustrated in Fig. F.5(a), it can be masked by right-clicking it to open the pop-up menu (Fig. F.5(a1)), selecting "Create Mask" from the menu to open the Mask Editor dialog box (Fig. F.5(b)), typing, say,

```
image(imread('Motor.png'))   or   disp('Motor')
```

in the Icon Drawing command field, and clicking OK, on the premise that an image file named "Motor.png" exists in the current directory or some folder on the MATLAB search path. Fig. F.5(c) shows the subsystem masked by an image of a motor. The mask can be edited through the pop-up menu (Fig. F.5(c1)) opened by right-clicking on the masked subsystem and selecting "Edit Mask" from the menu.

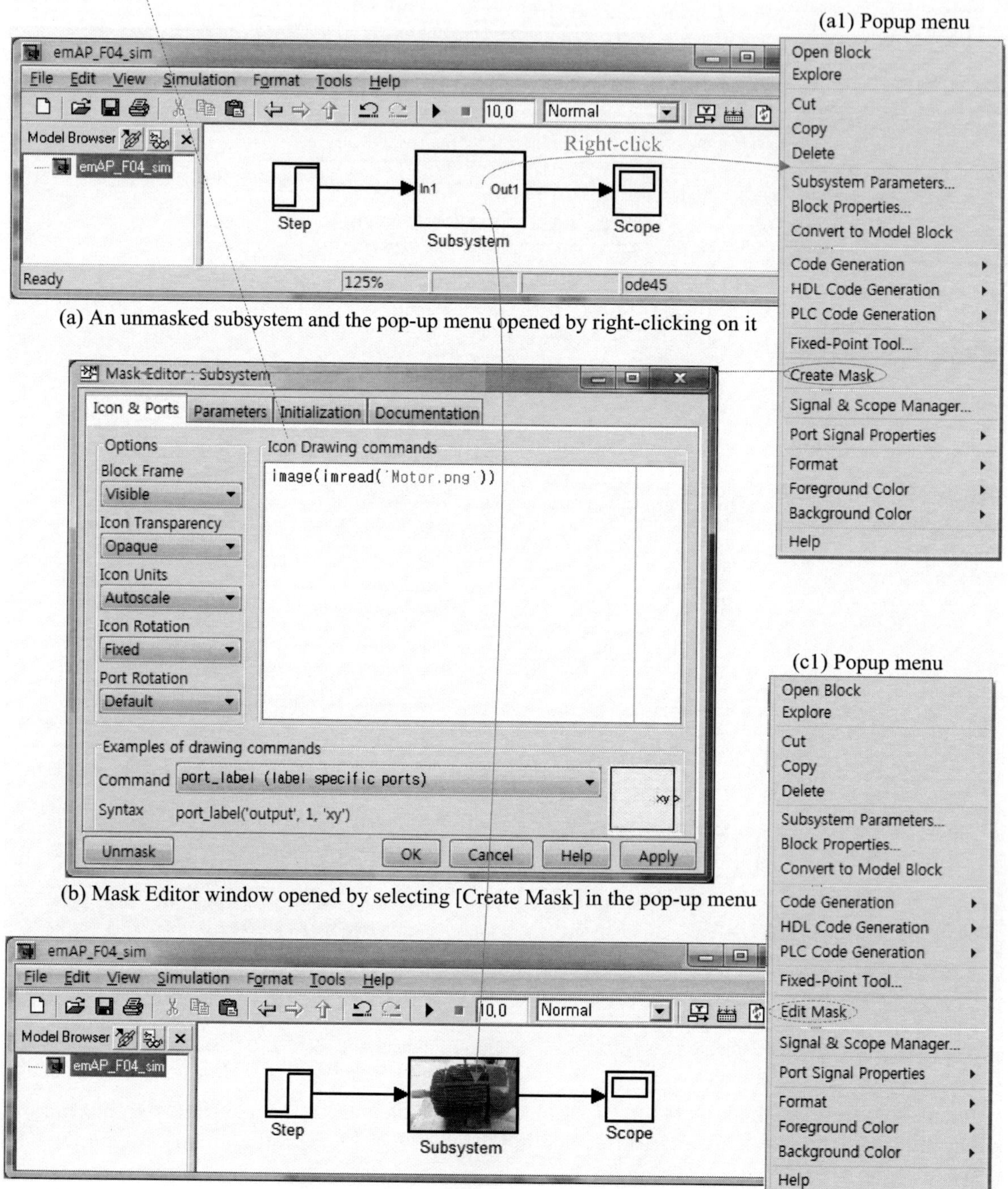

(a) An unmasked subsystem and the pop-up menu opened by right-clicking on it

(b) Mask Editor window opened by selecting [Create Mask] in the pop-up menu

(c) Simulink model with a masked subsystem

Figure F.5 Masking a subsystem ("emAP_F04_sim.mdl")

F.3 S-Function and MATLAB Function

An *S-function* is a special Simulink block whose behavior is described in MATLAB, C, C++, or Fortran. To practice using S-function for the simulation of the same system as above, write an S-function describing the continuous-time second-order system with transfer function

$$G(s)=\frac{Y(s)}{U(s)}=\frac{K}{s^2+a_1s+a_0}\overset{a_1=1}{\underset{a_0=0}{=}}\frac{K}{s^2+s} \tag{F.2}$$

in state equation form (Eq. (4.4.2)) as (see Sections 4.4 and 4.5)

$$\begin{aligned}\begin{bmatrix}x_1'(t)\\x_2'(t)\end{bmatrix}&=A\begin{bmatrix}x_1(t)\\x_2(t)\end{bmatrix}+Bu(t)\text{ with }A=\begin{bmatrix}0&1\\-a_0&-a_1\end{bmatrix}=\begin{bmatrix}0&1\\0&-1\end{bmatrix},\ B=\begin{bmatrix}0\\1\end{bmatrix}\\ y(t)&=C\begin{bmatrix}x_1(t)\\x_2(t)\end{bmatrix}+Du(t)\text{ with }C=[1\quad 0],\ D=0\end{aligned} \tag{F.3}$$

in an M-file named, say, "csfunc_my.m." Then, click and drag the S-function block from the User-

```
function [sys,x0,str,ts] = csfunc_my(t,x,u,flag,x0)
% See csfunc.m/dsfunc.m/sfuntmpl.m for other S-function templates.
% Copyright 1990-2009 The MathWorks, Inc. $Revision: 1.1.6.2 $
A=[0  1; 0  -1];  B=[0; 1];  C=[1  0]; D=0; % Eq.(F.3)
switch flag,
  case 0,  % Initialization
    [sys,x0,str,ts] = mdlInitializeSizes(A,B,C,D,x0);
  case 1,  % Derivatives in a continuous state equation
    sys = mdlDerivatives(t,x,u,A,B,C,D); % x'=Ax+Bu, y=Cx+Du
  case 3,  % Outputs
    sys = mdlOutputs(t,x,u,A,B,C,D);
  case {2,4,9},  % Unhandled flags
    sys = [];
  otherwise  % Unexpected flags
    DAStudio.error('Simulink:blocks:unhandledFlag', num2str(flag));
end
% ---------------------------------------------------------------------------
function [sys,x0,str,ts] = mdlInitializeSizes(A,B,C,D,x0)
% Return sizes, initial conditions, and sample times for the S-function.
sizes = simsizes;
sizes.NumContStates=2;  sizes.NumDiscStates=0;  sizes.NumOutputs=1;
sizes.NumInputs=1;  sizes.DirFeedthrough=1;  sizes.NumSampleTimes=1;
sys = simsizes(sizes);
%x0  = x0; %zeros(2,1); % Initial state vector
str = []; % Reserved for use in future versions of Simulink, set to null.
ts = [0 0]; % Sampling period and offset for continuous systems.
% ---------------------------------------------------------------------------
function sys = mdlDerivatives(t,x,u,A,B,C,D)
% Return the derivatives for the continuous states.
sys = A*x + B*u;
% ---------------------------------------------------------------------------
function sys = mdlOutputs(t,x,u,A,B,C,D)
% Return the block outputs.
sys = C*x + D*u;
```

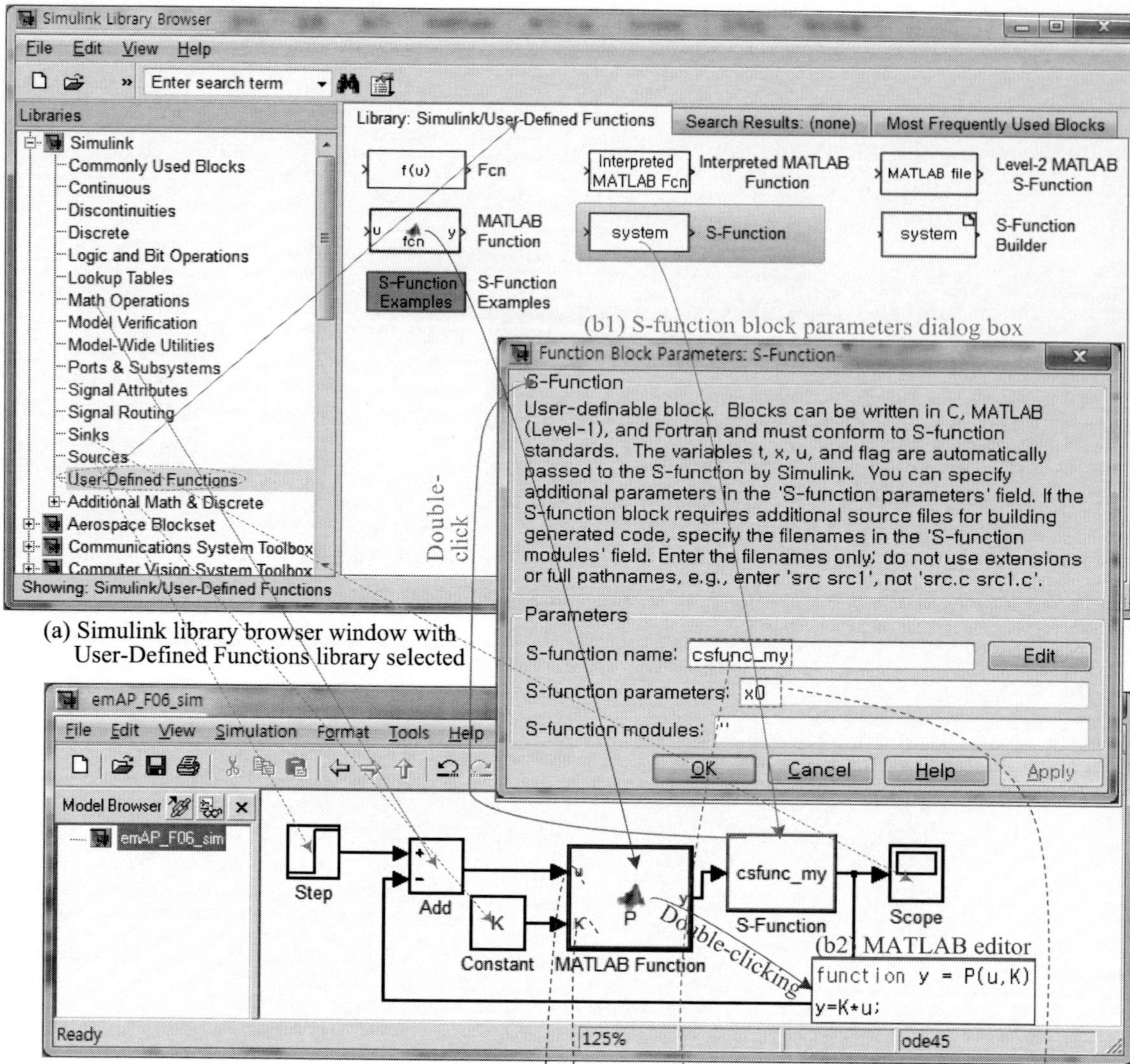

(b) Simulink model

Figure F.6 Using S-Function and MATLAB Function blocks ("emAP_F06_sim.mdl")

Defined Functions library to the model window (Fig. F.6), double-click on the S-function block to open the S-function block parameter dialog box (Fig. F.6(b1)), put "csfunc_my" and "x0" (with no quotation marks) into the fields of S-function name and parameters, respectively, and click OK. Also, to practice using the MATLAB Function block, click and drag the MATLAB Function block from the User-Defined Functions library to the model window, double-click on the MATLAB Function block to open the MATLAB Editor window, make a MATLAB function named, say, "`P()`,"

```
function y=P(u,K)
Y=K*u
```

and save it (in the MATLAB Function block). Since this function has two inputs, the MATLAB Function block will have two input terminals u and K where the K input terminal is connected to the K Constant block dragged from the Source library. Now that the model "emAP_F06_ sim.mdl" has been completed, type the following statements:

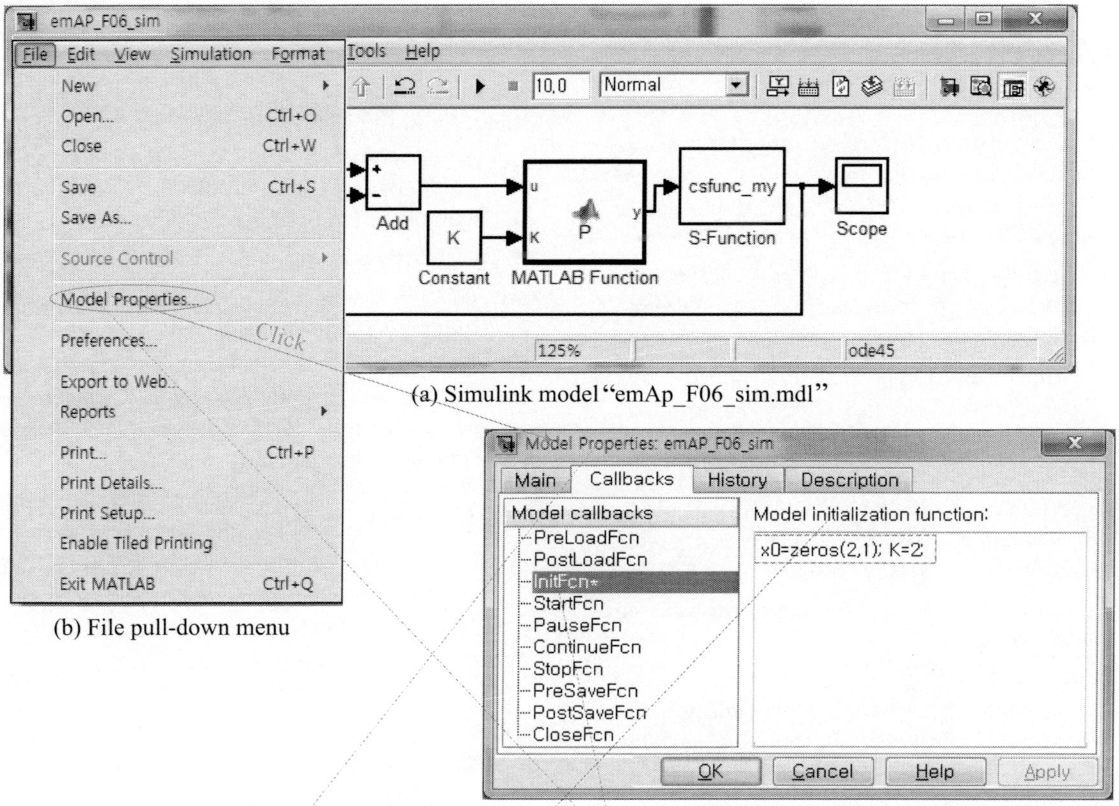

(a) Simulink model "emAp_F06_sim.mdl"

(b) File pull-down menu

(c) Model Properties dialog box

Figure F.7 Using callback function

```
>>x0=zeros(2,1); K=2;
```

at the MATLAB prompt, press the Enter key, and run the model. Note that x0 is the initial state to be given as the 5th input argument of the S-function "`csfunc_my()`" (Fig. F.6(b1)). An alternative to typing this statements at the MATLAB prompt is to select File>Model Properties… to open the Model Properties dialog box, select the Callbacks tab>InitFcn, type the MATLAB statements into the field of Model initialization function as shown in Fig. F.7, and click OK. Once the Simulink model has been completed, you can run it and see if the output signal waveform measured via the Scope block is similar to that shown in Fig. F.4(c2). Once you have saved the initial parameter values that are needed to run a Simulink model by using this callback function, you do not have to re-initialize the parameters when reopening the Simulink model even if all the variables in the MATLAB workspace have been overwritten or removed by the Clear command.

References

[C-1] Chakrabarty, A., *Elements of Ordinary Differential Equations and Special Functions*, 2nd ed., New Age International, New Delhi, 1996.

[G-1] Griffiths, David J. and Reed College Hill, *Introduction to Electrodynamics*, Prentice Hall, Inc., New Jersey, 1999.

[H-1] Helstrom, Carl W., *Probability and Stochastic Processes for Engineers*, Macmillan Publishing Company, New Jersey, 1984.

[K-1] Kolman, Bernard and David R. Hill, *Elementary Linear Algebra*, 7th ed., Prentice Hall, Inc., New Jersey, 2000.

[K-2] Kreyszig, Erwin, *Advanced Engineering Mathematics*, John Wiley & Sons, Inc., New York, N.Y., 1983.

[L-1]Leon-Garcia, Alberto, *Probability and Random Processes for Electrical Engineering*, Addison-Wesley Publishing Company, Boston, 1994.

[O-1] O'Neil, Peter V., *Advanced Engineering Mathematics*, 6th ed., Cengage Learning, Toronto, 2006.

[O-2] Oppenheim, A.V. and R. W. Schafer, *Digital Signal Processing*, Prentice Hall, Inc., Englewood Cliffs, N.J., 1975.

[P-1] Peaceman, D. W. and H. H. Rachford, "The Numerical Solution of Parabolic and Elliptic Differential Equations," *Journal of Society for Industrial & Applied Mathematics*, Vol. 3, pp. 28–41, 1955.

[P-2] Phillips, C.L. & H. T. Nagle, Jr., *Digital Control System Analysis and Design*, Prentice-Hall, Inc., New Jersey, 2002.

[S-1] Schilling, R.J. and S.L. Harris, *Applied Numerical Methods for Engineers using MATLAB and C*, Brooks/Cole Publishing Company, Pacific Grove, 2000.

[S-2] Stearns, S.D. and D.R. Hush, *Digital Signal Analysis*, 2nd ed., Prentice Hall, Inc., Englewood Cliffs, N.J., 1990.

[W-1] Weber, Hans J., Frank E. Harris, and George B. Arfken, *Essential Mathematical Methods for Physicists and Engineers*, Academic Press, California, 2003.

[W-2] Webpage <http://www.efunda.com/math/ode/linearode_undeterminedcoeff.cfm>

[W-3] Webpage <http://www.kth.se/polopoly_fs/1.220751!/Menu/general/column-content/attachment/labTutorialandexcercise5.pdf>

[W-4] Webpage <http://www.sosmath.com/calculus/sequence/stirling/stirling.html>

[W-5] Wilson, Howard B., Louis H. Turcotte, and David Halpern, *Advanced Mathematics and Mechanics Applications Using MATLAB*, 3rd ed., Chapman & Hall/CRC, 2003.

[W-6] Wylie, C. Ray and Louis C. Barrett, *Advanced Engineering Mathematics*, 6th ed., McGraw-Hill, Inc., 1995.

[Y-1] Yang, Won Y., Wenwu Cao, Tae-Sang Chung, and John Morris, *Applied Numerical Methods Using MATLAB*, John Wiley & Sons, Inc., New Jersey, 2005.

[Y-2] Yang, Won Y. et al., *Signals and Systems with MATLAB*, Springer-Verlag, Berlin Heidelberg, 2009.

[Z-1] Zill, Dennis G., Warren S. Wright, and Michael R. Cullen, *Advanced Engineering Mathematics*, 4th ed., Jones and Bartlett Publishers, Massachusetts, 2011.

INDEX

N

O

T

Index for MATLAB functions

‡rv stands for "random variable."

Note that the script files, named, say "em03e04.m," "em04p01.m," "em05f11.m," each having its filename in the first (comment) line, are related to Example 3.4, Problem 4.1, and Fig. 5.11, respectively. Note also that 'sim()' denotes the MATLAB function named "sim," which can be a MATLAB built-in function or a user-defined function, and it is supposed to be stored as an M-file named "sim.m" in a folder (that can be searched by MATLAB) on the computer.